Civil Engineering and Architecture

Donna Matteson, M.Ed.
State University of New York at Oswego

Deborah Kennedy, P.E.

Stuart Baur, Ph.D., AIA
Missouri University of Science and Technology

Eva Kultermann
Illinois Institute of Technology, College of Architecture

Australia • Brazil • Japan • Korea • Mexico • Singapore • Spain • United Kingdom • United States

Civil Engineering and Architecture
Donna Matteson, Deborah Kennedy, Stuart Baur, and Eva Kultermann

Vice President, Career, Education, and Training Editorial: Dave Garza
Director of Learning Solutions: Sandy Clark
Senior Acquisitions Editor: James DeVoe
Managing Editor: Larry Main
Product Manager: Mary Clyne
Development: iD8 Publishing Services
Editorial Assistant: Cris Savino
Vice President, Career, Education, and Training Marketing: Jennifer McAvey
Marketing Director: Deborah Yarnell
Marketing Manager: Katherine Hall
Production Director: Wendy Troeger
Production Manager: Mark Bernard
Content Project Manager: Mike Tubbert
Art Director: Casey Kirchmayer

Library of Congress Control Number: 2010930491

ISBN-13: 978-1-4354-4164-4

ISBN-10: 1-4354-4164-8

Delmar
5 Maxwell Drive
Clifton Park, NY 12065-2919
USA

Cengage Learning is a leading provider of customized learning solutions with office locations around the globe, including Singapore, the United Kingdom, Australia, Mexico, Brazil and Japan. Locate your local office at:
international.cengage.com/region

Cengage Learning products are represented in Canada by Nelson Education, Ltd.

For your lifelong learning solutions, visit **delmar.cengage.com**

Visit our corporate website at **cengage.com**

Printed in the U.S.A.
1 2 3 4 5 6 7 15 14 13 12 11

BRIEF CONTENTS

CONTENTS

PART III LOCATION, LOCATION, LOCATION

PART IV PLANNING FOR OCCUPANCY

PART V STRUCTURAL DESIGN: BUILDING FOR TOMORROW

PART VI MECHANICALS: THE BUILDING COMES ALIVE

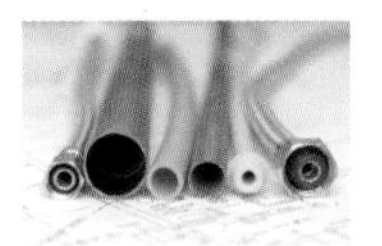

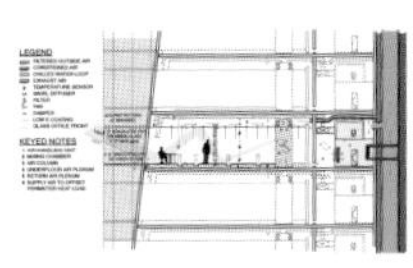

PART VII CURB APPEAL

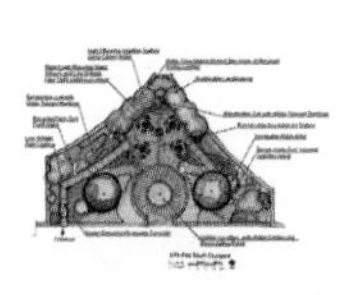

PART VIII SELLING THE PLAN

Chapter 19 Visual Communication of Design Intent 694

Chapter 20 Formal Communication and Analysis 710

PREFACE

Look out any window. What do you see? You probably see buildings and roads leading to more buildings. The buildings probably represent many different architectural styles. They employ many structural systems and serve a variety of functions in your community. You might pass these buildings every day and take them for granted, or you might wonder how they remain standing.

After reading this book, you will never take a building for granted again. As you travel through your community, you will understand the hundreds of decisions that went into the design and construction of your built environment. This text will take you on a journey through time and technology, from the earliest huts built from bones and animal skins to today's technologically sophisticated and environmentally sensitive structures. You will experience the process of designing a structure, from site discovery through landscape design and presentation drawings.

Civil Engineering and *Architecture* is a new textbook for teachers who want to inspire their students to explore career pathways in these challenging and exciting disciplines. By presenting the principles and concepts that engineers and architects use to shape today's built environment, *Civil Engineering and Architecture* will help students develop the problem-solving skills and technological literacy they need to embark on the journey.

CIVIL ENGINEERING AND ARCHITECTURE WITH PROJECT LEAD THE WAY, INC.

This text resulted from a partnership forged between Delmar Cengage Learning and Project Lead The Way® Inc. in February 2006. As a nonprofit foundation that develops curriculum for engineering, Project Lead The Way®, Inc. provides students with the rigorous, relevant, reality-based knowledge they need to pursue engineering or engineering technology programs in college.

The Project Lead The Way® curriculum developers strive to make math and science relevant for students by building hands-on, real-world projects into each course. To support Project Lead The Way's® curriculum goals, and to support all teachers who want to develop project/problem-based programs in engineering and engineering technology, Delmar Cengage Learning is developing a complete series of texts to complement all of Project Lead The Way's® nine courses:

Gateway To Technology

Introduction To Engineering Design

Principles Of Engineering

Digital Electronics

Aerospace Engineering

Biotechnical Engineering

Civil Engineering and Architecture

Computer Integrated Manufacturing

Engineering Design and Development

To learn more about Project Lead The Way's® ongoing initiatives in middle school and high school, please visit *http://www.pltw.org.*

HOW THIS TEXT WAS DEVELOPED

This book's development began with a focus group that brought together experienced teachers and curriculum developers from a broad range of engineering disciplines. Two important themes emerged from that discussion: (1) teachers need a single resource to help them teach a course that encompasses both civil engineering and architecture, and (2) teachers want an engaging, interactive resource to support project/problem-based learning.

For years, teachers have struggled to fit conventional textbooks to STEM-based curricula. *Civil Engineering and Architecture* addresses that need with an interactive text organized around the principles and applications essential to success in these disciplines. For the first time, teachers will be able to choose a single text that addresses the challenges and individuality of project/problem-based learning while presenting sound coverage of the essential concepts and techniques used in civil engineering and architectural design.

This book is unique in that it was written by a team of authors with pedagogy and engineering expertise to ensure a textbook that students can understand and one that contains valid engineering content. *Civil Engineering and Architecture* supports project/problem-based learning by:

- Creating an unconventional, show-don't-tell pedagogy that is driven by *concepts*, not traditional textbook content. Concepts are mapped at the beginning of each chapter and clearly identified as students navigate the chapter.
- Reinforcing major concepts with Applications, Projects, and Problems based on real-world examples.
- Providing a text rich in features designed to bring architecture to life in the real world. Case studies, Career Profiles, Boxed Articles highlighting human achievements, and resources for extended learning will show students how engineers and architects develop career pathways, work through failures, and innovate to continuously improve the success and quality of their projects.
- Reinforcing the text's interactivity with an exciting design that invites students to participate in a journey through the history and current practice of civil engineering and architecture.

Organization

Civil Engineering and Architecture presents concepts and applications in a flexible, multi-part format. This text begins by defining some of the roles of civil engineers and architects. Students will learn about some of the great milestones that these professions have created in the past as well as today. They will also learn the various processes involved in working with a client, finding and analyzing a site, and determining the appropriate design based on the needs of the client and site conditions. Along the way they will also learn the various aspects that go into a building design from structural analysis, building methods and materials of construction, and some of the codes that building designers need to consider. Student will learn technical terms and communicate design ideas commonly associated with the building industry. They can also learn the basics of how to design structural, mechanical, and electrical systems. Finally, students will discover the importance of these building systems and the means that we use to communicate the designs to other building professionals.

Features

Teachers want an interactive text that keeps students interested in the story behind the structures that shape our built environment. This text delivers that story with plentiful boxed articles and illustrations of current technology. Here are some examples of how this text is designed to keep students engaged in a journey through the design process.

Case Study

Throughout this textbook, we will take a close look at actual building projects. They will be referred to as case studies. One case study will examine the design and build of the Syracuse Center of Excellence, a 55,000-square-foot, five-story research facility (see Figure 1-47). The site was contaminated by previous industrial uses, but has been remediated, restoring the site for construction of the Center of Excellence. The building will house offices for staff, classrooms, public spaces, and research laboratories. The main laboratory is a total indoor environmental quality lab where controlled experiments on the human response to indoor environments will be conducted. Main sustainability features of the Syracuse Center of Excellence include ground-source heating and cooling, rainwater collection, a green roof, and photovoltaic panels.

A second case study will examine a large urban plan development: CityCenter in Las Vegas. CityCenter, a 67-acre urban development in Las Vegas, Nevada, includes condominiums, hotels, a casino, and a large retail and entertainment district (see Figure 1-48). The nine billion dollar venture includes six major buildings, along with a fire station, co-generation plant, parking garages, and a people-mover system. The goal of this urban plan is to create the diversity and vitality of a city, with a focus on sophistication. This requires that building designs offer timeless beauty and functionality, yet be different from any others. It was a challenging task to find architects to design such a look. Sven Van Assche, vice president of design, was up to the task. He began by clearly defining the term contemporary. Through his research, he determined that contemporary was not necessarily a pre-established design style using specific design elements but instead an innovative design that would be both timeless and inviting—a beauty people would enjoy for years to come. Van Assche searched the world for architects who would be able to design this "timeless" urban experience. Due to the project size and its mission—providing unique and distinctive choices that would attract diversity—several creative and experienced architects were needed. Although each would design a different building with its own vision, there was a need for meticulous leadership and collaboration between the professionals to create a dynamic and unified urban experience. Gensler, named the largest architectural firm in the United States by *Building Design+Construction* magazine for the past 24 consecutive years, was chosen as the executive architect of CityCenter. As the owner's representative, Gensler was responsible for collaborating with each of the architectural teams.

Figure 1-47: Syracuse Center of Excellence under construction in 2008.

32 Part I: Introduction to Civil Engineering and Architecture

searched world wide for modern architects to design CityCenter's one-of-a-kind buildings. When the search concluded, the result was a "dream team" of the world's foremost architects (Figure 1-49).

Figure 1-48: CityCenter ready for opening in December 2009.

- **Case Studies** let students explore successes, failures, and the process of architectural design.

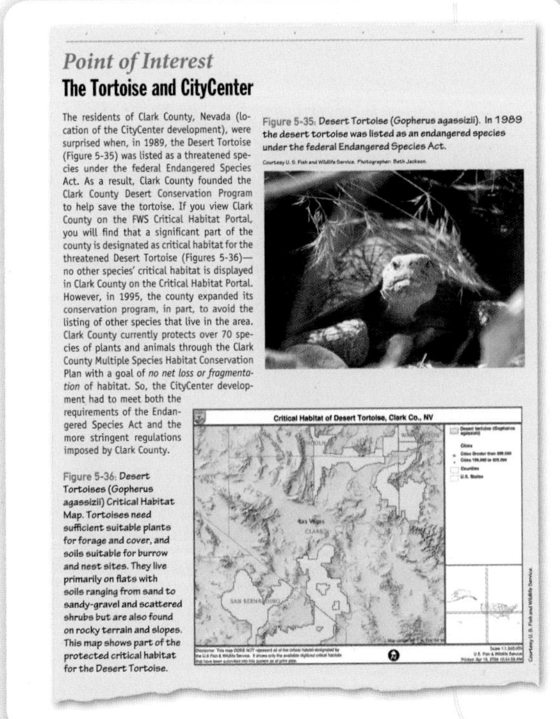

Point of Interest

The Tortoise and CityCenter

The residents of Clark County, Nevada (location of the CityCenter development), were surprised when, in 1989, the Desert Tortoise (Figure 5-35) was listed as a threatened species under the federal Endangered Species Act. As a result, Clark County founded the Clark County Desert Conservation Program to help save the tortoise. If you view Clark County on the FWS Critical Habitat Portal, you will find that a significant part of the county is designated as critical habitat for the threatened Desert Tortoise (Figures 5-36)—no other species' critical habitat is displayed in Clark County on the Critical Habitat Portal. However, in 1995, the county expanded its conservation program, in part, to avoid the listing of other species that live in the area. Clark County currently protects over 70 species of plants and animals through the Clark County Multiple Species Habitat Conservation Plan with a goal of *no net loss or fragmentation* of habitat. So, the CityCenter development had to meet both the requirements of the Endangered Species Act and the more stringent regulations imposed by Clark County.

Figure 5-35: Desert Tortoise (*Gopherus agassizii*). In 1989 the desert tortoise was listed as an endangered species under the federal Endangered Species Act.

Figure 5-36: Desert Tortoises (*Gopherus agassizii*) Critical Habitat Map. Tortoises need sufficient suitable plants for forage and cover, and soils suitable for burrow and nest sites. They live primarily on flats with soils ranging from sand to sandy-gravel and scattered shrubs but are also found on rocky terrain and slopes. This map shows part of the protected critical habitat for the Desert Tortoise.

- **Boxed Articles** highlight fun facts and points of interest on the road to new and better buildings and environments.

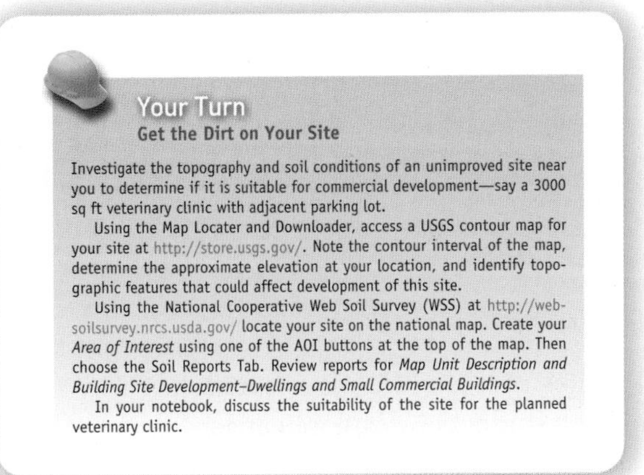

Your Turn

Get the Dirt on Your Site

Investigate the topography and soil conditions of an unimproved site near you to determine if it is suitable for commercial development—say a 3000 sq ft veterinary clinic with adjacent parking lot.

Using the Map Locater and Downloader, access a USGS contour map for your site at http://store.usgs.gov/. Note the contour interval of the map, determine the approximate elevation at your location, and identify topographic features that could affect development of this site.

Using the National Cooperative Web Soil Survey (WSS) at http://websoilsurvey.nrcs.usda.gov/ locate your site on the national map. Create your *Area of Interest* using one of the AOI buttons at the top of the map. Then choose the Soil Reports Tab. Review reports for *Map Unit Description and Building Site Development–Dwellings and Small Commercial Buildings*.

In your notebook, discuss the suitability of the site for the planned veterinary clinic.

- **Your Turn** activities reinforce text concepts with skill-building activities.

OFF-SITE EXPLORATION
Find out more about thermoluminescence and other dating technology from the Minnesota State University, Mankato, at http://www.mnsu.edu/emuseum/archaeology/dating/.

▶ **Off-Site Exploration** provides links to extended learning with resources for additional reading and research.

▶ **Key Terms** are defined throughout the text to help students develop a reliable lexicon for the study of engineering.

Built Environment: the human-made surroundings created to accommodate human activity, ranging from personal residences to large, urban developments.

Surveyors specify locations and describe spatial relationships using various representational systems, including metes and bounds descriptions and the rectangular survey system.

S T E M

STEM connections show examples of how science and math principles are used to solve problems in engineering and technology.

Careers in Civil Engineering and Architecture

PUTTING IT ALL TOGETHER

Life is never dull for architect Pamela Campbell. She works on multiple projects at once, all of them interesting. She sees them through various stages and meets with a wide range of people.

"Architects are coordinators to a great extent," Campbell says. "We pull different bodies of knowledge together to make a complete physical creation. It's very satisfying to put it all together for a project."

Pamela Campbell
Senior Associate, Cook + Fox Architects

On the Job

For six years, Campbell has worked on Henry Miller's Theatre in New York City's Broadway. She was part of the team that developed the building, and she has followed it through to construction. She works with the construction manager, the owner of the project, the tradespeople, and the theater company that will occupy the site. She also communicates with the press and others interested in the project.

"In this stage, I spend about a third of my week on the construction site for the theater," Campbell says. "I answer questions, do sketches, and try to resolve issues that come up during construction."

Campbell is also working on a house in Syracuse, New York, called the LiveWorkHome. It's a prototype intended to provide low-cost housing for single families. Part of the challenge is incorporating an environmentally sustainable element into a tight budget.

"It's been a fun process," Campbell says. "It's a lot smaller and a lot faster than the Henry Miller's Theatre project."

Inspirations

Campbell always liked to make things, even as a child playing in the yard or building sand castles.

"There's a cliché with architecture students that you start playing with Legos and you suddenly know you're meant to be an architect," she says.

Campbell enjoyed art classes in school, as well as math and science.

"Architecture is one of those professions that bring together lots of fields," she says. "It's a combination of the creative side of things and practical considerations, such as physics and measurement."

Campbell toured universities and found herself attracted to the atmosphere of architectural classes. "The creative aspect is what drew me to it," she says. "I liked all the little models and drawings. You are given a project and have to create something in physical form."

Education

Campbell enrolled in the Mackintosh School of Architecture in Glasgow, Scotland. She liked the fact that the program wasn't classroom-based.

"It was about self-motivation and discovering for yourself," she says.

In one of her first projects at school, Campbell had to build a children's playhouse. She measured her "clients," figuring out how big their hands were so she could create a structure that suited them. "It was a small project," she says, "but a fun starter place. As we went on, the projects became larger and more complex."

Advice for Students

Campbell suggests that high school students find an architect to shadow for a few days.

"I would try to get a sense of what the daily work is like before you start university and commit to it completely," she says.

Chapter 20: Formal Communication and Analysis 19

▶ **Career Profiles** provide role models and inspiration for students to explore career pathways in engineering.

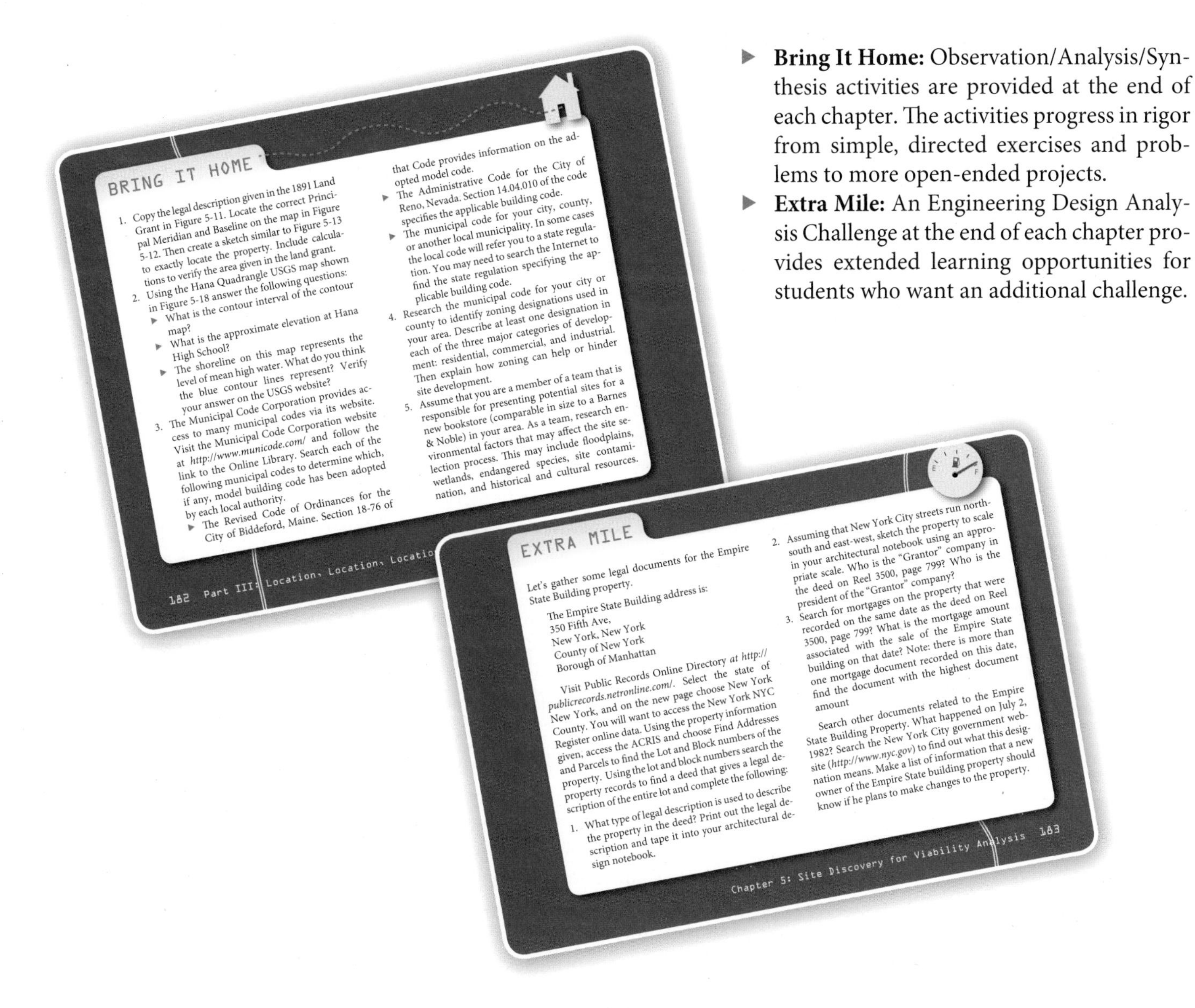

BRING IT HOME

1. Copy the legal description given in the 1891 Land Grant in Figure 5-11. Locate the correct Principal Meridian and Baseline on the map in Figure 5-12. Then create a sketch similar to Figure 5-13 to exactly locate the property. Include calculations to verify the area given in the land grant.
2. Using the Hana Quadrangle USGS map shown in Figure 5-18 answer the following questions:
   - What is the contour interval of the contour map?
   - What is the approximate elevation at Hana High School?
   - The shoreline on this map represents the level of mean high water. What do you think the blue contour lines represent? Verify your answer on the USGS website?
3. The Municipal Code Corporation provides access to many municipal codes via its website. Visit the Municipal Code Corporation website at *http://www.municode.com/* and follow the link to the Online Library. Search each of the following municipal codes to determine which, if any, model building code has been adopted by each local authority.
   - The Revised Code of Ordinances for the City of Biddeford, Maine. Section 18-76 of that Code provides information on the adopted model code.
   - The Administrative Code for the City of Reno, Nevada. Section 14.04.010 of the code specifies the applicable building code.
   - The municipal code for your city, county, or another local municipality. In some cases the local code will refer you to a state regulation. You may need to search the Internet to find the state regulation specifying the applicable building code.
4. Research the municipal code for your city or county to identify zoning designations used in your area. Describe at least one designation in each of the three major categories of development: residential, commercial, and industrial. Then explain how zoning can help or hinder site development.
5. Assume that you are a member of a team that is responsible for presenting potential sites for a new bookstore (comparable in size to a Barnes & Noble) in your area. As a team, research environmental factors that may affect the site selection process. This may include floodplains, wetlands, endangered species, site contamination, and historical and cultural resources.

182 Part III: Location, Location, Locatio

EXTRA MILE

Let's gather some legal documents for the Empire State Building property.

The Empire State Building address is:
350 Fifth Ave,
New York, New York
County of New York
Borough of Manhattan

Visit Public Records Online Directory *at http://publicrecords.netronline.com/*. Select the state of New York, and on the new page choose New York County. You will want to access the New York NYC Register online data. Using the property information given, access the ACRIS and choose Find Addresses and Parcels to find the Lot and Block numbers of the property. Using the lot and block numbers search the property records to find a deed that gives a legal description of the entire lot and complete the following:

1. What type of legal description is used to describe the property in the deed? Print out the legal description and tape it into your architectural design notebook.
2. Assuming that New York City streets run north-south and east-west, sketch the property to scale in your architectural notebook using an appropriate scale. Who is the "Grantor" company in the deed on Reel 3500, page 799? Who is the president of the "Grantor" company?
3. Search for mortgages on the property that were recorded on the same date as the deed on Reel 3500, page 799? What is the mortgage amount associated with the sale of the Empire State building on that date? Note: there is more than one mortgage document recorded on this date, find the document with the highest document amount

Search other documents related to the Empire State Building Property. What happened on July 2, 1982? Search the New York City government website (*http://www.nyc.gov*) to find out what this designation means. Make a list of information that a new owner of the Empire State building property should know if he plans to make changes to the property.

Chapter 5: Site Discovery for Viability Analysis 183

- **Bring It Home:** Observation/Analysis/Synthesis activities are provided at the end of each chapter. The activities progress in rigor from simple, directed exercises and problems to more open-ended projects.
- **Extra Mile:** An Engineering Design Analysis Challenge at the end of each chapter provides extended learning opportunities for students who want an additional challenge.

Supplements

A complete supplements package accompanies this text to help instructors implement twenty-first–century strategies for teaching engineering design:

- A **Student Workbook** reinforces text concepts with practice exercises and hands-on activities.
- An **Instructor's e-resource** includes solutions to text and workbook problems, instructional outlines and helpful teaching hints, a STEM mapping guide, PowerPoint presentations, and computerized testing options.

How *Civil Engineering* and *Architecture* **supports STEM Education**

Math and science are the languages we use to communicate ideas about engineering and technology. It would be difficult to find even a single paragraph in this text that does not discuss Science, Technology, Engineering, or Mathematics. The authors of this text have taken the extra step of showing the links that bind math and science to engineering and technology. The STEM icon shown here highlights passages throughout this text that explain how engineers and architects use math and science principles to support successful designs. In addition, the Instructor's e-resource contains a STEM mapping guide to this career cluster.

ACKNOWLEDGMENTS

A text like *Civil Engineering and Architecture* could not be produced without the patient support of family and friends and the valuable contributions of a dedicated educational community. The authors wish to acknowledge several individuals for their support and patience throughout this process:

The authors would like to thank CityCenter Las Vegas, Syracuse Center of Excellence, and Rowlee Construction, Inc., for case study contributions. The three major case studies offer real-life examples of residential, commercial land development and urban planning to increase student interest and understanding. Edward Bogucz Jr., Mark Lichtenstein, and Martin Wells of the Center of Excellence and Sven Van Assche, Ken Mize, Ken Karren Jr., and Natalie Mounier of the CityCenter Project all contributed valuable resources, and generously gave of their personal time and expertise to communicate the unique and inspiring stories behind the commercial building projects. A house designed and constructed by Rowlee Construction Inc. provided a common thread through the residential design chapters. Special thanks are given to designer/builder Taber Rowlee and homeowners Deb and Paul Foster for the in-depth look at the development of a charming residence from conception to completion. Rick Cook and Pam Campbell of Cook+Fox Architects in NYC are leaders in green, sustainable, and resilient architecture. Thanks to Rick Cook and Pam Campbell this book contains several examples of unique, innovative, and LEED (Leadership in Energy and Environmental Design) certified building designs.

Special thanks must be given to Tom White, nationally recognized and respected director of Project Lead The Way®, for his innovative ideas, support, and collaboration. The lead author Donna Matteson would like to thank the State University of New York at Oswego, for approval of a yearlong sabbatical to write this book. Throughout the process the book benefited from the authors' network of professional colleagues. Ms. Matteson would like to extend appreciation to daughter Star Matteson, a 13-year veteran of technology education and PLTW pre-engineering, for insightful manuscript contributions and for obtaining valuable student input on readability, interest, and relevancy.

Illustrations and photographs bring text to life. To that end the significant efforts of the following people are greatly appreciated. Ms. Matteson extends a special thanks to long-time colleague and friend Richard Kulibert Jr. for hundreds of page reviews, and the many ideas to expand and improve chapter content. In addition, Mr. Kulibert, NYS Technology teacher is to be recognized for his technical support and for creation of CAD drawings to enhance Chapter 9. Project Lead The Way® teachers Michael Elliott, Wendy Stearns, and Richard Salamone furnished Chapter 3 with a final project completed during summer training at RIT. Photographs of the drawings, models, and poster display illustrate expected student documentation and project outcomes. Dan Braun, graduate of SUNY at Oswego dedicated many hours to create CAD drawings of architectural elements featured in Chapter 4. Another SUNY Oswego graduate, Cole Moon, contributed his amazing photographs of the aftermath of Katrina. Throughout the lengthy book development process, Ms. Matteson was especially thankful for the continued support, encouragement, and inspiration from career-long colleague, Tom Frawley.

Ms. Kennedy would like to thank her children, Jennifer and Bradley (both PLTW students), for their support and patience during the development of this text.

Dr. Baur would like to acknowledge the drawings provided by Charles Berendzen and Donald Blocks' architectural drafting class at Rolla Technical Institute, as well as the electrical drawings from Greg Thiel and Leo Peirick. This work would not have been possible without the support of my wife Martina and sons Markus, Erich, and Wilhelm.

The publisher wishes to acknowledge the invaluable wisdom and experience brought to this project by our focus group and review panel:

Focus Group:

Connie Bertucci, Victor High School, Victor, NY

Omar Garcia, Kearny High School, San Diego, CA

Brett Handley, Wheatland-Chili Middle/High School, Scottsville, NY

Donna Matteson, State University of New York at Oswego

Curt Reichwein, North Penn High School, Lansdale, PA

George Reluzco, Mohonasen High School, Rotterdam, NY

Mark Schroll, Program Coordinator, The Kern Family Foundation

Lynne Williams, Coronado High School, Colorado Springs, CO

Review Panel:

Bob Bentley
Academy of Architecture and Engineering
Naples High School
Naples, FL

Todd Benz
Pittsford-Mendon HS
Pittsford, NY

Connie Bertucci
Victor Senior High School
Victor, NY

Dale Coalson
Science Academy of South Texas
Mercedes, TX

Eric Dunn
Sinclair Community College
Dayton, OH

Eric Fisher
Hamilton Heights High School
Arcadia, IN

Wendy Ku
Simsbury High School
Simsbury, CT

David Lynch
Memorial High School
Madison, WI

Dara M. Randerson
Oswego East High School
Oswego, IL

Greg Thiel
Lee's Summit High School
Lee's Summit, MO

Peter Tucker
Triad High School
Troy, IL

Technical Edit

Jean Uhl, P.E.
Instructor, Construction Management
Department of Construction Management & Civil Engineering
Georgia Southern University
Statesboro, GA

The publisher also wishes to thank our special consultant for this series:

Aaron Clark, North Carolina State University, Raleigh, NC

Finally, the publisher extends special thanks to Project Lead The Way's® curriculum directors Sam Cox and Wes Terrell for reviewing chapters at the manuscript stage.

ABOUT THE AUTHORS

Donna M. Matteson is an Associate Professor in the Department of Education at the State University at Oswego. Ms. Matteson has a total of 25 years of teaching experience and 8 years of experience in business and industry. Her area of expertise is Architecture and Computer Aided Design with AutoCAD certification through Autodesk Inc. During the summer months, Ms. Matteson is a Master Teacher for Project Lead The Way® and has provided graduate-level summer institutes at eight different universities. Ms. Matteson has a relationship more than a decade long with Project Lead The Way®, serving in the capacity of strategic planning, curriculum writing, and exam development. She was a lead writer and reviewer of the original PLTW Civil Engineering and Architecture curriculum and consultant and contributor to the latest edition.

Ms. Matteson is working on a dissertation of research to identify biophilia conditions that can improve the learning environment. During 2006–2008, she was project director for the current Introduction to Engineering Design curriculum and wrote the reverse engineering chapter for the first text on this series, *Engineering Design: An Introduction*. Ms. Matteson's research on reverse engineering and learning through inquiry has been published in professional journals and proceedings.

As one of the leaders in transitioning Industrial Arts to Technology Education in New York State, Ms. Matteson is well known in her profession and has presented at numerous state, national, and international conferences, including Hawaii and Japan. Currently she serves as scorer and reviewer of the NYS Teacher Certification

Content Specialty Exam. Her awards include the *USA Today* All Teacher Team for two consecutive years, IBM and Technology and Learning Teacher of the Year for New York and the northeast United States, and New York State Senate Woman of Distinction. Ms. Matteson's professional service includes more than 10 years of service to the New York State Technology Education Association, a three-year term on the Board of Directors for Technology Alliance of Central New York, and co-trustee of Epsilon Pi Tau, International Honor Society.

Deborah Kennedy is a professional engineer. She earned a B.S. in Civil Engineering from Purdue University and an M.S. in Civil Engineering from the University of New Hampshire. She holds an M.A. in Teaching Secondary Mathematics from The Citadel. Ms. Kennedy has practiced structural engineering with many consulting engineering firms including Bechtel, E. C. Jordan & Co., Rust International Inc., and Lockwood Greene. Ms. Kennedy also has 10 years of experience teaching high school mathematics and PLTW pre-engineering courses including Civil Engineering and Architecture. Since 2009 Ms. Kennedy has been a curriculum writer for PLTW Inc.

Stuart Baur is a registered architect and civil engineering doctorate. He joined the faculty of the Missouri University of Science and Technology in 2002. Dr. Baur has practiced with the architectural firms of Arquitectonica, The Nichols Partnership, and Bermello Ajamil and Partners in South Florida. His notable projects include the American Embassy in Lima, Peru, the Banque of Luxembourg, the Sawgrass Mills Mega, and the American Airlines Concourse A expansion at the Miami International Airport. A joint project between the University of Miami and Habitat for Humanity during the 1992 Hurricane Andrew aftermath piqued his interest in stability and sustainable buildings and inspired him to complete master's and doctoral degrees in civil engineering.

Dr. Baur has been involved with the UMR solar house project since 2002, acting as faculty advisor to the 2005, 2007, and 2009 solar house teams. Related endeavors include the development of a solar village that also serves as a student housing and research facility to educate the public about renewable energy technology and environmentally friendly landscaping.

Dr. Baur is a member of the U.S. Green Building Council and a founding member of the Heartland Chapter (Mid-Missouri). He is a 2002 Missouri Chamber of Commerce Leadership Missouri Graduate.

Dr. Baur holds civil engineering degrees from the University of Miami and the Missouri University of Science and Technology, and architecture degrees from the Ohio State University and the University of Miami.

Eva Kultermann is a licensed architect and assistant professor at the Illinois Institute of Technology, College of Architecture. She has a background in construction and is currently involved in research in the area of energy-efficient and resource-efficient construction technologies.

PART I
Introduction to Civil Engineering and Architecture

CHAPTER 1

Definitions and History of Civil Engineering and Architecture

GPS DELUXE

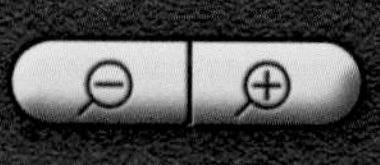

Menu

START LOCATION | DISTANCE | END LOCATION

Before You Begin

Think about these questions as you study the concepts in this chapter:

What are the differences and similarities between Civil Engineering and Architecture?

What were the earliest human-made structures? What structural systems were used?

How is a building's age determined?

How does a building maintain its architectural integrity when the building process extends past the lifetime of the architect?

Who were some of the distinguished architects and civil engineers in history and what were their accomplishments?

How did architecture and engineering achievements increase land use and urban development?

How did innovations in building materials contribute to the evolution of civil engineering and architecture?

Our human-made world is filled with examples of civil engineering and architecture. Just look out any window and you will see that civil engineering and architecture surround us. We travel roads, bridges, and rails to reach our home, school, and workplace. These structures represent the work of civil engineers and architects.

As you read this chapter, you will develop a better understanding of what civil engineers and architects do through study of their work. You will learn how to define the difference between civil engineering and architecture, as you read how these disciplines have contributed, separately and together, to our **built environment** (see Figure 1-1).

Built Environment: the human-made surroundings created to accommodate human activity, ranging from personal residences to large, urban developments.

As we examine a few examples of civil engineering and architecture in today's world, we will take an imaginary trip back in time to explore the history of civil engineering and architecture. Our journey will start in prehistoric times and make several stops on the way back to the present day. During this journey, you will read the stories behind history's most famous architects and civil engineers and their successful works.

Historical architecture provides clues about people, culture, environment, and concerns of the time. Just as architects and civil engineers have shaped our built environment, so has the natural environment shaped the course of architecture and civil engineering. As this chapter guides you on an imaginary journey through time, you'll learn how environmental and societal factors influenced the development of civil engineering and architecture. In turn, you'll discover how the achievements of architects and civil engineers throughout history have led to increased land use and urban development in today's built environment. Just look out any window.

Figure 1-1: *Miami skyline.*

DEFINING CIVIL ENGINEERING AND ARCHITECTURE

Before we begin our journey into the past, we need a clear understanding of the basic differences and similarities between the fields of civil engineering and architecture. Let's first take a look at civil engineering.

What Is Civil Engineering?

Civil engineering has a long history. The first engineer generally accepted by name and achievement is Imhotep, who lived from 2650 to 2600 BC. Imhotep is credited with overseeing the building of the Step Pyramid at Saqqarah in Egypt between 2630 and 2611 BC. We will make a stop during our journey to look at this structure.

The term *civil engineering* was first used in the eighteenth century to distinguish the newly recognized profession from military engineering. Civil engineering gained recognition as a discipline when the Institution of Civil Engineers was founded in London in 1818. This is how the Institution defined civil engineering in its original charter:

> ...the art of directing the great sources of power in nature for the use and convenience of man, as the means of production and of traffic in states, both for external and internal trade, as applied in the construction of roads, bridges, aqueducts, canals, river navigation and docks for internal intercourse and exchange, and in the construction of ports, harbors, moles, breakwaters and lighthouses, and in the art of navigation by artificial power for the purposes of commerce, and in the construction and application of machinery, and in the drainage of cities and towns.
>
> —*Reprinted from the Institution of Civil Engineers' original charter, 1828*

Point of Interest

The First Civil Engineer

Although the first engineers were military engineers making items for war such as catapults, rams, fortresses, and towers, civil engineering is the oldest of all other engineering disciplines. The first person actually to call himself a civil engineer was John Smeaton. Smeaton designed the first lighthouse to be located in the open sea. The lighthouse, called Eddystone, perched on a group of rocks 14 miles from the Southwest shore of England (see Figure 1-2). To build this structure, Smeaton created a form of cement that would harden in water.

Figure 1-2: The famous Eddystone lighthouse was made to withstand the pounding of waves on the reef using "hydraulic" cement and dovetailed stone blocks.

The American Society of Civil Engineers updated and expanded this definition in the twentieth century to describe how engineers apply their knowledge of math and science to meet human wants and needs:

> Civil Engineering is the profession in which a knowledge of the mathematical and physical sciences gained by study, experience, and practice is applied with judgment to develop ways to utilize, economically, the materials and forces of nature for the progressive well-being of humanity in creating, improving, and protecting the environment, in providing facilities for community living, industry and transportation, and in providing structures for the use of humanity.
>
> —*American Society of Civil Engineers, 1961*

Civil Engineering: the profession of designing and providing specifications for structural works that serve the general public, such as roads, bridges, water systems, and public buildings.

Today, we define **civil engineering** as the profession of designing and executing structural works that serve the general public. These structural works include bridges, canals, dams, roads, tunnels, harbors, water systems, and water treatment facilities (see Figure 1-3). Keep in mind that when researching the term *civil engineering,* you might find a variety of definitions.

Figure 1-3: **Wastewater treatment facility.**

Image copyright Wade H. Massey, 2008. Used under license from Shutterstock.com.

What Does a Civil Engineer Do?

Many people think of civil engineering as the design of bridges, but civil engineers provide structural design for a much wider variety of building projects. Today's civil engineers create structural designs for harbors, factories, power plants, transportation facilities, public meeting spaces, water treatment facilities, residential and commercial buildings, and others (see Figure 1-4). There are seven major interrelated branches of civil engineering:

BRANCHES OF CIVIL ENGINEERING

- Construction engineering
- Environmental engineering
- Geotechnical engineering
- Structural engineering
- Transportation engineering
- Urban and community planning
- Water resources engineering

Each branch of civil engineering requires specialized career training. You will read more about preparing for a career in civil engineering in Chapter 2.

Figure 1-4: **Civil engineers design complex transportation systems, such as the New York City subway.**

© iStock.com/Claudio Divizia.

Identifying Types of Structures

Civil engineers design structures to be safe, reliable, and resilient (see Figure 1-5). Their structures must withstand the vertical force of gravity and lateral (side) forces such as wind and earthquakes. Civil engineers ensure the safety and reliability of their structures by analyzing the bending, compression, and tension capacity of each building member.

As an example, consider the most basic structure of all: a simple wall. Structures built with walls include early stone and brick buildings, log cabins, concrete block, and preformed panels. By itself, a wall may not be very strong.

Figure 1-5: Bridges are designed to provide safe, reliable, and convenient means to cross water. The Brooklyn Bridge, NYC.

You may have noticed that the dug-out area around a new masonry block foundation wall is not backfilled with dirt until the house has been framed, or at the very least, a floor system has been added to the top of the foundation. Without vertical compression or connecting horizontal members, the wall cannot withstand the lateral or side loads from the backfill dirt.

Retaining walls are another structure that must resist the lateral movement of earth (see Figure 1-6). They are built to achieve an extreme change in grade and must withstand additional loads from stormwater.

Figure 1-6: The retaining wall in the foreground of this photo must withstand pressure from the lateral movement of the earth behind it and stormwater runoff.

Point of Interest

You may have seen a retaining wall that has failed or was about to fail. The Seattle.gov website provides a document highlighting an investigation conducted by the Seattle Public Utilities of a retaining wall that failed in December 2006 during a severe rainstorm. Visual inspection indicated that the concrete failed by torsion or twisting force. The final determination was that the failure was caused by increased water pressure from the rainstorm.

http://www.**seattle**.gov/util/stellent/groups/public/@spu/@esb/documents/webcontent/spu01_002567.pdf

Engineers have identified five basic structural systems based on geometric configurations (see Table 1-1). The reliability of each system can be analyzed by applying loads to its members. Using that information, civil engineers can identify advantages and disadvantages of each structural system, and select the best match for a given project. You will learn more about structural systems in Chapter 14. For the purpose of this chapter, you will identify the structural classification of several historical structures.

Your Turn

The PBS Building Big webpage provides several civil engineering activities.

Access the Public Broadcasting Service website at http://www.pbs.org/wgbh/buildingbig/ to learn about bridges, domes, skyscrapers, dams, and tunnels. Read about engineering wonders of the world and engineers and architects that "build big." Have some fun trying to solve the engineering challenges offered under each category.

You might want to copy the five classifications in your notebook and record the page number for future reference. As you study these classifications, you will begin to see them everywhere!

Table 1-1: **Structural Systems**

Category	Recognized by	Example	Illustration
Column and beam (post and lintel)	Distance between vertical members (columns/post) is determined by the spanning capacity of horizontal members (beam/lintel)	Post and beam construction	Column-and-beam construction © iStock.com/laughingmango
Corbel (pronounced *core-bull*)	Corbelling: objects laid in horizontal courses, each projecting slightly beyond the one below	Pyramids	Corbelled chimney © Cengage Learning 2012
Cantilever (pronounced *can't-ta-leave-r*)	Cantilevered: beams extend beyond their supports to form an overhang	Frank Lloyd Wright's Fallingwater	Cantilevered roof © iStock.com/Timothy Large
Arch and vault	Wedge-shaped members placed so that they press against each other to become self-supporting. A keystone is located at the top of the arch.	Arch	Archway © Cengage Learning 2012

(Continued)

Table 1-1: **Structural Systems (Continued)**

Category	Recognized by	Example	Illustration
Truss and space frame (pronounced *turr-us*)	Short members connected in triangular configurations	Railway bridge	Railway bridge © iStockphoto/MegapixelMedia
Tensile (pronounced *ten-sul*)	Use of suspended cables to support load	Suspension bridge	Suspension bridge © iStockphoto.com/Bill Grove

Your Turn

Take a minute to reread the definitions of civil engineering. Make entries in your notebook as to how the profession has evolved over time and the possible contributing factors. What global and human factors impact civil engineering today? Read your local newspaper or listen to a news station to find a local project related to civil engineering. Follow the project's progress and add articles and notes to your notebook. Determine why the project is being done. How will the completed project enhance the community? As you follow the progress of the project, list any problems that occur and how they are addressed. Determine when the project is expected to finish. How will the community be affected during the construction? For example, if a bridge were being replaced due to concerns of structural fatigue, where would the old bridge go? How would construction noise and dust affect the community? Would there be an increase in traffic congestion or accidents? Would traffic detours cause delays? And finally, what new material, design, or process will be used to enhance structure longevity and prevent a future failure? Throughout the review of your local project, continue to make entries in your notebook. When the project is complete determine if the project was finished on time and within budget. Make notes of things you would have done differently if you were "in charge."

What Is Architecture?

Architecture: responsible and purposeful design of functional, sustainable, and aesthetically pleasing buildings.

Unlike civil engineering, architecture is not clearly defined. If you ask a dozen architects to define architecture, you might get a dozen different answers. One might say that architecture exists where form and function come together; another might say that architecture achieves maximum comfort through minimal form; yet another might describe architecture as the form that emerges when we create harmony with the environment. For the purposes of this text, we will define **architecture** as the art and technique of building design through purposeful application of elements and principles of design. The visual elements of design are line, color, form, space, shape, texture, value, and tone. The principles of design are balance, contrast, emphasis, movement, pattern/repetition/rhythm, and unity/harmony. You will read more about the principles and elements of design and how architects apply them in Chapter 4.

Most architecture can be described by spatial relationships and purposeful combination of geometrical shapes. Prisms, pyramids, cones, cylinders, and other geometrical objects have attributes of size and proportion (see Figure 1-7). Their relationship as they touch, intersect, and combine is what creates a building's unique architectural form (see Figure 1-8).

Figure 1-7: An assortment of three-dimensional geometric shapes.

Beyond Form and Function

Architecture is not limited to a building's form and function. In addition to a visually pleasing and serviceable design, the building's architecture must *fit* within the surroundings. Recent architecture shows thoughtful consideration for the occupant's health, owner operation and maintenance cost, building longevity, future expansion, change of use, and the current and future impact on the environment. These factors are considerations of sustainable architecture. Sustainable architecture is achieved by addressing the impact of environmental, economical, and ethical factors. You will read more about green and sustainable architecture and building resiliency in future chapters.

Figure 1-8: How many geometric shapes do you recognize in Richard Meier's High Museum of Art in Atlanta, Georgia?

Point of Interest
An Architect's Definition

My area of study is architecture and its cultural manifestations, and such a quest for historical knowledge, especially of the arts, arises from my own need for a deeper understanding of the essence of architecture. Too frequently, the merely historical understanding of architecture is considered the complete and authentic understanding: whereas the sense of spiritual activity expressed in architectonic forms should be related to speculative interest in the nature of architectural forms more directly than to a historical interest in forms already realized, already claimed by the past. The one affords us the promise of architecture; the other, too often, ends in historicist ideology.

Richard Meier. *Richard Meier, Architect: Buildings and Projects, 1966–1976.* (New York, NY: Oxford University Press, 1976.)

THE HISTORY OF CIVIL ENGINEERING AND ARCHITECTURE: A JOURNEY THROUGH TIME

What were the earliest human-made structures? What types of structural systems were used? Let us begin our journey back in time to find out. Many of you have watched science-fiction movies or read stories of time travel where people are transported back in time through a portal or use of a time-travel device. Imagine that you found such a device similar to a Global Positioning System (GPS), which receives information from satellites to provide navigation. In addition to maps and directions, your time-travel device has a menu option for dates in time (see Figure 1-9).

Humans made dwellings by draping animal hides over bones and stakes.

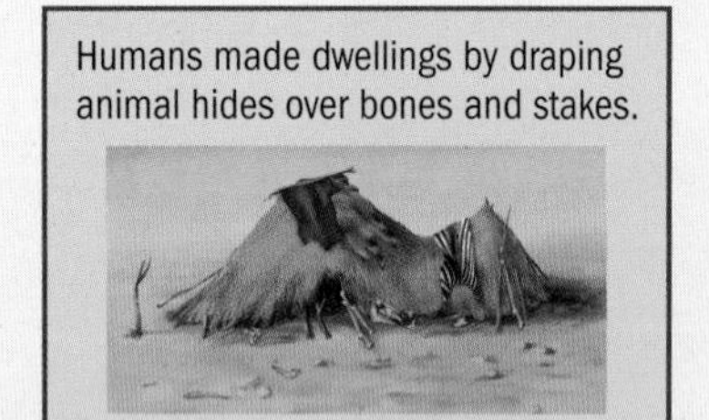

15,000–10,000 BC

Architecture evolved from the stone-age need to provide shelter toward the societal need to provide security and gathering spaces. **Stonehenge** was built near Salisbury, England, over a period of about a thousand years. Its purpose and means of construction remain a mystery.

3000 BC

Stone Age or Paleolithic | Mesolithic Age | Neolithic Age | Early Bronze Age

300,000 BC

Dwellings found in Nice, France, were made from stakes pushed into the sand and supported by a ring of stones.

8000 BC

The settlement of **Jericho** in the Jordan Valley consisted of circular shaped mud huts surrounded by a wall up to 27 feet thick enclosing almost 10 acres.

Building in Prehistoric Times: 300,000 BC

As you study civil engineering and architecture, you might wonder about the early buildings in history. What were the first human dwellings? What did they look like? What materials were used for construction? Armed with a notebook and pen, you decide to take your first trip back in time. You have always been interested in the prehistoric era, so you enter 300,000 BC, which is in the Paleolithic period. You are quickly transported to a sandy beach in Nice, a city on the Mediterranean coast of southern France. Oval huts ranging in length from 26 to 49 feet, and in width from 13 to 19 feet are built along the shore. These are the earliest known buildings on architectural record. **Thermoluminescence** dating of the hut floors indicate that they were built approximately 380,000 years ago! You explore further to discover central hearths and walls constructed with 3-inch diameter stakes set in sand and braced on the outside by a ring of stones.

Figure 1-9: **Imagine discovering a device that would allow you to travel back in time.**

During the Paleolithic and Mesolithic periods of the Stone Age, people were mobile. Mammoth bone huts were believed to exist in 20,000 BC in the Ukraine. The foundation was created of mammoth jaws and large bones, and the roof was constructed with tree branches and tusks.

After entering 15,000–10,000 BC into your time-travel device, you discover animal hide tents in France and in the severe climates of Russia and Switzerland pit houses, which were dug into the ground. These ancient pit houses might have been inspiration for today's earth bermed structures built partially underground for energy efficiency.

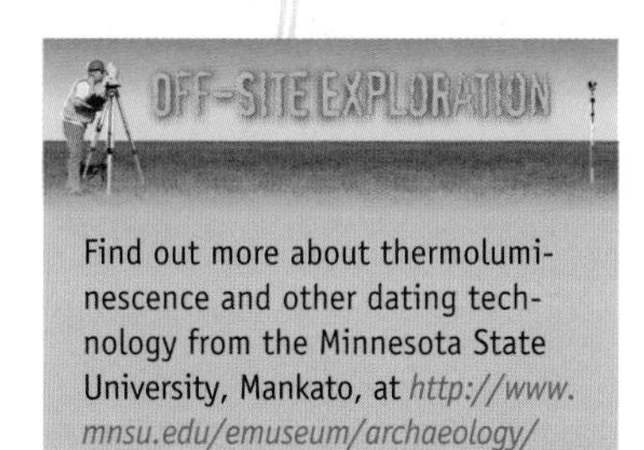

Find out more about thermoluminescence and other dating technology from the Minnesota State University, Mankato, at *http://www.mnsu.edu/emuseum/archaeology/dating/*.

2750 BC

The builder **Imhotep** constructed the step pyramid of King Djoser in **Ancient Egypt.** Many scholars consider Imhotep to be the first engineer and architect in history to be known by name.

2720–2560 BC

The straight-sided Great **Pyramid at Giza** originally covered more than 13 acres and stood 481 feet tall. Scholars continue to speculate on how it was built.

448–432 BC

Greek architecture displays classical design principles of symmetry and balance.

Hellenistic Period

250 BC–AD 1500

The Great Wall of China, stretching for more than 4,000 miles, was built to protect the borders of the Chinese Empire.

Roman Period

AD 36–50

The Nîmes aqueduct in France stretches about 30 miles from Uzès to Nîmes and includes the **Pont du Gard** bridge.

Point of Interest
Thermoluminescence Dating

Thermoluminescence dating, which is used on rocks, minerals, and pottery that existed between the years 300–10,000 BC, is based on the fact that almost all natural minerals are thermoluminescent. Energy absorbed from ionizing radiation frees electrons to move through the crystal lattice, and some are trapped at imperfections. Later heating releases the trapped electrons, producing light. Measurement of the intensity of the luminescence can be used to determine how much time has passed since the last time the object was heated. The light is proportional to the amount of radiation absorbed since the material was last heated. Natural radioactivity causes latent thermoluminescence to build up so the older an object is the more light is produced. Because thermoluminescence technology is still in a developmental stage, it is not considered accurate enough for archaeological standards.

Human populations use environmental resources in order to maintain and improve their existence.

After recording these findings in your notebook, you are off to Jericho in the Jordan Valley, around 8000 BC. The settlement of Jericho consisted of circular shaped mud huts, surrounded by a wall up to 27 feet thick enclosing almost 10 acres (see Figure 1-10). You are surprised to find a circular stone tower at the center of town. It is believed that Jericho reached a population of 2000 by 7000 BC.

Gothic architecture employs a skeletal system featuring arches, rib vaults, and flying buttresses

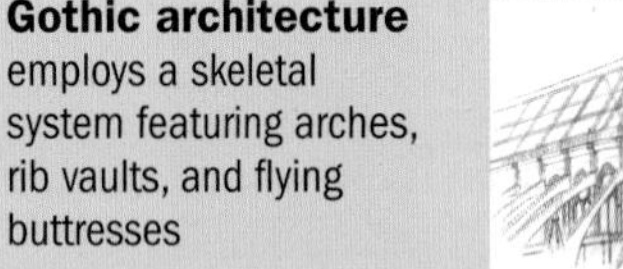

AD 1140–1520

Renaissance architecture valued formality and classical elements like pediments and circular forms.

AD 1400–1600

Roman Period — Medieval Period

27 BC–AD 25

The **Pantheon** displays **Romanesque** style, using a continuous wall to sustain the load

AD 1485–1547

Tudor architecture returned to a plainer, symmetrical style that emphasized horizontal elements; designed to provide luxurious interior comfort.

Before you leave the Stone Age, you make a quick stop at one of the oldest known Neolithic settlements in Europe, around 6220 BC. This settlement featured early timber frame houses approximately 25 feet by 25 feet in size. These mud-walled structures were built by inserting oak saplings into deep footings of mud. You may remember from history class that farming most likely led to the establishment of early urban societies. The agricultural surplus of these areas provided opportunity for some residents to take on roles unrelated to food, such as a religious leader, craftsperson, merchant, or builder.

Figure 1-10: **Circular houses built in ancient Jericho with corbelled mud bricks might have looked similar to these structures.**

You decide to make a note in your notebook that early hunters and gatherers needed temporary movable shelters, whereas early farmers required more permanent dwellings. Referring back to your structural systems chart, think about which systems may have been applied during the Stone Age. Tents may have employed a tensile system, whereas some of the early huts used a post-and-lintel framework system.

Early structures were not limited to living spaces. As you continue your time travel, you will discover tombs of stacked stone slabs across Europe and Asia. Architecture evolved from addressing the basic needs of shelter, security, or worship. Construction techniques were influenced by whatever materials were readily available.

Building in the Ancient World: 3200 BC to AD 337

Continue on your journey to what many consider the most famous and mysterious relic of prehistory in Europe: Stonehenge (see Figure 1-11). The cover story of the June 2008 *National Geographic* magazine was devoted to the "Secrets of Stonehenge."

Although the purpose of Stonehenge still remains a mystery, there are many theories. Cremation remains have led some to believe it was used as a cemetery. Others believe the stones were purposefully aligned with solar and lunar phenomena to create a temple or astronomical calendar. Whatever the purpose, the magnitude of this project has earned admiration from generations of scholars, tourists, and history buffs alike.

Stonehenge could date back as far as 3000 BC. Circular pits lining the inner edge of the circular ditch and bank of Stonehenge may indicate original wooden pole construction. As the construction of Stonehenge continued through several phases, extending over a period of a thousand years, it is likely that many changes and adaptations occurred during that time. The early bluestones weighing approximately four tons probably appeared around 2500 BC. The stones are thought to have been brought in from Wales, 250 miles away.

Figure 1-11: **Stonehenge, located near Salisbury, England.**

Many still wonder about the purpose of Stonehenge and its specific design and layout—a circle of stones 100 feet in diameter that towers approximately 16 feet above ground. The huge stones were shaped and placed in a very precise pattern. How were such large stones raised and placed into position? This very question still intrigues researchers. Some have surmised that construction methods employed a simple machine such as a lever and inclined plane, a sledge and greased track, wooden

Figure 1-12: A Stonehenge tenon is seen at the top of the tallest stone. Mortise or openings in the horizontal members were aligned with the vertical tenon, creating the interlocking construction.

© iStockphoto.com/James Robinson.

scaffolds, ropes, and the combined efforts of more than 100 people. We can see that the massive lintels, or horizontal members, are secured to their uprights by mortise-and-tenon joints (see Figure 1-12). This type of joinery is still used in woodworking and construction today (see Figure 1-13). Stonehenge's mortise-and-tenon interlocking construction has enabled this dry stone structure to withstand the elements for almost four thousand years.

Were these builders some of our first civil engineers? Looking back to your structural systems chart, which systems do you think were applied to the building of Stonehenge?

RECENT DISCOVERIES AND INVESTIGATIONS Recently, an investigation of Stonehenge has led researchers to a nearby Neolithic village at Durrington Walls, dating between 2600 and 2500 BC. Researchers believe that this village may have contained 300 houses, making it one of the largest Neolithic settlements to be found in Britain. It is believed the houses were constructed of wattle and daub, a building material of wooden strips or branches and a combination of wet soil, clay, sand, animal dung, and straw. This building technique has been used for more than 6,000 years, and can still be seen in developing countries. A similar process of lath and plaster, used extensively in the United States up until the late 1950s, might have evolved from the wattle and daub process.

In this text we describe what we currently know and accept as history. However, new discoveries can reveal the existence of other early shelters and settlements that change our perception of history. For example, archeologists recently unearthed a 4,000-year-old temple on the northern coast of Peru. The temple contains a mural of a deer caught in a net. Archeologists used **carbon dating** to confirm the age of the mural and temple. The temple was made of blocks built from river sediment. This primitive construction material created a stark contrast to the elaborate

Carbon Dating:
used by archeologists to estimate the age of organic remains and other substances rich in carbon. Carbon dating can be used on items as old as 60,000 years.

Figure 1-13: Mortise-and-tenon joints were used in this wooden post-and-beam construction.

© iStockphoto.com/troy wuelfing.

© Cengage Learning 2012

Fun Fact

A retired construction worker from Michigan, Wally Wallington, thinks he knows how Stonehenge was moved and erected. To prove his theory, he has made a cement replica of a 16-foot tall stone hedge arch and singlehandedly moved and stood the blocks by applying his knowledge of physics. Using two small stones and a long lever, he was able to move a 1-ton block 300 feet per hour. Gravity, weight, and blocking were then used to stand the 19,200 lb blocks. The feat was taped for the Discovery Channel and can be seen at http://www.theforgottentechnology.com.

mural and structure. Artifacts from the site included shells that might have come from coastal Ecuador, indicating the possibility that the region was a cultural exchange point. You can find out more about this discovery and see photos from the excavation at *http://www.reuters.com/article/newsOne/idUSN1018888320071111.*

Building in Ancient Egypt

The next stop on your journey will be ancient Egypt. During this period, Pharaohs believed they would need their worldly possessions in the afterlife, which led to the design and development of tombs. In your trip back in time, you discover that Mastabas were the first pyramids; however, the shape may surprise you. They were simple flat-roofed rectangular structures made of mud, brick, or stone. You begin to see the evolution of the pyramid in the design of step pyramid, which was comprised of six stacked mastabas, each progressively smaller. Look closely at Figure 1-14 and identify the structural system used to construct the step pyramid. You may be surprised to learn that this pyramid still stands today.

Figure 1-14: The step pyramid of King Djoser was constructed in 2750 BC.

Image copyright Svetlana Privezentseva, 2010. Used under license from Shutterstock.com.

The builder of the step pyramid was Imhotep, whom we mentioned earlier in this chapter as the first engineer to be recognized by name and achievement. Many also consider him to be the first architect in history known by name. Although he was not the first to build with stone, the immense size of the step pyramid, made entirely of stone, is considered a signficant innovation. Imhotep applied simple tools and mathematics of proportion and scale to develop the art of building with stone. Some believe that Imhotep may also be connected to the first use of columns in architecture.

Application of techniques, tools, and formulas to determine measurements and proportions.

When most of us think of a pyramid, we envision triangular forms. The Great Pyramid at Giza (Figure 1-15), built between 2720 and 2560 BC, is a good example. It covered more than 13 acres with sides 755 feet long angled at 51 degrees. Originally, these walls stood 481 feet high. Now standing at 450 feet, its features are so large it can be seen from the moon. Recent computer calculations estimate approximately 500,000 stone blocks, each weighing more than 2 tons, were used to build this pyramid. How were these huge stones moved and placed? Even today, there is much speculation about the construction of the pyramids. Tomb paintings depict huge blocks being moved on sledges over lubricated ground and the use of ramps to bring

Your Turn

Many questions concerning construction of the pyramids are still left unanswered. Even the mortar used remains a mystery. Some believe that analysis could lead to a more definitive dating of the pyramids. Use your Internet search engine to research pyramid mortar to read more about the carbon dating of organic material within the mortar. You may be surprised to learn that the mortar joints of the great pyramid are consistently only 1/50 of an inch.

the blocks to their position on the pyramid. Revisit your structural systems chart from this chapter. Into which category would you place the Great Pyramid at Giza? You are right if you said corbelling. You can see in Figure 1-15 how each new row of blocks was offset from the previous row.

Figure 1-15: The Great Pyramid at Giza is the sole remnant of the Seven Wonders of the World, with features so large they can be seen from the moon.

Image copyright javarman, 2010. Used under license from Shutterstock.com.

HISTORY INSPIRES MODERN-DAY DESIGNS Although materials and building construction are much different, the pyramid shape is still used in building design. The famous glass pyramid by Ieoh Ming Pei at the Louvre Museum in Paris (Figure 1-16) was inaugurated in 1989. The pyramid serves as main axis of circulation and provides access to the large reception hall beneath.

Another modern-day pyramid is the Luxor Hotel and Casino, built in Las Vegas, Nevada (see Figure 1-17). One of its interesting features is an inclinator that is similar

Figure 1-16: The famous glass pyramid at the Louvre Museum in Paris by architect Ieoh Ming Pei.

Image copyright Jozef Sedmak, 2010. Used under license from Shutterstock.com.

Figure 1-17: The Luxor Hotel and Casino built in Las Vegas, 1991–1993.

Image copyright Andy Z, 2010. Used under license from Shutterstock.com.

to an elevator but has the ability to move diagonally. The Luxor's inclinator travels 30 floors, moving people at a 39-degree angle from floor to floor. Another unique feature is its 42.3 billion candlepower beam, which projects outward from the upper point of the pyramid. It has been said to be the strongest beam of light in the world and is clearly visible from outer space.

Figure 1-18: *Great Wall of China.*

Building in Ancient Asia: 250 BC–AD 1500

The next stop on your journey is the Great Wall of China (see Figure 1-18). The Great Wall was built to protect the borders of the Chinese Empire and stretched across 4000 miles. It was originally built using materials readily available near the building site. Stone was used when building in the mountain areas, and compacted earth was used in the plains. Each year, thousands of visitors walk the long and often steep Great Wall of China.

Architecture in Ancient Greece: The Parthenon, 448–432 BC

A trip back in history isn't complete without a stop in Athens to visit one of the most famous examples of classical architecture—the Parthenon (Figure 1-19). Phidias, a famous sculptor, designed the Parthenon, and architects Ictinos and Callicrates supervised its construction between 448 and 438 BC. The structure is approximately 111 feet by 228 feet and stands on a hill called the Acropolis, meaning "high city." The Parthenon, made from 22,000 ton of marble, is a famous example of classical architecture and is easily identified by extensive detail and massive columns. Looking back at your structural systems chart, into which system category would you place the Parthenon? Did you choose post and beam? Why do you think the columns were spaced so close together? If a different material was used for the horizontal or beam pieces, could the columns have been spaced farther apart? What can you conclude about the properties of marble?

COLUMNS USED IN TODAY'S ARCHITECTURE There are many styles of columns in use today. Their function is both decorative and structural. You often see them used as structural members for a front porch or portico. Columns are one of the most distinctive features used to identify building styles. You will read more about architectural styles in Chapter 4: Architectural Design.

Figure 1-19: **The Parthenon stands in ruins from damage from a fire during ancient times and an explosion in 1687.**

Engineering in Ancient Times

On the next stop of your journey, you will discover the ancient Roman aqueducts. Aqueducts were one of the greatest engineering feats of the ancient world, allowing people to move water over long distances with minimal loss. They were constructed in Persia, India, and Rome, with some dating back to the bronze age of 3500 BC. These irrigation systems supplied water to settlements and agricultural areas in need of water.

Fun Fact

An exact replica of the Parthenon was built in Nashville, Tennessee, in 1897 for the city's 100th anniversary. The building serves as the city's art museum.

The re-creation of the 42-foot statue of Athena is the focus of Nashville's Parthenon, just as it was in ancient Greece. The building and Athena's statue are both full-scale replicas of the Athenian originals.

Originally built for Tennessee's 1897 Centennial Exposition, this replica serves as a monument to what is considered the pinnacle of Classical architecture (see Figure 1-20).

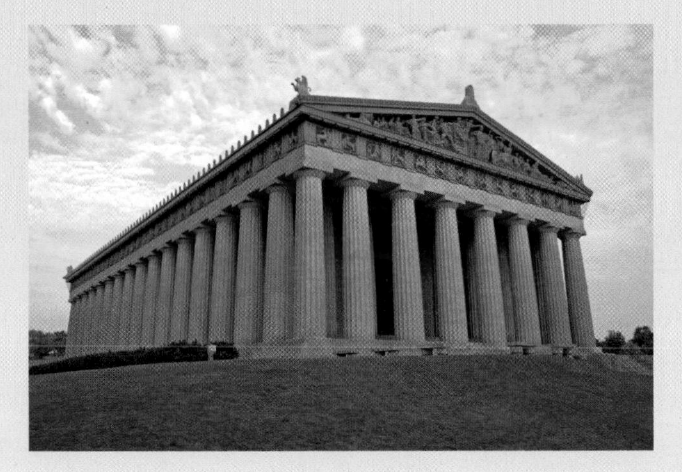

Figure 1-20: **The Parthenon replica stands proudly as the centerpiece of Centennial Park, Nashville's premier urban park.**

The ancient Romans were the most famous builders of aqueducts (see Figure 1-21). MatThey built eleven major aqueducts between 312 BC and AD 226. Builders used the principle of the arch to support these structures. (Revisit your chart on structural systems and reread the description of the arch.) Arches could span a wide distance and support weight. The early Roman arch used an odd number of bricks to form a semicircular pattern with a keystone or capstone at the very top.

The arch was used in doorways, bridges, gates, palaces, amphitheaters, and monuments. Many variations of the arch evolved, including the Gothic, jack, Tudor, and catenary arches. The catenary arch supports only its own weight. Can you think of a famous example of a catenary arch? You are right if you guessed the Gateway Arch in Saint Louis, Missouri. The Gateway Arch (Figure 1-22) was designed in 1947 and built between 1963 and 1968. The Gateway Arch is 630 feet wide at the base, and stands 630 feet tall.

Figure 1-21: **Pont du Gard.**

Architecture and Engineering of the Ancient Roman Empire

Your next stop will be the Pantheon in Rome, Figure 1-23. It is often difficult to determine the age of historical buildings like the Pantheon, but fortunately, important clues are often left behind by the architect or builder. The original Pantheon was a rectangular temple and included an inscription on the architrave indicating that it was built under the councilship of Marcus Vipsanius Agrippa. This inscription dates the construction of the original Pantheon between 27 and 25 BC. Building age is often determined by the building materials or construction techniques used. For example, the rebuild of the Pantheon following fires in AD 80 and 110 included bricks containing brickmaker stamps. These marks helped to determine that the restoration date was somewhere between AD 118 and 125. What other techniques are used to determine age of a structure? In this chapter, you have read about using thermoluminescence, carbon dating, and clues such as building materials, techniques, and inscriptions left by the architect or builder. Record this information in your notebook for future reference.

The Pantheon is easily identified by its Corinthian columns, set in a pattern of two rows of four columns each. These massive stone columns, which support the portico, are 39 feet tall by 5 feet in diameter.

Figure 1-22: The gateway in St Louis Missouri is an example catenary arch, and contains a tram system that transports visitors to an observation room at the top.

Gothic Architecture, Twelfth to Sixteenth Century AD 1140–1520

On next stop along your journey you will learn about Gothic architecture. Gothic architecture has been featured in many films and novels. For example, the famous Gloucester Cathedral, built in 1100 and rebuilt in 1331, was the setting for two Harry Potter films. Gothic architecture evolved from Romanesque architecture. The pointed Gothic arch allowed the building to be tall yet still have a large amount of window area for natural light (see Figures 1-25 and 1-26).

Because people's needs and wants change, new technologies are developed, and old ones are improved.

Gothic architecture flourished during the medieval period and is seen on many churches, cathedrals, castles, palaces, and university buildings. Although you can find many Gothic-style churches and universities, you will seldom see the style applied to residential structures.

Unlike Romanesque style, where a continuous wall sustains the load, a Gothic structure employs a skeletal system to transfer roof loads down to the ground at specific locations, thus affording huge uninterrupted spaces. The most recognizable features of Gothic architecture are pointed arches, rib vaults, and flying buttresses.

Originating in France, Gothic architecture soon spread to Britain, Germany, Italy, and Spain. In Italy, Gothic architecture was constructed with brick and marble instead of stone. Variations of Gothic architecture emerged, including Rayonnant and Flamboyant style. Rayonnant Gothic structures are known for their rose windows (see Figure 1-27). The more decorative

Figure 1-23: The Pantheon is considered to be the best preserved of all ancient Roman buildings.

Fun Fact

Did you know that the Roman Pantheon has the largest unreinforced solid concrete dome in the world? The dome has a diameter of 142 feet and a height of 142 feet from the floor. The dome has a hole 25.5 feet in diameter, which provides light to the interior. To reduce the weight of the dome, recessed panels and cement made with pumice were used in the upper levels (see Figure 1-24). What structural system is followed to create a dome? If you could slice a dome in half, what would you see?

Figure 1-24: **Looking from the inside of the Pantheon dome.**

Your Turn

To learn more about Roman architecture, use the Internet to search Roman villa, Roman Forum, amphitheater, Coliseum, Pantheon, Rome triumphal arch, and aqueduct. You might be surprised to learn of an ancient Roman system of central heating called a *hypocaust*, which created heated floors for public baths.

Flamboyant style is easy to recognize by its stone window tracings in the shape of an *S* or flame (see Figure 1-28).

Variations of styles are sometimes found on the same building. One famous example is the two contrasting towers on the west façade of the cathedral of Notre Dame de Chartres. One of the towers was damaged in the 1140s in a fire sparked by lightning. The new spire constructed in the sixteenth century featured a contrasting Gothic Flamboyant style. After examining Figure 1-29, can you tell which tower represents the Flamboyant style? (You are right if you chose the taller one.)

As you look at historical buildings in your own community, remember that there are many variations of architectural styles. Some are the result of availability of building materials, whereas other variations are based on design creativity. When creative designs became distinctively different from the original and gained popularity, the new variations of style were given a more specific, descriptive name, even if they still fell within the original, general style category. This was the case with Gothic architecture. Another such example is the British Perpendicular style of Gothic, which emphasized the vertical lines and elements (see Figure 1-30).

Figure 1-25: The Gothic Architecture of Santa Maria del Mar Church in Barcelona, Spain, provides large interior space and natural light.

Figure 1-26: The skeleton construction of Gothic architecture.

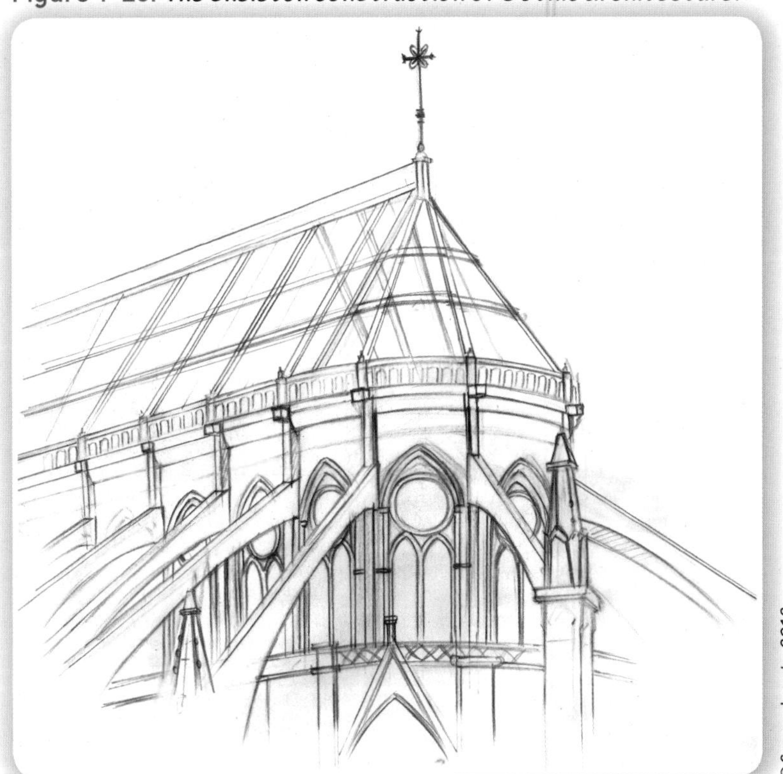

In historical perspective, science has been practiced by different individuals in different cultures. In looking at the history of many peoples, one finds that scientists and engineers of high achievement are considered to be among the most valued contributors to their culture.

Figure 1-27: A Rayonnant-style rose window in Notre Dame de Paris.

Figure 1-28: Flamboyant-style Gothic window.

Figure 1-29: The two contrasting towers of Notre Dame Cathedral in Chartres, France.

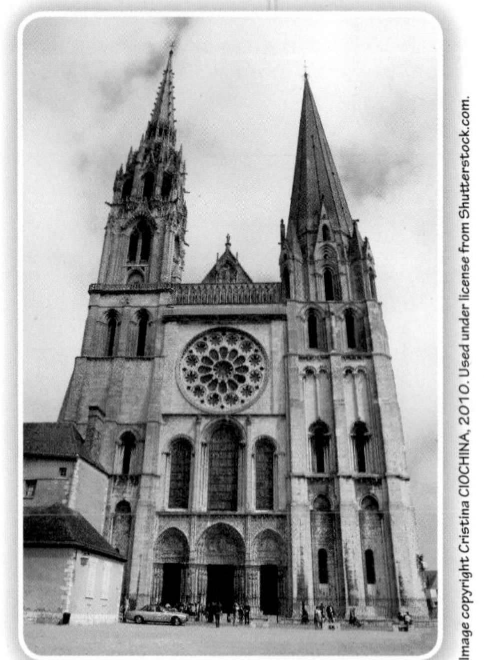

Figure 1-30: The Classic Perpendicular-style Lady Chapel was built between 1503 and 1512 at the eastern end of Westminster Abbey.

In the past, it took several years to build a structure. In some cases, it took several decades or even centuries! How did the builders maintain a structure's architectural integrity when the building process extended past the lifetime of several architects and designers? One solution was to match the form, materials, and details of the original style. A good example is Westminster Abby in London, England (Figures 1-32 and 1-33). Not much is left of the first structure, which opened in 1065. What you see now is the result of a major rebuild between 1245 and 1517 in the Gothic style. A chapel was added in the early 1500s. The two west front towers were added between 1722 and 1745. Finally, the north part of the building was completed in the nineteenth century. Although Westminster Abbey was built and rebuilt over several centuries, architects retained the building's Gothic style. Compare Figure 1-32 to Figure 1-33. What similarities and differences do you see? Notice the brick pattern? What problems do you think occurred during the process of rebuilding?

As the Gothic style evolved, the arches were flattened and windows were framed with rectangular trim. This evolution inspired the development of Tudor architecture between 1485 and 1547 (see Figure 1-34).

Renaissance Architecture 1400–1600

On the last stop of your historical journey, you will look at Renaissance architecture. This style of architecture was born in Italy following the Gothic period. The symmetrical and proportioned designs of the Renaissance were a stark contrast to the complex asymmetrical Gothic buildings. One of the most famous architects of this period was the Italian architect, Andrea Palladio (1508–1580). Andrea Palladio was known for classical design elements and mathematical order, which have influenced architectural styles for more than 500 years. The Villa Capra (Figure 1-35) is one of Palladio's best-known works.

Two hundred years after Palladio's time, Thomas Jefferson based the design of his famous home, Monticello, on Palladio's classical design principles (see Figure 1-36). Jefferson applied his understanding of geometric shapes

Your Turn

Use the Internet to research the Gloucester Cathedral, which was used in filming the Harry Potter films. Trace the timeline of the building and note the architectural features and changes that have occurred since 1072. Record the architectural style of the building, and print a picture of the fan-vaulted roof to include in your notebook (see Figure 1-31).

Historical buildings often have clues that uncover or document historical developments. See if you can locate a picture of the stained glass on the east wing of the cathedral, which dates back to 1350 and contains an image similar to the game of golf. There has been quite a debate as to where the game of golf originated; some believe it was in Scotland, while others think it was in the Netherlands.

Figure 1-31: Interior hallway of the Gloucester Cathedral.

Image copyright Ksenija Makejeva, 2010. Used under license from Shutterstock.com.

Figure 1-32: Westminster Abbey's west towers built between 1720 and 1745.

Image copyright Monkey Business Images, 2010. Used under license from Shutterstock.com.

Figure 1-33: The north entrance of Westminster Abbey completed in the nineteenth century.

Image copyright Stpehen Finn, 2010. Used under license from Shutterstock.com.

Figure 1-34: The history behind Hampton Court Palace in London dates back to 1236; however, the Tudor-style construction appeared in 1525.

Point of Interest

Did you know that Hampton Court Palace was once a prison for King Charles I? It has been reported to be haunted by ghosts. Go to http://www.cnn.com/2003/WORLD/europe/12/19/hampton.ghost/index.html?iref=allsearch to hear the story and learn more about the history of this building.

Your Turn

Use the Internet to research the four basic shapes of Gothic arches: Lancet, Equilateral, Flamboyant, and Depressed. Make sketches of them in your notebook and describe their features and the Gothic style they represent. Locate one historical example of each arch and print a copy for to your notebook.

and mathematical proportions to the Classical elements in Monticello. What geometric shapes and similarities do you see when comparing Figure 1-35 to Figure 1-36?

Thomas Jefferson was largely responsible for launching the Neo-Classical movement in architecture in the United States. He based his timeless, Classical design for the University of Virginia on mathematical rules and proportions

Figure 1-35: Construction of the Rotonda (Villa Capra) by Andrea Palladio began in 1550.

(Figure 1-37). In 1976, the American Institute of Architecture described Jefferson's "academical village" as "the proudest achievement of American architecture in the past 200 years."

You may be surprised to learn that the White House and the U.S. Capitol were also influenced by Palladio's Classical principles. Renaissance architecture is the last stop on your historical journey, at least for now. In Chapter 4, you will study

Figure 1-36: Thomas Jefferson began building his home at Monticello in Virginia in 1769 and continued to work on the structure for 40 years.

Figure 1-37: **Jefferson described his design for the University of Virginia as an "academical village." The central Rotunda (a) and extending pavilions and colonnades (b) are among the finest examples of Neo-Classical architecture in the United States.**

Courtesy of the University of Virginia Office of Public Affairs.

architectural design, beginning with American Colonial architecture around AD 1600 and continue to present day. In Chapter 4, you will also read about the "revival" of Greek and Gothic architecture.

Famous Buildings

Many buildings constructed in the 1500s and 1600s fell into disrepair due to wars, natural disasters, climate conditions, or lack of maintenance. Several of the survivors are quite famous. Saint Paul's Cathedral, Windsor Castle, and the Palace of Versailles were all built during this time period.

Structures require maintenance, alteration, or renovation periodically to improve them or to alter their intended use.

You have most likely heard of the Taj Mahal, which was built in India between 1632 and 1653. The lavish design of this tomb led to its name being used as a synonym for architecture that displays great wealth, extravagance, and detail (see Figure 1-38).

The building shown in Figure 1-39 is recognized worldwide for its unique shapes, detailing, and elaborate color scheme. Can you guess the name of this medieval Russian landmark built between 1555 and 1561?

FAMOUS ARCHITECTS

The ultimate aim of all creative activity is a building.

—*Walter Gropius, director of the Bauhaus School,*
Bauhaus Manifesto, *1919*

How do architects become famous? If you were to take a survey asking people to name a famous architect, many would name Frank Lloyd Wright. He was an architect, interior designer, writer, educator, and philosopher. Frank Lloyd Wright designed more than 1000 projects, but only 400 were built. He is most remembered

Figure 1- 38: Twenty-two thousand laborers and 1,000 elephants contributed to the construction of the Taj Mahal of India. It took them 22 years to build it, beginning in 1631.

Figure 1-39: Saint Basil's Cathedral was built in Moscow during the 1550s.

for Fallingwater, in Pennsylvania (see Figure 1-40), and the Guggenheim museum in New York City. When designing a new building, Wright considered even the smallest of details, including furniture and accessories.

During his career, Frank Lloyd Wright made many humorous and insightful remarks about architecture. In a 1953 interview for the *New York Times Magazine,* Wright said, "The physician can bury his mistakes, but the architect can only advise his clients to plant vines." What do you think he meant by that comment? As with any building, Wright's buildings were not perfect. There has been great controversy over the great cracks which formed on the cantilevered portions of Fallingwater. If you want to read about mistakes that plagued Fallingwater, use the Internet to search "Fallingwater cracks." Wright's designs ranged from a popular Prairie-style house made of natural materials to the expressionist modern style of Fallingwater. You might be surprised to learn that he also designed a nineteen story skyscraper in Bartlesville, Oklahoma. The Price Tower (Figure 1-41) was designated as a national historical landmark by the U.S. Department of the Interior in 2007.

Thousands of architects have left their marks on history. The table in Figure 1-46 lists several influential architects whom you might find interesting to research. Each artist has a unique story and a personal vision for architecture. For example, Ludwig Mies van der Rohe is one of the founding fathers of modern architecture. He was known for his simplified architectural designs, or "less is more"

Figure 1-40: Frank Lloyd Wright is one of the most influential architects in American history. Artist Walter DuBois Richards created this rendering of Wright's Fallingwater as part of a 16-stamp series honoring great accomplishments in American architecture. The stamps were issued in 1982.

Figure 1-41: **Frank Lloyd Wright's Price Tower was built in 1956 in Bartlesville, Oklahoma.**

approach to building. Mies van der Rohe created a "skin and bones" effect with steel framing and ground-to-ceiling plate glass windows. His 38-story Seagram building was erected in 1958 (see Figure 1-42). Notice the unique character achieved with the use of tinted glass, vertical bronze I-beams, and granite pillars.

Architects have made some interesting and thought-provoking comments. What do you think about this comment by architect Philip Johnson: "comfort is not a function of beauty…purpose is not necessary to make a building beautiful…sooner or later we will fit our buildings so that they can be used. Where form comes from I don't know, but it has nothing at all to do with the functional or sociological aspects of our architecture." In 1949, Philip Johnson designed the Johnson House, generally referred to as "the Glass House." Take a look at Figure 1-44. What statement do you think Johnson was trying to make?

One of the most talked about architects of today is Frank Gehry, a laureate of the Pritzker Architecture Prize. Many believe that he has created his own style of architecture, through his use of Deconstructivism, Expressionism, and Organic Modernism. His "architecture is art" philosophy has resulted in some of the most exciting architecture of our time. He has been described as "refreshingly original and totally American." His designs combine building elements into intriguing sculptures that seem to defy the laws of gravity (see Figure 1-45).

MAJOR DEVELOPMENTS IN CIVIL ENGINEERING IMPACT LAND DEVELOPMENT

Innovation and inventions in civil engineering play a major role in the growth of an area. For example, the arch was used to build aqueducts to distribute water over long distances. Aqueducts provided water and irrigation for crops to otherwise uninhabitable areas. The arch was also applied to bridge construction, making travel over large bodies of water possible. On the architectural side, the arch led to the development of large open spaces with high ceilings, ideal for cathedrals, palaces, and amphitheaters.

Even today, an area's growth and land development is controlled by the amount of resources available. Many places would be uninhabitable if they did not receive resources such as utilities from another area. What would it be like to live without electricity, communications, water, or sewage? The presence of roadways also affects land development. Would you buy a piece of land that was not accessible? In the past, new roads have fostered the development of suburban residential areas to alleviate city overpopulation. However, many suburban residents must travel daily back to the city to their place of employment. They depend on well-built and maintained roads and highways (see Figure 1-50).

Figure 1-42: **You can catch this view of the base of the Seagram Building at 375 Park Avenue in midtown Manhattan. The famous landmark was designed by Ludwig Mies van der Rohe with Philip Johnson and features a 2nd story roof garden.**

Point of Interest

If you visit New York City, you may want to see the Lever building (Figure 1-43). Built in 1952, it set a standard for office building design and was one of the first buildings to use a heat absorbing sealed glass.

In stark contrast to the glass office floors, the top floors are opaque to conceal mechanisms and mechanicals. The structure featured a 24-story building set on a two-story podium, which includes a roof garden. The base is a single-story pedestrian area supported by columns. Interesting features, still in service, include a window-washing mechanism and a mail conveyer to distribute mail to different floors.

Figure 1-43: The Lever building in New York City features a second-story roof garden.

From http://en.wikipedia.org/wiki/File:Lever_House_by_David_Shankbone.jpg

Figure 1-44: Philip Johnson built the Glass House as his personal residence in 1949. What statement do you think he was trying to make?

Courtesy of the Philip Johnson Museum.

Figure 1-45: Los Angeles Disney Concert Hall by architect Frank Gehry.

Image copyright Joel Shawn, 2010. Used under license from Shutterstock.com.

Fun Fact

The Los Angeles Disney Concert Hall initially had problems interacting with the environment. The highly reflective metal cladding directed sunlight onto the pavement, creating a fire hazard.

Point of Interest

Did you know that famous architect Frank Gehry appeared as a guest star to design a concert hall for Springfield in an episode of *The Simpsons*? If you enter "Frank Gehry in Simpsons" into your Internet search engine, you might be able to view the episode "The Seven-Beer Snitch" from season 16. For a more serious look at this architect and designer, search for the documentary *Sketches of Frank Gehry* by filmmaker Sydney Pollack. You will hear the architect's own interpretation of his famous buildings, including the home he designed for himself.

Your Turn

The table in Figure 1-46 lists a series of influential architects and important examples of their work. Select one architect from this list to research. Use the Internet to learn the architect's story. What was the vision or mission behind your architect's designs? Print examples of the individual's accomplishments and add notes of interesting details. If possible, identify the date of construction, building type, purpose, architectural style, and materials used. Identify which structural system may have influenced your architect's design.

(a)

(b)

(c)

Figure 1-46: (a) The Biltmore Estate represents the nineteenth-century architectural style of Richard Morris Hunt.

(b) Notre Dame du Haut was completed in Ronchamp, France, in 1955. The hilltop chapel represents Le Corbusier's vision of Postmodern Expressionism.

(c) The wings of the Milwaukee Art Museum's Quadracci Pavilion were designed by Spanish architect Santiago Calatrava to open and close in response to wind speed.

Figure 1-46: **Architects to research.**

Name	Birth/Death	Significant Work
Richard Morris Hunt	1827–1895	Biltmore Estate, Asheville, North Carolina
Henry Hobson Richardson	1838–1886	Ames Gate Lodge, North Easton, Massachusetts
Charles Greene	1868–1957	N. Bentz House Santa Barbara, California
Henry Greene	1870–1954	
Le Corbusier (Charles-Édouard Jeanneret-Gris)	1887–1965	Notre Dame du Haut, Ronchamp, France
Gerrit Rietveld	1888–1964	Schroder House, Utrecht, The Netherlands
Louis I. Kahn	1901–1974	Salk Institute, La Jolla, California
Marcel Breuer	1902–1981	Robinson House, Williamstown, Massachusetts
Charles Eames	1907–1978	Eames House, Pacific Palisades, California
Eero Saarinen	1910–1961	Washington Dulles International Airport, Chantilly, Virginia
John Lautner	1911–1994	Malin Residence (The Chemosphere), Los Angeles, California
Pierre Koenig	1925–2004	Stahl House, Los Angeles, California
Tadao Ando	1941–	Modern Art Museum, Fort Worth, Texas
Santiago Calatrava	1951–	Milwaukee Art Museum, Milwaukee, Wisconsin

Properties of Materials

In the past, materials of a given region often determined how things were built. It is probable our first builders and architects learned about the properties of these materials by trial and error and then passed along the knowledge to the next generation.

Case Study

Throughout this textbook, we will take a close look at actual building projects. They will be referred to as case studies. One case study will examine the design and build of the Syracuse Center of Excellence, a 55,000-square-foot, five-story research facility (see Figure 1-47). The site was contaminated by previous industrial uses, but has been remediated, restoring the site for construction of the Center of Excellence. The building will house offices for staff, classrooms, public spaces, and research laboratories. The main laboratory is a total indoor environmental quality lab where controlled experiments on the human response to indoor environments will be conducted. Main sustainability features of the Syracuse Center of Excellence include ground-source heating and cooling, rainwater collection, a green roof, and photovoltaic panels.

A second case study will examine a large urban plan development: CityCenter in Las Vegas. CityCenter, a 67-acre urban development in Las Vegas Nevada, includes condominiums, hotels, a casino, and a large retail and entertainment district (see Figure 1-48). The nine billion dollar venture includes six major buildings, along with a fire station, co-generation plant, parking garages, and a people-mover system. The goal of this urban plan is to create the diversity and vitality of a city, with a focus on sophistication. This requires that building designs offer timeless beauty and functionality, yet be different from any others. It was a challenging task to find architects to design such a look. Sven Van Assche, vice president of design, was up to the task. He began by clearly defining the term *contemporary*. Through his research, he determined that *contemporary* was not necessarily a preestablished design style using specific design elements but instead an innovative design that would be both timeless and inviting—a beauty people would enjoy for years to come. Van Assche searched the world for architects who would be able to design this "timeless" urban experience. Due to the project size and its mission—providing unique and distinctive choices that would attract diversity—several creative and experienced architects were needed. Although each would design a different building with its own vision, there was a need for meticulous leadership and collaboration between the professionals to create a dynamic and unified urban experience. Gensler, named the largest architectural firm in the United States by *Building Design+Construction* magazine for the past 24 consecutive years, was chosen as the executive architect of CityCenter. As the owner's representative, Gensler was responsible for collaborating with each of the architectural teams.

Figure 1-47: *Syracuse Center of Excellence under construction in 2008.*
Courtesy of the Syracuse Center of Excellence.

Figure 1-48: *CityCenter ready for opening in December 2009.*

Image copyright Scott Prokop, 2010. Used under license from Shutterstock.com.

Robert H. (Bobby) Baldwin, MGM MIRAGE's chief design and construction officer, and president and CEO of CityCenter, is credited as the project's visionary. After traveling the world to find the secrets of successful urban settings, he worked with New York's Ehrenkrantz, Eckstut, and Kuhn Architects to develop a master plan. Baldwin and Van Assche searched world wide for modern architects to design CityCenter's one-of-a kind buildings. When the search concluded, the result was a "dream team" of the world's foremost architects (Figure 1-49).

Pelli Clarke Pelli, architect of the World Financial Center in New York City, competed against Kohn Pedersen Fox to design the ARIA Resort & Casino in the heart of CityCenter. Pelli proposed a building of two intersecting arches of almost curvilinear form. Sunshades framed the glass curtain wall to provide sun protection as well as "make the whole building more graceful" stated Pelli (*Las Vegas Life*/November 2007). Kohn Pedersen Fox was chosen to design the Mandarin Oriental Hotel and condominium tower. Elegance and quality were essential to convey a luxurious feeling, which is signature of the internationally renowned resort brand Mandarin. Helmut Jahn Architects was chosen to design the two residential high rises known

Figure 1-49: **CityCenter's dream team of architects.**

Courtesy of MGM Mirage.

Gensler	Executive Architect	CityCenter
Pelli Clarke Pelli	ARIA Resort & Casino	ARIA/photo
Kohn Pedersen Fox Associates	Mandarin Oriental Las Vegas	Mandarin Oriental/photo
Helmut Jahn Architects	Veer Towers	Veer Towers/photo
Foster + Partners	The HarmonHotel, Spa & Residences	Harmon Hotel/photo
RV Architecture Architecture	Vdara Condo Hotel	Vdara Condo Hotel/photo
Studio Daniel Libeskind and Rockwell Group	Crystals	Crystals/photo

as Veer Towers. Jahn translated energy and excitement into physical form with the design of two angled 37-story glass towers that will shimmer both day and night. London's Foster + Partners, known for beautifully engineered intelligent and efficient buildings, was chosen to design the Harmon Hotel, Spa & Residences. This hotel will provide hip, exclusive living directly on The Strip for those that desire both privacy and profile. Rafael Viñoly was selected to design the soaring 57-story Vdara Condo Hotel deep within CityCenter. Vdara's distinctive crescent shape and skin of pattern glass will feature open floor plans and expansive views of the city and mountains. The retail and entertainment district known as Crystals was created by Studio Daniel Libeskind and Rockwell Group. The Crystals striking exterior features a dazzling and dynamic collection of angled roofs. Its luxurious interior provides a series of striking environments designed to invite, engage, intrigue, and relax.

Figure 1-50: Highway systems provide safe and efficient travel.

They learned that stone and brick were strong under compression but weak under tension, whereas wood, depending on the type, held up under both but was not as durable when exposed to the elements.

Early builders soon discovered that the material needed to be matched to its application. They also needed to use the materials in a way that would maximize their desirable qualities. One such desirable quality was strength. Although new techniques emerged, builders continued to search for stronger materials. A major development of its time was iron. Iron was strong, but brittleness limited how it could be used. In the nineteenth and twentieth centuries, iron was refined into steel, which became a popular building material and was soon applied to wire and cables. Can you imagine the advantages of replacing fiber rope with cables made of steel wire?

Materials were also combined to enhance performance. What do you think happened when concrete was reinforced with steel rebar? You probably guessed it would be stronger, but what else would change? Would the weight be the same? You will learn more about materials when you read Chapter 12: Building Materials and Components.

Innovations in Materials and Processes

Materials and applications change over time. Take road materials for example. Roads have gone from dirt, cobbles, brick, and stone to today's popular Macadam, often described as "black top." Most materials have both desirable and undesirable properties. For example, although Macadam provides a smooth traveling surface when new, it must be repeatedly coated or replaced in areas of harsh climates. During the winter months in northern United States, moisture can seep into small roadway cracks and freeze. When the water turns to ice it expands and causes larger cracks, letting even more moisture under the pavement, as shown in Figure 1-51. By spring, many of the roadways show signs of severe damage.

Macadam is designed to seal water out. Although this sounds like a good thing, it creates another problem. Under heavy rains, hydroplaning of automobiles can occur. When the water on the Macadam roadway builds beyond what the tire's tread can handle, cars literally drive on water, and without traction can easily slide off the road.

Another problem comes from rainwater run-off from large parking lots. Where does this water go so that it doesn't cause flooding? You will read more about run off in Chapter 6: Site Planning.

Figure 1-51: **Road cracks and pothole.**

Point of Interest

Today, civil engineers are working to develop and refine a porous road material to be used in warm climates. Use the Internet to search for the phrases "pervious concrete pavement" and "porous asphalt pavements."

Your Turn

You may be curious about engineering achievements and the engineers behind them. Table 1-2 provides a partial list of notable civil engineers identified by the American Society of Civil Engineers. Choose an engineer and use the Internet to research his or her work. What makes this engineer's life and work "notable?"

Table 1-2: **Notable civil engineers**

	Name	Major Accomplishment or Notability
1635–1703	ROBERT HOOKE	Hookes law, mechanics of materials
1724–1792	JOHN SMEATON	Eddystone Lighthouse 1759 Research in mechanical power watermills, windmills, and steam engines
1756–1836	JOHN LOUDON MACADAM	Process of roadway design and construction

(Continued)

Table 1-2: (*continued*)

	Name	Major Accomplishment or Notability
1764–1820	BENJAMIN HENRY LATROBE	Contracted by U.S. government to redesign the U.S. Capitol in 1815 Engineering projects included the Philadelphia Waterworks, plans for the Washington Canal, and the New Orleans Waterworks
1768–1849	MARC ISAMBARD BRUNEL	Thames Tunnel–underground tunnel
1772–1822	THEODORE BURR	Waterford style bridge The key feature was the arch that started below the deck at the abutments and ran near the top of the top chord at mid span. This was the first time in the United States that anyone used an arch in combination with a truss in order to provide both stiffness and strength.
1784–1864	STEPHEN H. LONG	Considered to be the first structural engineer in America. Patented Jackson Bridge, first engineer to write about the advantages of pre-loading a truss in order to achieve additional stiffness and the use of the parallelogram of forces
1789–1867	GRIDLEY BRYANT	Eight-wheel railroad car, the portable derrick, the rail switch, and the turntable
179 –1885	JOHN BLOOMFIELD JERVIS	Croton Aqueduct–brought fresh water to island of Manhattan
1798–1856	SIMEON BORDEN	Survey Instrument
1803–1858	HENRI PHILIBERT GASPARD DARCY	Darcy's Law and the Darcy-Weinbach equation
1806–1859	ISAMBARD KINGDOM BRUNEL	Great Western Railway, first tunnel under a navigable river, first propeller driven ocean going iron ship
1806–1869	JOHN AUGUSTUS ROEBLING	Idea of replacing the fiber rope with stronger wire rope for greater safety and economy; in 1841, filed a patent application for a cable made of parallel wires.
1811–1875	JAMES LAURIE	Design of the Warehouse Point Bridge across the Connecticut River on the New York-New Haven railroad line; replaced the older wooden bridge with an iron bridge without interruption of train service.
1813–1886	ELLIS SYLVESTER CHESBROUGH	Chesbrough's Chicago Water Supply
1814–1884	WENDALL BOLLMAN	Iron suspension-truss design railroad bridge
1820–1887	JAMES BUCHANAN EADS	Eads Bridge–Early steel bridge combined road and railway over the Mississippi at St Louis—when constructed it was the longest arch bridge in the world.
1826–1863	THEODORE DEHONE JUDAH	Central Pacific Railroad, Western terminus of America's first transcontinental railroad
1827–1897	ALBERT FINK	On May 9, 1854, received patent No. 10,887 for a truss bridge; received a second patent on April 9, 1867, No. 63,714, for a combination truss with wooden compression members and wrought iron tension members.

1832–1923	ALEXANDRE GUSTAVE EIFFEL	Eiffel Tower–worlds' tallest tower in 1889
1837–1926	WASHINGTON AUGUSTUS ROEBLING	Brooklyn Bridge–steel cable suspension bridge
1842–1911	ELLEN HENRIETTA SWALLOW RICHARDS	First woman to graduate from MIT; expert on water and sewage analysis
1843–1908	WILHELM HILDENBRAND	Suspension bridges and cables
1843–1903	EMILY WARREN ROEBLING	Brooklyn Bridge work; construction foreman at Montauk, Long Island, camp
1850–1935	GUSTAV LINDENTHAL	Smithfield Street Bridge–lenticular truss Queensboro Bridge and Hell Gate Bridge Upon completion the Hell Gate Bridge was the longest and heaviest steel arch bridge.
1855–1932	JOHN RIPLEY FREEMAN	Charles River Dam, Charles River Basin report
1858–1928	GEORGE WASHINGTON GOETHALS	Panama Canal–1914
1879–1965	OTHMAR HERMANN AMMANN	George Washington Bridge
1883–1924	CLIFFORD MILBURN HOLLAND	Holland Tunnel
1892–1989	ABEL WOLMAN	Combined engineering with public health and hygiene into the field that came to be known as sanitary engineering. Developed procedures for water and sewage chlorination and disinfection, and influenced federal policy concerning water pollution control and management.
1900–1989	STEPHEN D. BECHTEL, Sr.	Hoover Dam 1935 monolithic and the Trans Arabian pipeline
1902–1981	ARTHUR CASAGRANDE	Together with colleague Karl Terzaghi built the influential discipline of soil mechanics Earth fill dam design
1903–1994	ANTON TEDESKO	Thin shell concrete roofs, Hayden Planetarium dome, the Hershey Arena and NASA's Vehicle Assembly Building (VAB) at Kennedy Space Center, Florida. When completed in 1966, the VAB enclosed the greatest total volume of any building in the world.
1907–1997	MARIO SALVADORI	Salvadori Educational Center on the Built Environment *Why Buildings Stand Up* was the first of a series of popular works written for the general public.
1921–1996	GEORGE F. SOWERS	Soil Mechanics and Foundations
1922–1989	HARRY BOLTON SEED	Known internationally for his understanding of soil behavior during earthquakes

Source: American Society of Civil Engineers, http://live.asce.org/hh/index.mxml. Accessed 4/19/2010.

Careers in Civil Engineering and Architecture

Looking Back, to the Future

Edward A. Bogucz, Jr.: Executive Director
Syracuse Center of Excellence in Environmental and Energy Systems
Syracuse University

"When you're envisioning the future, the past can be a wonderful inspiration and guide," explains Ed Bogucz, executive director of the Syracuse Center of Excellence in Environmental and Energy Systems (SyracuseCoE). "Our organization develops next-generation energy and environmental systems. We trace our roots back to one of the greatest engineering accomplishments of all time—the design and construction of the Erie Canal. In the early 1800s, it was an audacious idea: connect the Atlantic Ocean to the Great Lakes via a 40-foot-wide canal across 360 miles of wilderness in upstate New York. U.S. President Thomas Jefferson called the idea 'a little short of madness.' Undaunted, New York's state government took on the project. The Erie Canal opened in 1825. It had a profound impact on the development of the U.S. It promoted the flow of people, products and ideas into, and out of, a young and growing country. During the Industrial Revolution, all kinds of new technologies and innovations were developed by companies along the route of the Erie Canal. One of the hubs was Syracuse, which was home to a wide variety of firms. In the early 1900s, Syracuse was known as the 'Typewriter City' because of a cluster of major manufacturers. L.C. Smith and Brothers (which later became Smith Corona) built a manufacturing plant along the Erie Canal in 1902. Simultaneously, L.C. Smith successfully encouraged Syracuse University to start an engineering college. Today, SyracuseCoE is led by Syracuse University and our new headquarters is being constructed on the site of the old Smith Corona plant."

"Without the Erie Canal, there would be no L.C. Smith, no College of Engineering at Syracuse University, and no SyracuseCoE. Our activities today were made possible by a sequence of historical events," says Bogucz. The historical significance of the site continues to challenge and intrigue Bogucz. During the building process, they uncovered mule shoes, plates, bottles, and many other artifacts dating back 200 years. Not all discoveries were welcome. One surprise was the discovery of an underground oil storage tank—made of wood—that dates from the days of the Erie Canal. Costs relating to the unexpected cleanup of this discovery had a major impact on the project's overall budget.

On the Job

Bogucz started his career in academia in 1985 as a junior faculty member at Syracuse University. He was attracted to Syracuse University because of its long history of encouraging faculty to work with industry. Building upon his childhood interest in air conditioning and his college study of heat transfer, the proximity to Carrier Corporation in Syracuse held the possibility of exciting collaboration opportunities. Bogucz explains, "all the planets aligned for me. As a junior faculty member I worked projects with Carrier to design software to help engineers design new air conditioners." Bogucz served eight years as Dean of the L.C. Smith College of Engineering and Computer Science when the school was in a period of rapid change. Bogucz led the development of a strategic plan to guide the future of the college. Energy and environmental systems was one of four different areas chosen for investments in faculty and facilities. Bogucz led teams that developed concepts for collaborations among firms and institutions throughout New York. These efforts culminated in 2002 with the establishment of SyracuseCoE as part of a statewide Centers of Excellence program that is intended to create economic benefits.

Inspirations

Bogucz grew up in northern New Jersey. His father was a mechanical engineer and executive in a company that made heating and cooling systems for homes, schools, and hospitals. He fondly remembers childhood visits to the plant where his father was employed. He remembers seeing molten iron poured to make cast iron boilers. He learned about engineering and how air conditioners work. From an early age, he envisioned following in his father's footsteps in the field of heating, ventilation, and air conditioning (HVAC) engineering. Responding to today's world challenges, HVAC engineers are developing innovations that will conserve energy and natural resources and improve indoor environmental quality for building occupants. As the executive director of SyracuseCoE, Bogucz plays a major role in collaborative efforts to improve the built environment and its interaction with the natural world.

Education

Bogucz recalls his interest in engineering developed in high school, "When I was in high school I had the privilege of taking courses similar to Project Lead The Way. I took two years of mechanical drawing and two years of architectural drawing. I had an interest in buildings and how things worked." As a senior he was torn between mechanical engineering and architecture. HVAC provided an interface between the two fields.

After receiving a Bachelors of Science degree in mechanical engineering from Lehigh University, Bogucz received a scholarship to study abroad at the University of London. It was during that period, working on his Masters degree, he envisioned an academic career. Upon return to the United States, Bogucz went back to Lehigh for his PhD, still interested in how things worked and the new and emerging technologies, which effect heating, cooling, and heat transfer.

Advice for Students

Bogucz challenges students to think about history. Whether the history of a building site, or an engineering marvel, history is important. Think about the Erie Canal, for example. Imagine what a challenge in 1817 it must have been to construct a bridge for the canal to pass over a river. The construction was done by hand, using locally available materials. The canal had to carry water over streams, rivers, and swamps. Bogucz recommends students question how things work; sometimes the knowledge can lead to innovation. Think about ways to conserve energy, or make things work more efficiently and use less energy. Challenge yourself to make a difference, just like innovators throughout history whose creativity inspires us today.

SUMMARY

Civil engineering and architecture contribute—separately and together—to the development of our built environment. Civil engineering, considered the largest engineering profession, includes seven major interrelated branches: construction, environmental, geotechnical, structural, transportation, water resources, and urban and community planning. The civil engineer is responsible for the design and implementation of safe, efficient, and resilient structural works that serve the general public. Structural works include buildings, bridges, canals, dams, roads, tunnels, harbors, water systems, and water treatment facilities. The civil engineer employs both math and physics to analyze the bending, compression, and tension capacity of building components. When these components are combined to create structural systems, they are analyzed again, this time by the geometric configuration of their members, and according to how they resist loads. A structure can be described based on five basic structural system categories; column and beam, corbel and cantilever, arch and vault, truss and space frame, and tensile.

The term *architecture* has been defined in various ways but is generally recognized as the art or practice of designing and building structures, especially habitable ones. While reading this chapter you progressed through an historical timeline of architecture to gain insight into the people, culture, environment, and concerns of the time. Early structures were need-driven. In the early Stone Age, tents made from bone and hides provided basic shelter and were easy to move. Once people began to farm and develop a sense of community, more stationary dwellings appeared made of materials such as mud, stone, and wood. The architectural timeline illustrated in this chapter progresses from ancient Egyptian to Greek, Roman, Gothic, and Renaissance architecture. During that time, it took several years to build a large structure and in some cases, it took several decades or even centuries. Some buildings deteriorated or were destroyed by wars and natural disasters before completion. Even though construction occurred over many years, the architectural integrity was maintained by matching the original building style's form, materials, and details. However, sometimes maintaining this architectural integrity was not possible because the materials were no longer available. In those cases, current materials or techniques resulted in a different appearance in the newer parts of a building. This makes it difficult to pinpoint a building's age, which is often determined by the materials and building techniques used during the construction process. Thermoluminescence and carbon dating have been used to date early structures and building materials. The architectural style of the building and inscriptions left by the architect or builder provide some of the best clues to a building's age. Architectural styles have recognizable elements, but many variations in style exist. Variations emerged because of materials local to a region, society and cultural influence, and the architect's creative style. One example is Gothic architecture. Although it originated in France, Gothic style soon spread to Britain, Germany, Italy, and Spain, where distinct variations emerged.

Today's architecture is not limited to a building's form and function. In addition to a visually pleasing and serviceable design, a building's architecture should *fit* within the surroundings. Responsible architecture shows thoughtful consideration for the occupants' health, the owners' operating and maintenance costs, building longevity, and impact on the environment. To address these considerations, architects and civil engineers come together to form our built environment, combining both art and design with structural integrity. Civil engineering advancements such as dams, water systems, and water treatment facilities have made otherwise uninhabitable areas viable for development. Expansion is also supported by roadways, bridges, tunnels, canals, and harbors, which allow people and materials to be transported great distances. The history of civil engineering and architecture plays an important role in today's world. Architectural and civil engineering achievements, developed over many centuries, provide a base of knowledge that continues to encourage innovation. Architects and engineers look to the past to learn from mistakes and build upon proven methods to develop new and better structures.

BRING IT HOME

1. Develop a chart to record the information you uncover during the research of a building's history. Include items that can be easily identified, such as the building style, structural system, building materials, and inscriptions. Leave space to include notes about the building's use over time and any renovations. Use the chart to record research on several historical buildings in your area.
2. Research 10 civil engineering accomplishments not listed in this chapter, which occurred between 3000 BC and the present. List them in sequential order and describe their historical impact. Identify connections to future developments.
3. Pick a local building and propose a small addition to the structure. Make a sketch of the construction and explain what you would do to insure the integrity of the buildings architectural style.
4. Make a list of utilities and resources that would be needed to build a large shopping center at the outskirts of your city or town.
5. Make a proposal to add a small pizza shop to a historical part of town. How could it be designed to "fit" into the area?
6. Working in a team, research several historical buildings in your area. Make a historical tour brochure that highlights the features and historical background of each building. Include interesting facts about the building's past use and events that occurred there.
7. Working in a team, research the history of a local civil engineering project, such as a bridge, roadway, watershed, or wastewater treatment facility. Develop a design for restaurant placemats that will provide interesting information about the project to visitors and area residents.

EXTRA MILE

CIVIL ENGINEERING AND ARCHITECTURE ANALYSIS CHALLENGE

In 1996, the American Society of Civil Engineers selected seven civil engineering wonders of the modern world. They were identified as the Channel Tunnel, the C N Tower, the Empire State Building, the Itaipú Dam, the Golden Gate Bridge, Netherlands North Sea Protection Works, and the Panama Canal. Form a team of four to six people and use the Internet to research each structure to determine why it was chosen. Record your findings in your engineer's notebook. As a team, research recent projects and choose your own list of civil engineering wonders of the last decade. Make a poster or presentation of your results.

CHAPTER 2
Careers

GPS DELUXE

Menu | START LOCATION | DISTANCE | END LOCATION

Before You Begin

Think about these questions as you study the concepts in this chapter:

1. What would it be like to be a civil engineer?
2. What education, training, and skills are required to become a civil engineer?
3. What are the stages of career development in becoming a licensed civil engineer?
4. What would it be like to be an architect?
5. What education, training, and skills are required to become an architect?
6. What are the stages of career development in becoming a licensed architect?
7. What other careers are related to the design and build of a structure?
8. How do civil engineers, architects, and professionals work together in the design and build of a structure?
9. What agencies and organizations are involved in the design and development of a structure?

INTRODUCTION

How many times have you been asked the question, "What do you want to be when you grow up?" It can be one of the most difficult questions to answer if you are not sure of the career options available, the amount of education required, or future job market opportunities. Some careers require four or more years of college, whereas others require only a two-year degree or hands-on training that can be acquired through an apprenticeship program. Training for your career will be one of the most substantial time and financial commitments affecting your future. Many students enter college without a major, or change majors, several times resulting in additional years to complete their college program. To prevent this from happening, it is important to research options that match your academic strengths and interests and interview people with careers that interest you. Overall, the most common advice you may hear is to choose a career you will enjoy. How do you determine which career path would be most enjoyable for you, without knowing what the career entails?

In this chapter we are going to examine two careers: civil engineering and architecture. You will read about job descriptions and the education they require. We will examine the stages of professional development and research the future job outlook. Finally, to expand your list of career options, we will take a brief look at other careers and agencies related to the design and construction of a structure.

Figure 2-1: *The number of civil engineers and architects in the workforce is expected to increase by 18 percent over the next decade. (Occupational Outlook Handbook, U.S. Department of Labor, Bureau of Labor Statistics.)*

CIVIL ENGINEERING AS A CAREER

Civil Engineer: an engineer responsible for the structural design and construction specification of public and private works such as buildings, roads, and harbors.

In Chapter 1, you read that **civil engineers** design structural works such as buildings, bridges, canals, dams, roads, tunnels, water systems, and water treatment facilities (see Figure 2-2). Structural engineers, a type of civil engineer, are responsible for the structural design of factories, power plants, processing plants, transportation facilities, and public meeting spaces. A structural engineer is often described as the one who designs *how* to build a structure. Architects depend on structural and civil engineers to insure that their residential and commercial buildings are strong, safe, and durable. During the building design process, a civil engineer may perform a site investigation as part of structural analysis. Civil engineers must adhere to strict codes when they specify requirements for a building project. These requirements include construction materials and procedures.

Civil engineers must continue learning about innovations in both building materials and construction techniques. Structures must be able to withstand live loads, such as occupants, and dead loads, such as building materials. A strong background in math and science is necessary to correctly apply the theory of hydraulics, thermodynamics, and physics. The civil engineer must plan for expected and unexpected natural forces. Civil engineers are innovative and try new designs, which sometimes fail. If a failure occurs, knowledge is gained from determining the cause of the failure. In some cases the civil engineer could be held liable. You will read more about loads in Chapter 9: Residential Space Planning. Civil engineers are often involved in some very exciting projects. Occasionally, investigations by civil engineers into the properties of materials, such as steel and concrete, have led to the creation and development of less expensive, stronger, and easier to manipulate materials with greater sustainability. A civil engineer could also be involved in environmental impact and site design, which will be explained in greater detail in Chapter 5: Site Discovery for Viability Analysis and Chapter 6: Site Planning.

Figure 2-2: **Civil engineers apply knowledge of hydraulics, thermodynamics, and physics to design safe and durable structures.**

Education, Training, and Skills Needed to Become a Civil Engineer

Let's imagine that your favorite classes in high school are math, science, and technology, and you have been successful at designing and constructing projects such as model bridges, towers, catapults, and cantilevers. You enjoy testing materials and performing the calculations required to determine how your structure will perform under loads. Civil engineering might be a career path for you (see Figure 2-3). However, before you decide, you will need to give it serious thought and research further to find the answers to the following questions:

- What is it like to be a civil engineer?
- Do I have an interest in this field?
- Do I have the academic skills necessary to succeed?
- What is the outlook for job availability after college?
- How much money can I expect to make?
- What are the opportunities for advancement?
- How many years of college will I need?
- What college courses are required?
- Which colleges offer this program?
- What additional certifications beyond college will I need?
- Will I enjoy this program?

Figure 2-3: A typical job posting for a civil engineer.

CIVIL ENGINEER

A local design firm is looking for a Civil Engineer with a concentration in structural engineering. Job responsibilities will include design and development of bridges and roadways including drainage and site planning. You must be an organized, independent worker who is able to communicate with clients and meet the federal and state regulations. Candidates must have a BS Degree in Civil Engineering, PE or be able to obtain a PE in 2 years and 2 years of experience is a must. Experience in structural engineering a plus.

Salary: 60–75K DOE.

Many of these answers can be found by accessing the Occupational Outlook Handbook on the U.S. Department of Labor Bureau of Labor Statistics website at *http://www.bls.gov*. This site provides valuable information that will assist you in making an informed career decision. For example, did you know that civil engineering in the United States has the highest employment of all the engineering specialty fields, and is expected to grow by 18 percent between 2006 to 2016? In 2006, there were over a quarter of a million civil engineers in the United States, ranking above the number of mechanical, industrial, electrical, electronics, aerospace, computer, environmental, and chemical engineers (see Figure 2-4)! You might also be interested to learn that the median earnings of a civil engineer in May of 2006 was $68,600, with the lowest 10 percent of civil engineers making $44,810 and the highest 10 percent earning $104,420. A survey conducted in 2007 by the National Association of Colleges and Employers reported the average starting salary for a civil engineer was $48,509. This information and much more can be found on the *http://www.bls.gov* website.

Figure 2-4: Civil engineering has the highest employment rate of all the engineering specialty fields.

A bachelor's degree is required to obtain an entry-level position as a civil engineer, and continuing education is necessary to keep current with new and emerging technology. College admission requirements for a civil engineering program require you to have a strong background in math and science. Once accepted into a four-year engineering program, your coursework would include general engineering, mathematics, physical and life science, design, computer applications, social sciences, and humanities. In addition, your program will incorporate a concentration of study and laboratory experiences specifically related to civil engineering.

Stages of Career Development to Become a Licensed Civil Engineer

The United States requires licensure for engineers to offer services to the public. It is important to choose an ABET accredited college or university if you plan to apply for licensure. The Accreditation Board for Engineering and Technology

(ABET) accredits many U.S. colleges and universities offering bachelor's degrees in engineering. Many civil engineers are licensed Professional Engineers, generally called PEs. PE licensure requires a degree from an ABET accredited engineering program, four years of relevant work experience, and successful completion of state exams. The first exam, the Initial Fundamentals of Engineering, can be taken late in the senior year or upon graduation. After passing the first exam, the engineer would be called an EIT, Engineer In Training, or an EI, Engineer Intern. After acquiring the necessary work experience, typically four or more years, the EIT will take a second exam called Principles and Practice of Engineering. Passing this exam will complete the licensure requirements. Most states recognize licensure from other states; however, many states have mandatory continuing education requirements for re-licensure. Continuing education addresses new and emerging technologies and practice and increase an engineer's opportunities for job advancement. Beyond fulfilling licensure requirements, some engineers choose to complete certification programs offered by professional organizations, whereas others pursue graduate and postgraduate degrees.

Website Resources

American Society of Civil Engineers, 1801 Alexander Bell Dr., Reston, VA 20191. Internet: http://www.asce.org

Information about careers in engineering is available from: JETS, 1420 King St., Suite 405, Alexandria, VA 22314. Internet: http://www.jets.org

Information on ABET-accredited engineering programs is available from: ABET, Inc., 111 Market Place, Suite 1050, Baltimore, MD 21202. Internet: http://www.abet.org

Those interested in information on the Professional Engineer licensure should contact: National Council of Examiners for Engineering and Surveying, P.O. Box 1686, Clemson, SC 29633. Internet: http://www.ncees.org

National Society of Professional Engineers, 1420 King St., Alexandria, VA 22314. Internet: http://www.nspe.org

Information on general engineering education and career resources is available from: American Society for Engineering Education, 1818 N St. NW., Suite 600, Washington, DC 20036. Internet: http://www.asee.org

ARCHITECTURE AS A CAREER

What would it be like to be an architect? Would a career in architecture match your interest and ability? It might, if you like looking at buildings and analyzing the room arrangement and exterior appearance and composition. Do you think about the building location and how it fits within its surroundings? Have you ever had an idea that would make a building look or perform better? Have you created sketches or drawings of these ideas? These may be clues that indicate you have an interest in architecture. As you read in Chapter 1, architects design residential and commercial habitable structures and built environments. Their designs are based on the principles and elements of design, spatial relationships, and environmental,

Figure 2-5: A typical job posting for an architect/project manager.

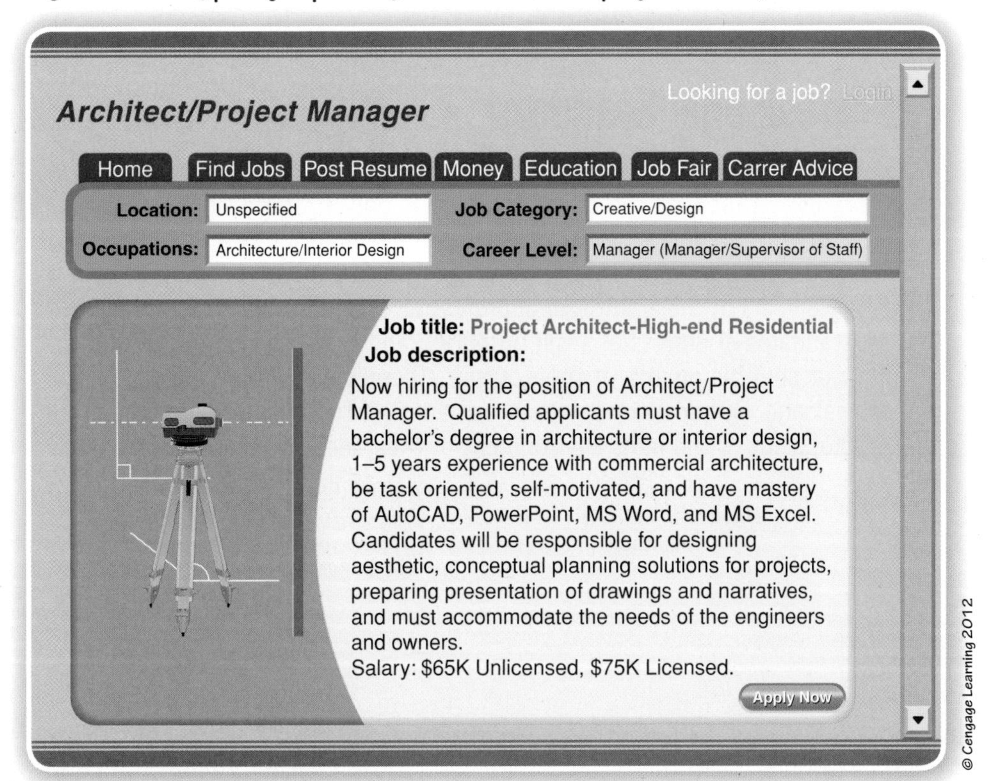

cultural, and functional considerations. **Architects** utilize math and science in harmony with artistic talents to create designs that are functional, aesthetically pleasing, and sustainable (see Figures 2-5 and 2-6).

> **"Architects:** licensed professionals trained in the art and science of building design."
> *U.S. Department of Labor Bureau of Labor Statistics Occupational Outlook Handbook.*

To research architecture as a career, we return to the U.S. Department of Labor Bureau of Labor Statistics website, where we learn that 132,000 architects were employed in 2006. That number is projected to increase by 18 percent to 155,000 over the next decade. The statistics also show that one in five United States architects is self-employed. An architect's job generally begins by meeting with a client to discuss the building objectives and budget. Following this meeting, the architect will begin the research and development of a suitable and sustainable plan. Architects are masters of blending form and function to create an aesthetically pleasing plan that fully addresses the client's needs. Many of today's architects are "designing for tomorrow," with conservative designs requiring less energy and maintenance.

In addition to preparing drawings, the architect often conducts research to support his or her building plan. This research could include a feasibility study, environmental impact study, land-use study, or cost analysis. The architect may follow a project throughout the construction process. Once the plans are complete, the architect will often oversee construction and collaborate with engineers, urban planners, interior designers, codes, and zoning officers to make sure their vision is attained.

The Bureau of Labor Statistics surveyed architect's earnings in May of 2006. The median annual earning for architects was $64,150. The lowest 10 percent earned

Figure 2-6: Architects must be creative, understand spatial relationships, have strong computer and communication skills, and be able to work both independently and within a team.

Figure 2-7: The path to become an architect includes an internship experience under the guidance of an experienced architect.

© Plus/iStockphoto.com.

Point of Interest

A Professional Standard

"It is important that the title 'architect' is only conferred upon individuals who can demonstrate the successful completion of a university level academic program and a period of assessed practical training.

Members of the architectural profession are dedicated to the highest standards of professionalism, integrity, and competence, and to the highest possible quality of their output. Thereby they bring to society special and unique knowledge, skills, and aptitudes essential to the development of the built environment of their societies and cultures."

Reprinted from UIA Accord on Recommended International Standards of Professionalism in Architectural Practice, June 1999, International Union of Architects. *http://www.aia.org/practicing/groups/international/uia/AIAP073960* (Accessed 4/2/2010)

around $39,400, and the highest 10 percent earned $104,940. Factors contributing to the wide range of earnings were experience, location, changing business conditions, and working for an architectural firm versus independent practice.

Education Training and Skills Required to Become an Architect

Like civil engineers, architects must complete a lengthy university study. The Bureau of Labor Statistics website identifies a five-year bachelor's degree in architecture as the most common degree for a student with no previous architectural training. In most states this degree must be from a college or university accredited by the National Architectural Accrediting Board (NAAB). A registered or licensed architect in the United States will have the designation initials RA. Most architects join the American Institution of Architects and use the initials AIA after their name. Many view architecture as a holistic discipline. Future architects learn to produce functional, attractive, and sustainable designs by collaborating with practicing architects and academia. An architect's preparation must include an internship at a structured, monitored, and assessed workplace. This is a valuable part of his or her preparation. It provides architecture students the opportunity to develop and demonstrate necessary critical reasoning, professional judgment, knowledge, and skill. Assessment of the future architect is comprehensive. It includes evaluation of case studies, review of the experiences recorded in his or her logbook, examinations, and interviews with experienced members of the profession. It is only after rigorous study, practice, and assessment that one will be considered for the title of architect.

Stages of Career Development to Become a Licensed Architect

Following graduation from a five-year architectural program, the graduate works as an intern for approximately three years to gain practical work experience preparing drawings and models and assisting in project design (see Figure 2-7).

Point of Interest

Recommended Training

In 1999, the International Union of Architects Assembly recommended a set of global guidelines for International Standards of Professionalism in Architectural Practice (the Accord). The following passages are from the UIA policies on Demonstration of Professional Knowledge and Ability and Practical Experience/Training/Internship:

Incremental Assessment

> "The scope and standard of competency at all stages of an architect's education and professional training should be subject to regular accreditation/validation by an objective panel. . . .
>
> "Architectural education and professional training must undergo continuous change and review if it is to keep pace with the changing nature of practice and expectations of the public. Concern with sustainability, health and safety, and access for the disabled are all examples of education and practice, which have changed significantly in a decade."

Recommended Guidelines for the UIA Accord on Recommended International Standards of Professionalism in Architectural Practice Policy on Demonstration of Professional Knowledge and Ability, International Union of Architects, June, 1999. http://www.aia.org/aiaucmp/groups/aia/documents/pdf/aias075206.pdf (Accessed 10/19/2010.)

3. Categories of Experience

An intern should receive practical experience and training under the direction of an architect in at least half of the areas of experience nominated under each of the following four categories:

3.1 Project and Office Management
Meeting with clients
Discussions with clients of the brief and the preliminary drawings
Formulation of client requirements
Pre-contract project management
Determination of contract conditions
Drafting of correspondence
Coordination of the work of consultant's office and project accounting systems
Personnel issues

3.2 Design and Design Documentation
Site investigation and evaluation
Meeting with relevant authorities
Assessment of the implications of relevant regulations
Preparation of schematic and design development drawings
Checking design proposals against statutory requirements
Preparation of budgets, estimates, cost plans, and feasibility studies

3.3 Construction Documents
Preparation of working drawings and specifications
Monitoring the documentation process against time and cost plans
Checking documents for compliance with statutory requirements
Coordination of subcontractor documentation
Coordination of contract drawings and specifications

3.4 Contract Administration
Site meetings
Inspection of works
Issuing instructions, notices, and certificates to the contractor
Client reports
Administration of variations and monetary allowances

From *Recommended Guidelines for the UIA Accord on Recommended International Standards of Professionalism in Architectural Practice, Policy on Practical Experience/Training/Internship*. International Union of Architects, 1999. *http://www.aia.org/practicing/groups/international/uia/AIAP073960* (Accessed 4/2/2010).

This experience helps the intern prepare for the challenging and comprehensive Architect Registration Examination (ARE). Following successful completion of all divisions of the ARE, many will take positions within an architectural firm and then advance as they gain experience. For example, Architect I would have 3 to 5 years of experience; Architect II, 6 to 8 years of experience; and Architect III, 8 to 10 years of experience. The next step might be advancement to Manager. Managers are generally licensed architects with more than 10 years of experience. Architects in senior management are generally referred to as Associates, and an owner or partner in an architectural firm holds the title of Principal. Architects must work effectively both independently and as part of a team, often working more than 40 hours a week. Computer-based productivity tools such as Computer Aided Design and Solid Modeling are essential for today's architect.

Website Resources

For information about education and careers in architecture, visit: The American Institute of Architects, 1735 New York Ave. NW, Washington, DC 20006. Internet: http://www.aia.org

Intern Development Program, National Council of Architectural Registration Boards, Suite 1100K, 1801 K St. NW, Washington, D.C. 20006. Internet: http://www.ncarb.org

Bureau of Labor Statistics, U.S. Department of Labor, Occupational Outlook Handbook, 2008–09 Edition, Architects, Except Landscape and Naval, on the Internet at http://www.bls.gov

Access the official website for America's Favorite Architecture provided by the American Institute of Architects located at *http://www.favoritearchitecture.org/#*. Explore and read interesting stories about structures, buildings, and built environments such as stadiums. For example, did you know that the corners of the famous Chrysler Building in New York City are adorned with replicas of Chrysler hood ornaments and radiator caps? You will also find some surprising facts about the Golden Gate and Brooklyn Bridges. While you are at this website, see if you can find the first baseball stadium to be covered by a roof of semi-transparent dome panels (see Figure 2-8).

Figure 2-8: Why was the original field of grass replaced with artificial grass in the first stadium to be covered by a roof? The answer may surprise you.

CAREERS RELATED TO CIVIL ENGINEERING AND ARCHITECTURE

Following intense research and discussions with your guidance counselor, family, and career professionals, you may have concluded that you do not want to pursue becoming a civil engineer or architect. There are many other related careers that might be of interest to you. For example, you might want to consider a career as a Computer Aided Design (CAD) technician (see Figure 2-9), model maker, interior designer, building inspector, landscaper, or one of the many jobs in the construction field.

There are many job areas related to the built environment (see Figure 2-10). Specific areas include the following; surveying, geotechnical, estimating and bidding, planning, sales, marketing, health and safety, electrical, plumbing, HVAC, hydraulic, pneumatic, and mechanical systems. Based on local and state requirements, jobs in these fields require varying degrees of training and experience.

Figure 2-9: **Job posting for CAD technician/drafter.**

CAD Technician/Drafter

A local company is looking for a CAD Technician/Drafter. Candidates must have a high school diploma or equivalent, be proficient in AutoCAD, and have 5 years of previous experience. Mechanical and civil drafting or construction experience is preferred. Candidates will be required to draft detail drawings using AutoCAD, create geometry, templates and symbol libraries, create record drawings and assist with the drawing archive process, plot check and submission sets, and update, develop, and maintain CAD software and hardware.

Salary: $23/hr.

Figure 2-10: **Each professional assesses the design and build of a structure from a unique perspective.**

CIVIL ENGINEERS, ARCHITECTS, AND RELATED AGENCIES WORK TOGETHER IN THE DESIGN AND BUILD OF A STRUCTURE

Today's architects and civil engineers do not work in isolation, but instead seek out information and advice from other professionals. Architects and civil engineers are acutely aware of climate change, natural disasters, energy consumption, and shortages of water and other resources, all of which necessitate thoughtful selection and the use of economically and environmentally friendly materials and practice. Their goal is to create a healthy, economical, and sustainable building which strikes harmony between humans and their environment. A common practice applied to achieve this goal is the application of **integrative design**.

Integrative Design: a process that begins at the design stage and involves the entire building team uniting building elements and construction disciplines in a whole-building approach.

Figure 2-11: Academy of Sciences building in Golden Gate Park, San Francisco, California is a LEED-certified building with a living roof and photovoltaic panels.

Kim Steele/The Image Bank/Getty Images.

Figure 2-12: Using the Whole Systems Design Approach, various professionals work together to develop a shared vision of project goals and the strategies for attaining them.

Image copyright Yuri Arcurs, 2010. Used under license from Shutterstock.com.

Using a Whole Systems Integrative Design approach, engineers and architects work together with other experts and stakeholders in the design and development of a building (see Figures 2-11 and 2-12). For example, a Whole Systems Integrated Design team might include the following: client, project manager, architect, interior designer, structural engineer, urban planner, cost estimator, environmental engineer, construction engineer, construction manager, carpenter, mason, plumber, electrician, landscape architect, building inspector, and community director. A *charrette* is a work session of about 12 to 30 people involved in the design, construction, and operation of a building project. This work session may last one day, or several days, as determined by the complexity of the building project. A charrette is often used to establish a shared vision and encourage creativity and resource efficiency within a cohesive design team.

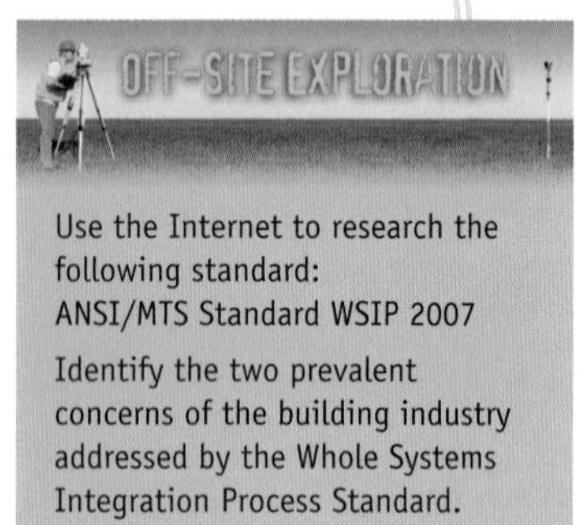

Use the Internet to research the following standard:
ANSI/MTS Standard WSIP 2007

Identify the two prevalent concerns of the building industry addressed by the Whole Systems Integration Process Standard.

So how does integrated design differ from traditional design? Whole Systems Integrated Design is focused on collaboration within a multidisciplinary team, compared to a traditional *linear* approach in which each area develops design solutions separately. In a linear approach the architect would work on the floor plan layout and exterior design of the building. Then the engineer would determine the structural requirements, and others would work on the site development and **mechanicals**. Although a project director oversees the entire process, the individual disciplines do not meet as a group to provide input in the design development.

Mechanicals:
plumbing, electrical, HVAC (heating, ventilation, and air conditioning), and protection systems.

During a Whole Systems Integration Process, a team works together from the beginning to understand and develop the design. This "group think" process and resourceful input results in effective solutions for cost efficient, green, sustainable buildings. Dozens of successful projects support the fact that integrated design is an effective approach for creating comprehensive green buildings on reasonable budgets. The Syracuse Center of Excellence (CoE) provides an example of how the approach can work. The rendering in Figure 2-13 shows the team's vision for the CoE's green roof.

Figure 2-13: A rendering of the Syracuse Center of Excellence, Syracuse, N.Y.

Courtesy of Toshiko Mori.

Think of it this way: if your teacher gave you one test to complete individually and a second test that could be completed together as class, which score would be higher? Whole System Integration works much in the same way. Because each individual has different experience and knowledge, the combined intelligence of the group is far greater and typically provides a better outcome.

There are many ways to approach the integrated design process, and different firms will adapt best practices to suit their projects and staff. The Rocky Mountain Institute is a nonprofit group that studies our use of energy and resources in the built environment. The Institute has defined the following four principles as the foundations of integrated design:

The CoE provides a shining example of the integrated design approach in practice. You can read about architect Toshiko Mori and the CoE team's process at *http://greensource.construction.com/features/0611mag_architects_Office.asp*.

1. **WHOLE-SYSTEMS THINKING:** taking interactions between elements and systems into account, and designing to exploit their synergies.
2. **FRONT-LOADED DESIGN:** thinking through a design early in the process, before too many decisions are locked in and opportunities for low-cost, high-value changes to major aspects of the design have dwindled.
3. **END-USE, LEAST-COST PLANNING:** considering the needs of a project in terms of the services (comfort, light, access) the end user will need, rather than in terms of the equipment required to meet those needs. For example, it's typically assumed that perimeter heating is essential to provide comfort at the edges of a building on cold mornings, but many green designers are now proving that high-performance buildings reduce energy use, noise, and maintenance demands.
4. **TEAMWORK:** Coming up with solutions, as a group, and collaborating closely on implementing those solutions aren't things that happen without strong facilitation and commitment.

Reprinted with permission from The Rocky Mountain Institute.

The Whole Systems Integrated Process Guide (WSIP) 2007 for Sustainable Buildings & Communities was approved in 2007 as ANSI/MTS Standard WSIP 2007. Research this standard to find examples of building projects that have used the process. For example, the Syracuse CoE design team held a series of charrettes with architect Toshiko Mori and a team of internationally recognized consultants.

Table 2-1: **Eight-Step Approach to Integration.** *Bunting Coady Architects, Vancouver*

1: **Shape and shadow:** Massing and orientation decisions that involve function, daylight, and structural considerations.	5: **How the building breathes:** Natural ventilation and passive heating and cooling.
2: **SITE Opportunities:** Where to locate the building on the property, and how it will relate to its immediate context.	6: **Comfort system:** With heating and cooling loads largely determined, the team can get into systems design.
3: **Envelope:** Types of walls and location of windows	7: **Materials:** Materials chosen for various surfaces
4: **Lighting design:** Look at both daylighting and electrical lighting throughout.	8: **Quality Assurance:** Review the building as a system.

Reprinted with permission from Bunting Coady Architects.

Alternatively, Bunting Coady Architects in Vancouver, Canada, takes an eight-step approach to integration that allows goals to emerge throughout the process, Table 2-1. Figure 2-14 compares the integrated design process to the traditional design process.

Figure 2-14: **Comparison of integrated design process to traditional design process.**

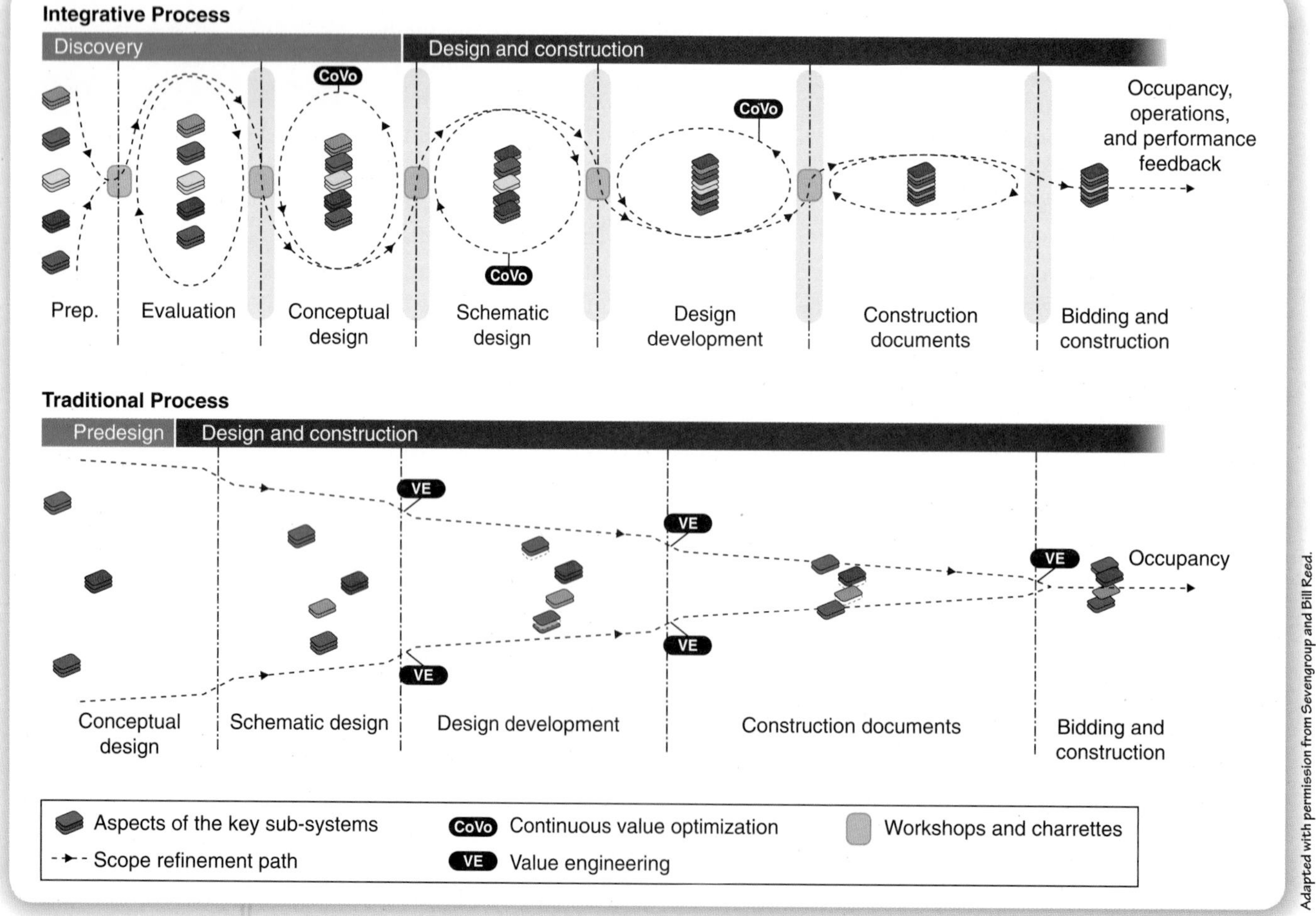

Adapted with permission from Sevengroup and Bill Reed.

AGENCIES AND ORGANIZATIONS INVOLVED IN THE DESIGN AND DEVELOPMENT OF A STRUCTURE

The Whole Systems Integration Process must include communication with many agencies during the design and build process to insure a safe, well-built, sustainable structure constructed without injury or negative impact to the environment. The specific agencies and organizations which must be contacted, and their level of involvement, is determined by the structure's location and purpose (see Figure 2-15). For example, if you wanted to build a small footbridge on a nature trail through **wetlands**, you would need to contact the Department of Environmental Conservation (DEC), the Environmental Protection Agency (EPA), and the Army Corps of Engineers.

In addition to complying with codes established by regulatory organizations, today's conscientious design teams plan to achieve **LEED** certification for high performance buildings. LEED stands for Leadership in Energy and Environmental Design, a voluntary national rating system developed by the U.S. Green Building Council for construction of high-performance, sustainable buildings that reduce negative environmental impact and improve occupant health and well-being. Building projects are rated on site sustainability, water efficiency, energy and atmosphere, materials and resources, and indoor environmental quality to earn Certified Silver, Gold, and Platinum LEED Certification.

As you read future chapters of this textbook, you will learn more about other agencies, such as the International Code Council (ICC), which is dedicated to building safety and fire prevention. Many cities, counties, and states adopt the ICC construction codes to guide the planning and construction of residential and commercial buildings. In addition to state and federal organizations, local groups, such as the Community Development Agency and the Industrial Development Agency, are contacted to determine concerns of area residents and the impact the structure may have on the local economy and community.

Wetlands:
lands where water saturation is the dominant factor. The resulting bogs, marshes, swans, and fens provide a habitat for many species of plants and animals.

LEED:
Leadership in Energy and Environmental Design, a voluntary national rating system for construction of high-performance, sustainable buildings.

The U.S. Environmental Protection Agency (EPA):
an independent regulatory agency responsible for establishing and enforcing environmental protection standards.

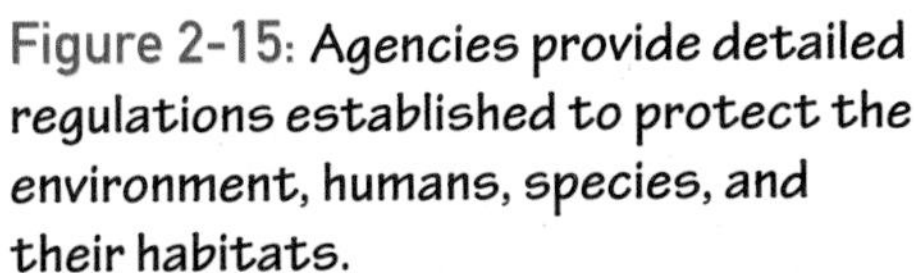

Figure 2-15: **Agencies provide detailed regulations established to protect the environment, humans, species, and their habitats.**

Using EPA Website Tools

The EPA website contains interactive tools that provide regulatory information to assist architects and engineers. For example, in Region 3, you can find a link to an article on beneficial landscaping. Beneficial landscaping is designed with environmental friendly practices and materials that require less energy to maintain.

REGIONAL REGULATORY INFORMATION Searching the EPA.gov website can help you determine how to comply with Region regulations,

Check out regulatory information for your region of the United States. Make notes in your notebook for future reference. *http://www.epa.gov/lawsregs/where/index.html*

Figure 2-16. The EPA serves as the on-the-ground staff enforcing the regulations. Select an EPA Region from the following drop-down box or from the map to find regulatory information for your Region.

You can choose from two options to learn more about state-specific regulatory information: state plans and programs required by federal law and state resource locators. The EPA website provides links to state-level laws, regulations, and administrative agencies and provides plans, programs, and designations as required by federal regulations (e.g., your state's implementation plan under the Clean Air Act).

Figure 2-16: **The EPA provides easy-to-access regulatory information for 10 regions of the United States at http://www.epa.gov/lawsregs/where/index.html.**

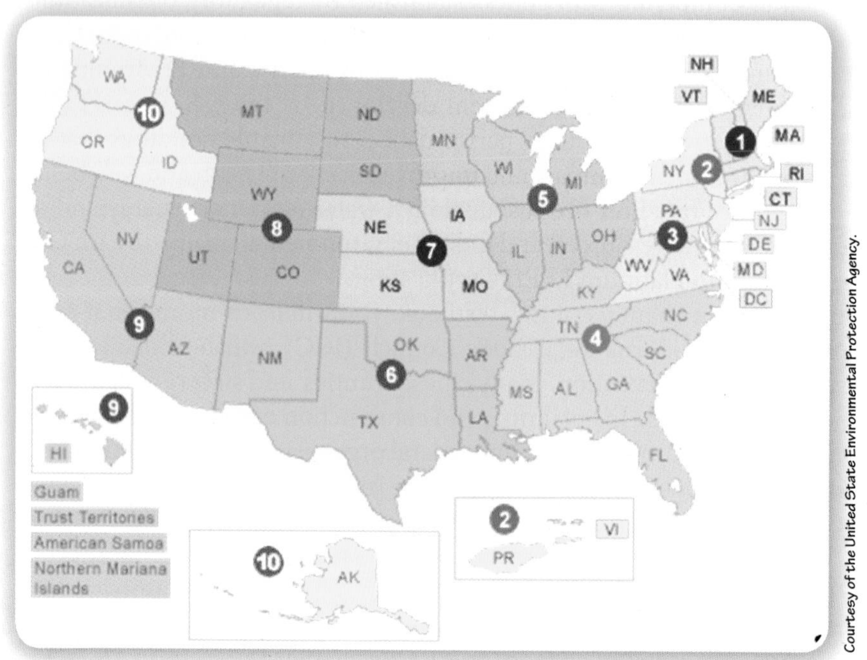

Courtesy of the United State Environmental Protection Agency.

Careers in Civil Engineering and Architecture

The Representative for Everything

JENNIFER WORKMAN, PROJECT ARCHITECT, GOOD FULTON AND FARRELL ARCHITECTS

Clients give Jennifer Workman an idea of what they want, and she takes it from there. As a project architect, she manages a team that turns the idea into a functional building. That means seeing the project through every stage, from the drawings to construction. She gets approval from city officials, picks a contractor, and determines the best materials to use given the available funds.

"You're responsible for the client's happiness on the project," Workman says. "You need to make sure that the best product is being built."

On the Job

For most of her career, Workman has developed retail structures. Lately, she's moved on to a different kind of project, larger in scale. Her most important assignment to date is developing a 150,000 square feet museum in Dallas, Texas, called the Museum of Nature and Science. Two architectural firms are involved, and Workman is the contact between the two. She also deals with the contractor, the consultants, the exhibit designer, and the budget. The building will take four years to complete from start to finish.

"To work on one project for so long is pretty intense," Workman says. "It means you're really the representative for everything, and you need to know about everything on the project. It's very exciting."

Inspirations

As a child, Workman was always interested in math and science, as well as art. Her stepfather is an architect, and at 15 she went to work for him building models. She got to watch him draw plans, and he also took her to construction sites. She decided to go into architecture herself after high school graduation—just in time.

"With architecture, it is better to decide immediately when you go to college," Workman says. "You have long studios from the beginning, and if you miss those you could easily be in school for an additional few years."

Education

Workman studied architecture at the University of Texas-Austin. It's a big place, but her architecture classes felt as intimate as her small high school.

"The architecture students were divided into 15-person studios, which is typical for most college studios," she says. "You build close relationships with people."

Workman's favorite part of college was traveling abroad. Her program took her to several cities in Western Europe, where she spent time looking at buildings. She focused on museums, sketching and photographing them.

"It's really important to travel and see other things," Workman says. "It helps you think outside the box. In the museum project I'm working on now, I'm able to say, 'At the Louvre in Paris they do this kind of thing.'"

Advice for Students

Workman worries that students who feel they aren't good at art or math will avoid an architecture career. "There are so many areas of architecture where you don't have to draw well to create buildings people can enjoy," she says.

Workman suggests that high school students work at an architectural firm in any capacity, just to get a feel for the profession. She herself worked at one firm as a receptionist.

"This will help you decide if you want to be an architect," she says. "You have to be really passionate about the work."

SUMMARY

In this chapter, you read about the differences and similarities between civil engineers and architects as well as the education, skills, and stages of development necessary to become a licensed professional. Other careers relating to the design and build of a structure were identified, and you learned how to research these occupations through use of the Occupational Outlook Handbook. Making a career decision is a difficult task. Only after thorough research of a profession's education and training requirements, job description, and future outlook can one begin to match their own academic strengths and interests with a career. During the career research process, one should visit with several professionals in the field, as these visits will provide a unique and personal insight into the career.

Most professionals in the architectural and civil engineering field do not work in isolation. Civil engineers, architects, stakeholders, and agencies work together from the very beginning of a project to create a relevant, safe, economical, and sustainable structure. This group is called Whole Building Integrative Design. The purpose of this approach is to develop a shared vision, encourage creativity, and use resources efficiency. Members of this team are acutely aware of climate change, natural disasters, energy consumption, and shortages of water and other resources, all of which necessitate thoughtful selection and the use of economically and environmentally friendly materials and practices. The goal is to create a healthy, economical, and sustainable building that strikes harmony between humans and their environment. The Integrative Design approach defines these goals and develops the strategies for attaining them.

BRING IT HOME

1. Select three careers involved in the design and build of a structure and use the Occupational Outlook Handbook, accessed from the U.S. Department of Labor Bureau of Labor Statistics website at *http://www.bls.gov* to determine the education requirements, job description, and future outlook of each.
2. Select one of the careers you researched in the previous question that matches your academic strengths and interests. Research four universities or colleges where the education or training is offered and compare cost, location, and program offerings.
3. With a group of classmates who have similar career interests, create a career questionnaire. Develop questions that will uncover what it's really like to have that career. With your teacher's permission, mail the questionnaire to several local or regional companies. Compare the responses and make a chart of similarities and differences.
4. Research local professionals in the civil engineering and architecture fields and suggest that your teacher invite them to discuss their careers with the class.

5. Make a list of the individuals and agencies that you would include in a Whole Building Integrated Design process for the development of one of the following projects: county airport, amphitheater, drive-thru credit union or bank, take-out pizza shop, secondhand clothing store, car wash, gas station, playground, nature center with hiking trails, campground, golf course and club house, small marina and bait shop, wastewater treatment plant, or skate park.
6. Develop a list of local concerns that would impact the design and construction of a sustainable structure.
7. Review the guide to becoming an architect provided by the American Institute of Architecture Students: *http://www.aias.org*

EXTRA MILE

Examine a new local construction and compare it to an older construction built for a similar purpose. Compile a list of differences and make a hypothetical guess as to why the design, construction technique, and/or materials were changed. Research to see if you can determine the actual reason.

PART II
Research and Design

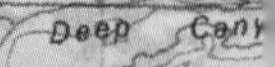

CHAPTER 3
Research, Documentation, and Communication

GPS DELUXE

Menu

START LOCATION | DISTANCE | END LOCATION

Before You Begin

Think about these questions as you study the concepts in this chapter:

1. What are three commonalities of all successful building projects?
2. Why is research vital to land development?
3. What documentation is created during the planning of commercial and residential development?
4. What is the purpose of a development program? What information is included and who is the audience?
5. How do working drawings and presentation drawings differ from one another?
6. Why is communication essential to achieving shared project vision and construction productivity?

Highly successful building projects have three things in common: They are the result of comprehensive research, documentation, and communication. Research, the first component, begins in the pre-project phase and continues through the building process as needed. Diligence in research will identify cost effective, sustainable resources, and current standards of **best practice.** Informed decisions based on thorough and thoughtful research will enhance owner and occupant satisfaction throughout the **building's lifecycle.**

Best Practice: proven performance or process of highest standard.

A building's lifecycle can be addressed through application of **Building Information Modeling (BIM).** Computer-based BIM programs allow for collaboration on building design, development, and test simulations to determine how the building will perform under projected conditions. Many events must occur before construction. Figure 3-1 shows a graph of a design process for

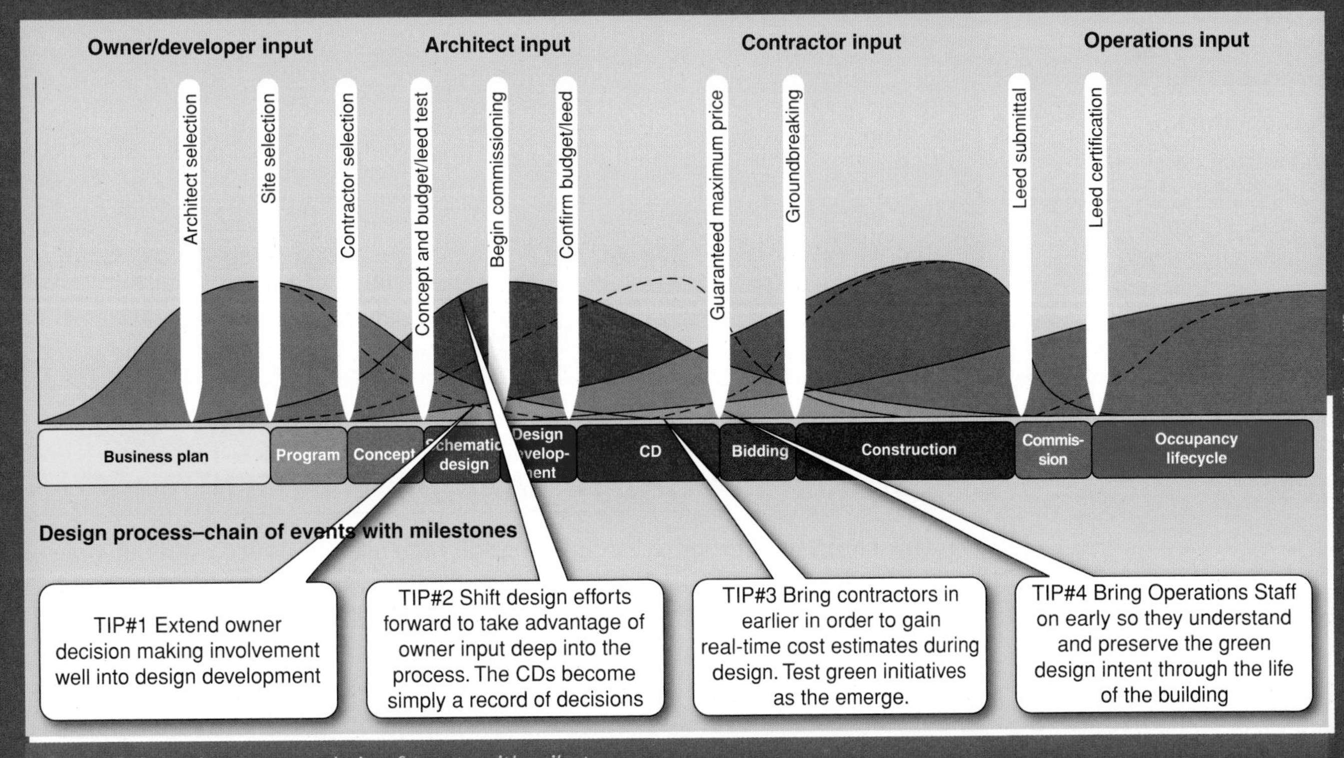

Figure 3-1: *Design process, chain of events with milestones.*

INTRODUCTION

Building Lifecycle:
all stages of a building over time, including construction, use, maintenance, preservation, revitalization, deconstruction, and demolition.

a green development. How many pre-construction events will require research and documentation?

The second component of successful building projects is documentation. Documentation begins at the project inception and provides a complete record of the building process. Initial concept drawings and communications are recorded and kept on file along with revisions and changes. When you think about project documentation, you may first think of the many sheets of technical drawings. Equally extensive are the supporting documents: studies, permits, and legal proceedings. In some locations, it may take years to complete the research and documentation necessary to build a commercial structure.

Technological tools, materials, and other resources should be selected on the basis of safety, cost, availability, appropriateness, and environmental impact.

Building Information Modeling (BIM):
collaborative building design based on sharing of information and test simulations of how the building will perform under projected conditions.

Program:
a written document highlighting the project goals, design objectives, constraints, and specifications.

The third component common to all land development and building projects is communication. Whether the project is small or large, residential or commercial, communication is key to a successful outcome. Think about all the stakeholders that are involved with a project. It is important that they are involved early in the planning process and kept informed throughout construction. Research and communication support the development of the comprehensive **program** required for permits and financing.

Finally, when the land development is approved and financed, continued communication between stakeholders will increase the likelihood of efficient construction progress and reduce the chance of a costly error or misunderstanding. This chapter will provide information to guide you through each of these three key areas as you complete your own land development project. Remember the famous quote by Thomas Edison: "Success is 10 percent inspiration and 90 percent perspiration"? Research, documentation, and communication are labor intensive, but essential for land development to become reality.

RESEARCH IS VITAL TO LAND DEVELOPMENT

A substantial amount of research is necessary to ensure the successful outcome of land development. Without deliberate and detailed research, one could make some very costly mistakes. Many people have discovered this fact the hard way. Failing to research a site's history, zoning, adjacent properties, and demographics before beginning a project can lead to financial ruin. Research determines building feasibility and identifies the **zoning, codes**, and **regulations** for the location.

Zoning, codes, and regulations are established to ensure the construction of safe, well-designed land developments that add value to the area. Building codes and regulations specify proven building practices that address building durability, sustainability, and occupant safety. Local governance determines the zoning codes and regulations that will pertain to your building project, however, many cities and towns follow a set of state or national codes (see Figure 3-2).

So where do you find general building code information? A good place to start is the International Code Council (ICC), developed in 1994 through the collaborative efforts of Building Officials and Code Administrators International, Inc. (BOCA), International Conference of Building Officials (ICBO), and the Southern Building Code Congress International (SBCCI). ICC's goal is to provide a single national building code for the United States.

Figure 3-2: Various code books for New York State.

ICC Publications

- International Building Code®
- International Energy Conservation Code®
- International Code Council Electrical Code Administrative Provisions®
- International Existing Building Code®
- International Fire Code®
- International Fuel Gas Code®
- International Mechanical Code®
- ICC Performance Code™
- International Plumbing Code®
- International Private Sewage Disposal Code®
- International Property Maintenance Code®
- International Residential Code®
- International Urban-Wildland Interface Code™
- International Zoning Code®

Another source for building code information is the National Conference of States on Building Codes and Standards (NCBCS). As you research building codes you will discover many code books available for purchase, including handy spiral bound Code Check reference books. Code books provide specific standards for specialties such as the NFPA 70 National Electrical Code® published by the National Fire Protection Association (NFPA), the Uniform Plumbing Code (UPC), Uniform Mechanical Code (UMC), and the Uniform Fire Code (UFC). Some codes can be obtained through the Internet, such as the Building Energy Codes provided by the U.S. Department of Energy. The U.S. government supports energy conservation by providing free software, RESCheck and COMCheck, to assist with the planning of energy efficient buildings (see Figure 3-3).

As you gather code information for your notebook, don't forget to locate guidelines for building and facility **accessibility** available through the American Disabilities Act Compliance Manuals. Due to the complexity of codes, websites such as The Building Oracle have been established to provide architects with code information that should appear on building plans.

Zoning:
local ordinances regulating the use and development of property by dividing the jurisdiction into land-use districts or zones represented on a map and specifying the uses and development standards within each zone.

Building Codes:
regulations, ordinances, or statutory requirements established or adopted by a governing unit relating to building construction and occupancy.

Building Regulations:
a set of guidelines designed to uphold standards of public safety, health, and construction and control the quality of buildings.

Accessible:
meets requirements of federal regulations for access for people with physical disabilities, including the Americans with Disabilities Act, the Architectural Barriers Act, and the Fair Housing Amendments Act.

Figure 3-3: Software programs RESCheck and COMCheck are provided by the U.S. Department of Energy to promote energy efficient building design.

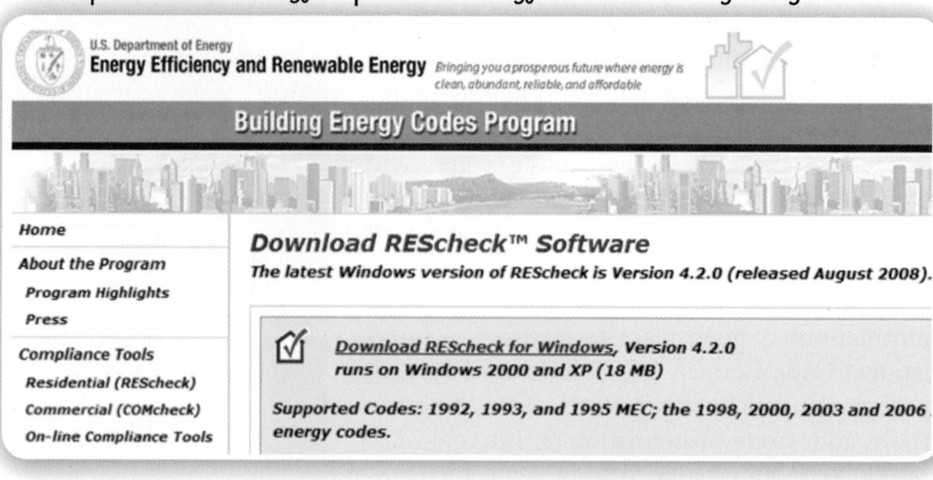

US Department of Energy

Your Turn

Research the following websites and record notes in your notebook about the information that is provided at each site. These notes will help you to quickly locate code information for your future development projects.

http://www.b4ubuild.com The Building Oracle

http://www.iccsafe.org International Code Council (ICC)

http://www.codecheck.com Code Check Reference books

http://www.nfpa.org National Fire Protection Association (NFPA)

http://www.iapmo.org International Association of Plumbing and Mechanical Officials (IAPMO)

http://www.access-board.gov/adaag ADA Accessibility Guidelines for Buildings and Facilities (ADAAG)

http://www.ncsbcs.org National Conference of States on Building Codes and Standards, Inc. (NCSBCS)

Setback:
the distance from the building to the property line.

Subcontractor:
a secondary person or company hired by the general contractor to do work.

Zoning refers to the local ordinances that regulate the use and development of a property, such as **setbacks** or height of a structure. Land is divided and zoned by building purpose and compatibility. Zoning prevents a factory from being constructed in the middle of a residential neighborhood. Building codes are specific rules established or adopted by the local governing unit and planning boards that identify safe and proven building practice the builder, or **subcontractor**, must follow. Often state and local governments adopt national or international codes.

Building regulations are designed to uphold standards of public safety, health, and construction and control the quality of buildings. Regulations are written by regulatory agencies. For example, the Environmental Protection Agency (EPA) is authorized by Congress to write regulations to address legislation about public health protection. You will learn more about zoning, codes, regulations, permits, and variances in Chapter 5: Site Discovery for Viability Analysis. For now, let us assume you have found a piece of property and have conducted thorough research. Your research confirmed that you are able to secure a clear **title**, but before you purchase the property, you want to make sure you will be allowed to build on the site.

Title:
a written document providing evidence of the right to legally possess or dispose of property.

A good place to start is at your local building department, planning board, or development services. Some may even have websites to guide you through the land development process, Figure 3-4.

Figure 3-4: The city of San Diego's Development Services Website provides information about building requirements and permit procedures.

Courtesy of San Diego Development Services, http://www.sandiego.gov/development-services.

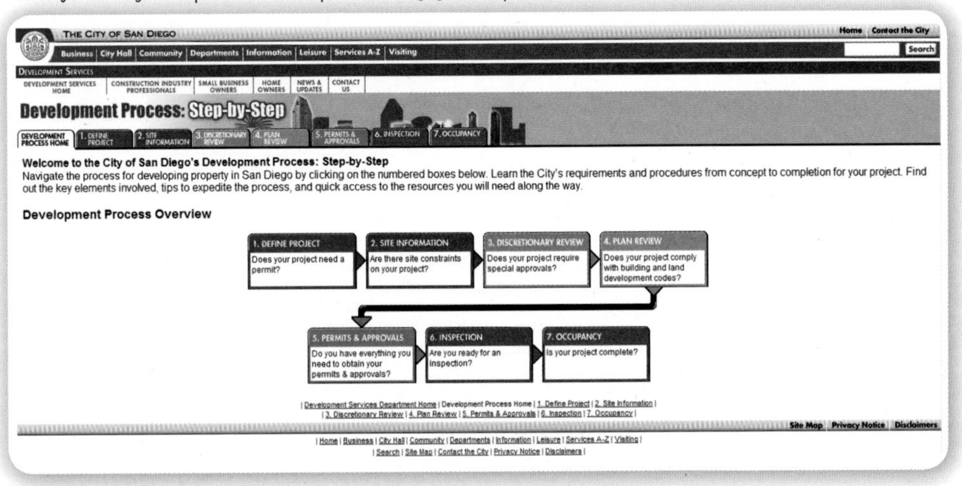

Most land development will require building permits and financial backing. To acquire these, you will need to communicate a clear, concise vision for your project. Just as you determine **viability** of the site, you will need to determine the viability of the development.

Viability:
practicality or feasibility.

For example, is there a need for your proposed facility? Will it be compatible with the adjacent properties or businesses? Will it add value to the surrounding area? Who will build your structure and how much will it cost? How long will it take to build? How will construction noise and debris affect the community? Each of these questions will require extensive research and study. Local community development agencies or civic groups may be able to provide you with some insight into community needs and concerns (see Figure 3-5).

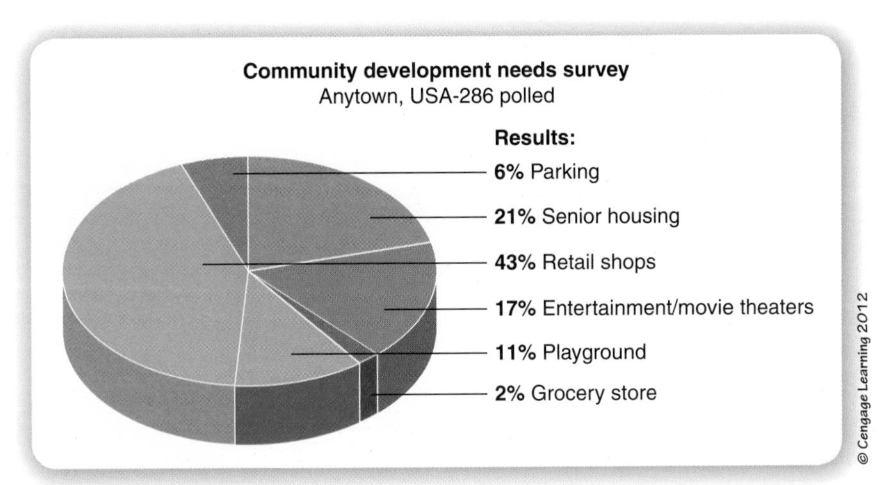

Figure 3-5: Local development agencies or civic groups can help you gather data about the planning concerns in your community. The data in this illustration identifies the development priorities of a fictional community.

Case Study

Let us take a look at our Syracuse Center of Excellence case study from Chapter 2. It took almost a year to complete the research and documentation needed to obtain their permit (see Figure 3-6).

The Center of Excellence has established collaborations with several universities to support learning through research, testing, and development. One of the collaborations involved students from Cornell University, Clarkson University, and the State University of New York College of Environmental Science and Forestry. Figure 3-7 shows some of the results of the students' study of traffic patterns near the Center of Excellence site.

Figure 3-6: *Syracuse Center of Excellence Building Permit dated October 2007.*

Courtesy of the Syracuse Center of Excellence.

City of Syracuse
DEPARTMENT OF COMMUNITY DEVELOPMENT
DIVISION OF CODE ENFORCEMENT
201 E. WASHINGTON ST.
SYRACUSE, NY 13202-1430

Permit Number: A. 84622 Date: 10.23.07

This notice, which must be prominently displayed on the property or premises to which it pertains. indicates that a

PERMIT

has been issued to LeChase Const.

permitting Const. New 5 Story Structure

at 727 E. Washington St

All work shall be completed in accordance with the conditions of the permit, plans and specifications, and all other applicable codes, laws, ordinances, rules and regulations. This permit does not constitute authority to work in violation of any federal, state or local law, rule of regulation.

CoE

The following inspections, checked (✓) and listed below, are required during the course of work. **Do not proceed beyond this points until initialed by our inspector.** The owner of his or her authorized agent shall be responsible for notifying this division at least 48 hours prior to a required inspection.

- [x] Sprinkler before closing ___
- [x] Footing before pouring concrete ___
- [x] Foundation before backfill ___
- [x] Framing before enclosing ___
- [x] Electric before enclosing ___
- [x] Elevator
- [x] Fire Alarm / Security ___
- [x] Plumbing before enclosing ___
- [x] Heating, ventilation, air conditioning before enclosing ___
- [x] Insulation before enclosing ___
- [x] Final Inspection ___

Permission has been granted to proceed with the work as set forth in the plans, specifications or statements now on file with this division. Any amendments made to the original plans, specifications or statements must be submitted to this division for approval prior to the commencement of any work.

FOR INSPECTION CALL:
448-8695

James Blakeman
Director of Codes

Figure 3-7: *This illustration shows the results of a study to determine traffic patterns at the Syracuse Center of Excellence location. The study was conducted by students at Cornell University, Clarkson University, and SUNY College of Environmental Science and Forestry.*

Cornell University, Clarkson University, and SUNY College of Environmental Science and Forestry. Reprinted with permission

2) Field campaign study

- Focus: Highway impacts on the air quality around CoE site
  - –Question 1: How intensive the highway traffic emissions will contribute to the air pollutants in various areas around CoE site?
  - –Question 2: Are there any other factors which impact the air quality around the CoE site?
- Approach
  - –Intensive study every quarter over two years (2008–2009) → Seasonal impact
  - –Field Measurements on seven locations around CoE → Spatial variations
  - –Measurements at different time of a day (morning and afternoon) → Diurnal variations
- Pollutants we measured
  - –CO: Carbon monoxide, mainly from gasoline vehicles
  - –BC: Black Carbon, mainly from diesel vehicles
  - –UFP: Ultrafine particles, with diameter less than 100 nm. They are from both gasoline and diesel vehicles.

Conclusion

Highway impacts could be clearly observed on black carbon (BC), UFP number concentrations and particle size distributions. We also identified the temporal and seasonal variations in the traffic impacts. Highway impacts, regional impacts and seasonal impacts need to be considered when designing the intelligent building control systems. Once two traffic imaging camera systems (shown in the pictures on right) are installed, we will be able to provide detailed information on the traffic patterns and their effects on the air quality.

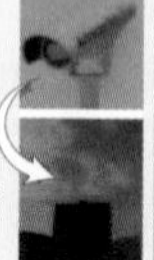

Diurnal variations

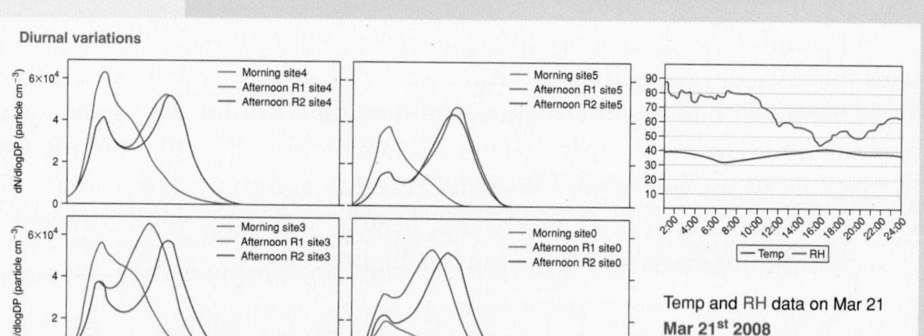

Both modes (10nm and 60~100nm) decayed as time went by, likely due to both traffic and meteorological conditions

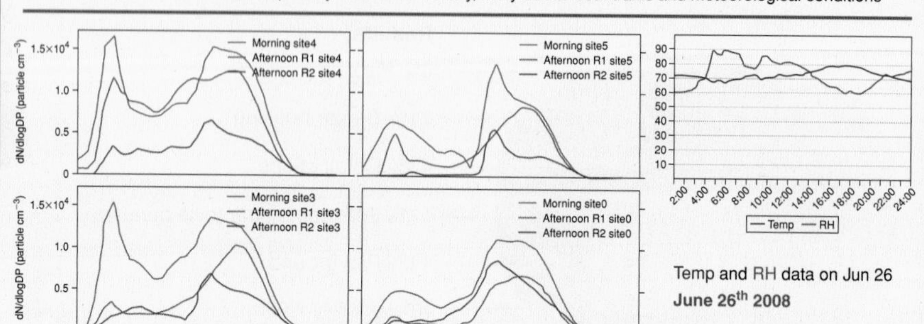

The second mode (~40nm) was observed in the afternoon, but without obvious spatial variation, suggesting a possible regional particale growth event.

Point of Interest
A Code for Urban Planning

Wouldn't it be nice to have all your daily needs, such as work, services, and entertainment, within walking distance of your residence? This is much different from the philosophy of zoning where areas are segregated. Form Based Codes (FCBs) support the preservation or revitalization of old buildings. This often results in a building having a second life or new purpose, such as an old school being converted into an apartment complex. FCBs encourage the development of shared spaces that foster a sense of community. The documents include a regulation plan, public space standards, and building form standards. This increasingly popular type of urban planning can save energy, cut pollution, and encourage a healthy, convenient, and enjoyable lifestyle.

DOCUMENTATION FOR LAND DEVELOPMENT

A project's purpose, size, and location affect the type and amount of documentation that must be completed for land development to occur. The same building built in three different states will most likely require different documentation to address specific state and local governance. A single family residence will require less documentation than a large commercial office building. In addition to study results, a **program** or business plan is generally required for commercial permits and approvals. The program provides detailed drawings and schedules necessary for the accurate and timely completion of the development. Documentation such as the architectural sketchbook and journal entries are not part of public record, but are essential for project development. Using the latest computer technology, today's architects and engineers work collaboratively utilizing Building Information Modeling (BIM) for **building optimization**.

Building Optimization: a building designed to function at its best or most effective.

Building models created with BIM, a 3D modeling software, can simulate building performance under real-world conditions. This valuable information, uncovered during the planning process, supports the design of buildings better able to withstand natural disasters such as hurricanes and earthquakes. BIM also allows architects and engineers to generate and exchange digital representations at all stages of the building process to support informed, collaborative decision-making. BIM can be used to uncover possible conflicts prior to construction.

Engineering design is an iterative process involving *modeling* and *optimization* used to develop technological solutions to problems within given constraints.

Architectural Brief and Program for Non-Residential Development

When land development is desired, it is usually described in terms of a problem or need. The architect and his or her team address this problem as an architectural brief. At this stage the architect may call together a group of consultants to identify

Figure 3-8: Autodesk® Revit® Structure software provides structural analysis, design, and documentation.

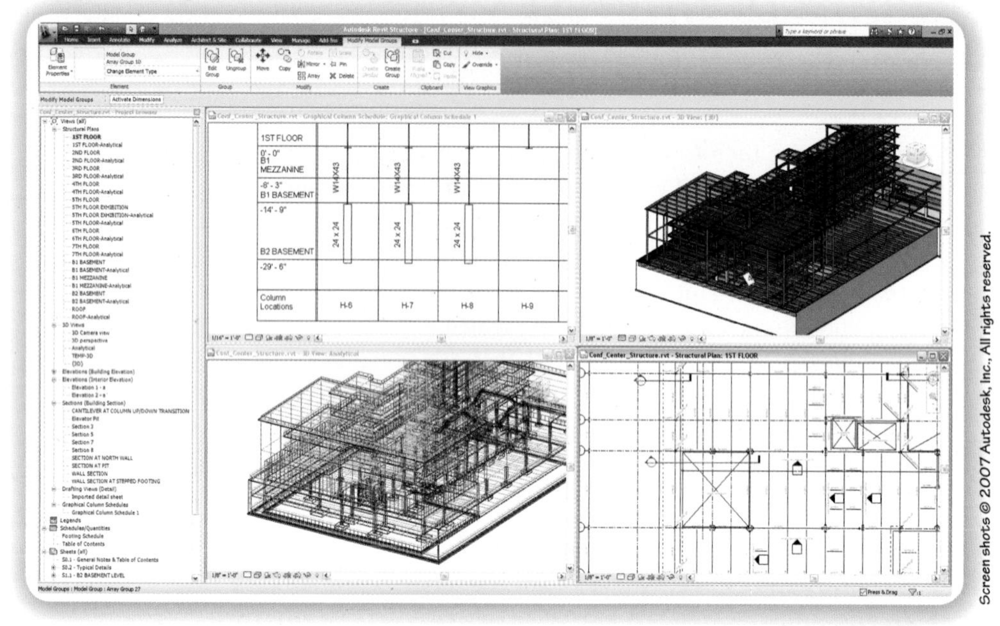

Your Turn

See how BIM can help an engineer plan a building that can survive an earthquake. Use your Internet search engine keywords Autodesk® BIM Earthquake (see Figure 3-8).

View videos on BIM, starting with a one minute introduction on the Autodesk® website http://www.usa.autodesk.com. Once on the Autodesk® site, enter the following in the search box: Building Information Modeling—Experience BIM http://usa.autodesk.com/adsk/servlet/index?siteID=123112&id=9970899

Navigate about the Autodesk® site to learn more about BIM http://usa.autodesk.com/adsk/servlet/index?siteID=123112&id=9976276

constraints, research requirements, and develop a unified vision for the project. This process is called **Integrated Design.** Most commercial developments require a business plan and program.

As the goals and requirements for the development become clear, space-planning is discussed. The number, size, and type of rooms are determined along with their desired proximity to adjacent spaces. Many times you will hear the term *program* used in conjunction with architectural brief. The program explains the space requirements and relationships, building flexibility and expandability, and site requirements. Special equipment, furnishings, and mechanical systems needed to support the building's function may also be listed in the program. The main purpose of a program is to provide governing agents, architects, and other stakeholders with clear and accurate details of the project's vision and requirements. This information is necessary to make informed project viability decisions such as approval for funding.

The main categories of a program include a summary, historical background, analysis and study results, scope of work, utility requirements, a schedule, and a detailed cost estimate. Developing a program for a large development is labor intensive and may take several months or even years to complete.

> Technological designs have constraints. Some constraints are unavoidable, such as properties of materials, or effects of weather and friction. Other constraints are realized through regulations to insure environmental protection and human safety. Constraints can limit design choices.

Components of a Program

- Title sheet
- Signature page
- Executive summary
  (essential information about the program for convenient reference, including highlights of the most important aspects: project justification, space requirements, cost model, and project schedule)
- Purpose and scope
- Site analysis
  (physical characteristics, orientation, function, relationship, and codes)
- Building requirements
- Space utilization
  (detailed room and space descriptions, purpose, and area)
- Building systems performance criteria
- Cost analysis

Development of a Single-Family Residence

As mentioned earlier, less documentation is generally required for a single-family residence. Let us assume you are building a single-family home for yourself. Most locations will require a building permit. Local governance will identify the building standards and specifications you must follow. In areas with zoning, you will need to confirm that the property is zoned for residential development. You will then need to verify that the property is suitable for building. Soil composition and percolation tests will determine land stability and drainage. A local civil engineer can perform these tests and assist with septic, drainage, and grading plans. You will read more about site analysis and development in Chapter 5: Site Discovery for Viability Analysis. Local governance generally requires a current survey of your property to prove you have street access and can accommodate necessary setbacks. You will also need to verify that you have access to necessary utilities, such as gas, water, and electricity. Wastewater and storm water will need to be removed, through a septic system or through connections to a sewer system and storm water drains.

Ideally, it is best to choose the site that matches your vision for the residence. For example, if you want a walkout basement, you would select land with a steep slope. There are many ways to acquire plans for your residence. You could contact a local builder, who may have plans to choose from, or you could contact an architect to create a custom plan. Many people research residential plans available online. Let us say you have searched online and have found a couple of plans that you like. They have the exterior style that will fit in the neighborhood, but you are not pleased with the interior layout. Maybe you are unsure if the plan will fit the property providing both solar orientation and curb appeal. Or you

Figure 3-9: **Builders can provide clients floor plans to choose from and the option of a turn-key building solution.**

may have specific needs, such as accommodations for a handicapped relative who visits you each month. These are all reasons why people employ an architect.

Your architect will review your site, research your local governance, prepare your custom plans, and possibly recommend a general contractor. A general contractor will oversee the entire construction and hire subcontractors, such as electricians and plumbers. Another option, if you do not already have property, is to locate a builder that has several locations and house plans from which to select. Many of these builders work in conjunction with a real estate agent. Working with this agent and your financial institution, a contract could be formed for your specific construction. This type of new construction is called *turn-key:* when the house is complete, you put your key in the door and move in (see Figure 3-9).

Keeping a Journal

Most architects and engineers keep a journal documenting research findings and communications that occur throughout the project. Keeping track of research, e-mails, addresses, and phone numbers is important, but notes from meetings, phone conversations, including personal reflections and thoughts may also prove valuable. Successful architects and engineers must be able to multitask, sometimes overseeing several projects at one time. Keeping a journal for each project helps them stay organized, and they use the notes and documentation as a quick refresher before returning a phone call or attending a meeting. Each journal entry should include a date, time, contact name and information, where the communication took place, and a detailed description of what was discussed.

Friday, May 3, 2009

Met with John Smith (123-456-7890), Codes Enforcement Officer, at 314 Maple Street project. Foundation passed inspection, Good location of foundation drainage tile. Site was wet. Discussed improving the below grade treatment of concrete wall exterior. John suggested adding Mel-Rol® and Platon exterior membrane materials. Research those and others on Monday—determine cost, availability, and installation procedure. Contact owner (Mary Jones) about benefit. Will need signed approval for additional time and cost.

Architectural Sketchbooks

Architects use sketchbooks to document ideas and stimulate design evolution and improvements. They serve as a place to record thoughts and observations derived from their research and meetings with the client. The portability of the sketchbook encourages spontaneous sketches and drawings at any time or place. Think about the many times you had an idea or saw a particularly interesting architectural feature. If you did not record it promptly, you may have forgotten some of the details. If you had your architectural sketchbook handy, you could have quickly sketched your idea for future reference.

Images can also be captured using a convenient camera phone or digital camera and included in your sketchbook. The architect's sketchbook is not meant to be shared, but instead serves as a personal workbook to enhance project ideation and

Figure 3-10: Architectural sketch for a proposed university clubhouse.

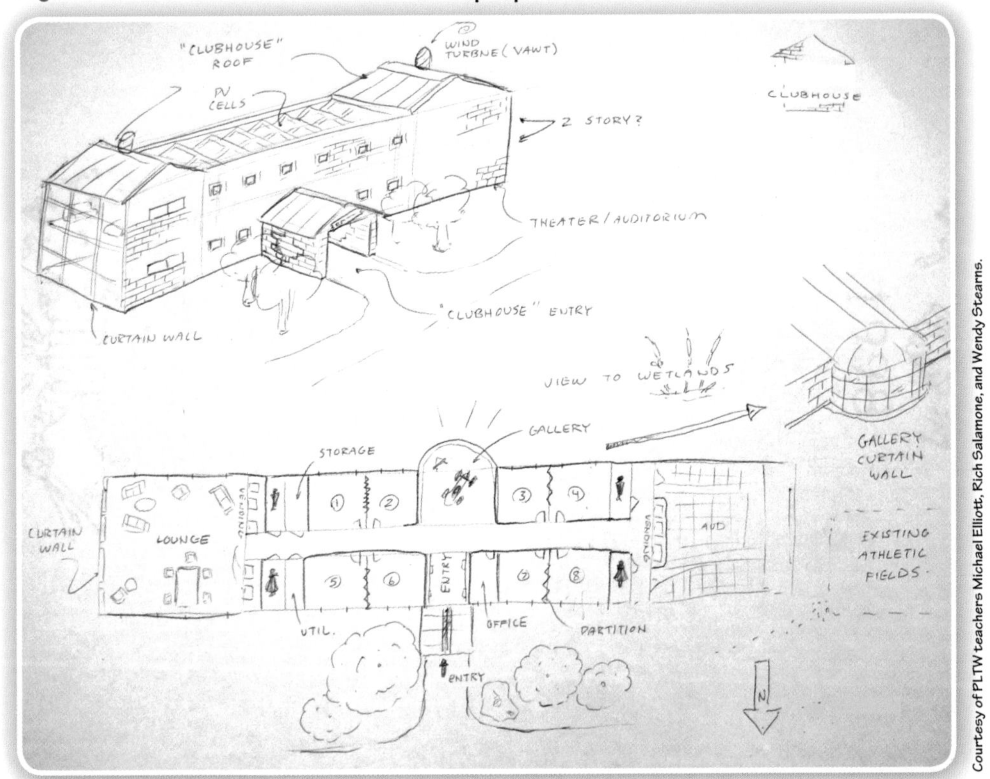

Courtesy of PLTW teachers Michael Elliott, Rich Salamone, and Wendy Stearns.

problem-solving. Architectural sketchbook entries are sometimes messy and incomplete. The freedom of forming quick sketches helps the architect to creatively develop an idea or concept (see Figure 3-10).

Architectural sketches go beyond documenting the surface appearance of an object to capturing an unseen quality that stirs emotion and attains harmony. By learning to "see" and capture that unseen quality on paper, you can replicate them, or better yet, use them to inspire new concepts for building design. Next time you notice a really fascinating building detail, stop and sketch the image. While sketching, think about what it is about the object that brings about the emotion (see Figure 3-11 a and b).

Figure 3-11 a and b: The annotated sketch (b) assesses the emotional impact of the architectural details shown in photo (a).

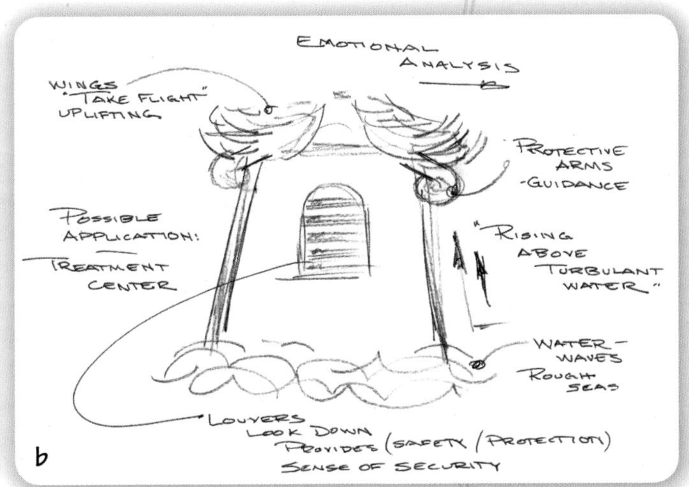

Figure 3-12: A bubble diagram for a proposed university clubhouse.

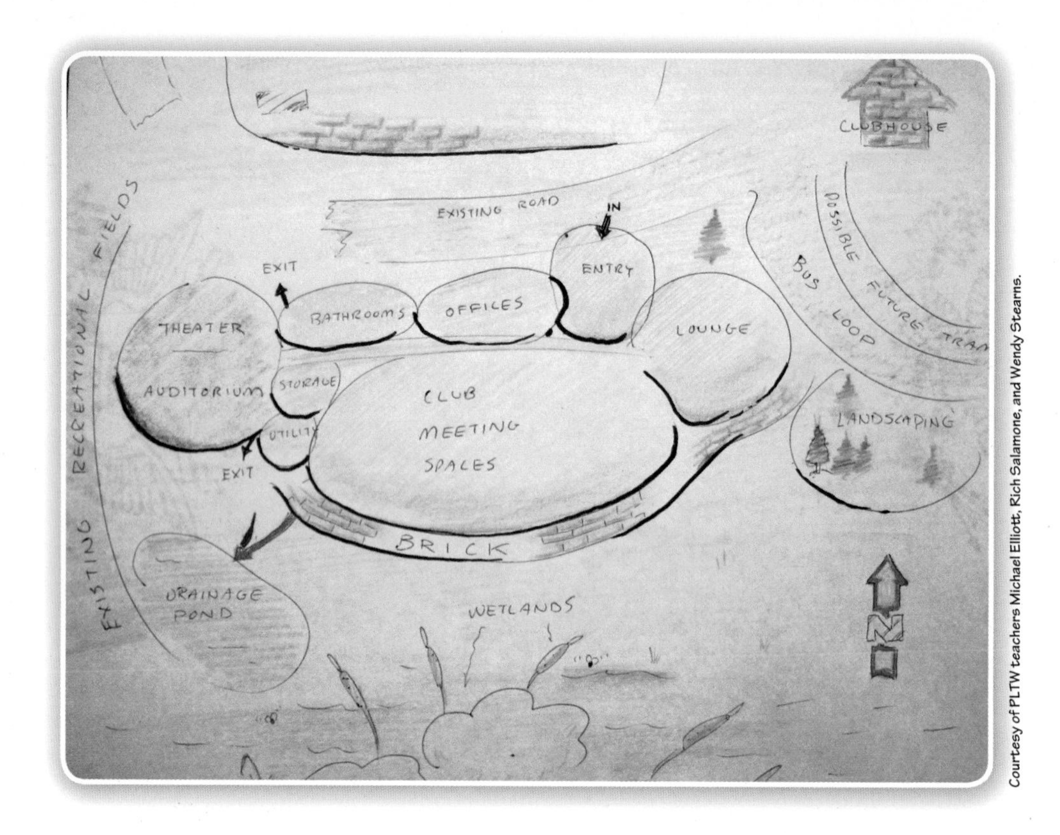

Courtesy of PLTW teachers Michael Elliott, Rich Salamone, and Wendy Stearns.

Bubble Diagrams

An architect's notebook often includes many bubble diagrams. A bubble diagram is a quick and easy way to define a building's relationship to a site or to develop a concept for a building layout by showing approximate space allotments and room placement in the overall plan.

Take a close look at Figure 3-12. What advantages and disadvantages do you see in the proposed layout of spaces?

The Difference between Working and Presentation Drawings

Working Drawings: technical drawings drawn to scale for the purpose of communicating specific information necessary for construction.

Section Views: views showing a slice of the structure, with detailed internal components and information at a virtual cut location.

Drawings are essential to the development process. In addition to supporting a building permit, drawings are used to specify building construction and obtain construction estimates, and they are often required by the bank or lending institution. Architectural drawings can be grouped into two general categories as determined by their purpose: working or presentation. Drawings that are needed to communicate the specific details of construction are referred to as **working drawings**.

Figure 3-14 shows examples of working and presentation drawings which would generally be placed on separate drawing sheets.

Working drawings are technical drawings, drawn to scale, for the purpose of communicating essential information to the civil engineer, contractors, excavators, builders, electricians, plumbers, building inspectors, and other professionals involved in the construction process. Several sets of working drawings are required to provide each professional with the information they need to complete their task. A set of working drawings is often referred to as *plans*. Plans must contain enough information and **section views** to allow accurate and complete construction.

Your Turn

Develop an idea for a small cabin, creating a bubble diagram and architectural sketch similar to the ones shown in Figure 3-13 a and b.

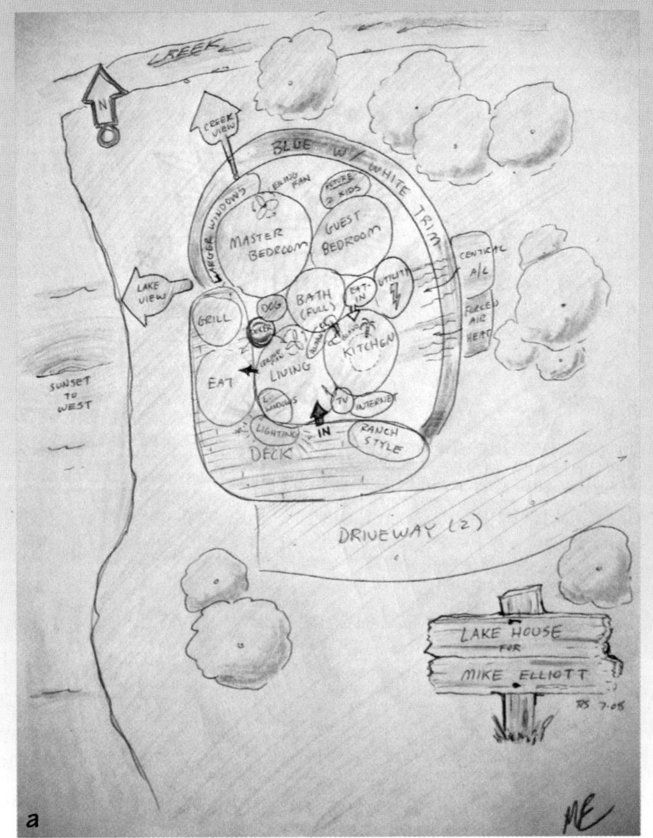

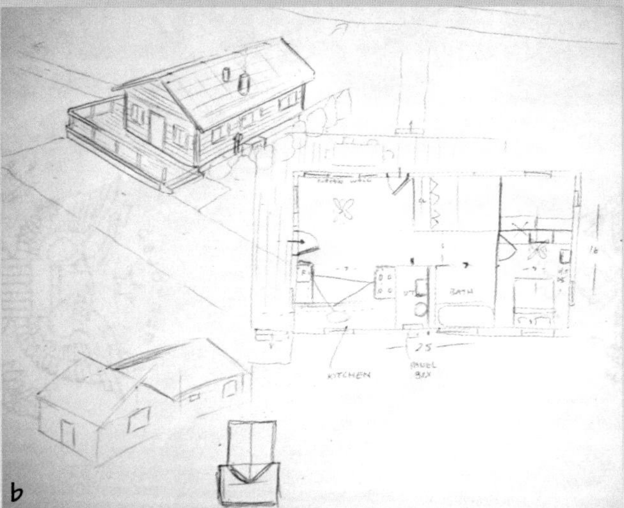

Figure 3-13 a and b: Architectural sketchbook entries for a small residence.

Courtesy of PLTW teachers Michael Elliott, Rich Salamone, and Wendy Stearns.

Figure 3-14: Working and presentation drawings for a proposed lake house.

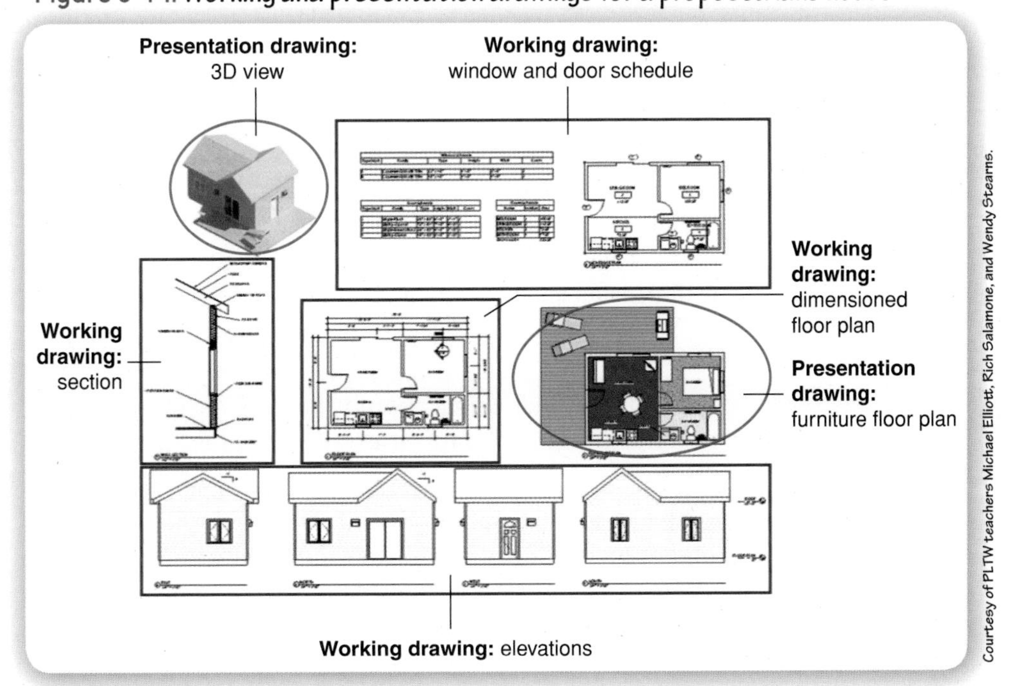

Courtesy of PLTW teachers Michael Elliott, Rich Salamone, and Wendy Stearns.

DRAWING SCALE AND SHEET SIZE Working and presentation drawings are usually placed on ANSI Standard sheets up to 34 × 44 sq in in size. Drawing sheets in a set of plans are generally all the same size. Most architectural drawing sheets are C size or larger.

Your Turn

Copy the sheet chart in Figure 3-15 into your notebook.

	Sheet Sizes in inches
A	8½ × 11
B	11 × 17
C	17 × 22
D	22 × 34
E	34 × 44

Figure 3-15: Common ANSI Standard sheet sizes.

Working drawings must be precisely scaled to fit onto the sheet and provide all geometry with a scaled reference. For example, on a scaled drawing you can measure any portion of the drawing and determine its actual size. A wall length that measures 2 in on a drawing, scaled at ¼ in = 1 ft 0 in is actually 8 ft in length. A window that measures ¾ in on the drawing sheet is actually 3 ft or 36 in wide. And a wall that measures 1 5/8 ft would be 5 ½ ft in length. On a scaled architectural drawing, you can take measurements from a floor-plan drawing and determine the actual distance between rooms, such as how far away the bathroom facilities are from a conference room.

To determine the scale of a drawing, you must compare the relationship of the actual size of the land or building to the size of the drawing sheet. Architectural and engineering firms generally specify a standard size sheet they use for their projects. Let us say that you work for a company that uses standard C size drawing sheets and you have a building with a floor plan of 80 ft × 100 ft If you scale it to ¼ in = 1 ft 0 in, it would take up a drawing space of 20 in × 25 in exceeding the 17 in × 22 in C size sheet. Remember that you must also account for the title block, border, **annotations,** and dimensions. Civil, site, landscape, and excavation drawings use engineering scale instead of the fractional scale of architectural drawings. See Figure 3-16.

Annotations: explanatory or essential comments or notes used to provide more information.

Scaling your 80 ft by 100 ft building to a scale of 1/8 ft = 1 ft 0 in would reduce it to 10 in by 12 3/8 in, leaving plenty of space for dimensions and annotations. Remember dimensions and annotations must be *clear and easy to read* and *be consistent in size and style* throughout the drawing set. As with sheet size, the architectural or engineering firm has most likely established a standardized font and text height. The range of acceptable dimension and text height is from 0.1 in to 0.125 in *Remember: Once a font and height is chosen, it should be consistent throughout all drawings.*

Documents drawn to scale, through application of proportional reasoning and mathematical process, provide accurate measurement information.

PLANS Both residential and commercial developments require plans. A set of plans for a residence might include only 8 to12 drawing sheets, including a site plan, elevations, floor plan, section drawing, and plans for plumbing, electrical, HVAC, and building performance. Commercial projects, on the other hand, may require more than a 100 drawings (see Figure 3-17).

Your Turn

Copy the drawing scale chart, Figure 3-16, in your notebook and use it to help you determine the best scale for the following to fit on C size sheets:

- A site plan of a plot 600 ft by 400 ft
- An elevation of a 200 ft long building with an overall height of 60 ft
- A section showing interior details of a 2 ft wide by 8 ft tall kitchen cabinet

Figure 3-16: **Commonly used drawing scales used for architectural drawings.**

Commonly Used Drawing Scale			
Type of Drawing	Smaller Sites and Buildings ↓		Larger Sites and Buildings ↓
Civil, site, landscape, excavation	1″ = 20′	1″ = 50′	1″ = 100′
Structural, architectural floor plans, elevations, mechanicals, plumbing, electrical, fire protection, security	1/4″ = 1′-0″	1/8″ = 1′-0″	1/16″ = 1′-0″
Section, details, and drawings required greater clarity	1″ = 1′-0″	3/4″ = 1′-0″	1/2″ = 1′-0″

Due to the number of drawings required, most architectural firms letter and number each sheet based on the type of drawing that appears on the sheet. Drawings are then grouped by category, before bound into a set. Drawings are easy to locate by referencing the drawing list on the title sheet, see Figures 3-18 and 3-19.

A = Architectural
C = Civil, site, or environmental
EX = Excavation
L = Landscape
LS = Life safety

Figure 3-17: **Plans for the Syracuse Center of Excellence comprise two volumes, each containing more than 100 drawings.**

Courtesy of Ashley McGraw Architects, PC and Toshiko Mori, Architect.

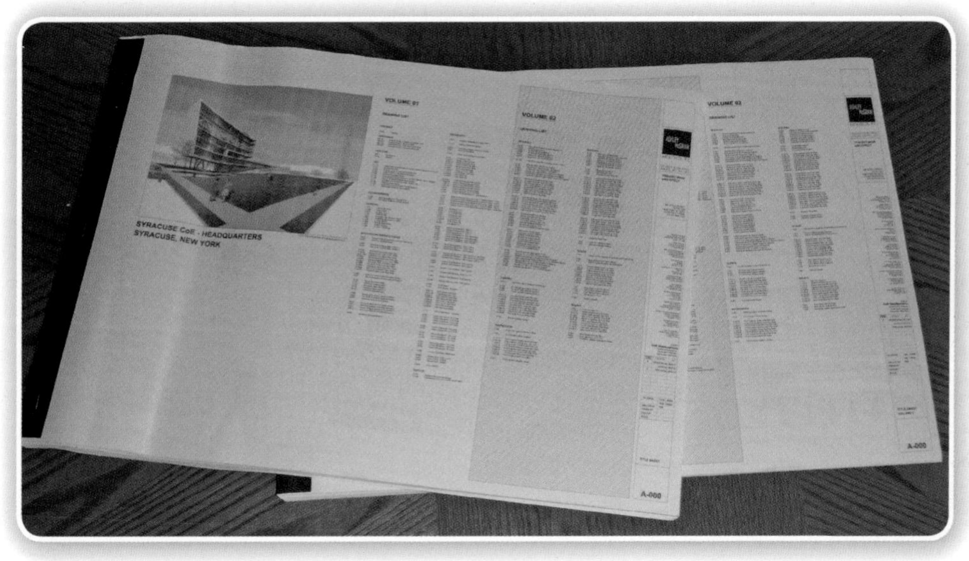

Figure 3-18: A partial list of numbered working drawings from the title sheet of the Syracuse Center of Excellence Plans.

ARCHITECTURAL

A-001	Symbols, Materials, and Abbreviations
A010	Partition Types and Details
A-025	Door/Frame/Interior Glazing Schedule
A-026	Door/Frame Details
A-030	Finish Room Schedule and Notes
A-100	Overall Site Plan
A-101-E	Street Level Plan East
A-101-W	Street Level Plan West
A-102-E	Second Level Plan East
A-102-W	Second Level Plan West
A-103-E	Third Level Plan East
A-103-W	Third Level Plan West
A-104-W	Fourth Level Plan West
A-105-W	Fifth Level Plan
A-106-W	Roof Level Plan
A-200-E	Exterior Building Elevation North
A-200-W	Exterior Building Elevation North
A-201-E	Exterior Building Elevation South
A-201-W	Exterior Building Elevation South
A-202	Exterior Building Elevation East
A-203	Exterior Building Elevation West
A-204	Exterior Building Elevation-Miscellaneous

Courtesy of Syracuse Center of Excellence.

M = Mechanical
P = Plumbing
F = Fire Protection
E = Electrical
T =Telecom
S = Security
S = Structural

All changes along the way must be indicated on the plans as revisions. Changes are often noted in red and are referred to as red-lined drawings. When changes to the plan occur during the construction process, a revised set of plans is provided upon completion. These plans are called *as built* or *just as built* and show all changes, exact dimensions, and location of all elements.

Presentation Drawings: drawings that communicate the general ideas for the development in a colorful, easy-to-understand format.

Presentation drawings, on the other hand, have a different purpose: to communicate the general plans in an easy-to-understand format during presentations to clients, community members, financial backers, and other interested parties. The presentation drawings are often colorful and provide a good visual, but they are seldom to scale (see Figure 3-20).

Figure 3-19: **Sheet A-201W drawn at 1/8 in = 1 ft scale showing the south elevation of the Syracuse Center of Excellence.**

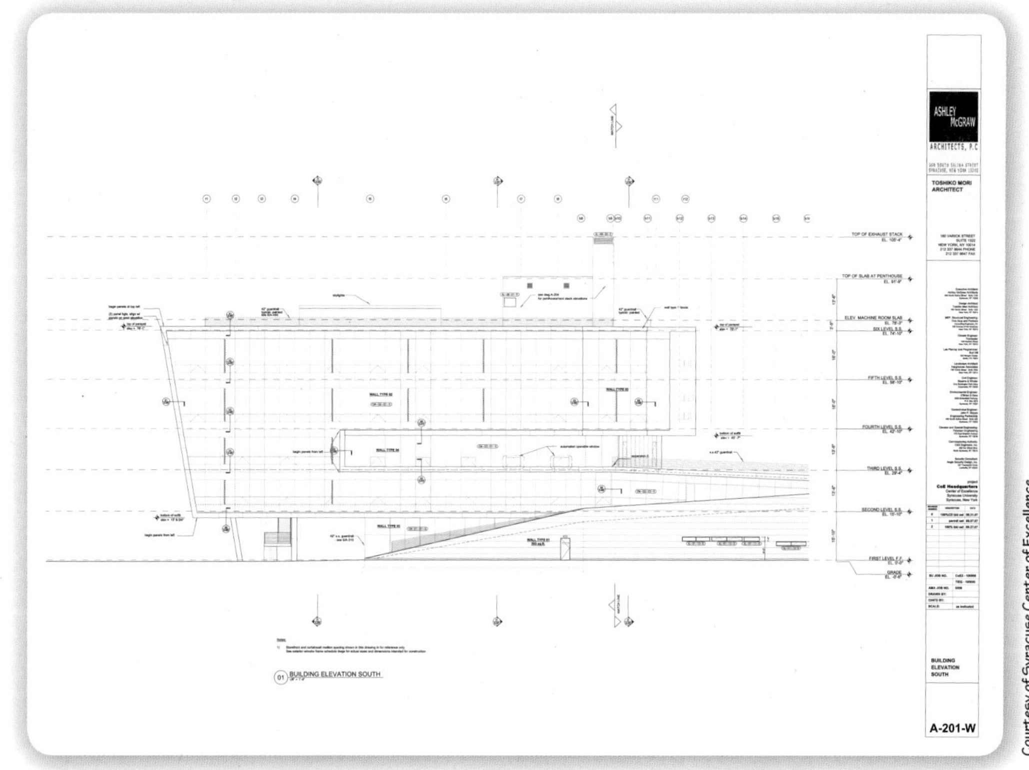

Courtesy of Syracuse Center of Excellence.

Floor plans are different in a presentation drawing format. For example, presentation drawings showing the floor-plan layout show how the room might appear when finished, accommodating various pieces of furniture. Illustrations that show three-dimensional views may accompany floor plans to enhance design communication. In comparison, a working drawing of a floor plan excludes furniture, but instead has many construction dimensions and annotations.

Figure 3-20: **A presentation drawing showing the south and east elevations of the Syracuse Center of Excellence.**

Courtesy of Syracuse Center of Excellence.

Case Study

At CityCenter in Las Vegas, presentation drawings are used to communicate floor plan options to prospective buyers (see Figure 3-21). Look at the two penthouse options for the Harmon, Figure 3-22 a and b.

Notice the small key at the lower left of the sheet showing where the condominium unit is located in reference to the building. They are both similar in square footage, but offer different floor plans. Which would you choose and why? Did having the furniture included in the presentation drawings help you decide?

Figure 3-21: *Model of the proposed Harmon at CityCenter, Las Vegas, 2007.*
Courtesy of CityCenter.

Figure 3-22 a and b: *Floor plans options for the Harmon at CityCenter, Las Vegas.*
Courtesy of CityCenter.

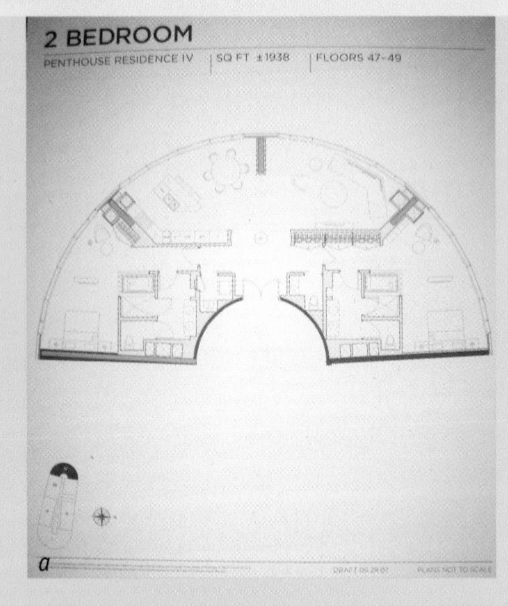

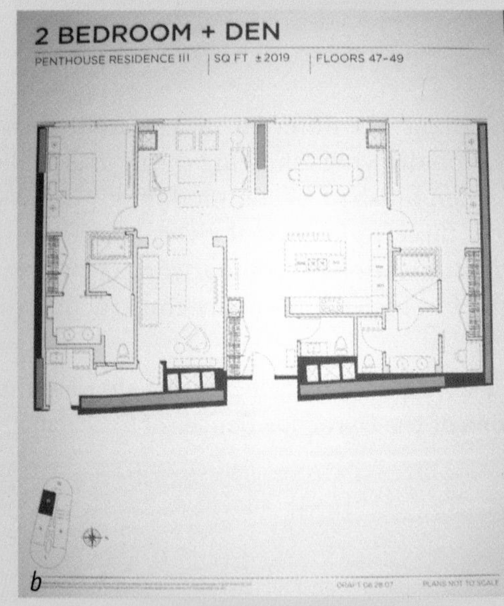

Your Turn

Create a drawing similar to Figure 3-23 to communicate your idea for a small cabin. Why would an architect place these drawings on separate sheets? What information is provided on this combined sheet? Could this building be constructed from the information provided? Make a list of missing information.

Figure 3-23: *Combined drawing of a proposed lake house.*

Courtesy of PLTW teachers Michael Elliott, Rich Salamone, and Wendy Stearns.

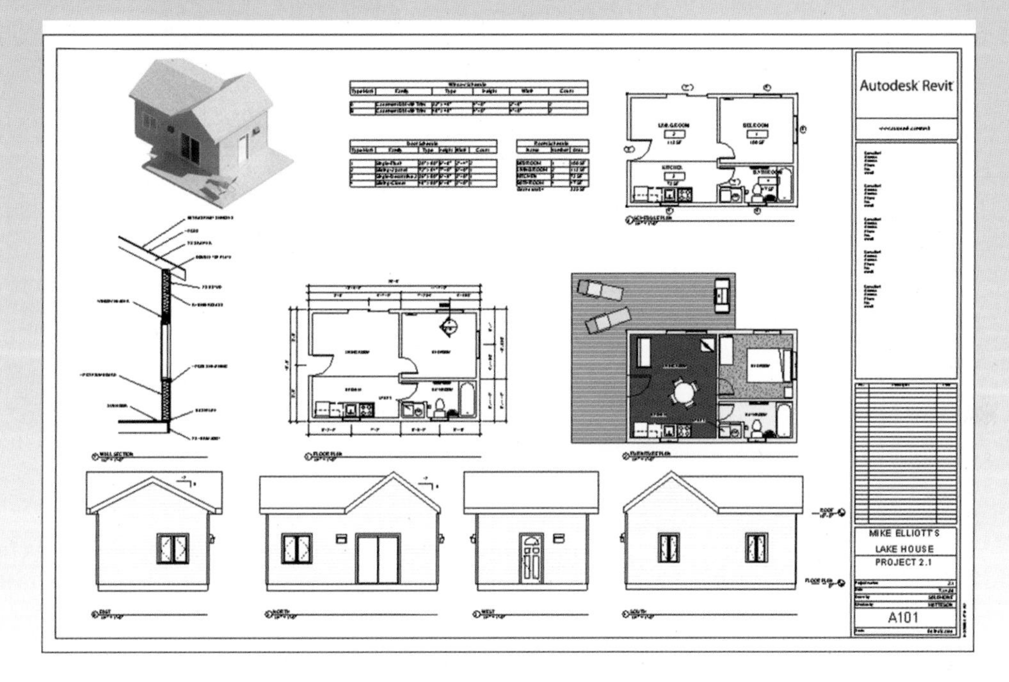

COMMUNICATION IS ESSENTIAL TO ACHIEVING SHARED PROJECT VISION AND CONSTRUCTION PRODUCTIVITY

Think about all the stakeholders that participate in a land development. It is important that these key players are kept current throughout the project. Sharing of vital information and maintaining a common vision will reduce the chance of misunderstanding or error. Remember reading about Whole Systems Integrative Design in Chapter 2: Careers? Today's architects and civil engineers understand that in order to create healthy, sustainable, and resilient buildings, they cannot work in isolation. Many choose to form a Whole Systems Integrated Design team where communication is the key to creativity and achieving a shared vision. The team might include a client, project manager, interior designer, structural engineer, urban planner, cost estimator, environmental engineer, construction engineer, construction manager, carpenter, mason, plumber, electrician, landscape architect, building inspector, financial backer, and community director. During the charrette, discussed in Chapter 2, a facilitator may be selected to insure that effective communications flow between the team members. The facilitator will listen and repeat what has been said back to the group for further discussion and confirmation.

Figure 3-24: Architect Toshiko Mori experiments with several building shapes during a charrette for the design of the Syracuse Center of Excellence.

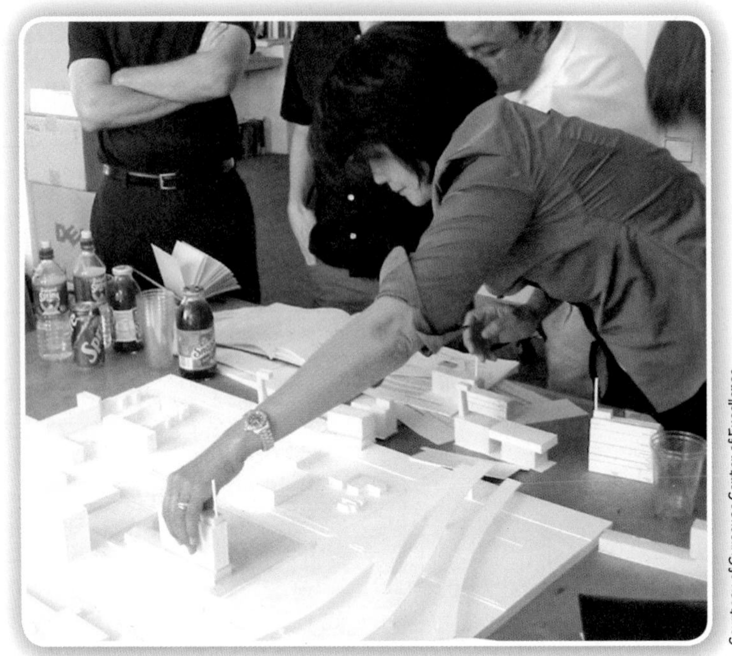

Courtesy of Syracuse Center of Excellence.

After fine-tuning the project's vision, the architectural team goes to work on the design, Figure 3-24. They may need to consult with engineers and other specialists along the way, but eventually they will come up with a proposal. Financial support and approval of the project is often determined by the success of the proposal. The architect must be able to effectively communicate a convincing proposal to justify all aspects of the plan. A formal proposal is often a combination of presentation slides, display boards, virtual and actual models, working and presentation drawings, schedules, and other supporting documents (see Figure 3-24). Colorful charts, graphs, and matrixes highlight and summarize important information. You will read more about presenting and justifying your proposal in Chapter 20: Formal Communication and Analysis.

Project management is essential to ensuring that technological endeavors are profitable and that products and systems are of high quality, built safely, on schedule, and within budget.

Just as positive communications enhance the success of a project, poor communication will do the opposite. An unconvincing proposal to a financing institution, planning board, or public forum could delay or even end plans for development. Once the development has been approved and the construction is under way, failure to complete verbal communications or written documentation in a timely fashion can result in major delays or a stop work order. If your building receives a stop-work order, all activity must cease until the problem is corrected and the stop work order has been lifted.

Careers in Civil Engineering and Architecture

DARE TO DARE

Who would ever guess they might someday walk a narrow steel I- beam of a bridge 90 ft high? Well that is just what happened to Mary Lou Noel. After obtaining her two-year degree in surveying, she joined the Operating Engineer's Union and took a job at a nuclear plant. When the job was over, she and a co-worker started their own surveying business. That business, and several successful bids, led to re-construction of many bridges along the NYS Thruway. It was during that time she was faced with the high walk across the steel I- Beam. Harnesses weren't required back then. In her own words, she sums up the experience with " ...you just had to dare to dare! I wouldn't back down, especially with all the men watching." But that wasn't the only daring challenge of her career. Every Monday through Friday she drove two hours each way to attend a college that offered a program in surveying. "I had an old Volkswagen Beetle. It was great on gas, but there was no heat in the winter. "Mary Lou was a non-traditional student in a non-traditional field. Some professors were biased against women entering the profession, but she forged ahead. "I wouldn't let them get me down, I just kept coming back. I knew I could do it." The day before she started college, she took a test to confirm her faith in her intellectual ability. She had read about Mensa, an international society that recognizes people who score at the 98th or higher percentile on a standardized IQ test. She wanted to "give it a try." Her score earned her an invitation to join. "I guess it was just something that I wanted to prove to myself."

Mary Lou Noel
Licensed Surveyor and Building Inspector

On the Job

Today Mary Lou is a building inspector and codes enforcement officer. She guides people on their journey through the building process. "Most people are unaware of the paperwork required. They need to submit plans and show documentation that the property is acceptable for development. They must adhere to the local ordinances such as the 50 ft setback from the road, and 25 ft side setbacks." Mary Lou visits the property several times during the building process to inspect the building to make sure it meets NYS code. She inspects everything except the electrical, which is done by a certified electrical inspector. She explains that safety, property value, and durability are main reasons behind codes. Take snow load for instance. Buildings in her town must have trusses that can withstand a 60 psf snow load. And because of cold winter climate, foundation walls must be at least four foot deep to prevent frost and ice from heaving the foundation. "There are no shortcuts. If they don't follow the code, I make them do it over."

Inspirations

Mary Lou recognizes her mother as the one person who inspired her the most. Growing up they built many things together, including several masonry projects. Mary Lou's skill with masonry was so well known that before going to college she was commissioned to build a large stone fireplace at Fort Ontario in Oswego New York, in honor of the women who served during the civil war. "People ask me how I know things like plumbing. I tell them I was born knowing, which is just what it feels like."

Education

Mary Lou remembers the day she decided to go to college for surveying. She arrived home and on her counter was an ad to study surveying. "The magazine was turned to that page, and it jumped out at me. I thought it sounded like a fun career. The classes required a lot of math; mostly trigonometry, but some of the classes were fun. I enjoyed learning about soils and masonry." Mary Lou's education didn't end upon graduation. While working on the NYS Thruway bridges, she learned about a high expanding cement mixture from Japan. Instead of blasting away the old concrete, small holes were drilled about 4 ft apart, and were filled with the special cement. When they arrived the next day, the old cement had cracked off the structure due to the high pressure of the expanding cement. This method proved to be much safer than blasting.

Advice to Students

"The best advice I can give to students is to go to college. Pick something that sounds fun and go on to school. If you don't, in two or four years from now you will look back and wish you had. You may think that two or four years is a long time, but it goes by quickly. Just make up your mind and make it happen. Don't let anyone discourage you."

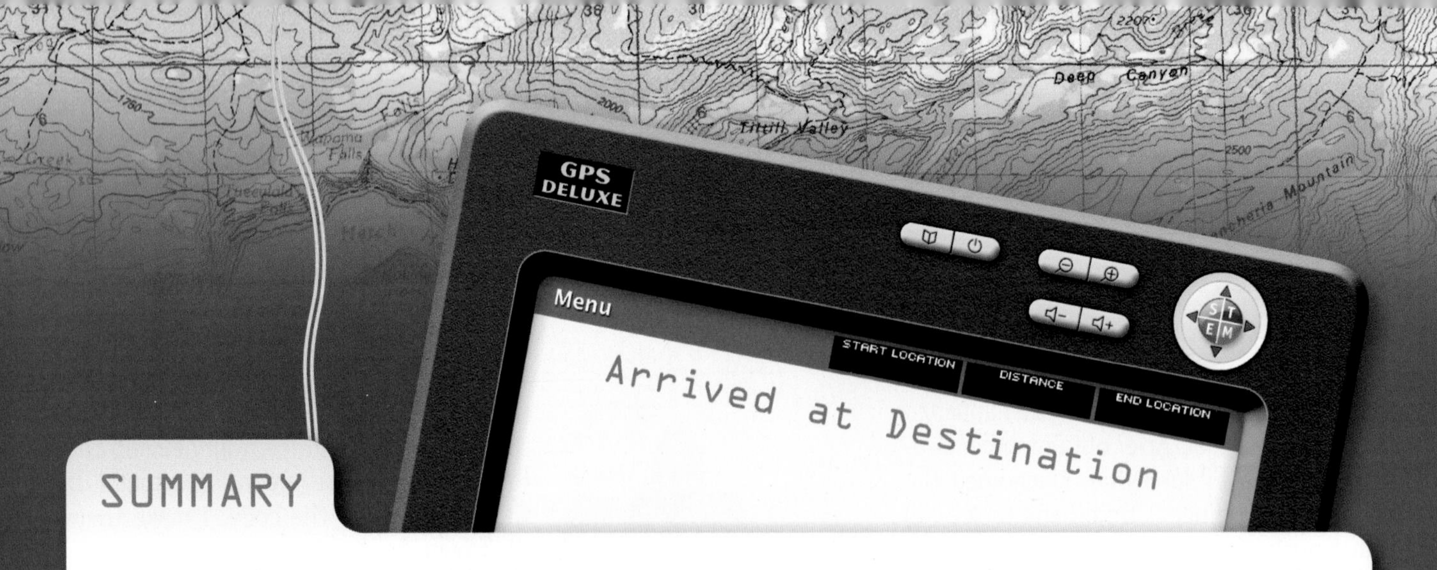

SUMMARY

All land developments have three things in common: They all require accurate and detailed research, documentation, and communication for the best possible outcome. Research for large development projects may take several months or even years to complete. An architectural sketchbook is used to record and develop ideas. Depending on the location and type of project, several studies and documents may be required to obtain necessary permits and approvals. Architects and civil engineers must research and select the best building materials and practices. In today's world of limited resources, intensive research is necessary to plan cost-efficient, safe, sustainable, and adaptable buildings that will be aesthetically pleasing and functional for many years to come. Architects and engineers are constantly challenged to identify choices that will provide maximum benefit to the client and end-user while having minimal negative impact on the environment. Documentation for land development takes many forms, from journals and sketches to detailed programs and plans.

Drawings fall into two general categories based on their purpose. Working drawings provide all necessary information required for complete and accurate construction, whereas presentation drawings are used to communicate a general vision for the project in a clear, easy-to-understand manner. Dimensioned floor plans, schedules, sections, elevations, and plot plans are all examples of working drawings. Working drawings are drawn, printed, or plotted to scale and placed on standard size sheets. These drawing sheets are grouped by type and easily located by a letter and number that is generally located on the lower right section of the title block. In comparison to working drawings, presentation drawings include colorful exterior and interior renderings and general concept floor plans showing room and furniture arrangement. They are not necessarily drawn to scale.

Communication with project stakeholders and governing agencies begins early in the planning stage and continues throughout the project. Communication can take many forms such as BIM collaborations, meetings, telephone calls, memos, and e-mails. Documentation of all communications will support productivity and accountability and avoid misunderstanding. The architect or manager responsible for overseeing the project is responsible for ensuring that required documents are submitted in a timely fashion. Maintaining communications between contributing parties will make optimal use of time and resources and keep the development on schedule.

BRING IT HOME

1. Use Google Earth, Map Live, or similar website to locate a map of your area. Select a location where you might like to build a video-gaming facility and check to see if it is zoned for commercial use. List the reasons why you selected this location.
2. Research to determine the setbacks required for the land selected for your video-gaming facility. Record a list of advantages and disadvantages of the setbacks in your notebook.
3. Design a survey instrument or list of questions that you could use to determine if your development is needed or wanted in your community.
4. Compile a list of eight professionals whom you would contact to participate in your charrette to help design your video gaming facility. Explain why you selected them and what information you expect they will contribute to the process. Identify who will facilitate your charrette and make a list of their responsibilities.
5. Draw a bubble diagram and architectural sketch of a 2500 sq ft single story video gaming facility. Make a list of design considerations.
6. Describe the information and presentation drawings you would include on a PowerPoint presentation to attract an investor.
7. Make a list of working drawings you would need for the construction of your video-gaming facility.
8. Calculate the scale you would use for your video-gaming facility floor plan to fit on a C size drawing sheet.
9. Create a proposed timeline for your video gaming facility. Label each part of the timeline identifying the research, documentation, and communication that will be required at each stage of development.

EXTRA MILE

Observe the governance of a land-development project by attending a local planning-board meeting or view a video archives offered on a government website. One such website is the City of San Diego Planning Commission, which posts live webcasts and video archives of their meetings at *http://www.sandiego.gov/planning-commission/*.

Select Live Webcasting and Video Archives of Meetings and watch a Land Use and Housing Committee meeting.

CHAPTER 4
Architectural Design

GPS DELUXE

Menu

START LOCATION | DISTANCE | END LOCATION

Before You Begin

Think about these questions as you study the concepts in this chapter:

1. What are the similarities and differences between residential and commercial buildings?
2. What are some design considerations of urban design?
3. What are the benefits of whole-building design?
4. Who are the essential members of the whole-building design team?
5. How do the features of a building help you to identify its architectural style?
6. How do roof, window, and door styles and exterior building materials impact form and function?
7. How do the principles and elements of design affect architectural styles?
8. What factors must an architect research and consider when planning a new structure?
9. How can a building's age help identify its architectural style?
10. How does Vernacular Architecture reflect the cultural and environmental conditions at the time of building design and construction?

INTRODUCTION

Have you ever wondered why buildings have so many different looks? When giving directions, you describe buildings by their shape, color, size, or materials. For example, you might tell someone to "turn right at the two-story brick house with green shutters." Other landmarks are identified by name and their unique appearance or style. We can all recognize the familiar style of a particular fast-food restaurant or mini-market. Famous buildings are also easy to recall. If someone mentions the Empire State Building, you mentally picture the tall sleek tower. The look or appearance of a building indicates its architectural style.

In this chapter, you will read about the difference between residential and commercial buildings and discover the many elements that influence a building's appearance. You will learn about the whole-building approach to architectural design, and you will see how design teams address issues of economics, environment, form, and function when developing a viable building proposal.

Not surprisingly, the same principles and elements of design that are fundamental to product design are also key to successful architectural designs. This chapter will provide architectural examples of these principles and elements of design with line drawings showing how they may be illustrated.

Most cities have buildings of various architectural styles. When you have finished this chapter, you will look at buildings differently. You will be able to recognize distinctive features and overall architectural styles. You might even be able to ascertain the age of the building or other factors that influenced its design. This chapter will become a foundation for planning your residential and commercial buildings (see Figure 4-1). Several tables in this chapter will provide quick references to key terminology of building features and details.

Figure 4-1: *Do you know the architectural style of the Empire State Building? At the end of this chapter, you will easily recognize the style of this and other famous buildings. Empire State Building, New York, New York, Shreve, Lamb, and Harmon, 1931.*

RESIDENTIAL AND COMMERCIAL ARCHITECTURE

For the purpose of this textbook, we will divide architecture into two basic groups: Residential and commercial. We will define *residential architecture* as a building used as a home, such as a single-family home or townhouse. *Commercial architecture* describes buildings constructed for the purpose of business or service. However, for this textbook we will also include industrial, public/private, assembly buildings, plan unit development, and urban design in the commercial category.

In reality, architecture is defined and divided into several categories based on commonalities and specific needs addressed during design of a particular type of building. For example, commercial architecture includes small businesses, stores, and hotels. The term *industrial architecture* describes buildings and factories designed for the manufacturing, assembly, or construction of products. Assembly

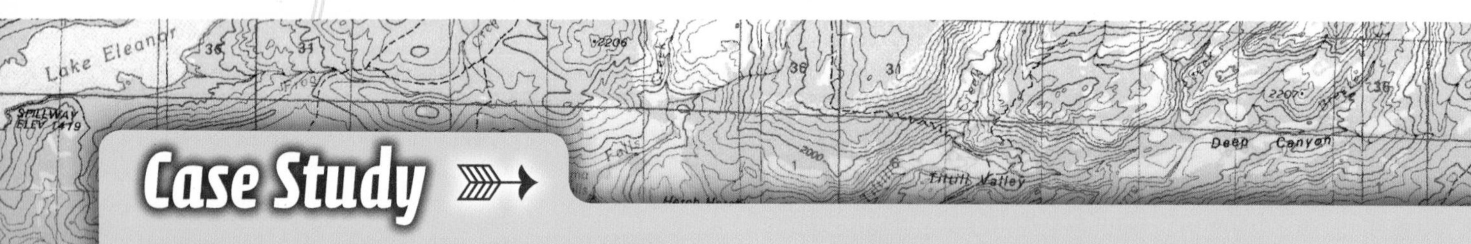

Case Study

URBAN DESIGN AT CITYCENTER, LASVEGAS

CityCenter in Las Vegas is an exciting example of contemporary urban design. The project is often called a city within a city. In addition to condominiums, a hotel, casino, retail shops, and entertainment, CityCenter's 76-acre setting has its own parks, power plant, fire station, and people mover (see Figure 4-2).

Three teams of urban planners spent six months researching options to maximize the property's value. Once the board of directors approved the project, several prominent architects were invited to propose possible building designs. Eight of those architects were ultimately chosen to design the major structures of the development. Urban development is a long, labor-intensive and detail-laden process. You will read more about the CityCenter project in other chapters of this textbook.

Figure 4-2: *CityCenter in Las Vegas is one of the largest environmentally sustainable urban communities in the world.*

Courtesy of MGM Mirage.

buildings are yet another category, which includes gathering places such as churches and community centers. Another category of commercial architecture is urban design. Urban design includes research, design, and management of public environments, including parks, streets, and parking facilities. Urban designers must have a clear understanding of how the space is to be experienced. They commit many months to researching property history, proposed use, environmental factors, and the surrounding community. This process is called *site discovery.* During site discovery, urban designers will consider the future impact their work will have on adjacent property and neighborhoods as they study current construction methods, materials, regulations, and codes. Their research will support a viability analysis for the proposed land development. You will learn more about site discovery and viability analysis when you read Chapter 5: Site Discovery for Viability Analysis.

WHOLE-BUILDING DESIGN

The whole-building design approach starts at the very beginning of the planning process and includes architects, engineers, building occupants, owners, building officials, contractors, and specialists in areas such as materials, energy, air quality, and acoustics. Synergy is one of the main goals of the team approach.

During the whole-building design process, the design team will research the **seven resources of technology,** which include people, capital, tools and machines, time, information, materials, and energy. The research will uncover the most reliable and available materials and building practices to maximize a building's energy, economic, and environmental performance. The goal of the whole-building design team is to design a building that is aesthetically pleasing, energy efficient, sustainable, well suited to the environment, and a healthy place to live or work (see Figure 4-3).

Synergy:
a result achieved when the unit or team becomes stronger than the sum of the individual members.

Structural design is influenced by the surrounding environment, natural resources, available materials, acceptable building practices, style preferences, building costs, lifecycle costs, and sustainability factors.

You will read more about energy, sustainability, and green building in Chapter 8: Energy Conservation and Design. Whole-building design can provide a healthier and more comfortable environment while reducing maintenance, energy use, and environmental impact. A plan using the whole-building philosophy would maximize the use of natural sunlight, energy-efficient windows and insulation, and environmentally friendly and renewable building materials. The plan would also take maximum advantage of the site's natural resources. For example, planners might try to locate the building for maximum southern solar exposure or wind protection.

When designing with the whole-building design philosophy, the team must consider factors specific to location such as natural resources. Using the Internet to research whole-building design, you will discover several whole-building design efforts, such as Building America. Building America is a private/public partnership sponsored by the U.S. Department of Energy. According to Building America's website, the partnership "conducts research to find energy-efficient solutions for new and existing housing that can be implemented on a production basis." For more information, visit their website at http://www.eere.energy.gov/buildings/building_america/about.html.

Sustainable:
building performance, operation, and maintenance that saves both money and natural resources. The Leadership in Energy and Environmental Design (LEED) recognizes this performance in five key areas of human and environmental health: sustainable site development, water savings, energy efficiency, materials selection, and indoor environmental quality.

Figure 4-3: Solar Energy Research Facility, Golden, Colorado.

Courtesy U.S. Department of Energy.

The Building America Partnership's goal is to work with builders to design and produce homes that have on-site power systems, reduce pollution by using less energy, and are constructed in less time using innovative energy and material-saving technologies.

Your Turn

Building America projects are appearing all over the United States. There are currently more than 40,000 Building America projects. Use the following Internet site to research project locations in your state and others: http://www.eere.energy.gov/buildings/building_america/cfm/project_locations.cfm.

Compare a building project in your state to one in another climate. What are the similarities and differences? How does the project incorporate whole-building design?

Look at both residential and commercial applications of the whole-building design approach at http://www.nrel.gov/buildings/whole_building_research.html. What did you discover? Make notes in your notebook for consideration when planning your next building.

If time allows, read more about building design at http://www.nrel.gov/buildings/.

IDENTIFYING BUILDING FEATURES

All buildings, whether residential or commerical, have specific characteristics and details. We recognize them by their size, shape, color, material, or building components. These details will help you to determine a building's architectural style. Just for fun, let's pretend you are a treasure hunter and you have received a tip that a treasure is hidden in the woods behind a Georgian-style house in your neighborhood. Unfortunately, you are not familiar with architectural styles and there are woods behind dozens of houses that line your street. Your best chance is to search for clues to help you find the Georgian-style building. Clues will include the overall shape and proportion of the building; roof type; building materials; size, style and placement of windows and doors; and other defining details.

You decide to begin your search by using the Internet and type in the words "Georgian architecture." You discover that this architectural style was named after the British monarchs George I through George IV, who ruled from 1714 until 1830, when this style of architecture was popular. You scroll further to find that the materials used in Georgian buildings include brick and stone. Georgian-style buildings traditionally featured reddish brick and white window trimming and cornices. You record this information in your architectural notebook along with the word *cornice* to research further. Next, you discover that although many Georgian buildings are commonly tan, red, or white, modern-day Georgian style homes might use a variety of other colors, and due to the cost of bricks, some Georgian homes are now built of wood.

At this point, your research has not been very helpful, but undaunted, you continue your quest. Reading further, you discover that Georgian-style buildings are symmetrical in appearance, often with chimneys located on both sides of the house. From your previous study of engineering design, you recall that symmetry is achieved with shapes that correspond, or mirror, the opposite sides of a virtual centerline. This would mean that the building would have a balanced appearance with opposite sides having matching components, such as windows or other building features. You continue your research by looking for the building features of the Georgian Architecture, and discover that windows are generally six-paned **sash windows** or larger, Figure 4-4. Some windows feature nine or twelve panes, or pieces of glass (see Figure 4-4).

Before you leave the website, you read about Georgian-style entrances, which are emphasized by a **portico** with **hipped** or minimally pitched **gable** roof. Georgian-style buildings often include a stone **parapet.** See figure 4-5 for examples of a portico and parapet.

Figure 4-4: The top and bottom sections of this six-over-six, double-hung window each contain six panes of glass.

Sash window:
single- or double-hung windows where one or both of the glass panels (sashes) overlap and slide up and down inside the frame.

Hipped:
a hipped roof slopes down to the eaves on all sides.

Figure 4-5: The buildings featured in this table are useful in identifying architectural style.

Name	Description	Illustration
Architrave *AR ka trave*	The principal beam and the lowest member of the entablature, resting directly on the capitals of the columns.	
Baluster *Bail-us-ter*	A small, decorative post that supports the upper rail. A row of balusters is called a *balustrade*.	*http://architecture.about.com/library/blgloss-balustrade.htm*

(Continued)

Figure 4-5: The buildings featured in this table are useful in identifying architectural style. (Continued)

Name	Description	Illustration
Capital	The crowning member of a column or a pilaster.	
Clerestory *CLEAR-storee*	A vertical extension beyond the single-story height of a room, generally with windows.	
Column	A pillar or post consisting of a base, shaft, and capital.	
Cornice *KOR niss*	A decorative projection or crown along the underside of the roof where the roof "hangs over" the wall. Originally, cornices were brackets that supported the overhanging roof.	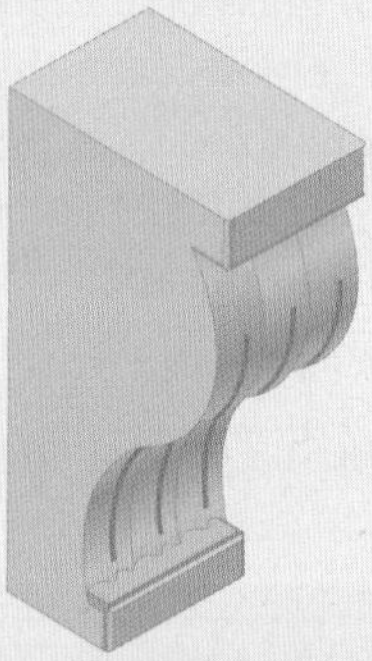
Cupola *KEWP oh la*	A small structure located above the roof for the purpose of adding light and air, or for ornamentation. Historically used as a lookout.	
Dentil *DEN till*	A small, rectangular block used in series under the cornice. From the Latin word *dentes* meaning "teeth." Dentils were thought to resemble a row of teeth.	

Name	Description	Illustration
Dormer *Door mer*	A structure with a window, added to a sloping roof.	
Entablature *en TAB la chur*	An order of horizontal moldings and bands that rest directly on the capitals and columns.	
Façade *fa SOD*	An architectural front or "face" of a building.	
Frieze *FREEZE*	The middle section of the order, below the cornice of a wall and above the Architrave.	
Jack arch	A brick or stone element supporting openings in the masonry. Decorative patterns and elements are often incorporated to visually frame architectural features. Also known as "flat arch" or "straight arch."	
Keystone	The central stone generally placed at the top profile of arched elements.	
Lintel *LIN til*	A supporting wood or stone beam generally placed across the top of window or door opening. (Also called a *header*.)	

(Continued)

Figure 4-5: The buildings features in this table are useful in identifying architectural style. (Continued)

Line illustration courtesy of Dan Braun, SUNY Oswego.

Name	Description	Illustration
Modillion *moh DILL yun*	An ornamental bracket, placed under the eaves, in series, for support or decoration. Their S, or scroll, shape easily identifies classic modillions. Later modillions are seen in the form of a plain block.	
Palladian window *pa LAY dee n*	A large, multi-paned window unit with a large center arched section and two short, narrow side windows. It was named after Andrea Palladio, who invented them in the sixteenth century. Independence Hall in Philadelphia has a Palladian window located over its rear entrance.	
Parapet *PAIR a pit, -pet*	A low wall along the edge of a roof, balcony, platform, or terrace.	
Pediment *PED a ment*	A decorative gable end extension, which emphasizes a building's width giving it a sturdy appearance. Often a pediment is visually supported by columns.	
Pilaster *pi LAS ter*	A rectangular or half-round column attached to a wall to provide the appearance of columns without the expense of actual columns.	
Portico *POOR ti coh*	A porch-type structure, with a roof supported by columns or walls, that leads to the entrance of a building.	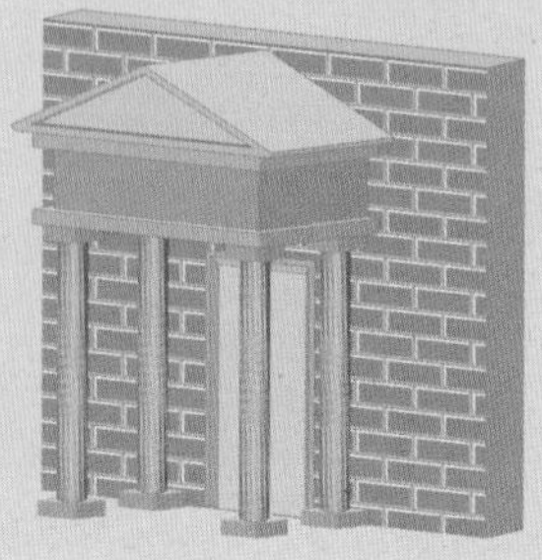

Name	Description	Illustration
Quoins *Koinz* *or* *Coins*	Brick-like designs placed at the corners of a building.	
Transom *TRAN sum*	A framed glass typically placed above doors to allow light into the entranceway.	
Turret *TUR et*	A small tower attached to a larger building.	
Water table	A horizontal row of specially molded bricks that extend out from the rest of the wall marking the ceiling of the basement.	
Widow's walk also known as: **Captain's walk**	A railed walkway built on a roof, traditionally for viewing the sea while waiting for fishing boats to return.	

Your research describes Georgian buildings in America as having **pilasters, water tables,** and complemented horizontal elements such as **pediments.** Now you have more unknown words to search. Without pictures or definitions, this is not as easy as you first thought. However, still enticed by the possibility of finding a treasure and armed with your list of words, you expand your search for the words *portico, parapet, hipped roof, pilaster, water table,* and *pediment.* As you search each one, you make sketches and record notes. The table in figure 4-5 will provide graphics and descriptions of several building features.

Fun Fact

The Longest Portico

The longest **portico** in the world consists of 666 arches and leads from the edge of the city of Bologna, Italy, up to the Sanctuary of the Madonna di San Luca (Figure 4-6)—a journey of a little more than two miles!

Figure 4-6: The portico of the Sanctuary of the Madonna di San Luca contains 666 arches and covers more than two miles.

Point of Interest

What's on Your Pediment?

Pediments date back to the architecture of classical Greece. Do you see the pediment on the Parthenon in Athens in Figure 4-7? The triangular sections of pediments were often adorned with sculptures depicting scenes from mythology. The theme on the two pediments on the Parthenon in Athens are Zeus's presentation of Athena to the Gods of Olympus and Athena's strife with Poseidon for the land of Attica.

Figure 4-7: Look closely at the remains of the Parthenon's pediment. Can you see a portion of the scene depicted?

So, what treasure did you find in the woods behind the Georgian-style house? With the help of a borrowed metal detector, you quickly found a small wax sealed tin with a few silver coins and a single page from the 1737 *Poor Richard's Almanac,* with the following quote from Benjamin Franklin: "The noblest question in the World is What good may I do in it." How many architects of Franklin's time read that very quote? Could it have been one of their driving forces to make a difference in the world through the design of great and noble buildings? Can you imagine that someday your career path might lead you to design a prominent building?

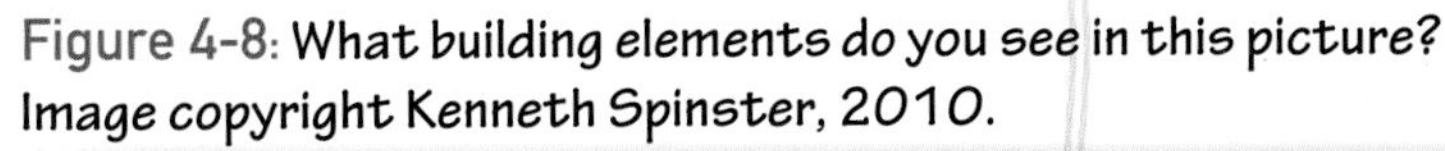
Figure 4-8: **What building elements do you see in this picture? Image copyright Kenneth Spinster, 2010.**

Today's treasure is that, during your quest to locate a Georgian-style house, you discovered many new words and their meanings. These words are part of the architect's vocabulary. Practice using them to describe the buildings you see in your own city or neighborhood. For example, you might see a house like the one in figure 4-8. Notice how the building is symmetrical with chimneys at both ends. What other features do you recognize?

Architects today continue to incorporate Georgian influences into commercial buildings to provide the visual appeal of symmetry, formality, luxury, and old-world charm. Architect and Urban Planner Jaquelin T. Robertson received the 2007 Richard H. Driehaus Prize for Classical Architecture based on his success in incorporating the principles of traditional and classical architecture into modern urban development. Which Georgian-style building elements can you identify in figure 4-9?

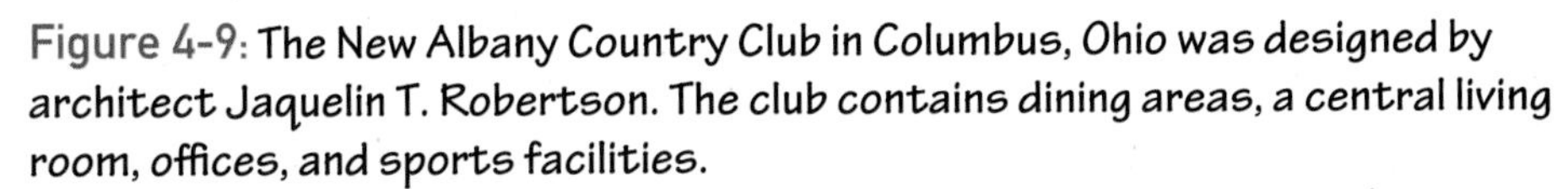
Figure 4-9: **The New Albany Country Club in Columbus, Ohio was designed by architect Jaquelin T. Robertson. The club contains dining areas, a central living room, offices, and sports facilities.**

(Courtesy to come. Photo: Robert Bensen ArchitectureWeek.com) Mary needs to pay fee

Roof Styles

Roof style is often a major clue to determining a building's architectural style. Beyond the basic function of protecting a building from the elements, the roof attains aesthetic quality from its size, shape, and finish. Structural roof vocabulary will be covered in

Figure 4-10: *Common roof styles.*

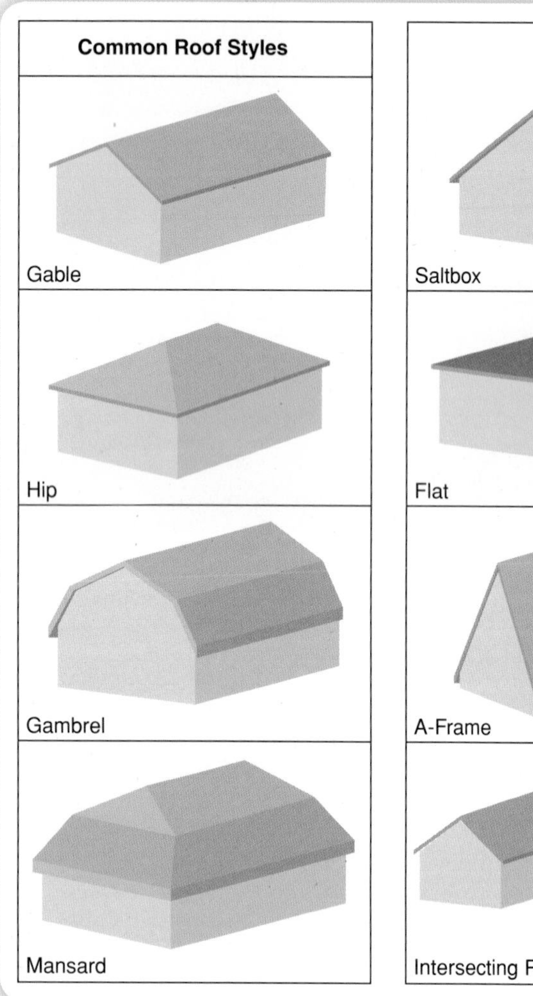

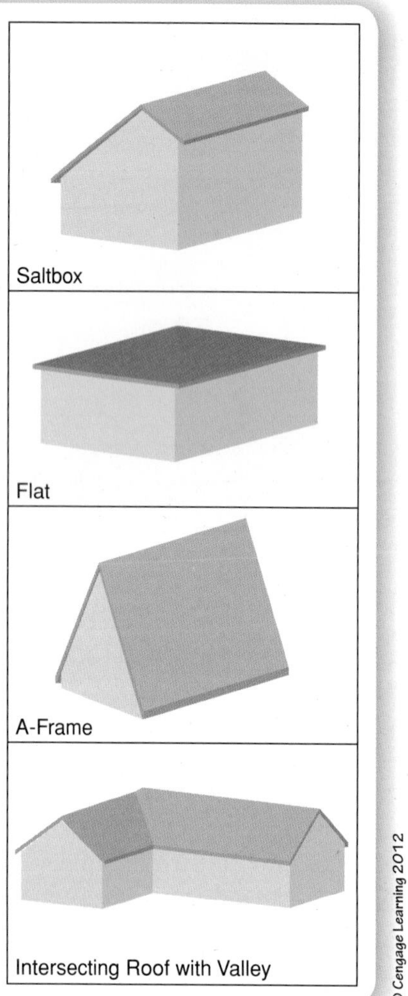

Chapter 13: Framing Systems: Residential and Commercial Applications and Chapter 14: Structural Systems: What Makes a Building Stand? To help you identify architectural styles, you will need to assess the roof pitch, or slope. Roof style and materials vary by region. For example, a steep roof pitch can reduce snow load, and in some locations, a minimum roof pitch is specified by code. Metal roofs are popular in the northeast United States, whereas clay tile roofs are popular in the South and Southwest. Figure 4-10 illustrates the most commonly used roof styles. When a **building footprint** is not rectangular, two roofs often come together to form a valley.

Window styles

Window styles play a major role in identifying a building's architectural style. Each style has specific properties and characteristics. Historically, windows were built using wood and small single panes of glass, whereas today's energy efficient double- and triple-layered windows achieve their multi-paned appearance with an overlaid grid section. These new, easy-to-clean-and-maintain windows have frames made of solid vinyl or wood clad with aluminum or vinyl. You will read more about energy-efficient windows in Chapter 8: Energy Conservation and Design.

Window styles vary in the way they function. Double-hung or sash windows slide up and down within their frame. Most historical buildings still have original sash windows with weights built into the frames to ease the operation of the heavy window. Casement and awning windows usually crank open or are pushed open with a bar that holds them in place. Picture windows are large, fixed windows that provide for a broad, uninterrupted viewing area or "picture." These windows are often used in conjunction with small windows on either side that open to provide airflow. The table in figure 4-11 will help you to identify many of the most common window styles.

ELEMENTS AND PRINCIPLES OF DESIGN APPLIED TO ARCHITECTURE

Developer John Jacob Raskob is said to have inspired the shape of the Empire State Building when he held up a pencil in front of architect William Lamb and asked, "Bill, how tall can you make it so that it won't fall down." Shape is but one of the elements used in the design of a building.

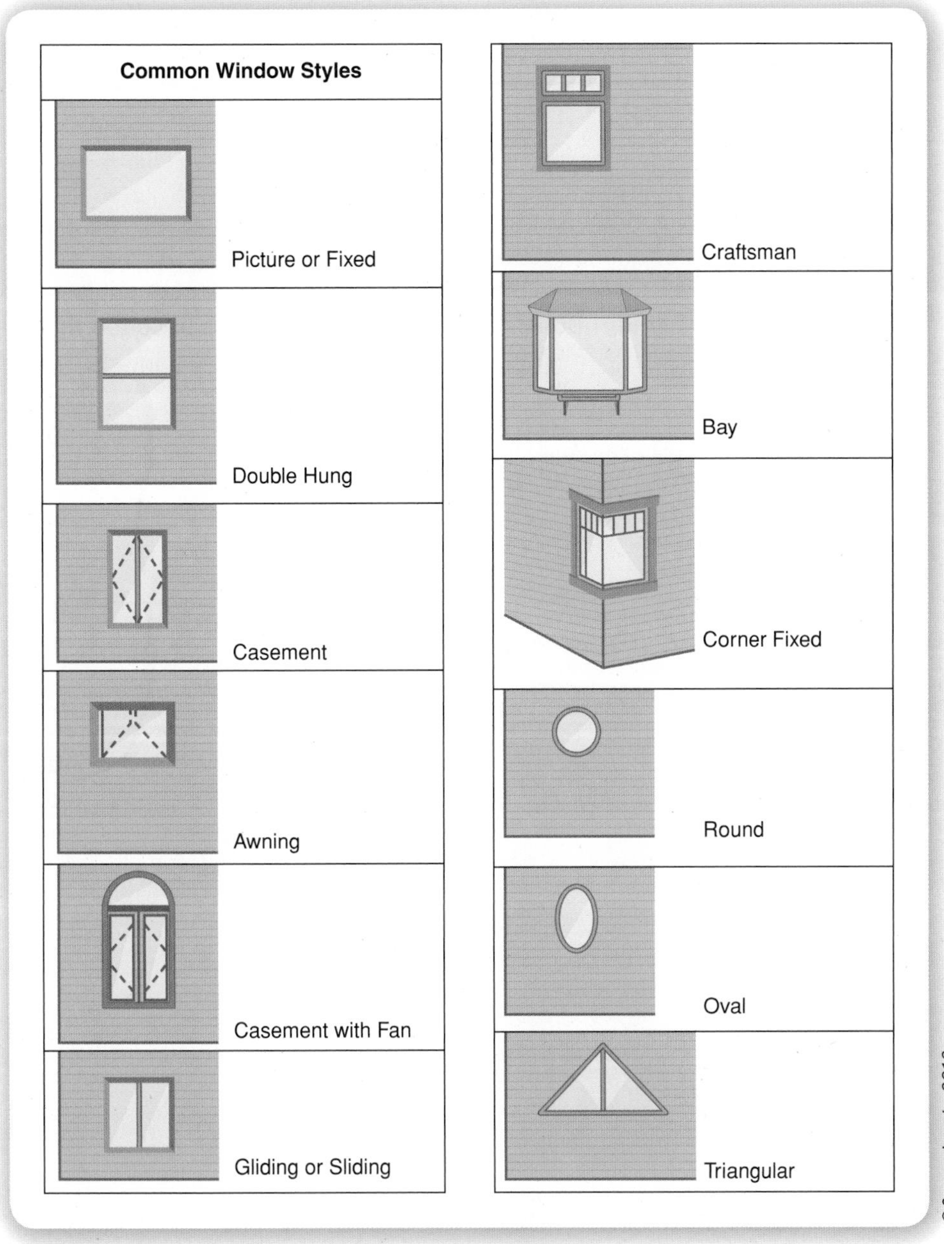

Common Window Styles

Picture or Fixed

Double Hung

Casement

Awning

Casement with Fan

Gliding or Sliding

Craftsman

Bay

Corner Fixed

Round

Oval

Triangular

Figure 4-11: *Common window styles.*

Buildings of the same architectural style have individual and unique characteristics. For example, the Empire State Building is easily identified by its shape. Architects design uniqueness into a style by applying **elements and principles of design.**

Elements and principles of design are applied in all areas of design, including architectural building design. Visual elements of design include line, color, form, space, shape, texture, value, and tone. The principles of design are balance, contrast, emphasis, movement, pattern/repetition/rhythm, and unity/harmony. Take a few minutes to review the elements and principles of design as defined in the tables provided (see Figures 4-13 and 4-14).

The design and construction of structures have evolved from the development of techniques for measurement, controlling systems, and the understanding of spatial relationships.

Fun Fact

Tallest of the Tall

The Empire State Building, constructed between 1930 and 1931 and finished in the Art Deco style, was the tallest building for 40 years at a height of one-fifth of a mile (1056 ft).

The Taipei 101 Tower, with its 60 ft spire, eclipsed the Empire State Building in 2004 at a total height of 1670 ft. The tallest structure in the world today is the Burj Dubai, standing at 2716 ft tall and containing 162 stories (see Figure 4-12).

Figure 4-12: **The tallest building in the world is currently the Burj Dubai at 2716 ft tall. From HACKER.**

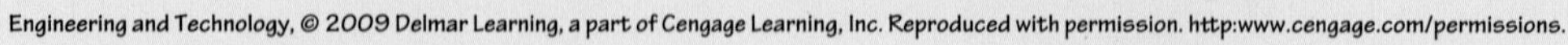

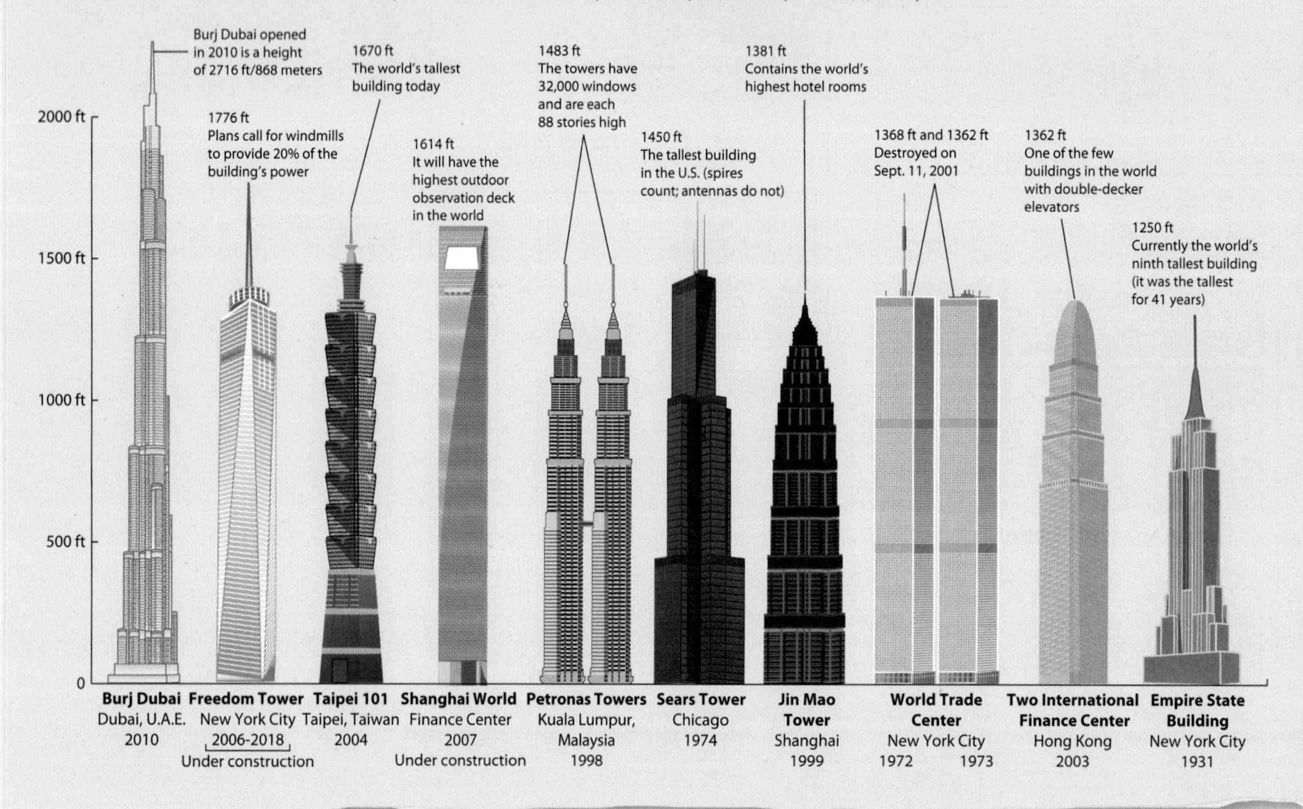

DETERMINING ARCHITECTURAL STYLE

The form, materials, and building techniques of an architectural style often define the culture, climate, and social concerns of the time period. As Ludwig Mies van der Rohe said, "True architecture is always objective and is the expression of the inner structure of our time from which it stems" (quoted in Werner Basel, *Mies van der Rohe,* Basel, Boston, Berlin: Birkhauser, 1997, p. 8). This timeless observation holds true today. How do culture, society, environment, and economy affect the needs and desires of a building's future occupant? If architecture is a true blend of both form and function, there is room for many variables. What is attractive and functional

Figure 4-13: *These architectural examples show the elements of design at work.*

Elements of Design Applied to Architecture

Element	Definition	Architectural Application Example	Illustration
Line	A marked path that is vertical, horizontal, diagonal, or curved.	Lines can define the design of a building. Lines can also provide visual effects. Vertical lines are used to make a building appear taller. The vertical lines of the pilasters add a sense of strength and significance to this building's entrance.	 The basilica of Sant'Ambrogio, Milan. Image copyright gallo23145, 2010. Used under license from Shutterstock.com.
Color	A particular hue seen when light is reflected off a surface.	Color helps define architectural style and distinguish exterior materials and accents shapes. Warm colors are reds, yellows, and oranges whereas cool colors are blues, greens, and violets. In architecture, neutral colors such as shades of white, ivory, tan, and gray are often used. Colors often support a building's architectural style such as the Victorian-style **Painted Ladies.**	 The term ***Painted Ladies*** is used to describe certain Victorian-style houses, usually painted in three or more colors to enhance their architectural details. The famous Painted Ladies occupying 712–720 Steiner Street in San Francisco were built between 1892 and 1896. © iStockphoto.com/Bill Storage.
Form	A shape, outline, or structure, as described by lines and geometric shapes.	Most buildings can be described based on their composition of geometric shapes. A round or hexagonal turret is often found on **Victorian**-style buildings.	 Queen-Anne–style Victorian. © iStockphoto.com/ jim plumb.

(Continued)

Figure 4-13: *These architectural examples show the elements of design at work. (continued)*

Elements of Design Applied to Architecture

Element	Definition	Architectural Application Example	Illustration
Space	A given area; a positive or negative visual space.	Architectural interest is achieved through a building's positive and negative visual space. Contrast of positive and negative space can be created by a change in shape, color, or materials.	 Contrast of positive and negative space can be created by a change in shape, color, or materials. Courtesy of Northern Arizona University, Thomas Paradis.
Shape	The contour, profile, or silhouette of an object.	The silhouette of the building will help you quickly identify many architectural styles. For example, the **saltbox-style** house was named after the boxes used to store salt during colonial times. The shape of a saltbox roof forms a lopsided triangle created by one-story rooms, often sheds, added to the rear of taller homes.	 Saltbox Colonial. © iStockphoto.com/ Richard Stouffer.
Texture	The roughness or smoothness, including reflective properties, of a material or object and its related look or feel.	Rough surfaces create an impression of stability and substance, such as concrete masonry or stone often chosen for Georgian-style buildings. The smooth, reflective surface of glass and metal provide a comparatively lighter, refined, modern effect for contemporary buildings. The texture of stucco and large heavy clay roof tiles are distinctive features of **Spanish**-style buildings.	 Spanish Mission. Image copyright MalibuBooks, 2010. Used under license from Shutterstock.com.
Value	The relative lightness or darkness of a color.	Lighter colors can make a building appear larger.	 Georgian Colonial. ©iStockphoto.com/Jeff Morse.
Tone	The degree and variation of lightness and darkness of an area.	Some buildings use light and materials to create a desired tone.	 Frank Gehry's Guggenheim Bilbao. Image copyright Jamo Gonzalez Zarraonandia, 2010. Used under license from Shutterstock.com.

Figure 4-14: **These architectural examples illustrate the principles of design.**

Principles of Design Applied to Architecture			
Balance	Parts of the design are equally distributed to create a sense of stability. Often referred to as *symmetry*.	Balance and symmetry are achieved in this Cape Cod by the placement of matching features (windows and dormers) being placed at equal distance from the virtual centerline of the building.	 Image copyright Gregory James Van Raalte, 2010. Used under license from Shutterstock.com.
Contrast	Two or more parts are noticeably different.	The decorative half timbers on this Tudor home show great contrast to the light stucco. Contrast is achieved by placing different colors, materials, shapes, or textures side-by-side.	 Image copyright Kevin Penhallow, 2010. Used under license from Shutterstock.com.
Emphasis	The point of attention in a design; the feature in a design that attracts one's eye.	The dome of the Massachusetts State house, designed by architect Charles Bulfinch, was gilded in gold leaf in 1874 to create emphasis. Emphasis can be achieved by using a unique color, material, texture, shape, or building component as a focal point.	 Image copyright Mary Lane, 2010. Used under license from Shutterstock.com.
Movement	A feeling of action or flow when viewing a structure's form.	Walt Disney Concert Hall, designed by Frank Gehry as the home of the Los Angeles Philharmonic, conveys a feeling of motion though its graceful curves and smooth, reflective exterior.	 Image copyright Artifan, 2010. Used under license from Shutterstock.com.

(Continued)

Figure 4-14: *These architectural examples illustrate the principles of design. (continued)*

Principles of Design Applied to Architecture			
Pattern/ repetition/ rhythm	The regularly repeated arrangement of lines, shapes, color, texture, or pattern. Regular, graduated, random, and gradated are four types of rhythm.	Art Deco designers achieve rhythm through the use of stepped forms, such as the repeatedly offset and rounded arcs of the Chrysler Building. Architect William Van Alen used the vertical, triple-striped decorative elements and black decoration to communicate rhythm through simple, geometrical repetition of pattern.	 Image copyright Steve Baker, 2010. Used under license from Shutterstock.com.
Unity/harmony	The consistent use of lines, color, and texture to achieve a pleasing effect, balance, symmetry, and a composed appearance.	Repeating horizontal bands of color, light, and texture follow the gentle curves of the proposed Harmon Hotel, Spa, and Residences designed by architects Norman Foster and Partners for CityCenter, Las Vegas.	 Courtesy of MGM Mirage.

to one person might not to another. Therefore, an architect must have a clear understanding of the many factors that affect building design. A long, labor-intensive research effort will address environmental concerns, building structure, function, and client needs and wants. The result is a cohesive plan for maximizing the property's potential while satisfying the client, or occupants, for many years to come. Building trends and architectural styles might provide a base for the design, but as variables are considered, the resulting building plan will be unique and different from all others. With this in mind, as you look at existing buildings, you will discover many variations within each architectural style. Although this chapter will provide photographs and illustrations of common styles, they provide only limited examples of a given style. Because so many variations exist, it is best to look for clues, design elements, and features, and then match them to an architectural style category.

To determine a building's architectural style, you will first need to identify the following information:

- roof shape and pitch, chimney placement
- building shape, size, and number of stories
- window and door style, size, and placement
- decorative details, such as brackets and cornice trim
- construction materials and finish
- footprint and floor plan
- building features, such as a cupola, dormer, portico, or turret

Learn to recognize these visual clues and make connections to the architectural styles in which they are used. Some elements are applied to more than one style, but with practice, you will be able to look at almost any building and determine its architectural style. To help you remember this information, you might want to create an entry in your architectural design book for each architectural style shown in Figures 4-15 and 4-16.

Figure 4-15: **After you research Georgian architectural style on the Internet, your notebook entries might look something like this.**

Architectural Style: Georgian

Roof shape and pitch, chimney placement and eave details	Building shape, size, and number of stories	Window and door style, size and placement	Decorative details such as brackets and cornice trim	Constitution materials and finish	Footprint and floor plan	Historical and other information
Usually gambrel or hipped Medium pitch Minimal overhang Balanced paired chimneys	Rectangular: symmetrical building 2-3 stories	Pediment, central door framed by a portico and transom, 9/9 or 12/12 pane double hung windows, 5 on 2nd floor. Dormers and Palladian window	Quoins Dentil cornice Pilasters	Brick Wood	Central hall with 1 or 2 rooms on other side	1720–1800 Revival: 1895–1930

Another clue to determining the architectural style of a building might be to identify the historical period in which the building was designed and constructed. Remember, however, that historical styles are revisited and built in more recent years. These buildings are identified as *revivals* or *neo*. The term *neo* is added to the historical building style term, such as Neo-Georgian. Sometimes a neo or revival has simplified details, newer materials, or finishes that require less maintenance. You will notice that some buildings are difficult to identify because they have elements and features from many different styles. These buildings are referred to as *Eclectic* or *Neo-Eclectic*. Before you begin this next section, you should

Figure 4-16: These notebook sketches illustrate several building elements that can help identify Georgian architectural style.

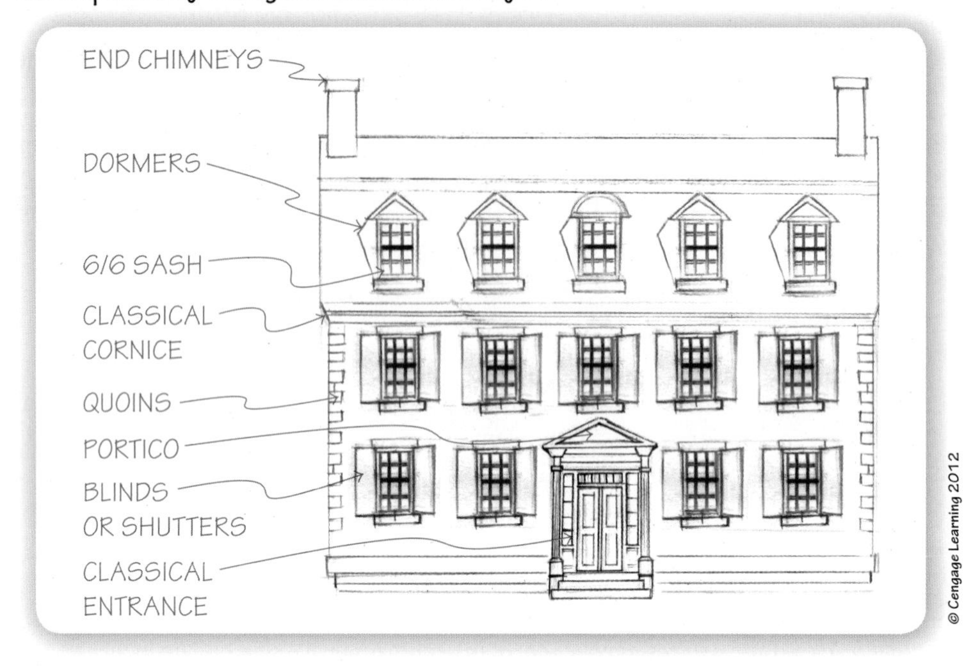

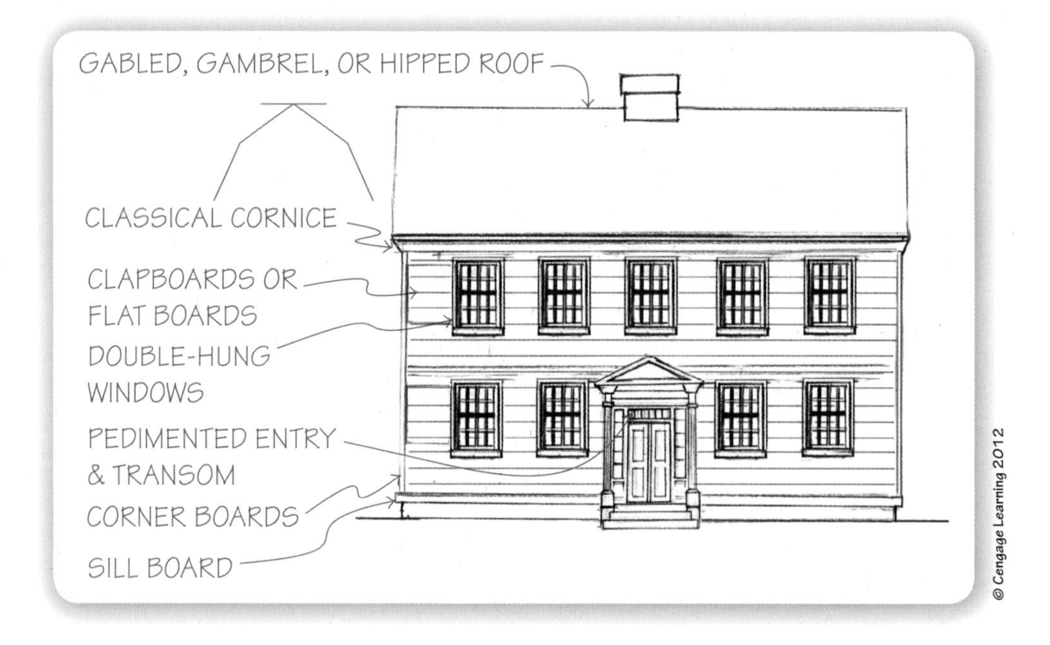

spend some time reviewing the building features and terminology in Figure 4-5. For example, in this section, you will read that Greek revival entablature was architecturally detailed with wide frieze panels. If you do not know the term *entablature,* the sentence will not make sense to you. Take a look back at the picture illustrating entablature in Figure 4-5. You can now picture the wide frieze panel and its location above a column. As you read this section, remember that there are more architectural styles than we can cover in this chapter. If you decide to pursue a career in architecture, you will undoubtedly learn about other architectural styles.

A basic principle followed by a design team, whether it is a design for a poster, clothing, bicycle, or building, is to begin by investigating similar items.

American Colonial Architecture (1600 to 1780)

European settlers in the New World often used the architectural styles from their homelands when designing their American homes. The homes built during this period were called *American Colonial Architecture* and became known as the *Colonial, Cape Cod, Dutch Colonial, Georgian,* and *Saltbox* styles (see Figure 4-17). Single-story Cape Cod cottages were some of the first homes built in the United States. Most Colonial homes were symmetrical, rectangular buildings with matching small-paned, double-hung windows placed on both sides of the front door. Colonials were built with locally available materials. Most Colonials had second-floor bedrooms and featured shed and gable roofs. Later Colonials had gambrel roofs, which provided more space on the second floor. Colonial revival homes were often sided with white clapboard with shutters and detailed cornices.

SALTBOX The Saltbox is an easy style to recognize because of its roofline. These square or rectangular two-story homes have a one-story shed attached at the back. The shed was usually oriented north to provide a windbreak. The entire building is covered by a single roof that drops down severely at the back. This simple, New England Colonial was given its name based on its resemblance to the boxes colonials used to store salt.

The Saltbox usually features a clapboard shingle exterior, large central chimney, and double-hung windows with shutters.

Figure 4-17: *A Saltbox-style home is easily recognized by its boxy shape and extended shed roofline.*

DUTCH COLONIAL The Dutch Colonial has a broad gambrel roof with flaring eaves that often extend over a porch. Although early Colonial homes were often a single room, the Dutch Colonial could receive additions to each end. Distinctive features included stone walls, chimneys at one or both ends, dormers, and a

Fun Fact

The Campbell-Christie House was home to John Walter Christie, born in the house on May 6, 1865. In the early 1900s, Christie built and raced cars against Louis Chevrolet, Henry Ford, and Barney Oldfield. Barney Oldfield was the first man to drive a car at 60 miles an hour; however, Christie beat Oldfield in a 50-mile race and held the world speed record. In addition to inventing the first automotive front-wheel drive, Christie invented a design for a high-speed tank that could move on either wheels or tracks. You can find more history from the Campbell-*Christie House at* http://www.bergen-countyhistory.org/Pages/cchousehistory.html

Figure 4-18: **The Dutch Colonial Campbell-Christie House was built by Jacob Campbell 1774 in northern New Jersey.**

Image courtesy of Bergen County Historical Society.

central Dutch double doorway. A Dutch door is split horizontally, allowing the occupant to open and close each half individually. In colonial days, the bottom section was closed to keep out animals, and the top half was left open to let in sunlight and air (see Figure 4-18).

GEORGIAN ARCHITECTURE (1720 TO 1800) The Georgian style was named after the four King Georges of England, who ruled during the time these buildings were popular. The rectangular, symmetrical Georgian structure features a central entrance with a pediment, or decorative crown, over the door. The buildings are two to three stories in height with symmetrically placed double-hung windows. Take a close look at the two Georgian Colonials, built in the 1700s, shown in Figures 4-19 and 4-20. What similarities and differences do you see? Notice that they both have a balanced row of five windows across the second story. Both pictures show flattened columns or **pilasters** on either side of the door. You will also see matching chimneys, pedimented dormers, multi-paned windows, and a transom window above the door. Now look closely at the windows and count the number of panes in each. One building features nine-over-nine windows and the other has twelve-over-twelve windows. Both window styles are common to the Georgian style of architecture. Other differences are more noticeable. For example, the Derby House has a gambrel roof style, whereas the Wentworth-Gardner House has a hipped roof. As you research this style further, you will discover that Georgian homes are generally built of brick or wood, and may have dentil molding along the eaves and quoins at the corners. Georgian revivals may also include other details, such as a cupola or a palladian window, to add decorative appeal.

Figure 4-19: The Georgian-style Derby House was built in 1762 in Salem, Massachusetts.

Figure 4-20: The Georgian-style Wentworth-Gardner House was built in 1760 in Portsmouth, New Hampshire.

Neo-Classical/Federalist/Idealist/ Rationalist (1750 to 1880)

Federal architecture emerged during the same period as the birth of the United States. Although Federal and Georgian styles are similar, Federal-style buildings feature elegant moldings and tall, slender chimneys, resulting in a more delicate look. A balustrade often caps the low-pitched roof (Figure 4-21). Federal style emphasizes the central entrance with semicircle fanlights above the door and narrow windows, called sidelights, on either side. Curved lines appear in Federal-style buildings in details such as recessed wall arches, circular or elliptical windows, or oval porticoes. Other unique features include triple-sash windows at the floor level and smaller windows on the top floors. The windows have large panes of glass and narrow mullions, or dividers. Particularly ornate Federal-style buildings will have a Palladian window, which is sometimes set into a recessed wall arch. Decorative flourishes in Federal architecture include dentil moldings, shutters, swags, and garlands. Federal-style architecture changed more than the exterior appearance of structures. The interior layout of a Federal building is asymmetrical and often includes an oval room, in contrast to the Georgian's stark arrangement of four square rooms around the central hallway.

Can you think of a famous Federal-style building? You are right if you thought of the White House! Look closely at Figure 4-22 and compare the size of the lower windows to the windows of the upper floor. What other design elements can you identify? Remember the Oval Office in the west wing of the White House?

Figure 4-21: This sketch shows a balustrade that caps the low-pitched roof of this Federal-style building.

Figure 4-22: The White House.

Figure 4-23: The Second Bank of the United States was designed by architect William Strickland in Greek Revival style. Can you tell what famous building inspired the architect?

Greek Revival (1790 to 1850)

You will not be surprised to learn that the ornamentation of Greek Revival architecture is inspired by the architecture of Ancient Greece (Figure 4-23). These buildings feature gabled pediments and ionic columns. The entablature is detailed with a wide frieze panel. The Greek revival style is easy to recognize by its multistory, full-width portico, which leads to recessed Greek revival doors. These doors are often flanked by narrow rectangular windows. Roofs are generally either gabled or hipped (Figure 4-24).

Victorian Architecture (1840 to 1900)

The Victorian period brought about a major change in building and design. Advancing technologies following the Industrial Revolution led to woodworking machines that could produce intricate details. Victorian buildings appeared with a variety of ornate moldings, spindles, columns, and brackets. Buildings were fanciful and featured towers, turrets, and bays. The buildings and trim were painted with contrasting vibrant colors. Victorian styles include the Queen Ann, Italianate, Gothic Revival, Eastlake, Romanesque, Second Empire, Shore Victorian, Richardsonian, and Shingle. Building materials were wood, cut stone, pressed brick, and plate glass. Victorian houses have distinct features that are easy to recognize. One example is the scrolled wooden bargeboard, used to decorate the many steep gables (see Figure 4-25). Fish-scale patterns are also distinctive decorative features of Victorian style. Victorians usually have **corbelled** chimneys and several large, spindled porches or verandas.

Corbelling:
stone or wood projecting from a wall or chimney for support or decoration.

Figure 4-24: This sketch of a Greek Revival–style House shows two story columns and roof pediment.

Figure 4-25: Bargeboard is common on Victorian-style buildings.

Figures 4-26 and 4-27 show two different Queen Anne buildings, one as a photograph and the other as an illustration. Architects often use sketches or illustrations, similar to the one shown, to communicate their plan for a building's exterior. You might want to sketch the Queen Anne in your notebook and label the defining elements. Can you see similarities and differences between two buildings?

ITALIANATE Italianate buildings have a strong, square form and shallow, hipped roofs with overhanging eaves and ornate brackets. Tall, narrow windows often extend from the floor and the tops are adorned with hood molding. Many Italianates have bays, porches, balconies, and a cupola. Multiple corbelled chimneys often appear in irregular locations and heights (Figures 4-28 and 4-29).

GOTHIC REVIVAL The Gothic Revival is marked by steeply pitched roof lines and ornate wooden detailing at the pedimented gable ends (Figure 4-30). The Gothic Revival shows its symmetry through balanced roof peaks, columns, and pilasters. Elaborate Gothic Revival buildings usually include towers or verandas and arched "Gothic" windows, as shown in Figure 4-31.

Arts and Crafts (1860 to 1930)

The Arts-and-Crafts Movement, in contradiction to the industrialization of the time, revived an interest in application of natural materials to manmade environments. Arts-and-Crafts houses

Figure 4-26: An elaborately painted Queen-Anne-style home.

Figure 4-27: A sketch of a Queen-Anne-style building might look like this.

Figure 4-28: This colorful Italianate buildings have a strong square form, front bays with tall windows, and decorative brackets beneath the eaves.

Figure 4-29: A sketch of an Italianate building might look like this.

Point of Interest

An Exacting Process

Gasson Hall has recently undergone exterior renovations to replace 95 percent of the exterior decorative cast stones that have deteriorated during the past 100 years. Before exact replicas could be constructed, professionals rappelled down the sides of the building to analyze and measure the stones. The information was documented using tablet PCs. The tower's four prominent spires were removed and used to create the mold for the replica towers. Use the Internet to research Gasson Hall at Boston College. A slide show of this large stone restoration project may be viewed at http://bcm.bc.edu/issues/fall_2007/features/heart-of-stone.html. A virtual 360-degree panorama from the Gasson Bell Tower can be viewed at http://at.bc.edu/slideshows/insideview/gasson.html.

have a rustic appearance achieved through use of earthy building materials including wooden shingles, stucco, and fieldstones (Figure 4-32). Exposed roof beams and rafters, stone chimneys, overhanging eaves, and various-shaped dormers are other features found on an Arts-and-Crafts-style building. The roofs are usually low-pitched gabled, shed, or hipped. The small, rectangular, one-and-a-half-story homes typical of this style often included a large front

Figure 4-30: *Gothic Revival home.*

Figure 4-31: *Balance and symmetry are apparent in the Gothic Revival-style Gasson Hall, built in 1908 for Boston College in Chestnut Hill, Massachusetts. The building, designed by architect Charles Donagh Maginnis, led the Collegiate Gothic style to popularity on other university campuses including Yale, Princeton, and Duke.*

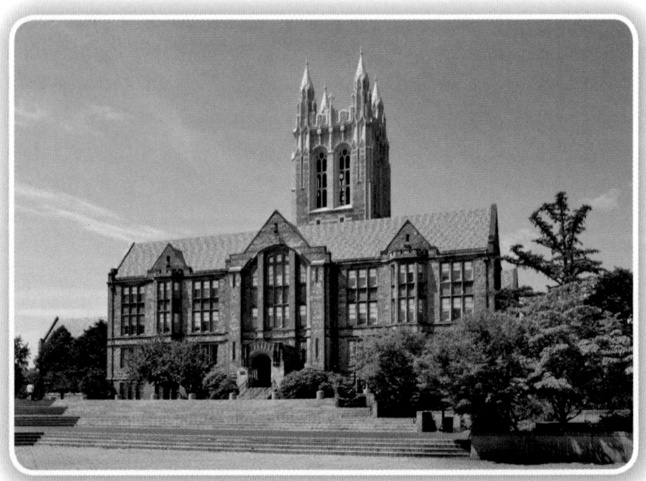

Figure 4-32: The Arts-and-Crafts Movement led to simple unadorned homes that blended with the landscape.

Figure 4-33: The Prairie-style boxy and symmetrical American "Foursquare" was named for its four-room over four-room floor plan.

porch. The Arts-and-Crafts Movement, and desire for a simpler lifestyle, influenced the development of the Prairie and Bungalow "Craftsman" styles. These simple and unadorned homes were designed to blend with the landscape.

The Prairie Style was created by the famous architect Frank Lloyd Wright and featured natural materials, low-pitched roofs with large overhangs, and clerestory windows. Prairie houses were one of two styles: either boxy and symmetrical (Figure 4-33) or low, spread out, and asymmetrical (Figure 4-34). Other details included casement windows and porches with large, square support posts. Brick and clapboard were the most common building materials.

Figure 4-34: Prairie-style architecture: The Frederic C. Robie House, 1909, Chicago by Frank Lloyd Wright.

Twentieth Century (1901 to 2000)

Neo-Georgian, Spanish Colonial, Colonial Revival, Dutch Colonial, English Tudor, Garrison, Cape Cod, French Normandy, and French Provincial were all popular in twentieth century. Many architectural designs emerged that were specific to location, such as the Spanish Mission architecture of the Southwest and Florida and the distinctive Pueblo-style homes in Arizona and New Mexico. The single-story ranch house came on the scene in 1932 and is still very prevalent today in almost every part of the United States. Many fine examples of twentieth-century architectural styles might be present in your own neighborhood. Look at the photographs provided in Figures 4-35 through 4-43 to see if you recognize these architectural styles.

French Provincial houses appearing between 1900 and 1930 featured two or more stories, brick, cut stone or stucco exteriors, steep, complicated roofs, massive chimneys, and a central turret.

Figure 4-35: Cape-Cod-style homes, popular in the 1940s and 1950s, are typically one-and-a-half stories with a gabled roof. Cape Cods often have small dormers to provide light and ventilation to the second floor.

Figure 4-36: This sketch of a Neo-Dutch Colonial shows a gambrel roof with flaring eaves, central doorways, and small dormers.

Figure 4-37: The Tudor Revival has a half timbering over stucco, steep gabled roofs, stone chimneys, narrow multi-paned casement windows, and a rounded doorway.

Your Turn

The Spanish Mission often has a quatrefoil window. Use the Internet to research this window style and sketch several examples in your architectural notebook.

ART DECO ARCHITECTURE (1925 TO 1935) Art Deco echoed the machine age with vertically oriented geometric shapes and use of hard materials, including metal and glass. Art Deco homes are usually two-story homes and have a smooth, light-colored, stucco finish, rectangular cutouts, and glass blocks (Figure 4-44). Most Art Deco buildings are commercial in nature (Figure 4-45).

INTERNATIONAL STYLE (1929 TO 1970) The International-style house has a horizontal, boxy look, a flat roof, and cantilevered rooms (Figure 4-46). A steel skeleton typically supports this type of building. Materials can include concrete, glass, and steel. Casement windows are frequently

Figure 4-38: This sketch shows the most recognizable feature of a Garrison Colonial, the cantilevered second story, traditionally with pineapple or acorn shape ornamentation below the overhang. Garrisons usually have an exterior chimney at one end and smaller second-story windows. Windows are either double-hung or casement with small panes of glass.

Fun Fact

The Origins of the Garrison

Some historians have surmised that the Garrison style has military origins, mimicking forts built with an upper projection to make them difficult to scale. However, another possibility of origin may be Elizabethan houses, which had second-story extensions over the streets. The upper projection increased their second-story living space and to shelter the shops and pedestrians below from the rain. Whatever the origin, the projection provides an architectural horizontal break.

Figure 4-39: Spanish Mission style usually has low-pitched reddish tiled roofs, light stucco walls, and rounded windows and doors. Other elements may include balconies with elaborate ironwork, decorative tiles around doorways and windows, arched dormers, roof parapets, or a bell tower.

Figure 4-40: Adobe architecture is characterized by flat roofs, parapet walls with round edges, earth-colored stucco or adobe-brick walls, straight-edge window frames, and roof beams that project through the wall. The interior typically features corner fireplaces, unpainted wood columns, and tile or brick floors.

Figure 4-41: Ranch homes are single-story, rectangular, L-, or U-shaped buildings with low-pitched roofs. They often have attached garages. Windows are generally large and include double-hung, picture, and sliding. Variations of the ranch include split-level and raised ranch.

Figure 4-42: A-frame homes are designed with a steep roof that extends to the ground to reduce snow loads. Most of the windows and doors are placed on the front and back gable ends, protected by large overhangs. This style has limited interior space but is quite popular in vacation homes.

Figure 4-43: This sketch of a French Provincial home suggests formality with high-pitched hip roofs, arched openings, and double French windows with shutters.

© Cengage Learning 2012

© Cengage Learning 2012

Figure 4-44: This is a sketch of the Earl Butler House in Des Moines, Iowa, designed by Kraetsch and Kraetsch and completed in 1937 in the Art Deco style.

© iStockphoto.com/Natalia Bratslavsky

Figure 4-45: The Chrysler Building is a widely recognized Art Deco–style building. It was designed by architect William Van Allen and built between 1928 and 1930 in Manhattan.

placed at corners. You can recognize International-style homes by their lack of ornamentation. The International style was generally used for commercial buildings (Figure 4-47).

CONTEMPORARY ARCHITECTURE (1950 TO PRESENT) Contemporary buildings are usually recognized for their lack of ornamentation (see Figure 4-48). Various window sizes and shapes are combined for optimum sunlight or view. The exterior building materials of contemporary residential architecture often combine vertical, horizontal, and diagonal patterns of wood mixed with **masonry.** Figure 4-49

Figure 4-46: This unique home features several international style features.

Figure 4-47: Widely considered the first international modernist skyscraper, the PSFS Building in Philadelphia is a National Historic Landmark. Built in 1932, by architects G. Howe and Wm. Lescaze, the building now houses the Loews Philadelphia Hotel.

shows a pre-fabricated contemporary home. This design features a central clerestory and is built at a factory to minimize the environmental impact on the home site.

POSTMODERN ARCHITECTURE (1950 TO PRESENT) Postmodern style reflects originality and uniqueness through unusual combinations of two or more different elements. A Postmodern building may be exaggerated or abstract, with forms that reflect irony, humor, and contradiction. They are often described as bizarre, surprising, or shocking (see Figure 4-51).

Architectural design is influenced by personal characteristics, such as creativity, resourcefulness, and the ability to visualize and think abstractly.

Figure 4-48: A contemporary house by Silver Ridge Design, Inc.

Figure 4-49: Prefab contemporary Breezehouse by John Swain is intended to be both ecologically and aesthetically pleasing through its use of natural sunlight and recyclable building materials.

Case Study

CONTEMPORARY ARCHITECTURE AT CITYCENTER

Sven Van Assche, Vice President of Design for the CityCenter in Las Vegas, traveled the world to examine contemporary architecture and to observe various melting pots of cultural diversity. The goal was to discover how to replicate the diversity and vitality of an urban setting in a high-quality environment that offers a variety of choices for discriminating residents and visitors. Robert H. Baldwin, MGM's Chief of Design and Construction, wanted unique buildings, unlike any other buildings in the world. The resulting designs for the contemporary-style buildings offered timeless simplicity a harmonious arrangement. The building designers have avoided the cold, hard edges and white walls of signature contemporary buildings and instead have chosen luxurious building materials including mosaic marble, wood, and patterned metals for a warm and inviting feel. The lower center of the photograph in figure 4-50 shows architect Helmet Jahn's contemporary 37-story glass Veer Towers, which are inclined 5 degrees from center in opposite directions. According to CityCenter the Veer Towers have been described as a "masterful translation of energy and excitement into physical form."

Figure 4-50: *Contemporary style was chosen for the buildings of CityCenter to communicate quality and timeless design.*

Courtesy of MGM Mirage.

NEO-ECLECTIC (1950 TO PRESENT) Neo Eclectic homes use modern materials like vinyl or imitation stone to replicate a variety of historic architectural styles. The roofs, windows, doors, and details draw upon several architectural styles (see Figure 4-52).

Residential home styles often reflect individuality and uniqueness as with the Venice Beach House by Frank Gehry, Figure 4-53.

Your Turn

Use the Internet to research buildings based on Buckminster Fuller's Geodesic Dome. Expand your knowledge of Architectural styles by researching definitions and examples of Art Moderne, the Bauhaus school (coined by Walter Gropius), Deconstructivism, Formalism, Modernism, Structuralism, and Postmodernism. Record the definitions and sketches in your architectural design notebook.

Figure 4-51: Postmodern architecture often has contradicting and surprising form.

Figure 4-52: This is a sketch of Neo-Eclectic building that shows a mix of building styles, materials, and components.

Figure 4-53: Venice Beach House by Frank Gehry.

VERNACULAR ARCHITECTURE Vernacular architecture describes a method of construction that uses locally available resources. This fundamental approach, often dismissed asunrefined, is commonly based on local traditions and building techniques passed down through generations. Proponents of vernacular architecture describe concern for cultural and economic sustainability as the basis for this approach (Figure 4-54).

Buildings That Address Specific Concerns

Some residential styles address specific concerns and needs. Although they have not become popular, several designs for modular dwellings have emerged to meet the needs of young families. The design starts with a basic three- to four-room piece and allows additional spaces to be added as the family's needs or finances change.

Other residential buildings are designed for maximum solar gain. These unique solar-powered homes often have south-facing glass and a large shed or flat roof line to accommodate photovoltaic cells or other solar-gathering devices. Occasionally, homes are built partially underground to save energy and provide protection from the elements. These are called *bermed* houses.

Figure 4-54: Vernacular architecture is a culturally and environmental influenced blend of traditional building methods and local materials.

Figure 4-55: This is a sketch of Katrina Cottage design. Durable Katrina cottages feature steel studs, a steel roof, rot- and termite- resistant siding, and moisture- and mold-resistant drywall to meet international building and hurricane codes.

Recently, many buildings have emerged to address specific economic and environmental factors. Katrina Cottages, like the one shown in Figure 4-55, were developed for families who lost their homes because of the August 29, 2005 hurricane Katrina and the floods that followed. There are more than two dozen versions of the prefabricated Katrina Cottage. The styles of such structures can be based on one or more traditional architectual styles, but are driven by a need for a quick, inexpensive, and durable shelter.

Careers in Civil Engineering and Architecture

LIVING LARGE

Sean Stadler is part of a large architectural firm, which means he usually works on large-scale commercial projects rather than single-family residences. His buildings can take several years to complete, and Stadler helps keep them on track. His responsibilities include making public presentations for his projects and speaking to neighborhoods, commissions, regulators, and clients.

Stadler is an associate principal in his firm—in other words, one of the senior leaders. He is the design principal on his projects, which means he is responsible for how the buildings look. The technical side of the projects is resolved by other architects at the firm.

"A lot of students who think about going into architecture have the misconception that you have to be super smart in math," he says. "My particular expertise is more on the artistic side."

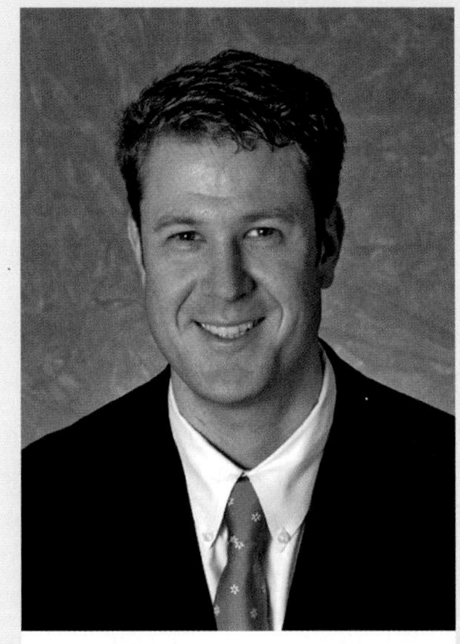

Sean Stadler, AIA, LEED AP, associate principal at WDG Architecture, PLLC

On the Job

One of Stadler's most interesting projects is Maryland's National Harbor, a mix of residential, retail, and hotel spaces on a waterfront. It's one of the largest projects on the East Coast, and Stadler's firm designed four of the buildings.

"In today's urban development, there's a concept called 'new urbanism,' which is intended to create a vibrant town atmosphere," he says. "National Harbor tried to achieve this."

Prior to beginning National Harbor, Milt Peterson, the project's owner, went to Barcelona, Spain, to draw inspiration from an area called Las Ramblas. "It's an urban town that has an active street life, mixing retail with residential units above," Stadler says. "The owner of National Harbor wanted to re-create this feeling in Maryland. We took that idea and designed the buildings for him."

Inspirations

Stadler never wanted to be anything other than an architect. As a child, he loved playing with Legos.

"I would play with them for hours," he says. "That's when everybody said, 'Oh, you're going to be an architect when you grow up.'"

Stadler could draw, paint, and sculpt. "I would get an image in my mind and want to create something," he says. "That's what you're able to do as an architect—create something from nothing."

Education

Stadler went to college at Ohio's Kent State University without knowing a lot about its architecture program.

"I got very lucky with some of the choices I made, not really having a whole lot of guidance and being the only college graduate in my family," he says. "I was fortunate that a local school had a very good architecture program."

Stadler appreciated Kent State's diverse approach to the architecture curriculum.

"Some schools teach from a more theoretical basis, some from a more technical basis," he says. "Our university married the two fairly well."

Stadler particularly enjoyed a student project for the Ohio Edison Company that integrated everything he'd learned in school. "It put you in a situation you might encounter in the real world, with a real client," he says.

Advice for Students

Stadler recommends that students interested in architecture simply go out and look at buildings. "Just understanding the scale of a building is really helpful," he says.

Stadler also suggests that students talk with practicing architects. A good place to find a local firm, he says, is the website of the American Institute of Architects: http://www.aia.org. The site also has a section on an architect's career development path.

SUMMARY

In this chapter, you read about the difference between residential and commercial architecture and discovered that most cities and towns have several architectural styles. These various styles are the result of variables in culture, economy, climate, human needs, and aesthetic perspective. Today, these variables are addressed with a whole-building design approach. Whole-building design involves architects, engineers, building occupants, owners, building officials, contractors, and specialists in areas such as materials, energy, air quality, or acoustics. They work together from the beginning of the project ideation to create a building that is aesthetically pleasing, energy-efficient, sustainable, well suited to the environment, and a healthy place to live or work.

This chapter introduced CityCenter in Las Vegas as a case study in whole-building design. The project has often been described as a city within a city. Several architectural firms and urban planners were involved in the planning process. The project required a long and labor-intensive research phase to successfully meet the needs of the client and maximize property potential in an ethical manner. It was necessary to select an architectural style that would complement the culture, society, environment, and economy. Contemporary architectural style was chosen for its timeless quality and aesthetic potential to attract diversity and promote an exciting city atmosphere.

Through a mini-quest to find a Georgian-style house, you searched for building shape, roof type, materials, windows, doors, chimneys, and trim as visual clues to help determine architectural style. You discovered that these elements are often used in more than one style, and when combined, they form an Eclectic- or Neo-Eclectic–style building. Tables were included within the chapter to provide illustrations of building features and show applications of principles and elements of design. Additional photographs and brief descriptions highlighted architectural styles in the United States, beginning with American Colonial architecture. During this chronological timeline, historical styles were revisited and built with newer materials, and identified as revivals or neo. Because so many variations exist within each architectural style, you learned to recognize clues, design elements, and features, and to match them to an architectural style category.

This chapter revealed the value of architectural terminology for communicating successfully within a whole-building design team to plan highly functional, aesthetically pleasing, sustainable buildings. Although building trends and styles might provide a base for building design, when all variables are considered, the resulting plan can be unique and different from all others.

BRING IT HOME

1. Research Monticello to identify the architectural features that support the building's Colonial Georgian style. Why is this building famous?
2. Research the building feature "buttresses" and add a sketch, description, and architectural style notes to your notebook.
3. Survey your neighborhood to determine the most popular roof style. After a survey of at least 20 neighboring houses, create a pie chart of your results.
4. Compare the buildings in your area to those in a different part of the country with a different climate. What are the differences in architectural styles, building materials, or building processes?
5. Conduct research to find the oldest residential and commercial buildings in your area. Using a print or digital photograph of each building, identify and label all the building components you recognize. Make a determination of each building's architectural style based on your analysis.

EXTRA MILE

1. Research a local historical building and identify the date of construction. Taking the role of the original building owner, make a list of concerns and needs addressed by the building's design. Determine how the building's current use compares to the original purpose. Research and make a list of renovations and updates that have occurred. Record your thoughts of what prompted the renovations.

PART III
Location, Location, Location

CHAPTER 5
Site Discovery for Viability Analysis

GPS DELUXE

Menu

START LOCATION | DISTANCE | END LOCATION

Before You Begin

Think about these questions as you study the concepts in this chapter:

1. What information is important to consider when you weigh the risks and benefits of developing a property?
2. How is land described?
3. How do you know what rules and regulations apply to land development?
4. How does zoning help or hinder site development?
5. What environmental regulations affect the usage of a site?
6. What climate data is important when researching a site for development?
7. Why should you consider utilities and traffic in your site analysis?
8. How does demographic data affect site selection?
9. What should you look for during a site visit?
10. How can you determine if a development is appropriate for a given site?

INTRODUCTION

Would you ever buy a used car without actually looking at it first? To make sure that you are getting what you pay for, and what you expect, you have to get an up-close-and-personal look at the vehicle. To avoid future problems you should take the car for a test drive, have a mechanic check the engine and car systems to make sure it is safe to drive and meets all regulations, and research its history to get accident, mileage and title information. You must be an informed buyer in order to avoid unpleasant surprises later. Developing land is an even larger investment than purchasing a car. Just as you would take precautions to discover everything you could about a used car before you sign on the dotted line, you should take reasonable steps to investigate a property, or area of land known as a **site.** You don't want to buy a lemon, whether it is in the form of a car or a piece of property (Figure 5-1).

Site: an area of land, typically one plot or lot in size, on which a project is to be located.

What makes a site "right" for a development? Aren't there many sites that could be used for any given development? The answer is that, yes, there are probably many sites on which a given development, say a pizza restaurant, could be built. But, just because you *can* build on a site does not mean you should. After all, does it make sense to build a pizza restaurant in an industrial park, surrounded by warehouses, and away from a major artery of traffic? Or on a lot that is home to a fragile ecosystem or could hide toxic contamination? You may never be sure that you have chosen the *best* site, but you can investigate the site to be reasonably sure that it does not pose problems that can jeopardize

Figure 5-1: *Site discovery helps avoid purchasing a "lemon" property.* © Gina Luck/iStockphoto.com.

INTRODUCTION

Figure 5-2: *Would this dry lake bed make a good location for your next project?* © silvrshootr/iStockphoto.com.

the success of your project or significantly increase the time and cost of construction. We refer to this investigation process as **site discovery**. In this chapter, we will use various resources in the site discovery process to gather important information about a site that will affect the construction and usability of a planned development. Once you have completed site discovery, you will be able to better evaluate a site to determine whether or not your proposed development is a "good fit" for the site; this evaluation is called **viability analysis**.

Design problems must be researched before they can be solved.

HOW LAND IS DESCRIBED

In order to research a site for potential development, you must first discover specifically what land is included in the property. This involves gathering legal documents and understanding the language of land description.

Property Research

How do you begin site discovery? What documents do you need? One of the first things that you should investigate is the legal ownership of the property and any legal cirumstances that could directly affect you as the purchaser of the property. For instance, a utility company may own rights to access a 50 ft wide strip through the middle of the property to maintain an underground supply line. Perhaps a prior owner has stipulated that the property remain undisturbed to provide a sanctuary for brown pelicans. If the previous owner has not paid his income taxes for several years, a new owner may be liable for the debt. You will save yourself much time and effort if these kinds of situations are identified before you purchase the property.

The best place to begin your search is the county office of the Recorder of Deeds. The Recorder of Deeds goes by many names: the Registrar-General, Registrar of Deeds, Registrar of Titles, Deeds Registry, Deeds Office, County Recorder, and Register Mesne Conveyance (RMC), among others. In some locations, the county clerk's office performs the responsibilities of the Recorder of Deeds. Whatever the name, this office maintains records of real estate ownership and other documents pertaining to rights over property.

The first document to obtain as part of your research should be the **deed** (Figure 5-3) and a copy of the **plat** (Figure 5-4) for the property from the Recorder of Deeds. These two documents will provide the legal description of the property and identify areas of the property to which others may have rights, such as easements or rights-of-way. An **easement** refers to the privilege to pass over the property of another for a specific purpose. For example, an **ingress** and **egress** easement might give a neighbor, who has no direct access to a public road, the right to cross your property in order to enter (ingress) and exit (egress) his land. Although often used interchangably with the term easement, **right-of-way** (or easement-of-way) often refers to the actual strip of land granted through an easement. For example, a municipality may hold a right-of-way along a public road for maintenance or expansion. Often public utilities are also located within the right-of-way.

Many properties are purchased by means of a **mortgage,** which is a document that gives a lender rights to a property as security for a loan. The mortgage is considered a property lien against the lot. In this case the **property lien** would give the company holding the mortgage the right to collect payment of a debt from the sale of the property. Property liens can also result from other debts such as unpaid taxes, nonpayment of contractors, or a court judgment. If you were to buy a property owned by a person named on a property lien, you could be forced to pay the debt. Fortunately, most **lending institutions** require a title search to establish a "clear title," which means that the property is free of any liens or judgments, before they will lend money to purchase a property. The Recorder of Deeds can also help you with the title search.

Deed:
a written document that provides the transfer of ownership of real property from one person(s) to another and must include a description of the property.

Plat:
a map of an area of land that shows property lines, streets, buildings, some easements, and rights of use over the land. Be aware that not all easements are shown on the plat.

Figure 5-3: *Partial deed to a property in Mount Pleasant, South Carolina, and recorded in Charleston County. The deed provides the legal description of the property in metes and bounds.*

STATE OF SOUTH CAROLINA)
) TITLE TO REAL ESTATE
COUNTY OF CHARLESTON)

KNOW ALL MEN BY THESE PRESENTS, THAT SOUTH CAROLINA ELECTRIC & GAS COMPANY ("SCE&G"), in the State aforesaid, for and in consideration of the sum of One and No/100 ($1.00) Dollars cash to it in hand paid at and before the sealing of these Presents by TOWN OF MT. PLEASANT, SOUTH CAROLINA, a body politic ("Grantee"), in the State aforesaid, (the receipt of which is hereby acknowledged) has granted, bargained, sold and released, and by these Presents, does grant, bargain, sell and release unto Grantee the following described property (the "Property"), to wit:

All that certain piece, parcel or lot of land, situate, lying and being in the Town of Mt. Pleasant, County of Charleston, State of South Carolina, containing 0.585 acres, more or less, and being more fully shown as Lot B on "Subdivision Plat, Lots A & B, SCE&G Bayview Substation" prepared by Absolute Surveying, Inc., James Kelly Davis, R.L.S. No. 9758, dated October 25, 2001, and having the following metes and bounds, to-wit:

Beginning at the southwestern intersection of Palm Street and Pineview Drive, running thence with the southern right of way of Palm Street approximately 160 feet to an iron pipe found at the corner of property of South Carolina Electric & Gas Company; thence running with the southern right of way of Palm Street S 73-03-39 W 89.76 feet to the POINT OF BEGINNING; thence turning and running S 16-49-10 E 150.00 feet to a point; thence turning and running S 73-03-39 W 170.00 to a point in property owned now or formerly by the Town of Mt. Pleasant; thence turning and running N 16-49-10 W 150.00 feet to a point in the southern right of way of Palm Street; thence turning and running with the southern right of way of Palm Street N 73-03-39 E 170.00 feet to the POINT OF BEGINNING.

Courtesy the Register Mesne Conveyance of Charleston County South Carolina.

Figure 5-4: Partial reprint of Plat Book DE, page 007, Charleston County, South Carolina. Lot B corresponds to the property conveyed in the deed shown in part in Figure 5-3. The Point of Beginning of the property description has been highlighted.

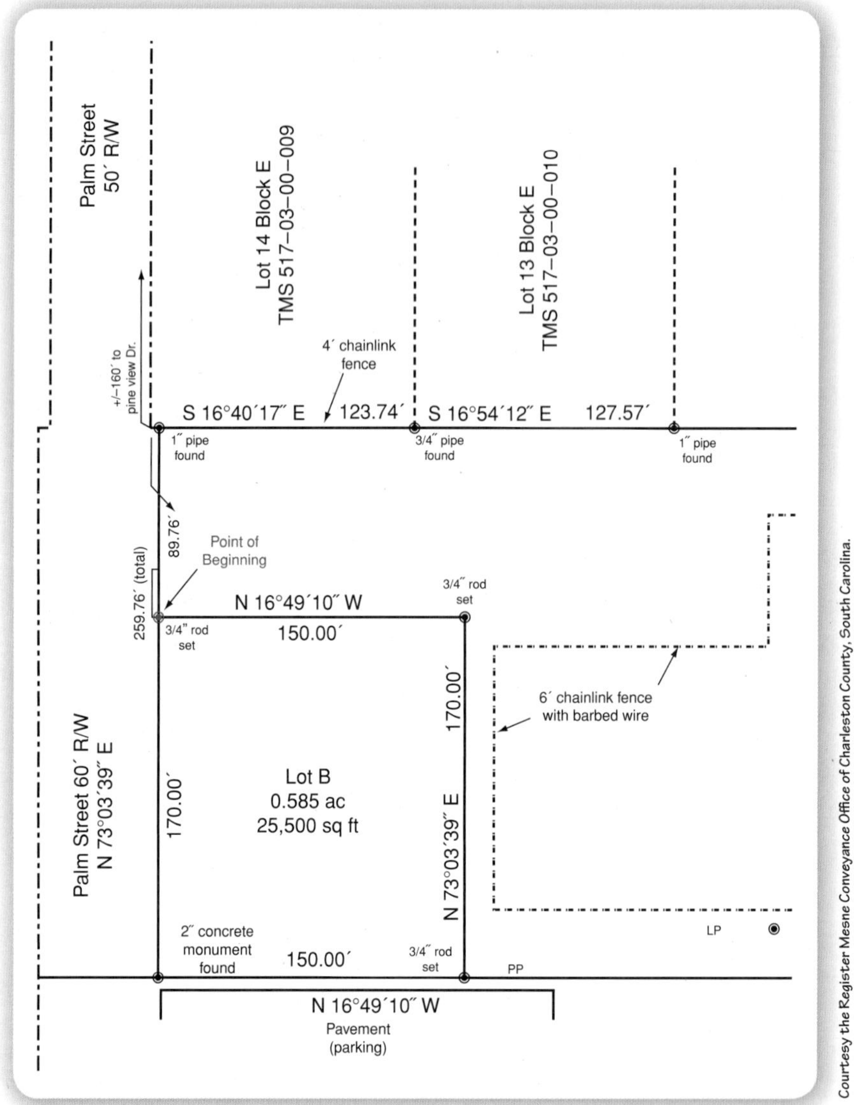

Courtesy the Register Mesne Conveyance Office of Charleston County, South Carolina.

You may visit the appropriate county office to perform your research, or in many counties, the Recorder of Deeds will provide Internet access to these records; the *Public Records Online Directory* may be helpful in locating county offices or online resources. Another way that property information is provided is through a Geographic Information System or a Land Information System. A **Geographic Information System (GIS)** is a computer-based mapping system that integrates geographic data with other relevant information such as transportation networks, hydrography, population characteristics, economic activity, political jurisdictions, environmental characteristics, and so on. All information is referenced to a common spatial coordinate system. A **Land Information System (LIS)** is a special type of GIS most often used by municipal agencies that manages information related to land ownership such as parcels, land use, zoning, and infrastructure (see Figure 5-5). In practice, an LIS is often referred to as a GIS.

Figure 5-5: Screen shot of the Charleston County, South Carolina, GIS map. A parcel in downtown Charleston is highlighted and property information is provided in the right column. The left column lists the available layers that can be included in the map view.

Courtesy of Charleston County, South Carolina.

OFF-SITE EXPLORATION

Identify a restaurant in Charleston County, South Carolina, that you would like to visit and find the address online. Visit the Charleston County website at *http://www.charlestoncounty.org/* and access the GIS online service. Search the GIS for the restaurant property by address. Print the map and insert it into your architectural notebook. Record the plat book and deed book numbers in your notebook. Then search for the deed and plat for the property—you will need to exit the GIS and visit the Register Mesne Conveyance (RMC) page. Note, in some cases, the deed is not available. If you cannot access the plat, choose another restaurant to research. Print out the deed, if available, and the plat and insert them into your notebook. Highlight any easements on the property.

Engineers and scientists rely on technology, such as GIS and LIS, to enhance the gathering and manipulation of data. New techniques and tools contribute to the advancement of science and engineering.

Information systems, such as GIS/LIS, can be used to manage information and inform and educate people.

Legal Descriptions

How do you know exactly what land is included with a property? An address or a parcel number, although helpful, does not provide enough information alone to adequately investigate the viability of the site. However, as we discussed earlier, the **legal description** of a property must be included in the deed. This description will provide the exact location and position of the property lines so that you can decide if the lot will be suitable in size and shape for your proposed project. In addition, once you know exactly where the property lies, you have boundaries for further investigation of the physical characteristics of the site. In order to interpret the language of legal descriptions you will need a little background information on the three types of legal descriptions: metes and bounds, rectangular survey system, and lot and block.

Legal Description: a written passage or statement that defines property and is descriptive enough so that the property can be differentiated from other properties and located without other evidence.

METES AND BOUNDS When the United States was colonized, land was described using the accepted English system of metes and bounds. The term **metes** refers to a boundary described by a distance and direction between property corners. The term **bounds** indicates a boundary that is described less accurately—for

Figure 5-6: The direction angle of a property line is given using bearings, an angle in degrees, minutes, and seconds. The angle is measured from due north or due south toward either east or west.

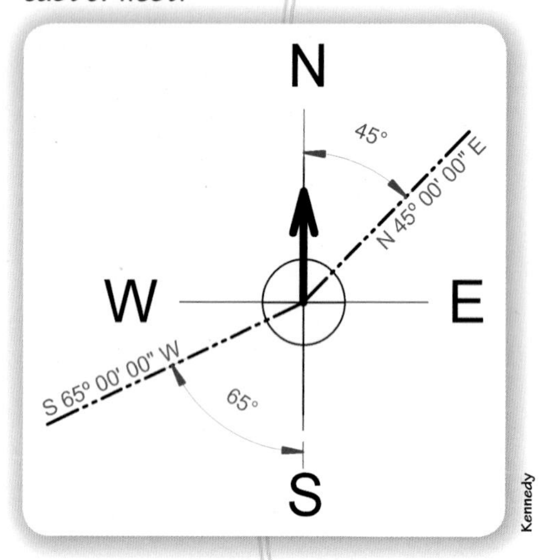

example, "down the center of Eagle Creek" or "along the stone wall at the edge of Issac Porcher's property." A metes and bounds survey generally describes a point of beginning (POB), which designates one of the property corners. Then each property line is described individually in sequence to complete a loop so that the final property line ends at the POB.

The length of each property line is given by a distance, typcially measured in feet today. The direction of a property line is indicated by a **bearing**. A bearing is an angle measure, less than 90 degrees, from either due north or due south. For example, you can determine the direction of a property line indicated by the bearing north 45 degrees east (N 45° E) by standing at the property corner, facing due north, and then turning 45 degrees to the east, or clockwise. The direction you now face is north 45 degrees east (see Figure 5-6). Likewise, south 65 degrees west (S 65° W) indicates a direction that is 65 degrees to the west, or clockwise, from due south.

Today, surveyors typically describe distances in decimal feet and bearing angles in degrees (°), minutes ('), and seconds ("). Remember from your math classes that there are 360 degrees in one full revolution. Sixty minutes are equivalent to one degree, and sixty seconds constitute one minute. So, 30 minutes is one-half of a degree. The bearing S 25° 56' 51" E (or S 25-56-51 E) indicates an angle of 25 degrees, 56 minutes, and 51 seconds to the east of due south.

It is important to realize that the bearing callout for a property line can refer to either endpoint of the line (property corner) regardless of the sequence used to describe the property lines. For example, the bearing N 35° E can represent the same property line orientation as the bearing S 35° W. Look at the property in Figure 5-7. The property lines are annotated with two alternate bearings that can be used interchangably. Notice that the two bearings for each property line are similar, the difference being that N and S are interchanged, and E and W are interchanged.

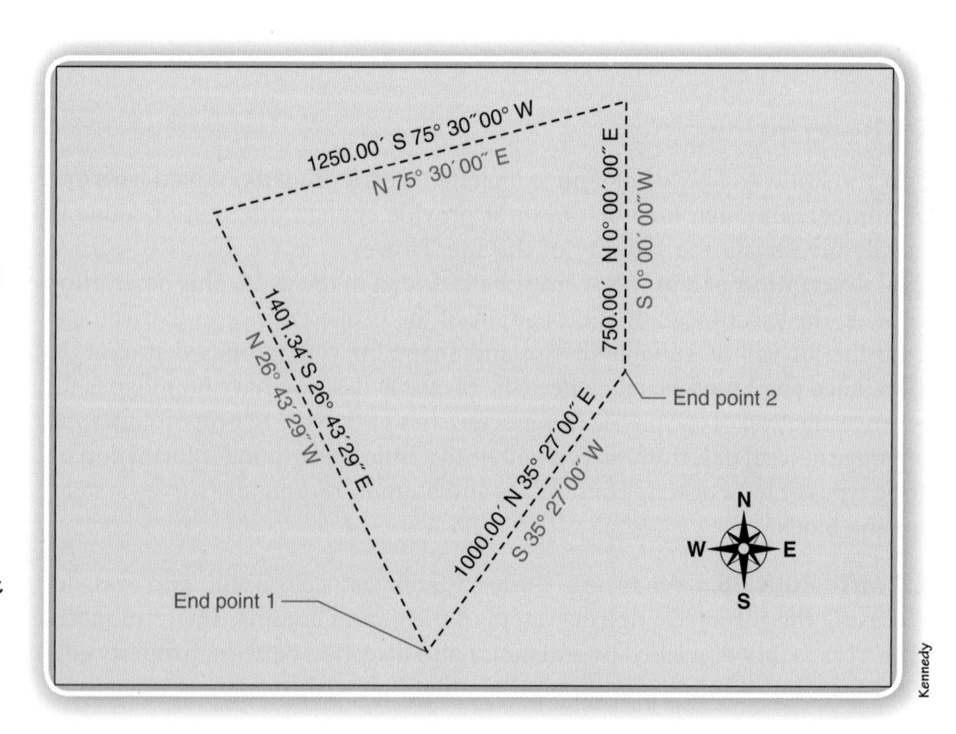

Figure 5-7: The direction of a property line may be indicated by two different bearings. The bearing N 35° 27' 00" E (in black) results when End point 1 is used as the origin of the bearing, and S 35° 27' 00" W (in blue) gives the direction when beginning at End point 2. Either bearing is acceptable on the plat. Both alternate bearings are given for each property line here, but you should include only one bearing designation on a plat or site plan.

Example: An Example of Metes and Bounds Description

The legal description given in the deed in Figure 5-3 is a metes and bounds description. Using the property description in the deed, trace out the property lines shown on the plat. The numbers provided in the description below correspond to the numbers inserted on the plat in Figure 5-8.

Figure 5-8: **Partial reprint of Plat Book DE, page 007, Charleston County, South Carolina. Lot B is the property conveyed in the deed shown in Figure 5-3. The numbered points correspond to points indicated by the metes and bounds description in the deed.**

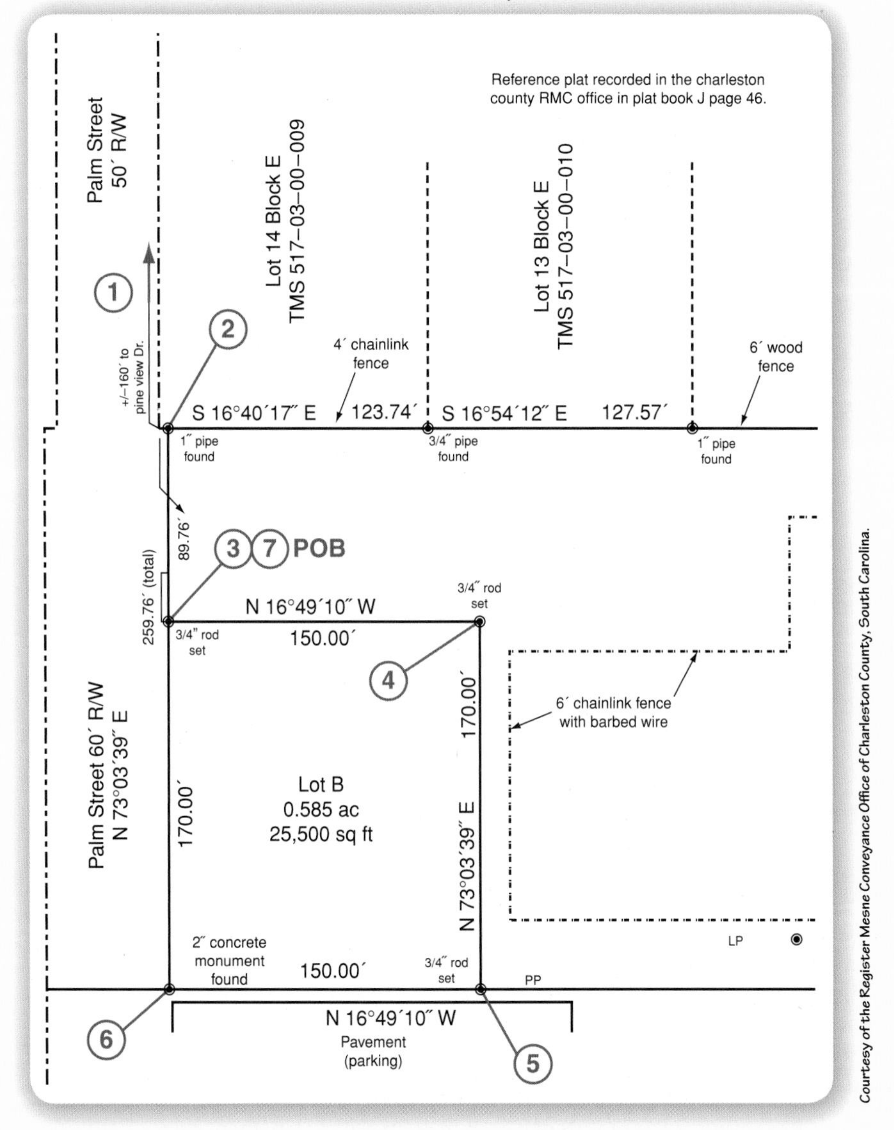

Courtesy of the Register Mesne Conveyance Office of Charleston County, South Carolina.

1. "Beginning at the southwestern intersection of Palm Street and Pineview Drive"

   Note: this point, called the Point of Commencement (POC), is not shown on the plat, but is indicated by an arrow pointing north

2. "running thence with the southern right of way of Palm Street approximately 160 ft to an iron pipe found at the corner of property of South Carolina Electric & Gas Company"
3. "thence running with the southern right of way of Palm Street S 73-03-39 W 89.76 ft to the POINT OF BEGINNING" (POB)

   Note that the bearing angle given on the plat is N 73° 03' 39" E, which indicates the same orientation as the bearing given in the deed
4. "thence turning and running S 16-49-10 E 150.00 ft to a point"

   Note that the bearing angle given on the plat is N 16° 49' 10" W, which indicates the same orientation as the bearing given in the deed
5. "thence turning and running S 73-03-39 W 170.00 to a point in property owned now or formerly by the Town of Mt. Pleasant"
6. "thence turning and running N 16-49-10 W 150.00 ft to a point in the southern right of way of Palm Street"
7. "thence turning and running with the southern right of way of Palm Street N 73-03-39 E 170.00 ft to the POINT OF BEGINNING."

Early metes and bounds descriptions often sound like instructions on a pirate's treasure map. Most were written in a running prose style and used rocks, trees, or creeks to describe the property. Distances were often recorded using units of measure that are no longer familiar such as chains, rods, or poles. A rod or pole is equivalent to 16.5 ft. A chain is equivalent to 4 rods or 66 ft. For longer distances, 10 chains is a furlong, and 80 chains is one statue mile. Ten square chains make an acre.

Fun Fact

The clergyman Edmund Gunter developed a simple method of surveying land accurately using a chain that was 66 ft long, referred to as *Gunter's chain*. As a result of the use of his chain, the word *chain* became synonymous with the measured length of 66 ft. And since the chain had 100 links, a *link* was one one-hundredth of a chain.

Example: Sketching Plot Plans from Metes and Bounds Descriptions

The following metes and bounds description was used on a deed dated February 13, 1810, and is recorded in Mercer County, Kentucky (Deed Book 7, p. 417).

> Beginning at the mouth of a branch at an ash stump, thence up the creek S 20 poles to 2 beech, thence east 41 poles to a small walnut in Arnett's line, thence north 50 east 80 poles to a linn hickory dogwood in said line, thence north 38 poles to an ash, thence west 296 poles with Potts' line till it intersects with Tolly's line, thence south 30 west 80 poles to a whiteoak and sugar thence east 223 poles to beginning…

Can you sketch the property lines of the described property? You will need a protractor and an engineer's scale (or ruler). Record the sketch in your architectural notebook at an appropriate scale. Be sure to label each property line with the distance and direction and each property corner with the appropriate landmark. Include a title and note the scale of your sketch. Figure 5-9 provides a seriers of photographs showing the progression used to sketch the property.

Step 1: *"Beginning at the mouth of a branch at an ash stump"*

Place a North arrow at the top of the paper. Start at the point of beginning and label this point POB.

Figure 5-9: **Sketching a plot plan from a metes and bounds description. Early metes and bounds descriptions often did not describe a closed loop.** Kennedy

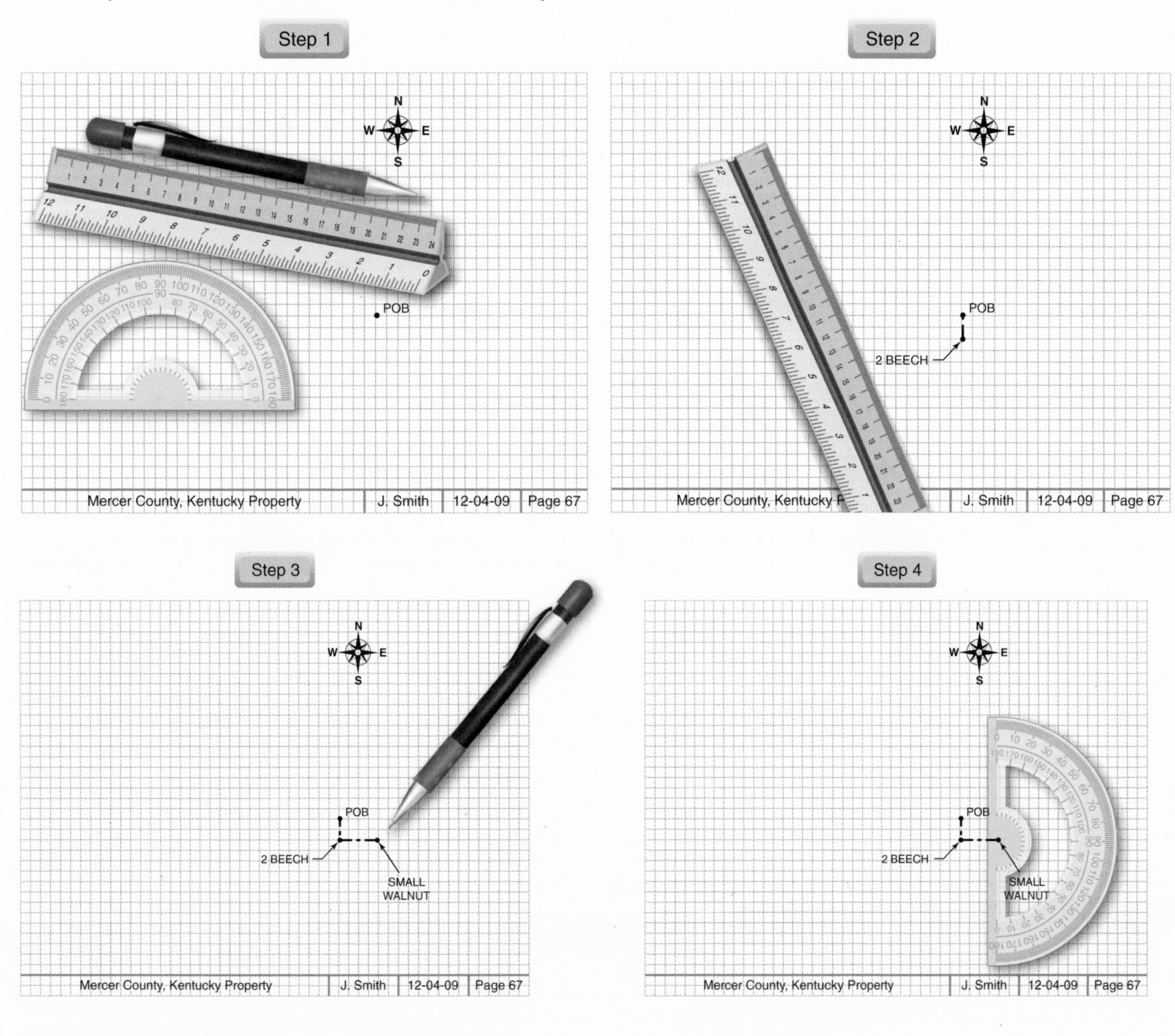

Figure 5-9: (*continued*)

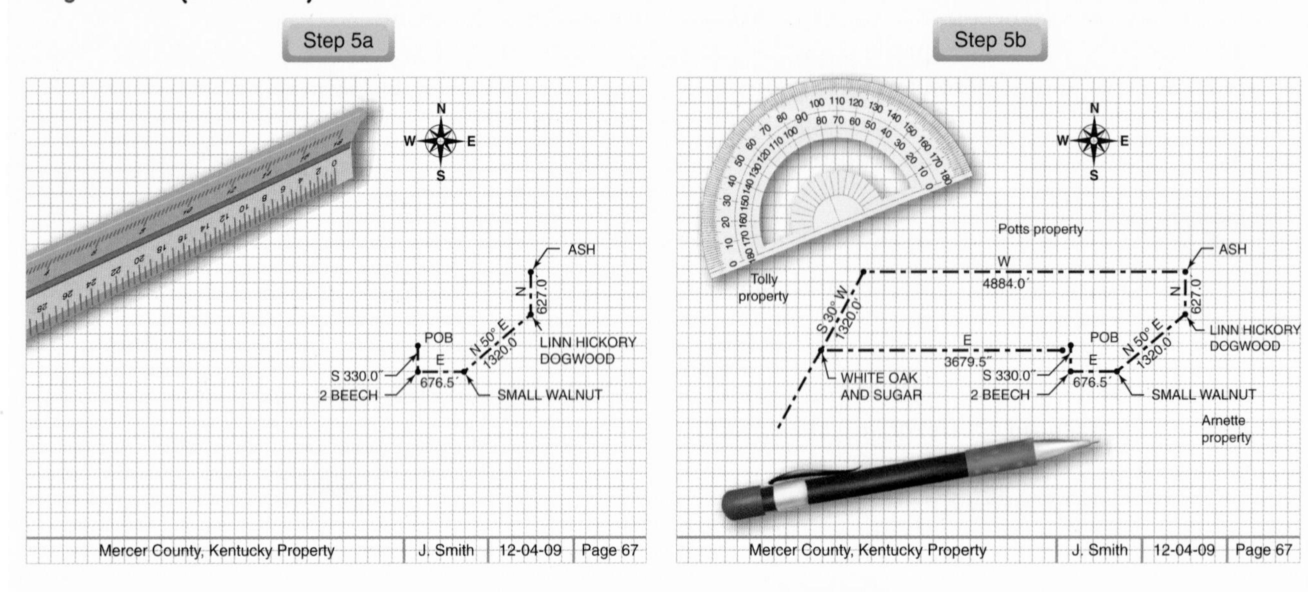

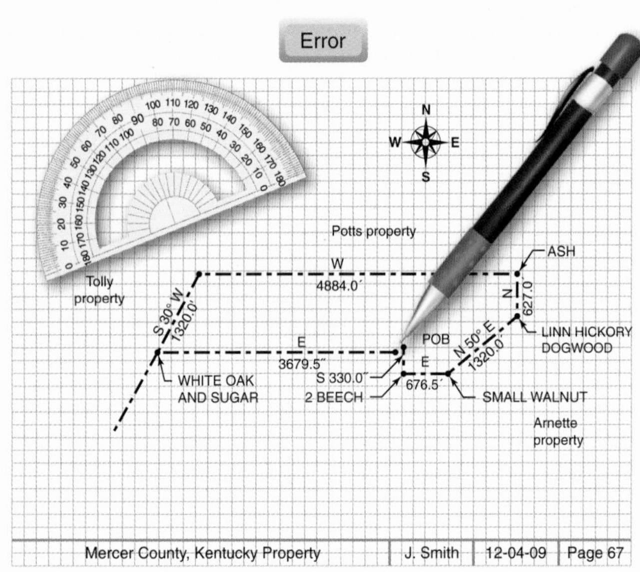

Step 2: *"thence up the creek S 20 poles to 2 beech"*

Sketch the first property line by determining the direction and distance. The first property line follows the creek south for 20 poles. Since a pole is 16.5 ft the length of the property line can be converted to feet.

$$Distance = 20\ poles \cdot \frac{16.5\ \text{ft}}{1\ pole} = 330\ \text{ft}$$

Using an appropriate scale, draw a line 330 ft long due south from the POB and label the end of the line (property corner) "2 BEECH."

Step 3: *"thence east 41 poles to a small walnut in Arnett's line"*

Sketch the next property line using the distance and direction beginning from the end point of the previous property line.

$$Distance = 41\ poles \cdot \frac{16.5\ \text{ft}}{1\ pole} = 676.5\ \text{ft}$$

Beginning at the previous end point, sketch a 676.5 ft long line that runs directly east. This will be a horizontal line. Label the endpoint "SMALL WALNUT."

Step 4: *"thence north 50 east 80 poles to a linn hickory dogwood in said line"*

Begin this property line at the small walnut tree. In this case the direction is given by the bearing north 50 degrees east (N 50° E), which means that the property line will run 50 degrees east of due north. To sketch the line, begin by sketching a north-south construction line through the end point of the previous line. Use your protractor to measure an angle of 50 degrees from due north toward the east, in a clockwise direction.

$$Distance = 80\ poles \cdot \frac{16.5\text{ ft}}{1\ pole} = 1320\text{ ft}$$

The line should be 80 poles or 1320 ft long. Label the endpoint "LINN HICKORY DOGWOOD."

Step 5:

Continue sketching property lines in this manner. Although the property line descriptions should bring you back to the POB, don't be surprised if the endpoint of the final property line does not coincide exactly with the POB. Many early surveys were imprecise and did not describe a closed loop.

The same property is represented in an architectural software CAD program in Figure 5-10. Notice that, because the property lines do not complete a closed loop, the final property line bearing and distance (row 8 in the table) do not match the legal description. The final property line was automatically created in the program to close the loop.

Figure 5-10: A property described by metes and bounds can be created in CAD software such as Autodesk Revit using a table of property lines. Notice that an extra property line has been added. This line was necessary to close the loop because the property description in the Mercer County Kentucky deed book resulted in a gap between the endpoint of the last property line and the POB.

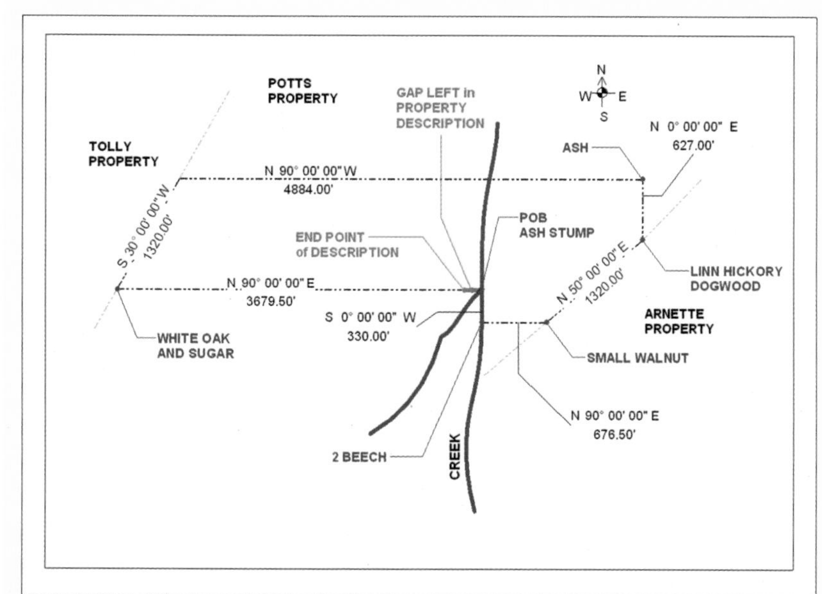

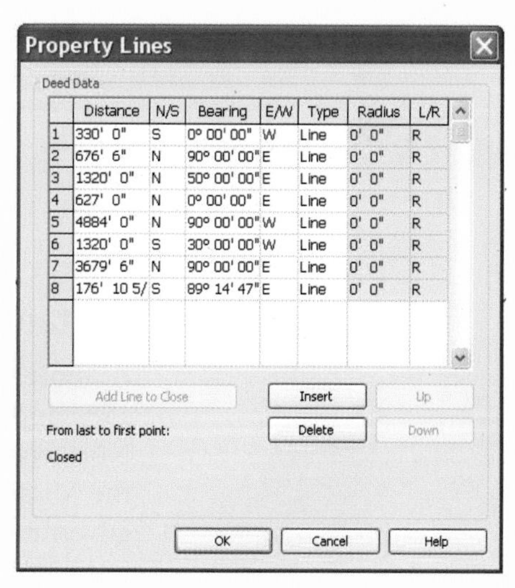

	Distance	N/S	Bearing	E/W	Type	Radius	L/R
1	330' 0"	S	0° 00' 00"	W	Line	0' 0"	R
2	676' 6"	N	90° 00' 00"	E	Line	0' 0"	R
3	1320' 0"	N	50° 00' 00"	E	Line	0' 0"	R
4	627' 0"	N	0° 00' 00"	E	Line	0' 0"	R
5	4884' 0"	N	90° 00' 00"	W	Line	0' 0"	R
6	1320' 0"	S	30° 00' 00"	W	Line	0' 0"	R
7	3679' 6"	N	90° 00' 00"	E	Line	0' 0"	R
8	176' 10 5/	S	89° 14' 47"	E	Line	0' 0"	R

The original 13 colonies as well as a few other areas of the (now) United States used metes and bounds legal descriptions to convey land. However, the land described under the metes and bounds system resulted in tracts of varying shapes and sizes that were described independently of each other and not connected to any universally accepted system. As you might imagine, this "...system led to many overlapping claims, boundary disputes, and clouded titles which the courts were swamped with" (White, 1983). When America won her independence from Britain, it became apparent that a more uniform system of land description was needed to precisely define tracts that would be sold to settlers and land speculators in areas that had not yet been divided.

Figure 5-11: Partial copy of a Land Patent issued to John Smith in 1891 containing a rectangular survey system legal description. Can you locate the property?

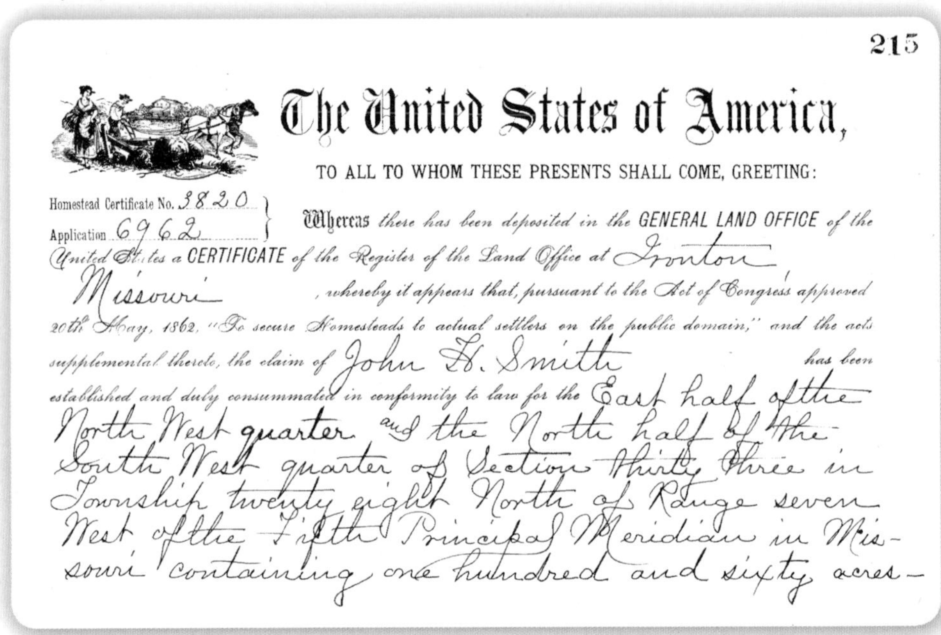

215

The United States of America,

TO ALL TO WHOM THESE PRESENTS SHALL COME, GREETING:

Homestead Certificate No. 3820
Application 6962

Whereas there has been deposited in the GENERAL LAND OFFICE of the United States a CERTIFICATE of the Register of the Land Office at Ironton Missouri, whereby it appears that, pursuant to the Act of Congress approved 20th May, 1862, "To secure Homesteads to actual settlers on the public domain," and the acts supplemental thereto, the claim of John H. Smith has been established and duly consummated in conformity to law for the East half of the North West quarter and the North half of the South West quarter of Section thirty three in Township twenty eight North of Range seven West of the Fifth Principal Meridian in Missouri containing one hundred and sixty acres—

Courtesy U.S. Department of the Interior, Bureau of Land Management.

Rectangular Survey System

A more uniform rectangular system was suggested by Thomas Jefferson in 1784 and adopted by the United States in 1805. The system was first used in the Northwest Territories (west of Pennsylvania, north of the Ohio River and east of the Mississippi River). The revenue from the sale of these lands helped the new government pay the tremendous war debt it had accumulated during the Revolutionary War.

The United States originally used a document called a *land patent* to transfer ownership of land. Figure 5-11 provides a partial copy of an 1891 land patent which conveyed 160 acres of land to John Smith. Today the rectangular survey system is the most widely used system in the United States. The original 13 colonies and areas of the country where land had been conveyed prior to the adoption of the new system still use metes and bounds.

Before you can locate a tract of land described by the rectangular survey system, you need a few definitions. The rectangular survey system is based on 6 mi by 6 mi tracts of land called **townships.** Townships are described with respect to

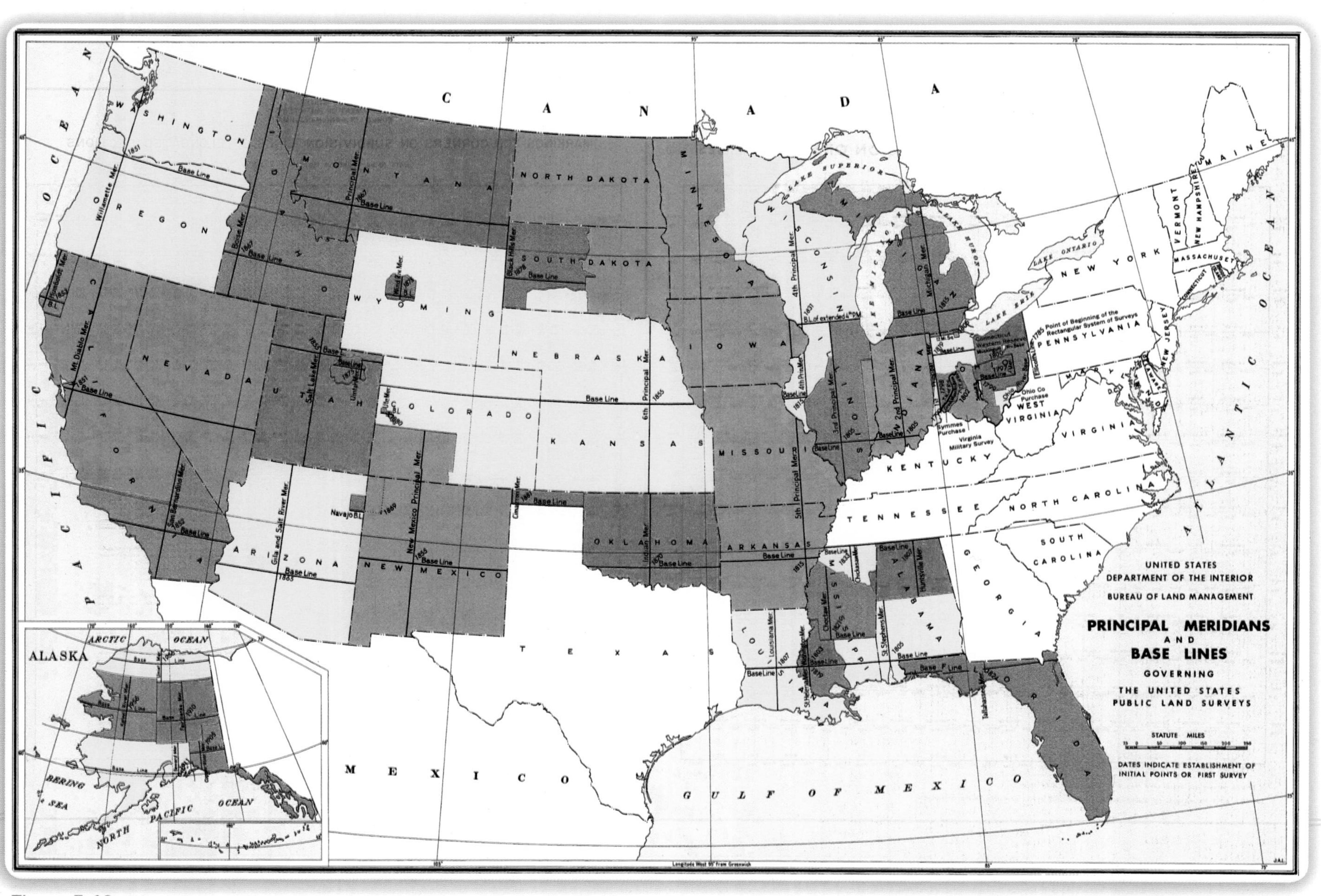

Figure 5-12: Principle meridians and baselines of the *great* land surveys.

Courtesy U. S. Department of the Interior, Bureau of Land Management.

Figure 5-13: In the Rectangular Survey System (or Public Land Survey System), Townships are divided into 36 Sections which can then be divided into fractions of sections. The shaded parcel is the north half of the southwest quarter of Section 14 of Township 2 South, Range 3 West.

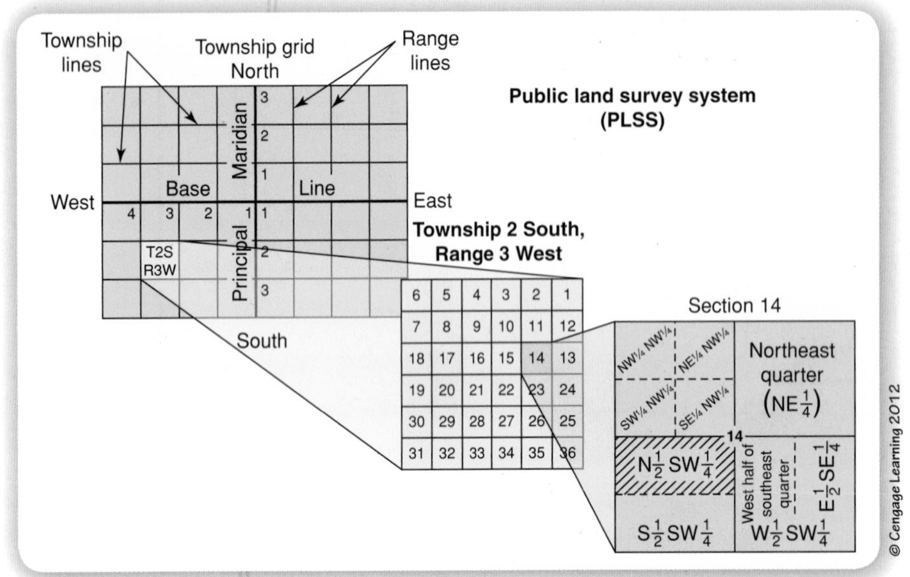

a north-south survey line called a **principal meridian** and a corresponding east-west survey line called a **baseline**. There are 31 pairs of these lines in the United States which are located on the map in Figure 5-12. The intersection of each principal meridian and baseline is the point of beginning for a great land survey covering a specific area of the country and is typically marked with a brass monument.

Townships are laid out from the point of origin in a grid pattern. The gridlines running north-south are called **township lines,** and the gridlines running east-west are referred to as **range lines**. Township numbers are assigned based on the location of the townships in the grid similar to cartesian coordinates in mathematics. For example, the township described as Township 2 South, Range 3 West, abbreviated T. 2 S., R. 3 W., is located two townships south and three ranges west of the POB as shown in Figure 5-13.

Each township is divided into 36 equal sections that each measure 1 sq mi and contains 640 ac. The sections are numbered consecutively in a surpentine pattern beginning with the most northeastern section in a township (see Figure 5-13).

Each section is further subdivided into halves or quarters, which can then, in turn, be subdivided further into halves or quarters, and so on. Figure 5-13 shows subdivisions of Section 14. The shaded portion of Section 14 can be described as the N ½, SW ¼, Sec. 14, T. 2 S., R. 3 W. The complete legal description would also refer to the specific principle meridian on which the description is based.

To determine the area of a tract described using the rectangular survey system, begin with the total section area of 640 ac. You can then find the area of each fraction of a section by multiplying the fraction by 640 ac. For fractions of fractions, continue to multiply by each fraction applied. Therefore, the area of the parcel identified in Figure 5-13 is 80 ac (= 640 ac × ½ × ¼).

Fun Fact

John Strickland has undertaken a project to photograph all of the Principal Meridians at their corresponding Base Lines. He has recorded his travels and pictures of the initial points on his Principle Meridian Project website at http://www.pmproject.org/ (see Figure 5-14). You can participate in this project by submitting your photographs of an initial point and a story of your visit.

Figure 5-14: John Shankland has documented initial points of several great land surveys on his Principle Meridian Project website. The webpage shown displays images taken at the initial point of the Fifth Principle Meridian survey in Louisiana. The initial point is marked by a monument that sits in a swamp in what is now The Louisiana Purchase Historic State Park. This principle meridian was established in 1815 to facilitate the survey of the Louisiana Purchase in order to convey land to men who had fought in the War of 1812.

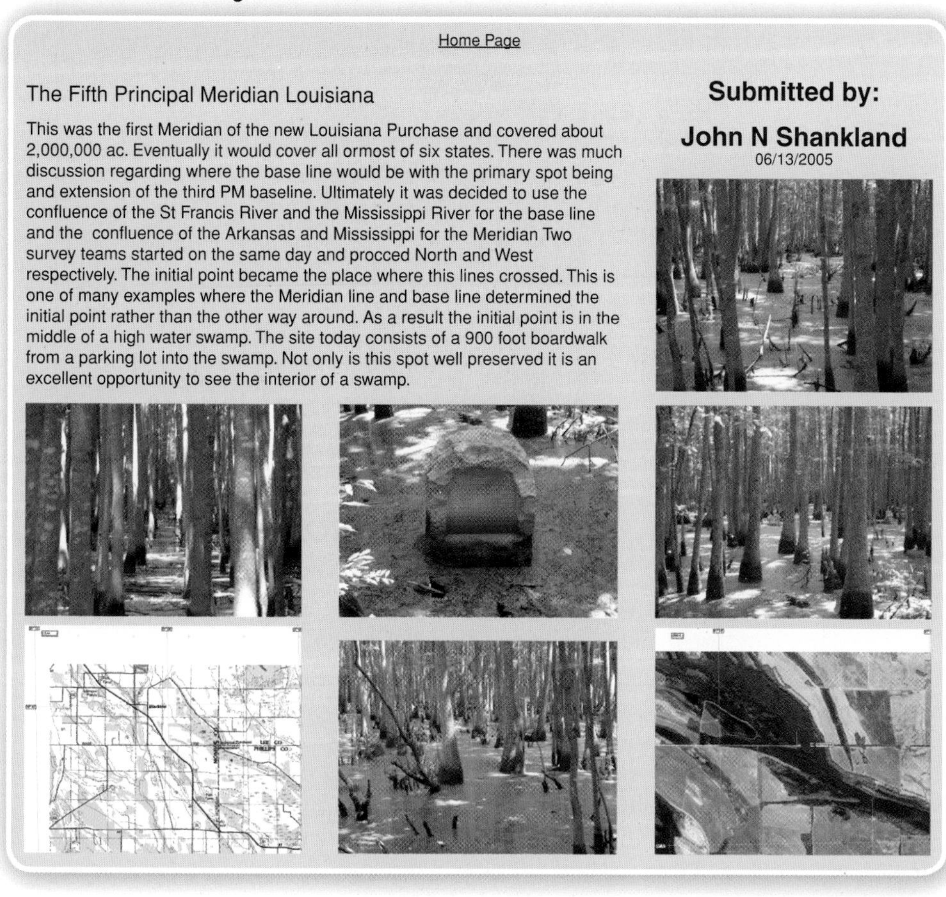

Home Page
The Fifth Principal Meridian Louisiana
This was the first Meridian of the new Louisiana Purchase and covered about 2,000,000 ac. Eventually it would cover all ormost of six states. There was much discussion regarding where the base line would be with the primary spot being and extension of the third PM baseline. Ultimately it was decided to use the confluence of the St Francis River and the Mississippi River for the base line and the confluence of the Arkansas and Mississippi for the Meridian Two survey teams started on the same day and procced North and West respectively. The initial point became the place where this lines crossed. This is one of many examples where the Meridian line and base line determined the initial point rather than the other way around. As a result the initial point is in the middle of a high water swamp. The site today consists of a 900 foot boardwalk from a parking lot into the swamp. Not only is this spot well preserved it is an excellent opportunity to see the interior of a swamp.
Submitted by:
John N Shankland
06/13/2005

Courtesy of John N. Shankland.

Surveyors specify locations and describe spatial relationships using various representational systems, including metes and bounds descriptions and the rectangular survey system.

Lot and Block

Eventually it became necessary to devise a system of land description that was not restricted to the rectangular parcels that resulted from the rectangular system. The **lot and block** system provides for the subdivision of a tract of land into lots of various shapes and sizes thereby inspiring the term subdivision. Today we refer to a neighborhood created by breaking up property into smaller lots as a **subdivision.** This system can be used to subdivide a parcel described by metes and bounds or a rectagular survey. The large tract is given a name, divided into blocks consisting of a series of individual lots or building sites. The legal description is then recorded as the lot number and block number of the named

Figure 5-15: Partial plat showing lot and block designations for property in "The Ponderosa" neighborhood. The legal description for the shaded lot is "Lot 34, Block R, Ponderosa Subdivision, County of Charleston, South Carolina."

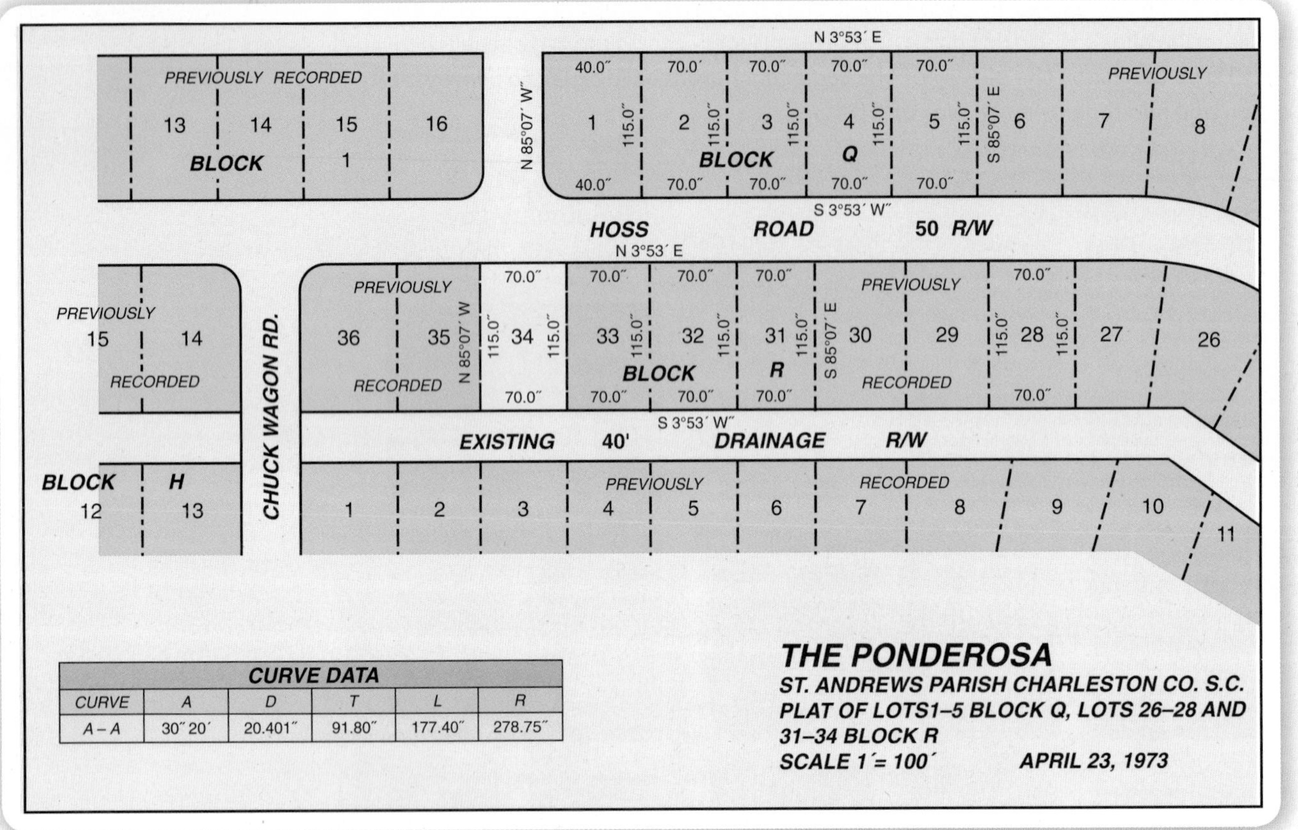

CURVE DATA					
CURVE	A	D	T	L	R
A – A	30″ 20′	20.401″	91.80″	177.40″	278.75″

Plat courtesy of Charleston County, South Carolina Office of Register Mesne Conveyance.

subdivision. Figure 5-15 shows a portion of plat recorded in Charleston County, South Carolina. The legal description in the deed corresponding to the shaded lot reads:

> All that piece, parcel or lot of land, situate, lying and being in the County of Charleston, State of South Carolina and known and designated as Lot 34, Block R, Ponderosa Subdivision, as shown on a plat…dated April 23,1973 and recorded in the RMC Office for Charleston County in Plat Book R at Page 64…

TOPOGRAPHY AND CONTOUR MAPS

Topography: the configuration of a surface including its relief and the position of its natural and cultural features.

Once you discover the specific location of property lines, the topography of the site can be investigated. **Topography**, also called relief, refers to the lay of the land or the rise and fall of the ground surface. How does the topography of the property affect its desirability for development? Certain topographical features, such as bodies of water, mountains, and valleys, lend themselves to desirable views or can provide protection from wind, sun, and noise. On the other hand, the relief can also result in land that drains poorly or too quickly, or a ground surface that does not provide a desirable level surface on which to build a structure. We will discuss how these topological features impact the site layout in the next chapter. During the site discovery process, you will research the important topographic

Case Study

The deed and plat for the CityCenter development are filed with the Clark County, Nevada, Recorder. Because of the complexity of the site, the recorded plat is five pages long (partially shown in Figure 5-16) and includes the legal description and two separate plat maps. Notice the legal description is very long and includes a combination of rectangular survey and metes and bounds descriptions. Because of the complexity of the description, tables are used to provide property line (and curve) data. Also note the many easements and restrictions on the property located on the map on sheet 2 of the document and denoted with numbered tags. The tags correspond to the numbered list on sheets 4 and 5.

Figure 5-16: *Plat document for CityCenter development filed with the Clark County Nevada Recorder. The document includes the property legal description, location and description of easements, and a plot plan of the property.*

Courtesy of Clark County, Nevada Assessor.

CITY CENTER

A COMMERCIAL SUBDIVISION

A MERGER AND RESUBDIVISION OF LOT 1 AND 2 AS SHOWN BY THAT MAP ON FILE 93 OF PARCEL MAPS, PAGE 27 AND PARCEL 2 AS SHOWN BY THAT MAP ON FILE 113 OF PARCEL MAPS, PAGE 88 IN THE CLARK COUNTY RECORDER'S OFFICE, CLARK COUNTY, NEVADA, TOGETHER WITH A PORTION OF THE EAST HALF (E 1/2) OF SECTION 20, TOWNSHIP 21 SOUTH, RANGE 61 EAST, M.D.M., CLARK COUNTY, NEVADA.

LEGAL DESCRIPTION

A MERGER AND RESUBDIVISION OF LOT 1 AND 2 AS SHOWN BY THAT MAP ON FILE 93 OF PARCEL MAPS, PAGE 27 AND PARCEL 2 AS SHOWN BY THAT MAP ON FILE 113 OF PARCEL MAPS, PAGE 88 IN THE CLARK COUNTY RECORDER'S OFFICE, CLARK COUNTY, NEVADA, LYING WITHIN THE EAST HALF (E 1/2) OF SECTION 20, TOWNSHIP 21 SOUTH, RANGE 61 EAST, M.D.M., CLARK COUNTY, NEVADA, TOGETHER WITH A PORTION OF THE NORTHEAST QUARTER (NE 1/4) OF THE SOUTHEAST QUARTER (SE 1/4) OF SECTION 20, TOWNSHIP 21 SOUTH, RANGE 61 EAST, M.D.M., CLARK COUNTY, NEVADA, DESCRIBED AS FOLLOWS:

COMMENCING AT THE NORTHWEST CORNER OF THE SOUTHEAST QUARTER (SE 1/4) OF SAID SECTION 20;
THENCE ALONG THE NORTH LINE THEREOF, SOUTH 89°00'39" EAST, 295.96 FEET TO A POINT ON THE EASTERLY RIGHT-OF-WAY LINE OF INTERSTATE NO. 15 AS DEDICATED BY THAT CERTAIN "FINAL ORDER OF CONDEMNATION" RECORDED IN BOOK 719 OF OFFICIAL RECORDS AS INSTRUMENT NO. 578203 IN THE CLARK COUNTY RECORDER'S OFFICE, CLARK COUNTY, NEVADA;
THENCE ALONG SAID RIGHT-OF-WAY LINE, SOUTH 00°12'57" WEST, 40.91 FEET TO THE **POINT OF BEGINNING**;
THENCE DEPARTING SAID RIGHT-OF-WAY LINE, AND ALONG THE SOUTHERLY RIGHT-OF-WAY LINE OF HARMON AVENUE AND THE WESTERLY PROLONGATION THEREOF AS DEDICATED BY THOSE CERTAIN DOCUMENTS RECORDED IN BOOK 20021113 OF OFFICIAL RECORDS AS INSTRUMENT NO. 00876 AND INSTRUMENT NO. 00877 IN SAID COUNTY RECORDER'S OFFICE;
THENCE ALONG SAID SOUTHERLY LINE, SOUTH 89°00'29" EAST, 272.77 FEET TO THE BEGINNING OF A CURVE HAVING A RADIUS OF 816.00 FEET;
THENCE CURVING TO THE LEFT ALONG THE ARC OF SAID CURVE, CONCAVE NORTHWESTERLY, THROUGH A CENTRAL ANGLE OF 43°29'29", AN ARC LENGTH OF 619.40 FEET TO A POINT THROUGH WHICH A RADIAL LINE BEARS SOUTH 42°29'58" EAST;
THENCE SOUTH 42°10'00" EAST, 5.01 FEET TO THE BEGINNING OF A CURVE HAVING A RADIUS OF 757.00 FEET;
THENCE CURVING TO THE RIGHT ALONG THE ARC OF SAID CURVE, CONCAVE SOUTHEASTERLY, THROUGH A CENTRAL ANGLE OF 43°22'37", AN ARC LENGTH OF 573.10 FEET;
THENCE SOUTH 88°47'23" EAST, 204.53 FEET;
THENCE SOUTH 86°15'50" EAST, 300.71 FEET TO THE BEGINNING OF A CURVE HAVING A RADIUS OF 193.50 FEET;
THENCE CURVING TO THE RIGHT ALONG THE ARC OF SAID CURVE, CONCAVE SOUTHWESTERLY, THROUGH A CENTRAL ANGLE OF 39°56'11", AN ARC LENGTH OF 134.87 FEET;
THENCE SOUTH 46°19'39" EAST, 35.21 FEET TO A POINT ON THE WESTERLY RIGHT-OF-WAY LINE OF LAS VEGAS BOULEVARD (FORMERLY OLD HIGHWAY 91) AS IT NOW EXISTS;
THENCE ALONG SAID RIGHT-OF-WAY LINE, SOUTH 00°02'31" EAST, 294.11 FEET;
THENCE SOUTH 00°02'00" EAST, 801.69 FEET;
THENCE DEPARTING SAID RIGHT-OF-WAY LINE, SOUTH 89°58'34" WEST, 200.00 FEET;
THENCE SOUTH 00°02'00" EAST, 179.88 FEET TO A POINT ON THE BOUNDARY OF PARCEL 1 AS SHOWN BY THAT CERTAIN MAP ON FILE IN FILE 113 OF PARCEL MAPS, PAGE 88 IN SAID COUNTY RECORDER'S OFFICE;
THENCE SOUTH 89°58'00" WEST, 399.91 FEET;
THENCE NORTH 00°02'00" WEST, 385.00 FEET;
THENCE SOUTH 89°58'00" WEST, 427.61 FEET;
THENCE SOUTH 00°02'25" WEST, 354.64 FEET;
THENCE NORTH 89°57'35" WEST, 122.61 FEET;
THENCE SOUTH 00°10'36" EAST, 564.88 FEET;
THENCE SOUTH 89°34'07" WEST, 403.08 FEET;
THENCE SOUTH 44°58'00" WEST, 50.68 FEET;
THENCE SOUTH 89°58'00" WEST, 341.40 FEET TO THE BEGINNING OF A NON-TANGENT CURVE HAVING A RADIUS OF 1060.00 FEET, A RADIAL LINE TO SAID POINT BEARS NORTH 81°10'52" EAST;
THENCE DEPARTING SAID BOUNDARY OF PARCEL 1 AND CURVING TO THE RIGHT ALONG THE ARC OF SAID CURVE, CONCAVE NORTHEASTERLY, THROUGH A CENTRAL ANGLE OF 09°02'09", AN ARC LENGTH OF 167.17 FEET TO A POINT THROUGH WHICH A RADIAL LINE BEARS NORTH 89°46'59" WEST, SAID LAST DESCRIBED COURSE BEING ALONG THE EASTERLY LINE OF THOSE LANDS AS CONVEYED BY THAT CERTAIN DOCUMENT RECORDED IN BOOK 20000131 OF OFFICIAL RECORDS AS INSTRUMENT NO. 00727 IN SAID COUNTY RECORDER'S OFFICE;
THENCE ALONG THE NORTHERLY LINE OF SAID LANDS, NORTH 89°46'59" WEST, 81.00 FEET TO A POINT ON THE AFOREMENTIONED EASTERLY RIGHT-OF-WAY LINE OF INTERSTATE NO. 15;
THENCE ALONG SAID RIGHT-OF-WAY LINE, NORTH 00°12'57" EAST, 1382.76 FEET TO THE **POINT OF BEGINNING**.

CONTAINING: 61.97 ACRES OF LAND.

Case Study

(continued)

Figure 5-16: *(continued)*

RESERVATIONS AND EASEMENTS

THIS SURVEY DOES NOT CONSTITUTE A TITLE SEARCH BY LOCHSA SURVEYING. TO DETERMINE OWNERSHIP OR EASEMENTS OF RECORD. FOR ALL INFORMATION REGARDING EASEMENTS, RIGHTS-OF-WAY OR TITLE OF RECORD, LOCHSA SURVEYING RELIED ON NEVADA TITLE COMPANY'S COMMITMENT FOR TITLE INSURANCE NO. 04-12-0244-DTL DATED APRIL 7, 2005 (1ST AMENDMENT).

ONLY THOSE EXCEPTIONS WHICH ARE NON-FINANCIAL IN NATURE ARE AS THEY APPEAR IN THE ABOVE REFERENCED TITLE REPORT AND LISTED BELOW.

(27) AN EASEMENT AFFECTING THAT PORTION OF SAID LAND AND FOR THE PURPOSES THEREIN AND INCIDENTAL PURPOSES THERETO, IN FAVOR OF NEVADA POWER COMPANY, FOR ELECTRICAL LINES, RECORDED JUNE 11, 1957, IN BOOK 131 AS DOCUMENT NO. 107719 OF OFFICIAL RECORDS.

[AFFECTS A PORTION OF THE SUBJECT PROPERTY AS SHOWN ON SHEET 2 OF 5.]

(28) AN EASEMENT AFFECTING THAT PORTION OF SAID LAND AND FOR THE PURPOSES THEREIN AND INCIDENTAL PURPOSES THERETO, IN FAVOR OF SOUTHERN NEVADA POWER COMPANY, FOR ELECTRICAL AND COMMUNICATION LINES, RECORDED JUNE 27, 1957, IN BOOK 133 AS DOCUMENT NO. 109240 OF OFFICIAL RECORDS.

A PARTIAL RELINQUISHMENT OF RIGHT OF WAY GRANT, RECORDED JUNE 29, 1998 IN BOOK 980629 AS DOCUMENT NO. 01253 OF OFFICIAL RECORDS.

[AFFECTS A PORTION OF THE SUBJECT PROPERTY AS SHOWN ON SHEET 2 OF 5.]

(31) AN EASEMENT AFFECTING THAT PORTION OF SAID LAND AND FOR THE PURPOSES THEREIN AND INCIDENTAL PURPOSES THERETO, IN FAVOR OF NEVADA POWER COMPANY, FOR ELECTRICAL LINES, RECORDED APRIL 17, 1961, IN BOOK 293 AS DOCUMENT NO. 237071 OF OFFICIAL RECORDS.

[AFFECTS A PORTION OF THE SUBJECT PROPERTY AS SHOWN ON SHEET 2 OF 5.]

(32) AN EASEMENT AFFECTING THAT PORTION OF SAID LAND AND FOR THE PURPOSES THEREIN AND INCIDENTAL PURPOSES THERETO, IN FAVOR OF NEVADA POWER COMPANY, FOR ELECTRICAL LINES, RECORDED APRIL 18, 1961, IN BOOK 293 AS DOCUMENT NO. 237215 OF OFFICIAL RECORDS.

[AFFECTS A PORTION OF THE SUBJECT PROPERTY AS SHOWN ON SHEET 2 OF 5.]

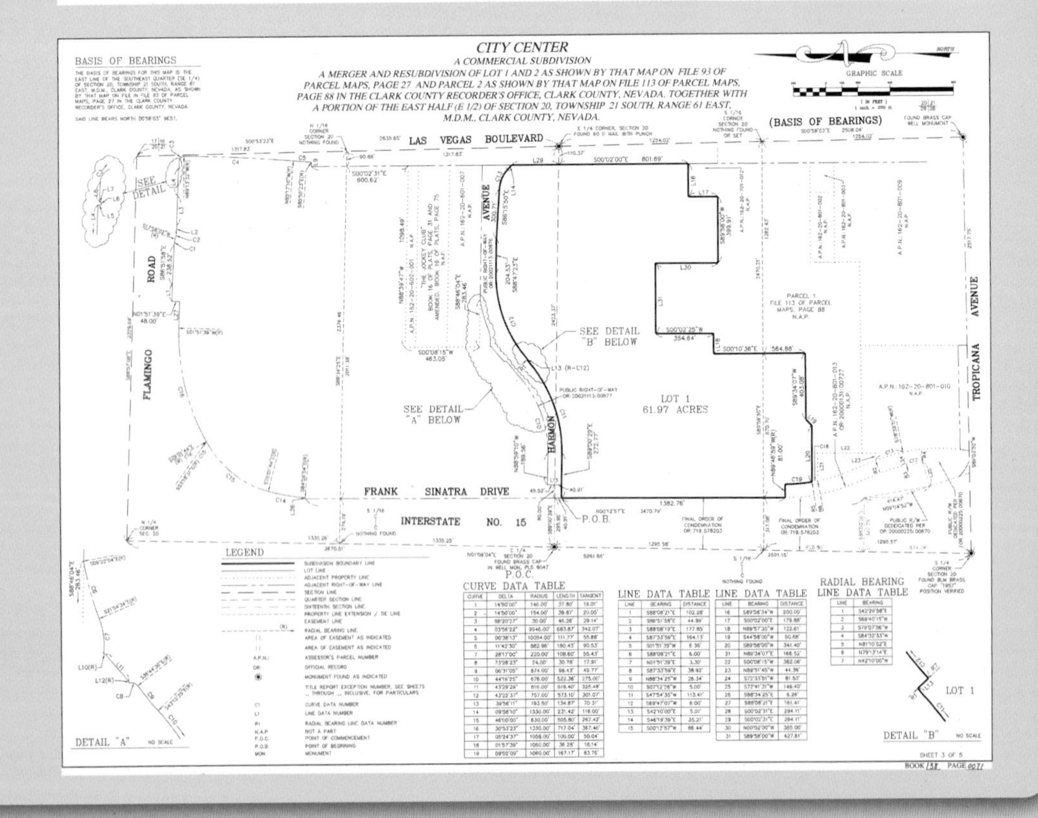

features, including the slope of the ground, and consider the soil characteristics of the site to help determine whether the property is a suitable building site for your proposed development.

You can get information about the site topography using a contour map. A contour map shows surface relief with curved **contour lines** which indicate points of equal elevation. To understand contour lines, look at Figure 5-17. The contour map shows an island with several contour lines. The thicker contour lines, called index contours, are labeled with a number indicating the **elevation,** or height, of the line above mean sea level. Most contour maps in the United States use ft as the unit for elevation, as is true in this example. Every point on the index contour marked 50 is at the same elevation of 50 ft above sea level. If you were to hike the island and follow the 50 ft contour line you would walk a level path—neither climbing nor desending. If you were to hike perpendicular to the contour lines, you would change elevation as you moved—traveling downhill when the elevation decreases or climbing uphill when the elevation increases.

Elevation: the height of a point above an adopted datum, such as mean sea level (MSL).

A cross-section of the island more clearly shows the rise and fall of the ground surface. The horizontal lines on the section view correspond to contour lines on the map. Notice that contour lines that are close together indicate a steeper slope and contour lines that are further apart indicate a flatter slope. The slope toward the beach on the east side of the island is much flatter, (indicted by the more widely spaced contour lines) than the slope toward the west side of the island (indicated by the more closely spaced contour lines). The three-dimensional view may help you understand how contour lines show the rise and fall of the land.

Figure 5-17: *Contour map of an island with corresponding section view and 3D view. Contour lines that are closer together indicate a steeper slope, and contour lines that are farther apart indicate a flatter slope.*

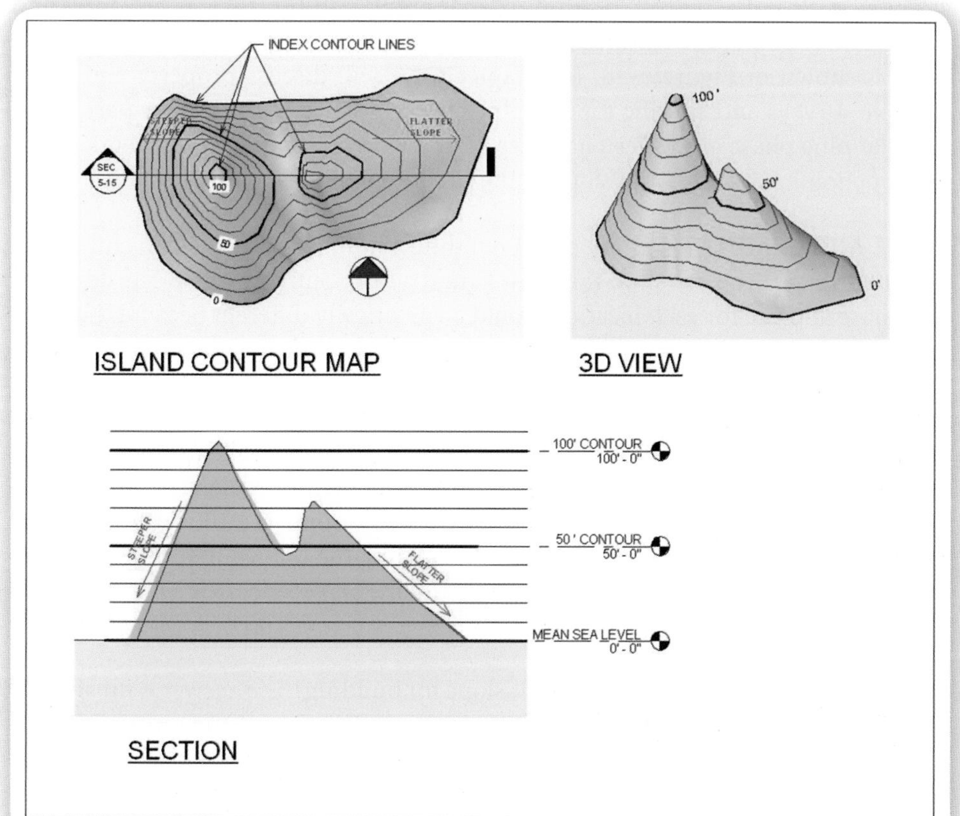

Contour Interval

The change in elevation between adjacent contour lines is called the **contour interval** of the map. The contour interval can be determined by dividing the change in elevation between two index contour lines by the number of *spaces* (not lines) between the two index contour lines.

$$\textit{Contour interval} = \frac{\textit{change in elevation between index contour lines}}{\textit{number of contour spaces between the index contour lines}}$$

Example: Calculating a Contour Interval

In order to find the contour interval of the island contour map in Figure 5-17, we will use the 100 ft and 50 ft index contour lines.

$$\begin{aligned}\textit{Contour interval} &= \frac{\textit{change in elevation between the contour lines}}{\textit{number of contour spaces between the contour lines}}\\ &= \frac{100\text{ ft} - 50\text{ ft}}{5\ \textit{spaces}}\\ &= 10\text{ ft}\ \textit{per space}\end{aligned}$$

Therefore the contour interval for the island contour map is 10 ft.

You can get preliminary topographic information from United States Geological Survey (USGS) contour maps. In addition to contour lines, the USGS topographic maps show general locations of manmade features, such as buildings and roads, and topographic features. A portion of a USGS contour map of Hana, Hawaii, is shown in Figure 5-18. In Chapter 6 we will discuss property and topographic surveys and specific topographic characteristics of the site, as part of the site planning phase of the design process. For now, you should get a general idea of the topographic features and slope of the site for viability analysis.

SLOPE AND GRADIENT Think about building a home on the side of a cliff in San Franciso or near the marsh in the "Low Country" of South Carolina. The design of the house and site for each location would be drastically different because the slope of the sites are very different. We will represent the steepness of the slope, or rate of change of the elevation by a **slope gradient,** the difference in elevation between two points, given as a percentage of the distance between those points. If you have ever driven through mountains, you may have noticed road signs, like the one shown in Figure 5-19, that show slope gradient. A 12 percent slope gradient (often referred to simply as slope) means that the ground rises 12 ft for every 100 ft of horizontal distance. You can calculate the slope gradient using the following formula:

$$\textit{Slope gradient} = \frac{\textit{change in elevation}}{\textit{length}} \cdot 100\%$$

So, is a slope of 3.8 percent a "good" slope for building? The answer is most likely yes. Slopes between 1 percent and 5 percent typically provide the most economical topographic conditions for commercial facilities because they accommodate large structures without requiring special designs. In addition, these gentle slopes provide good drainage of the site. Figure 5-20 gives general site considerations

Figure 5-18: **Example of a USGS 1:24000 quadrangle map – Hana Quadrangle, Hana, Hawaii.**

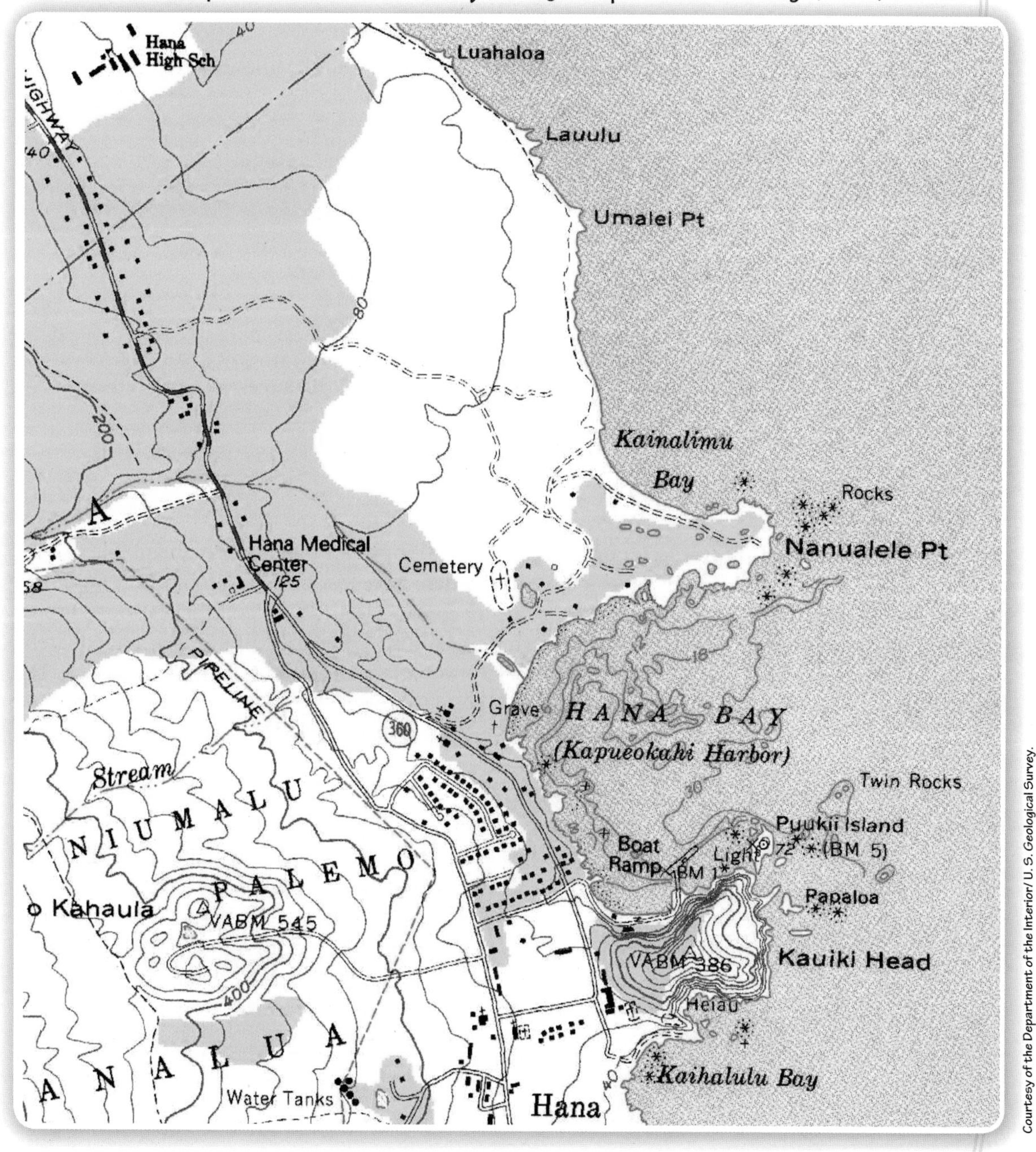

Courtesy of the Department of the Interior/ U. S. Geological Survey.

for various slopes. As illustrated, sites with slopes of less than 1 percent may pose drainage problems and require more expensive drainage systems. Slopes between 5 percent and 10 percent may require **regrading,** or moving soil around the site, and can result in large post development stormwater runoff and erosion. Slopes greater than 10 percent require significant regrading or design accomodations and should generally be avoided, if possible.

Engineers and architects use mathematics to represent and analyze rate of change in various contexts including the rate of change of the elevation of the surface of the earth expressed as slope or slope gradient.

Example: Calculating Slope Gradient

Find the slope gradient for a property on which the elevation rises 11.5 ft across the 300 ft width of the lot.

$$\begin{aligned} \textit{Percent gradient} &= \frac{\textit{change in elevation}}{\textit{length}} \cdot 100\% \\ &= \frac{11.5 \text{ ft}}{300 \text{ ft}} \cdot 100\% \\ &= 0.038 \cdot 100\% \\ &= 3.8\% \end{aligned}$$

Therefore, the slope gradient for this lot is 3.8 percent. In other words, on average, the ground rises 3.8 ft per 100 ft of horizontal distance.

Figure 5-19: This road sign indicates a steep 12 percent slope gradient.

Soils

In addition to the topography, soil conditions can make a site unsuitable for development. Many developers and builders have lost time and money because they encountered undesirable soils during construction. Weak soil that cannot support the weight of a building, soils that expand when wet or do not drain adequately, and a high water table can all adversely affect construction time and costs. The Natural Resources Conservation Service (NRCS) provides an abundance of online soils data via the Web Soil Survey (WSS) which can be accessed at http://websoilsurvey.nrcs.usda.gov/ (see Figure 5-21).

Although an in-depth soils study may be necessary later in the design process, at this point researching the general soil conditions may be adequate. The WSS rating for *Suitabilities and Limitations for Use* will indicate whether the soil conditions are acceptable or pose potential problems for development. You can choose to rate your selected area of interest for dwellings (residential development), landscaping, roads, or commercial development. The rating takes into consideration many soil characteristics, such as slope, load bearing capacity, depth to a water table, and compressibility of the soil—soil characteristics that we will more closely consider when we discuss site development and foundation design. Based

Figure 5-20: The slope of a property can indicate potential site drainage problems or the need for significant regarding. Slopes between 1 percent and 5 percent are typically most economical for commercial development. (Department of the Army, TM-5-803-6, Site Planning and Design, 14 Oct 1994.)

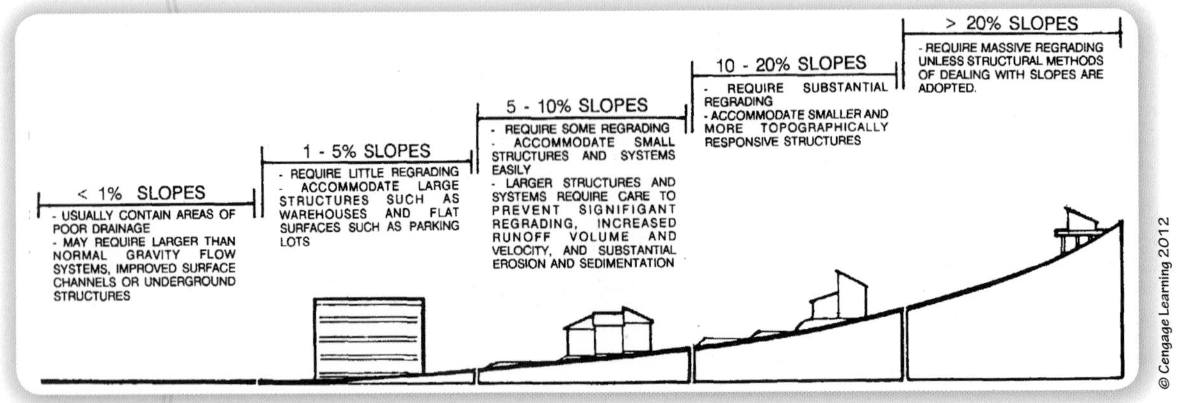

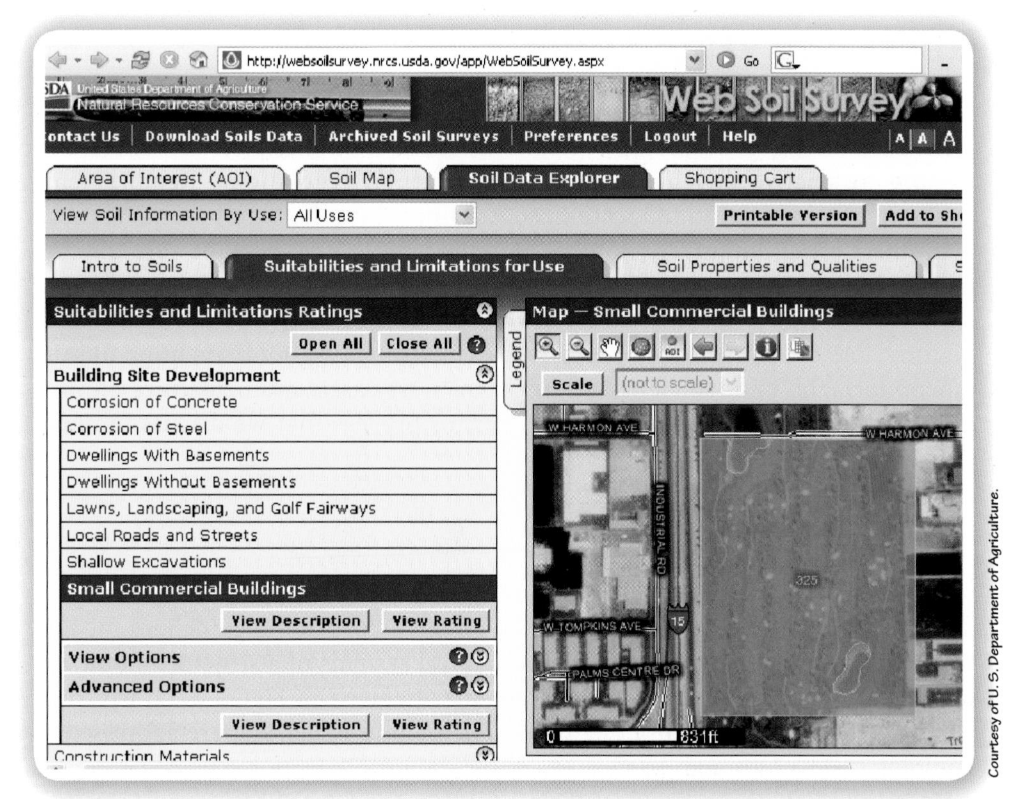

Courtesy of U. S. Department of Agriculture.

Figure 5-21: Screen capture of the National Conservation Service Web Soil Survey map for the vicinity of the CityCenter site. The website provides a vast amount of information pertaining to the soils of the United States including soil classifications, profiles, properties, and suitability for various uses. It is an excellent resource for preliminary soils information.

on the information you gather from the National Cooperative Web Soil Survey (WSS), you should be able to determine whether the soil is acceptable in its natural state or can be improved with special designs. In some cases you may decide that more investigation is necessary or that the site cannot be economically used for your development.

Your Turn

Get the Dirt on Your Site

Investigate the topography and soil conditions of an unimproved site near you to determine if it is suitable for commercial development—say a 3000 sq ft veterinary clinic with adjacent parking lot.

Using the Map Locater and Downloader, access a USGS contour map for your site at http://store.usgs.gov/. Note the contour interval of the map, determine the approximate elevation at your location, and identify topographic features that could affect development of this site.

Using the National Cooperative Web Soil Survey (WSS) at http://websoilsurvey.nrcs.usda.gov/ locate your site on the national map. Create your *Area of Interest* using one of the AOI buttons at the top of the map. Then choose the Soil Reports Tab. Review reports for *Map Unit Description and Building Site Development–Dwellings and Small Commercial Buildings*.

In your notebook, discuss the suitability of the site for the planned veterinary clinic.

RULES AND REGULATIONS

Your development will be controled by many rules and **regulations**, which can include building codes, state regulations, local or municipal ordinances and zoning, restrictive covenents, and environmental regulations. Requirements will differ from location to location, so it is important that you become familiar with these requirements for your proposed site so that you can be sure that the project you envision is legally allowable. Compliance with these rules and regulations can help ensure the success of your project. Failure to comply may result in fines, loss of licensure, or other adverse consequences, not to mention unsafe, uncomfortable, or inefficient buildings. In most cases, municipalities enforce these rules and regulations through a permit process that we discuss in detail in Chapter 6.

Regulation:
a rule adopted by a state regulatory agency to implement, interpret, or make specific the law enforced or administered by it, or to govern its procedure.

Building Codes

As discussed in Chapter 3, building codes are legal requirements designed to protect the safety, health, and welfare of the public by providing guidelines for fire, structural, electrical, plumbing, and mechanical systems of a structure. Some examples of model building code requirements are given in Figure 5-22, which illustrates the wide variety of design and construction issues that are addressed in building codes. Although some areas of the country have not enacted or do not enforce building codes, most structures built in the United States must, by law, comply with some set of minimal building standards.

According to the International Code Council (ICC), most states have adopted some or all of the International Building Codes. You may research your state and local jurisdictional adoptions of the International codes at http://www.iccsafe.org, however, you should check with your local building department to verify the building codes that will apply to your project. In many cases, local jurisdictions will revise or add requirements to the adopted model code. You may obtain copies of the International Building Codes and several of the earlier legacy codes from the ICC.

Point of Interest

The Code of Hammurabi

Building codes have been used for 4000 years to protect the public from poor construction practices and unsafe conditions. One of the first known building codes was The Code of Hammurabi, a code of law dating to the eighteenth century BCE. The code included 282 laws that were written on stone tablets and displayed in Babylon's temple of Marduk. Several of the laws dealt with construction including Law 229, which states: If a builder has built a house for a man and his work is not strong, and if the house he has built falls in and kills the householder, that builder shall be slain.

Figure 5-22: Building codes give requirements for many different aspects of building construction. Although building codes may vary from location to location, this diagram gives examples of building code requirements that may apply to your project. You should research local code requirements for each project.

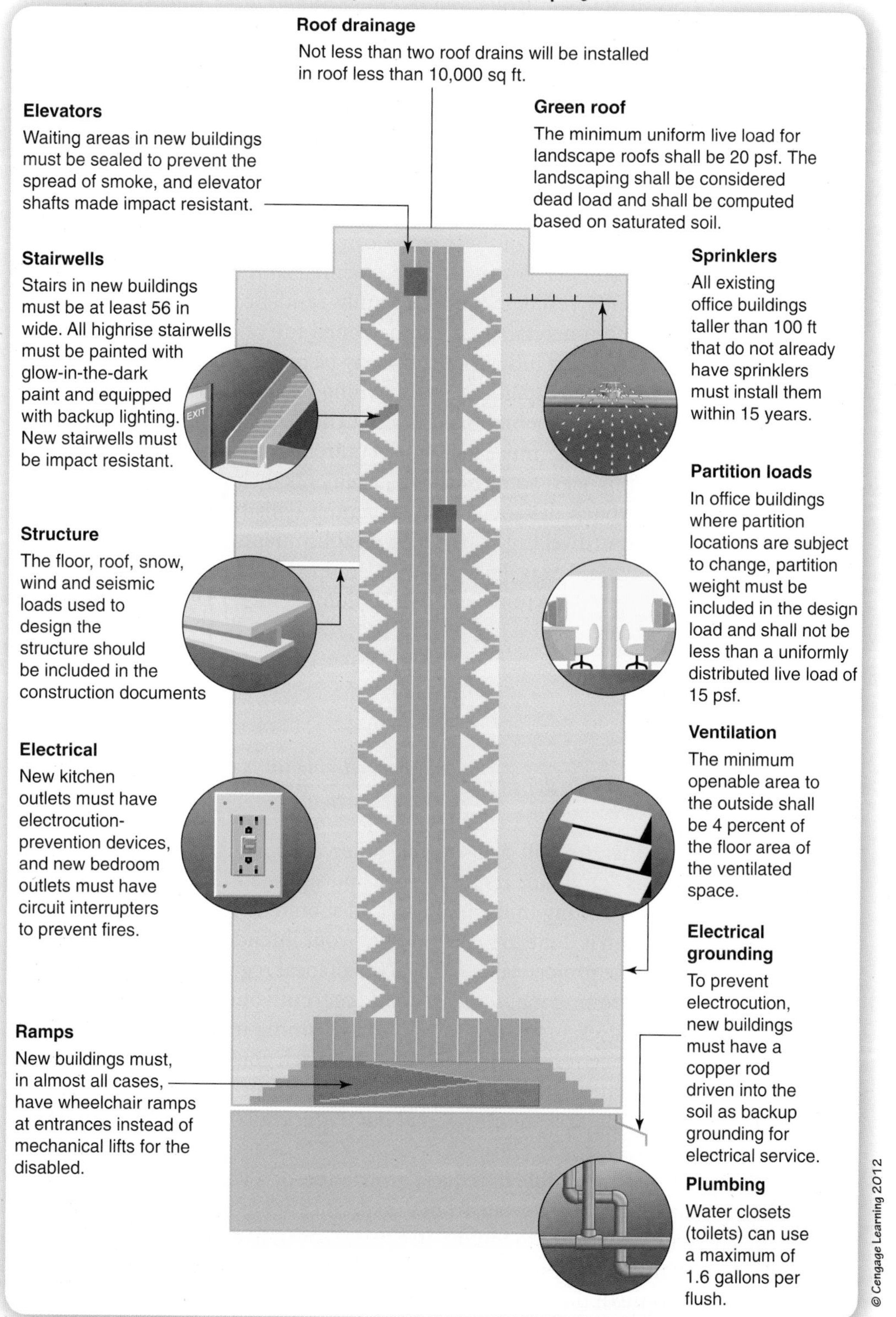

Roof drainage
Not less than two roof drains will be installed in roof less than 10,000 sq ft.
Elevators
Waiting areas in new buildings must be sealed to prevent the spread of smoke, and elevator shafts made impact resistant.
Green roof
The minimum uniform live load for landscape roofs shall be 20 psf. The landscaping shall be considered dead load and shall be computed based on saturated soil.
Stairwells
Stairs in new buildings must be at least 56 in wide. All highrise stairwells must be painted with glow-in-the-dark paint and equipped with backup lighting. New stairwells must be impact resistant.
EXIT
Sprinklers
All existing office buildings taller than 100 ft that do not already have sprinklers must install them within 15 years.
Partition loads
In office buildings where partition locations are subject to change, partition weight must be included in the design load and shall not be less than a uniformly distributed live load of 15 psf.
Structure
The floor, roof, snow, wind and seismic loads used to design the structure should be included in the construction documents
Electrical
New kitchen outlets must have electrocution-prevention devices, and new bedroom outlets must have circuit interrupters to prevent fires.
Ventilation
The minimum openable area to the outside shall be 4 percent of the floor area of the ventilated space.
Electrical grounding
To prevent electrocution, new buildings must have a copper rod driven into the soil as backup grounding for electrical service.
Ramps
New buildings must, in almost all cases, have wheelchair ramps at entrances instead of mechanical lifts for the disabled.
Plumbing
Water closets (toilets) can use a maximum of 1.6 gallons per flush.

4

Local Ordinances and Zoning

Local municipalities (cities, towns, and counties) pass ordinances, or laws, to ensure public safety and convenience and to protect or enhance the environment. These ordinances address a wide range of activities, such as pet waste, loud noises, and speeding within the municipal limits. Specific sections of the ordinances govern development and building construction. In addition, **zoning ordinances** designate "zones" within the municipality and provide rules for land use, type of occupancy, allowed building types, and population densities in the designated zone.

Most of the time a municipality is divided into three broad zones: residential, commercial, and industrial—but other zone types, such as agricultural and open space, can also be used. Each of these zones can be further subdivided to provide more specific requirements. For instance, residential zoning may include separate zones for single-family residences, multi-family residences, and apartment-type housing. Likewise, commercial areas can be zoned for very specific uses, such as schools or retail stores, and industrial areas may be broken down into light industrial and heavy industrial facilities. Other zoning designation may also apply to areas with special environmental conditions such as coastal areas or floodplains. Development in each zone must meet the requirements for the designated zone specified in the ordinances. Some of the requirements include maximum building area and height, setback distances, and maximum density. Density refers to the number of dwellings (dwelling density) or establishments per unit area. For example, the Jacksonville, Texas, Planning and Zoning Ordinances establish a maximum density for Zone D (Multi-Family (Medium Density) Dwelling District) of 10 living units per acre.

Setback: the minimum legal distance from a property line or street where improvements to a site can be built or the minimum distance from the property lines to the front, rear, and sides of a structure.

Dwelling Density

$$\textit{Dwelling density} = \frac{\textit{number of living units on site}}{\textit{area of site}}$$

The local ordinances will also dictate other construction elements, such as parking allocations, landscape requirements, and signage.

Ordinances vary among municipalities, and should be reviewed before a site is selected or any design is done to make sure that your intended use is allowed on the property and that the project can comply with all local regulations.

You can obtain zoning maps and ordinances from your local building department, or, in some cases, online. A portion of the zoning map for a part (Grid 21) of the City of San Diego is shown in Figure 5-23. Each zone designation is subject to different requirements that are given in the San Diego Municipal Code. Both the city zoning maps and the municipal code are available on the City of San Diego website.

In some cases it is possible to request a **variance,** or a waiver, from the governing body that allows you to deviate from the specifics of an ordinance. For instance, you may request a variance to build a structure that is 40 ft tall even though town ordinances restrict building heights to 35 ft. If you are granted the variance, the waiver applies only to the specific situation for which it was requested. Your local building or planning department can provide you with a copy of the local ordinances and information on how the ordinances apply to your project or how to apply for a variance.

Figure 5-23: A portion of the Official Zoning Map, Grid Tile 21, of San Diego, California dated 6/10/2008. According to the San Diego Municipal Code; CC is Commercial-Community (-3-5 indicates high intensity pedestrian orientation), OP is Open Space – Parks (-1-1 allows developed active parks), RM is Residential – Multiple Unit (the numbers correspond to maximum density requirements), and RS is Residential – Single Unit (the numbers correspond to minimum lot square footage requirements). LJPD is La Jolla Planned District and is regulated by the La Jolla Planned District Ordinance.

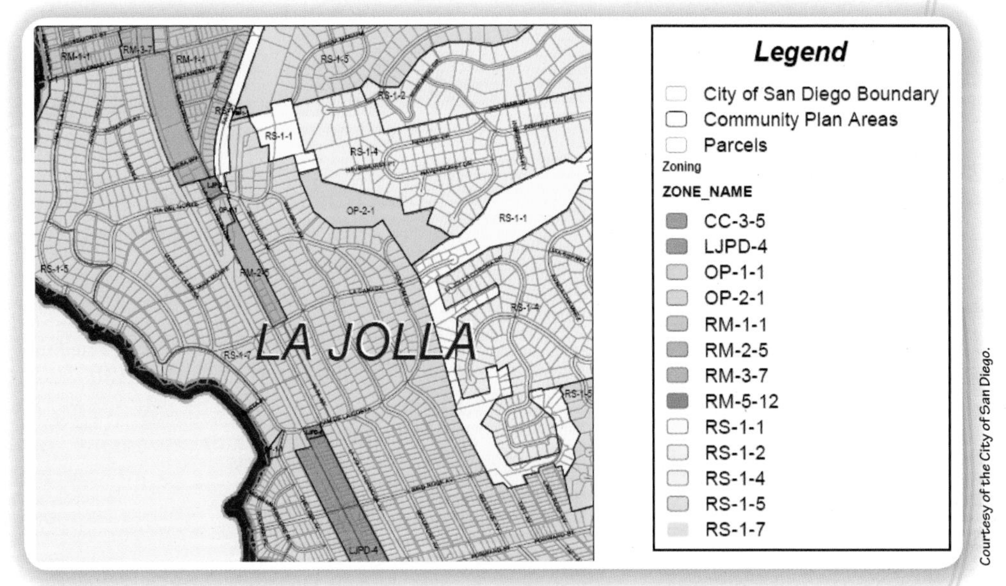

Courtesy of the City of San Diego.

Your Turn
A Code to Build By

The Municipal Code Corporation website at http://www.municode.com/ provides access to many municipal codes. Visit this site and follow the link to the Online Library. Search the municipal code for your city, county, or another local municipality to determine which, if any, model building code has been adopted by the municipality for non-residential construction (for example, the 2003 IBC). In some cases the local code will refer you to a state regulation. You may need to search the Internet to find the state regulation specifying the applicable building code. Then research the zoning ordinances to identify specific zone designations and their definitions.

Restrictive Covenants

In many new planned developments and residential communities the original developer, a neighborhood association, or an association of owners will control the appearance of the area and limit use of the properties with **restrictive covenants.** A restrictive covenant differs from a zoning regulation in that it is a contract between the landowners whose properties are affected by it, rather than an exercise of the governmental police power. These rules can apply to almost anything as long as the controling group agrees to the restriction. If development could potentially affect protected areas such as floodplains or wetlands, the restrictive covenants may require review and approval of government agencies.

Case Study

The CityCenter project in Las Vegas is actually outside the Las Vegas city limits—it is located in an unincorporated area of Clark County, Nevada. Clark County provides public access to its LIS/GIS system at http://gisgate.co.clark.nv.us/openweb/. The system includes zoning layers for several municipalities within the county including Clark County zoning. A screen shot of the zoning in the area of CityCenter and the legend is shown in Figure 5-24. As you can see, the CityCenter property is located within an H-1 zone, which is defined as Limited Resort and Apartment District (Section 30.36.010) in the Clark County Code. The following excerpt from Section 30.40.320 of the code describes the purpose of the H-1 zone:

> **30.40.320 H-1 Limited Resort and Apartment District.**
>
> A. Purpose. The H-1 Limited Resort and Apartment District is established to provide for the development of gaming enterprises, compatible commercial, and mixed commercial and residential uses, and to prohibit the development of incompatible uses that are detrimental to gaming enterprises.

The Clark County Code (also available online via the Clark County, Nevada, website) prescribes the allowable uses and building requirements for the CityCenter project. For example the height of a building in an H-1 zone is limited to 100 ft except that most buildings "over one (1) story or fourteen (14) ft shall be set back from any adjacent single family residential use a distance of three hundred percent (300 percent) of the height of the building or structure," per the code. This means that a building 100 ft tall must be at least 300 ft away from any nearby houses. Another code requirement for development within H-1 is that the density of dwellings cannot exceed 50 units per gross acre.

Figure 5-24: *Zoning for a portion of Clark County Nevada is shown using the county GIS system, accessed July 8, 2008. The location of the CityCenter project has been added. The CityCenter site is zoned H-1 which indicates Limited Resort and Apartment District.*

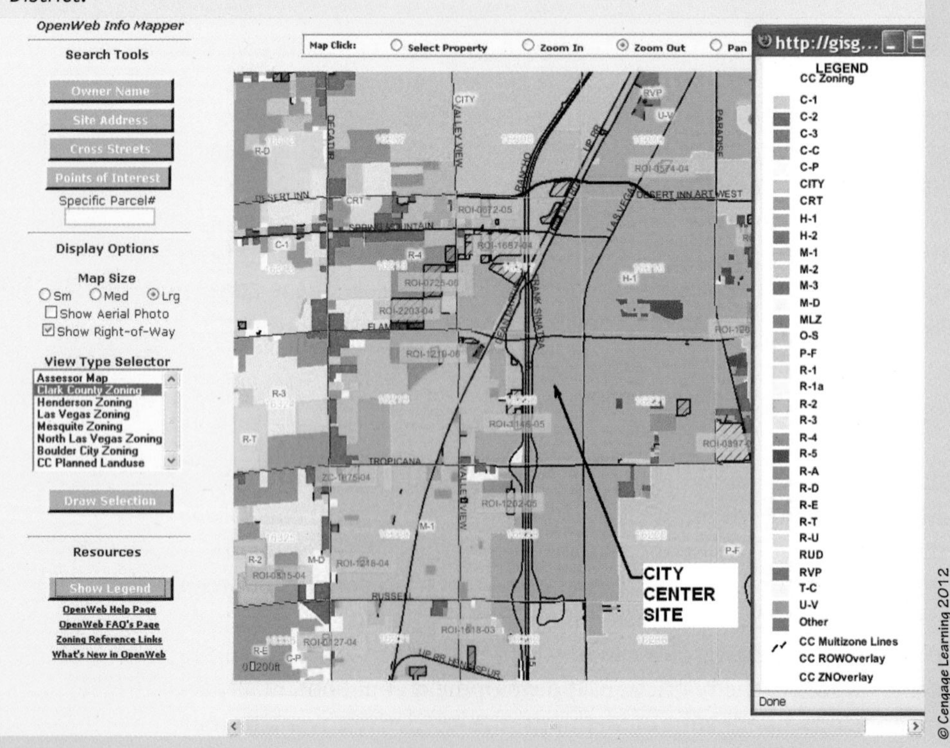

Most restrictive covenants will stipulate the size and types of construction of buildings but may also include other stipulations that would affect construction, such as exterior finish materials or colors, types of fencing, outdoor storage, and location of a garage for residences. Commercial covenants may restrict property usage to specific businesses or limit the parking of vehicles on the property. Some restrictions may reduce the desirablility of a lot for development and should be researched prior to purchasing the property. Your real estate agent, the current property owner, the developer, or the neighborhood association should be able to provide a copy of the restrictive covenants. However, most restictive covenants are a matter of public record, and you can obtain a copy at the county courthouse.

Another means of restricting the use of property is a **deed restriction.** While a restrictive covenant is typcially a document separate from the deed, a deed restriction is a clause in the deed to the property that limits how the owner may use the land. For example, a deed restirction may restrict the building of structure, paving of a parking lot, or the cutting and planting of trees on the land.

Identification of the criteria and constraints for a project includes research (of building codes, regulations, municipal ordinances, covenants, deed restrictions, and zoning laws) and determination of how these requirements affect the final design.

ENVIRONMENTAL REGULATIONS

The nation behaves well if it treats the natural resources as assets which it must turn over to the next generation increased, and not impaired, in value. —Theodore Roosevelt, 1907.

Perhaps one of the most controversial aspects of land development is the impact that it has on the natural environment. A multitude of environmental organizations have become increasingly important in shaping the national (and international) philosophy for the conservation of our natural world, and the government has responded with increased regulation of land development. A variety of federal, state, and local environmental regulations may apply to your development activity, requiring permits, inspections, and documentation throughout the design and construction process. It is important to understand the environmental requirements and gather information about your site that will help you factor in the expenses associated with compliance. Ignoring potential environmental issues can be financially devastating to your project and may lead to personal liability.

Environmental Impact

The first major environmental legislation in the United States was the National Environmental Policy Act (NEPA). NEPA was enacted in December 1969 to encourage consideration of the environmental impact of "major federal actions significantly affecting the human environment." "Major federal actions" has been interpreted to mean any project that is initiated by the federal government, benefits from federal funding, or can be prohibited or regulated by the federal government. According to the Council on Environmental Quality (CEQ), about a quarter of the states have also adopted similar rules to regulate state projects (2007). You may access a list of these states at http://ceq.hss.doe.gov/nepa/states.html. Since most commercial projects are somehow regulated by federal and/or state governmental agencies, many construction projects are governed by NEPA.

Some types of projects fall under a categorical exclusion and an environmental study is not required. However, if not specifically excluded, the developer must perform an environmental study of a proposed project. Depending on the project, an Environmental Assessment (EA) and/or a more in-depth Environmental Impact Statement (EIS) may be required, which may include consideration of ecological, aesthetic, historic, cultural, economic, social, and health effects. These studies are typically completed by environmental professionals with experience in assessing potential impact. The environmental studies are made available for public review, and public input is encouraged throughout the process. As a citizen, you have the right to make comments on any environmental study, and such comments must be addressed by the lead agency for the project. The entire NEPA process is illustrated in Figure 5-25.

You should understand that NEPA does not require that you avoid environmental impact, but it does require that you consider the environmental consequences of a development and alternatives that may reduce the impact.

Decisions regarding development involve the weighting of trade-offs between predicted positive and negative effects on the environment.

Floodplains

The natural beauty and recreational opportunities provided by coastal environments, rivers, and lakes have created a demand for property in areas near bodies of water. But what are the risks of developing a property near water? Often, these desirable areas are prone to flooding (see Figure 5-26). Congress created the National Flood Insurance Program (NFIP) in 1968 in an effort to reduce losses of life and property in these floodplains. As a result, communities have adopted floodplain management plans that include specific requirements for construction of new structures in flood areas. The requirements are enforced through regulations such as zoning ordinances and building codes. In exchange for participating in the program, a community may take advantage of federal government-backed flood insurance that is normally required in order to obtain a mortgage for a building in a flood prone area.

The NFIP is administered by the Federal Emergency Management Agency (FEMA). FEMA is required to create Flood Insurance Rate Maps (FIRMs) that show, among other things, the Special Flood Hazard Area (SFHA) and the Base Flood Elevations (BFE) for a **100-year flood.** The SFHA is the area predicted to flood during a 100-year storm—in other words, the area which has a one percent chance of flooding in any given year. The SFHAs are designated as A-zone or V-zone depending on the predicted severity of the flooding. V-zones are the more hazardous because structures within this zone are subject to high velocity wave action. A-zones are predicted to experience wave action of less than 3 ft in height. New construction built within an A- or V-zone requires flood insurance and must meet specific requirements that can increase the cost of a building project.

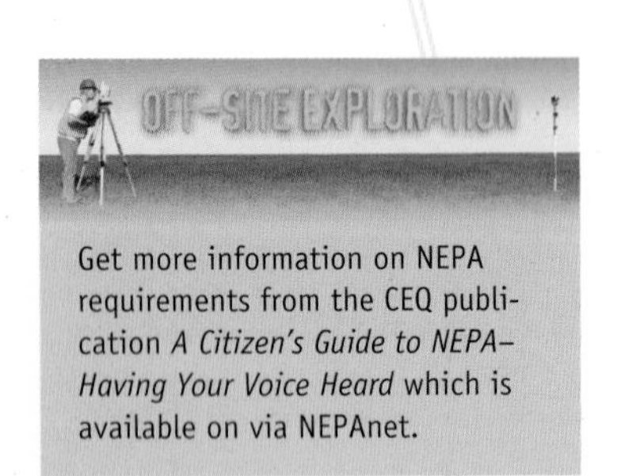

Get more information on NEPA requirements from the CEQ publication *A Citizen's Guide to NEPA–Having Your Voice Heard* which is available on via NEPAnet.

If your community participates in the NFIP, you can use the FIRMs for your area to determine which flood zone, if any, your property lies within. Your local planning department should have copies of these maps. FEMA also provides

Figure 5-25: The NEPA process.

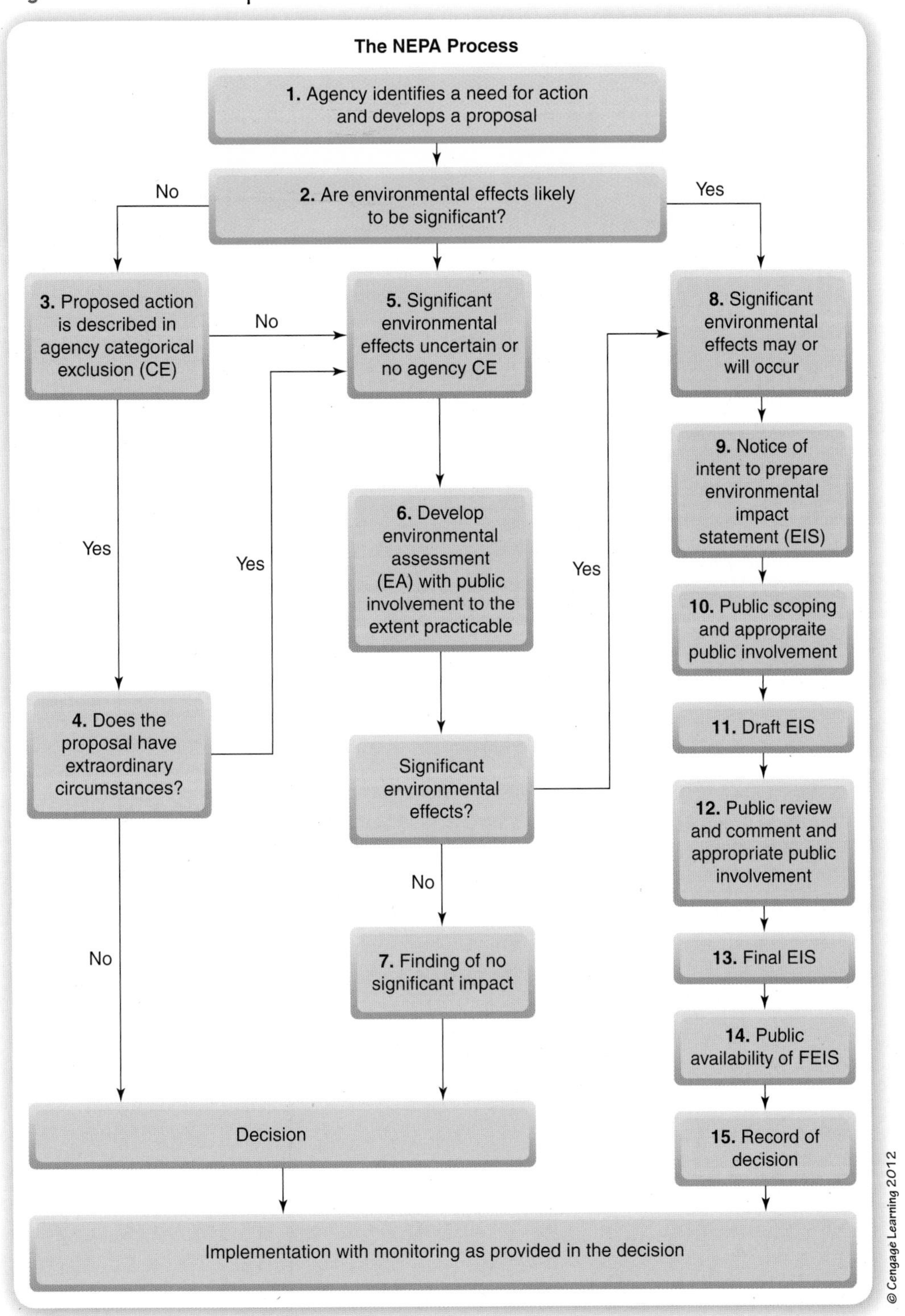

access to FIRMs for the United States via its online Map Service Center (*http://msc.fema.gov*). The Map Service Center was used to create the FIRMette, or portion of a FIRM, shown in Figure 5-27. For further information, you may contact your state's floodplain manager. A list of contact information for state floodplain managers is provided on the Association of State Floodplain Managers website.

Figure 5-26: New Hartford, IA, June 9, 2008 — The convenience store on York Street is closed until further notice due to the flooding of Beaver Creek. Photo by Greg Henshall / FEMA.

Figure 5-27: A portion of a Flood Insurance Rate Map, also called a FIRMette, created using the FEMA online Map Service Center at http://msc.fema.gov. Note the Base Flood Elevations in these zones are shown using contour lines.

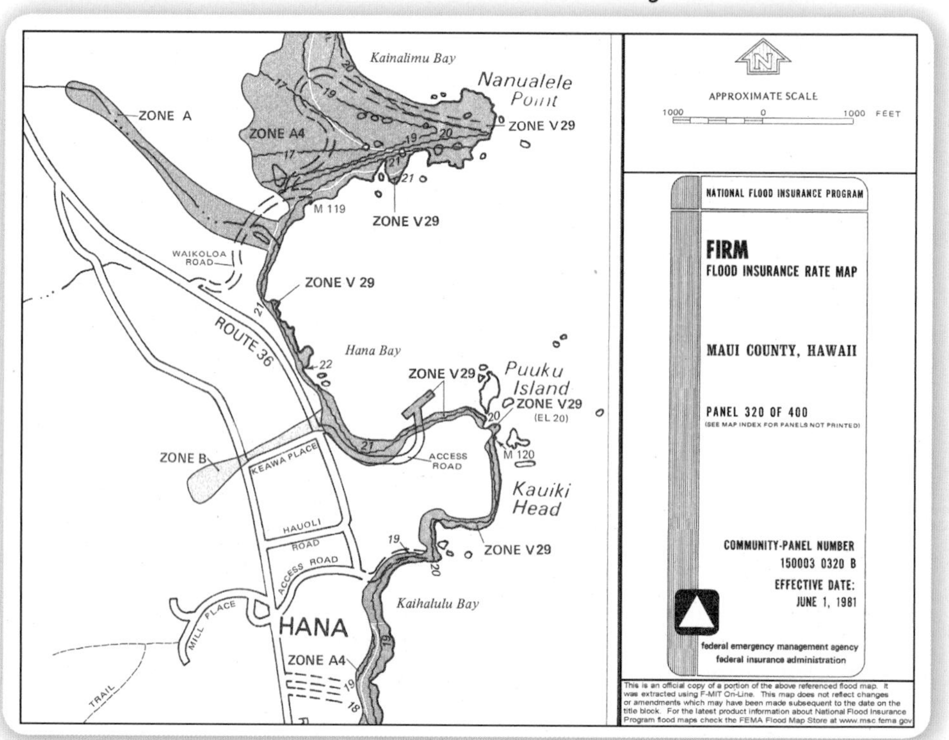

Courtesy of Federal Emergency Management Agency.

Your Turn
Are You Flood Prone?

Research the potential for flooding at your home, school, or a project site. Visit the FEMA Map Service website at http://msc.fema.gov and view the FIRMette Tutorial. Then create a FIRMette for your area of interest. If you determine that no FIRM exists for your area, use another area of interest. Insert the map into your architectural notebook, note the BFE, and define the applicable Flood Zone Designation. If your property is not in a Flood Insurance rate zone, create a second FIRMette for property near a local body of water that is in a SFHA.

Wetlands

Wetlands are transitional zones between the land and the water. They are "wet" for periods of time during the year and support plant life that thrives in saturated soils. Sometimes wetlands are easy to recognize—marshes and swamps are covered in water most of the year. However, other wetland areas are not always wet. In fact, wetlands are a fairly common land feature and occur in almost every county and climatic zone in the United States (EPA 2006). Because of regional differences in soils, topography, climate, and other factors, they take many forms, such as marshes, swamps, bogs, and fens (see Figures 5-28 and 5-29.) Wetlands provide habitat for thousands of species of wildlife, absorb and slow floodwaters, absorb nutrients, and remove toxins and sediment from water.

In the past, wetlands were often filled or drained. However, the federal government now protects wetlands through regulations, like Section 404 of the Clean Water Act. If your project will require filling a waterway—including any wetland—Section 404 of The Clean Water Act requires that you apply for a permit. In order to obtain a permit, you must show that you have made every effort to avoid or minimize the impact of your development on waterways. And in keeping with the national policy of "no net loss," if you cannot avoid impact on wetlands, you must provide compensation for the damage. The permit may require **mitigation** in the form of restoring, establishing, or enhancing other aquatic resources. The EPA and the Corp of Engineers each have some responsibility for the administration and enforcement of Section 404 and can provide information on permits.

Figure 5-28: Scarborough Marsh (saltwater), Maine. Many of the salt marshes along the Maine coast have been degraded or destroyed. The Scarborough Marsh accounts for 15 percent of the Maine's total tidal marsh area, making it the largest contiguous marsh system in the State.

Photo Courtesy U. S. Fish and Wildlife Service.

Figure 5-29: Simeonof Island bog in Alaska Maritime National Wildlife Refuge.

Photo courtesy U. S. Fish and Wildlife Service.

Fun Fact

Where Did All the Wetlands Go?

The bottomland hardwood-riparian wetlands along the Mississippi River once stored at least 60 days of floodwater. Now they store only 12 days because most have been filled or drained.

Natural ecosystems (such as wetlands and floodplains) provide an array of basic processes that affect humans including control of the hydrologic cycle, disposal of wastes, and recycling of nutrients. Humans are changing many of these basic processes, and the changes may be detrimental to humans.

Find out about wetland programs in your region by visiting the *EPA Wetlands Across the Country* webpage at *http://www.epa.gov/owow/wetlands/regions.html*.

In addition to federal regulations, many state and local governments have adopted wetland protection measures. Because avoiding wetland impact or mitigating the impact can increase construction time and cost for the development, it makes sense to investigate your site for wetlands as early as possible in the process. The U.S. Fish and Wildlife Service (FWS) provides access to national Wetland Inventory Maps that can be used to identify wetlands near your property (see Figure 5-30). Individual states may also provide a wetland identification program or local wetland maps.

An environmental consultant can help locate the boundaries of wetlands, identify project impacts, and assist with permitting in order to comply with federal

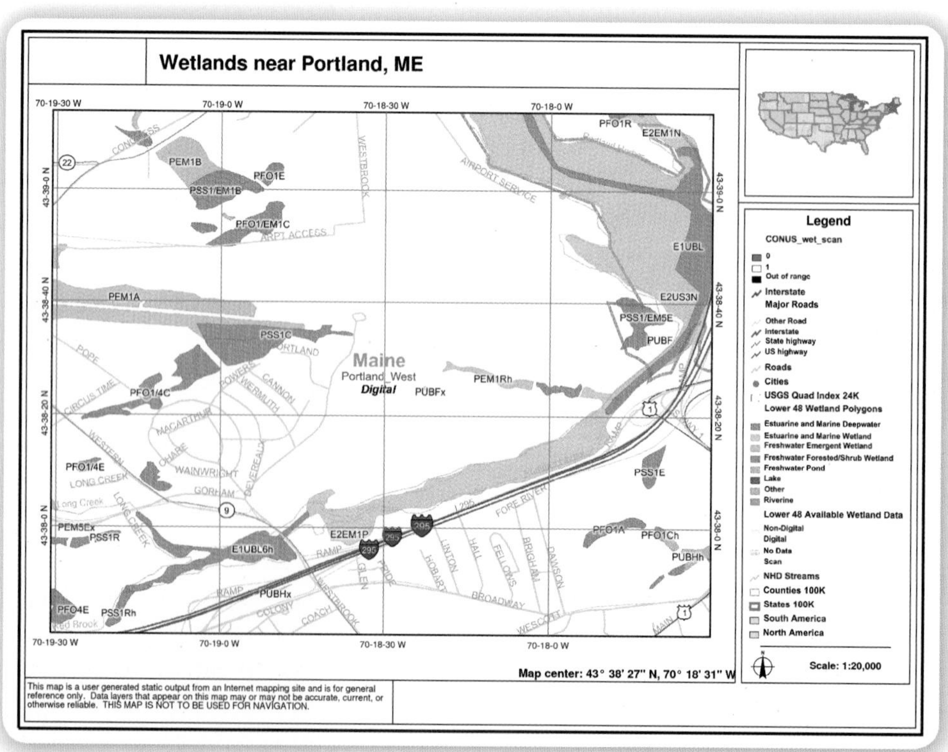

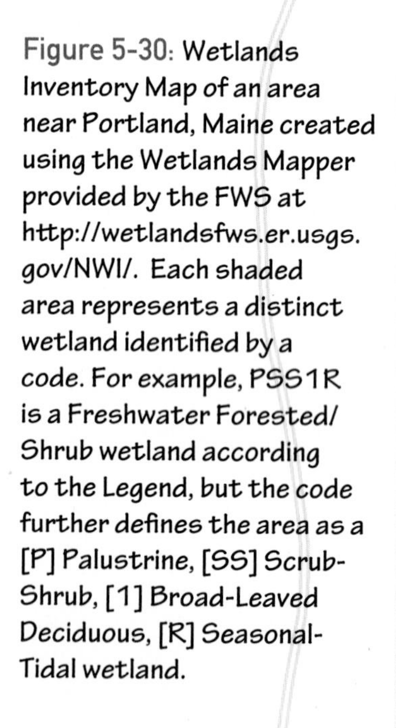

Figure 5-30: Wetlands Inventory Map of an area near Portland, Maine created using the Wetlands Mapper provided by the FWS at http://wetlandsfws.er.usgs.gov/NWI/. Each shaded area represents a distinct wetland identified by a code. For example, PSS1R is a Freshwater Forested/Shrub wetland according to the Legend, but the code further defines the area as a [P] Palustrine, [SS] Scrub-Shrub, [1] Broad-Leaved Deciduous, [R] Seasonal-Tidal wetland.

and state regulations. Figure 5-31 shows a map of wetland impacts prepared by an environmental consultant for the Environmental Impact Statement of a proposed highway project in the state of Washington.

Figure 5-31: **Map of identified wetland impacts attributed to alternatives for a highway project documented in the SOUTHEAST ISSAQUAH BYPASS, Final Environmental Impact Statement, and Section 4(f) Evaluation, December 2007.**

Courtesy of the City of Issaquah, Washington, The Washington State Department of Transportation, and the U. S. Department of Highways, Federal Highway Administration.

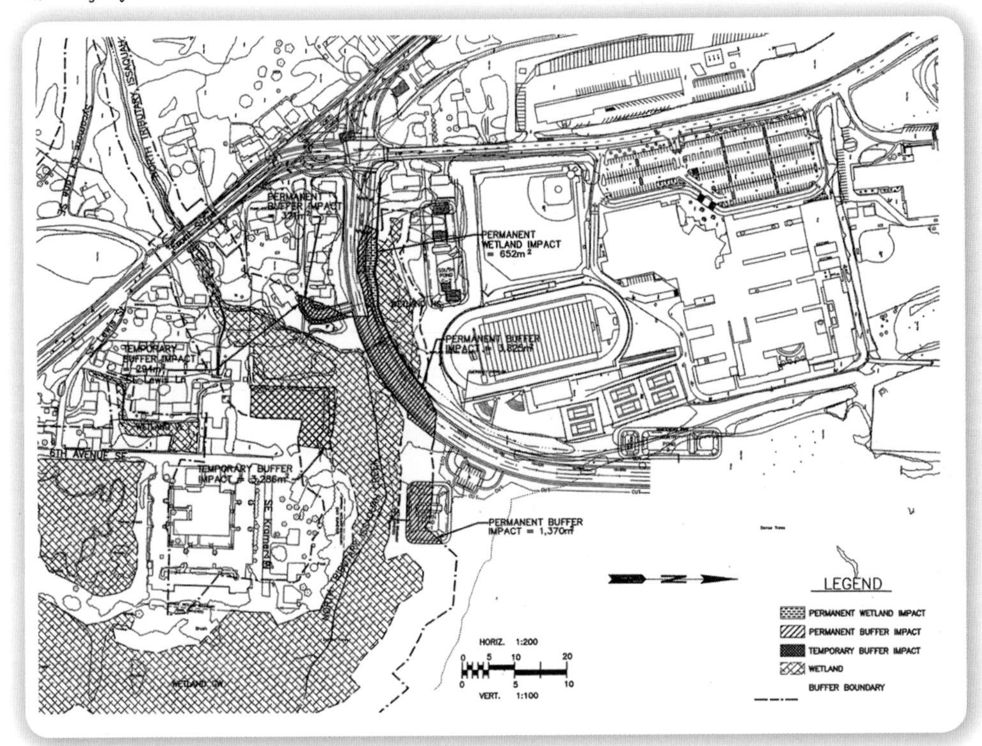

Stormwater Permits

The Clean Water Act includes another regulation, the National Pollution Discharge Elimination System (NPDES), which also directly affects land development and construction projects. Because construction activities disturb the land, stormwater that runs across a construction site often picks up soil and other contaminants. This runoff eventually drains into waterways and wetlands. If your development will disturb (clearing, grading, or excavating) one or more acres of land, the NPDES requires that you obtain a permit. Most states are authorized to issue NPDES stormwater permits and typically require that you have a written plan, called a Stormwater Pollution Prevention Plan (SWPPP). You must present your plan to avoid stormwater pollution, protect existing vegetation, and stabilize the soil before you will be issued a general construction permit (See Figure 5-32). Local governments may also have construction site requirements that differ from NPDES.

View the online module on Wetlands Functions and Values at *http://www.epa.gov/watertrain/wetlands/module11.htm*. When you are finished reading the information, take the self-test.

Coastal Zones

Additional approval and permitting may be required if your site falls within an identified coastal zone or coastal area of concern. In general, every state that borders an ocean or great lake has identified coastal zones and imposes regulations on development within the zone through the Coastal Zone Management Act. The Texas coastal zone is shown in Figure 5-33. Information on your state's coastal regulations can be

Figure 5-32: Some BMP's (Best Management Practices) for preventing storm water pollution are shown on this portion of a poster on Planning and Implementing Erosion and Sediment Control Practices.

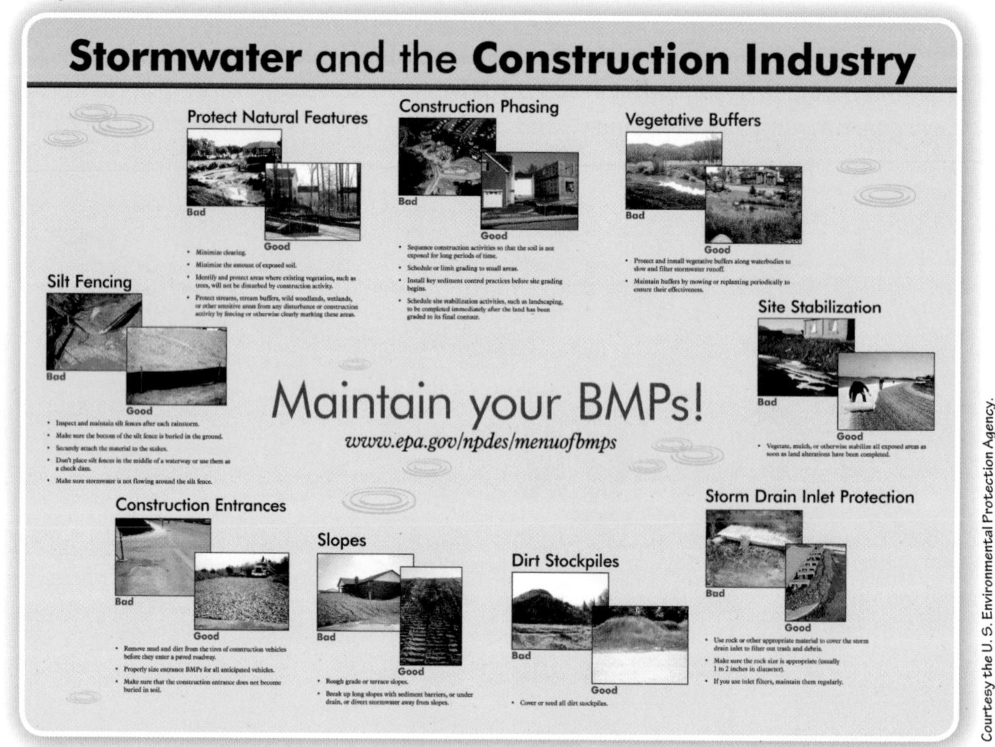

Courtesy the U. S. Environmental Protection Agency.

OFF-SITE EXPLORATION

Preventing Stormwater Pollution

The EPA has published a document entitled *Developing Your Stormwater Pollution Prevention Plan–A Guide for Construction Sites,* which is intended to help construction site managers write an effective SWPPP. Review an online copy of this publication at http://www.epa.gov/npdes/pubs/sw_swppp_guide.pdf. In your project notebook describe the difference between erosion and sedimentation. Then list the Ten Keys to Effective Erosion and Sediment Control and make notes that could help you reduce stormwater pollution at your construction site.

found on the National Oceanic and Atmospheric Administration (NOAA) Office of Ocean and Coastal Resource Management (OCRM) website at http://coastalmanagement.noaa.gov/mystate/welcome.html. Because of the many regulations that govern waterways, wetlands, and coastal areas an environmental consultant is often hired to perform studies and complete the necessary reports and permit applications.

Your Turn
Wetlands in Your Neighborhood

The U.S. Fish and Wildlife Service provides access to a Wetlands Online Mapper at http://www.fws.gov/wetlands/Data/mapper.html. Visit this site and work through the Quick Tutorial to become familiar with the GIS (Geographic Information System) and how to use the map. Be sure that your anti-pop-up software is disabled, and then launch the map by clicking on the area of the country in which you live. When the map opens, use the Locate tool to locate your zip code and then zoom in to enlarge your project site. Click on the Legend tool to display the legend. Choose at least two wetlands near your location and identify them using the Identify tool. Using the Print PDF command, print a copy of your wetlands map. Insert the map and record the geographic coordinates, wetland type, and number of acres of each wetland in your notebook.

Then visit the NOAA Ocean and Coastal Resource Management website at http://coastalmanagement.noaa.gov/mystate/welcome.html. If your state or territory is one of the highlighted states and works with NOAA to manage coastal areas, research your state's coastal management plans. Sketch a map of your state in your notebook and note counties that are included in the coastal zone. Make notes of some of the regulations that govern coastal development in your state.

Figure 5-33: The Texas Coastal Zone. Special regulations are imposed on development in these coastal areas as well as other coastal areas throughout the United States.

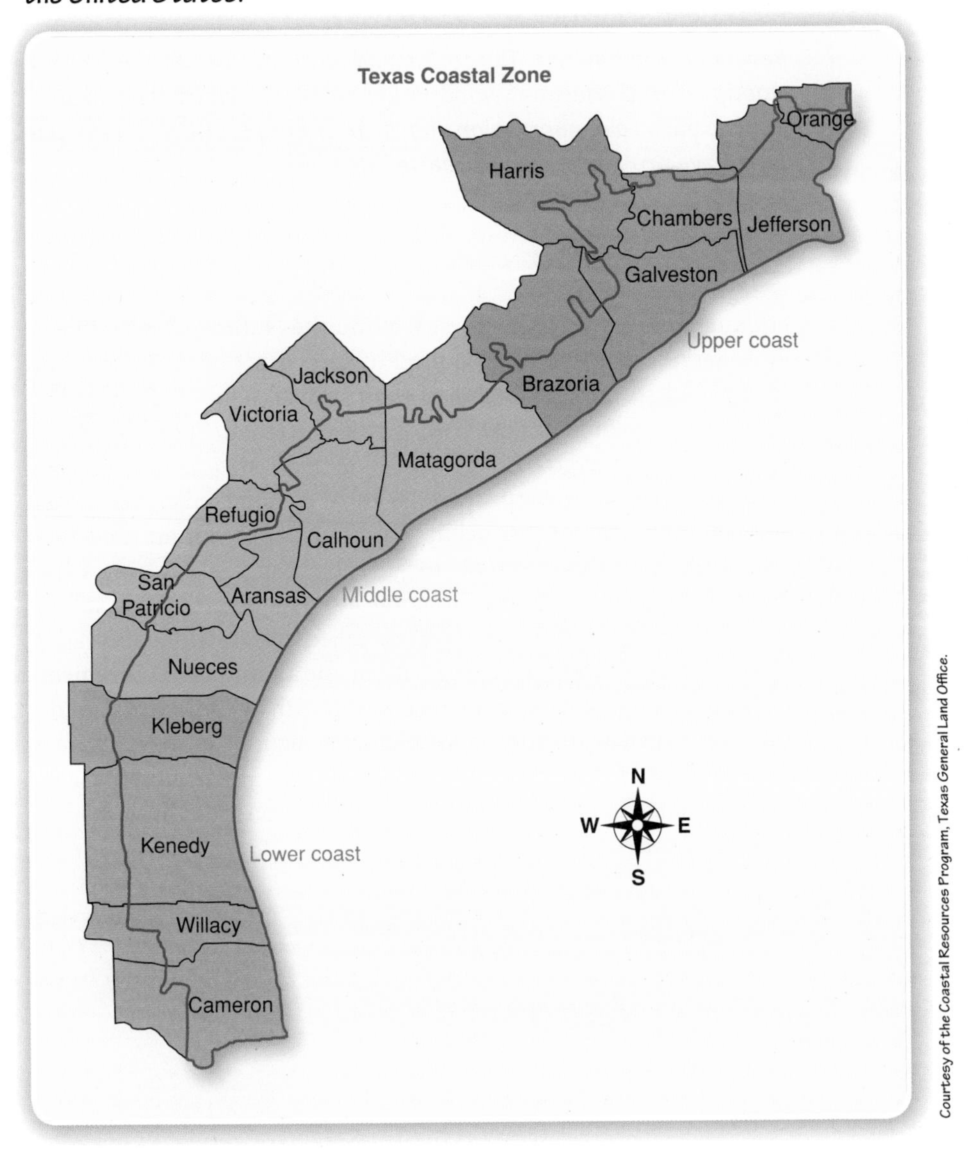

Courtesy of the Coastal Resources Program, Texas General Land Office.

Endangered Species

Scientists estimate that the natural rate of extinction on earth is one species lost every 100 years. However, the FWS reports that the United States has lost more than 500 species to extinction in only 400 years, more than one per year. Why should you care if the world loses a few more species, especially if conservation restricts the use of your property? Congress addressed this question in the preamble to the Endangered Species Act, which states that endangered and threatened species "are of esthetic, ecological, educational, historical, recreational, and scientific value to the Nation and its people." And Congress has made an effort to protect species from extinction with the Endangered Species Act (ESA).

The ESA protects both endangered and threatened species of plants and animals including fish, mammals, birds, reptiles, amphibians, and even insects. **Endangered** means a species is in danger of extinction such as the Ivory-billed woodpecker shown in Figure 5-34. **Threatened** means a species is likely to become

Figure 5-34: The Ivory-billed Woodpecker, once thought to be extinct, has been rediscovered in the Big Woods of Arkansas. Prior to the recent discovery of the Ivory-billed woodpecker, the last confirmed sighting was in 1940 after the species suffered a severe loss of habitat. The continuing existence of the Ivory-billed woodpecker, the largest woodpecker in the United States, is due in large part to conservation efforts over the last several decades that have protected large areas of land from fragmentation and development.

endangered. Per the ESA, the FWS and the National Marine Fisheries Service (NMFS) are required to keep lists of endangered and threatened species and maps of their critical habitats. This information is available on the FWS Critical Habitat Portal at http://criticalhabitat.fws.gov/.

If your development poses a risk to an endangered or threatened species or will damage its habitat, you must apply for an Incidental Take Permit. In this case, *take* includes harassing or harming a species or degrading its habitat. The application requires that you submit a plan for protection of the species, called a Habitat Conservation Plan. Be sure to check local regulations for additional requirements associated with protection of species other than those listed under the ESA. You can get applications and instructions for permits from your Regional FWS Endangered Species Office. Environmental consultants can help complete Habitat Conservation Plans and Incidental Take Permits.

Humans have a major effect on other species through land use—which decreases space available to other species—and pollution—which changes the chemical composition of air, soil, and water.

OFF-SITE EXPLORATION

Endangered Species in Your Backyard

Find out about endangered species where you live. Visit the U.S. Fish and Wildlife Service Critical Habitat Portal at *http://criticalhabitat.fws.gov/*. Investigate the critical habitat of endangered and threatened species in your area by expanding the Critical Habitat Data folder on the left and choosing By State/County. Once you have recorded the species for your county, view the Mapper for each species and study the boundaries of each critical area. Document your findings in your architectural notebook.

Tree Protection

Many municipalities control the destruction or cutting of trees with tree ordinances. For instance, the Mt. Pleasant, South Carolina, Code of Ordinances protects trees with diameters of 8 in or greater (measured at 4.5 ft above the ground) by requiring replacement trees to be planted if protected trees are lost. In addition, Mount Pleasant requires approval to remove most "significant" trees (16 in or greater measured at 4.5 ft above the ground) and "historic" trees (24 in or greater measured at 4.5 ft above the ground). If removal of a larger tree is approved, the town enforces even more stringent replacement requirements than for smaller protected trees. To show that you have complied with the requirements, you may be required to have a tree survey performed by a professional land surveyor, urban forester, or other professional to demonstrate compliance. A tree survey shows the size, location, and species of trees on a property Figure 5-37 illustrates the tree species protected by the City of Los Angeles. Because tree protection or replacement can significantly increase the cost of construction, you should be familiar with your local tree regulations and consider the size, species, and location of the trees when determining the viability of a site.

Point of Interest

The Tortoise and CityCenter

The residents of Clark County, Nevada (location of the CityCenter development), were surprised when, in 1989, the Desert Tortoise (Figure 5-35) was listed as a threatened species under the federal Endangered Species Act. As a result, Clark County founded the Clark County Desert Conservation Program to help save the tortoise. If you view Clark County on the FWS Critical Habitat Portal, you will find that a significant part of the county is designated as critical habitat for the threatened Desert Tortoise (Figures 5-36)—no other species' critical habitat is displayed in Clark County on the Critical Habitat Portal. However, in 1995, the county expanded its conservation program, in part, to avoid the listing of other species that live in the area. Clark County currently protects over 70 species of plants and animals through the Clark County Multiple Species Habitat Conservation Plan with a goal of *no net loss or fragmentation* of habitat. So, the CityCenter development had to meet both the requirements of the Endangered Species Act and the more stringent regulations imposed by Clark County.

Figure 5-35: Desert Tortoise (Gopherus agassizii). In 1989 the desert tortoise was listed as an endangered species under the federal Endangered Species Act.

Courtesy U. S. Fish and Wildlife Service. Photographer: Beth Jackson.

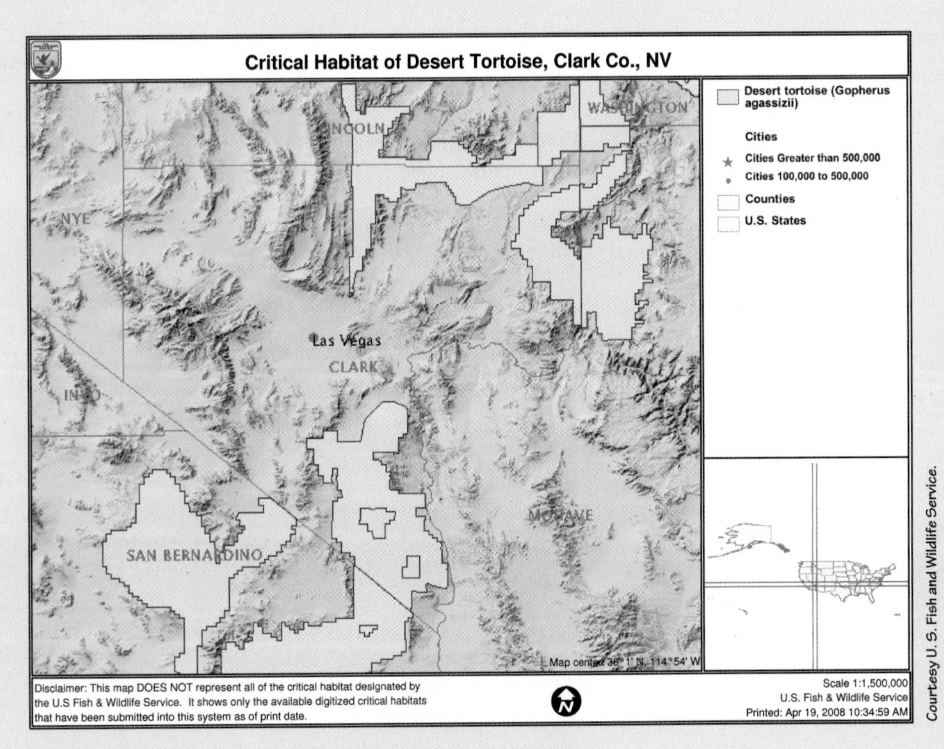

Courtesy U. S. Fish and Wildlife Service.

Figure 5-36: Desert Tortoises (Gopherus agassizii) Critical Habitat Map. Tortoises need sufficient suitable plants for forage and cover, and soils suitable for burrow and nest sites. They live primarily on flats with soils ranging from sand to sandy-gravel and scattered shrubs but are also found on rocky terrain and slopes. This map shows part of the protected critical habitat for the Desert Tortoise.

Figure 5-37: The City of Los Angeles has adopted a Native Tree Protection Ordinance that protects the species of trees shown.

Courtesy the City of Los Angeles, Department of Public Works, Bureau of Street Services, Urban Forestry Division.

The City of Los Angeles is home to one of the nation's largest urban forests.

In recognition of the aesthetic, environmental, ecological and economic benefits and the historical legacy that native trees provide the community, the City of Los Angeles has adopted the Native Tree Protection Ordinance, which will protect several of our most common native trees.

It is estimated that there are nearly twelve million trees within the city limits. Los Angeles' urban forest exists largely due to intensive tree planting starting as far back as the Pueblo de Los Angeles founding. Native trees are a significant component of this vast urban forest. The most prevalent native trees are the Oaks (Quercus spp.), Western Sycamore (Platanus racemosa), California Bay (Umbelullaria californica) and California Black Walnut (Juglans californica). The dominant native species is the Coast Live Oak (Quercus agrifolia).

Coast Live Oak (Quercus agrifolia) is long-lived, grows both single and multi-trunked, obtains 60 feet in height and usually grows in groups. The Valley Oak (Quercus lobata) is an extremely large and long-lived tree. The tree is deciduous, can reach heights of one hundred feet and have a one hundred foot canopy. They are tolerant of droughts and continued dry conditions. These two oak species are predominate but all oak species are protected except Scrub Oak.

California Black Walnut (Juglans califonica) is usually multi-trunk, deciduous, and most often does not reach a height of more than 25 feet. Their fruit was utilized by both native Indians and wildlife. The Western Sycamore (Platanus racemosa) is a very large tree that may reach heights of one hundred feet in ideal conditions. Typically, the Western Sycamore is single trunk but often grows with bends and twists. They may survive on little moisture. The California Bay (Umbelullaria californica) is usually single trunk, may reach heights of seventy-five feet, and cover large areas. They have edible fruit and are evergreen.

Figure 5-38: Love Canal homes after more than 900 families were evacuated from the site. As part of the cleanup, 229 homes and one elementary school surrounding the original 16-ac landfill were later demolished and the debris buried underground.

Courtesy the U. S. Environmental Protection Agency, Region 2.

Contamination and Containment

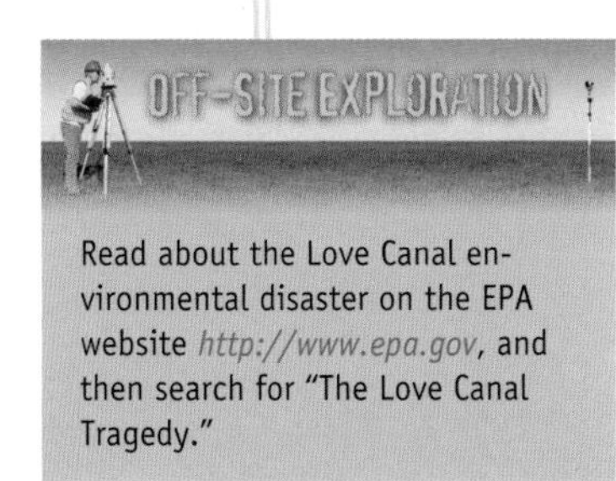

OFF-SITE EXPLORATION
Read about the Love Canal environmental disaster on the EPA website http://www.epa.gov, and then search for "The Love Canal Tragedy."

During the twentieth century, the United States suffered several major environmental disasters. One of the most devastating disasters, Love Canal (see Off Site Exploration and Figure 5-38) is often identified as the impetus for strict environmental regulations in the United States. The Comprehensive Environmental Response, Compensation, and Liability Act (CERCLA), also known as Superfund, was enacted "to reduce and eliminate threats to human health and the environment posed by uncontrolled hazardous waste sites." CERCLA gives the government authority to respond to the release of hazardous substances and provides for funding of cleanup activities. For developers and property buyers, perhaps the most important part of this is that the EPA has the power to hold a "potentially responsible party" (PRP) liable for the cost of cleanup. If you buy a contaminated property, you can be considered a PRP, even if you took no part in the contamination and had no knowledge of the contamination when you purchased the property. Therefore, it is important to discover contamination before you buy. The EPA has identified hundreds of contaminated sites throughout the country that have possible human health and environmental risk factors. These sites have been placed on the National Priority List (NPL). Be aware that even though your site is not on the NPL, it may still be polluted with undiscovered or undisclosed contamination. An environmental consultant can help locate contamination in the area of your project.

OFF-SITE EXPLORATION
Review the National Priority List on the EPA's NPL webpage at http://www.epa.gov/superfund/sites/npl/locate.htm.

Fortunately, amendments to CERCLA have provided some protection for property buyers who have made a diligent attempt to find out if the property is contaminated. Under the All Appropriate Inquiries Rule, if the property owner researches previous ownership and uses of the property *before* acquiring the property, he can be exempted from the cost of cleanup.

The alignment of development with natural processes maximizes performance and reduces negative impacts on the environment.

BROWNFIELDS Have you ever driven past an unused and rundown property that has been fenced with "No Trespassing" or "Warning" signs posted? Many times these sites were formerly commercial properties (for example dry cleaners, gas stations, or farms) or industrial facilities that are thought to have a potential for environmental contamination. Sometimes these sites, referred to as **brownfields,** are polluted; other times there is no significant hazard. In the past, developers avoided brownfields because of the potential liability. However federal and state government programs now encourage developers to cleanup and redevelop these unused, unsightly properties, which have encouraged revitalization of many blighted areas. Incentives vary from state to state and can include tax incentives, grants, or special loans and should be investigated through your state environmental protection agency. Figure 5-39 gives examples of brownfield redevelopments.

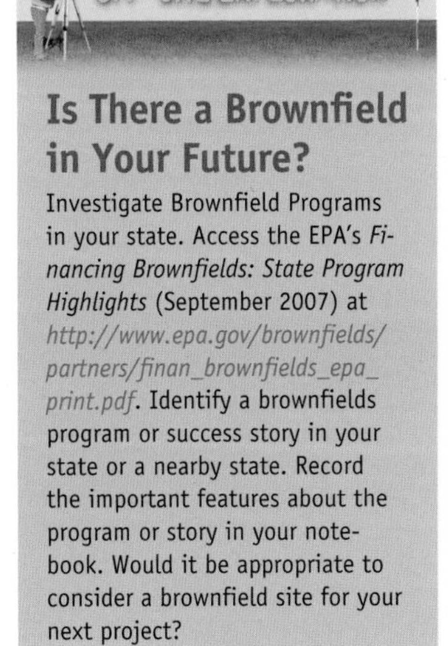

OFF-SITE EXPLORATION
Is There a Brownfield in Your Future?
Investigate Brownfield Programs in your state. Access the EPA's Financing Brownfields: State Program Highlights (September 2007) at http://www.epa.gov/brownfields/partners/finan_brownfields_epa_print.pdf. Identify a brownfields program or success story in your state or a nearby state. Record the important features about the program or story in your notebook. Would it be appropriate to consider a brownfield site for your next project?

Historic and Cultural Resources

Have you ever wondered who might have used the land that your home or school now occupies? Did any of the famous explorers, like La Salle, Marquette, or Lewis and Clark, cross the land? Which Native American tribes may have lived, hunted, or buried their dead where you now stand? Did colonial settlers use the land to grow crops, raise livestock, or build a well? Or are there structures on the property that are examples of historic architecture? Often historical and cultural artifacts are

Figure 5-39: A map showing successful brownfield redevelopment projects.

This map is used with permission of the National Association of Local Government Environmental Professionals (NALGEP)

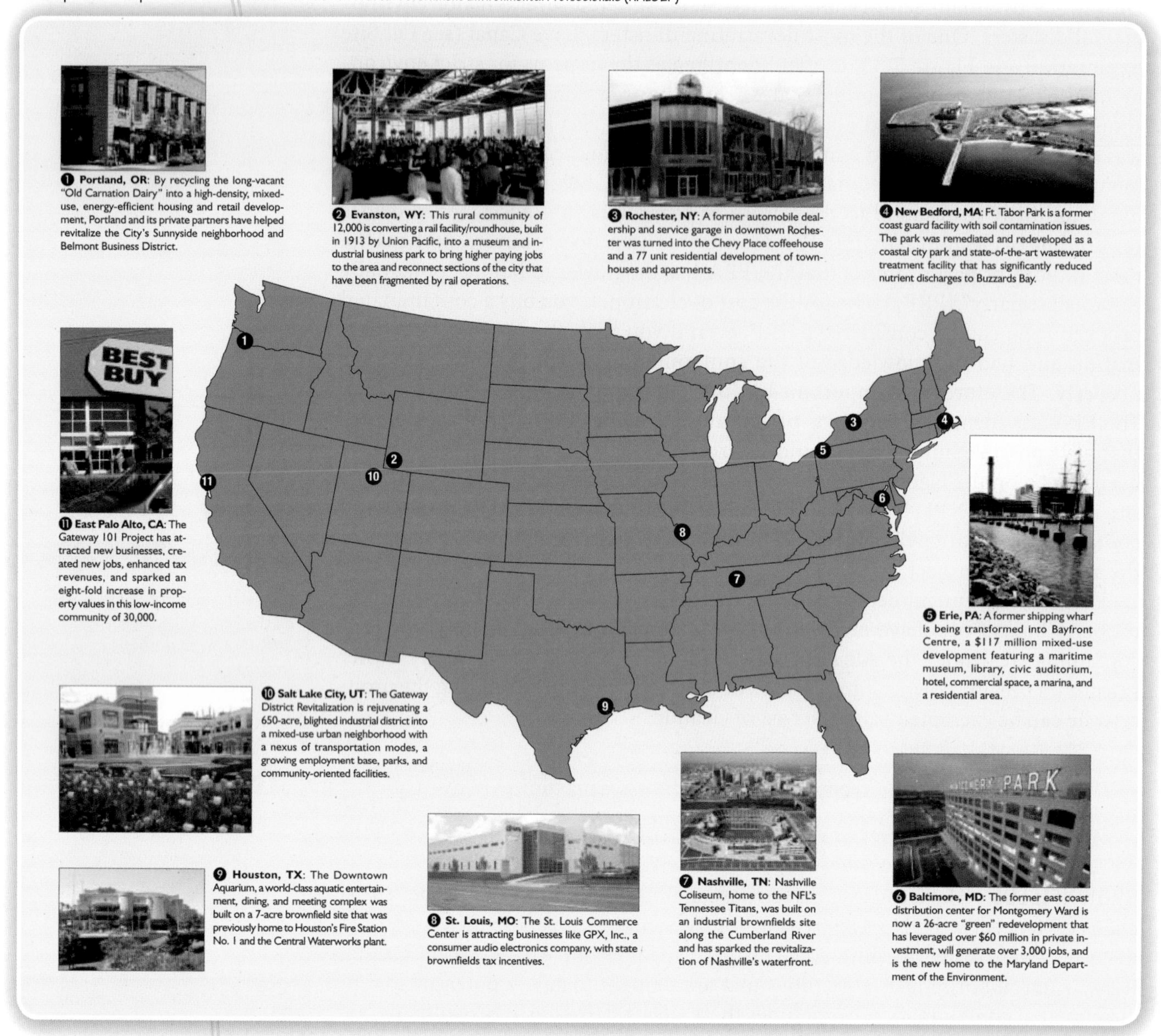

difficult to identify and are therefore lost when property is developed. In an effort to help preserve our nation's historic artifacts, Congress passed the National Historic Preservation Act in 1966. As amended, the act requires that federal agencies consider the impact of their actions on historic properties. As we have previously discussed, even developments that are not under the direct control of a federal agency are affected by such rules because they require permits or approval by a federal agency. Currently, the National Register of Historic Places includes over 85,000 historical places and over 13,000 designated historical districts which can be researched on the National Register of Historical Places website at http://www.nationalregisterofhistoricplaces.com/state.html. However, there are many more historical places that have not yet been identified. Undeveloped property should be investigated for Native American, colonial, and more recent artifacts that have been left behind by our predecessors.

❶ Portland, OR: By recycling the long-vacant "Old Carnation Dairy" into a high-density, mixed-use, energy-efficient housing and retail development, Portland and its private partners have helped revitalize the City's Sunnyside neighborhood and Belmont Business District.
❷ Evanston, WY: This rural community of 12,000 is converting a rail facility/roundhouse, built in 1913 by Union Pacific, into a museum and industrial business park to bring higher paying jobs to the area and reconnect sections of the city that have been fragmented by rail operations.
❸ Rochester, NY: A former automobile dealership and service garage in downtown Rochester was turned into the Chevy Place coffeehouse and a 77 unit residential development of townhouses and apartments.
❹ New Bedford, MA: Ft. Tabor Park is a former coast guard facility with soil contamination issues. The park was remediated and redeveloped as a coastal city park and state-of-the-art wastewater treatment facility that has significantly reduced nutrient discharges to Buzzards Bay.
BEST BUY
⓫ East Palo Alto, CA: The Gateway 101 Project has attracted new businesses, created new jobs, enhanced tax revenues, and sparked an eight-fold increase in property values in this low-income community of 30,000.
❺ Erie, PA: A former shipping wharf is being transformed into Bayfront Centre, a $117 million mixed-use development featuring a maritime museum, library, civic auditorium, hotel, commercial space, a marina, and a residential area.
❿ Salt Lake City, UT: The Gateway District Revitalization is rejuvenating a 650-acre, blighted industrial district into a mixed-use urban neighborhood with a nexus of transportation modes, a growing employment base, parks, and community-oriented facilities.
❾ Houston, TX: The Downtown Aquarium, a world-class aquatic entertainment, dining, and meeting complex was built on a 7-acre brownfield site that was previously home to Houston's Fire Station No. 1 and the Central Waterworks plant.
❽ St. Louis, MO: The St. Louis Commerce Center is attracting businesses like GPX, Inc., a consumer audio electronics company, with state brownfields tax incentives.
❼ Nashville, TN: Nashville Coliseum, home to the NFL's Tennessee Titans, was built on an industrial brownfields site along the Cumberland River and has sparked the revitalization of Nashville's waterfront.
PARK
❻ Baltimore, MD: The former east coast distribution center for Montgomery Ward is now a 26-acre "green" redevelopment that has leveraged over $60 million in private investment, will generate over 3,000 jobs, and is the new home to the Maryland Department of the Environment.

Simply developing property near an historic property, historic district, or archaeological site can cause harm. Harm to historic properties can include physical damage or destruction, or less obvious impacts such as altering the appearance of the property or surroundings; introducing noise, pollution, or unsightly elements to the site; or allowing the property to naturally deteriorate through neglect. If you will impact an historic property and the National Historic Preservation Act applies to your development, you must investigate possible impacts and explore alternatives to avoid or reduce harm to the historic properties. Other federal regulations may also apply, such as the Native American Graves Protection and Repatriation Act of 1990 or the Archaeological Resources Protection Act. As is true with most other regulations we have discussed, in addition to federal regulations, state and local governments may designate additional historic sites and enforce more stringent requirements, so you need to check state and local regulations as well.

Fun Fact

The Big Duck

Some of the sites listed as National Historic Sites are surprising. For example, the Big Duck on Long Island, New York (Figure 5-40) was added to the national register in 1997 for its architectural significance. The building was built in the 1930s and served as a poultry store.

Figure 5-40: The Big Duck, Suffolk County, New York. The Big Duck was built in 1930-31 in Riverhead on Long Island, New York as a poultry store and is included in the national register of historical places. The structure was constructed with a wood frame and surfaced with concrete on wire mesh.

Photo courtesy of the National Park Service.

Your Turn

Identify historically significant sites near CityCenter in Clark County Nevada via the National Register of Historical Places website at http://www.nationalregisterofhistoricplaces.com/state.html. Follow the links to Nevada and then to Clark County. Identify at least three sites in the county of architectural/engineering significance that have listed a familiar architectural style and at least two sites significant to Native American culture.

Point of Interest

An Historic Parking Lot?

The All Star Bowling lanes and parking lot (Figure 5-41) was added to the National Historic Sites list in 1996 because of its role in the civil rights movement. Due to the bowling alley owner's segregationist policy and blatant disregard for the Civil Rights Act of 1964, local college students held protests in the parking lot on January 29, February 5, and February 6, 1964. On February 8 the protests resulted in the shooting of two college students and one high school student on the South Carolina State University campus.

Figure 5-41: All Star Bowling Lanes in Orangeburg, South Carolina was listed in the National Register for Historical Places in 1996 based on its significant role in the Orangeburg Massacre (1968).

Courtesy of the South Carolina Department of Archives and History.

OTHER SITE CONSIDERATIONS

Climate

The local climate will affect many aspects of your project. The grading and drainage design, landscaping, and the placement of buildings on the site are affected by the wind, rain, and sun. The heating and air conditioning systems of the building will be designed based on the temperature and humidity of the area. Wind, snow, flood, earthquake, and hurricane forces will be used in the structural design of the building. Climate affects can be enhanced or reduced by architectural features to improve human comfort and energy efficiency. We will discuss the climatic effects on each of these aspects of your project design in later chapters. The information you learn will help you understand the importance of climate in your viability analysis. For now, you should obtain basic climate information: average monthly temperature ranges, quantity and frequency of precipitation, orientation and angle of the sun at sunrise and sunset at midwinter and midsummer, and prevailing wind direction.

There are many sources of temperature and precipitation data available. Two national online sources are provided by the NOAA.

- National Climactic Data Center provides climate normals for over 4000 weather stations throughout the United States at the U.S. Climate Normals website: http://cdo.ncdc.noaa.gov/cgi-bin/climatenormals/climatenormals.pl.
- U.S. Climate Map at http://www.cdc.noaa.gov/USclimate/states.fast.html provides links to graphical climatology data for several locations within each state. Figure 5-42 provides an example of the graphical data available.

Wind data is difficult to find. A national climate map showing annual prevailing wind speed and direction for the lower 48 states is shown in Figure 5-43. More precise wind roses, which are graphs that show speed and direction of wind at a specific location, are available on the National Water and Climate Center website at http://www.wcc.nrcs.usda.gov/climate/windrose.html. See Figure 5-44 for an example of a wind rose.

Although we know that the earth revolves around the sun, as an observer on earth you have surely noticed that the sun's path changes with the seasons. The sun appears to rise and set in a slightly different position on the horizon each day and rises higher in the sky during the summer than during the winter. The exact position of the sun from a position on earth at any given time is represented by two angles, an altitude and an azimuth. **Altitude** is the angle up from the horizon—zero degrees altitude means exactly on your local horizon, and 90 degrees is "straight up." **Azimuth** refers to the angle along the horizon typically measured clockwise from due north (see Figure 5-45).

Figure 5-46 shows a comparison of the apparent path of the sun on December 21 (winter solstice) and June 21 (summer solstice) from a location in the northern hemisphere. The U.S. Naval Observatory Sun or Moon Altitude/Azimuth Table at http://aa.usno.navy.mil/data/docs/AltAz.php provides the sun's apparent location in the sky from anywhere in the world at any given time. You should obtain sun data for the location of your project on both the summer solstice (June 21) and the winter solstice (December 21)—this information will be used later when locating your building on the site and when you consider energy efficiency and HVAC design for your building.

Figure 5-42: Thirty year (1961–1990) mean climate data for Kodiak, Alaska.

Image provided by the NOAA-ESRL Physical Sciences Division, Boulder Colorado from their Web site at http://www.cdc.noaa.gov/.

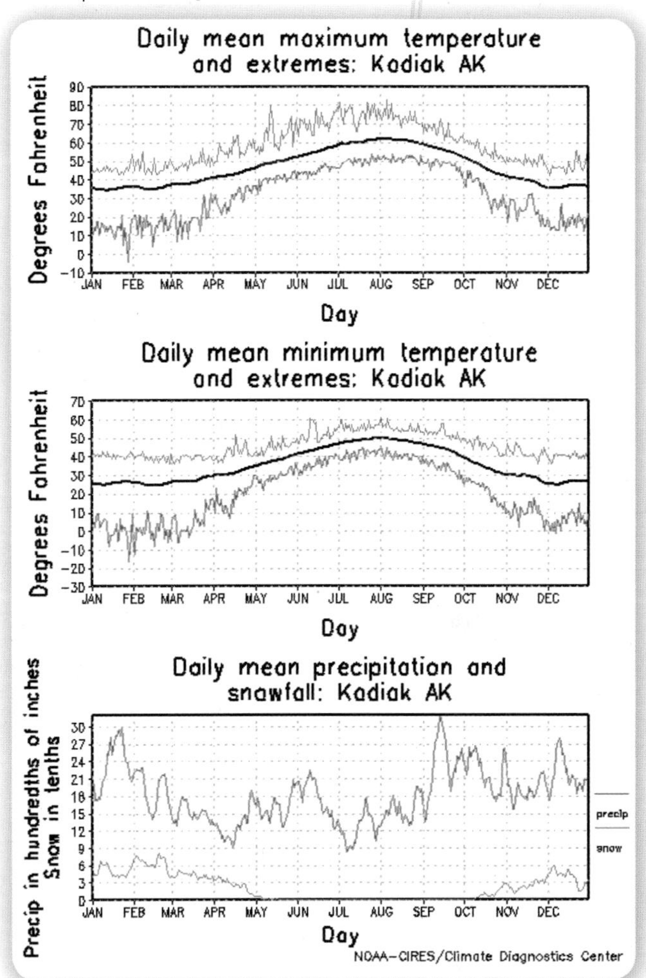

Charts and graphs, such as wind roses and line graphs of temperature normals, are used to communicate data or the results of an investigation.

MICROCLIMATE Have you ever noticed a change in climate between two locations that are relatively close together? Perhaps you were walking outside in the sun on a hot summer day but as you entered a shaded area next to a gentle stream the temperature dropped significantly. Or, on a cold windy day, you discovered that you were much warmer on the downwind side of a dense grove of trees that blocked the wind. We refer to weather that depends on local site characteristics as *microclimate.* Site features such as topography, vegetation, and bodies of water can alter the general weather conditions locally by affecting the temperature, humidity, air movement, and radiant heat energy. The microclimate can have a positive or negative affect on the cost of development, so any specific conditions that can change the climate of a site should be documented and taken into consideration during site selection.

Figure 5-43: Mean wind speed and prevailing direction for the lower 48 states. Similar information is available for Hawaii and Alaska from the NOAA Climate Maps of the United States (http://cdo.ncdc.noaa.gov/cgi-bin/climaps/climaps.pl).

Courtesy of the National Oceanic and Atmospheric Administration, National Climatic Data Center.

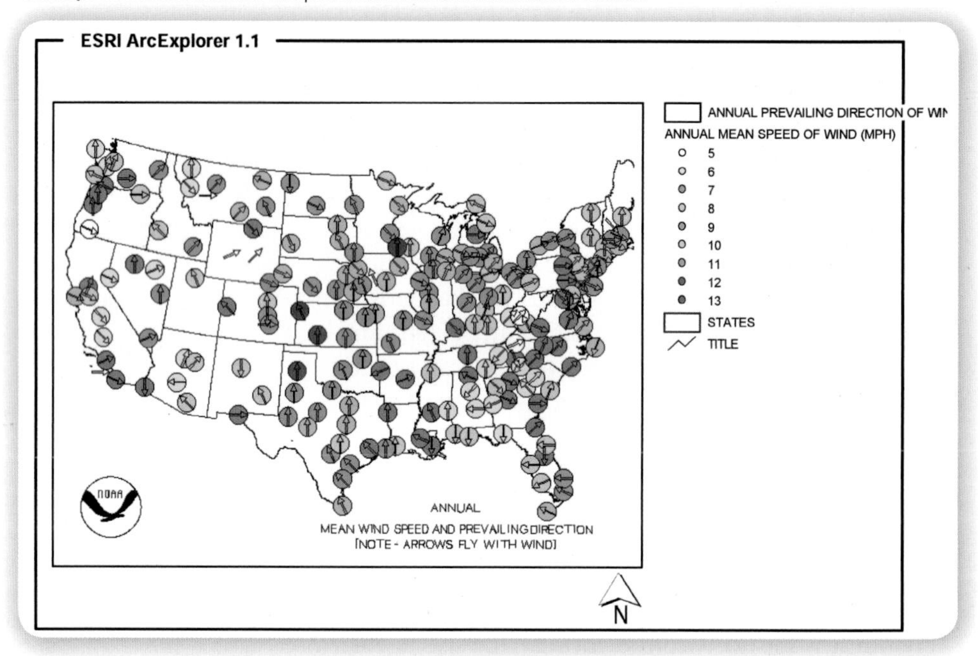

Utilities and Municipal Services

A successful project requires that you carefully consider the post-construction needs of the facility and the people who will use it. Most developments will need a source of energy, a supply of clean water, telephone and cable connections, and both wastewater and solid waste disposal. You will also want to be sure that your buildings will be protected by adequate police and fire fighting service. And, the facility will need access to roads so that people and products can safely travel to and from the new site. Unfortunately, you are not guaranteed these services just because you decide to develop a piece of land. You should always contact your local utility companies and municipality to find out if services are easily accessible from the property, how service can be established, and what fees will apply. If services are not available, or are very costly, you may need to consider building utility systems on your property. In either case, whether you use services provided by a municipality or private company, or you establish your own systems, the costs and time requirements should be considered when assessing the viability of a project. We will discuss utility planning and systems in greater detail in Chapter 6.

OFF-SITE EXPLORATION

The Land of the Midnight Sun

Is Prudhoe Bay, Alaska in the Land of the Midnight Sun? Investigate the sun's apparent position in the sky from Prudhoe Bay on the summer solstice using the U.S. Naval Observatory altitude/azimuth tables at *http://aa.usno.navy.mil/data/docs/AltAz.php*. Does the sun shine at midnight there? What area(s) of the world comprise the Land of the Midnight Sun?

Traffic Flow

Have you ever had to wait through several traffic light changes to enter or exit a shopping center parking lot or a neighborhood? Did you eventually resort to darting across traffic to get to where you were going? Or, have you been frustrated while trying to find a business because the building wasn't on the main road, but "hidden" on a side street or behind other buildings? Insufficient traffic access, poor

Figure 5-44: Wind Rose for Kodiak, Alaska based on 30 years of hourly wind data. Presented in a circular format, a wind rose shows the frequency of winds blowing FROM particular directions. The length of each "spoke" around the circle represents the frequency of time that the wind blows from that direction. For instance, based on the concentric scale, this wind rose shows that the wind blew from the northwest (to the southeast) roughly 22 percent of the time on average during February in Kodiak. The colors on the spokes also indicate the percentage of time that the wind blew within each particular speed range. In this case, the wind from the northwest blew at a speed of between 5.40 and 8.49 m/s for approximately 7 percent of February. For the month of February, the prevailing wind direction is west to northwest.

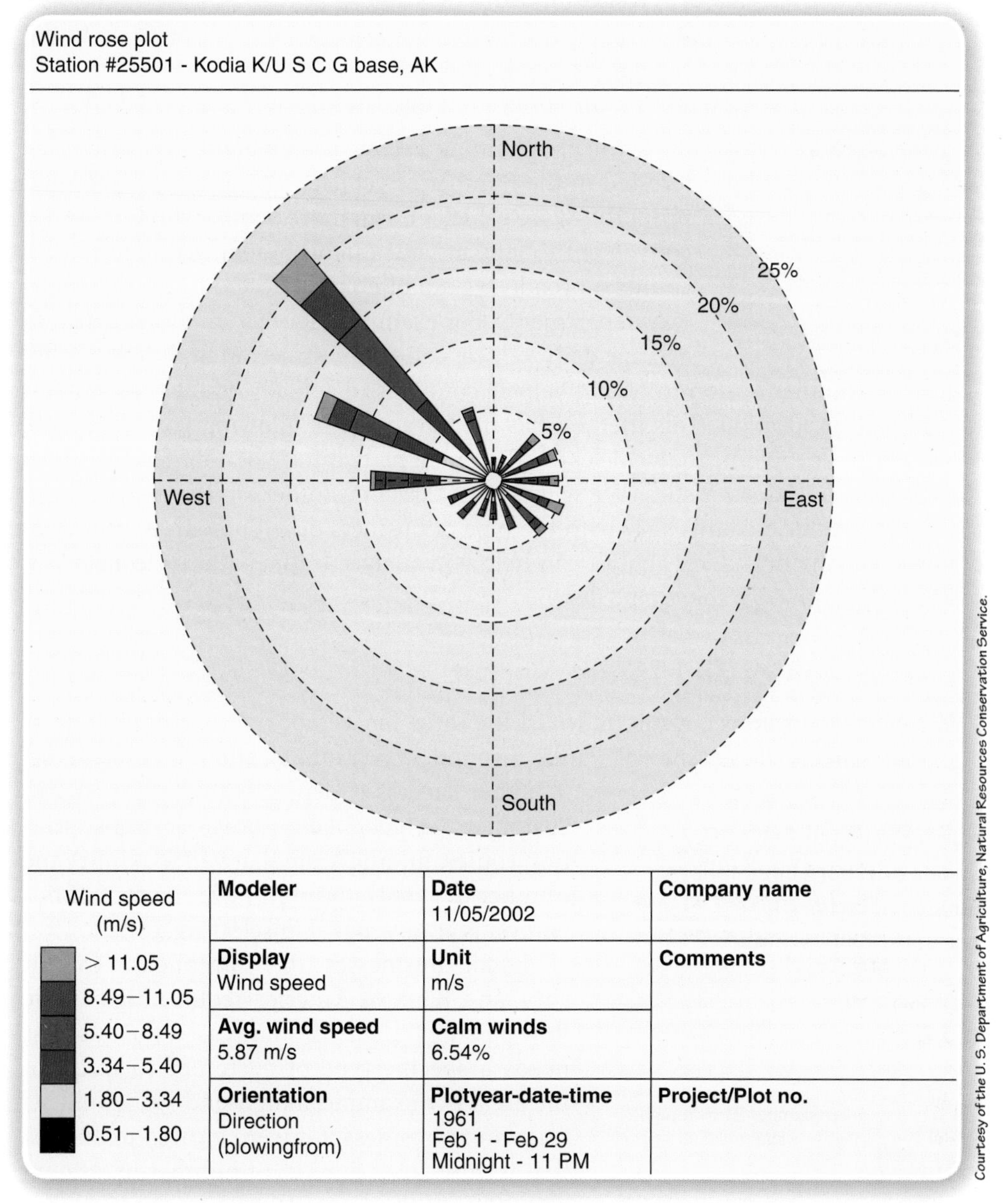

Figure 5-45: Using altitude and azimuth angles, you can describe the apparent position of the sun at any given time. Here the altitude is 40 degrees up from the horizon. The azimuth is 210 degrees (clockwise along the horizon from due north) or S 30 degree W.

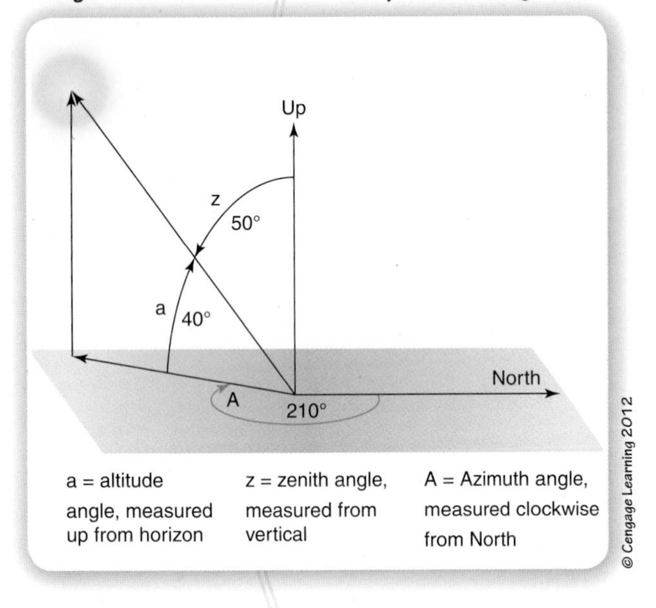

visibility, or poor vehicle flow around a property can reduce the desirability of a site for development. Depending on the type of development planned, easy access may be a key to the success of the facility. Site-specific situations that can create poor traffic conditions should be investigated during site discovery and include:

- No left turn lanes on busy adjacent roadways
- Poor **sight distance** at probable entrances and exits (vegetation, parking, or other obstructions that block a driver's view)
- Existing driveways or intersections near proposed entrances or exits
- Heavy traffic flow on adjacent roadways with no traffic lights to control flow
- Traffic congestion that severely restricts vehicle flow

If you are planning a retail development, you will be interested in the number of vehicles that pass by a potential property. More vehicles mean more people who may find it convenient to stop at your establishment. Average vehicle counts on roadways can often be collected from state highway departments. For example, Figure 5-47 shows a map of average daily vehicle counts for streets in St. Francis, Wisconsin. Another option to obtain vehicle counts is to actually count the cars that pass a given location. This method has the advantage of allowing you to visually classify the people in the vehicles—which may be helpful when targeting people with specific characteristics. We will discuss target customers in the next section. Desirable vehicle counts vary with the type of business development. A site selection company or market analyst can help you analyze traffic patterns to help you find a site with desirable traffic flow.

Demographics and Target Market

If you ask a business consultant to list the three most important factors in selecting a commercial site, she will surely respond, "Location, location, and location." Whether success is dependent on proximity to customers, a good labor pool, or transportation routes, finding a site that will maximize the number of clients and minimize the cost of operation is always a goal in selecting a site.

Retail businesses like Starbucks, Circuit City, and Bank of America require high traffic flow passing the site. In some cases, the business needs a specific type of customer to frequent the area. For example, a high-end spa would attract more business if it were located in an area that is home to a significant number of affluent women over the age of 18 and conveniently located in a well-traveled part of town. Other service providers that provide service away from the business location, like architects and carpet cleaners, may not need high traffic but should be located near their customers. Wholesalers

Figure 5-46: The apparent position of the sun in the sky changes throughout the year. In the northern hemisphere, the sun rises highest on the summer solstice (June 21) and is lowest in the sky on the winter solstice (December 21).

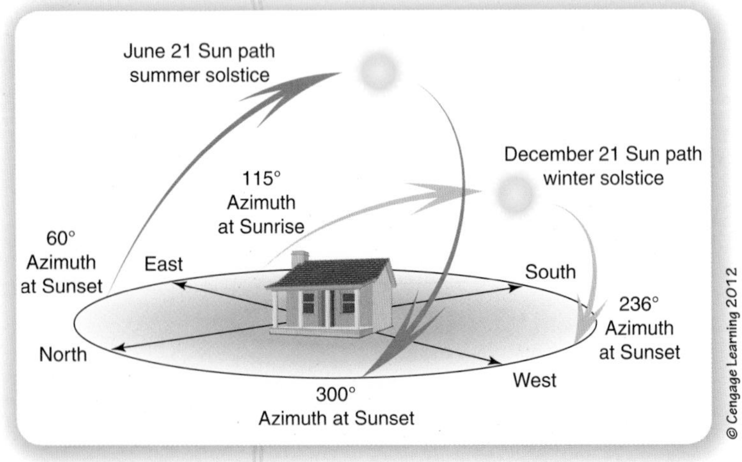

Figure 5-47: *Vehicle counts per day for a portion of the City of St. Francis, Wisconsin. The number of vehicles that pass a site can impact the success of a retail business.*

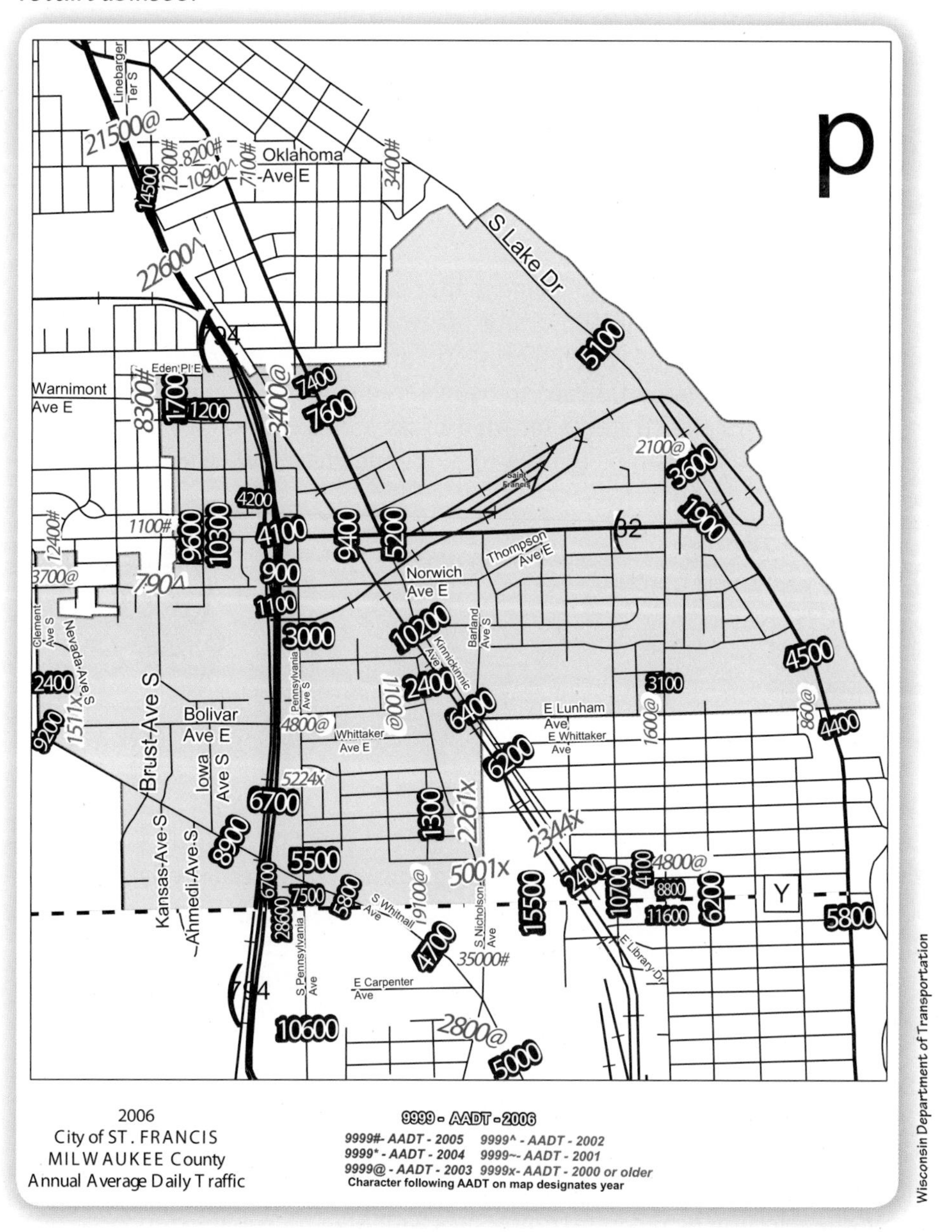

Wisconsin Department of Transportation

and distributors will probably be more concerned with finding adequate shipping and receiving facilities and being near major transportation routes. Manufacturers must take into account a host of factors including availability of raw materials, primary market location, transportation routes, and the availability of a (skilled) labor force. All of these factors must be researched and considered in the site selection process.

Because the success of a business is highly dependent on the people who live and work in the vicinity of the property and the businesses that have located nearby, you should gather **demographic data** to get an objective view of the characteristics of your potential customers, employees, and competition. In order to effectively use demographics, you need to know your target customer or employee,

Demographic Data or Demographics:
population characteristics such as race, age, household income levels, mobility (in terms of travel time to work or number of vehicles available), educational levels, home ownership, and employment status.

that is, what type of person or business would you like to hire or expect to use your product or service? Are there enough of these people or businesses in the area to support the proposed business.

Fun Fact

Success is in the Demographics

Locos Deli and General Store opened in 1988 as a tiny establishment that sold sandwiches, milk, toilet paper, and other general supplies to the students of the University of Georgia (http://www.locosgrill.com/). After 20 years, Locos has reinvented itself and boasts 22 Locos Grill and Pub locations throughout the southeast. Today, the company offers franchise opportunities to people interested in opening a Locos location. The success of Locos is due, in part, to the strategic location of its stores. Based on research, Locos has established the following traffic and demographic criteria to guide franchise site selection.

- Population: at least 50,000
- Traffic count: at least 25,000 vehicles per day
- Percentage of population under 54 years (70 percent)
- Median age range: 30–45 years
- Average minimum household income: $50,000
- Population growth of at least 5 percent

Target Market: a specific group of consumers at which a company aims its products and services.

Once you can identify desirable demographics, you can compare your **target market** to the demographics of the site being considered. Alternatively, if you are just beginning the site selection process, you can use demographic data to identify potential sites that would provide a good base of customers for your business. You can usually obtain demographic data from the city, town, or county in which the site is located. Many times local demographic data is taken from the United States Census Bureau data base available online at the U.S. Census Bureau site: http://factfinder.census.gov/.

An alternative to researching the local market yourself is to hire a site-selection service or a market analysis professional. These service providers can prepare a market analysis of a particular site or analyze and compare many sites in order to assist in site selection by recommending sites with high potential for success.

SITE VISIT

At this point, you have thoroughly researched your potential site. According to the information you have gathered during site discovery, everything is in good order. It seems that the site is a good choice for your development. But wait, when was the last time you visited the site? When the real estate agent showed you half a dozen properties? If you haven't looked at the site up close and personal since you began site discovery, you need to schedule a site visit. A site visit will give you the opportunity to gather site-specific information not available through other sources. It will also allow you to verify and reassess the data that you have gathered and look for any potential problems that you might have missed. The Site Inspection Checklist (Figure 5-48) will help you analyze the site. To make the most of your

Figure 5-48: *Site Inspection Checklist.*

SITE INSPECTION CHECKLIST

A site visit should be well documented with sketches, notes, and photographs. Be sure to take the following supplies when you visit the site:

- a site map,
- contour map (if available),
- writing utensil,
- note paper,
- tape measure,
- flags or stakes,
- a shovel and buckets (for soil samples).

□ Locate property corners, as accurately as possible, with flags or stakes.

□ Take photographs. Be sure to record the location and direction of all photos taken and locate all important information on the site map. You should photograph (at least) the following.

- Road frontage from the centerline of all adjacent roadways.
- Roadways–document the width and composition.
- Existing utilities, structures, and pavement – locate on map.
- Existing trees, bodies of water, and vegetation of interest.
- Potential building and parking locations.
- Views from potential building locations in all directions.
- Other areas of interest as noted below

□ Identify Potential Problems. Consider each of the following while walking the site. If a potential problem is identified, locate areas of concern on the site map, photograph each, and take notes.

- Inadequate size and shape of property for proposed use
- Poor physical location of site
- Incompatible noise, smells, or surrounding land use
- Extremely flat or steep terrain
- Unstable soils, excessive erosion, or sedimentation
- Wetlands, floodplains, or standing water
- Environmental contamination or hazards
- Evidence of historical artifacts or endangered species
- Undesirable microclimate

□ Record Traffic Conditions. Identify adverse traffic conditions as discussed earlier in the chapter.

- No left turn lanes or traffic lights on busy adjacent roadways
- Traffic congestion that severely restricts vehicle flow
- Poor sight distance at probable entrances and exits
- Existing driveways or intersections near proposed entrances or exits
- Traffic count, time permitting

□ Perform Percolation Test, if necessary. If a sanitary sewer system is not located near the site, perform a percolation test to assess the potential for a septic system. Record the location of the test(s) on the site map.

□ Gather Soil Samples. Dig to the anticipated level of foundation bearing and take soil samples at several locations. Record the location of the samples on the site map.

Research the demographics of your area. Visit the U.S. Census Bureau website at *http://factfinder.census.gov/*. Under the American Community Survey, choose "get data", then "data profiles." Select the county or urban area in which your project will be built, or, if information is not available for your project area, select another county or urban area near you. Document the social, economic, housing, demographic, and narrative profiles for your area of interest.

visit, consider possible site layouts and document your ideas with bubble diagrams before you go. With potential options in hand, you can more efficiently analyze the site to consider pros and cons of each option.

CONSTRUCTION COST ESTIMATE

How much will your project cost? Now that you have researched the site, you have a better appreciation for what is necessary to develop your plan. At this point in the process, you do not have enough information to get a detailed estimate of construction costs for your project because it has not yet been designed. However, you should have a plan for some of the basic parameters of the project: the floor area of the building and parking area, the height of the building, the probable type of foundation and building system, and the types of interior finishes. A cost estimate will include all costs associated with the project including preliminary expenditures (such as preliminary design, permits, and fees), engineering and architectural design fees, costs to prepare the site (such as clearing and grading), required infrastructure improvements, and the material and labor costs to construct the facility. An engineering or construction firm that has experience with similar projects in the vicinity of your planned project should be able to give you a ball park estimate of the construction cost for the project.

In lieu of professional services, you can perform a less reliable sq ft estimate of the construction cost. Many sources provide construction cost information including the well-known RS Means, which publishes extensive construction cost data. RS Means also provides a very simple online sq ft estimating tool via the RS Means Quick Cost Estimator at *http://www.rsmeans.com/calculator/index.asp*. This estimator requires only the type of building (i.e., single family home, hospital, etc.), the building floor area in sq ft, and the zip code of the project. Because many aspects of a project are not taken into consideration, you should not rely on this estimate as a final cost, but it will give you an idea of the potential cost of the building. Remember, there are many other costs involved with development—we have discussed many of them in this chapter. Be sure to consider all known costs in your estimate, as well as a *contingency* to cover unexpected expenses.

A good example of costs exceeding pre-construction estimates is the CityCenter Project. Initially a detailed estimate projected the cost of construction for the project to be 7.4 billion dollars, estimates rose to 9 billion within a year after construction began and continued to rise as construction progressed.

VIABILITY ANALYSIS

As you have progressed through the site discovery process, you have gathered various documents and learned a great deal about your site and the rules and regulations that control development. How can you use this information to help you decide whether or not your planned development is viable? Is spending more time and money on the design phase of your project a good decision or should you go back to the drawing board? A **viability analysis** requires you to consider the potential for success of the development by answering three important questions.

- Is the project legally permissible?
- Is the project physically possible?
- Is the project financially feasible?

Throughout the chapter we have discussed site conditions that affect the answers to these questions. However, one important consideration that has not been addressed is the cost of construction. It should be emphasized that project cost is a critical aspect of project viability. Although you are now familiar with many aspects of development that may increase the cost (such as complying with regulations, protecting the environment, and addressing site conditions) you need an estimate of the cost to physically build the development in order to decide if the project will be affordable and profitable, and therefore financially feasible. In addition, if you will not personally finance the project, you will need to involve a financial institution or investor to ensure a source of financing before proceeding with the project design.

Take some time to review the information gathered during site discovery to perform a viability analysis. Make sure you are confident that your proposed project is

- *legally allowable* because it meets all zoning ordinances, deed restrictions and covenants, and can comply with all other government regulations;
- *physically possible* because the size, shape, topography, available resources, infrastructure, and services are sufficient to successfully build and support the project; and
- *financially feasible* because, during the site discovery process, you have considered the cost of construction and other key aspects of development and are confident that the development will provide a reasonable profit.

We have discussed all three of these considerations throughout this chapter. If you are reasonably sure that your project meets all three criteria, and that your plans are viable, you may proceed with the design phase of the project with a measure of confidence that you have avoided many of the potential problems that can doom a development to failure.

OFF-SITE EXPLORATION

Value Engineering is a process using a team from a variety of disciplines to improve the value of a project through the analysis of its functions. Find out about Value Engineering on the U.S. Army Corp of Engineers webpage at *http://www.nab.usace.army.mil/* and search the site for "Value Engineering Program."

Resources such as information, energy, materials, machines, energy, people, time, and capital are needed to get a job done.

BEST AND HIGHEST USE

In some cases your goal in site discovery is not to find a viable site for a given project but to determine the best use of a given site. Let's say that you inherited a piece of property from your favorite rich uncle. Although the property was part of a farm when your uncle bought it 40 years ago, the town has grown and the old farm is now zoned for commercial use. You would like to make the most of the property. Should you simply sell the undeveloped property? Should you develop the property? If development is the answer, what should you build to bring the highest profit?

In real estate, a property is appraised at its **highest and best use,** which means, the property value is based on the use that would produce the most profit. In order for a potential use, say a video store, to be the highest and best use, it must meet each of the three criteria discussed in the previous section and it must be the most productive use. In order to find the most productive use of the property, you should compare several potential uses.

Consider an example. First, consider all legally allowable uses for your property.

> The 38-ac property is zoned for apartments, recreational, or commercial retail. No deed restrictions apply and all uses can meet government regulations.

Second, determine whether each of the uses is physically possible on the site.

> The site is rectangular and has a gently sloping terrain that contains a 4-ac pond toward the northeast corner of the property. Because of the relatively flat slope, rectangular shape of the property, proximity to the road, and accessibility to utilities, all permitted uses are physically possible.

Next, the financial feasibility of each possible use is considered. Only uses that have the potential to produce enough income to justify the cost of construction and produce a profit are considered further.

> Based on moderate growth in this area, demand for apartments and recreational facilities is expected to increase. Apartment housing has a minimal vacancy of 3 percent. There is one small gym in the area that is consistently crowded and busy, and few other recreational opportunities. However, there is an overabundance of retail commercial strip centers in the area resulting in 30 percent vacancy in retail space. Therefore, commercial retail space is not recommended for this site. Feasible economic uses for this site include apartments, and recreational facilities.

Finally, the maximally productive use is determined by comparing the estimated profit for each use. If you plan to develop the property and then sell it at market value, this requires a comparison of estimated cost to fair market value of each option. A simplified comparison for the property follows

Recreational Facility

Cost to construct a 10,000 sq ft recreational facility: $1,200,000.

Market value of 10,000 sq ft recreational facility: $1,350,000.

Profit for recreational facility = $1,350,000 − $1,200,000 = $150,000.

$$\textit{Percent profit} = \frac{\textit{profit}}{\textit{cost}} \cdot 100\% = \frac{\$150{,}000}{\$1{,}200{,}000} \times 100\% = 12.5\%$$

Apartment Complex

Cost to construct 10,000 sq ft apartment complex: $1,500,000.

Market value of 10,000 sq ft apartment complex: $1,750,000.

Profit for apartment complex = $1,750,000 − $1,500,000 = $250,000.

$$\textit{Percent profit} = \frac{\textit{profit}}{\textit{cost}} \times 100\% = \frac{\$250{,}000}{\$1{,}500{,}000} \times 100\% = 16.7\%$$

Mathematics can be used to solve a variety of problems such as determining the best and highest use of a property.

Even though the recreational facility will cost less to build, the potential profit for the apartment complex is greater than that for the recreational facility. The highest and best use of the property is an apartment complex. Because every property is different in characteristics and location, determining the highest and best use of a property will generally require the expertise of a licensed or certified real estate appraiser—we will not estimate the potential profit for each building type. However, the cost of construction alone is a valuable tool in site selection.

SUMMARY

After reading this chapter, you know that finding a viable site for a new development will require some time and detective work. You also know that the effort involved in the site discovery process can help ensure a successful project. Investigating the physical characteristics of the land, the natural environment (including plants, animals, and natural resources), the climate and weather, and the infrastructure available to the site will help you estimate the cost of construction and the usability of the completed project as well as provide information that will help you protect the health and safety of the public and the environment. In addition, researching the rules and regulations that govern the development will provide information necessary to determine if the proposed development is feasible. The Site Discovery Checklist provided in Figure 5-49 can help you gather the information you need to make an informed decision.

You may never be sure that you have chosen the *best* site, but you can be reasonably sure that the site you chose is a "good fit" for your project and does not pose problems that can jeopardize the success of your project. Without sufficient site discovery you put your project and your reputation at risk.

Figure 5-49: **Site Discovery Checklist.**

SITE DISCOVERY CHECKLIST

Note: The following list includes sources of general information for a site; however, site-specific information is preferable and may be necessary for accurate site analysis.

□ Deed – Recorder of Deeds or *Public Records Online Directory website*
□ Plat – Recorder of Deeds or *Public Records Online Directory website*
□ Tax map–Recorder of Deeds or *Public Records Online Directory website*
□ Topo map (free USGS topo Maps via Map Locator at http://store.usgs.gov/)
□ Soils map and soil characteristics data, soil boring data (if available)
WSS at http://websoilsurvey.nrcs.usda.gov/)
□ Building Code
□ Municipal Ordinances (Online Library at http://www.municode.com/)
□ Zoning Map (municipal ordinances)
□ FIRM (FEMA Map Service website at http://msc.fema.gov)
□ Wetlands Inventory (National Wetlands Inventory at http://www.fws.gov/wetlands/Data/Mapper.html)
□ Coastal Zone regulations if applicable (http://coastalmanagement.noaa.gov/mystate/welcome.html)
□ Endangered Species Critical Habitat Map (FWS Critical Habitat Portal at http://criticalhabitat.fws.gov/)
□ Tree Survey (site specific)

- □ Superfund sites (EPA National Priorities List webpage at http://www.epa.gov/superfund/sites/npl/locate.htm)
- □ Historically and Culturally significant sites (http://www.nationalregisterofhistoricplaces.com/state.html)
- □ Climate Normals (http://www.cdc.noaa.gov/USclimate/states.fast.html)
- □ Wind Data (Prevailing Wind Map or Wind Roses available at http://www.wcc.nrcs.usda.gov/climate/windrose.html)
- □ Solar Orientation (U. S. Naval Observatory Sun or Moon Altitude/Azimuth Table at http://aa.usno.navy.mil/data/docs/AltAz.php)
- □ Utilities and Municipal Services (contact local municipality and/or service providers)
- □ Traffic (state highway department or local municipality)
- □ Demographics (U. S. Census Bureau site http://factfinder.census.gov/)
- □ Site Visit
- □ Cost Estimate (RSMeans Quick Cost Estimator at http://www.rsmeans.com/calculator/index.asp)

BRING IT HOME

1. Copy the legal description given in the 1891 Land Grant in Figure 5-11. Locate the correct Principal Meridian and Baseline on the map in Figure 5-12. Then create a sketch similar to Figure 5-13 to exactly locate the property. Include calculations to verify the area given in the land grant.
2. Using the Hana Quadrangle USGS map shown in Figure 5-18 answer the following questions:
   - What is the contour interval of the contour map?
   - What is the approximate elevation at Hana High School?
   - The shoreline on this map represents the level of mean high water. What do you think the blue contour lines represent? Verify your answer on the USGS website?
3. The Municipal Code Corporation provides access to many municipal codes via its website. Visit the Municipal Code Corporation website at *http://www.municode.com/* and follow the link to the Online Library. Search each of the following municipal codes to determine which, if any, model building code has been adopted by each local authority.
   - The Revised Code of Ordinances for the City of Biddeford, Maine. Section 18-76 of that Code provides information on the adopted model code.
   - The Administrative Code for the City of Reno, Nevada. Section 14.04.010 of the code specifies the applicable building code.
   - The municipal code for your city, county, or another local municipality. In some cases the local code will refer you to a state regulation. You may need to search the Internet to find the state regulation specifying the applicable building code.
4. Research the municipal code for your city or county to identify zoning designations used in your area. Describe at least one designation in each of the three major categories of development: residential, commercial, and industrial. Then explain how zoning can help or hinder site development.
5. Assume that you are a member of a team that is responsible for presenting potential sites for a new bookstore (comparable in size to a Barnes & Noble) in your area. As a team, research environmental factors that may affect the site selection process. This may include floodplains, wetlands, endangered species, site contamination, and historical and cultural resources.

Draft a letter to your client summarizing your findings and the possible impacts associated with developing the site.

6. Research your local municipal code to identify tree ordinances and document the important requirements in your notebook. Using a plat map as a basis, prepare a tree survey drawing of your home property (or a project site) according to your local code. If your local ordinances do not require a tree survey drawing, create a drawing of the site to include a footprint of any existing buildings and all trees larger than 8 in in diameter. For each tree indicate location, size (diameter at a height of 4.5 ft), and species of tree.
7. Investigate the climate conditions near your school including information on temperature, precipitation, wind, and sun. How would each of these climatic conditions affect the design of a new commercial building located near your school?
8. Identify at least two businesses in your community that suffer from poor traffic conditions. How do these conditions affect the success of the businesses?
9. Research the demographics of your area using the U.S. Census Bureau website at *http://factfinder.census.gov/*. Under the American Community Survey, choose "get data," then "data profiles." Select a county or urban area near you. Document the social, economic, housing, demographic, and narrative profiles for your area of interest. Why is this information important when researching a site for development?
10. Using the Site Inspection Checklist included in this chapter, perform a site inspection of a potential project site. Be sure to obtain permission to assess the site before the inspection.
11. Pretend that you are investigating a particular site for development of a retirement community. List-specific information that you could potentially obtain during site discovery that could provide justification for not locating the development on the site.

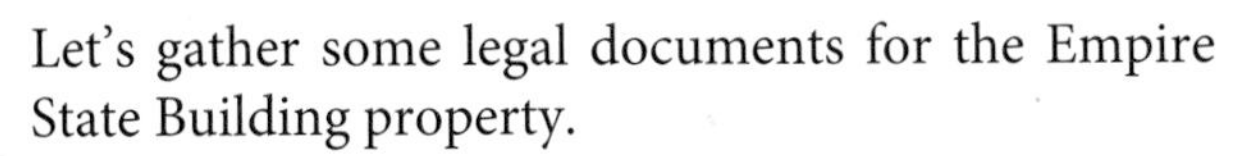

EXTRA MILE

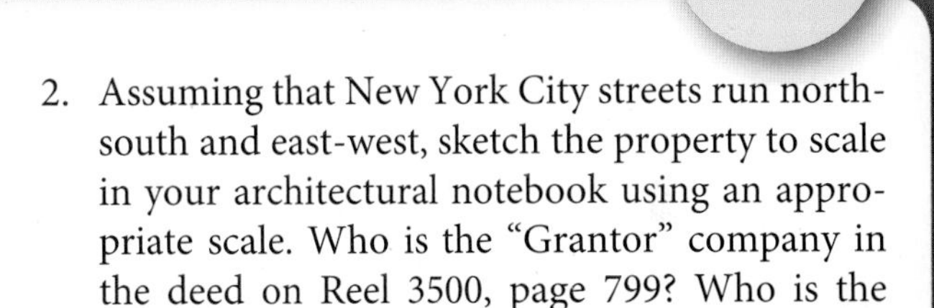

Let's gather some legal documents for the Empire State Building property.

The Empire State Building address is:
350 Fifth Ave,
New York, New York
County of New York
Borough of Manhattan

Visit Public Records Online Directory *at http://publicrecords.netronline.com/*. Select the state of New York, and on the new page choose New York County. You will want to access the New York NYC Register online data. Using the property information given, access the ACRIS and choose Find Addresses and Parcels to find the Lot and Block numbers of the property. Using the lot and block numbers search the property records to find a deed that gives a legal description of the entire lot and complete the following:

1. What type of legal description is used to describe the property in the deed? Print out the legal description and tape it into your architectural design notebook.
2. Assuming that New York City streets run north-south and east-west, sketch the property to scale in your architectural notebook using an appropriate scale. Who is the "Grantor" company in the deed on Reel 3500, page 799? Who is the president of the "Grantor" company?
3. Search for mortgages on the property that were recorded on the same date as the deed on Reel 3500, page 799? What is the mortgage amount associated with the sale of the Empire State building on that date? Note: there is more than one mortgage document recorded on this date, find the document with the highest document amount

Search other documents related to the Empire State Building Property. What happened on July 2, 1982? Search the New York City government website (*http://www.nyc.gov*) to find out what this designation means. Make a list of information that a new owner of the Empire State building property should know if he plans to make changes to the property.

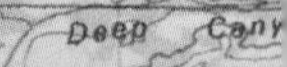

CHAPTER 6
Site Planning

GPS DELUXE

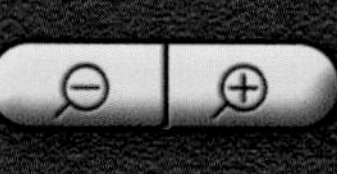

Menu | START LOCATION | DISTANCE | END LOCATION

Before You Begin

Think about these questions as you study the concepts in this chapter:

1. How do you determine the specific requirements and constraints for the project design?
2. What information can be obtained from a land survey and how does the information affect the design of a project?

3. How do you find out about the type of soil present on a site? How does the type of soil present on a property affect the project design?

4. What factors affect the "buildable" area of your site?

5. How does the building orientation on the site affect the site opportunities available and constraints to the design?

INTRODUCTION

Have you ever begun a project without planning the outcome? Perhaps you are building a DVD rack to store your video collection under your bed. You know how big the DVD cases are and the opening size that will accommodate the cases. So you visit your local building supply store and buy supplies. After a couple of hours of cutting, gluing and nailing you discover that you didn't buy enough wood, so you head back to the store. When you return, you cut the final piece and put the final touches on the rack—you are finished, at last. But wait, the DVDs do not fit in the rack—you forgot to consider the thickness of the wood during assembly. Should you disassemble the rack and rebuild or start over? In frustration, you decide it is too late to do anything and try to place the rack under your bed. Snap! You did not plan for a clearance between the rack and the bed, so the wood scrapes the bottom of the bed. And you really hadn't given much thought to making the rack easy to use by including some sort of handle that allows you to access the rack without crawling under the bed. Oh, and you forgot that you already store your guitar, sleeping bag and tent, and old Lego bricks under your bed. So the DVD rack would not fit anyway. Because of your lack of careful planning in the beginning, the project has turned into a minor disaster that will take more time, money, and effort to fix. The same thing can happen when developing a site—without careful planning, the project can result in material waste, damage to the surroundings, congestion of features, and a facility that is not user-friendly.

The goal of site design is to enhance the project with the least disruption to people and the environment. To make the best use of a site, civil/site engineers must consider many factors and balance the use of man-made elements with natural features. The best designs are achieved by following a site design process that involves the entire design team. First the goals of the project must be identified and detailed site information gathered. This information should then be evaluated with respect to its impact on the project. Next, site design concepts should be generated for comparison. Once a site concept is agreed upon, a site plan can be developed. The preliminary plan can be adjusted and revised in order to create a final site plan that accomplishes the project goals and optimizes the use of the site. The people involved in the project (clients, users, employees, and design professionals) are sometimes called **stakeholders** and should be involved throughout the design process to ensure that the site design meets the needs of the user and client in an attractive and environmentally friendly way.

In certain circumstances, when the project will significantly impact surrounding property owners or the larger community, it is advisable to include representatives from these groups in the process. By soliciting feedback, you can address potential complaints and provide solutions that may alleviate the concerns of those who will be impacted.

CONCEPT- Enlarged Partial Site Plan

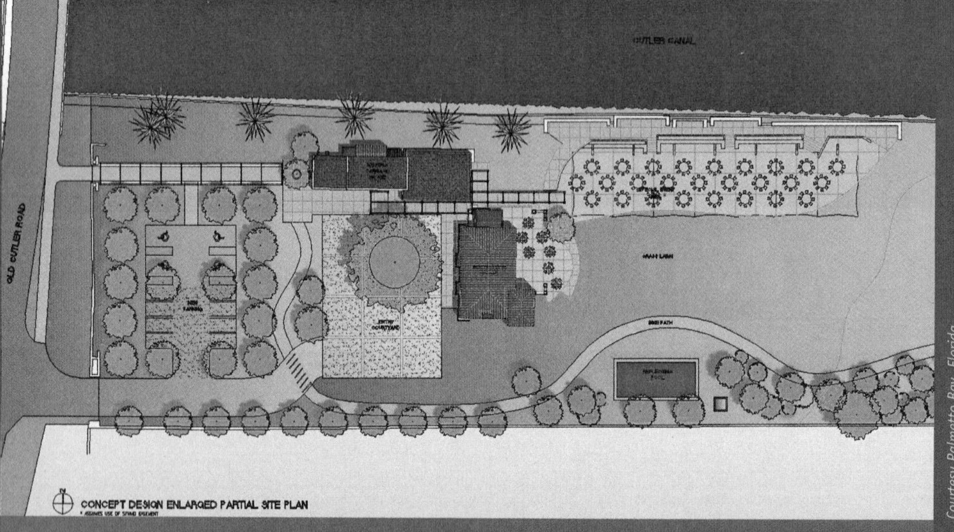

Courtesy Palmetto Bay, Florida.

Figure 6-1: *The best site designs address the needs of the users in an attractive and environmentally friendly way. This property on Biscayne Bay surrounds a mansion built in 1926 that will be conserved as a public park.*

PROBLEM IDENTIFICATION—PROGRAMMING

Before you can solve a problem, you must first know *exactly* what the problem is. The specific problem will be different for every project. A "good" design will meet the unique needs of the client and the users while responding to the characteristics of the specific site, meeting all regulations, and minimizing the cost of construction, maintenance, and operation. So, the first step in the site design process is to identify the specific design problem and then translate the problem into constraints and criteria for the project. Architects typically prepare a written document called a **program** to describe the design objectives, constraints, and criteria of a project.

Needs Analysis

Who better to tell you what features the facility needs than the people who will use it? Often, the client is the end-user or is a representative of the end-user and can provide much of the information that you will need, but you should have input from everyone who will use the final project: the owner, employees, customers, service personnel, and surrounding community members when determining the project criteria. Gathering information on the goals, needs and function of the project, design expectations and available budget is often called a **needs analysis** and is used to prepare the project program.

Figure 6-2: A needs analysis should include input from the client as well as employees, customers, service personnel, and other stakeholders.

Every aspect of the project design will be based on the needs analysis, so it is important to have a complete understanding of the project goals before design begins. Although in this chapter we will concentrate on needs that will influence site design, the initial needs analysis should provide information to guide the whole building design. In many cases it may not be possible to fully meet every need, but the needs analysis provides you with a basis for your design—a goal. Survey your client and other stakeholders. Ask a lot of questions and research other similar facilities to make sure you have the important information you need to create a "good" design. We will revisit the needs analysis in Chapter 8 when we discuss space planning.

The decision to develop a project is influenced by corporate culture and financial considerations as well as societal opinions and demands.

ADJACENCY MATRIX The needs analysis should provide information about which types of activities, both interior and exterior, will occur on the site and what features are needed to make those activities possible. Sometimes activities

Point of Interest

In 2004, the city of Scottsdale, Arizona, purchased the 42-acre site of an abandoned mall with the intent of revitalizing the area. The city's goal was to create "an urban, mixed-use knowledge-based center, which includes high-tech business incubation, education, research, office, and possibly create options to incorporate appropriate residential, commercial, and supportive retail uses." Scottsdale hired Urban Design Associates (UDA) of Pittsburgh, Pennsylvania, to perform a needs analysis and to prepare design guidelines for the project. To provide input from the citizens of Scottsdale (the end-users) the city council created a Citizens Advisory Working Group to work with the city and UDA to develop the guidelines. The result was a 53-page document titled "Design Guidelines and Development Framework for the ASU-Scottsdale Center for New Technology and Innovation and the Surrounding Area." The document details the needs to be addressed and the vision of the city for the development. For example, one of the guiding principles for the development was to create "meaningful open space and public uses," which included creating indoor and outdoor community gathering places. A resulting design guideline (criteria) states that the outdoor space "must be accessible and welcoming to the residents of the adjacent neighborhoods and be of sufficient size (1.0 to 1.5 acres) and designed for outdoor performances and festivities, with public art, benches, trees and shelters for shade, water features, and appropriate lighting."

Courtesy Pei Cobb Freed & Partners Architects LLP.

Figure 6-3: Perspective rendering of the ASU-Scottsdale Center showing where people gather under the SkySong shade structure.

Your Turn
A Perfect School

Assume that you are a member of the design team assigned to design a new high school for your district. What do you need to know in order to create a good design? Think about educational, technical, recreational, medical, and personal needs. Consider the needs of the community and the administration. What about energy efficiency, aesthetics, and durability? Create a list of questions that you would want answered before beginning the design of the school.

Answer the questions from a student's perspective. What guiding principles and specific features would you include in the design of a new high school? Record these ideas in your project notebook.

Ask a teacher for input in the design. Are the needs and wants of your teacher different from those of students? Record the teacher's responses in your project notebook.

Invite a principal or administrator from your school to your classroom to answer your questions. Record the answers in your project notebook. How did the administrator's ideas differ from yours? How would the ideas of a cafeteria worker or janitor differ? How about a nurse, coach, guidance counselor, or school resource officer?

Ask the opinions of several community members who don't have children. Will their views differ from people who have children?

Create a list of design criteria or guidelines for the design of a high school.

or uses, in a building or on a site, are incompatible and should not be located near each other—for example a playground and vehicular traffic. Other times site activities or features should be located adjacent to (that is, next to) each other for ease of use—a kitchen should be near a dining room. An **adjacency matrix** is a special type of table that can be used to rate the desired relationship of the activities or uses and help determine the best spatial relationships for a project's features. The adjacency matrix in Figure 6-4 uses four categories for desired proximity of site design elements for a proposed engineering office building. Each category is designated by a number as indicated by the key. For instance, to find the desired proximity of the septic field to the wetlands area, follow the septic field row to the wetland column to find a rating of 4 as shown in red in the figure. Because the wastewater dispersed by a septic system can contaminate the wetland, a 4 is used in the adjacency matrix to indicate that adjacency should be avoided, the septic field should not be located near the wetland area. And, since the desired proximity between the parking area and the building has a rating of 1 (as shown in blue), the parking area should be located adjacent to the building.

Figure 6-4: A site adjacency matrix will indicate the proximity of site features.

Adjacency Matrix Engineering Office Bldg Site	Building	Parking	Ingress/Egress	Detention Pond	Septic Field	Existing Trees	Wetlands
Building							
Parking	1						
Ingress/Egress	3	1					
Detention Pond	3	2	3				
Septic Field	4	3	3	4			
Existing Trees	2	2	3	3	4		
Wetland	4	4	3	2	4	3	

Key
1 = Adjacency is essential
2 = Adjacency is desirable
3 = Adjacency is unimportant
4 = Adjacency should be avoided

SITE CHARACTERISTICS

A good site design will take into consideration and enhance the natural characteristics, good and bad, of the site. This demands a detailed knowledge of the site. Although you will have gathered an extensive amount of information during site discovery, you will need to investigate the site in more detail in order to be able to create an effective site plan.

So, how do you get an intimate knowledge of your site? Surveying, in the broad sense of the word, refers to gathering and processing information about the earth and the environment. You should survey all aspects of the site. If required, you should initiate an environmental study (discussed in Chapter 5) to delineate areas of the site, species, and cultural resources that should be protected. If not already complete, a property survey should be performed in order to establish property boundaries, utility locations, and critical lines. A topographic survey of the property will establish the ground contours and drainage patterns of the property that can affect your site layout. If the site is forested, a tree survey may be required in order to locate trees that might be protected by local codes. A soils survey and geotechnical analysis are often needed because the drainage system, pavement design, and structural design may all be affected by the soil characteristics.

Land Surveys

Although surveying can refer to many different types of studies and investigations, it is most often used to refer to **land surveying.** Land surveying is the science of determining the relative positions of points on or near the earth's surface and can be separated into two major categories. **Geodetic surveys** factor in the curve of the earth's surface and are typically performed with great precision. Geodetic surveys are used to establish highly accurate control networks, which are used as a basis for maps and locating features. The curve of the earth will affect the design of larger projects such as highway systems. **Plane surveys** ignore the curvature of the earth and assume a flat planar surface. This approach simplifies position calculations because distances can be measured as straight lines. Plane surveys can be used within small areas in which the earth's curvature will have little effect. Most architectural projects can be based on plane surveys.

Many types of surveys can be performed depending on the purpose of the survey. Some, but not all, of the more common types of surveys are listed below:

1. **Control survey.** Control surveys are used to establish precise horizontal and vertical positions of points that serve as a reference framework for other types of surveys.
2. **Construction survey.** A construction survey locates points and elevations that can be used to establish correct locations and elevations for civil engineering and architectural projects and is often called an *engineering survey.* For instance, a construction survey might stakeout a line and grade for a foundation, fence, or road.
3. **Topographic survey.** A topographic survey is used to gather data to prepare topographic maps that show the location of natural and man-made features and the shape and elevation of the ground. Engineers and architects often use a topographic survey when designing improvements to a site.
4. **Property survey.** A property survey establishes property lines and is sometimes called a boundary survey, land survey, or cadastral survey. A **plat** is typically created from the survey. Professional registration is required in order to complete an official property survey in the United States. Sometime the terms *site survey, plot survey,* or *lot survey* refer to the combination of a property survey and a topographic survey and may be required in order to receive a construction permit. A plat is shown in Figure 6-5.
5. **Hydrographic survey.** A **hydrographic survey** provides data for mapping shorelines and the bottom of bodies of water. Hydrographic surveys are often conducted using **SONAR** (Sound Navigation and Ranging), which uses sound waves to determine depth and find objects in the water. Sometimes a survey that combines a topographic survey and a hydrographic survey is called a **cartographic survey.**
6. **Route survey.** A route survey is used to map existing routes or lay out new projects with long horizontal extents such as highways, pipelines, canals, and so forth.

HORIZONTAL AND VERTICAL DATUM As you can imagine, there have been millions of surveys conducted across the United States for many different reasons. In order for all of the resulting maps, charts, and construction drawings to align with each other, a set of reference points was needed so that locations and elevations were measured from common references. In the United States, the National Spatial Reference System (NSRS) provides horizontal (latitude and longitude) and vertical (elevation) reference points that can be used for any activity that requires accurate location information. This collection of reference points with known latitude and longitude is called the **horizontal datum.** Likewise, the collection of points with known elevation, called **benchmarks,** is referred to as the **vertical datum.** The National Geodetic Survey has placed markers throughout the country to identify these reference points of known horizontal or vertical position. Each marker is typically set by embedding a brass, bronze, or aluminum disk in concrete (see Figure 6-6) or bedrock but may be a deeply-driven metal rod, water tower, church spire, or other identifiable object. In addition, many local municipalities have established local coordinate systems that provide horizontal and vertical reference points for use within the municipality.

Benchmark:

A benchmark is a permanent mark that establishes the exact elevation of a point. Benchmarks are used by surveyors as a starting point for surveys to establish elevations of other points.

Figure 6-5: A plat is created from a property survey.

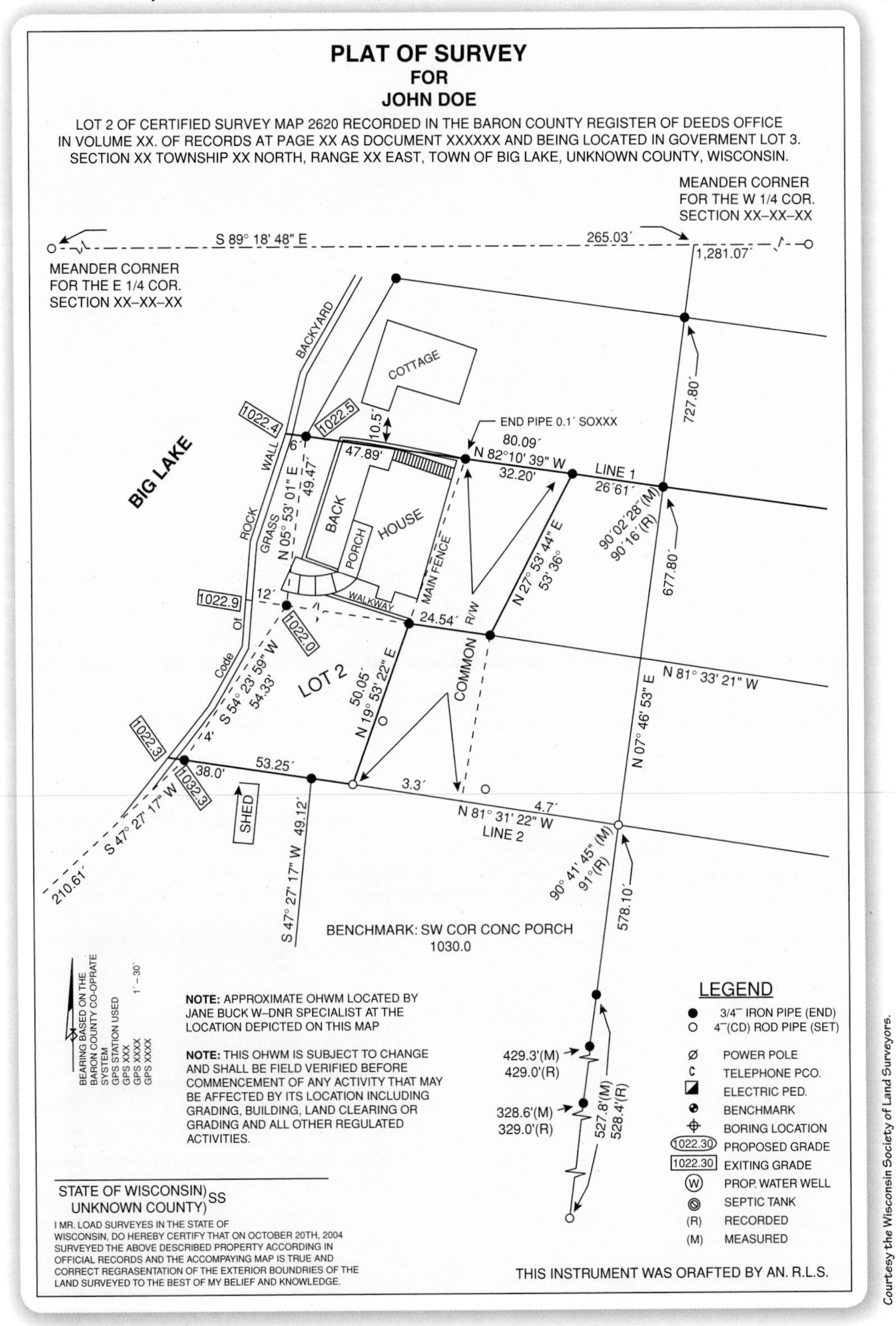

Courtesy the Wisconsin Society of Land Surveyors.

We depend on the horizontal and vertical data to help us accurately locate construction projects. Imagine that you are designing an extension to a city transit system that must meet the existing system at specific points. You need very accurate location information on the existing system, and you need to know the exact location of the extents of the property you are allowed to use. Even more importantly, this information

Figure 6-6: A survey team sets a new survey disk, in this case, a horizontal control mark. Reference marks are typically created by setting a bronze or brass disk in concrete.

Courtesy National Oceanic and Atmospheric Administration.

Your Turn

Mark Recovery

You probably pass at least one reference survey mark every day on your way to school, you just don't notice. Even though the survey marks are often intentionally placed in plain sight, many of the 1.5 million benchmarks in the United States have not been used recently. In fact, some have been lost—no one has seen them for a long time. But you can help recover a survey mark and have its location recorded in the national database. Visit the National Oceanic and Atmospheric Administration's education website at http://oceanservice.noaa.gov/education/welcome.html and search for "survey mark hunting" to find an activity on the subject. Follow the instruction in the activity to locate a survey mark in your area that has not been documented for at least five years, take a picture, and log your findings.

Science / SERVICE / Stewardship UNDERSTAND THE EARTH

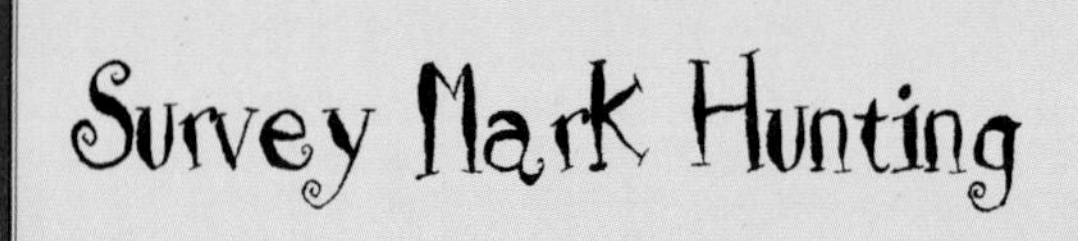

National Geodetic Survey Satellite Triangualtion Program, illustrating the idea of modeling the Earth. Courtesy NOAA Geodesy

Imagine bridges not meeting in the middle...
Airplanes landing next to runways instead of on them...
Ships frequently running aground...
This is just a glimpse of life without geodesy.

What's geodesy? It's the science of measuring the size and shape of the Earth and accurately locating points on the Earth's surface (and is pronounced "gee - ODD - ess - ee").

READ ON, and find out how geodesy can be a lot of fun!

Another way to think about geodesy is to imagine a world globe with a lot of pins stuck in it. Geodesy is about giving each of those pins its own "address" written as latitude and longitude. Why is this important? Because each of those pins can serve as a starting point for describing the location of any other point on Earth; just like when you want to tell someone how to get to your house, you give them a starting point that they know, like a road or a building. In the United States, these reference points are developed and maintained by NOAA's National Geodetic Survey (NGS).

Hang on, we're almost to the fun stuff!

So where are all those pins stuck in the globe? They are everywhere—more than 1,200,000 in the United States!

Of course, they really aren't pins. Instead, NGS uses permanent marks called "survey marks" (you may hear survey marks called "benchmarks," but benchmarks are only one type of survey mark). Often, survey marks are marked with a metal disk like the photo below, set in concrete or bedrock:

What You Will Do

Get information on the location and description of survey marks in your geographic area, and find out how to share your survey marking discoveries with the rest of the world!

Survey marks can also be stainless steel rods driven into the ground, drill holes in bedrock, bottles, pots, or landmarks visible from a long distance, such as a water tower, a radio mast, or a church steeple.

Courtesy NOAA Geodesy Collection

93

National Oceanic and Atmospheric Administration (NOAA)

Figure 6-7: Survey mark hunting has become a popular hobby. The National Oceanic and Atmospheric Administration provides easy instructions for finding and reporting lost survey marks. Visit http://oceanservice.noaa.gov/education/for_fun/SurveyMarkHunting.pdf for full instructions.

Courtesy NOAA National Ocean Service.

Fun Fact

One interesting application of the horizontal datum is monitoring the movement of the San Andreas Fault in California—the source of many earthquakes. Two of the Earth's tectonic plates meet at the 800 mi long fault that extends 10 mi below the ground surface. Geologists can determine how much the upper layers of the earth's crust move during earthquakes by monitoring the movement of monuments in the horizontal datum (see Figure 6-8).

Figure 6-8: **Survey reference marks have many uses including tracking the movement of tectonic plates. Geologists determine how much movement occurs in the upper layers of the earth's crust at the San Andreas Fault by monitoring the movement of reference monuments in the horizontal datum.**

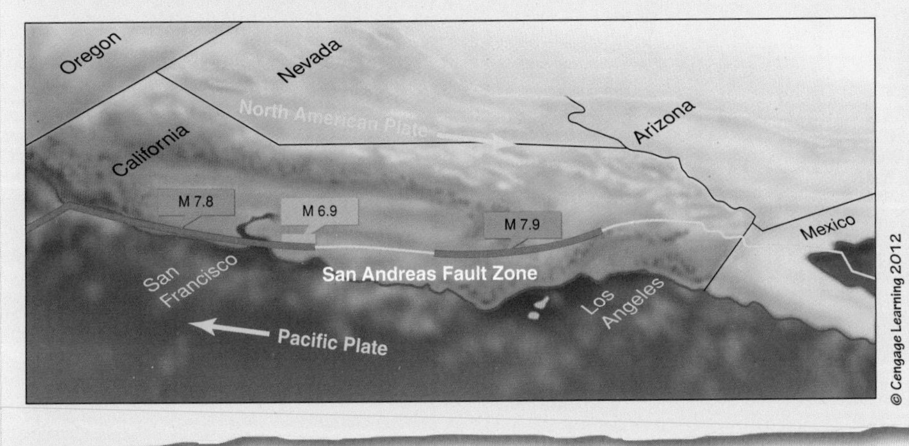

must be based on the same information that the designer of the existing system and all of the property owners along the route used during design and construction. Otherwise your project could end up in litigation when you infringe on another's property or when your design results in rails that don't line up with the existing system.

Conventional Surveys

Traditionally surveyors performed conventional surveys using a tape measure, for measuring distances; a level, to determine elevation; and a transit or **theodolite** to measure vertical and horizontal angles as well as distances. The level, transit, and theodolite are considered optical equipment, which means they rely on line-of-sight observations. In line-of-sight observations a surveyor looks through a telescope, sites a target, and reads a measurement as illustrated in Figure 6-9. An automatic level, or auto level, is commonly used on building sites to determine elevation because it is easy to set up and use. The auto level houses an internal compensator that automatically levels the instrument when approximately level.

Figure 6-9: **This autolevel is an optical surveying instrument—the surveyor looks through a telescope to sight a target and read a measurement.**

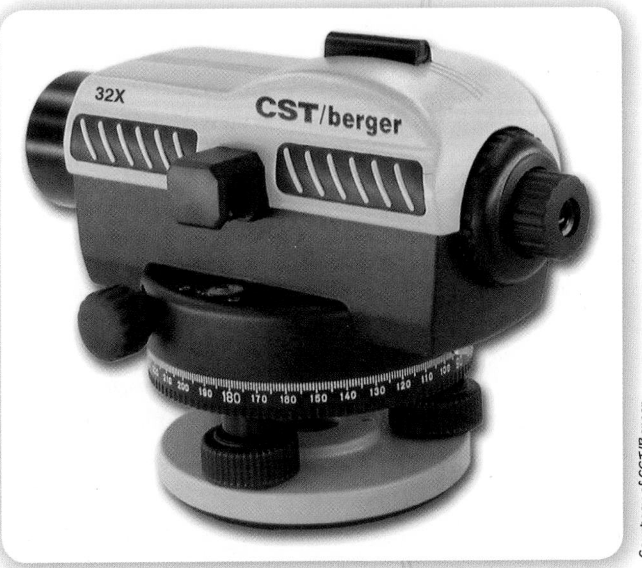

Figure 6-10: Optical surveying equipment is mounted on a tripod to stabilize the equipment and position the equipment for easy sighting through the telescope.

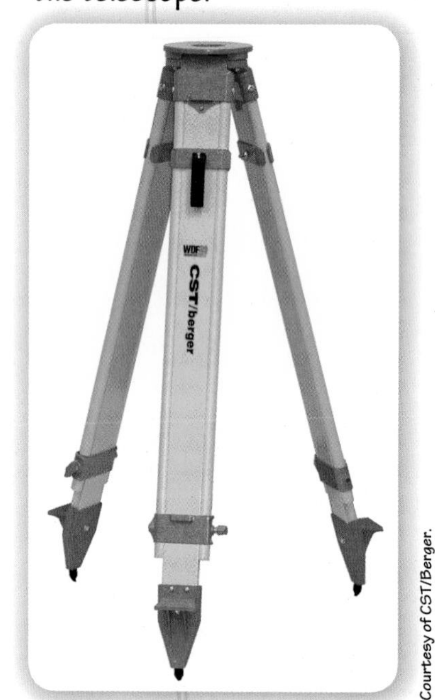

Courtesy of CST/Berger.

Figure 6-11: English mathematician and astronomer Edmund Gunter developed this type of metal chain in about 1820 for measuring length and determining the area of plots of land. The chain is 20 m (66 ft) long, and is made of 100 links and was frequently used as a unit of measure in early surveys.

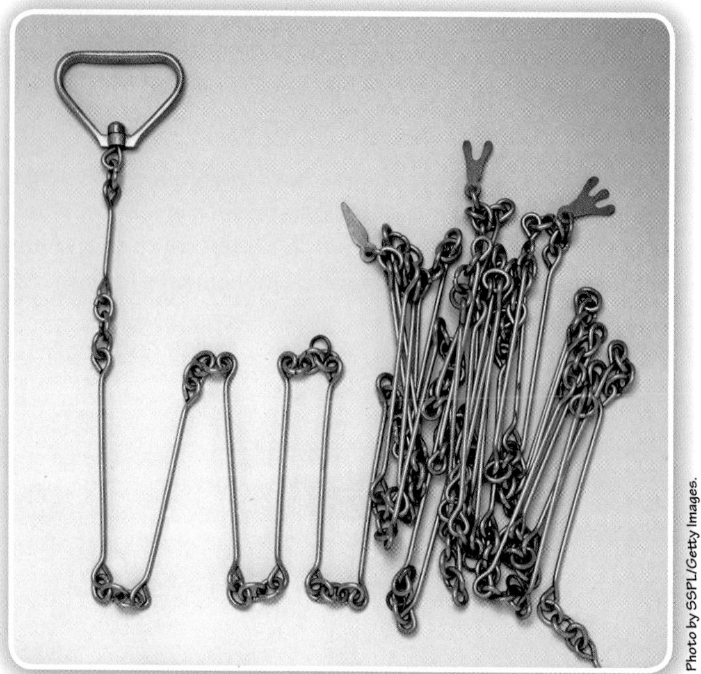

Photo by SSPL/Getty Images.

Theodolites and transits are very similar and the terms are often used interchangeably although they possess differences in the means of measuring angles. Levels, theodolites, and transits are mounted on a tripod to stabilize the equipment at a height at which the surveyor can look through the telescope (see Figure 6-10).

VERTICAL CONTROL—DIFFERENTIAL LEVELING The vertical position of a point can be established using **differential leveling,** a method that can be performed using a *level* and a *leveling rod.* A level essentially establishes a plane on which all points are at the same elevation. This is accomplished by setting the telescope of the instrument (which has a set of crosshairs) in a perfectly horizontal position that is perpendicular to the earth's surface. When the telescope is perfectly level, every point viewed through the telescope on which the horizontal crosshair falls is at the same elevation (see Figure 6-12). The vertical position of this equal elevation is referred to as the **height of instrument** (HI). The HI elevation can be determined by measuring

Fun Fact

In early surveys length was frequently recorded using units of chains and links as measured with a metal chain often referred to as a surveyor's chain (see Figure 6-9).

Figure 6-12: When a leveling instrument is set level, every point that corresponds with the horizontal crosshairs when looking through the telescope falls on a horizontal plane at the height of instrument. Every point on this plane is at a constant elevation.

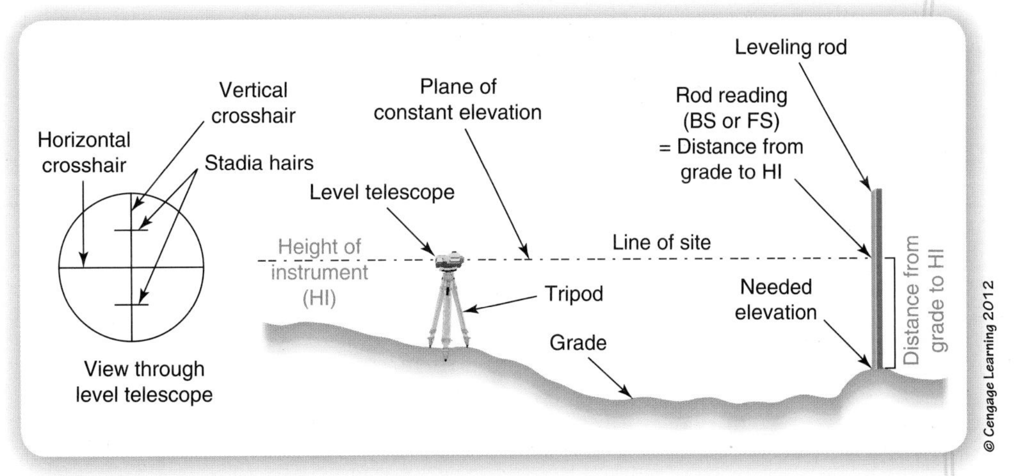

the *vertical* distance from a point of known elevation (a reference point, for example) to the HI. By measuring the vertical distance from the horizontal HI plane to a point of interest (which may be a spot on the ground, the top of a concrete curb, the top of a finished floor, or any other point), the elevation of that point can be determined (see Figure 6-13).

A **leveling rod** is used to measure the distance from the ground (or any other point of interest) to the HI elevation. A leveling rod is a long "stick" marked with graduations indicating accurate length measurements—it is really an extra long measuring stick (see Figure 6-14). By using the level to sight on the leveling rod (held in a vertical position) that is resting on the point of interest, the surveyor can determine the vertical distance from the point to the HI by reading the rod at the horizontal crosshair.

Figure 6-13: Differential leveling involves establishing the elevation of the horizontal plane of the telescope and using rod readings to determine the height of a point of unknown elevation using the height of a point of known elevation.

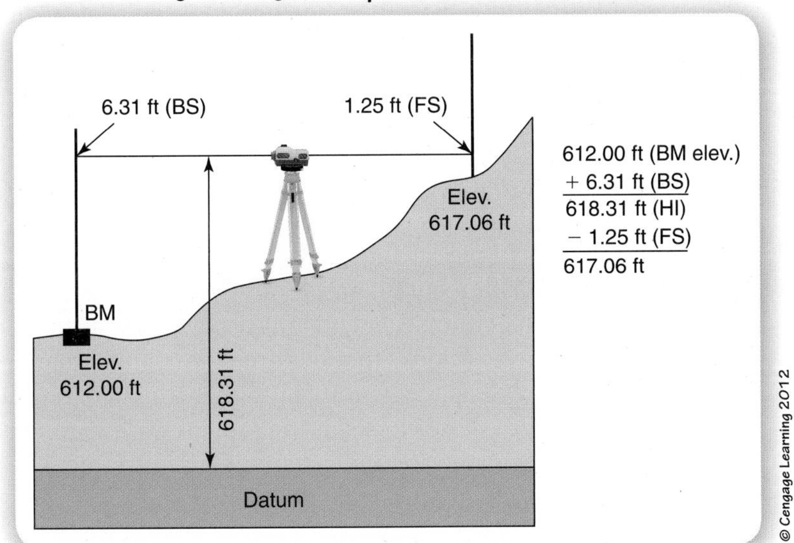

Figure 6-14: Leveling rod.

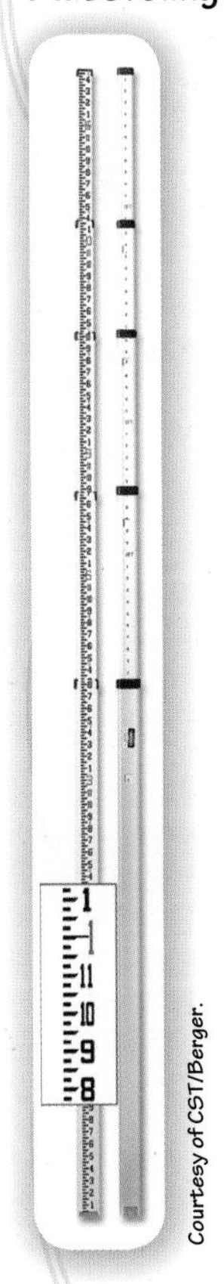

Figure 6-15: A total station has a theodolite and an electronic distance-measuring (EDM) instrument that allows distance and angle measurements to be taken with just one piece of equipment. The total station shown here is a Topcon GPT-3002LW Total Station.

Courtesy of Topcon Positioning Systems, Inc.

If you begin your rod readings by positioning the leveling rod on a point of known elevation (a bench mark (BM)), you can calculate the HI by adding the rod reading to the known elevation. A rod reading taken at a point of known elevation is called a **backsight (BS)** or plus sight. Once the HI is established, you can rotate the level telescope to sight in any direction and identify points of equal elevation. A rod reading taken at a point of unknown elevation is referred to as a **foresight (FS).** Because the FS is the vertical distance measurement from the point of unknown elevation to the HI, you can find the elevation of any visible point of interest by subtracting the FS reading from the HI.

Because the curvature of the earth is always neglected in plane surveys, elevations determined by plane surveys can be inaccurate. However, you can minimize, if not eliminate, these errors by taking care to equalize the distances of the FS and BS for every level location. This will balance the error in the plus sight and the minus sight so that the error is canceled. You can use the stadia readings to help ensure that your FS and BS are approximately the same distance from the level.

HORIZONTAL CONTROL Horizontal control establishes points of known horizontal position (longitude and latitude) on the earth. Today, conventional horizontal control is often established using a theodolite and an electronic distance-measuring device (EDM) or a total station (see Figure 6-15), which combines the features of a theodolite, an EDM, and a computer that collects data. EDMs and total stations typically emit infrared or microwave signals that are reflected back to the device by a reflector prism at a far point. The pattern in the returning signal is read and the computer calculates the distance to the prism. If the instrument is set up over a point of known location (for instance, a horizontal datum) such that the HI is known and is referenced to another known point, the three-dimensional coordinates (latitude, longitude, and elevation) of a distant point can be determined.

GLOBAL POSITIONING SYSTEMS In recent years the U.S. Department of Defense's **Global Positioning System** (GPS) has rapidly become a popular tool in surveying and mapping because it can provide highly accurate three-dimensional location information and can drastically reduce the amount of equipment and man-hours required by conventional surveying. The GPS is a constellation of satellites (see Figure 6-16) that orbit the earth and continuously beam radio waves. You can use a GPS receiver to intercept a satellite signal and measure the time it takes for the signal to reach your receiver. The computer in the receiver can calculate the distance to the satellite based on the time measurement. You must know the distance from three satellites to get a precise three-dimensional location. Three spheres intersect at two points, but one of the points is in outer space and cannot be a position on earth. The outer space location is thrown out, which leaves one possible location based on the three satellite readings (see Figure 6-17).

Practically, you need to intercept signals from four satellites. This is because accurate distance calculations depend on very precise time measurements. The satellites have precise clocks that tell time to within 40 nanoseconds (0.000000040 seconds). Most GPS receivers are not nearly this accurate and may not be perfectly synchronized to the clocks in the satellites. To compensate for the resulting timing error a fourth satellite, distance is used to calculate the receiver's correct position. More satellite signals mean even more accuracy in the position calculations.

OFF-SITE EXPLORATION

The accuracy of the position determined by a GPS unit depends on the type of receiver. Many handheld GPS units are only accurate to within several meters. To improve accuracy, the National Geodetic Survey developed a network of hundreds of accurate GPS receivers that operate 24 hours a day. This system is called the Continuously Operating Reference Stations (CORS) network. The data received by these stations is used to update the National Spatial Reference System and is published on the Internet. By using the CORS website at *http://www.ngs.noaa.gov/CORS/cors-data.html*, you can improve the accuracy of your coordinates to the centimeter.

Figure 6-16: The Global Positioning System (GPS) is a constellation of satellites that orbit the earth 11,000 mi above the surface.

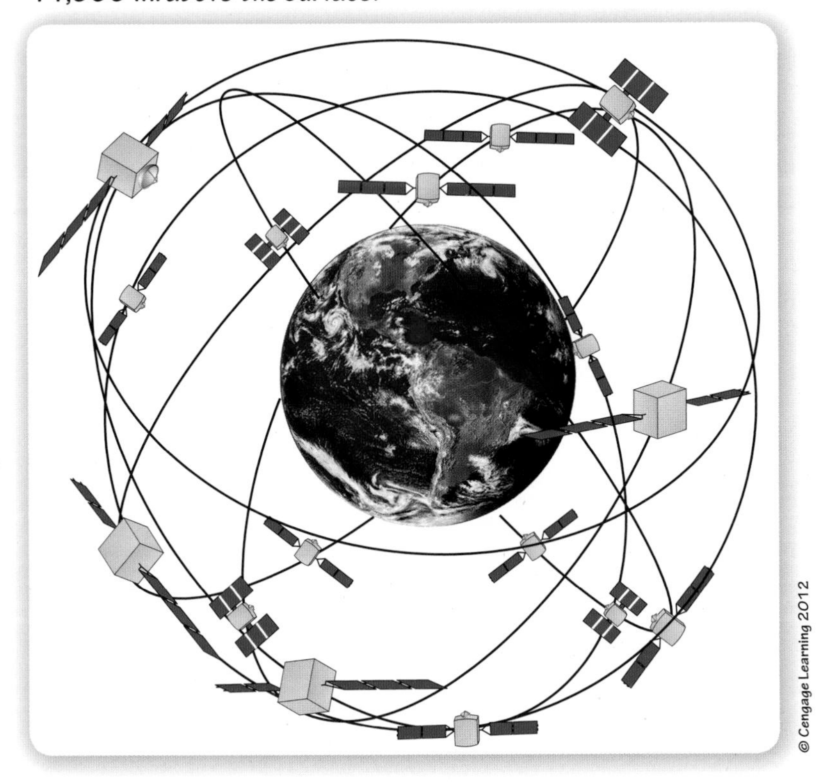

© Cengage Learning 2012

Figure 6-17: The collection of points that are equidistant from a satellite form a sphere. Distance measurements from three satellites represent three different spheres that will intersect at two points. One of these points will not be on the earth's surface and can be thrown out. However, to precisely define a position on earth, the distance from four satellites is needed to correct for inaccuracy in time measurements in the GPS receiver.

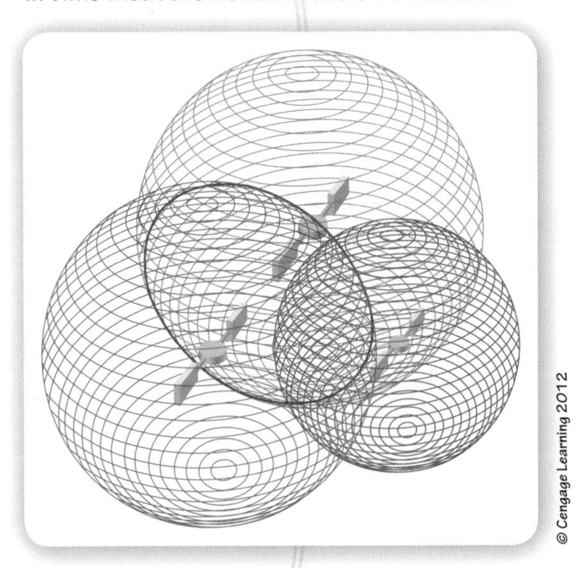

© Cengage Learning 2012

Technology is an integral part of the practice of engineering and architecture. A variety of technology tools, such as surveying equipment, GPS receivers, and computers, are used to collect, analyze and display data.

Fun Fact

When the GPS system was first created, timing errors were inserted into GPS transmissions to limit the accuracy for non-military uses to about 300 ft (100 m). U.S. president Bill Clinton ordered this practice, known as Selective Availability, eliminated on May 1, 2000.

Topographic Survey

During site discovery, you gathered existing information about the topography of the site area from existing resources, such as USGS topographic maps, but since topography can have a major affect on the project planning, more detailed information is often necessary. A topographic survey is conducted to locate natural and man-made features and to document the relief, or change in elevation, of the site and can be conducted using conventional surveying techniques, GPS, or aerial photography. Aerial photography is especially useful when topography is needed for a large area. The data collected from a topographic survey is used to create a contour map of the site.

The first thing that you need in order to gather topographic data is good control—a benchmark or temporary benchmark—to use as a reference for the survey data. Once the benchmark is located and the elevation is known, the elevation of other points on the site can be determined.

Although more sophisticated equipment is commonly used for topographic surveys (including GPS), one method of conducting a *conventional* topographic survey is the grid method. The grid method involves laying out a grid on the site using a transit or theodolite to achieve right angles. Alternatively, you can use a tape measure to lay out the grid if you check diagonal lengths to make sure the grid lines intersect at 90-degree angles. Once the grid is marked, you can use differential leveling to find grade elevations at the grid intersection points. A grid spacing of 10, 20, 50, or 100 ft can be used depending on the needed accuracy of the survey and the topography of the site. If the site is relatively flat or gently sloping, a wide spacing may be acceptable. If the terrain is rugged with many topographic features, a close spacing should be used.

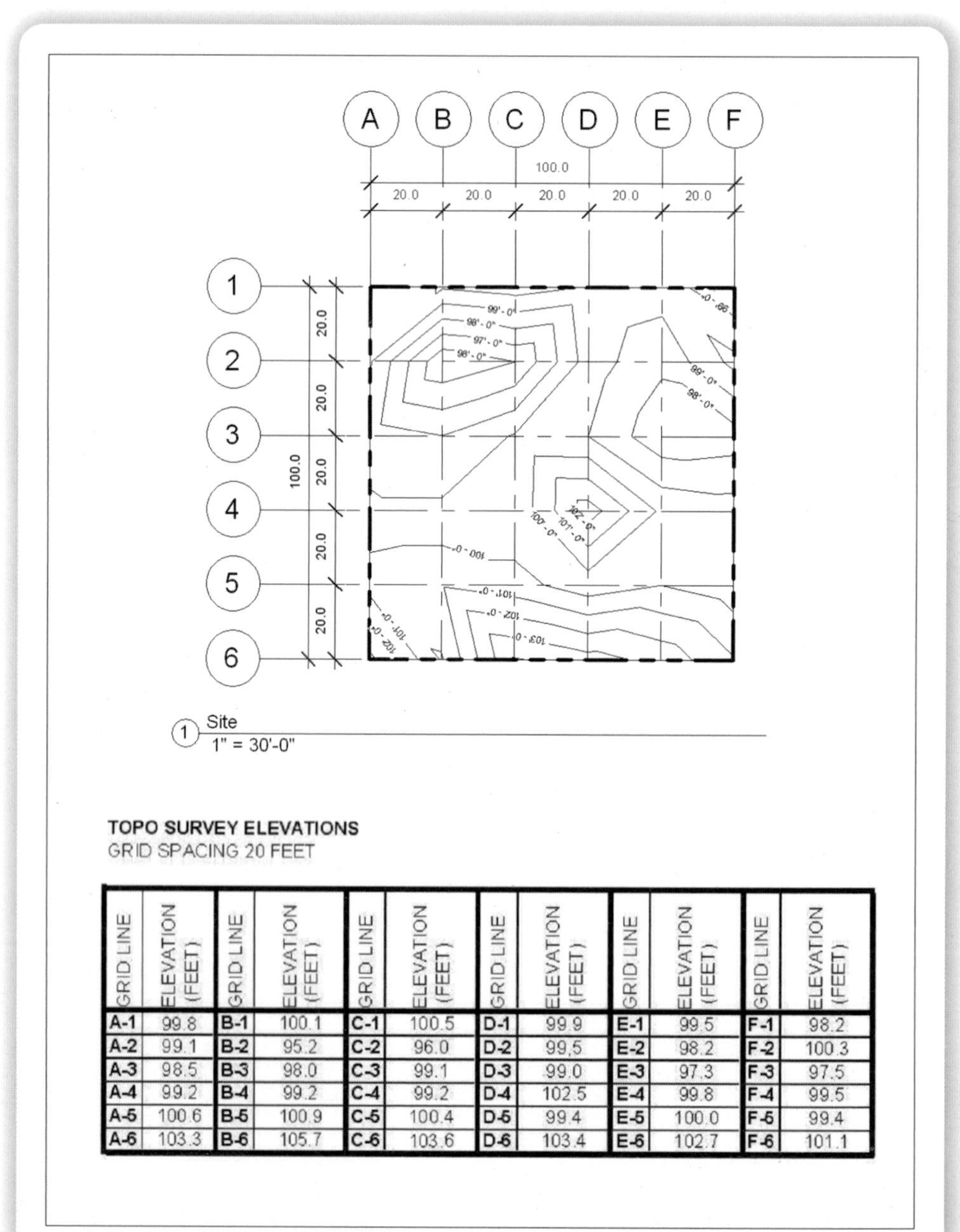

TOPO SURVEY ELEVATIONS
GRID SPACING 20 FEET

GRID LINE	ELEVATION (FEET)	GRID LINE	ELEVATION (FEET)	GRID LINE	ELEVATION (FEET)	GRID LINE	ELEVATION (FEET)	GRID LINE	ELEVATION (FEET)	GRID LINE	ELEVATION (FEET)
A-1	99.8	B-1	100.1	C-1	100.5	D-1	99.9	E-1	99.5	F-1	98.2
A-2	99.1	B-2	95.2	C-2	96.0	D-2	99.5	E-2	98.2	F-2	100.3
A-3	98.5	B-3	98.0	C-3	99.1	D-3	99.0	E-3	97.3	F-3	97.5
A-4	99.2	B-4	99.2	C-4	99.2	D-4	102.5	E-4	99.8	F-4	99.5
A-5	100.6	B-5	100.9	C-5	100.4	D-5	99.4	E-5	100.0	F-5	99.4
A-6	103.3	B-6	105.7	C-6	103.6	D-6	103.4	E-6	102.7	F-6	101.1

Figure 6-18: **Elevation data from a topographic survey can be used to create a topographic map.**

Once a topographic survey has been conducted the elevations of specific points on the ground can be found. These elevations are then used to create contour lines on a topographic map. A table of elevations representing topographic survey data for a 100 ft × 100 ft square lot and the corresponding contour map created using CAD software is shown in Figure 6-18. If you want to try topographic surveying, check out the Extra Mile Activity at the end of the chapter.

3

SOILS INVESTIGATION

Have you ever seen a brick building with long vertical cracks in a wall or next to a chimney? Or maybe you have noticed embankments where the soil seems to be falling off the slope (see Figure 6-19)? These types of failures can often be explained by the characteristics of the soil. A famous example of soil failure is the Tower of Pisa.

If we had a choice, we probably would not use soil as a structural material, but unfortunately we must depend on soil to carry the loads of our buildings, bridges, and roads. Soil also acts as a sponge to absorb and filter rainwater or wastewater and as a conveyor to funnel water into drainage systems. Some soil types are perfect for providing nutrients

Figure 6-19: Slope failure adjacent to a highway.

Courtesy California Department of Transportation.

Point of Interest

The Tower of Pisa—A Product of Flawed Geotechnical Engineering

Construction of the bell tower of the Pisa Cathedral (Figure 6-19) began in 1173. The tower was built on a 10 ft deep foundation situated on a bed of dry stones. Shortly after construction, the tower began to lean—it had been built on an old riverbed composed of layers of sand and clay interspersed with pockets of water. The soil could not support the tremendous weight of the tower and responded by compressing unevenly—and the tower began to lean. Over the next 800 years, construction progressed sporadically as the tower lean increased. Several attempts to halt the lean only made the situation worse. In 1990, the lean had increased to 15 ft from vertical at the top. The tower was closed to the public and remedial measures resulted in reversing the lean. In 2008, the tower had a lean of approximately 13 ft and is considered an engineering failure, although, due to its popularity with tourists, no one wants to return the tower to a vertical position.

Figure 6-20: The bell tower of the Pisa Cathedral became known as the Leaning Tower of Pisa due to a failure of the supporting soil.

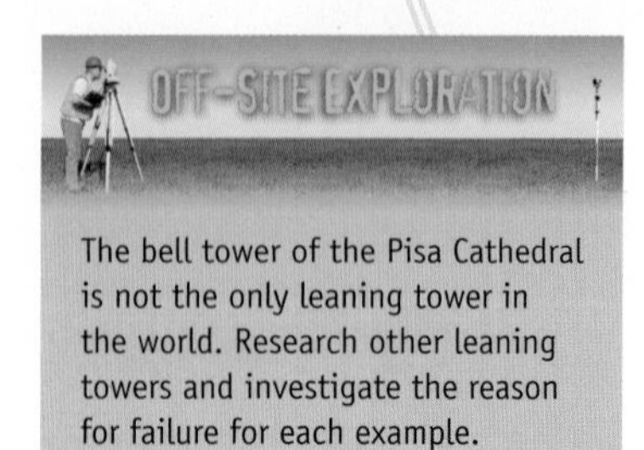

The bell tower of the Pisa Cathedral is not the only leaning tower in the world. Research other leaning towers and investigate the reason for failure for each example.

for plants but cannot support a building (organic soil). Other types of soils provide excellent strength to support a foundation, but easily wash away if exposed to moving water (sand). Still other soil types make good dams because they can provide a watertight seal, but would ooze and compact if used under a roadway (clay). Without detailed soils information, engineers cannot confidently design safe and functional projects. Fortunately, we can predict the behavior of soil based on the soil characteristics.

Geotechnical engineers are trained in analyzing soils and designing systems that involve soils and are often hired to perform **geotechnical investigations** of a site. A geotechnical investigation involves evaluating the physical and chemical properties of soils in order to predict soil behavior and determine compressibility, strength, and other characteristics that can influence a construction project. The geotechnical engineer will plan the investigation based on the type of structure being constructed and the characteristics of the site.

Because soil generally occurs in distinct layers that have been deposited over millions of years, the testing often includes boring into the ground and taking soil samples at various depths. The samples are then tested to determine the engineering characteristics of each soil layer. Geotechnical engineers analyze the test results so that they can make design recommendations in an engineering report.

Soil Classification

Do you remember making mud "pies" as a kid? You probably figured out that the best pies are made with soils that stick together. You didn't know it, but you were performing a soils analysis by testing the soils for cohesion. Soils characteristics vary in other aspects as well: strength, permeability, and compressibility, to name a few. These characteristics are important to design engineers because they can have a drastic impact on many aspects of a project.

We will identify the type of soil contained in a soil sample using the Unified Soil Classification System (USCS). This system distinguishes among 15 soil types based on the size of the soil particles (grain size), the amounts of the various sizes (size distribution), and the characteristics of the very smallest grains.

GRAIN SIZE The first step in classifying a soil is to find the approximate sizes of individual pieces, or grains, of the soil and the distribution of grain sizes. A **sieve analysis,** also known as a *gradation test,* is used to separate the soil into ranges of grain sizes. To perform a sieve analysis the soil sample is dried and then passed through a series of screens, called *sieves,* which have progressively smaller and smaller openings (see Figure 6-21). This process separates the soil by retaining soil particles larger than the opening size on each screen—smaller particles pass through the sieve. Of course, some objects mixed in with the soil are just too large to handle and are not included in a sieve analysis. Material that is larger than 3 in in diameter but less than 12 in is referred to as cobbles. Boulders are larger than 12 in and can be as large as 20 ft in diameter. Material larger than boulders is referred to as rock.

The openings in the screens are sized such that they will separate the sample into specific types of soil—gravel, sand, silt, and clay. The U.S. sieve sizes are indicated in one of two ways. A dimension in inches indicates the opening size of the screen, such as ¾ in. Otherwise, a number is used to indicate the quantity of screen openings per inch, such as a no. 4 sieve, which has 4 openings per inch. Figure 6-22 indicates that gravels pass the 3-in sieve but are retained on the no. 4 sieve. Sands pass the no. 4 sieve but are retained on the no. 200 sieve. Silt and clay pass the no. 200 sieve (which actually looks more like a tightly woven fabric).

Gravels are further separated into coarse and fine, and sand is divided into coarse, medium, and fine. Figure 6-22 shows the soil designation given to a soil based on the sieve on which it is retained and shows both the U.S. sieve sizes and

Figure 6-21: Sieves.

Courtesy of Avery Fox.

Figure 6-22: A sieve analysis separates soil samples into soil types based on the size of soil particles. Each sieve retains particles that are larger than the openings in the sieve. The types of soils listed to the left of a specific sieve in the figure will pass that sieve. For example, sand, silt, and clay will pass the no. 4 sieve. Only silt and clay will pass the no. 200 sieve. Comparable International Standards Organization sieve sizes are also shown.

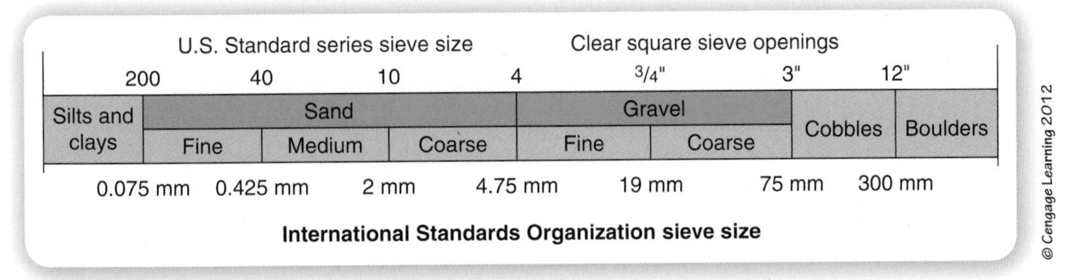

© Cengage Learning 2012

Figure 6-23: Examples of coarse-grained soils. All of these soils are clean and uniformly graded. The soil on the left was retained on the no. 40 sieve and is a medium sand. The soil in the center was retained on the no. 12 sieve and is a course sand. The soil on the right is a fine gravel.

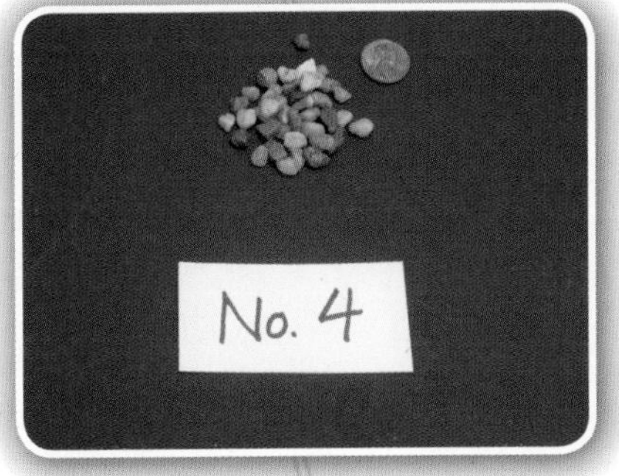

Kennedy

Example: Determine the grain size distribution of a soil sample

Assume that you are responsible for analyzing the results of a sieve analysis of a soil sample collected from a job site. The sample was dried, weighed, and sieved. You have been given the set of sieves with retained soil. Determine the grain size distribution of the sample so that the soil can be classified.

Step 1: ***Determine the mass of soil retained on each sieve and the mass of soil that passed through all the sieves and was collected in the bottom pan. Remember that the mass retained does not include the weight of the sieve. Create a data table similar to the following and enter these measurements.***

Table 6-1

U.S. Sieve Size	Mass Retained (g)	Percentage of Sample Retained	Cumulative Percentage Retained	% Passing
¾"	21			
No. 4	35			
No. 10	69			
No. 40	119			
No. 200	42			
PAN	34			
TOTAL	320			

Step 2: ***Calculate the percentage of soil retained on each sieve and in the pan.***

$$\% \ retained = \frac{mass\ retained}{total\ mass} \cdot 100\%$$

So, for the ¾-in sieve the percent of soil retained is:

$$\% \ retained = \frac{21g}{320g} \cdot 100\% = 6.6\%$$

Add the calculated values to the table.

Table 6-2

U.S. Sieve Size	Mass Retained (g)	Percentage of Sample Retained	Cumulative Percentage Retained	% Passing
¾"	21	6.6		
No. 4	35	10.9		
No. 10	69	21.6		
No. 40	119	37.2		
No. 200	42	13.1		
PAN	34	10.6		
TOTAL	320	100.0		

Step 3: *Calculate the cumulative percentage for each sieve. The cumulative percentage is the percentage of soil retained to that point. For example, the cumulative percentage of soil for the no. 40 sieve is the total of the percentages retained on the ¾-in, no. 4, no. 10, and no. 40 sieves together. Include the cumulative percentages in the table.*

Table 6-3

U.S. Sieve Size	Mass Retained (g)	Percentage of Sample Retained	Cumulative Percentage Retained	% Passing
¾"	21	6.6	6.6	
No. 4	35	10.9	17.5	
No. 10	69	21.6	39.1	
No. 40	119	37.2	76.3	
No. 200	42	13.1	89.4	
PAN	34	10.6	100.0	
TOTAL	320	100.0		

Step 4: *Calculate the percent of soil passing each sieve.*

% passing = 100% − cumulative percentage retained

% passing = 100% − *cumulative percentage retained*

Therefore, the percent finer than the no. 10 sieve is

% passing = 100% − 39.1% = 60.9%

% passing = 100% − 39.1% = 60.9%

Table 6-4

U.S. Sieve Size	Mass Retained (g)	Percentage of Sample Retained	Cumulative Percentage Retained	% Passing
¾"	21	6.6	6.6	93.4
No. 4	35	10.9	17.5	82.5
No. 10	69	21.6	39.1	60.9
No. 40	119	37.2	76.3	23.7
No. 200	42	13.1	89.4	10.6
PAN	34	10.6	100.0	0
TOTAL	320	100.0		

Step 5: *Create a Gradation Chart. The grain size distribution is often displayed on a gradation chart that shows the percentage of soil passing each sieve. For each sieve size (listed on the horizontal axis) plot the percent passing from the table above.*

A gradation chart shows the grain size distribution of a soil sample.

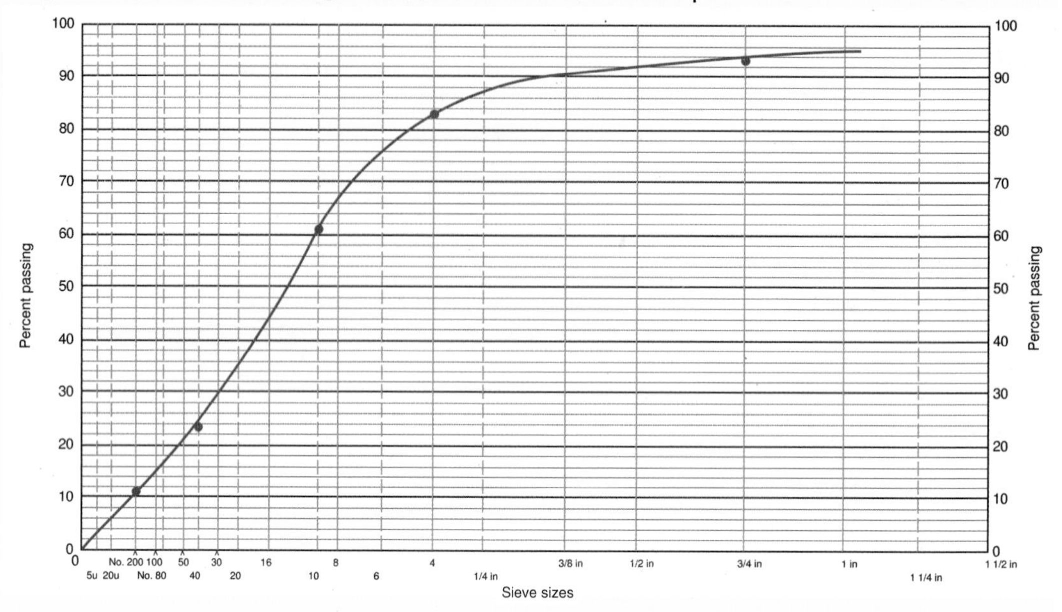

the International Standards Organization (ISO) sieve sizes. For example, a fine gravel would pass the ¾-in sieve but be retained on the no. 4 sieve. The ISO sizes indicate the screen opening size in millimeters. Once the soil is separated by grain size, the percentage of each soil type is determined.

Engineers use a variety of representations for relations among quantities that vary including symbolic expressions, tables of data, charts, and graphs. These representations help engineers analyze information, such as soil characteristics, in order to solve problems.

Coarse-Grained Soils: soils in which 50 percent or more, by weight, of the soil is retained on the no. 200 sieve.

Fine-Grained Soils: soils in which more than 50 percent, by weight, of the soil passes the no. 200 sieve. That is, more than 50 percent of the soil is composed of silt and clay.

Moisture Content: refers to the ratio of the weight of water to the weight of the dried soil in a sample expressed as a percent. Values of greater than 100 are possible.

If less than 50 percent of the soil passes the no. 200 sieve and therefore contains mostly sand and gravel, the soil is considered a **coarse-grained soil**. If more than 50 percent of the soil passes the no. 200 sieve and is therefore composed mostly of silt or clay, the soil is referred to as a **fine-grained soil**. Silt and clay may also be called *cohesive soils* because the particles tend to be cohesive, or stick together, when wet. Generally, the behavior of coarse-grained soils is fairly consistent and predictable. However, the behavior of fine-grained soils can vary drastically especially when the amount of water in the soil is changed. More information is needed to predict the behavior of fine-grained soil.

PLASTICITY Silt and clay can exist in three different states depending on the moisture content—a viscous liquid, a plastic substance that can be molded, or a non-plastic solid state. The amount of water present in the soil, called the **moisture content**, determines which state the cohesive soil takes. At a moisture content of zero, the soil will be solid and non-plastic, but as the moisture content is increased, the soil may eventually take on a plastic state. The moisture content at which the soil transforms from a non-plastic solid to plastic is called the **plastic limit (PL).** As the moisture content continues to increase, the soil will begin to act like a thick liquid—the moisture content at this transition is referred to as the **liquid limit (LL).** (see Figure 6-24).

The liquid limit and plastic limit are sometimes called *Atterberg limits* and are determined in a laboratory by performing tests on the portion of the soil that passes the no. 40 sieve. The mathematical difference between the liquid limit and the plastic limit is called the **plasticity index.** The plasticity index (PI) is an important predictor of the soil's behavior and is used in the classification of fine-grained soil.

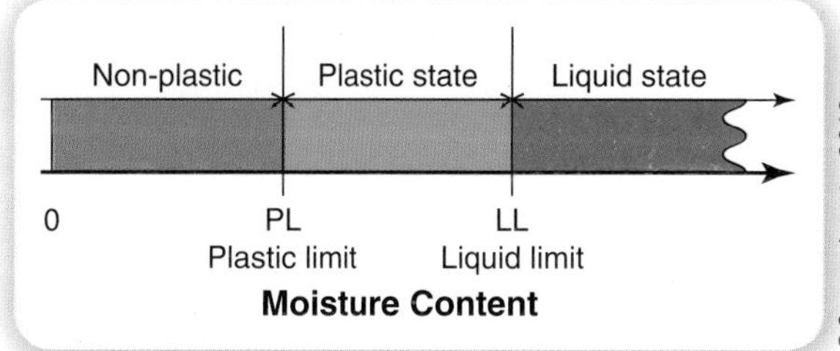

Figure 6-24: As the moisture content increases from zero a fine-grained soil will transition from a non-plastic to a plastic state at the plastic limit and from a plastic to a liquid state at the liquid limit.

Plasticity Index

$$PI = LL - PL$$

Where PI = plasticity index
LL = liquid limit
PL = plastic limit

UNIFIED SOIL CLASSIFICATION SYSTEM Once the grain size distribution and, if necessary, the Atterberg limits are determined, a soil sample can be classified according to the USCS. The system distinguishes among 15 different soil classifications, each of which is identified by a symbol consisting of letters from the following table.

Table 6-5

Soil Type	Gradation	Liquid Limit
G = gravel		
S = sand	W = well graded	H = high plasticity (LL over 50)
C = clay	P = poorly graded	L = low plasticity (LL under 50)
M = silt		
O = organic		
Pt = peat		

The first letter of the symbol indicates the type of soil that makes up the majority of the sample. Typically a second letter provides additional information about the soil. A course-grained soil that has very little (less than 5 percent) fine-grained particles is referred to as *clean.* The second letter in the symbol for a clean gravel or sand gives an indication of the variety of sizes of grains in the soil. The letter W indicates that the soil is **well graded** and will contain a wide range of grain sizes that are well distributed across the range. For example, a soil classified as GW is well-graded gravel, which means it is composed mostly of gravel and contains a wide variety of gravel sizes. A **poorly graded** soil is indicated by the letter P and will *not* contain a well-distributed range of particle sizes. For example, a *uniformly graded* soil is poorly graded because it contains particles that are mostly within a narrow range of sizes. A gap-graded soil is also poorly graded but will contain a large percent of particles in more than one narrow range of sizes. Figure 6-25 shows grain size distributions of a well-graded soil and two poorly graded soils. Notice that the uniform distribution is narrower and steeper than the well-graded soil. The gap-graded soil contains grains that mostly fall into two narrow ranges of sizes as indicated by the two relatively steep positive slopes separated by a fairly flat section in the distribution. Look back at the gradation chart for the last example. Do you think this soil sample represents a well-graded or poorly graded soil?

Figure 6-25: A coarse-grained soil is classified as well graded if it contains a wide variety of grain sizes that are well distributed across grain sizes. Poorly graded soils contain grains that are mostly limited to a narrow range or ranges of grain sizes. This grain size distribution chart shows a representation of a well-graded soil as well as two examples of poorly graded soils: a uniform soil and a gap-graded soil.

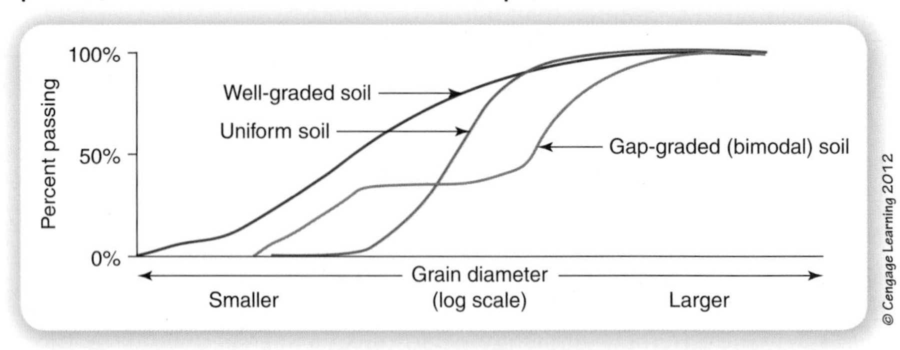

Figure 6-26: Classification of fine-grained soils can be accomplished with a plasticity chart. To determine the soil classification, plot the plasticity index and liquid limit of the portion of a soil sample that passes the no. 40 sieve and identify the corresponding region. The A-line (equation: PI = 0.73 (LL − 20)) represents the division between silt and clay. Points plotted above the A-line indicate clay, and points plotted below the A-line indicate silt or an organic soil. For example, a soil with LL = 84 and PI = 61 would be plotted in the CH region of the chart (as indicated by the orange dot) and would be classified as a clay with high plasticity.

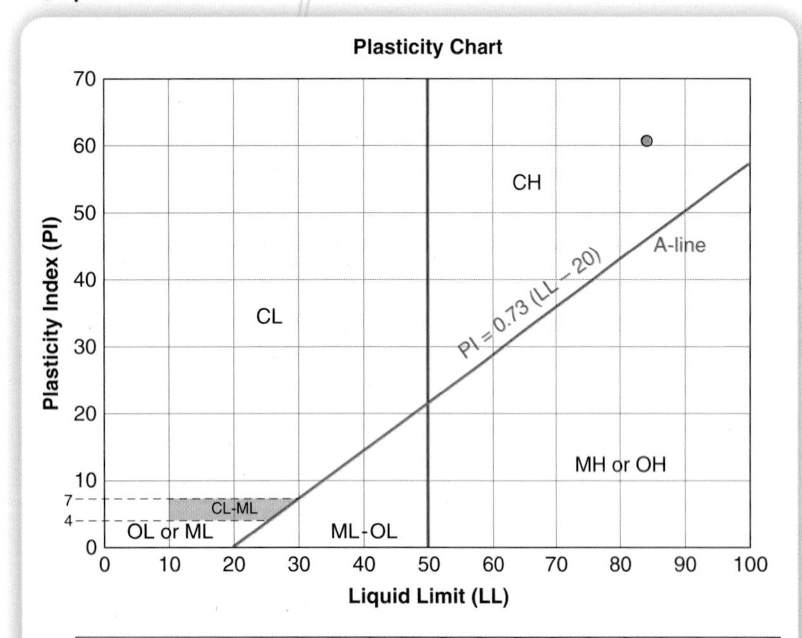

If the soil is made up of mostly gravel or sand but contains a significant amount of fines (more than 12 percent), the second letter in the classification symbol will identify the fines as clay (C) or silt (M). So, for example a soil classified as SM is primarily sand but has a significant amount of silt. Coarse-grained soils that have more than 12 percent fines are said to be *dirty*.

If the soil is made up mostly of fine-grained soils, the soil classification will be based on the plasticity of the soil as indicated by the plastic index and the liquid limit. Soils with a high liquid limit (greater than 50) are considered to be highly (H) plastic. Soils with a liquid limit less than 50 have low (L) plasticity. A plasticity chart (see Figure 6-26) can be used to identify the soil as silt or clay based on its Atterberg limits.

Organic soils and peat differ from gravel, sand, silt, and clay in that they contain a high percentage of organic matter (such as pieces of leaves and sticks). Organic soil and peat can often be identified by an "earthy" odor or a dark brown, dark gray, or black color. They will also have a fibrous texture and typically have high moisture content. Organic soil is a poor engineering material because it will break down in time and compress. In most cases, organic soils occur at the surface and are removed from areas of the site on which structures and pavement will be built. Often this soil can be used elsewhere on the site and is stored in a pile on the site for later landscaping.

Example:

Classify a soil using the USCS. Once the grain size distribution and the plasticity index are determined, a soil can be classified using the flow chart in Figure 6-27. Enter the chart at the top and use the analysis data to determine the flow path to the soil classification at the bottom.

For this example we will use the grain size analysis data from the previous example. The Atterberg limits of the soil that passed the no. 40 sieve were determined to be: PL = 23 and LL = 84.

Step 1: *Visual Analysis*

Because a sieve analysis was performed, we can assume that the soil was not identified as organic based on color or odor.

Step 2: *Percent Retained on No. 200 Sieve*

Based on the analysis data above, 89.4 percent of the sample was retained on (or before) the no. 200 sieve. Therefore, move to the left branch as indicted by the blue arrow A in the soil classification chart.

Step 3: *Determine the Percents*

We must compare the percentage of gravel to the percentage of sand.
% G = % coarse gravel + % fine gravel = 6.6% + 10.9% = 17.5%
% S = % coarse sand + % medium sand + % fine sand = 21.6% + 37.2% + 13.1% = 71.9%
Therefore, % S > % G. Move to the right on the chart (B).

Step 4: *Percent Passing No. 200 Sieve*

From the table, 10.6 percent of the sample passed the no. 200 sieve, which is between 5 and 12 percent. Therefore, follow the middle branch down (C). The soil will require a dual symbol.

Step 5: *Plot Atterberg Limits*

To determine the dual symbol, we will analyze the sample in two ways: as if it had less than 5 percent fines, and as if it had more than 12 percent fines. The resulting symbols will be combined to determine the dual designation.

1. Assuming less than 5 percent fines, follow the green C1 arrow to the left. Based on the shape of the grain size distribution plot from the previous example, the sand is well graded. Therefore, a designation of SW applies.
2. Assuming more than 12 percent fines, follow the orange C2 arrow to the right. Since LL = 84 and PL = 23, the Plastic Index is $PI = LL - PL = 84 - 23 = 61$.

The point (84, 61) is plotted on the plasticity chart (Figure 6-26) and falls above the diagonal line, which indicates clay. Note that the point actually falls within the CH region, but for a dual designation we are concerned only with the fact that it can be classified as clay. Therefore, according to the classification chart, a designation of SC (clayey sand) applies.

Figure 6-27: *Soil classification chart.*

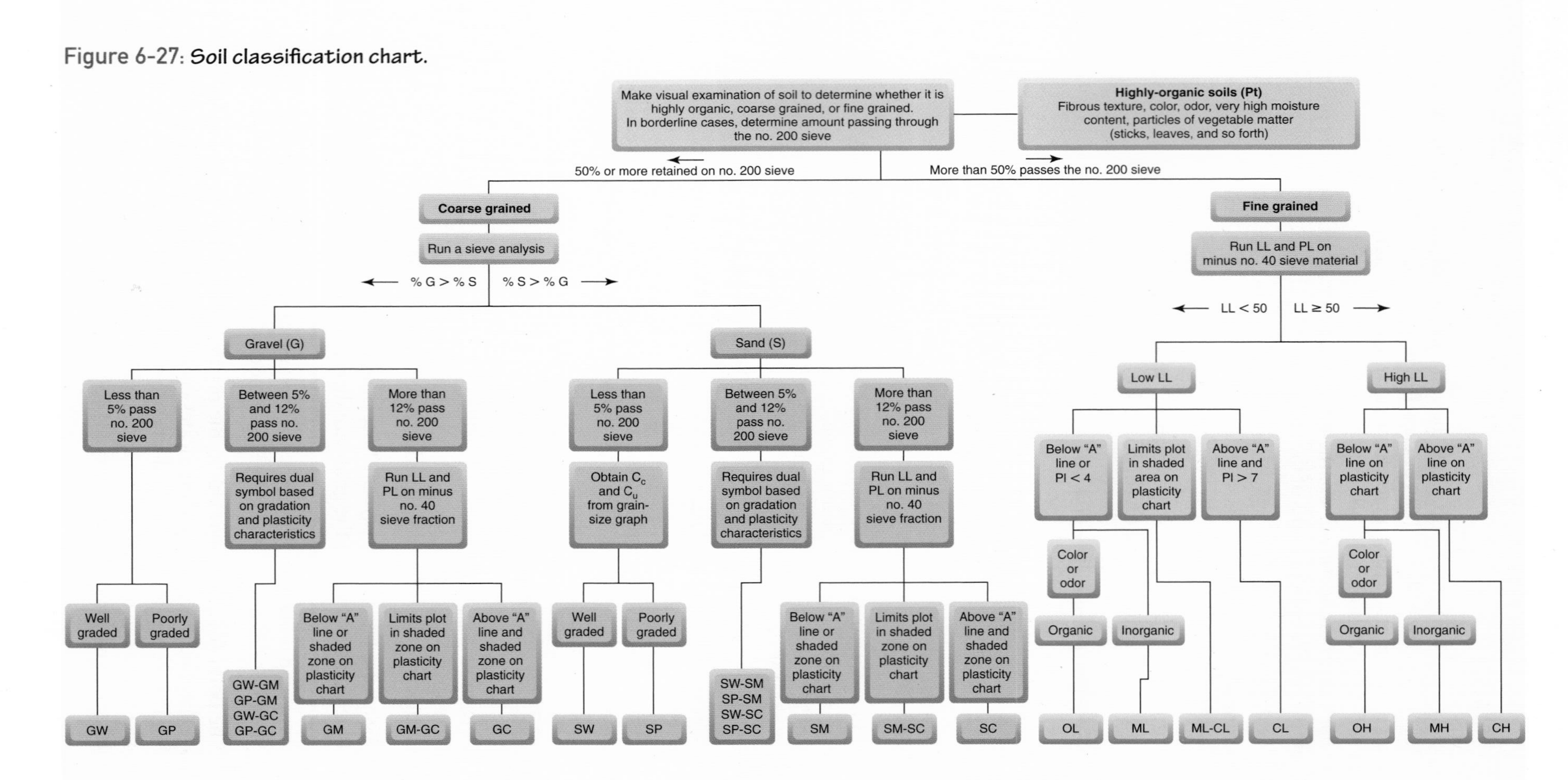

Soil classification chart.

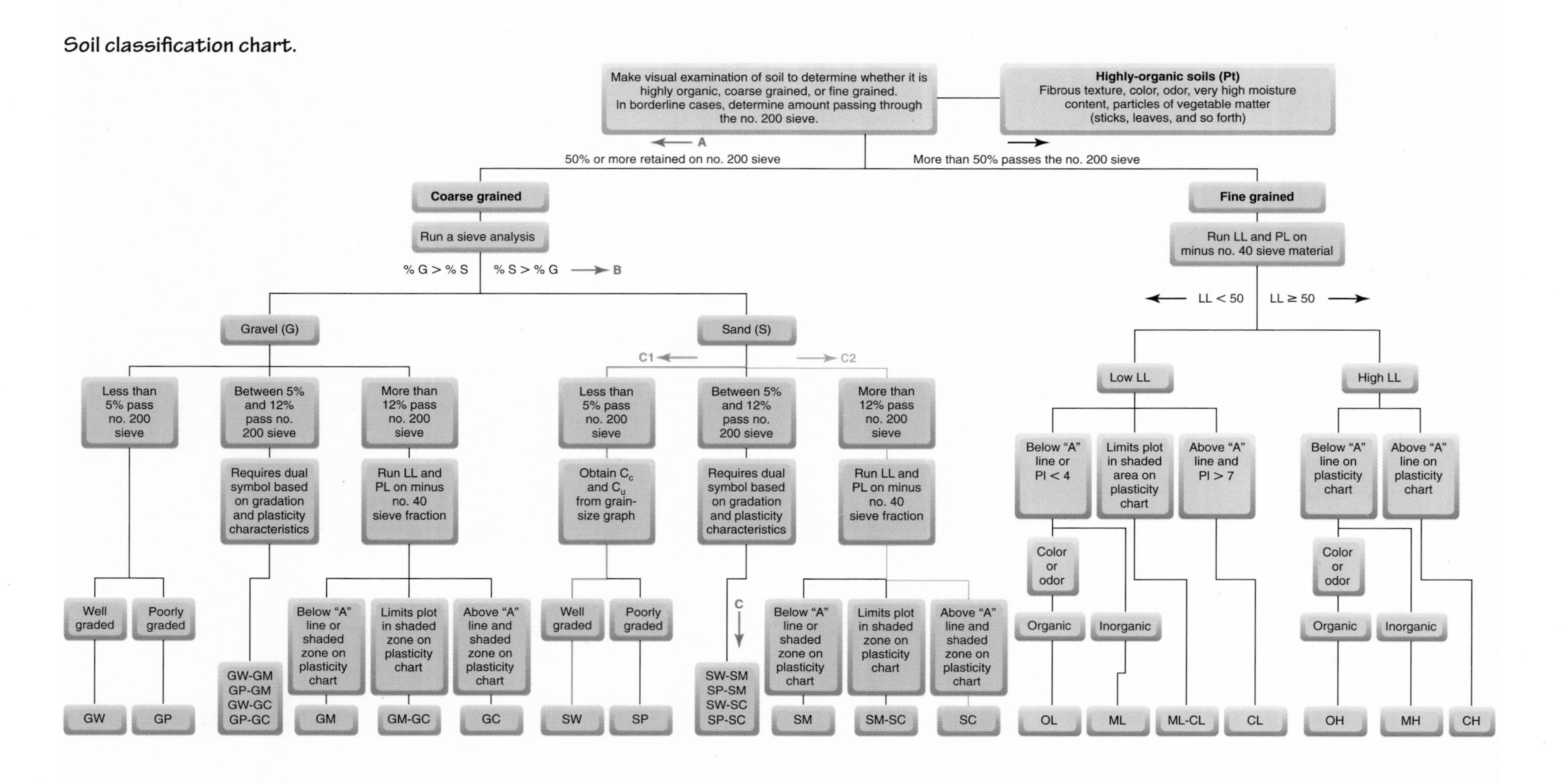

Step 6: *Classification of Soil*

Because this course-grained soil contains between 5 and 12 percent fines, it requires a dual designation. Based on the soil analysis the soil can be classified as a SW-SC, which indicates that it is well-graded sand with a significant amount of clay as shown in the chart.

Your Turn

Visit The University of the West of England website on Soil Classification at http://environment.uwe.ac.uk/geocal/SoilMech/classification/default.htm. Although the British Soil Classification is slightly different than the USCS, try the Fine Soils Simulation, complete the table, and use the plasticity chart (Figure 6-26) to classify the soil based on the USCS classification. Then, try the Coarse Soil Simulation on the Glacial Till sample (but assume that the 5.00 mm sieve is equivalent to USCS no. 4 sieve and that the 0.063 mm sieve is equivalent to USCS no. 200 sieve), complete the table, and classify the soil based on USCS.

Differential Settlement occurs when the structure experiences different rates of settlement at different locations.

Permeability: refers to the speed at which water travels through the soil.

THE GEOTECHNICAL REPORT In addition to the sieve analysis and Atterberg limit test, the soils investigation may include other tests to address specific site situations or project requirements. A geotechnical engineer will analyze the results of the soils investigation and present predictions about the behavior of the soil in an engineering or geotechnical report. The report will generally include a description of the existing site conditions such as terrain, drainage, and groundwater conditions as well as the soils data collected and soils test results. The engineering properties of the soils and engineering recommendations should also be presented in the engineering report and may include:

- Allowable soil bearing pressure, used to design foundations
- Recommended foundation type
- Evidence of expansive soils and recommendations
- Lateral (horizontal) earth design pressure
- Potential for **differential settlement** (different rates of settlement across a structure)
- Frost depth (dictated by code)
- Ground water table depth
- Maximum safe slope of soil
- Discussion of potential to use existing soils as engineered fill material
- Recommended pavement design
- Soil **permeability**

Geotechnical investigations are not normally required for small residential projects because the building loads are small. For larger projects, geotechnical investigations should be performed before the project design is started. The soil properties and recommendations made in the geotechnical report are important to the design of the site as well as the design of structures. Failure to properly investigate the soil conditions and take into consideration the information presented in a geotechnical report can result in costly problems during construction as well as throughout the life cycle of the project.

Case Study

WHY DID THE LEVEES FAIL?

Hurricane Katrina (August 28, 2005) was the most costly hurricane (at a cost of over $84 billion dollars) and the third deadliest (claiming 1500 lives) to hit the U.S. mainland in history. Katrina was recorded as a category 3 hurricane at landfall, which means wind speed was in the range of 111 to 130 mph. Although category 3 hurricanes typically cause extensive damage, the damage is generally much less severe than category 5 hurricanes, which will often cause catastrophic damage. Why did this category 3 storm result in more damage than any previous category 4 or 5 storm? Much of the reason for Katrina's devastating effects is the fact that much of New Orleans is built below sea level. A system of levees was in place to protect the city from inundation of water.

Hurricane damage can typically be traced to a combination of two factors: high wind speeds and a storm surge. A storm surge results when the hurricane pushes a large volume of water onto the coast. Category 3 storms often cause storm surges of 9 to 12 ft of water above normal water levels. Category 5 storms can cause surges of over 18 ft above normal water levels. Unfortunately, many of the New Orleans levees failed when exposed to the 20 to 30 ft storm surge (FEMA).

A team of engineers known as the Independent Levee Investigation Team (ILIT) investigated the New Orleans levee failures that occurred as a result of Hurricane Katrina in 2005. The final report resulting from this investigation concluded that soils, and not overtopping, were responsible for several levee failures. As reported, large sections of the levees protecting the Lower Ninth Ward (see map) were constructed of "unacceptably erodeable materials (which) included sands and lightweight shell-sands, and the massive and catastrophic erosion of these materials caused the rapid failure of great lengths of levees. . . ." Other Ninth Ward levee failures along the Industrial Canal were attributed to seepage of water through permeable layers of "marsh" that ran under the levee walls. Once the storm surge water level rose outside the levee walls, the increased water pressure forced water to flow through the marsh layer causing a "blowout" of the soil on the inside of the levee due to the pressure. According the ILIT report, "The boring data was far too sparse along this section for the importance of the design. . . and for the complexity of the local geology."

The soil profile shows a cross section of soils in New Orleans near one of the levees that failed during Hurricane Katrina. The long vertical black lines indicate soil borings that were performed to provide information on the soil types shown in the cross section. The soil symbols indicate the USCS soil classifications of the soils found during the investigations.

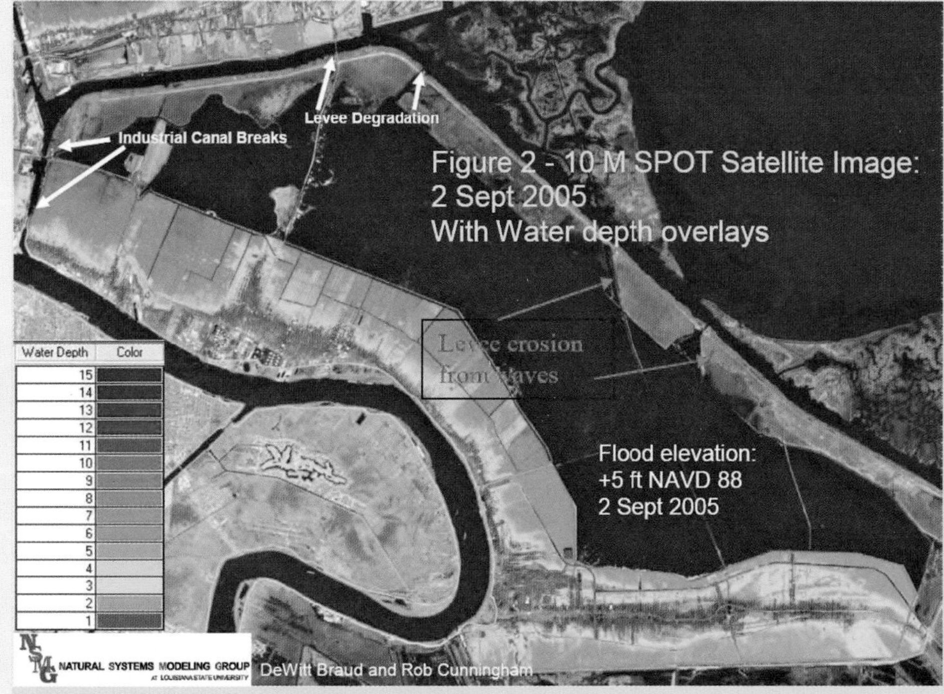

Depth of flooding of St. Bernard Parish and the Lower Ninth Ward on September 2, 2005 (four days after Hurricane Katrina). An independent team of engineers investigated the levee failures and concluded that the levee erosion from waves shown on the map resulted from the use of highly erodible soils in the levee construction. The two breaks along the Industrial Canal were found to have been caused by underseepage. From the Investigation of the Performance of the New Orleans Flood Protection Systems in Hurricane Katrina on August 29, 2005, completed by the Independent Levee Investigation Team (ILIT). Original map (without annotations) courtesy Louisiana State University Hurricane Center.

Case Study

(continued)

Soil profile showing the location of soil borings near the Lower Ninth Ward levee in New Orleans. From the Investigation of the Performance of the New Orleans Flood Protection Systems in Hurricane Katrina on August 29, 2005, completed by the ILIL.

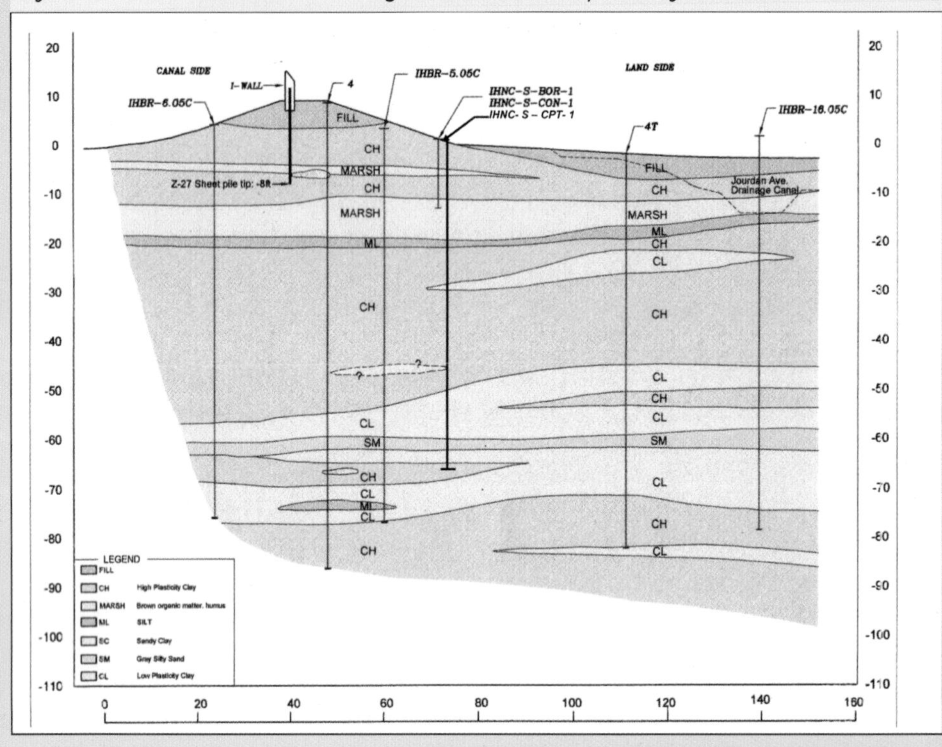

This boring log records the findings of a geotechnical boring performed near the Katrina levee failure in the Lower Ninth Ward in New Orleans after the hurricane. The boring number corresponds to one of the borings shown on the soil profile. The purpose of this boring was to retrieve soil samples for testing in order to determine the engineering properties of the soil. Notice that the fine-grained soils (CH) were tested to determine liquid limit, plastic limit, and the in situ moisture content. From the Investigation of the Performance of the New Orleans Flood Protection Systems in Hurricane Katrina on August 29, 2005, completed by the ILIL.

BORING NUMBER IHNC-S-BOR-1

PAGE 1 OF 1

UC Berkeley
Davis Hall
Berkeley, California

CLIENT ILIT (Independent Levee Investigation Team) **PROJECT NAME** Lower Ninth Ward
PROJECT NUMBER **PROJECT LOCATION** Lower Ninth Ward, New Orleans, Louisiana
DATE STARTED 2/17/06 **COMPLETED** 2/17/06 **GROUND ELEVATION** .93 ft N.A.V.D. **HOLE SIZE** 4"
DRILLING CONTRACTOR STE **GROUND WATER LEVELS:**
DRILLING METHOD Hollow Stem Auger **AT TIME OF DRILLING** N/A
LOGGED BY A. Athanasopoulos **CHECKED BY** D. Cobos-Roa **AT END OF DRILLING** ---
NOTES South end of South breach (Claiborne) **AFTER DRILLING** ---

DEPTH (ft)	GRAPHIC LOG	MATERIAL DESCRIPTION	SAMPLE TYPE NUMBER	RECOVERY %	BLOW COUNTS (N VALUE)	Su, Strength (tsf)	Dry Unit Weight (tsf)	▲ SPT N VALUE ▲ 20 40 60 80 / PL MC LL 60 120 180 240 / ☐ FINES CONTENT (%) ☐ 20 40 60 80
0		FILL: Augered through the first 6' to go through placed fill by USACE (fill came from old levee and was dumped).						
10		CH: Medium, gray clay.	ST 1	37 (100)		0.42		
		WOOD	ST 2	68 (100)				
		CH: Soft, gray clay with organics and wood.	ST 3	63 (100)		0.13 0.19		
		Bottom of hole at 14.0 feet.						

Case Study

(continued)

Katrina is the costliest hurricane to strike the U.S. mainland. Extensive damage resulted from the high winds and flooding that was exacerbated by the failure of many New Orleans levees.

Photo credit: U.S. Army Corps of Engineers.

CONCEPT DEVELOPMENT

Once you have analyzed the existing site characteristics that will affect the design of your project, you may begin the preliminary design phase to develop project concepts. Concept development involves generating ideas for site development that will creatively address the project requirements documented in the needs analysis while simultaneously taking into consideration the site characteristics. You will explore alternatives for site development that satisfy the needs of your client, comply with the rules and regulations that govern your project, and minimize the negative impacts of the project on the surrounding people and environment in a cost effective way. Because the site design will affect many aspects of the project, you should take a multidisciplinary approach and involve the entire design team and representatives of the people who will use the facility in the concept development. A design **charette** can provide valuable information at this stage of the design.

Buildable Area

4

Since the building is typically the main feature of a project, the building location on the lot is one of the most important decision in site design. It is sometimes difficult to resist the initial tendency to place the building in the center of the property, but many factors can influence the best building location.

As you learned in the previous chapter, local ordinances often dictate many aspects of site design including how parts of the site can be used. Before locating the building and other major elements of your project, you should review applicable regulations to identify restrictions for use and begin to define the area of the site on which you can build your structure—the **buildable area.**

On a base map of the property, which should include property lines and contours, sketch required setbacks and easements. In addition, many codes require

Figure 6-28: The building location is perhaps the most important decision in site design. Orientation of the building on the property will depend on terrain, views, solar orientation, wind direction, and nearby noise and sounds.

buffer zones between new development and adjacent uses. **Buffer zones** are areas, typically at property lines, that separate different land uses or adjacent properties. Screening materials such as fences and landscaping may be required in the buffer to provide a visual and sound break. The buildable area is the area resulting from the most restrictive setback, easement, and buffer lines or other restrictions. An example site map showing the buildable area is shown in Figure 6-29.

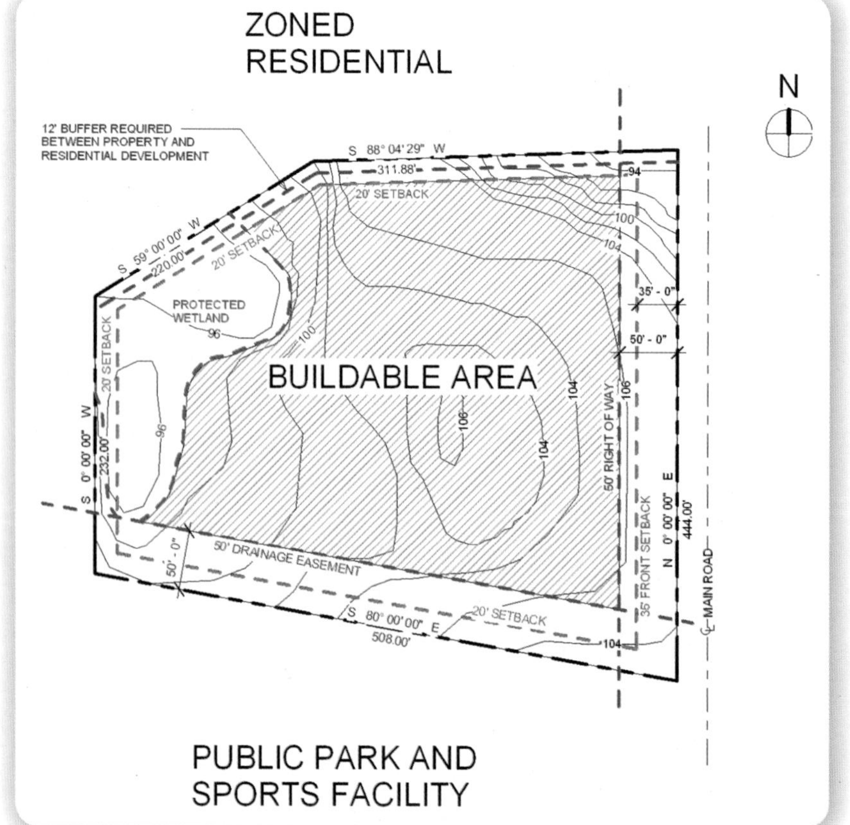

Figure 6-29: The buildable area is the area resulting from the most restrictive setback, easement buffer lines, and other site restrictions such as protected wetlands.

Site Orientation

Once you are familiar with the regulations that control construction on the site and you have defined the buildable area, you should consider the physical and environmental features of the site itself. Sometimes you will have little choice in the location of the building(s) on a property because the lot is small, street frontage dictates the front of the building, and property line setbacks restrict building placement. If this is the case, site planning is a matter of squeezing the necessary project elements within the buildable portion of the lot. But in many cases, you will be able to locate the building in order to take advantage of positive site features and to minimize undesirable features. The physical and environmental elements that will affect the building location include terrain, solar orientation, wind, sound, and views from the site. Together these elements will help determine the **site orientation**.

Site Orientation: the placement of a structure on a site based on physical and environmental factors such as terrain, wind, sound, solar orientation, and views from the site.

SOLAR ORIENTATION The sun has the potential to directly provide both heat and light energy for a building. If appropriate equipment is provided, solar energy can also be converted to electricity to provide energy for other building systems. The best solar orientation allows a long wall of the building to receive continuous sun throughout the day during winter. As you learned in Chapter 5, the azimuth and altitude of the sun will vary throughout the year. The azimuth and altitude of the sun on the winter solstice can provide you with important information that will help you position your building for the best winter solar orientation. Generally, on a clear day, the sun will shine on the southern face of a building for much of the day if nothing blocks the rays (Figure 6-30). In North America, we refer to a building with this ideal orientation—a long exterior wall facing true south—as having a *southern exposure.*

If you use a compass, you can determine *magnetic* north on your lot. A compass points to magnetic north, which is different than true north—you must adjust your compass reading to determine *true* north. Magnetic declination is the difference between true north and magnetic north and varies depending on your location (see Figure 6-31). In order to determine true north, rotate the compass face so that the needle points to the declination angle for your location. The N on the compass face indicates true north, and the S indicates true south.

For best solar orientation, position your building on the site so that as much exterior surface as possible will receive winter sun based on the winter solstice azimuth. Avoid the shadow of large obstructions that will limit sunlight exposure in the winter such as evergreen trees, mountains, and other structures. Of course, in most U.S. climates, we will want to protect the building from the intense summer sun. This is conveniently handled by strategically positioning deciduous trees, which lose their leaves in the winter, to provide summer shade to the building. We will discuss the use of trees and other landscaping techniques to address site orientation issues in Chapter 18.

Figure 6-30: A southern exposure, which orients a long wall of the building facing true south, is the ideal solar orientation for North America.

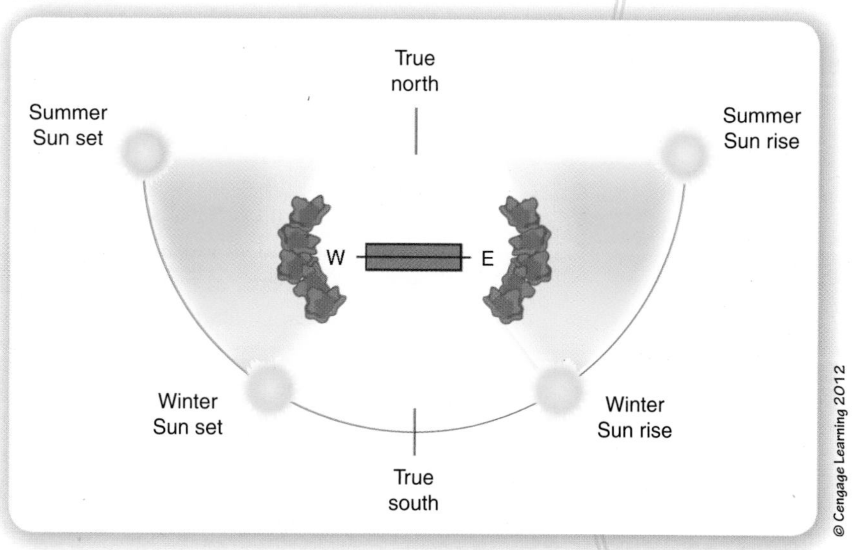

Figure 6-31: Magnetic declination for the contiguous United States. The contour lines indicate the direction of magnetic north (the direction a compass points) from true north. For example, for any location that falls along the 10°E contour, magnetic north is 10° east (counterclockwise) of true north. As a result, magnetic south is 10° west (clockwise) of true south.

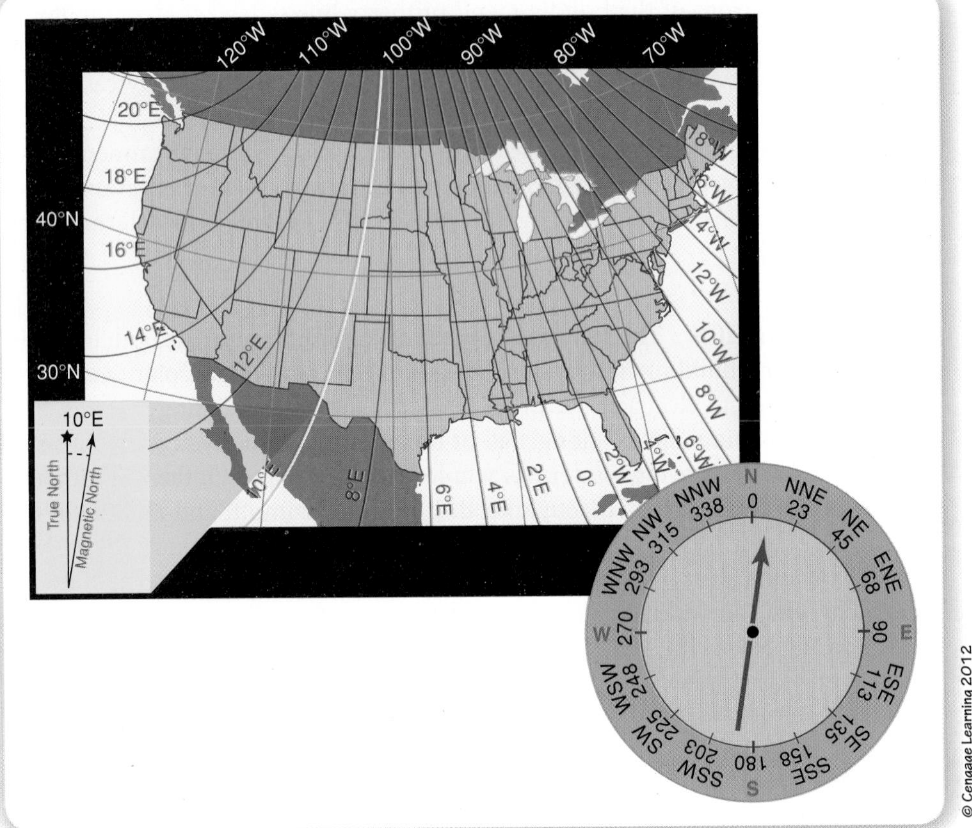

Technological processes can be aligned with natural processes to maximize performance and reduce negative impacts on the environment. For example, buildings can be strategically oriented to the sun to maximize solar gain thereby reducing energy demands.

WIND ORIENTATION The prevailing winds that blow across your site can improve the comfort and energy efficiency of your building. You should have gathered prevailing wind data during site discovery (see Chapter 5). If you carefully plan the building orientation, your site design elements can capture cool summer breezes and block cold winter winds.

For best wind orientation, the building should be positioned to reduce exposure to the winds and shield the habitable space from prevailing cold winter winds (see Figure 6-32). Natural features such as forest and hills can provide protection from wind. Landscaping elements such as evergreen trees and berms can also be positioned to serve as a windbreak, as we will learn in Chapter 18. Alternatively, all or part of the building can be built underground, allowing the earth to insulate the structure from the elements. An example of an earth-sheltered building is shown in Figure 6-33.

You should also consider the possibility of capturing cool summer breezes and directing the air flow through the building by taking advantage of *cross ventilation*. Cross ventilation is achieved by providing openings on the wall that will receive

Figure 6-32: For maximum cooling benefits during hot summers, orient the long axis of the building to receive the summer prevailing wind. To reduce the negative effects of cold winter winds, reduce the building exposure to the prevailing winter wind.

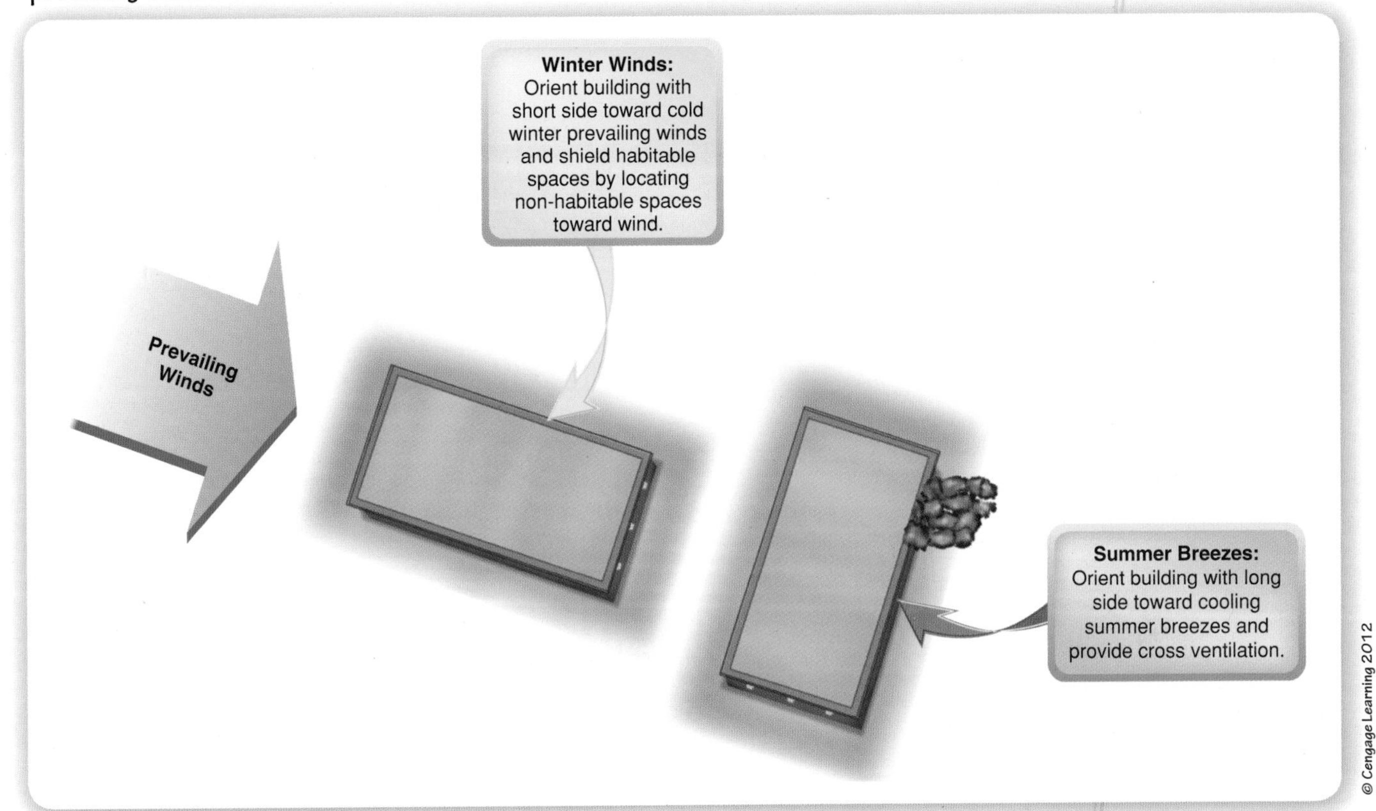

Winter Winds:
Orient building with short side toward cold winter prevailing winds and shield habitable spaces by locating non-habitable spaces toward wind.
Prevailing Winds
Summer Breezes:
Orient building with long side toward cooling summer breezes and provide cross ventilation.

the prevailing wind and openings on opposite walls as shown in Figure 6-34. The air pressure on the windward face is higher than the air pressure on the leeward face causing the air to flow through the building. Landscaping can also be used to funnel the wind into the building, as we will discuss in Chapter 18.

Figure 6-33: Earth-sheltered buildings are protected from the cooling effects of cold winter winds.

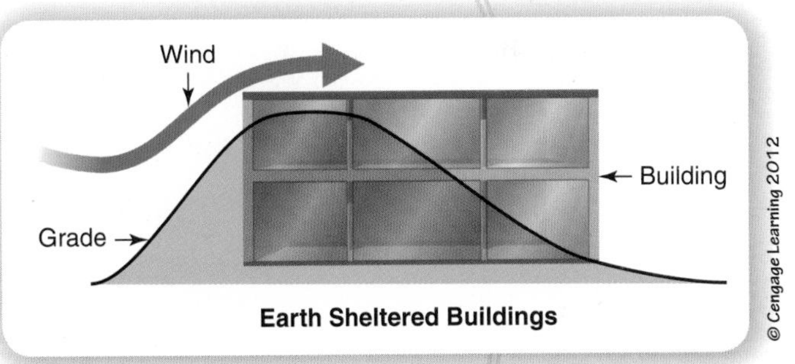

Wind
Building
Grade
Earth Sheltered Buildings

SOUND ORIENTATION Unfortunately you cannot always locate a project in a quiet, rural setting. You may need to contend with noise from adjacent businesses and neighbors or nearby roadways. If your site will be exposed to excessive noise pollution, the best sound orientation is a building position away from the noise. Since sound waves tend to travel up and away from the source, a position that is level with or below the elevation of the source may help reduce the noise. In addition, natural or man-made sound insulation such as staggered rows of evergreen trees, hedges, hills, berms, fences, and walls constructed of heavy material can help absorb and redirect the sound (see Figure 6-35).

VIEW ORIENTATION If you or your client purchased the project site at a premium because of the potential views from the property, *view orientation* may take precedence over all other orientations. Orienting the building to take best advantage of the view may result in exposing the building to cold winter winds or may require that the structure be elevated above a nearby roadway allowing more noise

Figure 6-34: The Syracuse Center of Excellence incorporates operable windows on opposite walls to take advantage of cool breezes during warm weather by cross ventilation.

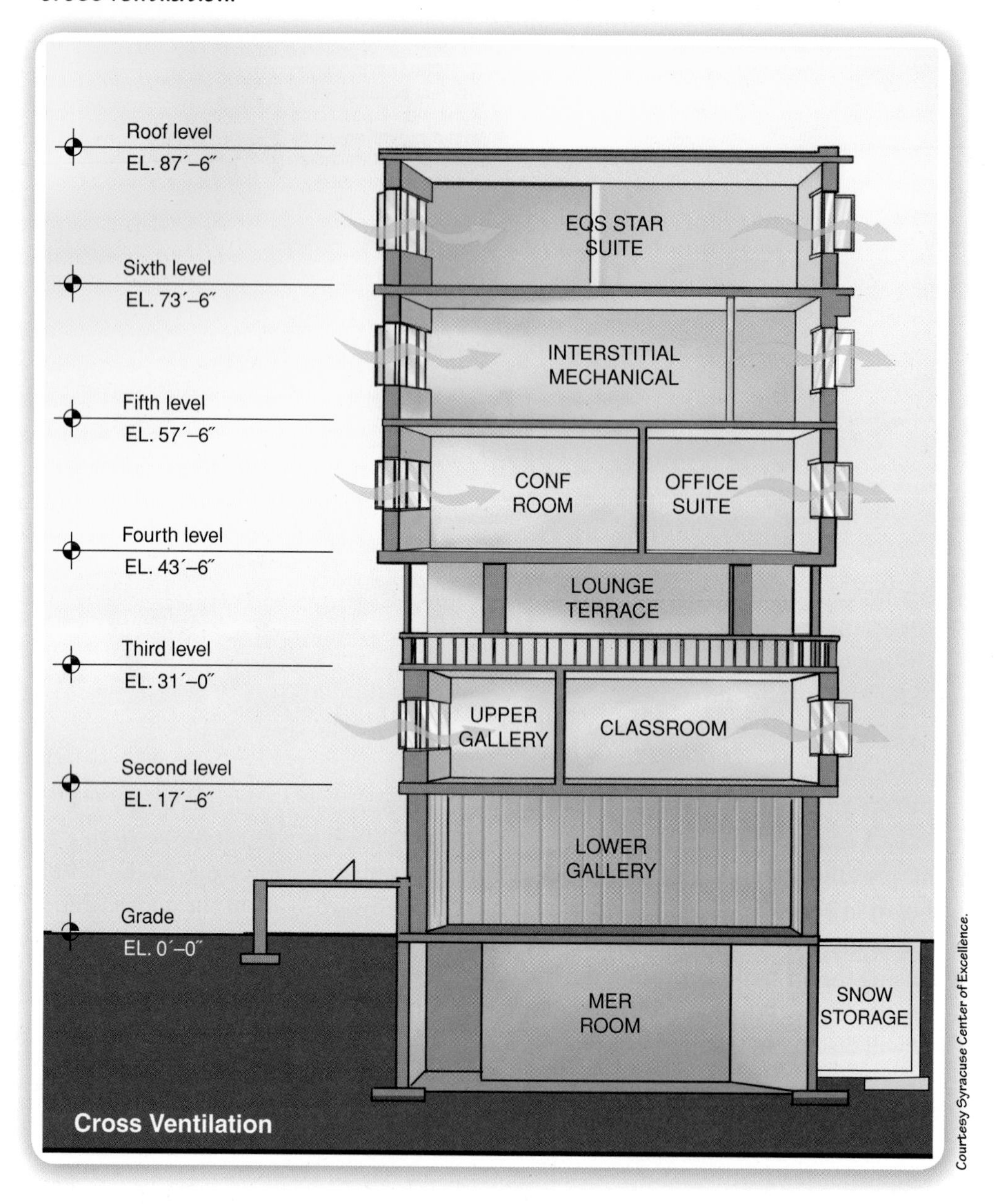

Courtesy Syracuse Center of Excellence.

to reach the building. In many cases, the property owner "bought" the view and is willing to sacrifice other advantages to capitalize on his investment. To achieve the best view orientation, the designer must carefully consider both the exterior and interior space design to avoid obstructions and capitalize on the view.

TERRAIN ORIENTATION The physical characteristics of the land on which the building will be constructed can affect the placement of the structure on the property and the type of building that will be constructed. We discussed desirable slopes for development in Chapter 5 and discovered that sites slopes between 1 and 5 percent provide the most economical sites because they accommodate large structures and provide good site drainage without requiring special designs. So what terrain factors should you consider when locating your builidng on the property?

Figure 6-35: Dense foliage, including trees and a hedge, and a fence provide sound insulation for the house on the left. Staggered rows of evergreen trees and a berm insulate the house on the right from the road noise. From JEFFERIS AND MADSEN. Architectural Drafting and Design, 5e.

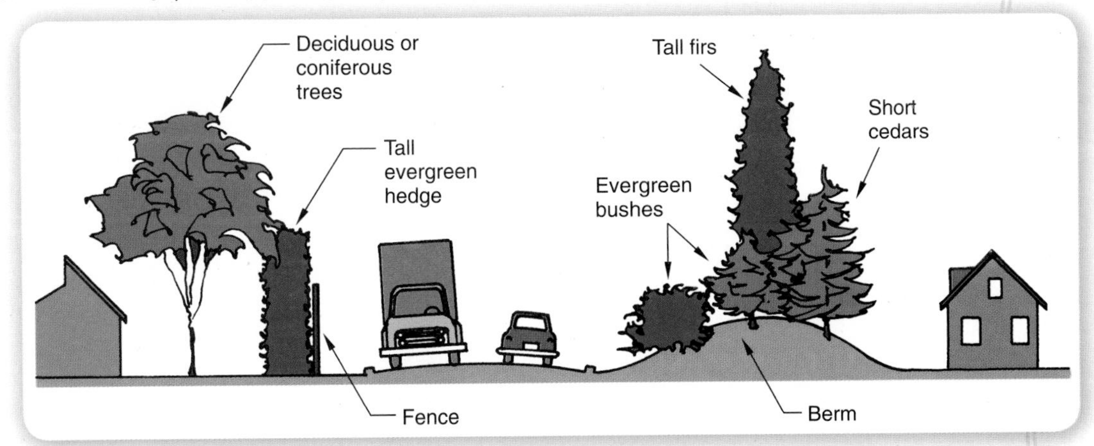

1. Locate buildings on high ground. Try to locate your structure in a relatively flat area of the property with natural drainage away from the building location to keep storm water from inundating your building. Later in this chapter we will discuss specific code requirements for ground slope and the height of floors above grade.
2. Avoid steep slopes. Keep all structures away from steep slopes (more than 33.3 percent). Steep slopes are prone to slope failure and can cause excessive stormwater runoff. If your structure is at the base of the slope, erosion and shallow foundation failure may occur. If your building is situated at the top of a steep slope that fails, the building foundation could lose support from the soil resulting in excessive settlement or foundation failure. The International Building Code (IBC) specifies required setbacks from steep slopes (see Figure 6-36).

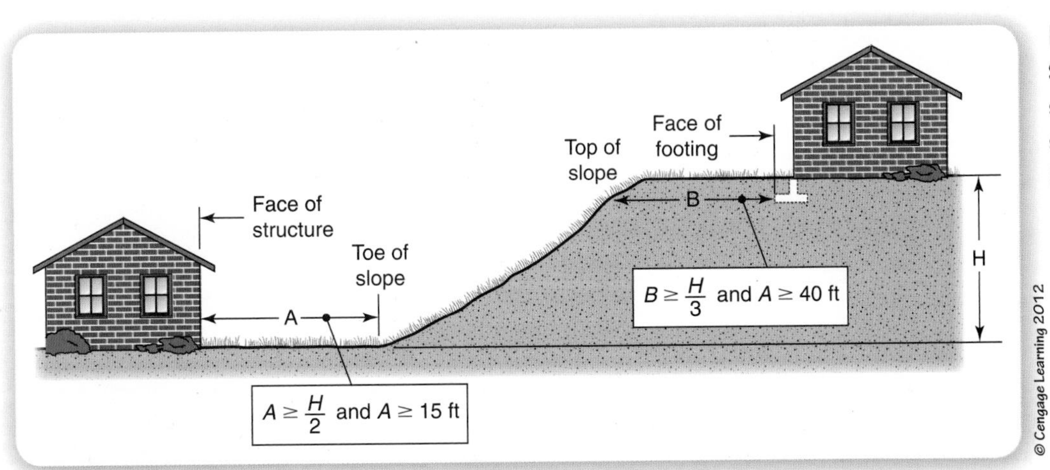

Figure 6-36: Structures must be set back from slopes steeper than 33.3 percent.

3. Select a building location that will minimize the amount of soil that will have to be moved during construction. It is expensive to move, purchase, and dispose of soil, so the amount of earthwork necessary during construction can greatly affect the cost of construction, as we will discuss later in this chapter.

Of course, your site may not have an ideal slope—it may be very flat (less than 1 percent slope) or steep (more than 10 percent slope). As discussed in Chapter 5, undesirable slopes will impact the architectural and structural design of your building and the design of the site drainage that will typcially increase the cost of construction. Sections of some building designs that are responsive to steep slopes are shown in Figure 6-37. Multilevel buildings are contoured to the ground slope by incorporating a "stepped" design. Fall-away designs include one or more floor levels that are open on one side but not obvious from the opposite side because they are built partially below grade. In residential construction, these floors are often called *daylight basements.* Cantilever building designs provide a consistent floor elevation but require taller foundations than multistory buildings or cantilevered structural designs that allow the building to extend beyond the foundation. Earth sheltered structures are built partially or fully below the ground surface, thereby requiring more *excavation,* or digging, in order to construct the building. However, earth sheltered structures may provide wind and sound orientation advantages.

SITE OPPORTUNITIES MAP A site opportunities and constraints map, like that shown in Figure 6-39, provides an overview of the site orientation and environmental considerations for a property. A site opportunities and constraints map should be created for each potential site in order to verify that the site can adequately meet the client/user needs and comply with all applicable regulations. The maps can be used to compare alternate sites and to determine the best building location and orientation.

Figure 6-37: **Designs that accommodate steep grades include multilevel, fall-away, cantilever, and earth-sheltered designs.**

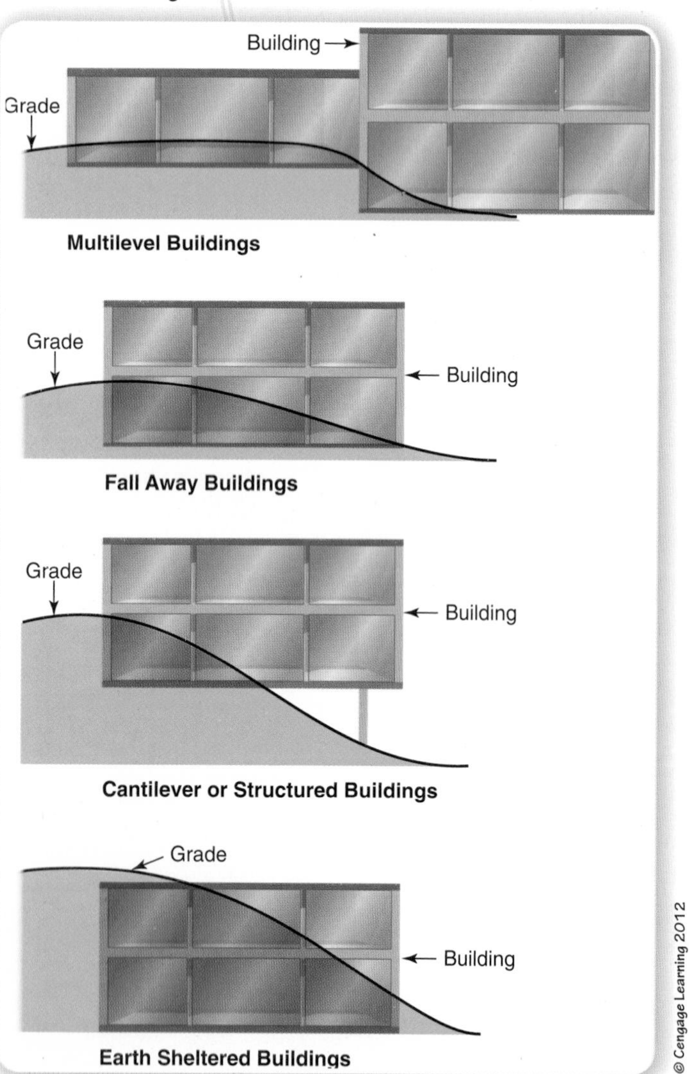

Figure 6-38: **The terrain of a building site may drastically affect the building design. The Ostrog Monastery in Montenegro is earth sheltered and is truly integrated into the site topography.**

Point of Interest

Bill Gates Goes Underground

Bill Gates, former CEO of Microsoft, Inc. and one of the richest people on earth, and his wife Melinda own a large earth sheltered mansion that was designed as a joint venture between Bohlin Cywinski Jackson and Cutler Anderson Architects in the Pacific Rim–style. Much of the home is embedded into the hillside of Lake Washington in Medina, Washington. The home design features an underground 6300 sq ft garage space. To access the garage, you must drive over the roof and onto a wooden trestle that extends past the open face of the house. The retaining walls used in the design created a drainage problem on the site. To reduce runoff, storm water is collected in underground pipes, which drain into a man-made stream and wetland estuary. The home won a 1997 Honor Award from the American Institute of Architects.

Bill and Melinda Gates' home is an earth sheltered mansion built into the hillside overlooking Lake Washington in Medina, Washington.

Figure 6-39: **A site opportunities and constraints map, like the one shown, here can help you record the site orientation opportunities and constraints when selecting a building location on the site.**

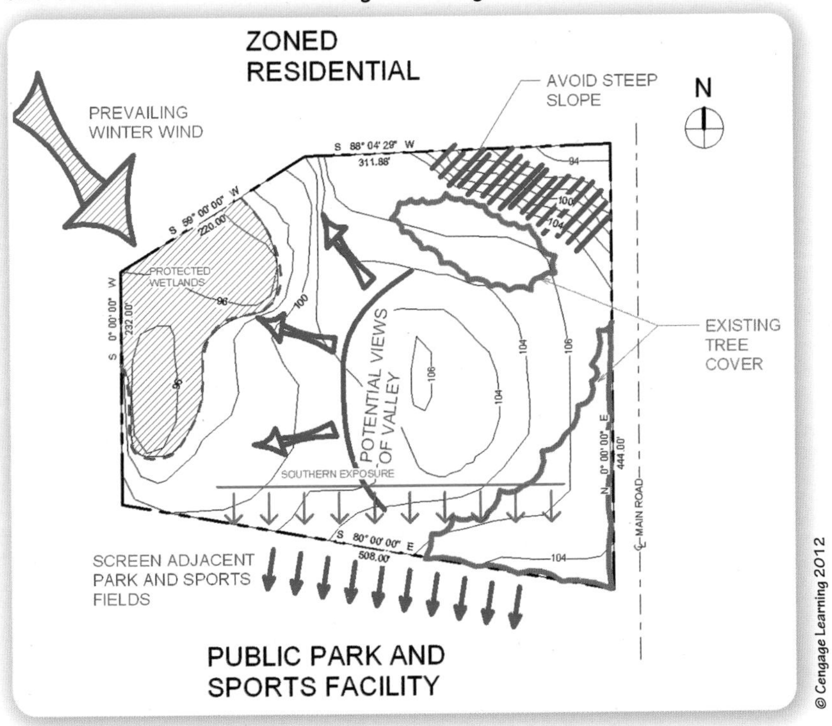

Engineers and architects use a wide variety of sources to identify criteria and constraints and determine how these will affect the design. When conflicts arise among competing criteria and constraints, they must identify and consider trade-offs in order to select the best design solution.

Point of Interest

Center of Excellence Orientation

The Syracuse Center of Excellence in Environmental and Energy Systems designers incorporated good site orientation practices in the design of the facility. The building is relatively thin with the largest walls of the building facing north and south (see Figure 6-40) providing excellent solar orientation. Extensive windows provide natural light to the occupants in most indoor spaces and opportunities for excellent views and natural ventilation. The southern face of the building is a wall of insulated glass with integrated electronically controlled blinds that allow occupants to control solar gain and glare. The window openings on the north and south faces are operable and allow cross ventilation to cool the building during warm weather.

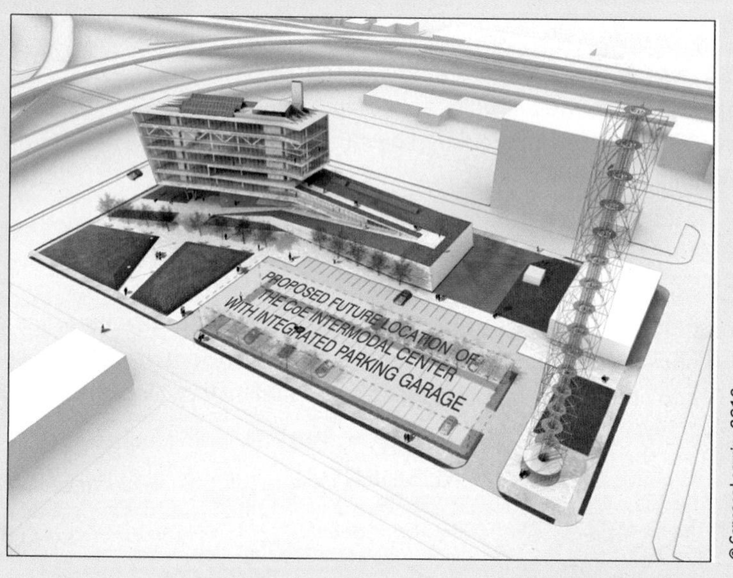

Figure 6-40: The Syracuse Center of Excellence has excellent site orientation. The building is oriented with its longest wall (almost completely composed of glass) facing south providing good solar orientation and providing good views of the green area of the site. In addition, the building was oriented to take advantage of cooling summer breezes with operable windows on the north and south elevations to allow cross ventilation. Courtesy of the Syracuse Center of Excellence.

Careers in Civil Engineering and Architecture

Sewers for Fort Lauderdale

As a project construction manager for the large engineering firm CH2M HILL, Theo Melo works on a project to build a sewer system for Fort Lauderdale, Florida. Almost half the city was without sewers before the 10-year, $691 million program began.

Though Melo works solely on this program, he's never bored. "It's something new every day," he says.

Theo Melo
Project Construction Manager, CH2M HILL

On the Job

Melo manages the day-to-day construction of sewers and pump stations and the renovation of water and wastewater treatment plants. He coordinates with the city of Fort Lauderdale, manages permits, and makes sure the contractor gets the work done on time and on budget. He also deals with residents' complaints. "If there's a problem," he says, "you try to fix it."

Melo is motivated by the fact that his work in Fort Lauderdale will help the environment. The sewers are replacing aging septic tanks, which had posed a threat to the city's water system.

"We're cleaning up the environment for the residents," he says. "That's the biggest problem we've solved here."

Inspirations

Melo comes from a family of civil engineers on both sides, and he'd wanted to be one himself ever since childhood. "I was always into structures, bridges, and tall buildings," he says.

Melo went to a magnet high school that specialized in engineering. He built robots and programmed them, and also learned to draw via computer.

Melo liked the flexibility of civil engineering. "The beautiful thing about it is that there's so much you can do," he says. "There are seven main areas, and you can go into any one of them, or mix and match."

Education

Melo received his civil engineering degree from Auburn University. He took classes in all seven areas of civil engineering and enjoyed creating projects in each one. He designed a bridge made from balsa wood, an interstate system, an airport, a traffic system, and a wastewater treatment plant. He also engaged in competitions, using concrete to build canoes and horseshoes.

"The good thing about all those projects," Melo says, "is that you came out of college with real-life experience."

Advice for Students

Melo sees a lot of job opportunities in the future. "We'll need young engineers down the road," he says.

Melo recommends that students interested in engineering take as much math and science as they can in high school.

He also suggests that they talk to people at engineering firms, as well as college students majoring in engineering.

"Get familiar with what they do and what they like to do, and see if that's what you like to do," he says. "That way you'll have a better sense of what interests you, and you won't have to change your major after you start your program. Ask questions—you have nothing to lose."

SUMMARY

A construction project is influenced by a wide variety of conditions and circumstances. Without proper investigation and planning, a project can veer off a successful path. Once a site has been investigated and deemed potentially adequate for a project a conceptual plan for the construction of the project on that site can be undertaken. Initially, it is imperative that the criteria and constraints for the project are clearly defined and documented. A needs analysis should be performed and an architectural program written so that all stake holders are working toward the same clearly defined goal.

It is also important to continue to gather important information about the site that will affect the design of the project. Land surveys, including topographic, property, construction, and control surveys will provide information that is critical during the design and construction phases of the project. Surveys are performed with a wide variety of instruments that range from conventional optical equipment to GPS equipment, which uses signals from a constellation of satellites to calculate exact position.

Because the existing soils on a site can greatly influence the design and construction of a project, a geotechnical investigation is often necessary for commercial development. A geotechnical report provides important information on the physical and chemical properties of the soils. In addition, the report provides the specific characteristics of the existing soils and typically includes the USCS classification of the soils based on grain size and distribution. This classification, along with other soils test results, allow geotechnical engineers to estimate lateral soil pressure, soil strength, and soil permeability as well as to make recommendations for pavement, foundation, and drainage design.

Once an architectural program has been approved and you have gathered sufficient site information, you may begin the preliminary design phase to develop a conceptual design of the project. A conceptual design must address the constraints and criteria in the architectural program and consider the physical and environmental features of the site. A key consideration is the orientation of the project components on the site. Depending on the site topography, building function, and location, the site terrain and potential views may be an important consideration. In addition, the solar, wind and sound orientation should inform the design. A site opportunities and constraints map can be created to reflect all of the considerations important to the design to provide a basis for the conceptual design.

BRING IT HOME

1. Create a list of the major use areas inside your high school (cafeteria, library, front office, attendance, guidance, gym, etc.). Then create an adjacency matrix using these areas for the perfect high school (which will not necessarily correspond to the existing adjacencies in your school).
2. What types of surveys might be useful during the development of a commercial facility? List each survey type and explain how the survey information could be beneficial to the project.
3. As a surveyor, you have been asked to set a temporary benchmark on a development site in order to provide a point of known elevation on the property. You have located a documented benchmark with an elevation of 642.09 ft less than a block from the site and installed a temporary bench marker on the site. Using optical equipment, you obtain a rod reading of 5.93 ft at the benchmark (foresight) and a rod reading of 3.65 ft at the temporary benchmark. Create a sketch similar to Figure 6-12 showing the instrument set-up and the line-of-sight to both the permanent and temporary benchmark. On the sketch indicate the rod readings. Then calculate the HI and the elevation of the temporary benchmark.
4. As a geotechnical engineer, your firm has been contracted to provide a soils report for the site of a new development. The results of a sieve analysis and the Atterberg limits for sample 6C are presented below. Classify this soil using the Unified Soil Classification System (USCS).
   a. Total sample mass = 738.23 g
   b. Retained on no. 4 sieve = 247.60 g
   c. Retained on no. 40 sieve = 179.39 g
   d. Retained on no. 200 sieve = 98.71
   e. Liquid limit = 43
   f. Plastic limit = 31
5. You have been hired to design a residence on a property in a small town in central Washington State. The map shown below was developed from a site survey.

   The state claims an easement of 40 ft along Highway 19. The town ordinances require a setback of 25 ft along public roadways, and side and rear setbacks of 20 ft. There is a drainage easement of 30 ft along the east property line. Additional research has indicated that Highway 19 supports heavy traffic and produces continuous road noise. In addition, there are streetlights along Highway 19 that illuminate much of the south portion of the lot. However, there are excellent views of a large lake to the west of the property. Perform the necessary research to establish the best site orientation and create two maps for the property.
   a. A map showing the buildable area of the property.
   b. An opportunities and constraints map.

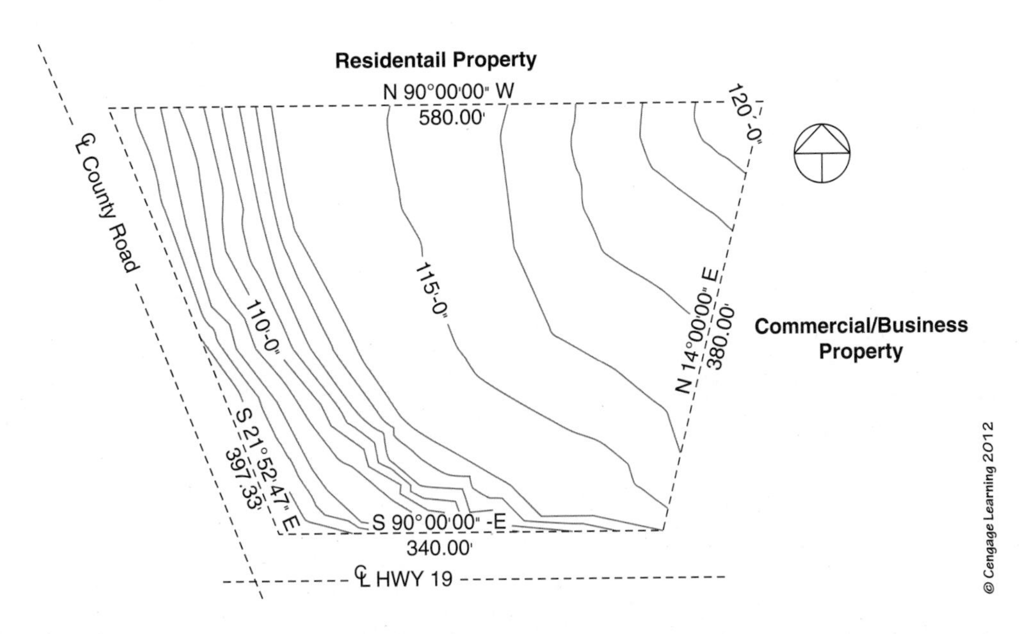

EXTRA MILE

CREATE A TOPOGRAPHIC MAP

A topographic survey will determine the elevations of specific points on the ground. These elevations are then used to create contour lines on a topographic map. A topographic survey was conducted using the grid method for a 100 × 100 sq ft lot and a table of the elevations of the grid intersection points was prepared (see below). A 20 ft grid spacing was used. The lettered grid lines run north-south where Grid Line A coincides with the west property line. The numbered grid lines run east-west and Grid Line 1 coincides with the north property line. Use this data to create a topographic map of the property, either by hand or using software.

GRID LINE	ELEVATION (FT)	GRID LINE	ELEVATION (FT)	GRID LINE	ELEVATION (FT)	GRID LINE	ELEVATION (FT)	GRID LINE	ELEVATION (FT)	GRID LINE	ELEVATION (FT)
A-1	650.7	A-1	650.7	A-1	650.7	A-1	650.7	A-1	650.7	A-1	650.7
A-2	649.0	A-2	649.0	A-2	649.0	A-2	649.0	A-2	649.0	A-2	649.0
A-3	648.1	A-3	648.1	A-3	648.1	A-3	648.1	A-3	648.1	A-3	648.1
A-4	649.2	A-4	649.2	A-4	649.2	A-4	649.2	A-4	649.2	A-4	649.2
A-5	648.2	A-5	648.2	A-5	648.2	A-5	648.2	A-5	648.2	A-5	648.2
A-6	644.7	A-6	644.7	A-6	644.7	A-6	644.7	A-6	644.7	A-6	644.7

CHAPTER 7
Site Design

GPS DELUXE

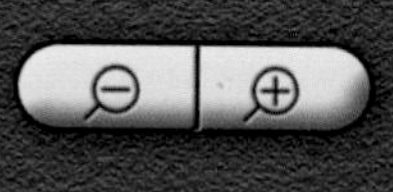

Menu

START LOCATION | DISTANCE | END LOCATION

Before You Begin

Think about these questions as you study the concepts in this chapter:

1. What accommodations for persons with disabilities must be incorporated into your site design?
2. What factors should be considered when designing the parking and circulation of vehicles and pedestrians on a site?
3. How does the grading of a site affect the cost of construction?
4. How do you design a site to adequately control stormwater runoff?
5. What information do you need to design an adequate supply (or disposal) of necessary utilities?
6. What information is typically shown on a site plan?

INTRODUCTION

You may not notice the design of a site unless there are problems. You may take for granted that there will be parking spaces of adequate size, sidewalks where they are needed, and handicapped access to buildings. Most likely you just assume that rainwater will not inundate a building or form rivers across a parking lot when it rains. You are accustomed to having clean water run into a drinking fountain when you push a button and, although you may not know where the used water goes, you are confident that it is being carried away from the building to be treated. And, although you may not notice the attractive appearance of many facilities, you probably will note unattractive properties. The goal of site design is to enhance the project with the least disruption to people and the environment, and therefore, the site is often overlooked (see Figure 7-1).

To make the best use of a site, civil/site engineers must consider many factors and balance the use of man-made elements with natural features. Essentially, the overriding consideration in site design is to create positive flows. Traffic should flow smoothly onto and off of the site and through the parking areas. Pedestrians should be provided with a safe passage to and from the building with access for handicapped people. Stormwater should flow away from buildings and parking areas and not pool in areas to be used by people. Clean water, power, and other utilities should flow freely to the site. On demand, used water and solid waste should be easily discarded and removed. If all of these systems flow smoothly and the needs of the users and client are addressed in an attractive and environmentally friendly way, the site design is successful. If the systems do not operate effectively, the site will display negative conditions that draw attention and indicate poor site design.

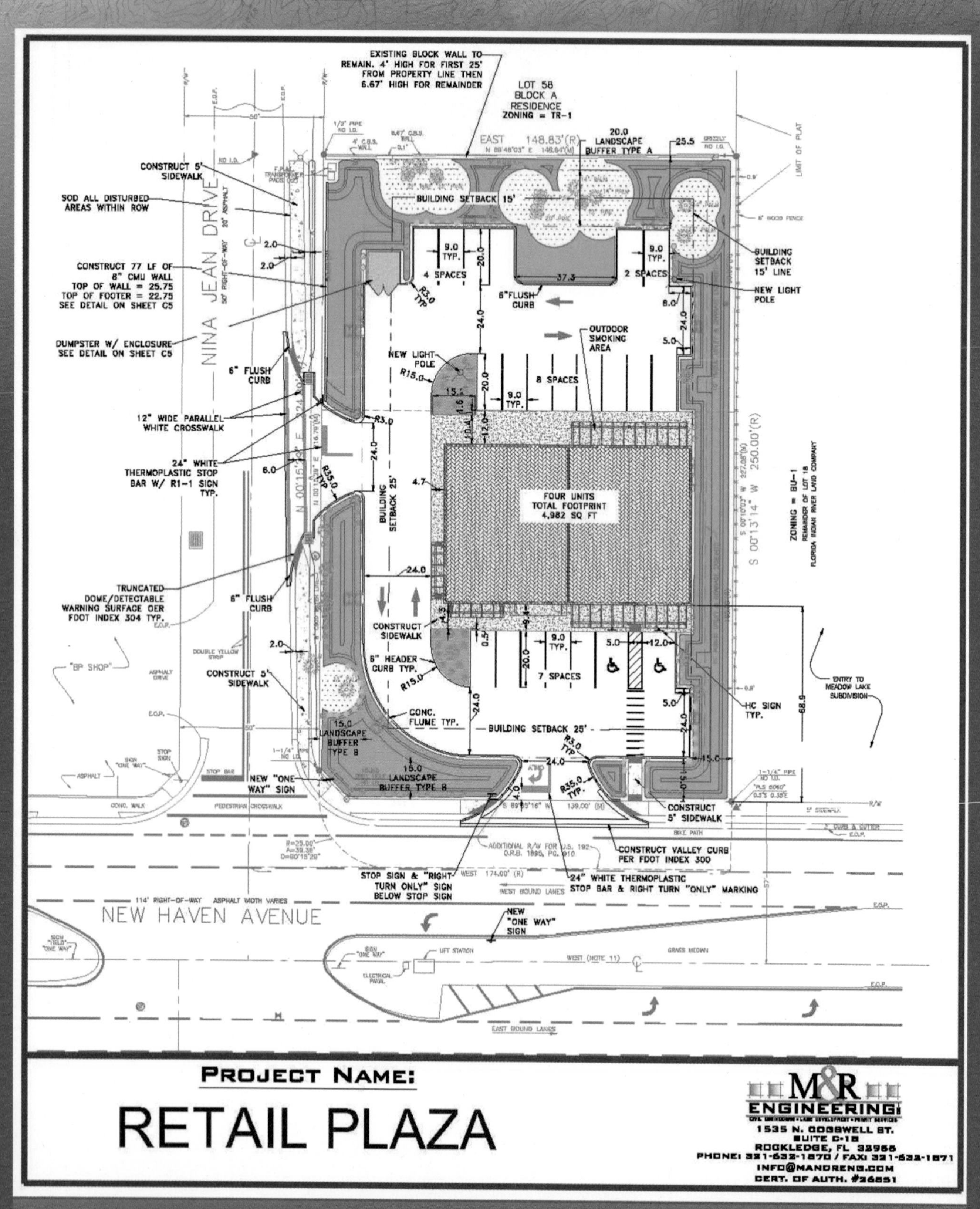

Figure 7-1: *Site plan example. Drafting and Civil Engineering by M&R Engineering, Inc.*

ACCESSIBILITY

Do you think your life would change if you broke your leg playing soccer and were confined to a wheelchair? What if a friend was in a traffic accident that resulted in him losing his sight? Would he be able to maneuver through the maze of life without injury? Many of us take for granted the ability to perform everyday tasks such as climbing steps, opening a door, or turning on a light switch. And, we don't realize the hazards posed by common features of buildings. For instance, if your broken leg has left you wheelchair-bound for several weeks, could you retrieve a box of your favorite cereal from the top shelf at your local grocery store without assistance? A wall-mounted metal box storing a fire extinguisher seems like a good feature to have in a building, but what if your blind friend is using a walking cane and the box projects too far into the walkway? If your friend walks close to the wall he could easily collide with the box and be injured (see Figure 7-2).

Figure 7-2: A temporary condition, like a broken leg, as well as chronic disabilities can negatively affect your ability to travel from place to place and to access public and private sites, buildings, and facilities.

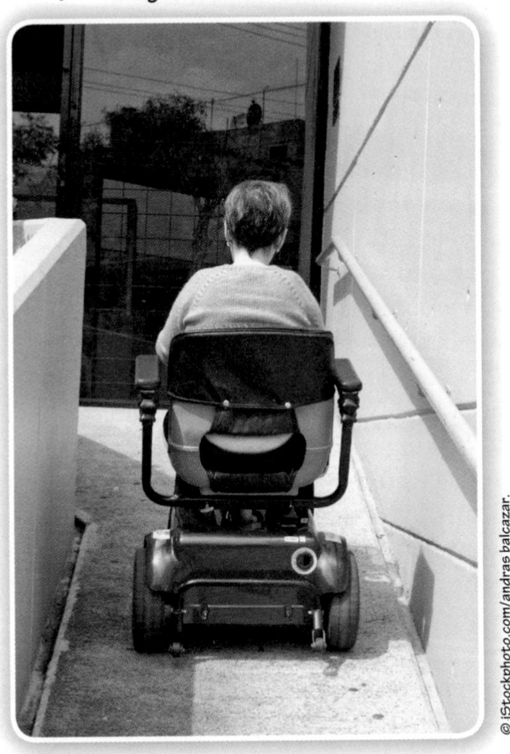

© iStockphoto.com/andras balcazar.

Accessible: describes a site, building, or facility that meets the requirements of the applicable federal regulations for access (Americans with Disabilities Act or ADA, the Architectural Barriers Act or ABA, and/or the Fair Housing Amendments Act or FHAA).

Accessibility: is the ease and convenience with which customers, tenants, employees, and other users can enter a property. Accessibility also refers to building design and alterations that enable people with physical disabilities to enter and maneuver in the building.

There are millions of people who, due to temporary injuries or lifelong disabilities, struggle to complete day-to-day activities. Although many common tasks remain difficult for people with disabilities, the federal government has enacted many laws that prohibit discrimination against people with disabilities related to building construction. Specific requirements for accessible design depend on the type of facility you are building. In general accessibility is not mandated for residential construction with less than four units.

The purpose of the accessibility laws is to give everyone the opportunity to safely participate in everyday activities, such as eating in a restaurant, attending a football game, taking a class at the local community college, or working at the zoo.

The Americans with Disabilities Act of 1997 (ADA) prohibits excluding, on the basis of disability, access to public accommodations, employment, transportation, state and local government services, and telecommunications. ADA Title II covers activities of state and local governments. ADA Title III covers commercial facilities and businesses and nonprofit organizations that are public accommodations. A **public accommodation** is any private entity that offers products or services to the public.

Construction or building alterations for the federal government and federally funded projects are subject to the Architectural Barriers Act of 1968 (ABA) which requires compliance with the Uniform Federal Accessibility Standards (UFAS). The Americans with Disabilities Act and Architectural Barriers Act Accessibility Guidelines (ADA-ABA Guidelines) specify requirements for new and altered construction that improve accessibility for everyone.

The Fair Housing Amendments Act (FHAA) as amended in 1988 requires new housing units with four or more units to be accessible to persons with disabilities. The ADA, ABA, and FHAA are federal legislation and are enforced through the Department of Justice. If you do not comply with these laws, you could be sued in federal court for failure to comply.

A Standard: is a level of quality or excellence that determines a level of attainment that is acceptable. It is not a law or an enforceable code unless it is adopted by a governmental entity.

In addition to federal regulations, the American National Standards Institute (ANSI) standard A117.1 is a national **standard** that provides guidance for barrier-free design upon which the ADA-ABA Guidelines, FHAA, and UFAS access requirements are based, although there are differences between the three. The International Building Code (IBC) references ICC/ANSI A117.1, which essentially makes it part of the code. Therefore, if your state has adopted the IBC, then the ANSI A117.1 standards are enforceable. However, many states and local governments have adopted ADA-ABA Guidelines or have developed their own special regulations related to accessibility that will control your design. Your local building inspector will enforce the adopted regulations, therefore, you should be clear as to the accessibility accommodations that you are required to incorporate into your design.

Two important accessibility requirements that affect site design are accessible routes and accessible parking spaces. *Accessible routes* are paths of travel that are designed to be easily navigated by persons in a wheelchair and cannot include stairs, steps, or escalators. Accessible parking spaces are sized to allow a disabled person to park a car (or van) while providing enough additional space to maneuver a wheelchair next to the car and allow unconfined transfer of the person between the vehicle and wheelchair.

The IBC and the ADA-ABA Guidelines (2004) have similar accessibility requirements—both typically require accessible routes within the site from public transportation stops, accessible parking, accessible passenger loading zones, and public streets or sidewalks to the accessible building entrance unless the only means of building access does not provide for pedestrians. For our discussion in this textbook, we will use the Americans with Disabilities Act Accessibility Guidelines of 2004 (ADAAG), which will apply to most commercial facilities, to illustrate some of the accessibility requirements. Be aware that the discussion here represents only a small portion of the accessibility design elements required by law. Please review the accessibility requirements that apply to your project. The following list provides some of the basic ADAAG requirements:

1. The width of the accessible route (a solid surface) should be a minimum of 36 in, but if the route is less than 60 in wide, a passing space should be provided every 200 ft.

2. Every effort should be made to create a level route, however a vertical level change of ¼ in is allowed, and a bevel can be used for level changes of up to ½ in, as shown in Figure 7-3.
3. The running slope of a walking surface included in an accessible route should not exceed 1:20, which means the increase in height (rise) cannot exceed 1 in for every 20 in in horizontal distance (run).
4. Any walking surface that has a slope greater than 1:20 is considered a ramp. Ramps cannot exceed a slope of 1:12 along accessible routes and cannot have a rise of more than 30 in (see Figure 7-4).
5. Each ramp must have a landing at the top and bottom, as specified in Figure 7-5.

Figure 7-3: Changes in level. Vertical changes in level of ¼ in (6.4 mm) high maximum are permitted along accessible routes. Changes in level between ¼ in and ½ in (13 mm) must be beveled with a slope not steeper than 1:2. From the US Access Board, Americans with Disabilities Act (ADA) Accessibility Guidelines for Buildings and Facilities, 36 CFR Part 1191.

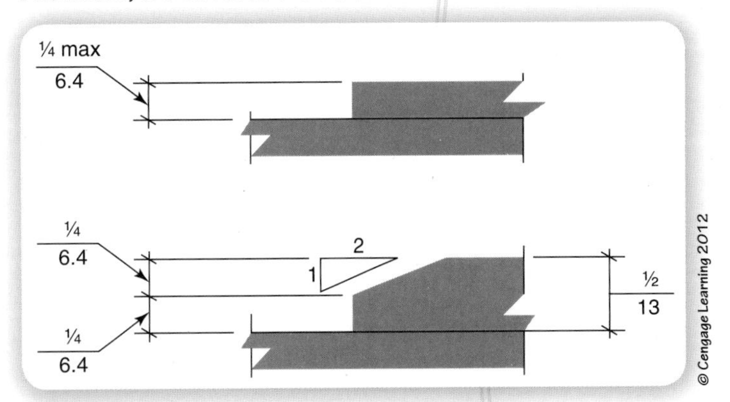

Figure 7-4: Ramps. Changes in level greater that ½ in (13 mm) high must be ramped with a maximum slope of 1:12. The maximum rise for any ramp run shall be 30 in maximum, and every ramp must have landings at the top and bottom of each run.

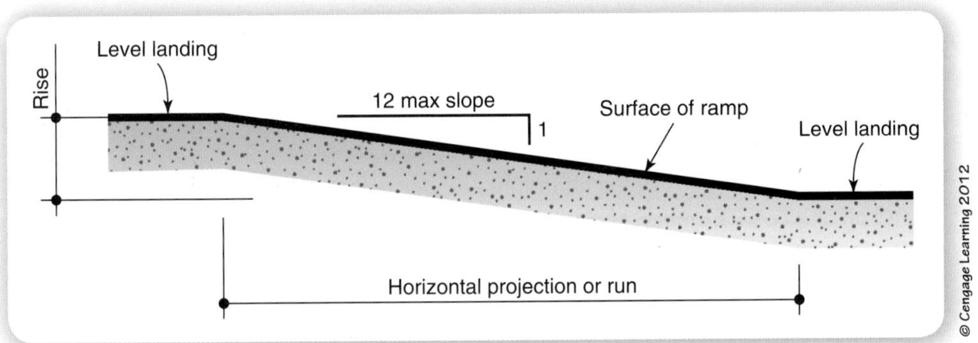

Figure 7-5: Landings are required at the top and bottom of each ramp run. The landing width should be at least as wide as the widest ramp run leading to it, and the landing length shall be at least 60 in (1525 mm) long. From US Access Board, Americans with Disabilities Act (ADA) Accessibility Guidelines for Buildings and Facilities, 36 CFR Part 1191.

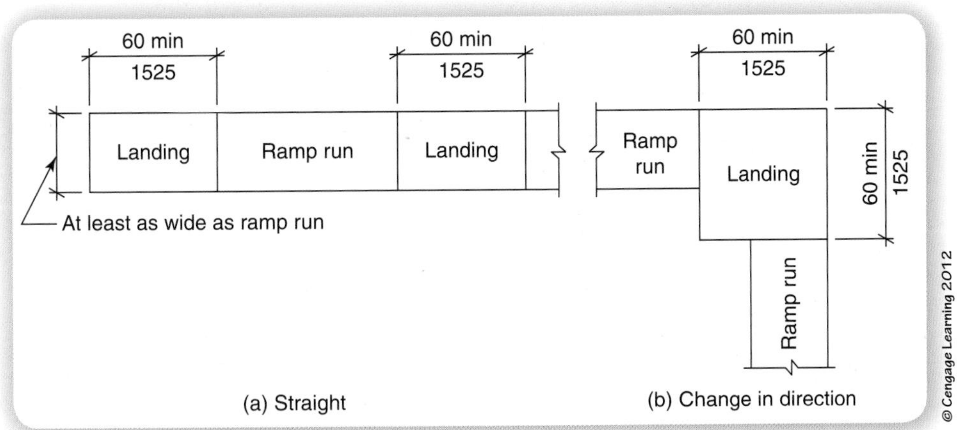

Figure 7-6: Curb ramp. Curb ramp flares shall not be steeper than 1:10 and landings shall be provided at the tops of the ramp. From US Access Board, Americans with Disabilities Act (ADA) Accessibility Guidelines for Buildings and Facilities, 36 CFR Part 1191.

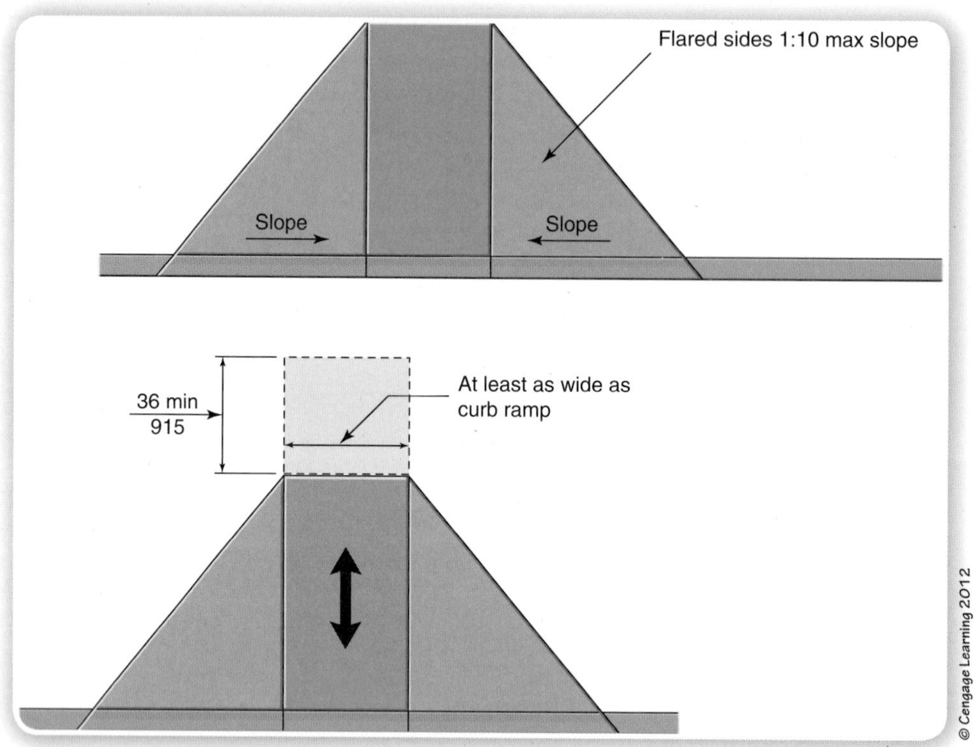

Figure 7-7: ADAAG requires a minimum number of accessible parking spaces based on the total number of parking spaces provided according to this table. At least one of every six accessible spaces must be van accessible.

Source: US Access Board, Americans with Disabilities Act (ADA) Accessibility Guidelines for Buildings and Facilities, 36 CFR Part 1191.

Total Number of Parking Spaces Provided in Parking Facility	Minimum Number of Required Accessible Parking Spaces
1 to 25	1
26 to 50	2
51 to 75	3
76 to 100	4
101 to 150	5
151 to 200	6
201 to 300	7
301 to 400	8
401 to 500	9
501 to 1000	2 percent of total
1001 and over	20, plus 1 for each 100, or fraction thereof, over 1000

6. If the ramp rise exceeds 6 in, handrails should be provided on both sides of the ramp according to the accessibility requirements, but are not required at curb ramps.
7. Accessible ramps are required to provide an accessible route at curbs. Curb ramp details are shown in Figure 7-6.
8. If parking spaces are provided for the facility, ADAAG requires accessible parking spaces according to Figure 7-7 and should be located at the shortest possible distance along the accessible route.
9. At least one of every six accessible spaces must be van accessible.
10. Car accessible and van accessible spaces must meet the dimensional requirements shown in Figure 7-8. Other parking space configurations are possible depending on the applicable code.

Figure 7-8: Accessible car parking spaces should be 96 in (2440 mm) minimum width, and van accessible spaces should be a minimum of 132 in (3350 mm) wide. Each accessible space must have an adjacent aisle that adjoins an accessible route. An aisle must be a minimum of 60 in (1525 mm) wide but two spaces may share a common aisle. From US Access Board, Americans with Disabilities Act (ADA) Accessibility. Guidelines for Buildings and Facilities, 36 CFR Part 1191.

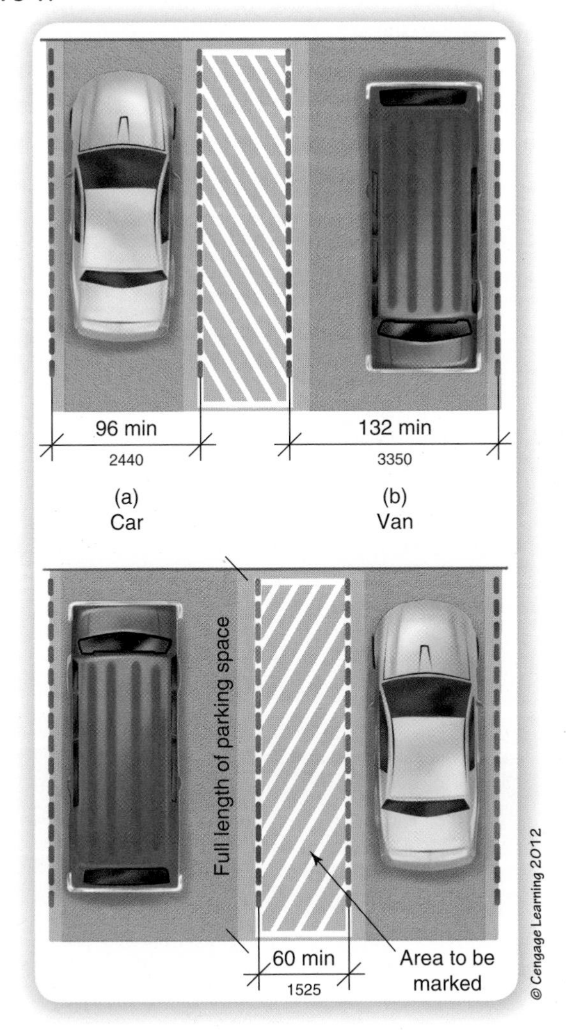

PARKING AND CIRCULATION

The parking lot design and the circulation of people and vehicles on a site can often have a significant effect on the safety, efficiency, and the **aesthetics** of your development. Is there anything more frustrating than arriving to the theater with only a few minutes to spare before your movie begins only to find that there are no parking spaces or that cars are idling in the travel lanes because the parking lot is grid locked? Have you ever had to cross busy lanes of traffic or wade through deep puddles of water to get from your car to a building? Careful site design can avoid unpleasant and dangerous experiences for employees, customers, and clients as well as enhance your project.

What do you need to know in order to create "good" circulation and an efficient parking lot design? Knowledge of local regulations, familiarity with engineering design processes, and common sense can provide a good basis for your design. The following list gives general considerations to circulation and parking design. Many of these recommendations are illustrated in Figure 7-9. Please note that parking requirements are often addressed in local regulations—you should check the local ordinances to verify your design.

Ingress and Egress: refers to entrance and departure, respectively, the means of entering and leaving, and the right to do so.

1. *Ingress and Egress.* How will people and vehicles enter and exit your site? Access to your site should be controlled—that is, the number and size of entrances and exits should be carefully planned to reduce conflict between normal street traffic and traffic entering and exiting your site (see Figure 7-10). State and local governments will regulate access to public roadways often enforcing restrictions on the width and slope of the access drive, radius or curvature of the entry drive, and the composition of the pavement (depth of stone base, concrete, asphalt, etc.) within the right of way. State and local municipalities

Figure 7-9: Parking and circulation design.

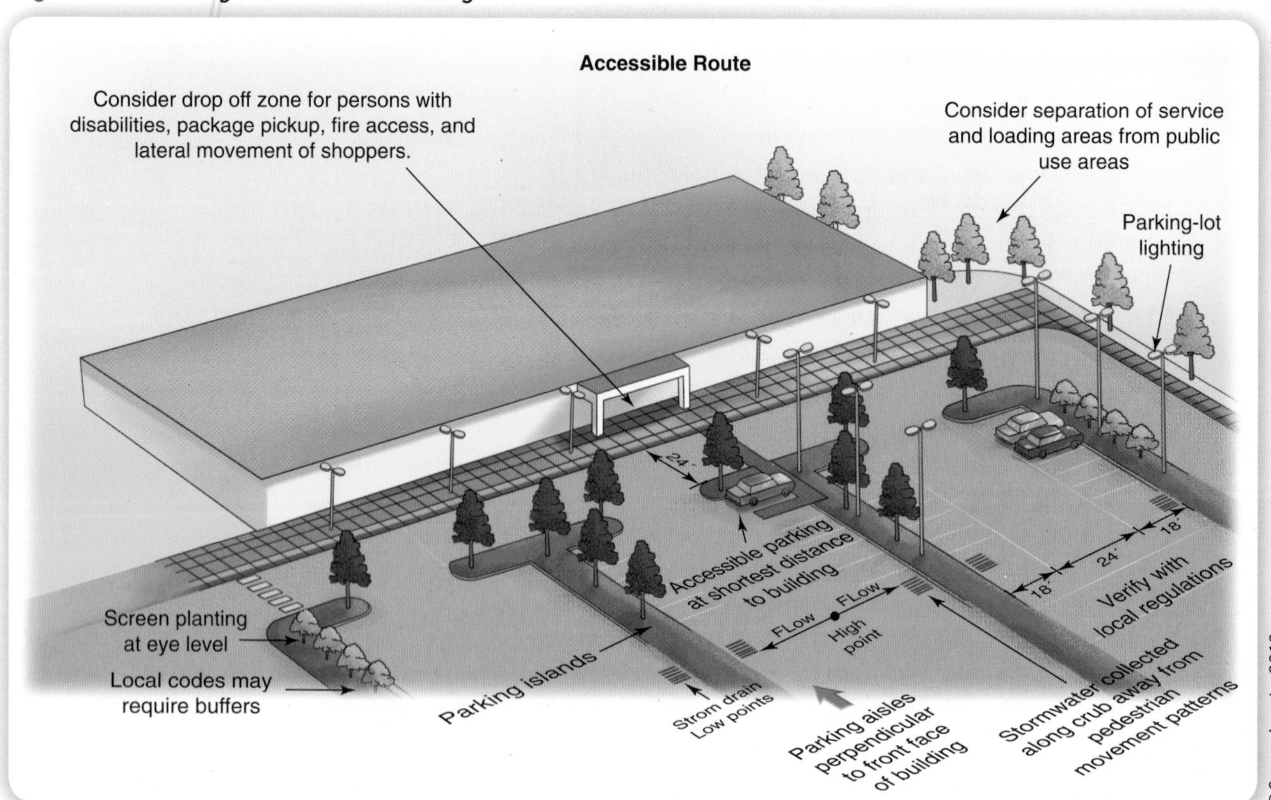

Figure 7-10: An example of driveway spacing criteria for state highways. The spacing requirements take into consideration the time it takes a driver to react to an obstacle and the distance that a car travels after the brakes are applied. The spacing requirements apply whether the adjacent driveways are private access drives or public streets.

Source: Georgia Department of Transportation Regulations for Driveway and Encroachment Control. http://www.dot.ga.gov/doingbusiness/PoliciesManuals/roads/Documents/DesignPolicies/DrivewayFull.pdf.

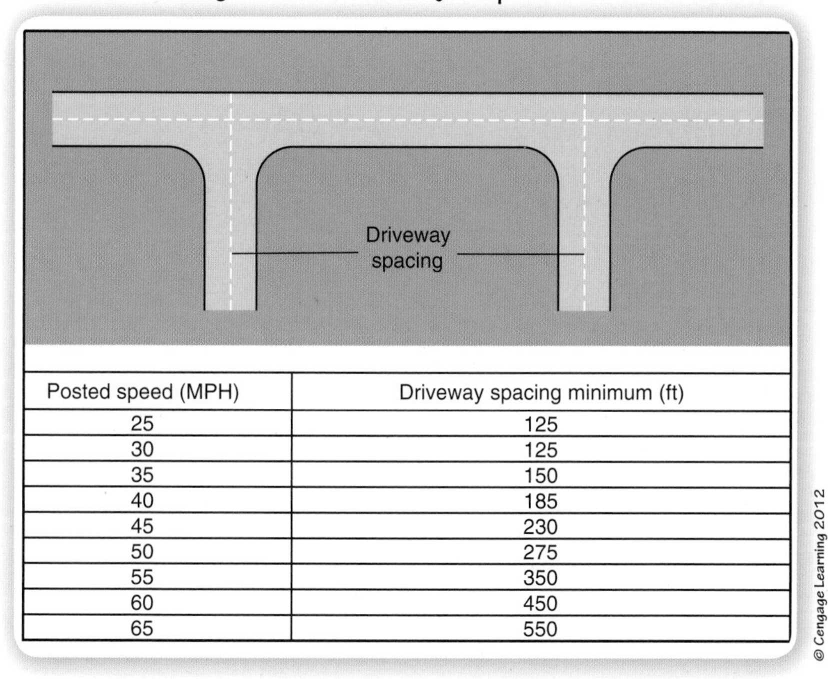

Posted speed (MPH)	Driveway spacing minimum (ft)
25	125
30	125
35	150
40	185
45	230
50	275
55	350
60	450
65	550

typically require that you apply for an *encroachment permit* to construct an access driveway (or to perform any construction activity) within the road right-of-way.

Always try to design your site circulation to discourage through traffic. A general rule of thumb that may be used for commercial facilities, if it does not conflict with local codes, is that if the daily traffic to the site is less than 5000 vehicles per day, you should try to limit access to a single two-way access drive or two one-way drives. When selecting the location for entrances and exits, try to align new drives with access drives that occur directly across the street—this will reduce the number of points that vehicles are slowing and turning.

2. ***Number of Parking Spaces.*** Local ordinances typically require a minimum number of parking spaces depending on the use of the building. If the local municipality does not have parking regulations, the recommendations in Figure 7-11 may be helpful.
3. ***Size of Parking Spaces.*** Ninety-degree parking provides the greatest number of spaces for a given area, but 90-degree spaces are more difficult to enter and exit than angled parking. So, 90-degree parking is not recommended for lots where drivers are parked only a short time, such as fast food restaurants or shops. Parking space size requirements will vary with the angle of parking (see the examples in Figure 7-12). Sixty-degree spaces are the most utilized due to the ease of use of spaces at this angle. Parking space size requirements are often included in local regulations, but a standard perpendicular (90 degree) parking space is 9 ft wide by 18 to 19 ft long.

Figure 7-11: Recommended parking requirements.

Source: *Asphalt Paving Design Guide*, The Asphalt Paving Association of Iowa. http://www.apai.net/cmdocs/apai/designguide/AsphaltCompositeSmFst.pdf.

Land Use	Spaces/Unit
Residential	
Single-Family	2.0/Dwelling
Multifamily	
Efficiency	1.0/Dwelling
1-2 Bedroom	1.5/Dwelling
Larger	2.0/Dwelling
Hospital	1.2/Bed
Auditorium/Theater/Stadium	0.3/Seat
Restaurant	0.3/Seat
Industrial	0.6/Employee
Church	0.3/Seat
College/University	0.5/Student
Retail	4.0/1000 GFA
Office	3.3/1000 GFA
Shopping Center	5.5/1000 GLA
Hotels/Motel	1.0/Room
	0.5/Employee
Senior High Schools	0.2/Student
	1.0/Staff
Other Schools	1.0/Classroom

GFA = Gross Floor Area in square feet
GLA = Gross Leasable Area in square feet

4. ***Accessibility.*** You must provide accessibility for disabled persons: accessible parking spaces and barrier free access to the building, as discussed earlier in this chapter. Figures 7-7 and 7-8 provide ADAAG requirements for accessible parking spaces.
5. ***Aisle Width.*** Generally, two-way traffic is preferable in parking areas for ease of circulation, but may require more paved area. The required aisle width in a parking lot between rows of parking spaces is generally indicated in local codes, but a common two-way aisle width is 24 ft. You may be able to reduce aisle width for one-way traffic if your local code allows.
6. ***Pedestrian Circulation.*** You will want to separate pedestrian and vehicular traffic whenever possible to promote safety. Provide an accessible route with a safe walking surface to the entrance of the building from the parking lot, from public transportation stops, and from adjacent buildings as dictated by the applicable regulations. Generally, a walkway should be 3 ft wide to accommodate one-way pedestrian travel and 5 ft wide to accommodate two-way pedestrian traffic. People also tend to use the aisles in a parking lot when walking from a parked car to the entrance of a building. Try to accommodate this tendency by aligning the parking spaces so that the aisles are perpendicular to the front face of the building, and pedestrians do not walk through the narrow spaces between parked cars (see Figure 7-13).

Figure 7-12: An example of regulations controlling the size of parking spaces and aisle width. Notice the required aisle width for two-way traffic is 24 ft for all angles.

Source: Land Development Code for Osceola County, Florida.

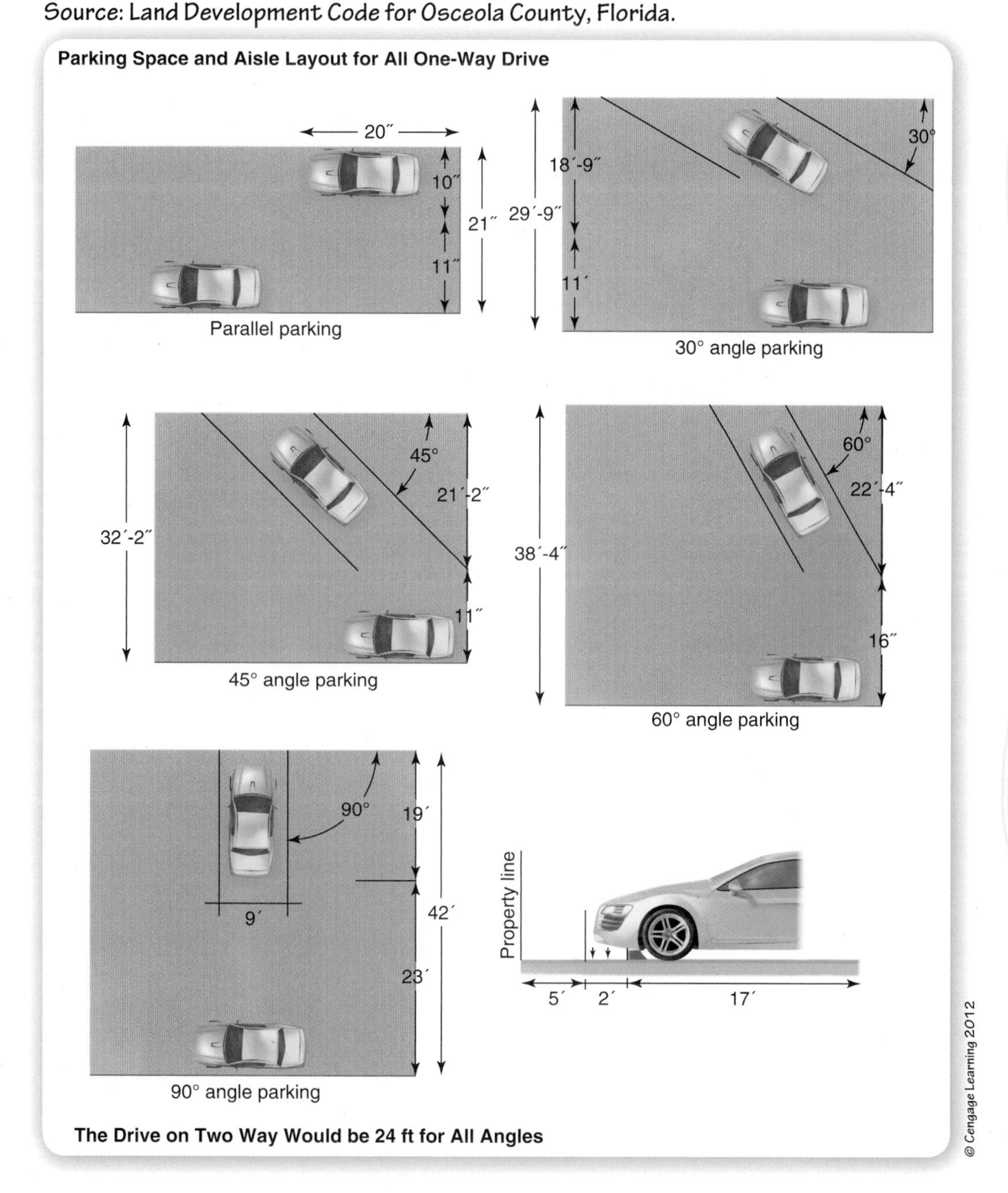

7. ***Special Vehicle Access.*** Will the facility need access for large vehicles—fire trucks, city buses, tractor trailers, school buses, garbage trucks? Research the roadway requirements for vehicles that will need access to your site and provide adequate drive widths, turn radii, and parking for vehicles that will need access to your site. Specific requirements may be specified in your local municipal code.
8. ***Off-Street Loading Area.*** Municipal regulations may require off-street loading areas for commercial facilities. The number and size of the loading area(s) is often based on the type of usage and size of the facility. It is a good idea (and often required by regulations) to locate loading areas away from vehicular and pedestrian traffic and out of view. Check your local code for requirements.

Figure 7-13: The preferred circulation allows pedestrians to walk along the aisles of the parking lot as shown on the right, not between parked cars as shown on the left.

Source: TM 5-803, Site Planning and Design, courtesy Department of the Army.

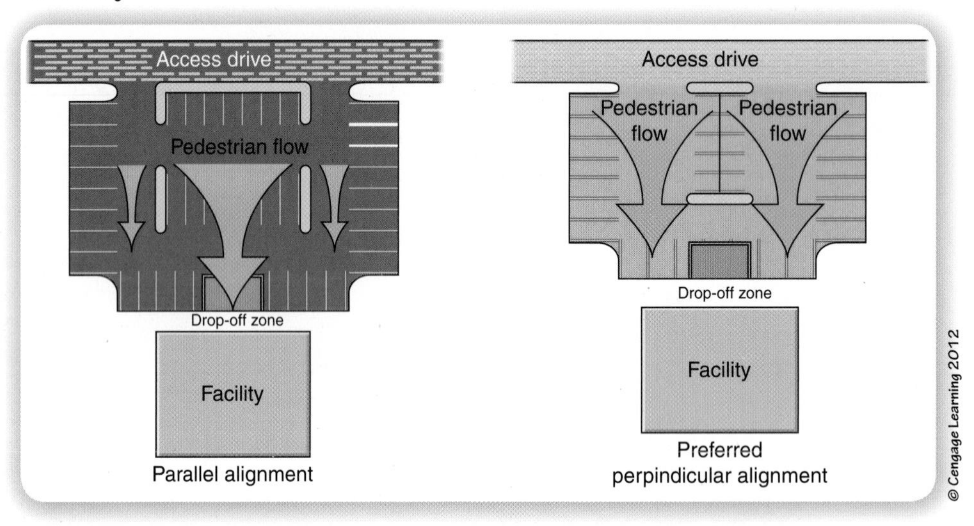

Figure 7-14: Typical dumpster pad plan. Local codes often require that waste disposal pads be screened from view using walls, fences, and/or landscaped screens.

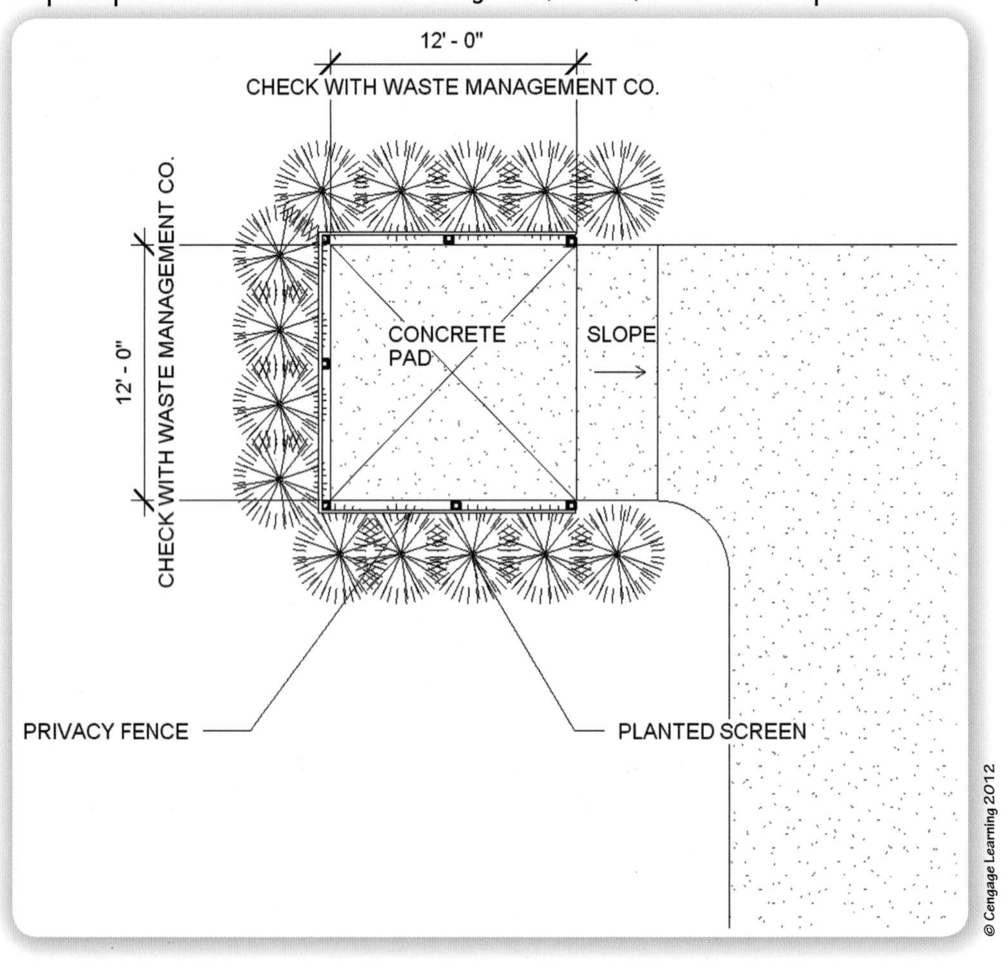

9. ***Waste Disposal.*** Provide pads for required dumpsters and adequate access for waste disposal trucks. It is a good idea (and often required by regulations) to screen the dumpsters from view by locating them behind buildings and providing screening material such as walls, fences, and landscaping. An example of a dumpster pad plan is shown in Figure 7-14. Check with the waste management company for size requirements.
10. ***Drainage.*** The parking lot should be designed so that water does not collect in spaces, aisles, or walkways. Traditional drainage techniques would slope the pavement surface toward storm drains where water would be collected then conveyed to a municipal storm drainage system. We will discuss storm drainage considerations later in this chapter.
11. ***Landscaping.*** Landscaping can drastically improve the aesthetics and environment of your project. A combination of tall canopy trees that can provide shade, and low branching shrubs with dense foliage to screen the expanse of pavement and vehicles can enhance the visual appeal of any facility. Often local regulations contain specific requirements for parking islands, **buffers**, and landscaping. We will discuss landscaping in more detail in Chapter 18.
12. ***Lighting.*** Parking lot and pedestrian access lighting will make your site safer and is often required by local regulations. Lighting should provide adequate light to your property but should not illuminate or cast glare onto adjacent property. Lighting design will be discussed in Chapter 15.

Buffer:
a vegetated zone adjacent to a development used to minimize the effects of the development.

Your Turn

Create a list of parking and site circulation design criteria for your project. Research your local municipal code and determine specific design criteria based on the preceding list of parking and circulation considerations. Record your criteria, with references to specific code requirements, in your project notebook.

GRADING AND DRAINAGE

The topography of a site can affect the design of structures, parking, and circulation and will impact the cost of development. Many times, a site that is not naturally level must be recontoured to provide a relatively flat area on which to construct a building or parking lot and to ensure adequate drainage. We refer to the act of changing the topography of a site as **grading** or *regrading.* Grading is typically accomplished using heavy construction equipment—some examples are shown in Figure 7-15.

Grading:
is the process of changing the topography of a property for a purpose.

When regrading a site it is important to ensure that the ground slopes away from structures and activities so that stormwater is carried away from these areas. You want to be sure to use slopes that are neither too shallow or too steep to minimize design and construction challenges. Some grading is almost always required during construction, but it is best to maintain exisiting topography whenever possible because less grading means less impact on the environment and a lower cost of construction. At the very least, you will probably have to remove organic topsoil where structures will be located to ensure adequater bearing capacity. You may also need to excavate for ponds, level the site to provide a better building surface, or create slopes to promote drainage.

Figure 7-15: *Construction equipment used for site preparation.*

Grading Plan

A grading plan is often required in order to obtain a building or construction permit. The grading plan is prepared by a civil engineer and shows the proposed grading and all related work. Remember from Chapter 5 that you will likely be required to prepare a Storm Water Pollution Prevention Plan (SWPPP) detailing the measures you will take to avoid stormwater pollution, protect existing vegetation, and stabilize the soil. Your proposed erosion prevention and sediment control measures should also be included on the grading plan.

When preparing a grading plan, consider the intended use of each area of the site and how the ground slope will affect it. Different uses have different ideal slopes. For instance, it is very difficult and dangerous to mow grass on a steep slope, and a relatively flat slope (as long as the area is drained well) would be appropriate for a picnic area—otherwise it would be a challenge to place the picnic tables so that your lunch would not roll off the table. The table in Figure 7-16 provides recommended slopes that you should incorporate into your site design. Slope gradient (percent slope) is discussed in Chapter 5.

Finish Grade: is the final elevation of ground surface after excavating or filling.

To prevent stormwater from entering buildings, finished floor elevations should be set above the outdoor grade per the local code (usually a minimum of 6 in) at the perimeter of the building, and the **finish grade** should slope away from a building at a minimum of 5 percent slope for approximately 10 ft, as shown in Figure 7-17. Always take care that any grade change you make to your property does not cause increased stormwater discharge to nearby properties or waterways. We will discuss stormwater management later in this chapter.

Figure 7-16: Recommended grade slopes.

Use	Minimum Slope	Maximum Desirable Slope
Parking Areas	1%	5%
Lawn Areas	1% (2% is better)	25%
Handicap Ramps	5%	8.33% (1 unit rise per 12 units run)
Walkways	0.5%	5% (anything over 5% is a ramp)
Swales	1%	10%
Stabilized Slope (Groundcover, Rocks, etc.)	1%	50%

Figure 7-17: Typical grading near a building. Grade should be set below the finished floor elevation according to the local code (typically 6 in minimum) and slope away from the building. Pavement can be sloped to a catch basin or toward ditches or swales where the water will be collected and transported to a stormwater system or an on site storage and treatment facility.

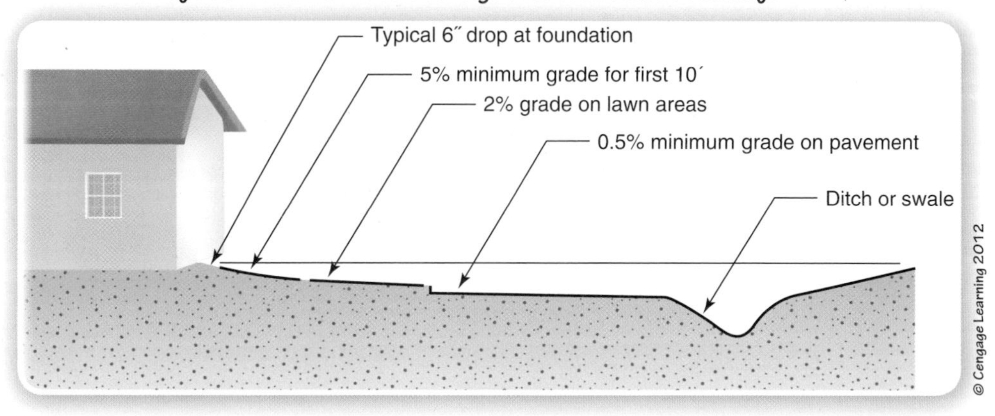

A grading plan is shown in Figure 7-18. When creating a grading plan, it is accepted practice to show the existing grade using dashed contour lines. The new grade is shown with solid contour lines where it is different from the existing grade. For projects with minimal regrading, the grading plan is often incorporated into the site plan.

Cut and Fill

During construction and final grading, soil may be removed from or added to different areas on the site. Removing and adding soil are referred to as **cut and fill**, respectively (see Figure 7-19). The costs associated with purchasing soil, disposing of excess soil, and hiring people and equipment to move the soil can be high, and increased soil disturbance means increased environmental impact. For those reasons, engineers usually try to minimize the amount of earth that must be moved. Regardless, cut and fill is almost always necessary, so the goal is to balance the amount of cut with the amount of fill. That way, all of the cut material is simply moved to another location on the site and used as fill.

Cut (or Excavation): the removal of naturally occurring earth materials.

Fill: the deposit of soil, rock, or other materials. Fill can also refer to the material being deposited.

As discussed in Chapter 5, both steeply sloped sites and very flat sites provide design and construction challenges that can require a considerable amount of earth moving. Very flat slopes may require significant regrading in order to provide

Figure 7-18: Grading plan. From JEFFERIS AND MADSEN. Architectural Drafting and Design, 5e.

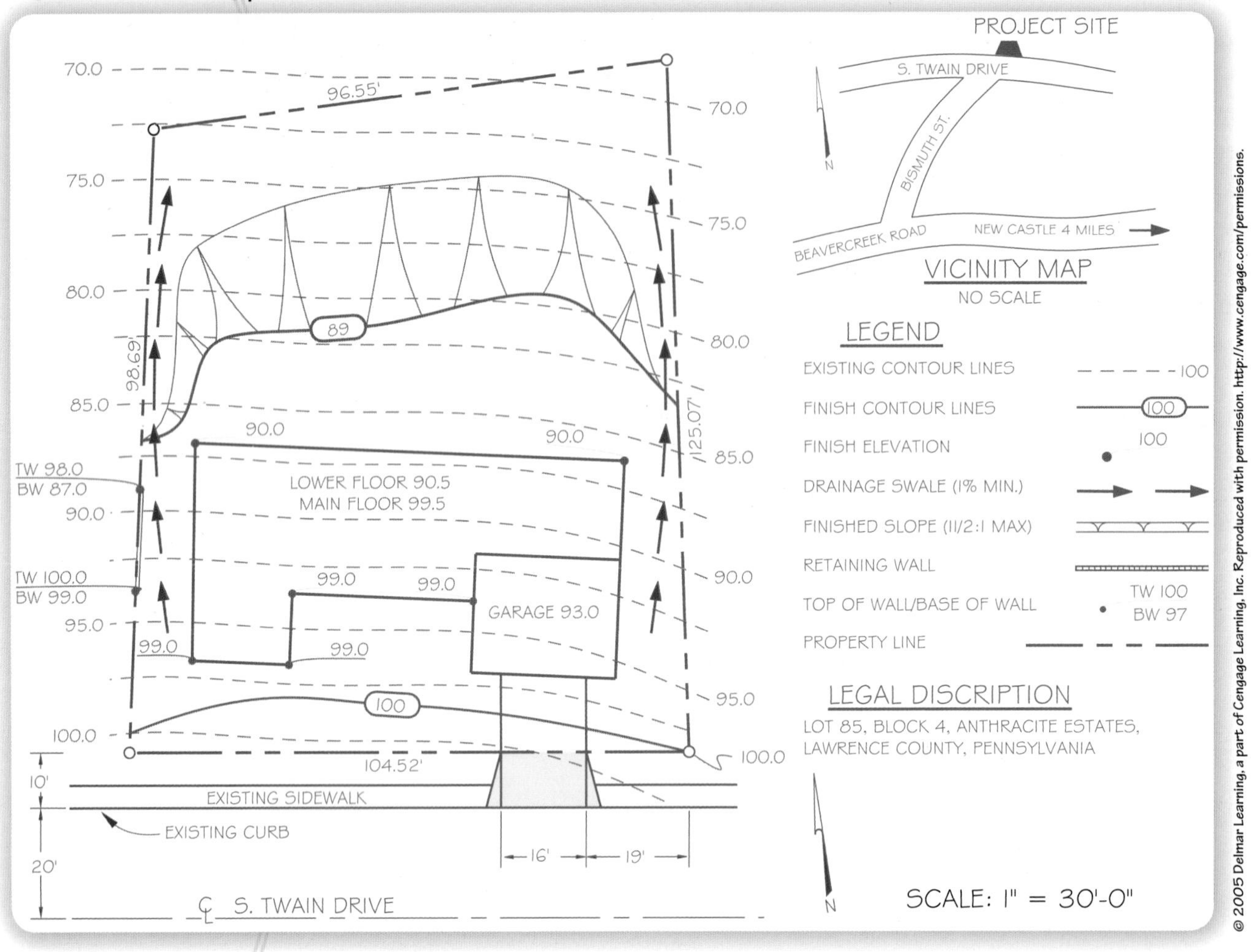

Figure 7-19: Cut and fill is often required when a property is regraded to improve drainage patterns or to create a level surface on which to build.

adequate drainage and may increase the cost of wastewater disposal. Steep slopes can require substantial cut and fill and may limit the feasible size of standard one-story buildings.

The cost of cut and fill is based on the type of soils and the volume of soil removed or added. The volume depends on the area of the site that requires grading, and the depth of the cut or fill. In order to calculate the volume of soil, you need to know the topography of both the existing site and the proposed finished site. The difference in the existing ground surface and **finish grade** is referred to as the depth of cut (if the proposed grade is lower than the existing grade) or fill (if the proposed grade is higher than the existing grade). Figure 7-18 illustrates the use of cut on a site that was regraded to provide a relatively flat area to accommodate the construction of a two-story building. The isometric pictorial shows a three dimensional view of the finished grade on the site. The **profile** shows that a large volume of soil will be removed from the site to accommodate the building, but no areas of fill. In this case, the material removed will have to be disposed of offsite.

Of course, the cut or fill depth is rarely constant across the site and so finding the volume of soil to be added or removed may seem like a complicated task. To simplify this calculation we will use a grid, similar to the one set up for the topographic survey, to break the site into several smaller (mostly square) areas. If we know the average depth of cut or fill in the square area, we can then multiply the surface area of the square by the average depth—assigning fill volumes negative values and cut volumes positive values. The average depth of cut or fill is found by averaging the depths at the corners of each area. By adding the cut or fill volumes for all of the square areas, we can approximate the volume of excess soil (if the total is positive) or required fill (if the total is negative). Let's look at an example of cut and fill calculations for a simple site.

Profile:
a vertical section showing the ground surface. A profile may show exising and new grade lines and/or soil strata.

Example: Cut and Fill Calculations

As a member of the project design team you are responsible for estimating the amount of cut and fill required to regrade the construction site. From survey data, and the grading plan, you have approximated the existing and the new grade elevations at the grid intersection points as shown.

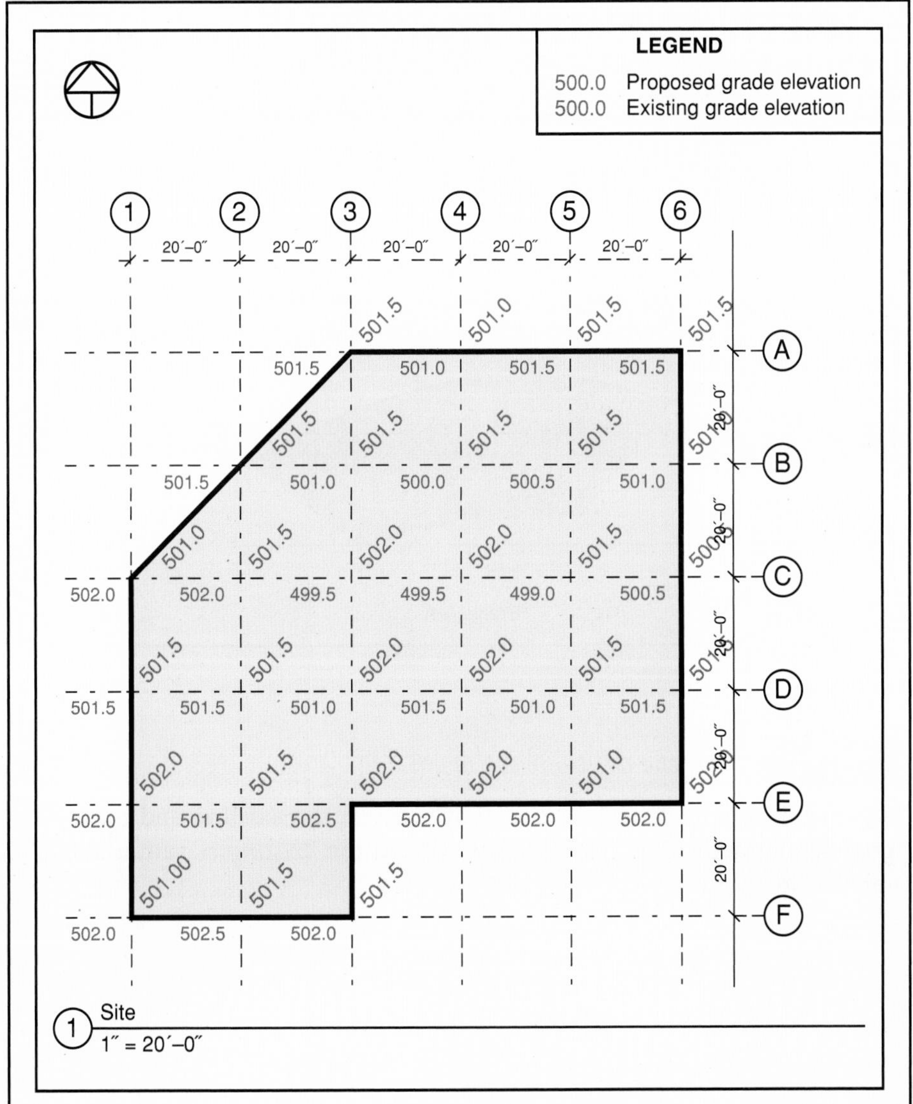

© Cengage Learning 2012

An existing contour map and profile showing existing and proposed grade are also provided.

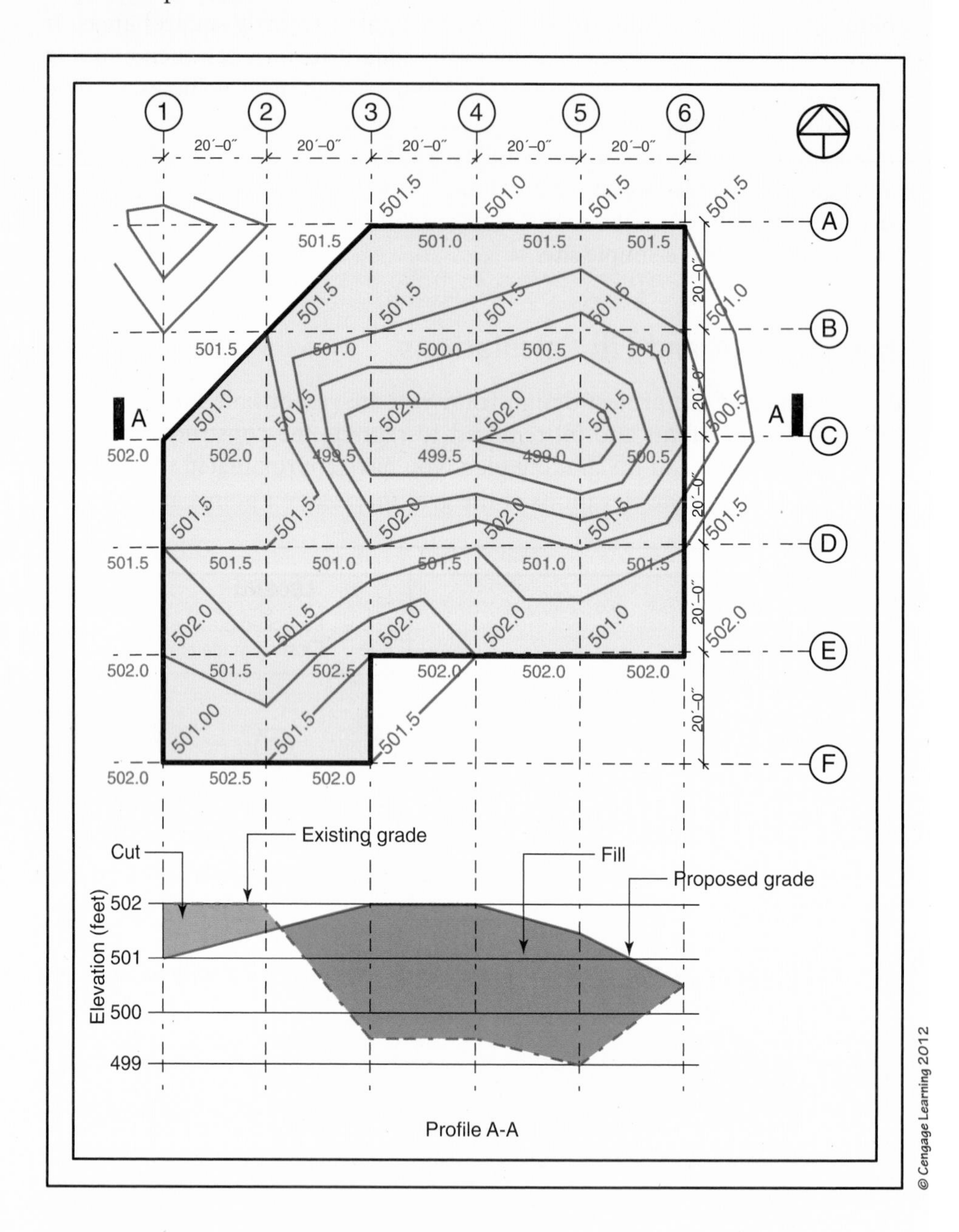

Step 1: *Find the depth of cut or fill.*

To find the change in grade elevation between the existing and proposed grades, subtract the *proposed* elevation from the *existing* elevation at each intersection point.

Change in elevation = existing elevation − proposed elevation

If the change in elevation is negative, fill is required.

If the change in elevation is positive, cut is required.

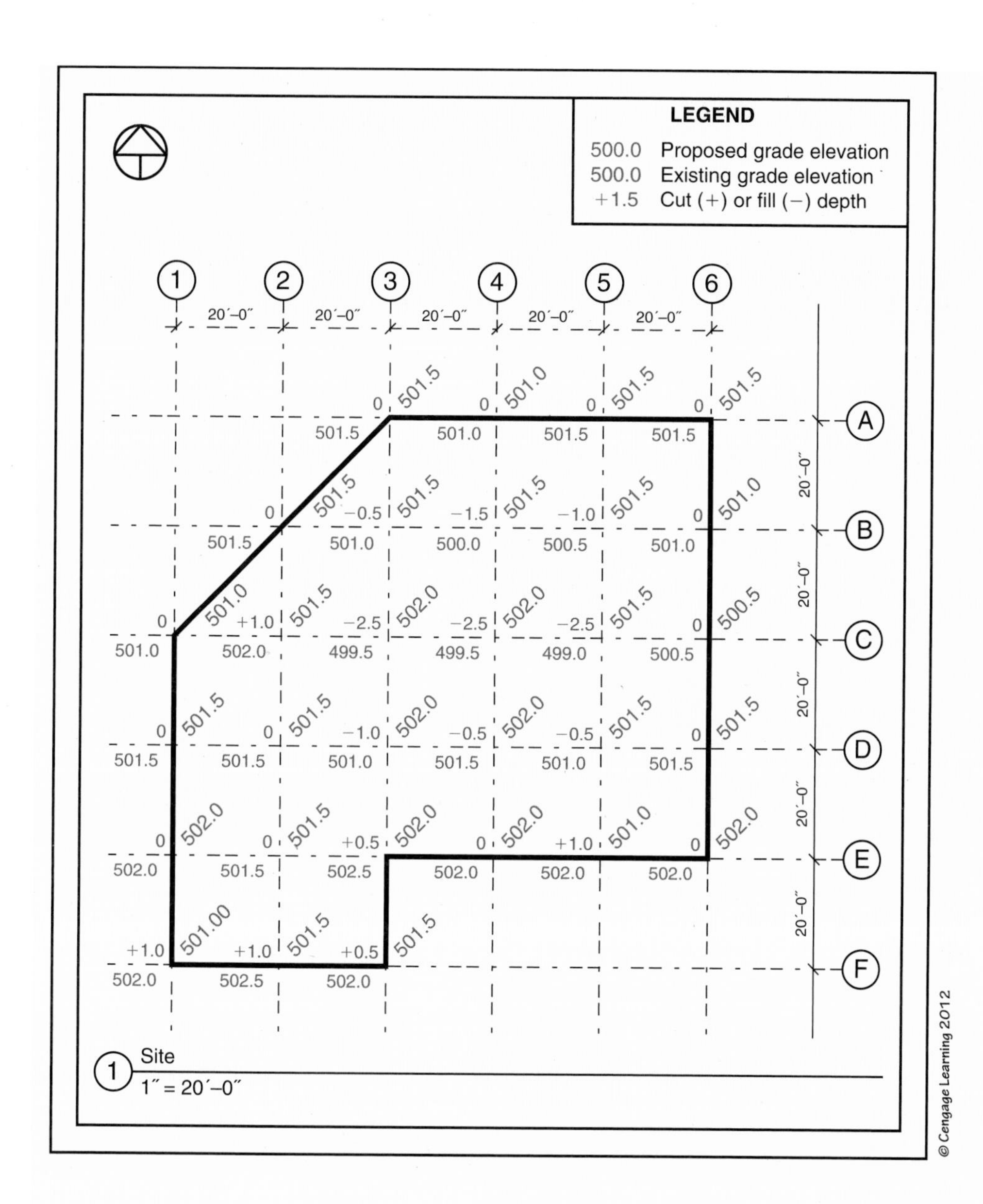

For example, at the point C-4

$$\textit{Change in elevation} = \textit{existing elevation} - \textit{proposed elevation}$$
$$= 499.5 - 502.0 = -2.5 \textit{ ft}$$

Since the depth change is negative, 2.5 ft of fill is required at C-4.

Likewise, the change in elevation for each grid intersection point has been calculated and is shown on the map below.

Step 2: *Find average depth of cut or fill.*

Average the change in elevation for the corners of each area of the grid by summing the depth values on each corner and dividing by the number of values. For example, using the shaded square on the map below, the average depth is

$$\textit{Average depth} = \frac{0 + 1.0 - 1.0 - 2.5}{4} = -0.625 \text{ ft}$$

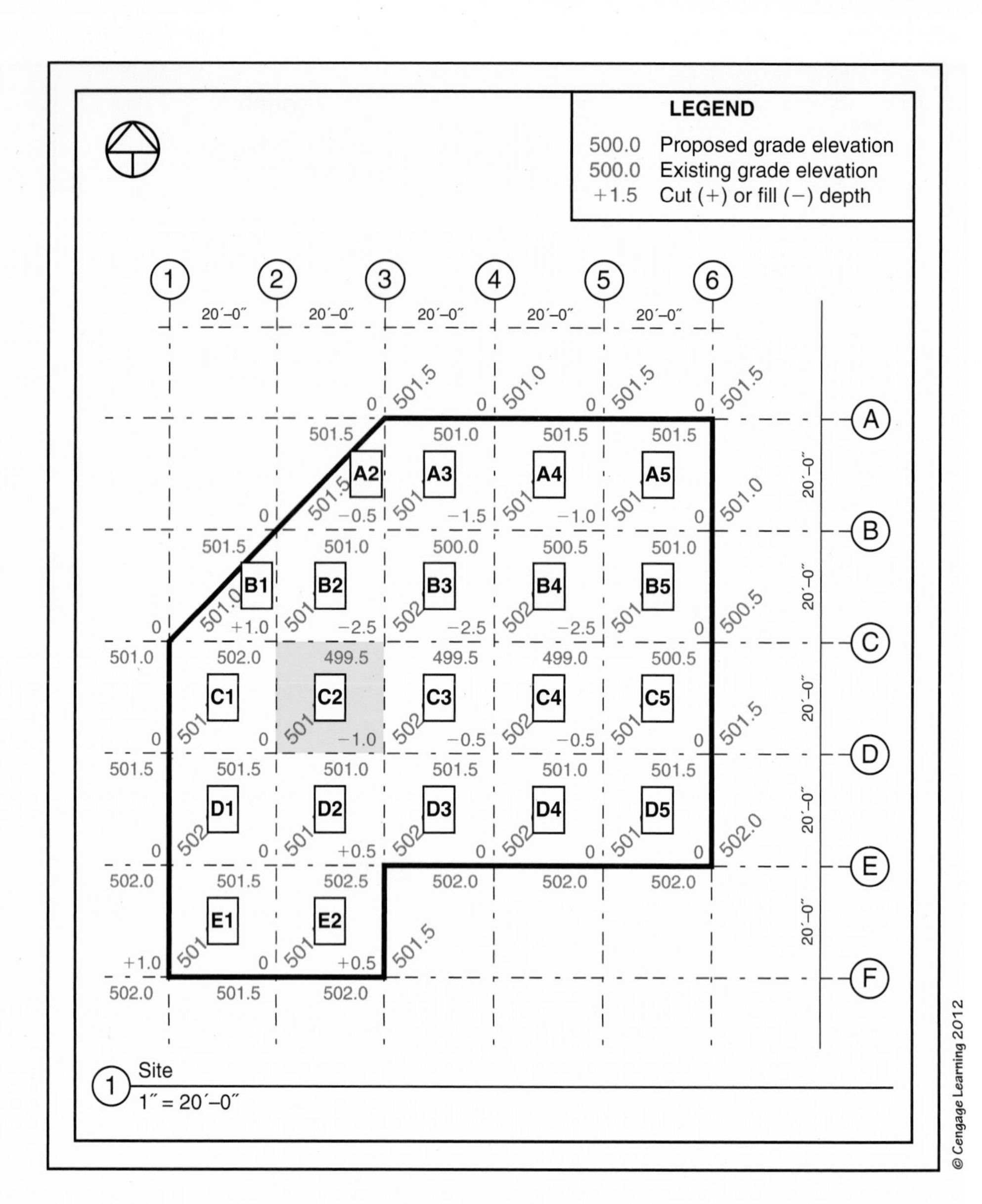

The average change of elevation is shown for several of the grid areas in the table below.

Step 3: ***Find the volume.***

The volume of soil added or subtracted for each square area is simply the area of the square multiplied by the change in elevation of the grade.

Volume of cut and fill

$$V = \Delta h \cdot A$$

where V = volume of soil

Δh = change in elevation

A = area of grid space

For this example, the area of each full square is 400 sq ft (= 20 ft length × 20 ft width). Therefore, the area shaded in the map above would produce

Grid Area	Change in Elevation (ft) Corner 1 / Corner 3	Corner 2 / Corner 4	Average Change in Elevation (ft)	Area (sq ft)	Volume of Cut (+) or Fill (−) (cu ft)	Volume of Cut (+) or Fill (−) (cu yd)
A2		0	−0.17	200		
	0	−0.5				
A3	0	0	−0.50	400		
	−0.5	−1.5				
A4	0	0	−0.63	400		
	−1.5	−1				
A5	0	0	−0.25	400		
	−1	0				
B1		0	0.33	200		
	0	1				
B2	0	−0.5	−0.50	400		
	1	−2.5				
B3	−0.5	−1.5	−1.75	400		
	−2.5	−2.5				
B4 to D5						
E−1	0	0	0.50	400		
	1	1				
E−2	0	0.5	0.50	400		
	1	0.5				

$$
\begin{aligned}
V &= \Delta h \cdot A \\
&= -0.625\, ft \cdot 400\, ft^2 \\
&= 250\, ft^3
\end{aligned}
$$

Most of the time, cut and fill volumes are expressed in cubic yards. Therefore, you need to convert cubic feet to cubic yards. Since there are 3 ft in a yard, the conversion factor is

Conversion factor; − cubic yards and cubic feet

$$\left(\frac{1\, yd}{3\, ft}\right)^3 = \frac{1}{27}\frac{yd^3}{ft^3}$$

Therefore, 250 cu ft is equivalent to 9.3 cu yd.

$$250\, ft^3 \left(\frac{1\, yd^3}{27\, ft^3}\right) = 9.3\, yd^3$$

The volume calculations and conversions are recorded in the table below. Cut and fill calculations can be performed quickly by civil engineering computer programs and CAD software. Spreadsheet programs can also reduce calculation time for repetitive calculations like these—the table below was created using a spreadsheet application. The goal of equalizing cut and fill was not realized in this case. Additional material must be brought to the site. Alternate grading plans can be created if the cost becomes prohibitive.

Grid Area	Change in Elevation (ft) Corner 1 / Corner 3	Corner 2 / Corner 4	Average Change in Elevation (ft)	Area (sq ft)	Volume of Cut (+) or Fill (−) (cu ft)	Volume of Cut (+) or Fill (−) (cu yd)
A2		0	−0.17	200	−34.0	−1.26
	0	−0.5				
A3	0	0	−0.50	400	−200.0	−7.41
	−0.5	−1.5				
A4	0	0	−0.63	400	−252.0	−9.33
	−1.5	−1				
A5	0	0	−0.25	400	−100.0	−3.70
	−1	0				
B1		0	−0.33	200	66.3	2.44
	0	1				
B2	0	−0.5	−0.50	400	−200.0	−7.41
	1	−2.5				
B3	−0.5	−1.5	−1.75	400	−700.0	−25.93
	−2.5	−2.5				
B4 to D5		Not included for brevity		4800	−3008.0	−111.41
E1	0	0	0.50	400	200.0	7.41
	1	1				
E2	0	0.5	0.50	400	200.0	7.41
	1	0.5				
					TOTAL =	−149.19

Stormwater Management

As you know from middle school science class, the water cycle (also called the hydrologic cycle) describes the journey of water as it circulates throughout the earth and its atmosphere (see Figure 7-20). Water falls in the form of rain or snow onto the surface of the earth. Some of this water is absorbed by plants. Some of the water infiltrates into the ground and is trapped under the surface—we call this **groundwater**. The rest of the water, which we call **surface water**, runs off (either above or below the surface) into a stream, river, lake, or ocean. Although the surface and subsurface water may be used by people, animals, and plants, the water is eventually returned to the atmosphere in vapor form—surface water evaporates and plants release water through transpiration—where it condenses into clouds. When the clouds encounter cool air, the water is released in the form of precipitation (rain or snow). If the precipitation falls on land, the water infiltrates, runs off, or is absorbed by plants—and so the water cycle continues.

Figure 7-20: **The water cycle includes precipitation, accumulation through surface and subsurface runoff, evaporation, transpiration, and condensation.**

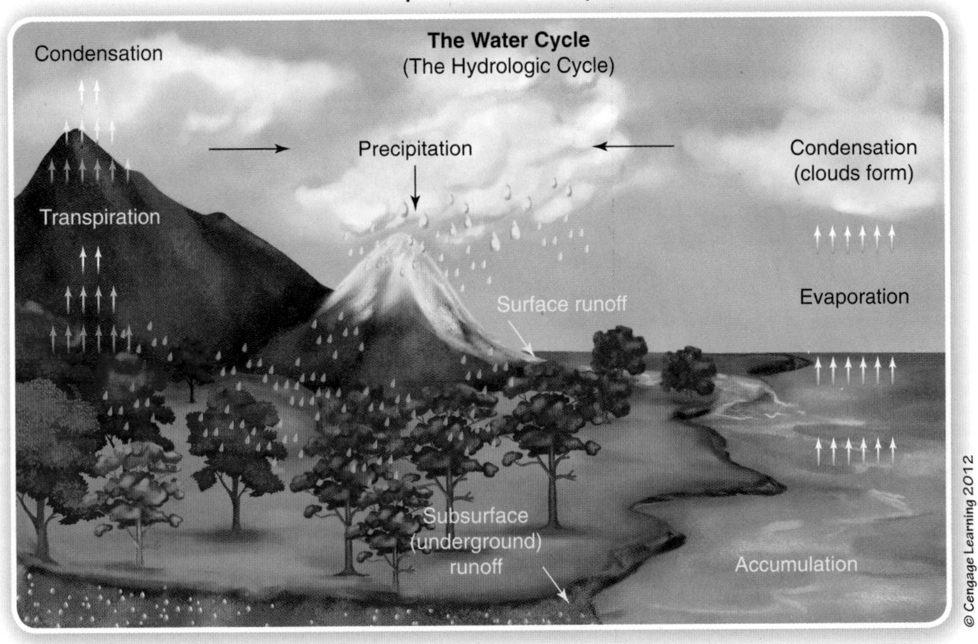

When land is developed, the journey of water is often altered. When you construct a building or pave parking lots and sidewalks on a site that was previously covered with forest, grass, or farm fields, the water can no longer infiltrate into the ground or be absorbed by the plants. The result is that much more stormwater runs off of the land after a property is developed as illustrated in Figure 7-21. This can result in flooding, erosion, or sedimentation on downstream property or in natural

Figure 7-21: **Land development alters the natural water balance. Typically, development includes covering parts of the ground surface with impervious materials. Therefore, less water infiltrates the ground and less water is absorbed by plants, both of which greatly increase the amount of surface runoff.**

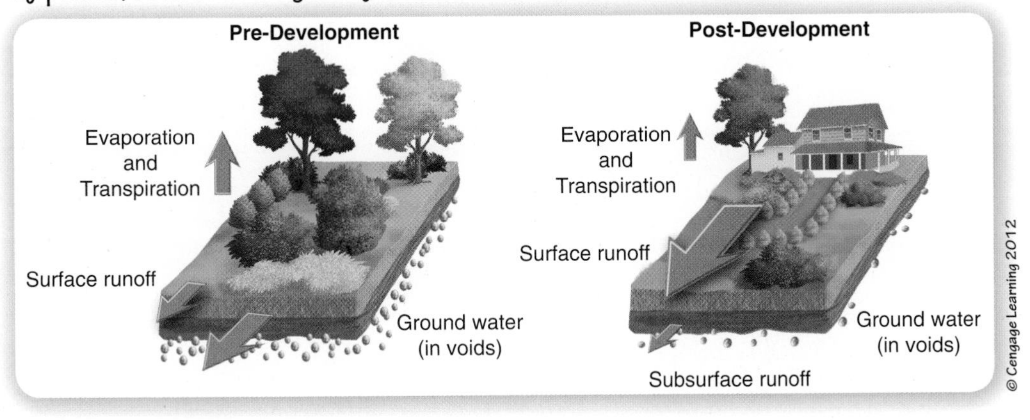

Figure 7-22: Development within a watershed area increases runoff and overloads natural waterways. Increased flow during a flood caused severe bank erosion of this river. Several houses were lost because the soil eroded from beneath them.

Source: FEMA.

© Cengage Learning 2012

waterways (see Figure 7-22). Human activity that occurs on a developed site may leave toxins such as gasoline, fertilizer, pesticides, and pet waste that are easily carried away by stormwater. Depending on the topography, the rainwater runoff can pass over many other properties before it reaches a body of water. In Chapter 5 we discussed the NPDES stormwater permit requirements and a StormWater Pollution Prevention Plan (SWPPP) that delineates steps you plan to take to prevent stormwater pollution during construction. You must also consider the future impact of the project on the quality and volume of stormwater leaving the site after construction.

The earth is a system that contains a fixed amount of water. Water can exist in several different reservoirs and move between solid earth, oceans, atmosphere, and organisms as part of a geochemical cycle.

In order to protect property owners and the environment, and in order to avoid overloading stormwater systems, many municipalities have adopted regulations that control the quality and restrict the quantity of stormwater leaving a developed

OFF-SITE EXPLORATION

Stormwater pollution from factories, farmland, or solid waste treatment facilities is fairly easy to identify and pinpoint. However, much of the water pollution that reaches our streams, ponds, rivers, and oceans comes from sources that are not so easily identified—it occurs when rain or snowmelt runs over the ground and picks up pollution from many small sources on many different properties. We call this **nonpoint source pollution (NPS).** Visit the EPA webpage on Nonpoint Source Pointers at *http://www.epa.gov/owow/nps/facts/*. Read the Pointer No. 1 Factsheet and record the three principle sources of Water Quality Impairment (pollution) in our rivers, lakes and estuaries in your notebook. Also list common NPS pollutants, and record the progress that the United States has made in addressing NPS pollution in the last 10 years.

Nonpoint source pollution results when stormwater runs over land and picks up contaminates from many small sources that are difficult to pinpoint. Pesticides, fertilizer, animal waste, oil, and grease are examples of toxins that can be carried by stormwater to water bodies.

© Cengage Learning 2012

site. If you do not comply with the regulations, you risk fines. For example, the county of Roanoke, Virginia, can impose a fine of up to $32,500 or imprisonment for each violation. Although the regulations often do not apply to individual single-family residential properties that are not part of a larger development, it is important for everyone—homeowners, developers, business owners, and large corporations—to attempt to reduce the negative impact of his land use by minimizing stormwater runoff and by reducing the amount of toxins deposited on the ground. Check your local codes and ordinances for stormwater regulations.

PEAK STORMWATER RUNOFF Local codes typically require that you limit the stormwater runoff from your development to the amount of runoff that occurred before the development was built. The amount of rainfall associated with a single storm can be described by three characteristics: duration, depth, and intensity. **Duration** (minutes or hours) refers to the length of time a storm lasts; **rainfall depth** (in) indicates the total amount of rainfall during a storm; and **intensity** (in per h) is determined by dividing depth by duration. A **return period**, or the average length of time between storms, can be estimated for a storm of a given duration and depth. For example, a storm that dumps 12 in of rain in 1 h may have a return period of 100 years for a given location. If a storm has a return period of 100 years there is a 1 percent chance of a storm of that magnitude occurring in any given year. If a storm has a 10-year return period, it has a 10 percent chance of occurring in any given year. We will use these storm characteristics to estimate the amount of stormwater runoff from a site.

There are many methods to calculate stormwater peak flow; we will use the Rational Method, which is commonly used to estimates the peak flow of stormwater runoff for small drainage areas. The general procedure for determining peak runoff using the Rational Method is as follows:

1. Estimate the drainage area (in ac).
2. Determine the **runoff coefficient** (C) and the correction factor (C_f).
3. Estimate the hydraulic length or flow path that will be used to determine the **time of concentration**.
4. Determine the time of concentration (T_c) for the drainage area.
5. Determine the rainfall intensity using the time of concentration.
6. Substitute the drainage area, C_f, C, and intensity into the Rational Formula to determine the peak rate of runoff.

DRAINAGE AREA When determining the change in runoff on a given property, we are concerned only with the area of land that will be altered. The quantity of stormwater that enters the site from other property will not change due to the alteration and therefore can be ignored when calculating the change in stormwater runoff. Therefore, use the area of the property on which you are building for this calculation.

RUNOFF COEFFICIENT The **runoff coefficient** indicates the percentage of rainwater that is not absorbed, but runs off of the surface, and ranges from 0 to 1.0. For instance, a runoff coefficient of 0.95 indicates that approximately 95 percent of the rainwater that falls on the surface will run off and 5 percent will be stored or absorbed by the surface. The higher the coefficient, the more rainwater runoff resulting in higher peak flows. Vegetated cover, such as forest, pasture, and lawn will generally have relatively small runoff coefficients, but **impervious** surfaces, such as concrete, asphalt, and roofs, will have large runoff coefficients.

Impervious: describes a surface that does not allow water to infiltrate or penetrate

There are a number of sources that provide runoff coefficients, and sources may differ slightly on recommended coefficients. We will use the runoff coefficients shown in Figure 7-23. You may want to conservatively use the higher coefficient unless you have specific information about the conditions at your site that justify lower coefficients. The U.S. Department of Agriculture's Web Soil Survey (*http://websoilsurvey.nrcs.usda.gov/app/*) may be helpful to investigate the drainage characteristics of the soil on your property.

Figure 7-23: **Recommended runoff coefficients, C, for use with the Rational Formula. Reprinted with permission from Michael R. Lindeburg, PE, Civil Engineering Reference Manual 9th ed., copyright**

Rational Method Runoff Coefficients

Categorized by Surface	
Forested	0.059–0.2
Asphalt	0.7–0.95
Brick	0.7–0.85
Concrete	0.8–0.95
Roof	0.75–0.95
Lawns, well drained (sandy soil)	
Up to 2% slope	0.05–0.1
2% to 7% slope	0.10–0.15
Over 7% slope	0.15–0.2
Lawns, poor drainage (clay soil)	
Up to 2% slope	0.13–0.17
2% to 7% slope	0.18–0.22
Over 7% slope	0.25–0.35
Driveways, walkways	0.75–0.85
Categorized by use	
Farmland	0.05–0.3
Pasture	0.05–0.3
Unimproved	0.1–0.3
Parks	0.1–0.25
Cemeteries	0.1–0.25
Railroad yard	0.2–0.40
Playgrounds (except asphalt or concrete)	0.2–0.35
Business districts	
Neighborhood	0.5–0.7
City (downtown)	0.7–0.95
Residential	
Single-family	0.3–0.5
Multi-plexes, detached	0.4–0.6
Multi-plexes, attached	0.6–0.75
Suburban	0.25–0.4
Apartments, condominiums	0.5–0.7
Industrial	
Light	0.5–0.8
Heavy	0.6–0.9

Notice that the table provides runoff coefficients for specific surfaces. Each coefficient applies only to the area covered by the indicated surface. The table also gives runoff coefficients for categories of land use, such as parks and single family residential areas, which usually have many different types of surfaces, such as roofs, driveways, walkways, lawns, and so forth. The runoff coefficients that are categorized by use can be considered **composite runoff coefficients**. In this case, composite means that the runoff coefficients of all of the different types of surfaces have been weighted—that is, the coefficient is adjusted to take into consideration the surface area of each type of surface—to produce a runoff coefficient that can be used for the overall area.

Composite runoff coefficient

$$C = \frac{C_1A_1 + C_2A_2 + \cdots + C_nA_n}{A}$$

where

C = composite runoff coefficient (dimensionless)
A = total drainage area in acs
n = the number of separate surfaces contributing to the drainage area
Therefore,

$C_1, C_2, \ldots, C_n$ represent runoff coefficients for surface areas 1, 2,...,n

$A_1, A_2, \ldots, A_n$ represent the area (in sq ft) for surface areas 1,2,...,n

Example: Find the composite runoff coefficient for a 15 ac site composed of 5 ac of brick paving, 50,000 sq ft of retail space, and the remainder unimproved pasture.

Step 1: *Determine the area of each different surface in acre.*

A (brick) = 5 ac

$$A(building) = 50{,}000\, ft^2 \cdot \frac{1ac}{43{,}560\, ft^2} = 1.15\, ac$$

A (pasture) = 15 ac − 5ac − 1.15 ac = 8.85 ac

Step 2: *Find the runoff coefficient for each surface.*

From Figure 7-23,
C (brick) = 0.85, C (roof) = 0.95, C (pasture) = 0.3

Step 3: *Calculate the composite.*

$$C = \frac{C_{brick}A_{brick} + C_{roof}A_{roof} + C_{pasture}A_{pasture}}{A}$$

$$= \frac{(0.85 \cdot 5) + (0.95 \cdot 1.15) + (8.85 \cdot 0.3)}{15} = 0.53$$

HYDRAULIC LENGTH For design purposes, to find peak flow at a point, engineers typically use a duration that is equal to the **time of concentration**. The time of concentration is the time that it takes for water to drain from the most distant point (when considering the drainage path) to a given point. Because the slope of the ground and the type of soil and cover (pavement, forest, grass, etc.) on the property affects how quickly water will drain on the site, you will need a contour map and information about the existing site conditions—information you should have gathered during site discovery.

Rainwater will drain down a slope toward the lowest elevation and will seek the quickest (or steepest) path. The **hydraulic length** is the length of the longest drainage path within the drainage area to the point at which the water exits the site. To estimate the path of a drop of water on your site, begin at the point of splash down and sketch the shortest line segment to the next lowest contour line. Continue drawing the shortest line segment from one contour to the next lowest until you can no longer decrease in elevation. Sketch the drainage path from several points on your contour map as demonstrated in Figure 7-24 and determine the length of the shortest path.

Figure 7-24: Rainwater will drain down a slope toward the lowest elevation and will seek the quickest (or steepest) path. On this site, the rainwater that is not absorbed or detained will run off toward the southwest corner of the site. To estimate the path of a drop of water on your site, begin at the point of splash down and sketch the shortest line segment to the next lowest contour line. Continue drawing the shortest line segment from one contour to the next lowest until you can no longer decrease in elevation. The length of the longest path is shown by the heavy blue line and has an estimated length of 215 ft. This image was created in AutoDesk Revit. The length of the line was estimated using the tape measure tool.

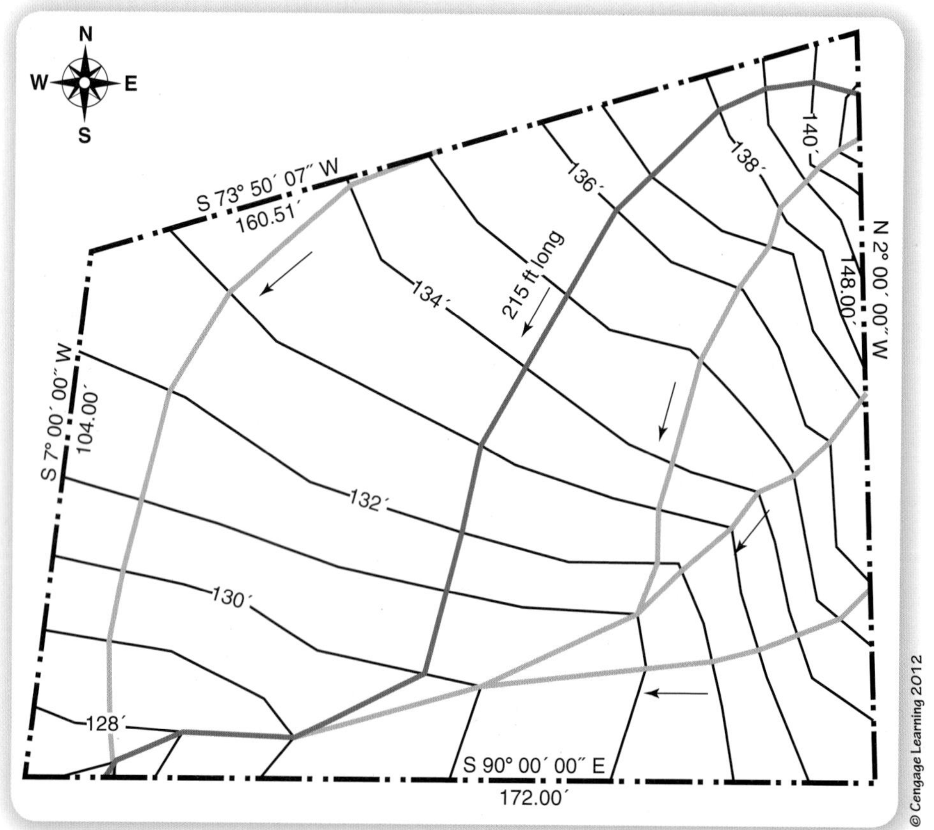

TIME OF CONCENTRATION The severity of a storm can be expressed by the **precipitation intensity**, which indicates the average amount of rain that falls per hour of a storm and is calculated by dividing the amount of rainfall that falls during a storm (in inches) by the duration of the storm (in hours). The time of concentration is based on three factors including the hydraulic length, the ground slope, and the runoff coefficient of the land. Once these three quantities are known, the time of concentration can be estimated using a Seeyle Chart (see Figure 7-25). If you determine a time of concentration of less than 10 minutes, use a minimum of 10 minutes.

Figure 7-25: The Seeyle Chart is used to estimate time of concentration. Adapted from the Seeyle Chart in the Town of East Hartford Manual of Technical Design.

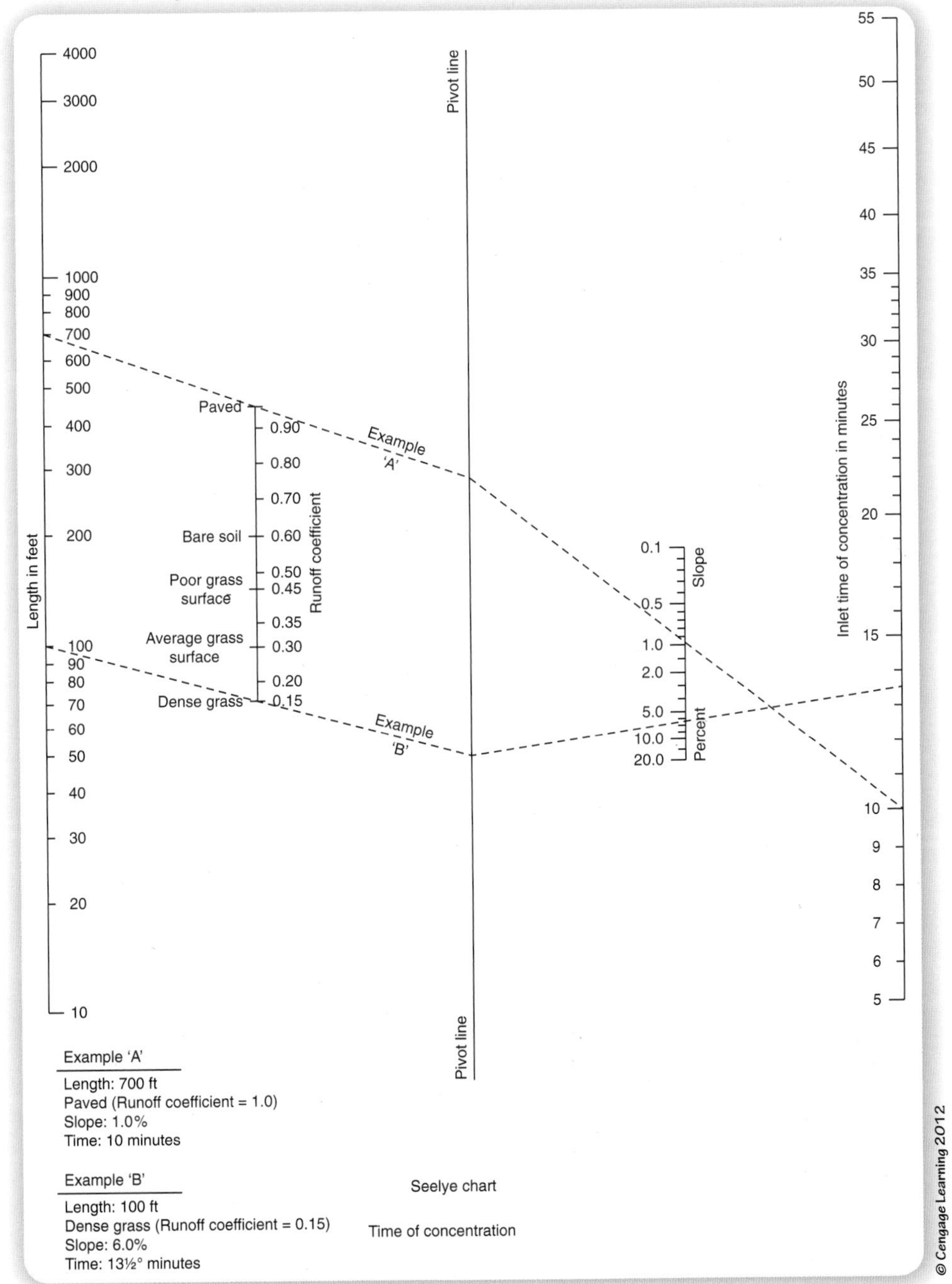

Example: Calculate Time of Concentration

Find the time of concentration for the site shown in Figure 7-24. Assume the site is completely forested. For simplicity, we will assume that the water flows overland for the entire distance and does not concentrate into rivulets or channels and therefore use the Seeyle Chart to estimate time of concentration.

Step 1: ***Find the hydraulic length. The hydraulic length is 215 ft (from the previous example).***

UNF 7-9. Intensity-Duration-Frequency curve for Roanoke, Virginia. Courtesy NOAA's National Weather Service.

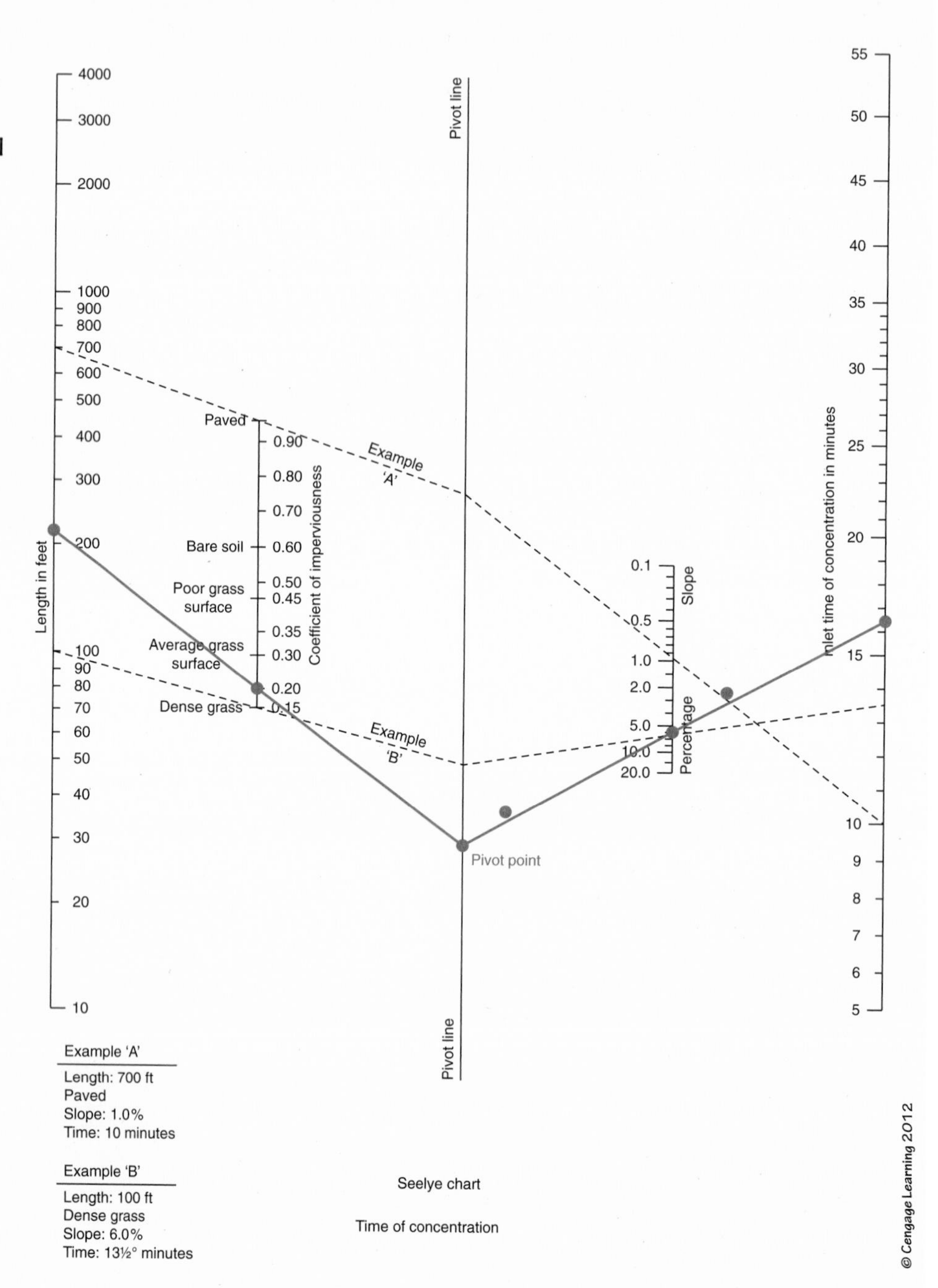

Step 2: ***Find the slope.***

Since the slope on the site is fairly consistent along the flow path, use the average slope along the path.

$$Average\ slope = \frac{change\ in\ elevation}{length\ of\ drainage\ path} \cdot 100\% = \frac{141\ ft - 127\ ft}{215\ ft} \cdot 100\% = 6.5\%$$

Step 3: ***Find the runoff coefficient.***

We will conservatively use $C = 0.2$, which is the maximum runoff coefficient for forested land given in Figure 7-23. Note that if more than one type of cover is present, you must calculate a composite runoff coefficient.

Step 4: ***Find the time of concentration.***

Use the Seeyle Chart to determine the time of concentration.

1. Locate the point on the distance scale that corresponds to the hydraulic length (215 ft).
2. Locate a point on the Coefficient of Imperviousness that corresponds to the runoff coefficient ($C = 0.2$).
3. Draw a line through these two points and extend it to the pivot line.
4. Locate a point on the percentage slope scale that corresponds to the ground slope (6.5 percent).
5. Draw a line beginning at the pivot point through the slope point and extend to the time of concentration scale. Read the value for the time of concentration, which is approximately 16 minutes in this example.

RAINFALL INTENSITY Once you have estimated the time of concentration (t_c) you can find rainfall intensity for a given storm return period using an Intensity-Duration-Frequency (IDF) chart for your site location. You can often get IDF charts from your state highway department, or use the NOAA Precipitation Frequency Data Server at *http://hdsc.nws.noaa.gov/hdsc/pfds/index.html.* Be sure to select *Precipitation Intensity* for data type. A typical IDF chart is shown in Figure 7-26. For example, the rainfall intensity for a 10-year storm with a time of concentration of 30 minutes is 3.5 in/h.

Your Turn
This is Intense

Search your state highway department website for IDF charts or visit the NOAA Precipitation Frequency Data Server online. Locate an IDF chart for your project site (or your school site) that gives Rainfall (Precipitation) Intensity in in per h. Print out the IDF chart and include it in your project notebook. Then use the chart to estimate the Precipitation Intensity (in/h) for a commercial development during a 100-year storm if the time of concentration is 20 minutes.

Figure 7-26: Example of Intensity-Duration-Frequency (IDF) chart. In order to determine rainfall intensity, enter the chart with a duration equal to the time of concentration. Next, move up to the desired storm-return period, then move horizontally to read the rainfall intensity. Rainfall intensity is read in either mm/h (on the left axis) or in/h (on the right axis). Courtesy Federal Aviation Administration.

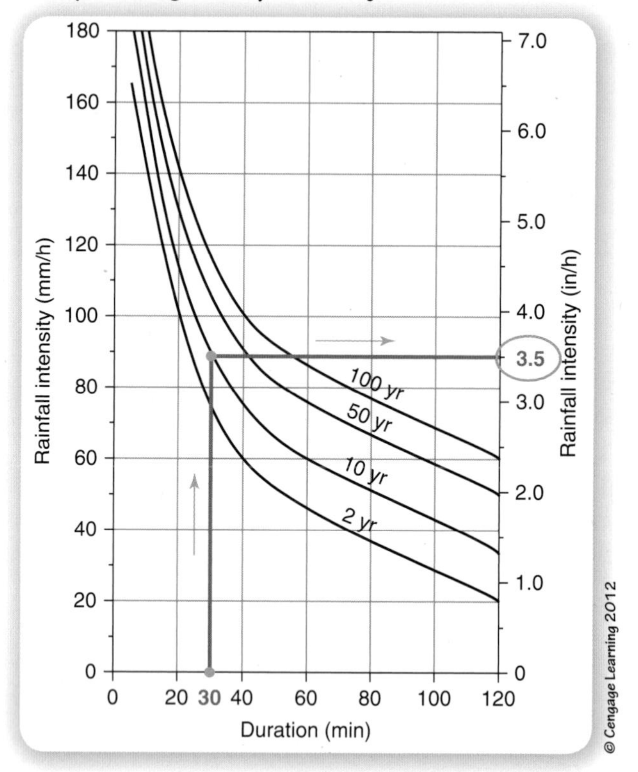

THE RATIONAL FORMULA Many methods are available to estimate peak runoff; we will use the Rational Formula, which is commonly used to estimate runoff for small drainage areas.

The Rational Formula
$Q = C_f CiA$

where:

Q = peak rate of runoff in cubic ft per second (cfs)
C = composite runoff coefficient, indicates the percentage of rainwater that runs off of a surface
i = average Precipitation Intensity for the time of concentration, t_c, for a selected return period in in per h (in/h)
A = watershed area in acre (ac)
t_c = the time of concentration equal to the time required for water to flow from the hydraulically most distant point in the watershed to the point of design in minutes.
C_f = runoff coefficient correction factor (per Figure 7-27) depends on the return period of the storm and accounts for the fact that more severe storms will eventually saturate the soil and result in more surface runoff.

Figure 7-27: Runoff coefficient correction factors for Rational Method.

Recurrent Intervals (years)	C_f
2, 5, and 10	1.0
25	1.1
50	1.2
100	1.25

Note: C_f multiplied by C must be less than or equal to 1.0
Source: Virginia Department of Transportation, Drainage Manual, 2001. pp. 6–17.

Many regulations require that you limit post-development runoff to predevelopment levels for a specific **design storm**, usually stated in terms of return frequency (e.g., a 25-year storm). In order to determine if your development will cause an increase in stormwater runoff, you must estimate both the pre-development runoff and the post-development runoff. If the post-development runoff is greater, most codes will require that you handle the excess water on-site. We will discuss methods to handle excess stormwater later in this chapter.

Example: Calculate the Change in Peak Runoff

As an example, we will calculate the change in peak stormwater runoff for a site in Roanoke County, Virginia. The property is 5.7 ac and is currently forested, but will be developed into a retail center. The plan includes 100,000 sq ft of retail space, 80,000 sq ft of parking, and 20,000 sq ft of well-drained lawn at a slope of 4 percent. The remainder of the property will remain forested. A drainage analysis has determined the time of concentration to be 30 minutes pre-development and 10 minutes post-development.

The Roanoke County ordinances state that "[t]he 25-year post-developed peak rate of runoff from the land-disturbing activity shall not exceed the 10-year

Intensity-Duration-Frequency curve for Roanoke, Virginia. Courtesy NOAA's National Weather Service.

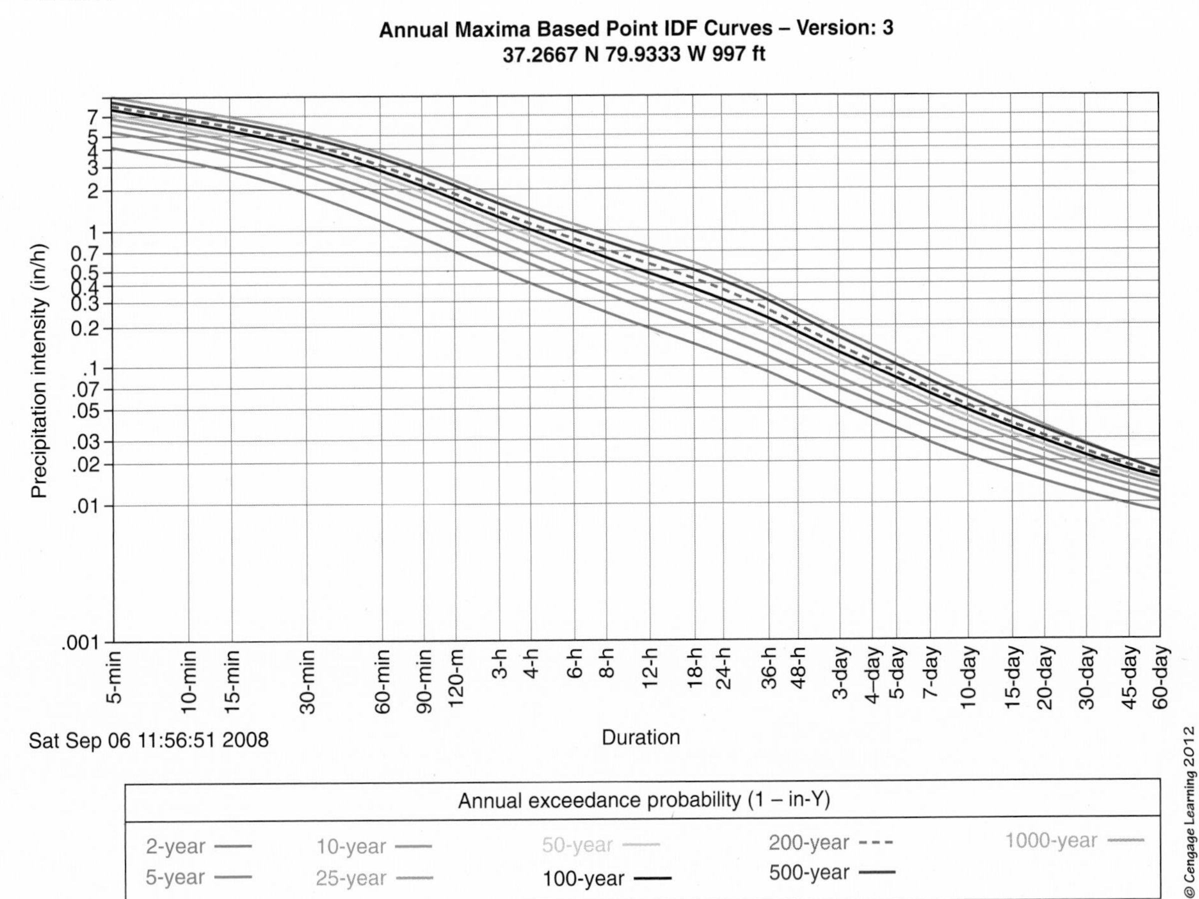

pre-developed peak rate of runoff." We will calculate the pre-development runoff for a 10-year storm and compare the rates to a post-development runoff for a 25-year storm. If stormwater regulations are not available, assume that you must limit the post-development runoff to pre-development flow for a 25-year storm.

Step 1: ***Determine the pre-development runoff.***

$C_f = 1.0$ (Figure 7-27 for 10 — year return period)
$C = 0.2$ (Figure 7-23 for forested land)
$i = 3.0\ in/h$ (*Roanoke IDF with* a 10 − year storm and a time of concentration of 30 minutes)
$A = 5.7\ ac$

Therefore,

$Q = C_f CiA$
$Q(pre) = (1.0)(0.2)(3.0)(5.7) = 3.4\ cfs$

Step 2: ***Determine post-development runoff.***

First, convert the areas of retail space (roof), parking, and landscaping to acres.

$$A_{roof} = 100{,}000\ ft^2 \cdot \frac{1\ ac}{43{,}560\ ft^2} = 2.30\ ac \text{ and } C_{roof} = 0.95$$

$$A_{parking} = 80{,}000\ ft^2 \cdot \frac{1\ ac}{43{,}560\ ft^2} = 1.84\ ac \text{ and } C_{parking} = 0.95$$

$$A_{lawn} = 20{,}000\ ft^2 \cdot \frac{1\ ac}{43{,}560\ ft^2} = 0.46\ ac \text{ and } C_{lawn} = 0.15$$

$$A_{forested} = 5.7\ ac - 2.30\ ac - 1.84\ ac - 0.46\ ac$$
$$= 1.10\ ac \text{ and } C_{forested} = 0.2$$

Next, calculate the composite runoff coefficient.

$$C_{composit} = \frac{A_{roof} \cdot C_{roof} + A_{parking} \cdot C_{parking} + A_{lawn} \cdot C_{lawn} + A_{forested} \cdot C_{forested}}{A_{total}}$$

$$C_{composit} = \frac{(2.30 \cdot 0.95) + (1.84 \cdot 0.95) + (0.46 \cdot 0.15) + (1.10 \cdot 0.2)}{5.7} = 0.74$$

$C_f = 1.1$ (Figure 7-27 for a 25 − year return period)
$C_{composit} = 0.74$
$i = 5.5\ in/h$ (Roanoke IDF using a 25 − year storm and a time of concentration of 10 minutes)
$A = 5.7\ ac$

Therefore
$Q = C_v CiA$
$Q_{post} = (1.1)(0.74)(5.5)(5.7) = 25.5\ cfs$

Step 3: ***Find the change in runoff.***

$\Delta Q = Q_{post} - Q_{pre}$
$\Delta Q = 25.5 - 3.4 - 22.1\ cfs$

STORMWATER STORAGE AND TREATMENT Stormwater management systems are often required by codes to handle the excess stormwater when the post-development peak stormwater runoff exceeds pre-development levels. But more and more often, municipalities are enacting laws that also address the effect of development on water quality. Holding water on-site and/or allowing it to filter through vegetation and soil reduces peak runoff and helps remove contaminants. Typically stormwater is collected (from roof drains, parking lots, etc.) and transported to a storm management structure through drains, pipes, ditches, or swales. Today, stormwater ponds (see example in Figure 7-28) are frequently used on commercial sites or in residential developments to store and treat stormwater, but other structures may be used for stormwater management. Some common stormwater management devices include:

Figure 7-28: In addition to reducing stormwater runoff and improving water quality, stormwater ponds can be used as a landscape feature.

1. A **retention pond**, also known as a wet pond, collects stormwater in a permanent on-site pond. These ponds treat stormwater by allowing pollutants to settle out or be removed by biological activity. This is one of the most cost-effective stormwater management systems and is widely used today. In some cases the retained water is re-used for irrigation and can provide a significant savings for sites that have a high demand for irrigation, like golf courses. In order to maintain a permanent pool, the drainage area feeding the pond should be at least 25 ac. However, these ponds are difficult to maintain in arid and semi-arid climates.

Figure 7-29: A detention pond (dry pond) is designed to temporarily hold stormwater but slowly release the water at pre-development levels. A retention pond is designed to capture stormwater in a permanent on-site pond. From Army Corp of Engineers. Area Planning, Site Planning and Design. TI 804-01 SITE PLAN pdf pp. 4-21.

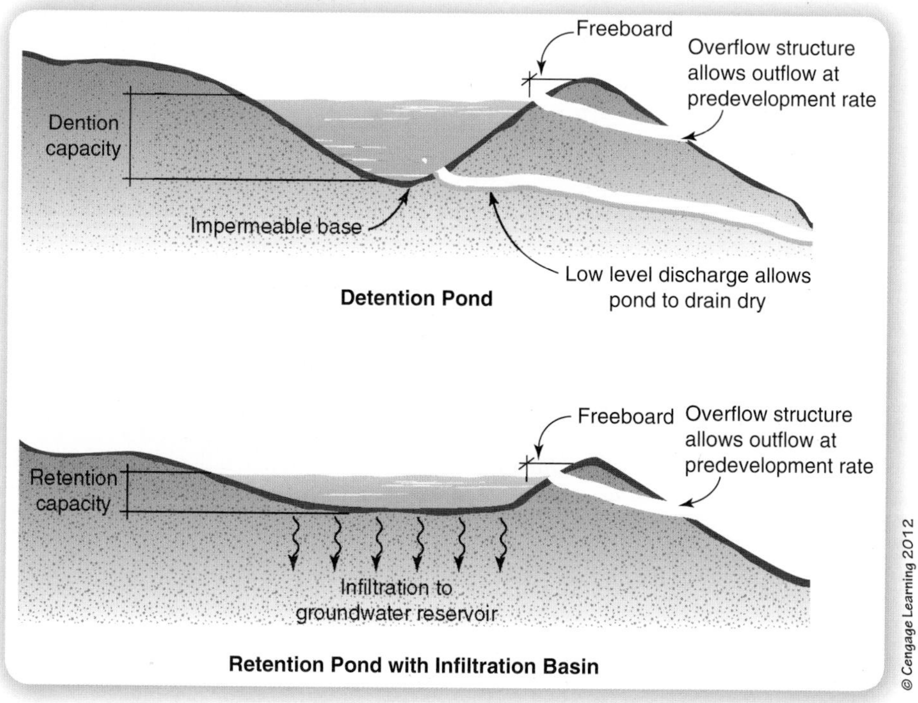

2. A **detention pond**, also known as a dry pond or an extended detention basin, collects stormwater and then slowly releases the water into the municipal stormwater system. These ponds are designed to detain water for a minimum duration (for example 24 h) which allows pollutants to settle out, but are not intented to maintain a permanent pool. Dry ponds should be used only for sites with a minimum drainage area of 10 ac because the smaller pipes needed to restrict water flow with less rainfall on smaller sites tend to clog. A variation of the dry detention pond that can be used for smaller sites uses large underground tanks to detain the stormwater. However, underground tank storage also tends to be expensive and is generally only used when there is not enough room on the site for surface storage (see Figure 7-29).
3. **Stormwater wetlands**, also known as constructed wetlands, are similar to wet ponds in that they include a permanent pool. In this system, stormwater is diverted to a shallow pool incorporating wetland plants (see Figure 7-30). Pollutants are removed through settling and biological activity. Although wetlands can be used to manage stormwater in most areas of the United States, a drainage area of at least 25 ac is needed to maintain the permanent pool—possibly more, depending on the average rainfall of the area.

Figure 7-30: **The image on the left shows a newly constructed wetland. The image on the right shows the same constructed wetland 2 years later.**

Photographer: Lloyd Rozema.

ESTIMATING THE SIZE OF STORMWATER PONDS Good site design requires that stormwater management be considered early in the design process. A good rule of thumb for large residential developments and commercial site design is to set aside 10 percent of the development area for stormwater management structures—ponds, constructed wetlands, or bioretention areas. Often, if the available area is limited, with careful design the required area can be reduced to a minimum of 5 percent of the total site area. If, because of other site requirements, the property will not provide sufficient area for surface stormwater management, you may need to consider other methods such as underground storage tanks or permeable pavement to reduce runoff.

Civil engineers must consider many contributing factors when deciding what type of stormwater management system that will best control the quantity and improve the quality of the stormwater runoff from a particular site. The volume of stormwater, the amount of contamination, the site soil characteristics, the depth to the water table, and the topography of the site will all impact the design. Although the design of stormwater management systems is beyond the scope of this textbook, we will approximate the volume of water that must be handled by the system using a simple formula.

Volume of stormwater

$$V = \frac{0.2\ ac \cdot ft}{ac} \cdot A$$

where V = volume of stormwater in ac-ft
A = drainage area (area of site) in acres

The volume calculation gives you an idea of how much water you will be required to store, but it doesn't tell you how big (length, width, and depth) the structure needs to be. In order to determine how much land you will need for the detention structure (pond, wetlands, bioretention area), you will need to estimate the depth of the pond. The recommended depth varies and depends on the type of structure you will use. For safety reason, you should design stormwater ponds so that the bottom of the pond is very shallow and slopes gradually at the edges. This will minimize the danger of drowning. Although not visually desirable, local regulations may require fences around the ponds.

Once you have determined a water depth, you can approximate the surface area of the detention facility using the following formula:

$$A_D = \frac{V}{d}$$

where A_D = area of detention facility
V = volume of stormwater in ac-ft
d = depth of detention facility

The volume of a rectangular pond and circular pond are illustrated in Figure 7-31.

Be aware that if you plan to use a wet pond to control stormwater, you must add the required stormwater volume to the volume of the permanent pool of water before calculating the area. You can use the calculated area for preliminary site design before the final stormwater management system has been formally designed.

Figure 7-31: **Rectangular pond and circular pond.**

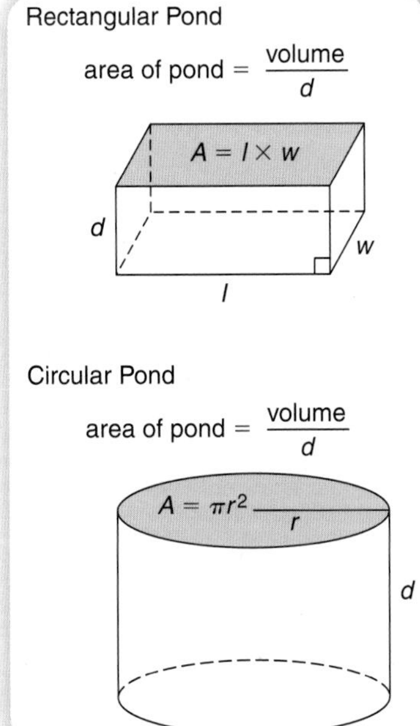

The calculation of quantities such as the area, surface area, and volume of geometric figures is an important part of many engineering problems such as the design of stormwater storage facilities.

Example: Estimate the Size of a Detention (or Retention) Pond

As part of the site design for a retail shopping complex, you must design a stormwater management system for the 15 ac site. Due to the size of the site (less than 25 ac) you propose to construct a detention pond to store the excess rainwater. Approximate the volume of water that must be stored. Then, assume an average pond depth of 4 ft to estimate the site area needed for the pond.

$$V = \frac{0.2\ ac \cdot ft}{ac} \times A$$

$$V = \frac{0.2\ ac \cdot ft}{ac} \cdot 15\ ac = 3\ ac \cdot ft$$

Therefore,

$$A_D = \frac{V}{d} = \frac{3\ ac \cdot ft}{4\ ft} = 0.75\ ac$$

Now convert the area to sq ft,

$$A_D = 0.75\ ac \cdot \frac{43{,}560\ ft^2}{1\ ac} = 32{,}670\ ft^2$$

Your site design should allow 0.75 ac or 32,670 sq ft for a detention pond with an average depth of 4 ft.

You can then estimate the plan size of the pond. If the pond will be approximately circular you can use the Area formula for a circular pond shown in Figure 7-31 to find the required pond radius:

By manipulating the equation you can solve for the radius,

$$A = \pi \cdot r^2$$

$$r^2 = \frac{A}{\pi}$$ divide both sides by π

$$r = \sqrt{\frac{A}{\pi}}$$ take the square root of both sides

$$r = \frac{R32{,}670\ ft^2}{\pi}$$ substitute the calculated area, and solve

$$r = 102\ ft$$

Therefore, a circular pond should have a radius of approximately 102 ft or a diameter of 204 ft.

Figure 7-32: The key elements of LID.

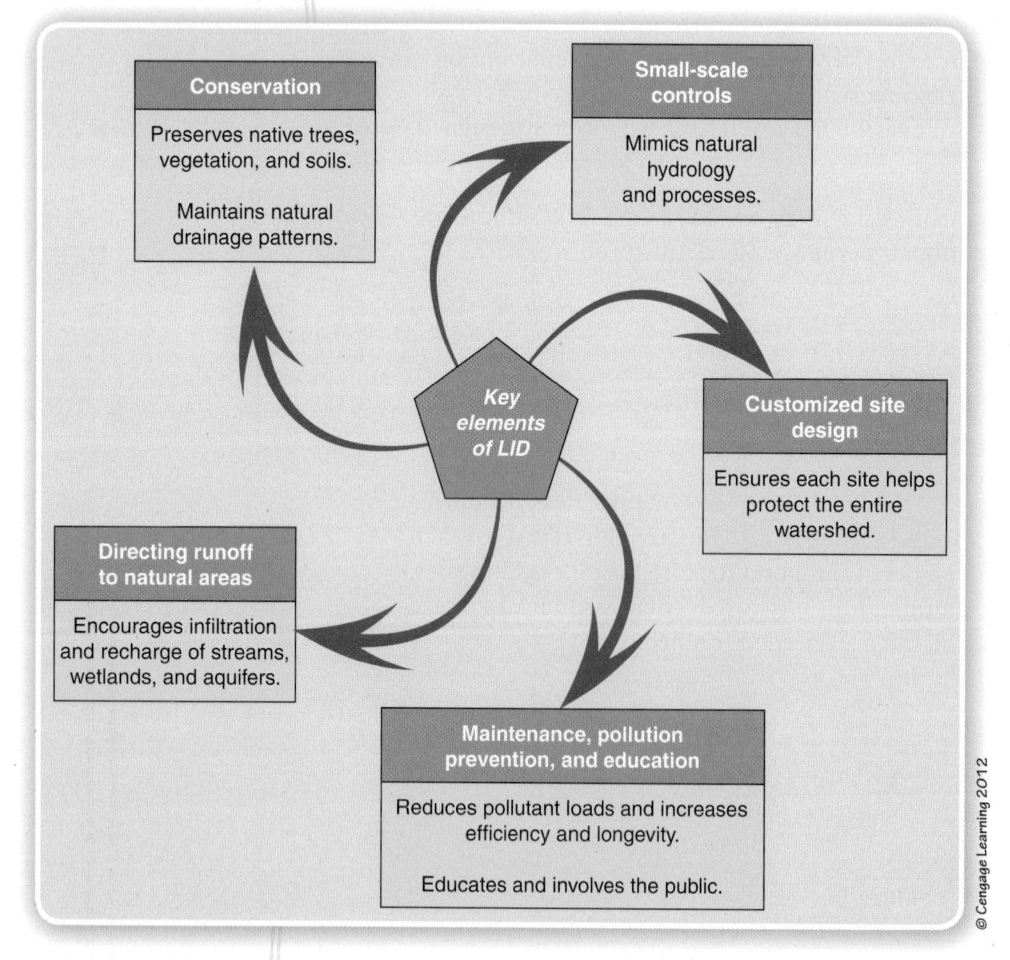

Low Impact Development

Low Impact Development (LID) is a stormwater management approach that uses green space, native landscaping, and techniques that mimic a site's pre-development water cycle: water infiltrates the ground, is stored, or evaporates naturally (see Figure 7-32). The approach results in a smaller volume of cleaner stormwater runoff, which means less expensive infrastructure for stormwater management and less damage to lakes, streams, and oceans. An added bonus is that LID enhances "quality of life" features by creating a greener design.

Some of the LID techniques that you may want to consider for your site include the following:

1. Rain gardens and bioretention cells – divert stormwater to low-lying landscaped areas that can store rainwater until it infiltrates the ground or is evaporated.
2. Impervious surface reduction and disconnection – minimize the area covered with imperious materials (such as concrete and asphalt) and add frequent pervious "breaks" like planter islands to allow water to infiltrate the ground.
3. Green roof – use plant material on a specially constructed flat roof to reduce stormwater runoff. Green roofs also save energy by keeping roofs cool in the summer and insulated in the winter. We will discuss green roofs in Chapter 18.
4. Rain barrels and cisterns – store water collected by roof gutters and drains in rain barrels or underground tanks (called cisterns) and use the rainwater to irrigate landscaping during dry months.
5. Permeable pavers – use gravel-filled or soil and grass-filled plastic cells, interlocking concrete paver blocks, or porous concrete for walkways and parking areas.

An example of a site plan that incorporates LID techniques is shown in Figure 7-33.

Figure 7-33: Site plan of a commercial Low Impact Development.

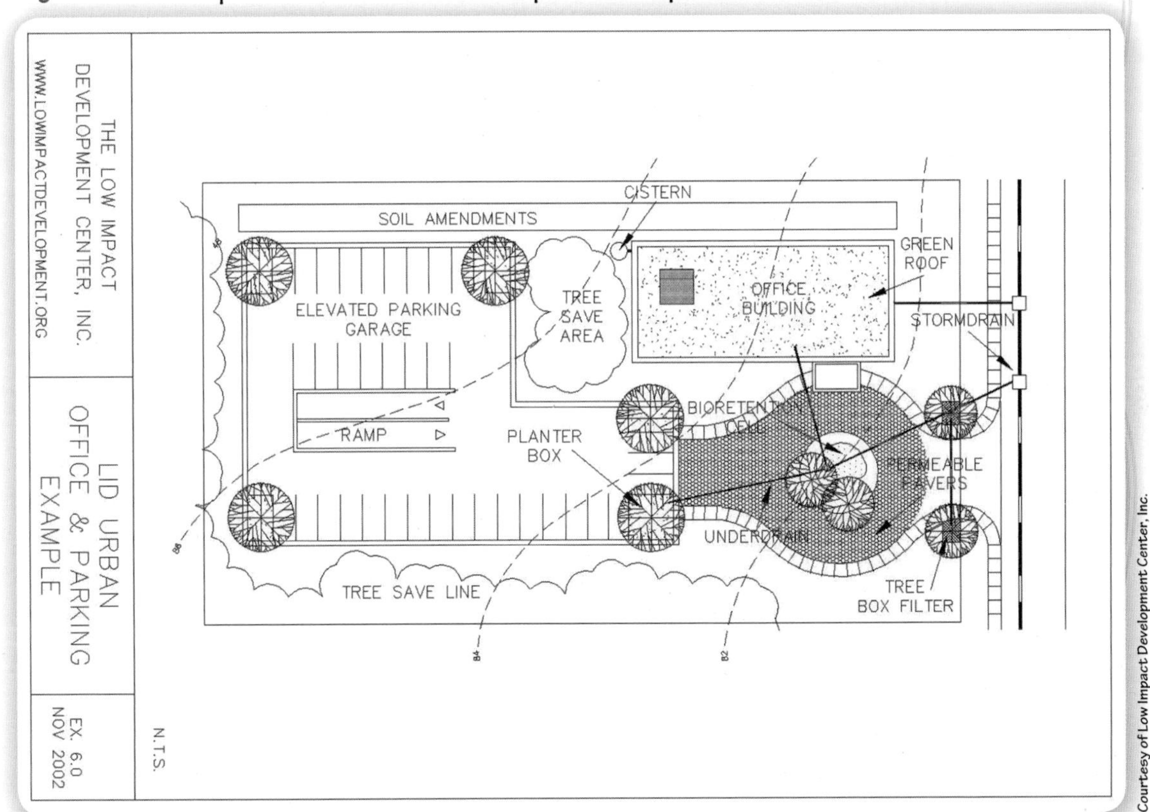

Courtesy of Low Impact Development Center, Inc.

Engineers and architects design and implement technologies to reduce the negative consequences of their designs to the environment. For example, effective stormwater management systems can be designed to reduce the quantity and improve the quality of stormwater discharge from a site.

Your Turn

Many cities and towns across the country have begun to encourage LID design in order to manage urban runoff. One example is the city of Sequim, Washington, which has established *Design Standards and Guidelines for Large Retail Establishments* to serve as a standard of excellence for development within the city. The document includes standards, which are mandatory, and guidelines, which are not mandatory but are intended to guide designers and planners in meeting the goals of the community. Find the document online at http://www.ci.sequim.wa.us/planning/designguidelines/index.cfm. Review sections on site planning, landscaping, parking lot design, and environmentally conscience development. Identify at least four LID techniques incorporated into the guidelines.

UTILITIES

Have you ever turned on a faucet expecting a stream of water but you get only a trickle? Are you frustrated when you lose your cable connection during your favorite show or your Internet connection while performing research for a school project? Does your home frequently lose electrical power so that you can't operate the microware and the heating and cooling units don't run? Convenient and reliable utilities are essential to American life and business. No development can be successful without careful consideration and design of water, sewer, power, and communication services.

You should have contacted all utility service providers during site discovery to find out if adequate services are easily accessible from the project property, how service can be established, if permits are required, and what fees will apply. If available, municipal or community services are often distributed through service lines within the right-of-way or easements of roads. Figure 7-35 shows a cross-section

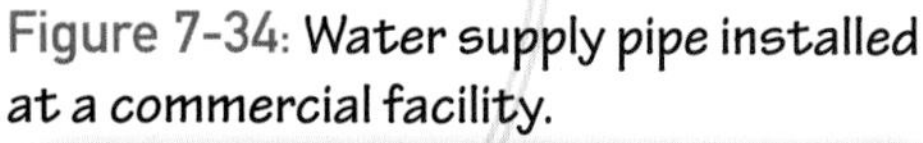

Figure 7-34: **Water supply pipe installed at a commercial facility.**

Photo courtesy of constructionphotographs.com

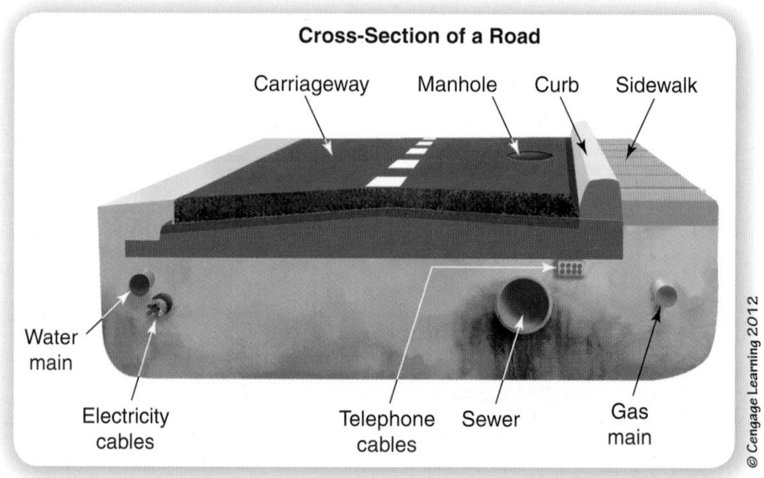

Figure 7-35: ***Cross-section of a road showing typical placement of utilities in the right-of-way. Notice the manhole access to the sewer line. Utility locations in your area may be specified by local regulations that differ from the locations shown here.***

of a roadway with underground utilities lines. Be aware that the arrangement of utilities here may be different in your area. Often local regulations specify where each utility may be placed in the right-of-way.

If services that meet the requirements for the project are not available, or very costly, you may need to consider building utility systems on the property. For example, if cable is not available to the site, installing a satellite dish may be a good alternative. Many property owners have installed solar panels or wind turbines to produce electrical power as an alternative to purchasing power from a utility company. Wells may be drilled and pumps used to provide water at acceptable pressures, and wastewater can be handled with a septic system if a municipal sanitary sewer system is not available.

We have already discussed stormwater management systems and practices, which are typically designed by civil engineers. In this section we will concentrate on two more utility services in which civil engineers are involved: the design and construction of water supply systems and wastewater management systems. A water supply system provides clean water to users. A wastewater system is designed to treat water that has been "used" by homeowners, businesses, or industry and then discarded.

Water Supply

Remember the disappointing trickle that drips out of some water faucets when you open the valve? An adequate quantity and pressure are important characteristics of a good water supply. Acceptable quality is also necessary—the water should have no unpleasant odors, tastes, or discoloration (see Figure 7-34).

Civil engineers consider both quantity and quality when designing a water supply system, which typically involves obtaining raw water from a water source (wells, rivers, lakes, or reservoirs), treating the water to insure that it is safe to use, conveying the water to a storage facility (usually a tank of some kind), and distributing the **potable water** to users (see Figure 7-36).

Potable Water: is raw or treated water that is considered safe to drink.

Figure 7-36: A typical potable water supply system will obtain raw water from a surface water (such as lakes, rivers, and reservoirs) or groundwater (as shown) source. The raw water is treated to remove or neutralize solids and contaminants and then pumped to a storage tank. One storage option is an elevated tank (as shown in the figure) which allows gravity flow of the water and applies a static head which maintains pressure within the system.

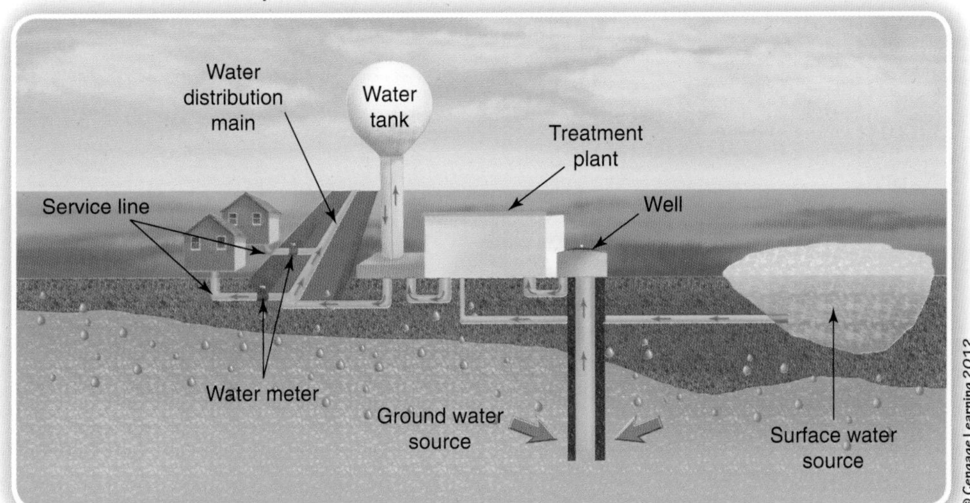

Pressure:

is the force applied to a unit area of surface. Water pressure at a points acts in all directions with the same force.

The design of a water supply systems is beyond the scope of this text; however, understanding the water needs of your individual project and the basic principles of hydraulics that apply to the movement of water in water supply and wastewater collection systems will help you provide a better site design.

WATER PRESSURE If you dive down to the bottom of a swimming pool, you may notice that the force of the water on your body (and your eardrums) changes as you descend—the deeper you dive, the more force you feel. If you were to measure the force you would discover that the force is 0.433 pounds on every square inch or lb per sq in (psi) of your skin when you are at a depth of 1 ft below the surface—this is exactly the weight of the 1 in × 1 in × 1 ft column of water that is directly above the 1 in sq area (see Figure 7-37). This force per area is referred to as **pressure**—so the water pressure at a 1 ft depth is 0.433 psi. And the force on a square inch of skin would increase 0.433 lb for every additional foot that you descend—0.866 psi at 2 ft, 1.299 psi at 3 ft, and so on. If you are a scuba diver, you are familiar with this principle and, hopefully, take precautions to avoid serious consequences when spending extended time at high pressure. In other terms:

$$1\ \textit{foot of water} = 0.433\ \textit{psi}$$

or

$$2.31\ \textit{ft of water} = 1\ \textit{psi}$$

Figure 7-37: **Water pressure on a given area at a given depth is equal to the weight of water above the area.**

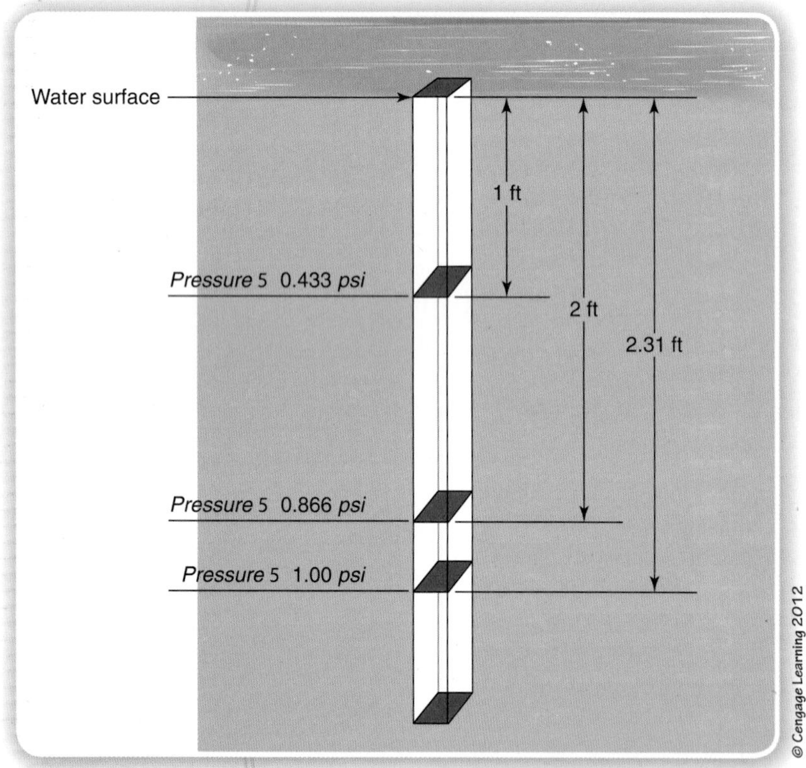

STATIC HEAD Water pressure is often described by the depth of water required to create the pressure and is referred to as **static head**. The word *static* indicates that the water is at rest, not moving through the pipes. Pressure and head are two ways to describe the same thing. The static head of water at a point is simply depth to that point. So, we can describe the pressure at the bottom of the 10 ft deep pool as having a static head of 10 ft or having a static pressure of 4.33 psi as the following calculation indicates:

$$10\,ft \cdot \frac{0.433\ psi}{1\ ft} = 4.33\ psi$$

or

$$10\,ft \cdot \frac{1\ psi}{2.31\ ft} = 4.33\ psi$$

Static Head:

(also called elevation head): at a point, the vertical distance between the point and the free water surface. It is also the level to which a liquid would rise in a piezometer or open-ended tube connected to the liquid at the point of interest.

The relationship between static head and pressure is:

Water Pressure

$$pressure = static\ head \cdot 0.433\ psi/ft$$

or

$$pressure = \frac{static\ head}{2.31\ ft/psi}$$

It is important to understand that static water pressure at a point depends only on the height of water above that point, not the shape or size of the container holding the water. Therefore, if you plugged the end of a 10 ft long straw, held it vertically, and filled it with water, the pressure at the bottom would be the same as the pressure at the bottom of a 10 ft deep pool. Also, the water column does not have to be directly above the point of interest. If you filled a long crazy straw (you know, one of those straws with lots of loops and spirals), the pressure in the crazy straw at a point that is 8 ft below the water surface would be the same as the pressure at an 8 ft depth in the straight straw or an 8 ft depth in the pool. See Figure 7-38. Similarly, the static water pressure in a pipe in a distribution system fed from an elevated tank depends only on the height of the water surface in the elevated tank above the height of the pipe itself, even if the pipe is miles away from the tank.

Figure 7-38: **The static water pressure depends only on the depth from the water surface. Therefore, the water pressure is the same at an 8 ft depth in a straight pipe, a crazy straw, or a swimming pool.**

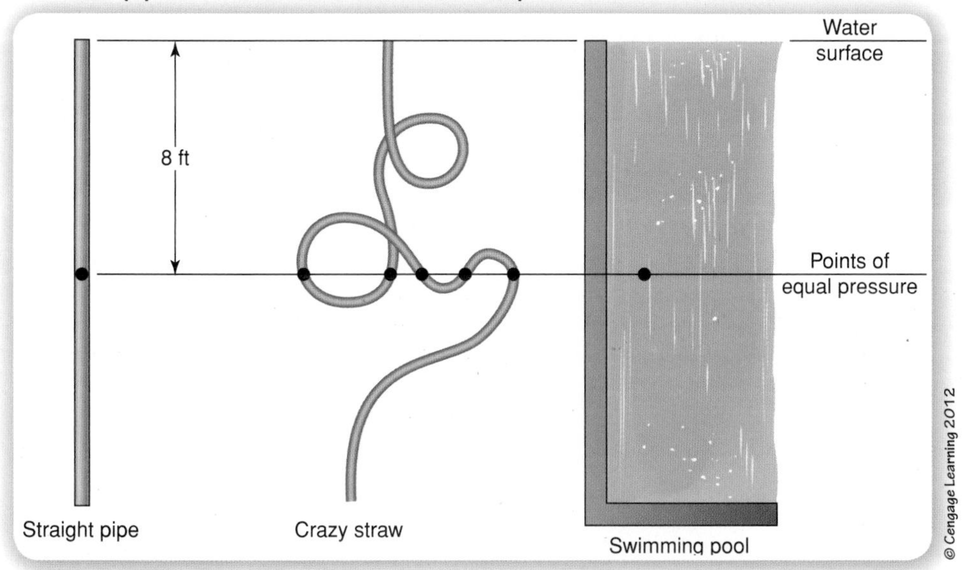

To demonstrate the calculations involved in estimating supply water pressure, let's look at an example in the town of Paradise. Assume that you are planning to construct a two-story apartment complex on the outskirts of town. Of course, since you are in Paradise, every building must be ideal in every way—including the water pressure. The Paradise Code of Ordinances includes the following regulation:

> Every residential unit within Paradise must be supplied potable water at a pressure of 55 psi measured at the supply side of the service meter. And every plumbing fixture within the residence must be supplied water at a minimum pressure of 40 psi.

We'll use this project throughout this section to demonstrate the calculations involved.

Example: Static Water Pressure

The town of Paradise obtains raw water from Utopia Lake, treats the water at the municipal treatment facility, and then pumps the water into an elevated storage tank. The town pumps water only during the nighttime hours when the cost of electricity is less, so the water level is typically highest in the morning at an elevation of 535 ft and continually drops during the day as the people of Paradise go about their daily business. At 11 p.m. the water level is generally at its lowest level of 519 ft when the pumps start up. What is the minimum static head and static pressure at the meter for the apartment complex if the service pipe elevation is 300 ft?

Because water pressure at a point depends only on the height of water above that point, the static head is the change in elevation between the water level in the tank and the point of interest in the pipe. The minimum water pressure will occur when the water level in the tank is lowest. Therefore the minimum static head is the difference in elevation between the lowest tank water level and the point of interest in the pipe.

$$static\ head = 519' - 300' = 219\ ft$$

Then convert static head to static pressure.

$$\begin{aligned} static\ pressure &= static\ head \cdot \frac{0.433\ psi}{1\ ft} \\ &= 219\ ft \cdot \frac{0.433\ psi}{1\ ft} \\ &= 94.8\ \text{psi} \end{aligned}$$

Alternatively

$$\begin{aligned} static\ pressure &= static\ head \cdot \frac{1\ ft}{2.31\ psi} \\ &= 219\ ft \cdot \frac{1\ ft}{2.31\ psi} \\ &= 94.8\ psi \end{aligned}$$

HEAD LOSS Water is rarely static in a water supply system. If you, or anyone connected to the system, opens a water valve by turning a faucet, starting a dishwasher, or operating a sprinkler system, the water will move through the pipes. To understand what happens when water flows through a pipe, imagine that you have connected a soaker hose (a flat hose with small openings at regular intervals along its length) to your water faucet. What will happen if you slowly increase the water flow just enough so that you see water flowing smoothly out of the last hole? Will the water flow equally through all of the holes?

If the hose is laid out on a flat surface you should notice that the stream of water flowing from the hole closest to the water faucet is taller than the rest. In fact, the height of each stream decreases linearly as the distance from the faucet increases (see Figure 7-39). When water flows through a pipe (or hose) friction is generated between the water molecules and the walls of the pipe—energy is lost along the entire length. As a result, the water pressure decrease as the water travels further and further along the pipe. We refer to the frictional loss in pressure as *head loss.*

The amount of pressure lost in a given pipe as water flows through it depends on several things: the diameter of the pipe, the length of pipe, the type of pipe, the age and/or condition of the pipe, the flow rate of the water, and **pipe fittings** (such as elbows, tees, and valves) that the water must pass through. Figure 7-40 illustrates several types of pipefittings. Although there are several methods to determine the loss of pressure due to friction, we will use the Hazen–Williams formula to estimate head loss, h_f.

Figure 7-39: Water that passes through a soaker hose (or pipe) loses energy due to friction between the water molecules and the walls of the hose, as demonstrated by the fact that the height of the jets of water from a soaker hose decrease in the direction of flow.

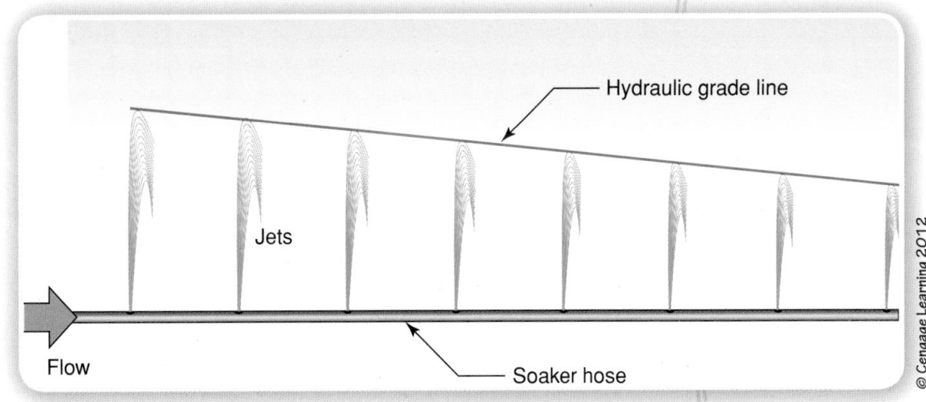

Figure 7-40: Some common fitting shapes. These steel fittings have screw connections.

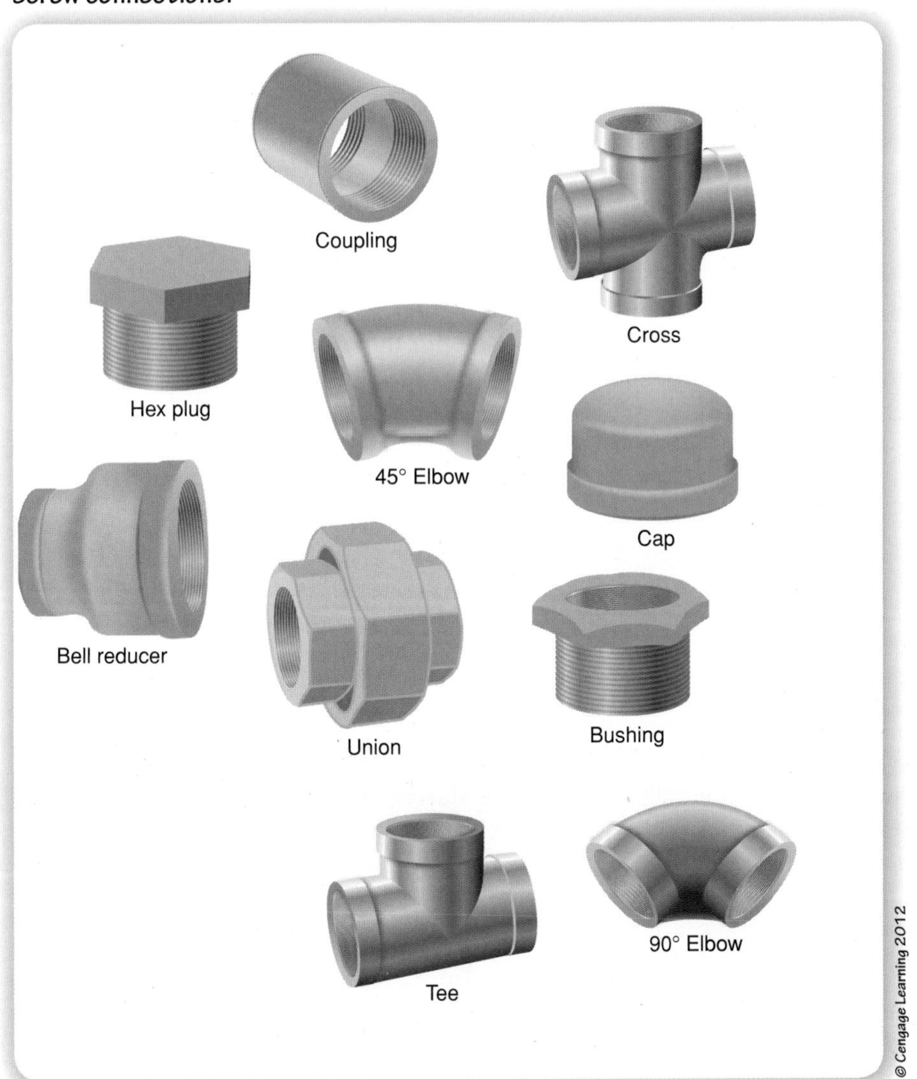

Hazen–Williams Formula

$$h_f = \frac{10.44 \cdot L \cdot Q^{1.85}}{C^{1.85} \cdot d^{4.8655}}$$ where h_f = head loss in ft of water

L = length of pipe in ft

Q = flow of water in gallons per minute (gpm)

C = Hazen–Williams coefficient (see Figure 7-41)

d = diameter of the pipe in in

Be careful to use the correct units when applying the Hazen–Williams formula: length in ft, flow in gallons per minutes (gpm), and diameter in in.

Figure 7-41: **Hazen Williams Coefficient, *C*.**

Pipe Material	Typical Range	Clean, New Pipe	Typical Design Value
Cast Iron and Wrought Iron	80–150	130	100
Copper, Glass or Brass	120–150	140	130
Cement lined Steel or Iron		150	140
Plastic PVC or ABS	120–150	140	130
Steel, welded, and seamless or interior riveted	80–150	140	100

The value of the Hazen–Williams coefficient, C, is based on the pipe material and condition. Generally accepted coefficients are shown in Figure 7-41. Notice that the *design* value of the coefficient is lower than the coefficient for *clean* (new) pipe. This is because the inside of a pipe generally becomes rougher as it ages due to pitting and the buildup of scale and rust. As the roughness increases, friction increases causing more head loss. Because *C* is in the denominator of the Hazen–Williams formula, a smaller *C* value will increase the head loss.

Engineers typically use the design value for the coefficient, C, when designing a water system to ensure that the water pressure will be adequate for as long as the system is in use—20, 30, or 50 years.

Engineers use symbolic expressions, such as the Hazen–Williams formula to represent relationships among quantities related to the problem to be solved.

Example: Calculate Head Loss

The apartment complex in Paradise (from the previous example) is 5 mi from the elevated water storage tank. Use the Hazen–Williams formula to estimate the head loss due to friction if water travels through a 10 in diameter cast iron pipe at a flow rate of 120 gpm.

Step 1: ***Convert length of pipe to ft. Note: 1 mi = 5280 ft.***

$$5\ mi \cdot \frac{5280\ ft}{mile} = 26{,}400\ ft$$

Step 2: *Choose the appropriate Hazen–Williams coefficient from Figure 7-41.*

Use the design value for cast iron pipe, $C = 100$.

Step 3: *Use the Hazen–Williams formula to estimate head loss.*

$$h_f = \frac{10.44 \cdot L \cdot Q^{1.85}}{C^{1.85} \cdot d^{4.8655}}$$ Note: L in ft, Q in gpm, and d in in

$$h_f = \frac{10.44 \cdot 26{,}400 \cdot (120)^{1.85}}{(100)^{1.85} \cdot (10)^{4.8655}}$$

$$h_f = 5.26\,ft$$

Thus, water flowing through the pipe will lose 5.26 ft of head as it moves through the 5 mi of pipe to the apartment complex.

The Hazen–Williams formula is an empirical formula, which means that it is based on experimentation. In the 1920s Mr. Hazen and Mr. Williams measured head loss through pipes of various diameters and materials at different flow rates in order to devise their formula and coefficients. Experimentation has also shown that pressure is lost when pipefittings, such as elbows and tees, force the flowing water to change directions inside a pipe. Fittings that block or disrupt the flow, such as valves and reducers, will also cause energy loss. The head loss caused by a pipefittings is referred to as a *minor loss.*

Your Turn
A Valve for Every Job

Many types of valves exist to control the flow of fluids, but in general valves can be divided into two basic types: stop valves and check valves. Stop valves are used to shut off the fluid flow by an outside force (like a person turning a knob) and can be one of four general types: gate, globe, butterfly, and ball valves. Check valves are used to keep fluids from flowing in the wrong direction in a system and are operated by the internal flow of fluid. Check valves allow fluid to flow in one direction, but close when the fluid flow reverses thereby protecting against backflow and contamination. Check valves may be of the swing, lift, or ball type. Research the various types of valves and sketch each in your project notebook.

A simple way to include minor losses in the head loss calculation is to determine the **equivalent length** of each fitting installed in the pipe run. The equivalent length of a fitting is the straight length of pipe that causes the same head loss as the fitting. For example, a regular 90-degree elbow in a 10 in, flanged steel pipe causes the same head loss as 12 ft of straight 10 in steel pipe. Therefore we say that this elbow has an *equivalent length* of 12 ft Figure 7-42 gives the equivalent length of steel flanged and screwed connections. Notice that the equivalent lengths vary greatly among fittings. Most valves, for example, have a large equivalent length. A gate valve is an exception since the gate in the valve moves out of the water stream when the valve is open. On the other hand, a long radius elbow has a relatively small equivalent length. So, a long radius elbow causes much less resistance to water flow than does a valve. Although equivalent length will vary depending on the pipe material, manufacturer and condition, we will use the generic values in Figure 7-42 for all head loss calculations.

Figure 7-42: Approximate equivalent length of pipefittings for flanged and screwed connections. Although the equivalent pipe length of a fitting varies with the pipe material and condition of the pipe, we will use these approximations for all head loss calculations.

Equivalent Lenght of Straight Pipe for Valves and Fittings (ft)												
Screwed Fittings		Pipe Size										
		1/4	3/8	1/2	3/4	1	11/4	11/2	2	21/2	3	4
Elbows	Regular 90 deg	2.3	3.1	3.6	4.4	5.2	6.6	7.4	8.5	9.3	11.0	13.0
	Long radius 90 deg	1.5	2.0	2.2	2.3	2.7	3.2	3.4	3.6	3.6	4.0	4.6
	Regular 45 deg	0.3	0.5	0.7	0.9	1.3	1.7	2.1	2.7	3.2	4.0	5.5
Tees	Line flow	0.8	1.2	1.7	2.4	3.2	4.6	5.6	7.7	9.3	12.0	17.0
	Branch flow	2.4	3.5	4.2	5.3	6.6	8.7	9.9	12.0	13.0	17.0	21.0
Return Bends	Regular 180 deg	2.3	3.1	3.6	4.4	5.2	6.6	7.4	8.5	9.3	11.0	13.0
Valves	Globe	21.0	22.0	22.0	24.0	29.0	37.0	42.0	54.0	62.0	79.0	110.0
	Gate	0.3	0.5	0.6	0.7	0.8	1.1	1.2	1.5	1.7	1.9	2.5
	Angle	12.8	15.0	15.0	15.0	17.0	18.0	18.0	18.0	18.0	18.0	18.0
	Swing Check	7.2	7.3	8.0	8.8	11.0	13.0	15.0	19.0	22.0	27.0	38.0
Strainer			4.6	5.0	6.6	7.7	18.0	20.0	27.0	29.0	34.0	42.0

Equivalent Lenght of Straight Pipe for Valves and Fittings (ft)																		
Flanged Fittings		Pipe Size																
		1/2	3/4	1	11/4	1 1/2	2	21/2	3	4	5	6	8	10	12	14	16	18
Elbows	Regular 90 deg	0.9	1.2	1.6	2.1	2.4	3.1	3.6	4.4	5.9	7.3	8.9	12	14	17	18	21	23
	Long radius 90 deg	1.1	1.3	1.6	2.0	2.3	2.7	2.9	3.4	4.2	5	5.7	7	8	9	9.4	10	11
	Regular 45 deg	0.5	0.6	0.8	1.1	1.3	1.7	2.0	2.6	3.5	4.5	5.6	7.7	9	11	13	15	16
Tees	Line flow	0.7	0.8	1.0	1.3	1.5	1.8	1.9	2.2	2.8	3.3	3.8	4.7	5.2	6	6.4	7.2	7.6
	Branch flow	2.0	2.6	3.3	4.4	5.2	6.6	7.5	9.4	12.0	15	18	24	30	34	37	43	47
Return Bends	Regular 180 deg	0.9	1.2	1.6	2.1	2.4	3.1	3.6	4.4	5.9	7.3	8.9	12	14	17	18	21	23
	Long radius 180 deg	1.1	1.3	1.6	2.0	2.3	2.7	2.9	3.4	4.2	5	5.7	7	8	9	9.4	10	11
Valves	Globe	38.0	40.0	45.0	54.0	59.0	70.0	77.0	94.0	120.0	150	190	260	310	390			
	Gate						2.6	2.7	2.8	2.9	3.1	3.2	3.2	3.2	3.2	3.2	3.2	3.2
	Angle	15.0	15.0	17.0	18.0	18.0	21.0	22.0	28.0	38.0	50	63	90	120	140	160	190	210
	Swing Check	3.8	5.3	7.2	10	12	17	21	27	38	50	63	90	120	140			

In order to get a more accurate estimate for the head loss in a pipeline, use a pipe length that takes into account the affect of every fitting installed in the line by adding the equivalent length of each fitting to the actual length of the pipe. This total equivalent length, L, can then be used in the Hazen–Williams formula.

Example: Equivalent Length and Head Loss

Estimate the head loss in the water distribution supply line from the town's elevated tank to the Paradise apartment complex including the minor losses. Assume that the water supplied to the project will pass through 2 regular elbows, 16 line flow tees, three branch flow tees, one angle valve (in which the water makes a 90-degree turn within the valve), and five gate valves before reaching the meter for the complex. Remember that the distribution lines are 10 in flanged pipe.

Step 1: ***Create a table to organize your calculation of total equivalent length.***

Fitting	Quantity	Equivalent Length (ft)	Total (ft)
Regular 90 elbow	2		
Line flow tee	16		
Branch flow tee	3		
Angle valve	1		
Gate valve	5		
		EQUIVALENT LENGTH for Minor Losses	

Step 2: ***Enter the equivalent length for each fitting using Figure 7-42.***

Fitting	Quantity	Equivalent Length (ft)	Total (ft)
Regular 90 elbow	2	14.0	
Line flow tee	16	5.2	
Branch flow tee	3	30.0	
Angle valve	1	120	
Gate valve	5	3.2	
		EQUIVALENT LENGTH for Minor Losses	

Step 3: ***Find the total length for each type of fitting.***

Multiply the quantity by the equivalent length in each row. Record the product in the right column.

Fitting	Quantity	Equivalent Length (ft)	Total (ft)
Regular 90 elbow	2	14.0	**28.0**
Line flow tee	16	5.2	**83.2**
Branch flow tee	3	30.0	**90.0**
Angle valve	1	120	**120.0**
Gate valve	5	3.2	**16.0**
		EQUIVALENT LENGTH for Minor Losses	

Step 4: ***Sum equivalent length totals.***

Fitting	Quantity	Equivalent Length (ft)	Total (ft)
Regular 90 elbow	2	14.0	28.0
Line flow tee	16	5.2	83.2
Branch flow tee	3	30.0	90.0
Angle valve	1	120	120
Gate valve	5	3.2	16.0
		EQUIVALENT LENGTH for Minor Losses	**337.2**

Step 5: ***Calculate the total equivalent length by adding the equivalent length for minor losses to the actual pipe length and***

$$\begin{aligned} \textit{total equivalent length} &= \textit{actual pipe length} + \textit{equivalent length for minor losses} \\ \textit{total equivalent length} &= 26,400\,ft + 357.2\,ft \\ &= 26,757.2\,ft \end{aligned}$$

Step 6: ***Calculate the total head loss using the Hazen–Williams formula.***

$$h_f = \frac{10.44 \cdot L \cdot Q^{1.85}}{C^{1.85} \cdot d^{4.8655}}$$

$$h_f = \frac{10.44 \cdot 26,757.2 \cdot (120)^{1.85}}{(100)^{1.85} \cdot (10)^{4.8655}}$$

$$h_f = 5.33\,ft$$

If you compare this total head loss with the head loss neglecting minor losses from the prior example, you will notice that the difference is small, a mere 0.07 ft (= 5.33 ft – 5.26 ft). Although minor losses can be large and should be considered in some cases, often minor losses are negligible. In general, if the ratio of actual pipe length to pipe diameter is greater than 1000, the affect of minor losses will be insignificant and can be ignored.

If $\frac{L}{d} > 1000$, then you may ignore minor losses.

Be careful to make sure that the units on the length and diameter are the same. In the case of the Paradise apartment complex the pipe diameter is 10 in, which is equivalent to 0.833 ft (= 10 in/12 in/ft). The ratio of length to diameter (both in ft) is

$$\frac{L}{d} > \frac{26{,}400\ ft}{0.833\ ft} = 31{,}693 > 1000.$$

Therefore, minor losses can be considered insignificant for the Paradise apartment complex water supply and can be ignored.

Total Dynamic Head (TDH): is the pressure within a pipe system when the water in motion.

TOTAL DYNAMIC HEAD Water that travels through many miles of pipe from the storage tank to a home or business can lose a significant amount of pressure. The actual pressure you feel at the service connection when you turn on a faucet will be less than the static pressure because the water will lose energy due to the frictional head loss. The actual pressure is referred to as *dynamic pressure,* in psi, or **total dynamic head (TDH),** in ft of water. Here *dynamic* refers to the fact that the water is in motion.

Total Dynamic Head: the pressure within a pipe system when the water is in motion, measured in ft of water.

Total Dynamic Head (TDH)

$$TDH = static\ head - head\ loss$$

$$dynamic\ pressure = TDH \cdot \frac{0.433\ psi}{ft}\ OR\ dynamic\ pressure = TDH \cdot \frac{psi}{2.31\ ft}$$

If you have experienced "low" water pressure, you know how frustrating it can be when water just trickles from a faucet or showerhead. What level of water pressure is sufficient to satisfy the normal consumer? For residential customers, the recommended water pressure is 55 psi, but the minimum pressure in the distribution main should not drop below 40 psi. A water pressure of 65 to 75 psi is often recommended for a water distribution system. This is normally sufficient to provide adequate pressure to buildings up to ten stories tall. Most state codes give 80 psi as the maximum safe water pressure for any building. Be aware that pipes and fittings in most water systems are designed for pressures less than 150 psi—higher pressures can cause leaks in the system. However, if water is supplied to your project at a pressure that is too high, a pressure-reducing valve can be installed in the supply line so that water entering the building can be adjusted to a desirable pressure.

Example: Calculate Dynamic Pressure

What is the actual (dynamic) pressure at the water meter for the Paradise apartment complex? Will the water be at an ideal pressure? If not, what can you do to regulate the pressure?

Step 1: ***Find the Total dynamic head (TDH).***

From previous examples the static head in the system is 219 ft and the head loss (including minor losses) is 5.33 ft.

$$TDH = static\ head - head\ loss$$

$$TDH = 219' - 5.33' = 213.67\ ft$$

Step 2: ***Convert head to pressure.***

$$dynamic\ pressure = TDH \cdot \frac{0.433\ psi}{ft}\ OR\ dynamic\ pressure = TDH \cdot \frac{psi}{2.31\ ft}$$

$$dynamic\ pressure = 213.267\ ft \cdot \frac{0.433\ psi}{ft} = 92.3\ psi$$

This pressure is greater than the maximum recommended pressure of 80 psi and is not the ideal of 55 psi. Therefore, a pressure-reducing valve may be installed to reduce the pressure.

Fun Fact

Water, Water Everywhere but Not a Drop to Drink

The oceans hold 97 percent of earth's water, but ocean water is too salty to drink. More than 2 percent of our water is frozen or trapped underground and not easily accessible. That means that less than 1 percent of the water on our planet is available for us to drink without expensive desalination or drilling processes.

Wastewater Management

Americans use a lot of water. The USGS reports that the daily amount of water withdrawn from surface and groundwater supplies in the United States in the year 2000 was over 43 billion gallons per day (gpd). The typical residential service draws 600 gpd. What happens to all of that water? You don't *drink* that much water, and your family pet is not a camel. Well, the amount of water that your family uses at home depends on many factors including the climate you live in, the types of appliances you own, how much you water your lawn, and your water conservation efforts.

Any water that you use outside your home, for irrigation or washing your car, goes the way of stormwater—some of the water will evaporate, infiltrate, or be absorbed by plants. The rest will run into the stormwater system (which is a waste of perfectly good water). But what about the water you use inside your home? Most of that water is used for baths, showers, washing clothes and dishes, brushing teeth, washing hands, and flushing toilets. After you use it, this "used" water, referred to as wastewater, usually ends up running down a drain, but where does it go from there?

Wastewater is normally not safe to drink and is often not clean enough to be discharged into lakes or streams (although conservation efforts have increased the reuse of cleaner "used" water, referred to as *greywater*). In most cases, wastewater is treated to improve its quality and protect the environment before it is discharged into a waterway. There are two different approaches to treating wastewater: **Publically Owned Treatment Works (POTW)** and onsite treatment facilities.

> **Publically Owned Treatment Works (POTW):** a system that is owned by a state or municipality that is involved in storage, treatment, recycling, and reclamation of municipal sewage or industrial wastes of a liquid nature. It includes sewers, pipes, and other conveyances if they carry wastewater to the POTW treatment plant.

PUBLICALLY OWNED TREATMENT WORKS Wastewater that comes from residences, commercial businesses, and institutions is called domestic or **sanitary wastewater**. Sanitary wastewater usually contains human waste, cleaning solutions, food particles, oil, and grease. In many communities with areas of concentrated population, the sanitary wastewater is collected by pipes and treated in a municipal wastewater plant to remove solid, chemical, and biological pollutants. Wastewater from industrial processes can be polluted with toxins not normally found in sanitary wastewater. Codes may require that it be pretreated to remove some of the pollutants before it is discharged into the sanitary sewer system.

If your facility has easy access to a municipal sanitary sewer system, you will most likely connect to the existing system because this is the easiest and most environmentally friendly alternative. Connecting to a municipal system, or Publically Owned Treatment Works (POTW) will involve transporting wastewater from the building through an underground sewer pipe, called a sewer lateral, to a branch or main line (pipe) of the municipal system. The branches convey the wastewater to larger sewer mains, or trunks, that carry the wastewater to a treatment facility. Manholes are placed at intervals along the sewer lines and at branch intersections to provide access to the pipes. A simplified illustration of a municipal sanitary sewer system is shown in Figure 7-43.

> **Sanitary Wastewater:** the spent or used water from a residential, institutional, or commercial facility that contains dissolved and suspended matter that is harmful to human health and the environment.

Figure 7-43: **Municipal sanitary sewer systems collect wastewater from homes and businesses and convey the wastewater to a Publically Owned Treatment Works (POTW) where the wastewater is treated and then discharged.**

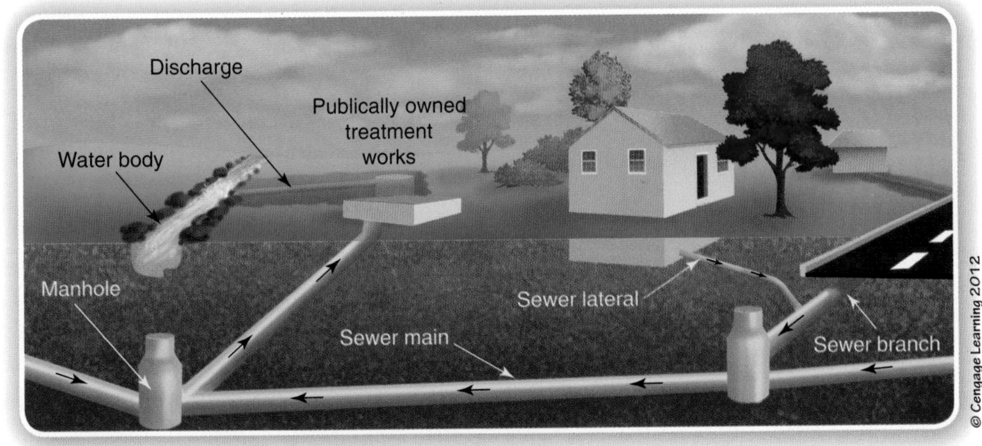

Although small municipalities with low flows may use alternate systems, most large municipalities use treatment systems that include primary treatment, secondary treatment, and disinfection. Primary treatment includes screening, grit removal, and settling to remove matter that floats to the surface or falls to the bottom. Secondary treatment uses biological treatment in which masses of organisms, called activated sludge, "eat" pollutants in the water and are then separated out by gravity. Finally, the wastewater is disinfected to kill bacteria, viruses, and protozoa before it is reused in the treatment process or discharged into a watercourse. The solid sludge waste that is removed may be incinerated or spread over farmland.

Fun Fact

Yuck!

Although ancient Rome is well known to have had an impressive water supply and sewer system with public baths and latrines, there were few rules about sanitation. Most human waste was collected in pots and thrown out of windows into the street. However, the Dejecti Effusive Act was created to protect pedestrians who passed by open windows at inopportune moments. If an unsuspecting citizen was hit and injured by projectile waste, the thrower of the waste was required to pay damages, but only if the incident occurred during daytime hours. No damages were paid for clothing.

You should have already contacted the local water and sewer departments to locate nearby branch and main sewer lines and determined connection fees during site discovery. Local codes will specify sewer line requirements. In most cases, to connect your project to a POTW, you must provide a lateral sewer line from each building to the nearest branch or main sewer line (see Figure 7-44). A sewer lateral is generally a **gravity sewer**, which means the pipe slopes down away from the building toward the main sewer line so that the wastewater flows by means of gravity. A properly designed lateral will be large enough to provide adequate flow and installed at an angle that is steep enough to keep the sewage moving in the pipe. It is also important to keep sanitary sewer lines away from other utility lines, especially water supply lines. If a sewer line leaks or breaks, the sewage can infiltrate the water supply line and contaminate the potable water.

Figure 7-44: A sewer lateral transports wastewater from a building to the sanitary sewer system.

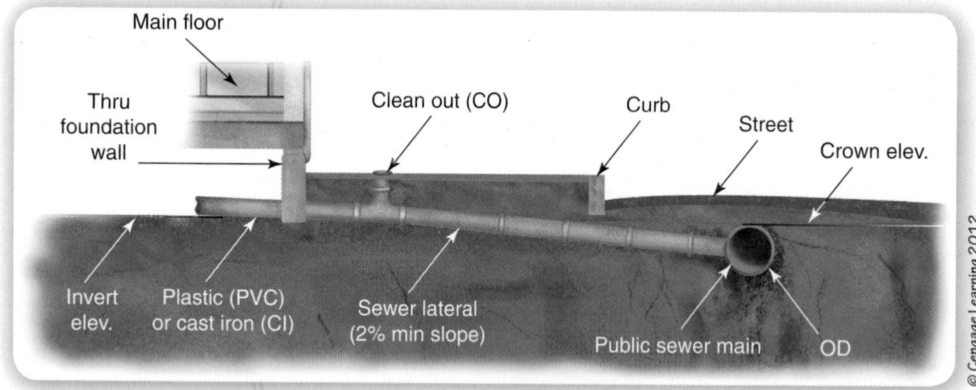

The following recommendations are common code requirements for sewer line installations:

1. *Pipe size.* A good rule of thumb for sanitary sewer pipe size is to use a 4 in pipe for single family residential structures, a 6 in pipe for multi-family residential, retail, or commercial facilities, and at least an 8 in pipe for an industrial facility.

2. *Depth.* A standard dimension to provide the adequate clearance for construction of the sewer line under a basement floor slab is 2 ft below the lowest service elevation. That is, the sewer line should be at least 2 ft lower than the elevation of the lowest floor requiring sanitary sewer service. In addition, sanitary sewers should be installed completely below the frost depth to reduce the possibility of freezing. Figure 7-45 shows a map of frost depth in the United States; however, local conditions may vary. Check with your local building department to get local requirements.

Figure 7-45: Frost depths may vary due to local conditions. Check with your local building department for more specific information.

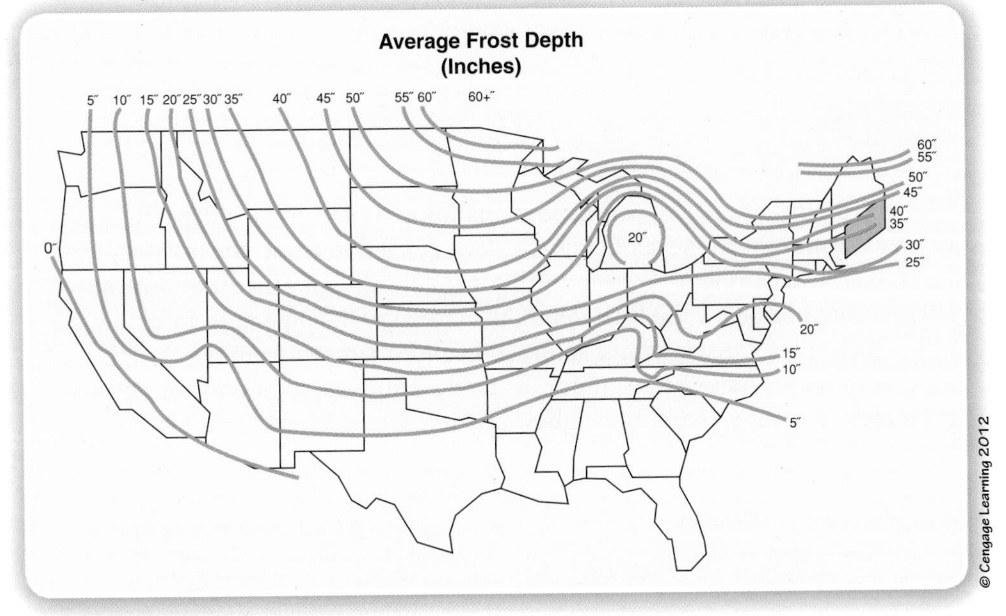

3. *Slope.* Ideally, sewer laterals should be installed with a minimum slope of 2 percent, equivalent to ¼ in drop per foot of pipe length (although shallower slopes may be allowed by local codes). Be aware that steeper sewer line slopes will cause the wastewater to flow more quickly and, if the slope is too steep, may result in damage to connections when the wastewater slams into pipe walls. If your lateral slope will be more that 10 percent, check with your local water and sewer department for guidance.
4. *Separation.* In general, maintain a horizontal distance of 10 ft between water and sewer lines when they run parallel. Sewer lines should also be installed at least 18 in below the elevation of water supply lines. In some cases exceptions can be made if it is not feasible to meet the separation requirements—check with your local building department.

Sewer Lateral Slope

$$\text{Sewer Lateral Slope} = \frac{\text{invert elve. of lateral at building} - \text{crown elev. of main} + \frac{1}{2}OD}{\text{distance from building to sewer main}} \cdot 100\%$$

where $\frac{1}{2}OD$ = *half the outside diameter of the sewer branch or main*

Note that this formula assumes that the invert of the lateral at the main falls at approximately the horizontal centerline of the main.

Example: Calculate Sewer Lateral Slope

Your company is designing a new pizza restaurant in Jackson, Wyoming, just south of Yellowstone National Park. As the civil engineer assigned to the job, you are in charge of the site design and must therefore be sure that the sanitary sewer line meets code requirements. The restaurant is 160 ft from the service connection at the 10 in sewer main which has a crown elevation, or the elevation of the top of the pipe, of 6213.65 ft. The elevation of the lowest (ground) floor of the building is 6221.52 ft. Using the recommendations provided in the previous section, determine the highest possible lateral invert elevation at the building foundation, and then calculate the resulting pipe slope. Will it meet the 2 percent minimum slope requirement? What is the lowest invert elevation of the lateral at the building foundation that would provide adequate slope?

Step 1: ***Find the highest lateral invert elevation at the building foundation.***

Use a 6 in pipe size per recommendations for commercial facilities. The minimum clearance for the sewer below the floor is 2 ft. However, the frost depth in Jackson is 30 in per Figure 7-45 and will control the design. Since the entire sewer lateral should be below the frost depth, the crown of the pipe must be below the frost depth. Therefore, you must subtract the pipe diameter from the elevation as well. In addition, assume that grade outside the building is 6 in below the ground floor elevation. As a result, the highest possible lateral elevation at the building foundation will be

$$\begin{aligned}\textit{Highest lateral elev. at building} &= \textit{floor elev.} - \textit{distance to grade} - \textit{frost depth} - \textit{pipe diameter}\\ &= 6221.52 - 6\ in \cdot \frac{1\ ft}{12\ in} - 30\ in \cdot \frac{1\ ft}{12\ in} - 6\ in \cdot \frac{1\ ft}{12\ in}\\ &= 6218.52\ ft\end{aligned}$$

Step 2: ***Calculate the slope.***

$$\textit{Sewer lateral slope} = \frac{\textit{invert elev. of lateral at building} - \textit{crown elev. of main} + \frac{1}{2}OD}{\textit{Distance from building to sewer main}} \cdot 100\%$$

$$= \frac{6218.02\ ft - 6213.65\ ft + \frac{1}{2}(10\ in)\left(\frac{1\ ft}{12\ in}\right)}{160\ ft} \cdot 100\%$$

$$= 3.0\%$$

The sewer lateral has a slope of 3.0 percent, which exceeds the minimum slope requirement of 2 percent. Therefore, using an invert elevation at the foundation of 6218.52 ft is acceptable.

Step 3: ***Calculate the lowest lateral invert elevation at foundation.***

The lowest invert elevation at the building foundation that would still provide adequate slope can be found by setting the slope gradient to 2 percent and solving for the invert elevation.

$$\textit{Sewer lateral slope} = \frac{\textit{invert elev. of lateral at building} - \textit{crown elev. of main} + \frac{1}{2}OD}{\textit{Distance from building to sewer main}} \cdot 100\%$$

$$\frac{\textit{Sewer lateral slope}}{100\%} \cdot \text{distance} + \text{crown elev. of main} - \frac{1}{2}\,\text{OD} = \begin{array}{c}\text{invert elev. of lateral}\\ \text{at building}\end{array}$$

$$\frac{2\%}{100\%} \cdot 160\,ft + 6213.65\,ft - \frac{1}{2}\,(10\,in)\left(\frac{1\,ft}{12\,in}\right) = 6216.43\,ft$$

So, the sewer lateral could be installed at an invert elevation as low as 6216.43 ft at the foundation and still maintain a 2 percent slope.

ONSITE AND DECENTRALIZED WASTEWATER TREATMENT SYSTEMS It is preferable to discharge your wastewater into a municipal system if it is available. However, sometimes a building site will not have easy access to a municipal system, so wastewater will have to be handled on the site. Approximately 25 percent of households in the United States and 33 percent of new developments use on-site and clustered (decentralized) wastewater treatment systems as an alternative to a POTW (EPA, 2003). Unfortunately many of these systems do not properly treat wastewater or are improperly located too close to ground or surface waters. The result is environmental degradation of land and water sources. Careful design and maintenance of on-site systems are necessary in order to provide adequate water treatment.

A conventional on-site treatment system has three components: a **septic tank**, a drainage (leach) field, and soil (see Figure 7-46). A single home may have an individual septic system. A large development may have a larger septic system that collects wastewater from many buildings and treats it in a central location. A septic tank is a large underground concrete box that is designed to hold wastewater

Decentralized Wastewater Treatment Systems: are individual onsite systems that collect, treat, and disperse or reclaim wastewater from individual residences, businesses, or small clusters of buildings. These systems are commonly referred to as septic systems, private sewage systems, individual sewage treatment systems, onsite sewage disposal systems, or "package" plants.

Figure 7-46: A conventional decentralized wastewater treatment system, or septic system, has three components—a septic tank, a drainage (leach) field that includes a distribution box and perforated pipe, and soil.

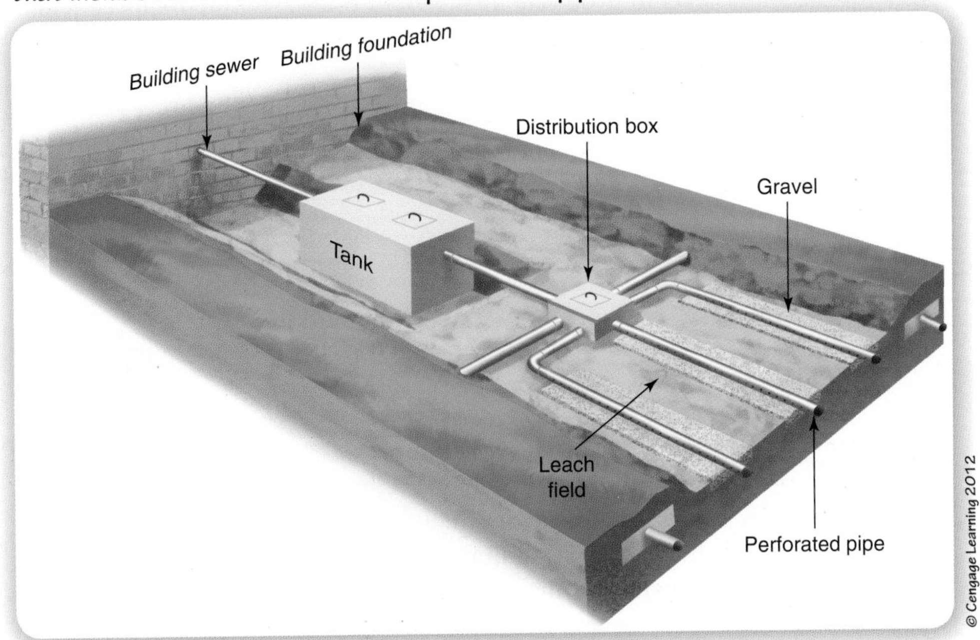

Figure 7-47: Section view of a septic tank. The scum and sludge are retained in the tank and should be pumped out when the volume of solids exceeds 30 percent of the tank volume.

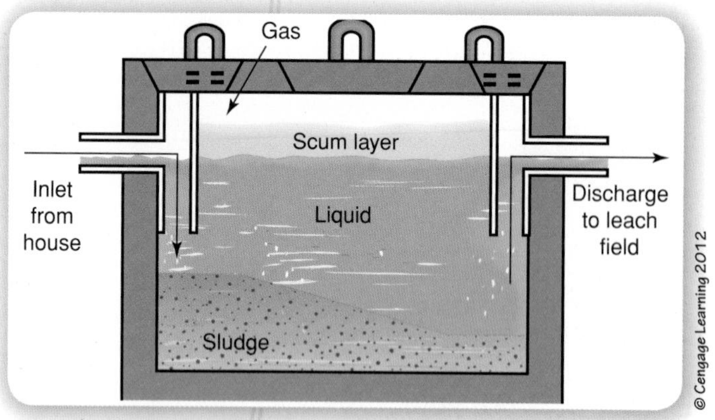

for about 2 days. This detention time allows sludge (heavy solids) to settle out and scum (grease, oil, and floating debris) to float to the surface of the water. The solids are retained in the tank while the liquid is discharged into a distribution box as shown in Figure 7-47. When the system is operating properly, anaerobic decomposition will significantly reduce the amount of solids in the tank, but the tank should be pumped out when the solids exceed 30 percent of the tank volume.

The liquid in the septic tank is discharged into a **distribution box** that disperses the wastewater to perforated pipes buried under the ground. The pipes are placed on a bed of gravel in permeable soil so that the water easily infiltrates the soil. The area onto which the wastewater is distributed is called an infiltration field or **leach field**. The design of the leach field depends on the characteristics of the native soil. As you can imagine, if the soil is granular so that water infiltrates relatively quickly into the soil, you will need a much smaller leach field than if the soil is cohesive and resists infiltration.

Not all soil is capable of accepting and adequately treating wastewater discharge. Evaluation of the soil in the area of the leach field is required before a septic system can by installed and used. The required method of evaluation varies among local jurisdiction, so you should check with your local building department before conducting soil tests. Some states or local jurisdictions required a **percolation** (or perc) **test** to estimate the rate of water infiltration into the soil. Many different test methods have been developed for percolation tests, but basically the test involves digging several holes in the area of the proposed leach field. The holes are filled with water and the water is left to saturate the surrounding soil. The holes are again filled with water and then the water level is measured periodically to determine the rate at which the water percolates or infiltrates into the soil. The rate is recorded in minutes per inch—that is, how many minutes it takes for the water level to drop one inch. Codes often base the required length of perforated pipe in the leach field on the percolation rate and the flow rate of wastewater from the home or facility.

Because percolation tests can be unreliable, especially during a rainy season or draught, a soil evaluation by a professional engineer, geologists, or environmental specialist may be required in order to obtain a permit to install a septic system.

Engineers and architects can devise technologies to conserve water, soil, and energy through such technologies as reusing, reducing, and recycling.

SITE PLAN

Engineers and architects document the final design of a building site on a **site plan**. A site plan is a drawing, or set of drawings, representing a view of the project site, looking down from above, which shows important site elements. In order to obtain a building permit, you are often required to provide a site plan that includes

Your Turn
To Perc or Not to Perc

Use the U.S. Department of Agriculture Web Soil Survey (keyword: Web Soil Survey) to research the suitability of soil on your site or near your school for an on-site wastewater treatment system. Begin by finding your Area of Interest (AOI) by navigating to your state and county. Find your location on the map using the zoom and pan tools and then indicate the AOI using the Define AOI tool. Then click on the Soil Data Explorer tab and then the Suitability and Limitations for Use sub-tab. Investigate the information on Septic Tank Absorption Fields under the Sanitary Facilities category. What is the rating for the soil in your area? What are the reasons for this rating? Would an on-site wastewater treatment system be a good choice for this location? Record you findings in your Project Notebook.

Next, find out if your local building department requires a percolation test in order to get a permit to install a septic system. If so, obtain the specific test procedures. If not, research methods to perform a percolation test. Perform a perc test on your project site or school property and record your results. Be sure to sketch a map showing the location of your test hole(s) and details of the procedure you used. Calculate the percolation rate for each hole and show your work.

Do your percolation test results support the Septic Tank Absorption Field rating you found using the Web Soil Survey?

New York City Underground

A civil engineer's work supplying humans with life's necessities and the modern conveniences that support our society is often hidden below our ft. Few places in the world can rival the underground network of utilities and transportation lines that are buried beneath New York City. Power, steam, gas, water, and cable supply lines form a complicated maze above subway and vehicular tunnels. Sewer lines that carry away wastewater are found even deeper below the surface at depths of up to 200 ft Far, far below the sewer system, workers called *sandhogs* continue to excavate City Tunnel No. 3, a Department of Environmental Protection project that is designed to guarantee a safe water supply for the city. Learn more about the complexity of NYC's underground infrastructure at the National Geographic website *http://www.nationalgeographic.com/nyunderground/docs/nymain.html*. Take the tour of each underground level, and view the photo tour. Although mostly hidden from view, New York City could not exist as it is today without the infrastructure designed by civil engineers.

specific elements as dictated by your local building department. Site plan requirements vary by jurisdiction but may include:

- Plan scale
- North arrow
- Property description (legal description, address, or other required by local jurisdiction)
- Property lines with bearing and dimensions
- Area of lot
- Existing and proposed roads with center line and curb elevations
- Impervious surfaces—driveways, patios, walks, decks, and parking areas, with indications of type of material
- Existing and proposed structures (including area of roof and overhang) with finished floor elevation
- Utility locations and easements—overhead and underground
- Setbacks
- Existing trees—outline of forested areas; location size and species of large trees
- Proposed trees (size and species)
- Site contours and/or corner elevations
- Stormwater drainage—erosion control, catch basins, ditches, ponds, etc.
- Critical areas and buffers—wetlands, waterways, flood area, critical habitat, etc.
- Septic system—location of tank and leach field

A sample residential site plan for the City of Portland, Oregon, is shown in Figure 7-48. A partial example of a commercial site plan is shown in Figure 7-49.

PERMITS

Compliance: is a state of being in accordance with established guidelines, specifications, rules, codes, or regulations

The needs of your client are not your only concern. As you have learned in previous chapters, there are rules and regulations that will govern every aspect of your project including the design of the site. Codes, ordinances, and other regulations should be reviewed at every step of the design process to make sure the design is in **compliance**. You will likely be required to create a set of working drawings of your proposed project and submit specific plans with applications for various permits. We have already discussed the need for several types of permits that may be required based on the location of your project and the type of project being proposed. Some of these permits include stormwater permits, Incidental Take Permits, coastal construction permits, and wetland permits (see Chapter 5). Additional permits will likely be needed. For example, you may need a water permit and sewer permit to connect to local water and sewer systems. An encroachment permit may be required to connect entrance drives to state or local roads. A building permit is usually required when you construct new buildings or make additions to existing buildings. Often the requirements for obtaining other permits are included in the building permit requirements. Noncompliance can result in fines, construction delays, redesign, and possibly demolition of noncompliant features, none of which are good for your project budget or schedule.

Figure 7-48: **Sample residential site plan for Portland, Oregon.**

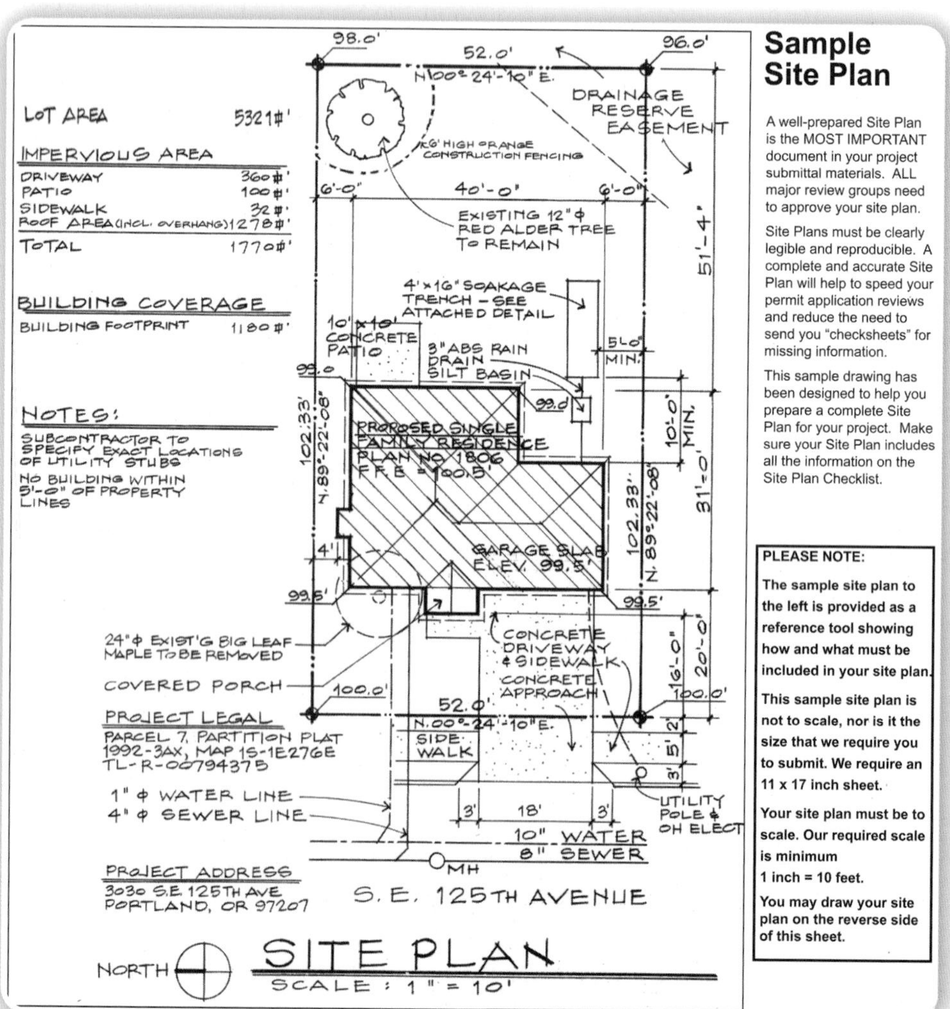

Courtesy of Bureau of Development Services, City of Portland.

Figure 7-49: Partial commercial site plan with grading.

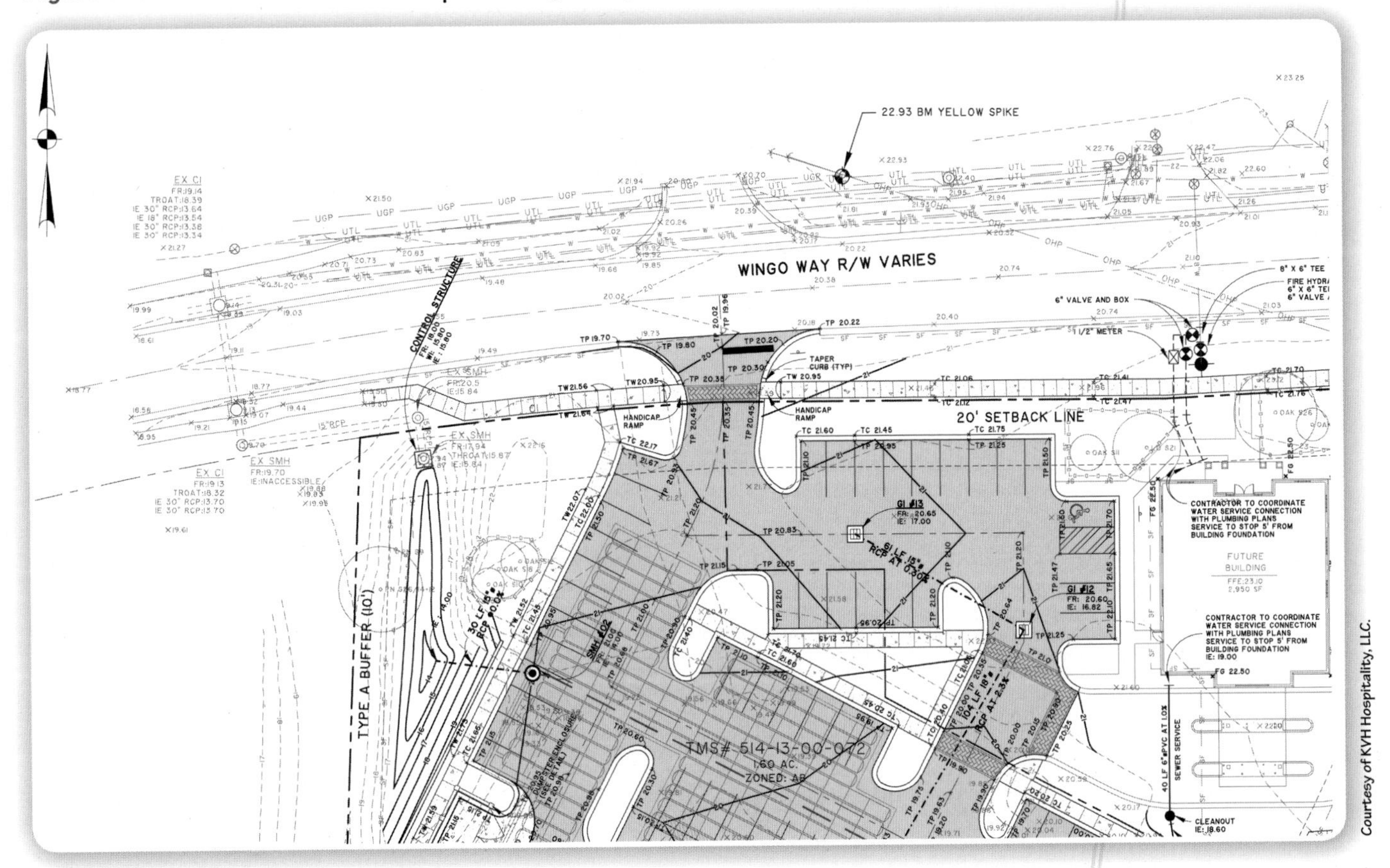

Courtesy of KYH Hospitality, LLC.

SUMMARY

Civil engineers are typically responsible for the design of a site and must take into consideration a wide variety of factors that affect the design of the property. Providing adequate access to the property for pedestrians and vehicles, including handicapped people and their vehicles, is an important consideration. Other factors that will influence the site design and affect usability include accessibility to disabled persons, site circulation and parking, grading and drainage, and the availability of utilities. In addition, many of these site elements must comply with codes and regulations that will control the project design

Civil engineers document site design with a site plan that will locate the structure on the property. In addition, a site plan provides information necessary to construct other site features such as stormwater drainage systems, utility connections, ingress and egress routes, parking facilities, and possibly landscaping.

Although the civil engineer must incorporate an extensive array of testing, analysis, and design techniques throughout the site design process, with careful consideration of all contributing factors, the final design will provide an efficient, cost-effective, environmentally friendly, and visually appealing environment that will enhance the finished project.

BRING IT HOME

1. As a member of the design team for a 10,000 sq. ft retail electronics store, you have been given the responsibility for creating a preliminary site plan for a 400 ft × 400 ft flat property. Create a sketch to show your preliminary plan for the major site elements (building, parking, walkways, ingress and egress, landscaping). Include a detailed parking lot layout. Then indicate on the sketch the design details that you will incorporate to make the site accessible. Use the following criteria for the design:
   - Use 90-degree parking.
   - Provide parking spaces with a minimum size of 9 ft × 18 ft
   - Provide two-way parking aisles.
   - Provide 4 ft wide sidewalks along both adjacent roads.
   - Provide a 12 ft (minimum) landscaped buffer (including the sidewalk) between the parking area and the roadways.
   - Provide a 10 ft buffer along the property line of adjacent properties.
   - Provide an enclosed dumpster pad.
   - Landscaped parking islands must be incorporated such that no more than eight parking spaces are aligned side-by-side.
   - Parking islands must be a minimum of 4 ft wide and the length of adjacent parking spaces.

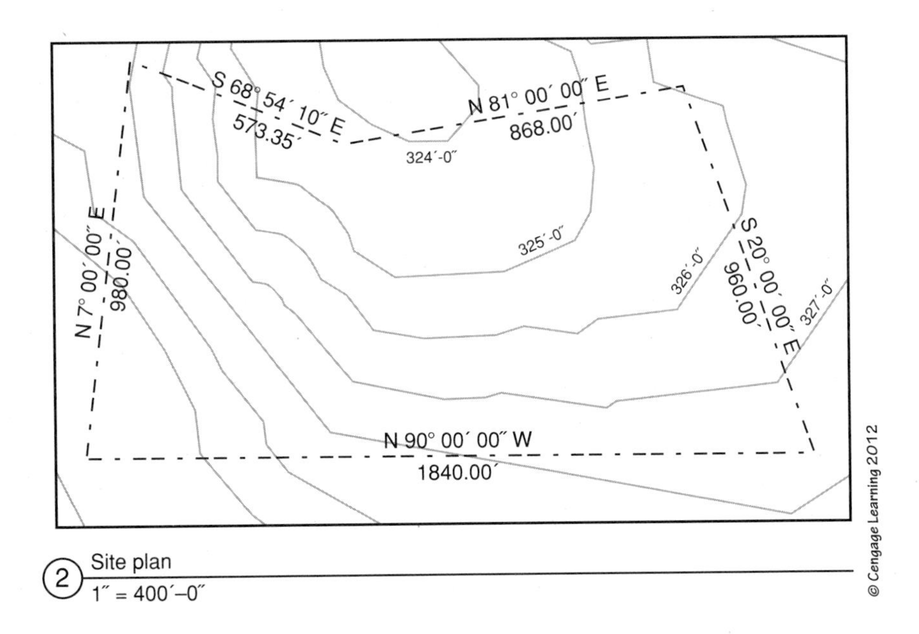

(2) Site plan
1″ = 400′–0″

2. As the civil engineer on the design team for a 30 ac multi-building industrial complex, you must determine the change in stormwater runoff due to development of the site. The existing site is pastureland but will be developed into 550,000 sq ft of building space and 225,000 sq ft of asphalt parking area and roadways. In addition, landscaping will include 3 ac of poorly drained grassed area. The remainder of the site will remain unimproved pasture. Assume that the local codes require that the post-development peak runoff not exceed the pre-development levels for a 25-year storm. Assume the pre-development time of concentration is 15 minutes.
   a. Find the change in runoff once the site is developed.
   b. Approximate the size of a detention pond to collect the excess runoff.
   c. Draw a profile view of the pond. Calculate cut and fill volumes for the detention pond.
3. Select an older small commercial facility near you and create a sketch to show the existing site elements. Include major site components such as the building, parking, sidewalks, landscaped areas, and natural areas. Note the treatments in each area. For example, if applicable, annotate the parking lot as "ASPHALT PARKING LOT," the sidewalk as "POURED CONCRETE WALKWAY," and the landscaped area as "LANDSCAPED AREA WITH WOOD MULCH." Also indicate any stormwater facilities present on the property such as swales or ponds. Create a list of LID recommendations for the owner that would improve the stormwater management system of the property by reducing the quantity of runoff and improving the quality of runoff. Using a colored pencil, show how you would incorporate these recommendations on the sketch of the site.
4. A new condominium complex is planned on the outskirts of Newtown, USA. The water level in the storage tank that will supply the development with water fluctuates from 1381 ft to 1407 ft The water will travel through approximately 5.2 mi of 10 in flanged ductile iron distribution main which includes eight regular 90-degree elbows, 10 branch flow tees, 3 line flow tees, 16 gate valves, and a swing check valve before reaching the water meter at the complex. The elevation at the service line is 1207 ft Use the Hazen–Williams method to determine each of the following values. Then determine if the water pressure is sufficient for the complex.
   a. Static head in the supply line
   b. Static pressure in the supply line

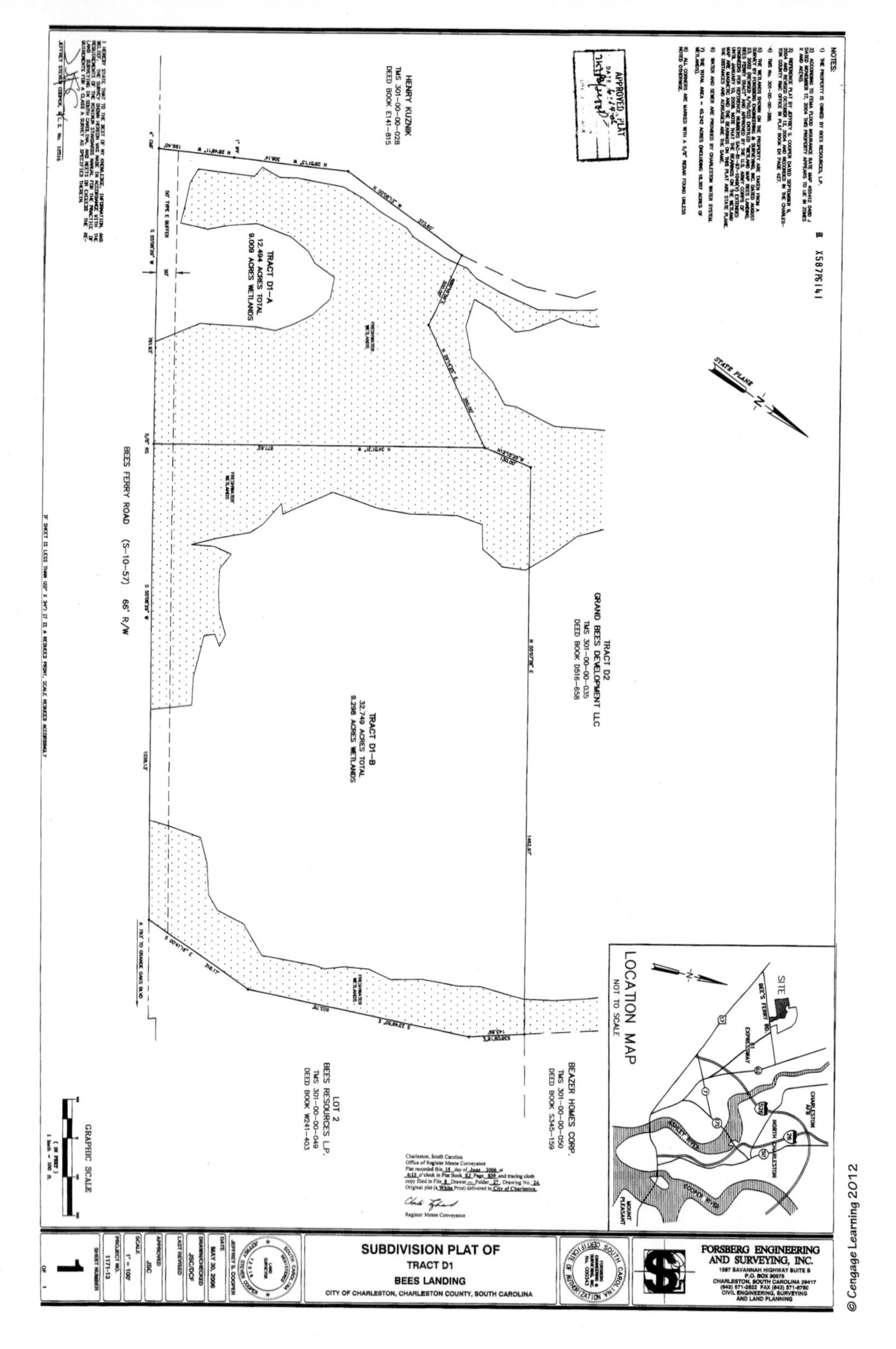

c. Head loss in the distribution main between the storage tank and the development
d. Dynamic head at the supply line
e. Dynamic pressure at the supply line

5. The existing 3 in sewer lateral for an existing commercial facility in northern California has never functioned adequately and will be replaced with a new sanitary sewer lateral. The lowest

floor elevation in the building is 356.1 ft and the grade surrounding the building is 355.2 ft. The building is 249 ft away from the 12 in sewer main which has a crown elevation of 346.9 ft at the point of connection. The construction contractor proposes to install the sewer lateral pipe such that it passes through the foundation wall closest to the sewer main at an invert elevation of 353.5 ft Answer the following questions.

a. What size sewer lateral would you recommend for the replacement? Why?
b. Is the proposed invert elevation for the sewer lateral acceptable in terms of frost depth and distance below the floor elevation? Explain.
c. Will the proposed elevation meet the minimum slope requirements for the sewer lateral? Show your work.
d. What is the minimum invert elevation at the foundation wall that will provide adequate lateral slope? Show your work.

EXTRA MILE

A 50-bed community hospital will be built on the property shown on the plat. Based on what you have learned in this chapter, create an efficient site plan for the project to include the building, adequate parking and circulation (including accessibility requirements), a screened dumpster with vehicle access, appropriate walkways, water and wastewater connections, and a stormwater management system (including a stormwater pond). Incorporate as many LID techniques as possible in the design. Write a project narrative that could be used as a Case Study in this textbook to illustrate the site design process.

PART IV
Planning for Occupancy

CHAPTER 8
Energy Conservation and Design

GPS DELUXE

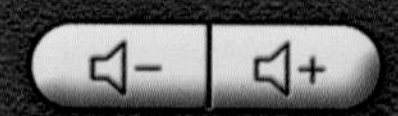

Menu

START LOCATION | DISTANCE | END LOCATION

Before You Begin

Think about these questions as you study the concepts in this chapter:

1. Where does the majority of energy that we use come from today?
2. When did building systems change?
3. What happened as a result of the energy crisis of 1972?
4. What is energy conservation?
5. How much energy do lighting systems in buildings consume in the United States?
6. What is the difference between passive and active solar designs?
7. How does solar geometry impact building design?
8. What is environmentally conscious design?

INTRODUCTION

Today we live in a global economy that is greatly dependent on energy from fossil fuels such as wood, coal, and petroleum. They not only move our cars and provide us with power for our buildings but also allow us to enjoy products from various parts of the world.

What are the biggest users of energy? It is estimated that the three biggest consumers of energy are transportation, buildings, and industry, with each consuming approximately one-third of the world's energy (Figures 8-1 through 8-3). In 2005 the United States consumed approximately 60 percent of the world's energy. While future estimates indicate that as other developing countries begin to grow their economy, the demand for the same resources will not only decrease the supply faster but also increase the costs to supply the energy needs of both the developed and developing countries. This is leading to certain instability in the global economy. How do we prevent this? The simple answer is

Figure 8-1: *Various modes of transportation consume about one third of the world's energy supply.*
http://www.istockphoto.com/stock-illustration-7942046-world-transport-globe.php

Figure 8-2: *Buildings are consumers of energy.*

http://www.istockphoto.com/stock-illustration-2372825-cityscape-detailed-urban-scene-with-cafes-and-shops.php

to conserve current energy resources and develop alternative resources that will provide for continued growth and prosperity.

The building construction industry has a major role in this partnership. In 2005 the U.S. Department of Energy estimated that 80 percent of all electricity was produced by coal-burning plants, Figure 8-4. While certain estimates indicate that we may have as much as a 100 years' supply of coal at its current state, other studies have shown that with the addition of new coal burning plants every day, we may only have as little as 25 years. It is one of many types of scenarios

Figure 8-3: *Industry accounts for about one-third of global energy consumption.*

http://www.istockphoto.com/stock-illustration-8795695-cartoon-industry-banners.php

that necessitates the development of new types of building systems and will ultimately lead to greater energy conservation and sustainability efforts.

http://www.istockphoto.com/stock-illustration-4472682-coal-burning.php

http://www.istockphoto.com/stock-illustration-7933526-urban-factory.php

Figure 8-4: *In 2005, the U.S. Department of Energy estimated that 80 percent of all electricity was produced by coal-burning plants.*

EARLY BUILDING DESIGN

Before the turn of the nineteenth century, building design and energy usage was usually left to the control of the architect or the individual homeowner (Figure 8-5). The building design centered upon the thermo comfort that was provided by a fireplace, (Figure 8-6 and 8-7) which also happened to provide lighting for the residents. Lighting was also provided by windows and oil lamps at night. At this time, cooking on a stove was also becoming central in providing not only food preparation needs but also warmth in the kitchen area. This was about to change with the dawn of the Industrial Revolution.

By the twentieth century, thermo comfort (Figure 8-8) and lighting (Figure 8-9) had been engineered through new types of building systems and technologies. These systems included mechanical, electrical, and lighting that provided new types of livable space. Such systems as air conditioning, the electric lightbulb, and radiant heating systems were engineered to provide a comfortable living environment with little regard to available resources (Figure 8-10 and 8-11).

The energy crisis of 1972 raised a new public consciousness (Figures 8-12, 8-13 and 8-14), and the industry began to realize the importance of providing a livable space and doing so with minimal impact on resources. According to a study, about

Figure 8-5: Thomas Jefferson's architectural masterpiece, Monticello, included many of fascinating innovations, but still relied on fireplaces for warmth and lighting.

© Image copyright n4 Photovideo, 2010. Used under license from Shutterstock.com

35 percent of all energy produced in the United States is consumed by buildings. In response, the building industry has looked at new measures in reducing energy consumption while providing a balance between building design using green building systems. Some of the noticeable changes in energy-conscious design include the use of energy efficient appliances, as provided by EPA's Energy Star Designation, (Figure 8-15 and 8-16) the use of double glazed windows to provide daylighting, and greater resistance to thermal transmission to minimize air conditioning and lighting needs.

Additionally, some buildings are designed to consider passive thermal and cooling systems with the use of shading devices as well as direct and indirect solar gains (Figure 8-17 and 8-18). These passive designs also incorporate landscaping around the surrounding building as a means to reduce thermal losses. The importance of energy conscious design is also discussed in the following chapters.

Figure 8-6: Log cabins relied on stoves for cooking and providing warmth.

© iStockphoto.com/Jello5700

Figure 8-7: The log cabin fireplace used wood to produce energy.

© iStockphoto.com/Todd Bates.

Figure 8-8: Willis Carrier, an early pioneer in air conditioning technology, with the first chiller.

Courtesy of Carrier Corporation.

Figure 8-9: Thomas Edison is credited with inventing the incandescent bulb.

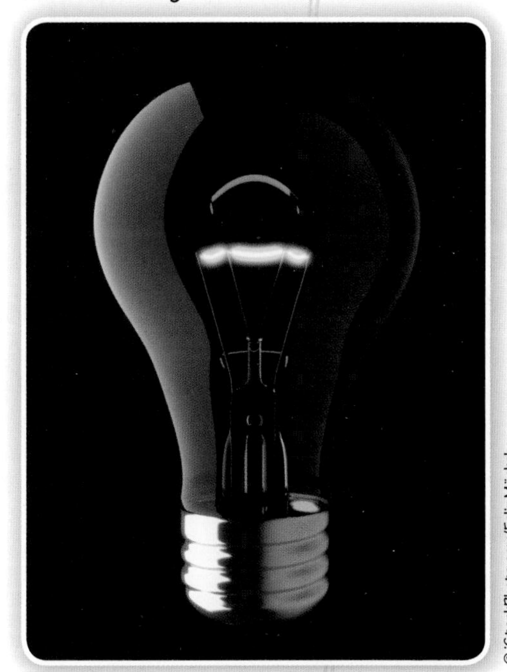

© iStockPhoto.com/Felix Möckel.

BUILDING DESIGN AND ENERGY

Building design before the nineteenth century was regionally based and depended on local trades and available building materials in the general area. **Architectural style homes** were found in various parts of the country such as Cape Code homes in the Northeast (Figure 8-19), Spanish Revival in parts of the South and the West Coast (Figure 8-20), adobe in the desert regions of the Southwest (Figure 8-21), and bungalows along the West Coast (Figure 8-22). The designs of these buildings appeal to our creativity and respond appropriately to our environment. These environmentally conscious designs incorporated passive solar techniques through solar geometry, land topography, and natural resources. In the case of homes built in

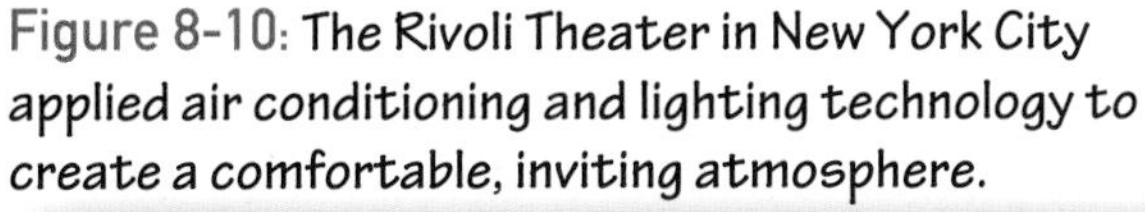

Figure 8-10: The Rivoli Theater in New York City applied air conditioning and lighting technology to create a comfortable, inviting atmosphere.

Courtesy of Carrier Corporation.

Figure 8-11: This 1950s office environment was designed for ample lighting with little regard for consumption of natural resources.

Getty Images, Hulton Archives.

Figure 8-12: The energy crisis of 1972 raised public consciousness about energy consumption. Industry responded by finding ways to design livable spaces with minimal impact on resources.

© iStockphoto.com/Lilli Day.

Figure 8-13: Shortages in supply increased the price of traditional energy resources, including oil.

© iStockphoto.com/Alex Slobodkin.

Between 1908 and 1940 you could order a variety of prefabricated homes from Sears Roebuck and Company. Those same homes can be found throughout the country and with several different options. Learn more about the Sears homes by going to *http://www.searsarchives.com/homes/1908-1914.htm*. Answer the following questions:

1. What types of Sears homes are there?
2. How much did they cost then and now?
3. Where are these homes in the United States?

the North, roofs were designed steep to shed the snow. Overhangs were short to provide for direct sunlight, and windows were small to prevent minimal heat loss. In the case of the Southeast region of the country, roofs on homes were low pitch with large overhangs to provide protection from direct sunlight. Windows were large to allow cross ventilation during hot summers.

During the twentieth century, just as mass production took hold within the auto industry, a similar movement was beginning to take shape in the home building industry. The dawn of the Industrial Revolution introduced new types of technology that would permit the mass production of homes and slowly diminish the importance of regionally based architecture. Production homes were becoming the norm, and the building styles were becoming more uniform (Figure 8-23). The inherent benefits of mass-produced structures included simplicity of construction, lower cost of building and purchase, efficient design, and quality assurance (Figure 8-24).

Figure 8-14: Shortages in supply increased the price of electricity.

© iStockphoto.com/Bill Grove.

Figure 8-15: Energy Star.

Source: US Environmental Protection.

Figure 8-16: Energy-efficient appliances.

Figure 8-17: The designers of the Fort Worth modern museum in Texas used ample overhangs shading to create energy-conserving building shading.

Figure 8-18: Double-glazed windows can produce energy-saving solar gain.

Figure 8-19: The Cape Cod style home is well suited to the northeast U.S. environment, with a steep-pitched roof to shed snow and small windows to prevent heat loss.

Figure 8-20: This Spanish-revival style home has large windows to admit cooling cross breezes in summer.

Figure 8-21: Adobe-style homes are common in the Southwest region of the United States.

Figure 8-22: The bungalow-style home's large roof overhangs help keep the structures cool in warmer environments.

Figure 8-23: Production homes provided many benefits: simplicity of construction; lower cost of building and purchase; efficient design; and quality assurance.

Figure 8-24: An engineered home under construction.

Figure 8-25: Articulated plan showing light well configurations.

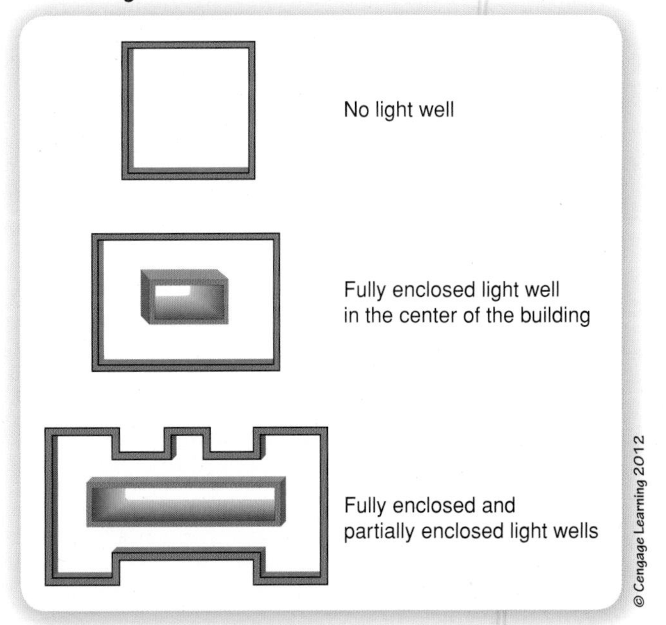

Similarly, larger structures were being designed and constructed with the advances in new materials and building components. Some of the new structures included the early years of the skyscrapers. These buildings while reaching multiple stories high, were limited by their length and width due to the need to provide natural ventilation and daylighting. Modern air conditioning for large office buildings was not available until the end of the 1920s, and electric lighting was typically limited to incandescent lighting systems that offered minimum illumination. To solve these challenges, architects placed large light wells in the building, giving every office external light and air (see Figure 8-25). **Light wells** were enclosed in the interior of the building or adjacent to the exterior of the building (see Figure 8-26). Architects were acutely aware of the proportions of a room's height to its width were considered when natural lighting was needed to illuminate a room as indicated in the figure below. This design also had the advantage of providing additional external corner offices, offices that could be rented at higher rates. With the dawn of air conditioning and artificial lighting, buildings became less dependent on their environments and energy seemed cheap and unlimited (see Figure 8-27).

Light Wells: openings or shafts that allowed for light and natural ventilation in buildings large or small. Early twentieth-century skyscrapers often employed light wells prior to the days of air conditioning and lighting. Today, enclosed atriums employ light wells to provide daylighting for interior court spaces as found in some shopping malls.

After the Second World War, the country experienced a building boom and growth in all economic sectors. With the country's growth came new and expanded industries. The economic boom saw also the greatest increase in demand for energy. In 1971 and 1972, 69 percent of the total increase in demand for energy was supplied by oil. New supplies of crude oil were coming from the Eastern Hemisphere in an effort to keep up with demand. The United States became increasingly vulnerable to foreign supplies of oil. Then the energy crisis of 1972 struck as an oil embargo was declared, causing a reduction in oil supplies. Shortages were quickly felt at the pump, transportation was stifled, the cost of goods and services began to rise, and the country's economy was coming to a standstill. The embargo lasted until the beginning of January in 1974 when countries in the Eastern Hemisphere agreed to supply the United States with crude oil once again.

Figure 8-26: A building well used for light and natural ventilation.

Figure 8-27: Sears Tower without any light wells.

Learn more about the Energy Crisis of 1972 by searching the Internet. Find out what were some of the conditions that led the United States to this crisis. What were the challenges? How did our government respond during and after the crisis and how does it impact us today?

In the years that followed, the government began to reform multiple sectors of the country including the building industry. To help reduce consumption, the federal government initiated the following measures: in 1975, a national maximum speed limit of 55 mph was imposed along with the creation of the Strategic Petroleum Reserves through the Emergency Energy Conservation Act; in 1978, the Department of Energy was created, followed by the National Energy Act, and that same year saw the Public Utility Regulatory Policy Act was passed to promote the use of renewable energy systems. Then in 1992 Congress passed the Energy Policy Act (EPACT) to address energy efficiency, energy conservation, and energy management, reforming the Public Utility Holding Company Act and amended parts of the Federal Power Act of 1935.

ENERGY CONSERVATION

As a result of the energy crisis, changes in the building industry, with support from government and the private sector, became increasingly more evident in the establishment of energy guidelines. The result was the adoption of the Energy Policy Act (EPACT), which mandated that buildings be designed to conserve energy through regulatory guidelines.

As a professional, you will need to create and implement project plans considering available resources and requirements of a project/problem to accomplish realistic planning in design and construction situations.

In turn, EPACT also provided a direct impact on active systems such as improving energy efficient appliances and the development of new types of lighting systems such as compact fluorescents (Figure 8-28) and light-emitting diodes (Figure 8-29),

Figure 8-28: Compact fluorescent bulbs use a fraction of the energy their incandescent counterparts need.

© iStockphoto.com/craftvision.

Figure 8-29: Light-emitting diodes save energy.

© iStockphoto.com/Brian Hitch

which use a fraction of the power of traditional incandescent bulbs with the same amount of lumen output. EPACT also accelerated advances in lighting controls. Dimmers are one such device (Figure 8-30), allowing us to set lighting levels manually for various activities. Occupancy sensors turn the lights on or off when a room is not in use, Figure 8-31. A photometric sensor (Figure 8-32) may be used to balance the lighting system with the natural light to ensure an even distribution of light level throughout the room. Indirectly, EPACT also increased awareness of pre-Industrial Revolution designs utilizing low-tech systems such as natural ventilation, (Figure 8-33) daylighting, passive cooling, Figure 8-34 and 8-35 and heating systems.

Electrical and Lighting

In today's building design, there is a complex system of electrical power throughout the building. In the case of a residential design, some of the most common electrical loads include appliances such as stoves, refrigerators, dishwashers, washer and dryers, hot water heaters, and electronics such as computers, televisions, stereos, clocks, and printers.

In the case of commercial buildings, lighting, heating, ventilating, and air conditioning (HVAC) make up about 70 percent of the energy consumed. Computers, electronics, and other miscellaneous equipment make up the remainder. That is why there is a focus on lighting system, appliances, and other electronics.

Figure 8-30: Dimmer switches can allow consumers to control their energy consumption.

© iStockphoto.com/Samuel Kessler.

Energy Conservation through Lighting Design

Over the last few decades, significant energy savings have been achieved in lighting systems through the development of energy-efficient sources. Lighting controls, however, is one area where unclaimed energy savings can be improved. If you walk around most office buildings, you will likely see that a fraction of the offices are empty yet lights are still on. If the lighting in or above these unoccupied work areas is either dimmed or turned off, then the lighting power is reduced, resulting in energy savings over the course a day, month, and year.

In a high school or university environment, where classrooms are used throughout a day, but the occupants change almost hourly, classrooms are often vacated without the instructor or the last person out

Figure 8-31: This ceiling-mounted occupancy sensor can help control energy use in buildings.

©2010 Lutron Electronics, Inc.

the door turning off the lights. If no class occupies the space immediately thereafter, the lights likely will remain on, resulting in significant amount of wasted energy. Similarly, in spaces where people are working, energy savings can be achieved if the lights are switched or dimmed to satisfy occupant preference or due to the availability of daylight.

The amount of energy consumed in commercial office buildings for lighting generally ranges from 20 to 50 percent of the total building electrical load. The actual percentage consumed depends not only on the magnitude of the lighting load, but also on the volume of the non-lighting loads in a building (such as the load for computers, refrigeration, and other mechanical equipment). In an effort to reduce the amount of lighting energy consumed in buildings, most states have adopted an energy code that limits the amount of lighting power that you may install in a building. The Illuminating Engineering Society of North America (IESNA) and the American Society of Heating, Refrigeration, and Air-Conditioning Engineers (ASHRAE) have developed a joint standard for new commercial buildings that includes lighting power limits.

ASHRAE/IESNA Standard 90.1 was released and addressed lighting power limits. It offers the advantage of being written in a code format, and it is much easier for designers to apply, and for regulating agencies to enforce, because of its simplicity.

Comply with regulations and applicable codes to establish a legal and safe workplace/jobsite. As professionals, engineers regard safety and codes with the utmost care in order to protect the public in all aspects of the project.

Figure 8-32: A photometric Sensor.

Courtesy of Delta OHM S.R.L.

Figure 8-33: Natural ventilation is an important aspect of energy-efficient design.

© Cengage Learning 2012

Electric Utility Costs

Electric utilities bill large consumers of electric power (building owners) using two simultaneous measures of consumption. Typically, a utility bill first applies an **energy charge** to its customers.

A building owner is charged for each **kilowatt-hour** of energy consumed. A kilowatt-hour is 1000 W-h, which is the amount of energy required to burn a 100-W lamp for 10 h. In many cases, for commercial customers, this energy charge is time-of-day dependent. Under such a plan, higher rates are charged during daytime hours, while lower rates are charged during nighttime hours. These rates may also be seasonal, with higher amounts being charged during the peak season for energy consumption than at other times of the year. Some utilities are now even charging time-of-day pricing, where the price of energy varies with the hour of the day. Higher prices are again charged during the peak hours of the day to encourage savings at these times.

In addition to the basic energy charge, most large consumers of electricity are also billed through a **demand charge**. In this case, a demand meter samples the average power (in watts, or more correctly in kW) consumed within a building over consecutive 15- or 20-min intervals (see Figure 8-36). The maximum value measured in one of these intervals during each billing period is then used to determine the demand charge for that month (see Figure 8-37). This peak power consumption is important because it more closely affects the capacity that an electric utility must be able to provide. You may have heard that some areas of the country are experiencing problems where the electrical grid approaches capacity on peak days, often requiring large users to curtail operations. Some utilities are forced to employ rolling blackouts to reduce peak demand. **Rolling blackouts** are where power is turned off to certain sections of a utility's service area for short periods of time.

Figure 8-34: This shading technique helps this building save on cooling costs.

Energy Charge:

An **energy charge** is the amount on a customer's billing which reflects the actual energy they use over the billing period.

Demand Charge:

A **demand charge** covers the costs associated with maintaining sufficient electrical facilities at all times to meet each customer's highest demand for energy. It is based on the greatest amount of electricity used by the customer in any half-hour period during the billing period.

The demand charge is expressed as a dollar per kilowatt (kW) rate and is applied to the customer's maximum kW demand, or the highest rate at which the customer required energy during the month.

Figure 8-35: An overhang can provide shading.

Figure 8-36: You can check the electric meter at your home to track the kilowatt hours your family uses.

Figure 8-37: Kilowatt hours translate into your family's utility bill.

It may surprise you to learn that the demand charge can exceed a utility's monthly energy charge for some buildings that experience high peak electrical demand. In addition, some utilities apply a **ratchet clause**, which means that the demand charge billed in subsequent months cannot be less than a specific percentage of the maximum demand charge over the previous 12 months (for example, the minimum demand charge may not be less than 50 percent of the peak demand over the previous 12 months). Because of the costs associated with high peak building demand, some modern building management systems are designed to monitor a building's electrical demand and attempt to limit the power consumed at peak load conditions. This type of load management will become more important as time-of-day billing rates become more commonplace.

Ratchet Clause: refers to a provision whereby a customer is never charged less than some percent of the maximum demand established during a previous time period—frequently the past year or the past summer months. This can be a severe and controlling penalty if, for example, a water utility uses an extra pump during a high water demand, but never or seldom uses it again. This ratchet demand can control the entire year's charges.

The savings associated with the above energy and demand charges can often pay for energy efficient design solutions (lamps, ballasts, luminaires, and controls). These energy-saving measures are likely to cost more initially, but save building owner's money over the long run. In particular, if the operating hours are long, such as in corridors and lobbies, the payback period for energy saving technologies is likely to be much shorter. To determine the approximate payback period, one needs to know the energy and demand rates that are in effect for a building.

Use appropriate units to determine accurate measures, spaces, and building systems.

Example: Determine cost savings for converting your home's lighting system from incandescent to fluorescent over a course of 1 year given the following:

20 incandescent bulbs at 100 W/h
20 compact fluorescent bulbs (equivalent of 100 W incandescent) at 23 W/h

Assume lights are on 8 h a day 7 days a week.
Determine the W/h of all lights used in the house.

Incandescent system:

20 bulbs × 100 W/h/bulb × 8 h/day × 7 days/week × 52 weeks/year
= 5,824,000 W or 5824 kW

Convert watts to kilowatts you need to know 1kW = 1000 W

Compact fluorescent system:

20 bulbs × 23 W/h/bulb × 8 h/day × 7 days/week × 52 weeks/year
= 1,339,529 W or 1340 kW

If we say the utilities cost is approximately $0.09/kW

Then the costs of the systems are as follows:

Incandescent system:

5824 kW × $0.09/kW = $524.16

Compact fluorescent system:

1340 kW × $0.09/kW = $120.60

A savings of $524.16 − $120.60 = $403.56 over the course of a year

ASHRAE/IESNA 90.1 1999 LIGHTING REQUIREMENTS

This new lighting energy standard is relatively short and simple, as you will see. It has a few key features that the design of building lighting systems must address. Some of these features are application specific, while others pertain to general lighting and control issues. The following is a summary of these requirements.

Application Areas

First, it is important to consider the types of spaces to which the standard does not pertain, including the following:

- Emergency lighting that operates only in the case of an emergency
- Lighting within living units
- Lighting that is designed to fulfill health or safety requirements
- Gas lighting systems

The standard applies to most other general architectural lighting and installed task lighting. The fact that it addresses task lighting is important. Most designers will not consider the power that is supplied to task lights in the calculations of a room power density (watts/sq ft).

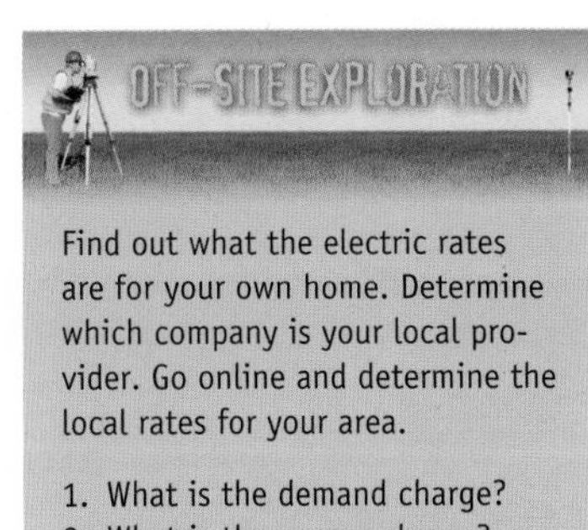

Find out what the electric rates are for your own home. Determine which company is your local provider. Go online and determine the local rates for your area.

1. What is the demand charge?
2. What is the energy charge?
3. Is there a service charge, and if yes, what is it?
4. Do you receive a reduced rate if you consume more power?

Automatic Lighting Shutoff

Interior lighting in buildings larger than 5000 sq ft must have an automatic control device in all spaces. This control device can be a time-of-day device, but the area controlled must be no more than 25,000 sq ft and must be on a single floor. One can fulfill this requirement by using an occupancy sensor that turns off the lights within 30 minutes of the space being vacated. Occupant intervention is, of course, permitted. Lighting that is intended for 24-h operation, such as in a building lobby or 24-h retail environment, does not require an automatic control device. In most large buildings, the automatic lighting control requirement can be served by placing control of the lighting system on a central building management system (BMS). Such a system would likely also control the HVAC equipment in the building.

Space Lighting Control

Each space that is enclosed by ceiling-height partitions must contain at least one control device for the general lighting in that space. This device can be either a manual or an automatic device (such as an occupancy sensor). The maximum floor area that may be controlled by a single device is 2500 sq ft in spaces less than 10,000 sq ft and 10,000 sq ft in enclosed areas larger than 10,000 sq ft In remote areas where remote control is required for safety or security, the control device must include an indicator light to signal when the system is in operation and a label that lists the area that the device controls.

Separate Control

The following applications are special situations where a separate control device is required for the lighting hardware.

- Display/accent lighting
- Lighting in display case
- Task lighting, which must have a wall-mounted switch that is accessible from a point within view of the controlled luminaire(s)
- Lighting for non-visual applications, such as plant growth and food-warming
- Lighting that is for sale and on display in a retail environment or is for demonstration purposes as in an education setting
- Hotel and motel guestrooms, which must have a master control device at the main entrance to the space that controls all fixed luminaries and switched receptacles

Determining Building Power Limits

The ASHRAE/IESNA standard generally limits the total lighting power that may be used within a building in the design of a lighting system. This building lighting power limit can be determined in one of two ways. You can consider the entire square footage of a building and a single average power density (watts/sq ft) based on the building type as specified in the standard. We will refer to this method as the **building area method.** In this case, the power density (watts/sq ft) for the building system is multiplied by the total building square footage to determine the total lighting power that you can design for a building to use.

Table 8-1 shows the lighting power density for a variety of common building types, which are applied to the total building square footage using the total building approach.

The second option, the **space-by-space method,** is a more time-consuming way to arrive at the interior lighting power allowance. Here, each space is considered individually. Multiply the power density listed in the standard for each space by its square footage, and then sum these values to determine the total interior lighting power allowance.

In both methods, the square footage to apply is the gross square footage, taken to the exterior of the building in the case of the building area method, or to the center of a wall in the space-by-space method. Once you obtain the building power allowance, you are permitted to use it anywhere within the building, as long as you do not exceed the total across all interior lighting systems.

Table 8-2 contains representative power densities from the standard for a number of common spaces found in most building types. These values are for use in the space-by-space method of determining a building's interior power allowance.

The standard also provides details on how the *installed lighting power* is to be determined for a designed lighting system. In the case of luminaires with screw base sockets and no ballast (incandescent and tungsten halogen luminaires), the wattage considered must be the maximum labeled wattage of the luminaire. If you are using a 75 W lamp in a luminaire that can accept a 150 W lamp, you must use 150 W as the connected wattage unless the luminaire will specifically be labeled with maximum lamp wattage of 75 W. Luminaires with transformers, such as low-voltage track lighting, must be based on the maximum input wattage to the transformer. This is different from line-voltage track lighting, where the wattage is a minimum of 30 W per linear foot of track, or the actual of the luminaries attached to the track, whichever is greater.

Table 8-1: Power Densities for Different Building Types for Use in the Building Area Method (from ASHRAE/IESNA Standard 90.1, 1999)

Building Type	Power Density for Entire Building (watts/sq ft)
Dining bar lounge/leisure	1.5
Dining: cafeteria/fast food	1.8
Dining: family	1.9
Hospital/health care	1.6
Hotel	1.5
Library	1.7
Manufacturing	2.2
Museum	1.6
Office	1.3
Performing arts theater	1.5
Religious	2.2
Retail	1.9
School/university	1.5
Warehouse	1.2

Additional Lighting Power

Under certain conditions, you are able to exceed the lighting power specified above, but only if this lighting power is used for a specific purpose in that space. The situations under which this added power can be applied are addressed in the list below.

1. For decorative lighting, such as sconces, chandeliers, and the lighting of art, 1.0 W/sq ft may be added.
2. For the lighting of spaces with visual display terminals (VDT), provided the lighting system meets the specific VDT lighting requirements listed by the IESNA, 0.35 W/sq ft may be added.
3. For retail lighting used to highlight merchandise, you may apply an additional 1.6 W/sq ft to light general merchandise or 3.9 W/sq ft to light fine merchandise (jewelry, fine clothing and accessories, china and silver) and art galleries.

These special power allowances can only be used for the designated purpose and cannot be added to the overall building power allowance. It can only be applied for the designated purpose in that space.

Exterior Lighting Control

Exterior lighting must be controlled by a photo sensor or astronomical time switch, except for areas such as covered vehicle entrances or building exits, or where lighting

Table 8-2: Power Densities for Different Space Types for Use in the Space-by-Space Method (from ASHRAE/IESNA Standard 90.1, 199 9)

Space Type	Power Density for Space (watts/sq ft)
Office enclosed	1.5
Office open planned	1.3
Classroom	1.6
Conference meeting room	1.5
Lobby	1.8
Dining area	1.4
Stairs-active	0.9
Corridor	0.7

Find out more about ASHRAE Standard 90.1 1999/2001 by visiting the following website *http://www.cooperlighting.com/content/source/energy_legislation/ashrae.cfm*. Answer the following questions:

1. Two of the three types lighting power calculations have been described. What is the third type?
2. Name at least three key changes made to ASHRAE 90.1 Standard from the 1989 version to the 1999/2001.
3. Which states have adopted energy codes prior to ASHRAE 90.1 Standard 1999?

is required for safety and security, or where the purpose of the lighting is to allow people to adapt to large changes in lighting level between the daylight exterior and an interior or covered space.

Exterior Lighting Efficacy Requirements

In exterior environments, the minimum source efficacy for luminaires operating at greater than 100 W is 60 lm/W. This effectively rules out the use of incandescent or mercury sources. No minimum value is applied if the luminaire is controlled by a motion sensor, is used for security/safety, is integral to signage, or is part of a historic structure.

Exterior Building Lighting Power

The exterior power allowance is the sum of the power allowance provided at all entrances and exits as described below. As in interior spaces, the total exterior power allowance may then be divided among the various areas in any proportion, provided the total allowance is not exceeded. Façade lighting must be kept separate from entrance and exit lighting (it is not included in the exterior building power allowance and cannot be shared with other applications). The allowable power density you can apply in each of these applications is provided in Table 8-3 below.

Table 8-3: **Allowable Power for Exterior Applications**

Exterior Space	Power Allowed
Building entrance with canopy	3 W/sq ft^2 of canopy area
Building entrance without canopy	33 W/lin ft of door width
Building exit	20 W/lin ft of door width
Building facades	0.25 W/sq ft of illuminated façade area

Signage and transportation signal as well as directional and marker lighting are exempt if the luminaires have an independent control device.

PASSIVE SOLAR DESIGN

Passive solar design of heating, cooling, lighting, and ventilation systems relies on sunlight, wind, vegetation, and other naturally available resources on the site (Figure 8-38). Passive solar design can provide greatest impact and cost-effectiveness when applied early in the design and development phase of a building's process or product.

A successful passive solar building needs to be very well insulated in order to make best use of the sun's energy. The result is a quiet and comfortable space, free of drafts and cold spots. Today, many buildings are designed to take advantage of this natural resource through the use of passive solar heating and daylighting.

The advantages of passive solar design:

<u>H</u> igh energy performance: lower energy bills all year round.
<u>I</u> nvestment: independent from future rises in fuel costs, continues to save money long after initial cost recovery.

V alue: high owner satisfaction, high resale value
A ttractive living environment: large windows and views, sunny interiors, open floor plans
L ow Maintenance: durable, reduced operation and repair
U nwavering comfort: quiet (no operating noise), warmer in winter, cooler in summer (even during a power failure)
E nvironmentally friendly: clean, renewable energy doesn't contribute to global warming, acid rain, or air pollution.

Figure 8-38: The five elements of passive solar design.

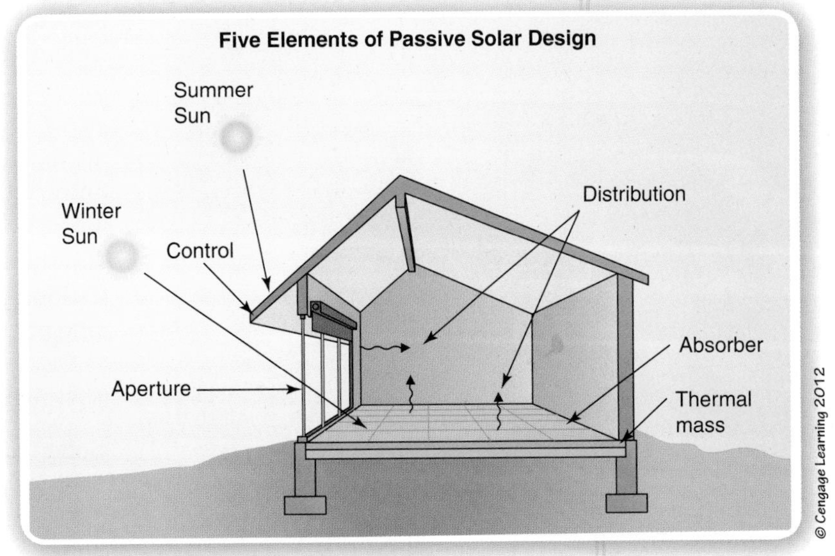

The basic natural processes that are used in passive solar energy are the thermal energy flows associated with radiation, conduction, and natural convection. When sunlight strikes a building, the building materials can reflect, transmit, or absorb the solar radiation. In the winter, the sun is low in the sky that allows the heat to penetrate into windows on the south face of a structure.

During hot summer months, passive solar design can be achieved by making use of convective air currents, which are created by the natural tendency of hot air to rise. In the summer, south-facing windows can be shaded by an overhanging roof or awning to keep out the high hot summer sun. Because much of a building's heat is lost through its windows, the majority of windows in a passive solar building are located on the south wall. Depending on the climate and the design, as much as 100 percent of a building's heating needs can be provided by the sun. Given a balanced design, in a climate such as Albuquerque, 80 percent of a building's heating needs can be done with solar power. Even if 50 percent or 30 percent power is sun-generated, heating bills will reduce significantly. Additionally, the heat produced by the sun causes air movement that can be predictable in designed spaces. Thus summer comfort can often be achieved without the need for air conditioning by employing shading and natural cooling techniques.

ACTIVE SOLAR DESIGN

Similar to passive solar design, active solar design systems rely on sunlight to provide energy to a building. Unlike passive solar design, active solar design uses growing technological advances to convert energy to usable power. The more common active solar systems are photovoltaics, solar collectors, and solar thermal systems.

Photovoltaics

Photovoltaics are made up of many solar cells. Solar cells are made of positive and negative semiconductors. Sunlight strikes a solar cell (Figure 8-39) and the "p" and "n" semiconductors act as a filter separating the electrons from the sunlight. The electrons are then carried across and become part of an electric current also known as the "photovoltaic (PV) effect." A PV cell typically produces 1 to 2 W, hardly enough power for any applications. A collection of connected

Figure 8-39: Solar cell.

© iStockphoto.com/Lena Andersson.

cells however forms a module (Figure 8-40). When modules are connected together they form an array (Figure 8-41). In this way one can built a PV system to meet almost any power need. Alone a module or array is not a complete system. Additional components include the structure to support and point them toward the direction of the sun and components that take the direct current (DC) from the PV and conditions the electricity to be used for specific applications. The system is often referred to as building integrated photovoltaic (BIPV).

Solar Collectors

Solar collectors are the most common of all solar energy systems. They are designed to capture the sun's light energy and convert it to heat energy. The collectors are typically used to provide heated water for use in both residential and commercial applications. Several types of solar collectors exist, including flat plate collectors, evacuated tube collectors, concentrating collectors, and transpired-air collectors.

Flat plate collectors (Figure 8-42 and 8-43) typically use an insulated metal box with a glass or plate cover. Encased in the box is a series of tubes or dark absorber plate. As the sun heats the absorber plate the heat is transferred to air or a liquid passing through the collector. The heat is then transferred for use as part of the domestic hot water supply or heating the home itself during winter.

Evacuated tubes (Figure 8-44 and 8-45) are similar in purpose as a flat plate collector but they convert the sun's energy more efficiently. Sunlight enters through the outer glass tube and strikes the absorber, where energy is converted to heat. The heat is transferred top the liquid or gas and then transferred to the manifold at the top of the tube. The manifold transfers the heat to the passing liquid to be used in residential or commercial applications.

Figure 8-40: Solar Solar module.

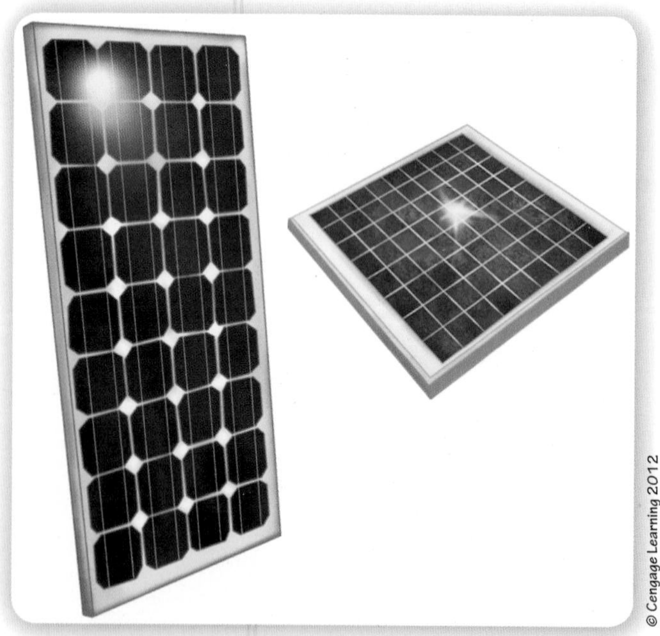

© Cengage Learning 2012

Figure 8-41: Solar array.

© iStockphoto.com/Franck Boston.

Figure 8-42: Flat plate collector.

© iStockphoto.com/Pavlo Vakhrushev.

Evacuated tubes (Figure 8-46 and 8-47) are more efficient then flat plate collectors for a couple of reasons. First, the chambers inside the tubes are vacuum-sealed during production. This reduces the amount of heat loss. Second, because the tubes are circular, the sunlight is perpendicular to the sun for most of the day, unlike a flat plate collector where the sunlight is perpendicular to the collector at noon. These characteristics—the vacuum minimizes heat losses to the outdoors and the extended time of optimal performance—make these collectors particularly useful in cold and cloudy environments.

Figure 8-43: Flat plate collector diagram.

Reprinted from the U.S. Department of Energy fact sheet (DOE/GO-10096-051) titled "Residential Solar Heating Collectors" (March 1996) by NREL, http://www.nrel.gov/docs/legosti/fy96/17460.pdf. Accessed February 25, 2010.

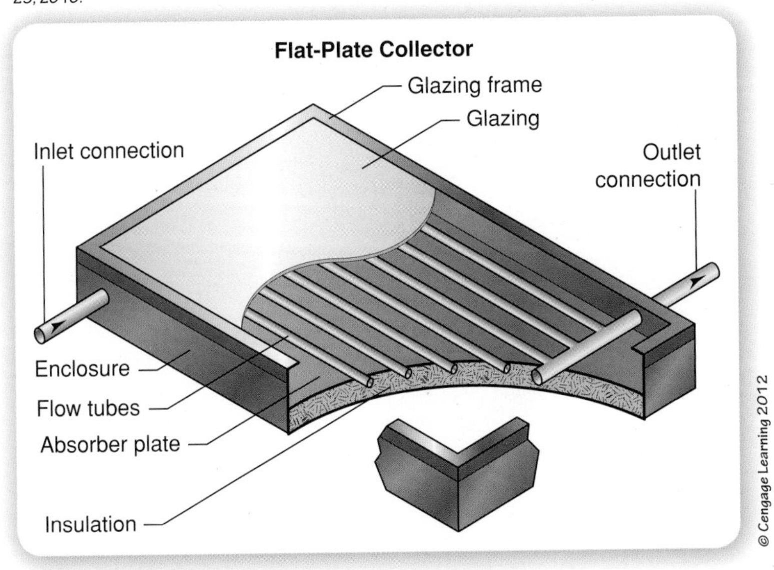

© Cengage Learning 2012

Figure 8-44: Evacuated tubes.

© iStockphoto.com/Ben Whittle.

Figure 8-45: Evacuated tubes diagram.

Reprinted from the U.S. Department of Energy fact sheet (DOE/GO-10096-051) titled "Residential Solar Heating Collectors" (March 1996) by NREL, http://www.nrel.gov/docs/legosti/fy96/17460.pdf. Accessed February 25, 2010.

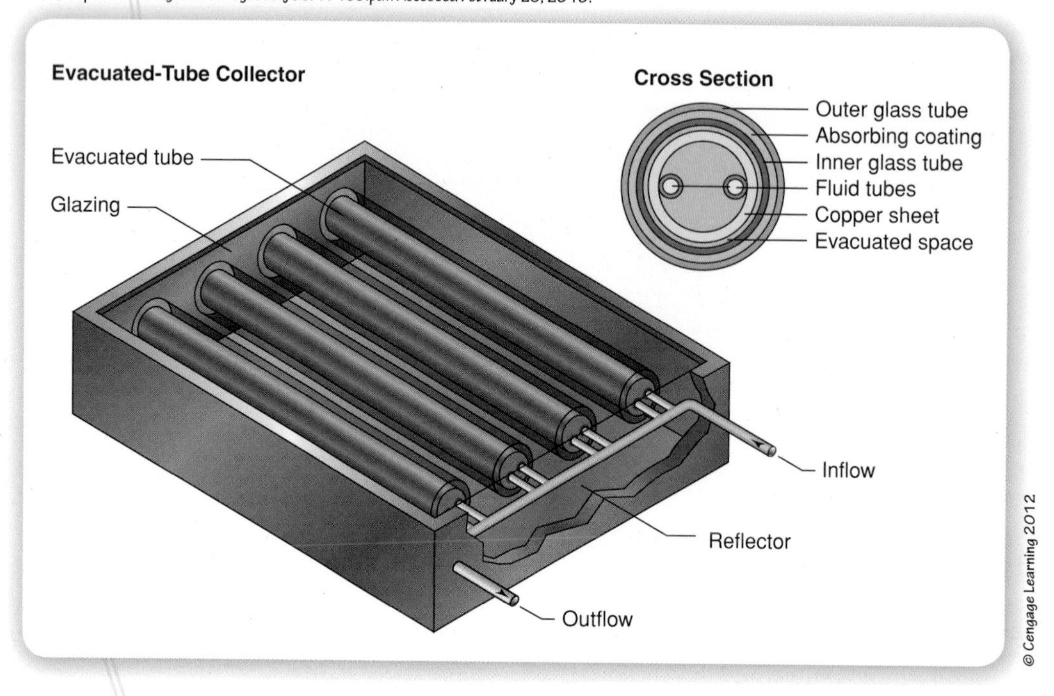

Figure 8-46: Solar concentrator.

Figure 8-47: Solar concentrator diagram.

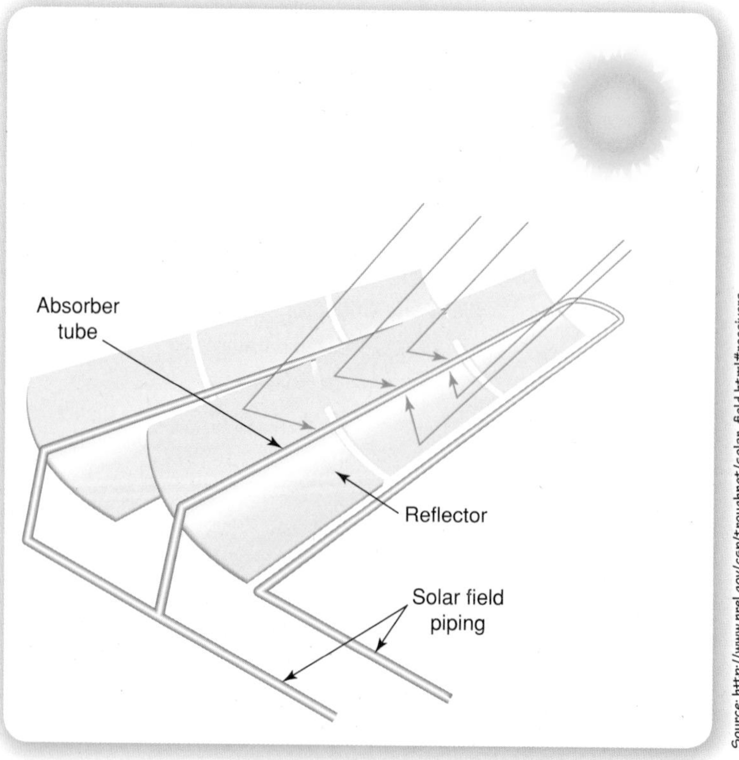

Source: http://www.nrel.gov/csp/troughnet/solar_field.html#receivers.

Concentrating collectors use a parabolic mirror or dish to focus the sun's energy on the receiver to a focus area known as the absorber. Because the magnification effect of concentrating collectors is approximately 60 times greater than flat plat collectors, concentrating collectors are typically used in commercial or industrial applications. Unlike the evacuated tubes the concentrating collectors only work in direct sunlight and require single or dual axis trackers to increase service time. Because concentrators require direct sunlight and perform poorly on hazy or cloudy days, they are typically found in locations near the equator in dry arid climates such as the U.S. Southwest.

Figure 8-48: Transpired-air collectors at a Federal Express facility in Littleton, Colorado.

Courtesy of DOE/NREL; credit—Warren Gretz.

Transpired-air collectors (Figure 8-48 and 8-49) are constructed using a thin metal plate perforated with holes. The sun heats the metal and a fan is used to force air over the heated metal. At about 70 percent efficiency, the transpired-air collectors have been used in pre-heating ventilation systems and crop drying. Because it requires no glazing and insulation, the transpired-air collector is considered one of the most economical systems of its kind.

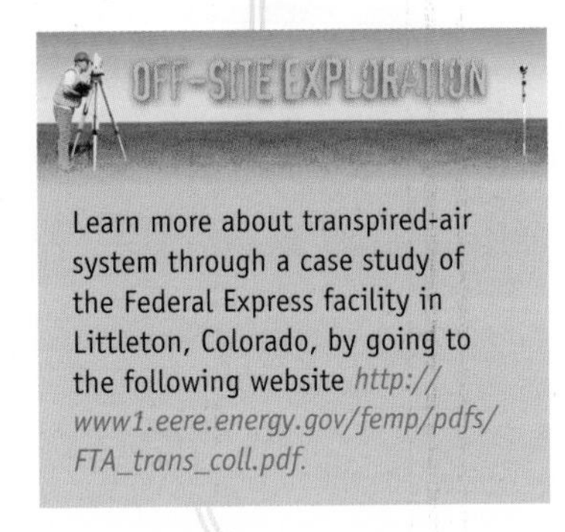

Learn more about transpired-air system through a case study of the Federal Express facility in Littleton, Colorado, by going to the following website *http://www1.eere.energy.gov/femp/pdfs/FTA_trans_coll.pdf.*

To understand how solar energy is optimized for a particular site one needs to understand the basics in solar geometry (see Figure 8-50). The sun emits solar radiation to the earth's surface at a relative constant amount also known as the **solar constant.** The amount of radiation reaching the surface of the earth varies with sun angles and composition. This is also impacted by earth's elliptical orbit around the sun and the tilt of the earth's axis at 23.5 degrees. Due to its tilt, the earth has seasonal changes—fall, winter, spring, and summer. In the northern hemisphere, hot and cold season typically start on June 21 for summer and December 21 for winter, also known as our summer solstice and winter solstice, respectively. Spring and fall typically start March 21 and September 21 and are commonly referred to as the vernal and autumnal equinox. On June 21, the summer solstice indicates that the sun's rays will be perpendicular to the earth's surface at latitude of 23.5 degrees north of the equator or along the Tropic of Cancer. The same can be said 6 months later: on December 21 the sun is at a 23.5 degrees south of the equator or along the Tropic of Capricorn in the southern hemisphere. How does this affect the solar gain on a building then? The angle at which the sun strikes the earth surface is a function of the geographic location, the time of day and the day of year. Given a location and the three previous mentioned factors, the solar altitude angle can be determined. For example, if you were located on the equator at fall or spring equinox, the solar altitude angle would be considered at 90 degrees above you at solar noon.

As you can see in Figure 8-51, the effect of the sun's energy is strongest when the sunlight is perpendicular to the earth; however, it diminishes at lower angles due to the amount of atmospheric layers being passed through. This is particularly noticeable at sunrise and sunset as we are able to view the sun directly without any harmful effects to our eyes.

Figure 8-49: Transpired-air collector diagram.

Reprinted from the U.S. Department of Energy fact sheet (DOE/GO-10096-051) titled "Residential Solar Heating Collectors" (March 1996) by NREL, http://www.nrel.gov/docs/legosti/fy96/17460.pdf. Accessed February 25, 2010.

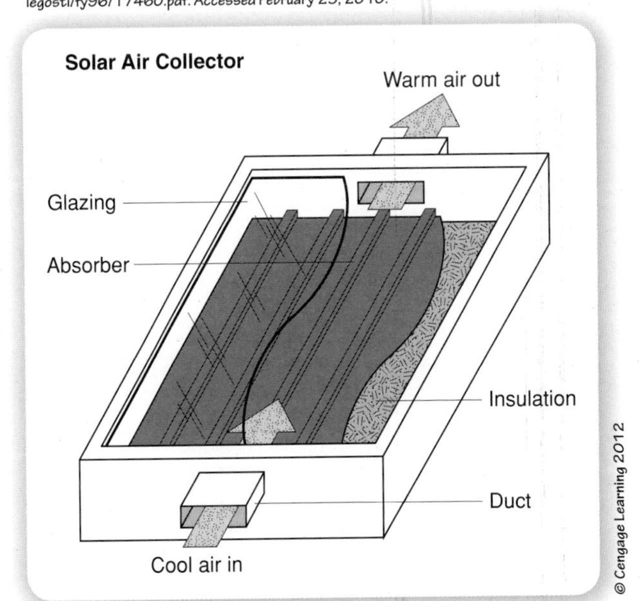

Figure 8-50: Solar geometry.

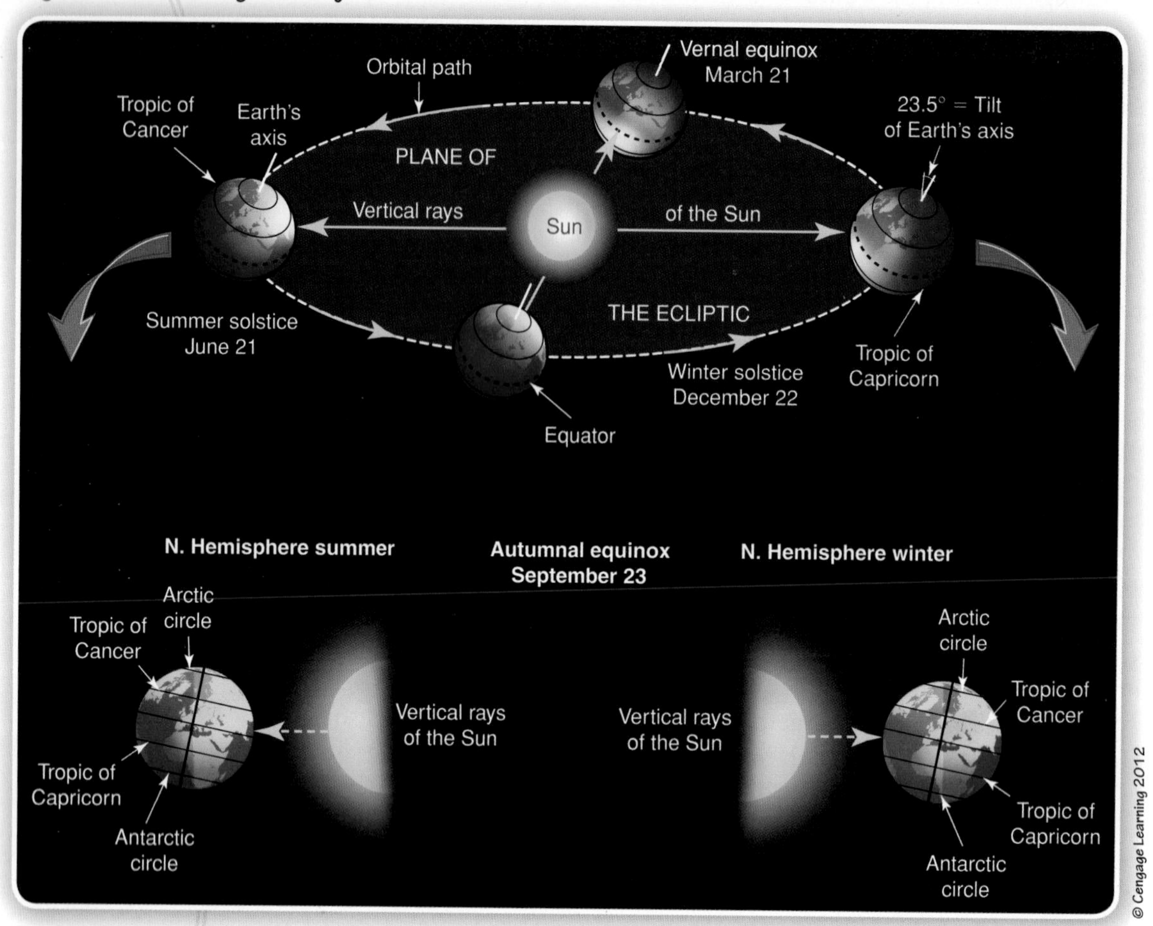

Figure 8-51: Solar position consideration. Reprinted from the U.S. Department of Energy fact sheet (DOE/GO102000-0790) titled "Passive Solar Design" (December 2000) by NREL, http://www.nrel.gov/docs/fy01osti/29236.pdf. Accessed February 25, 2010.

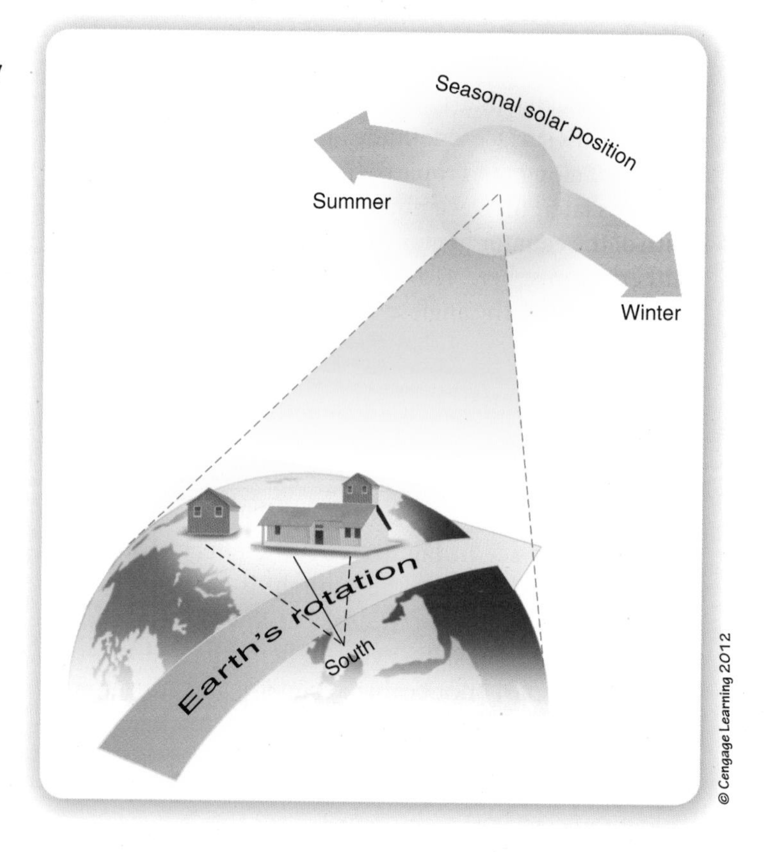

To understand the path of the sun in our location, we may want to develop a solar window using a sky dome (see Figure 8-52). While we have the solar altitude angle, an azimuth angle is needed to plot the sun's path throughout the seasons. The **azimuth angle** is simply the horizontal angle of the sun measured from the north-south line. If we determine the solar altitude angles and the azimuth angles for each hour of the sun for summer solstice (June 21), spring and fall equinoxes (March 21 and September 21), and winter solstice (December 21) between sunrise and sunset, we can plot a three-dimensional sky dome showing the sun's path for a particular location. Consider the solar window is when the sun's rays have the most potential in providing solar power to your location. This typically occurs between 9:00 a.m. and 3:00 p.m. Plotting the solar window on a sky dome will provide a graphic representation of the solar potential for a particular location as illustrated below.

Figure 8-52: A sky dome.

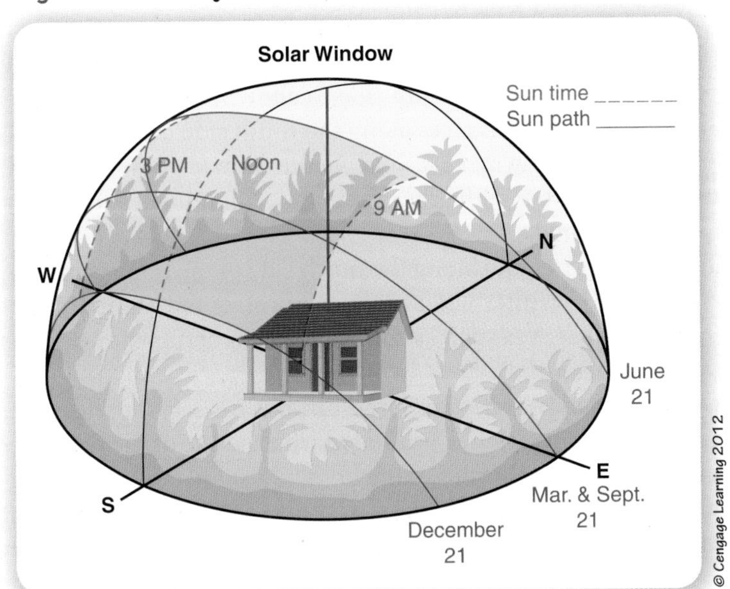

Your Turn

Determine the solar altitude angle and azimuth angle for your home for the given calendar days and times using the following website http://www.sunposition.info/sunposition/spc/locations.php#1:

Jan 21	Feb 21	March 21	April 21	May 21	June 21	July 21	August 21	Sept 21	Oct 21	Nov 21	Dec 21

From the hours of 8:00 a.m. to 5:00 p.m.

Print out the results of each of your dates and overlap to the best of your capabilities your results. Discuss your responses to the following questions:

1. Are there any similarities to the solar angle and why?
2. What was the shortest day of the year and why?
3. What was the longest day of the year and why?
4. Does the solar altitude angle equate to zero for your location? If so when? If not, why not?

Knowing the angle of the sun throughout the course of a year for a particular location allows a building design to use both passive and active system to incorporate a more environmental conscious design while conserving our energy needs. It is essential to understand the site's impact on a building and how a building's orientation and landscaping can capture the full potential of solar access and other natural resources to conserving the resources. In climates where hot summers prevail, shade and ventilation are more desirable. The opposite may be said for cold climates, where sunlight and minimal ventilation is desired due to efforts to reduce heat loss. In areas containing slopes from hills or mountains, location on the southern slopes is usually preferable for two reasons: first, the south slopes receive the most sunlight and the potential for harnessing solar energy is greater than on the remaining sides of the slope. Second, the effects of shading are less dominant as objects' cast their shortest shadows on the south side of the mountain.

Type of vegetation can also be used to influence the amount of heating and cooling loads on a building. Assume a home in a cold climate of the country has the greatest amount of glazing on the south side to allow for direct solar gain during the winter. The use of deciduous trees along the south side of the home allows for proper shading during the warmer seasons. As the climate goes from mild to cold, the deciduous trees begin to lose their leaves, allowing for direct solar gain at a time of year when that is optimal.

On the northwest corner and along the north side of the home, a series of evergreen trees can provide a winter wind barrier to protect the home from heat loss. Other types of natural barriers such as rocks and earth berms may be constructed to the home from winter winds.

In warm climates, homes rely on large canopy trees to provide year-round shading from the sun. Ventilation is promoted with the use of the same trees, large overhangs, and windows on all sides of the home.

For urban environments, a building's relation to its site provides challenges stemming from dense urban living that causes what is known as the **heat-island effect;** the close proximity of multiple buildings in an urban area stores heat energy from the sun during the daylight hours, resulting in larger cooling loads. Several alternatives in passive cooling design include the use of terrace and roof gardens, which provide shading. Additionally, the use of lighter color materials for roofs, walls, and other flat surfaces will reduce the absorption of heat energy. As a rule of thumb, a white roof has less heat absorption as compared to a black roof due to its reflectance coefficient or albedo. Ongoing research studies on double glass curtain wall systems for skyscrapers have shown a 30 to 50 percent reduction on cooling and heating loads.

Although using geometric formulas to determine areas and volumes of various structures is important, for both passive and active solar design applications, it also is important in understanding the relationship of a building's orientation to its site.

Point of Interest

Department of Energy – Solar Decathlon

Imagine you are given the task of design and build a green home that is 100 percent solar powered. Keep in mind that you are required to abide by the following requirements:

1. Footprint size of livable conditioned space has to be a minimum of 450 sq ft and a maximum of 800 sq ft.
2. The maximum height of the home cannot exceed 18 ft from the ground to the highest point of the structure itself.
3. The home must be designed to provide power for the following areas:
   a. Home office – computer and printer
   b. Kitchen – cooking and refrigeration appliances
   c. Home Entertainment – Digital TV
   d. Washer and dryer
   e. A means to provide for domestic hot water needs
   f. Conditioning of space
4. The home is to be transported to the National Mall in Washington, DC, and rebuilt on-site.
5. The home is in a competition against 19 other teams in 10 categories, including: Architecture, Market Viability, Engineering, Lighting Design, Communications, Hot Water, Comfort Zone, Appliances, Net Metering, and Home Entertainment.

Numerous challenges exist for such a project to design an attractive zero-energy home that is also transportable. The designs incorporate both passive and active systems. The total building systems integration is based on the latest and greatest technology available today. Every two years, 20 universities from around the world come together to compete. While past solar decathlons have had a different set of individual competitions, the overall size of the home has not changed.

Learn more about the solar decathlon by visiting http://www.solardecathlon.org.

Point of Interest

CityCenter – A Planned Urban Community

When completed, CityCenter will be one of the largest planned urban communities in the world. Valued at $9 billion dollars, the development will be a city within itself located in Las Vegas, Nevada. When open, the 67 ac site will contain a 61-story, 4000-room resort casino, three hotels, approximately 2600 luxury residences, and a 500,000 sq ft retail and entertainment district. Some of the world's most renowned architects—including Pelli Clarke Pelli, Rockwell Group, Studio Daniel Libeskind, Kohn Pederson Fox, Helmuth Jahn, RV Architecture, Rafael Vinoly, Foster + Partners, and Gensler—are developing various parts of the project.

CityCenter's dramatic design and exciting amenities utilize natural light and innovative strategies to conserve power, and much of the power consumed at CityCenter will be generated on-site. CityCenter's cogeneration solution will utilize the excess heat generated by the on-site power plant to heat water for use by guests. This reduces demand on the power grid and benefits the entire community.

Other energy-saving measures include special incentives afforded to both guests and residents doing their part to support our sustainable community. Preferred parking will be provided for employees, visitors, and residents driving low-emitting or alternative-fuel vehicles. Additionally, to encourage the use of high-occupancy vehicles (HOV), preferred parking spaces will be provided for employee vehicles with an occupancy of two or more passengers. Guests and residents will have access to designated carpool pick-up and drop-off locations near building entrances, as well as access to an online Ridership Board for organizing and coordinating carpool opportunities. For guests traveling by bicycle, complimentary valet parking will be available as well as separate secure bicycle storage for residents.

Figure 8-53: CityCenter.

Photograph courtesy of MGM Mirage.

SUMMARY

The need to conserve energy is increasing as the world's resources are limited and developing countries are finding themselves competing for the same resources. The building industry accounts for approximately a third of all energy consumption. As architects and engineers, it is our responsibility to become better advocates not only by designing buildings that conserve our resources but also by developing alternative resources such as renewable energy systems.

Energy conservation measures were illustrated through lighting design as a result of the energy crisis of 1972. Some of the outcomes of the energy crisis was the development of various controls and advances in lightbulbs to help conserve energy by as much as one-third to two-thirds of energy consumed. In addition to energy conservation measures, the Energy Star rating system and why it was developed by the U.S. Environmental Protection Agency was discussed.

This chapter also discussed both passive and active systems. Understanding how solar geometry impacts the environment will assist building professionals in designing a building that maximizes its use of passive and active solar designs. The final result will not only create a building that is more environmentally conscious but provide a step in the right direction for a sustainable future for generations to come.

BRING IT HOME

1. Perform an energy analysis of your lighting system at home and determine what amount of energy is consumed by the lighting system in your home compared to your overall electric bill.
2. Building upon the previous problem, propose an upgrade to the same lighting system and determine what the energy savings and cost savings would be per year?
3. Determine the energy consumed by your electric appliances per year for the following: refrigerator, water heater, and washer and dryer. Go to the EPA's Energy Star Program and compare the list of recommended Energy Star appliances to the ones you have at home. Also compare energy savings.
4. If you were asked to design a solar thermal system for Phoenix, Arizona, at what angle to the sun would you optimize the system for and why?
5. A photovoltaic system is attached to the roof of a house in St. Louis, Missouri, with a slope of 4:12 (4 in in height to every 12 in in length). What season is the angle of the photovoltaic optimized for?
6. Define the solar window for the Portland, Oregon, and San Diego, California. At what angles would you design a solar powered system and why?

EXTRA MILE

The Federal Energy Management Program lists several buildings as case studies to demonstrate various techniques used to develop high-performance buildings. Choose a case study and prepare a presentation to your class as to the types of systems and designs that were used for your particular building design. Be prepared to answer questions, such as how were the buildings LEED rating acquired in the area of energy? Go to the following web address to see the choices of buildings as possible case studies: *http://femp.buildinggreen.com/mtxview.cfm?CFID=47767840&CFTOKEN=5658606*

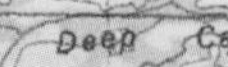

CHAPTER 9
Residential Space Planning

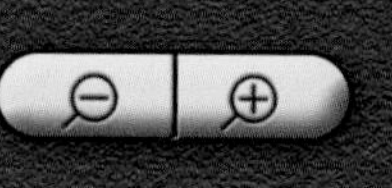

GPS DELUXE

Menu

START LOCATION | DISTANCE | END LOCATION

Before You Begin

Think about these questions as you study the concepts in this chapter:

1. What are the parameters, specifications, and limitations of space planning?
2. How does the client's budget define building options?
3. How do the architect and client reach a shared project vision?
4. Why are space standards important to the development of floor plans?
5. How does the process of investigation, analysis, and development result in efficient use of space?
6. What are some major considerations when planning functional and appealing spaces?

Space planning for residential construction dates back to the pyramids. Ancient tombs were designed with specific space requirements and purpose. The pyramids, the Parthenon, and other early structures provide historical evidence of a uniform system of measurement and proportion. The size and shape of most structures is determined by precise measurements and spatial relationships to human dimensions. The documentation of these dimensions and the range of human motion have led to the development of space standards. Architects rely on these standards to help them design efficient living spaces (see Figure 9-1). These spaces must allow specific tasks to be performed in a safe, comfortable, and efficient manner. Whether designing a pedestrian bridge or a simple staircase, the architect must accommodate for various human sizes and capabilities.

©iStockphoto.com/Fertnig.

Figure 9-1: *A simple staircase is designed using standards based on human size and capability.*

The term **space planning** took hold in the 1930s in response to the need for high-performance office environments. These carefully planned work environments proved to be productive, and space standards gained in popularity. However, over the years it became apparent that space-planning standards needed to evolve and change with the times. For example, the office of the 1930s did not anticipate the technology needs brought about by the computer. When typewriters were replaced with computers and other electronic devices, different floor plan layouts and **infrastructure** became necessary.

In addition, functional spaces of the 1930s did not provide for changes in the way business is conducted. Today, the process of conducting business in a productive and convenient manner requires different **space adjacencies**.

Space Planning: the design of safe, comfortable, and efficient spaces.

Space Adjacencies: proximity to other rooms.

Space planning for a residence is similar. There are certain adjacencies that enhance the occupants' comfort and efficiency. The process of space planning involves detailed and purposeful research/investigation, analysis, and development (see Figure 9-2). This process is not linear, but instead

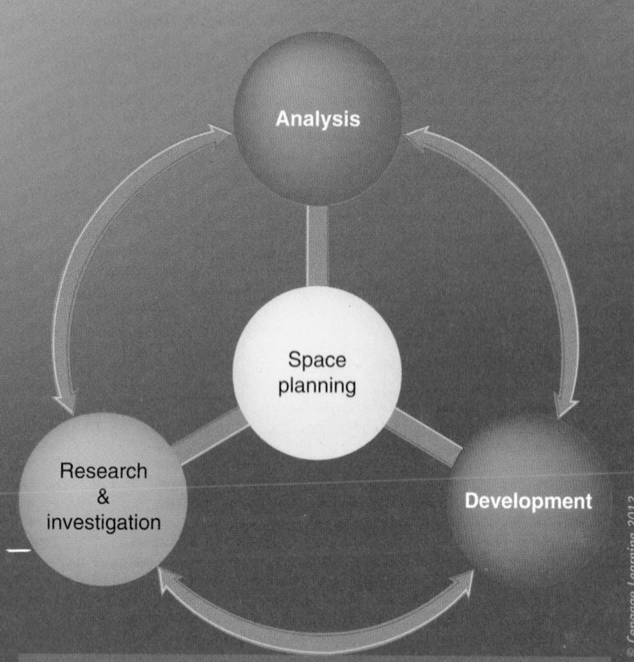

Figure 9-2: *Most architects and space planners move back and forth between research, analysis, and development.*

the architect or space planner moves back and forth through the stages. For example, during analysis, additional research may be necessary to identify a more cost effective material. During development of the floor plan, the architect may step back to analyze room adjacencies and traffic patterns.

Generally, space planning begins by gathering as much information as possible about the project. When designing a residence, the architect or space planner will research the neighborhood, site characteristics, and local governance such as codes and zoning. In addition, he or she must develop a clear picture of the needs, desires, and financial position of the owner. This information must then be analyzed to determine reasonable parameters and goals. The architect or space planner will then apply standards, creative spatial visualization, and personal expertise to brainstorm design options. These options must address the client's needs and wants while keeping within the financial budget.

Space planning goes beyond the design of an efficient floor plan that meets the client's needs. It involves thoughtful arrangement of both interior and exterior features that maximize site characteristics such as solar orientation, **topography,** and view (see Figure 9-3). Space planning applies elements and principles of design to enhance the desired architectural style of the building. In addition, optimal space planning provides for future flexibility, expandability, and resale value.

Infrastructure: when used in reference to buildings, infrastructure is the internal systems such as electricity and telecommunications.

Topography: features on the surface of an area of land.

Photograph courtesy of Donna Matteson.

Figure 9-3: *When planning spaces, the architect considers the scenic views of the site.*

RESEARCH AND INVESTIGATION DEFINE PROJECT PARAMETERS, SPECIFICATIONS, AND LIMITATIONS

Residential space planning begins with the gathering of information. This information will help the architect determine parameters, specifications, and limitations of the project. The architect must understand the client's needs and wants and uncover the client's vision for the project. One of the early tasks is to identify a general budget. The cost of labor and building material vary by region. Other cost factors include the complexity of the plan and site considerations. You might be curious how your residential development compares to the national average. According to the U.S. Census Bureau, the average size of a single-family house constructed in 2009 was 2438 sq ft. Fifty-three percent featured two or more stories and three bedrooms. Thirty-four percent had four or more bedrooms. The 2009 U.S. Census identified the average lot for a new single-family residence was 17,462 sq ft (0.4 ac). However, lot size varied by region; lots in the northeast United States averaged 35,176 sq ft (0.8 ac) whereas lots in the West averaged 10,081 sq ft (0.23 ac).

Engineers and architects identify constraints and external factors that may impact the outcome.

During the investigation process, the architect will review characteristics of the site and accessibility options. Each project is unique in the amount of research required to develop an optimal plan. Research often includes specific code requirements, space standards, and the cost, availability, and features of building materials.

Budget Defines Building Options

During one of the first meetings, the architect must uncover the client's general budget for the project. This budget determines the amount of square footage, building complexity, and level of quality that can be constructed. Building complexity refers to the shape and details of construction. Obviously, a simple rectangular building with minimal details will be least expensive and will take the least amount of time to build. The level of quality is determined by the building material choices, such as siding, roofing, windows, flooring, cabinets, countertops, and trim.

One of the most important considerations of space planning is cost. It is important that the client have a realistic expectation of the total cost of the project. General estimating websites may provide a general square footage cost factor, but they are not accurate enough for most builders. The architect does rough budget, but the final estimates are generally provided by the general contractor based on final plans and specifications (see Figure 9-4). General contractors and builders use spreadsheets to estimate a building project. A detailed spreadsheet is provided in Chapter 11: Dimensioning and Specifications.

When costing a building there are many factors to consider. Building type, complexity, size, shape, and the quality of materials will all impact the construction cost. Standard size components such as windows, doors, and cabinets are less expensive than custom components. Due to the many customer choices available, even standard size cabinets and

Figure 9-4: Sample estimate summary.

Microsoft Excel - Estimating_Form-House_ Sample

	A	B	C	D
1		TOTAL BY DIVISION		
2	Sample Estimate Summary			
3				
4	Division 1	General		$3,930.00
5	Division 2	Sitework		$32,940.00
6	Division 3	Concrete		$11,560.00
7	Division 4	Masonry		$9,620.00
8	Division 5	Steel		$1,230.00
9	Division 6	Wood & Plastics		$54,639.00
10	Division 7	Thermal Moisture		$14,780.00
11	Division 8	Doors & Windows		$14,230.00
12	Division 9	Finishes		$26,800.00
13	Division 10	Specialties		$1,060.00
14	Division 11	Equipment		$0.00
15	Division 12	Furnishings		$0.00
16	Division 13	Special Construction		$0.00
17	Division 14	Conveying Systems		$0.00
18	Division 15	Mechanical		$21,320.00
19	Division 16	Electrical		$7,870.00
20	Division 17	Demolition		$0.00
21				
22		TOTAL W/O OH&P		$199,979.00
23				
24		OVERHEAD	10.0%	$19,997.90
25		TOTAL W/ OVERHEAD		$219,976.90
26		PROFIT	5.0%	$10,998.85
27				
28		TOTAL W/ OH& P		$230,975.75
29		Lot		$0.00
30		Other		$0.00
31				
32		GRAND TOTAL		$230,975.75

Courtesy of Rowlee Construction, Fulton, NY.

windows may take four to 6 weeks for delivery. A contractor must factor estimated delivery dates into the construction schedule, so that materials are available when needed. The time of year, weather conditions, and labor costs all impact the construction estimate. Prices of building materials tend to fluctuate between the time of a quote and the actual build. For example, when the price of oil goes up, so does the price of related products, such as asphalt shingles and vinyl siding. Material shortages caused by increase in demand can cause a price increase, such as when large quantities of plywood are purchased to cover windows in preparation for a forecasted hurricane. Even rising fuel and transportation costs impact the total cost of a building project. Because of these many variables, builders give cost estimates that are limited to a specific timeframe.

Optimal solutions to building design address desired needs within realistic constraints.

Often a builder will provide price allowances for certain items. For example, the price of hardwood floors varies greatly, depending on hardwood choice, size, style, and finish. A cost allowance per square foot would be placed in the contract to prevent the client from exceeding the budget. Clearing, contouring, and excavation of the site and providing for water and septic will add to the total land development cost. Other factors impacting total cost are not as apparent. The owner is typically responsible for providing builders' risk insurance during construction. This covers material or property loss resulting from unforeseen occurrences such as theft or natural disaster. Reputable general contractors provide worker's compensation and general liability, and require their subcontractors to carry the same. This insurance provides coverage should injuries occur on the jobsite.

Covenants: building specifications required by a planned development.

Site Investigation Determines Building Potential

An architect commissioned to design a new residence may begin gathering site information about the neighborhood, adjacent properties, climate conditions, **covenants**, and local governance.

Chapter 5: Site Discovery for Viability Analysis and Chapter 6: Site Planning explain how topography, size, shape, soil type, road frontage, setbacks, and utilities impact a site's building potential. An experienced architect maximizes building potential by designing a structure that enhances the positive features of the site. The corner site in Figure 9-5 offered a choice of road access. The southern exposure at the back of the property provided an ideal location for a scenic and private 10 ft × 12 ft sunroom.

Figure 9-5: Map of a building site with construction footprint.

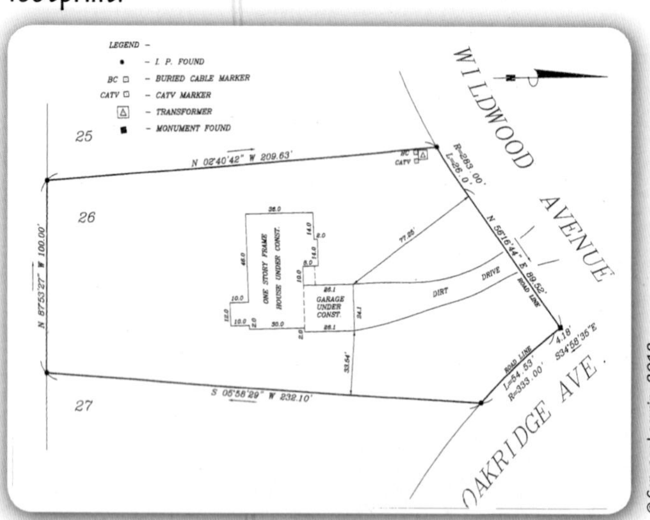

Client Survey and Interview Lead to Project Vision

After uncovering the client's general budget, the architect must obtain detailed information about the future occupant. Most architects use a client survey and interview to gather this information. The client is often encouraged to bring in pictures and ideas that he or she find desirable. An architect will not copy these images, but instead will use them to learn what appeals to the client. A client survey instrument helps to gather information about the client's lifestyle and interests (see Figure 9-6). It may uncover specific needs, such as existing pieces of furniture that must be accommodated in the new residence. With this information, the architect will develop a profile of the client, the client's family and guests, and even future owners.

Figure 9-6: Example of topics addressed in a client survey.

Client Survey Topics

Family Information

Names and description
Special physical needs
Hobbies and interests
Pets (number/type/size)

Architectural Design

Style
Roof type and pitch
Square footage
Type of windows
Exterior colors/textures
Exterior materials
Interior materials
Site considerations
Security issues
Ceiling height
Specialized:
i.e., fireplace

Accommodations

Furniture
Electronics
Size, type, location
New, existing

Work Needs

How many offices
Size
Equipment
Furniture
Files/storage
Internet, phone
Public access

Kitchen

Appliances
Cabinetry
Island/peninsula
Special cooking
Storage
Dining
Location/adjacencies

Bathroom

Number and type
Full, ¾, ½, 1 ½
Special fixtures
Storage
Location/adjacencies

Bedrooms

Number
Type; master, guest
Size
Electronics
Reading

Bedrooms

Number
Type; master, guest
Size
Electronics
Reading
Special needs
Location/adjacencies

Outdoors

Landscaping, gardens, walkways
Decks/porches
Hot tub
Grading desired

Garages

Size
Attached/detached
Storage
Multipurpose

Utilities

Type of heat
Air conditioning
Central vacuum
Internet
Cable
Phone
Intercom
Security systems
Energy conservation

Future Planning

Projected resale date
Children, more or less
Elderly parents
Changes: income, entertainment, hobbies
Work at home

Your Turn

Develop a client survey to gather information from a prospective client. Have a student in your class assume the role as your client. Ask him or her to complete your survey before the interview. During the interview, discern the client's budget for the project. Revisit Chapter 4 on Architectural Design when you interview your client to determine his or her architectural style preference. Refer to Chapter 12: Building Materials and Components when you discuss material options. Discuss material cost, care, and maintenance preferences. Ask about accessibility and future needs. To help determine future needs, ask your client to visualize the building 10, 20, and 30 years into the future. Discuss how their needs might change. How will space requirements differ? Will occupancy increase or decrease? Will activities or hobbies change? How might personal income and operating expenses change? Add the results from the client survey and interview to your notebook.

Figure 9-7: **Human measurement charts.**

		Men	Women
A	Elbow rest height	7.7/9.4/11.2	7.4/9.4/10.6
B	Shoulder height	21.6/23.8/25.9	20.3/22.4/24.6
C	Eye height	29.3/31.7/38	27.2/29.4/31.5
D	Sitting height	33.7/36/38.3	31.3/33.6/35.9
E	Thigh clearance	5/6/7	4.1/5.5/6.9
F	Buttock-knee length	21.6/23.5/25.4	20.7/22.7/24.7
G	Knee height	19.7/21.5/23.3	18.9/19.8/21.4
H	Stool height	14.6/16.3/18.1	13.4/15.0/16.7

Dimensions are in inches and represent the 5/50/95 percentile

		Men	Women
I	Stature	64.6/68.9/73.3	59.8/64/68.1
J	Shoulder height	52.1/56.2/60.2	48/52.1/56.3
K	Eye height	60.1/64.5/68.9	55.1/59.4/63.6
L	Elbow height	39.6/42.5/46.1	35.9/39.1/42.1
M	Fingertip height	23.7/26.0/28.3	21.3/23.6/25.9

Dimensions are in inches and represent the 5/50/95 percentile

Source: National Institute for Occupational Safety and Health.

Architects plan for the future. Although the client may not be thinking about selling a building that has not yet been built, the architect knows that most buildings are eventually placed on the market. Future marketability is important consideration of the design process. When that time comes, the client will want to recover his or her investment. It will be difficult to sell a house for $500,000 if it is located in a development of homes that sell for $250,000. Location is not the only factor impacting future marketability. The number of bedrooms, floor-plan layout, architectural style, and sustainability also contribute to a home's resale value.

Space Standards Guide the Development of Comfortable and Effective Spaces

The architect and client work together to develop a realistic project vision. Once a project vision is determined, the architect will begin the development process. The architect must utilize their knowledge of space standards. These standards identity the dimensions and clearances needed for a safe, efficient, and enjoyable living environment. The architect first considers the number of occupants to use each space, their size, and capabilities. The size and shape of humans vary by population groups, ages, and gender. Architects look to **anthropometrics**, the study of human body measurements for guidance (see Figure 9-7). Anthropometrics is used to design safe, comfortable, and functional working and living spaces.

Anthropometrics:
(from the Latin *anthro* meaning "man" + *metric* meaning measurements) the study of human body measurements.

So how does the study of human measurements impact space standards, building guidelines, and codes? Let's look at one example: bedrooms. Bedrooms must provide an escape window of a specific size. In New York State, the bedroom window must have a minimum net opening of 24 in high by 20 in wide and be no more than 44 in off the floor. The size and placement of this window escape was determined by human body measurements and capabilities.

Engineers and architects work within universal standards and codes to insure comfort and safety.

Habitable Space: a space used for living, sleeping, cooking, and eating, such as a bedroom, living room, dining room, kitchen, or family room.

In space planning, **ergonomics** refers to the science of designing to meet the physical needs of the occupant. Space-planning standards specify things like optional counter height, wall cabinet height, and kitchen work zones. The field of ergonomics is most apparent in the layout of stairs. You may have experienced stairs, especially in older buildings, that are so shallow or uneven in height they cause you to trip. Ergonomically designed stairs are both safe and comfortable to navigate. Although the minimum width dimension of stairs and hallways is 3 ft, some architects provide additional width to enhance client comfort. Stairways are wider in commercial applications to provide accessibility, greater capacity, and safe passage.

Residential building plans allow a clear width of three or more feet in all hallways, stairways, or other passageways.

Other standards that impact residential space planning are identified in state codebooks. Accommodation of these standards can impact the room's layout. In New York State, windows in **habitable spaces** must be a minimum 8 percent of the square feet of the room for light and 4 percent of the square feet for ventilation. A double-hung window provides only half the square footage of ventilation compared to room light, because only one half of the area of window can be open. Consider furniture size and placement when choosing the size and location of a window (see Figure 9-8).

Figure 9-8: Windows placed to accommodate the client's existing furniture.

Courtesy of Paul and Deb Foster

Your Turn

What size double-hung window would you specify to meet the minimum square footage window requirement for a 9 ft × 12 ft bedroom? Record the minimum window light and ventilation specifications in your notebook. Add the net minimum size of a bedroom escape window. Turn to Chapter 4: Architectural Design, Figure 4-11, and sketch the common window styles chart into your notebook.

Non-Habitable Space: a space such as a bathroom, laundry, closets, pantry, utility room, or kitchenette.

PLANNING FOR SPECIAL NEEDS Accessibility is generally thought of as the accommodation of people confined to a wheelchair. However, there are many other special needs to consider when planning a residence, such as the elderly or people with limited mobility, vision, or hearing. Some people might have environmental illnesses and must avoid all but natural materials. Other clients may desire custom heights for cabinets and countertops. Considerations such as these will impact space planning on an individual basis. Accommodation of special needs will generally require additional research to locate alternative fixtures, cabinets, and building materials (see Figure 9-9).

Figure 9-9: Dimensions to guide accessibility planning.

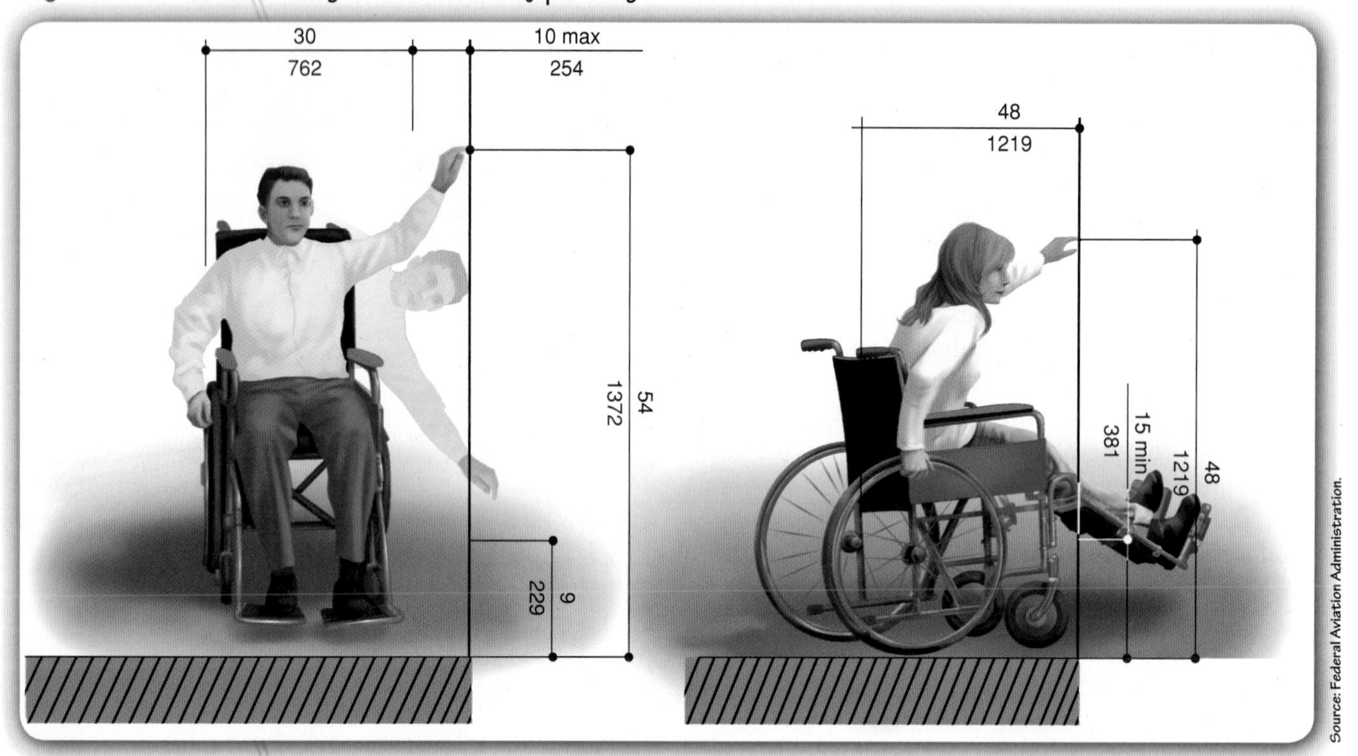

If your residential plan must accommodate a wheelchair, you will need to follow specific guidelines. An adult-size wheelchair is approximately 26 in wide by 42 in long and will need 32 in clear wide doorway and 36 in clear wide hallway. ADA code 4.13.5 specifies a clear width of 32 in, with the door open at 90 degrees. Most architects will use a 34 in hinged door to provide the clearance necessary. If a swinging door is at the end of a hallway, then allow an additional 12 in of hallway space at the opening side of the door for the wheelchair to maneuver when opening the door. **Space planners allocate a 5 ft diameter turn around space for wheelchairs.** Architects designing for special needs will strive to keep the layout in proportion. In a handicapped bathroom the architect will provide a minimum wheelchair turnaround space, handicapped accessible lavatory sink, and special height water closet. When placing wall cabinets or towel bars an architect must also consider that the maximum reach from a wheelchair is 24 in horizontal and 48 in vertical. Additional accessibility guidelines for residences are available at the American Disabilities Act website, *http://www.access-board.gov/adaag.* You will read more about handicapped accessibility in Chapter 10: Commercial Space Planning.

PLANNING BUILDING COMPONENT SPECIFICATIONS AND OPTIONS Architects must provide specifications and options for windows, doors, cabinets, and other building supplies. Architectural software programs normally provide libraries of components and may offer the ability to incorporate specific manufacturer components. Architects usually keep a library of several manufacturer catalogs. Many use Sweets© catalog files. The Sweets Network© provides access to products, catalogs, specifications, CAD, Building Information Modeling/BIM, and green building. The Sweets BIM© collection offers Revit® models from several manufacturers.

Contemporary engineering tools such as Computer Aided Design and Building Information Modeling are used to analyze and solve problems.

Your Turn

Use the Internet to download manufacturers' building catalogs from major catalog listing websites. Search for the keywords *building catalogs* or *products construction* to find websites such as http://products.construction.com/residential, http://www.ebuild.com, and http://www.4specs.com.

Locate and download catalogs for windows, doors, and cabinetry to determine size and style options during residential space planning (see Figure 9-10).

Figure 9-10: Many building supply catalogs are available online.

MAIN | IDEAS | PRODUCTS | LEARN | SERVICE | ABOUT ANDERSEN

OVERVIEW
ALL WINDOWS
REPLACEMENT WINDOWS
DOORS
HELP ME CHOOSE
OPTIONS & ACCESSORIES
WHAT'S NEW
ABOUT PRICE

Windows

WINDOW DESIGN TOOL
See the difference Andersen windows and doors can make in your home. Use our interactive design tool to create window and door combinations that you can save and share.
Start Designing

Casement Windows
Have a hinge at the side and crank out to open.

Awning Windows
Have a hinge at the top and open outward.

Double-Hung Windows
Open by sliding one sash vertically past another.

Gliding Windows
Open by sliding one sash horizontally past the other.

Picture & Transom Windows
Picture & transom windows are stationary windows designed to match double-hung or casement windows.

Specialty Windows
Are stationary windows with curved shapes or angles other than 90°.

ANALYSIS LEADS TO OPTIONAL RESIDENTIAL SPACE PLANNING

Analysis occurs during both the research and development process. Readily accessible, detailed documentation will enhance the efficiency and quality of the analysis. The goal is to determine the most viable choice. This could involve comparison of features, prioritizing, and making trade-offs. Notes taken during the investigation should question the relevance and impact on the project. For example, how will the solar orientation of the site impact the floor plan? How will the client's style and quality preferences impact construction cost? What is the payback time needed to offset of additional cost of energy efficient options? Remember, the space planning process is not linear. You should perform continual research and analysis throughout the development process.

The Ability to Visualize Space Is Key to Space Planning

Architects and space planners must visualize a room's size, shape, and ability to accommodate desired fixtures, appliances, and furnishings. Each space must allow convenient access and traffic flow with minimum impact upon the activities taking place in the room. When designing a building, you must understand the difference between gross and rentable space. When houses are built, they are described by their overall construction size. The exterior walls of cost efficient buildings are generally laid out in 2 ft even increments to reduce construction time and material waste and allow for use of stock items such as trusses. One might describe their new home as a 26 ft × 48 ft single story. **Garages, porches, basements and attics are not included in square foot calculations**. To find the gross square footage of this house, you multiply the 26 ft width by the 48 ft length, which equals 1248 sq ft. The gross square footage is generally the figure used if you are listing a house for sale, paying for a building permit, or buying homeowners' insurance. Gross square footage, based on exterior dimensions, is not actual living space. Rental space is square footage based on the interior dimensions. If you are inexperienced in space planning, you might not allow enough interior space when planning the layout a floor plan. Walls, chimneys, and closets greatly reduce the amount of usable interior space. Take a look at a 24 ft × 36 ft two-story colonial. If you wanted two bedrooms at the 24 ft end with a full bath in between, what size bedrooms could you have? Initially you might think the bedrooms could each be 9 ft × 12 ft, because after you subtract 6 in for each of the exterior walls, the 24 ft wide colonial has an interior space of 23 ft. Adding the dimensions of 5 ft, required for a full bathroom, and 9 ft width dimension for each of the two bedrooms you get 23 ft. But what about the interior walls of the bathroom? (See Figure 9-11.)

Plumbing walls containing the large waste pipes coming down from the second floor are built with 2 × 6 lumber. A 2 × 6 wall with ½ in sheetrock on both sides can reduce your interior space by over 6½ in. So why are these dimensions so important to consider? Let's take a look at planning for furniture in these two bedrooms (see Figure 9-12). A standard walkway space of 2 ft should be provided on *both* sides of the

Figure 9-11: Most 2 × 4 interior walls reduce your space by almost 5 in, depending on the thickness of the wallboard or sheetrock.

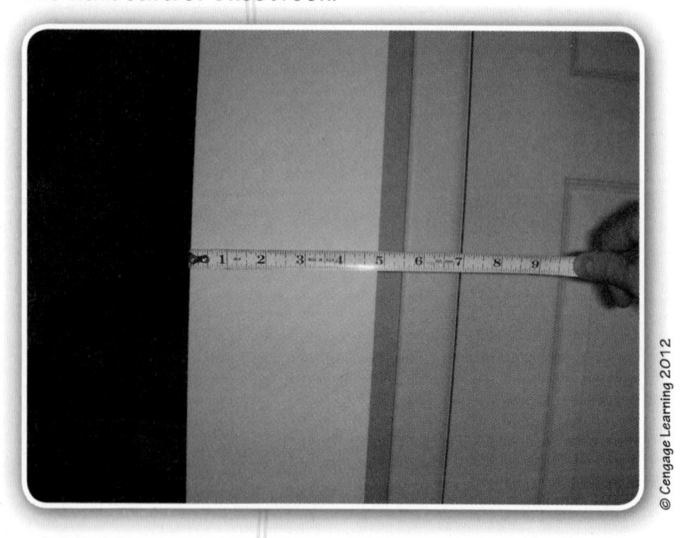

Figure 9-12: **Provide a 2 ft minimum walkway on both sides of a bed.**

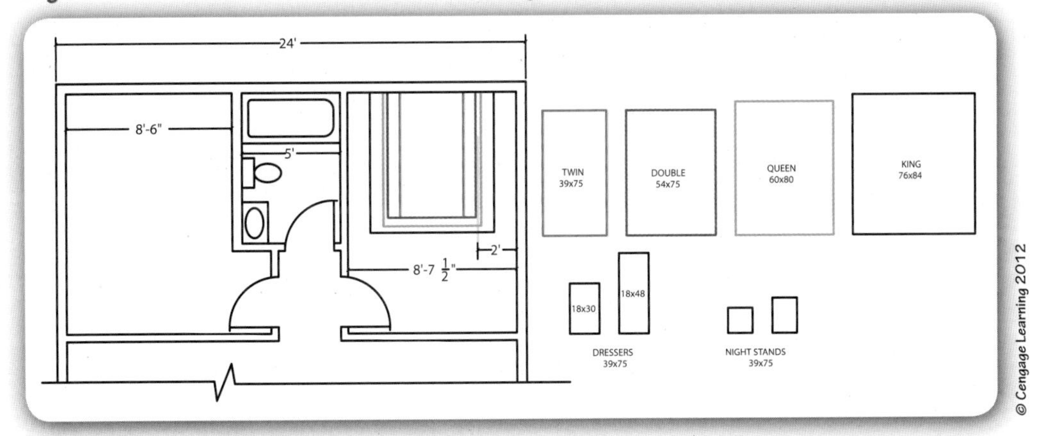

bed, as shown by yellow dashed lines. How does that limit the bed options? Think about where you could place the dresser while keeping a front clearance of 3 ft to allow convenient opening of dresser drawers. What other bedroom accommodations are needed in this plan? Is there space for a closet? How do you think this plan could be improved?

Your Turn

Record in your notebook the amount of usable space lost by interior walls; include both standard and plumbing walls. Sketch and note the dimensions of the standard bed sizes shown in Figure 9-12. Record the clearance requirements of 2 ft minimum walkway space on both sides of the bed and 3 ft at the front of a dresser. Include your own thoughts about bedroom-planning considerations, such as room function, existing furniture, closet and storage needs, and window placement.

Many space planners design spaces that exceed minimum standards. It is generally recognized that bedrooms cannot be less than 7 ft in any horizontal direction, and must have headroom of at least 7 ft. Codebooks often specify a minimum of 70 sq ft with a minimum horizontal dimension of 7 ft. Bedrooms are identified as habitable spaces and must have an emergency exit, usually a window. Now think about furniture needed in a bedroom and its placement. As you can see in Figure 9-12, minimum room size limits furniture size and placement. Is it possible to design a space that is too large for safety, comfort, and efficiency? Absolutely. Overly large or out of proportion spaces are difficult to furnish and are uncomfortable for the occupant. To avoid this, the architect must create spaces based on the activities planned for that room. A general chart of room sizes will provide a basic understanding of proportions; however, room size is determined by the client's unique needs and how the rooms fit within the overall plan. Most space planners agree that rectangular rooms offer ease and flexibility of furniture placement.

ARCHITECTS VISUALIZE SPACES IN THREE DIMENSIONS Architects and space planners think in three dimensions. In addition to the width and depth of a room, they must consider the wall height.

Architects visualize three-dimensional spaces and envision the relationship of humans and objects residing within the spaces.

Your Turn

Copy the chart of general room dimensions, shown in Figure 9-13, into your notebook. Before going home, make a list of the rooms in your house and estimate whether they are small, medium, or large. When you arrive home, measure the rooms and compare them to your size chart. Did you estimate the sizes correctly? Tour each room and note the features that make the space efficient and comfortable.

Figure 9-13: **Table of general room sizes.**

General Room Dimensions			
Room	**Small**	**Medium**	**Large**
Kitchen	10 ft × 14 ft, 12 ft × 12 ft	12 ft × 14 ft, 13 ft × 15 ft	14 ft × 16 ft, 12 ft × 22 ft
Dining	10 ft × 12 ft	12 ft × 14 ft, 12 ft × 16 ft	14 ft × 16 ft
Bedroom	7 ft × 10 ft, 8 ft × 10 ft 9 ft × 11 ft	10 ft × 12 ft, 12 ft × 12 ft, 12 ft × 14 ft	12 ft × 16 ft, 14 ft × 20 ft
Living room	12 ft × 12 ft, 12 ft × 14 ft	13 ft × 15 ft, 14 ft × 16 ft	14 ft × 18 ft, 16 ft × 20 ft, 16 ft × 22 ft
Family room	13 ft × 16 ft	16 ft × 18 ft	18 ft × 22 ft
Full bathroom	5 ft × 7 ft-6 in	8 × 10 ft, 7 ft × 11 ft	9 ft × 12 ft, 10 ft × 12 ft
Garage	12 ft × 24 ft single car	22 ft × 22 ft, 22 ft × 24 ft 24 ft × 24 ft double car	28 ft × 26 ft, 30 ft × 30 ft

Figure 9-14: **Notebook entry of initial space planning.**

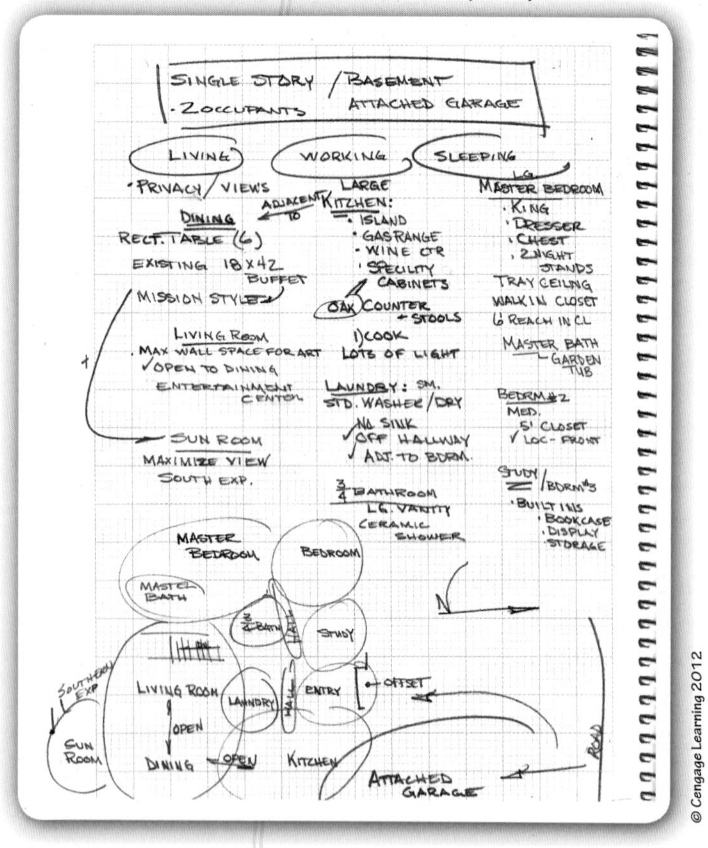

A ceiling cannot be less than 7 ft in height because standard doors with door casing are 6 ft 8 in. Most people would not feel comfortable with a 7 ft ceiling height. Also this minimum height might not provide the space for desired options such as a lighted ceilings fan. Therefore, the ceiling height of most residences is between 8 ft and 9 ft. Ceiling height can reflect housing style, such as the high ceilings of Victorian homes. Other times the size and type of windows might suggest a suitable ceiling height. Windows with a special feature, such as a transom or half round, may require higher ceilings. When locating windows they are generally placed at least 18 in off the floor for safety.

Space Adjacencies Are Analyzed for Convenience and Comfort

Interior spaces fall into three distinct categories; **living, working,** and **sleeping.** Each area has a distinct purpose. The architect or space planner must analyze the activities to be performed in each area and review the client's preferences and specific requests. Thoughtful consideration and analysis of this information will determine the approximate size and adjacency of the rooms. The architect or space planner may document this brainstorming process with a notebook entry similar to Figure 9-14.

An adjacency matrix is often used to plan the proximity of each room. Examine the adjacencies desired for each room shown in Figure 9-15. Look closely to determine the two primary adjacencies for the living room. A secondary adjacency indicates a desire for convenient location that is close, but not necessarily directly adjacent, to another room.

Efficiency Is Measured by Projecting Traffic Flow

During the development of a floor plan, the architect thoughtfully considers placement of entrances, stairs, and hallways in relationship to the overall plan. Wisely placed entrances and stairs can reduce the amount of hallway space and add to the convenience of the owner and guests. A well-designed floor plan will allow efficient **traffic patterns** (see Figure 9-16).

Traffic Pattern: the projected movement of people for a given floor plan.

MAJOR DEVELOPMENT CONSIDERATIONS WHEN PLANNING FUNCTIONAL AND APPEALING SPACES

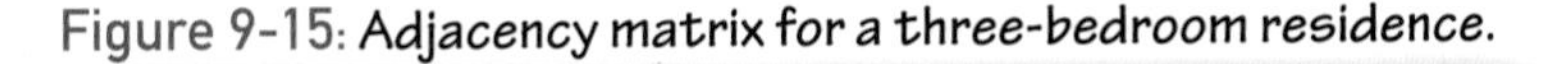

Remember the principles and elements of design you read about in Chapter 4: Architectural Design? Visual elements of design include line, color, form, space, shape, texture, value, and tone, whereas the principles of design are balance, contrast, emphasis, movement, pattern/repetition/rhythm, and unity/harmony. Successful architects and space planners apply these principles and elements during the development process to create a desirable and appealing plan.

Figure 9-15: Adjacency matrix for a three-bedroom residence.

● Primary adjacency
○ Secondary adjacency

	Kitchen	Dining	Sunroom	Laundry	Full bath	Living room	Master bedroom	Master bathroom	Bedroom #2	Bedroom #3/study	Garage	Service entrance	Main entrance/foyer
Kitchen		●		○	○						○	●	
Dining			●			●							
Sunroom						○							
Laundry							○		○	○		●	
Full bath								○	●	●			
Living room							○						●
Master bedroom								●	○	○			
Master bathroom										○			
Bedroom #2													
Bedroom #3/study													
Garage												●	
Service entrance													
Foyer													

Figure 9-16: This traffic pattern shows the anticipated movement of the residents and guests and identifies congested areas.

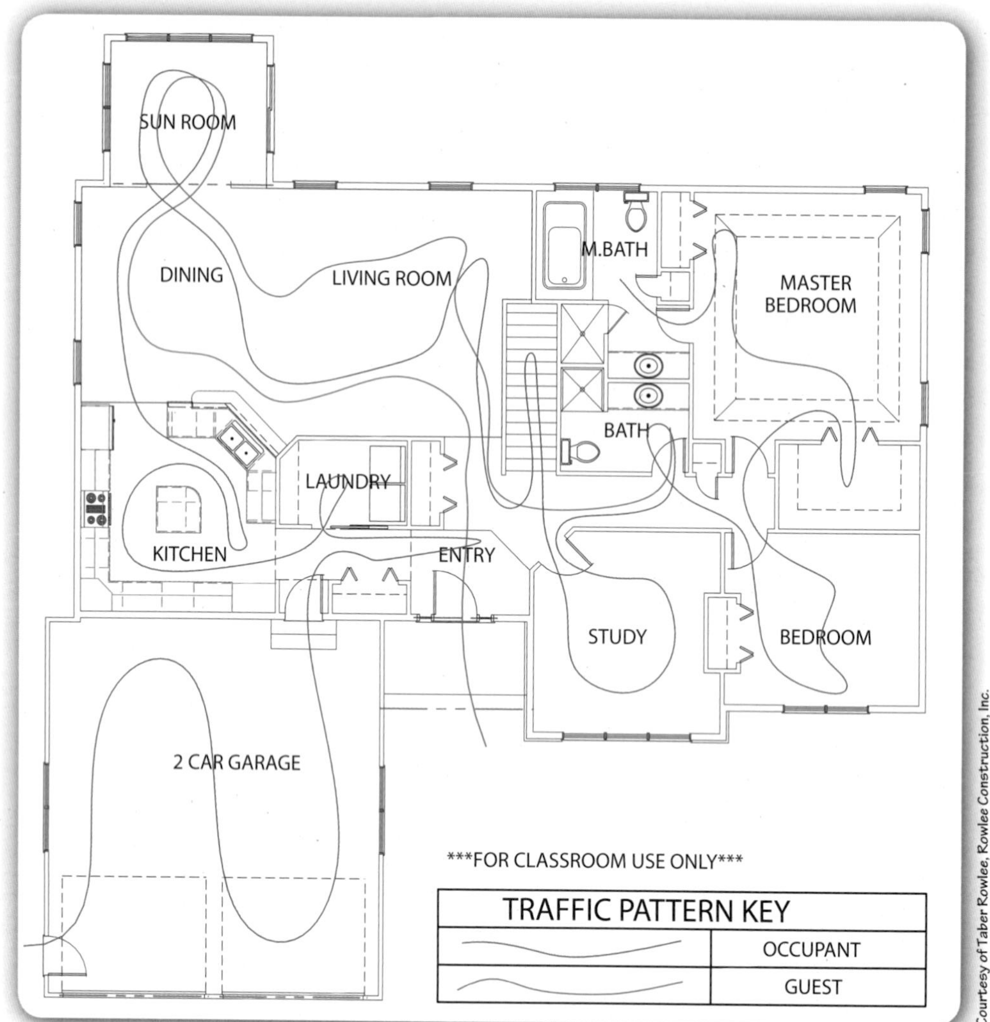

Courtesy of Taber Rowlee, Rowlee Construction, Inc.

It takes great skill to plan spaces that will be functional and appealing from both an inside and outside perspective. The size, type, and placement of every window and exterior door must be carefully chosen to achieve a pleasing design. Each room is thoughtfully placed to maximize the site features and enhance the comfort of the occupants and guests. Rooms are grouped and located based on a variety of considerations, from maximizing natural light to providing for cost-effective plumbing. As architects brainstorm several layout options, they strive to achieve harmony between areas designed for working, living, and sleeping. Room placement and window locations can impact a building's exterior appearance. An architect must strike a balance between interior functionality and exterior beauty.

These uniquely different spaces must be positioned so that they can be utilized without impacting negatively on each other. Rooms that accommodate noisy activities are carefully positioned away from quiet areas (see Figure 9-17). Interior walls can be sound insulated when rooms require close proximity, but have different noise considerations. Storage is another consideration when planning spaces. An architect must provide for convenient storage throughout the residence to meet a variety of specific needs.

Figure 9-17: Living, working, and sleeping areas of a house.

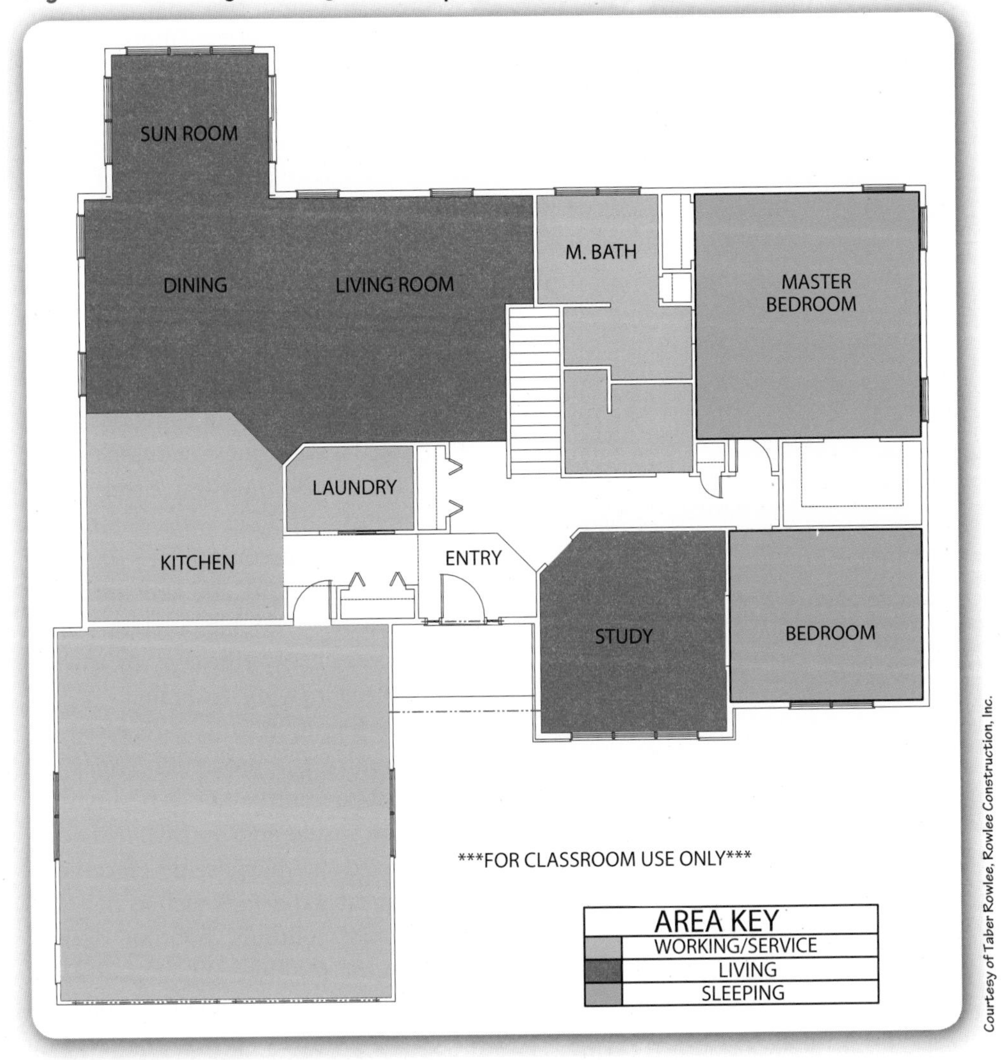

Courtesy of Taber Rowlee, Rowlee Construction, Inc.

Planning within a Budget

Planning within a budget is not as easy as you might think. Many clients have unrealistic expectations of building cost and desire more square footage or building quality than they can afford. This challenges the architect to make the most of every square foot (see figure 9-18). Open living spaces with views of the outside can provide a spacious feeling without additional square footage (see Figure 9-18).

Often the staircase is placed in a central location to provide convenient access to rooms on the second floor (see Figure 9-19).

Another space-saving strategy is to place the second floor staircase above the first floor staircase, as in Figure 9-20.

What other space-saving strategies are proposed to clients? Architects often suggest dual-purpose rooms, such as a study that can be used as an occasional guest room. Room placement on the floor plan and state codes often impact space options. In some states, homes with attached garages built with attic trusses provide clients with a bonus room accessible from a second floor hallway. Research of fire codes would be necessary before suggesting this feature.

Figure 9-18: An open staircase to the lower level provides additional visual space to the area.

Courtesy of Deb and Paul Foster.

Figure 9-19: Centrally located stairs provide convenient access to rooms and avoid long hallways. This drawing was completed by Richard Kulibert Jr.

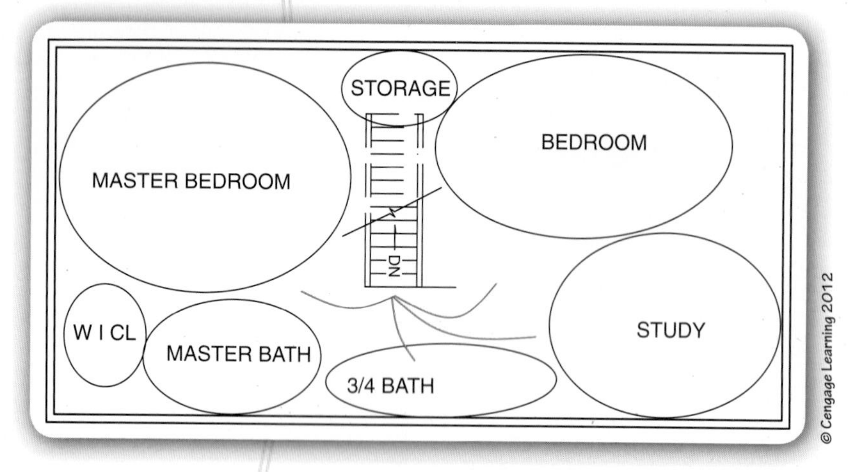

Figure 9-20: Stairs are often stacked to save space.

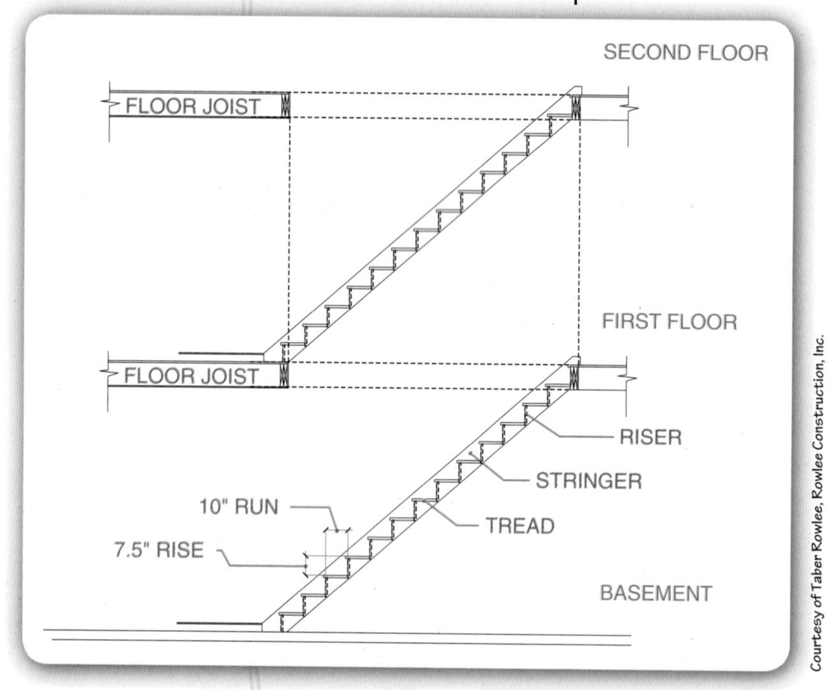

Planning Living Areas

Living areas are where the residents and their guests play and relax (see Figure 9-21). Various lifestyles warrant different rooms and layouts. Living area choices include great rooms, living rooms, family rooms, dining rooms, dens, libraries, sunrooms, and home theaters. A formal entrance or foyer is generally placed in or near a main living space, usually the living room. The foyer should provide an inviting entrance with and visual for smooth transition to the living areas. In other words, your guest shouldn't need a map to navigate their way. Front entry doors are usually 36 in wide × 6 ft 8 in tall, which will accommodate the passage of large furniture and appliances. In harsh climates, a double entry or porch can increase energy efficiency by shielding out the cold, wind, and rain.

Many homes feature a recessed or covered entry for protection. Keep in mind that the entry should be inviting from both inside and outside. This can be achieved by purposeful placement and architectural details such as detailed entry doors, columns, lighting, accent materials, or roofline. When guests arrive at the house, they should not have to search for the entry. What architectural details can you identify in Figure 9-22?

When planning the interior, the architect anticipates the traffic patterns of both owner and occupant. They consider entries, door swings, and various furniture placements for different types of activities. Natural light and scenic views are desirable for dining rooms, living rooms, and sunrooms, but not for home theaters. When planning living areas, a guest bathroom is generally placed in a convenient location, separate from the bedroom area.

PLANNING CONSIDERATIONS FOR THE DINING AND LIVING ROOM When planning adjacent spaces, consider the convenience of the design for the occupants. Room adjacencies should provide both visual appeal and accessibility. Noise is another consideration. For example, the noise of the living room television may have a negative impact on guests in the dining room.

When planning the individual dining and living room spaces, consider the size of furniture and the safe and comfortable navigation about the room. Placement of furniture should accommodate the activities planned for the room. A living room might require an arrangement conducive to entertaining and conversation and the flexibility to view television, a fireplace, or the outside. During consideration of room size and shape, plan ahead for extra space requirements, such as space needed to open a sofa bed. Consider furniture that must move to accomodate the open sofa bed. For example, where will the coffee table be moved? Identify the the function and desired occupancy of each room. How many people can be comfortably seated or participate in the planned activity. Will a recreation room with a pool table have the same space requirements as room with exercise equipment? Flexibility is highly desirable in space planning. Think about how a client might rearrange or repurpose space to accommodate new interests. Determine how furniture may be repositioned for maximum capacity during an occasional party. These are all selling points when presenting a house plan to a client. (See Figure 9-23.)

The space planner will recommend size, style, and placement of windows to maximize interior and exterior aesthetics, function, and natural light. A chart of window types is shown in Chapter 4: Architectural Design. You will read more about windows in Chapter 8: Energy Conservation and Design and Chapter 12: Building Materials and Components.

When planning a dining room, one should allow enough space at and around the table (see Figure 9-24). Each person needs 18 in × 30 in of table space. A distance of at least 36 in from the table to any adjacent surface will allow someone to walk behind the person seated. If possible, one should consider the size and type of chairs. Large chairs with arms will require extra space. Other dining room furniture to consider may include china cabinets, curios, hutches, buffet tables, extra chairs, or a serving cart.

Working and Service Areas

The working or service areas would include the kitchen, laundry, bathroom, utility room, and garage. The working areas are often grouped together to reduce plumbing cost. These areas can be noisy, which is another main consideration when choosing the location of these spaces. An exterior entrance for occupants is usually located in the service area, providing easy access from the garage or backyard. This service entrance provides for convenient delivery of groceries and goods. A small half bathroom near the service entrance will provide convenient accessibility from the garage or outside.

Figure 9-21: The living area of a house includes the dining room.

Courtesy of Deb and Paul Foster.

Figure 9-22: A sheltered entry provides protection from snow, wind, and rain.

Courtesy of Deb and Paul Foster.

Figure 9-23: An open floor plan allows for several furniture arrangements.

Courtesy of Deb and Paul Foster.

Figure 9-24: Space standards for dining are based on human measurements and range of movement.

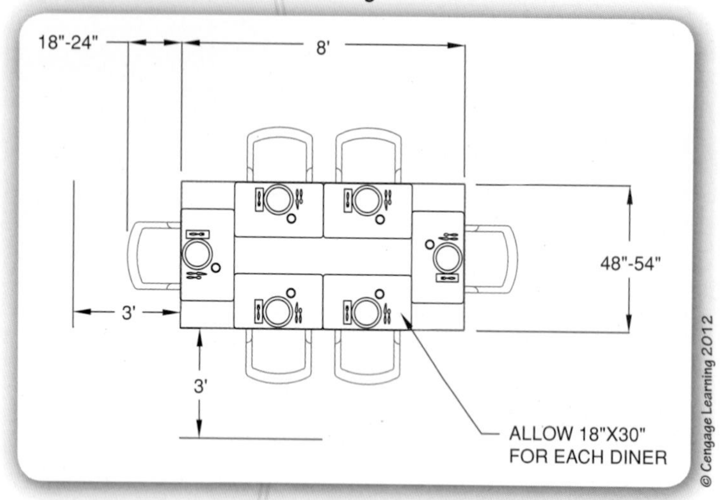

Figure 9-25: The kitchen is a challenging space to design.

Courtesy of Paul and Deb Foster.

When designing a two-story home, it is recommended to have a bathroom on each level. In some older homes, the only bathroom is on the second level near the bedrooms. This arrangement forces guests into the private areas of the home and limits handicapped accessibility.

KITCHEN PLANNING The kitchen is one of the most challenging spaces to design, as well as the most expensive to build and furnish (see Figure 9-25). Although the architect may provide a general layout, a kitchen planner may be employed to determine a detailed layout with cabinet and appliance specifications. Generally, the builder will work with the client to establish an allowance to cover custom choices of kitchen cabinetry, countertops, appliances, and flooring. These choices are often made when the owner can physically walk through the rough framed house. There are many options allowing for endless design possibilities (See Figure 9-26). It is imperative for the kitchen planner to have a thorough understanding of the client's needs. What level of food preparation is desired? How many will participate in the preparation? What appliances are desired? How much cookware and bakeware must be stored? Will the kitchen include space for a table or a counter with stools? These questions and many more must be answered.

Figure 9-26: A client may desire a special feature such as a wine center.

Courtesy of Paul and Deb Foster.

A kitchen is often viewed as having three work areas; preparation, cleaning, and storage. To remember this easily, think of the preparation area as the stove, the cleaning area as the sink, and the storage area as the refrigerator. These three areas are carefully planned for maximum convenience. Each area will need cabinets to serve each purpose (see Figure 9-27). For example, cooking utensils should be stored near the range top and baking dishes near the oven.

The preparation area is where the food is prepared and cooked. If there is only one cook, allow a counter space of at least 48 in adjacent to the cooking and cleaning areas. Two cooks will need 72 in The preparation space is often located between the range and sink. Allow a counter space of 18 in or greater at the opening side of the refrigerator. Think how difficult it would be to pour a glass of milk

Figure 9-27: A pot-and-pan drawer conveniently located in the kitchen's preparation area.

Courtesy of Paul and Deb Foster.

Figure 9-28: The kitchen cleaning area.

Courtesy of Paul and Deb Foster.

without an adjacent countertop. Another 18 in or more of counter should be provided next to the cooking unit. Remember, most cooks will claim that you can never have too much counter space.

The kitchen cleaning area includes the sink, dishwasher, and garbage disposal (see Figure 9-28). As with the preparation area, it should be near the recycling station or wastebaskets. Think about the location of the dishwasher in relation to the sink. A convenient location is directly to the right or left of the sink. Not only does the close proximity reduce plumbing cost, but many people scrape their plates over a garbage disposal and rinse them before placing them in the dishwasher. A sink located on an exterior wall is commonly placed under a window to provide natural light and an outside view during cleanup. Other times the sink is located in an island or peninsula with a view toward the home's interior space. Provide counter space on both sides of the sink; a minimum of 36 in on one side for stacking dirty dishes and 24 in on the other side for clean dishes.

Kitchen Work Triangle An efficient kitchen plan provides a safe, comfortable, and convenient experience for the user. The kitchen layout can be analyzed for efficiency by drawing a work triangle and measuring the distance between the refrigerator (storage), stove (preparation), and sink (cleaning) (see Figure 9-29).

Ergonomics and anthropometrics, studies of human measurement, specify that an efficient work pattern has total distance of 15 ft to 22 ft, with no one side being less than 4 ft.

Figure 9-29: The kitchen triangle shows the users' work pattern.

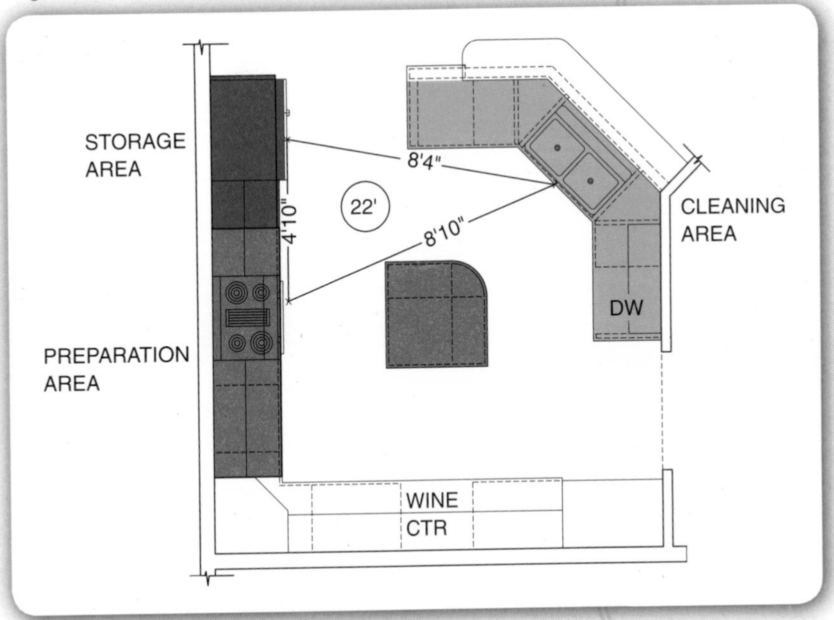

Courtesy of Taber Rowlee, Rowlee Construction, Inc.

Ideally, the work triangle should not be disrupted by people walking through the kitchen to gain access to another room. When planning a kitchen consider how cabinet, refrigerator, and oven doors obstruct the floor space and prohibit movement when in the open position.

Your Turn

Sketch Figure 9-29 into your notebook. Make a separate note of acceptable work triangle dimensions. Add notes indicating the recommended counter space for each area, as identified in the paragraph on kitchen planning.

Kitchen Arrangements There are several standard arrangements for kitchens: single wall, corridor, L-shape, U-shape, peninsula, and island (see Figure 9-30).

Each arrangement has unique characteristics. The single wall has the storage, preparation, and cleaning areas along the same wall. This arrangement serves a small space with one cook. In a corridor arrangement, sometimes called a galley kitchen, the storage preparation and cleaning areas are placed along two opposing

Figure 9-30: Standard kitchen arrangements.

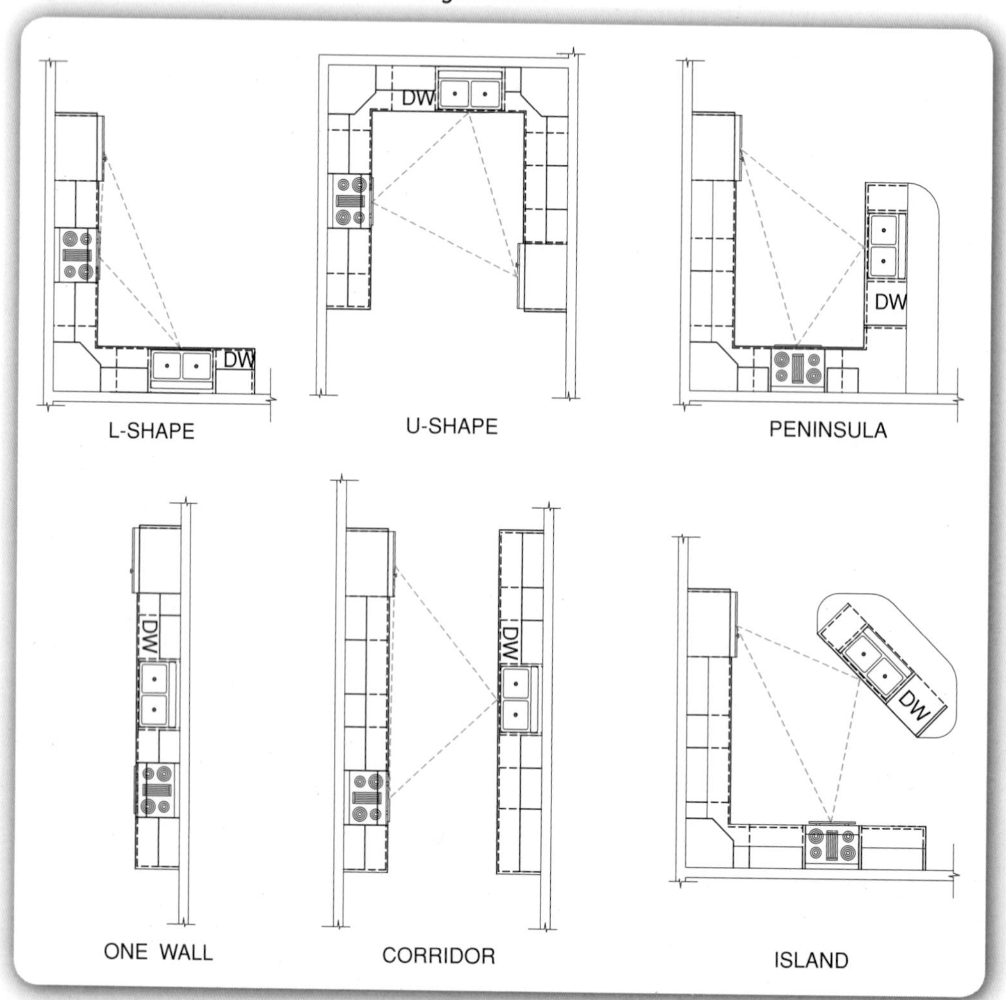

walls. Allow a minimum corridor width of 48 ft between the faces of the base cabinets. The L- and U-shape kitchens are popular for their convenient layout and comfortable accommodation of more than one cook. Both of these arrangements offer an uninterrupted work triangle. Large kitchen plans often include a peninsula or island. Peninsulas and islands can be of almost any shape. A peninsula plan has one counter that extends from the wall into the room, which may feature a range top, a sink, additional counter space, or an eating area. The island kitchen features a stand-alone counter height unit with all of the same options as a peninsula. The peninsulas and island kitchens generally require the most space, because of the floor clearance required. An island must be carefully sized and placed so that it does not hinder productivity. Think about the kitchen triangle and the movement to and from the storage, preparation, and cleaning stations. An island should not impede this movement.

Figure 9-31: Cabinet options include slide out recycling bins.

Courtesy of Paul and Deb Foster.

Cabinets and Pantries The client and general contractor often contact a kitchen planner or kitchen supplier to develop a detailed kitchen plan from the architect's plans. In the detailed plans the kitchen cabinets are described specifically by their size, type, and location. Cabinet types fall into four main categories; base, wall, utility, and specialty. Base cabinets serve as a base for the countertop. They are generally 24 in deep and come in several styles. Base cabinets can be ordered as single door, double door, corner, or drawer units. Cabinet options include slide out shelves, recycle bins, corner carousels, and many others. (See Figure 9-31.)

Wall cabinets are generally 12 in deep and range from 12 in to 42 in high. However, the standard height of wall cabinets is 30 in and 36 in. Depths greater than 12 in are available for locations over the refrigerator or when a staggered layout is planned. When a staggered layout is used, some cabinets must be deeper to accommodate crown molding. (See Figure 9-32.)

Cabinets are identified on the plan by type and size. The letters identify the type of cabinet followed by the size, (see Figure 9-33). When planning kitchen cabinetry, one must consider cabinet door swing and the clearance needed to open drawers. Fillers strips and corners that match cabinetry provide the necessary

Figure 9-32: Wall cabinet depths vary to accommodate crown molding.

Courtesy of Paul and Deb Foster.

Figure 9-33: Cabinets are labeled by type and size.

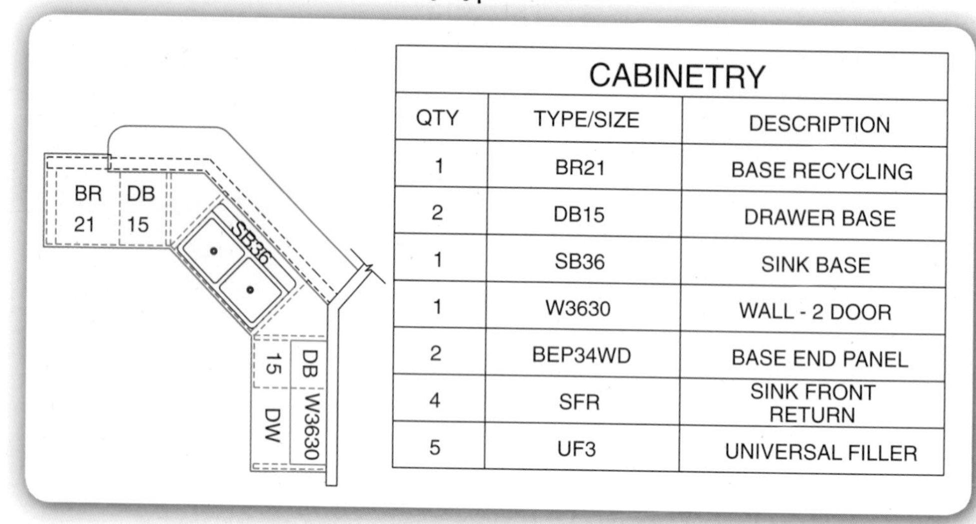

CABINETRY		
QTY	TYPE/SIZE	DESCRIPTION
1	BR21	BASE RECYCLING
2	DB15	DRAWER BASE
1	SB36	SINK BASE
1	W3630	WALL - 2 DOOR
2	BEP34WD	BASE END PANEL
4	SFR	SINK FRONT RETURN
5	UF3	UNIVERSAL FILLER

Water Closet: a toilet.

Lavatory: a bathroom sink.

Vanity: a bathroom base cabinet.

space clearance to open drawers and doors. To create a unified look, appliance fronts that match cabinetry can be added to refrigerators and dishwashers. Base end panels are used to give a finished look to end cabinets, which are not butted against a wall.

Full-height utility cabinets can be ordered with or without shelves to accommodate either dry goods or tall items such as a step stool or broom.

BATHROOMS Bathrooms come in many sizes and have specific terminology.

The smallest is a half bath, which contains a **water closet (toilet)** and **lavatory (sink)**. A **vanity** is a cabinet that supports the lavatory. (See Figure 9-34.)

Another option for a small bathroom is a pedestal sink. It requires only minimal space but does not provide the storage of a vanity. As mentioned earlier in this chapter, half baths are often placed close to work areas, with easy access from the garage, outside, or kitchen. Half baths are also popular near living areas for the convenience of visitors and residents. The next larger bath is a three-quarter bath, which includes a water closet, lavatory, and shower. The very common full bath contains a water closet, lavatory, and a bathtub or shower/tub combination. In some locations, local governance and lending institutions will require a residence to have at least one full bath, to accommodate young children or the elderly. When a bathroom has a separate tub and shower, it is called a bath-and-a-half. There is no limit to the amount of bathroom features that can be included in a bathroom plan. Bathroom suites may have separate water closet rooms, a grooming area, a sauna, a steam room, a whirlpool tub, or even a lap pool (see Figure 9-35).

Figure 9-34: Bathroom with vanity base and a lavatory top.

Residential planners follow basic standards of bathroom design. First, bathrooms are usually grouped together whenever possible to reduce the cost of plumbing (see Figure 9-36).

Second, each fixture is placed in a position for safety and ease of use. Specific guidelines include an allowance of 18 in from the center line of the water closet to an adjacent wall or tub. When a water closet is placed alongside a lavatory, allow 15 in from the water closet centerline to the side of the lavatory.

A minimum of 21 in of clearance is required in front of a water closet and a floor space of at least 30 in × 48 in is needed in front of the lavatory, tub, and shower (see Figure 9-37).

Your Turn

Sketch the diagram in Figure 9-37 into your notebook. Include all clearance dimensions. Label the water closet, lavatory, and bathtub. List the four types of bathroom and the features of each.

LAUNDRY ROOM Most house plans include a laundry area (see Figure 9-38). It can be as small as 36 in × 66 in. A laundry area is commonly accessible from a hallway or integrated within a large bathroom. A convenient laundry location

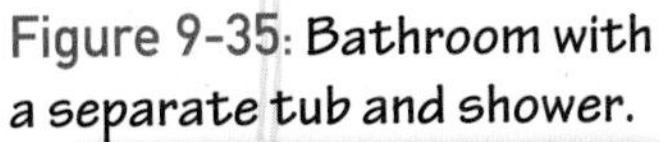

Figure 9-35: Bathroom with a separate tub and shower.

Courtesy of Paul and Deb Foster.

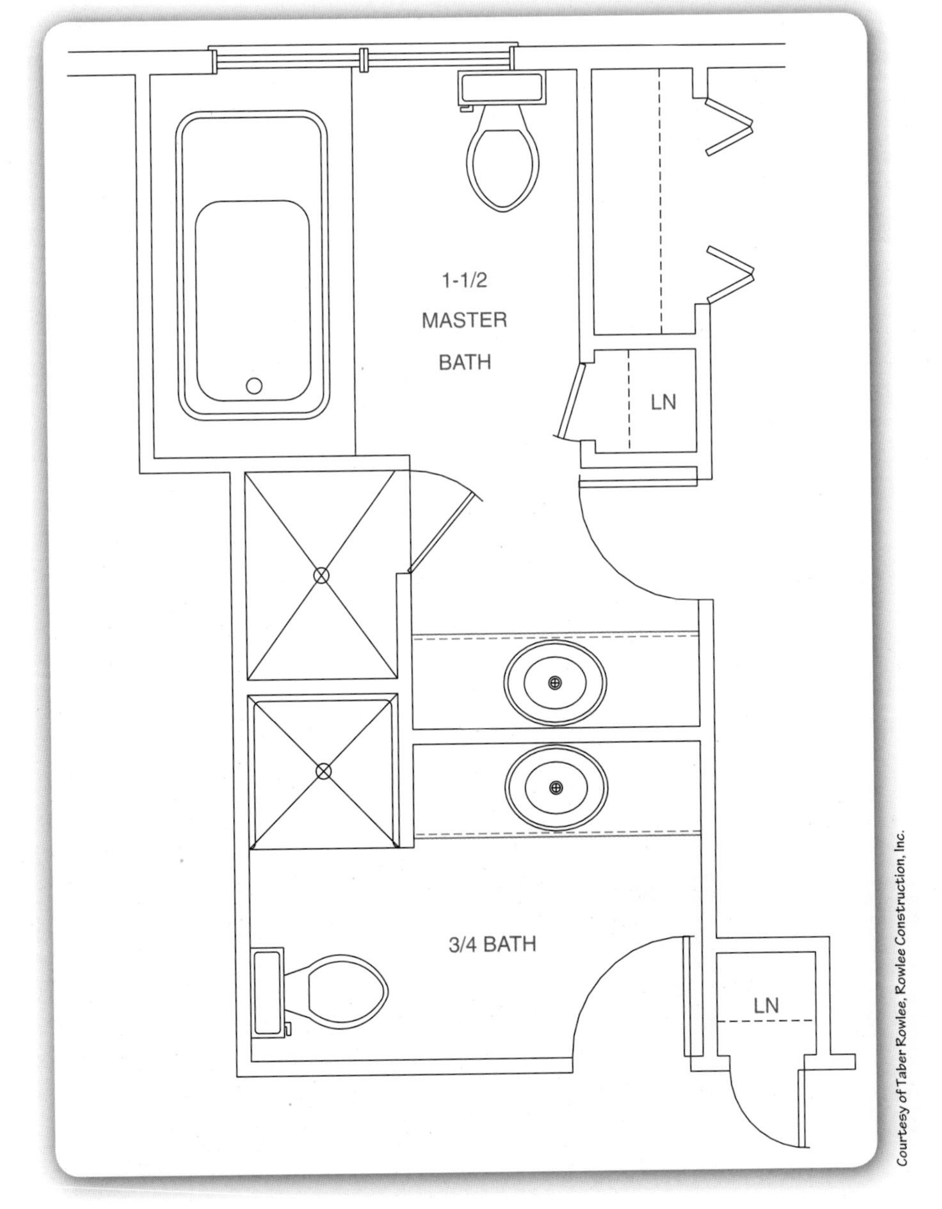

Courtesy of Taber Rowlee, Rowlee Construction, Inc.

Figure 9-36: Bathrooms placed back to back, or one above the other, will reduce plumbing cost.

Figure 9-37: Minimum bathroom clearances.

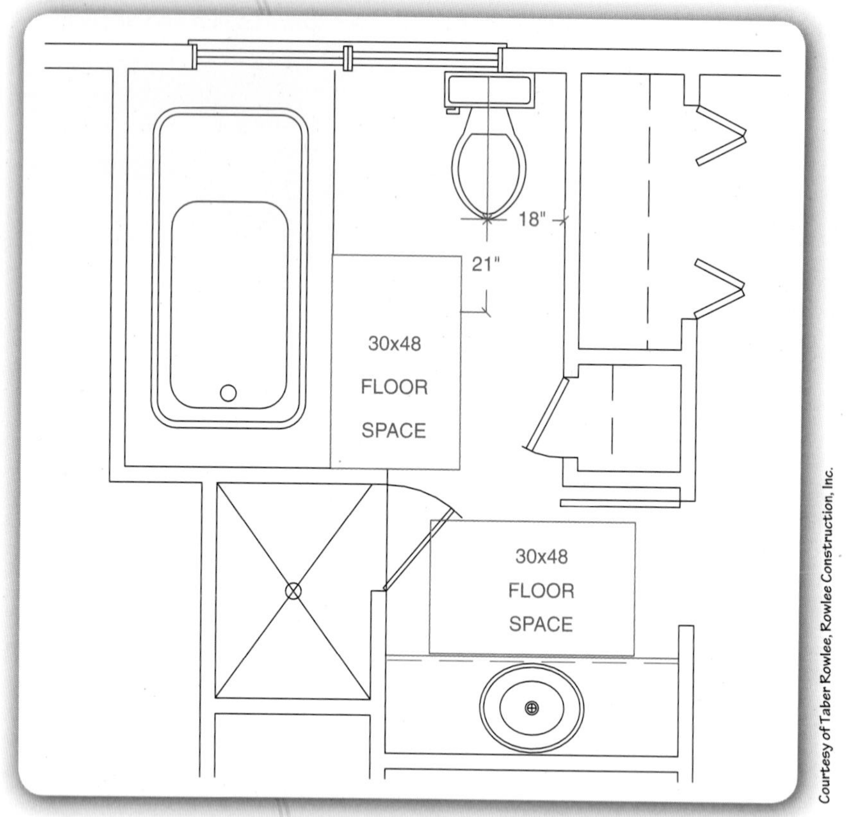

Courtesy of Taber Rowlee, Rowlee Construction, Inc.

does not require the owner to carry clothes up and down stairs. Local codes and appliance manufacturers provide guidelines for proper ventilation, which may impact laundry location options. Sometimes a laundry is planned as a separate room containing a sink, folding table, clothes rack, or ironing board. The door of a laundry room must be 2 ft 8 in or 2 ft 10 in to allow the passage of appliances. To reduce building costs, the laundry should be placed near other plumbing areas. Some clients may request that the laundry room include outside access for clothesline drying. Wherever the location, remember to provide adequate storage for laundry supplies.

Sleeping Area

Bedrooms are the sleeping areas of the home (see Figure 9-39). The client will specify the number and approximate size of bedrooms they desire. When planning bedroom locations, the architect will consider privacy, noise levels, activity, and room adjacencies. The client's master bedroom may include a sitting area for reading, a master bath, and large walk-in closets. The master bedroom is often located at the back of the house to provide the homeowner privacy, peace, and quiet (see Figure 9-40). Other bedrooms locations are based on the client's personal situation. A client with a young family may want the other bedrooms placed in close proximity to their master bedroom, whereas a client with teenagers may desire the additional bedrooms placed in a more distant location. Clients with elderly parents may need one bedroom and bath located on the main floor. A

Figure 9-38: This laundry room is accessible from a hallway and features a pocket door to maximize interior space.

Courtesy of Paul and Deb Foster.

Figure 9-39: Bedrooms are located to insure quiet and privacy.

Courtesy of Paul and Deb Foster.

guest bedroom can be designed to serve a dual purpose of a study or den. When planning sleeping areas, it is important to consider adjacency to bathroom facilities. Remember, you should never have to walk through a bedroom to get to another room. *Bedrooms should be private.*

MECHANICAL ROOMS Mechanical rooms house items such as a hot water tank, water filter, and heating unit. They are often installed in the basement; however, not all homes have basements. In places such as south Florida, where the water table is too high for basements, homes have mechanical rooms on the main floor. Their size is determined by code, number, and type of mechanical system, and floor space needed for control, repair, or replacement.

Figure 9-40: This master bedroom, located in the back of the house, features a tray ceiling, a master bath, and a walk-in closet.

Courtesy of Paul and Deb Foster.

Storage and Closets

Most people claim they never have enough storage. It is usually one the main concerns a client will bring up when planning a new home. So how does the architect determine the adequate amount of storage? Each case is based on individual needs of the client and his or her family. You will need to refer to space planning guidelines for closet and storage sizes. For example, bedroom reach-in closets need 24 in of interior depth to accommodate clothes hangers. It is common practice to provide each occupant a minimum of 4 ft to 6 ft of closet space. Climate can impact the amount of storage needed. Someone living in the upper United States may need clothes and accessories for both warm and cold seasons. Walk-in closets are very popular storage solution. The minimum depth and width of walk in closets is 5 ft, which allows for the swing of the door and a rod of clothes along one side and across the back. A common size for walk-in closets is 6 ft × 6 ft; however, many clients prefer larger walk-in closets equipped with closet systems. A closet system provides a custom arrangement of racks, shelves, drawers, and rods. A walk-in closet can also be used as a dressing room (see Figure 9-41).

A coat closet should be provided near the entry door for visitor use. A closet for occupant use may be located near the service entrance. Additional storage should be provided for cleaning supplies, linens,

Figure 9-41: A 6 ft x 10 ft walk-in closet provides space for both storage and dressing.

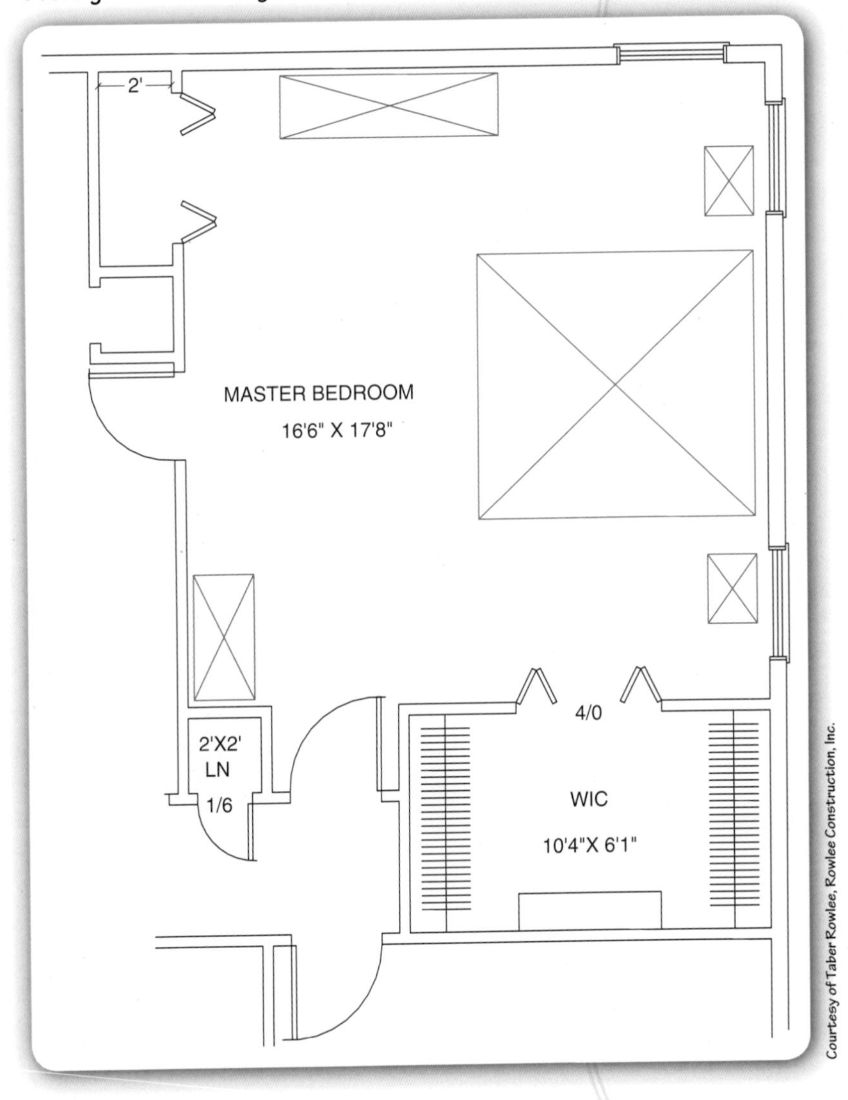

Courtesy of Taber Rowlee, Rowlee Construction, Inc.

Figure 9-42: Linen closet storage for sheets and towels should be located near the bath and bedroom area.

Courtesy of Paul and Deb Foster.

Figure 9-43: The arched ceiling of this hallway adds light and interest.

Courtesy of Paul and Deb Foster.

and seasonal items. These closets are often referred to as linen or utility closets. Linen closets are generally 12 in to 18 in deep and feature several shelves (see Figure 9-42).

Utility closets are designed based on the items they store. It may be helpful to revisit the client survey and interview notes to identify other storage needs. You may discover the client has hobby equipment and supplies that require storage. Don't forget things like books, DVDs, and electronic game equipment. Home theaters are popular in today's homes and require many electronic components. Well-designed storage offers organization, accessibility, and expandability.

Stairs and Hallways

Once the living, working, and sleeping needs are determined, the architect considers room adjacency and accessibility. As mentioned earlier, stairways are often centrally located to avoid excessive hallways. Based on room needs, the architect will determine a stairway shape and location that will accommodate both floors (see Figure 9-43).

Several math calculations will be used to determine the stairway space allowance (see Figure 9-44). First, the architect will determine how many steps will be needed to reach the next floor and then use that number to calculate staircase area. To begin these calculations, the architect first needs to know the vertical distance in inches from finished floor to finished floor. Finished floor materials such as carpet, hardwood, or tile must first be determined because they vary in thickness.

Once the architect has determined the vertical distance in inches from finished floor to finished floor, he or she divides the distance by 7.5 in, the ideal step **rise**.

Rise:
vertical distance.

After dividing by 7.5 in, the architect rounds the number up or down to the nearest whole number. This identifies how many vertical steps one must take to reach the next floor. *It is important that all steps are equal in height.* The total

Figure 9-44: Common stair layouts.

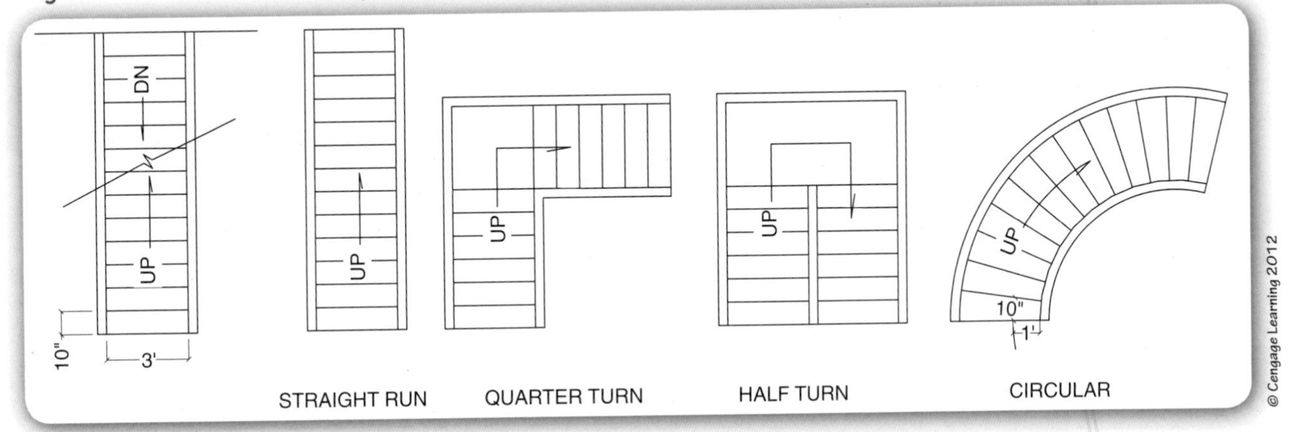

vertical distance in inches, from finish floor to finish floor, is now divided by this number of steps. This will provide the equal distance in rise between steps. This number should not be less than 7 in or greater than 8 in.

We can now calculate the linear space needed for the stairs. Each step has a **tread**, or foot space, from the front the back. The tread of basement steps is often more shallow than the tread of a main staircase because they take up less space. The depth of common tread is generally 9, 9.5, 10, or 10.5 in. The size of the tread will impact the total staircase **run**.

To determine the staircase run, or horizontal distance needed, take the number of steps, minus one, and multiply that number by the tread size. You may wonder why we subtracted one step. Look at figure 9-45 to see that the top step is actually the floor. It is a vertical step up, but it does not need to be included in our calculation. You should now have the total horizontal distance of the staircase. The area needed for the staircase is determined by the horizontal distance and the width of the stairs. It is recommended that stairs be at least 36 in wide. Many clients prefer a slightly wider staircase of 42 in or 48 in. Similar calculations are used to for alternative layouts. Some require an addition of a **landing**. An L-shaped staircase will require a landing at least the width of the stairs in both directions where it turns 90 degrees.

In the design of a circular staircase, the stair tread cannot be less than 6 in at the innermost side. The tread must be at least 10 in at a distance of 12 in from the inner side (see Figure 9-44). If a window is placed in a staircase for light, it must be high enough so that a person would not fall into the window if he or she accidentally tripped coming down the stairs.

Staircases must provide for a minimum headroom clearance of 6 ft 8 in or 80 in (see Figure 9-45). Using the step rise you calculated earlier, you can quickly calculate how many steps you

Tread:
the width of the horizontal part of a step, measured from front to back.

Run:
horizontal distance.

Landing:
a level area at entry doors or between flights of stairs.

Figure 9-45: Staircases must provide for a minimum headroom clearance of 6 ft 8 in or 80 in.

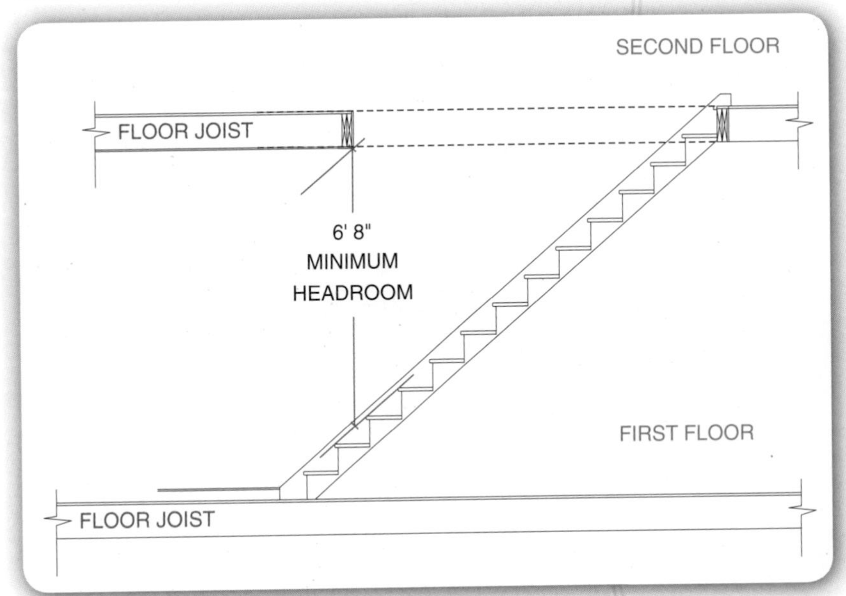

would have to take to equal this height. For example, if the step rise were 7.25 in, it would take 11 steps. Then multiply the number of steps by the tread distance to find the horizontal distance you will need to provide for the stair opening.

When a staircase, such as stairs leading to the basement, is closed off by a door, the door must not swing over the stairs. The best solution is to provide a square landing equal to width as the stairs. However, if space is limited, the door may swing away from the staircase (see Figure 9-46).

Figure 9-46: **Door swing options for closed staircases.**

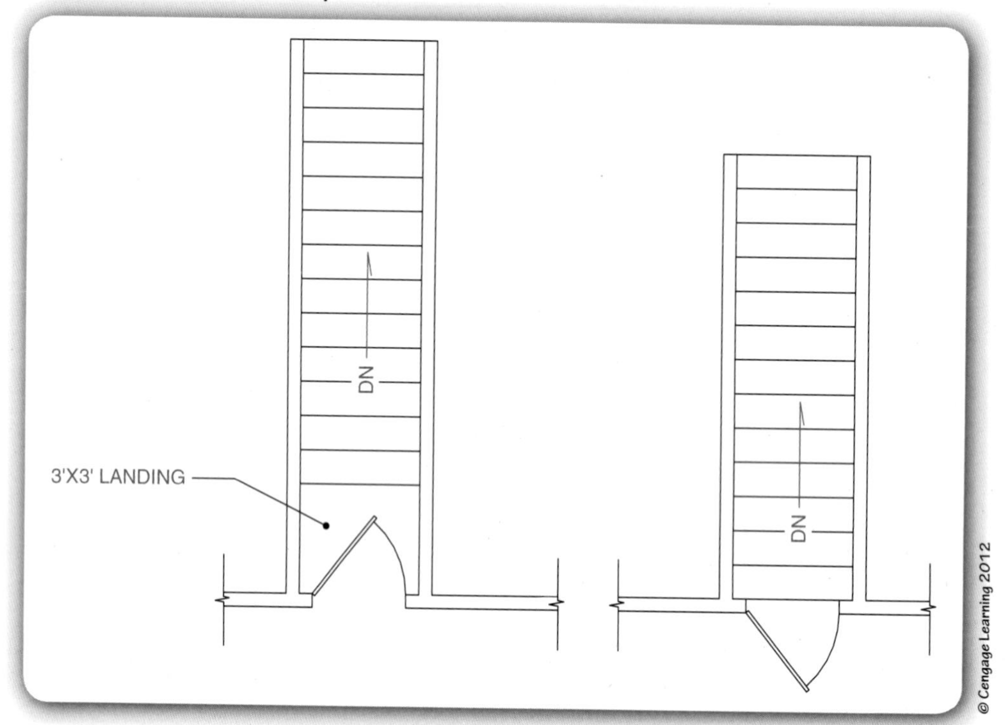

Your Turn

Try your hand at calculating a set of stairs. The vertical distance from the top of the floor on one level to the top of floor on the next level is 103.75 in.

To find how many steps you will take to reach the next floor, divide the total vertical distance (103.75) by the ideal step rise. Your calculation of 103.75 divided by 7.5 should equal 13.8333. Round the number to the nearest whole number, which will give you 14. Divide 103.75 by 14 to determine the exact rise of each step. In this scenario each step rise would be 7.4 in.

Now determine the staircase run using a tread of 9.5 in. Don't forget to subtract one step for the top floor. Thirteen steps times 9.5 equals 123.5 in of horizontal distance, or run. A 42 in wide staircase will need a space of 42 in × 10 ft 3.5 in. Try another scenario. The distance between floors is now 112 in. How many vertical steps will you take to reach the top? What will be the distance between steps? If you plan to use 10.5 in tread, what horizontal distance will you provide? If time allows, draw a plan view of an L-shaped set of stairs using the information above, including a 36 in × 36 in landing as the fourth step from the bottom.

Spiral Staircases Spiral staircases are used as design feature or for supplemental access (see Figure 9-47). If a spiral staircase is your only access to a second floor, it will be difficult to move furniture to that level. Keep in mind that although spiral stairs take up only small footprint, some people may feel they are uncomfortable or difficult to navigate.

© iStockphoto.com/Jeff Morse.

Figure 9-47: Spiral staircases may be difficult to navigate.

HALLWAYS Hallways provide convenient access and privacy to certain areas of a residence. Bedrooms, bathrooms, and laundry facilities are generally all accessed from a hallway. This helps to separate the private sleeping areas from the often noisy service areas. Hallways also provide bedrooms with privacy from the living areas. Another consideration is visual privacy. For example, when planning an entry to a bathroom, avoid placing it where it can be seen from a living space. When you design a hallway, allow a minimum of 36 in. Consider the comfort of the client. A wider hallway of 44 in to 48 in will provide better lighting and more comfortable passage. Most architects try to keep the length of hallways to a minimum. One solution is to locate room entrances where they maximize space. Compare the two floor plans shown in Figure 9-48.

When laying out a floor plan, remember to avoid long, dead-end, or maze-like hallways.

Your Turn

Record the hallway, stairs, and entry minimum dimensions into your notebook for future reference.

Hallways: 36 in width minimum. Locate bathrooms and bedrooms off hallways for privacy.

Stairs: 36 in width minimum. Calculate space by calculating steps closest to 7.5 in rise × 10 in tread. A stair landing should be as long as the stairs are wide. A door may not swing over stairs.

One entry door should be 36 in minimum, with a landing on both sides.

Planning a Garage

There are many factors to consider when planning a garage (see Figure 9-49). As you read in Chapter 6: Site Planning, the garage must adhere to all setbacks required by local governance. In addition to investigating material and labor cost, site, and local building codes, the architect must question the client concerning the number and type of vehicles, storage requirements, and other possible uses. Several questions must be answered. Will the client need direct access to the house, backyard, or bathroom facility? Will the garage need a workbench or utility sink? What items will be stored? Yard care equipment and recreational vehicles may take up a large amount of floor space, whereas seasonal items and hobby supplies could be stored on shelves or hooks. A roof constructed with rafters or attic trussed can offer additional storage. If this is the case, how will the client access the space? Will the plan include fold down attic stairs or a full staircase?

The basic purpose of a garage is vehicle storage. Vehicles range from small compact cars to extended cab pickups. The architect or space planner must accommodate for the future needs of the client. Although they may have a sports car today, they

Figure 9-48: Which plan reflects the most efficient use of space?

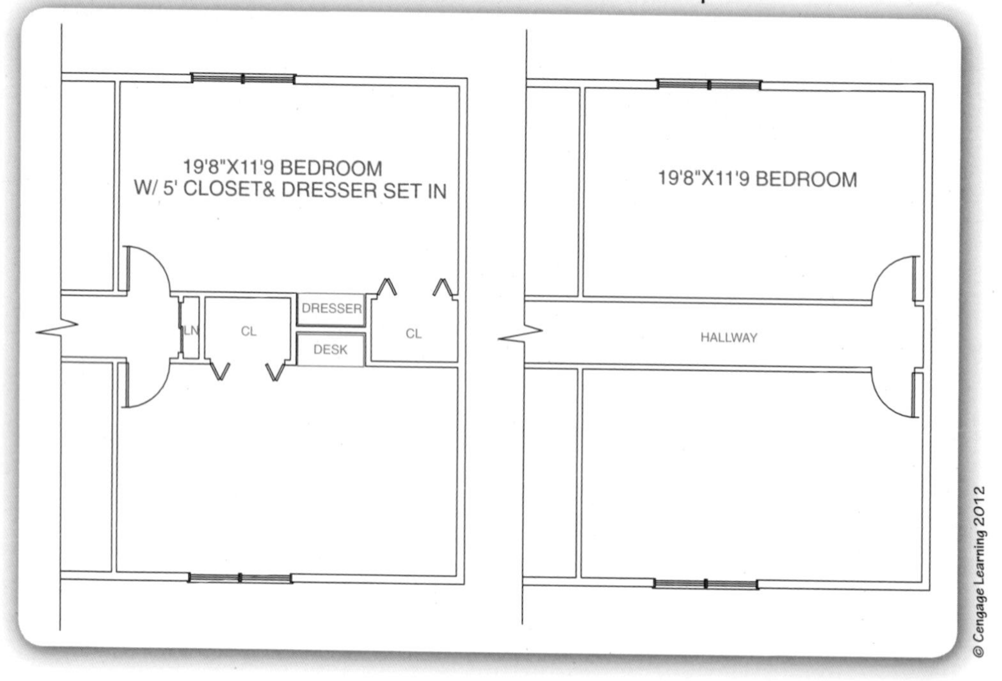

could have a sport utility vehicle tomorrow. Garage standards for vehicle parking include a 3 ft wheel stop distance to the back wall and 2 ft 6 in side clearance and space between cars.

The standard minimum interior dimensions of a one-car garage are 11 ft 8 in × 22 ft and 20 ft 10 in × 22 ft for a two-car garage. To reduce building costs, garages are generally designed using whole foot *exterior* dimensions. Thus, a 12 ft × 24 ft garage would minimally accommodate one car, and a 22 ft × 24 ft garage would accommodate two (see Figure 9-50). A larger garage will allow storage of yard equipment and tools, or provide workspace. Garage doors come in several sizes. A standard 8 ft × 7 ft door will accommodate most cars; however, a width of 9 ft would be needed for a truck with large side mirrors. A taller door can be ordered to allow for storage of a motor home. In harsh climates, the architect will place roof peak at the front of the garage to direct the snow and rain off to the side. In this case, an additional roof or portico may be placed over a walk through door located on the side.

Figure 9-49: A garage should complement the style of the house.

FIREWALL A firewall of a multi-tenant duplex or apartment may require a continuous wall from the foundation to the underside or beyond the roof structure. Non-combustible materials such as masonry or gypsum board over metal studding are often used to construct firewalls. Firewalls are designed to provide additional protection for the home and its occupants.

A firewall is constructed between an attached garage and residence and between multi-tenant dwellings such townhouses or duplexes. The firewall

Figure 9-50: A 22 ft × 24 ft garage would minimally accommodate two cars.

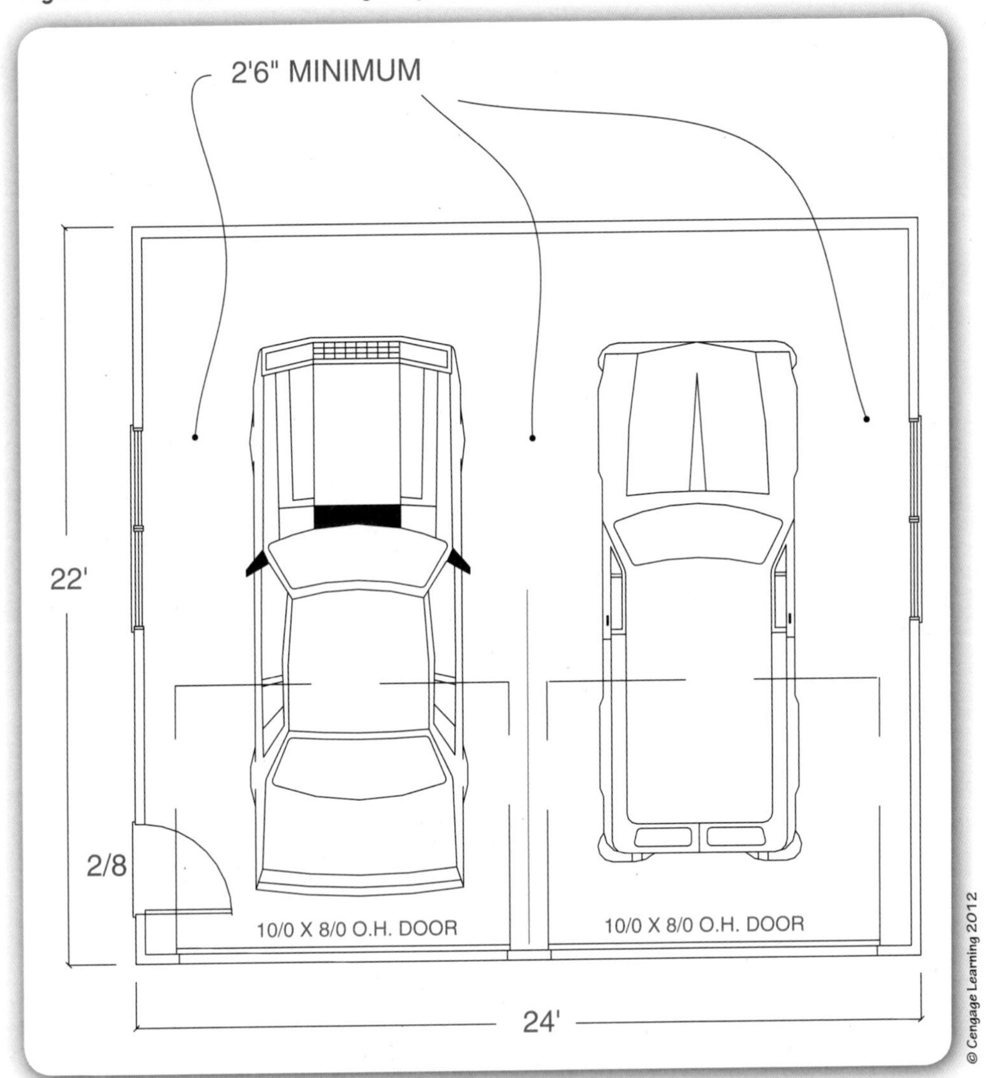

assembly is governed by national, state, and local codes specific to the type of building. An attached garage may require a firewall assembly of 5/8 in Type X gypsum board on both sides of the studded wall. Codes may specify a C labeled door with closure be used in the firewall if access to the residence is desired. Codes may also require 5/8 in Type X gypsum board on ceilings to slow the progression of a fire. Materials used for construction of firewall are coded based on fire rating, or the length of time it would take for fire to penetrate the material.

DESIGN EMPHASIS Garage design elements of line, color, form, space, texture, value, and tone can add emphasis to a residence. Additional rooflines or wall offsets can enhance visual interest and appeal. The roof pitch of the garage generally complements the roof pitch of the house. Depending on the building's architectural style, the garage can support a symmetrical or asymmetrical appearance (see Figures 9-51 and 9-52). It should be apparent that the garage belongs to the property. Unity can be achieved through application of similar materials or colors.

Figure 9-51: The side-entrance garage on this home allowed the architect to maintain symmetry.

Figure 9-52: The asymmetrical garage can achieve unity through the use of similar building materials and features.

Porches, Sunrooms, Decks, Patios, and Balconies

The living space of a home may be enhanced by the addition of a porch, deck, sunroom, patio, or balcony (see Figure 9-53). The architect designs these spaces by blending exterior and interior elements. A covered porch or sunroom may offer a pleasant place to rest by combining outside views with inside benefits of a floor, roof, and screened windows. These amenities provide shelter from sun, rain, and insects.

Decks and patios accommodate outside activities without much shelter. They can be elaborate spaces with fire pits, hot tubs, and built in barbeque grills, which are sometimes designed around a pool or garden. Balconies are small upper-level cantilevered platforms that offer residents the opportunity to step outside. The architectural design of a house can be enhanced by a porch, sunroom, deck, patio, or balcony, as in the case of the ornate front porch of a Victorian home.

Figure 9-53: A small sunroom can expand and enhance living space.

Courtesy of Paul and Deb Foster.

Elevation Options: Internal and External Spaces Working in Harmony

As discussed in Chapter 4: Architectural Design, architectural style is achieved through the application of principles and elements of design. Even though the architect has determined the desired architectural style at the beginning of the space-planning process, it is possible to create a variety of exterior looks for the same floor plan. Many different looks can be achieved through creative selection and manipulation of roof type and pitch, exterior materials, details, doors, and windows. An Arts-and-Crafts style was chosen for the home shown in Figure 9-54. The goal was a home of simple beauty and elegance that would fit comfortably within the natural surroundings.

Figure 9-54: **An Arts-and-Crafts inspired residence, designed by Taber Rowlee.**

Photo courtesy of Paul and Deb Foster.

Figure 9-55: **Repetitive patterns of three appear in timbering and window grids.**

Courtesy of Paul and Deb Foster.

To create a unique style within the Arts-and-Crafts realm, the architect cut back the gable ends of the roof and applied repetitive patterns of three and natural stone. The focal point is a front arched window emphasized by the roofline and surrounding stonework (see Figure 9-55).

The arch is repeated in the sunroom and the garage door windows. A horizontal line on the top window section creates unity by following across all windows regardless of size. Can you see the horizontal line in Figure 9-56?

The choice of window, door, roof, and exterior materials can greatly impact a building style. A contemporary feel could be achieved with tall, thin, unadorned casement windows, vertical wood or masonry exterior, and a shed roof. The same floor plan could take on a more colonial look with double hung windows, shutters, gable or gambrel roof, horizontal siding, and a front porch.

Figure 9-56: **A horizontal window grid line provides unity to multi-sized windows.**

Courtesy of Paul and Deb Foster.

Careers in Civil Engineering and Architecture

"FIGURE IT OUT"

Taber L. Rowlee: Estimator and Architectural Designer

Taber L. Rowlee remembers the day he was working on the job site of one of his father's building projects, when he was given the chance to prove himself. His dad was headed for vacation, and left Taber with a set of building plans in need of an estimate. His final words were, "here, you figure it out." "That moment was pretty important," explains Taber, "it showed that I had earned the respect to do it on my own." That successful estimate and others that followed led to a promotion from fieldwork to the office, where he now estimates and designs buildings full time. "It is both challenging and rewarding to create plans for buildings that blend design with value engineering. Today's economy requires an architectural designer to work collaboratively to maximize building potential at a realistic cost. It often entails a 'figure it out' process of negotiation between the client and designer to determine a win-win solution."

On the Job

Taber L. Rowlee has a passion for sports that dates back to his childhood. Growing up he enjoyed playing hockey, lacrosse, and other sports. He developed an acute appreciation for competition and working in as a team. Today he continues to play hockey and now coaches his son's team. His commitment to teamwork and "hate to lose" attitude drives this passion to figure out solutions to on the job challenges. Over the past decade, Taber has worked with his father and brother to sustain and grow the family construction business. What once was a residential construction company, now designs, builds, and develops commercial structures for chains like Dunkin Donuts, Pizza Hut, Kinney Drugs, and many others. When asked how this change occurred Taber explains that the construction business is impacted by building trends, "what is a lucrative job today, is gone tomorrow." So when the residential market waned, the company took on repairs, maintenance, and construction of industrial facilities. When those jobs diminished, they started looking at commercial construction. "It is important to be flexible, and look for ways to maximize every opportunity." Through the successful outcome of early commercial projects, the company's reputation for excellence and commitment spread. "We take on projects that are of hands-on size which allows us to offer a personal connection with our clients, such as there is Rowlee involved on every job in some way shape or form that can either answer any questions that may arise or to perform a certain portion of the work that needs to be accomplished. Our clients know we can get the job done, on time, and within budget."

Inspirations

When asked about inspirations, Taber explains that he has always had a strong sense of family and community. Taber and his wife Melissa, have three children, each with unique talents and abilities of their own. "My family is very important to me. They are all very driven. They inspire me to make a positive impact on our community and our future."

Education

Taber's educational path is anything but traditional. He had a passion for drawing and design from an early age and in high school was confident that he wanted to become an architect. In the summer months from high school through college he received an "on the job" education as he worked for the family construction business performing many labor intensive tasks from land clearing & building excavations to building houses & working on jobs in industrial settings. During college he couldn't deny his growing interest in the sports or health and fitness related field. All these career possibilities posed a problem that Taber had to "figure out." The process resulted in a few changes during the college years that sidetracked his goal of becoming an architect. After earning his degree at University at Buffalo, The State University of New York, he entered a health- and fitness-related field, his passion for architectural design however led him back to the family construction business. That is when he created his niche and was able to combine his drawing and design capabilities with his on the job experiences and use that with his estimating skills to not only create work for his family business but to

Careers in Civil Engineering and Architecture

sustain it in a ever increasingly changing world. Taber describes his life as an ongoing education. "Each day you learn something new."

Advice to Students

"Everyone has a different situation, but everyone has potential. Maximize what you have; interests, abilities, or relationships. Work on developing a new skill or relationship within a field that interests you, either paid or volunteer. Learn as much as you can about how things work. Life is a process which requires risk and trial and error. Don't be afraid of failure. If you head in the wrong direction, take another one. Sometimes you must take a risk. If you don't shoot, you can't score!"

SUMMARY

Residential space planning requires detailed and purposeful investigation, analysis, and development. The process is not linear, but instead may require several transitions between investigation, analysis, and development. The goal of an architect or space planner is to design a functional and aesthetically pleasing plan that can meet the client's current and future needs and wants. Research and investigation is needed to gather information about the client, neighborhood, building site, materials, local governance, and space standards. The acquired information is then analyzed to determine project parameters, specifications, and limitations. The architect must have a clear picture of the needs, desires, and financial position of the owner to determine a budget. The client's budget impacts the square footage and level of complexity and quality of building components. Trade-offs and revisions are common during the investigation and analysis process. The goal is to agree on a realistic project vision that will meet client's needs while keeping within the budget. Architects often propose creative solutions to maximize the efficiency of spaces, such as dual-purpose rooms and built-in cabinetry (See Figure 9-57 A and B). Current and future activities of the occupants are major considerations when planning spaces. During the development process, the architect will analyze room adjacencies and traffic patterns.

Standards account for human dimensions and assure mobility and adequate space for safety and comfort. Standards identify distances and clearances to guide the placement of cabinetry, furniture, fixtures, and appliances. Architects apply these standards along with principles and elements of design to plan spaces that will be functional and appealing from both an interior and exterior perspective. When planning interior spaces, the architect or space planner identifies each room by purpose; working, living, or sleeping. Rooms are then thoughtfully arranged to enhance the comfort of the occupants and guests, maximize the site features, and achieve harmony. Architects and space planners understand that optimal space planning provides for future flexibility, expandability, and resale value. Diligent efforts in space planning will enhance owner satisfaction over the building's lifecycle.

Figure 9-57 A and B: Built-in bookcases enhance a library's appearance and function. (A)

Courtesy of Taber Rowlee, Rowlee Construction Inc.

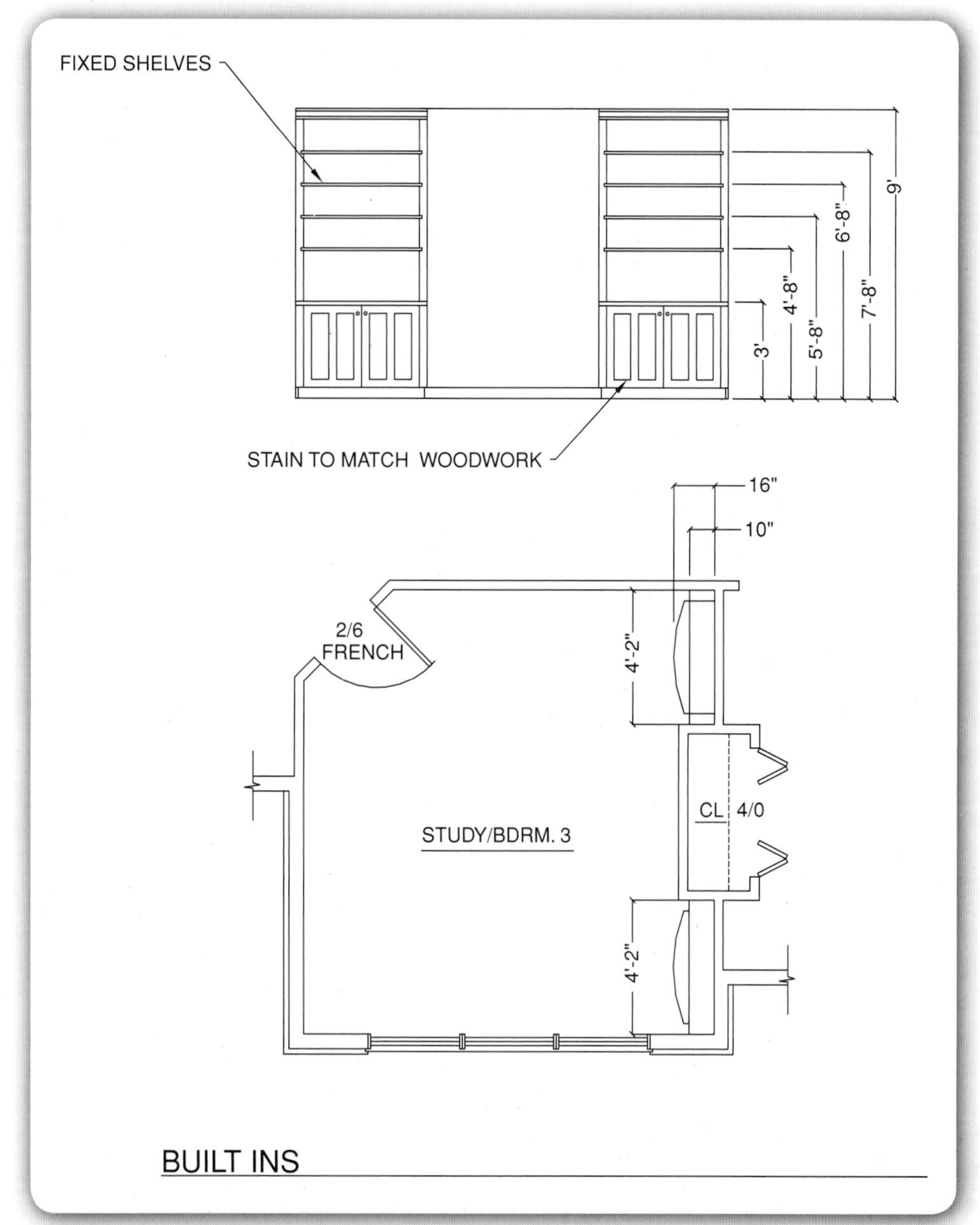

Figure 9-57 A and B: (*continued*) (B)

Courtesy of Paul and Deb Foster.

BRING IT HOME

1. Create an adjacency matrix to guide the floor-plan development of a two-story urban rowhouse (see Figure 9-58) for the client described below. Show both primary and secondary adjacencies. The building will be located between two brick rowhouses. The building site will accommodate a 16 ft × 48 ft, 2.5-story structure. The single client desires a floor-plan layout where he or she could entertain up to five guests on the main floor. The main floor must include a large kitchen, a medium dining area and a living room, and a half bath. On the second floor, the client has requested a large master bedroom, full bath, and laundry facilities. The client works from home as a game designer and would like to have the third floor serve as a computer room.
2. Using ¼ in graph paper, create ¼ in = 1 ft scale representations of the rooms requested. Refer to chart in Figure 9-13. Label and group the rooms into living, working/service, and sleeping areas. Arrange several floor plans using graph paper cutouts allowing space for a centralized entry, hallway, and staircase. Create bubble diagrams of two or three options, and label all rooms and entry locations.
3. Analyze the bubble diagrams and choose one to convert into a ¼ in = 1 ft floor-plan sketch. You may use graph paper. Draw all walls a thickness of 1/8 in to represent the six in walls in scale. Show storage, hallway, stairway, and entry locations. Label each room.
4. Consider both the interior and exterior appearance and sketch a front and back elevation showing window and door style and placement. Identify and label the elements that will support the desired architectural style.
5. Using standard furniture charts or ¼ in scale templates, sketch the location of all furniture, making sure to allow adequate space clearance.
6. Plan a kitchen layout by sketching the locations of cabinetry, appliances, and fixtures. Label all work areas, and include a kitchen triangle that

meets efficiency guidelines. Label the actual distance on each side of the work triangle, and record the total distance in the center.

7. Plan layouts for the two bathrooms by sketching fixture locations. Include dimensions and note required clearances.
8. Using two different colored pencils draw the traffic patterns of both the occupant and guests. Analyze the plan for efficiency. Are there any congested areas? Does the plan provide convenient and comfortable movement during activities?

Figure 9-58: Rowhouses are generally long and narrow.

EXTRA MILE

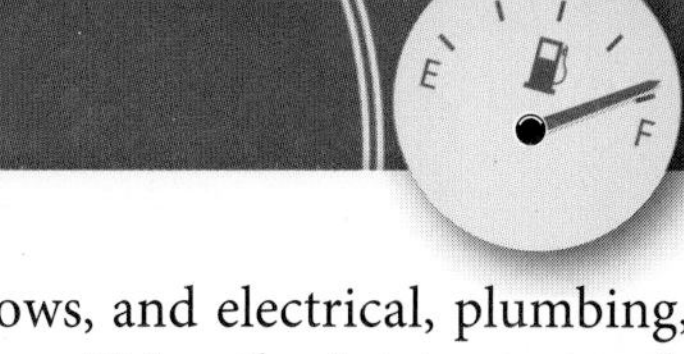

1. Working in a small group, locate a vacant building that is still structurally intact. Research to determine the building's exterior dimensions and footprint. Expand the investigation to include site features, neighborhood, and local governance. Working as a team, brainstorm design options to renovate or convert the structure to a residence or townhouse.
2. ISO shipping containers are used to ship goods worldwide, but after two or three years they are replaced with newer models. What happens to these durable corrugated steel and tubular steel frame structures with thick marine grade plywood floors? Some believe they can be a good source of housing. They can equipped with doors, windows, and electrical, plumbing, and HVAC systems. Using the Internet, search "shipping container housing." You will discover that a common 20 ft to 24 ft container size is 8 ft wide × 8.5 ft high. International shipping uses containers of 40 ft to 53 ft in length. Plan an arrangement using three 40 ft × 8 ft shipping containers and design the interior space to accommodate a family with two children. Shipping containers can be combined, stacked, and insulated.

To view a video showing the construction of a shipping container house, go to *http://www.bobvila.com/BVTV/Bob_Vila/Video-0201-05-1.html*

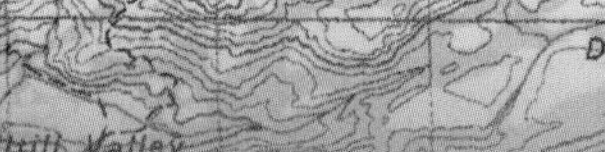

CHAPTER 10
Commercial Space Planning

GPS DELUXE

Menu | START LOCATION | DISTANCE | END LOCATION

Before You Begin

Think about these questions as you study the concepts in this chapter:

1. How does commercial space planning differ from residential space planning?
2. How does purposeful space planning contribute to a building's sustainability?
3. How do space planners provide for safe and efficient occupant circulation?
4. What are a building's public spaces and how are they incorporated into the space plan?
5. What are the factors to consider when placing entrances and exits?
6. How do workplace environments impact individual and group performance?
7. Why are mechanicals and infrastructure considered early in the space-planning process?
8. What are the building support spaces?
9. How are the areas of a commercial building measured and described?

➢ INTRODUCTION

As you may recall, for the purpose of this textbook we divided buildings into two major categories: residential and commercial. Although commercial architecture specifically defines private nonresidential facilities that affect commerce, such as factories, early in this textbook, we broadened our definition to include places of public accommodations. Places of public accommodations include lodging, restaurants, theaters, libraries, stadiums, daycare centers, stores, office buildings, and health, recreation, and transportation facilities. So how does the planning of commercial space differ from residential space? The most obvious difference is that they are handicapped accessible (see Figure 10-1). You undoubtedly have noticed a difference between the restroom facilities in your school and the bathrooms in your home. The restrooms in your school are most likely wheelchair accessible and designed to accommodate several people at one time. Public restrooms incorporate commercial toilets, urinals, sinks, hand dryers, privacy partitions, and dispensers. Usually a custodial closet is nearby for service of the restroom. Restrooms are just one example of the many differences between commercial and residential spaces. In the hallways of commercial buildings you may see drinking fountains, fire alarms, sprinklers, exit signs, Automatic External Defibrillators (AEDs), and other safety equipment.

Commercial buildings have restricted access to specific rooms and spaces. In your school you do not have access to file rooms, teacher break areas, custodial closets, or spaces filled with mechanical, electrical, and communications equipment. Every building space is thoroughly researched to determine

Figure 10-1: *Commercial spaces provide accessibility for people with disabilities.*

the most functional, aesthetically pleasing, and cost-effective location. Locations impact building performance, ease of construction, and occupant safety, comfort, convenience, and accessibility. Why do you think the main office, gymnasium, auditorium, music rooms, and technology areas of your school are usually on the ground floor? A ground floor location allows a quick exit. Some rooms are placed in a ground floor because of noise, odors, or structural considerations. Commercial buildings must also comply with occupancy limits, which specify how many people can safely occupy and circulate about a space. Circulation studies determine the space necessary for efficient movement about the building. What would happen if the hallways in your school were too narrow? Another feature of commercial buildings is a main entrance. The main access doors of a commercial building are usually centrally located in a highly visible location and provide convenient access to the main lobby. In your school, the main entrance is most likely placed near the main office.

Today's commercial buildings are planned for sustainability. A building might be positioned to maximize natural lighting or solar gain (see Figure 10-2). Building construction, operation, maintenance cost, building durability, and lifecycle are considered during the planning phase. When designing school buildings, for example, the designer knows that student populations, subjects, and teaching methods change over time. These changes may require the building to be expanded or reconfigured to adapt to new technologies. Consider the renovations that were needed in schools built around 1980, before computers were commonplace. Some schools didn't have enough electrical outlets, or space to run cables for networking. How will wireless technology make some building features obsolete? Today commercial space planners work together with collaborative partners to gather information, analyze, and strategize to create a sustainable plan that will serve occupants and owners for years to come. Think again about your school. Is the student population always the same year after year?

Figure 10-2: *Spaces are often positioned to maximize natural lighting or solar gain.*

Architects, property owners, and other stakeholders have increased interest in green and sustainable buildings. A sustainable building is considered more affordable to operate and maintain. New research has linked building environment to occupant performance. Air quality, natural light, aesthetics, and other physical characteristics are believed to effect attendance, mood, and even test scores! A commercial space planner performs many tasks. In addition to adhering to building codes and regulations for occupant safety and comfort, space planners often research to locate materials, mechanical equipment, windows, doors, and building components to enhance building performance, aesthetics, and sustainability.

COMMERCIAL SPACES ARE DESIGNED FOR SAFETY, ACCESSIBILITY, FLEXIBILITY, AND SUSTAINABILITY

Commercial buildings must follow a distinct set of building codes. Building codes are developed for safety and accessibility of building occupants, and to enhance the longevity and structural integrity of the building. You may want to review the information on codes in Chapter 3: Research Documentation and Communication. The space planner is well versed in the local building codes and common practice. For example the designer may follow the general practice of providing space for one water cooler or drinking fountain for every 75 employees. In addition, the building code may require that each floor have a drinking fountain. Codes may also stipulate that each floor have both male and female restroom facilities. A public restroom with several stalls will include privacy dividers and have privacy offset at the entrance so that people cannot see into the restroom from the corridor. Codes for restroom facilities are usually based on occupancy and use. For example, depending on the size of an adjacent assembly area, code may require a minimum of one male toilet stall per 75 to 125 people, and one female toilet for every 65 to 75 people. It is not uncommon to provide three toilets in a women's restroom for every two provided in men's restrooms. For assembly areas, codes may specify a minimum of one sink or lavatory for every 200 people. An office building might require one toilet stall for every 50 people and one sink for every 80. Check the codes in your area to determine specific requirements.

In addition to occupant comfort and health, codes promote occupant safety. For example, corridors for egress, or exit, are specified by multiplying the total occupancy load of the floor, multiplied by a width variable, divided by the number of exit directions (see Figure 10-4). Space planners must consider the increase in corridor and stairway load if one exit becomes blocked. Codes establish corridor minimums based on building use, occupancy, and number of exit directions. You must use the corridor minimums even if your space calculations come up less. ADA recommends that a corridor should be wide enough for both a wheelchair and a person walking to pass comfortably (see Figure 10-5).

Example: A common formula to calculate corridor and stair width for safe egress:

Minimum width for safe egress = occupancy (load) × width variable ÷ number of exit directions

Width variable	
0.2	level exit
0.3	exit stair

Example: If a floor has an occupancy or load of 750 with 3 exit directions, the formula for calculating the corridor width would appear as follows:

750 (occupancy) × 0.2 (variable) = 150 in ÷ 3 (exits) = 50 in minimum width for safe egress

For safety, the space planner may calculate the corridor load with two exits, allowing for increased load should one of the three exits become blocked.

750 × 0.2 = 150 in ÷ 2 (exits) = 75 in minimum width

Case Study

In our case study of the Center of Excellence (CoE) in Syracuse, New York, the plans include classrooms, laboratories and testing facilities, offices, meeting rooms, restrooms, and storage areas. The CoE will provide a facility for collaborative research projects to improve indoor environments though advances in lighting, sound, air quality, thermal comfort, water quality, and renewable and clean energy. Renewable and green energy include wind, solar, and geothermal. The new CoE building in Syracuse, designed to use half the energy of an average building the same size, will serve as a "test bed" to monitor and measure employee performance based on innovative features such as individually controlled ventilation systems. (See Figure 10-3.)

Figure 10-3: *The fifth-level Syracuse Center of Excellence building will provide spaces for research, testing, and development of sustainable and green technology.*

Level 5
There will be office space available on this floor for our visiting partners, or those who may like to maintain space in our building. The Carrier TIEQ Lab will be located on this floor as well as a conference room and a lounge.

Level 4
There will be office space available on this floor for our visiting partners, or those who may like to maintain space in our building. There will also be a monitoring lab and a lab preparation and equipment storage space.

Level 3
Aside from an indoor and terrace lounge, Level 3 will also have a conference room and a space that could be used as a lab, preparation space or office space depending on the needs of our partners.

Level 2
This level will have more than 1000 sq ft of gallery space to display the work of our partners. It will also include office space available to partners who would like to maintain a desk for use when they are in town, a conference room and classroom; all available for partner use.

Level 1
There will be moderate-sized reception area on level one as guests enter the building. There will also be more than 12,600 sq ft of laboratory space equipped to provide our partners with adequate space to test and process biofuels.

Courtesy of the Syracuse Center of Excellence.

Your Turn

Calculate the stair width for a floor occupancy of 500 with two stair exits. Remember to use the 0.3 width variable.

Record the formula and width variable chart in your notebook for future use.

Figure 10-4: The formula to calculate corridor and stair width for safe egress includes variables for occupancy and number of exit directions.

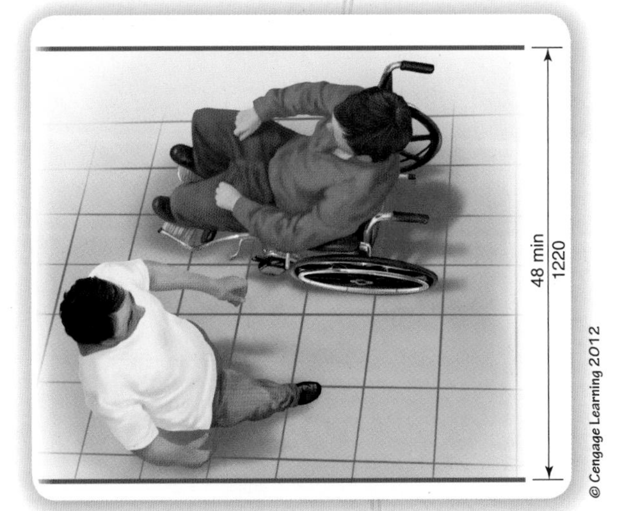

Figure 10-5: Minimum passage width for one wheelchair and one ambulatory person. Adapted from the ADA Standards for Accessible Design, U.S. Department of Justice, July 1, 1994.

Healthcare facilities, hospitals, schools, and other buildings with unique circulation needs are required to provide wider hallways to allow safe and efficient movement of occupants and equipment.

Corridor and stair width is usually specified by building use. For example, school corridors serving 4 classrooms, which total 100 person occupancy, will require a minimum of 6 ft wide corridors. School corridors serving more than four classrooms may need a 10 ft to 12 ft wide corridor to insure safe circulation.

Local governments often adopt national building codes such as those developed by the International Code Council (ICC). Conscientious space planners work diligently to meet or exceed codes while economically maximizing building value. Green, sustainable, and resilient features are often proposed. Leadership in Energy and Environmental Design (LEED) recognizes these areas as sustainable sites, water efficiency, energy and atmosphere, materials and resources, indoor environmental quality, and innovation in design (see Figure 10-6 a and b).

When you designed your residential building, you began with assessment of the site and the future homeowner's needs and wants. Commercial space planning begins much the same way, but instead of research to meet the needs of a single family, you will assess the needs of many stakeholders who will own, utilize, and maintain the building. In addition to gathering site information, a commercial space planner will seek out information from the building owner, manager, occupants, and maintenance, custodial, and security personnel. This information will form a starting point to identify the public, private, and support spaces of the building. Space allotments are determined based on occupancy, purpose, amenities, how business is conducted, and individual style or preference. In office environments, 160 to 250 of usable square feet per person is generally allotted. Benchmarks and space calculators may be used as a starting point to determine the amount of useable space needed (see Figure 10-7).

Once the space planner has a general listing of spaces, the approximate square footage is assessed based on function and activities planned for the space. This document is referred to as the Program for Space Types (see Figure 10-8).

Figure 10-6 a and b: The Hearst Tower, a high-rise green office building, was the first building in New York to achieve LEED Certification.

(a) Image copyright Michel Stevelmans. Used under license from Shutterstock.com. (b) AFP/Getty Images.

Figure 10-7: Typical business space estimates.

	Typical Business Space Estimates
President's office or Chairman of the Board	250 to 400 sq ft (4 to 5 windows in length)
Vice-President's office	150 to 250 sq ft (3 to 4 windows in length)
Executive's office	100 to 150 sq ft (2 windows in length)
Employee office	100 to 125 sq ft (desk, lateral file, 1 visitor chair)
Partitioned open space	80 to 110 sq ft
Open space	60 to 110 sq ft
Workstation area	50 to 100 sq ft for clerical (depending on file and equipment needs) 64 to 80 sq ft for technical personnel
Conference rooms	15 sq ft per person: theater style 25 to 30 sq ft per person: conference seating
Mail room	8 to 9 ft wide with 30 in counters on either side. Length depends upon usage
Reception area	125 to 200 sq ft Receptionist and 2–4 people 200 to 300 sq ft Receptionist and 6–8 people
File room	7 sq ft per file with a 3 ft to 4 ft aisle width
Library	Allow 12 in for bookshelf width 175 to 450 sq ft with seating for 4–6
Lunch rooms	15 sq ft per person, not including kitchen. Kitchen should be one third to one half of seating area
Coat closets	1 lineal ft accomodates 4 coats, calculate 3 coats per person

Source: http://www.officefinder.com.

Figure 10-8: The Syracuse Center of Excellence Program of Space Types.

SyracuseCoE HQ Program Space Types

Location: 727 East Washington St.
Syracuse, NY 13244

		Sq Ft[1]	√ Available by Appt (A) License (L)
Ground Floor (cement)			
Unit 101	Mezzanine/Lobby	1,500	
Common	Elevator lobby rest rooms, lockers, showers	1,000	
Unit 102	Building mechanical rooms	2,500	
Lab Wing (design and fit out required):			
Unit 108	Bio Fuels Lab, SUNY ESF	2,415	
Unit 106	High Bay Flex	2,150	√ (L)
Units 103-112		5,700	√ (L)
Second Floor (raised)			
Unit 201	Office	630	√ (L)
Unit 202	Office hotel (6 cubes + collaboration space)	1,100	√ (L,A)
Unit 203	Classroom	980	√ (A)
Unit 204	Conference room	450	√ (A)
Common	Gallery	1,000	√ (A)
Common	Elevator lobby, kitchenette	600	
Third Floor (raised)			
Unit 301	Office, SyracuseCoE	600	
Unit 301a	Office/storage, SyracuseCoE	250	
Unit 302/303	Office, SyracuseCoE	2,000	
Unit 304a	Collaboration area	180	√ (A)
Common	Lounge, kitchen, terrace, rest rooms	1,500	
Fourth Floor (cement)			
Unit 401	Monitoring/controls, SyracuseCoE	1,200	
Unit 402	Interstitial testing, SyracuseCoE	2,000	
Unit 403	Office or lab	700	√ (L)
Unit 404	Office or lab	1,000	√ (L)
Common	Elevator lobby, rest rooms	600	
Fifth Floor (raised)			
Unit 501	Office, TIEQ PI	1,100	
Unit 502	Carrier TIEQ Lab I	720	√ (A)
Unit 503	Carrier TIEQ Lab II	720	√ (A)
Unit 504	TIEQ Prep	460	√ (A)
Unit 506/507	Fanger Room	1,000	√ (A)
Common	TIEQ Reception, elevator lobby, rest rooms	700	

[1] Square footage notations are net approximate.

Rev. 4/1/09

Courtesy of the Syracuse Center of Excellence.

Your Turn

Use the online calculator at http://www.officefinder.com/officespacecalc.html to approximate the square footage estimate for a spacious office plan to accommodate:
10 employees
1 president
2 executives
Partitioned open space for 7 employees
1 conference room for 40 people
1 mailroom
1 file room
1 library
Reception area for 6 people

Record the square foot estimate in your notebook. Recalculate using an efficient plan and compare the square footage savings. How much square footage was saved with efficient space planning? Now change the partitioned area for the 7 employees to an open (bullpen) plan. How does this change impact the square footage? What are some possible advantages and disadvantages of an open plan? Record your answers in your notebook.

A space planner will research the occupancy, furniture, equipment, and activities planned for each space. Some activities will require a specific amount or type of light, whereas others may require a quiet, private, or secure space.

Your Turn

Assess the list of spaces, planned for your commercial building, to determine the building occupancy of each public, private, and shared space. Think about the room relationships. Which spaces will complement each other and which should be placed a distance apart? This information will be helpful when you begin to plan adjacencies. Copy the following analysis list into your notebook and add other considerations you think will be relevant to effective space planning. Use this list to plan your commercial spaces.

Individual Space Analysis Considerations:
Function, activities, occupancy, ADA
Square footage desired
Equipment type and size
Furniture and movable partitions, type and size
Window types, views, and solar directions
Electrical and communication needs
Plumbing needs
Natural and artificial lighting—specify general, task, decorative
Acoustical needs
Air quality and thermal needs
Security needs
Storage facilities required
Ceiling heights
Public and internal access
Relationship to other spaces, stairs, elevators, restrooms

Adjacency Matrix

Once the general square footage of each space has been determined by occupancy and use, the spaces are assessed for adjacency. An **adjacency matrix** is a common technique, used by space planners, to determine where the rooms should be placed. When developing an adjacency matrix, the space planner considers the type of space, function, furnishing, equipment, commonalities, and connectivity to other spaces. The matrix helps identify essential space relationships, compatible zones, and shared space options (see Figure 10-9 a and b).

Your Turn

Take a closer look at the program document outlining space recommendations and planning of handicapped accessibility for the Madison County, Montana County Courthouse proposed by Schlenker & McKittrick Architects. Search Montana County Courthouse Program using the Internet, or use the PDF below: http://madison.mt.gov/aboutus/PB/Final ProgramDoc2.pdf

Figure 10-9 a: An adjacency matrix of Madison Courthouse in Virginia City Montana.

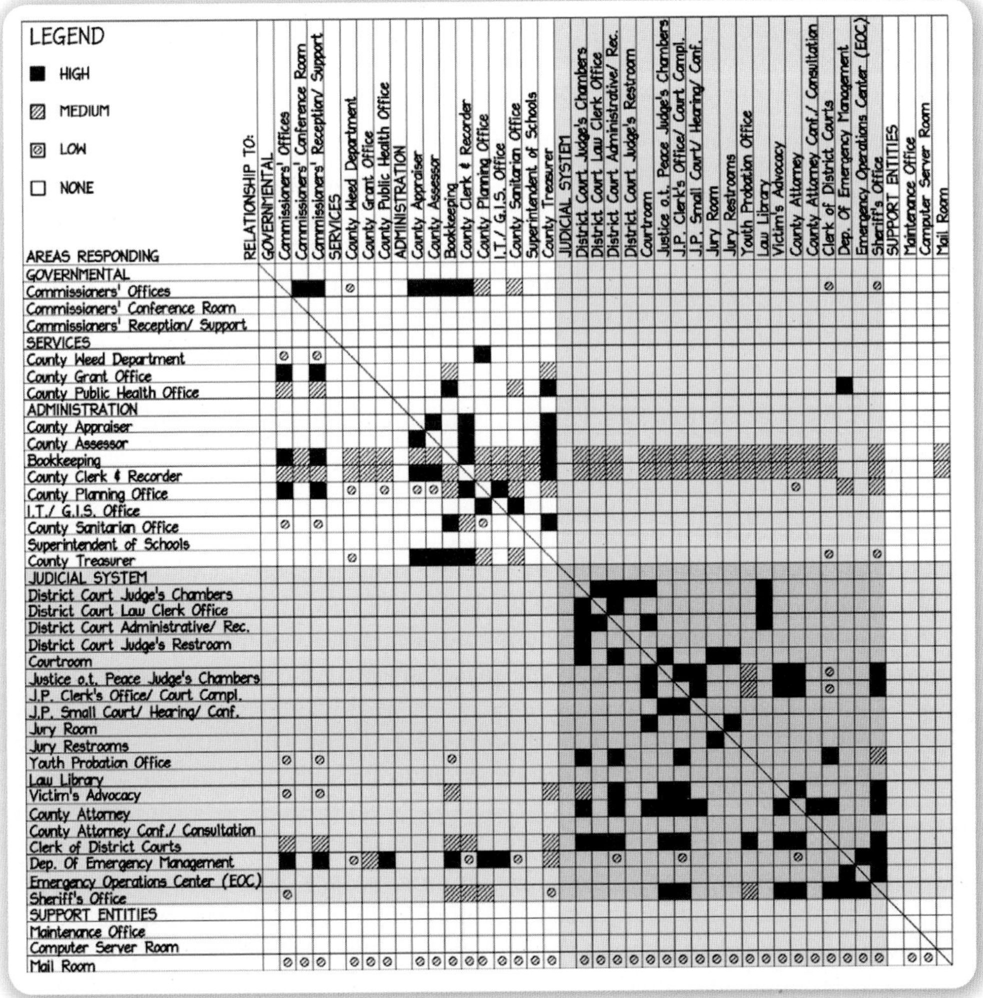

Courtesy of Schlenker & McKittrick Architects, P.C.

Figure 10-9 b: *Space analysis and building program for the Madison County, Montana County Courthouse.*

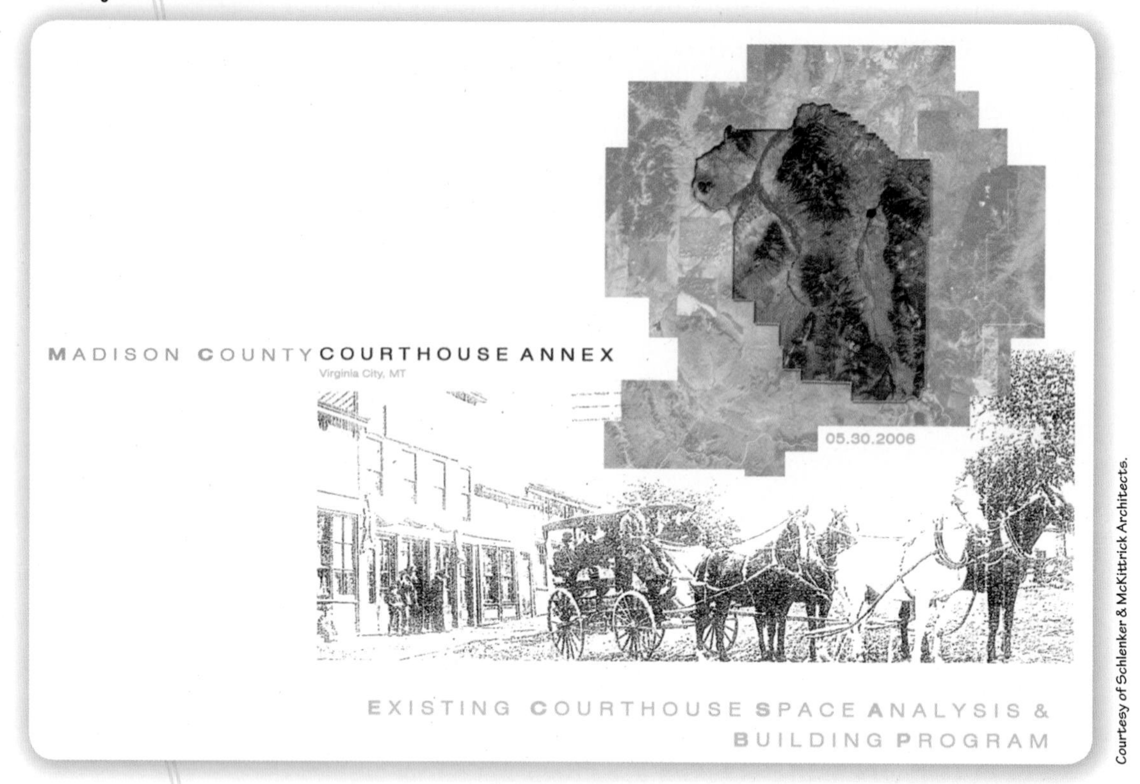

Following the development of an adjacency matrix, the space planner begins brainstorming for groupings and develops bubble diagrams of possible arrangements. Spaces are located to maximize site features, enhance building performance, reduce construction costs, and minimize flow of traffic through the building. The location of spaces and adjacencies can enhance the productivity of staff, maintenance, custodial, and security personnel.

THE SPACE PLAN ALLOWS SAFE AND EFFICIENT CIRCULATION OF BUILDING OCCUPANTS

Commercial spaces are designed with consideration for mobility and accessibility for all people, including those that are physically disabled. Commercial space planners look to the Americans with Disabilities Act Accessibility Guidelines (ADAAG) for minimum dimensions and clearances for access and mobility. Technical requirements of accessible spaces are posted on government websites (see Figure 10-10).

In addition to 48 in of corridor space needed for one wheelchair and one walking person to pass, the space planner must allow enough space for the wheelchair to turn around. A wheelchair will need 78 in × 60 in to make a smooth U-turn (see Figure 10-11).

Interior doorways must have a minimum *clearance* of 32 in for a wheelchair to pass through. If you look closely at Figure 10-12, you will see that the door is larger than 32 in. Space planners specify 34 in or 36 in internal doors and 36 in to 40 in exterior doors to provide adequate access.

Figure 10-10: *Government websites are a good source for ADA information.*

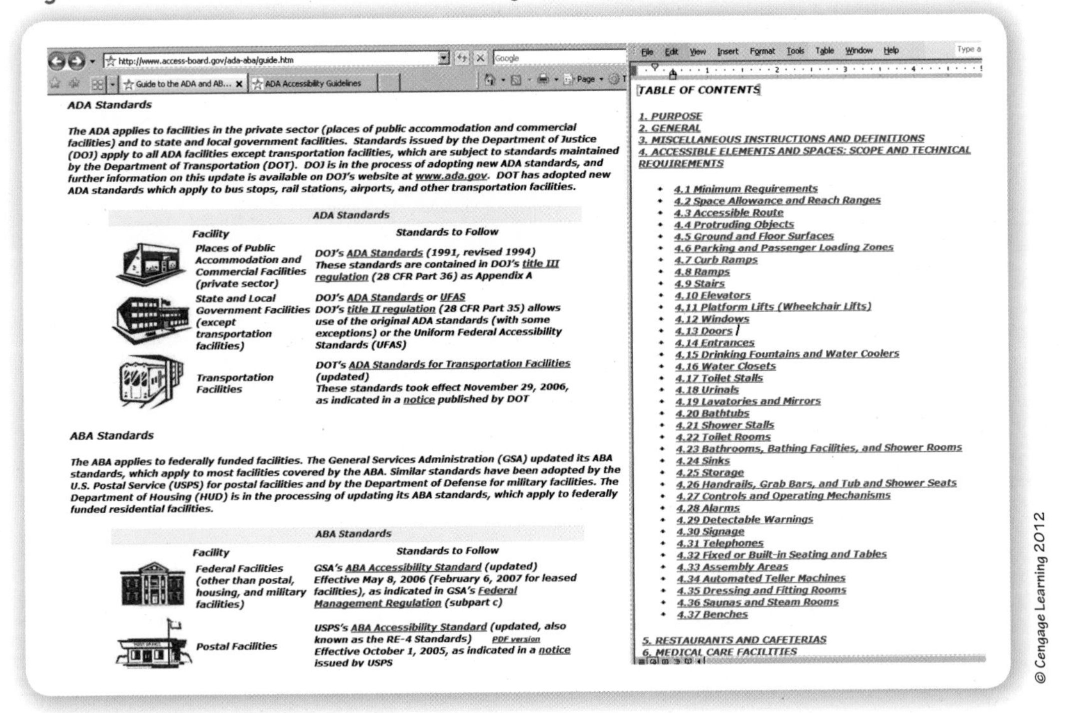

http://www.access-board.gov/ada-aba/guide.htm

Guide to the ADA and AB... | ADA Accessibility Guidelines

ADA Standards

The ADA applies to facilities in the private sector (places of public accommodation and commercial facilities) and to state and local government facilities. Standards issued by the Department of Justice (DOJ) apply to all ADA facilities except transportation facilities, which are subject to standards maintained by the Department of Transportation (DOT). DOJ is in the process of adopting new ADA standards, and further information on this update is available on DOJ's website at www.ada.gov. DOT has adopted new ADA standards which apply to bus stops, rail stations, airports, and other transportation facilities.

ADA Standards

Facility	Standards to Follow
Places of Public Accommodation and Commercial Facilities (private sector)	DOJ's ADA Standards (1991, revised 1994) These standards are contained in DOJ's title III regulation (28 CFR Part 36) as Appendix A
State and Local Government Facilities (except transportation facilities)	DOJ's ADA Standards or UFAS DOJ's title II regulation (28 CFR Part 35) allows use of the original ADA standards (with some exceptions) or the Uniform Federal Accessibility Standards (UFAS)
Transportation Facilities	DOT's ADA Standards for Transportation Facilities (updated) These standards took effect November 29, 2006, as indicated in a notice published by DOT

ABA Standards

The ABA applies to federally funded facilities. The General Services Administration (GSA) updated its ABA standards, which apply to most facilities covered by the ABA. Similar standards have been adopted by the U.S. Postal Service (USPS) for postal facilities and by the Department of Defense for military facilities. The Department of Housing (HUD) is in the processing of updating its ABA standards, which apply to federally funded residential facilities.

ABA Standards

Facility	Standards to Follow
Federal Facilities (other than postal, housing, and military facilities)	GSA's ABA Accessibility Standard (updated) Effective May 8, 2006 (February 6, 2007 for leased facilities), as indicated in GSA's Federal Management Regulation (subpart c)
Postal Facilities	USPS's ABA Accessibility Standard (updated, also known as the RE-4 Standards) PDF version Effective October 1, 2005, as indicated in a notice issued by USPS

TABLE OF CONTENTS

Figure 10-11: Space needed for wheelchair turn around. Adapted from the ADA Standards for Accessible Design, U.S. Department of Justice, July 1, 1994.

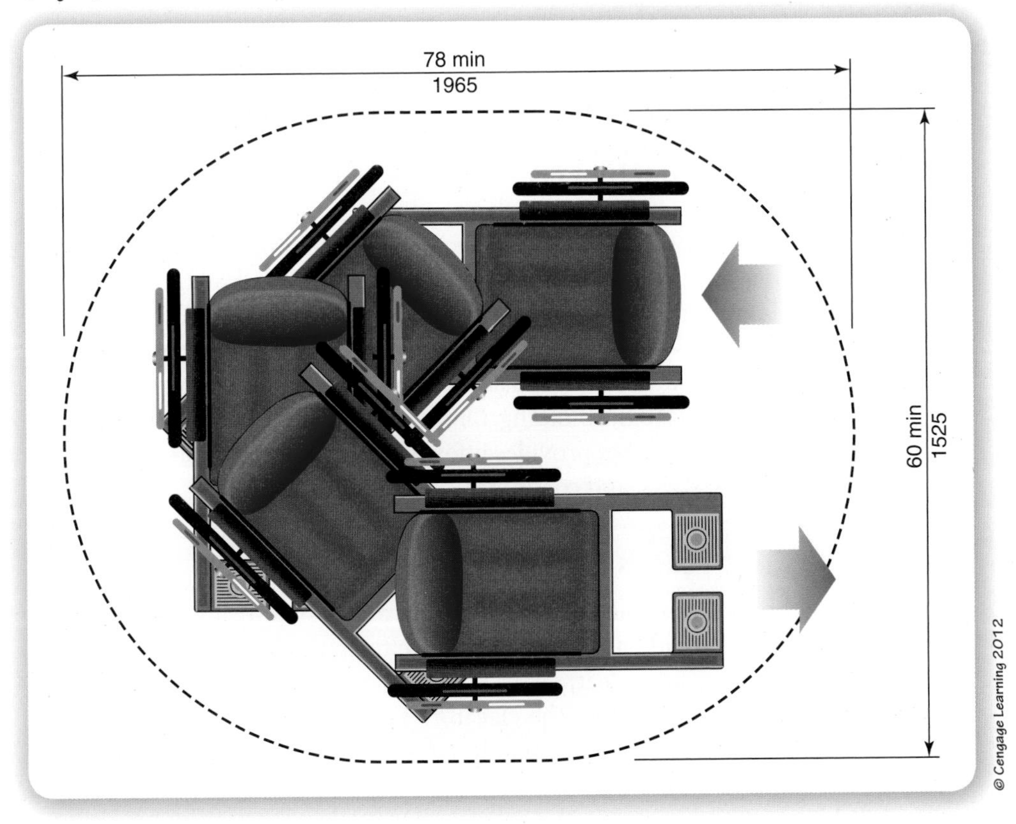

Figure 10-12: Clear doorway width for wheelchair passage. Adapted from the ADA Standards for Accessible Design, U.S. Department of Justice, July 1, 1994.

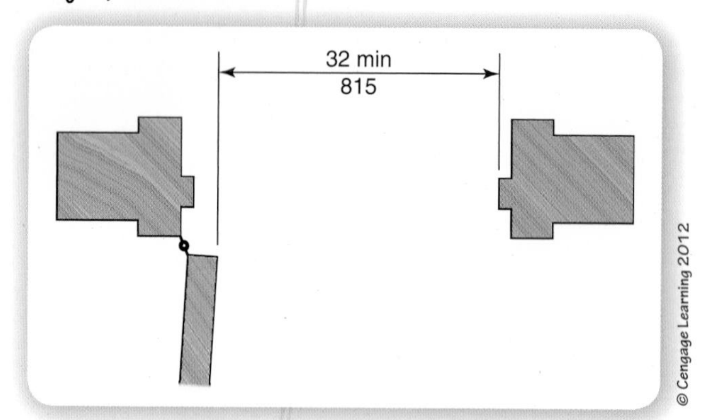

Figure 10-13: A building's main entrance should be highly visible.

PUBLIC SPACES ARE LOCATED FOR VISIBILITY AND ACCESSIBILITY

Public spaces such as lobbies or reception areas should be in highly visible locations (see Figure 10-13). Visitors to the building should not have to search for the main entrance. The central lobby acts as a hub and provides close access to the elevator, stairway, general restroom, and other public areas, such as auditoriums, cafeterias, or gift shops. The lobby is designed to accommodate a large amount of occupant traffic and is usually brightly lit and monitored by security.

Exits and Entrances Provide Safety and Accessibility

The main entrances of large buildings often feature double glass doors or revolving doors called *turnstiles.* If a turnstile is used, an ADA-accessible door must be provided adjacent to the revolving door. Entrances and exits are located to provide safe and efficient **ingress** and **egress** to the building.

Ingress: the way in.

Egress: the way out.

In addition to the central public entrance, many buildings have separate entrances for employees. Fire exits are located at far ends of the building or ground level of stairs. Main entrances should be visible and conveniently accessible from public sidewalks, public streets, parking, and loading zones for public transportation, if available. A building's entrance must balance aesthetics with function, and provide safe and convenient access to the building. Once inside, it is common to find a reception area, lobby, or vestibule that provides information and clear visible connections to elevators, escalators, stairs, and corridors.

Access to Other Floors

Building elevators are generally centrally located and visible from the lobby (see Figure 10-14). Stairs provide a healthy alternative to elevators and are required by code for egress during a fire. Clearly marked stairs are generally located at the ends of the building and centrally near the elevator. The number of elevators is determined by occupancy, traffic analysis, building type, number of floors, and capacity and speed of the elevator units.

For general occupant use, a minimum of two elevators are usually placed together, providing a back up in case one requires repair or maintenance. Average elevators have a 3500 to 4000 lb capacity and will transport 18 to 20 people at a time. A general calculation used to determine elevator population uses 150 sq ft per person. Larger buildings like the Empire State Building have several elevators that are designated to serve specific floors (see Figure 10-15). Service elevators are designated for a specific purpose, such as transportation of large furniture or supplies. Often service elevators are located near a shipping dock or outside access.

Escalators are used in shopping malls, airports, and convention centers to transport a large volume of people in a short amount of time. A 32 in wide escalator from one floor to the next can transport 3000 people per hour (pph). A 48 in wide escalator can move 4000 pph!

Figure 10-14: **Elevators are usually located near the lobby.**

Public Restrooms

Most public bathrooms provide one large handicapped accessible stall that is usually located at the far end for efficient use of space and convenience of wheelchair turn around. Sometimes a handicapped accessible sink, or lavatory, is included in the stall, but if not, one of the lavatories must be positioned lower and provide 27 in of knee clearance (see Figure 10-16).

Your Turn

Look closely at the features of public restroom. In your notebook, list eight to ten features that are unique to a commercial restroom.

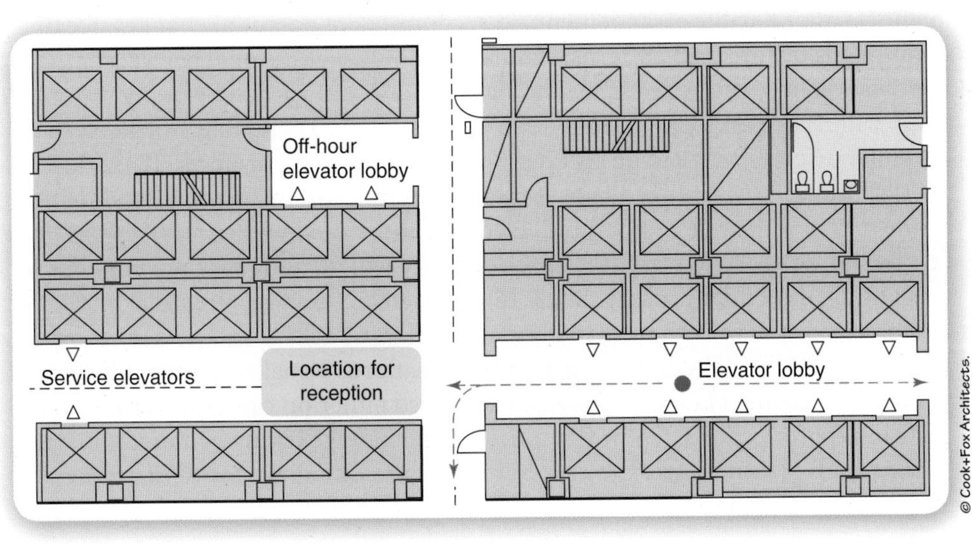

Figure 10-15: The 24,382 sq ft of office space designed for Skanska, USA, on the thirty-second floor of the Empire State Building in New York is served by 10 general, 2 service, and 2 off-hour elevators.

Figure 10-16: ADA guidelines for accessibility in public bathrooms. Adapted from the ADA Standards for Accessible Design, U.S. Department of Justice, July 1, 1994.

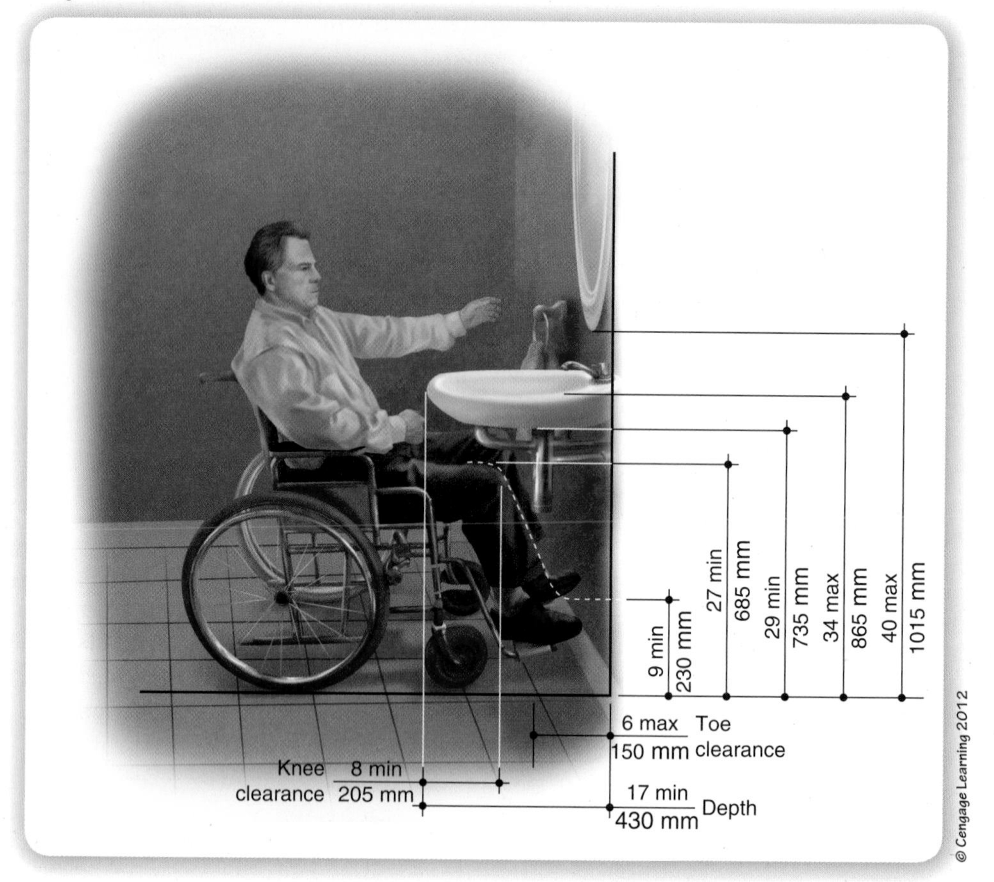

A SUSTAINABLE SPACE PLAN PROVIDES FLEXIBLY, ADAPTABILITY, AND AFFORDABILITY

A building's purpose may change many times throughout its lifecycle. Some changes are driven by changes in technology and the way businesses are conducted. To maintain a competitive advantage, companies must stay current with proven methods that foster employee creativity and productivity. Most space planners offer flexible arrangements to accommodate potential changes. Flexible space plans feature grid patterns, standard ceiling heights, movable walls, modular components, and wireless technologies. Work settings can be easily reconfigured with space dividers and modular furnishings. Expandable worktables and modular display areas can be set up for different tasks and projects. These flexible workspaces can be arranged to accommodate group brainstorming sessions and teamwork and then be easily subdivided into individual work areas.

To support flexible plans, the space designer will plan for easy reconfiguration of lighting and outlet arrangements. In commercial buildings this is achieved by use of a grid system with plug in floor and ceiling panels. A grid system provides uniformity and sustainability. The grid system affords visual alignment of window mullions to ceiling systems, which can be placed either on center or offset 50 percent in both directions. Columns and partitions are generally centered on grids. Floor panels placed by grid are easy to reconfigure and allow convenient access to power, telephone, and data outlets.

Figure 10-17: Net-Zero definitions developed by the National Renewable Energy Laboratory, one of the U.S. Department of Energy's national laboratories.

	Net-Zero Definitions
Net-zero site energy	A building that produces at least as much energy as it uses in a year, when accounted for at the site. The measurement time frame is annual.
Net-zero source energy	A building that produces at least as much energy as it uses in a year, when accounted for at the source. "Source energy" refers to the primary energy required to generate and deliver the energy to the site. To calculate a building's total source energy, multiply imported and exported energy by the appropriate site-to-source conversion multipliers.
Net-zero energy costs	A building where the amount of money a utility pays the building's owner for the energy the building exports to the grid is at least equal to the amount the owner pays the utility for the energy services and energy used over the year.
Net-zero energy emissions	A building that produces at least as much emissions-free renewable energy as it uses from emission-producing energy sources annually. Carbon, nitrogen oxides, and sulfur oxides are common emissions that NZEBs offset.
Near zero energy	A building that produces at least 75 percent of its required energy through the use of on-site renewable energy. Off-grid buildings that use some nonrenewable energy generation for backup are considered near zero energy because they typically cannot export excess renewable generation to account for fossil fuel energy use.

Ceiling Heights

Although ceiling height specifications for government buildings vary, office spaces larger than 150 sq ft require a minimum of 9 ft ceiling height and smaller offices under 150 sq ft require a minimum of 8 ft. Government guidelines recommend using a uniform ceiling height to provide flexibility for future floor-plan changes (United States General Services Administration Public Building Service).

Space Planning and Energy Consumption

Efficient space planning of a commercial building can impact energy consumption and enhance sustainability. Building size, shape, envelope, solar orientation, building material, and energy choices all affect the buildings energy efficiency and affordability (see Figure 10-19). Although the initial cost of an energy-efficient building may be greater, once built, such buildings cost less to operate, use fewer non-renewable resources, and have less impact on the environment. Often products are chosen based on lifecycle cost analysis. Lifecycle cost analysis considers initial operation, and maintenance, cost and dependability of product and company, length and terms of warranty, update and adaptation, and recommended industry standards. For more information, you may want to revisit Chapter 8: Energy Conservation and Design.

Figure 10-18: The Aldo Leopold Legacy Center qualifies as a site ZEB, source ZEB, and emissions ZEB.

ZERO ENERGY BUILDINGS
Learn more about Zero Energy Buildings

The 11,900 sq ft Aldo Leopold Legacy Center in Baraboo, Wisconsin, is a carbon-neutral, net-zero energy building (ZEB) (see Figure 10-18). It produces roughly 10 percent more than the energy needed to operate over the course of a year.

Use the Internet to search the U.S. Department of Energy website to learn more about the Energy Efficiency and Renewable Energy Building Technologies Program and Net-Zero Commercial Building Initiative: http://www1.eere.energy.gov/buildings/commercial_initiative/

The U.S. Department of Energy has identified 16 commercial building types that represent approximately 70 percent of the commercial buildings in the United States. This benchmark information can give you a perspective on commercial buildings and annual energy use per square foot in various climates. The whole-building energy analysis of these benchmark models was achieved through use of EnergyPlus simulation software.

Figure 10-19: *Average building sizes.*

Building Type Name	Floor Area (ft²)	Number of Floors
Large office	498,590	12
Medium office	53,630	3
Small office	5500	1
Warehouse	52,050	1
Stand-alone retail	24,690	1
Strip mall	22,500	1
Primary school	73,960	2
Secondary school	204,170	3
Supermarket	45,000	1
Fast food	2500	1
Restaurant	5500	1
Hospital	241,350	5
Outpatient health care	10,000	2
Small hotel	43,200	2
Large hotel	122,116	6
Midrise apartment	33,600	4

Source: U.S. Department of Energy, Energy Efficiency and Renewable Energy, http://www.eere.energy.gov.

The U.S. Department of Energy reports that commercial and residential buildings use almost 40 percent of the primary energy and approximately 70 percent of the electricity in the United States. The Department of Energy predicts that without change in building design electricity consumption of commercial buildings is expected to increase 50 percent by 2025. To address this issue, the U.S. Department of Energy has established the Net-Zero Commercial Building Incentive to promote the development of technology and practice of cost-effective zero-energy commercial buildings (ZEBs) by 2025. A Net-Zero energy building (NZEB) typically uses on-site renewable energy and produces as much energy as it uses over the course of a year (see Figure 10-17).

WORKPLACE ENVIRONMENTS INFLUENCE WORKER PERFORMANCE

Today's workspaces are designed to maximize worker productivity and provide the flexibility to accommodate growth, change in practice, and new and emerging technologies. Research has shown that the physical environment can affect a person's physical and psychological well-being and, ultimately, impact performance (see Figure 10-20).

Indoor air quality, sunlight, shadows, and views have been proven to stimulate and evoke feelings of comfort or well-being. Biophilia is the term used to define human's innate emotional attachment to nature. This challenges the architect to apply his or her knowledge of elements and principles of design to elicit positive employee feelings. The location and adjacencies of workspaces and the use of visuals, color, materials, form, and shape are believed have the power to inspire teaming, creativity, and encourage company pride. An integrated workplace considers the nature and impact of physical settings, work patterns and efficiencies, and the organizations culture and management style.

When Cook+Fox designed a 25,000 space for fashion designer Elie Tahari on the fiftieth floor of One Bryant Park, the goal was to provide a high-fashion showroom with a natural world feel. The showroom space needed to be able to transform to match a desired look and feel. A creative collaboration between the architect and Elie Tahari identified theatrical and cinematic themes to guide the design process. Designers looked to the building for clues. The glass-skinned building allowed breathtaking views of the skyline, weather patterns, and natural elements of the upper atmosphere. It was agreed that the concept for the showroom was to reinforce these **biophilic** connections while being careful not to upstage the fashion design on display.

Figure 10-20: **Space planners manipulate the physical environment to promote positive feelings.**

When Cook+Fox began deliberating to envision the showroom for One Bryant Park they assessed the views available from every side of the building: Central Park, the Hudson River, the Statue of Liberty, the Empire State Building, the East River, the Chrysler Building, Bryant Park, and the downtown skyline (see Figure 10-21).

Biophilic:
in tune with nature.

The designers at Cook+Fox determined the solar effect on the space in various seasons and how the natural light and resulting shadows would impact the space, fashion displays, and occupants. Functionality, aesthetics, and emotions were considered when planning the arrangement and placement of spaces. You may enjoy reading some of the designer's notes for planning of this space (see Figures 10-22 a and b).

You may wonder how all the furniture and equipment will comfortably fit the plan. Figure 10-23 shows a detailed space plan with equipment and furniture placed in each area. The inclusion of furniture assists design visualization and affirms that enough space has been allotted.

Design visualization is enhanced by the construction of models. You may remember reading about models in Chapter 3: Research, Documentation, and Communication. Models provide viewers a three-dimensional understanding of the spaces. The model of the fiftieth floor shown in Figure 10-24 includes scaled human models to enhance the viewers comprehension of proportion.

Your Turn

Take a few minutes to look at Figures 10-21 to 10-24. They illustrate the space plan for the fiftieth floor of One Bryant Park in New York City. Why do you think the designers selected glass walls for the five 137 sq ft executive offices? What exterior views are offered from the showroom? What advantage does the showroom location offer?

Figure 10-21: *Views and natural lighting helped determine the optimal location for the fashion showroom.*

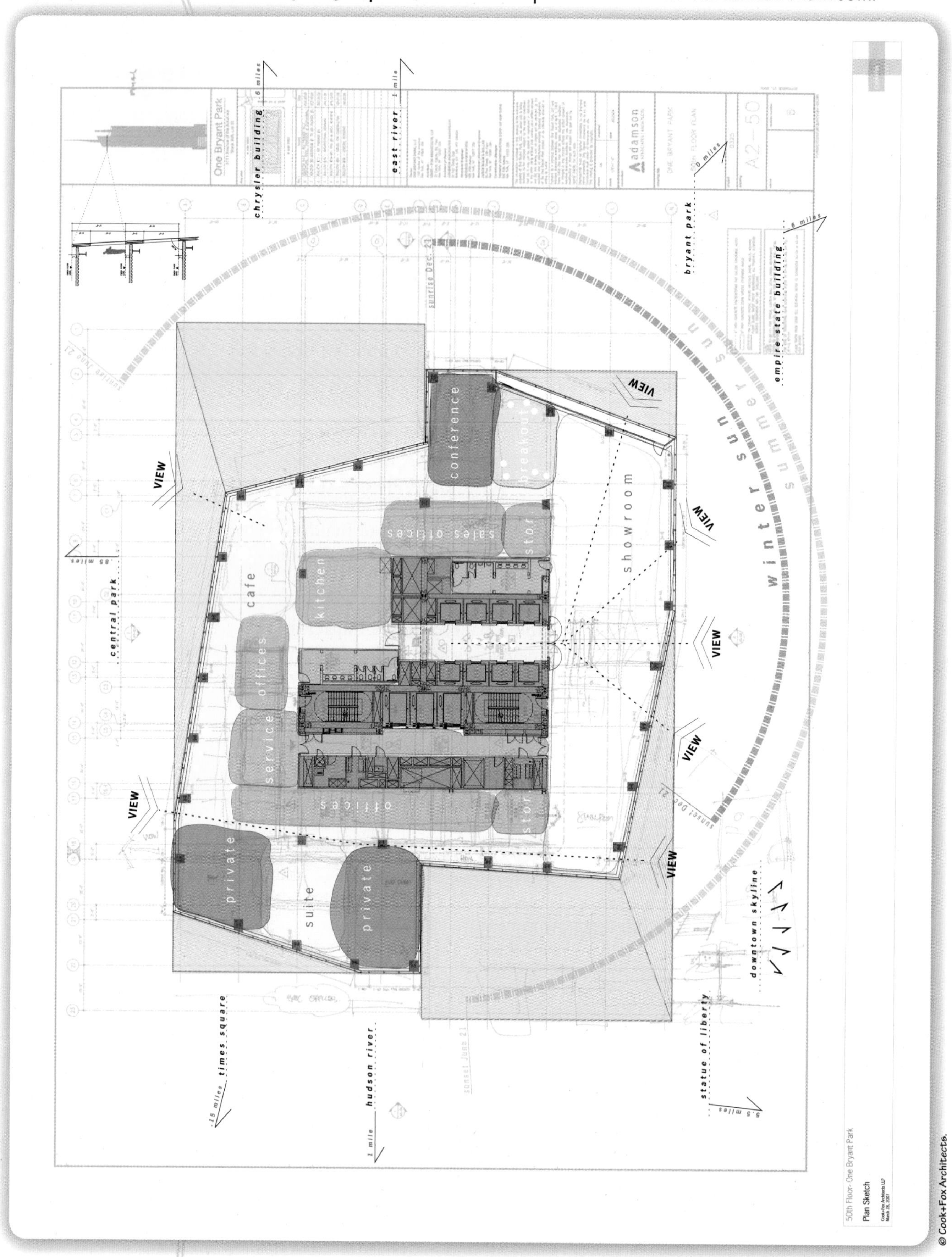

© Cook+Fox Architects.

An Open Office Plan versus a Closed Office Plan

When designing an office, a space planner will research the organization's mission, purpose, and business functions. This information will identify and assess patterns of daily workflow (how work is conducted and how information is gathered, processed, and communicated). Research may include how the workplace has changed over time and how it may continue to change to keep a competitive advantage. These profiles help the space planner design an environment that will enhance workflow and increase productivity.

One of the major decisions in office planning is whether to use an open or closed office plan. You may wonder how they differ and what advantage and disadvantages they offer. A traditional or **closed plan** generally features individual offices and separate work and break areas with full high walls. An **open plan** features informal workspaces, sometimes referred to as a bullpen, grouped by commonality with low or no partitions. Workers have access to shared and unassigned spaces. Advocates of the open plan say that the open plan is a better use of space, provides better overall lighting, and fosters teamwork, collaboration, communication, and a sense of belonging. Opponents say the open plan limits privacy, encourages interruptions, impedes confidential communications, and is distracting (see Figure 10-25).

Many space planners design integrated workplaces to enhance workflow. An integrated workplace accommodates different work styles by providing space for individual work, teamwork, socialization, and offers areas for private meetings and conversations. Spaces are designed to be visually inspiring, provide encouragement, and promote company pride, corroboration, networking, and mentoring. Some companies offer their employees amenities such as in house coffee shops or fitness facilities. Integrated workplaces may include centrally located unassigned offices, conference rooms, and workrooms with convenient access to supplies and materials. These shared spaces may be used for individual work, group work, training sessions, and client presentations. Individuals are encouraged to manipulate the spaces to meet their project needs. Companies often request an open plan because of versatility and the ability to be economically reconfigured to adapt to future changes in business procedures.

Figure 10-22a: The designer considers the function and aesthetics of each space.

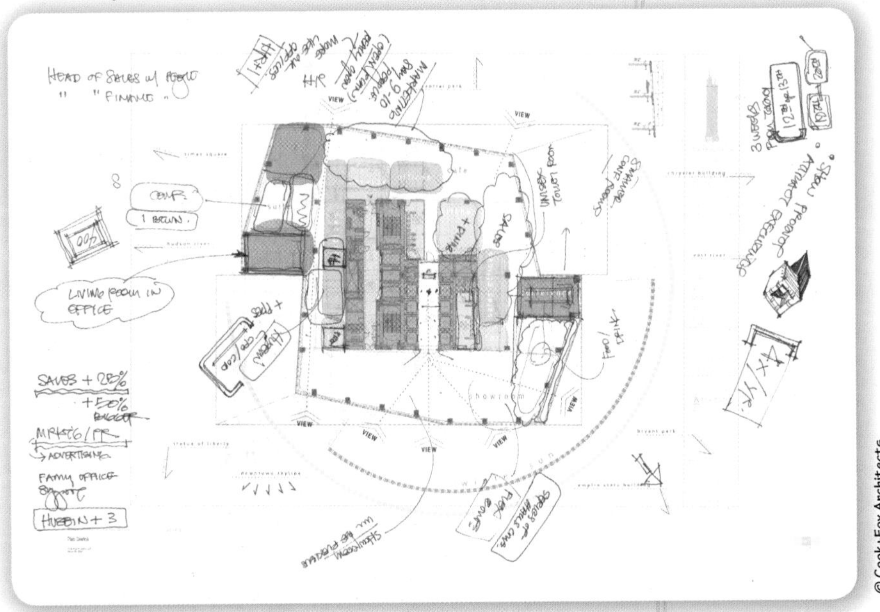

Figure 10-22b: Notes include feelings the designers want to evoke with color, materials, lighting, and view.

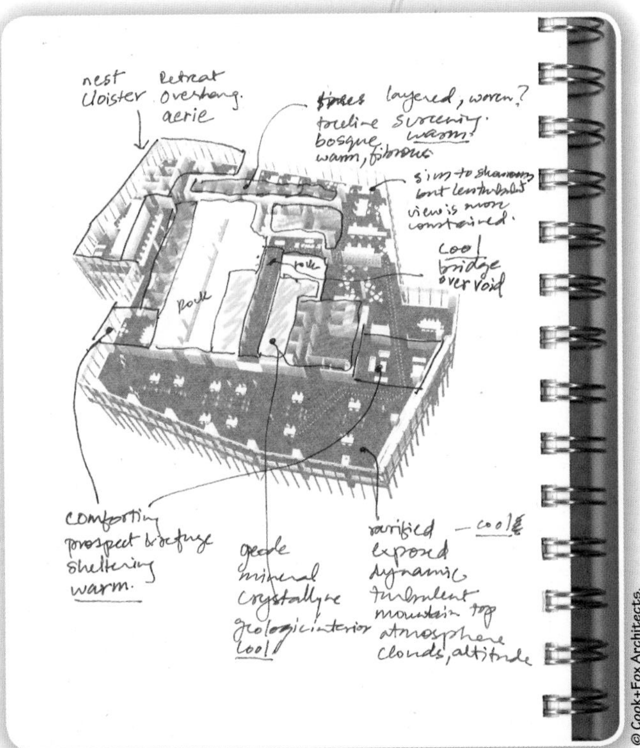

Figure 10-23: The proposed plan for the fiftieth floor of One Bryant Park.

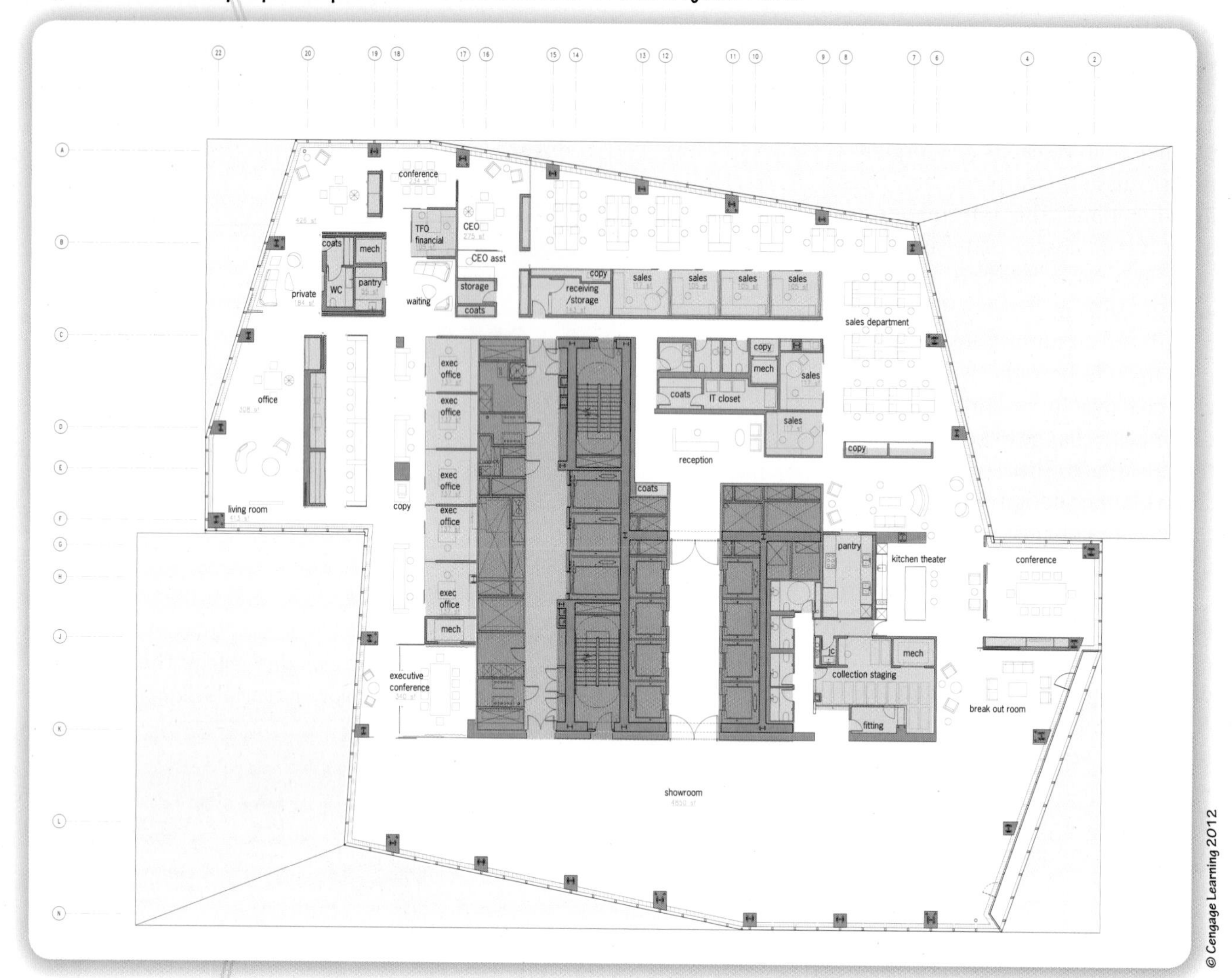

Figure 10-24: A three-dimensional visual of the fiftieth floor of One Bryant Park.

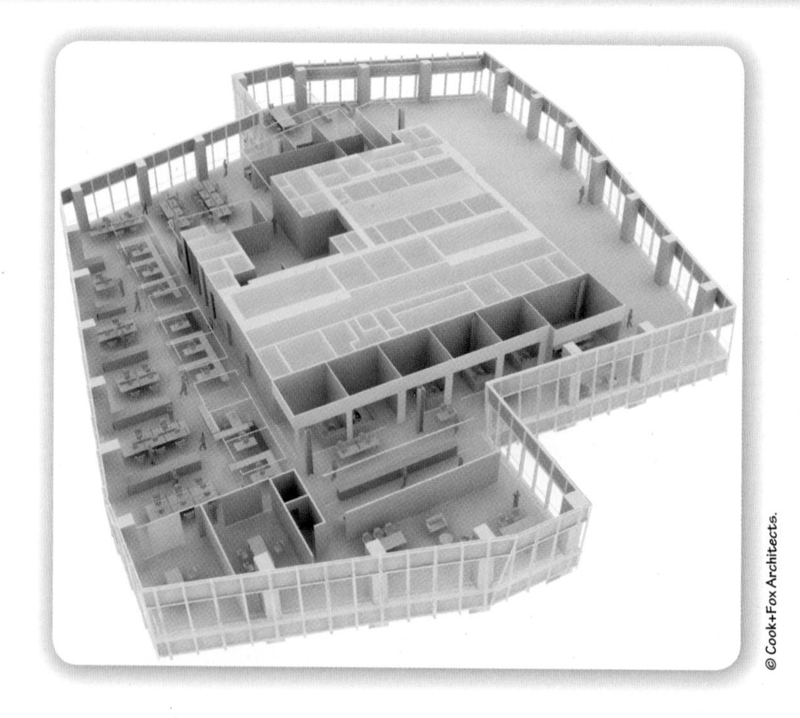

When Cook+Fox Architects created a plan for a 24,382 sq ft office space to be located on the thirty-second floor of the Empire State Building, they used a bubble diagram to group spaces and plan the adjacencies of 17 private offices and 41 workstations (see Figure 10-26).

The design intent was to create a healthy, comfortable, modern workplace that would exemplify the company's dedication to sustainability and environmentally responsible design. The resulting open plan emphasized daylight and individually controllable under-floor air distribution (see Figure 10-27). Almost every workstation was within 15 ft of operable perimeter windows to provide both daylight and view. Glass partitions allowed daylight to reach the interior spaces.

Figure 10-25: **Open office plan.**

Figure 10-26: **Bubble diagrams can be used to group spaces.**

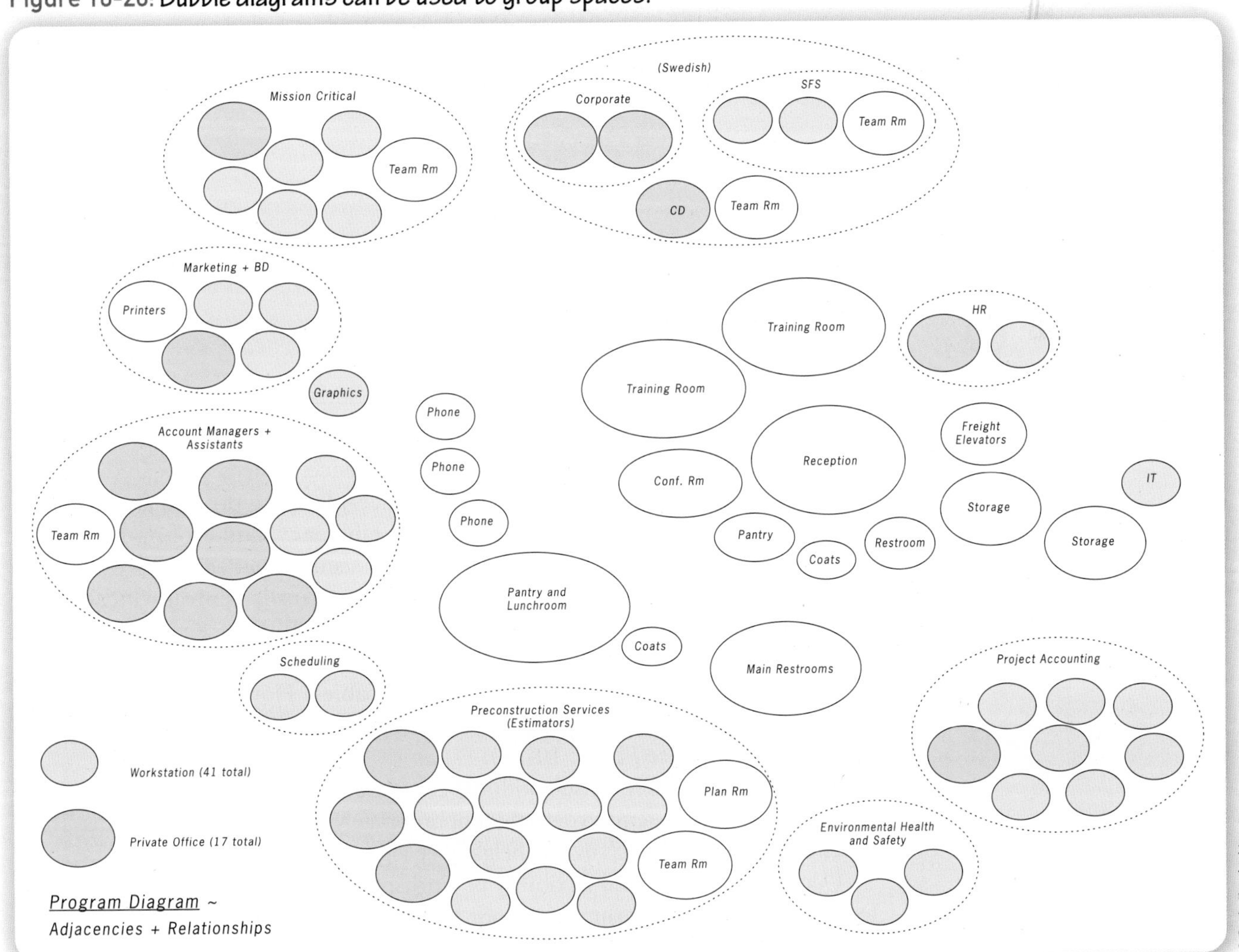

Figure 10-27: The office plan for the thirty-second floor of the Empire State Building emphasized daylight and view.

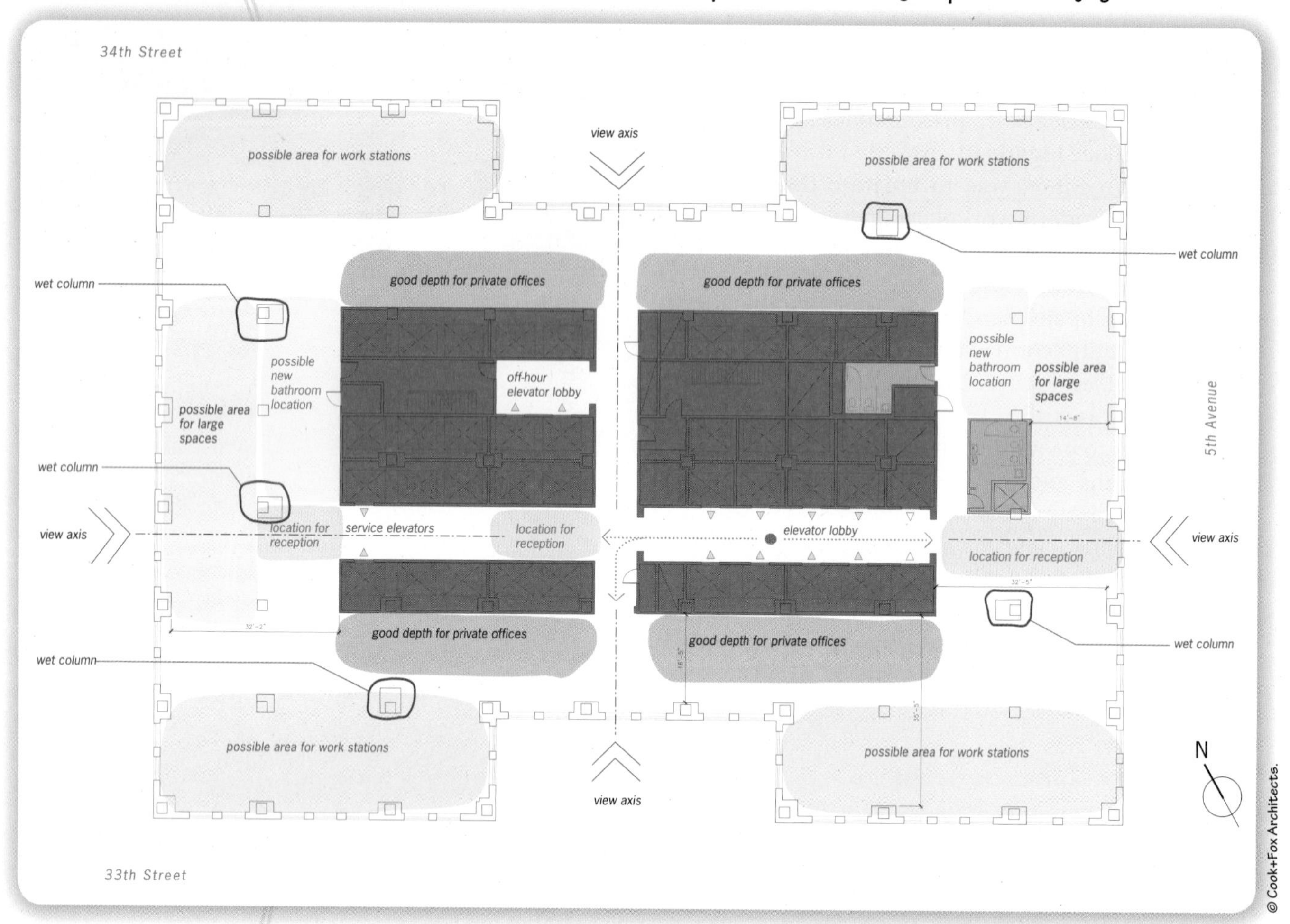

PLANNING SPACE FOR BUILDING EQUIPMENT AND UTILITIES

Plumbing, mechanical, electrical, and communications components are necessary for a building to function and require space. Spaces dedicated to these functions must be considered early in the space-planning process and are often aligned and grouped for safety, accessibility, efficiency, and economic reasons. Equipment access and control panels must be secured yet easily accessible for repair and maintenance. Equipment rooms are generally centrally located and stacked vertically for convenience, accessibility, and to reduce material and installation cost. Horizontal pathway systems and ductwork provide space for electrical wire, HVAC, and communication cables. Horizontal pathways feature cable trays and wire ways, with separate channels for power and communication systems (see Figure 10-28).

Space planners often consult with experts on indoor environmental control and intelligent building systems to determine appropriately sized and placed systems. These systems affect the building's overall height (see Figure 10-29).

Occupant comfort and flexibility of workspace are two considerations that affect system choice. For example, a floor air **plenum** distribution system can be easily reconfigured and allows workers to control their own air and create a personally comfortable environment.

Plenum:
a closed chamber with higher pressure than the surrounding atmosphere.

The choice of mechanical systems and their location can enhance a buildings flexibility and adaptability. For example, gridded components and removable panels provide ease of repair, maintenance, and reconfiguration. Other considerations include lighting, sprinklers, and vents, which all require space. Earlier in this chapter you read about the office space for fashion designer Elie Tahari on the fiftieth floor of One Bryant Park. The architectural firm Cook+Fox devised a creative way to disguise mechanicals so they would not inhibit the high fashion showroom. The architectural design team created a "smart cloud" that would disguise the high ceiling, irregular beams, and ductwork and provide service points for lighting, sprinklers, and vents.

The logic for the smart cloud was found in the Voronoi tessellation, a mathematically rigorous pattern that is often found in nature. The team used advanced computer modeling techniques to create the tessellated pattern that became the framework for the ceiling (see Figure 10-31 a and b). The ceiling design provided a solution to cover the ductwork and house lighting, sprinklers, and vents. Visually the smart cloud provided the organic feel of an outdoor environment (see Figure 10-30 a and b).

Figure 10-28: Space in ceiling and floors can be used for ductwork, horizontal pathways, and plenum systems.

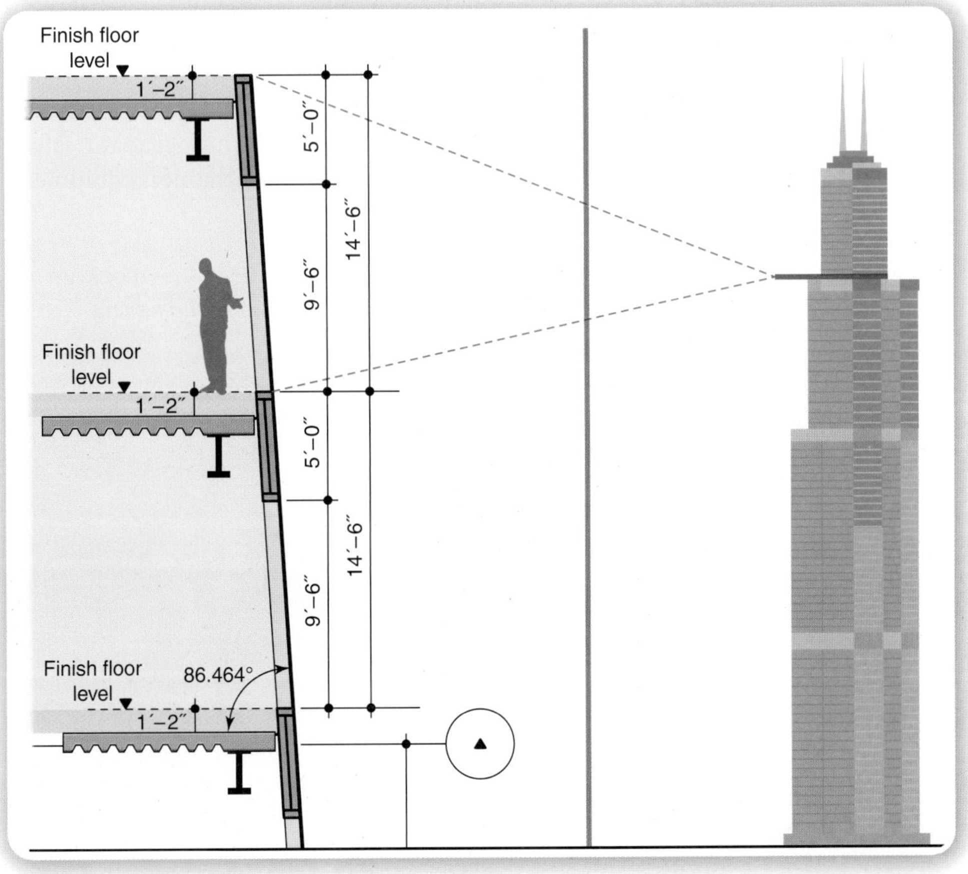

Figure 10-29: Five feet of structural and mechanical space adds to the floor-to-floor height.

Figure 10-30 a: The office and showroom design for fiftieth floor of One Bryant Park.

© dbox for Cook+Fox Architects.

Figure 10-30 b: The cells of the "smart cloud" were extruded and fabricated from Mylar to create a visually organic and harmonious cloud ceiling.

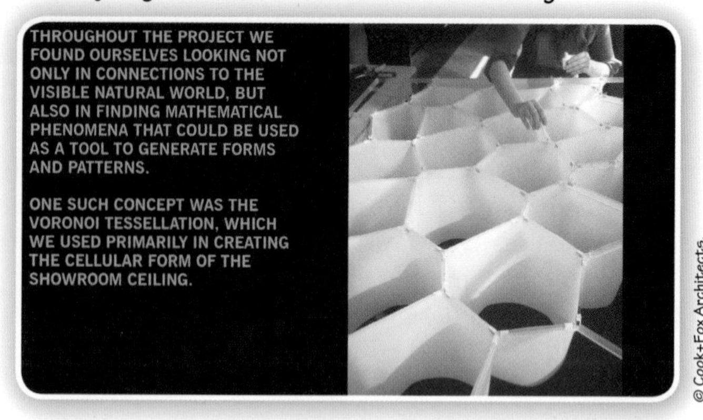

© Cook+Fox Architects.

Building Support Spaces

Sometimes when we think of the spaces of a commercial building, we only recall areas we experience as building occupants. However, a building design must include several support spaces. These spaces include loading, staging, and storage areas, security and fire control centers, and trash, maintenance, custodial, and equipment rooms. Many buildings have service or equipment closets, at least 2 ft in depth, located throughout the building. The locations of building spaces are planned for cost efficiency, access, security, and convenience. For example, lockable custodian closets are usually accessible from the corridor and located on each floor near the toilet facilities. On a multilevel building, the support spaces are often stacked (located above one another). Equipment rooms must be easily accessible to building support personnel and have adequate space around each piece of equipment for maintenance, repair, or replacement. In addition to addressing all codes and local governance, designers plan support adjacencies based on cost, function, efficiency, convenience of maintenance, and occupant safety and comfort. The designer considers the noise levels from mechanical equipment

Figure 10-31 a: Computer modeling was used to create the tessellated pattern for the smart cloud ceiling. © Cook+Fox Architects. © Cook+Fox Architects.

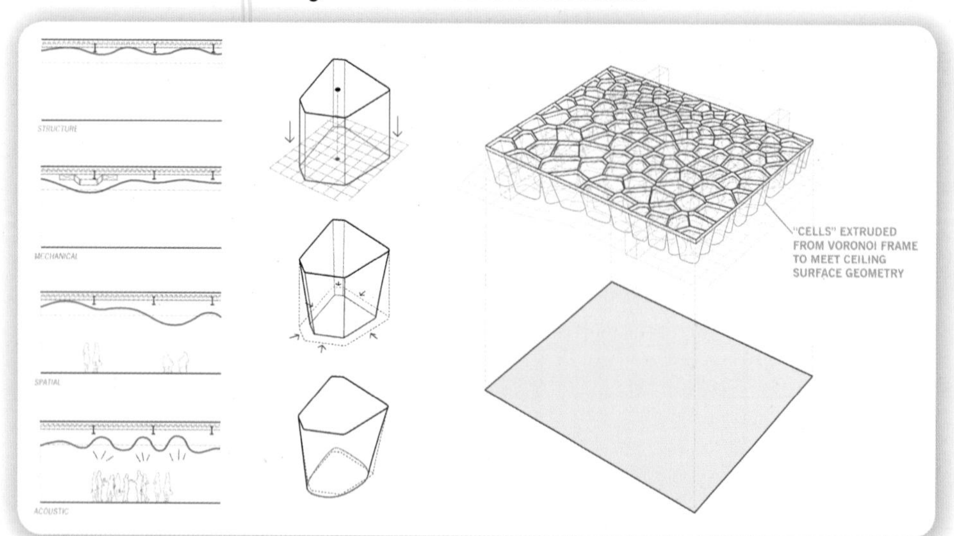

Figure 10-31 b: A mock-up of the smart cloud ceiling.

© Cook+Fox Architects.

and selects a location that provides both functionality and occupant comfort. For building safety, electrical rooms and closets are separate from areas where water could possibly overflow, such as the toilet or custodial closet areas. For example, in a multilevel building, a main electrical switchgear would not be located below a toilet area.

If a building includes a security control center, it is generally located adjacent to the main lobby and elevators for quick response. Today's **intelligent buildings** may have a separate room for monitoring and control of the building automation system.

Intelligent Building: a self-regulating, computer-controlled building.

An intelligent building is computer monitored and controlled to provide optimum comfort and safety. The building's computer gathers and records data, and adjusts multiple systems to create an optimum environment. You will read more about intelligent buildings in Chapter 17: Indoor Environmental Quality and Security.

SPACES ARE IDENTIFIED BY NAME, NUMBER, AND SQUARE FOOTAGE

After you have finalized your floor plan, describe each room by name, number, and square footage (see Figure 10-32). The contractors that bid construction will use the gross building area to calculate construction costs. You will learn in Chapter 11: Dimensioning and Specifications that the gross area is based on the exterior measurements of the structural members. A single-story building with a structural exterior of 42 ft × 96 ft would have a gross building area of 4032 sq ft. The usable or occupiable area is the gross area minus the exterior walls. Rentable area is a little trickier. Generally, rentable area is gross area minus the exterior walls and major vertical penetrations, and may include a pro-rated share of the building common areas to calculate rental cost. An office area might be described as the actual floor space or the distance from interior finished side of the exterior wall to the center of interior partitions separating adjacent offices.

Room Numbers

Rooms are numbered following common practice developed for clarity and ease of movement about the building. Generally the main floor, ground level, or major pedestrian entrance level is considered the main level, and rooms are numbered

Figure 10-32: **Areas are calculated based on purpose.**

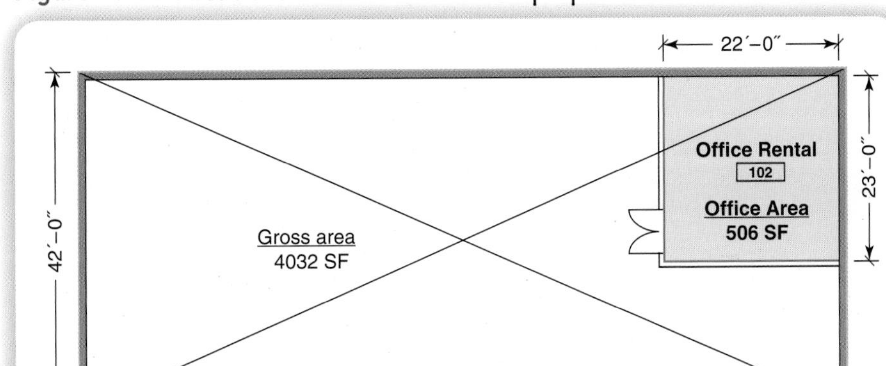

beginning with 101, meaning first floor, room one. The second floor would start with 201, third floor would begin with 301, and so on. Rooms located on a basement level or sub-basement may be labeled beginning with B-01 and SB-01. When buildings such as an apartment have similar floor plan on consecutive floors, the room numbers generally align vertically. For example, room 201 is located directly above room 101. The numbers 100, 200, 300, and so on are reserved for circulation spaces such as stairways, corridors, open lobbies, elevators, and lobbies. These numbers are followed by an alpha character, or letter, to represent the type of space. For example, the lobby on the first floor would be labeled 100L. When there is more than one staircase or corridor, a number is added following the letter. For example, the fourth corridor on the second floor would be labeled 200C4. Larger buildings, such as hospitals with multiple wings, may include a numbering system that identifies the floor number, wing number, sequential ID/room number. The number 3401 could be interpreted as the first room of the fourth wing of the third floor.

A room with two entrances and a movable partition may have two room numbers. If there is a possibility of future subdivision of large rooms, room numbers can be reserved. You may have noticed in your school or at a hotel that odd numbers are placed on one side of a corridor and even numbers on the opposite side. This technique makes it easier for building occupants to locate rooms.

Rooms within a main room, such as a filing room at the back of an office, will be identified by the primary room number followed by a letter. For example, a small office located at the back of classroom 202 would be labeled 202A. Facility service rooms such as restrooms, custodial closets, and mechanical rooms are usually numbered in sequence with the other rooms, but the number will be followed by a letter that represents the service. Using this numbering system, the mechanical room on the fourth floor located between rooms 406 and 410 would be labeled 408M.

SUMMARY

Commercial space planners are architectural designers who adhere to a separate set of codes than residential planning to provide safety and accessibility. Codes are developed to insure structural integrity and provide physical and environmental guidelines for occupant safety and comfort. Architectural designers arrange spaces for safe and convenient ingress, egress, and circulation about the building. The ADA provides guidelines and dimensions to accommodate wheelchair mobility. Space planners work with a variety of systems professionals to gather the specifications that will impact building layout, cost, functionality, and sustainability. Spaces are also designed to provide occupant safety, comfort, and convenience. Studies have been conducted linking building environment to occupant well-being. Air quality, natural light, aesthetics, and other physical characteristics are believed to affect health, performance, and mood. Before floor plans are developed, the architectural designer researches adjacent properties and considers potential views. After a list of public, private, and support spaces developed, the square footage of each space is assessed based on function, occupancy, accommodation of equipment, and activities planned for the space. For example, some activities will require a specific amount or type of light, whereas others may require a quiet, private, or secure space.

Sustainable plans include flexible features that will easily accommodate changes in workforce, type of business, or how business is conducted. Shared spaces, such as workrooms and open office plans, provide economical and efficient use of space. The architectural designer will group rooms and plan adjacencies that will complement each other to enhance functionality. Solar orientation, prevailing winds, and exterior landscape are also considered. The buildings main entrance and lobby is generally centrally located in a highly visible location. The lobby acts as a hub and provides convenient access to other parts of the building via corridors, elevators, and stairs. The traffic pattern should promote safe and easy navigation of the building. Commercial buildings may be divided into public, private, and support spaces. A commercial building plan must allow for spaces to accommodate plumbing, electrical, indoor environmental quality, communications, and protection systems. Building support spaces include loading, staging, and storage areas, security, and fire control centers, and trash, maintenance, custodial, and equipment rooms.

On the floor plan, all rooms are identified by name, number, and square footage. Rooms are usually numbered following commonly accepted guidelines to provide clarity and ease of movement about the building. For example, room 101 would be the first room on the first, or main floor. It is also common to place odd numbers on one side of a corridor and even numbers on the opposite side.

BRING IT HOME

1. Have your teacher select a piece of local property that would be suitable for a small, 10-unit strip mall. For this activity you can assume an existing adjacent parking area is available for patrons. Discuss the property in detail to determine the solar advantages, prevailing winds, existing views, and landscape and road accessibility. Make a sketch of the property with details.
2. As a class, determine a total of 10 units—stores, service, restaurant, or entertainment facilities—that you will include in your mall. As a class, divide up the units to be researched by small teams. Two groups may select to design of the building support spaces and public restrooms. Research your unit or space to determine the approximate overall square footage based on function, occupancy, supplies, furniture, equipment, and activity planned. Make a list to justify your square footage.
3. As a class, debate the location for each space of the strip mall, including the support spaces and public restrooms. When all students have had a chance to present their unit's needs and desired location, come to an agreement on the general location of each unit. Create a bubble diagram that shows the adjacencies of each unit. Determine the exact length and width of your space that will fit together with the other units. Sketch the exterior walls of entire strip mall, label the divided spaces, and identify the dimensions of each unit. In your notebook explain the advantages and disadvantages of your location.
4. Working in your team, plan a strip mall unit that promotes the business mission and provides adaptability, flexibly, and sustainability. Design an ADA-accessible unit that is functional, aesthetically pleasing, and sustainable. Make a list of features that will enhance worker and patron comfort, safety, and mood.
5. As a class, place your completed plans together to create the finished strip mall. Discuss the final outcome as a class and determine the strengths and weaknesses of the plan. In your notebook describe your ideas to enhance the lifecycle of the mall.

EXTRA MILE

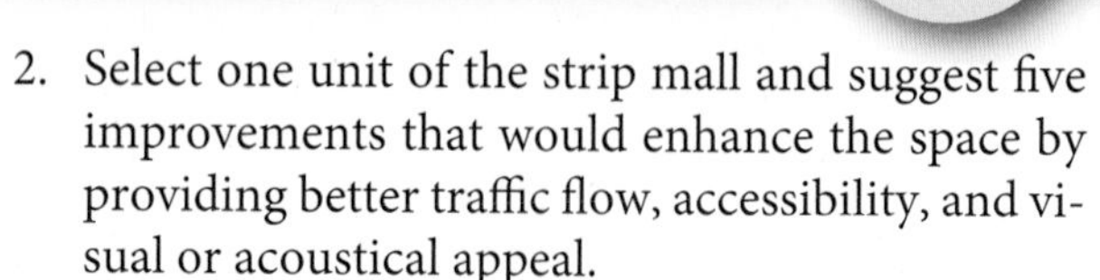

1. Using the unit plan you created for the strip mall, redesign the space into a totally different business. If it was a store, try turning it into a restaurant. If the original plan was a theater, think about converting it into indoor mini-golf or skate park.
2. Select one unit of the strip mall and suggest five improvements that would enhance the space by providing better traffic flow, accessibility, and visual or acoustical appeal.
3. Using one unit of the strip mall, research and recommend five changes that would make the unit more green, sustainable, or resilient.

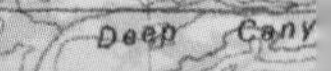

CHAPTER 11
Dimensioning and Specifications

GPS DELUXE

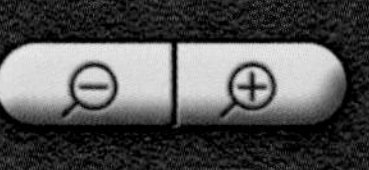

Menu

START LOCATION | DISTANCE | END LOCATION

Before You Begin

Think about these questions as you study the concepts in this chapter:

1. Why are working drawing dimensions and annotations universally understood?
2. How does one clearly and accurately communicate specifications necessary for construction?
3. How is dimension placement influenced by the material and process of construction?
4. What information is placed on elevation views?
5. How are specifications for hidden building materials and components communicated?
6. How does one identify component and finish specifications without complicating a drawing?

INTRODUCTION

What would it be like to bake a cake without a recipe or play a new game without knowing the rules? In order to achieve a desired outcome, either a good cake or to win the game, you will need specific information. The building process is similar. In order to reach a desired building outcome, contractors and builders need detailed information about the building's design and composition. This vital information is communicated through a set of plans or **working drawings** (see Figure 11-1).

The size and type of project will influence the number of working drawings needed. For example, a small residential construction project may require only a dozen drawings, whereas a large commercial project may need several hundred. Drawings sets are organized by main categories, which include civil, architectural, structural, mechanical, plumbing, electrical, fire protection, security, excavation, site environmental, and landscape. But drawings alone do not communicate all of the information necessary for construction. Drawings must be supplemented by purposeful dimensioning and annotation. Within this chapter you will view examples of accepted dimensioning and annotation practice for both residential and commercial building projects. Starting with floor plans and continuing through supporting elevations, sections, and details, you will see how each drawing is dimensioned and annotated for clear, accurate, and complete communication.

Figure 11-1: *Working drawings provide detailed building information.*

DIMENSIONING AND ANNOTATIONS FOLLOW DRAWING CONVENTIONS FOR UNIVERSAL UNDERSTANDING

How do you communicate the size, shape, material, and location of building components to contractors, subcontractors, and builders? Effective communication is achieved through application of a standardized format of dimensioning and annotation. This standardized format is based upon **drawing conventions**.

Drawing Conventions: accepted practices to speed the drawing process and provide universal communication.

In the case of architectural drawings, conventions refer to standards for dimensioning, graphic symbols, abbreviations, notes, and legends. Architectural drawings use some of the same conventions as engineering drawings that specify line type, weight, and application (see Figure 11-2). Well-know standards, set forth by the American National Standards Institute (ANSI), include ANSI Y 14.2 Line Conventions and Lettering, ANSI Y 14.3 Multiview and Sectional View Drawings, and ANSI Y 14.5 Dimensioning and Tolerances. Remember standards are not actually written by ANSI. They are developed by accredited standards developing organizations such as American Society of Mechanical Engineers. Standards developed by accredited organizations may be accepted and designated as American National Standards.

Engineers and architects use symbolic languages of communication such as drafting standards and math equations to develop documentation within accepted standards of precision.

Figure 11-2: Universally recognized line conventions of architectural drawings.

LINE CONVENTIONS			
TYPE	EXAMPLE	PROPERTIES	APPLICATION
BORDER LINES		THICK .5MM	SHEET BORDER AND TITLE BLOCK
OBJECT LINE		MEDIUM .35MM	WALLS IN PLAN VIEW
CENTER LINE		THIN .25MM	CENTER POINTS DIMENSIONING, GRID LINES
BREAK LINE		THIN .25MM	SHORTEN A DRAWING
HIDDEN LINE		THIN .25MM	OBJECTS HIDDEN FROM VIEW
PHANTOM LINE		THICK .5MM	CUTTING PLANE LINE
CONTINUOUS LINE		THIN .25MM	EXTENSION LINES, DIMENSION LINES, LEADERS
SECTION LINES OR HATCH		THIN .25MM	MATERIALS IN SECTION DRAWINGS

Symbols

Symbols are used on working drawings to quickly identify drawings, rooms, levels, building orientation, components, and materials. Common symbols for this purpose include reference numbers or letters (see Figure 11-3). The title symbol, located in the upper right of Figure 11-3, is commonly found at the lower left of all drawings.

Symbols reduce the amount of text necessary to communicate the desired information. For example, imagine how much text would be needed to communicate the sight direction of the elevation, the view identification number, and the sheet number where the elevation is located. All of this is communicated through one very small elevation symbol shown on the floor plan of Figure 11-4.

Figure 11-3: Symbols are used to quickly communicate information using minimum space.

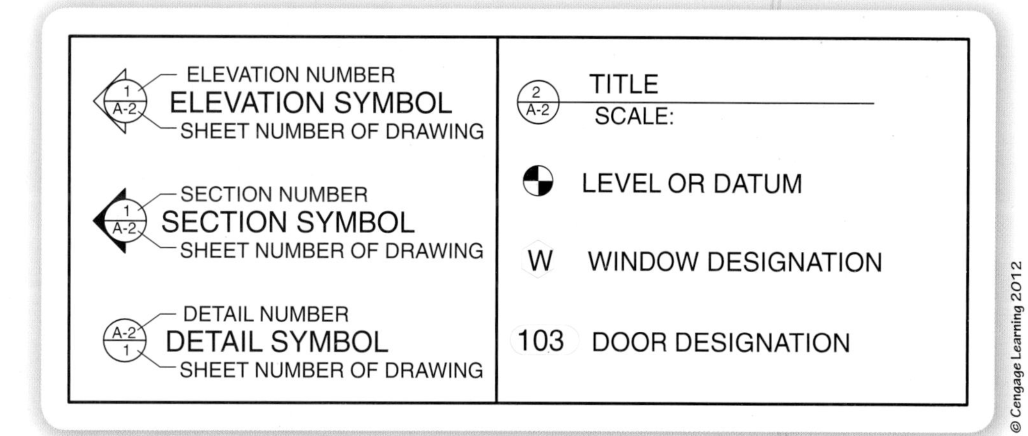

© Cengage Learning 2012

Figure 11-4: Elevation symbols on the floor plan identify line of sight, elevation drawing number, and sheet number where the elevation drawing is located.

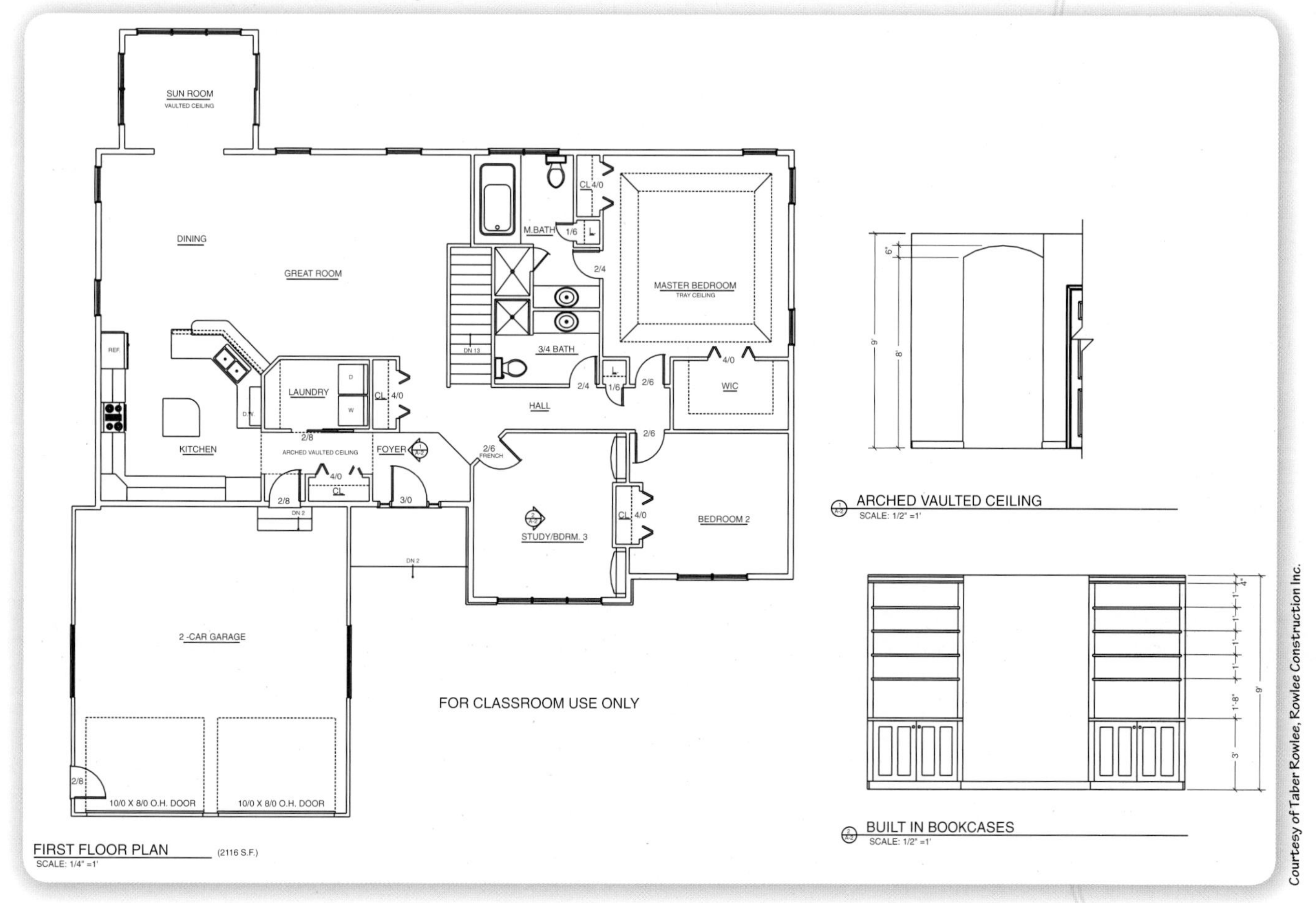

Courtesy of Taber Rowlee, Rowlee Construction Inc.

Building components such as windows and doors are marked using symbols that correspond to a schedule of information. This detailed information is needed for ordering, framing, and installation. Other drawings such as mechanical, plumbing, electrical, fire protection, and security plans employ symbols to identify components and placement. Symbols are usually accompanied by a legend or key. Another common symbol is the elevation symbol which is used to identify the height of various building levels (see Figure 11-5).

Figure 11-5: Elevation symbols act as level indicators.

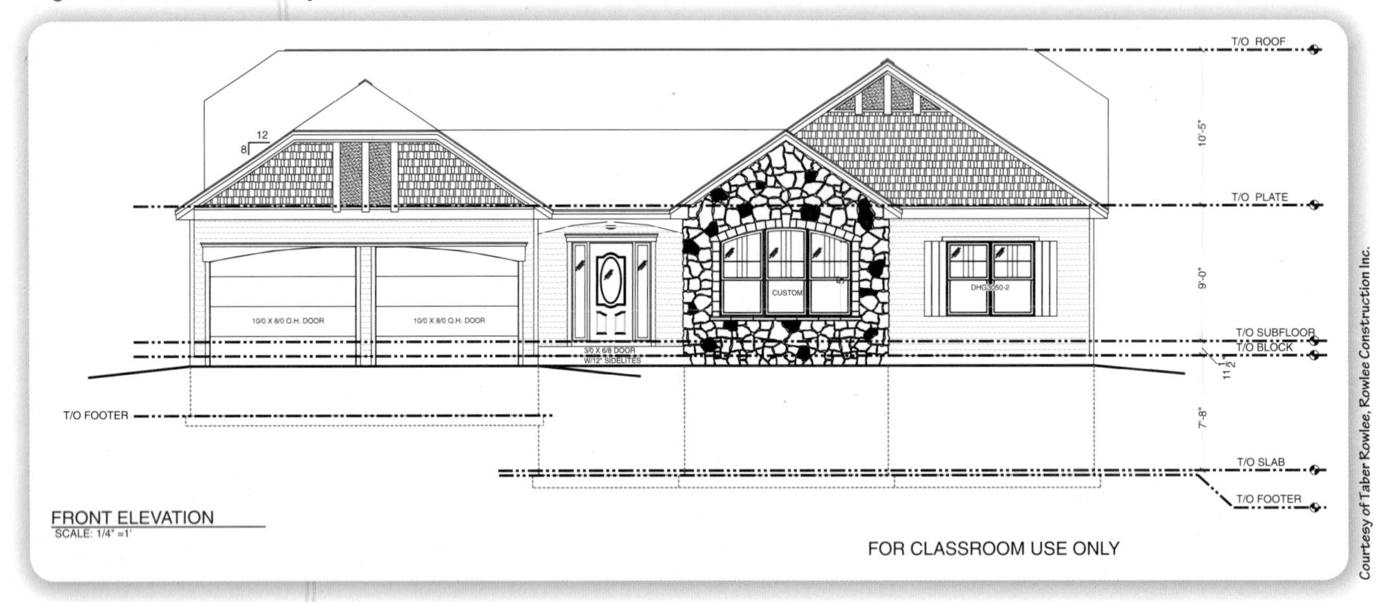

Courtesy of Taber Rowlee, Rowlee Construction Inc.

Your Turn

Sketch the line conventions chart and symbols chart, shown in Figures 11-2 and 11-3, into your notebook. Write a brief explanation as to why line conventions and symbols are used.

Text and Dimensions

The lettering or application of text on architectural drawings generally shows more flair than what is seen on standard technical drawings. Fonts such as City Blueprint or Country Blueprint are two examples commonly chosen when using Computer Aided Design or three-dimensional modeling software. The height of most text and dimensions is usually 0.125 in (1/8 in) or 0.2 in (3/16 in). Titles underneath individual drawings and subtitles such as room names are slightly larger at 0.2 in (3/16 in) or 0.25 in (1/4 in) in height, respectively. The goal of standard sizes is to achieve maximum clarity and consistency by using a text that is small enough to provide adequate room for notes and dimensions, yet large enough to be legible. Text font and dimension style are often dictated by region, company policy, and standards. It is important to be consistent with text font and dimension style throughout a set of plans.

Dimensions define distances for measurement between specific points. The dimension style of most architectural drawings use ticks, small slant marks, to indicate end points of measurement. Architectural ticks provide a clean alternative to arrowheads that are commonly used on a mechanical drawing. The process of dimensioning architectural drawings is similar in some ways to dimensioning mechanical drawings. When adding dimensions to a drawing, extension lines are placed a minimum of 1/16 in away from the object and extend 1/8 in beyond the dimension line (see Figure 11-6).

Figure 11-6: Architectural dimensioning spacing and format.

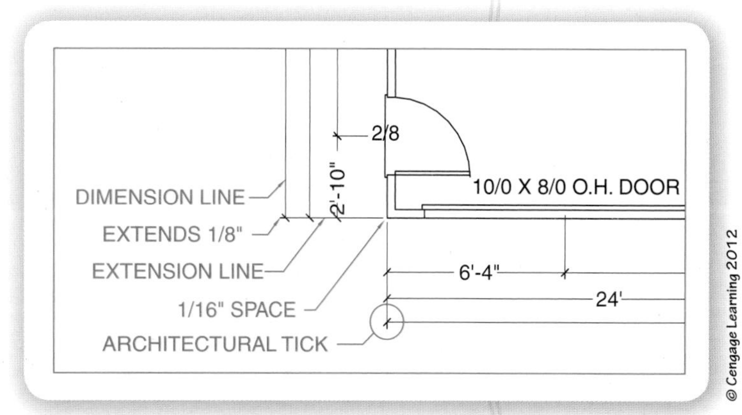

The first set of dimensions is placed approximately 3/8 in away from the projected roof overhang of the floor plan. Subsequent lines of dimensions are placed approximately ¼ in apart. A major difference in dimensioning architectural drawings is dimension format. For example, a floor plan is dimensioned using feet and inches, 0 ft 0 in. However, not all drawings for a building project use the same dimension format. For example, site plans and surveys show a format of feet and decimal parts of a foot, generally carried two places to the right of the decimal. Dimension settings can be formatted into your CAD or modeling software. To enhance drawing clarity, plan the dimension placement carefully to avoid crossing dimensions whenever possible. This usually requires the placement of smaller dimensions closer to the object.

Computer-aided drawing and modeling tools are used to enhance the accuracy and clarity of communication.

Your Turn

Sketch the diagram shown in Figure 11-6 in your notebook. Use architectural ticks, recommended dimension height, and note all spacing.

Notes

Notes provide information about the intended construction which cannot be clearly explained by drawing or symbol (see Figure 11-7).

Notes on construction drawings fall into two categories: general and specific. **General notes** for a set of drawings are often placed on the first sheet of the drawing set. Thoughtful organization and placement of notes avoid confusion or the possibility of a note being overlooked. **Specific notes** are placed in close proximity to the item they reference. Specific notes are often accompanied by a leader, as shown in Figure 11-8. Short, direct leader lines connect a note to the item of reference. Leader lines should point away from the beginning or end of a note, but never from the middle. Careful planning will prevent the crossing of other leaders or dimension lines.

Abbreviations

Abbreviations of words or phrases are common to working drawings. To avoid any confusion, a legend is provided on the title sheet of the plan set (see Figure 11-9).

Figure 11-7: Multiple lines of notes on the same sheet are uniformly spaced and numbered for clarity.

GENERAL NOTES:

1) ALL EXTERIOR WALLS OF HEATED STRUCTURE SHALL BE 2×6 STUDS @ 16" O.C. W/ HEADERS AS FOLLOWS:
3'-0"-5'-0' (2)-1 3/4" × 9 1/2" MICROLAM BEAMS U.N.O.
5'-0"-10'-0" (2)-1 3/4" × 11 7/8" MICROLAM BEAMS U.N.O.
10'-0" PLUS (REFER TO PLAN)
2) ALL FRAMING MEMBERS, BEAMS, FLOOR JOIST, AND RAFTERS SHALL HAVE A MIN FIBER BENDING STRESS OF 1,200 PSI LAMINATED MEMBERS TO HAVE A MIN FIBER BENDING STRESS OF 2900 PSI.
3) PROVIDE DOUBLE JOIST UNDER BEARING PARTITIONS RUNNING PARALLEL W/ JOIST.
4) PROVIDE DOUBLE STUD BEARING FOR OPENINGS 6'-0" OR LARGER.
5) ALL FLOOR JOIST SHALL BE 2×10 @ 16" O.C. W/ BRIDGING @ SPANS GREATER THAN 8'-0" U.N.O.
6) SUBFLOOR SHALL BE 3/4" T&G O.S.B. GLUED &NAILED TO JOIST AS PER APA GLUED SYSTEM.
7) PROVIDE 5/8 " TYPE × G.W.B.@ BOTH SIDES OF COMMON HOUSE/ GARAGE WALLS MIN. 4'-0" BEYOND HEATED SPACE (U305 1 HR) &5/8" TYPE × G.W.B. (FM FC 172 1 HR) @ ENTIRE GARAGE CEILING.
8) PROVIDE "C" LABEL DOOR W/ CLOSER BETWEEN HOUSE AND GARAGE.
9) ALL BASEMENT FOUNDATION WALLS SHOWN ARE STD. LOAD BEARING HOLLOW CORE W/ HORIZONTAL JOINT REINFORCING 3RD, 5TH, &7TH COURSE U.N.O.
10) PROVIDE #4 RE-BAR FULL HEIGHT FROM TOP OF FOOTING TO TOP OF BLOCK, GROUTED SOLID, @ 4'-0" O.C. MAX FOR STRAIGHT WALLS 30'-0" OR LONGER.
11) GROUT BLOCK CORES SOLID AT ALL STL. BEAM AND COLUMN BEARING CONDITIONS.
12) ALL FILL UNDER CONC. SLABS SHALL BE COMPACTED TO 95% OF MAX. DRY DENSITY.
13) FOOTINGS SHALL BE 3000 PSI CONC. AND SIZES AS FOLLOWS:
1'-4" MIN × 8" DEEP FOR 8" C.M.U. WALLS.
1'-8" MIN × 10" DEEP FOR 10" C.M.U. WALLS.
14) FOOTINGS HAVE BEEN DESIGNED FOR A SOIL BEARING PRESSURE OF 2,500 PSF CONTRACTOR TO VERIFY SOIL CONDITIONS AND REPORT ANY DEVIATIONS TO THE ARCHITECT.
15) BOTTOM OF FTGS. SHALL BE 4'-0" BELOW FINISHED GRADE U.N.O.
16) INSTALL 4" DRAIN PIPE OUTSIDE W/ 24" COVER OF #2 WASHED STONE AND/OR INSIDE W/ SUMP PIT PER SITE CONDITIONS.
17) PROVIDE 1/2" ANCHOR BOLTS 6'-0" O.C. AND AT CORNERS W/ 2×6 P.T.SILL ON SILL SEALER.
18) SAWCUT BASEMENT, AND OTHER INT. EXPOSED SLABS, IN ACCORDANCE W/ GENERALLY ACCEPTED STDS.
19) ALL COLUMNS SHALL BE 3" DIA. STEEL PIPE COLUMNS W/ A 6" × 6" × 1/2" BASE PL ON BTM. OF ALL COLUMNS & A 4" × 6" × 1/2" PL ON TOP OF COL. ON 2'-0" × 2'-0" × 1'-0" CONCRETE FOOTINGS U.N.O.
20) ALL WORK TO COMPLY WITH THE LATEST ADDITION OF THE NYS BUILDING CODE AND ANY APPLICABLE STATE, COUNTY OR LOCAL CODES, CODES GOVERN OVER DRAWINGS.

TRUSSES

TRUSSES SHALL BE ENGINEERED BY A LICENSED, N.Y. STATE STRUCTURAL ENGINEER FOR A 40# PER S.F. LIVE LOAD AND 15# PER S.F. DEAD LOAD, SUBMIT SHOP DRAWINGS SHOWING SPECIES, SIZES AND STRESS GRADES OF LUMBER TO BE USED; PITCH, SPAN, CAMBER CONFIGURATION AND SPACING FOR EACH TYPE OF TRUSS REQUIRED; TYPE SIZE, MATERIAL, FINISH, DESIGN VALUE AND LOCATION OF METAL CONNECTOR PLATES, BEARING, LATERAL BRACING AND ANCHORING DETAILS.

FOSTER RESIDENCE
LOT 26 GLEN 2 FULTON, N.Y. 13069
ROWLEE CONSTRUCTION
81 PIERCE DRIVE FULTON, NEW YORK 13069

JOB NO:
DATE: 06/23/05
REV. DATE:07/18/05
DRAWN BY: TAB
CKD. BY:
SCALE: AS SHOWN

A-1

Courtesy of Taber Rowlee, Rowlee Construction Inc.

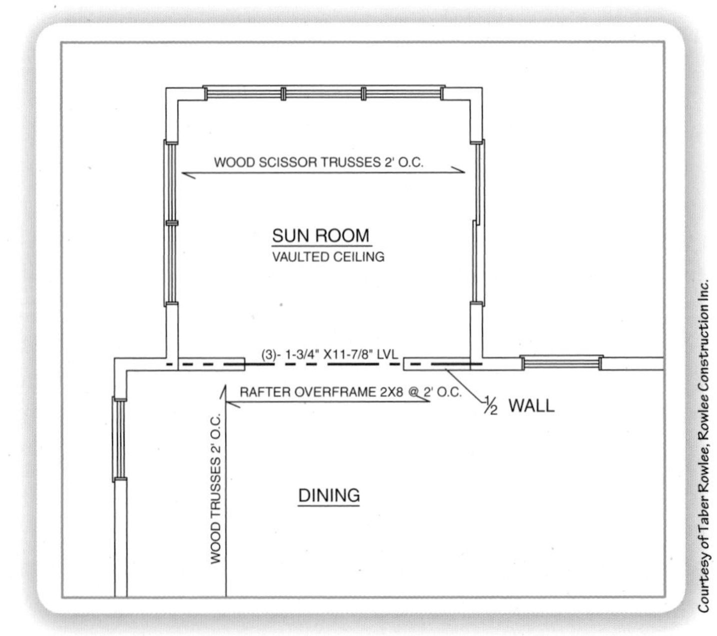

Figure 11-8: A leader is used to connect a note to the item it explains.

Courtesy of Taber Rowlee, Rowlee Construction Inc.

Figure 11-9: **Abbreviations commonly found on architectural drawings.**

BLDG	Building
BR	Bedroom
BRK	Brick
BSMT	Basement
CAB	Cabinet
CEM	Cement
CFT	Ceramic Floor Tile
CIP	Cast in Place
CLO	Closet
CONC	Concrete
CMU	Concrete Masonry Unit
DH	Double Hung
DN	Down
DWG	Drawing
EQ	Equal
EXIST	Existing
EXT	Exterior
FLR	Floor
FT	Feet/Foot
GL	Glass
GWB	Gypsum Wall Board
HVAC	Heating Ventilation and Air Conditioning
INSUL	Insulation
INT	Interior
MFR	Manufacturer
MTL	Metal
NO	Number
NTS	Not to Scale
OC	On Center
PLYWD	Plywood
RO	Rough Opening
SQ	Square
SS	Stainless Steel
STD	Standard
T&G	Tongue and Groove
TYP	Typical
UL	Underwriter Laboratories
VR	Vapor Retarder
WC	Water Closet
WD	Wood
W/O	Without

DIMENSIONS AND ANNOTATIONS ARE USED TO COMMUNICATE SPECIFICATIONS NECESSARY FOR CONSTRUCTION

A set of working plans includes precisely scaled, detailed drawings. Drawings alone do not communicate the specifications required to complete a successful construction. Dimensions and annotations are necessary to communicate building specifications in a clear, concise, easy-to-read format.

Building design and documentation require precise measurement, analysis, and understanding of space relationships.

Let's look at the process of dimensioning of a floor plan. Most architectural drawings are based upon the floor plan, which is a view of a building from above, with the cutting plane at 4 ft above the floor. Objects that fall above the cutting plane are shown with hidden lines, such as upper kitchen cabinets.

Before you begin dimensioning, recheck the drawing for accuracy. Are walls, windows, and doors correctly sized and placed? How does the placement look from both an inside and outside perspective? On residential plans, doors generally swing inward to the rooms, whereas in commercial buildings fire codes require doors to swing outward from assembly rooms for a safe exit. Once the accuracy of the floor plan has been confirmed, the dimensioning process can begin.

Figure 11-10 a and b: *Construction methods are based on building material. (a) Steel frame. (b) Wood Frame.*

Dimension Placement Is Influenced by Wall Construction and Material

During the dimensioning process it will help to think like the person who must read the drawing, to accurately construct and place the walls. A mason will expect to see dimensions in a format specific for concrete, masonry, and stonework. The rough framer of a wood construction will expect a different placement of dimensions. A steel frame construction drawing shows yet another dimensioning method (see Figure 11-10 a and b). In the following example, we will look at a typical residential wood construction on a masonry foundation. Many foundations are built with concrete masonry units (CMU), poured concrete, or precast foundation walls. A mason will need dimensions placed at the face of walls and rough openings of windows and doors (see Figure 11-11).

The face-to-face dimension technique can be seen on most concrete or masonry foundation plans. The foundation plan shows a top view of the footings, foundation walls, columns, beams, girders, and openings (see Figure 11-12). In areas prone to flooding, a foundation plan may include piers. Dimensions and notes provide detailed information

Figure 11-11: Dimensions for masonry construction are provided at the face of all openings.

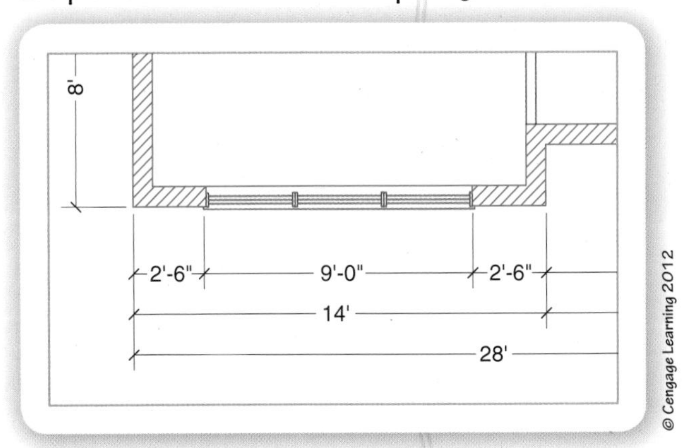

Figure 11-12: Dimensioned foundation plan shows concrete block (CMU) foundation walls constructed on a poured concrete footing shown by red hidden lines.

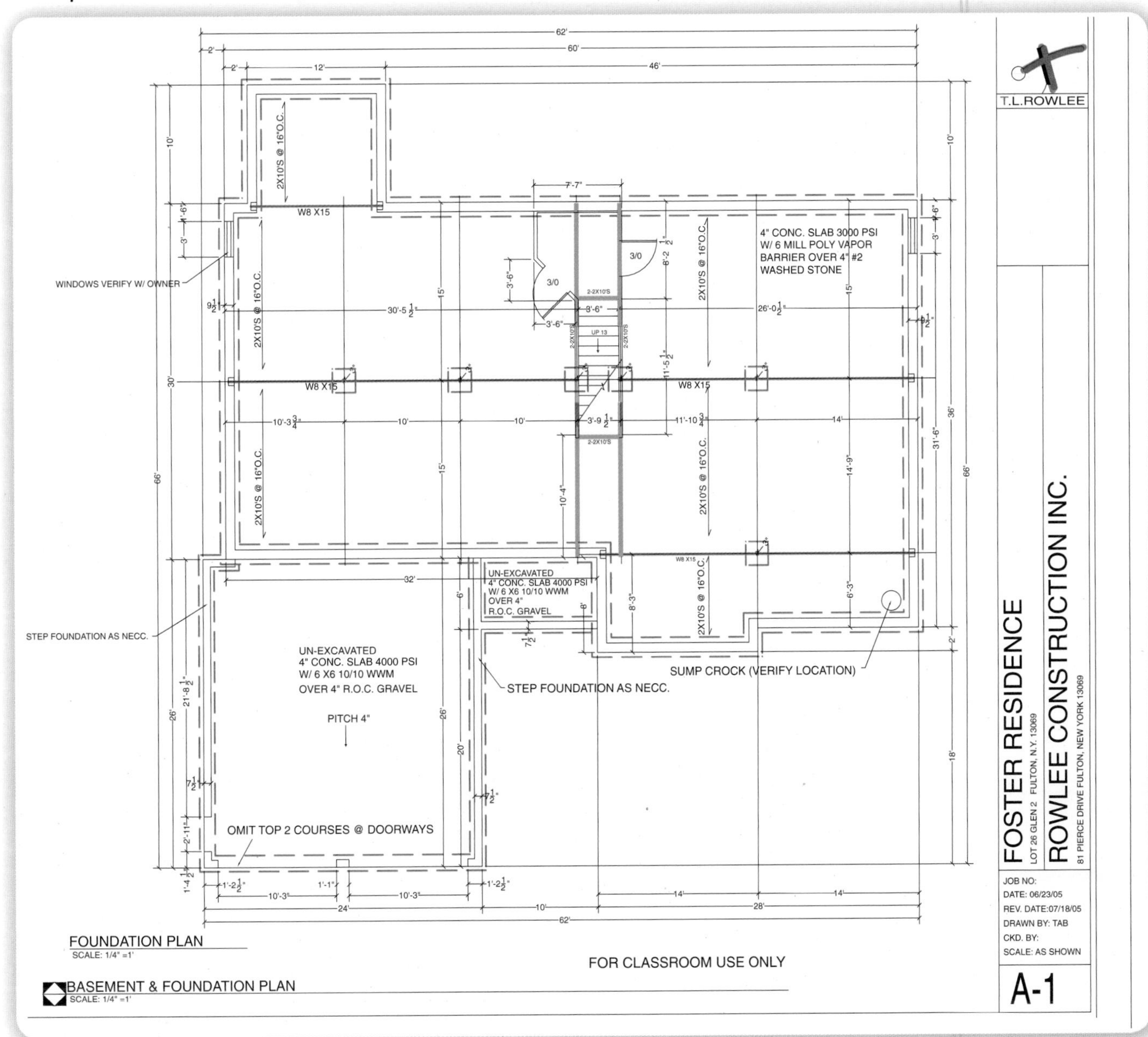

to guide construction of the foundation. You will learn more about structural systems in Chapter 13: Framing Systems: Residential and Commercial Applications.

Once the foundation is in place and the floor system is installed, the floor plan is used to guide wall construction. Keep in mind that the exterior stud wall is usually built in vertical alignment, or flush with the foundation wall. The floor plan of a wood construction will include dimensions for the overall building size from exterior stud face to exterior stud face. Window and door placement is generally shown with dimensions to centerlines. Rough opening sizes and other important information of the windows and doors is located on a separate schedule. Although some states may accept a different format, it is generally common practice to place dimensions using a hierarchy format. Smaller dimensions such a location of windows and doors are placed closest to the floor plan. The next set of dimension lines provide the major wall dimensions, while the outermost dimensions provide the overall dimension of the exterior wall (see Figure 11-13). Although not always the case, interior walls, dimensioned outside the floor plan, are dimensioned to the interior wall centerline. Most importantly, you should confirm that dimensions of each set have the same total (see Figure 11-14).

Figure 11-13: Dimensions are placed from smallest to largest to avoid crossing.

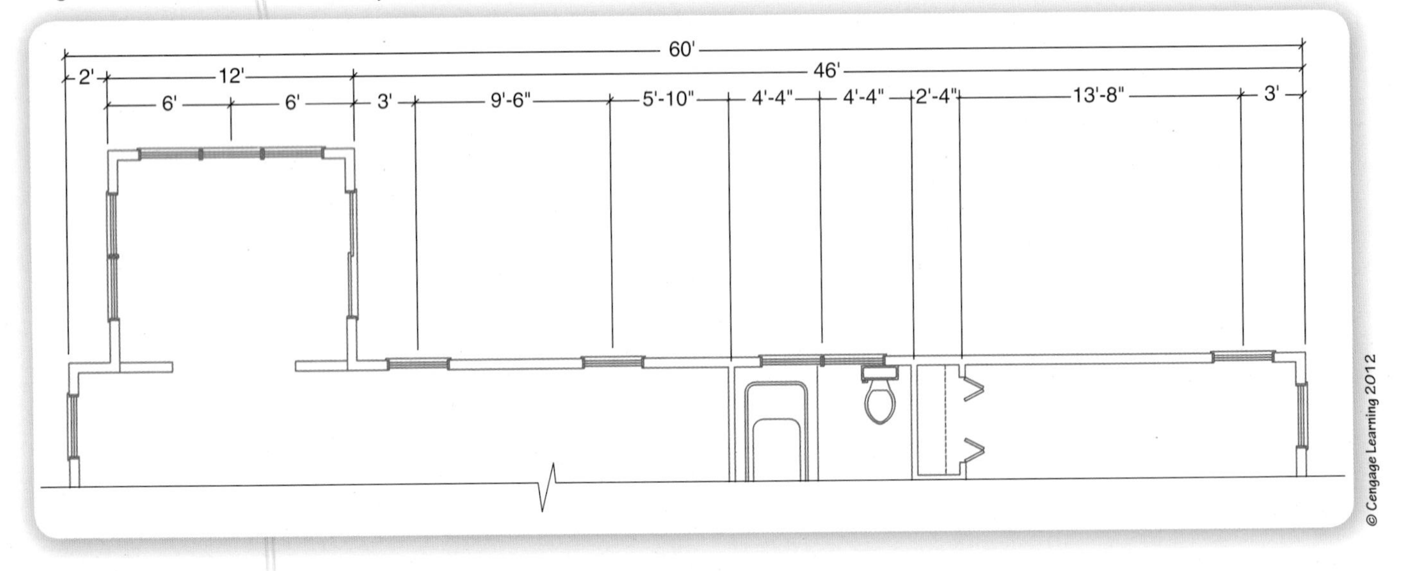

Interior dimensions are placed in consideration of the builder and efficient building practice. In a well dimensioned plan, the builder should not have to stop and add or subtract dimensions to determine wall placement. However, the amount of dimensions on a working plan may range depending upon the amount of freedom the architect allows the builder. Obvious dimensions, such as the centering of a doorway in a small space or closet depth under the stairs, are often omitted.

When a wood frame construction has a stone or brick veneer such as in Figure 11-15, the dimensions are placed to the outside face of wood stud and the veneer information can be provided as a separate note.

Dimensions for steel construction are shown using a structural grid, which aligns with the centerline of the steel columns (see Figure 11-16).

Curtain walls are commonly used in commercial building and are dimensioned on a separate drawing, Figure 11-17.

Curtain Walls:
metal framework, typically filled with glass, spanning several floors

Figure 11-14: Fully dimensioned floor plan for a wood frame construction.

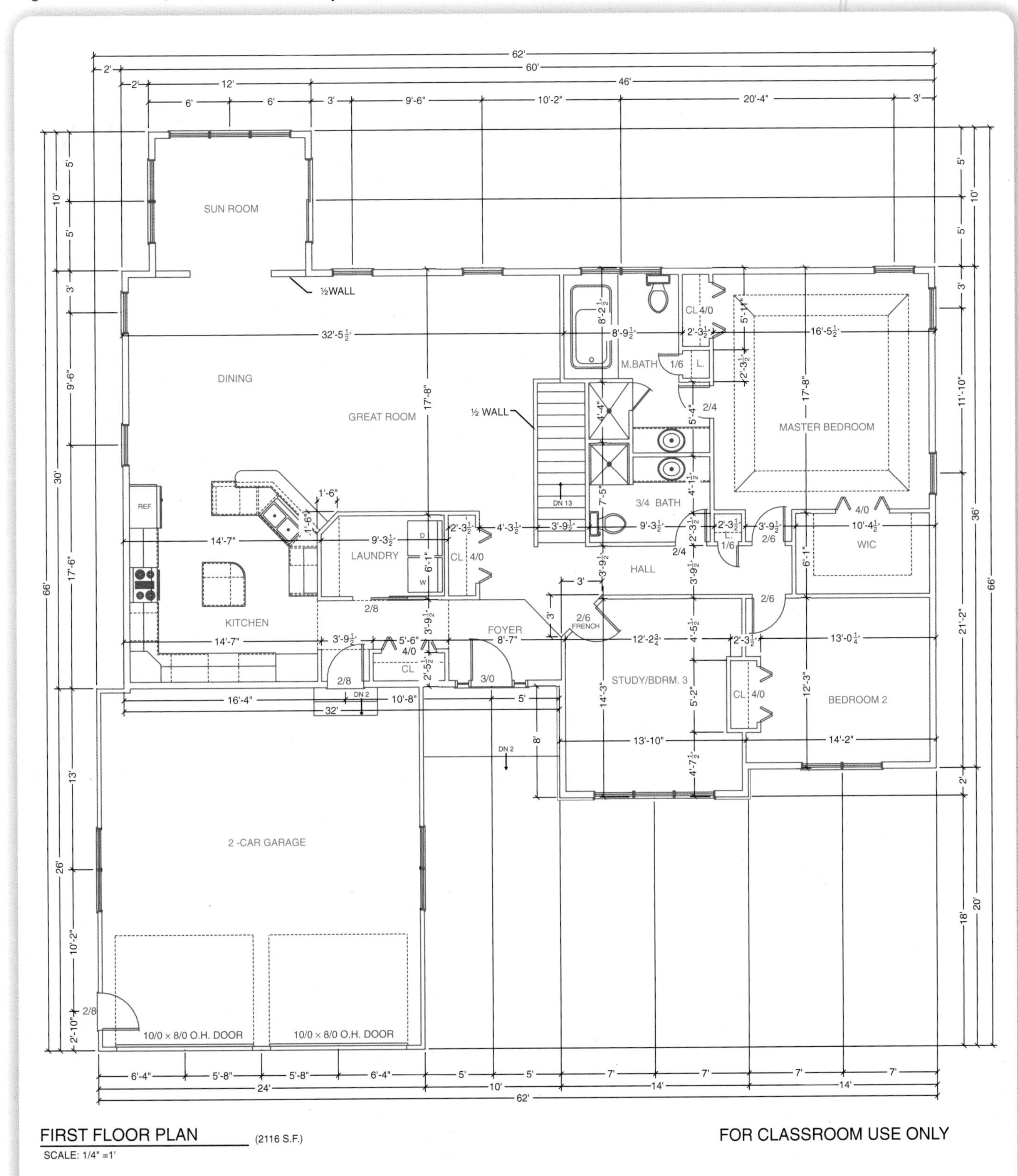

Courtesy of Taber Rowlee, Rowlee Construction Inc.

ELEVATION VIEWS PROVIDE A PREVIEW OF THE FINISHED CONSTRUCTION

Elevation drawings show vertical dimensions, materials, architectural design, and construction details not apparent in the floor plan. Exterior elevations give you an idea of how the front, right side, rear, and left side of the building will appear when finished. Vertical dimensions identify the location of finished grade, bottom of footing, and height of windows, doors, roof, floor, ceiling, and building features such as chimneys. Notes may be added to reference materials, finishes, roof pitch, and other construction details. Generally, exterior elevations are printed or drawn at the same scale as the floor plan (see Figure 11-18). The size of the building and size of the sheet determines which scale is chosen. When using CAD, buildings are drawn in actual size and printed in an accepted architectural scale to fit the drawing sheets. Common scales for exterior elevations are ¼ in = 1 ft 0 in, 1/8 in = 1ft 0 in, or 1/16 in = 1ft 0 in. Interior elevations are generally printed at ½ in = 1 ft 0 in scale. A ½ in =1 ft 0 in scale shows small components, such as cabinetry shelves. Interior elevations provide a vertical view, which can be dimensioned and annotated to communicate desired construction specifications. If space allows, the interior elevation might be placed on

Figure 11-15: Dimensioning of wood frame construction with stone or brick veneer does not include the veneer.

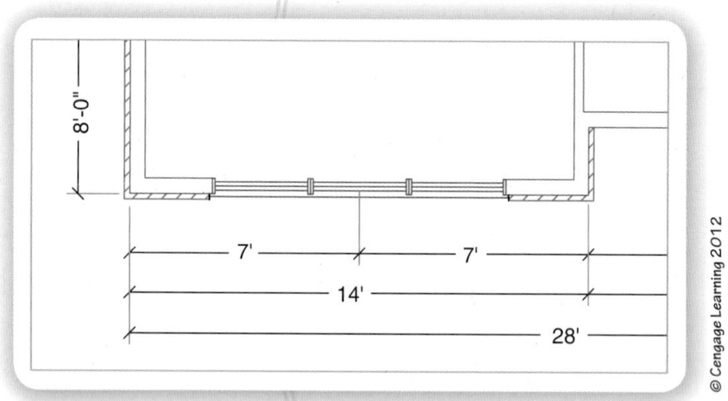

© Cengage Learning 2012

Figure 11-16: Steel construction is dimensioned using a grid.

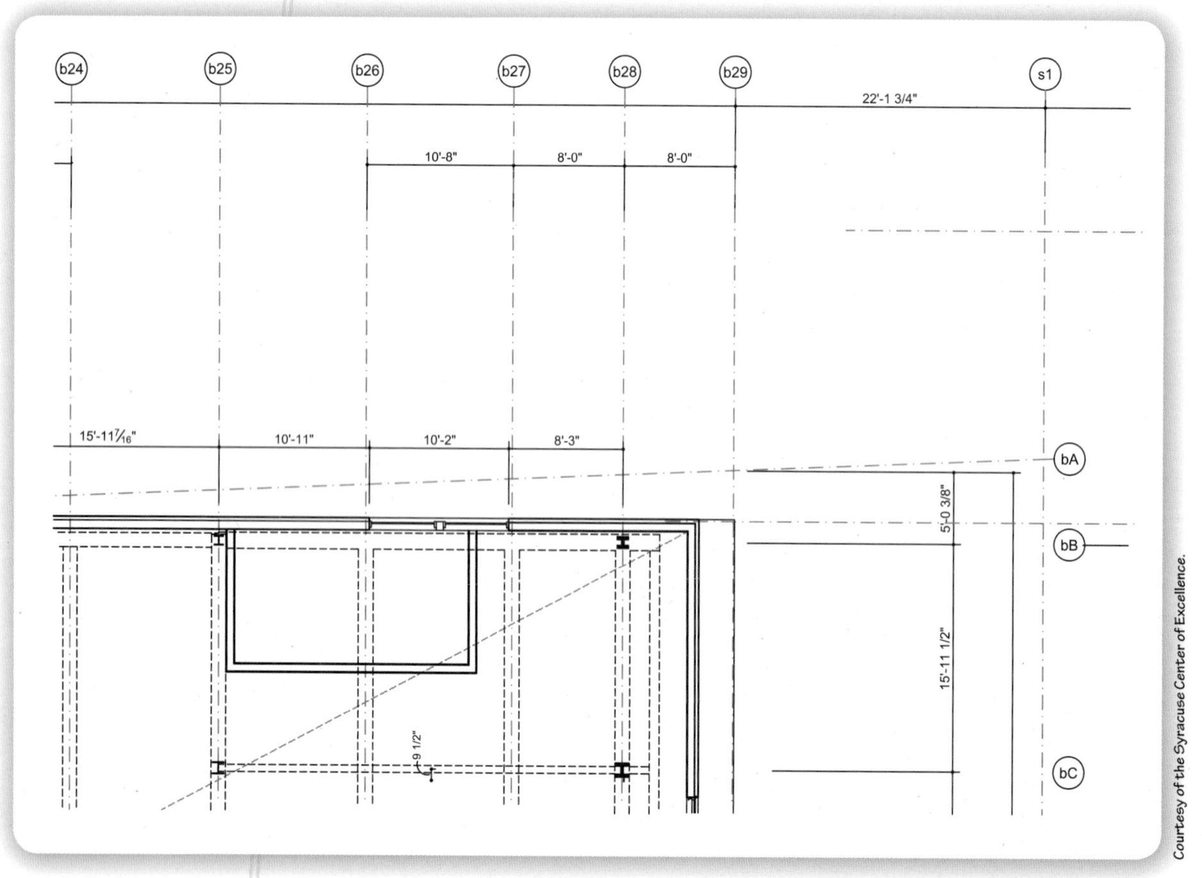

Courtesy of the Syracuse Center of Excellence.

Figure 11-17: Dimensioning of a curtain wall.

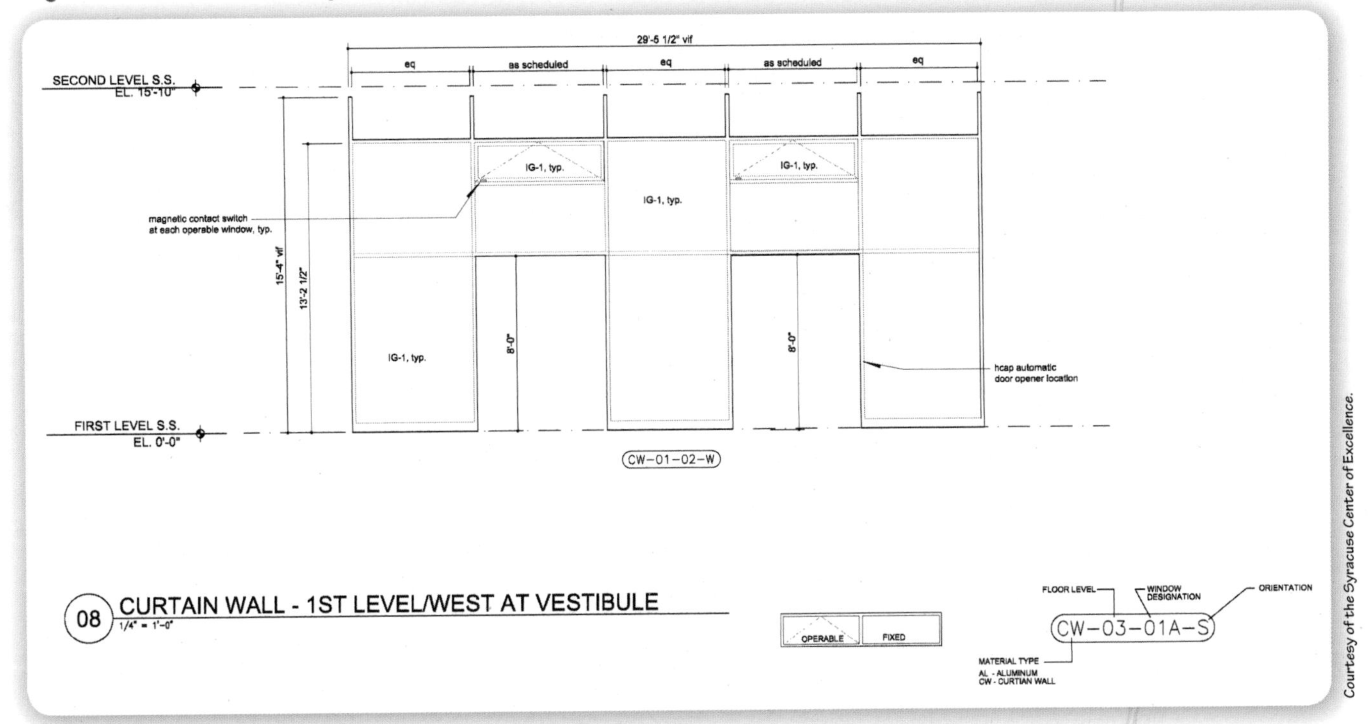

Courtesy of the Syracuse Center of Excellence.

Figure 11-18: Exterior elevation drawings of Foster Residence.

Courtesy of Taber Rowlee, Rowlee Construction Inc.

the floor plan. In a large project, a separate sheet is used to accommodate several interior elevations. All elevations are referenced on the floor plan with an elevation symbol that indicates the direction of view, the elevation number, and the page number where the elevation view can be located. For example, the elevation symbol on Figure 11-19 shows the direction of sight, elevation number 04, and page A723. Figure 11-20 shows the corresponding elevation located on page A723.

SECTION VIEWS ALLOW DIMENSIONING AND ANNOTATION OF HIDDEN MATERIALS AND COMPONENTS

Figure 11-19: Section of floor plan showing symbol that directs the viewer to sheet A723 for kitchenette elevation drawing, see Figure 11-20.

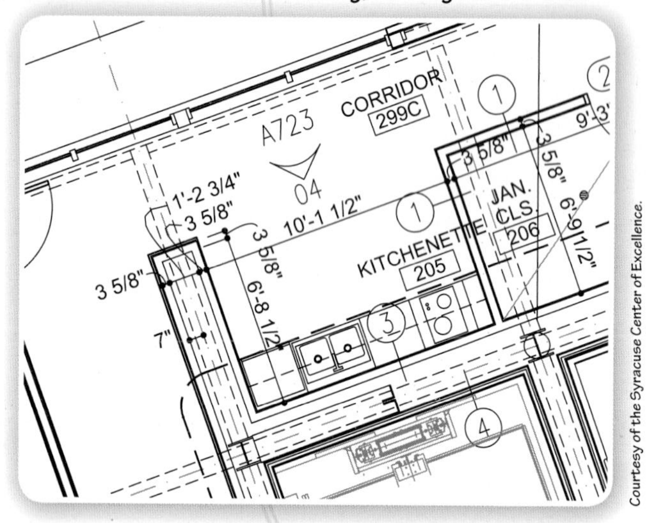

Courtesy of the Syracuse Center of Excellence.

As with interior elevations, sections are usually drawn or printed at a scale of ½ in = 1 ft 0 in. However, a **section view** is quite different from an elevation view. A section view allows you to see what is inside. For example, a section through an exterior wall shows what you would see if you cut a wall vertically with a saw and moved half of it away. You would see everything from the exterior siding to the interior sheetrock or gypsum board. An exterior wall section drawing shows vertical dimensions and size and type of building materials (see Figure 11-21).

Details

Details are magnified drawings enhanced with additional graphics to clarify atypical or unusual construction. If space allows, details may be placed directly on the drawing they reference. A separate sheet is used

Figure 11-20: Interior elevation view of kitchenette from page A723 Elevation 04.

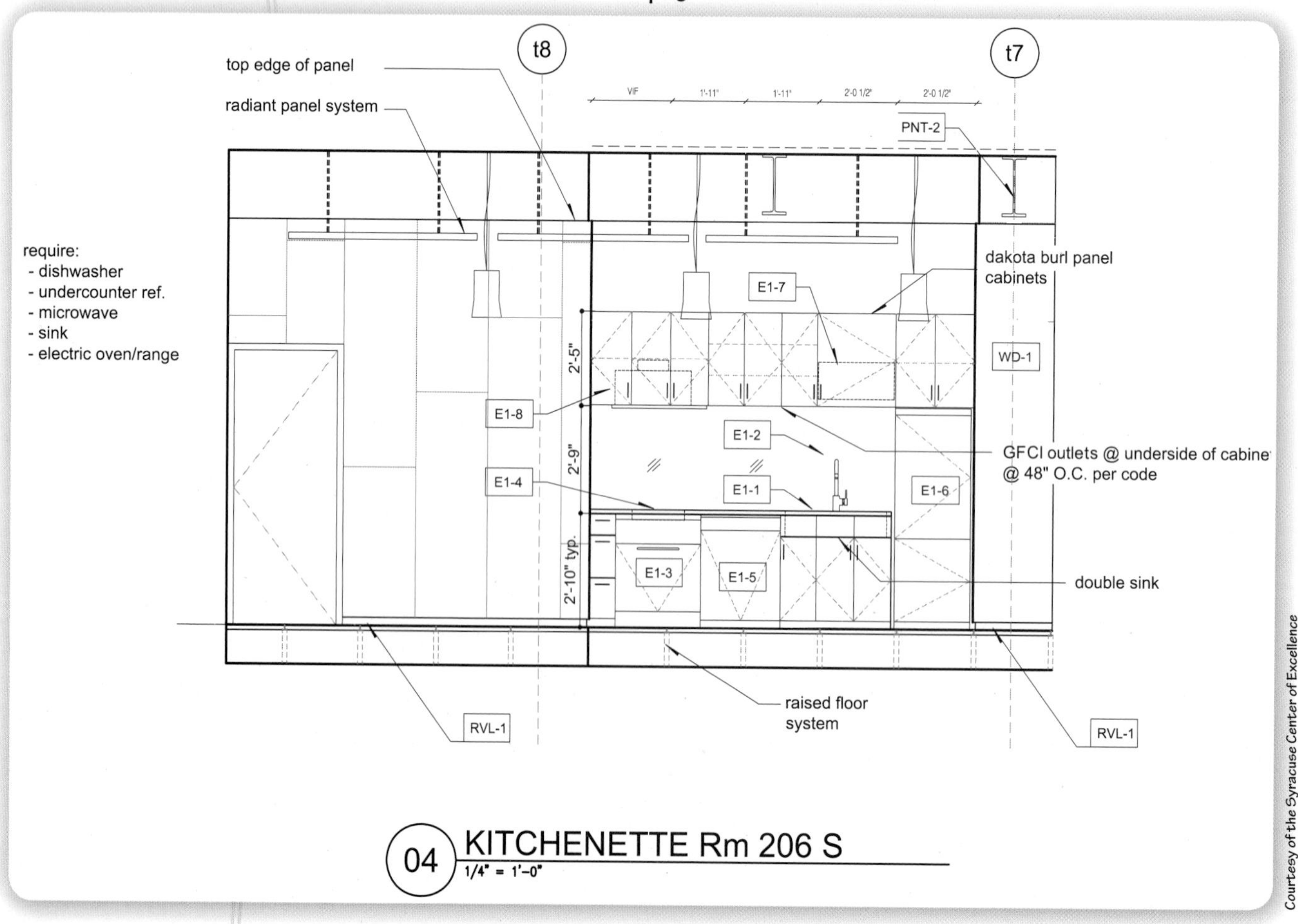

Courtesy of the Syracuse Center of Excellence

Figure 11-21: A section shows interior details.

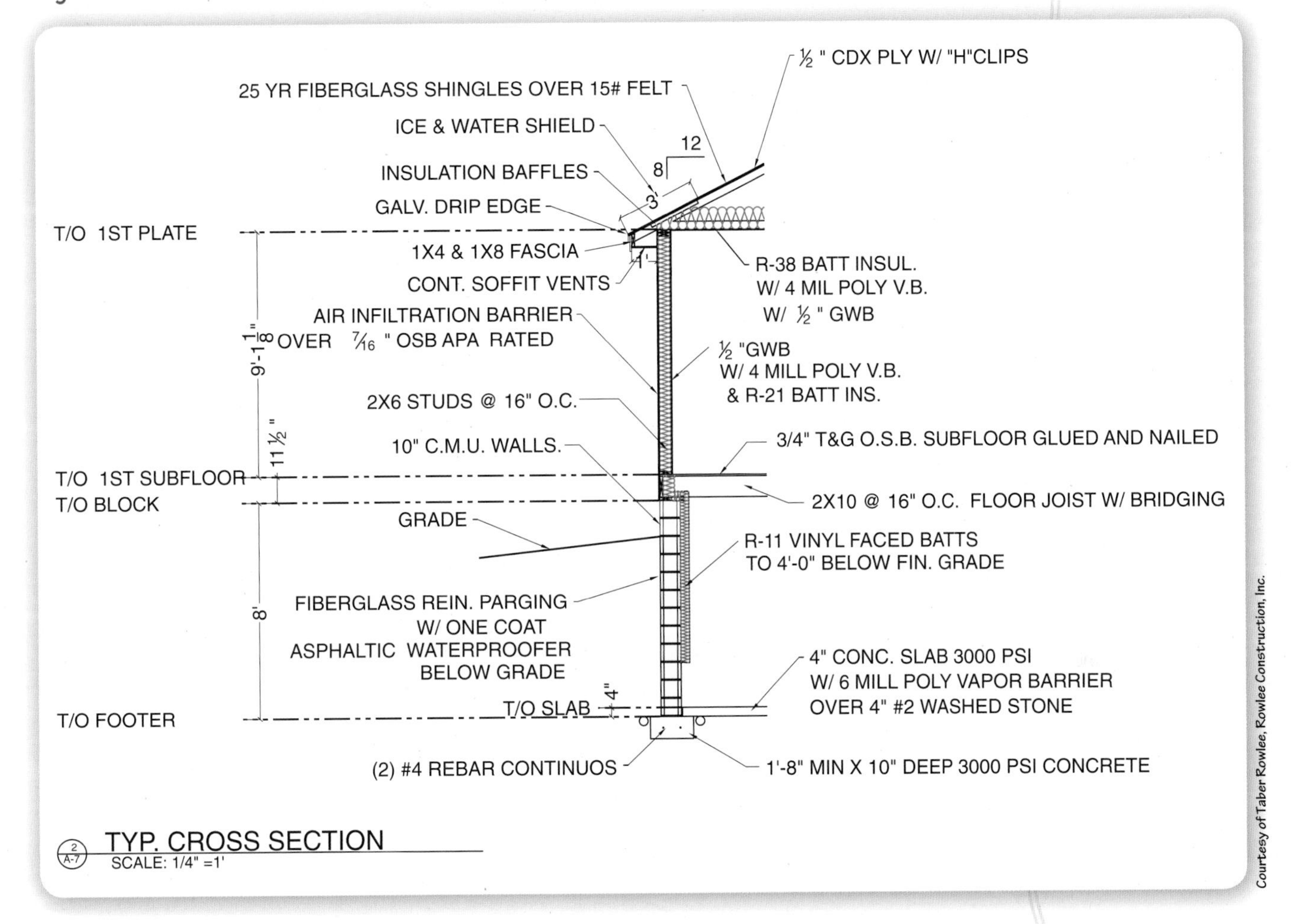

Courtesy of Taber Rowlee, Rowlee Construction, Inc.

in large complex building projects that require many details. Details are referenced by a symbol placed on the floor plan showing the detail number and sheet number where the detail is located. Because the purpose of a detail is to provide clarity to complex construction components or assemblies, they are commonly drawn or printed in 1 in = 1 ft 0 in or 1½ in = 1 ft 0 in scale. A larger scale of 3 in = 1 ft 0 in may be used to magnify very small areas. Figure 11-22 a shows a portion of a floor plan that has a detail view on sheet A-406. Figure 11-22 b shows the magnified detail of the typical southwest corner of the building. The word *typical* communicates that the southwest corner will be the same on all levels unless otherwise noted. Figure 11-23 shows the southwest corner of the finished building.

Figure 11-22a: Detail of South West Exterior Corner, Detail 01, Sheet A406.

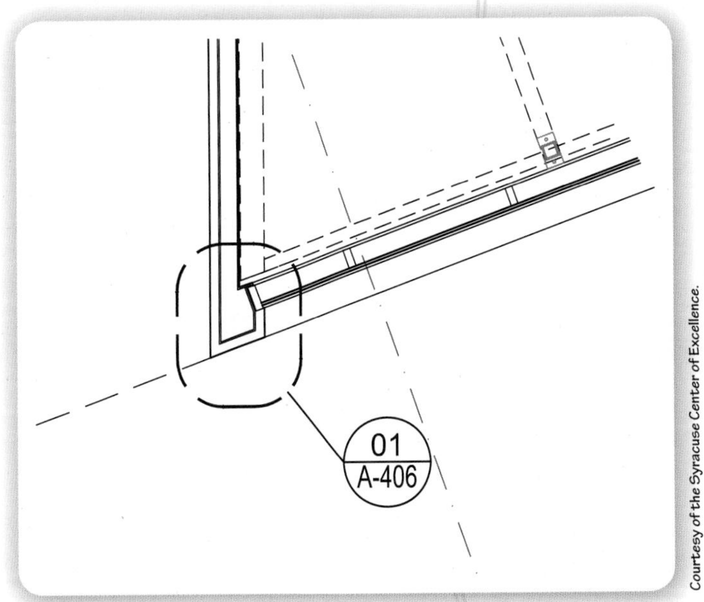

Courtesy of the Syracuse Center of Excellence.

Figure 11-22 b: (*continued*)

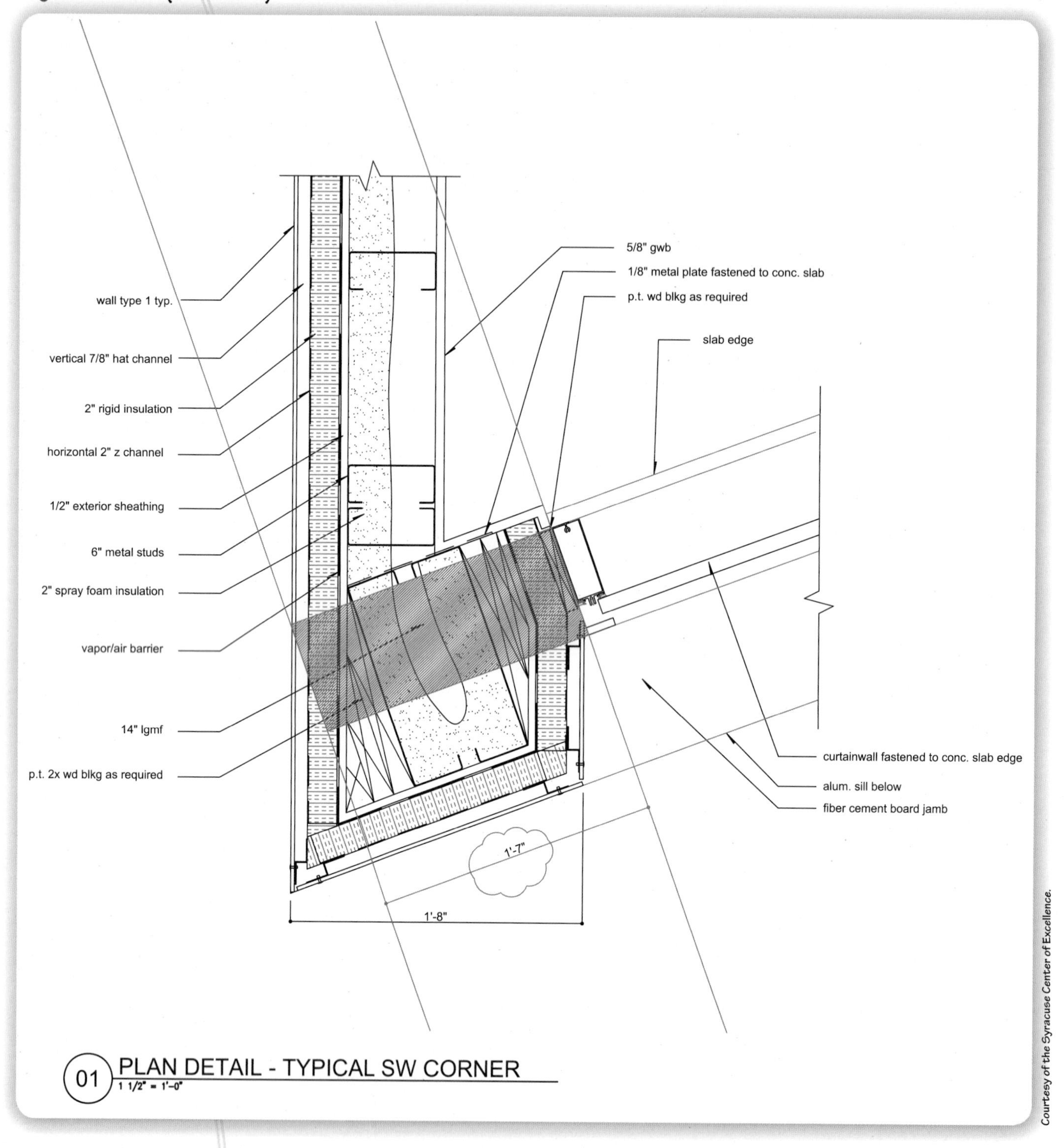

Courtesy of the Syracuse Center of Excellence.

SCHEDULES PROVIDE NUMEROUS SPECIFICATIONS WITHOUT COMPLICATING A DRAWING

6

Schedules offer a clean, concise way to itemize the features of building components or applications. A set of working drawings may include many schedules depending on the size and scope of the project. Schedules are used to identify specifications for room finishes, windows, doors, and mechanical and electrical components. They are also used in structural plans to itemize column and beams. Symbols, sometimes called tags, are used to show the location of the room or component. For example, symbols for exterior windows and doors are placed on the floor plan at the exterior side of the component (see Figure 11-24). Symbols that represent interior doors can be placed near the center of the component. A corresponding window or door schedule will list parameters such as type, size, rough opening, composition, material, finish, and other desired information. A highly detailed schedule may even specify the manufacturer and catalog number. In addition to providing detailed information to the builder, schedules are helpful to contractors when preparing a cost estimate.

Figure 11-23: The southwest corner after construction.

Courtesy of the Syracuse Center of Excellence.

Figure 11-24: Schedules provide necessary information without over crowding the drawing.

WINDOW SCHEDULE

MARK	QTY	TYPE	SIZE	MFGR	UNIT NUMBER	ROUGH OPENING	REMARKS
A	7	DOUBLE-HUNG	2'-8" X 4'-9"	TALON	DHG2849	2'-8 1/2" X 4'-9 1/2"	CHERRY- UNFINISHED FLAT BETWEEN THE GLASS MUNTIN BARS
B	1	DOUBLE-HUNG COMBINATION	6'-0" X 3'-0"	TALON	MDH 3030	6'-0 1/2" X 3'-0 1/2"	CHERRY UNFINISHED FLAT BETWEEN THE GLASS MUNTIN BARS

1 / A-6 WINDOW SCHEDULE
SCALE: 1/4" =1'

B A CL 3 A M.BATH 4 L. 1 MASTER BEDROOM FOR CLASSROOM USE ONLY A DN 13 3/4 BATH 3 4 WIC HALL 1 2

Your Turn

Analyze a Small Area of the Syracuse Center of Excellence Commercial Case Study

The Syracuse Center of Excellence steel construction and curtain walls were dimensioned using the grid approach. Examine the floor plan shown in Figure 11-25 a, sheet A102W, and locate the office suite 201. Would Figure 11-25 b be described as an elevation, section, detail, or schedule?

Now look at the exterior elevation A200W shown in Figure 11-26. Record in your notebook the additional information provided in the exterior elevation. Compare the exterior elevation shown in Figure 11-26 to the section view shown in Figure 11-27. List the drawing differences in your notebook.

Now look at the interior elevations for the second floor. Figure 11-28 shows the interior elevation A721 of the office suite. Read the construction notes and add a brief description of the construction in your notebook. Do you remember the abbreviations for GL, WD, and GWB shown earlier in the Figure 11-9 chart?

Let's see what other information we can find about the second floor office suite. Use the schedule and index shown in Figure 11-29 (a and b) to identify the interior finishes planned for the 201 Office Suite. Record this information in your notebook.

During this analysis you have examined an exterior and interior elevation, a section, a detail, a schedule, and an index. What purpose does each of these drawings serve? Record your answers in your notebook. Add dimensions and specifications to your own plans. Use the checklist shown in Figure 11-30 to help you annotate your drawings.

Figure 11-25 a: *(a) Second Floor Plan Sheet 102W.*

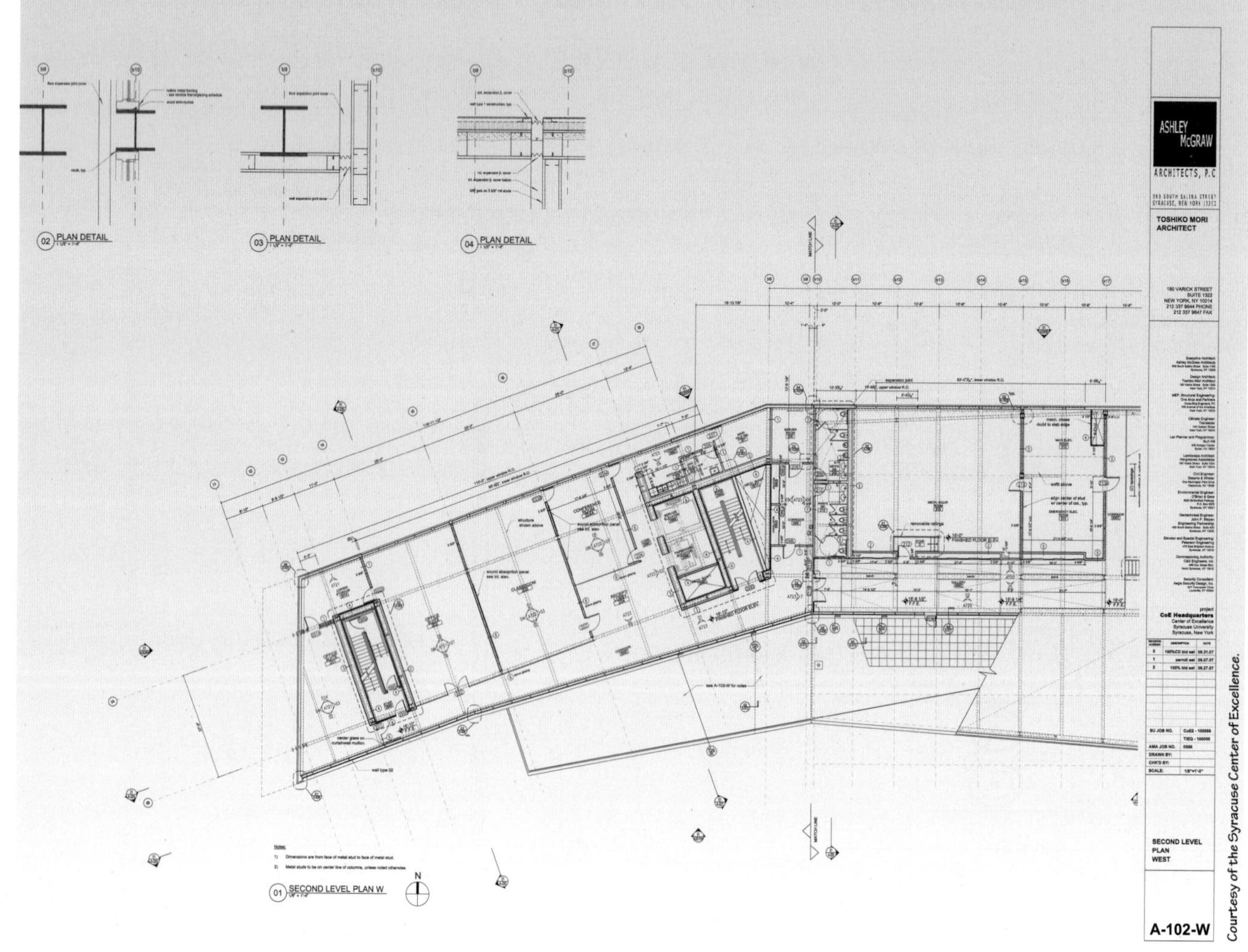

Courtesy of the Syracuse Center of Excellence.

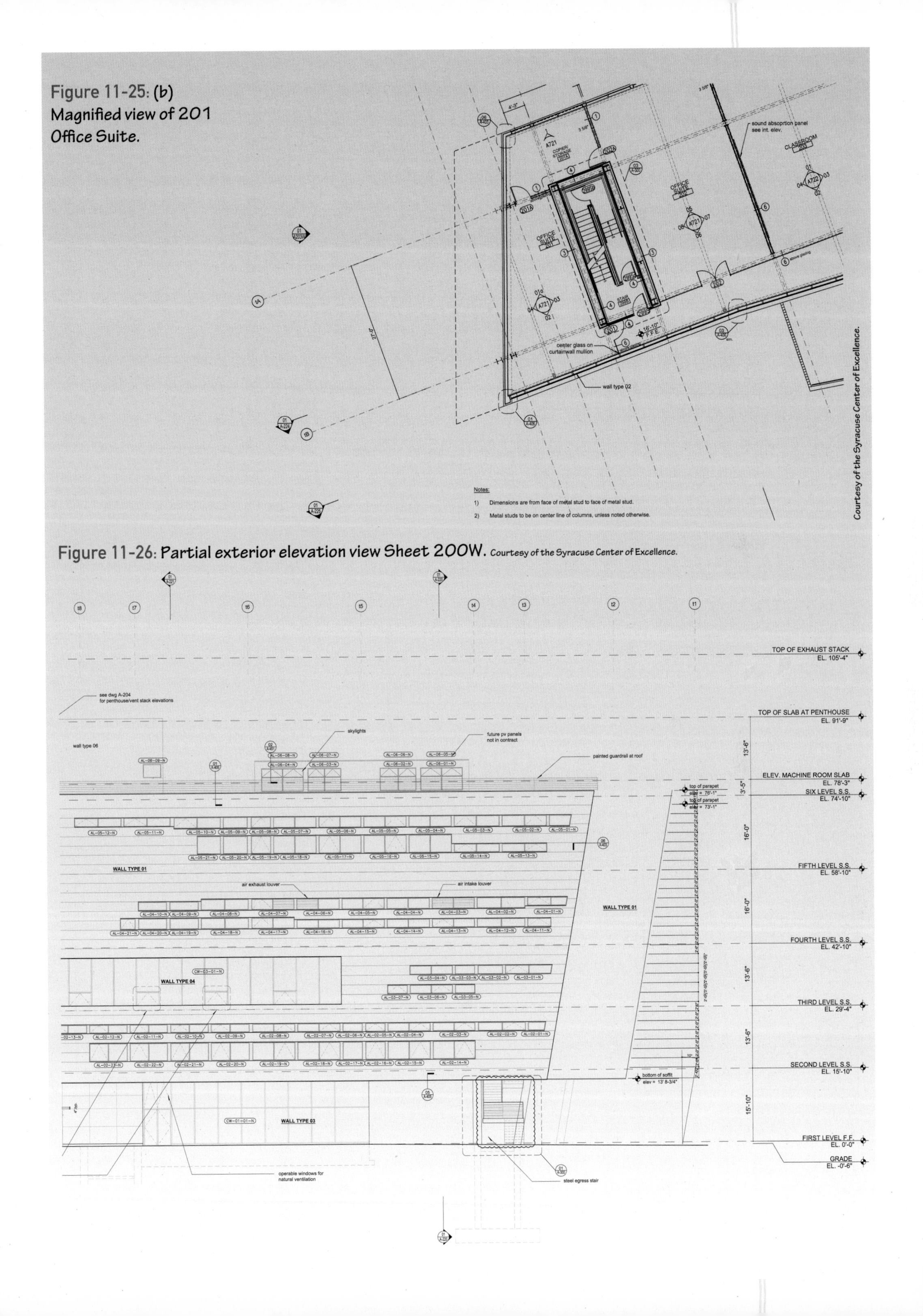

Figure 11-25: (b) Magnified view of 201 Office Suite.

Figure 11-26: Partial exterior elevation view Sheet 200W. Courtesy of the Syracuse Center of Excellence.

Figure 11-27: Partial section view Sheet A224 W. Courtesy of the Syracuse Center of Excellence.

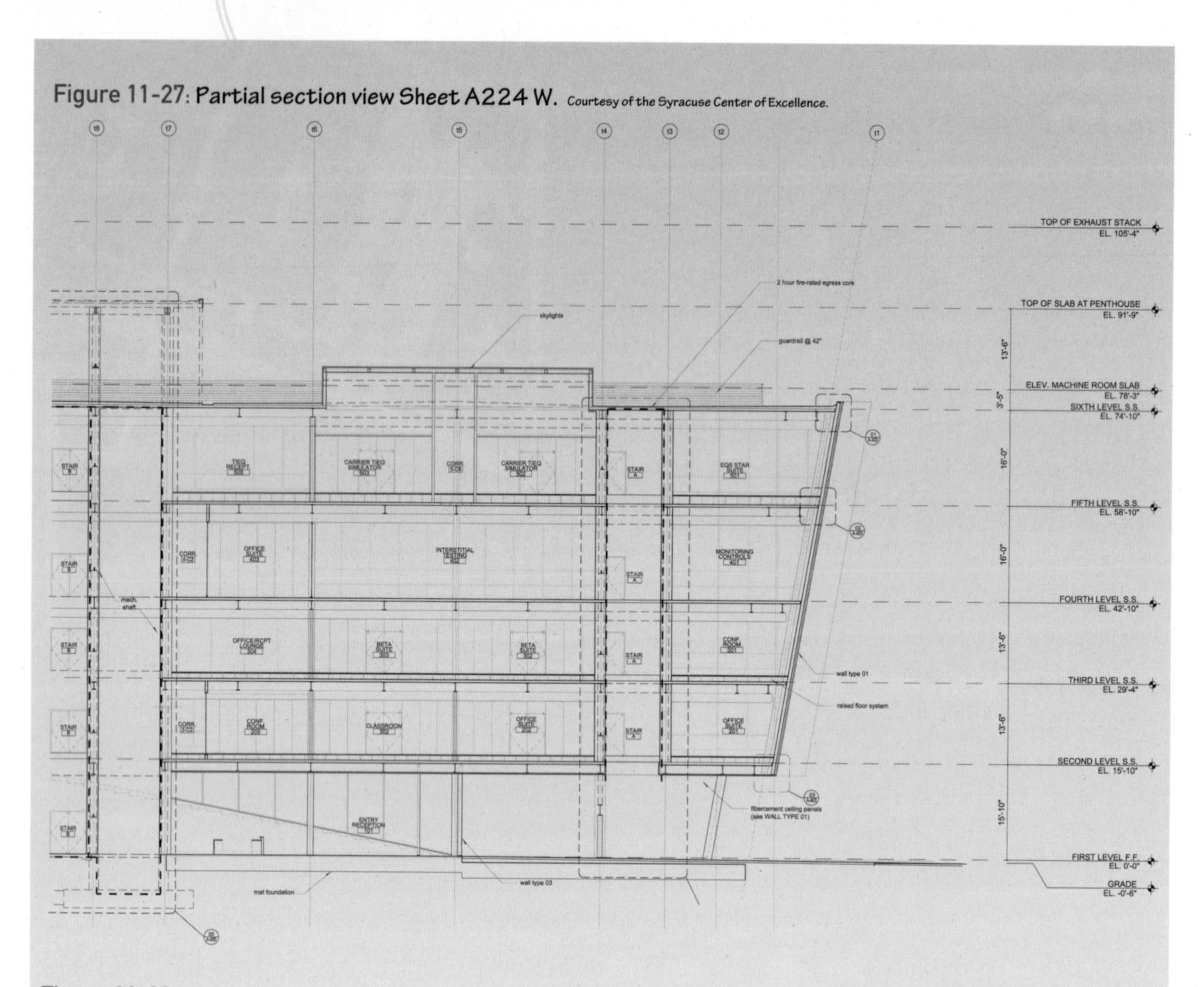

Figure 11-28: Interior elevation 3 from Sheet A 721 showing interior wall of office suite 102.

Courtesy of the Syracuse Center of Excellence.

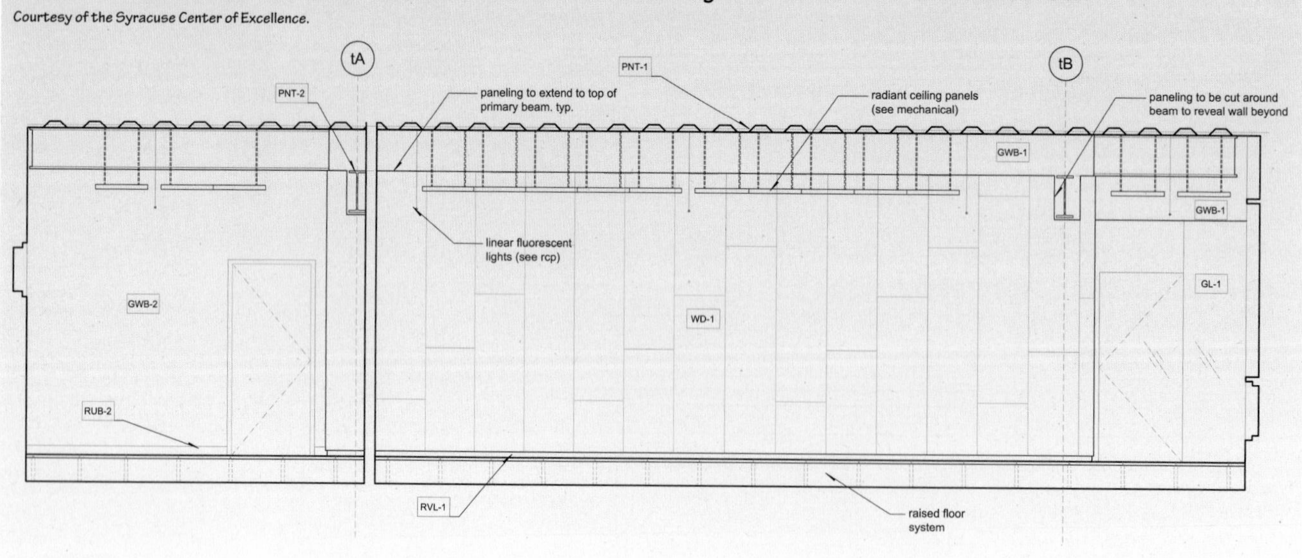

Figure 11-29a, b: Finish Schedule and Material Index for the Syracuse Center of Excellence.

CoE HQ- FINISH SCHEDULE

		FLOOR	BASE	WALL NORTH	SOUTH	WEST	EAST	CEILING	NOTES
STREET LEVEL									
100	VESTIBULE	TER-1/GRILLE	-	-	-	-	-	FCMT-1	③
101	ENTRY RECEPTION	TER-1	RVL-1	-	GWB	-	WD-1	FCMT-1	②③
102	MECH. EQUIP. ROOM	SEALER	RUB-2	GWB/PNT	GWB/PNT	GWB/PNT	GWB/PNT	-	-
102B	FIRE PUMP	SEALER	RUB-2	GWB/PNT	GWB/PNT	GWB/PNT	GWB/PNT	-	-
102C	INTER. TELE. RM	SEALER	RUB-2	GWB/PNT	GWB/PNT	GWB/PNT	GWB/PNT	-	-
-		-	-	-	-	-	-	-	-
104	TEMP. CONSTRUCTION	-	-	PRIMER ONLY	-	-	-	-	-
105	TEMP. CONSTRUCTION	-	-	-	PRIMER ONLY	-	-	-	-
106	CHEM. WASTE STORAGE	SEALER	RUB-2	GWB/PNT	GWB/PNT	GWB/PNT	GWB/PNT	GWB/PNT	-
107	BIOFUELS STORAGE	-	-	PRIMER ONLY	PRIMER ONLY	PRIMER ONLY	PRIMER ONLY	-	-
108	BIOFUELS PILOT PLANT	-	-	CMU/PRIMER ONLY	-	CMU/PRIMER ONLY	-	-	-
108A	BIO R&D			-	CMU/PRIMER ONLY	CMU/PRIMER ONLY	-		
114	MEN'S RR	CT-1	CT-1	CT-1/PNT/MIR-1	CT-1/PNT	CT-1/PNT	CT-1/PNT	APC-2/PNT	①⑦
115	WOMEN'S RR	CT-1	CT-1	CT-1/PNT	CT-1/PNT/MIR-1	CT-1/PNT	CT-1/PNT	APC-2/PNT	①⑦
116	VESTIBULE	GRILLE	RVL-1	GWB/PNT	GWB/PNT	GWB/PNT	GWB/PNT	EXPOSED/PNT	-
-	-	-	-	-	-	-	-	-	-
1-C1	CORRIDOR	CON-1	RVL-1	GWB/PNT	GWB/PNT	-	-	EXPOSED/PNT	-
1-C3	CORRIDOR	CON-1	RVL-1	GWB/PNT	GWB/PNT	GWB/PNT	GWB/PNT	EXPOSED/PNT	④
1-C4	CORRIDOR	TER-1	RVL-1	WD-1	GWB/PNT	-	GWB/PNT	FCMT-1	②③
1-C5	CORRIDOR	CON-1	RVL-1	-	GWB/PNT	GWB/PNT	CMU/PNT	EXPOSED/PNT	④
1-C6	CORRIDOR	CON-1	RVL-1	-	-	GWB/PNT	CMU/PNT	EXPOSED/PNT	④
-	-	-	-	-	-	-	-	-	
B	STAIR	SEALER	RUB-2	GWB/PNT	GWB/PNT	GWB/PNT	GWB/PNT	EXPOSED/PNT	⑥
C	STAIR	SEALER	RUB-2	GWB/PNT	GWB/PNT	GWB/PNT	GWB/PNT	EXPOSED/PNT	⑥
		-		-					
SECOND LEVEL									
201	OFFICE SUITE	CPT-1	RVL-1	GWB/PNT	-	GWB/PNT	WD-1	EXPOSED/PNT	②④
201A	COPIER/STORAGE	CPT-1	RVL-1	-	GWB/PNT/WD-1	GWB/PNT	GWB/PNT	EXPOSED/PNT	④
202	OFFICE SUITE	CPT-1	RVL-1	-	-	WD-1	GWB/PNT	EXPOSED/PNT	②④
203	CLASSROOM	CPT-1	RVL-1	-	-	GWB/PNT/FSAP	GWB/PNT	EXPOSED/PNT	④
204	CONFERENCE ROOM	CPT-1	RVL-1	-	-	GWB/PNT/FSAP	-	EXPOSED/PNT	④
205	RECEPT/LOBBY	RUB-1	RVL-1	-	-	GWB/PNT	-	EXPOSED/PNT	④
-	-	-	-	-	-	-	-	-	
206	KITCHENETTE	RUB-1	RVL-1	-	WD-1	WD-1	WD-1	EXPOSED/PNT	②④
207	JAN. CLOSET	SEALER	RUB-2	GWB/PNT	GWB/PNT	GWB/PNT	GWB/PNT	EXPOSED/PNT	
208	SERVER ROOM	SEALER	RUB-2	GWB/PNT	GWB/PNT	GWB/PNT	GWB/PNT	EXPOSED/PNT	
208A	MAIN TELE. RM	SEALER	RUB-2	GWB/PNT	GWB/PNT	GWB/PNT	GWB/PNT	EXPOSED/PNT	
209	MENS RR	CT-1	CT-1	CT-1/PNT	CT-1/PNT	CT-1/PNT	CT-1/PNT/MIR-1	APC-2	
210	WOMENS RR	CT-1	CT-1	CT-1/PNT	CT-1/PNT	CT-1/PNT	CT-1/PNT/MIR-1	APC-2	
211	GALLERY	RUB-1	RVL-1	GWB/PNT	GWB/PNT	-	-	EXPOSED/PNT	④
212	MECH. EQUIP. ROOM	SEALER	RUB-2	GWB/PNT	GWB/PNT	GWB/PNT	GWB/PNT	EXPOSED/PNT	
213	MAIN ELEC. ROOM	SEALER	RUB-2	GWB/PNT	GWB/PNT	GWB/PNT	GWB/PNT	EXPOSED/PNT	
214	EMERGENCY ELEC. ROOM	SEALER	RUB-2	GWB/PNT	GWB/PNT	GWB/PNT	GWB/PNT	EXPOSED/PNT	
-	-	-	-	-	-	-	-	-	
2-C1	CORRIDOR	RUB-1	RVL-1	WD-1	-	-	-	EXPOSED/PNT	④
2-C2	CORRIDOR	RUB-1	RVL-1	-	-	-	WD-1	EXPOSED/PNT	②④
2-C3	CORRIDOR	RUB-1	RVL-1	GWB/PNT	GWB/PNT	GWB/PNT	GWB/PNT	EXPOSED/PNT	④
2-C4	CORRIDOR	RUB-1	RVL-1	GWB/PNT	-	GWB/PNT	-	EXPOSED/PNT	④
2-C5	CORRIDOR	RUB-1	RVL-1	GWB/PNT	WD-1	-	GWB/PNT	EXPOSED/PNT	②④
-	-	-	-	-	-	-	-	-	
A	STAIR	SEALER	RUB-2	GWB/PNT	GWB/PNT	GWB/PNT	GWB/PNT	EXPOSED/PNT	⑥
B	STAIR	SEALER	RUB-2	GWB/PNT	GWB/PNT	GWB/PNT	GWB/PNT	EXPOSED/PNT	⑥
C	STAIR	SEALER	RUB-2	-	-	-	-	EXPOSED/PNT	⑥
-	-	-	-	-	-	-	-	-	

Syracuse CoE- Headquarters
MATERIALS INDEX

Designation	Description	Manufacturer	Product Number	Contact Number	FINISH Color/Number	NOTES
FLOOR						
TER-1	TERR-CON Polished Concrete Floor	Floored, LLC		Jason Bye (585) 461-3060	White marble chips in white matrix	
CON-1	Retroplate polished Concrete topping slab	Floored, LLC	-	Jason Bye (585) 461-3060	Water-based concrete stain (see Paint Spec)	-
CPT-1	Carpet Tile 19.67" x 19.67"	Interface	138190250A	InterfaceFLOR Commercial (212) 994-9969	Color = #6274 Exact, Pattern = Clarity	-
RUB-1	Rubber Tile 36" x 36"	Roppe	-	N.R.F. Distributors (207) 622-4744	#195 Light Gray	PVC free rubber floor tiles
CT-1	Ceramic Mosaic Tile, 1" x 1"	Dal-Tile	#D-317	Rock Roberts, (315) 432-0295	White with white grout to match	-
SEALER	Concrete Sealer	Benjamin Moore		Henderson-Johnson Co. (315) 479-5561		
		-	-	-	-	-
BASE						
RVL-1	Aluminum base reveal extrusion, 5/8" x 2-1/2"	Fry Reglet	-	(800) 237-9773	-	-
CT-1	Ceramic Mosaic Tile, 1" x 1", built-up base	Dal-Tile	#D-317	Rock Roberts, (315) 432-0295	White with white grout to match	-
RUB-2	Pinnacle Rubber surface applied base, 4"	Roppe International	#170 white	Great Northern Assoc., Inc. (315) 431-1915	White, smooth, no toe	PVC free, rubber base
WALL						
GWB	Gypsum Wall Board		-	(800) 237-9773	Skim coat and paint	Non VOC
CMU-1	CMU	-	-	-	-	
WD-1	Dakota Burl (Alt. 01 = Bio Fiber Wheat), 3/4" Thick	Environ Biocomposites		(800) 324-8187	Stained and/or clear finish. Color TBD.	
CT-1	Ceramic Tile, 1" x 1"	Dal-Tile	#D-317	Rock Roberts, (315) 432-0295	White, with white grout to match	
MIR-1	1/4" Mirror	-	-	-	-	
PNT	Paint, see spec for type specific to each space type	Benjamin Moore	-	-	Decorator's White	
FSAP	Fixed Sound Absorptive panels	Decoustics	HIR-1	-	Guilford FR701 TBD	
CEILING						
PNT	Exposed ceiling with painted finish	-	-	-	-	
FCMT-1	Fiber Cement Board	Swiss Pearl	-	Paradigm Products Group, Inc. (856) 309-0102	-	
GWB-4	Gypsum wall board hung ceiling	-	-	-	Skim coat and paint	
APC-1	Acoustic Panel Ceiling, suspended, TIEQ spaces	-	-	-	-	
APC-2	Acoustic Panel Ceiling, suspended, Rest Rooms	-	-	-	-	
PNT	Paint, see spec for type specific to each space type	Benjamin Moore			Decorator's White	

Courtesy of the Syracuse Center of Excellence.

Courtesy of the Syracuse Center of Excellence.

Figure 11-30: Checklist for annotating architectural plans.

Annotation of Architectural Plan Checklist

- Legends, list of abbreviations, and index
- General notes
- Title and scale symbol at lower left of all drawings
- North arrow, datum, or reference points
- Building and spaces labeled by name or name and number
- Individual room and overall square footage
- Unique features, such as vaulted or tray ceilings and half walls
- Labels for closets, fixtures, appliances, attic access, and built-ins
- Staircase direction and number of stairs
- Handicapped accessibility: wheelchair turn-around ramps
- Direction and size of structural members: floor joists, rafters, and trusses
- Cross-reference symbols for all elevations, sections, and details
- Specific notes on sections, details, and elevations
  (size and type of material, construction method, and condition)
- Symbols for windows, doors, and other components with coordinating schedules
- Elevation symbols, grade, and roof pitch

SUMMARY

In order for a building plan to reach the desired outcome, the architect must provide plans that explain detailed information about the building design and composition. The majority of this information will be conveyed through universally recognized dimension and annotation practice. Drawing conventions increase the readability and understanding of vital building information. The goal is to provide comprehensive building information in an efficient, easy-to-understand format. Symbols and abbreviations reduce the amount of text needed on a drawing and are described in lists and legends located at the beginning of the drawing set. The beginning of a drawing set will also list general construction notes, which provide information about the intended construction that is not explained with the drawing alone.

Essential details about the building plan, materials, and procedures are communicated through the dimensioning and annotation of elevations, sections, and details. Elevations provide a visual of how the interior or exterior will appear when viewed from a specific direction. Elevation views provide the opportunity to place vertical dimensions and visualize the planned construction outcome. Internal details, such as the internal composition of a wall, are communicated through a section view. An exterior wall section view is labeled with construction notes and materials. When a building has unusual or complicated features, a detail view provides a magnified view. These magnified views explain the arrangement of specific building components to support accurate construction. Schedules refer to the itemized lists of components or finish specifications. They provide detailed information without complicating the drawing. Architects purposefully place dimensioning and annotation to enhance communication of essential information and increase the likelihood of the desired outcome.

BRING IT HOME

1. Place an elevation symbol on the floor plan to view an interior feature. Create a corresponding ½ in = 1 ft 0 in elevation view of the feature. Use leaders and notes to identify materials and construction specifications.
2. Explain the difference between a general and a specific note and provide an example of each.
3. Draw three examples that illustrate the differences between dimensioning for masonry, wood, and steel construction.
4. Completely dimension your own floor plan, using the conventions and methods shown in this chapter.
5. Place a section cutting plane line and symbol through an exterior wall of your floor plan. Complete a ½ in = 1 ft 0 in scale section of the exterior wall. Add elevation levels and use leaders and notes to identify materials and construction specifications.
6. Place door and window symbols on your floor plan and create a corresponding window and door schedule. Include categories for mark, quantity, type, size, manufacturer, unit number, rough opening, and finish
7. Create a legend showing all the symbols that appear on your floor plan.

EXTRA MILE

Working in a team, examine an actual set of working plans. Look at the floor plan and choose a symbol for an elevation, section, and detail. Find the corresponding sheets and analyze each supporting drawing. As a team, discuss the purpose of each drawing. Record your thoughts and make a list of the people you think will need this information.

PART V
Structural Design: Building for Tomorrow

CHAPTER 12
Building Materials and Components

GPS DELUXE

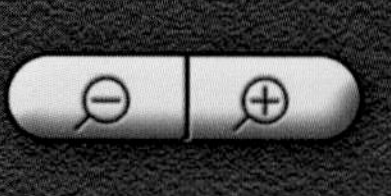

Menu

START LOCATION | DISTANCE | END LOCATION

Before You Begin

Think about these questions as you study the concepts in this chapter:

1. What are the ways in which construction materials are categorized?
2. What are the critical performance characteristics of construction materials?
3. How do the building codes affect the material selection process?
4. What is the format used to specify construction materials?
5. What are the main parameters that should be considered when selecting construction materials?
6. How are environmental issues related to the manufacturing and use of construction materials?
7. How do interior finish materials affect human health concerns?

Construction is a multifaceted undertaking that brings together architects, engineers, skilled workers, and a vast array of materials and equipment to create a finished building (see Figure 12-1). The construction industry is supported by a broad range of industries and manufacturers that produce the materials and components needed for a project. These include suppliers of stone and cement products; manufacturers of lumber, steel beams, windows, siding, roofing, and finish materials; and dealers of appliances and mechanical equipment. The production of building materials is a complex process that involves the extraction, processing, and refinement of a variety of raw materials. Many manufacturing endeavors involve the cutting, shaping, and assembling of these materials into useful products. Each product area uses a tremendous variety of materials, production, and installation methods that are continuously refined and updated. Materials have become more complex in recent decades, drawing from all naturally occurring elements in the periodic table, compared with a fraction in use in 1900. As materials have become more complex, so has the process of selecting them for construction. In order to make effective material selections for buildings, architects and engineers must be familiar with both the origins and latest developments of construction materials. This chapter introduces basic concepts related to the classification and selection of materials for construction.

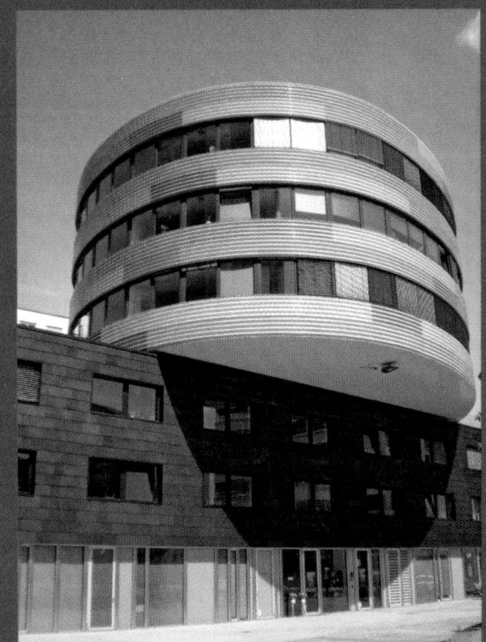

Courtesy of Eva Kultermann.

Courtesy of APA—The Engineered Wood Association.

Figure 12-1: *Materials, workers, and equipment come together in the building construction process.*

Figure 12-2: Brick construction technology developed over thousands of years to produce a highly refined architecture.

Courtesy of Eva Kultermann.

THE HISTORY OF BUILDING MATERIALS

Natural materials obtained from the earth have always been essential to our human development. Historically, buildings were built and occupied in close harmony with nature. The earliest settlements and their **vernacular** architecture demonstrated an integral fit with their local surroundings. Transportation was difficult so materials were extracted locally from available natural sources. Building styles and traditions arose according to what the local environment provided. For example, stone was used in rocky areas, while brick and adobe building dominated in the rich, clay-bearing plains (see Figure 12-2). Wood structures from crude log huts to complex timber frames were developed in heavily forested regions throughout the world (see Figure 12-3).

These early structures also displayed a close fit with their landscape and local climate. Buildings were oriented to take advantage of sunlight for both heating and lighting. Openings were arranged to admit cooling summer breezes while preventing the entry of cold winter winds. Wall tended to be thick and massive in both cold northern and hot arid climates, while reeds and grasses created pervious enclosures in hot humid ones. An innate knowledge of how to use a particular material in response to natural elements was perfected over thousands of years of experimentation and readjustment.

It was the Industrial Revolution that first changed these longstanding building traditions. Advances in engineering, new machinery, and the growth of larger communities began the move away from individually built, craft-inspired buildings. The rapid expansion of dense cities in close proximity to factories, mills, and mines mandated the use of uniform building types and materials. The period saw a rapid increase in the development and mass production of new products for buildings. The production of brick, for example, was first mechanized in the nineteenth century. The laborious process of hand-molding, which had been in use for more than 3000 years, was abandoned in favor of "pressed" bricks. These were mass produced by an extrusion process in which clay was squeezed in a continuous column through a rectangular die and sliced to size by a wire cutter (see Figure 12-4).

Figure 12-3: This timber frame uses large wood members in a complex structural frame.

Courtesy of Eva Kultermann.

Figure 12-4: The modern brick manufacturing process produces units that are uniform in shape, size, and color.

Courtesy of Eva Kultermann.

Timber technology also underwent rapid development during this same period. In North America the vast softwood forests were being harvested and processed by new industrial methods. Steam- and water-powered sawmills began producing standard dimensional lumber in large quantities to feed the westward expansion of the country. The production of inexpensive machine-made nails provided the other necessary ingredient that made possible a major innovation in building construction: the light wood frame, which is the common residential building skeleton that still dominates American residential construction today.

Courtesy of Eva Kultermann.

OFF-SITE EXPLORATION

Study examples of the development of vernacular building styles in different parts of the world by visiting the Green Home Building website at *http://www.greenhome-building.com/vernacular.htm*.

By the 1880s, the mass production of iron and steel had transformed building technology in ways never before imagined. In growing urban centers, the pressure of rising land values led property developers to demand taller buildings. Architects and engineers responded to this challenge with new high-rise buildings that utilized a lightweight, all-metal structure. The steel frame, together with the invention of the passenger elevator, allowed for an entirely new building type: the high-rise building. By 1895 a mature high-rise building technology had been developed consisting of steel I-beam frames with bolted or riveted connections, diagonal wind bracing, and lightweight infill materials.

Large amounts of steel were first introduced to the construction industry in the 1850s with the invention of the Bessemer process. By blowing air into the molten iron, impurities were burned out, resulting in economical and high-quality steel that could be produced quickly.

In the twentieth century, technological developments ushered in a vast new assortment of materials and products manufactured from previously unknown substances and material combinations. Traditional materials were reinvented through mechanical manipulations such as cutting, turning, and finishing. New products were formed by combining more than one substance and grinding, mixing, heating, and pressing them together. Several materials could be sandwiched with adhesives to develop strong and lightweight sheet materials including plywood and gypsum wallboard. Developments in chemistry brought forth innovative synthetic materials including plastics, fabrics, and fiberglass resins.

Today, a vast industry works to supply uniform products in standard sizes that are tested and certified by industry recognized standards. Material production is highly mechanized, resulting in strong, durable components whose performance characteristics can be confidently predicted. The industry is supported by a number of testing agencies and trade associations that conduct research and publish best practice guidelines on the proper use of materials. The evolution of building materials will undoubtedly continue in the years to come, as still unimagined materials and compounds are developed.

Point of Interest

A new type of smog-eating concrete can actually clean the air by dissolving pollutants. The cement is treated with titanium dioxide, which reacts with ultraviolet light to break down pollutants such as smog-forming nitrogen oxides. Using light and air, the photo-catalytic concrete helps to break down organic and inorganic substances responsible for air pollution. The chemical reaction prevents bacteria and dirt from accumulating on a building's surface, and also helps keep concrete clean and white.

CONSTRUCTION MATERIALS FUNDAMENTALS

Vernacular Architecture: a term used to categorize building traditions that address local needs and construction methods that use locally available resources.

Throughout the design and space-planning process, considering which material to utilize where, informs and refines design decisions. Each material carries with it distinct associations and capabilities. A large convention hall might employ steel for its lightweight and long spanning capacity, while a summer home could make use of natural wood to convey a warm and inviting feeling. Architects and engineers must be well informed on how materials are classified, what materials are available for use in a particular location, and what their performance characteristics are.

Material Families

For the purposes of classifying basic material composition, construction materials can be grouped into five broad categories:

Metals
Polymers
Ceramics
Composites
Organic/natural materials

Ferrous Metals: iron-based metallic materials.

Nonferrous Metals: metallic materials in which iron is not a principal element.

Metals are refined from minerals that have been extracted from the earth. They are generally pliable, strong, and are especially useful when tensile (pulling) forces are expected to occur. **Ferrous metals**, such as cast iron and steel, are those in which the chief ingredient is the chemical element iron. Iron is found in large quantities in the earth's crust where the ore is mixed with other minerals. To be useful, the iron must be extracted by mining, melted and refined to alter its properties, and then formed into useful products. All metals other than iron and steel are considered **nonferrous**, designating that they contain very little or no iron. Examples of nonferrous metals include aluminum, copper, and zinc.

Metals provide an efficient structural material, in that they can achieve very high strength with relatively low weight (Figure 12-5). They are generally tough and resistant to impact. Metals are used widely in applications ranging from structural steel frames, to siding and roofing systems, and building fasteners and hardware. Although the environmental impacts of both the mining and production of metals are high, they are readily recycled. According to the Steel Recycling Institute (*http://www.recycle-steel.org*), most structural steels are produced with up to 96 percent of both post-consumer and post-industrial scrap steel.

Figure 12-5: **These lightweight steel joists provide high-strength and long-spanning capabilities.**

Courtesy of Eva Kultermann.

Ceramics encompass such a vast array of materials that a concise definition is almost impossible. We can say that ceramics are a refractory (resistant to heat), inorganic, and nonmetallic material. The raw materials that make up ceramics are also extracted from the earth: silica for glass; clays for masonry; and lime for concrete. Most ceramics are formed by combining materials and preparing them either by baking as with brick and glass, or through the addition of water in concrete. Ceramics tend to be hard, rigid, brittle, and heavy, and are often used when compression forces are expected to occur.

Plastics and polymers include a number of natural and synthetic materials with numerous properties and purposes. Although the use of plastics in construction is relatively new, the material is used widely today for both structural and decorative purposes. The scientific definition of a plastic is a group of man-made polymers containing carbon atoms covalently bonded with other elements. Polymers are obtained by breaking down materials such as petroleum and coal under heat, and casting, pressing, or extruding them into useful shapes. Their inherent malleability allows for a variety of uses including sheets, films, and tubes among many others. Plastics are generally synthetic (not man-made) in nature, meaning they are produced by taking natural materials through a series of chemical manipulations. Recent decades have seen a greater increase in the use of plastics in construction when compared to all other types of building materials (see Figure 12-6).

Composite materials are engineered materials made from two or more constituent materials with significantly different physical or chemical properties. The most primitive composite materials were straw and mud combined to form bricks. These materials remain separate and distinct on a macroscopic level within the finished structure. Composite materials consist of a matrix (the mud in bricks for example) that surrounds a reinforcement material (the straw), which imparts its special mechanical and physical properties to support the matrix properties. Examples of composite materials include fiber-reinforced polymers, engineered wood products such as plywood and oriented strand board, wood plastic composite, asphaltic concrete, and carbon fiber.

Natural and organic materials can be defined as those that require little in the way of processing and refinement to be of use in construction. Natural materials such as wood, stone, jute, and straw have been used in construction for millennia (see Figure 12-7). They are usually derived from renewable sources such as living plants that have the ability to regenerate themselves quickly. Renewable materials are preferred because, when managed properly, their supply will not be depleted. Most result in biodegradable waste and tend to have reduced negative environmental

Ceramics:
a group of materials made from non-metallic minerals fired at high temperatures.

Plastic:
a group of man-made polymers that contain carbon atoms covalently bonded with other elements.

Composite Materials:
engineered materials made from two or more constituent materials with significantly different physical or chemical properties.

Figure 12-6: Polymer-based cladding materials are available in a variety of colors and patterns.

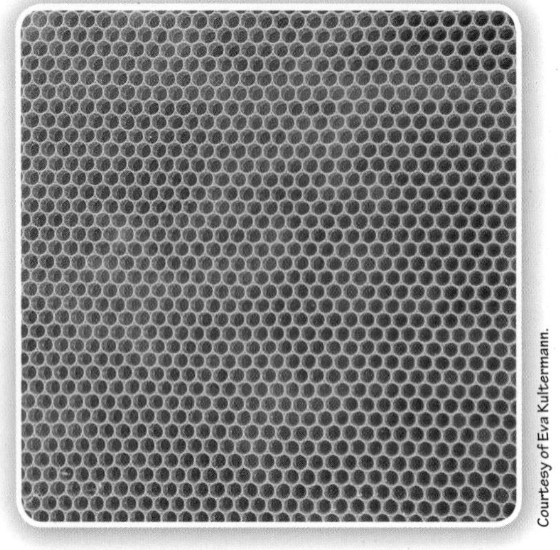

Courtesy of Eva Kultermann.

impacts across their lifecycle when compared to materials derived from fossil fuels.

Material Properties

Within each of the basic material families, materials are evaluated according to their technical performance characteristics. Each material has distinct quantitative properties that can be used to compare the benefits of one material with another and aid in the material selection process. All materials react to changes in applied loading, temperature fluctuations, and moisture content. To be useful, a material must be able to perform under a variety of loadings, stresses, and environmental conditions.

A material property may be a constant or may be a function of one or more independent variables, such as temperature. Materials properties can vary to some degree according to the direction within the material that they are measured in; a condition referred to as anisotropy (as opposed to isotropy, which implies identical properties in all directions).

Figure 12-7: Natural wood is one of the most common building materials.

Courtesy of Eva Kultermann.

As described in Chapter 4, buildings are subjected to both vertical and horizontal forces. Vertical or compressive forces, those which tend to push on a material, result from the weight of building walls, floors, and roofs. Horizontal forces are the result of wind, subsurface water, soil, and earthquakes. The mechanical response of a material to these various forces must be judged against the expected conditions for a given project.

Mechanical properties are a measure of a material's strength in resisting an applied force. They are related to the response of a material to both static (continuous) and dynamic (intermittent) loads. Mechanical properties include tensile strength (resistance to pulling), compressive strength (resistance to pushing), and shear strength (resistance to being split). Elasticity describes the ability of a material to deform under loading by bending or stretching, and return to its original form once the load is removed. When a material is loaded beyond its elastic limit it can no longer return to its pre-loaded form and will permanently deform or rupture. Materials that can tolerate deformation under load are considered ductile. Copper, for example, is a very **ductile material**. When tested under tension, it stretches for some distance without breaking. In contrast, brittle materials display low elastic limits and will fail without prior deformation.

Hardness is a measure of the ability of a material to resist indentation or surface scratching, and is related to the tensile strength of the material. Generally, a material that has a high hardness rating also will have a high tensile strength. Impact strength is the ability of a material to resist a rapidly applied load, such as the strike of a hammer. It is an indication of the toughness of the material. A material with high impact strength will absorb the energy of impact without fracturing. Impact strength is affected by strength and ductility. Metals that are strong and ductile also have high impact strength. Ceramics are strong in compression but are brittle (lack ductility) and break under impact. Plastics, however, are very ductile but low in strength and do not absorb impact well.

Figure 12-8: Building wrap is a material of high permeability that is designed to provide air sealing while still allowing moisture generated inside a building to escape to the exterior.

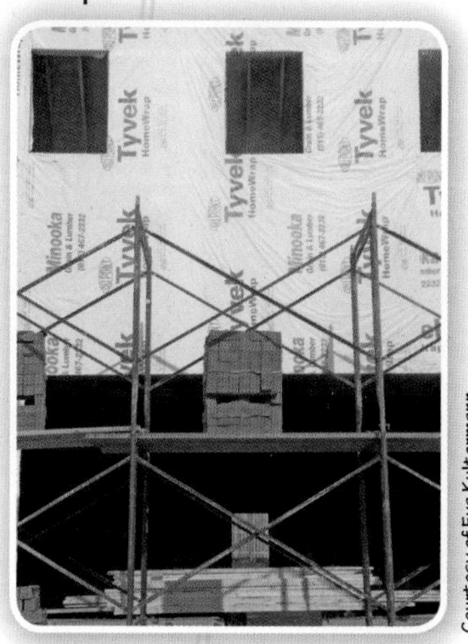

Courtesy of Eva Kultermann.

The rate that water flows through a material is a function of the material's permeability. **Permeability** is an important consideration for sheets and foils designed to restrict the passage of moisture through the floor, wall, and ceiling assemblies of a building (see Figure 12-8). The permeability of a material is measured in units of *permeance,* called *perms*, and is typically referred to as the *perm rating.* A vapor retarder for example, is defined as a material having a perm rating of 1.0 or less.

The thermal properties of a material are those that are related to the material's response to heat. When a material is subjected to a change in temperature it may expand, contract, conduct, or reflect heat. If a piece of metal is held in the hand and heated on one end, it will quickly conduct heat through its structure and become too hot to touch. Construction materials can be classified either as insulators, materials that resist the transfer of heat, or conductors, materials that encourage the transfer of heat. Fibrous materials, such as rock wool, fiberglass, and cotton are good insulators (see Figure 12-9), while most metals are good conductors.

Thermal conductivity is measured in the number of Btu's (British thermal units) that can pass through 1 sq ft of a material 1 in thick when the temperature difference between the two surfaces is 1 degree Fahrenheit.

The chemical properties of a material describe its tendency to undergo a chemical change or reaction due its composition and interaction with the environment. A chemical change can alter the original composition of a material and thereby affect its properties. Iron, for instance, has a tendency to oxidize or corrode under certain conditions. In addition to corrosion, other important chemical properties are ultraviolet degradation and fire resistance.

Materials and the Building Codes

A study of the materials of construction involves not only their technical aspects but also the influences on their use by the building codes and other standards. The use of a material for a particular application and its installation requirements are regulated by the building codes. Building codes control the design and construction of buildings by enforcing minimum standards established to safeguard the life, health, property, and general welfare of the public. The codes provide strict guidelines upon which architects and engineers design buildings, and material and equipment suppliers fabricate products. They ensure the stability and safety of buildings by ensuring that construction guidelines are adhered to through periodic site inspections.

Ductile (Metals): metals capable of being drawn into a wire or hammered thin.

Permeability: a measure of the rate that water or moisture will flow through a material.

Figure 12-9: *Cellulose insulation, made from recycled newspaper, provides an excellent thermal insulation material.*

Courtesy of Eva Kultermann.

In the United States, the International Building Code (IBC) regulates all structures within the jurisdictions where it has been adopted except one and two family dwellings. The International Residential Code for One- and Two-Family Dwellings is a separate residential code for smaller domestic structures and multiple single-family dwellings, such as townhouses.

Building codes specify construction requirements for a building based on the type of use, or occupancy. A home or school for example will have different requirements than would a hazardous manufacturing facility. Engineering requirements focus on both the structural elements of a building and its exterior **cladding** systems. In addition, requirements for elements such as building openings, architectural trim, and projections including balconies and canopies are outlined. Allowable design loads on the various components of the building are carefully detailed in the code, and structural components must meet minimum strength allowances. Criteria for calculating different types of structural loadings, structural tests and inspections, and the construction of foundation systems are given in great detail.

Cladding: the external finish covering the base material on a wall.

Your Turn

Contact your local department of building construction to find out which building code has been adopted for your town. Check for local amendments and changes that adjust code requirements for your area. Why were these amendments made?

Fire Resistance

An emphasis on the resistance to fire is a large component of building code constraints. Whether or not materials will readily burn, or their combustibility, is of prime importance in the selection of building materials. A material of low combustibility is said to be fire resistant. For example, a firebrick in a fireplace must withstand temperatures of up to 2000 degrees Fahrenheit. Building codes classify buildings into five basic types of construction based on their ability to resist fire.

Type I and II construction are considered inherently non-combustible and utilize materials such as concrete and steel. Part of their fire resistance comes from the fire resistance of the material itself, and some from applied fireproofing materials. Fireproofing refers to the act of making building assemblies more resistant to fire by covering structural elements in non-combustible materials. Steel frames for example can be sprayed with fire resistant paints (see Figure 12-10), or have main beams and columns surrounded by gypsum wallboard.

Type III construction utilizes exterior building materials that are non-combustible with load bearing materials that may be of varying degrees of combustibility. Type IV construction defines exterior walls that are of non-combustible materials such as brick, with interior structural elements built of exposed solid or laminated wood. This type is commonly used to designate heavy timber buildings. Large wood timbers are difficult to ignite and will withstand a fire much longer than small wood frame members.

Type V, often called wood frame construction, describes buildings in which the structural elements, exterior walls, and interior walls are made of combustible materials. Most small wood buildings will ignite and burn readily and must therefore be protected with fireproofing materials. Structural elements for example must be covered in a fire-resistant material that will withstand a fire for a minimum of one or more hours. This is usually achieved by cladding walls and roofs with gypsum board or plaster.

A building's maximum height and area allowance is limited according to the type of construction utilized. Buildings of Type I and II construction are generally allowed unlimited areas due to their intrinsic fire resistance. Greater allowable areas can be permitted if a building is equipped with an automatic sprinkler system or divided into smaller areas through the use of fire separation walls. Firewalls must extend from the foundation through the roof and have a fire rating of two to four hours.

Figure 12-10: A sprayed-on fire-resistant material protects this steel frame structure.

Courtesy of Eva Kultermann.

Specifying Materials – The MasterFormat

As outlined in Chapter 3, construction documents for a building project consist of two interdependent components: the drawings and the specifications. Construction drawings show the dimensional relationships between all aspects of the building; their forms, sizes, and locations. Since the drawings alone cannot give all of the details involved in describing the quality of specific materials or their construction methods, a written manual called the specifications is added. The specifications give more detailed information on the exact types of materials to be used and the ways in which installation processes are conducted.

Developed by the Construction Specifications Institute, the MasterFormat provides a standard for writing specifications using a system of numbered categories to organize construction activities, products, and requirements into a standard order that facilitates the retrieval of information. Since the design and completion of construction projects involves individuals in many technical fields, the ability to effectively communicate by having a consistent sequence for identifying and referring to construction information is essential. The format numbering system is also used by material manufacturers to organize technical literature pertaining to individual materials.

Divisions 1 through 23 of the MasterFormat are primarily concerned with the construction of buildings. The division ordering is loosely arranged to follow the sequence of the construction process itself, allowing for easy recall of where information is located. Division three, concrete, is the material from which most foundations are built. The divisions move on through major structural and cladding materials, to interior furnishings and components, and finally to mechanical and electrical systems. Table 12-1 outlines the initial divisions of the MasterFormat. Students should become familiar with the numbering of these first 23 divisions.

MasterFormat: the trademarked title of a uniform system for indexing construction specifications published by the Construction Specifications Institute and Construction Specifications Canada.

Visit the website of the Construction Specifications Institute website at http://www.csinet.org to see a full listing of the MasterFormat divisions. Within the listing, search for the location of roofing tiles, fabric structures, and elevators.

Table 12-1: The MasterFormat

Division 00	Procurement and Contracting Requirements
Division 01	General Requirements
Division 02	Exiting Conditions
Division 03	Concrete
Division 04	Masonry
Division 05	Metals
Division 06	Wood, Plastics, and Composites
Division 07	Thermal and Moisture Protection
Division 08	Openings
Division 09	Finishes
Division 10	Specialties
Division 11	Equipment
Division 12	Furniture
Division 13	Special Construction
Division 14	Conveying Equipment
Division 15– 21	*Reserved for future expansion*
Division 21	Fire Suppression
Division 22	Plumbing
Division 23	Heating, Venting, and Air Conditioning

MATERIAL SELECTION CRITERIA

The process of selecting materials for construction is a lot more complex than one might think. Each material, whether used in a structural frame or as finish material, has a profound impact on the success of the completed building. Simple measures or rules of thumb that would make the selection process easy would be welcome, but they are hard to obtain, if they exist at all. In practice, the designer is frequently forced into trading off a good result here for a not-so-desirable outcome there.

During the design process, basic material categories suitable for a particular purpose are identified. The designer must have a good understanding of current construction techniques in order to isolate initial choices among the diverse possibilities. As material properties are complex, technical data and other supporting materials are gathered that help to define the performance characteristics of the material. Reference standards and other recommendations for the use and care of the various material options are available from a number of sources.

The building industry relies heavily on independent standard writing groups that issue best practice guidelines for various materials and equipment. These organizations support the construction industry by conducting research and testing of materials and components and publishing the resulting information. An example of one of these organizations is the American Society for Testing and Materials (ASTM), a nonprofit corporation formed to develop standards on the characteristics and performance of materials, products, systems, and services. Standards developed by the society are published annually in a 48-volume publication, *The Book of ASTM Standards.* The standards outline procedures for material testing methods, recommended practices, and detailed specifications the designer can consult in order to make informed decisions.

Fun Fact

The American Society for Testing and Materials (ASTM) (http://www.astm.org) was formed in 1898 by a group of scientists and engineers led by Charles Benjamin Dudley. The aim was to address the frequent rail breaks plaguing the fast-growing railroad industry. The group developed the first standards for the steel used to fabricate rails.

Other organizations are devoted to the advancement of knowledge about individual materials and methods used in construction. Trade associations provide specific information on selected materials that influence the content of the building codes and establish industry-wide standards. A **trade association** is an organization made up of manufacturers involved in the production or supply of materials and services in a particular area. An example is the American Institute of Steel Construction. The goal of a trade association is to provide information on the proper use of the materials, products, and services of the material in question. Some trade associations support research into the proper use and improvement of construction materials and better construction methods. As they do this, they develop material specifications and performance procedures for a particular trade area. Sometimes trade associations administer programs of certification. Products made according to their specifications and standards are sometimes identified with a seal, label, or stamp that guarantees a material's performance. The American Plywood Association (APA) for example certifies that engineered sheet materials meet its specifications by allowing them to bear the APA stamp (Figure 12-11).

Trade Association: an organization of manufacturers and businesses involved in the production and supply of materials and services to various areas of the construction industry.

The designer may also choose to research existing buildings in a similar location or with the same use as the project in question. In this way, materials can be evaluated in service, and their weathering and durability characteristics directly observed. Specialty consultants may be required for some building types to consider the particularities of a material. For example, in the design of a concert hall, an acoustical consultant may be needed to judge the sound performance of interior finish materials.

Once all of the basic technical data has been collected, each material can be evaluated against a variety of criteria. The location of the project in question, its durability requirements, and budget issues must all be weighed against

Figure 12-11: This stamp on a piece of plywood certifies the span and exposure rating for the panel.

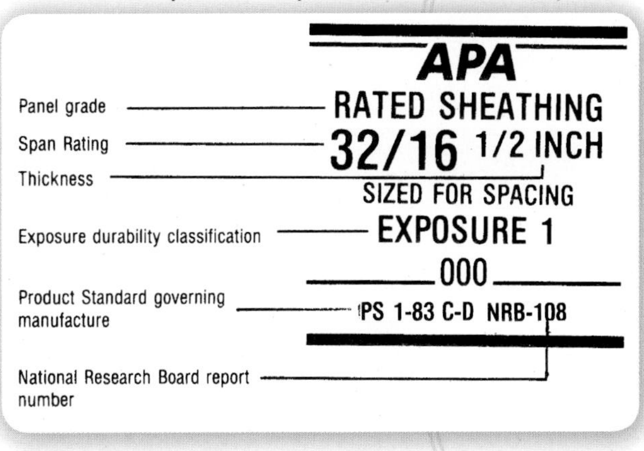

APA
Panel grade — RATED SHEATHING
Span Rating — 32/16
Thickness — 1/2 INCH
SIZED FOR SPACING
Exposure durability classification — EXPOSURE 1
000
Product Standard governing manufacture — PS 1-83 C-D NRB-108
National Research Board report number

Courtesy of APA—The Engineered Wood Association.

each other to arrive at the best material for a particular application. The following section lists selection criteria that are most commonly considered, regardless of the type of building in question.

Strength

Each material must be evaluated as to its ability to carry design loads without excessive deflection and deformation. The strength of a material is measured by the amount of loading it can carry without failing. This is known as the *ultimate strength* and varies widely between different materials. Strength characteristics are often the most critical of all selection criteria, particularly for structural materials.

Constructability

Constructability: the ease with which a material can be cut shaped and joined to other materials.

The ease with which a material can be manipulated is defined as its **constructability**. A material must be able to be easily cut and shaped with tools that are readily available. It must be able to be connected firmly to other building materials. Skilled craftsmen, trained in the installation of the material, must also be available.

Durability

The material must be able to perform well over time and under a variety of environmental conditions. A flooring material, for instance, must be considerably more durable than do products used for a ceiling. Durability concerns are considered in terms of the expected life span of a building, with more durable finishes resulting in longer periods between remodels. Measures of durability include hardness and impact strength.

Expense

The material must be evaluated in view of the proposed project budget. While a material may be deemed ideal for a particular application, its use can be impractical due to specified budget constraints. Cost considerations include the expected life of a material before replacement is required, regular maintenance, minor repairs, and the possibility of increased replacement costs years later because of inflation and increased labor costs. Using lifecycle costing, calculations can reveal if it would be less costly over the long run to use higher quality, more expensive materials.

Availability

The material must be readily available, preferably within close proximity to the project site. The transportation of materials consumes fossil fuel energy and many designers attempt to lessen this energy cost by specifying materials from within the region of the project site. Materials procured over very large distances can also have adverse effects on delivery and construction schedules.

6

Environmental Considerations

The selection of environmentally benign materials is a central issue for sustainable design and construction. According to the U.S. Green Building Council, each year between 30 and 40 percent of all raw materials extracted globally are used

Your Turn

Sweets Catalog is a resource used by architects and engineers for more than 100 years to aid in the material selection process. *Sweets* is published annually and includes hundreds of building material catalogs from leading product manufacturers. The catalogs are organized by material type, allowing you to easily browse and compare different types of products. Visit the Sweets Network website at http://products.construction.com/ and launch a search under the new products category.

in the building sector. The environmental impacts of extracting, processing, and transporting these materials are far-reaching. Fundamental questions arise as the construction industry attempts to define what materials are sustainable and how to evaluate them. **Environmentally Preferable Products (EPPs)** are defined by Federal Government Executive Order 13101 as materials that have "[a] lesser or reduced effect on human health and the environment when compared to competing products that serve the same purpose."

As manufacturers provide an ever-growing array of material options and new products with untested "green" performance claims, architects and engineers must be familiar with the factors that influence the environmental impact of a material. These factors include a material's embodied energy content, its impacts on natural resources and habitat degradation, and its potential for toxicity to humans and the environment.

Environmentally Preferred Products (EPPs): materials that have lesser or reduced effect on human health and the environment when compared to competing products that serve the same purpose.

EMBODIED ENERGY **Embodied energy** is defined as the energy consumed by all processes associated with the production and use of a material or assembly, beginning with the acquisition of natural resources and ending with their final reuse or demolition (Figure 12-12). Whereas the energy used in operating a building can be easily measured, the embodied energy contained within building materials is more difficult to calculate. A **Life Cycle Analysis (LCA)** is a detailed procedure for compiling and analyzing the inputs and outputs of resources and

Embodied Energy: the energy consumed by all processes associated with the production and use of a material, beginning with the acquisition of natural resources and ending with their final demolition.

Life Cycle Analysis (LCA): a detailed procedure for compiling and analyzing the inputs and outputs of resources and energy and their associated environmental impacts due to the use of a product throughout its life cycle.

Figure 12-12: **The stages considered when determining the embodied energy content of a building material or component.**

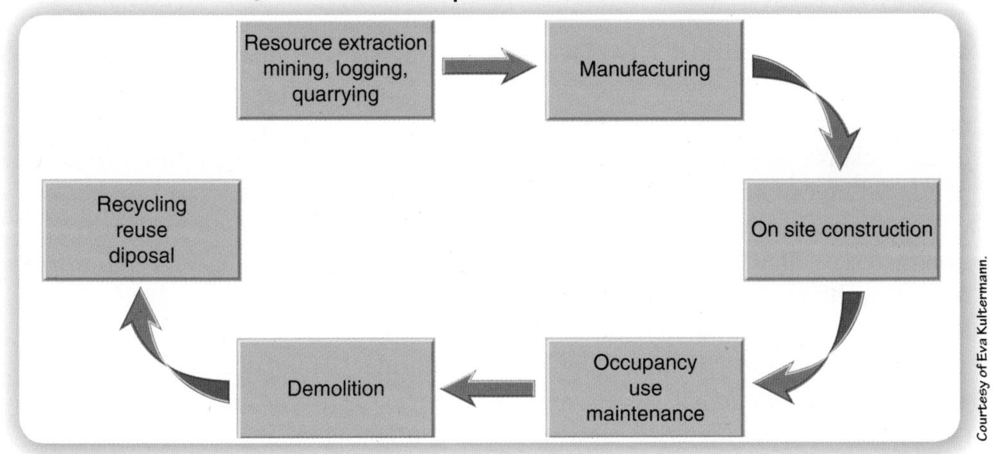

Courtesy of Eva Kultermann.

energy and their associated environmental impacts due to the use of a product throughout its lifecycle. Lifecycle assessment takes into account all of the resource and energy inputs that go into a material, while simultaneously calculating the resulting airborne emissions, solid and water borne wastes, and any other by-products and releases.

The lifecycle energy cost of most building products starts with the extraction of raw materials. Energy is consumed in the harvesting, mining, or quarrying of a material, the building of access roads, and the transportation of raw resources to the mill or plant. The manufacturing stage, where materials are formed and processed, accounts for the largest portion of embodied energy and emissions associated with the lifecycle of a building product.

More energy is consumed in the transportation of products and assemblies from manufacturing plants to regional distribution centers, and finally to the building site. The on-site construction is like an additional manufacturing step where individual materials and components come together in the assembly of an entire building. Once the building is complete and occupied, operating energy is calculated, taking into account functions like heating, cooling, lighting, and water use. Demolition marks the end of a building's life cycle, although it is not necessarily the end for all individual materials, some of which may be reused or recycled.

Typically, embodied energy is measured as a quantity of energy per unit of building material or component in megajoules (MJ) or gigajoules (GJ) of energy per unit of weight (ton) or area (square foot) of a material. Generally speaking, natural and organic materials have a much lower embodied energy content than do highly processed ones. Metals and plastics are among the highest embodied energy materials in use today. Published figures should be used with caution because values change depending on a variety of factors. For example, if a material is sourced locally, its embodied energy due to transportation will be lower than an identical one sourced from a great distance. In considering the environment, designers should strive to always select materials with the lowest embodied energy content.

NATURAL RESOURCES/HABITAT DEGRADATION A life-cycle analysis of materials also takes into account the impacts of material choices on the larger natural environment. The extraction, manufacture, and use of a construction material have profound effects on ecological systems and the supply of natural resources. To avoid habitat degradation, materials whose mining, harvesting, and use results in reduced effects on erosion, salinity, vegetation loss, changes in the nutrient characteristics of soil, and aesthetic damage to landscapes are preferred. Other considerations include the impacts on water systems, such as the release of salts, toxins, or suspended solids into aquatic systems.

MATERIALS AND HUMAN HEALTH Modern advances in material technologies have led to an increasing number of building-related health problems. The Environmental Protection Agency (EPA) has ranked indoor air quality (IAQ) as one of the most prominent environmental problem today. The quality of the air inside of a building can be compromised through the introduction of a vast number of new synthetic products developed to supply the increased demand for construction materials. **Sick Building Syndrome (SBS)** is a term used to describe building-related health complaints such as headaches; dizziness; nausea respiratory problems; eye, nose, or throat irritations; and skin problems.

Certain materials, including polyvinyl chloride (PVC), formaldehyde and arsenic are known to off-gas harmful substances. **"Off-gassing"** occurs when chemically unstable materials slowly release contaminants. Particularly harmful are **volatile organic compounds (VOCs)**. These organic solvents form vapors at room temperatures that easily evaporate into the air. Substances like plywood and fiber boards utilizing formaldehyde have been found to be a major culprit of toxic off-gassing. Additional chemical concentrations find their way into buildings in the form of paints, solvents, carpeting, office equipment as well as cleaning agents, and pesticides. Materials that off-gas not only create problems for building occupants, but also for workers during the construction process.

Material choices should be carefully evaluated their potential for adverse health consequences. Products such as engineered wood products containing formaldehyde oil based solvents and adhesives, and materials utilizing PVC should be avoided. A number of certification programs including Green Seal and the Carpet and Rug Institute now offer third-party certification programs that certify low-emitting materials and products.

Sick Building Syndrome (SBS):
a term used to describe building-related health complaints such as headaches; dizziness; nausea respiratory problems; eye, nose, or throat irritations; and skin problems.

Off-Gassing:
the process by which solid materials evaporate at room temperature, causing chemically unstable substances to slowly release contaminants.

Volatile Organic Compound (VOC):
organic solvents that at room temperature form potentially hazardous vapors that easily evaporate into the air.

Your Turn

GreenFormat provides architects and engineers with product information to help meet "green" requirements. Access the GreenFormat website at http://www.greenformat.com to search for construction materials that have been defined as environmentally preferable. Select green materials for the floors, walls, and ceiling materials, as well as furnishings for a remodel of your class room.

Summary of Material Selection Criteria

As new materials are developed each year, architects, engineers, and contractors must keep abreast with their properties in order to use them in the most effective and safe manner. For a material to be suitable for a particular application it must have predictable behaivor. When the properties of the material are defined, the designer can predict its behaivor and verify performance against that specified by codes and standards. This process requires that the design professional choose materials that adhere to certain requirements in a variety of considerations. Every effort should be made to choose materials that reduce environmental hazards for both the environment and human health.

MAJOR AND EMERGING CONSTRUCTION MATERIALS

A number of basic materials are used frequently in the construction industry. Construction technologies are constantly changing as new material installations and assemblies are developed. The fundamental principles that govern materials and construction however do not. The following section discusses the dominant materials used in construction today and new technologies that are being developed to improve their performance capabilities.

Concrete

Concrete is one of the most widely used construction materials with a long history of performance (Figure 12-13). Its constituent ingredients derive from a variety of naturally occurring materials that are readily available in most parts of

Figure 12-13: The unreinforced concrete dome of the Pantheon has survived for two millenia.

Courtesy of Eva Kultermann.

Figure 12-14: The basic ingredients of concrete are Portland cement, aggregates, and water.

Courtesy Portland Cement Association.

the world. Concrete provides an inexpensive construction material that is strong, hard, and durable with a high degree of fire resistance. The material has no form or tensile strength of its own. Forms must be built to shape it into useful structures, and reinforcing steel must be added to provide tensile strength. Concrete is one of the most versatile construction materials; it can be used to achieve almost any shape or form, and be finished with an unlimited variety of surface textures and patterns.

Concrete is produced by combining Portland cement, coarse and fine aggregates (sand and stone), and water in the proper proportions (Figure 12-14). The cement is a fine, pulverized material consisting of compounds of lime, iron, silica, and alumina. A chemical reaction between the cement and water produces heat that leads to a hardening of the mass. The exact composition of Portland cement varies, with different types providing a variety of properties. Some formulations are better suited for cold weather construction, while others possess sulfate-resisting properties for use in marine environments. Portland cement is a high embodied energy material that produces high concentrations of emissions during its manufacture. To reduce this environmental cost, new concrete mixes make use of fly ash, a fine, glass-like powder recovered from the gases created by coal-fired electric power generation. U.S. power plants produce millions of tons of fly ash annually, much of it previously discarded in landfills. Fly ash can provide an economical and environmentally preferable substitute for the Portland cement used in concrete, brick, and block production.

The strength and quality of concrete is determined by its water-cement ratio. If too much water is used, giving a high water-cement ratio, the paste is thin and will be porous and weak once it has hardened. If too little water is used, the concrete will be stronger but difficult to place during construction. As a general rule, a ratio of 45 to 60 percent of the weight of water to the weight of cement will result in strong and workable mixes. Concrete mixes are specially designed to provide the needed characteristics for a particular application. A mix designed for structural purposes will be higher in strength while one intended for architectural finishes will have superior surface qualities.

Well-formulated concrete is a naturally strong and durable material. It is dense, relatively watertight, and able to resist changes in temperature and wear and tear from weathering. The strength of concrete in the hardened state is usually measured in compressive strength. Both strength and durability are affected by the compaction process during placement of the wet concrete. Compaction refers to the process of removing air bubbles from the concrete during placement. Proper compaction results in concrete with an increased density that is stronger and more resilient.

Since concrete has no useful strength in tension, steel reinforcing is added to provide the required tensile strength (Figure 12-15). Concrete and steel react similarly to changes in temperature, so they work together to provide an efficient structural system. The concrete bonds to the steel, and the steel is protected from corrosion by the concrete. The two commonly used steel reinforcing materials are reinforcing bars called "rebars" and welded wire sheet reinforcement. The reinforcing steel is located so as to be covered by a minimum amount of concrete cover that varies depending on whether it is used in an interior or exterior application.

Cast-in-place concrete is produced by setting wood, metal, or molded plastic forms in place; placing reinforcing material in the forms; and pouring the concrete over the reinforcing, filling the form. After the concrete has been poured, it is consolidated and the excess is scraped off the top of the forms in a process called *screeding*. Once it has set for a short period, the final surface finish is completed with a trowel or float. Cast-in-place concrete members can be engineered in a wide range of sizes and shapes, with an unlimited variety of surface textures and colors. These include spread footings and foundation caissons, slabs on-grade, beams, columns, and wall and roof elements.

Cast-in-Place Concrete: concrete members formed and poured on the building site in the locations where they are needed.

Pre-cast concrete is cast in a factory under controlled conditions and then moved to the construction site for assembly. Units can be used to form both structural and nonstructural elements of a building, including columns, beams and girders, floor and roof slabs, and finished exterior and interior wall panels (Figure 12-16). The size of pre-cast units is limited only by what can fit on a truck and be transported to the site. Pre-cast concrete construction is used widely because it affords a number of advantages when compared with cast-in-place concrete. Casting takes place in automated facilities where experienced crews control the mixing and placing of concrete to produce higher quality units. The forms can be used repeatedly, thereby reducing costs. The erection process is faster, and inclement weather does not slow down precast construction as easily as it does cast-in-place jobs.

Pre-cast Concrete: concrete cast in a form and cured before it is lifted into its intended position.

Figure 12-15: Concrete columns with reinforcing steel exposed prior to the next pour.

Courtesy Portland Cement Association.

Figure 12-16: A pre-cast concrete wall panel being hoisted into position.

Courtesy Portland Cement Association.

Figure 12-17: A rammed earth wall showing the striations of subsequent layers of material.

Courtesy of Eva Kultermann.

New concrete formulations are finding increased use in the construction industry. Rammed earth for example can be classified as a green building material because it utilizes locally available materials with little embodied energy. Rammed earth walls are made by compressing a damp mixture of earth mixed with sand, gravel, clay and a small amount of Portland cement into conventional formwork. The damp material is poured into the forms in thin layers that are "rammed" (compacted) with a pneumatically powered tamper to around 50 percent of its original height. Subsequent layers of the material are added and the process is repeated until the wall has reached the desired height (Figure 12-17).

Masonry

The term **masonry** is used to describe either solid stone or modular units that are formed into building blocks. The modular units are manufactured from ceramic materials and hardened either by heat as in brick, clay tiles of terra cotta, or by chemical reaction in the case of concrete block. By themselves, masonry units are relatively weak and rely on mortar to join them together into structural assemblies.

Masonry:
construction made of either solid stone,or modular units that are formed into building blocks.

Mortar:
a plastic mixture of cementitious materials, water, and a fine aggregate used in masonry construction.

Mortar is the bonding agent used to join the individual masonry units into a solid structure. In addition to bonding the units together, it also seals the spaces between the units so they are not penetrated by air or moisture. The mortar holds the steel reinforcement required to connect multilayered walls and embed anchor bolts and other fasteners into the walls. The tooling of the mortar provides opportunities for design in lines of color and shadows. For general masonry applications, mortars may contain Portland cement or masonry cement, sand, hydrated lime or lime putty, and water. Clay brick is made from surface or deep mined clays that have the necessary plasticity when mixed with water to permit molding to the desired shape. The bricks are formed either by pressing them into molds or through an extrusion process. The clay must have enough stiffness and strength to hold the bricks' shape while wet and contain particles that will fuse together when subjected to high temperatures. The manufacture of clay bricks entails processing the raw materials; forming the bricks; allowing them to dry; and finally burning them in large automated kilns. Properly manufactured brick is strong, durable, and fire-resistant. The form of the brick in small units and their arrangement in an assembly provides the designer the opportunity to create a variety of designs and patterns (Figure 12-18).

Concrete masonry units are made of a relatively dry blend of Portland cement, aggregates, and water. The dry materials are mixed with water and consolidated into molds by pressure and vibration. Once formed, the wet units are allowed to dry before being cured in a steam powered kiln. Concrete masonry units are manufactured in a wide range of standard sizes and custom designed architectural units. They are one of the most widely used modern construction materials and are used in both structural and nonstructural applications. The physical properties of concrete, discussed previously, apply to concrete masonry units as well.

A relatively new masonry material is the autoclaved aerated concrete (AAC) block made of Portland cement mixed with lime, silica sand or recycled fly ash, water, and aluminum powder. The mix is poured into a mold where a chemical

Figure 12-18: A variety of brick positions, orientations, and types creates a unique wall pattern. From HAM. Residential Construction Academy Masonry: Brick and Block Construction.

Figure 12-19: A composite masonry wall utilizing an interior concrete block with a clay brick facing.

Courtesy of Eva Kultermann.

reaction between the aluminum and cement causes microscopic hydrogen bubbles to form, expanding the mix to about five times its original volume. The hydrogen is allowed to evaporate and the material is cut into blocks and formed by steam-curing in a pressurized chamber (an autoclave). The result is a lightweight, non-toxic, airtight material that can be used for wall, floor, and roof panels to provide structural capacity as well as excellent thermal, fire, and acoustical resistance properties.

Masonry walls are easy to design and construct, often resulting in a more economical building than do other construction methods. Masonry is a heavy material, limiting its use in high-rise buildings to that of an applied exterior facing material. Solid masonry bearing walls are generally used to construct low-rise buildings only. Masonry walls may be unreinforced or reinforced, solid masonry or cavity type. Many applications use a composite wall built up of several different types of units, such as clay brick over concrete blocks.

Cavity wall construction consists of two masonry walls separated by an air space. They are generally used for exterior walls because they control moisture penetration and can be insulated. Composite masonry walls have an exterior veneer of a quality masonry unit, such as brick, tile, or stone, while the hidden interior portion of the wall is built from a more economical unit, such as concrete block (Figure 12-19). When designing composite walls, the engineer designing the wall must consider how differences in thermal expansion, moisture absorption, and load-bearing capabilities between the two types of masonry will affect the wall.

To learn more about a science based approach to forest management, visit the Yale School of Forestry and Environmental Studies Global Institute of Sustainable Forestry at *http://research.yale.edu/gisf/*.

Wood

Wood in the form of natural boards and timbers is one of the most familiar construction materials (Figure 12-20). An enormous variety of components made from wood are also in use today including dimensional lumber, plywood, oriented strand board, and a variety of engineered wood products. Wood is unique among construction

Figure 12-20: Wood frame construction is the dominant and most familiar construction type in the United States. From VOGT. Carpentry, 4e.

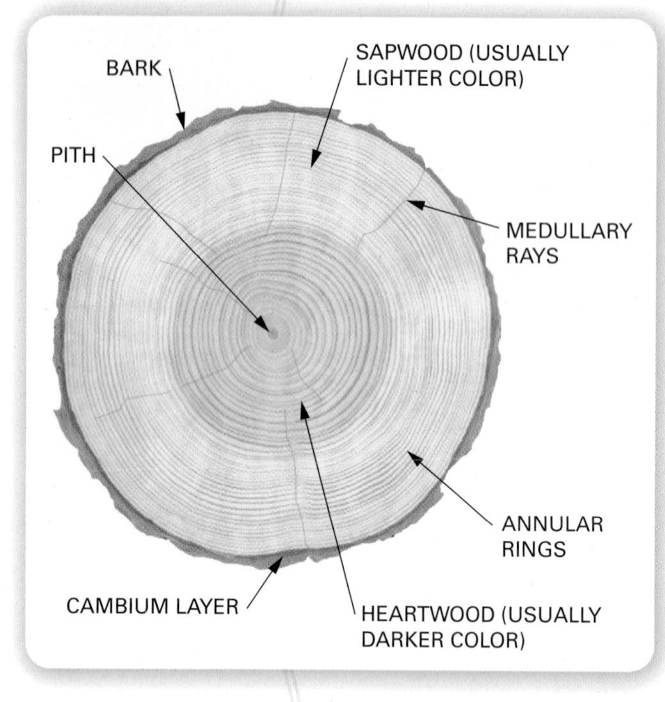

materials in that it is a natural and renewable resource. Carefully managed timber farms and natural wild growth provide a continuing and sustainable source of wood.

Woods are divided into two classes, *hardwoods* and *softwoods*. This division is based on botanical differences and not on their actual hardness or softness. Softwoods are referred to as *coniferous,* meaning that they bear cones and, with a few exceptions, have needlelike leaves that stay green all year long. Much of the production of wood for commercial use is in the class of softwoods that are used for structural framing, sheathing, roofing, subflooring, siding, trim, and millwork. Hardwood, or *deciduous,* trees are broad-leaved and shed their leaves in the winter. Hardwoods are more expensive than softwoods and find use in cabinets, furniture, paneling, interior trim, and flooring.

Since wood is a naturally occurring material, it has considerable variation in its physical properties including color, density, weight, and strength. While light in weight and easy to work, the material is quite durable and strong. The direction of the grain is a major consideration in the use of wood as a structural material. The strength of wood parallel to the grain is much greater than the strength perpendicular to the grain. Wood can withstand around one-third more load in compression than in tension when loaded parallel to the grain. Lumber is evaluated for strength during the manufacturing process by mechanical stress-rating equipment that subjects each piece to bending loads. The stress grade is electronically calculated and a grade designation is stamped on each piece.

A piece of wood may have natural defects that occurred while the tree was growing, or seasoning defects that are produced as the wood is dried for use. Natural defects include knots, shake, wane, insect holes, and pitch pockets (Figure 12-21). Knots occur where a branch that is imbedded in the trunk as the tree grows is cut. Knots weaken the wood and are one factor a lumber grader considers as wood is graded.

Figure 12-21: Examples of common wood defects. From SPENCE and KULTERMANN. Construction Materials, Methods, and Techniques, 3e.

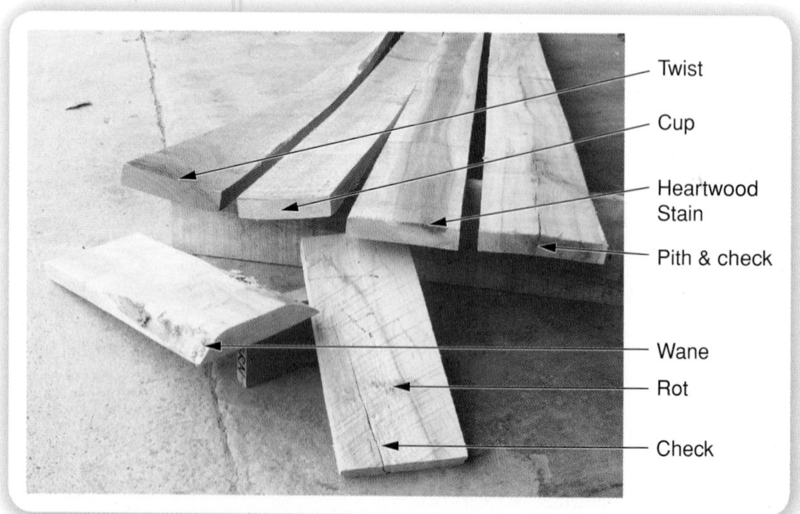

Solid wood lumber is classified according to size. Boards are less than 2 in in nominal thickness and 1 in or more in width. Narrow pieces of less than 6 in wide are classified as strips. Dimension lumber is from 2 in up to but not including 5 in thick and 2 in or more in width. The common 2 × 4 is an example of dimensional lumber. Timbers are 5 in or more in their least dimension and are subdivided into classes such as beams, posts, and girders.

Engineered wood is manufactured by bonding together wood strands, veneers, lumber, or other forms of wood fibers to produce larger composite units that are stronger than solid wood members. These products are manufactured with various resins, glues, and adhesives under high pressure, to produce solid wood substitutes. Engineered wood products provide higher strength and stiffness in smaller cross-sections than do solid lumber. These products are considered environmentally preferable because they utilize otherwise undervalued species in addition to wood chips and other forestry waste.

Engineered Wood: materials made by bonding wood strands, veneers, lumber, or fibers to produce larger composite units that are stronger than solid wood members.

One of the most widely used engineered wood products is plywood. Plywood panels are made by bonding together thin layers or plies of wood. Panels always have an odd number of plies, such as three, five, or seven, with the grain in each ply perpendicular to the ply adjacent to it. In addition to plywood, other products made from wood chips and bonded into panels under heat and pressure, include particle board, wafer board, and oriented strand board. Construction grade sheet materials find use as sheathing, decking, and flooring materials. A variety of select plywoods are used for interior paneling and millwork.

A number of other structural products enable the use of wood for building types beyond residential structures. Wood trusses combine smaller wood members in triangulated arrangements to enable wider spans for floor and roof construction. Most trusses are made from 2 in × 4 in and 2 in × 6 in lumber joined with wood or metal gusset plates (Figure 12-22). They are carefully engineered to carry known loads over specified distances. Glued laminated wood members or *glulams* are formed of solid sawn lumber glued end to end and then face bonded in laminations. Thinner wood strips are used for curved members and 1.5 in thick wood strips for straight members. A variety of new engineered wood products include structural components such as laminated veneer lumber (LVL), and parallel strand lumber (PSL). Most are produced by bonding thin wood veneers or wood chips with glues under pressure into larger components that are suitable for use as structural beams and columns. Both LVLs and PSLs are increasingly replacing solid lumber in applications like headers, beams, and columns.

Figure 12-22: Wood floor trusses made of 2 × 4 material connected by metal gusset plates.

Courtesy of Eva Kultermann.

Figure 12-23: This steel frame building achieves a column-free interior through the use of roof-bearing girders.

Courtesy of Eva Kultermann.

Steel and Other Metals

Since it became widely available in the late nineteenth century, steel has had tremendous influence on the evolution of architecture. Previous materials worked mostly in compression, restricting the spans that could be achieved. The stiffness of steel enables designers to achieve much greater spans and heights than is possible in either wood or masonry construction. In engineering terms there is almost no limit to what steel can achieve (Figure 12-23).

Steel is used widely in the construction industry for both light and heavy structural building frames. The material finds use in many other building components as well, including window and door frames, building

Figure 12-24: A steel skeleton frame with precast concrete floor and roof panels under construction.

Courtesy of H. H. Robertson.

hardware, and fasteners. In structural applications steel provides high strength and stiffness with low weight, making it one of the most economical materials available. The material steel is subject to corrosion and must be treated by coating, painting, or chemically treating it to prevent rusting. While steel will not readily burn, it does loose strength and eventually fails under very high temperature conditions. For this reason, steel structural elements must be encased or sprayed with fire-resistant materials.

Steel is produced by refining mined iron ore into a product called pig iron. Molten pig iron, metal scraps, and fluxes are mixed and heated to produce molten steel. Products are produced by rolling, extruding, cold-drawing, or casting the molten steel. Rolling is used for most structural shapes such as I-beams, wide-flange columns, channels, and angle iron. The properties of steel can be altered by applying a variety of heat treatments. Annealed steel is produced by subjecting the metal to an additional heating process that results in increased strength and stiffness. Many forms of high-grade steels, such as stainless, tool, heat resisting, and alloy, are used for special applications. Most steel structures utilize a "skeleton frame" of vertical steel columns and horizontal beams, constructed in a rectangular grid to support the floors, roof and walls of a building (Figure 12-24). The construction of a structural steel frame involves two principal operations: its fabrication and erection. Fabrication occurs off site and involves the cutting and processing of steel products to form the finished members of the structure. Beams and columns are cut to size, and holes are drilled to permit their being joined. Erection includes the hoisting or lifting of members to their proper place in the structure and making the finished connections between members. Steel members are connected to each other with bolts and threaded fasteners, or by welding. Metal decking covered with site cast concrete or pre-cast concrete slabs are commonly used to construct floors and roofs for steel structures.

Other nonferrous metals (those containing no iron) find extensive use in building construction. Among the most commonly used are aluminum, copper, leads, tin, and zinc.

Aluminum is a versatile material that is used widely for roofing, flashing, siding, and exterior trim as well as in door and window construction. The material is strong, very lightweight, corrosion resistant, and easily worked. Aluminum members can be joined by any number of fastening techniques including screws, bolts, and rivets as well as by welding, brazing, and soldering.

Copper is a non-magnetic reddish brown metal that displays excellent electrical and thermal conductivity. It has the highest conductivity properties of all commonly used metals and is used widely for electrical wiring. Copper is ductile, malleable, and easily worked. When combined with other metals, it offers a wide range of properties, making it a very valuable and widely used material in construction. Copper and copper alloys are excellent for outdoor uses such as siding, roofing, flashing, guttering, and screen wire. The alloys are used extensively for plumbing pipe in residential and commercial structures as well as in the manufacture of plumbing fittings, such as valves, drains, and faucets.

Point of Interest

The Tortoise and CityCenter

Corrosion between metals occurs when minute amounts of electricity in the atmosphere or soil flow from one metal (the anode) to a dissimilar metal (the cathode) through a current carrying medium (moisture) called an *electrolyte*. If two metals having different potentials are electrically coupled, the anode (more negative metal) will corrode.

The electrolyte is usually a water or gas, such as carbon dioxide or sulfur dioxide.

Metals are ranked by their tendency to be anodic or cathodic.

When a material near the top of the list, such as steel, is placed in contact with a material near the bottom of the list, such as copper, and an electrolyte is present, corrosion will occur. The metal that is destroyed (corrodes) is one that is high on the galvanic table.

Galvanic Series

↑	Magnesium alloys
	Zinc
Increasingly	Aluminum alloys
Anodic	Carbon steel
(Corrodes)	Stainless steel
	Lead
	Tin
Increasingly	Brass
Cathodic	Copper
(Protected)	Bronze
	Stainless steel
↓	Gold

Metals are one of the most widely recycled materials in the world. The steel industry for example has been engaged in recycling for more than a century. Steel loses none of its inherent physical properties during the recycling process, and has drastically reduced energy and material requirements compared with refinement from iron ore. Today 60 to 90 percent of all metal products contain recycled content and the energy saved by recycling substantially reduces the annual energy consumption of the industry

Figure 12-25: **Large glazed facades are used widely in high rise buildings.**

Courtesy of Eva Kultermann.

Glass

Glass, in the form of windows and larger glazed openings, is one of the most important materials used in building design (Figure 12-25). It is through glass that we admit natural light and provide views and a connection between the interior and exterior. Windows and glazed walls have historically been regarded as net energy drains on a building. Recent developments and the introduction of insulated windows, glass tinting, and selective film coatings are beginning to change this perception.

Glass has an unusual internal structure because it is rigid and has the characteristics of a solid, yet the atoms in glass are arranged in a random order similar to those in a liquid. Technically, glass is a super-cooled liquid. Of the thousands of workable glass compositions in use today, most are combinations of the same basic ingredients: sand, soda ash, or potash, mixed with stabilizers such as lime or alumina.

Most glass worldwide is produced by the float process. During production, furnaces create a molten glass which is conveyed to a bath of molten tin. The molten tin provides a very flat surface that supports the glass as it is polished by the application of heat from above. The ribbon of glass is moved to a cooling zone, annealed (reheated) to add strength, and finally cut into the desired size. Sheets of glass produced by this method have parallel surfaces; a smooth, clear finish; and high optical clarity.

Glass has great inherent strength that is weakened only by surface imperfections giving everyday glass its fragile reputation. The material is easy to cut, very durable, and more resistant to corrosion than many other materials. Since it is not porous, it does not absorb moisture or chemical elements in the ground or atmosphere. Glass is also able to withstand intense heat or cold as well as sudden temperature changes, making it an ideal exterior cladding material.

Most glass is utilized in window units that are installed in a wall or roof to admit light and air to an enclosure. Window frames are made from solid wood, wood clad with plastic or aluminum, solid plastic, steel, aluminum, and other composite materials. For larger glass openings, the glass is fixed in a variety of different installation methods directly on the construction site. Commercial entrances and storefronts fix glass within extruded aluminum members. The aluminum frame is typically filled with flat architectural glass products, such as tempered and heat-strengthened glass, laminated glass, and insulating glass. The glass is held either by clamping pressure plates, a structural adhesive tape, or a combination of the two.

Innovative new glazing systems utilize mechanical anchors at discrete locations near the glass edge, rather than continuous edge supports to fix the glass (Figure 12-26). Known as point supported glazing, the system allows for large glass areas with minimal interior framing and no exterior pressure plates.

When deciding whether or not to use glass in a specific application, its ability to withstand breakage is a major consideration. Several types of safety glass are produced for use in high hazard areas such as door glazing and windows in high-impact areas. Tempered glass is three to five times as resistant to damage as normal annealed glass. When the thin tempered skin on the glass is broken, the entire sheet disintegrates into small pebble-like particles rather than sharp slivers. Laminated glass is composed of multiple sheets of annealed glass with intermediate layers of plastic resins that hold the glass in place in case of breakage.

Figure 12-26: New systems for fixing glass include this point-supported spider fitting.

Courtesy of Eva Kultermann.

Single pane glass is highly conductive to the passage of heat and cold, resulting in energy loss through the material. Insulating glass is a manufactured glazing unit composed of two layers of glass with an airtight, vacuum air space between them. The units lower heating and cooling costs by reducing air-to-air heat transfer. Low-E glazing units have special coatings applied to the glass panes that are designed to further reduce a window unit's heat transfer. Solar radiation in the form of short wavelengths has the ability to pass through the glass of a window unit. Once admitted through the glass into a space, this radiation is absorbed by objects and re-radiated as long-wave radiation, or heat energy. The properties of low-E coatings are such that the transmission of solar radiation in the short wavelengths is high but drops off dramatically in the long wavelengths. The result is that solar radiation is allowed to enter a space but is prevented from re-radiating back though the glazing unit. This allows the window unit to act as a collector of solar energy within the space.

Research is now concentrating on the development of **building-integrated photo-voltaic (BIPV)** systems. This new technology will enable glazing to not only provide enclosure and daylight for a building but also to simultaneously generate electricity.

A typical photo-voltaic (PV) glazing unit uses two panes of clear or tinted glazing with the photo-voltaic modules adhered to the front of the window inside pane (Figure 12-27). Most PV glazing is available only by special order. The PV cells can be spaced in a variety of patterns and densities to aid in blocking sunlight from interior spaces.

Figure 12-27: This glazing panel integrates photo-voltaic modules sandwiched between two layers of glass.

Courtesy of Eva Kultermann.

Finish Materials

Interior finishes refers to the wall, ceiling, and floor finishes of a building's interior. In our modern society we spend the majority of our time indoors. The texture, colors, and pattern of interior finish materials provide the surroundings for much of what we do (Figure 12-28). Interior finish installation is highly regulated by the building codes, which vary depending on the proposed occupancy of the building. The architect and the owner work together to select materials that meet the requirements of the design intent and the building code.

Building-Integrated Photo-Voltaic (BIPV): photo-voltaic units that have been joined directly in building components.

Figure 12-28: This warm and inviting interior uses a variety of different finish materials.

Courtesy of Andersen® Windows and Doors.

Figure 12-29: Solid wood strip flooring being installed with a power nailer.

FLOORING Flooring materials are installed over substrates of various kinds and in a variety of applications. The designer must consider the characteristics of the substrate, the expected traffic loads on the floor, its required maintenance, and of course, it's appearance. Countless choices exist in materials for floors; the following discussion outlines just a few.

Both hardwoods and softwoods are used to form strip flooring (Figure 12-29) and various types of parquet. Some types have a factory applied finish, and others are sanded and finished after installation. Engineered and laminate flooring are manufactured products that use a multi-ply construction to create a solid product that is more dimensionally stable than many other flooring materials. Most consist of a laminate of three layers of solid wood with tongue-and-groove edges and ends. The top veneer is usually prefinished and is available in many wood species.

Resilient flooring materials provide a soft surface that is economical and durable. Linoleum and vinyl are available in either sheets or individual tiles. Rubber flooring is another form of resilient flooring that is particularly comfortable for walking. It wears well and resists damage from oils, solvents, alkalis, acids, and other chemicals. A number of new rubber flooring products are available that utilize rubber from recycled automobile tires.

Tile, brick, and stone flooring are probably the most durable and are commonly used for high traffic applications. Ceramic tile is used for both floor and wall applications. Often laid as mosaics, they are small, modular units of clay or natural porcelain that have a hard, glazed surface finish. Quarry tiles and pavers come in larger units and are almost impervious to dirt and moisture. Stone floorings, including granite, marble, and slate, provide a strong and highly durable surface.

Carpet is widely used as floor covering in both residential and commercial construction. It competes well with other finish floor products in terms of cost, durability, and maintenance. Carpet is available in a wide range of fibers including wool, acrylic, nylon, and polyester, as well as numerous textures and colors. The type of carpet used, and its potential effect on indoor air quality, must always be considered during the selection process. All carpet types trap dust, moisture, and pollutants to varying degrees depending on the depth of the pile, the carpet density, and the type of carpet. Synthetic fibers are traditionally made from petroleum and can off-gas. Seam sealants, carpet padding, and carpet treatments often contain volatile organic compounds, formaldehyde, and other pollutants. In response to these issues, the Carpet and Rug Institute (CRI) has developed the "Green Label" indoor air quality (IAQ) testing and labeling program for carpet and the adhesives used with them. CRI tests each carpet line four times a year for four categories of emissions. Those that contain no hazards to building occupants are certified by the Green Label.

WALLS AND CEILINGS Walls and ceilings provide the primary background for the furnishings and occupants within a space. Since they are not subject to weathering forces or excessive wear, they can be selected from a wider range of materials than other building components. Wall finishes may be integral to the building structure, as in an exposed brick wall, or be composed of surface layers that are attached to the wall framing. Additional finishes consist of very thin coatings and coverings that are adhered to a substrate material. The color and surface texture of wall coverings can have a major impact on the feel of a room and must be chosen with the intended use in mind. Light-colored walls will reflect light well and result in a bright interior. Dark-colored walls on the other hand tend to absorb light and will make a room more difficult to illuminate.

One of the most common finishes used for both walls and ceilings is gypsum, in the form of either plaster or wall board. Gypsum is a naturally occurring hydrated calcium sulfate mineral found in rock formations. A gypsum plaster finish is applied on a supporting base of either gypsum board or metal lath, usually in three coats of wet plaster. Gypsum board, often called drywall or wallboard, is the generic name for a series of panel products having a non-combustible core of calcined gypsum with paper surfacing on the face, back, and long edges (Figure 12-30). The board can be installed over a variety of framing types and the seams between panels are finished with a plaster like material that produces a smooth, continuous surface finish. Both gypsum wallboard and plaster are finished by painting or by applying a flexible wall covering such as wallpaper.

Figure 12-30: Gypsum wallboard is applied to both wall and ceiling surfaces.

Other common wall finish materials include solid wood, fabrics, and more recently, polymer-based wall panels. A wood wainscot is a horizontal band of wood often applied to the lower half of a wall surface. In this location it helps to protect the wall from damage in high-traffic areas. A wide range of fabrics and wallpapers are also available providing an unlimited palette of both color and textural variations.

The selection of interior finish materials should consider the environmental characteristics of products. Products are now available that are certified by independent organizations to meet stringent indoor air quality standards. The GREENGUARD Environmental Institute (GEI) is an example of a non-profit organization that oversees the GREENGUARD Certification Program. As an ANSI Accredited Standards Developer, GEI establishes acceptable indoor air standards for indoor products, environments, and buildings.

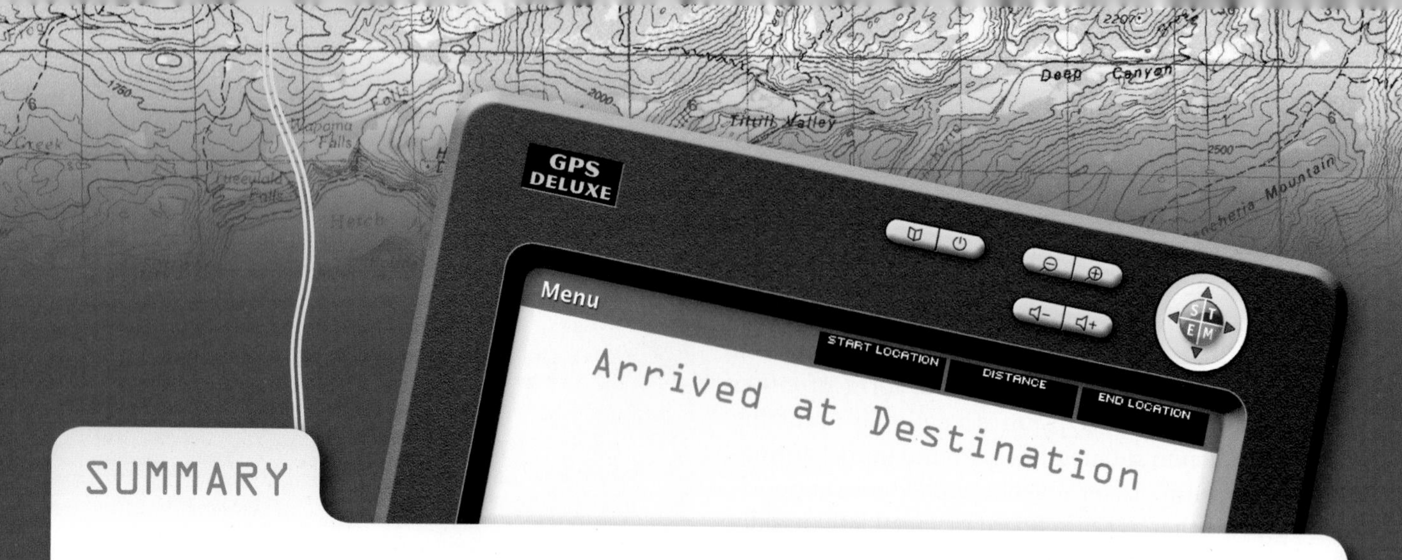

SUMMARY

Materials for construction have been constantly refined and developed since the beginning of human settlements. The last century alone has seen the introduction of countless new and innovative construction materials and components. Materials can be classified by their basic composition into five basic families: metals, polymers, ceramics, composite, and organic and natural materials. Within these basic families all materials are evaluated according to their performance properties. These distinctive and qualitative parameters are used during the material selection process. In addition to their performance characteristics, the selection of materials is governed by the building codes, which place restrictions on use based on strength and fire resistance.

The choice of materials for construction affects every aspect of the completed building, from its structural integrity to the quality of its interior surface finishes. While there may be no hard and fast rules of thumb, a few basic concepts should be adhered to. All material choices must be made within strict adherence to building code requirements. Each material must be able to support structural loadings and provide long lasting service. Architects and engineers should always strive to design for long life and adaptability, using durable low maintenance materials. Environmental considerations are increasingly entering the material selection process. Every attempt should be made to select low embodied energy materials, which may include materials with a high recycled content. Whenever possible, preference should be given to natural, organic, and renewable material choices.

BRING IT HOME

1. Develop a chart to compare the basic families of building materials. Include metals, ceramics, concrete, steel, wood, glass, and polymers. Compare each material for tensile and compressive strength, hardness and impact strength, as well as weight and density.
2. Have each member of your class write to a material trade association to request literature describing their construction material. Display, compare, and discuss their publications.
3. Have the same class members write to material suppliers to obtain product catalogs for the materials covered by the trade associations. Evaluate whether the materials reference and adhere to the standards of the trade association.
4. Visit as many local construction sites as possible and observe the materials being used for the structural frame, exterior cladding materials, and interior finish materials. How do the materials used differ according to the type and size of the building? Note and sketch the installation methods of various material assemblies.
5. Secure a copy of your local building code. What is the allowable building height and area for a commercial building utilizing light wood frame construction? How do the allowable height and area restrictions change if a steel frame is used?
6. Design and construct a reinforced concrete beam 6 to 8 ft in length. Consider the formwork, reinforcing, height-to-width ratio, and texture and color of the concrete. Produce a photographic record of the process.
7. Collect samples of various interior finish materials. Include wood, stone, tile, carpet, and wallboards among others. Label each, giving as much information about its properties as you can collect.
8. Conduct research on environmentally preferred building materials. Collect samples of alternatives for conventional and potentially harmful building materials.

EXTRA MILE

Contact local brick suppliers to secure a quantity of clay bricks for use in the class. Working in weekly teams, lay up a dry stacked wall that uses all of the available bricks, is structurally stable, is no more than 2 ft 4 in in height, incorporates at least one corner, and is beautifully made. Consider the coursing of the brick, the width of the wall, and its pattern and texture. At the beginning of each week, each team can present its process, objectives, and outcomes of the exercise to the class.

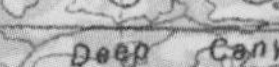

CHAPTER 13
Framing Systems Residential and Commercial

GPS DELUXE

Menu

START LOCATION

DISTANCE

END LOCATION

Before You Begin

Think about these questions as you study the concepts in this chapter:

1. What is a frame?
2. What are some of the basic framing systems in residential construction?
3. What are some of the basic framing systems in commercial construction?

INTRODUCTION

Think of a building as a human body (Figure 13-1). We can associate parts of the human body that function in similar capacities to similar parts of a building. The skeletal frame of the body could be considered the support and frame of the building. Working from the ground up, the human body has feet that help to distribute the above loads to a larger area on the ground much like the foundation system distributes the load of a building to supports on or in the ground. Legs are similar to a building's columns transferring vertical loads downward and our hip section would be similar to a floor transferring horizontal loads to the columns. The remaining torso provides the structural **framing system** for floors and walls. Ultimately, the head and shoulders of the human body is much like the roof framing system of a building.

Framing Systems:
basic **framing systems** (Figure 13-2) help support the building structure not only from the exterior but from the interior side of the building by transferring loads both horizontally and vertically down to the foundation of the building.

While there are similarities and differences between residential and commercial framing, a general understanding of the types of framing systems can be applied throughout the construction industry. This chapter covers some of the basic framing system types currently used in construction today.

Framing system for both residential and low-rise commercial buildings share common construction methods and materials. A low-rise building is considered a building that is three-stories or less. Some low-rise commercial building types include hotels, apartments, townhouses, offices, and restaurants, to name a few.

The other commercial building types typically fall under large or high-rise buildings. These types of buildings include skyscrapers, hotels, and office buildings. The use of a type of framing system is often based on several factors including the type of building, the type of loads to be supported, the building code requirements, the costs and the site conditions.

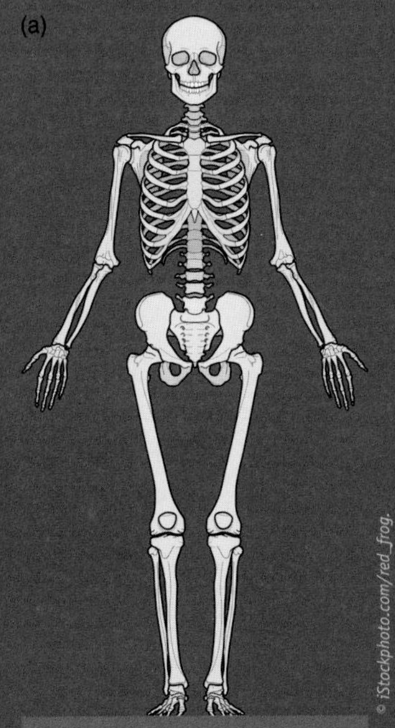

Figure 13-1: *Human skeletal frame as compared to a building section. (a) Skeletal. (b) Building section.*

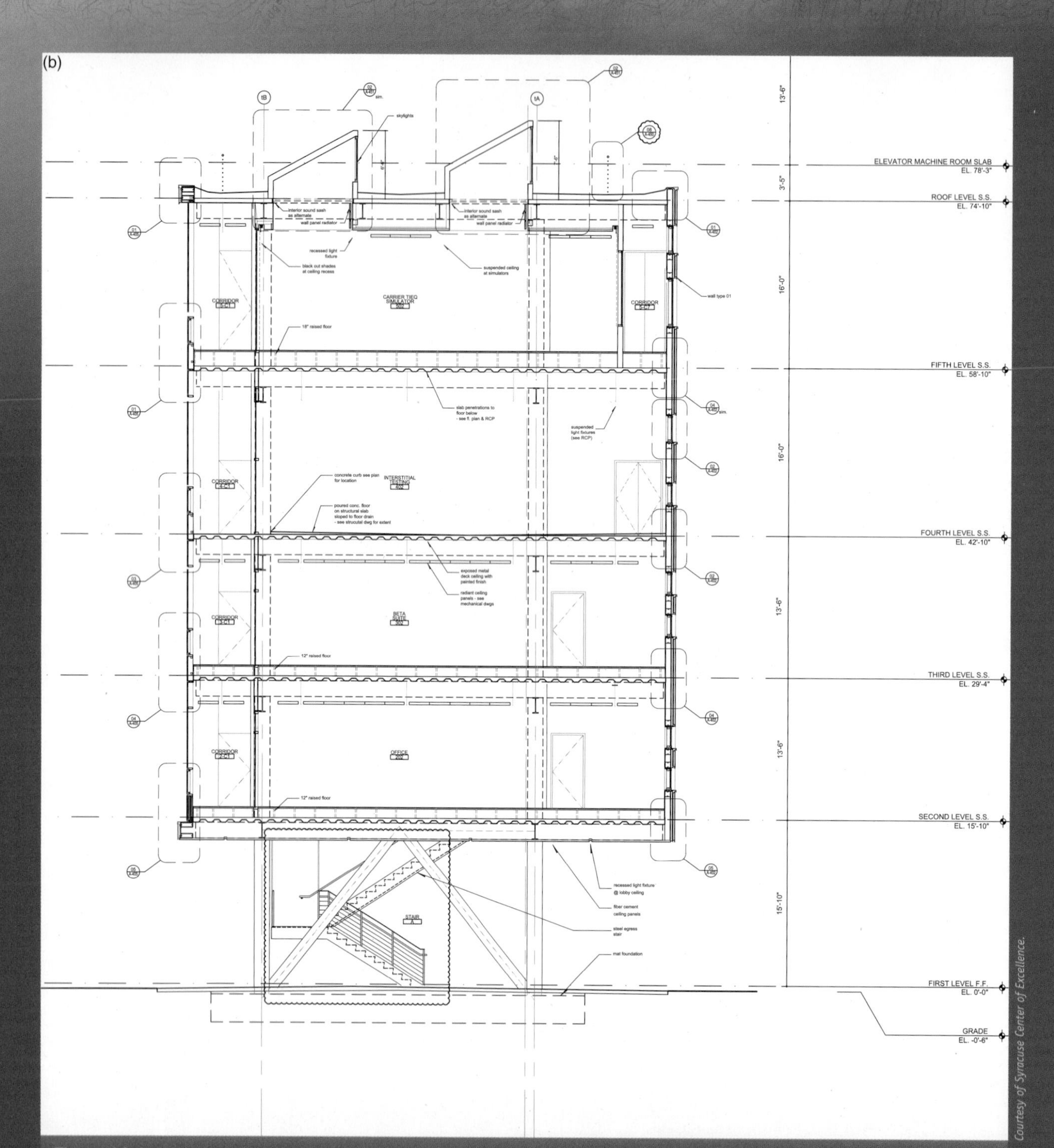

Figure 13-1. *(continued)*

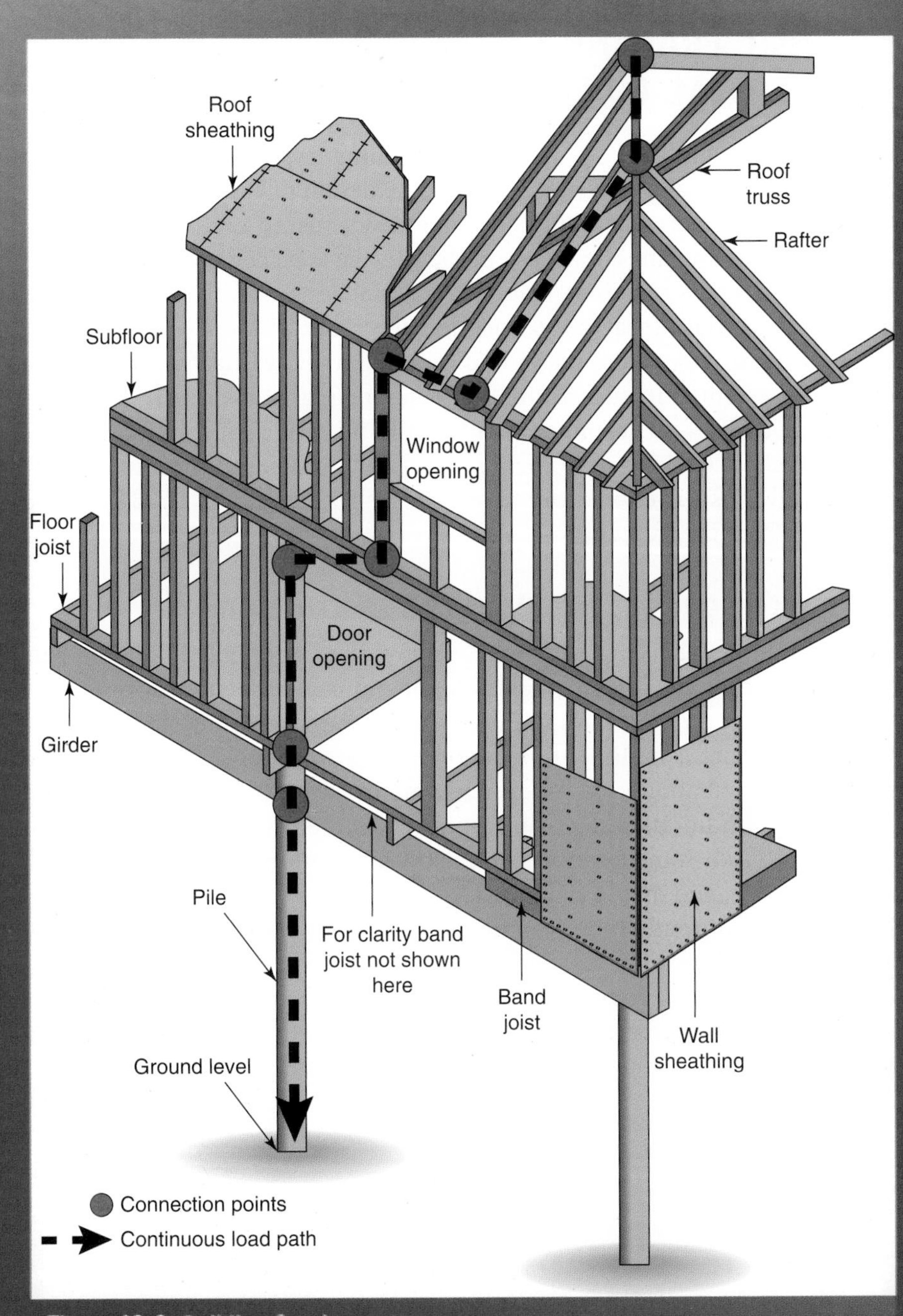

Source: FEMA.

Figure 13-2: *Building framing system.*

Resources are the things needed to get a job done, such as tools and machines, materials, information, energy, people, capital, and time.

SOME BASIC FRAMING SYSTEMS IN RESIDENTIAL CONSTRUCTION

Foundations:
foundations are a part of the framing system that transfers the load of a building to an area of the ground.

Starting from the ground up, residential framing systems typically include **foundations**, floors, walls, and roof systems. The design of each system is dependent upon several factors including the type of loads placed on the structure, the location of the building, the conditions of the site, the local codes and regulations, and the manufacturer's recommendations for installation. Additional factors like availability of materials; local trades such as carpenters, welders, and so on; and economics are also key to designing and building a successful framing system. Each system is essential in supporting the structure at any given site. A building on a weak foundation will crumble, as will a building with a solid foundation and poorly designed or constructed framing system. Each system needs to be carefully engineered and constructed in order to provide a shelter that will serve to protect the occupants from the elements.

Basic Foundations for Residential Construction

Basic foundation systems for residential and low-rise commercial building construction are typically classified as shallow foundations. As the term suggests, shallow foundations require minimal soil disturbance and, due to the smaller size of the structure, require a minimal foundation system to support the loads above. The type of the foundation used is dependent upon the loads being carried to the ground, the soil conditions and bearing capacity, weather conditions, and location of the site to be used. Some examples of shallow foundations include spread footings, slab-on-grade, and mat foundations.

A spread footing foundation (Figure 13-3) functions like a beam carrying a uniform load over a long distance. The soil is excavated in a trench-like fashion and is usually placed on top of compacted soil or undisturbed soil. Reinforcement such as steel bars or rebar may be used to give it greater strength. These reinforcements are typically cast in place due to ease of construction during assembly.

Figure 13-3: Spread footings.

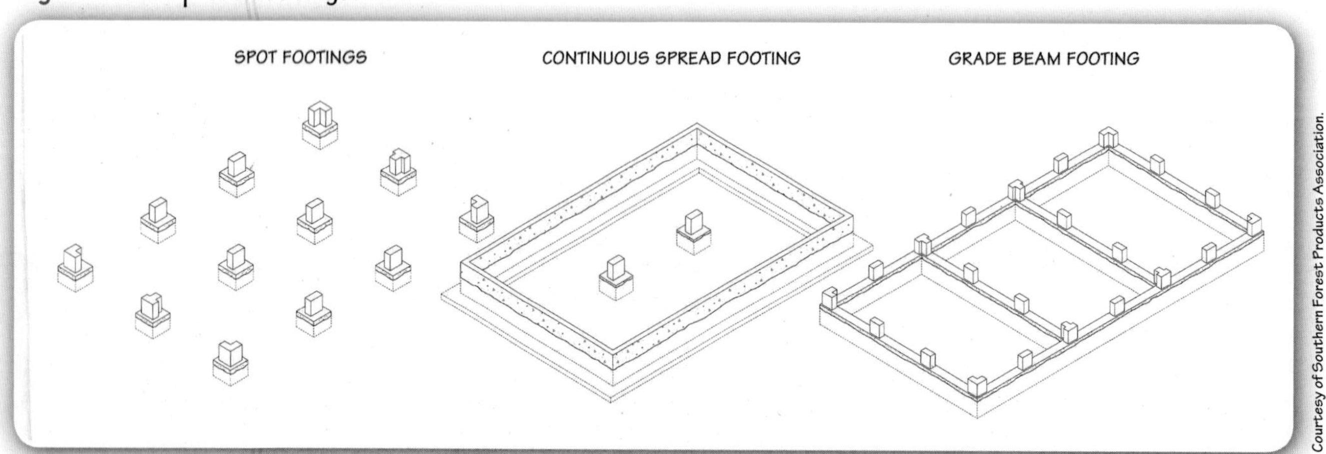

Figure 13-4: Foundation wall using masonry units.

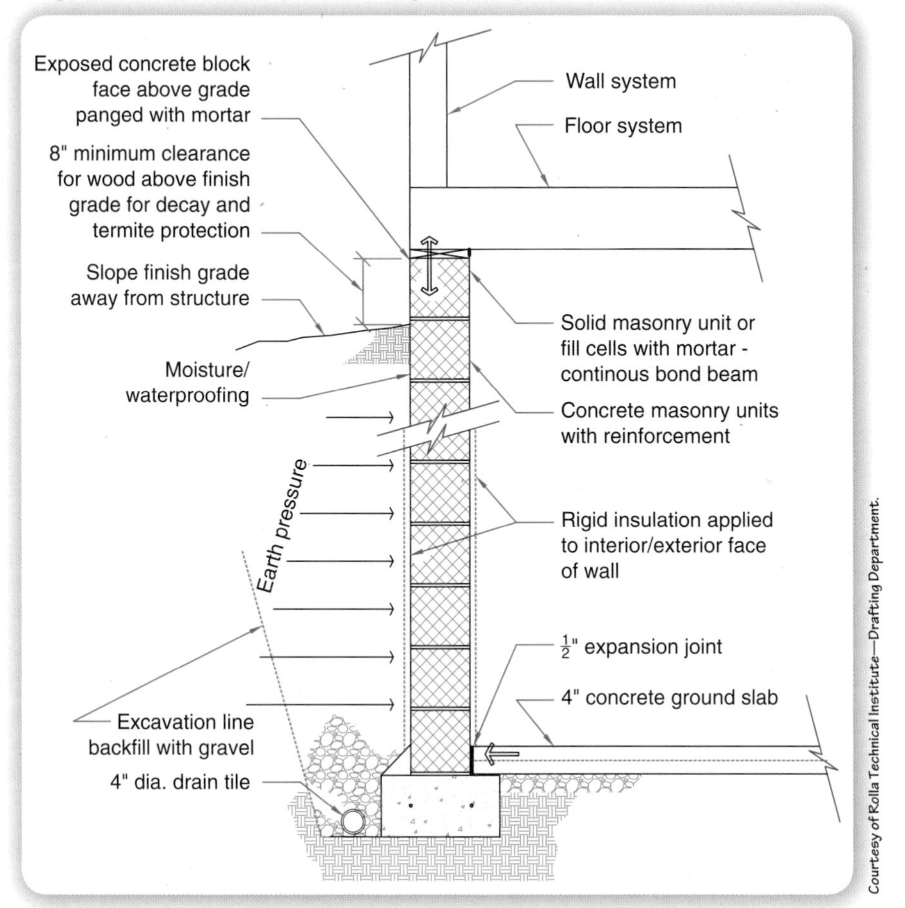

Courtesy of Rolla Technical Institute—Drafting Department.

Foundation walls (Figure 13-4) are typically used when crawl spaces or basements are designed as part of the structure. Foundation walls carry the weight of the structure to the footing.

These walls not only serve to carry vertical loads down to the footing but also act as retaining walls that withstand lateral forces from the soil. Foundation walls are usually constructed using **masonry units** or reinforced concrete.

Using math operations such as estimating and distributing materials and supplies to complete jobsite/workplace tasks.

Another type of foundation common in residential construction is called a column footing (Figure 13-5). Column footings distribute a load to the soil below much like a spread footing. These footings typically take on a square, round, or rectangular form dependent upon the column and the soil conditions.

In areas where soil conditions and frost depth is not a concern, a type of foundation system known as slab-on-grade (Figure 13-6) is typically used. It requires minimal excavation and the foundation is formed on site with poured-in-place reinforced concrete on the grade. Essentially, an area for the slab is staked on the ground, forms retain the poured concrete in its place, and reinforcements are added to increase the strength of the slab. Once finished, the structure is built over the slab.

Masonry Units: the dimensional properties of varying construction blocks. A cement masonry unit describes both the material and the dimensional properties of the construction unit.

Figure 13-5: Column footings.

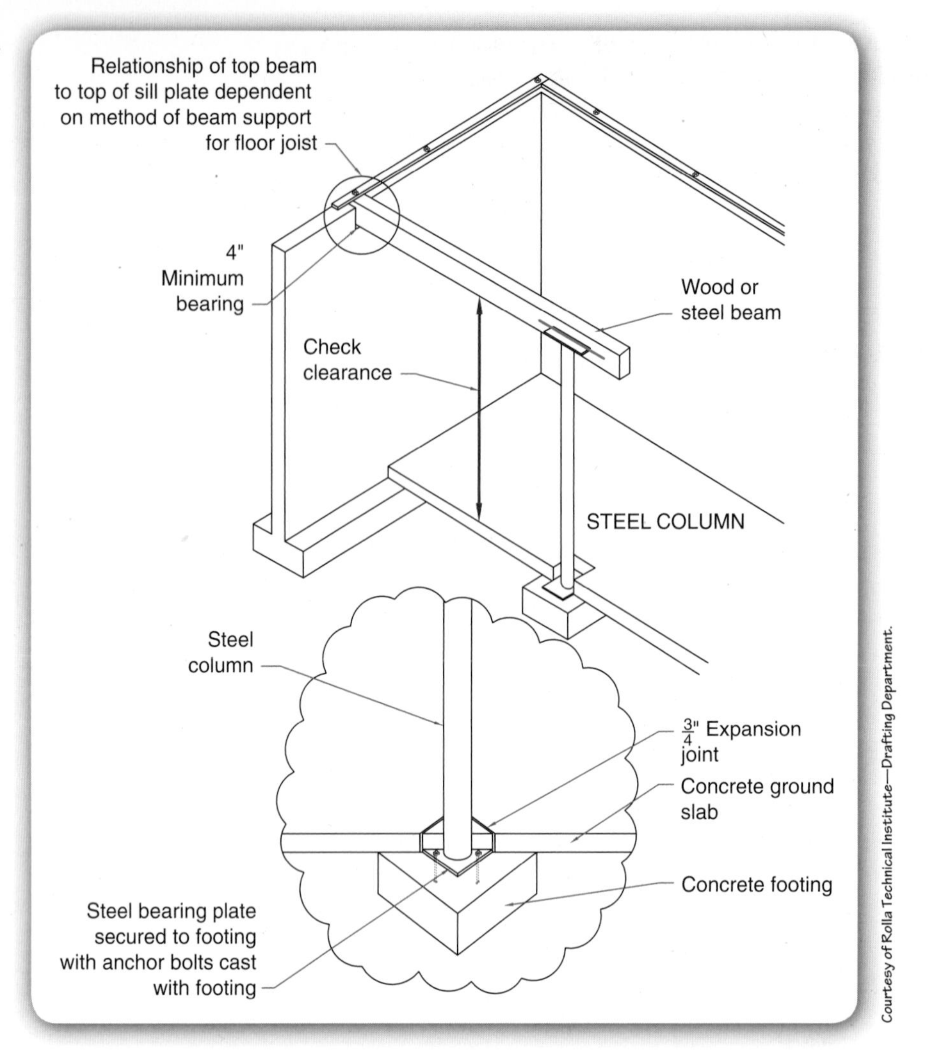

Courtesy of Rolla Technical Institute—Drafting Department.

A similar type of slab is referred to as *mat, raft,* or *floating* foundation. This type of system is used in areas where poor soil-bearing capacities exists and also where high water tables are prevalent. Like the slab-on-grade foundation, a mat, raft, or floating foundation helps to distribute structure load on a reinforced platform.

Figure 13-6: Slab-on-grade.

Courtesy of Donald Block.

Figure 13-7: Mat foundation.

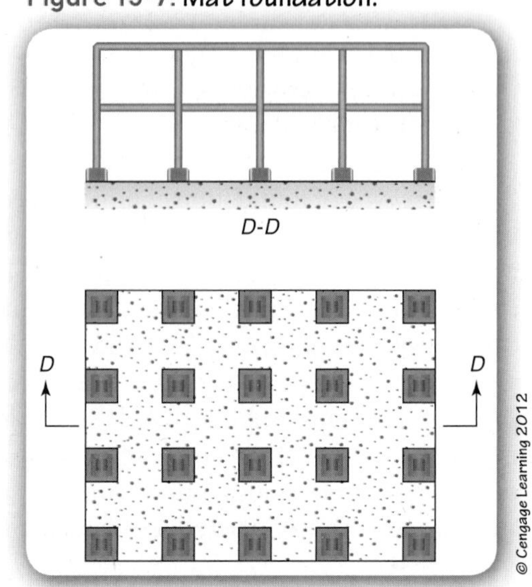

The floating foundation is typically larger than the building's footprint and allows the building to be supported on a continuous footing (Figure 13-7).

While dependent on location, some not so typical foundation systems for residential construction are used where soil conditions are poor. **Pile foundations** in the form of treated timber, reinforced concrete, or steel columns may be driven into the ground in coastal or low lying areas. These friction or bearing piles are driven far enough into the ground to reach a level of greater soil stability. Beams are placed under the load-bearing walls with the weight being transferred to the piles. Typically coastal residences use these types of framing systems to construct and secure their homes (Figure 13-8).

Figure 13-8: Coastal homes on piles.

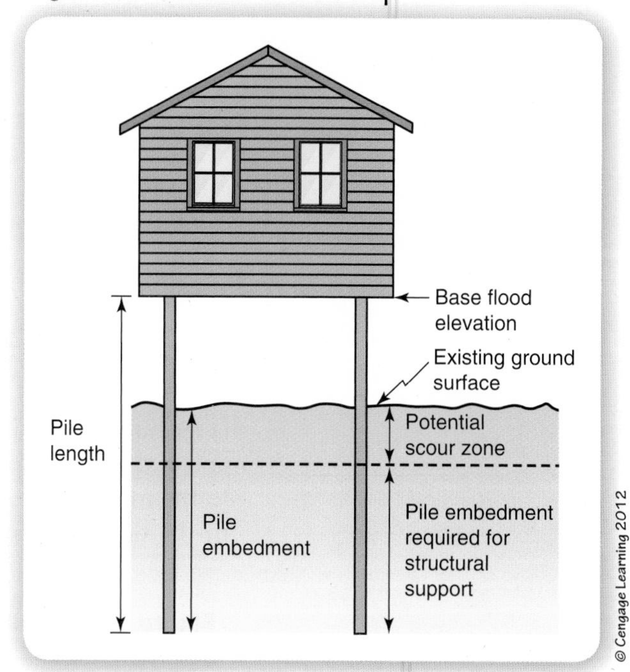

Basic Floor Framing for Residential Construction

Floor framing systems are the primary way that horizontal loads are carried to either the beams or columns. They also provide lateral support for the walls. A floor system may be composed of a series of beams or joists and decking with other flooring material. The sizing of members is dependent upon the span length between supports, load, and the strength of the materials being used. The type of connection of a floor system to both the foundation and wall system further impacts the strength of the structure.

A basic residential floor framing system, wood joist or wood plank and beam (Figure 13-9), combines a series of joists or beams spaced evenly apart and are supported by either beams or walls. The deck material consists of either sheathing or a series of planks. Wood joist systems are typically spaced 12 in, 16 in, or 24 in on center depending on the structural requirements. Wood beams may be spaced further apart depending on structural requirements.

When joists or beams span greater than 8 ft, **bridging** (Figure 13-10) is placed between members to provide resistance to twisting and overturning thus improving the structural capabilities of the flooring system.

Connecting the floor to the foundation system is typically referred to as *platform framing.* A series of sill plates are constructed using 2 × 4, 2 × 6, or 2 × 8 pressure treated lumber, and then sealed and anchored to the foundation wall

Floor Framing Systems: the primary way that horizontal loads are carried to the beams or columns. They also provide lateral support for walls.

Bridging: used when additional members are placed in between spanning floor members to prevent twisting and overturning. These are typically placed in members spanning more than 8 ft and the depth of the member is 6 or more times its width.

Figure 13-10: Bridging.

Source: FEMA.

Figure 13-9: Wood joist and wood plank and beam.

Courtesy of Rolla Technical Institute – Drafting Department.

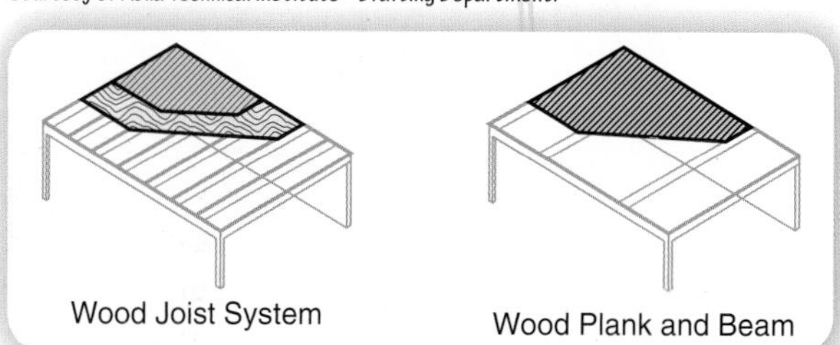

The American Wood Council has additional span tables for both joists and rafters for different species of woods. Go to the American Wood Council website and look under technical information and then span tables. The website provides an online span calculator to assist in designing a structural system based on the parameters you provide.

Fun Fact

According to the Department of Energy, construction of a typical "stick-built" wooden frame house of 1500 sq ft, with one story and a basement, takes roughly 10,000 board feet of lumber. If you assume 10,000 board feet of soft wood is needed to construct the house, at minimum it would take 24 trees that are 17 in across and 35 ft long, or 100 trees that are 8.5 in wide and 20 ft long.

(Figures 13-11 and 13-12). A series of joists or beams are secured to the sill plate and at the ends using a header board and fasteners. The subfloor is then fastened to the members to complete the floor framing system. Openings and overhangs in the floor framing system may require the use of two beams and two cross

Figure 13-12: Wall section of platform framing.

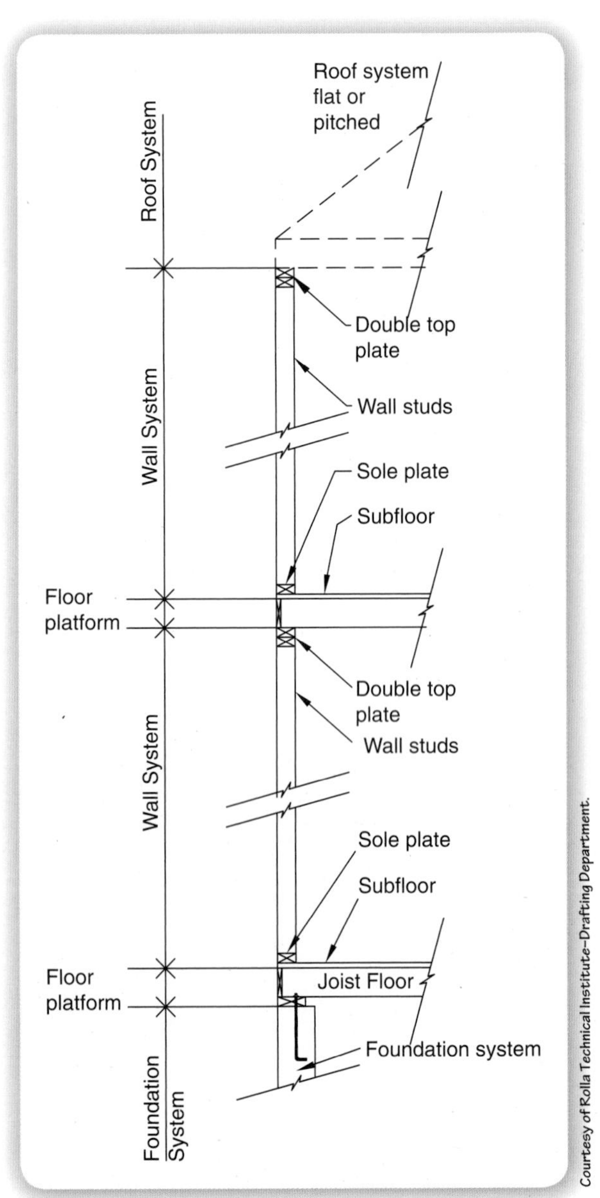

Figure 13-11: Platform framing.

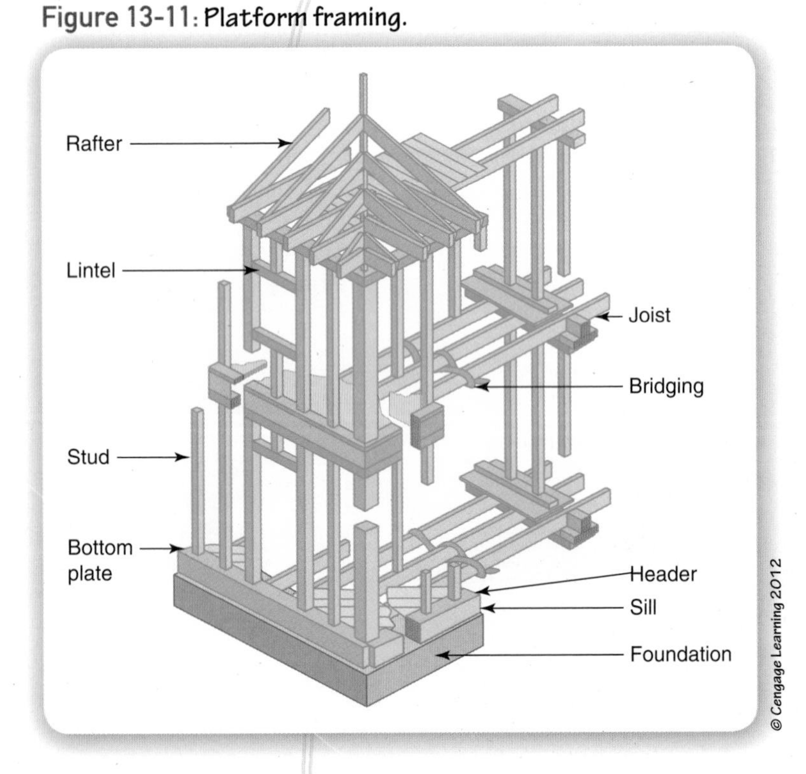

members on each side of the opening or overhang to support the additional span length. Below is a table indicating various types of floor joist and beam systems (Table 13-1).

Engineered wood flooring systems (Figure 13-13) have recently entered the market with new types of closed web (Figure 13-14) and open web (Figure 13-15) wood joist systems. They have many advantages compared to their predecessors in the form of reduced cost, increased strength capabilities, and longer spans. Open web joists also have the advantage of allowing plumbing, electrical, and HVAC systems to be installed and maintained through the flooring of the home. Table 13-2 is a set engineered floor joist and its span length based on their design loads.

Engineered wood members are composite materials which are manufactured by binding together the strands, particles, fibers, or veneers of wood together with adhesives. These engineered members are tested to meet national or international standards to ensure quality control.

Figure 13-13: **Engineered wood framing products.**

Courtesy of APA–The Engineered Wood Association.

Table 13.1: **Maximum Spans for Floor Joists: *General Cases***

Species Group	Grade	Joist Size	Maximum Span (ft-in)								
			With Strapping			With Bridging			With Strapping and Bridging		
			Joist Spacing (in)			Joist Spacing (in)			Joist Spacing (in)		
			12	16	24	12	16	24	12	16	24
DFir-L	No. 1 and No. 2	2 × 6	10-2	9-7	8-7	10-10	9-10	8-7	10-10	9-10	8-7
		2 × 8	12-2	11-7	11-0	13-1	12-4	11-3	13-9	12-10	11-3
		2 × 10	14-4	13-8	13-0	15-3	14-4	13-6	15-10	14-10	13-10
		2 × 12	16-5	15-7	14-10	17-2	16-2	15-3	17-10	16-7	15-6
Hem-Fir	No. 1 and No. 2	2 × 6	10-2	9-7	8-7	10-10	9-10	8-7	10-10	9-10	8-7
		2 × 8	12-2	11-7	11-0	13-1	12-4	11-3	13-9	12-10	11-3
		2 × 10	14-4	13-8	13-0	15-3	14-4	13-6	15-10	14-10	13-10
		2 × 12	16-5	15-7	14-10	17-2	16-2	15-3	17-10	16-7	15-6
S-P-F	No. 1 and No. 2	2 × 6	9-7	8-11	8-2	10-4	9-4	8-2	10-4	9-4	8-2
		2 × 8	11-7	11-0	10-6	12-5	11-9	10-9	13-1	12-2	10-9
		2 × 10	13-8	13-0	12-4	14-6	13-8	12-10	15-1	14-1	13-2
		2 × 12	15-7	14-10	14-1	16-4	15-5	14-6	17-0	15-10	14-9
Northern Species	No. 1 and No. 2	2 × 6	8-3	7-8	7-1	9-3	8-5	7-5	9-4	8-5	7-5
		2 × 8	10-6	10-0	9-4	11-3	10-7	9-8	11-10	11-0	9-8
		2 × 10	12-4	11-9	11-2	13-1	12-4	11-7	13-8	12-9	11-10
		2 × 12	14-1	13-5	12-9	14-9	13-11	13-1	15-4	14-4	13-4

Notes: (a) Nailed $^5/_8$" subfloor
(b) Live Load = 40 psf
(c) Dead Load = 10 psf
(d) Deflection = Span/360
(e) Spans include consideration of vibration criteria.

Source: North American Retail Hardware Association, 2004.

Fun Fact

Did you know the building industry in the United States uses a measuring system commonly referred to as the U.S. Customary Units? Developed initially by the British, the units commonly associated with building materials are typically in dimensions of inches, feet, and yards. In the case of framing systems, dimensional lumber is commonly referred to in nominal dimensions; for example, a common "2 × 4" in actual dimensions it measures approximately 1 ½ in by 3-½ in. Other common lumber sizes include:

Nominal	Actual	Nominal	Actual	Nominal	Actual
1 × 2	¾ in × 1 ½ in (19 mm × 38 mm)	2 × 2	1 ½ in × 1 ½ in (38 mm × 38 mm)	4 × 4	3 ½ in × 3 ½ in (89 mm × 89 mm)
1 × 3	¾ in × 2 ½ in (19 mm × 64 mm)	2 × 3	1 ½ in × 2 ½ in (38 mm × 64 mm)	4 × 6	3 ½ in × 5 ½ in (89 mm × 140 mm)
1 × 4	¾ in × 3 ½ in (19 mm × 89 mm)	2 × 4	1 ½ in × 3 ½ in (38 mm × 89 mm)	6 × 6	5 ½ in × 5 ½ in (140 mm × 140 mm)
1 × 6	¾ in × 5 ½ in (19 mm × 140 mm)	2 × 6	1 ½ in × 5 ½ in (38 mm × 140 mm)	8 × 8	7 ¼ in × 7 ¼ in (184 mm × 184 mm)
1 × 8	¾ in × 7 ¼ in (19 mm × 184 mm)	2 × 8	1 ½ in × 7 ¼ in (38 mm × 184 mm)	2 × 8	1 ½ in × 7 ¼ in (38 mm × 184 mm)
1 × 10	¾ in × 9 ¼ in (19 mm × 235 mm)	2 × 10	1 ½ in × 9 ¼ in (38 mm × 235 mm)		
1 × 12	¾ in × 11 ¼ in (19 mm × 286 mm)	2 × 12	1 ½ in × 11 ¼ in (38 mm × 286 mm)		

Today's framing materials are designed and constructed based on a modular system. If we consider sheets of plywood or drywall typically comes in sizes of 4 × 8 or 4 × 12 ft, the stick built framing system is designed and constructed to accommodate these size panels by having stud spacing at 12, 16, or 24 in on center. This allows for a panel to be fully supported. This type of system can also help reduce the time, cost, and resources needed for fabricating and constructing a facility when the design uses construction materials efficiently during planning stages.

Figure 13-14: Engineered wood floor framing systems: closed web.

Courtesy of Donald Block.

Figure 13-15: Engineered wood floor framing systems: open web.

Courtesy of Southern Forest Products Association.

Example: Problem

Design a floor framing system using a wood floor joist system (Douglas Fir) and an engineered wood joist system based on the following requirements:
Dead Load: 10 lb/ft^2
Live Load: 40 lb/ft^2
Total Load: 50 lb/ft^2
Span Length: 15 ft
Beam spacing is 16 in O.C.
Framing system to be built with strapping and bridging

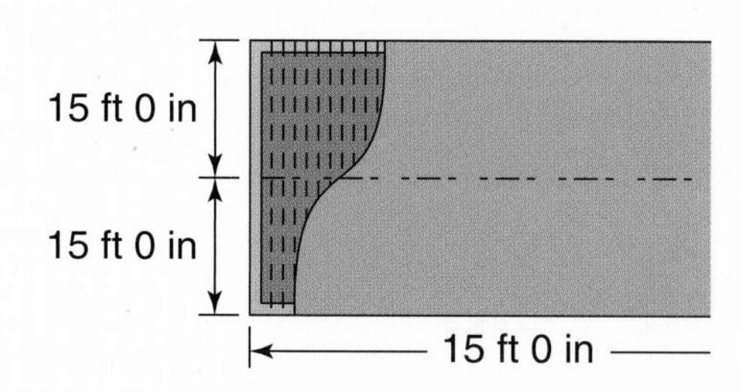

Answer:
Wood floor joist system using douglas fir:
From Table 13–1 indicates a 2 × 10 at 16 in O.C. would span a maximum length of 15 ft 10 in
Engineered wood system:
Determine the tributary for spacing at 16 in O.C.
Divide 16/12 = 1.333 ft
Determine factored load due to tributary area:
Live Load = 40 lb/ft^2 × 1.333 ft = 53.33 lb/ft
Dead Load = 10 lb/ft^2 × 1.333 ft = 13.33 lb/ft
Total Load = 66.66 lb/ft
From Table 13–2 indicates a 1 ½ × 9 ¼ at 16 in O.C. can support a total load of 104 lb/ft and 73 lb/ft live load.

The Internet has additional span tables for engineered wood joist and rafter systems. Select a search engine search on the Internet and look for "Engineered Wood Span Tables." Most companies provide their span tables for various manufactured joist and rafter systems.

Wood flooring in most residential construction comes in the form of three primary systems: oriented strand board (OSB), plywood, and wood plank. Oriented strand board and plywood sheets are typically produced in sizes of either 4 ft × 8 ft or 4 ft × 12 ft and may have **tongue and groove** joints for tighter fit. Wood planks come in varying thicknesses and widths depending on the span and finish (Figure 13-16 and 13-17).

Tongue and Groove: a type of joint that has one edge with a protruding thinner piece of wood that is inserted into a groove-cut edge of an adjoining piece to form a tight uniform fit.

Preparation of Floor Framing Plans

Floor framing plans (Figure 13-18) consist of floor plans, construction details, and notes. In a floor plan, structural members are indicated as to the orientation of the members and superimposed on an architectural background. The plan also indicates additional framing considerations needed for openings. Additional information such as construction details and member types and sizes and spacing between members are typically indicated on the plans.

Table 13-2: 1.75E LSL Uniform Floor Load (PLF) Tables: 1-½".

Courtesy of LP®Building Products

TO USE:

1. *Select the span required.*
2. *Compare the design total load to the Total Load column.*
3. *Compare the design live load to the appropriate Live Load column.*
4. *Select a product that exceeds both the design total and live loads.*

EXAMPLE:

Floor live load = 480 Plf, L/360 deflection limit
Floor total load = 660 Plf, L/240 deflection limit
Beam span center-to-center of supports = 16'- 6"

SOLUTION:

A 2-ply 1-½" × 16" can support 736 plf Total Load and 536 plf Live Load at L/360.

	1-½" × 5-½"			1-½" × 7-¼"			1-½" × 9-¼"			1-½" × 9-½"			1-½" × 11-¼"			
	Live Load		Total	Live Load		Total	Live Load		Total	Live Load		Total	Live Load		Total	
Span	L/480	L/360	Load	L/480	L/360	Load	L/480	L/360	Load	L/490	L/280	Load	L/480	L/360	Load	Span
5'	286	392	560	605	806	937	1127		1475	1203		1550	1797		2102	5'
6'	171	229	340	371	494	650	710	947	1023	761	1014	1075	1165		1472	6'
7'	110	147	218	242	323	476	472	630	750	507	677	789	791	1054	1080	7'
8'	75	100	147	166	221	329	328	439	573	353	471	602	558	744	825	8'
9'	53	71	103	118	158	234	237	316	452	255	340	475	406	542	651	9'
9'-6"	45	60	88	101	135	199	203	271	402	219	292	426	351	468	584	9'-6"
10'	39	52	75	97	116	171	176	235	347	190	253	375	304	406	526	10'
11'	—	—	—	66	88	129	134	179	264	145	193	285	233	311	434	11'
12'	—	—	—	51	68	99	104	139	204	113	150	221	183	244	360	12'
13'	—	—	—	40	54	77	83	110	161	89	119	174	145	194	286	13'
14'	—	—	—	32	43	62	66	99	129	72	96	140	117	157	230	14'
15'	—	—	—	—	—	—	54	73	104	59	78	113	96	128	187	15'
16'	—	—	—	—	—	—	45	60	86	49	65	93	80	107	154	16'
16'-6"	—	—	—	—	—	—	41	55	78	44	59	84	73	97	141	16'-6"
17'	—	—	—	—	—	—	37	50	71	41	54	77	67	89	128	17'
18'	—	—	—	—	—	—	32	42	59	34	46	64	56	75	108	18'
18'-6"	—	—	—	—	—	—	—	—	—	32	42	59	52	70	99	18'-6"
19'	—	—	—	—	—	—	—	—	—	—	—	—	48	64	91	19'
20'	—	—	—	—	—	—	—	—	—	—	—	—	41	55	78	20'
21'	—	—	—	—	—	—	—	—	—	—	—	—	36	48	66	21'
22'	—	—	—	—	—	—	—	—	—	—	—	—	31	42	57	22'

Figure 13-16: **Platform floor framing system.**

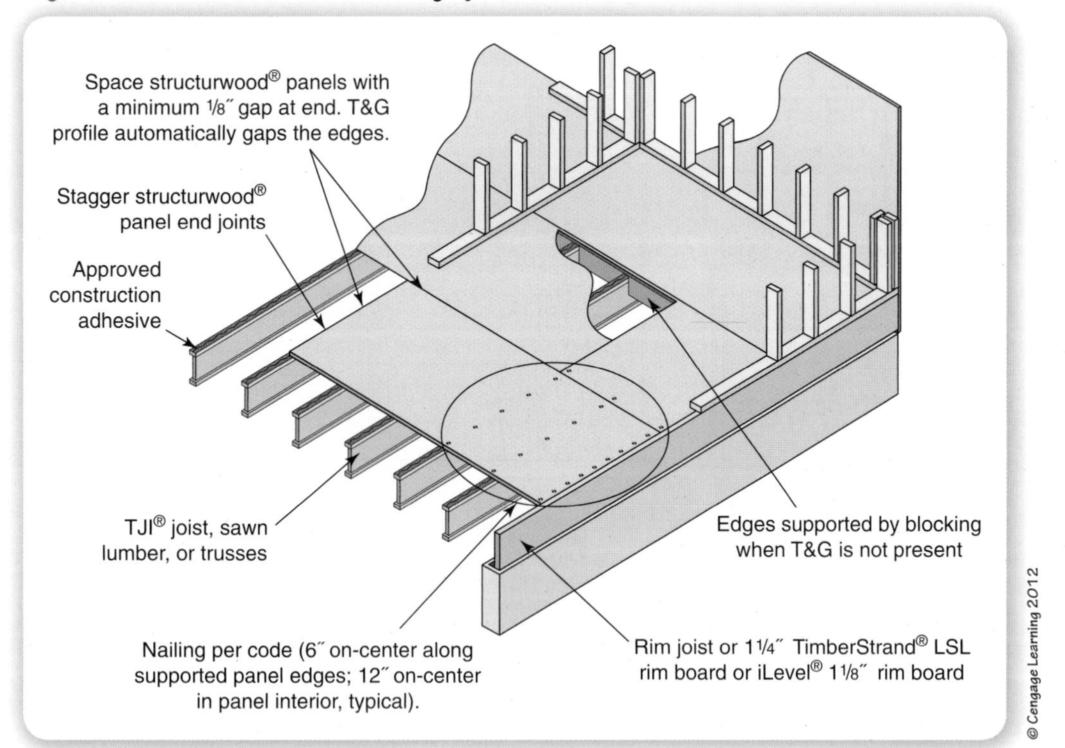

Basic Wall Framing Systems for Residential Construction

In most residential construction, exterior walls provide structural support for the floors and roof above. They also serve to protect the building's interior from exterior conditions. Interior walls may be load-bearing or non–load-bearing and serve as dividers between spaces. The sizes of door and window openings to allow access, ventilation, natural lighting, and views are all supported by wall systems (Figure 13-19).

A wall system for residential construction is typically built using three different types of framing systems. The most common is wood or light gauge steel stud framing. As the name implies a series of studs

Figure 13-17: **Tongue and groove floor panel.**

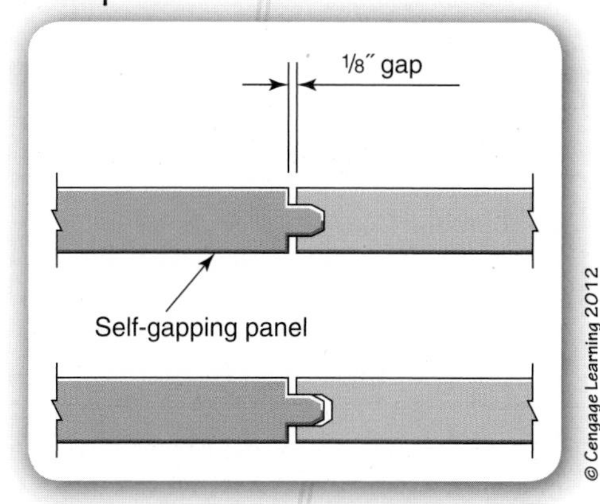

Figure 13-18: **Floor framing plan.**

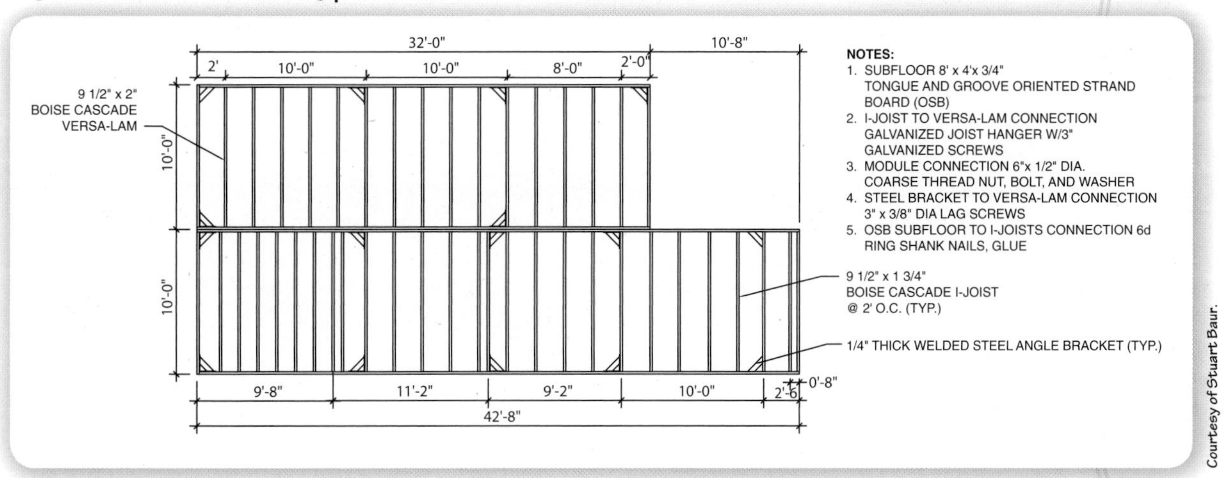

Courtesy of Stuart Baur.

Metal Track:
the horizontal framing member connecting the bottom or top of a metal stud wall assembly.

Bottom Plate:
(also known as a sole plate) the horizontal framing member connecting the bottom of a wood stud frame to the floor of a building's structure.

Header:
a structural member that spans the distance above a framed opening such as a door or a window.

Sill Plate:
a structural member that spans the distance below a framed opening such as a window.

Figure 13-19: Wall systems.

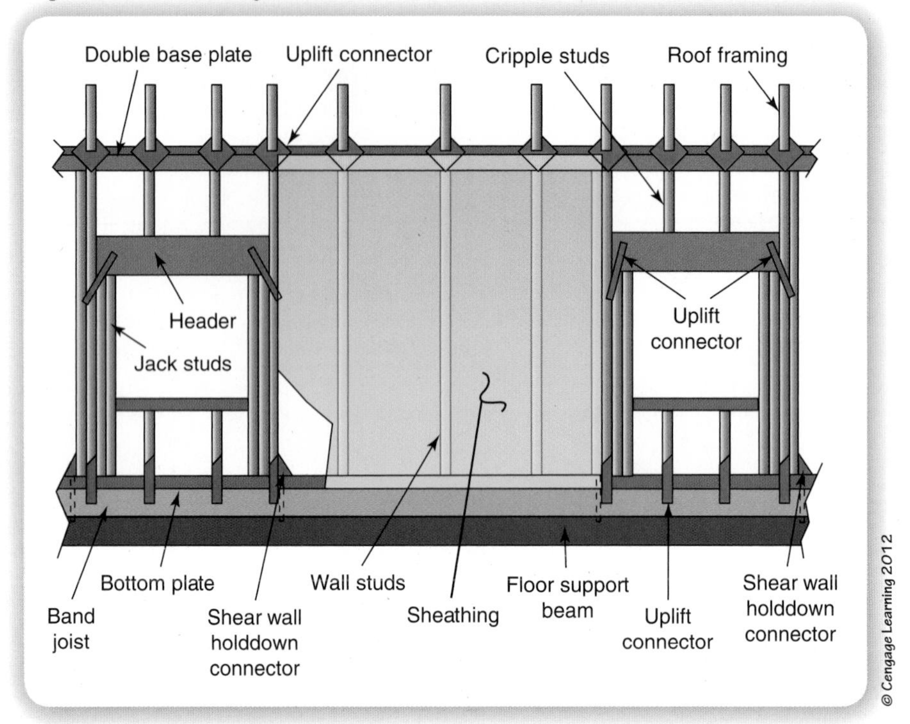

(2 × 4 or 2 × 6 nominal dimensions) are spaced 16 in or 24 in on center depending on load-bearing and non–load-bearing walls to a top and **bottom plate** or **metal track**. The height of the wall may vary, but due to pre-cut lumber heights, typically range from floor to ceiling height of 8, 10, or 12 ft. Insulation, vapor barrier, electrical, and plumbing may be accommodated in the stud wall. Exterior and interior finishes may be applied to wall sheathing or directly to the studs. The framing is connected to the flooring through a series of fasteners (see Figures 13-20 and 13-21).

Window or door openings in stud frame connections require an additional stud on each side of the opening. The additional stud surrounding the opening provides greater support. A **header** is placed at the top of the opening to provide additional support at the top and for windows a **sill plate** is also placed to complete the rough opening (Figure 13-22) for a window.

Point of Interest

The international residential code recommends wall construction for load-bearing applications space studs 16 in on center and for non–load-bearing applications at 24 in on center. Prior to the code, it was common practice to place wood studs 16 in on center regardless whether or not the wall panel was to be used as a load or non–load-bearing assembly. Two inherent benefits result from this application. First, 25 percent less wood is used in non–load-bearing walls, thus reducing the cost of construction while also reducing environmental impact. Second, the wall provides a greater thermal barrier due to the fact that there are fewer thermal breaks. A thermal break occurs at a point where thermal insulation is not continuous thus forming a gap between thermal barriers as in the case for wood framing systems the studs along the exterior wall prevent a continuous thermal

Figure 13-20: Axonometric view of a platform framing system.

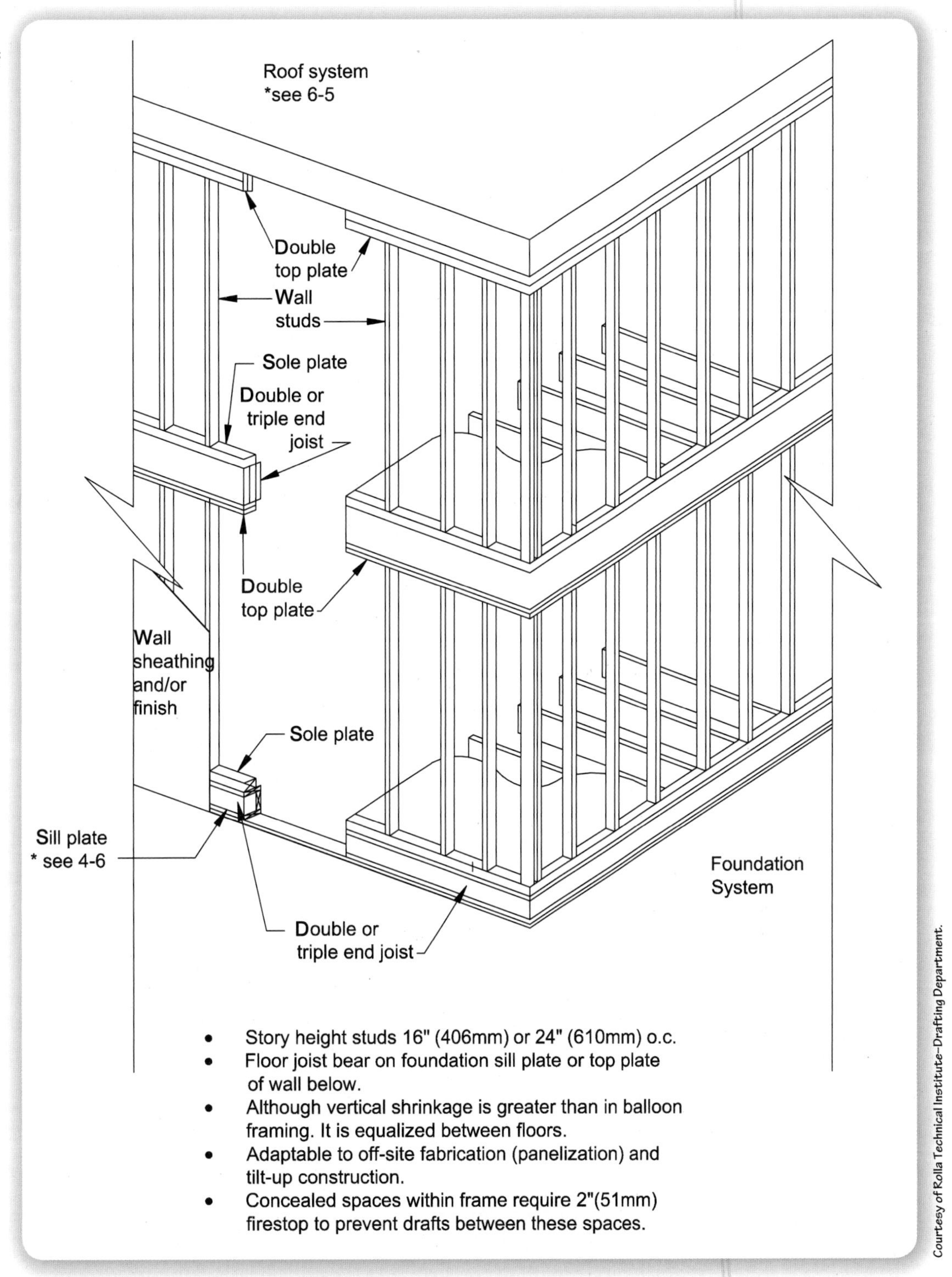

Roof system
*see 6-5
Double
top plate
Wall
studs
Sole plate
Double or
triple end
joist
Double
top plate
Wall
sheathing
and/or
finish
Sole plate
Sill plate
* see 4-6
Foundation
System
Double or
triple end joist
• Story height studs 16" (406mm) or 24" (610mm) o.c.
• Floor joist bear on foundation sill plate or top plate of wall below.
• Although vertical shrinkage is greater than in balloon framing. It is equalized between floors.
• Adaptable to off-site fabrication (panelization) and tilt-up construction.
• Concealed spaces within frame require 2"(51mm) firestop to prevent drafts between these spaces.

Point of Interest *(continued)*

insulated wall. To learn more about the importance of thermal insulation, perform a Internet search on "Energy Star qualified homes thermal bypass checklist." View thermal images of thermal losses in framing systems and how to minimize these losses during construction.

Figure 13-21: Partial wall section.

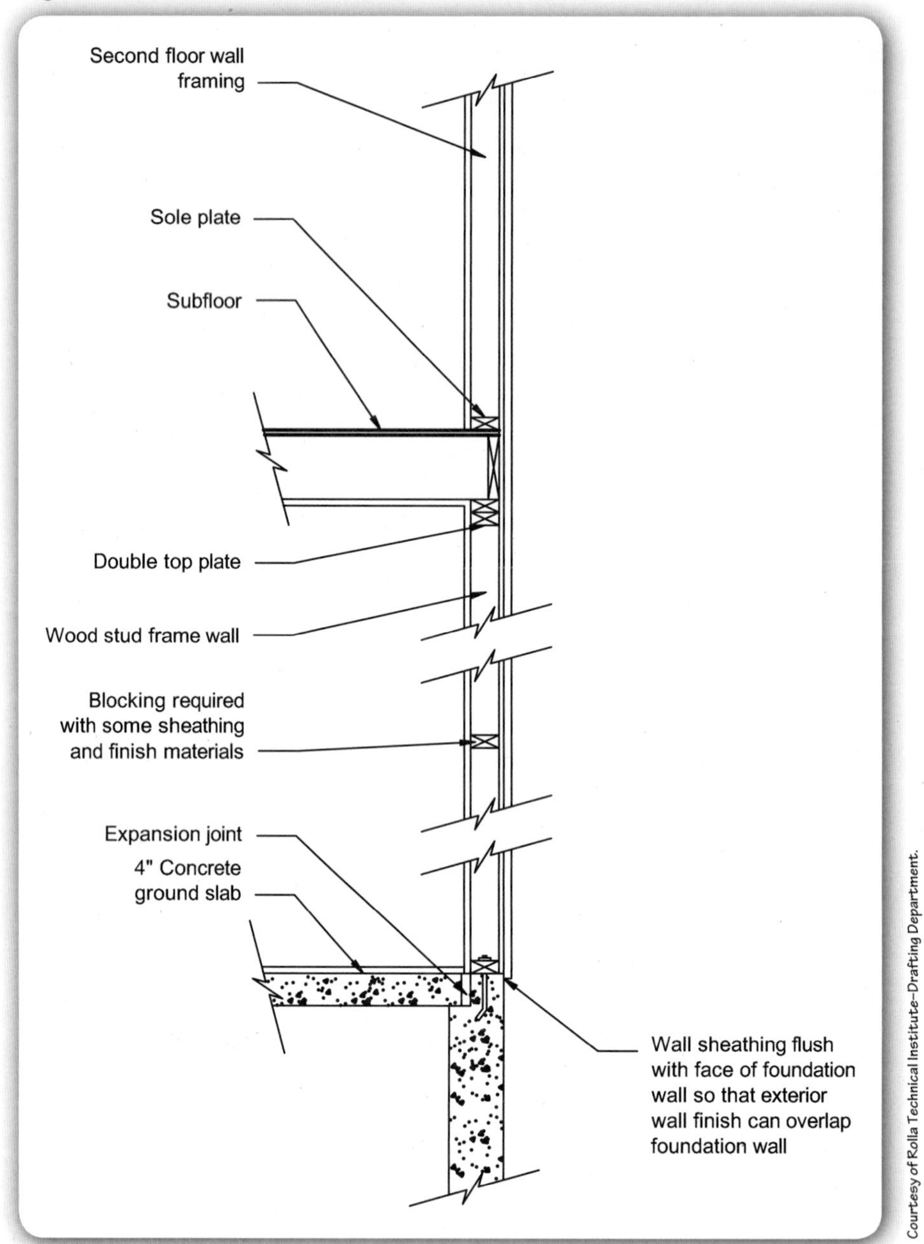

Courtesy of Rolla Technical Institute–Drafting Department.

When connecting two or more stud wall assembly corner joints, additional blocking and fastening systems are required (Figure 13-23). This is critical as the strength of the structure depends heavily on the corner connections.

Wall sheathing typically consists of either plywood or oriented strand board on the exterior side of the stud wall frame. This type of diaphragm system increases a building's structural integrity against varying loads such as wind or earthquake but also protect it from other natural elements.

A second type of system is commonly known as *wood post and beam framing.* The advantage of wood post and beam over stud wall systems is that it allows for larger openings. It is typically left exposed highlighting the workmanship and the detailing of its construction, and the walls are non–load-bearing. The structural

Figure 13-22: Opening in stud wall framing systems.

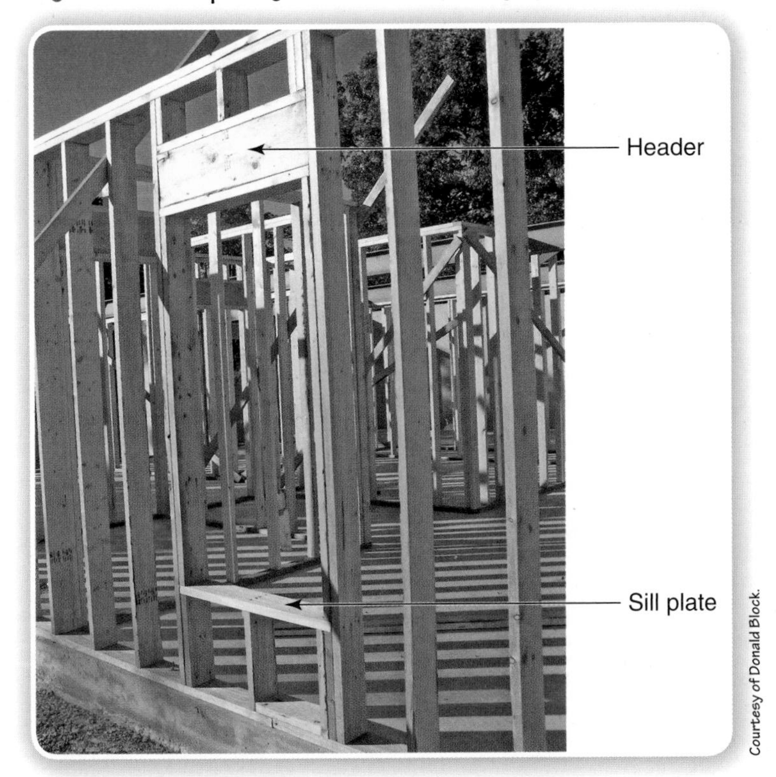

Figure 13-23: Stud wall corner connections.

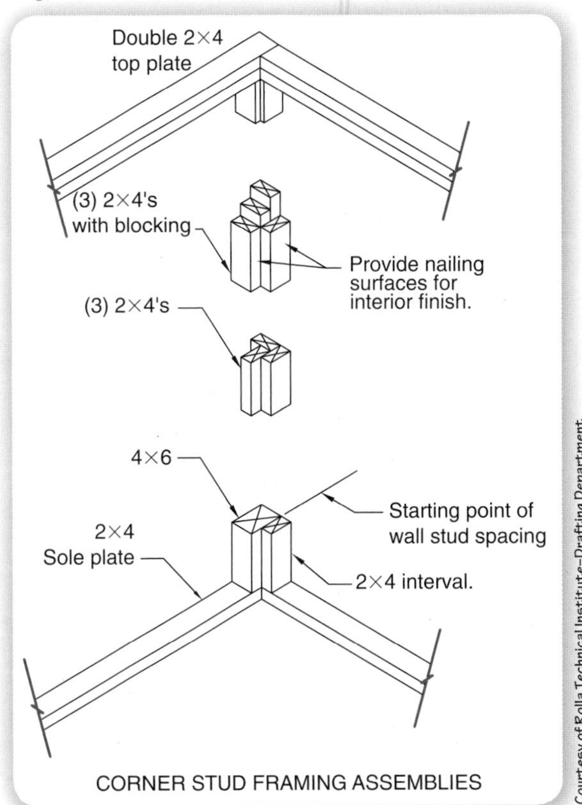

strength of post and beam construction is dependent upon the member's joints and the combination of the infill wall panels with its structural frame. Log cabins employ this type of construction (Figure 13-24).

The third type of framing system, the masonry wall, provides great strength yet has lateral load requirements. The wall finishes may be applied directly to the surface or as part of a structural system. Electrical, plumbing, and insulation

Figure 13-24: Today's log cabin.

Figure 13-25: Reinforced concrete block wall system.

Figure 13-26: Brick wall sections.

Courtesy of Rolla Technical Institute.

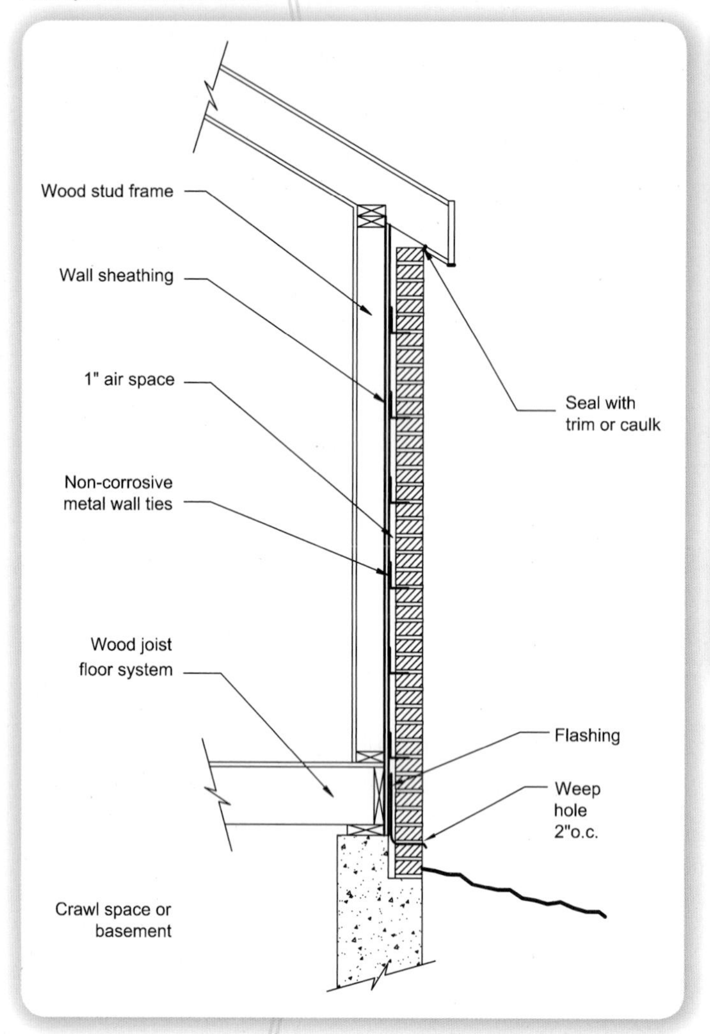

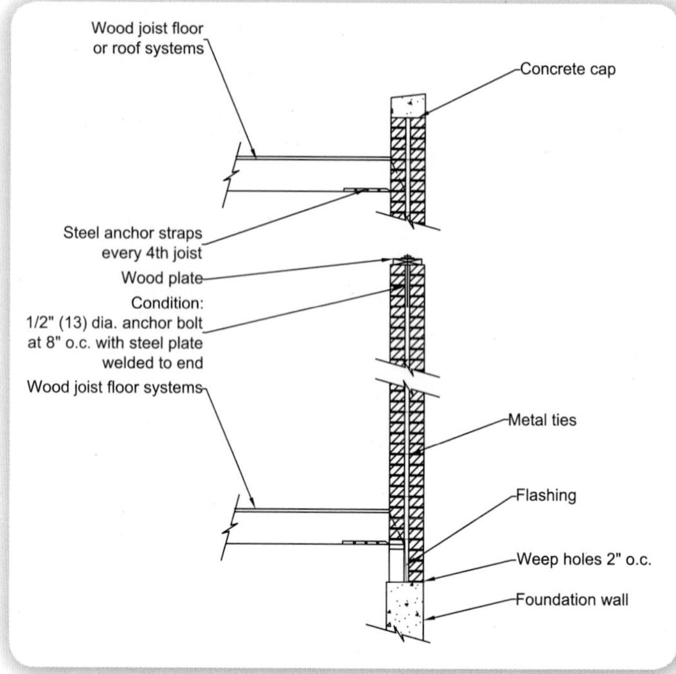

are typically placed between furring strips or in a separate structural framing system. Common materials consist of structural brick, concrete masonry, or stone. Reinforced concrete block walls are typically used in areas where homes undergo severe weather conditions such as earthquakes, hurricanes, and earth pressure below grade (Figure 13-25). Other masonry framing systems consist of solid brick, brick cavity with ties, block and brick cavity with ties, and brick veneer (Figure 13-26).

lintel: a **lintel** is a beam supporting the weight above.

Masonry openings are typically constructed using **lintels** or arches. A lintel is a type of beam that supports the weight above an opening such as a window or door. Lintels for brick wall openings are typically constructed using steel reinforced brick masonry or pre-cast concrete. Steel lintels are constructed by placing steel angle over the opening to act as a shelf and support the brick over the opening. A brick masonry lintel is formed by constructing a reinforced concrete beam within the brick wall. A pre-cast concrete lintel is a prefabricated beam placed over the opening and is typically visible (Figure 13-27).

An alternate way of spanning a masonry opening may be the use of arches. Based on the compressive properties of the masonry, the arch transcends the above load to a combination horizontal and vertical force. An equal but opposite force is then pushed from the adjacent walls (Figure 13-28).

Roof Framing Systems: the primary means of protecting the building and its occupants from the elements.

Basic Roof Framing Systems for Residential Construction

The **roof framing system** acts as the primary means of protecting the building from the elements. As such, the roof is also the primary generator of building loads to the overall structure carrying not only the live loads of the elements outside such

Figure 13-27: Masonry lintel over opening.

© iStockphoto.com/Linda Steward.

Figure 13-28: Arch lintel over opening.

© iStockphoto.com/Linda Steward.

as wind, snow, and rain but also the weight of the roof itself. The roof system could potentially be the most expensive component of the whole building when considering cost of installation, durability, and maintenance. Thermal resistance could impact the heating and cooling loads. The visual appeal of the building may be dependent on the roof's design. The overall finish, whether gravel, tile, or shingle may also impact the building's appeal (Figure 13-29).

Some of the most common roof systems are joist and rafter. Similar to a floor framing system for a flat roof, rafters span the length of the support (Figure 13-30). Unlike a flat roof, a low pitch roof typically has a slope of 2:12–4:12, where the first number 2 is the vertical distance and the second number 12 is the horizontal distance, and is made of two rafters spanning a larger area and supported at the middle by a ridge beam (Figure 13-31). A high pitch roof typically 4:12–12:12 slope would also include a collar beam to add additional support to the rafter (Figure 13-32). The following span table (Table 13-3) provides estimating and preliminary sizing of roof joists or rafters.

Roof support conditions are dependent upon the type of wall system and overhang conditions to which the wall is connected. Figure 13-33 provides several different types of roof-to-wall connections.

Another type of roof framing system includes the use of *roof trusses.* A roof truss system provides greater stability over longer spans. This allows for greater flexibility in the design of the building's interior spaces. Some of the most common

Figure 13-29: Roof designs.

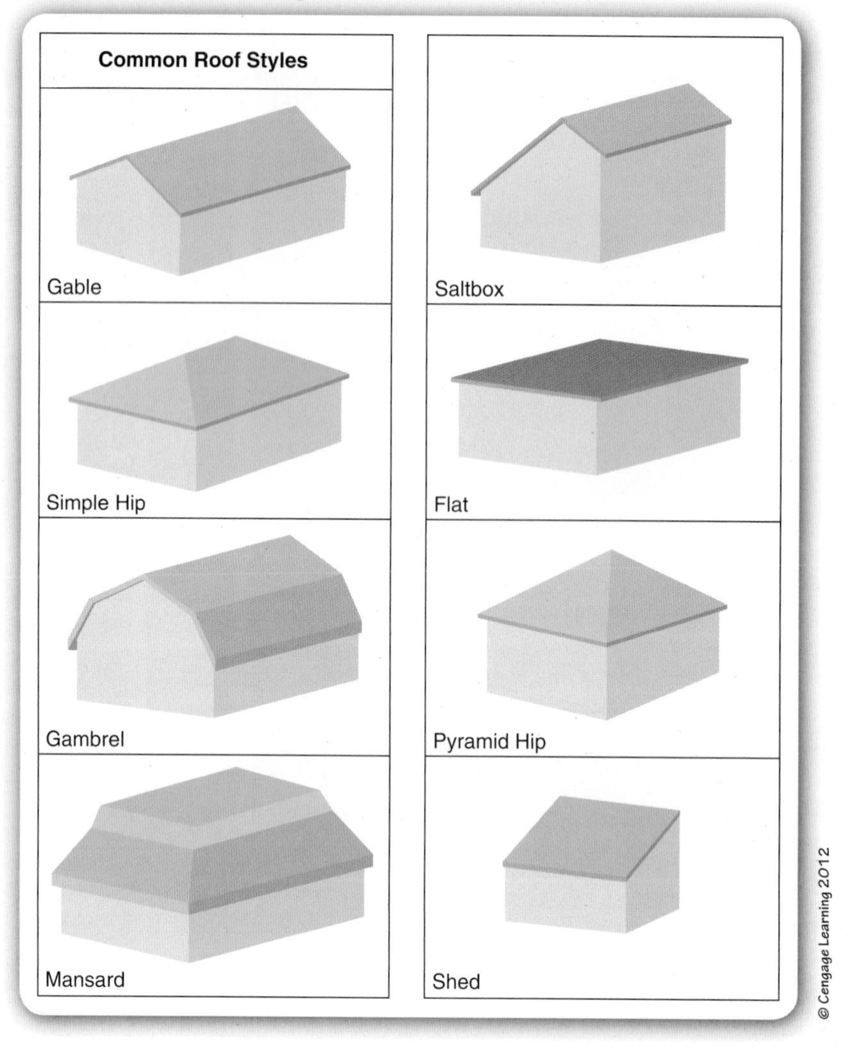

Figure 13-30: Flat roof.

© iStockphoto.com/bev hamer.

Figure 13-31: Gable roof with both high and low pitch roofs.

© iStockphoto.com/Anton Foltin.

roof truss designs include *Howe, Fink,* and *flat truss* (Figure 13-34). Roof trusses are normally engineered and fabricated at the manufacturer. They can also accommodate the building's mechanical and electrical within the frame itself.

Roof sheathing is typically constructed using oriented strand board or plywood sheets in sizes of typically 4 ft × 8 ft or 4 ft × 12 ft with a recommended ¾ in minimum thickness. Wood trusses are typically spaced between 2 ft and 4 ft on center depending on the location of structure and local building codes.

Figure 13-32: High pitch roof.

© iStockphoto.com/Majoros Laszlo.

Figure 13-33: Roof-to-wall connections.

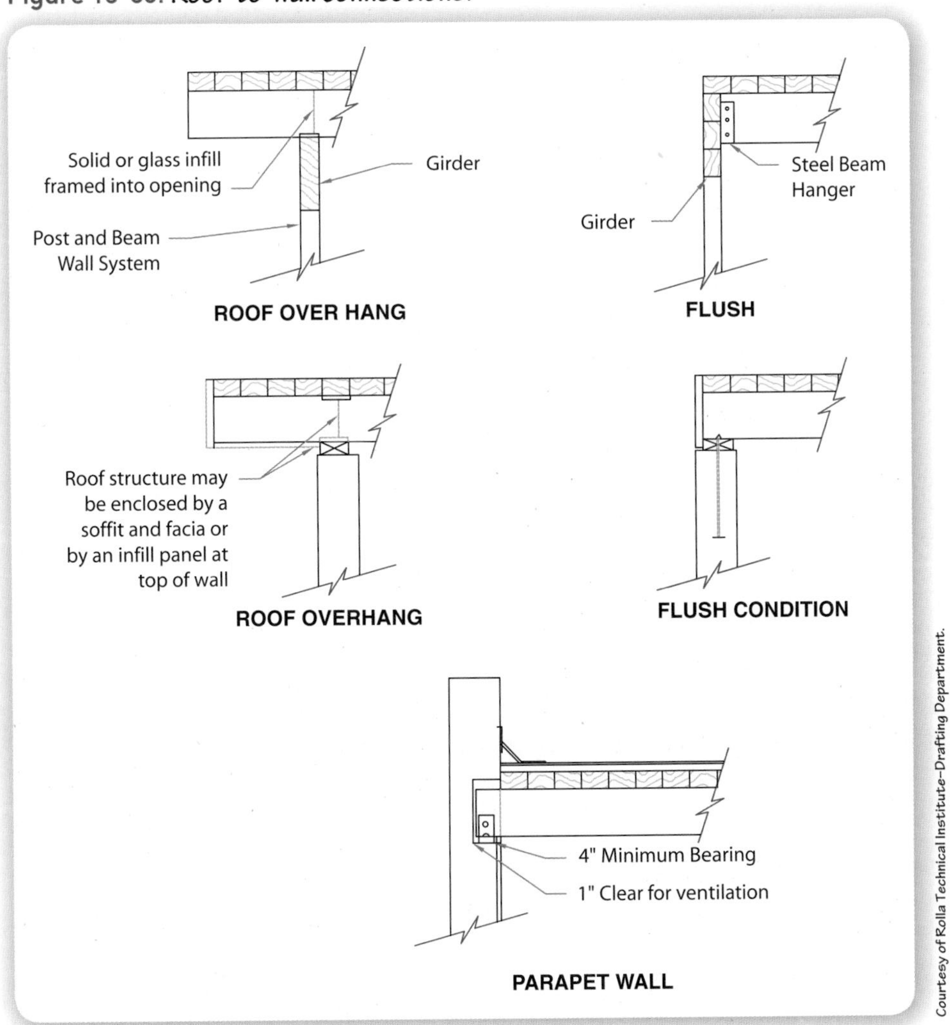

Courtesy of Rolla Technical Institute–Drafting Department.

Table 13-3:

Courtesy of Southern Forest Products Association.

Southern Pine Span Tables

Maximum spans given in feet and inches
Inside to inside of bearings

Table 21 Rafters — 40 psf live load, 10 psf dead load, 240 deflection, Cd = 1.15
Light Roofing; Drywall Ceiling; Snow Load

Size Inches	Spacing Inches on Center	Grade									
		Visually Graded				Machine Stress Rated (MSR)			Machine Evaluated Lumber (MEL)		
		ss	No. 1	No. 2	No. 3	2400f-2.0E	2250f-1.9E	1950f -1.7E	M23	M14	M29
2 × 6	12.0	12-9	12-6	12-3	10-0	13-3	13-0	12-6	12-9	12-6	12-6
	16.0	11-7	11-5	11-2	8-8	12-0	11-10	11-5	11-7	11-5	11-5
	19.2	10-11	10-8	10-2	7-11	11-4	11-1	10-8	10-11	10-8	10-8
	24.0	10-2	9-11	9-2	7-1	10-6	10-4	9-11	10-2	9-11	9-11
2 × 8	12.0	16-10	16-6	16-2	12-9	17-5	17-2	16-6	16-10	16-6	16-6
	16.0	15-3	15-0	14-5	11-0	15-10	15-7	15-0	15-3	15-0	15-0
	19.2	14-5	14-1	13-2	10-1	14-11	14-8	14-1	14-5	14-1	14-1
	24.0	13-4	13-1	11-9	9-0	13-10	13-7	13-1	13-4	13-1	13-1
2 × 10	12.0	21-6	21-1	19-11	15-1	22-3	21-10	21-1	21-6	21-1	21-1
	16.0	19-6	19-2	17-3	13-0	20-2	19-10	19-2	19-6	19-2	19-2
	19.2	18-4	17-6	15-9	11-11	19-0	18-8	18-0	18-4	18-0	18-0
	24.0	17-0	15-8	14-1	10-8	17-8	17-4	16-9	17-0	16-9	16-9
2 × 12	12.0	26-0*	25-7	23-4	17-11	26-0*	26-0*	25-7	26-0*	25-7	25-7
	16.0	23-9	22-10	20-2	15-6	24-7	24-2	23-3	23-9	23-3	23-3
	19.2	22-4	20-11	18-5	14-2	23-1	22-9	21-11	22-4	21-11	21-11
	24.0	20-9	18-8	16-6	12-8	21-6	21-1	20-4	20-9	20-4	20-4

These spans are intended for use in enclosed structures or where the moisture content in use does not exceed 19 percent for an extended period of time unless the table is labled Wet-Service. Applied loads are given in psf (pounds per square foot). Deflection is limited to the span in inches divided by 360,240, or 180 and is based on live load only. The load duration factor, C_D, is 1.0 unless shown as 1.15 or 1.25. An asterisk (*) indicates the listed span has been limited to 26'60" based on availability; check sources of supply for lumber longer than 20'. Highlighted sizes/grades are NOT commonly produced.

The Southern Pine Council does not grade or test lumber, and accordingly, does not assign design values to Southern Pine lumber. The design values contained herein are based on the *2002 SPIB Standard Grading Rules for Southern Pine Lumber,* published by the Southern Pine Inspection Bureau, and modified as required by the *2001 National Design Specification® (NDS®) for Wood Construction* published by the American Forest & Paper Association (AF&PA).

The primary purpose of this publication is to provide a convenient reference for joist and rafter spans for specific grades of Southern Pine lumber. The maximum spans provided herein were determined on the same basis as those in *Span Tables for Joists and Rafters,* published by AF&PA. Accordingly, the Southern Pine Council, its principals and/or members, do not warrant in any way that the design values on which the span tables for Southern Pine lumber contained herein are based are correct, and specifically disclaim any liability for injury or damage resulting from the use of such span tables.

The conditions under which lumber is used in construction may vary widely, as does the quality of the lumber and workmanship. Neither the Southern Pine Council, nor its principals and/or members, have any knowledge of the construction methods, quality of materials and workmanship used on any construction project; and accordingly, cannot and do not, warrant the performance of the lumber used in completed structures.

Figure 13-34: Roof truss configurations.

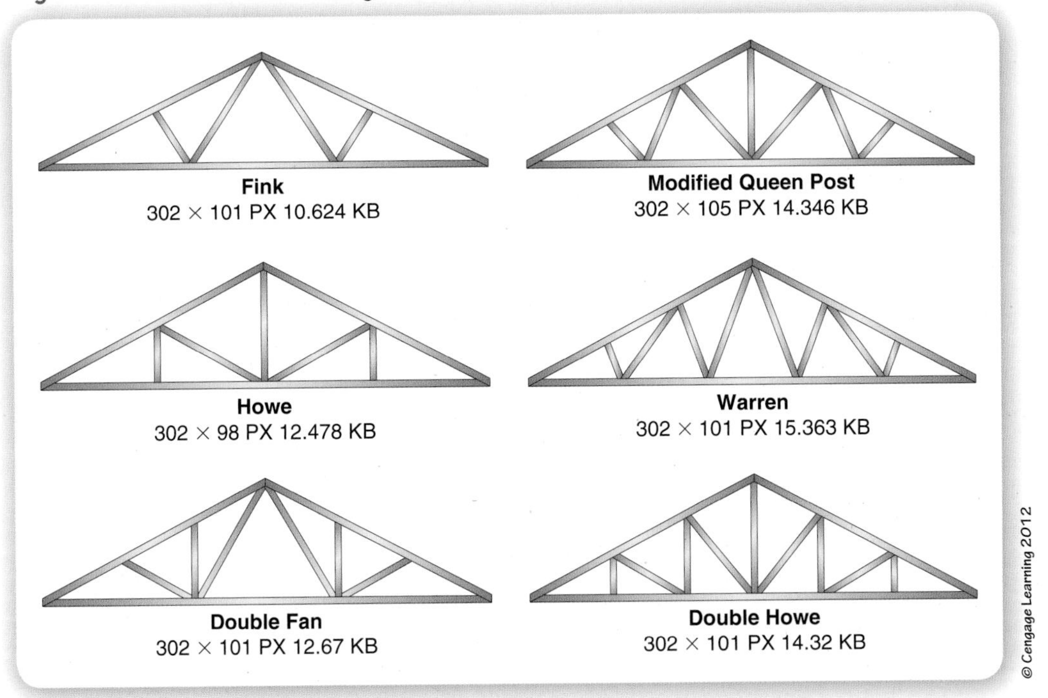

Preparation of Roof Framing Plans

Roof framing plans (Figures 13-35 and 13-36) are similar to floor framing plans. In a roof framing plans plan, structural members are indicated as to the orientation of the members and superimposed on an architectural background. The roof framing plan may vary depending on the type of construction methodology being used. For most residential construction roof systems are constructed using trusses or platform construction methods. The plan also indicates framing

Figure 13-35: Roof framing plan with hip roof. From Jefferis and Madsen, *Architectural Drafting and Design, 5e.*

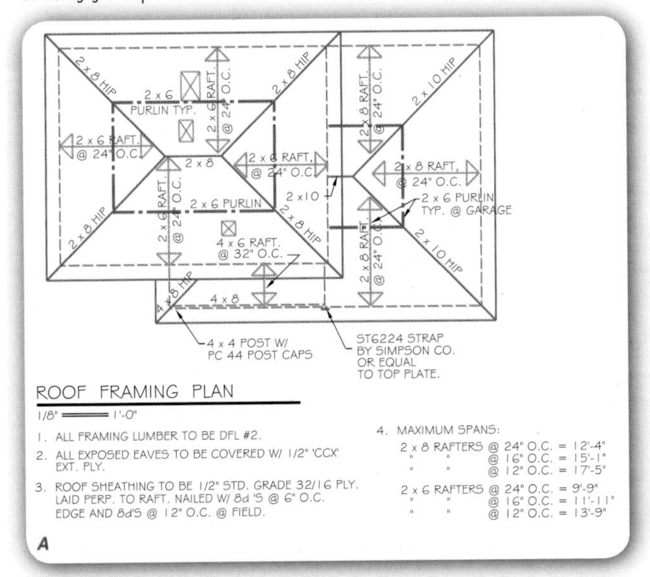

Figure 13-36: Roof framing plan with hip roof showing all framing members. From Jefferis and Madsen, *Architectural Drafting and Design.*

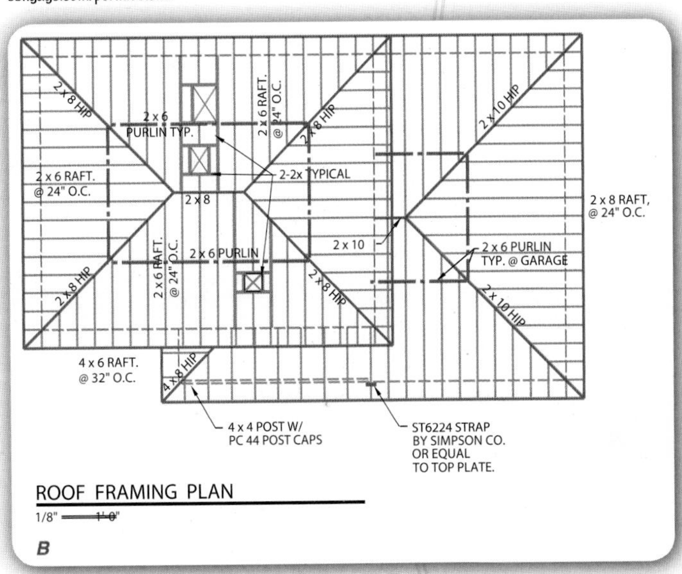

considerations needs such as shape, openings, and overhangs of the roof. Additional information such as construction details, types, sizes, and spacing are typically indicated as well.

Computer Aided Drawings (CAD) can generate several alternative solutions to potential problems. This is only going to improve with Building Information Modeling (BIM), a technology that will eventually provide better scheduling, material callouts, and communication of design to actual construction.

SOME BASIC FRAMING SYSTEMS IN COMMERCIAL CONSTRUCTION

Basic Foundation Systems for Commercial Construction

The difference between residential and commercial buildings is usually size and load. There are examples where a commercial building may be as small as a residential structure. In larger commercial structures, foundation systems are classified in two categories: shallow and deep. The foundation design is dependent upon several factors including the size of the building, the soils bearing capacity and the types of loads for which the building will be designed. Most low-rise buildings (three-stories or less) may be able to employ **shallow foundation** (Figure 13-37) such as spread footings, foundation walls, or mat foundation systems. With taller and heavier structures, typically deep foundations are needed. **Deep foundations** are usually built in order to support a heavier load from the structure above and come in the form of pile, caisson, and pile cap foundations (Figure 13-38). Depending on the structure and the soil conditions the foundation design may require a combination of deep foundation systems.

Shallow Foundation: helps support lighter loads and distributes them to the ground. They include spread footings for walls and columns.

A pile foundation transfers loads through poor soil conditions or water on to soils with better bearing conditions including rock. Piles are designed to anchor buildings not only against lateral loads but also against wind forces and waves. Concrete piles can be either pre-cast or cast-in-place piles. Pre-cast piles are cast to the desired length and cured before they are delivered. They are prepared using steel reinforcement and can be square or octagonal in cross-sectional shape. The reinforcement is added to allow the pile to resist the vertical loads and the bending moment caused by lateral loads such as wind. Once delivered to the jobsite, the piles are augered, driven, or jetted into the ground (Figure 13-39).

Figure 13-37: Shallow foundation systems.

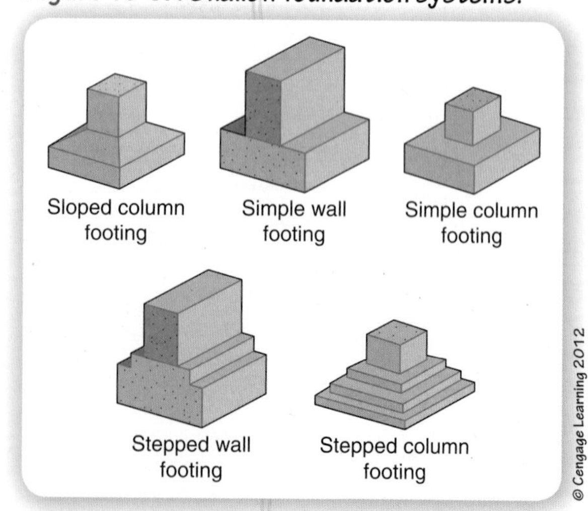

Cast-in-place piles are built by digging a hole in the ground and then filling it with concrete. The piles are constructed using two methods: cased and uncased. Cased piles are made by a steel casing drilled into the ground and then filling the casing with concrete. At the bottom of the pile a pedestal is placed which rests on the bedrock for stability

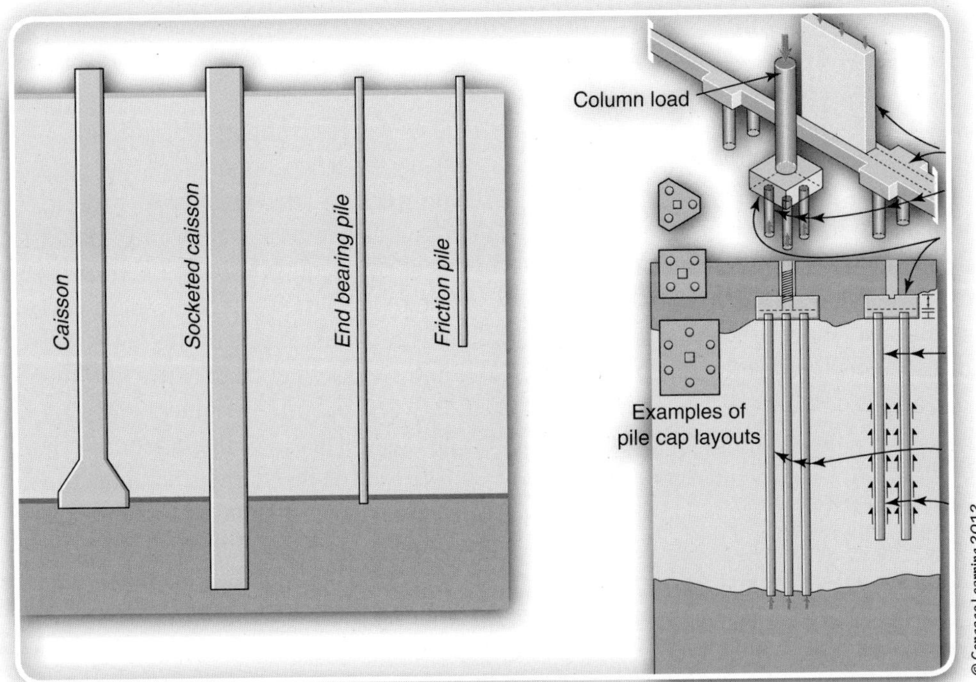

Figure 13-38: **Deep foundation systems.**

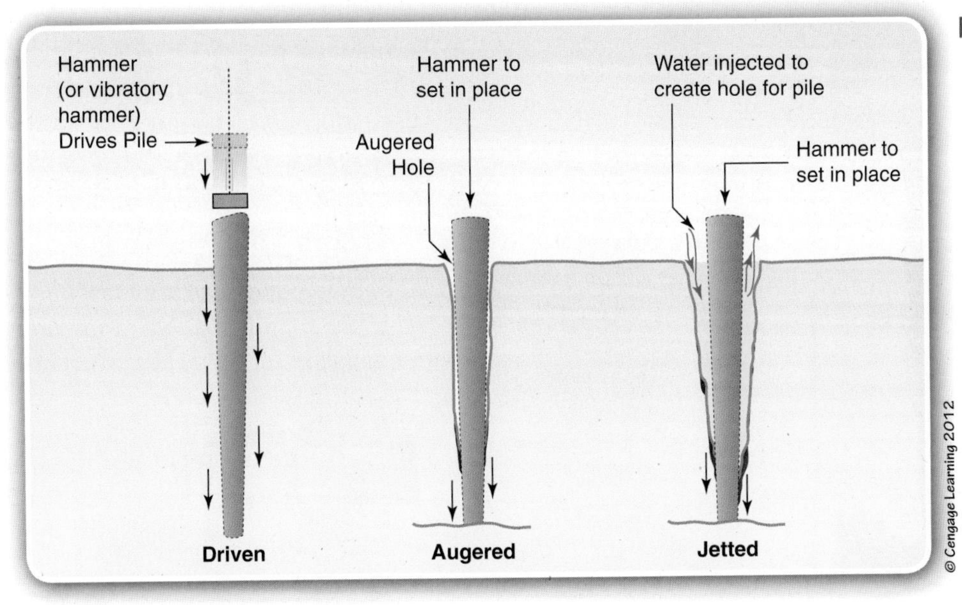

Figure 13-39: **Pre-cast piles.**

Deep Foundation:
helps support large loads and distributes them to the ground. They include pile, caissons, pile cap, or a combination of foundation systems.

Figure 13-40: **Cast-in-place piles.**

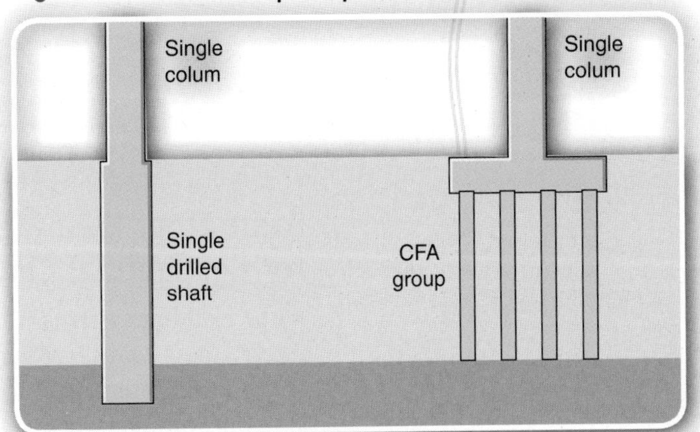

and additional bending moment support. Uncased piles are made by drilling a hole and then filling the hole with rebar and concrete. In the case where loads need to be distributed over a deep broad area a group of cast-in columns connected with a matt like foundation often referred to as a pile cap is used (Figure 13-40). Other types of pile foundations include the use of steel pipe or H-shaped steel or treated wood timber.

In the 1990s, engineers believed the Leaning Tower of Pisa was on the verge of collapse. About 10 years earlier, monitors noted that the tower was tilting by 0.04 in (1 mm) per year, raising the concerns of the structure's stability. In fact, British engineer Professor John Burland, who was overseeing the project to stabilize the tower, commented that it was hard to believe the tower had not collapsed. Computer models indicated the tower should have collapsed once it reached an angle of 5.44 degrees, but in 1990 it was at 5.5 degrees with the bell tower overhanging the base of the tower by 14.76 ft (4.5 m). What could engineers do to keep the Leaning Tower of Pisa (Figure 13-41) from leaning any further? The solution was to place lead weights on the north side of the tower and remove soil from the north side of the foundation area. Using a corkscrew drill, the tower began to settle into the dug area. Steel cables were placed not to pull it upright but to hold the tower from collapsing (Figure 13-42). By June of 2000, the tower corrected itself by 16 cm, thus returning to its original angle in 1870. On June 6, 2001, the tower had settled to its original angle in 1838. Why did the tower begin to tilt? It just so happens that the tower was built on soft sediments from a former riverbed, which explains why the tower was so unstable. Professor Burland described it like having a tower of bricks on a soft carpet: after it reaches a certain height it becomes shaky and will fall if taken any higher. Is the tower stable now? Professor Burland offers the following two scenarios: the first says the tower has been stabilized for good; the second says that the leaning process will start again—and that in two or three hundred years, the tower will be back where it was in 1990, on the brink of collapse.

Figure 13-41: **Leaning Tower of Pisa.**

Figure 13-42: **Saving the Leaning Tower.**

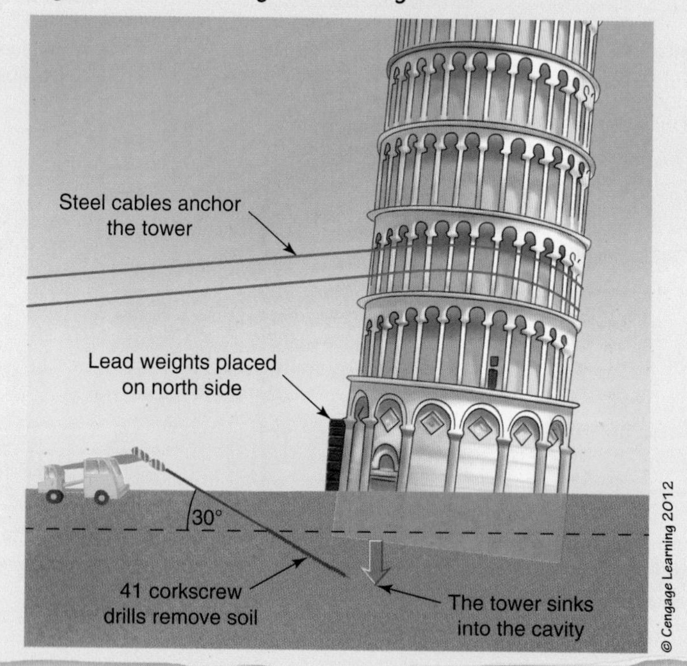
Steel cables anchor the tower
Lead weights placed on north side
30°
41 corkscrew drills remove soil
The tower sinks into the cavity

Basic Floor Framing Systems for Commercial Construction

Floor framing systems in commercial construction must take into consideration heavy traffic loads, resistance to wear and tear, greater loads over longer spans, and easy maintenance. Because they meet these needs, steel and concrete framing systems dominate the commercial building industry.

Steel framing systems include steel joist systems and steel beam and decking. A steel joist system is typically an open-web system. Two types of steel joist systems are used: structural steel joists consisting of hot-rolled members fabricated into a truss-like configuration and light-gauge steel joists fabricated from cold-formed steel members. The benefits of structural steel framing members such as joists systems generally include a high strength-to-weight ratio allowing for greater span; fabrication with standardized lengths, depths, and carrying capacities; and flexibility in installing mechanical and electrical systems. Because of its standardized shape and length, open-web steel joists lend themselves to a grid layout with the frame supporting a uniform load (Figure 13-43). See the table below for an estimate and preliminary sizing of structural steel joists members (Table 13-4).

Steel joist minimum bearing points to a wall or beam depend upon the type of support structure as illustrated in Figure 13-44.

Light-gauge steel joist floors provide a lightweight alternative for light loading conditions. They are able to have a clear span of up to 32 ft and come in varying shape and sizes. Connections are typically welded, bolted, or screwed and joists are spaced 16 in, 24 in, or 48 in on center based on structural design requirements. Additionally, punched holes in the web reduce the weight of the frame and allow for small electrical and plumbing lines. Some of the common connection details are typically to angles and C-shaped channels (Figure 13-45).

Unlike hot-rolled steel joist systems that are usually fabricated from standard steel members, cold-formed steel members are made from thin sheets of metal that are sent through a series of rollers to shape and form varying structural members. While there are some common shapes, such as C-shaped and angle shapes, other proprietary shapes have been formed by private companies. As a result, manufacturers of cold-formed steel trusses have conducted performance analysis on their steel truss design and provided span tables for their use. One such company, Trussteel, has listed its cold-formed steel span tables based on its truss designs, which are provided in Table 13-5.

Learn more about cold-formed steel by searching the Internet for a paper titled "An Overview of Cold-Formed Steel" by Helen Chen. Learn about how members are formed, what types of applications are there for cold-formed steel, and what the differences are between cold-formed steel versus hot-rolled steel.

Identify the physical properties present when using common construction materials in order to use the materials safely, effectively, and efficiently.

While some similarities between wood and steel skeletal frame construction, the steel frame is normally designed and constructed to carry heavier loads. Some of the loads may include uniform or distributed loads, meaning a load that is evenly distributed over a broad area such as the weight of the floor structure itself or a concentrated load such as a large air conditioning unit on top of an office building. In addition to these loads, lateral forces such as wind loads and live loads such as occupants in a building must be resisted by rigidly braced connections. Once the framing system is designed and engineered, the steel is normally cut, shaped, and drilled by fabricators prior to being shipped to the site. A common beam-to-beam or a beam-to-column connection is referred to as a framed (Figure 13-46) or moment connection (Figure 13-47). A wall supported connection typically consists of a beam pocket to allow a steel beam to be supported on the ends (Figures 13-48 and 13-49).

A complete floor system requires a subfloor, also known as *steel decking.* Basic floor types that are supported by steel beams include cast-in-place reinforced concrete, pre-cast concrete, and composite decking.

Figure 13-43: Structural steel joist flooring.

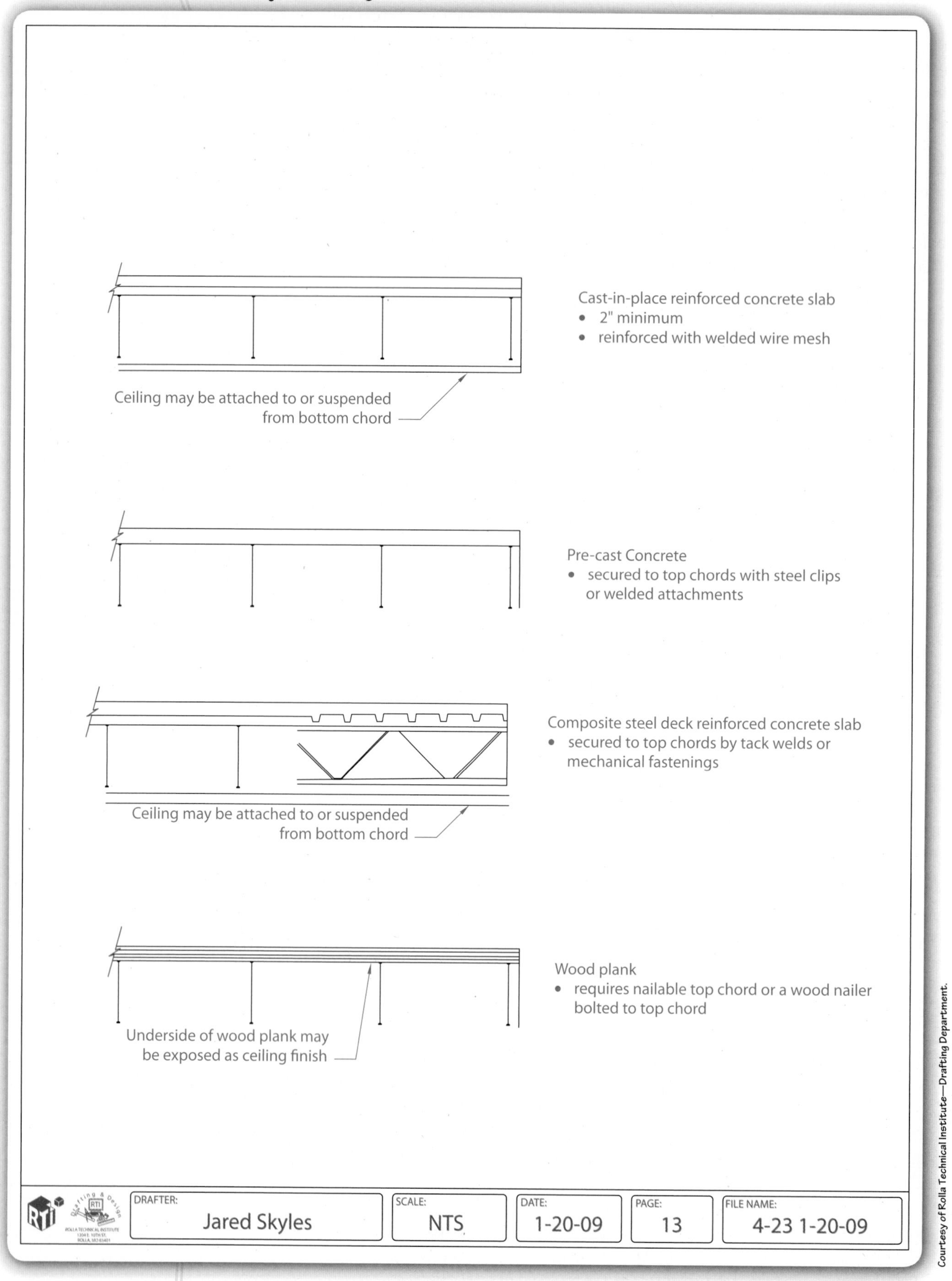

Courtesy of Rolla Technical Institute—Drafting Department.

Table 13-4: Steel Joist Span and Load (Steel Joist Institute)

STANDARD ASD LOAD TABLE
OPEN WEB STEEL JOISTS, K-SERIES

Based on a 50 ksi Maximum Yield Strength
Adopted by the Steel Joist Institute November 4, 1985
Revised to November 10, 2003 - Effective March 01, 2005

The black figures in the following table give the TOTAL safe uniformly distributed load-carrying capacities, in pounds per linear foot, of **ASD K-Series** Steel Joists. The weight of DEAD loads, including the joists, must be deducted to determine the LIVE load-carrying capacities of the joists. Sloped parallel-chord joists shall use span as defined by the length along the slope.

The figures shown in **RED** in this load table are the nominal LIVE loads per linear foot of joist which will produce an approximate deflection of 1/360 of the span. LIVE loads which will produce a deflection of 1/240 of the span may be obtained by multiplying the figures in **RED** by 1.5. In no case shall the TOTAL load capacity of the joists be exceeded.

The approximate joist weights per linear foot shown in these tables do not include accessories.

The approximate moment of inertia of the joist, in inches^4 is;

$I_j = 26.767(W_{LL})(L^3)(10^{-6})$, where W_{LL} = **RED** figure in the Load Table and L = (Span - 0.33) in feet.

For the proper handling of concentrated and/or varying loads, see Section 6.1 in the Code of Standard Practice for Steel Joists and Joist Girders.

Where the joist span exceeds the unshaded area of the Load Table, the row of bridging nearest the mid span shall be diagonal bridging with bolted connections at the chords and intersections.

ASD

STANDARD LOAD TABLE FOR OPEN WEB STEEL JOISTS, K-SERIES
Based on a 50 ksi Maximum Yield Strength - Loads Shown in Pounds per Linear Foot (plf)

Joist Designation	8K1	10K1	12K1	12K3	12K5	14K1	14K3	14K4	14K6	16K2	16K3	16K4	16K5	16K6	16K7	16K9
Depth (in)	8	10	12	12	12	14	14	14	14	16	16	16	16	16	16	16
Approx Wt (lbs/ft)	5.1	5.0	5.0	5.7	7.1	5.2	6.0	6.7	7.7	5.5	6.3	7.0	7.5	8.1	8.6	10.0
Span (ft) ↓																
8	550															
	550															
9	550															
	550															
10	550	550														
	480	550														
11	532	550														
	377	542														
12	444	550	550	550	550											
	288	455	550	550	550											
13	377	479	550	550	550											
	225	363	510	510	510											
14	324	412	500	550	550	550	550	550	550							
	179	289	425	463	463	550	550	550	550							
15	281	358	434	543	550	511	550	550	550							
	145	234	344	428	434	475	507	507	507							
16	246	313	380	476	550	448	550	550	550	550	550	550	550	550	550	550
	119	192	282	351	396	390	467	467	467	550	550	550	550	550	550	550
17		277	336	420	550	395	495	550	550	512	550	550	550	550	550	550
		159	234	291	366	324	404	443	443	488	526	526	526	526	526	526
18		246	299	374	507	352	441	530	550	456	508	550	550	550	550	550
		134	197	245	317	272	339	397	408	409	456	490	490	490	490	490
19		221	268	335	454	315	395	475	550	408	455	547	550	550	550	550
		113	167	207	269	230	287	336	383	347	386	452	455	455	455	455
20		199	241	302	409	284	356	428	525	368	410	493	550	550	550	550
		97	142	177	230	197	246	287	347	297	330	386	426	426	426	426
21			218	273	370	257	322	388	475	333	371	447	503	548	550	550
			123	153	198	170	212	248	299	255	285	333	373	405	406	406
22			199	249	337	234	293	353	432	303	337	406	458	498	550	550
			106	132	172	147	184	215	259	222	247	289	323	351	385	385
23			181	227	308	214	268	322	395	277	308	371	418	455	507	550
			93	116	150	128	160	188	226	194	216	252	282	307	339	363
24			166	208	282	196	245	295	362	254	283	340	384	418	465	550
			81	101	132	113	141	165	199	170	189	221	248	269	298	346
25						180	226	272	334	234	260	313	353	384	428	514
						100	124	145	175	150	167	195	219	238	263	311
26						166	209	251	308	216	240	289	326	355	395	474
						88	110	129	156	133	148	173	194	211	233	276
27						154	193	233	285	200	223	268	302	329	366	439
						79	98	115	139	119	132	155	173	188	208	246
28						143	180	216	265	186	207	249	281	306	340	408
						70	88	103	124	106	118	138	155	168	186	220
29										173	193	232	261	285	317	380
										95	106	124	139	151	167	198
30										161	180	216	244	266	296	355
										86	96	112	126	137	151	178
31										151	168	203	228	249	277	332
										78	87	101	114	124	137	161
32										142	158	190	214	233	259	311
										71	79	92	103	112	124	147

Figure 13-44: **Steel joist to wall or beam connection details.**

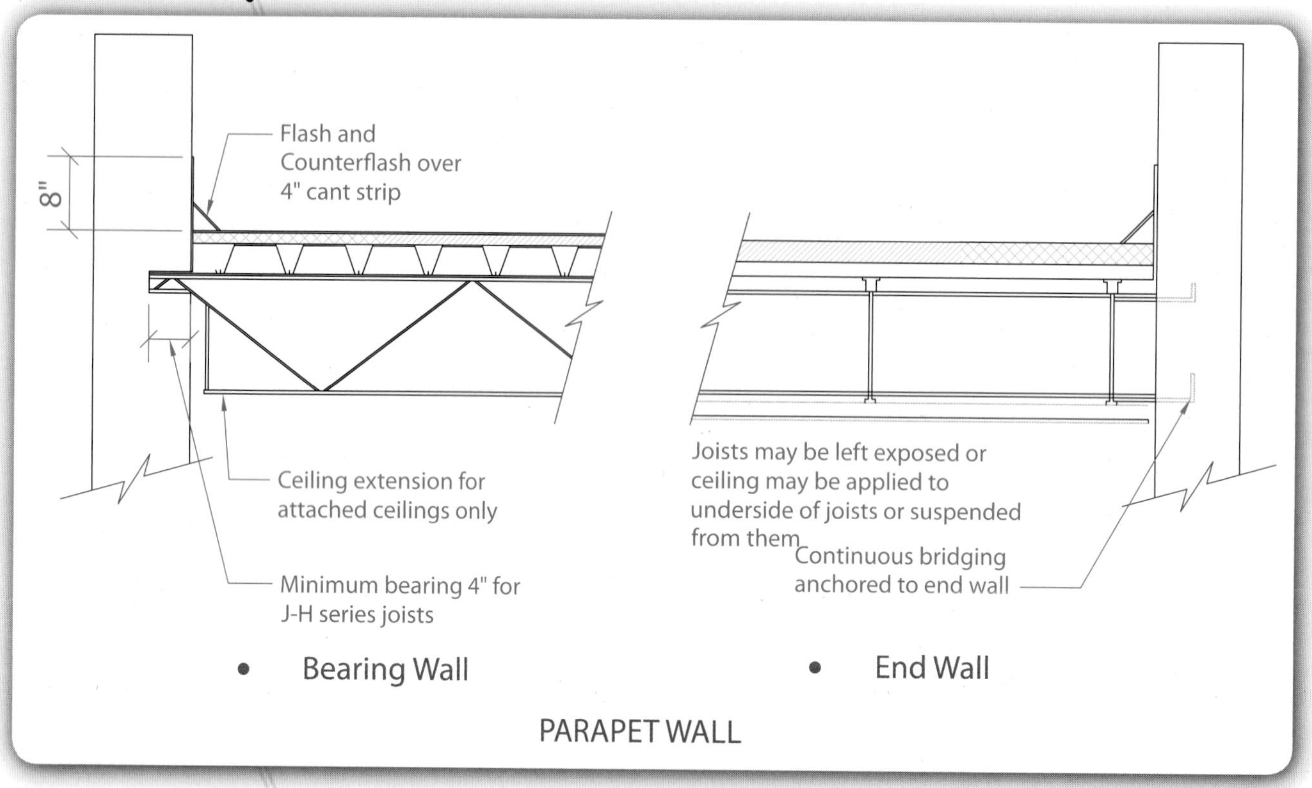

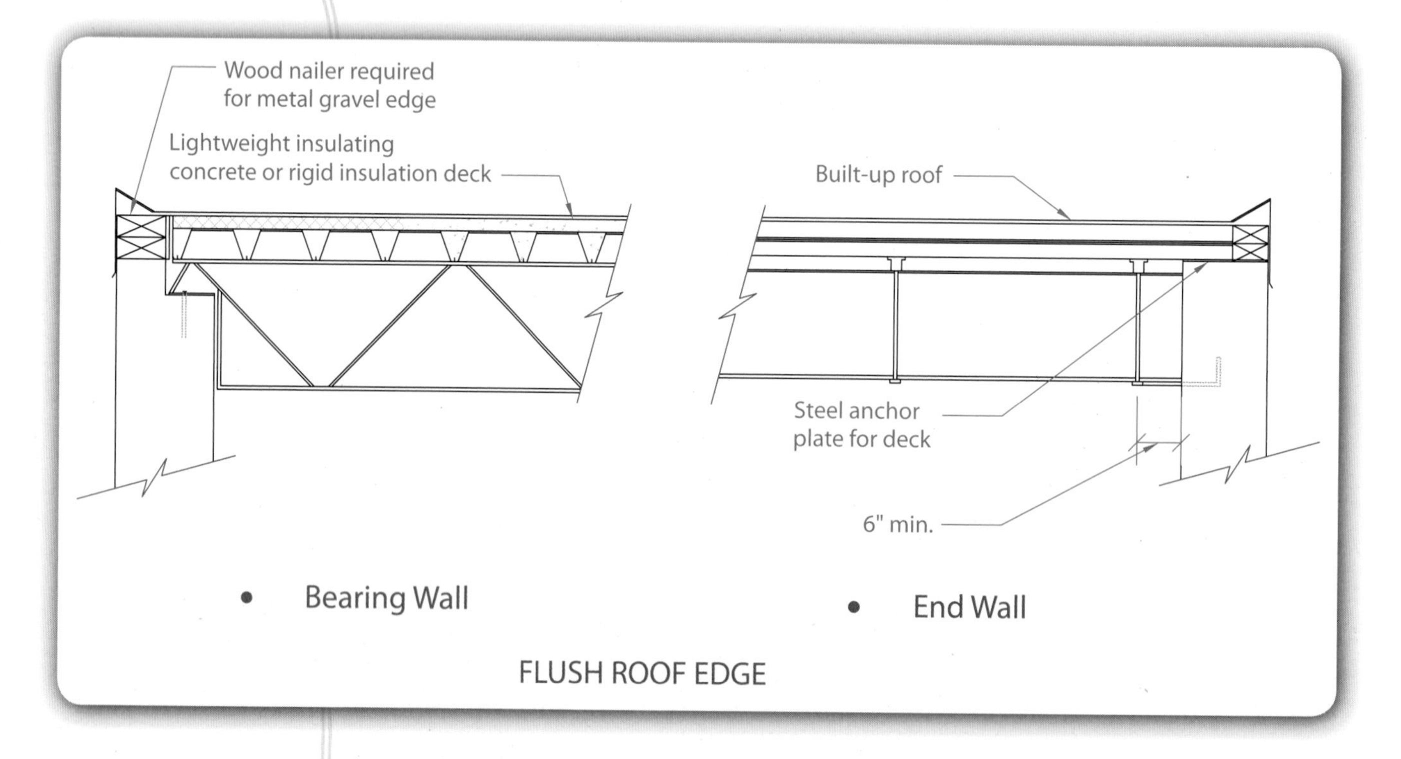

A pre-cast concrete deck is a series of pre-fabricated panels and comes in two types: solid core or hollow core. The span length for a solid panel 2 in thick is approximately 5 ft. The span length for a hollow core panel ranges from a 4 in thick panel reaching 12 ft in length to a 10 in thick panel reaching 32 ft in length. They are typically secured to the top **chords** using clips or welded attachments.

Figure 13-44: *(continued)*

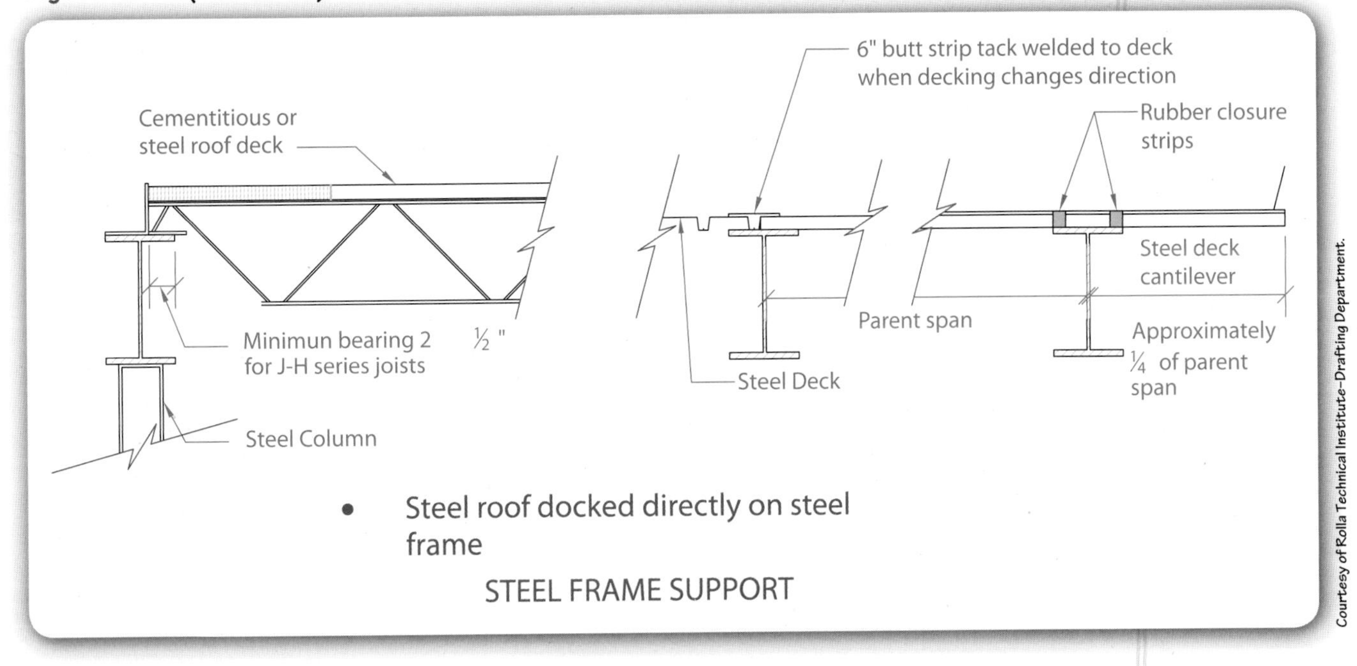

Figure 13-45: **Light-gauge steel joist to wall connection details.**

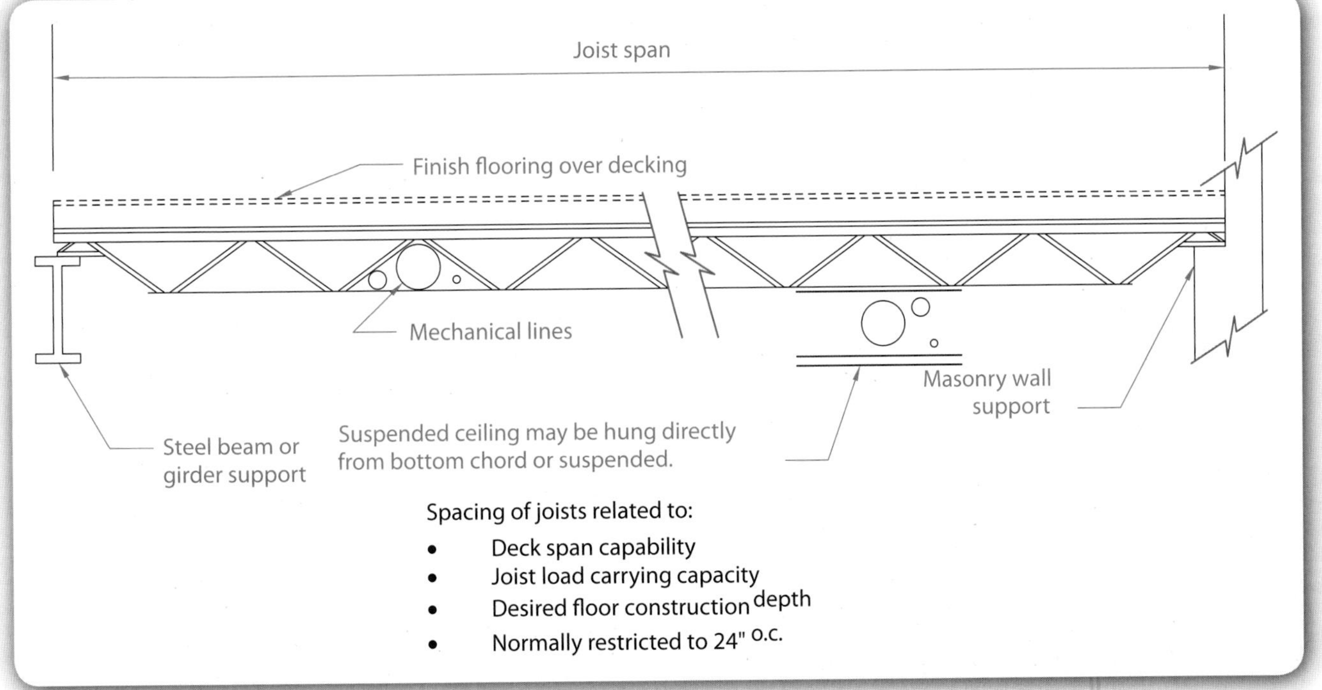

A **composite deck** is composed of corrugated metal deck, reinforcements, and concrete. With the frame in place, the corrugated metal deck is then secured to the frame by a series of tack welds or mechanical fasteners. Rebar or wire mesh is placed on top of the corrugated metal to add additional strength to the deck. Depending on the service of the floor, a lightweight concrete (110 lb/ft^3) as compared to normal weight concrete (150 lb/ft^3) is then poured and set in place. Once cured the composite deck works as a combined system (Figures 13-53 and 13-54).

Composite Deck: a combination floor framing system that is composed of a corrugated metal sheet fastening to the beams strengthened with steel reinforcing rods and concrete.

Table 13-5: **Cold-Formed Steel Floor Joist Span and Load (Trussteel)**

Chord Size / O.C. Truss Spacing	Load 1 40, 10, 5 psf								Load 2 80, 10, 5 psf							
	TSC2.75				TSC4.00				TSC2.75				TSC4.00			
Depth	12"	16"	19.2"	24"	12"	16"	19.2"	24"	12"	16"	19.2"	24"	12"	16"	19.2"	24"
12"	22	20	18	17	24	24	22	20	17	15	14	13	21	18	18	16
14"	25	22	21	19	28	27	25	23	19	17	16	15	24	21	20	18
16"	28	25	23	21	32	31	28	25	22	19	18	17	27	24	22	20
18"	31	27	26	23	36	34	31	29	24	21	20	18	29	26	24	22
20"	33	30	28	26	40	37	34	31	26	23	22	20	32	29	27	24
22"	36	32	30	27	44	39	37	34	28	25	23	22	34	31	29	26
24"	38	34	32	29	47	42	39	36	30	27	25	23	37	33	31	28

Chord Size / O.C. Truss Spacing	Load 3 125, 10, 5 psf							
	TSC2.75				TSC4.00			
Depth	12"	16"	19.2"	24"	12"	16"	19.2"	24"
12"	14	13	12	11	17	16	14	13
14"	16	14	14	13	20	18	17	15
16"	18	16	15	14	22	20	19	17
18"	20	18	17	15	25	22	21	19
20"	22	19	18	16	27	24	22	21
22"	23	21	20	18	29	26	24	22
24"	25	22	21	18	31	28	26	24

General Notes:

1) Spans Shown in charts are in feet.
2) Loads shown above are outlined as Top Chord Live Load (TCLL). Top Chord Dead Load (TCDL), and Bottom Chord Dead Load (BCDL).
3) Top and bottom chords designed assuming structural sheathing offers lateral restraint.
4) Deflection limits: Live Load - L/480
   Total Load - L/360.
5) Chases are to be located in center of span. Maximum chase width allowed is 24 inches.
6) Refer to TrusSteel Technical Bulletin TB971125 for TrusSteel floor truss design criteria.
7) Designs may include multiple gauges for top and bottom chords as determined by the designer using Alpine's steelVIEW engineering software. Maximum chord gauges are 43 mil (18 GA) for the TSC2.75 chord and 63 mil (16 GA) for the TSC4.00 chord.

Figure 13-46: **Steel beam framed connection.**

Courtesy of Donald Block.

Figure 13-47: Steel beam moment connection detail.

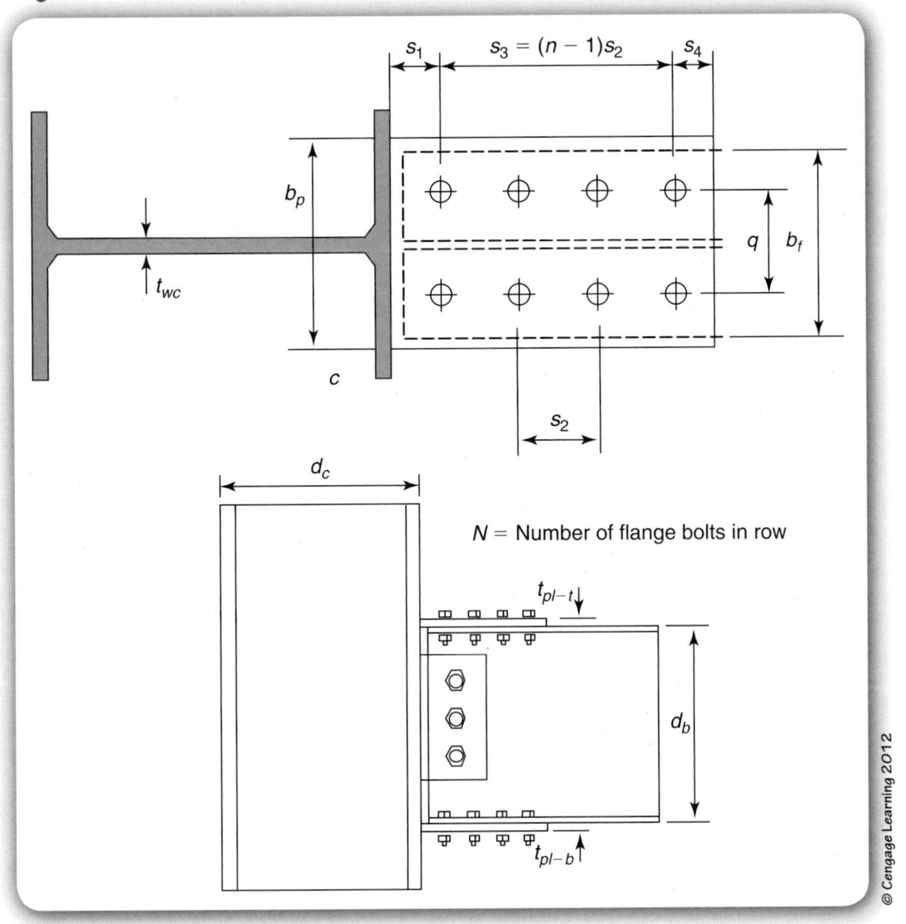

Figure 13-48: Beam pocket.

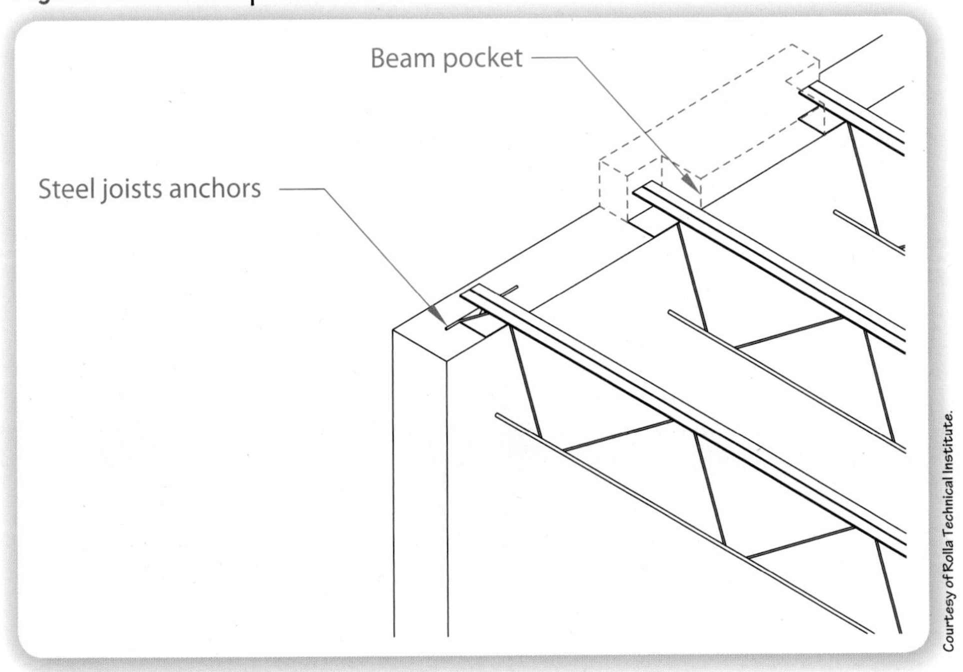

Figure 13-49: Sectional view of a steel joist to masonry connection using a beam pocket.

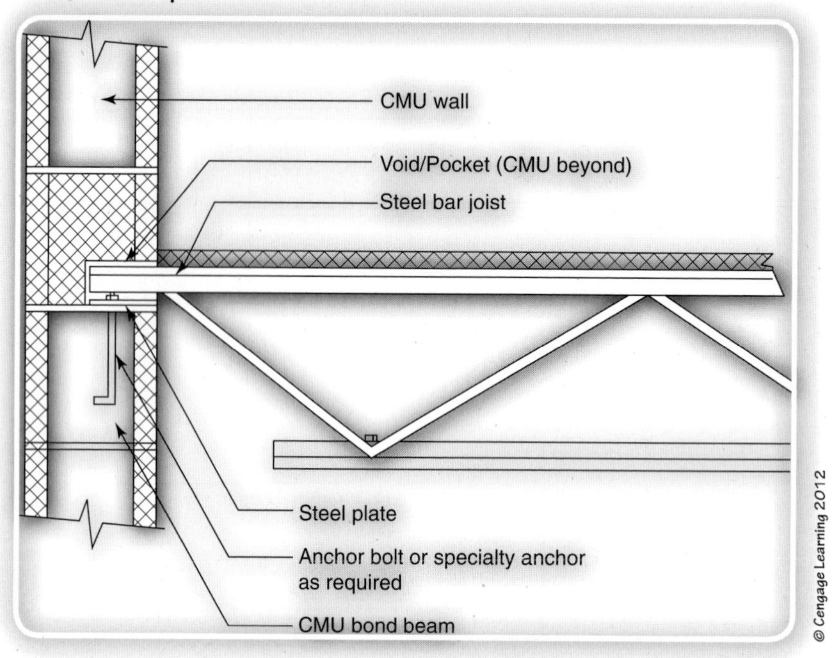

Point of Interest

Skyscrapers evolved in the United States as an American architectural icon in the early twentieth century thanks to several technological breakthroughs. One of the main advances was the development of steel framing. Prior to this technique, buildings relied on exterior and interior walls to support the floors. The taller the buildings were built, the thicker the walls had to be to support the massive weight of the building. A good example of a load-bearing masonry skyscraper is the Monadnock Building in Chicago (Figure 13-50). At the base of the building, the walls are reportedly 4 ft to 6 ft thick. Additionally, the load only allowed for small deep window and door openings. In contrast the Wainwright Building in St. Louis (Figure 13-51) uses a steel framing system that transfers the loads of the building to its beams and columns (Figure 13-52) and then to its foundation system. The exterior and interior walls are non–load-bearing allowing for large openings such as storefront windows and double swing doors. The framing system allows for modularity and future skyscrapers (as pictured) were allowed greater heights and larger openings.

Figure 13-50: Monadnock building.

Point of Interest (continued)

Figure 13-51: **Wainwright building.**

Courtesy of the Library of Congress, HABS MO, 96-SALU, 49-4.

Figure 13-52: **A drawing from William Le Baron Jenny's Chicago Fair Store showing steel framing clad in terra cotta tile.**

From *Industrial Chicago*, Volume 2, Chicago: Goodspeed Publishing Company, 1891.

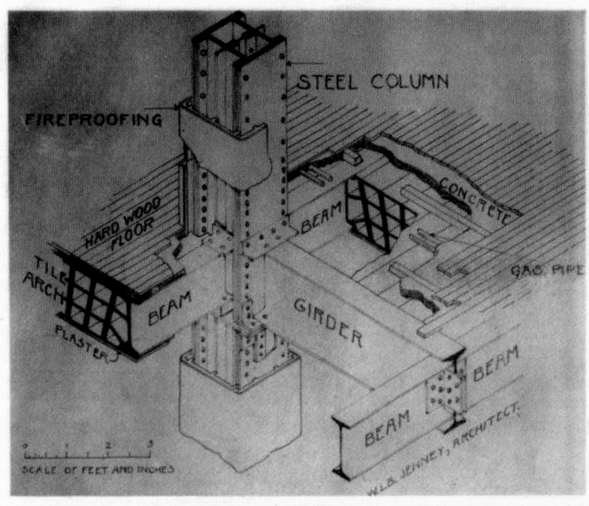

Besides steel, concrete framing systems typically fall into two categories: reinforced and pre-cast. Reinforced concrete is commonly cast in various size columns, beams, and slabs. Some of the most common floor framing systems include the one-way slab, two-way slab, one-way ribbed slab, two-way waffle slab, two-way flat plate, and two-way flat slab.

Figure 13-53: **Composite steel deck and framing.**

Courtesy of Donald Block.

Figure 13-54: **Composite floor deck.**

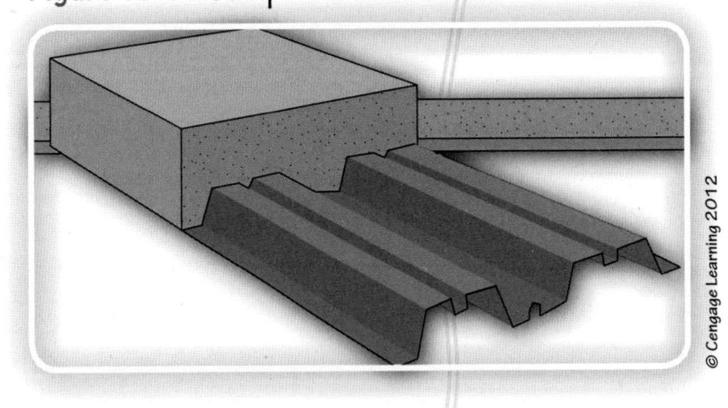

Point of Interest

Marina City (Figure 13-55) is a mixed-use residential/commercial use building complex built along the Chicago River in Chicago's Loop district. The complex consists of two 65-story residential towers, an auditorium building, a mid-rise hotel building, and a small marina. Once ranked as the tallest residential building constructed out of reinforced concrete, this complex was "a city within a city." It's radial design incorporated a core that contained the hall wrapped around the elevators with 16 pie-shaped apartments branching out from the hallway. The unique design allowed for living spaces to be farther out of the pie-shaped design with service areas such as kitchen and bath to be located near the core. During construction, the core of the building was formed first with each floor being constructed in place. Constructed at a rate of one floor per day, each core contains five elevators and two sets of stairs (Figure 13-56).

Figure 13-55: **Marina City Towers.**

Figure 13-56: **Marina City Tower under construction.**

A one-way slab is supported on two sides, usually by a beam, a wall, or a combination of both. This type of design is usually used for medium to heavy loads over short spans. For longer span support, a two-way slab provides support on all four sides. A one-way ribbed slab is used for light to medium loads over medium spans. A two-way waffle slab not only provides greater strength compared to its one-way ribbed system it allows for greater appeal. A two-way flat plate is a simplified slab on columns and no beam support and is typically designed for low to medium loads. For medium to heavy loads a two-way waffle slab (Figure 13-57) provides drop plates and column capitals as additional support.

Figure 13-57: Two-way waffle slab floor system.

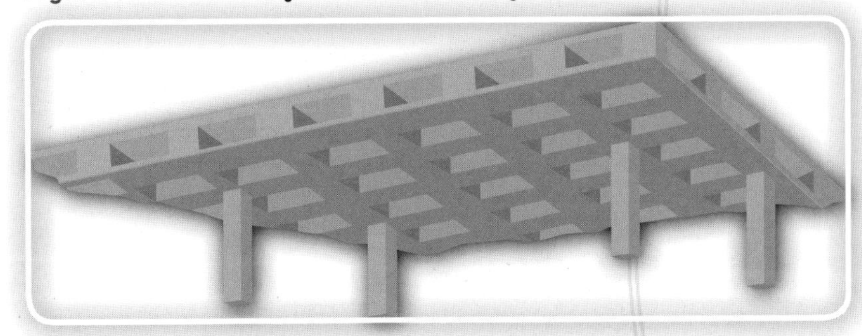

Pre-cast concrete plank floor systems offer several advantages compared to reinforced concrete floor systems, including pre-fabrication at the plant, consistent quality of strength and durability, shorter assembly time, and less labor. The disadvantages include limited flexibility in overall layout and limited span length. Connections between pre-cast floor systems for walls typically require the use of ties embedded in poured concrete or embedded fasteners designed and fitted at time of fabrication (Figure 13-58).

Other types of pre-cast concrete are formed for use in both one-way and two-way slabs (Figure 13-59). These systems often employ a process known as pre-stressing. During fabrication a series of tendons (cables) are strategically placed to provide tensile strength to the overall member. Tension is added to the tendons through the use of jacks and anchors. The concrete is then poured and allowed

Figure 13-58: Pre-cast concrete floor systems.

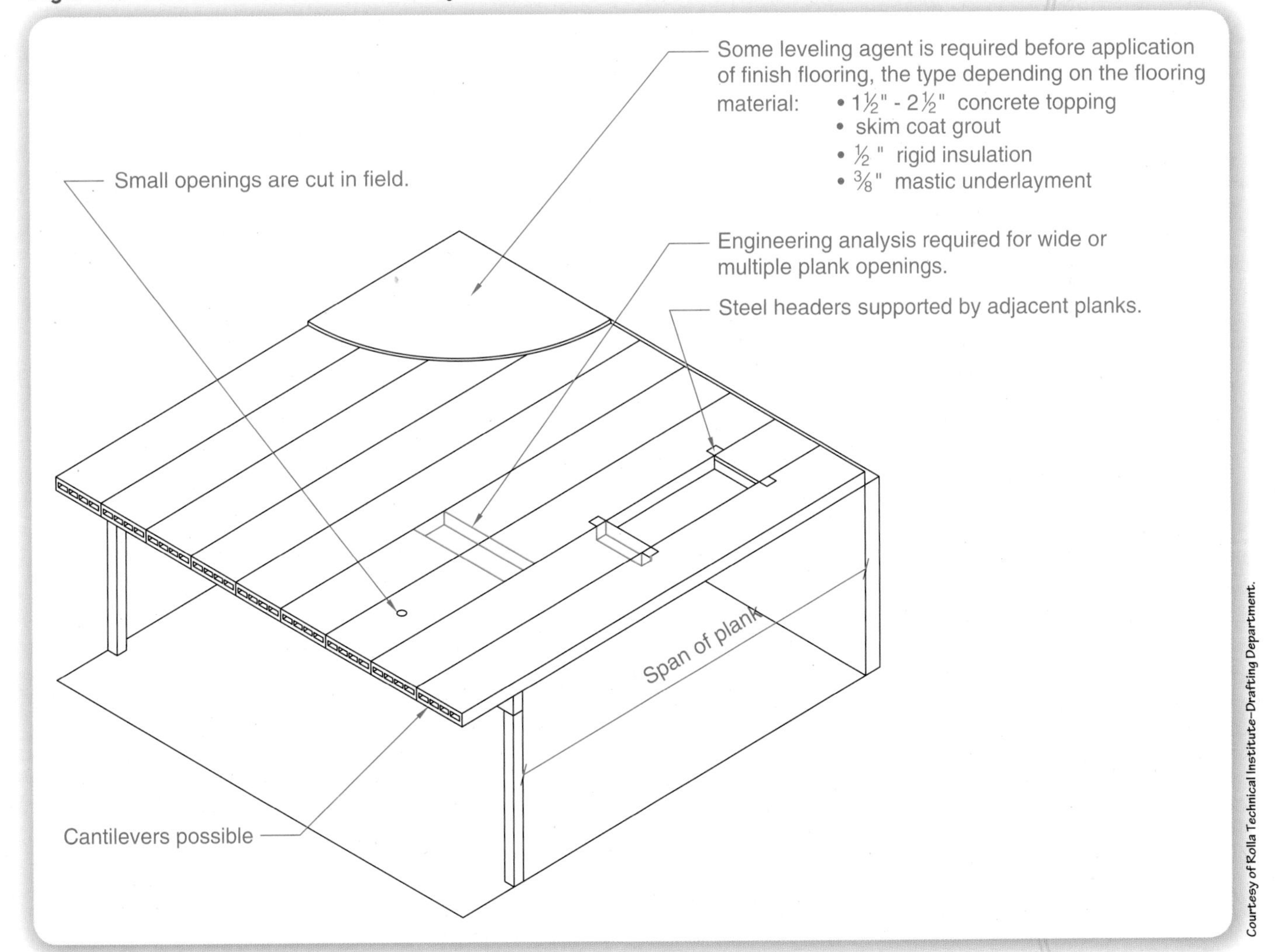
Some leveling agent is required before application of finish flooring, the type depending on the flooring material:
• 1½" - 2½" concrete topping
• skim coat grout
• ½" rigid insulation
• ⅜" mastic underlayment
Small openings are cut in field.
Engineering analysis required for wide or multiple plank openings.
Steel headers supported by adjacent planks.
Span of plank
Cantilevers possible

Figure 13-59: **Pre-stressed floor systems.**

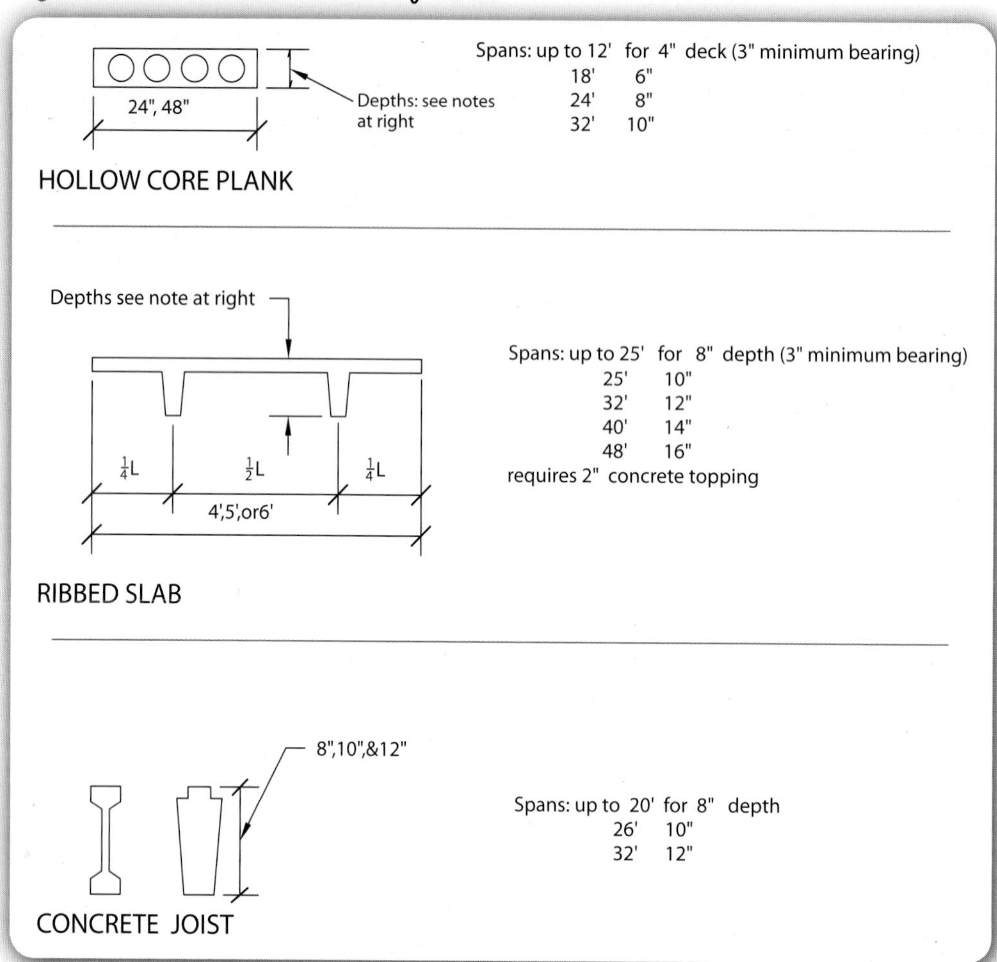

Courtesy of Rolla Technical Institute–Drafting Department.

to cure. Once complete, the formwork is removed, and the structural member is ready for use. As a result of pre-tensioning the tendons, the structural member has greater strength resulting in increased load-carrying capacity. This increase in turn allows for smaller and lighter members to carry the same load as a normal reinforced concrete member.

In a similar method, post-tensioning, a concrete beam (Figure 13-60) is set and cured during fabrication. Once hardened, a series of tendons (cables) are placed in various predetermined holes in the length of the member. With a anchor on one end and a mechanical jacking device on the other end, the tendon is pulled to a calculated tensile stress and anchored. The effect increases the load-carrying capacity in the structural element. Post-tensioning provides greater carrying capacity for samller structural elements and ineffect reducing amount of materials and reducing a structures overall weight.

Basic Wall Framing Systems for Commercial Construction

Wall framing systems for commercial buildings are typically classified as one of two types: load-bearing and non–load-bearing. As the term implies, load-bearing walls are walls that carry a load from the structure above. In low-rise commercial buildings, load-bearing walls may be constructed using metal stud walls, steel and reinforced concrete framing, or a masonry wall.

Similar to wood stud walls, metal stud walls can be constructed as a load-bearing assembly by using heavier gauge steel members at spaced at 12 in, 16 in, or 24 in on center. Wall sheathing and finish options are similar to wood stud construction. Corner connections between wall assemblies require additional blocking. Channel bridging and tension straps provide additional bracing to resist lateral loads (Figure 13-61).

Another type of wall framing system for commercial buildings is masonry. In low-rise commercial buildings, masonry is usually constructed with concrete masonry units and the cavities are filled with rebar and concrete. This type of system provides a solid wall framing system with extra reinforcements through the use of rebar and concrete.

Steel and reinforced concrete framing are two other load-bearing types of construction and are similar to post-and-beam wall systems. The difference lies in the strength of the material; stronger material allows for greater spans and larger loads. Panels in between columns and beams are non–load-bearing and may be assembled on-site using light-gauge metal studs and finishes or the panels may be pre-fabricated by a manufacturer and then placed in between supports with mechanical fasteners.

In high-rise structures, the use of a skeletal frame systems such as columns and slabs allows walls to serve as non–load-bearing partitions. In the case of where masonry wall systems are used, these typically serve two purposes: first to provide separation of spaces and second to provide protection from the elements within the building such as fire and weather (Figure 13-62). Areas that typically require masonry construction include mechanical rooms, paths of egress, and areas where safety from man-made and natural disasters are required.

Over the past century, wall systems have evolved thanks to advances in structural framing systems. These advances have led the way to more and larger openings. The result is a wall system that is lightweight and serves as a separation of spaces or a barrier to the

Figure 13-60: Post-tensioning.

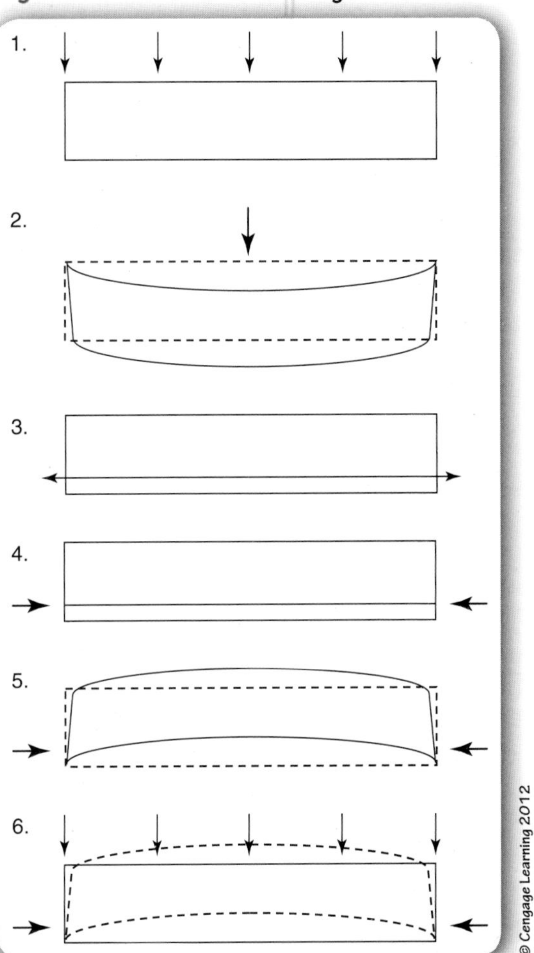

© Cengage Learning 2012

Figure 13-61: Metal stud wall system.

© iStockphoto.com/Phil Augustavo.

Figure 13-62: Masonry wall system.

© iStockphoto.com/Mel Stoutsenberger.

Figure 13-63: *Curtain wall system under construction (CityCenter, Las Vegas, Nevada).*

Courtesy of CityCenter.

Curtain Wall Systems: the exterior wall as a non-load-bearing wall that is adhered to the building's framing system to provide protection from the elements.

elements. This is commonly referred to as a **curtain wall system**. This system typically attaches to the frame of the structure with a series of mechanical fasteners. Several different types of curtain wall systems are shown below (Figure 13-63).

Basic Roof Framing Systems for Commercial Construction

Roof framing systems for commercial construction are similar to those for residential construction in a lot of ways, including roof designs: flat, gable, or shed roof. Another similarity is the type of framing system used, such as rafter roof, truss, and joist roof systems. The exceptions for the two types of construction include type of materials used to support the frame and the type of cover applied to the roof. Large scale-commercial buildings typically use steel and concrete for roofing systems (Figure 13-64). The strength of the materials allow for greater spans and heavier loads.

Figure 13-64: *Commercial roof framing systems.*

Courtesy of City Center, Las Vegas, Nevada.

Depending on the size and load to be carried, short span roofs are typically designed using light-gauge steel rafter or truss systems (Figure 13-65). The light gauge steel allows for ease of assembly, the members are placed across the supports and easily fasten to an anchoring system. Steel or wood decking is fastened to the top chord of the roofing members.

In the case where a roof system is required to support longer spans and medium loads, an open web steel joist system is preferred (Figure 13-66). These members are typically fabricated at the manufacturer using hot-rolled steel members. A standard open web steel joist 8 in to 30 in in depth can span up to 60 ft in length. A long span joist 18 in to 48 in in depth can

Figure 13-65: Light-gauge steel roof framing systems.

© iStockphoto.com/James Pauls.

span up to 96 ft in length while a long span joist 52 in to 72 in can span up to 144 ft. The open web allows for various electrical and mechanical systems to be run though the framing system. Because of the standard sizes and lengths of the open-web joist a grid layout is most economical. The structural system is designed to be efficient under uniform loads, but may be engineered to support concentrated loads over panel points.

Structural steel framing systems are normally designed to carry relatively heavy loads over long spans (Figures 13-67 through 13-70). In areas of high lateral and wind loads, the steel frame provides the necessary resistance due to its rigid braced connections. Since structural steel is difficult to handle on-site, members are typically prepared by fabricators off-site according to the designer's specification. A structural grid layout is commonly used for this type of framing system.

Figure 13-66: Open-web steel joist framing systems.

© iStockphoto.com/Robin Vondrak.

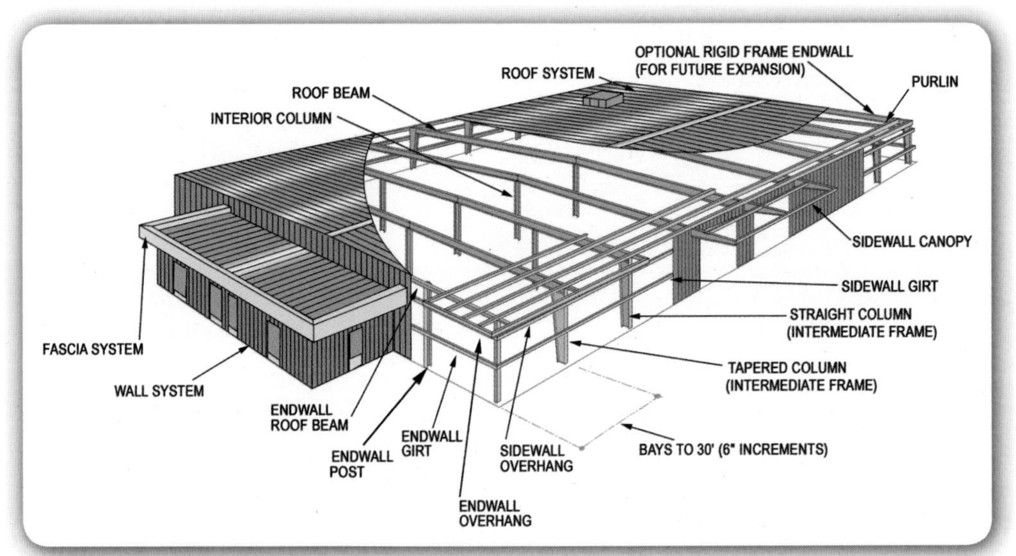

Figure 13-67: Wide-span structural steel roof framing systems.

Courtesy of Butler Manufacturing, a division of BlueScope Buildings North America, Inc.

Figure 13-68: Open web steel trusses and solid web steel beam roof framing system.

Courtesy of Butler Manufacturing, a division of BlueScope Buildings North America, Inc.

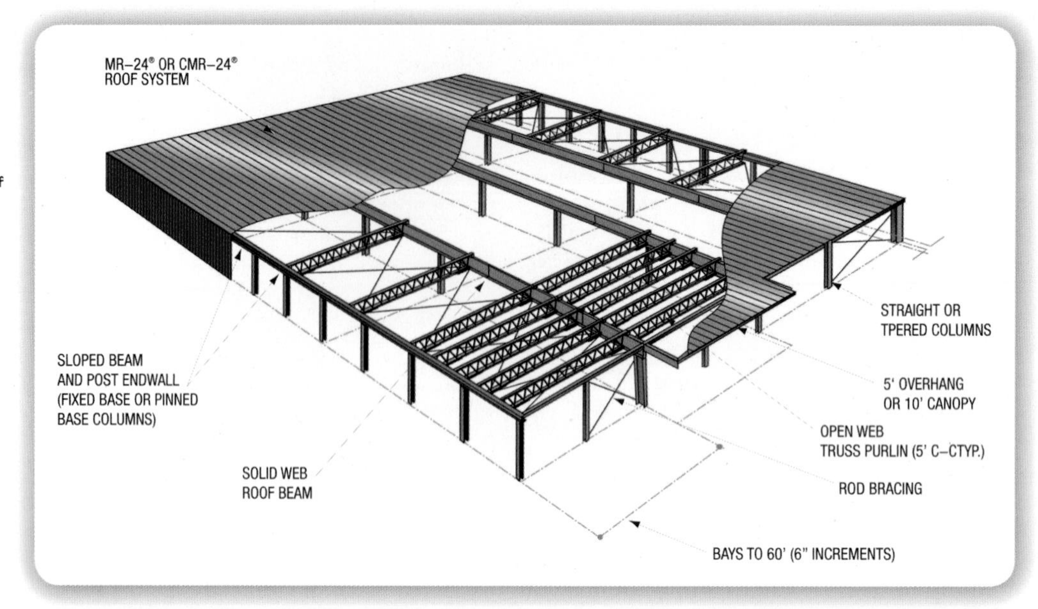

Roof decking is like floor decking in commercial buildings; the only exception is that in an effort to minimize the load, roof decks may use a plywood sheathing or lightweight concrete set to a corrugated metal deck. Another type of roof decking similar to floor decking is the use of pre-cast concrete planks. Pre-cast panels are placed above the roof framing system and secured to the top chords with steel clips.

Another type of roof framing system is called a composite roof deck. It combines a serious of corrugated steel sheets welded to the structural frame. A series of rebar is placed to provide reinforced with concrete slab to provide the roof greater strength and load-bearing capacity. The decks are typically welded to the top chord of the trusses and rebar for extra strength is placed prior to the concrete being poured and set in place (Figure 13-71).

Figure 13-69: Multi-story structural steel framing system.

Courtesy of Butler Manufacturing, a division of BlueScope Buildings North America, Inc.

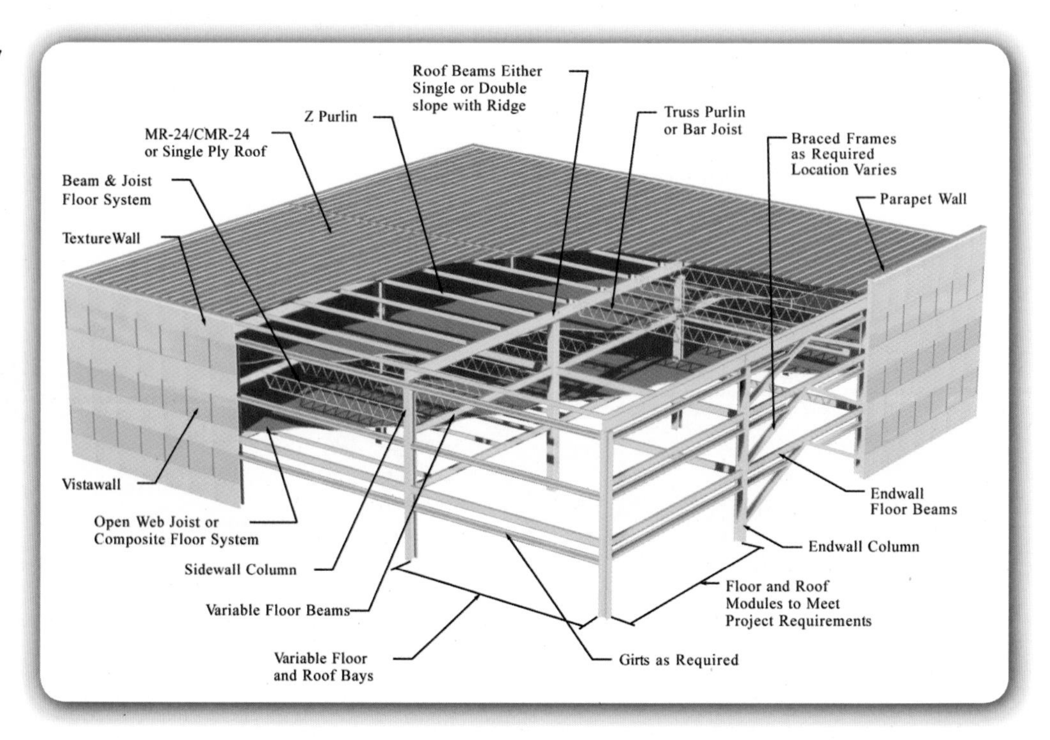

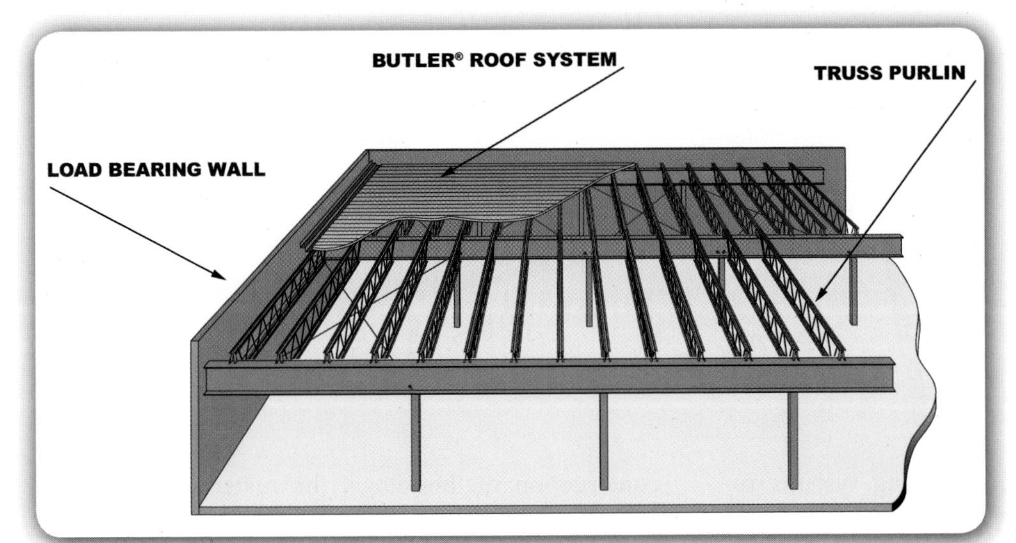

Figure 13-70: Hard wall structural steel framing systems.

Courtesy of Butler Manufacturing, a division of BlueScope Buildings North America, Inc.

Figure 13-71: Commercial roof decking systems.

Courtesy of Butler Manufacturing, a division of BlueScope Buildings North America, Inc.

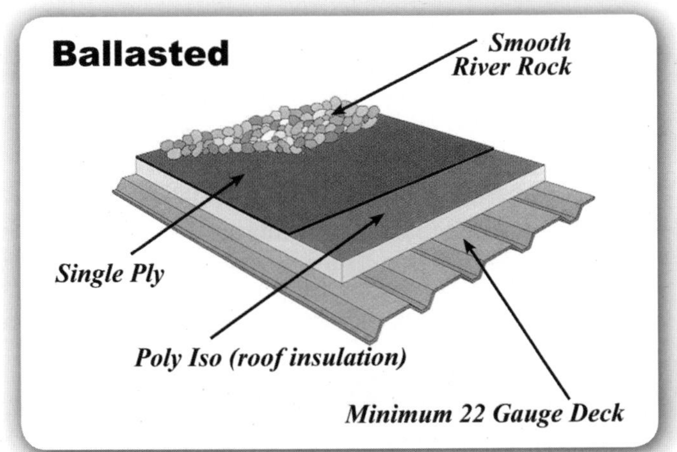

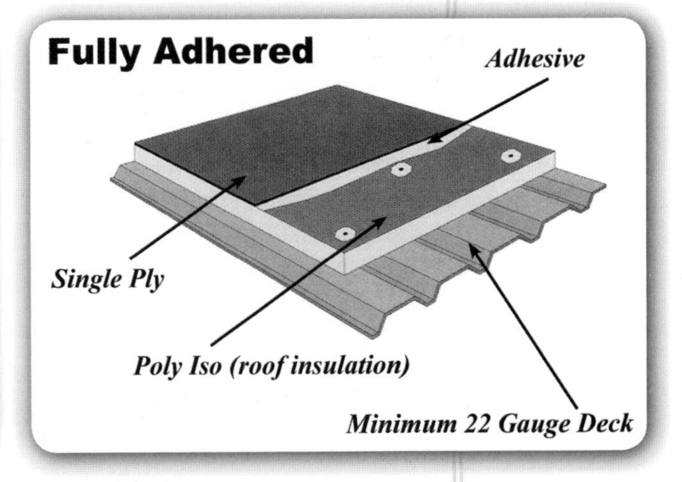

Fun Facts

The designers of the world's tallest building had to overcome several major obstacles including framing and materials systems. Learn more about the buildings structural design by searching the Internet for "Fun Facts about Burj Dubai Structural System" or visit http://gconnect.in/gc/lifestyle/get-ahead/burj-dubai-the-supertall-skyscraper.html.
Answer the following questions:

1. What type of foundation system did they use and why?
2. How much concrete was used?
3. What type of weather conditions was the building designed?

SUMMARY

Residential and commercial framing systems require a vast knowledge of what structural loads are being supported or resisted. Understanding the site conditions in terms of weather and soil will typically impact the building's design, ultimately influencing its structural framing system. As discussed, the framing system considers everything from the foundation to the roof.

In this chapter, we have learned the differences between shallow foundations and deep foundations, examined the use of various types of floor framing systems and the difference in sizes and strength of lumber to engineered members, and discussed the type of materials used in floor framing systems for commercial and residential applications.

While the wall framing systems for residential and commercial construction were similar in construction methodology, the materials and applications differ. Wood stud framing systems are typically reserved for residential construction. Metal stud framing and curtain wall construction are typically used in commercial construction.

Roof systems, like walls in general, share the same methodology when considering residential to commercial applications. The difference once again is the material used for constructing these assemblies. Wood roof framing systems are typically reserved for residential construction. Metal framing systems are typically used in commercial construction.

While the methodology and materials may differ between residential and commercial construction, the challenge is the same: design a building to withstand the forces and loads that are being applied to it.

BRING IT HOME

Visit the Solar Decathlon website at http://www.solardecathlon.org and from a previous decathlon pick at least two team's sets of drawings from the technical resources category. Review both sets of drawings by comparing both sets of framing plans for each home.

1. From the set of drawings of one house respond to the following questions:
   - What type of foundation system was the house using?
   - What type of floor framing system design was the team using?
   - What type of framing system was used to construct the walls and roof?
2. From the second set of drawings respond to the same set of questions.
3. Compare and list the difference between the two homes.

EXTRA MILE

Prior to the Burj Dubai being declared the tallest building in the world, Taipei 101 had the title. Compare the two skyscrapers in terms of the following:

1. Site conditions in terms of location, weather, and soil
2. What type of foundation system was employed by both buildings?
3. Why did they employ these foundation systems? List the key factors.
4. What type of skeletal framing did they employ for the walls and roof?

CHAPTER 14
Structural Systems: What Makes a Building Stand?

GPS DELUXE

Menu

START LOCATION | DISTANCE | END LOCATION

Before You Begin

Think about these questions as you study the concepts in this chapter:

1. What forces act on a building and must be taken into account when designing a structure?
2. How does a building's structural system resist the loads to which it is subjected?

3. How can you determine the forces applied to the individual structural components?

4. What is the difference between a material property and a section property, and how do they each affect the design of structural components?

5. What are Allowable Stress Design and Load Resistance Factor Design and what is the difference between them?

6. What factors must be considered when designing a tension member?

7. What factors must be considered when designing a compression member?

8. How do you size a beam?

9. How do you design a foundation?

INTRODUCTION

As with all aspects of building design, the goal of the structural design is to safely satisfy the client's needs while complying with the building codes and local regulations. You learned about many building systems in the previous chapter: floor, roof, wall, and foundation. These systems work together to provide the structural framework of the building. The structural framework acts as the skeleton—it provides the form, strength, and stability necessary to support all of the loads that will affect the building. When you consider the forces that act on a building, you will realize that a wide range of loads must be included in the structural design. The weight of the people, furniture, and equipment and the weight of the building itself are probably the first loads that come to mind. However, the building will also need to support all of the loads applied by natural phenomenon such as snow, wind, floods, earthquakes, and the pressure from soil. The goal of the structural design is to safely transfer all of these loads to the ground without excessive movement of the structure while also maintaining an environment that is serviceable for the people and equipment housed in the building. Of course, the second goal is to achieve all of this in the least expensive way (see Figures 14-1-14-3).

A structural engineer must carefully analyze the building to quantify the applied loads and then determine how the building systems will work together to transfer the loads to the ground. Most loads applied to the building will pass through multiple structural components on their path to the supporting soil, and each component through which they pass must be designed to safely carry and

©iStockphoto.com/rognar.

Figure 14-1: *Vimanmek Palace (The Palace in the Clouds) is a former royal palace. Built in 1900 from wood, it is the largest building in the world built of golden teak.*

transfer those loads. Of course, as with other aspects of the design, building codes and regulations provide design guidance and specify requirements for structural design that must be met.

In this chapter, you will learn about various types of loads that are applied to buildings and how to determine the magnitude of load applied to each component of the structure as the load follows its path to the ground. You will also be introduced to the basics of analysis and design of some common structural components including tension members, compression members, beams, and foundations. Although much of the content of this chapter centers around calculations, structural engineers must use their knowledge, experience, ethics, personal judgment, and creativity to find a comprehensive approach to safely support the building. Structural design is much more than just crunching numbers.

Figure 14-2: *Burj Dubai, the world's tallest tower, has a structure composed mainly of reinforced concrete. Special concrete mixes were used to withstand the extreme forces imposed by the massive building.*

Figure 14-3: *The John Hancock building in Chicago incorporates an innovative steel frame design with exterior x-bracing. Approximately 46,000 tons of steel were used in the construction.*

LOADS

To put it bluntly, you are a load—your mass creates a force on the surface on which you are standing or sitting. When you enter a building, the structure, in order to remain standing, must support your load and all of the other loads that are applied to the structure. The floor itself is a load, as are the walls, the heating and air conditioning ducts and equipment, the furniture, and anything else that has mass and is attached to or bears on the building. In addition, if the wind is blowing, the air will create forces, also called loads, on the outside of the building. All of these loads will make the building bend and move—just like a tree bends and sways in the wind. Bending and movement is a normal and expected reaction to loads. But if the skeleton of the building is not designed to safely withstand all of these forces and limit the movement to a reasonable (if not imperceptible) level, the building may not perform within acceptable limits. At the extreme, the building may break or collapse.

Obviously, a structural engineer needs to carefully evaluate all of the possible loads that will affect a building—failure to do so could be disastrous. Every anticipated force on a building must be considered when the structure is designed. The type and magnitude of loads that are included in the design of a structure depend on the location of the structure, the type of building being designed, and the size and shape of the building. Some of these loads are fairly easy to determine; for example, we can accurately estimate the weight of a floor or obtain the weight of an air conditioning unit from the manufacturer. But most loads are more difficult to pin down. It is difficult to precisely predict the wind speed during a future hurricane or the force created by dancing partygoers packed like sardines onto a balcony. Often, structural engineers must design structures to resist loads that have the potential to affect the structure but will never actually occur. For instance, a structural engineer may design a building to support 5000 people in a concert hall that has only 2500 seats or may design a building to resist a category 5 hurricane that will never occur during the life of the structure. Because the failure of a structure could have catastrophic effects on the health and safety of the people who use the building, the loads used to design structures are *conservative;* that is, the loads used for design are larger than the forces that are expected to occur in order to provide a level of confidence that the structure can withstand the unexpected.

Dead Load:
the weight of construction materials used in the building, such as the roof, walls, floors, cladding, and the weight of fixed service equipment, such as plumbing, electrical feeders, heating, ventilating and air conditioning systems, and fire sprinkler systems.

Dead and Live Loads

Loads that result from the force of gravity are referred to as **gravity loads**. Anything that has mass and is attached to or is supported by the building will apply a gravity load to the structure. The weights of the building components (floors, walls, roof, etc.) are considered **dead loads**; they do not change or move. Dead loads always act vertically downward. **Live loads**, on the other hand, are forces that can change in magnitude or can change position within the structure, such as people, furniture, and cars (Figure 14-4). However, live loads can also act horizontally. For instance, a person leaning against a railing can apply a horizontal live load to the railing; or, a car that turns a corner in a parking garage will apply a horizontal live load to the floor of the structure.

Live Load:
a load produced by the use and occupancy of the building including, but not limited to, the forces produced by people and movable equipment; but a live load is not an environmental load such as a wind load, snow load, rain load, earthquake load, flood load, or dead load.

The magnitude of a dead load is fairly easy to determine. If you know what materials and components will be used in construction of the structure, you can weigh the components, use accepted units weights of construction materials, or obtain the weights of specific components from the manufacturer. Of course, in the preliminary phases of the structural design, you may not know the weight of specific components because they have not been selected or because you have not

Figure 14-4: Dead loads include the weight of construction materials and fixed service equipment. Live loads include forces produced by people and moveable equipment. The loads shown are approximate loads that are commonly used in residential construction.

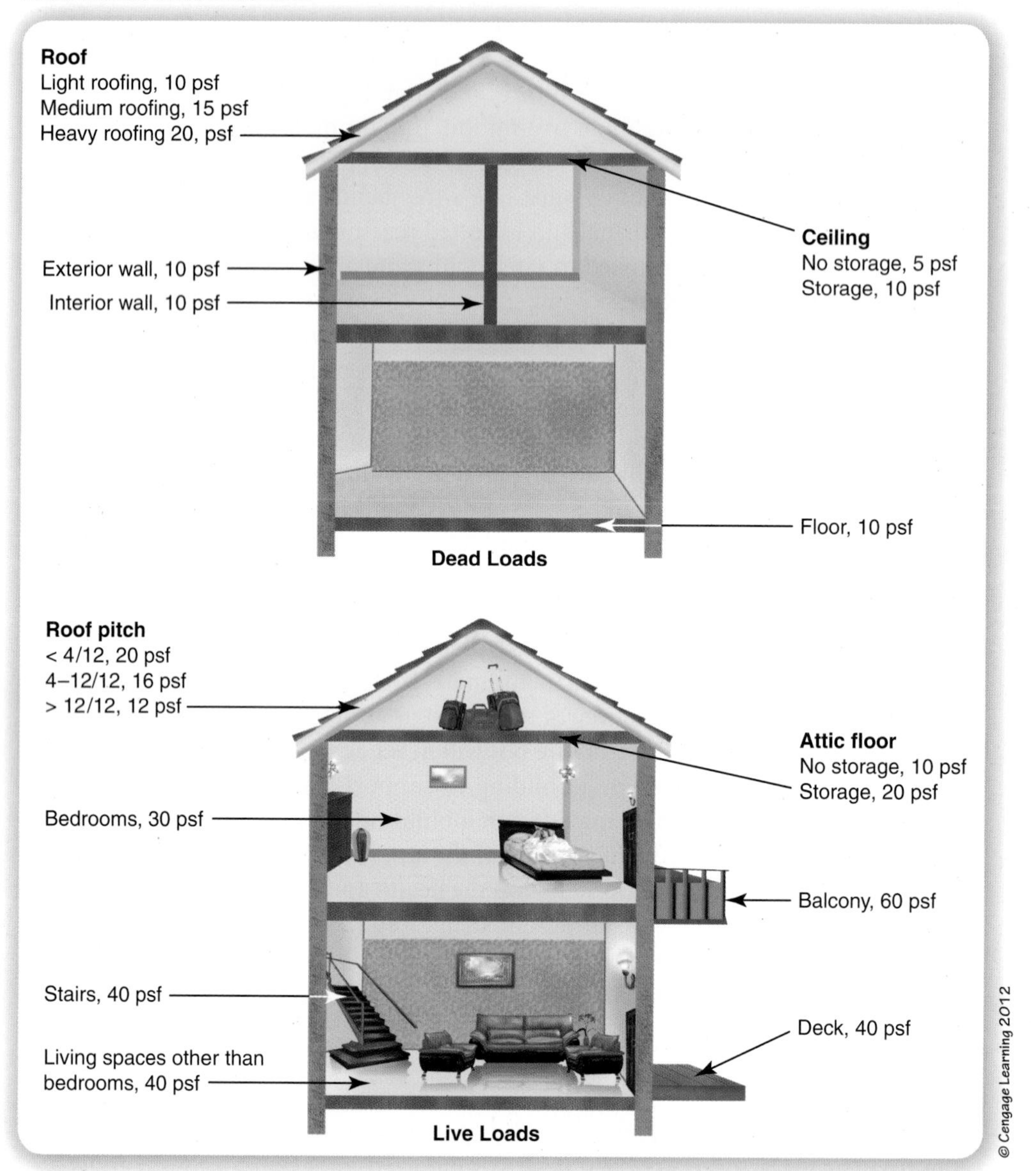

yet designed them. In this case, you must make your best guess as to what building components will be used in the design, and then when the final components have been chosen check that the final design can adequately support the actual loads.

Example:

Calculate the dead load (weight) of an elevated floor if the floor is a 5 in thick concrete slab on a steel deck that also supports a suspended ceiling, lights, and plumbing, electrical, and mechanical equipment for the floor below.

The accepted density of normal weight concrete is 150 lb per ft cube, which means that every cubic foot of concrete (a 1 ft × 1 ft × 1 ft cube) weighs about 150 lb.

So, if a floor slab is 5 in thick, the dead load of the slab will be

$$(5\ in)\left(\frac{12\ in}{1\ ft}\right)\left(150\ \frac{lb}{ft}\right) = 62.5\ \frac{lb}{ft}\ or\ 62.5\ psf \text{ of floor area}$$

Weights of common construction materials (including steel deck and suspended ceilings) are tabulated in Figure 14-5. Assume that the weight of the mechanical, electrical, and plumbing (MEP) equipment is 10 psf. Therefore, the uniform dead load applied to the floor can be estimated as follows:

Construction Material	Uniform Dead load (psf)
5 in concrete slab	62.5
Steel deck	3
Suspended ceiling	1.5
PEM	10
Total floor dead load	77 psf

Figure 14-5: Weight of materials (table). *Architectural Graphic Standards, Student Edition*, an Abridgment of the 9th edition, by Charles George Ramsey, Harold Reeve Sleeper, and John Ray Hoke.

BRICK AND BLOCK MASONRY		PSF
4" brickwork		40
4" concrete block, stone or gravel		34
4" concrete block, lightweight		22
4" concrete brick, stone or gravel		46
4" concrete brick, lightweight		33
6" concrete block, stone or gravel		50
6" concrete block, lightweight		31
8" concrete block, stone or gravel		55
8" concrete block, lightweight		35
12" concrete block, stone or gravel		85
12" concrete block, lightweight		55
CONCRETE		**PCF**
Plain	Cinder	108
	Expanded slag aggregate	100
	Expanded clay	90
	Slag	132
	Stone and cast stone	144
Reinforced	Cinder	111
	Slag	138
	Stone	150

FINISH MATERIALS	PSF
Acoustical tile unsupported per ½"	0.8
Building board, ½"	0.8
Cement finish, 1"	12
Fiberboard, ½"	0.75
Gypsum wallboard, ½"	2
Marble and setting bed	25–30
Plaster, ½"	4.5
Plaster on wood lath	8
Plaster suspended with lath	10
Plywood, ½"	1.5
Tile, glazed wall, ⅜"	3
Tile, ceramic mosaic, ¼"	2.5
Quarry tile, ½"	5.8
Quarry tile, ¾"	8.6
Terrazzo 1", 2" in stone concrete	25
Vinyl tile, ⅛"	1.33
Hardwood flooring, 25/32"	4
Wood block flooring, 3" on mastic	15
FLOOR AND ROOF (CONCRETE)	**PSF**
Flexicore 6" precast lightweight concrete	30

(continued)

Figure 14-5: *(continued)*

Flexicore 6" precast stone concrete		40
Plank, cinder concrete, 2"		15
Plank, gypsum, 2"		12
Concrete, reinforced, 1"	Stone	12.5
	Slag	11.5
	Lightweight	6–10
Concrete, plain, 1"	Stone	12
	Slag	11
	Lightweight	3–9
FUELS AND LIQUIDS		**PCF**
Coal, piled anthracite		47–58
Coal, piled bituminous		40–54
Ice		57.2
Gasoline		75
Snow		8
Water, fresh		62.4
Water, sea		64
GLASS		**PSF**
Polished plate, ¼"		3.28
Polished plate, ½"		6.56
Double strength, 1/3"		26 oz
Sheet A, B, 1/32"		45 oz
Sheet A, B, ¼"		52 oz
Insulating glass 5/8" plate with airspace		3.25
¼" wire glass		3.5
Glass block		18
INSULATION AND WATERPROOFING		**PSF**
Batt, blankets per 1" thickness		0.1–0.4
Corkboard per 1" thickness		0.58
Foamed board insulation per 1" thickness		2.6 oz
Five-ply membrane		5
Rigid insulation		0.75
LIGHTWEIGHT CONCRETE		**PSF**
Concrete, aerocrete		50–80
Concrete, cinder fill		60
Concrete, expanded clay		85–100
Concrete, expanded shale-sand		105–120
Concrete, perlite		35–50
Concrete, pumice		60–90
METALS		**PCF**
Aluminum, cast		165
Brass, cast, rolled		534
Bronze, commercial		552
Bronze, statuary		509
Copper, cast or rolled		556
Gold, cast, solid		1205
Gold coin in bags		509
Iron, cast gray, pig		450
Iron, wrought		480
Lead		710
Nickel		565
Silver, cast, solid		656
Silver coin in bags		590
Tin		459
Stainless steel, rolled		492-510
Steel, rolled, cold drawn		490
Zinc, rolled, cast or sheet		449
MORTAR AND PLASTER		**PCF**
Mortar, masonry		116
Plaster, gypsum, sand		104–120
PARTITIONS		**PSF**
2 x 4 wood stud, GWB, two sides		8
4" metal stud, GWB, two sides		6
4" concrete block, lightweight, GWB		26
6" concrete block, lightweight, GWB		35
2" solid plaster		20
4" solid plaster		32
ROOFING MATERIALS		**PSF**
Built up		6.5
Concrete roof tile		9.5
Copper		1.5–2.5
Corrugated iron		2
Deck, steel without roofing or insulation		2.2–3.6
Fiberglass panels (2½" corrugated)		5–8 oz

(continued)

Figure 14-5: *(continued)*

Galvanized iron	1.2–1.7
Lead, 1/8"	6–8
Plastic sandwich panel, 2½" thick	2.6
Shingles, asphalt	1.7–2.8
Shingles, wood	2–3
Slate, 3/16" to ¼"	7–9.5
Slate, ⅜" to ½"	14–18
Stainless steel	2.5
Tile, cement flat	13
Tile, cement ribbed	16
Tile, clay shingle type	8–16
Tile, clay flat with setting bed	15–20
Wood sheathing per inch	3
SOIL, SAND, AND GRAVEL	**PCF**
Ashes or cinder	40–50
Clay, damp and plastic	110
Clay, dry	63
Clay and gravel, dry	100
Earth, dry and loose	76
Earth, dry and packed	95
Earth, moist and loose	78
Earth, moist and packed	96
Earth, mud, packed	115
Sand or gravel, dry and loose	90–105
Sand or gravel, dry and packed	100–120
Sand or gravel, dry and wet	118–120
Silt, moist, loose	78
Silt, moist, packed	96
STONE [ASHLAR]	**PCF**
Granite, limestone, crystalline	165
Limestone, oolitic	135
Marble	173
Sandstone, bluestone	144
Slate	172
STONE VENEER	**PSF**
2" granite, ½" parging	30
4" granite, ½" parging	59
6" limestone facing, ½" parging	55
4" sandstone or bluestone, ½" parging	49
1" marble	13
1" slate	14
STRUCTURAL CLAY TILE	**PSF**
4" hollow	23
6" hollow	38
8" hollow	45
STRUCTURAL FACING TILE	**PSF**
2" facing tile	14
4" facing tile	24
6" facing tile	34
8" facing tile	44
SUSPENDED CEILINGS	**PSF**
Mineral fiber tile ¾", 12" x 12"	1.2–1.57
Mineral fiberboard 5/8 ", 24" x 24"	1.4
Acoustic plaster on gypsum lath base	10–11
WOOD	**PCF**
Ash, commercial white	40.5
Birch, red oak, sweet and yellow	44
Cedar, northern white	22.2
Cedar, western red	24.2
Cypress, southern	33.5
Douglas fir (coast region)	32.7
Fir, commercial white; Idaho white pine	27
Hemlock	28–29
Maple, hard (black and sugar)	44.5
Oak, white and red	47.3
Pine, northern white sugar	25
Pine, southern yellow	37.3
Pine, ponderosa, spruce: eastern and sitka	28.6
Poplar, yellow	29.4
Redwood	26
Walnut, black	38

NOTE:
To establish uniform practice among designers, it is desirable to present a list of materials generally used in building construction, together with their proper weights. Many building codes prescribe the minimum weights of only a few building materials. It should be noted that there is a difference of more than 25% in some cases.

Live loads are specified in codes and, based on testing results, are a best guess at the maximum load that could potentially be applied to the building. Because it is impossible to predict all of the possible conditions that may affect the magnitude and location of load applications, code-specified live loads are conservative; that is, the loads are larger than the actual loads expected. The International Building Code (IBC) live load requirements are given in Figure 14-6.

Notice that the table specifies both **uniform loads** and **concentrated loads**. A uniform (or distributed) load is a load that is applied uniformly, or evenly, over a given area or along a structural member. Figure 14-7 shows a scenario in which a uniform load might result from the even distribution of merchandise over a floor. The line drawing next to the illustration is called a **beam diagram** and shows a simplified depiction of the beam and loading. The IBC live loads requirements (Figure 14-6) are given in pounds per square foot. So, for example, the required floor live loading for a storage warehouse is 125 psf for a lightly loaded floor and 250 psf for a heavily loaded floor.

A concentrated load is a load that acts on a relatively small area of the floor. Figure 14-8 shows two examples of how a concentrated load might be applied to a structure. A concentrated load can be imposed by hanging a weight (perhaps HVAC ductwork or equipment) from a floor or by a wall or furniture supported by the floor. A concentrated load is generally shown as a single arrow on a beam diagram. The IBC gives concentrated live load requirements in pounds. This would account for any piece of heavy equipment, furniture, or file storage that could potentially load a small area of the floor. For example, a concentrated floor live load of 2000 lb must be applied when designing office buildings. The IBC also stipulates that the concentrated load be applied over a 2.5 ft by 2.5 ft area (see Figure 14-6); however, for design simplicity a concentrated load is often assumed to be applied to a single point and is shown as a single arrow on a beam diagram. It is important to note that the IBC stipulates that the floor or roof be designed to carry either the uniform load or the concentrated load, "whichever produces the greater load effects."

Another important consideration when designing nonresidential facilities is the potential for rearranging room configurations in the future. **Partition walls** are often relocated to accommodate changing company organization. Partition walls are walls that serve to separate spaces, but do not carry structural loads. According to the IBC, in office buildings and in other buildings where partition walls can be moved, the weight of partition walls must be included as a live load in the design, whether the walls are included on the working drawings or not. The code specifies a minimum uniformly distributed live loading of 15 psf to account for the partition wall's weight, except when the live load already exceeds 80 psf.

In many cases, the IBC allows the designer to reduce the required live loading to structural components depending on the area of the structure that contributes load to the component and how the component carries load. In this text, for simplicity, we will ignore live load reductions in our calculations.

Gravitation is a universal force that each mass exerts on any other mass, such as the force exerted on the building components and the occupants, furnishings, and equipment within a building and by the earth.

Figure 14-6: Minimum uniformly distributed live loads and minimum concentrated live loads. International Code Council, 2009 International Building Code. Table 1607.1, p. 310–11.

TABLE 1607.1
MINIMUM UNIFORMLY DISTRIBUTED LIVE LOADS, L_o, AND MINIMUM CONCENTRATED LIVE LOADS[g]

OCCUPANCY OR USE	UNIFORM (psf)	CONCENTRATED (lbs)
1. Apartments (see residential)	—	—
2. Access floor systems		
Office use	50	2,000
Computer use	100	2,000
3. Armories and drill rooms	150	—
4. Assembly areas and theaters		—
Fixed seats (fastened to floor)	60	
Follow spot, projections and control rooms	50	
Lobbies	100	
Movable seats	100	
Stages and platforms	125	
Other assembly areas	100	
5. Balconies (exterior) and decks[h]	Same as occupancy served	—
6. Bowling alleys	75	—
7. Catwalks	40	300
8. Cornices	60	—
9. Corridors, except as otherwise indicated	100	—
10. Dance halls and ballrooms	100	—
11. Dining rooms and restaurants	100	—
12. Dwellings (see residential)	—	—
13. Elevator machine room grating (on area of 4 in^2)	—	300
14. Finish light floor plate construction (on area of 1 in^2)	—	200
15. Fire escapes	100	—
On single-family dwellings only	40	
16. Garages (passenger vehicles only)	40	Note a
Trucks and buses	See Section 1607.6	
17. Grandstands (see stadium and arena bleachers)	—	—
18. Gymnasiums, main floors and balconies	100	—
19. Handrails, guards and grab bars	See Section 1607.7	
20. Hospitals		
Corridors above first floor	80	1,000
Operating rooms, laboratories	60	1,000
Patient rooms	40	1,000
21. Hotels (see residential)	—	—
22. Libraries		
Corridors above first floor	80	1,000
Reading rooms	60	1,000
Stack rooms	150[b]	1,000

continued

TABLE 1607.1—continued
MINIMUM UNIFORMLY DISTRIBUTED LIVE LOADS, L_o, AND MINIMUM CONCENTRATED LIVE LOADS[g]

OCCUPANCY OR USE	UNIFORM (psf)	CONCENTRATED (lbs)
23. Manufacturing		
Heavy	250	3,000
Light	125	2,000
24. Marquees	75	—
25. Office buildings		
Corridors above first floor	80	2,000
File and computer rooms shall be designed for heavier loads based on anticipated occupancy	—	—
Lobbies and first-floor corridors	100	2,000
Offices	50	2,000
26. Penal institutions		—
Cell blocks	40	
Corridors	100	
27. Residential		—
One- and two-family dwellings		
Uninhabitable attics without storage[i]	10	
Uninhabitable attics with limited storage[i, j, k]	20	
Habitable attics and sleeping areas	30	
All other areas	40	
Hotels and multifamily dwellings		
Private rooms and corridors serving them	40	
Public rooms and corridors serving them	100	
28. Reviewing stands, grandstands and bleachers	Note c	
29. Roofs		
All roof surfaces subject to maintenance workers		300
Awnings and canopies		
Fabric construction supported by a lightweight rigid skeleton structure	5 nonreducible	
All other construction	20	
Ordinary flat, pitched, and curved roofs	20	
Primary roof members, exposed to a work floor		
Single panel point of lower chord of roof trusses or any point along primary structural members supporting roofs:		
Over manufacturing, storage warehouses, and repair garages		2,000
All other occupancies		300
Roofs used for other special purposes	Note 1	Note 1
Roofs used for promenade purposes	60	
Roofs used for roof gardens or assembly purposes	100	
30. Schools		
Classrooms	40	1,000
Corridors above first floor	80	1,000
First-floor corridors	100	1,000
31. Scuttles, skylight ribs and accessible ceilings	—	200
32. Sidewalks, vehicular driveways and yards, subject to trucking	250[d]	8,000[e]
33. Skating rinks	100	—

continued

(continued)

Figure 14-6: *(continued)*

TABLE 1607.1—continued
MINIMUM UNIFORMLY DISTRIBUTED LIVE LOADS, L_o, AND MINIMUM CONCENTRATED LIVE LOADS[g]

OCCUPANCY OR USE	UNIFORM (psf)	CONCENTRATED (lbs)
34. Stadiums and arenas		
Bleachers	100[c]	—
Fixed seats (fastened to floor)	60[c]	
35. Stairs and exits		Note f
One- and two-family dwellings	40	
All other	100	
36. Storage warehouses (shall be designed for heavier loads if required for anticipated storage)		
Heavy	250	
Light	125	
37. Stores		
Retail		
First floor	100	1,000
Upper floors	75	1,000
Wholesale, all floors	125	1,000
38. Vehicle barrier systems	See Section 1607.7.3	
39. Walkways and elevated platforms (other than exitways)	60	—
40. Yards and terraces, pedestrians	100	—

For SI: 1 inch = 25.4 mm, 1 square inch = 645.16 mm², 1 square foot = 0.0929 m², 1 pound per square foot = 0.0479 kN/m², 1 pound = 0.004448 kN, 1 pound per cubic foot = 16 kg/m³

a. Floors in garages or portions of buildings used for the storage of motor vehicles shall be designed for the uniformly distributed live loads of Table 1607.1 or the following concentrated loads: (1) for garages restricted to passenger vehicles accommodating not more than nine passengers, 3,000 pounds acting on an area of 4.5 inches by 4.5 inches; (2) for mechanical parking structures without slab or deck which are used for storing passenger vehicles only, 2,250 pounds per wheel.

b. The loading applies to stack room floors that support nonmobile, double-faced library bookstacks, subject to the following limitations:

1. The nominal bookstack unit height shall not exceed 90 inches;
2. The nominal shelf depth shall not exceed 12 inches for each face; and
3. Parallel rows of double-faced bookstacks shall be separated by aisles not less than 36 inches wide.

c. Design in accordance with ICC 300.

d. Other uniform loads in accordance with an approved method which contains provisions for truck loadings shall also be considered where appropriate.

e. The concentrated wheel load shall be applied on an area of 4.5 inches by 4.5 inches.

f. Minimum concentrated load on stair treads (on area of 4 square inches) is 300 pounds.

g. Where snow loads occur that are in excess of the design conditions, the structure shall be designed to support the loads due to the increased loads caused by drift buildup or a greater snow design determined by the building official (see Section 1608). For special-purpose roofs, see Section 1607.11.2.2.

h. See Section 1604.8.3 for decks attached to exterior walls.

i. Attics without storage are those where the maximum clear height between the joist and rafter is less than 42 inches, or where there are not two or more adjacent trusses with the same web configuration capable of containing a rectangle 42 inches high by 2 feet wide, or greater, located within the plane of the truss. For attics without storage, this live load need not be assumed to act concurrently with any other live load requirements.

j. For attics with limited storage and constructed with trusses, this live load need only be applied to those portions of the bottom chord where there are two or more adjacent trusses with the same web configuration capable of containing a rectangle 42 inches high by 2 feet wide or greater, located within the plane of the truss. The rectangle shall fit between the top of the bottom chord and the bottom of any other truss member, provided that each of the following criteria is met:

i. The attic area is accessible by a pull-down stairway or framed opening in accordance with Section 1209.2, and
ii. The truss shall have a bottom chord pitch less than 2:12.
iii. Bottom chords of trusses shall be designed for the greater of actual imposed dead load or 10 psf, uniformly distributed over the entire span.

k. Attic spaces served by a fixed stair shall be designed to support the minimum live load specified for habitable attics and sleeping rooms.

l. Roofs used for other special purposes shall be designed for appropriate loads as approved by the building official.

1607.6 Truck and bus garages. Minimum live loads for garages having trucks or buses shall be as specified in Table 1607.6, but shall not be less than 50 psf (2.40 kN/m²), unless other loads are specifically justified and *approved* by the *building official*. Actual loads shall be used where they are greater than the loads specified in the table.

1607.6.1 Truck and bus garage live load application. The concentrated load and uniform load shall be uniformly distributed over a 10-foot (3048 mm) width on a line normal to the centerline of the lane placed within a 12-foot-wide (3658 mm) lane. The loads shall be placed within their individual lanes so as to produce the maximum stress in each structural member. Single spans shall be designed for the uniform load in Table 1607.6 and one simultaneous concentrated load positioned to produce the maximum effect. Multiple spans shall be designed for the uniform load in Table 1607.6 on the spans and two simultaneous concentrated loads in two spans positioned to produce the maximum negative moment effect. Multiple span design loads, for other effects, shall be the same as for single spans.

TABLE 1607.6
UNIFORM AND CONCENTRATED LOADS

LOADING CLASS[a]	UNIFORM LOAD (pounds/linear foot of lane)	CONCENTRATED LOAD (pounds)[b] For moment design	CONCENTRATED LOAD (pounds)[b] For shear design
H20-44 and HS20-44	640	18,000	26,000
H15-44 and HS15-44	480	13,500	19,500

For SI: 1 pound per linear foot = 0.01459 kN/m, 1 pound = 0.004448 kN, 1 ton = 8.90 kN.

a. An H loading class designates a two-axle truck with a semitrailer. An HS loading class designates a tractor truck with a semitrailer. The numbers following the letter classification indicate the gross weight in tons of the standard truck and the year the loadings were instituted.

b. See Section 1607.6.1 for the loading of multiple spans.

1607.7 Loads on handrails, guards, grab bars, seats and vehicle barrier systems. Handrails, *guards*, grab bars, accessible seats, accessible benches and vehicle barrier systems shall be designed and constructed to the structural loading conditions set forth in this section.

1607.7.1 Handrails and guards. Handrails and *guards* shall be designed to resist a load of 50 pounds per linear foot (plf) (0.73 kN/m) applied in any direction at the top and to transfer this load through the supports to the structure. Glass handrail assemblies and *guards* shall also comply with Section 2407.

Exceptions:

1. For one- and two-family dwellings, only the single concentrated load required by Section 1607.7.1.1 shall be applied.
2. In Group I-3, F, H and S occupancies, for areas that are not accessible to the general public and that have an *occupant load* less than 50, the minimum load shall be 20 pounds per foot (0.29 kN/m).

1607.7.1.1 Concentrated load. Handrails and *guards* shall be able to resist a single concentrated load of 200 pounds (0.89 kN), applied in any direction at any point.

Figure 14-7: Even distribution of merchandise on a floor is represented with a uniformly distributed load.

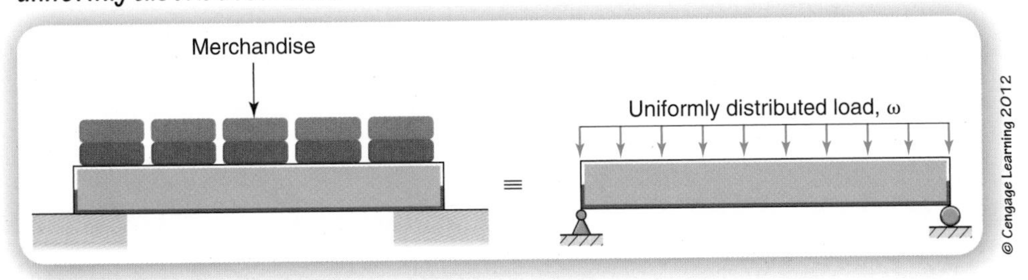

Figure 14-8: Loads that are applied over a small area are analyzed as concentrated loads and represented by a single arrow.

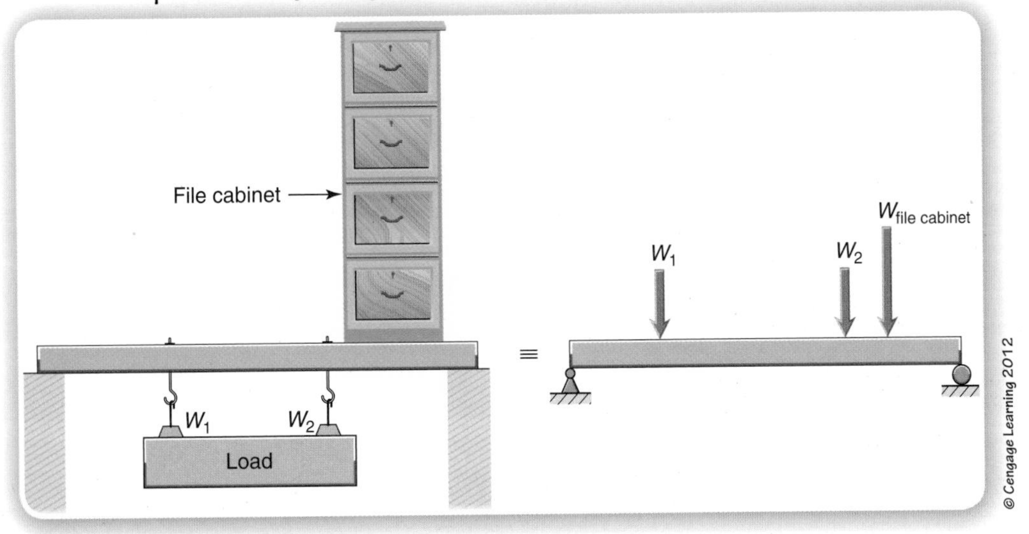

Snow Load

In most areas of the United States, heavy snow can accumulate on the roof of a building adding to the roof load. Consequently, the weight of the snow must be considered when designing a roof system and the structural components that will carry the roof load to the ground. Figure 14-9 shows the IBC ground snow load map. This map provides a guide for determining the weight of snow that accumulates on the ground during the winter in the contiguous United States. Snow loads in Alaska are provided in a separate table in the IBC. Snow loads in Hawaii are assumed to be zero except in mountainous areas; check with the local building official if you are designing a structure in Hawaii.

The ground snow load is used as the basis for the design snow load on the roof of a structure. The IBC dictates that ground snow load be adjusted per the publication *Minimum Design Loads for Buildings and Other Structures* (ASCE 7-05) by the American Society of Civil Engineers (Figure 14-11). The calculation of the design snow load should take into consideration several factors that can affect the accumulation and weight of snow on a roof, including exposure to wind, heat loss from the building, roof slope, and the importance of the structure to human health and safety. The required adjustments are expressed as multipliers, or coefficients, to the ground snow load.

Figure 14-9: *Ground snow loads for the United States (psf). International Code Council, (2009) International Building Code.* FIGURE 1608.2, p. 316–17.

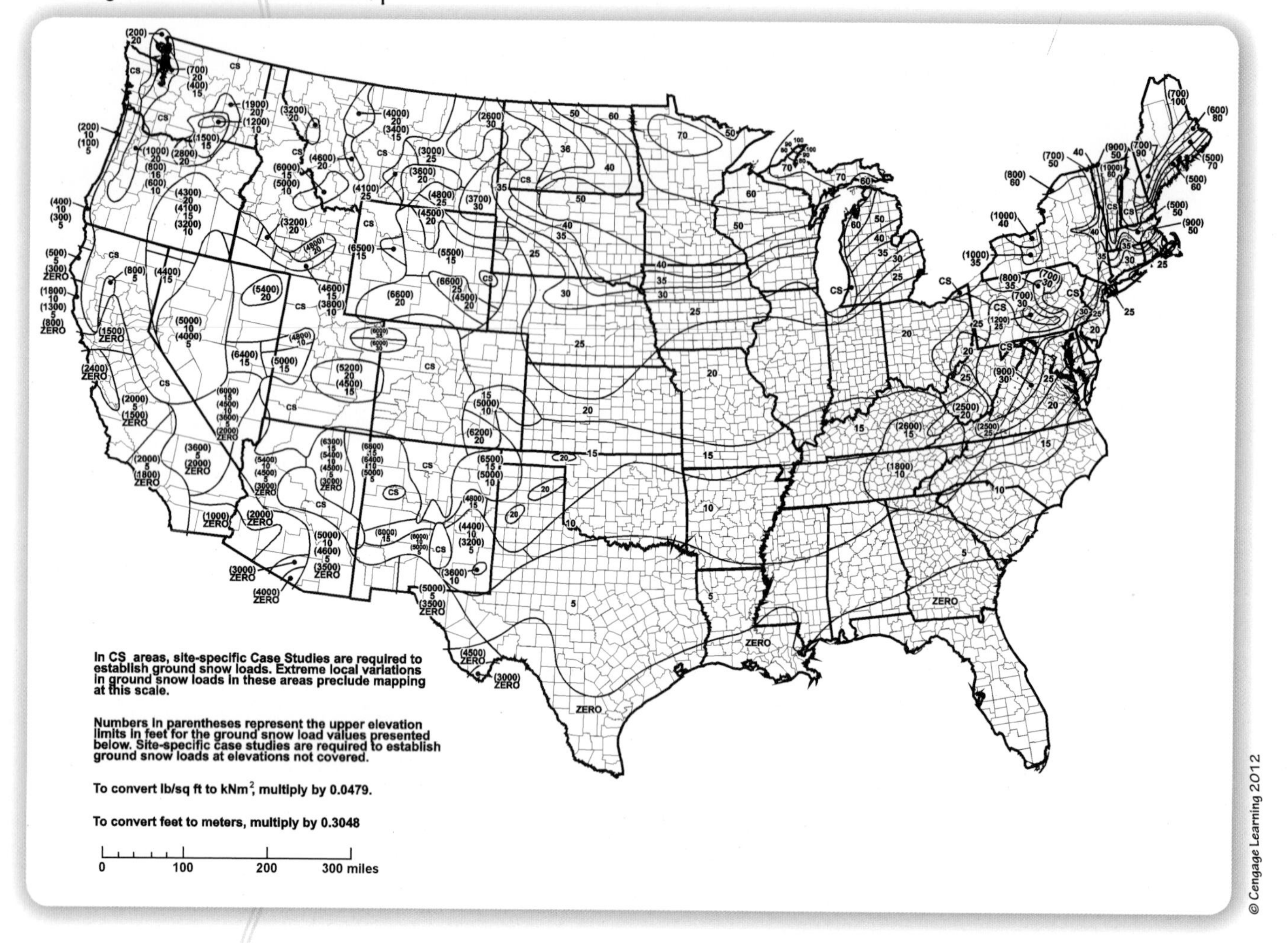

In CS areas, site-specific Case Studies are required to establish ground snow loads. Extreme local variations in ground snow loads in these areas preclude mapping at this scale.
Numbers in parentheses represent the upper elevation limits in feet for the ground snow load values presented below. Site-specific case studies are required to establish ground snow loads at elevations not covered.
To convert lb/sq ft to kNm², multiply by 0.0479.
To convert feet to meters, multiply by 0.3048
0
100
200
300 miles

The ASCE-7 formula for design snow load is

$$p_s = 0.7\ C_s C_e C_t I_s p_g$$

where P_s = design snow load
C_s = roof slope factor based on the roof slope and roofing materials
C_e = exposure facture based on the building's exposure to wind
C_t = thermal factor
I_s = occupancy importance factor (Figure 14-10)
P_g = ground snow load (Figure 14-9)

For flat roofs, the minimum design snow load is $p_s = I_s p_g$ or $p_s = 20I_s$ when p_g is greater than 20 psf.

Snow load in many locations within the United States is due not to a single snowfall but to the accumulation of snow over an entire season. The ground snow load takes into account this accumulation. However, statistically, a roof will lose more snow over time than is lost at the ground level. The 0.7 factor in the ground snow formula corrects for the fact that the roof will often support less snow than is on the ground after multiple snows.

Figure 14-10: *Occupancy of buildings and other structures. International Code Council, (2009) International Building Code. Table 1604.5, p. 307.*

TABLE 1604.5
OCCUPANCY CATEGORY OF BUILDINGS AND OTHER STRUCTURES

OCCUPANCY CATEGORY	NATURE OF OCCUPANCY
I	Buildings and other structures that represent a low hazard to human life in the event of failure, including but not limited to: • Agricultural facilities. • Certain temporary facilities. • Minor storage facilities.
II	Buildings and other structures except those listed in Occupancy Categories I, III and IV
III	Buildings and other structures that represent a substantial hazard to human life in the event of failure, including but not limited to: • Buildings and other structures whose primary occupancy is public assembly with an occupant load greater than 300. • Buildings and other structures containing elementary school, secondary school or day care facilities with an occupant load greater than 250. • Buildings and other structures containing adult education facilities, such as colleges and universities, with an occupant load greater than 500. • Group I-2 occupancies with an occupant load of 50 or more resident patients but not having surgery or emergency treatment facilities. • Group I-3 occupancies. • Any other occupancy with an occupant load greater than 5,000[a]. • Power-generating stations, water treatment facilities for potable water, waste water treatment facilities and other public utility facilities not included in Occupancy Category IV. • Buildings and other structures not included in Occupancy Category IV containing sufficient quantities of toxic or explosive substances to be dangerous to the public if released.
IV	Buildings and other structures designated as essential facilities, including but not limited to: • Group I-2 occupancies having surgery or emergency treatment facilities. • Fire, rescue, ambulance and police stations and emergency vehicle garages. • Designated earthquake, hurricane or other emergency shelters. • Designated emergency preparedness, communications and operations centers and other facilities required for emergency response. • Power-generating stations and other public utility facilities required as emergency backup facilities for Occupancy Category IV structures. • Structures containing highly toxic materials as defined by Section 307 where the quantity of the material exceeds the maximum allowable quantities of Table 307.1(2). • Aviation control towers, air traffic control centers and emergency aircraft hangars. • Buildings and other structures having critical national defense functions. • Water storage facilities and pump structures required to maintain water pressure for fire suppression.

a. For purposes of occupant load calculation, occupancies required by Table 1004.1.1 to use gross floor area calculations shall be permitted to use net floor areas to determine the total occupant load.

For the purpose of designing structural members in this textbook, we will make several simplifying assumptions. If the assumptions made here are not appropriate for your project or your project is located in Alaska or Hawaii, you may research the IBC and ASCE-7 to determine the contributing factors (C_s, C_e, and C_t) and the design snow load for your building.

First, we will assume that the building is in an urban or suburban area surrounded by other buildings and is partially exposed to the wind; therefore, we will use an exposure factor equal to 1.0. We will also assume that the structure is heated and loses heat through the roof and, therefore, will use a thermal factor equal to 1.0. And, we will ignore any reduction in design snow load resulting from the roof slope; therefore, we will conservatively use a roof slope factor of 1.0. We will, however, include the appropriate IBC snow importance factor in our calculations.

Example: Calculate Design Snow Load

Calculate the design snow load for an elementary school in Indianapolis, Indiana. The school is located in an urban environment, has a flat roof, and has an enrollment of 700 students with 42 faculty and staff members.

Step 1: *Find the ground snow load.*

The ground snow load for Indianapolis is 20 psf as per Figure 14-9.

Step 2: *Find the roof slope factor, C_s.*

For a flat roof assume $C_s = 1.0$.

Step 3: *Find the exposure factor.*

For a building located in an urban or suburban area surrounded by other buildings such that it is partially exposed to the wind, use an exposure factor equal to 1.0 (per ASCE–7).

Step 4: *Find the thermal factor, C_t.*

Since an elementary school will be heated, some heat will be lost through the roof, use $C_t = 1.0$ (per ASCE-7).

Step 5: *Determine the occupancy importance factor.*

Per Figure 14-10, an elementary school is classified as Occupancy Category III. Therefore, according to Figure 14-10, use $I_s = 1.1$.

Step 6: *Calculate the design snow load, p_s.*

$$P_s = 0.7\, C_s C_e C_t I_s p_g$$
$$= 0.7(1.0)(1.0)(1.0)(1.10)(20\ psf)$$
$$= 15.4\ psf$$

Step 7: *Check minimum snow load.*

$$P_s = I_s p_g = (1.10)(20\ psf) = 22.0\ psf,$$

which controls the design.

The design snow load is 22 psf.

Figure 14-11: Importance Factor (Snow Load). American Society of Civil Engineers, (2005) Minimum Design Loads for Buildings and Other Structures [ASCE/SEI 7-05]. Table 7-4, P. 93.

Category[a]	I
I	0.8
II	1.0
III	1.1
IV	1.2

[a]See Section 1.5 and Table 1-1.

Lateral Loads

If you have ever personally experienced a strong wind—like a tornado or hurricane—you know from experience that high winds can create forces that are devastating to man-made structures. Flood water, earthquakes, and soil pressure (including landslides) can cause similar destruction if not properly considered during the design of a structure (see Figures 14-13 and 14-14).

Figure 14-12: Lateral loads can result from an earthquake, wind, flood, or earth pressure against a structure.

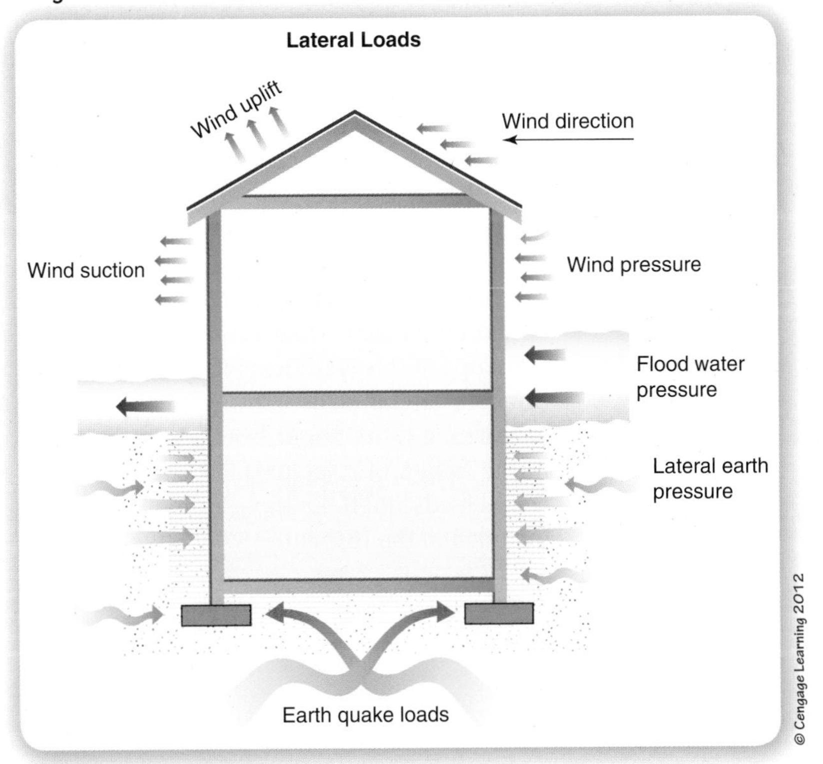

Figure 14-13: Northridge Earthquake, California, January 17, 1994. Buildings, cars, and personal property were all destroyed when the earthquake struck. Approximately 114,000 residential and commercial structures were damaged and 72 deaths were attributed to the earthquake. Damage costs were estimated at $ 25 billion. FEMA News Photo.

Although we know they can and do occur, these natural forces are variable and erratic—we can never be sure what forces they will exert on to a structure. In order to provide a safe building, however, an engineer must design a structure to withstand these unknown natural forces. So, we attempt to predict how these forces will

Figure 14-14: Orange Beach, Alabama. The Windemere Condominiums show the vast fury of Hurricane Ivan's 130 mph winds and 30 ft swells. The eye of Ivan passed directly over Orange Beach early morning on September 16, 2004. FEMA Photo/Butch Kinerney.

affect our buildings. For the most part, we assume that these natural forces apply **lateral loads** to a building. A lateral load is a force (or pressure) that is applied horizontally or perpendicular to the surface of a building (see Figure 14-12), as opposed to gravity loads, which act vertically downward.

Building codes specify the magnitude and direction of lateral forces that must be used in building design including wind, earthquake, soil, and flood loads. Calculating the magnitude of these lateral loads can become quite complicated, and the design of the structural elements that resist lateral loads is beyond the scope of this textbook; we will not include calculations for lateral loads. However, if you are interested in learning more about how these forces are applied in the design of structures, a detailed presentation of lateral loads and their application to building design is included in the IBC and ASCE-7.

Fun Fact

The Empire State Building was selected as one of the Seven Wonders of the Modern World by the American Society of Civil Engineers. The building is 1224 ft tall (1454 ft tall including the lightening rod) and encloses 102 floors. One of the challenges in designing the Empire State Building was making sure it could withstand the wind loads. The building is supported by 210 steel and concrete columns, 12 of which run from the foundation to the very top. It is said it would take a wind blowing at 4,500,000 lb pressure to knock the building over (National Society of Professional Engineers).

Empire State Building.

Point of Interest

Hurricane-Resistant House

Much research has been performed to identify construction methods that reduce structural damage due to a natural disaster, such as a hurricane. Rima Taher, an architecture professor at the New Jersey Institute of Technology, has written extensively on best practices for residential and commercial building design and construction to resist high winds and hurricanes. Her recommendations for residential design to provide hurricane resistance include the following:

- Raise structure above flood level
- Use a square, hexagonal, or octagonal floor plan
- Use a hip roof
- Use a roof slope of 30 degrees
- Limit roof overhangs to less than or equal to 20 in
- Connect roofs to walls using nails (not staples) and hurricane clips

This coastal South Carolina home exhibits many hurricane-resistant design features.

Photo courtesy of Deborah Kennedy.

Load Combinations

It would seem to make sense that if we design a structure to simultaneously resist the maximum load of every type (dead, live, snow, wind, flood, earthquake, etc.) that could possibly be applied to the building, we would surely end up with a strong and safe structure. But this is not necessarily true.

Of course, one argument against using all of the possible loads when designing is that the probability of all these forces acting at the same time is very low. A structure designed with these mega-load combinations would most likely require more structural materials than a design using only reasonable load combinations.

This would increase the cost of the building, and this added cost would most likely never be justified in the life of the building.

An even more important argument against simply designing a structure for the simultaneous application of the maximum loads is that the resulting structure may not be safe. In some cases, a worst-case scenario can occur with the application of one or two types of loads. For example, wind can cause a suction force on a wall or roof of a building, which means the wind pressure will actually pull outward on the surface, rather than push inward. If the wind is strong enough, the upward suction force on a roof can exceed the dead load of the roof, and the roof beams may bend upward rather than downward. If the beams are not designed for this bending force, they could fail. As a designer, you may not have recognized this possibility if you had included the maximum live load and snow load in your design thereby theoretically eliminating the upward bending of the roof beams. So, a design using more load does not necessarily result in a better or safer structure. It is better to carefully consider all of the possible combinations of loads that may be applied to a structure and make sure that your design can safely and efficiently resist every case.

In an effort to consider loading conditions that may create extreme conditions for various structural elements, building codes specify a variety of load combinations that must be considered when designing a building. The IBC states that "structures and portions thereof shall resist the most critical effects" resulting from specified load combinations that include various combinations of dead, live, flood, soil pressure, snow, thermal, and earthquake loads. For example, the following load combinations are included as alternate basic load combinations in the IBC:

Alternate Basic Load Combinations

$$D + L + (L_r \text{ or } S \text{ or } R)$$
$$D + L + (\omega W)$$
$$D + L + \omega W + S/2$$
$$D + L + S + \omega W/2$$
$$D + L + S + E/1.4$$
$$0.9D + E/1.4$$

where D = dead load

L = live load

L_r = roof live load

W = wind load

ω = wind coefficient (if wind loads calculated using Chapter 6 in ASCE 7 $\omega = 1.3$, else $\omega = 1$)

S = snow load

E = earthquake load

R = rain load

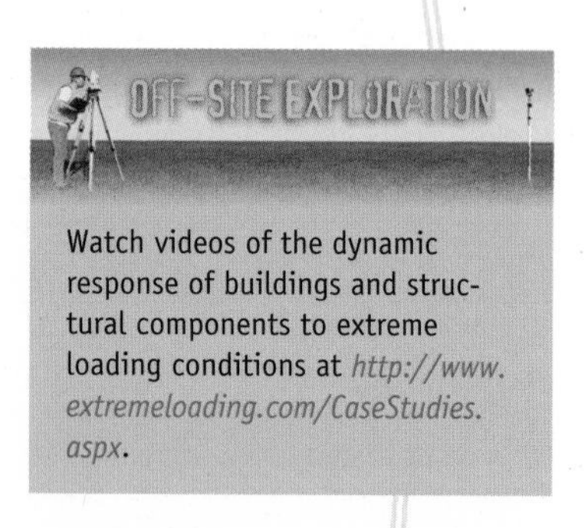

Watch videos of the dynamic response of buildings and structural components to extreme loading conditions at *http://www.extremeloading.com/CaseStudies.aspx*.

The IBC specifies many more load combinations not listed here. Engineers identify the load cases used for a design based on the structural design method used and the specifics of the building being designed.

As you have seen in previous chapters, there are a wide variety of materials and structural systems that can be employed for the structural design of a given building. In most cases, the final design choices are made based on input from all stakeholders and a structural investigation of several alternatives. Once consensus is reached, a preliminary design that includes the types of roof, floor, wall, and foundation systems to be used is laid out. Additional structural considerations are incorporated into the preliminary plan and may include lateral support systems (which we will discuss in the next section), additional structural components needed to support initial systems (for example, a steel or concrete frame to support floor and roof systems not supported by walls), and connections between systems. General dimensions and structural component spacing are estimated based on client needs and economic considerations. Once a comprehensive preliminary structural plan is accepted, the structural engineer can begin the analysis and design of the structure.

The design of structures includes a number of requirements.

Tributary Width and Tributary Area

In order to design any structural component, you must know the location and magnitude of the loads that must be resisted by the structure. Once the **design loads** have been determined based on the types of loads and load combination required to be resisted, you can begin to identify which structural components will initially receive the applied loads. In the case of concentrated loads, if the source of the load is static and will not move during the life of the structure, the point of the load application during analysis is fairly easy to locate. If the load can move, like a forklift truck, for example, the designer must consider all possible locations and determine the location that will produce the most critical load condition for each component that will eventually receive the load. For instance, assume that a forklift with its load weighs 5000 lb. The most critical location for the forklift in the design of a supporting floor beam would be at mid-span of the beam. However, the most critical location for the design of a column that supports that floor beam would be at the end of the beam, directly over the column. The design of each component would include consideration of the forklift in a location that is most critical for that component (Figure 14-15).

Figure 14-15 a, b: (a) A forklift will impose the most critical load for the beam at mid-span. (b) A forklift at the end of a beam, over the column, will create the most critical load for the column.

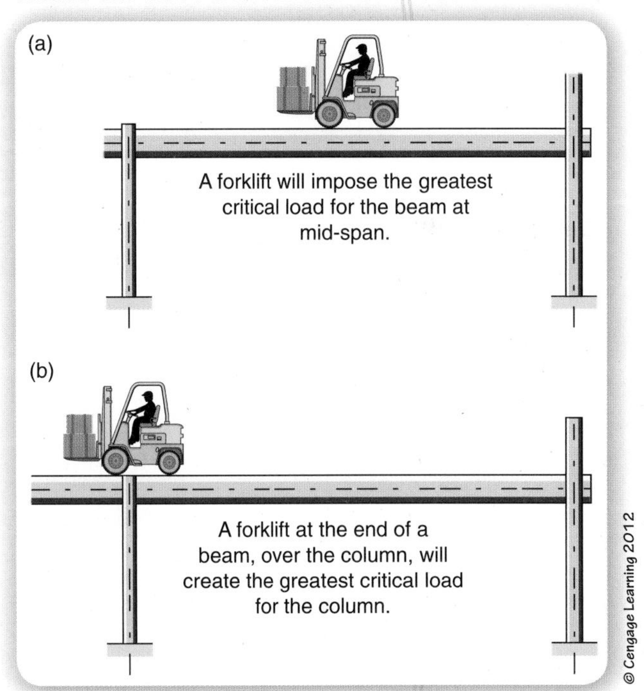

In the case of uniformly distributed loads, structural engineers use the concept of **tributary area** (or **tributary width**) to determine the distribution of load to individual components. The tributary area of a structural component refers to the loaded area of a building that will contribute load to that member. The tributary area considered is usually a floor or roof area, but can be a wall area in the case of lateral loads. So, for example, a floor may be supported by several beams as shown in the three-dimensional view of a structural steel frame shown in Figure 14-16. Obviously, each beam must carry only a portion of the dead and live loads applied to the floor.

Figure 14-16: A floor or roof may be supported by many beams. Each beam must only carry a portion of the floor load.

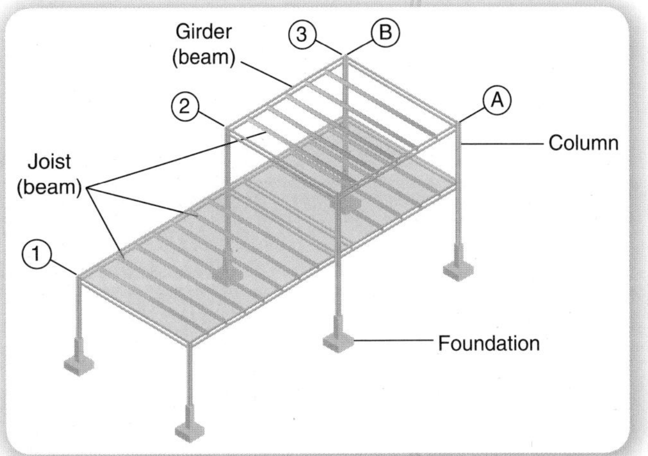

A partial framing plan for the first elevated level of the building frame is shown in Figure 14-17. The beam just to the right of column line 2 is highlighted blue. This beam must carry only the narrow portion of the floor for which this beam is the nearest support. Therefore, it will carry only half the area between itself and the next adjacent

Figure 14-17: The tributary width and tributary area identify the portion of the floor that will be supported by the beam.

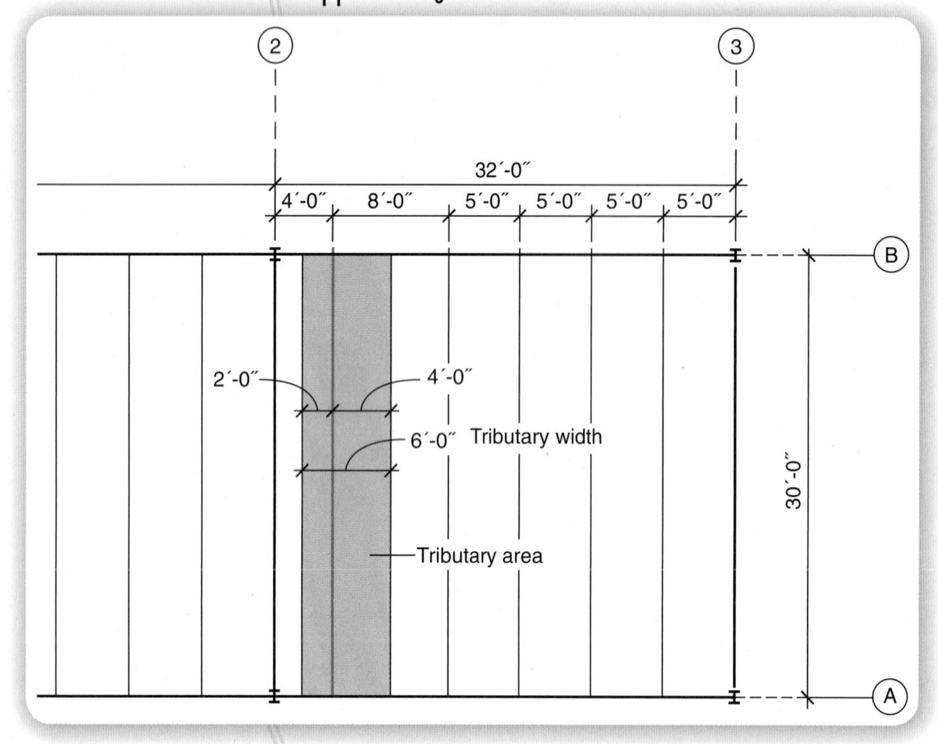

beam on each side. So, for this particular beam, it must carry 2 ft of floor to the left (half of the 4 ft distance to the beam on the left) and 4 ft of floor to the right (half of the 8 ft distance to the beam on the right). This gives a tributary width of 6 ft (2 ft on the left and 4 ft on the right).

The beam must carry the loads applied to the floor area represented by the shading—this is the tributary area. The tributary area can be calculated by multiplying the length of the beam by the tributary width. In this case, the tributary area for the beam under investigation is $180 \text{ ft}^2 = 30 \text{ ft} \cdot 6 \text{ ft}$. For beams that are loaded from only one side, like the beam on column line 3 in Figure 14-17, the tributary width is half the distance to next beam on the loaded side only (2 ft 6 in in this case).

Your Turn

Use the framing plan in Figure 14-17 to find the tributary width and tributary area of the interior beams between column lines 2 and 3.

Load Path

The ultimate job of a structure is to carry all of the loads applied to the building safely down to the ground. Every load that is applied to a building must be safely passed from one structural component to another until the load is finally transferred to the ground. But, most structural components do not directly support the applied loads—the loads are transferred from other structural components. For example, the live load on a floor may be transferred to beams, then to girders, then to columns, and finally to the foundation. How can you determine exactly what load each structural component must resist? The answer is that you must trace the loads along the paths that they take to the ground. Every load that is applied—horizontal and vertical—must have a specific path through which it moves. And every component along that path must be designed to carry every load that is transferred through it plus any additional loads applied directly to that component.

The chain of structural components through which a load is transferred is known as the **load path**. It is essential that the load path is continuous because a break in the path, even if the individual components are adequate, creates a weak link in the structure that can lead to failure. As a result, both the design of structural components and the design of connections between them are critical in the structural design of a building.

In order to determine the path of a load through a structure, you must first determine where the load will be applied. Loads applied to structural elements that are in direct contact with the ground, such as a concrete slab on grade, are transferred directly to the soil as illustrated by the floor loading on the ground floor in Figure 14-18.

Figure 14-18: Load path for loads placed on a slab on grade.

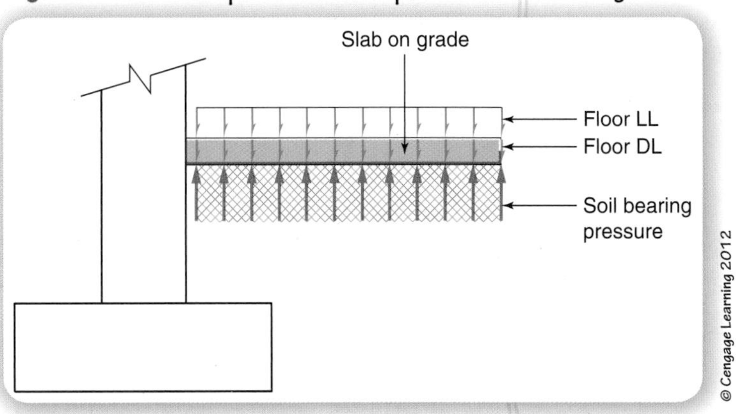

Gravity loads and other vertical loads that are supported above the ground on an elevated floor or roof typically follow a predictable load path. The load is applied to the floor (or roof) and carried by the floor (or roof) framing, which usually consists of **beams** (see Figure 14-16). A beam is a structural member that carries load that is applied transverse (perpendicular) to its length. The gravity load will typically be transferred through one or more beams (sometimes called joists, rafters, girders, or trusses) to a vertical structural element—a load-bearing wall or a column. A **load-bearing wall** is simply a wall that carries a load other than its own weight. A **column** is a thin vertical member that carries load through **compression** along its length. The bearing wall or column then transfers the load to the foundation that, finally, applies a bearing pressure on the supporting soil. One example of the load path of a gravity load through a structural frame is shown in Figure 14-19. Each structural member through which the load is transferred must be designed to safely carry every load that could potentially pass through it.

A similar process of load transfer occurs in other types of structures. Figure 14-20 shows the load path of several gravity loads applied to a wood framed house.

When the wind blows against a building or an earthquake shakes the earth below a structure, the resulting lateral forces will follow a different load path through the structure than that followed by a

Figure 14-20: Floor and roof loads applied to a residential structure typically travel through floor beams or roof members to load-bearing walls (or columns) and then to the foundation. The foundation transfers the loads to the supporting soil.

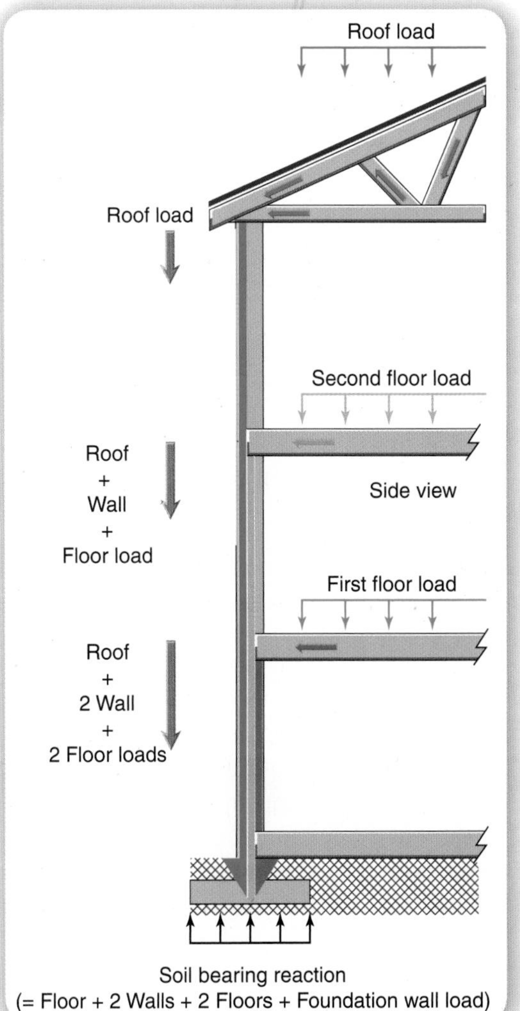

Figure 14-19: The load path of a concentrated load on the floor of this structure includes a floor beam, girders, columns, and foundations. The foundations transfer the load to the supporting soil. Note: TYP stands for "typical" and indicates that all similar objects are similarly labeled.

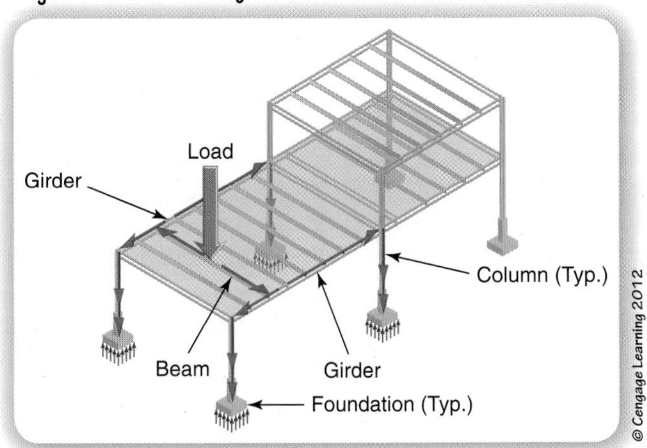

Figure 14-21: Resistance to lateral loads is provided by a rigid frame, diagonal bracing, and shear walls.

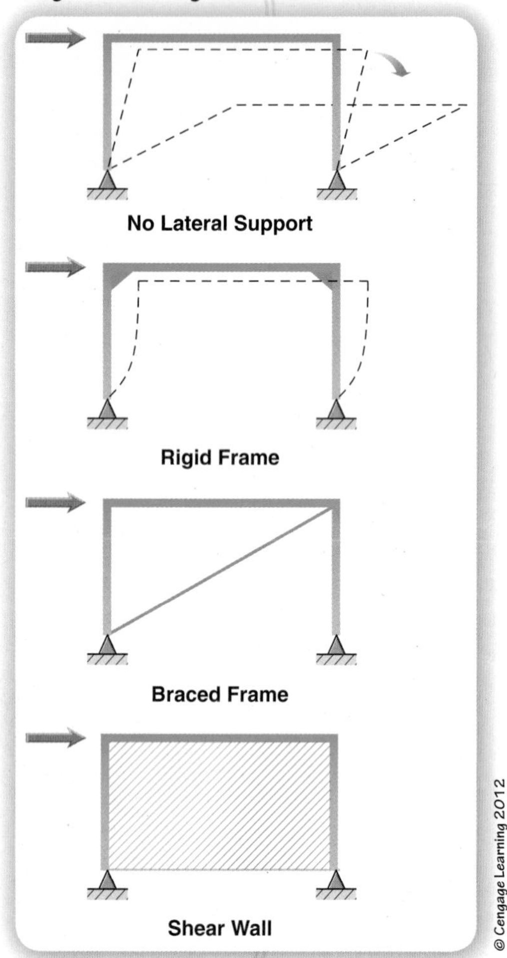

gravity load. Columns and beams are designed primarily to carry vertical loads and, by themselves, do not provide good lateral support. Think about building a "frame" out of plastic straws. If you attached the base of the frame to a table, it is easy to displace the box out of "square" by pushing on the side of the box with a lateral force: the frame just folds over.

There are several things you could do that would improve the lateral resistance of the frame. For instance, you could create stronger connections between the horizontal straws and the vertical straws (see Figure 14-21). In some cases, lateral support is provided by rigidly attaching the beams to columns so that the beams and columns become, more or less, a single element. We call the resulting structure a **rigid frame**.

Another option to horizontally stiffen the straw box would be to add diagonal straws from a bottom corner to an opposite top corner, as shown in Figure 14-21. These diagonal members are referred to as horizontal or **lateral bracing** and are structural elements that typically resist lateral forces by transferring an **axial load**—pushing or pulling along the length of the member. This structure would be called a **braced frame**.

A flat "plate"—like a floor, roof, or wall—can also transfer lateral load. If you were to fill in the top and sides of the straw frame with rigid cardboard, it would be much more difficult to displace the top of the box. A horizontal flat, thin, element that transfers horizontal forces to other structural elements is called a **diaphragm**. Often a diaphragm performs two functions, as is the case with a floor or roof—it supports vertical loads as well as lateral loads. Walls that transfer lateral loads, like the vertical cardboard pieces in our straw frame, are called **shear walls**.

Point of Interest

Hyatt Regency Walkway Collapse

On July 17, 1981, hundreds of guests had gathered for a dance party in the atrium of the new Hyatt Regency Hotel in Kansas City. Many of the partygoers had moved to walkways that were elevated above the atrium floor at the second, third, and fourth floors of the building. Little did they know that the walkways could not support the resulting live load and the evening would end in disaster.

The walkways were suspended from the roof structure by steel rods. The original walkway design drawings indicated that the second-floor walkway be constructed directly below the fourth-floor walkways and that the same hanger rods be used to support both walkways.

The original design was highly impractical because it required that the hanger rods be threaded along the entire distance between the second and fourth floors in order for the connection nuts to be installed according to the detail. Normal construction practices, in all likelihood, would have damaged the threads to the point that the fourth-floor connection would have been impossible to construct as per the plans. The construction contractor recognized this problem and proposed a new design, which had to be approved by the design engineer.

On July 17, 1981, a suspended walkway in the Kansas City Hyatt Regency hotel collapsed killing 114 people and injuring more than 200 others. At the time, it was the deadliest structural collapse in U.S. history.

Source: Wikimedia, author: Dr. Lee Lowery, Jr., P.E.

Original walkway support design.

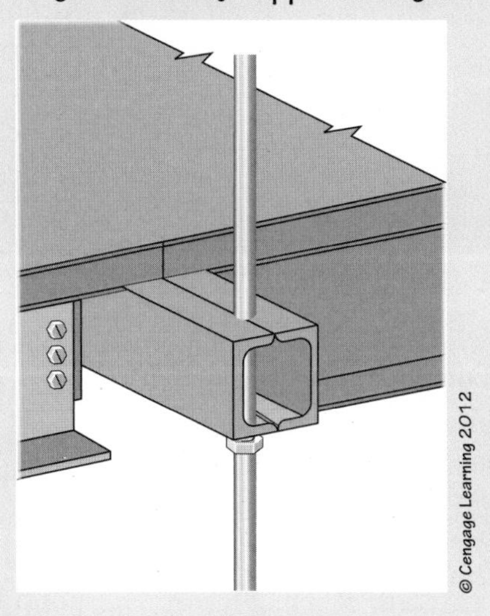
© Cengage Learning 2012

Through a series of miscommunication and negligent actions, the installed connection used two separate hanger rods (as opposed to one continuous rod) at each support location—one rod supported the fourth-floor walkway from the atrium roof, another rod supported the second-floor walkway from the fourth-floor walkway.

The installed design did not meet building codes. The load path for the original design transferred dead and live loads from the second-floor walkway into the support rods and then directly to the roof framing and did not affect the fourth-floor walkway framing. However, the load path traveled by the second-floor loads in the installed configuration included load transfer through the fourth-floor walkway framing before reaching the hanger rods. The fourth-floor beams were not designed to resist these additional forces. This load path change contributed to the collapse of the second- and fourth-floor walkways, caused the loss of 114 lives, and resulted in millions of dollars in costs.

Installed walkway support detail.

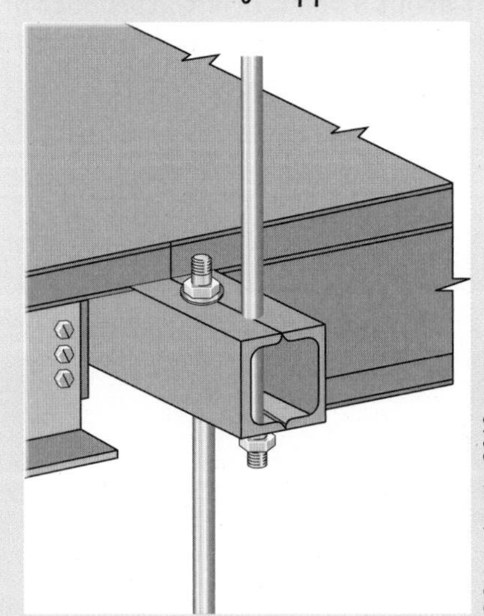
© Cengage Learning 2012

This picture of the failed fourth-floor walkway connection in the Kansas City Hyatt Regency was taken during the investigation of the disaster.

Source: Wikimedia, author: Dr. Lee Lowery Jr., P.E.

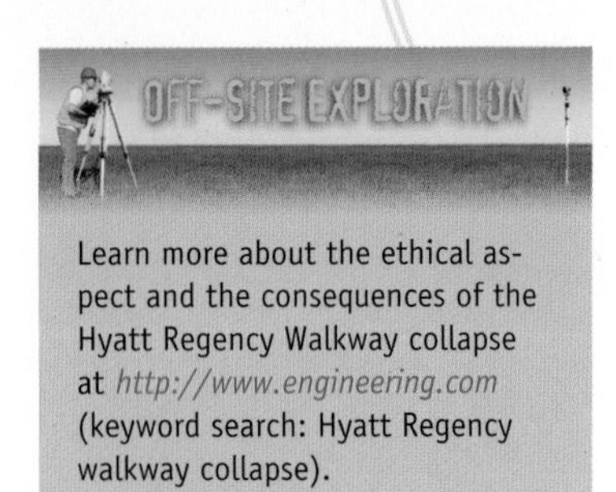

Learn more about the ethical aspect and the consequences of the Hyatt Regency Walkway collapse at http://www.engineering.com (keyword search: Hyatt Regency walkway collapse).

STATICS

Although we know, and accept, the fact that buildings move—floors bend when we walk on them and structures sway when the wind is strong—structural engineers typically assume a building will remain static, or at rest, when loaded. The branch of physical science that deals with forces on bodies at rest is referred to as **statics**. In order to design a safe structure, we must first understand how structural members resist the loads to which they are subjected while simultaneously remaining stationary.

Forces

Equilibrium: exists when the forces acting on an object counteract each other so that the object is in a state of rest.

In order for an object to remain static, the object must be in **equilibrium**—the forces acting on the object must counteract each other. In other words, if you push an object to the right, there must be an equal force pushing the object to the left in order for the object to be in equilibrium and therefore static (Figure 14-22). The forces that act in a straight line are referred to as linear forces and have units of pounds or kilograms in the Imperial system. A kip is equivalent to 1000 lb, and a ton is equivalent to 2000 lb.

Moment of a Force: the tendency of the force to cause rotation of a body about a point and is equal to the product of the magnitude of the force and the perpendicular distance to the force.

Sometimes a force that acts on an object will cause the object to turn or twist. For example, a merry-go-round is a piece of playground equipment with a circular riding surface that rotates around a central vertical axis (see Figure 14-23). You can rotate a merry-go-round by pushing on the outside edge of the circular surface. We refer to the tendency of a force to cause rotation of an object as a **moment of the force**. However, if someone wanted to stop you from rotating the merry-go-round, he would push the circular base in a direction that would oppose your effort. If you were forcing the equipment to move in a counterclockwise direction, the person would push in a clockwise direction. He would not have to stand directly across from you; he could stand anywhere around the circumference of the equipment and resist the rotation with the same force at the same distance from the center. Now imagine that the other person got a long pole and laced it between the rails on

Figure 14-22: An object is in equilibrium if the forces acting on the object counteract each other.

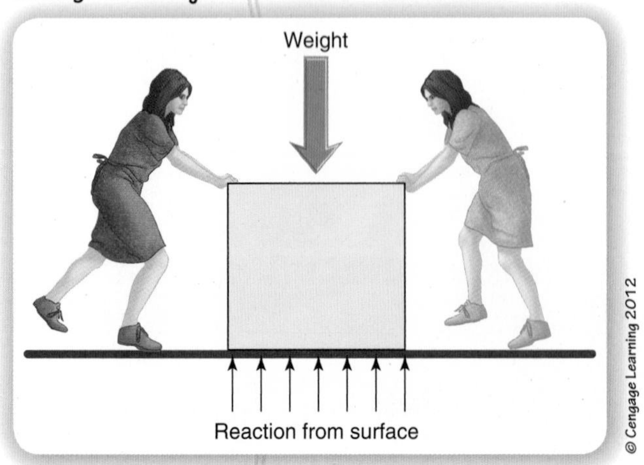

Figure 14-23: The tendency of a force to cause rotation of an object is called the moment of the force. The moment about a point, C, is the product of the magnitude of the force, F, and the perpendicular distance from the point to the force, d.

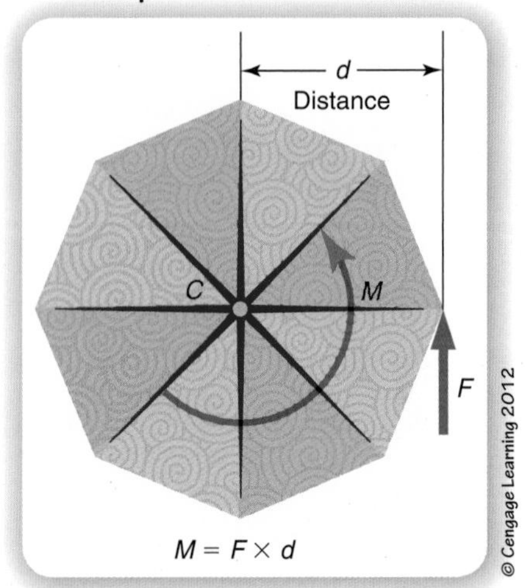

the merry-go-round so that he could apply a force to the merry-go-round from a point further away from the center (see Figure 14-24). You would intuitively agree that, in order to stop you from rotating the equipment, he would have to push with less force clockwise than you exert in a counterclockwise direction—how much less depends on his distance and your distance from the center point. But the result is that the moment that you create by pushing counterclockwise is opposed by an equal moment created by the other person pushing clockwise.

In order for a body to be static and therefore in equilibrium, the following must be true:

Fundamental Principles of Equilibrium

- $\Sigma F_v = 0$ The sum of all vertical forces acting on a body must equal zero
- $\Sigma F_h = 0$ The sum of all horizontal forces acting on a body must equal zero
- $\Sigma M_p = 0$ The sum of all moments (about any point) acting on a body must equal zero

We will use the fundamental principles of equilibrium to analyze structures, to determine the forces that must be resisted by each component, and to calculate the forces transferred from one component to another.

Figure 14-24: The moment about point C of a force F applied at a distance, d, from C can be counteracted with a force with a magnitude of 1/2F if the force is applied at a distance of 2d.

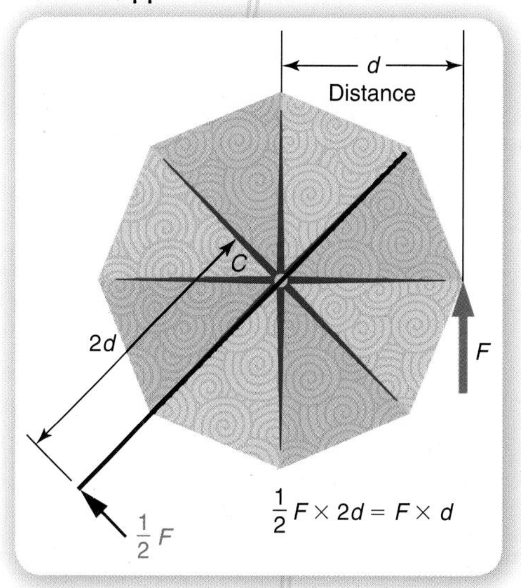

FREE-BODY DIAGRAMS When analyzing structural members, it is convenient to isolate each individual component from the structural system and identify the forces acting on that member. This is generally done using a **free-body diagram**. A free-body diagram is a graphical representation of an object and the forces (indicated by arrows) that are acting on the object. Two types of forces are shown on a free-body diagram: applied forces and reaction forces. An **applied force** is an external force that loads the structure, like a dead, live, or wind load. A **reaction force** is a force that resists an applied force. If we know the applied force(s), we can often calculate the reaction force(s) in simple structural members like axially loaded columns and simple beams.

A free-body diagram of a rigid brace is shown in Figure 14-25. The applied load, W, is resisted by the reaction forces of F_{xA}, F_{yA}, F_{xB}, and F_{yB}.

Figure 14-25: Free-body diagram of a rigid brace.

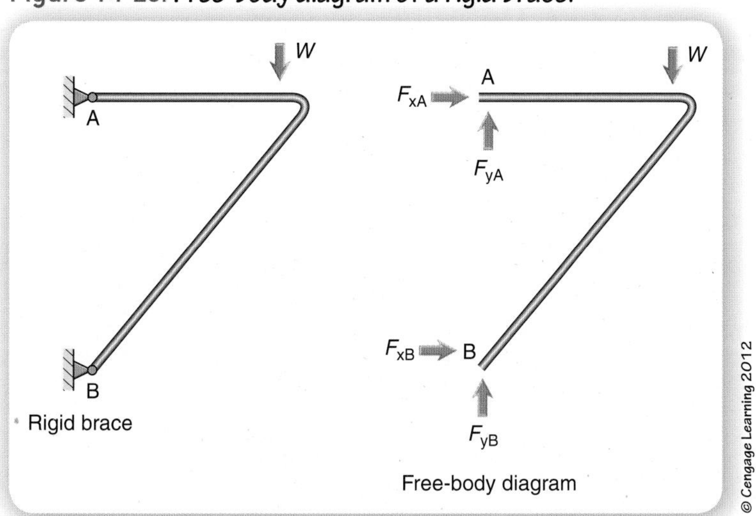

Figure 14-26: **Free-body diagram of a building column.**

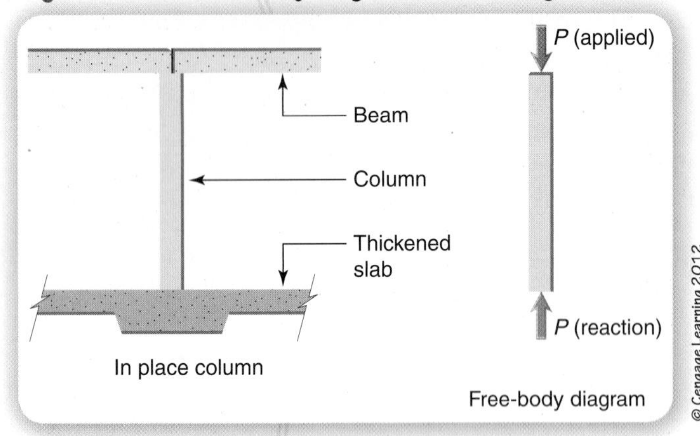

The free-body diagram of a **compression** (or **tension**) member is fairly simple to draw. Figure 14-26 illustrates a building column that rests on a thickened slab and is loaded by gravity loads transferred from beams above. The gravity loads are represented by the arrow, sometimes called a *force vector,* that points in the direction of the load (downward) at the point of load application (the top of the column). The **P** represents the magnitude of the gravity loads. Because the column is at rest, it must satisfy the fundamental principles of equilibrium. If we assume that the applied load is an axial load through the center of the column (so that the load doesn't cause any bending in the column), we only have to consider one of the equations of equilibrium in this case—the sum of the vertical forces must equal zero. Therefore, the resulting force on the bottom of the column must be equal to and opposite from the applied force acting at the top—it must have a magnitude of P and act upward. So, for example, if a column carries a gravity load of 50 k (= 50,000 lb), the resulting force at the base of the column must be a 50 k force upward. There are no horizontal forces or moments acting on this column, so we can ignore the remaining principles of equilibrium.

Tension:
is a force that tends to stretch or elongate an object.

Compression:
is a force that tends to press together or shorten an object.

Translation:
the motion of a body in a straight line.

Rotation:
the act or process of turning around a center or an axis.

The free-body diagram for a structural member that carries a transverse load (called a beam) is somewhat more complicated than that of an axially loaded member because there is always more than one reaction involved. Sometimes, depending on the way that the beam is supported, a moment reaction, in addition to force reactions, may result from the applied loading. There are three basic types of beam support: roller, pinned, and fixed. Each type will generate a different set of reactions. A **pinned support** will not allow translation in any direction (no vertical or horizontal movement), but it will allow rotation about the support. An example of a pinned support is shown in Figure 14-27. A **roller support** allows **translation** in one direction and will allow **rotation** of the beam about the support—it resists loads perpendicular the support surface; this also may be referred

Figure 14-27: **Pinned support. The support between the brace and the awning beam is assumed to provide resistance to translation movement but no rotational resistance.**

Figure 14-28 a, b, c: Rocker supports (a) and roller supports (b) were commonly used to support bridges prior to the 1970s to accommodate the large thermal expansion and contraction that results from exposure of the structure to the weather. Today bridges are often supported by elastomeric bearing pads (c) that allow the bridge girder to translate and rotate at the support.

to as a *rocker support.* You can see rocker and roller supports (Figure 14-28) on older highway bridges, but newer connection designs use elastomeric material to allow translation and rotation at the support. A **fixed support** restricts translation and rotation so that the end of the beam is completely immobile like the support shown in Figure 14-29. The table in Figure 14-30 shows idealized symbols and the reaction forces that are possible at each type of support.

A beam is classified based on how it is supported. For example, a beam that is supported on one end by a roller and on the other end by a pinned support is classified as a **simple beam** (see Figure 14-31). A simple beam will have three reaction forces—two at the pin and one at the roller. Since there are three unknown reactions, we can use the fundamental principles of equilibrium to calculate the support reactions for a simply supported beam;

Figure 14-29: Fixed connection. The connection between the cantilevered window shade beams and the wall resists both translational and rotational movement.

Figure 14-30: Beam supports and corresponding reactions.

Type of connection	Idealized symbol	Reaction	Number of unknowns
Pinned		F_x, F_y	Two unknowns. The reactions are two force components.
Roller or rocker		F	One unknown. The reaction is a force that acts perpendicular to the surface at the point of contact.
Fixed		M, F_x, F_y	Three unknowns. The reactions are the moment and two force components.

Figure 14-31: Simply supported beam diagram and corresponding free-body diagram.

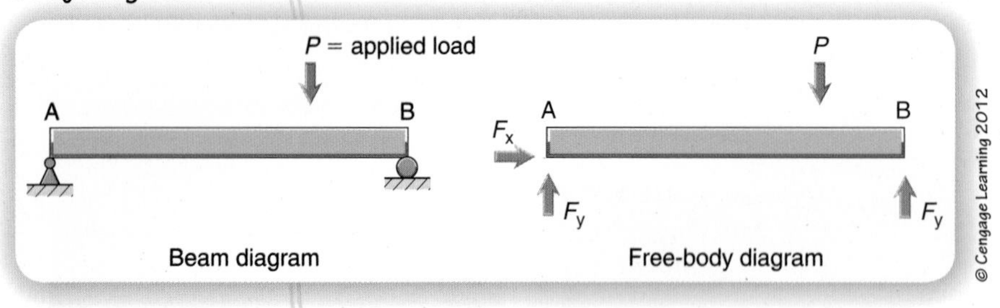

Figure 14-32: A statically indeterminate beam has more than three unknown reactions.

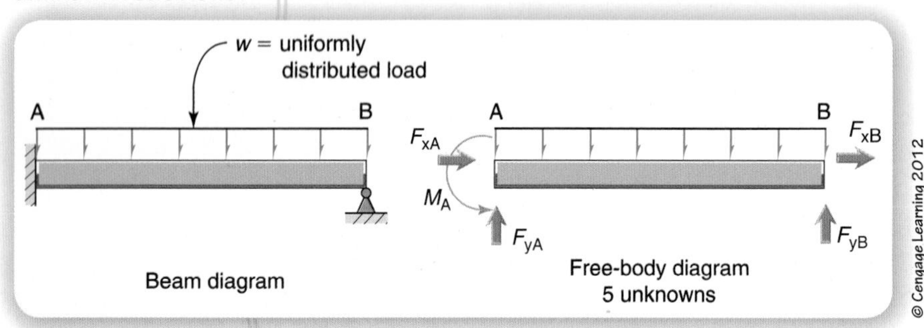

a system of three equations with three unknowns can be solved with algebra. Beams for which the reactions can be calculated using the equations of equilibrium are called **statically determinate beams**. As long as there are at most three reaction forces on any one beam, the beam will be statically determinate.

In some cases, due to the configuration of the supports, there are more than three reaction forces. For example, a beam with a fixed end and a pinned end would result in five unknown reactions (see Figure 14-32). You may remember from algebra that you can solve for multiple variables if you have enough equations that relate those variable—you need one equation for each unknown. Therefore, if there are more than three unknown reaction forces, it is impossible to determine the reactions of this beam using the equations of equilibrium because there are only three equilibrium equations. Beams with more than three unknown reaction forces are called **statically indeterminate beams** and require advanced methods and/or computer software to find the resulting forces. We will limit our discussion in this text to statically determinate beams.

Some statically determinate beam configurations are shown in Figure 14-33 with the corresponding free-body diagrams for a uniformly distributed load.

Figure 14-33: Examples of statically determinate beams. Each beam has only three unknown reactions.

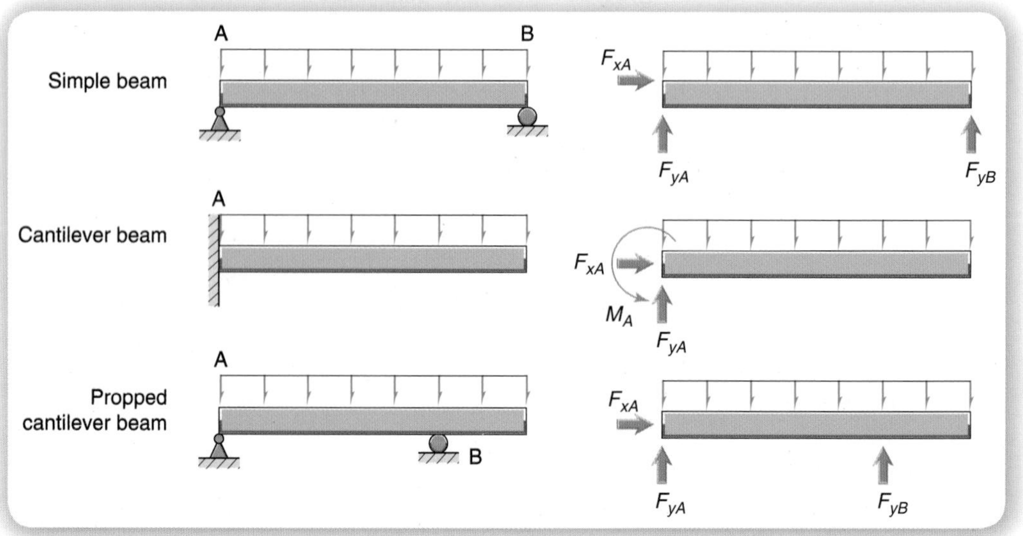

Your Turn

A diving board is an example of the use of a propped cantilever beam. Find at least one example of each of the beam types listed in Figure 14-33 in your town. Document your research with photographs.

A diving board is an example of the propped cantilever statically determinate beam.

A

B

Example:

Use the principles of equilibrium to find beam end reactions for the determinate beam shown below.

Step 1: *Sum forces horizontally.*

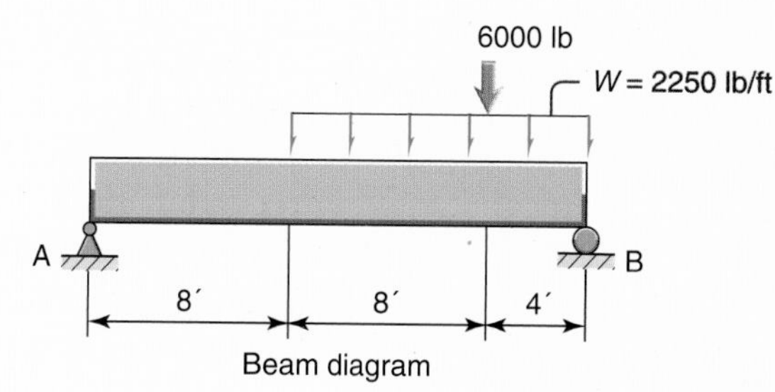

Beam diagram

$\Sigma F_x = 0$ Assume right is positive $\rightarrow +$

$F_{xA} = 0$ Since there are no horizontal applied loads, there will be no horizontal reaction

Step 2: *Sum forces vertically.*

$\Sigma F_y = 0$ Assume up is positive $\uparrow +$

$F_{yA} + F_{yB} - 6000\ lb - \left(2250\ \frac{lb}{ft}\right)(12\ ft) = 0$

$F_{yA} + F_{yB} - 6000\ lb - 27{,}000\ lb = 0$

$F_{yA} + F_{yB} = 33{,}000\ lb$

We need another equation to solve for the two unknowns.

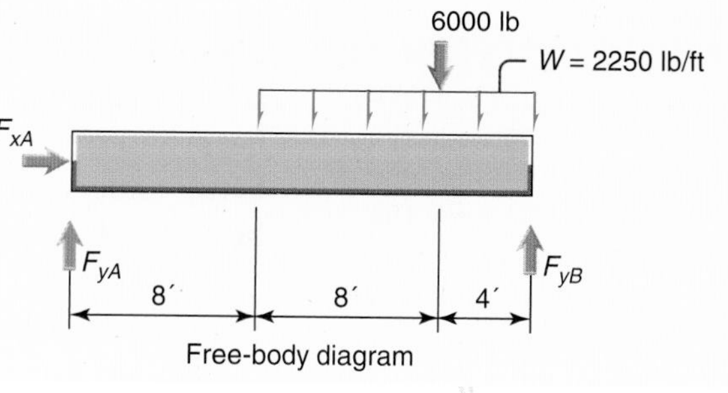

Free-body diagram

Step 3: ***Sum moments about a point.***

Strategically choose either point A or B.

$\Sigma M_B = 0$ ↺+ Assume that a counterclockwise moment is positive

$$-F_{yA}\,(20\,ft) + (6000\,lb)\,(4\,ft) + \left(2250\,\tfrac{lb}{ft}\right)(12\,ft)\left(\tfrac{12\,ft}{2}\right) - F_{yB}\,(0) = 0$$

Note that a uniformly distributed load is applied at the center of the load

$$-F_{yA}\,(20\,ft) + 24{,}000\,ft \cdot lb + 162{,}000\,ft \cdot lb + 0 = 0$$

$$F_{yA}\,(20\,ft) = 186{,}000\,ft \cdot lb$$

$$F_{yA} = 9300\,lb$$

Step 4: ***Substitute the known reaction into previous equation.***

$$F_{yA} + F_{yB} = 33{,}000\,lb$$

$$9300\,lb + F_{yB} = 33{,}000\,lb$$

$$F_{yB} = 23{,}700\,lb$$

CHECK:

Forces downward: $6000\,lb + \left(2250\,\tfrac{lb}{ft}\right)(12\,ft) = 33{,}000\,lb$

Forces upward: $9300\,lb + 23{,}700\,lb = 33{,}000\,lb$

Forces downward = Forces upward OK

Shear Stress:
a stress in which the material on one side of a plane pushes on the material on the other side of the plane with a force that is parallel to the plane and opposite in direction.

Moment about a Point (P):
the tendency of a force to rotate an object about point P. It is equal to the product of the magnitude of the force acting upon the object and the perpendicular distance from the point P to the force.

Shear Force and Bending Moment

Placing a transverse load on a beam creates two separate responses from the beam. First, the load causes internal fibers to displace in the direction of the load. These fibers slide against the adjacent fibers, which then slide against the next fibers and so on until the load is transferred through **shear stress** to the end support of the beam. Second, the transverse load causes the beam to bend and creates a **bending moment** within the beam. A bending moment is an internal response that involves a combination of tension stress and compression stress. When a beam is loaded with a gravity load, the top region of the beam will be compressed and the bottom region will be stretched as the beam bends downward—a combination of compression stress and tension stress that we often refer to as **bending stress** or flexural stress. This flexural stress is illustrated in Figure 14-34.

We will look at a specific example to help explain shear and bending moment. Figure 14-35a gives a beam diagram for a 20 ft long simply supported beam with a 35 k load applied at 8 ft from the left end. If we create an imaginary break in the beam and create a free-body diagram for the left portion of the beam only, we can better understand the forces, both internal and external, that are acting upon the beam (see Figure 14-35b).

Although we are using only a portion of the beam in the free-body diagram, you must remember that the remainder of the beam does exist and can apply forces to the diagrammed portion of the beam. Once the load has been applied and the

Figure 14-34: When a beam is loaded with a gravity load, the top of the beam is placed in compression and the bottom of the beam is placed in tension. This creates an internal moment, called a *bending moment.*

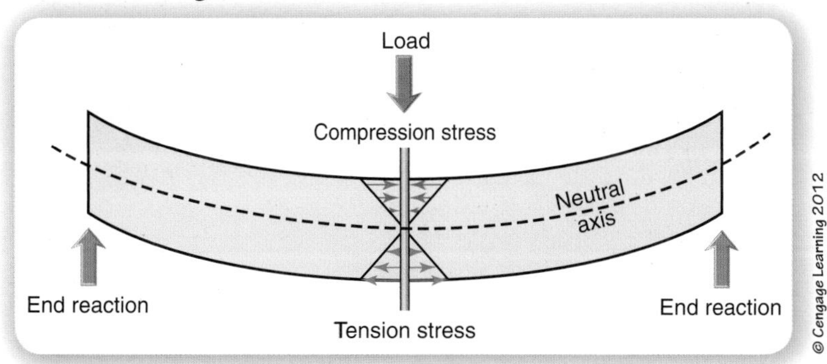

Bending Stress: tensile and compressive stress developed in a beam to resist the bending moment.

beam is static, the equilibrium equations must hold true, and we can make the following general observations.

- $\Sigma F_x = 0$ There are no horizontal applied forces and therefore $F_{xA} = 0$ and is not shown. However, we know that tension and compression stresses are present at an imaginary cut, therefore, the tension and compression stresses must be equal and opposite to each other in order to maintain equilibrium.
- $\Sigma F_y = 0$ A vertical force (shear force) must exist that is equal in magnitude and opposite in direction to the end reaction, F_{xB}. Because there are no external elements to deliver a load, the shear force must be internal and be delivered from the removed part of the beam.
- $\Sigma M_p = 0$ Strategically selecting the point P for our summation of moments, we observe that the moment, M, caused by the tension and compression stresses (called a bending moment) is counterclockwise, and must counteract the sum of the moments about P caused by the applied load and the end reaction as shown in the example below.

You can calculate the internal shear force and bending moment at any point along the beam using a free-body diagram of the portion of the beam to the left of the point of interest. We can then use the calculated shear force and bending moment to find the shear and bending stresses at that point.

Figure 14-35: Internal shear and bending moment at a point can be determined by considering only the portion of the beam to the left of the point and applying the equations of equilibrium.

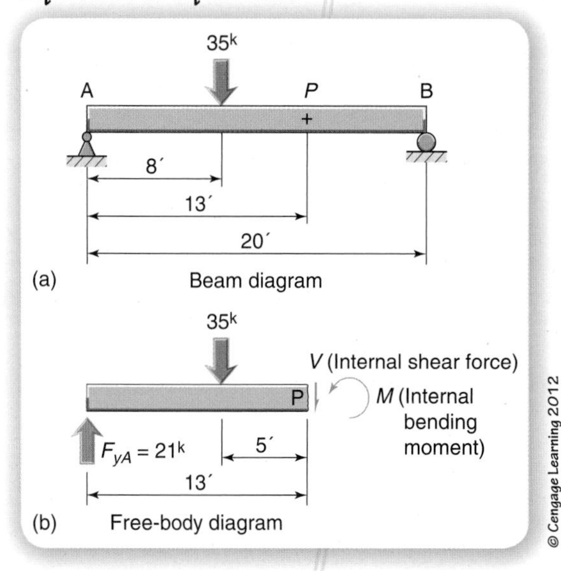

Example: Calculate Shear Force and Bending Moment at a Point

Find the shear force and bending moment at a point that is 13 ft from the left end of the beam shown in Figure 14-35a.

Step 1: *Sketch a free-body diagram for the entire beam.*

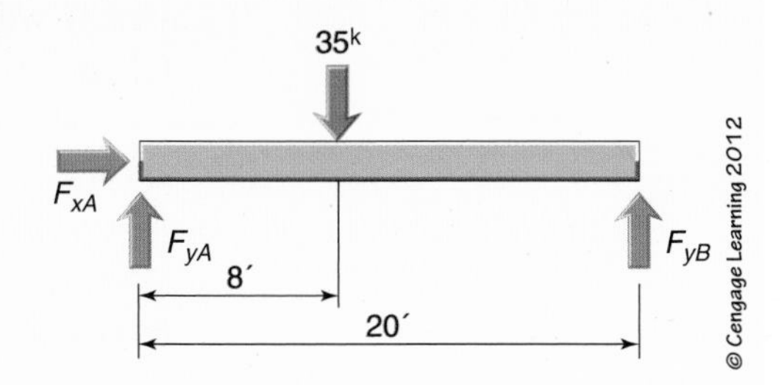

Step 2: ***Find the reaction forces, F_{xA}, F_{yA}, and F_{yB}.***

Use the equations of equilibrium to determine that $F_{xA} = 0$, $F_{yA} = 21\ k$, and $F_{yB} = 14\ k$.

Step 3: ***Sketch a free-body diagram for the portion of the beam to the left of the point of interest, P.***

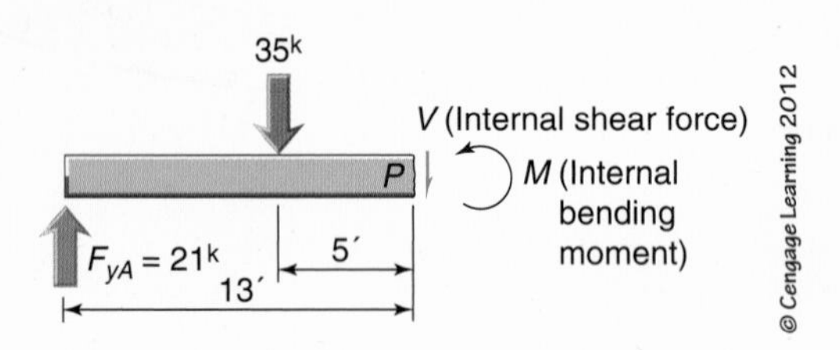

Step 4: ***Calculate the shear force, V, at P using the equilibrium equation (assuming a sign convention where down on the right is positive).***

$$\sum F_y = 0$$
$$F_{yA} - 35\ k - V = 0$$
$$V = F_{yA} - 35\ k$$
$$V = 21\,k - 35\ k$$
$$V = -14\,k$$

Step 5: ***Calculate the bending moment at P using the equilibrium equation (assuming a sign convention of counterclockwise rotation about P is positive).***

$$\sum M_p = 0$$
$$-F_{yA}(13\,ft) + (35\ k)\,(5\,ft) + M = 0$$
$$M = (21\,k)\,(13\,ft) - (35\ k)\,(5\,ft)$$
$$M = 273\,ft \cdot k - 175\,ft \cdot k$$
$$M = 98\,ft \cdot k \text{ counterclockwise}$$

Note that M is positive because we have drawn M counterclockwise. We have assumed the direction to be counterclockwise, a positive value of M will indicate that the bending moment is indeed counterclockwise. A negative M will indicate that the bending moment is in the opposite direction (clockwise).

Therefore, the internal reactions to the applied load at point P are a shear force of 21 k up and a bending moment of 98 ft · k in a counterclockwise direction.

SHEAR AND MOMENT DIAGRAMS In practice, engineers are often most interested in the maximum shear and maximum bending moment in a beam because these internal forces will create the most critical condition within the member. Engineers will often use diagrams that graphically represent the shear and bending moment at every point along a beam. These diagrams are referred to as **shear and bending moment diagrams** and are typically shown as a pair under the corresponding beam diagram. The shear and moment diagrams from the last example are shown in Figure 14-36.

Shear and bending moment diagrams can be approximated by calculating the internal shear force and bending moment at multiple points along the beam, plotting the values at the corresponding horizontal beam locations, and connecting the points with a smooth curve.

Think of a shear diagram as a method of keeping track of all of the vertical loads that affect the beam, including applied loads and resulting reactions. We will use a sign convention such that an upward external force to the left of the point of interest creates a positive change in shear at the point (see Figure 14-37). When creating a shear diagram, start at the left support and move to the right, adjusting the shear value for each load that you encounter. Each load will create a change in the shear value according to the following:

- **Concentrated loads** (and reactions) will result in an instantaneous change in the shear value equal to the magnitude of the force. An upward force results in an increase in the shear; a downward force results in a decrease in the shear. Notice the shear magnitude drops at both the 12 k load and the 9 k load and increases at each end reaction in Figure 14-38.
- **Uniformly distributed loads** will result in a linear change in the shear value such that the shear will change at a rate equal to the uniform load. The slope of

Figure 14-36: A shear and bending moment diagram for a simply supported beam with a single concentrated load. These diagrams correspond to the previous example. Notice that the magnitude of the shear and the bending moment at 13 ft from the left end of the beam agree with the values calculated.

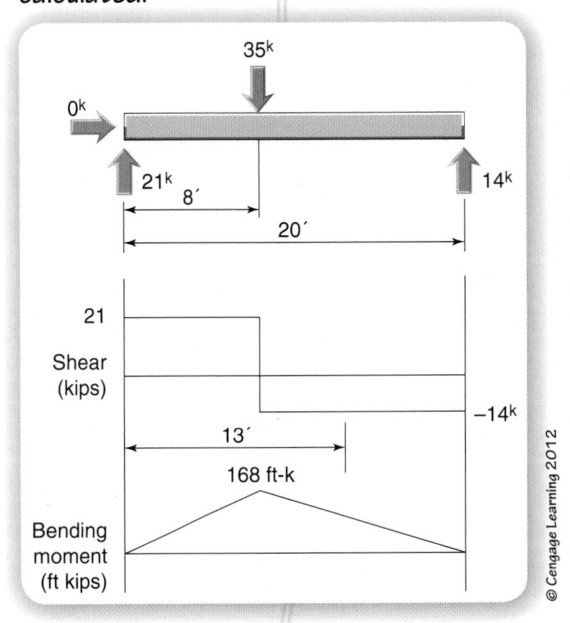

Figure 14-38: Shear diagram.

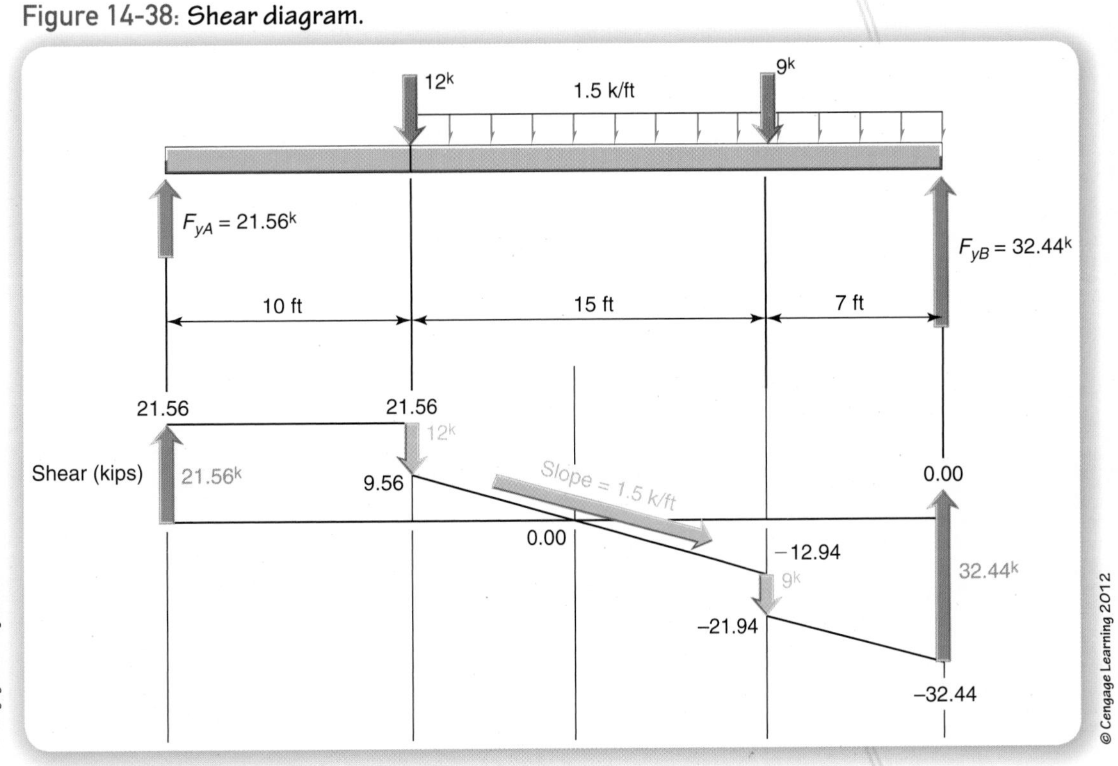

Figure 14-37: Sign conventions for shear and moment diagrams.

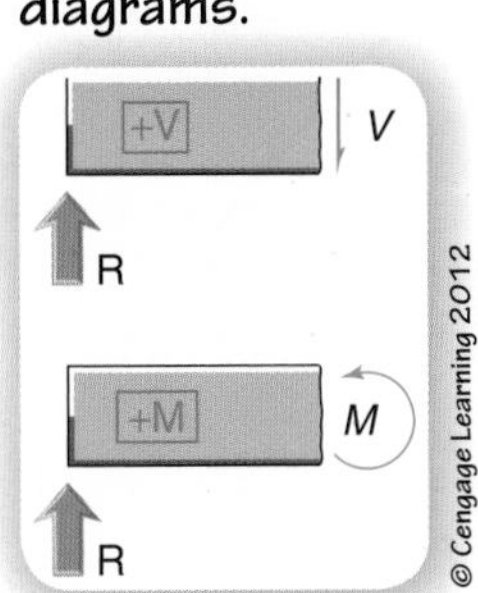

the shear diagram is equal to the magnitude of the uniform load. For instance, if the uniform load is 1.5 k/ft downward, the shear diagram will show a drop in the shear value at a rate of change of 1.5 k/ft of length of the beam. The slope of the shear is negative (down to the right) because the uniform load is downward. If the uniform load is upward, the slope of the shear is positive. See Figure 14-38.

Because bending moment at a point is a function of both the magnitude of the loads to the left of the point and the distance to each load, and because the shear diagram is simply a representation of load magnitudes and their location along a beam, you can use the shear diagram to help create the moment diagram.

For instance, if we calculated the bending moment for the beam shown in Figure 14-38 at a point located 10 ft from the left support only the left end reaction would be involved in the calculation. The 12 k load would not apply a moment at this point because the load acts through point *P* and therefore has a perpendicular distance of zero from *P*. The magnitude of the bending moment at this point, *P*, is the product of the magnitude of the left end reaction, 21.56 k, and the distance from *P* to the support, 10 ft, which is 215.6 ft · k (= 21.56 k · 10 ft). This bending moment magnitude is exactly the area under the shear diagram curve to the left of the point of interest indicated by the shading in Figure 14-39.

Figure 14-39: The bending moment at a point is equivalent to the area under the shear diagram to the left of the point. Therefore, the area under the shear diagram to the left of point *P* is equivalent to the bending moment at point *P*.

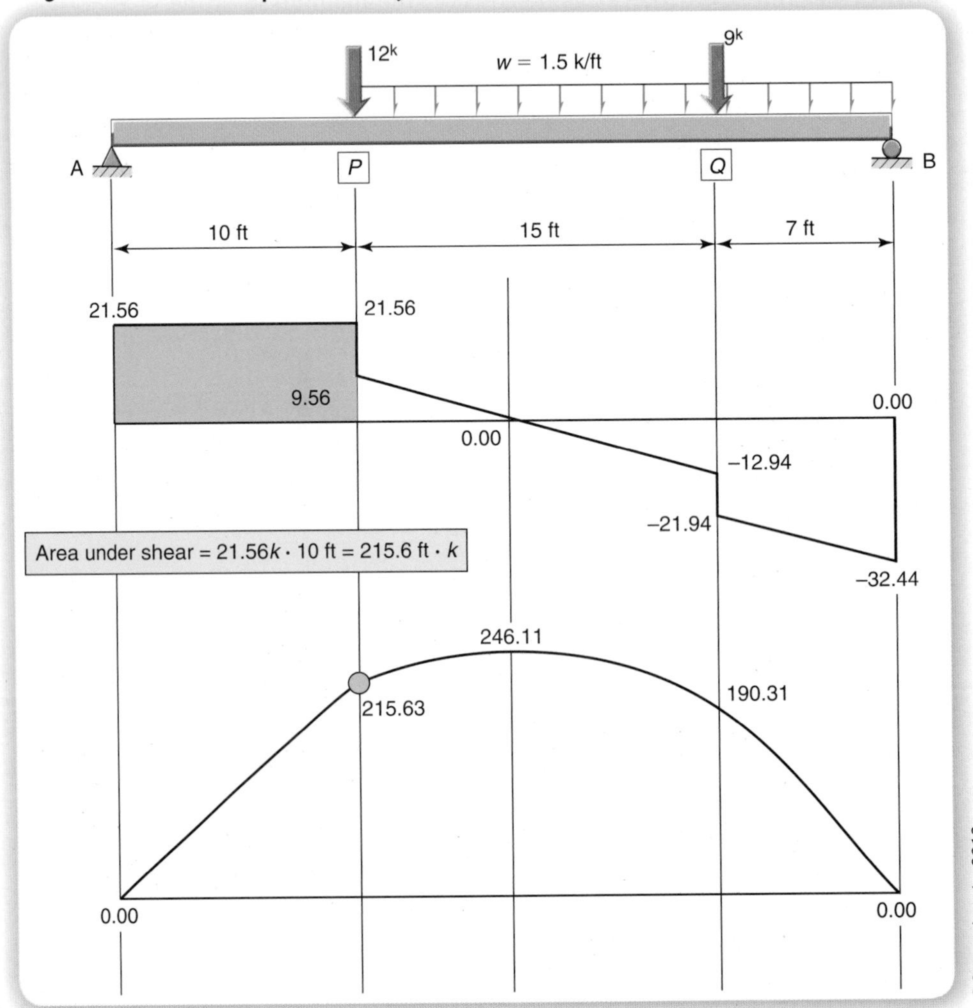

Likewise, in order to calculate the bending moment at point Q, find the area under the shear diagram to the left of Q. Note that a shear area below the horizontal axis contributes a negative area to the sum of areas (see Figure 14-40).

By calculating the bending moments at critical points (where changes in slope or magnitude occur in the shear diagram or where the shear is equal to zero), you can achieve a good approximation of the moment diagram for a beam. The following hints will help you create an accurate moment diagram:

- The bending moment at any point on a beam is equal to the area under the shear diagram to the left of the point.
- The maximum bending moment always occurs at a point of zero shear.
- The slope of the moment diagram at a given point is equal to the magnitude of the shear force at that point. As a result,
    - If the shear diagram has a constant magnitude (the line is horizontal) the moment diagram has a constant slope (and is therefore a straight line). If the shear magnitude is positive, the moment diagram slope is positive (up to the right), and if the shear magnitude is negative, the moment diagram slope is negative (down to the right) at that point.
    - If the shear diagram magnitude has a constant rate of change (that is the line is straight with a constant slope), the moment diagram will be parabolic.

Figure 14-40: **The area under the shear diagram to the left of point Q is equivalent to the bending moment at point Q.**

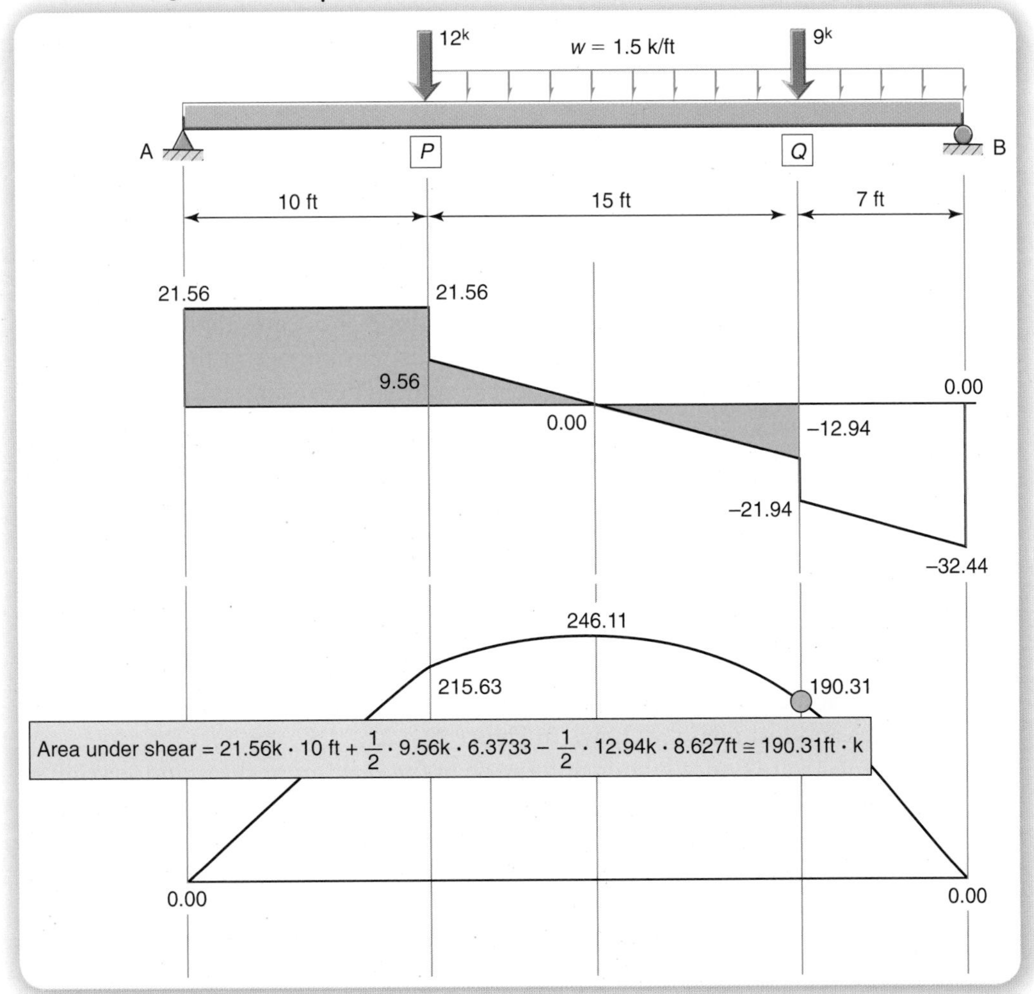

Figure 14-41: Shear and bending moment diagrams for a simply supported beam with an imposed uniformly distributed load.

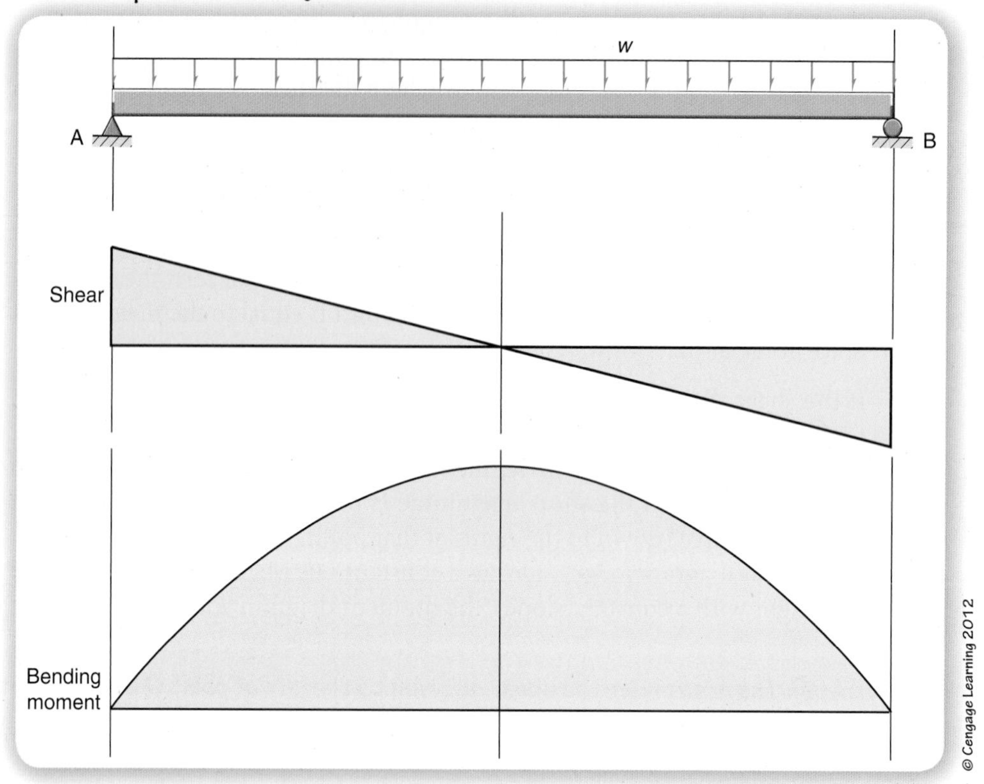

- A positive shear slope indicates that the moment diagram is concave up (shaped like a bowl holding water). A negative shear slope indicates that the moment diagram is concave down (shaped like an upside-down bowl).

If you have studied calculus, you may recognize that the shear diagram is the derivative of the moment diagram.

Once you understand the process of sketching shear and moment diagrams, you may begin to see similarities between the diagrams of beams that are loaded in a similar way. For instance, every simply supported beam subjected to a uniformly distributed load across its span will have shear and moment diagrams shaped like those shown in Figure 14-41. The magnitude of the shear and bending moment at any point depends on the magnitude of the distributed load and the length of the beam. This can be expressed by an algebraic function. As a result, if the loading condition is known, it is possible to create standard formulas to express the magnitude of the shear and bending moment at any point on a beam. Therefore, in lieu of creating a shear and bending moment diagram for each beam loading condition, an engineer can use standard formulas to calculate maximum shear and bending moment. Some standard beam diagrams with corresponding shear and moment diagrams and formulas are shown in Figure 14-42.

Engineers are typically most interested in the maximum shear and maximum bending moment because the beam must be designed to resist these maximums. For all symmetrically loaded beams—that is, beams subjected to a loading condition that is symmetric about the mid-span of the beam, the maximum moment (and zero shear) will occur at mid-span.

Notice that the third beam case is simply the sum of the first two loading conditions. Also, notice that the magnitude of the maximum shear and maximum

Figure 14-42: Beam diagrams with corresponding formula for maximum shear and bending moment.

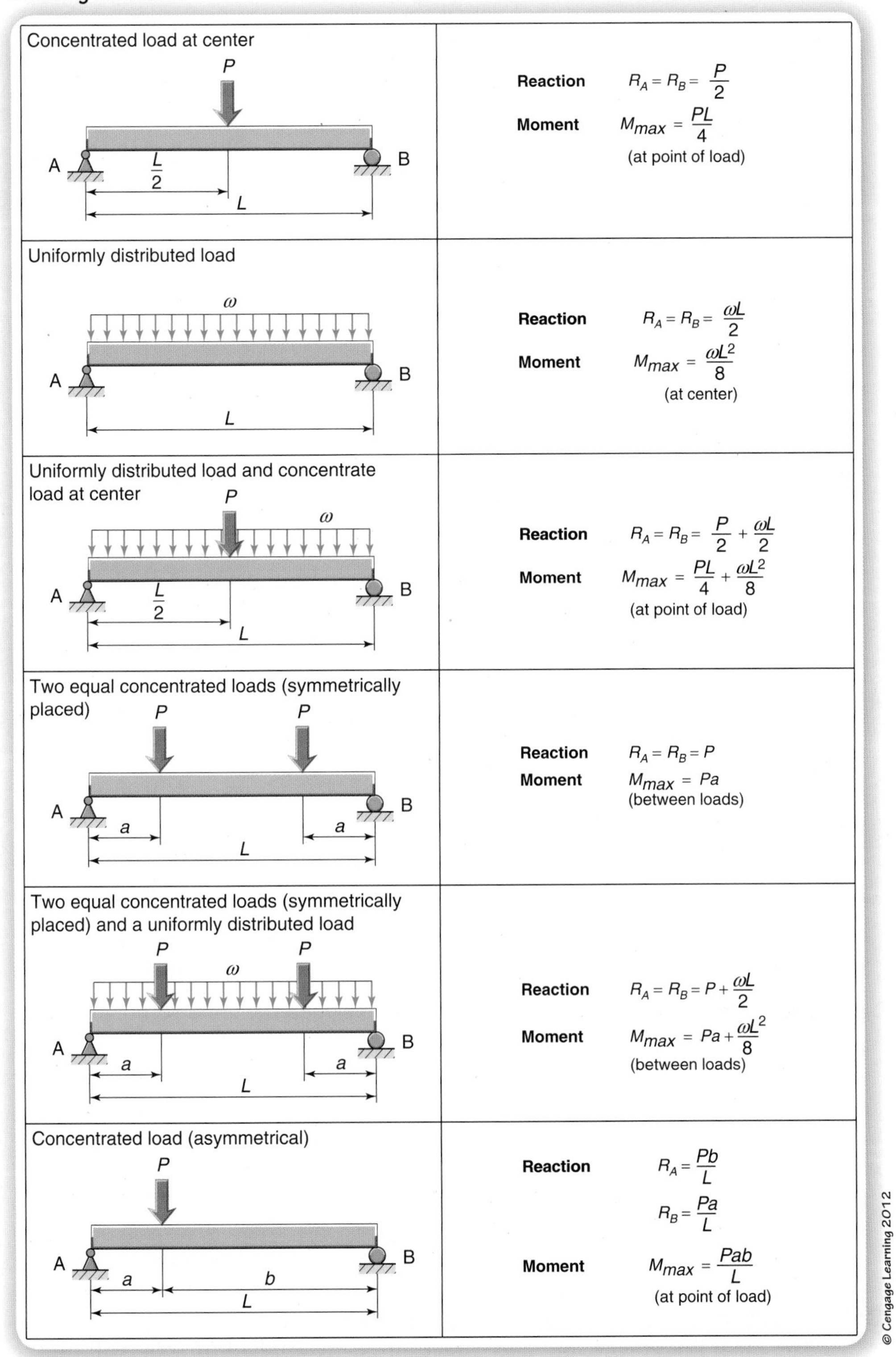

bending moment for the third case are also the sum of the shear and moment from the first two cases. This is true for all beams. If a beam is subjected to two distinct loading conditions, the shear or bending moment at any point along the beam is equal to the sum of the shear or bending moment resulting from the individual loading conditions at that point.

STRENGTH OF MATERIALS

As you have learned, when a structural member is loaded, the member responds with resisting forces. If the member cannot resist the forces, it will fail. How does a structural member resist applied forces and transfer them safely to the next element in the load path? The answer lies within each member, on a microscopic level. One important factor that engineers must consider is the material from which the member is made. As you learned in Chapter 12, every material has unique characteristics that influence its use as a building material. Structural engineers choose materials based, in part, on their strength and capacity to carry loads.

Stress:
a measure of the magnitude of the internal forces acting between the particles of the body exerted as a reaction to external forces on the body. Stress is expressed as the average amount of force exerted per unit area of the surface on which internal forces act.

Stress

Stress is an internal reaction to the external loads that are placed on the member; the particles that directly experience the load will push, pull, or slide against adjacent particles, which then push, pull, or slide against other particles, and so on and so forth, until the forces are finally transferred to a point of support or the next component in the load path.

We typically analyze stress with respect to the cross-sectional area of a structural member. The cross-sectional area can be found by imagining a cutting plane perpendicular to the length of the member (transverse) and finding the area of the cut surface (see Figure 14-43).

Figure 14-43: *Cross-sectional area.*

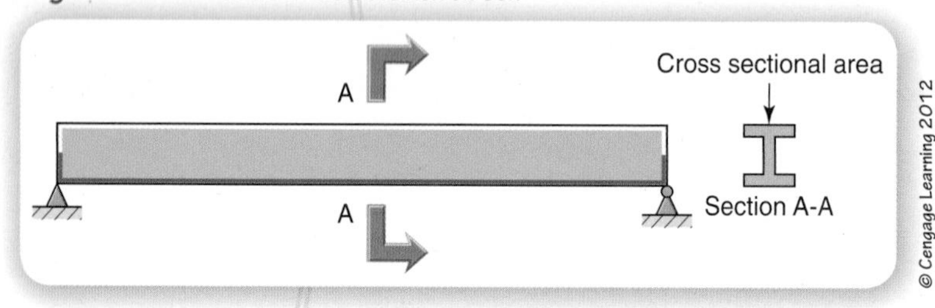

We quantify stress as the force per area over which an internal force acts. Therefore stress is often expressed in pounds per square inch (psi) or kips per square inch (ksi) in the imperial system. In general, there are five types of stress that can be created within a structural member: tension, compression, bending, shear, and torsion. Each is described and illustrated in Figure 14-44.

Tension and compression forces act along the length of the member and are sometimes referred to as *axial forces.* These axial forces create internal stresses that act perpendicular to the cross sectional area of the member. A stress that acts perpendicular to a surface is called a **normal stress**. Axial stress, including stress caused by bending, is often represented by the Greek letter sigma, σ.

Shear and torsion loads are resisted by internal stresses that act parallel to the resisting area; for our discussion, the resisting area is the cross-sectional area of the member as shown in Figure 14-43. Shear stress is commonly represented by the Greek letter tau, τ.

Flexure is, as previously discussed, a combination of normal tension and compression stresses. The plane along the length of the member on which there is no tension or compression in referred to as the **neutral axis**.

The magnitude, location, and direction of the applied force(s) will determine the type and magnitude of the internal stress. Often a structural member will experience more than one type of stress under a particular load combination. In order to design a safe structure, the engineer must ensure that the internal stresses do not exceed the maximum stresses that the material is allowed, by code, to carry. We will discuss maximum allowable stresses later in this chapter.

Figure 14-44: Types of stress.

Stress type	Description	Illustration
Tension	Pulling, stretching, elongating Stress acts perpendicular to cross-sectional area	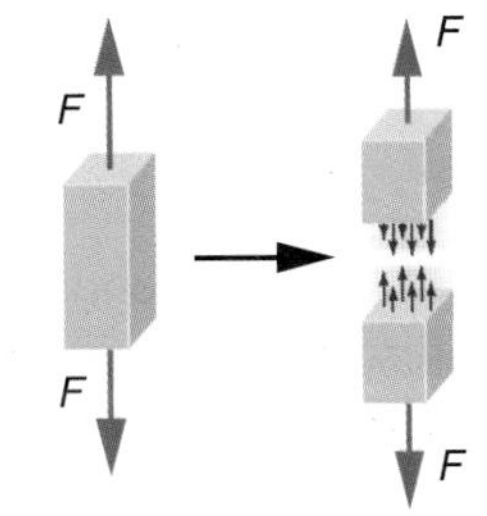 F F → F F
Compression	Pushing, squeezing, shortening Stress acts perpendicular to cross-sectional area	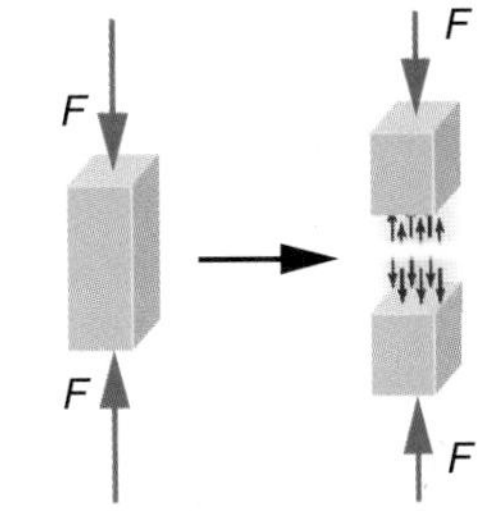 F F → F F
Shear	Tearing or sliding Stress acts parallel to cross-sectional area	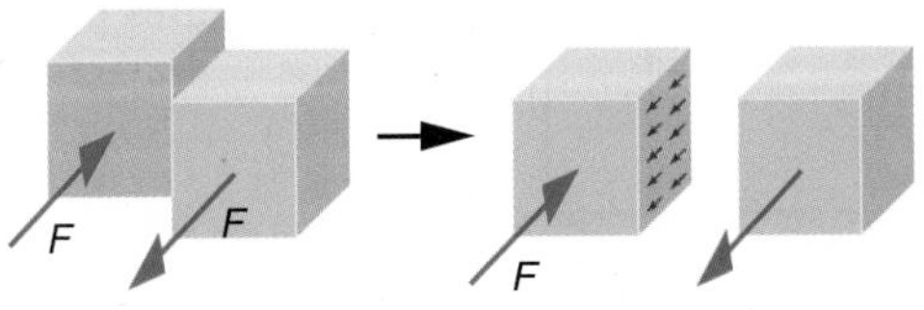 F F → F
Flexure	Bending Stress acts perpendicular to cross-sectional area	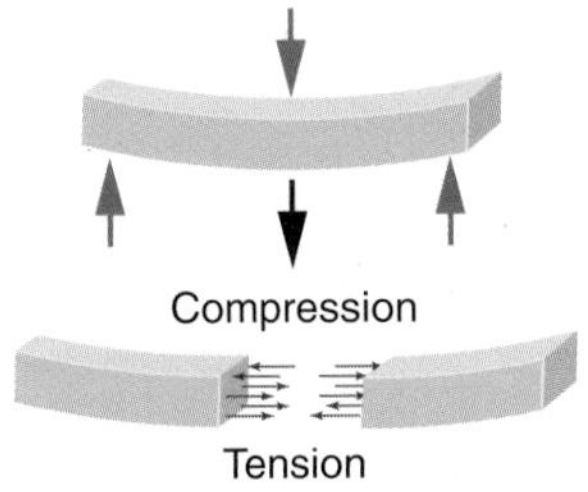 Compression Tension
Torsion	Twisting Stress acts parallel to cross-sectional area	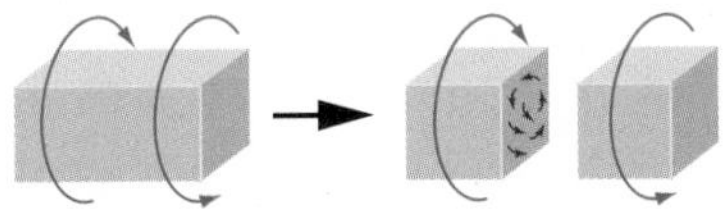

Stress

$$Stress = \frac{F}{A}$$

where

F = applied force

A = resisting area

Calculate Axial Stress

Let's assume that a braced frame includes a **diagonal brace** that carries a tension force of 17 k (17,000 lb). The brace is a 1 in diameter circular rod. What is the tension stress in the brace?

$$Stress(\sigma) = \frac{F}{A}$$

For a circular cross section the area is $A = \pi r^2$

$$\text{Stress}(\sigma) = \frac{17\ k}{\pi(.5\ in)^2} = 21.6\ ksi$$

Strain:
the change in size or shape of a material caused by the application of external forces, also referred to as *deformation.*

Strain

When a material, such as wood, steel, rubber, or an elastic band is subjected to loading (and therefore internal stress), something else happens within the material—it changes shape. **Strain**, or deformation, is the change in the size or shape of an object that results from the application of a load and will result from all types of loads—tension, compression, shear, bending, and torsion. In some cases, the deformation is obvious, like the elongation of a rubber band when you stretch it, or the bending of a wooden ruler when you load it transversely. In other situations deformation is not obvious, like the minute shortening of a column when snow accumulates on a roof.

Axial strain is typically calculated as a ratio of the change in length of the material to its original (unloaded) length and is expressed in the units of inches per inch (in/in). See Figure 14-45.

Figure 14-45: Strain.

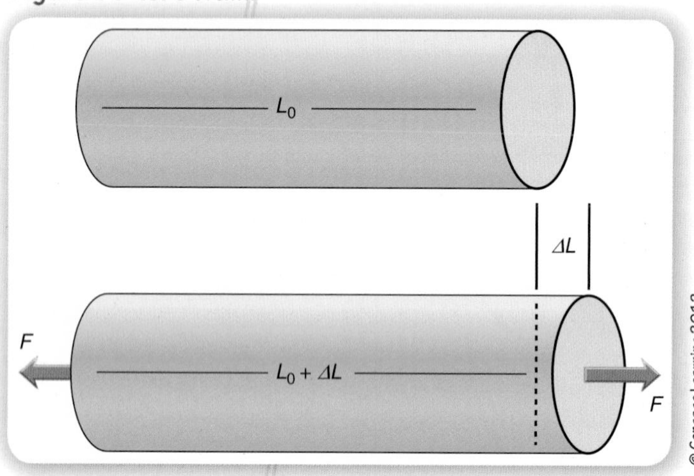

Axial Strain

$$\varepsilon = \frac{\Delta L}{L}$$

where

ε = unit strain (in/in or unitless)
ΔL = total deformation, elongation (in)
L = original length (in)

Example: Calculate Strain

Assume that a tension brace from the previous example was installed in a structural steel building with a column spacing of 25 ft and a floor-to-floor height of 18 ft. Under maximum loading, the brace increased in length by 0.27 in.

Calculate the strain in the brace.

$$\varepsilon = \frac{\Delta L}{L}$$

The length of the brace can be calculated using the Pythagorean theorem:

$$L^2 = H^2 + S^2$$

$$L = \sqrt{(18\ ft \cdot \tfrac{12\ in}{ft})^2 + (25\ ft \cdot \tfrac{12\ in}{ft})^2} = 369.7\ in$$

$$\varepsilon = \frac{0.27\ in}{369.7\ in} = 0.00073\ \frac{in}{in} = 7.3x10^{-4}\ \frac{in}{in}$$

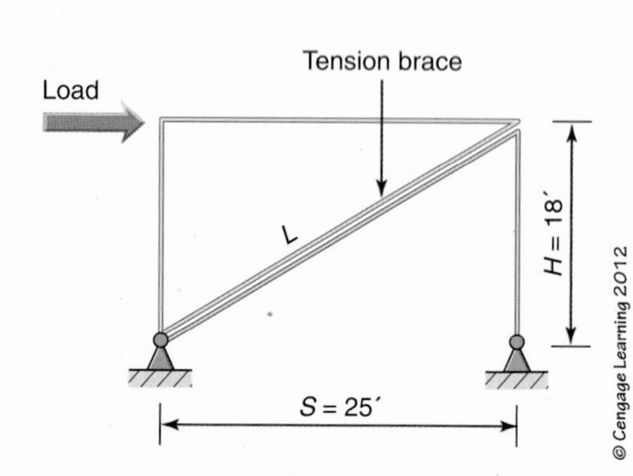

Elastic and Plastic Behavior

Figure 14-46: Stress-strain curve.

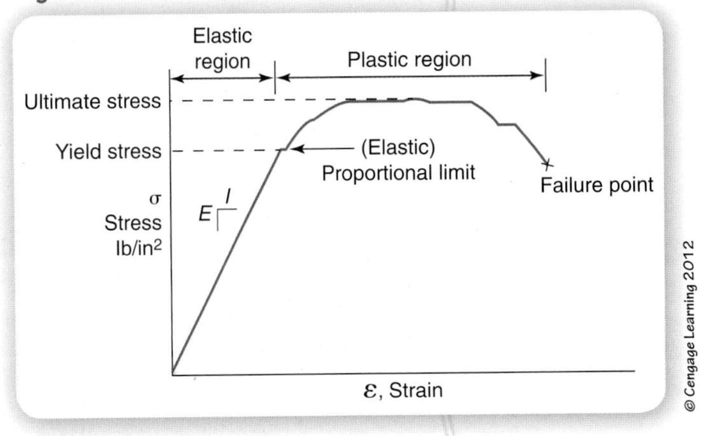

As you can imagine, structural engineers would prefer to minimize the deformation of the structures they design, but deformation is inevitable. If you load a structure, it will deform. In fact, in many materials, strain is directly proportional to the stress, up to a certain point. That means that there is a linear relationship between stress and strain, until you reach a specific stress level called the **yield stress.** The yield stress is the stress at which the deformation will begin to change more quickly per unit of stress. So, if you were to load an object with a 10 lb load which caused a deformation of 0.5 in, doubling the load to 20 lb would also double the deformation to 1.0 in, as long as the resulting stress did not exceed the yield stress. The stress–strain relationship is different for each material, so wood behaves differently than steel, which behaves differently than concrete.

Perhaps the simplest example to demonstrate stress and strain is a *tensile load test,* in which samples are loaded in tension and the resulting strain is recorded. Figure 14-46 gives a general stress-strain curve for structural steel in tension.

The initial straight portion of the curve illustrates the linear relationship between stress and strain that exists in the initial stages of tensile loading, before the stress reaches the yield stress. Up to a certain point, called the **elastic limit,** a member under load will return to its original size and shape when the load is removed. In other words, even though a structural member will deform when it is loaded, it will return to its original size and shape when the load is removed as long as the stress does not exceed the elastic limit. The elastic limit is very close to the yield stress, and for the purposes of this discussion, we will assume that they are the same stress point.

The part of the curve to the left of the yield point (elastic limit) is referred to as the **elastic region** of the curve. The ratio of stress to strain in the elastic region is constant and is often referred to as the **elastic modulus**, although it is also called the *modulus of elasticity* or *Young's modulus.* The elastic modulus depends on the material. For structural steel the accepted value for the elastic modulus is 29×10^6 psi.

The yield stress, elastic limit, and the elastic modulus (E) are referred to as *material properties* because the values depend only on the material from which a structural component is made.

Elastic Modulus:

(also called the modulus of elasticity or Young's modulus): the ratio of axial stress to strain within the elastic range of a material. It is a fundamental **material property** and is an indication of the stiffness of the material.

Elastic Modulus

$$E = \frac{\text{axial stress}}{\text{strain}} = \frac{\sigma}{\varepsilon}$$

where

E = elastic modulus

σ = tensile stress

ε = strain (elongation)

If load is added so that the resulting stress exceeds the elastic limit, the material will continue to deform at a much higher rate. The material will also experience plastic deformation, which means it will not completely return to its original size and shape when the load is removed—some deformation will be permanent. The portion of the curve between the elastic limit and fracture is referred to as the **plastic region**. Structural engineers try to avoid allowing structural members to enter the plastic region so that deformation is not permanent.

Students will understand measurable attributes of objects, the units, and processes of measurement, such as for stress and strain, and apply appropriate techniques, tools, and formulas to determine measurements.

Like axial stress, shear stress results in deformation. Shear deformation is more of an offset in the cross section of the object (as opposed to the shrinking or stretching which results from axial stress). A ratio similar to the elastic modulus, called the **shear modulus**, represents the linear relationship between shear stress and strain before plastic deformation occurs, and is often referred to as rigidity.

Shear Modulus

$$G = \frac{shear\ stress}{shear\ strain} = \frac{\tau}{\gamma}$$

where
G = shear modulus
τ = shear stress
γ = shear strain

SECTION PROPERTIES

The stress that results when a structural member is loaded depends on two things: the loading condition and the shape of the cross section of the member. The material from which the member is manufactured does not affect the magnitude of the stresses induced.

We have already discussed loads and load combinations, but in order to analyze and design structures, you must be familiar with **section properties**. Section properties reflect characteristics of a member that are dependent only on the cross-sectional size and shape of the member. Some of the section properties that will be used in the design of structural members include:

- A_g, gross area of the cross section.
- I, area moment of inertia. The area moment of inertia is a measure of the stiffness of the cross section of a member and is used to predict deflection in a bending member.
- r, radius of gyration. The radius of gyration is used to predict the susceptibility of a compression member to buckling.
- Z, plastic section modulus. The plastic section modulus is used to predict stress in a bending member.

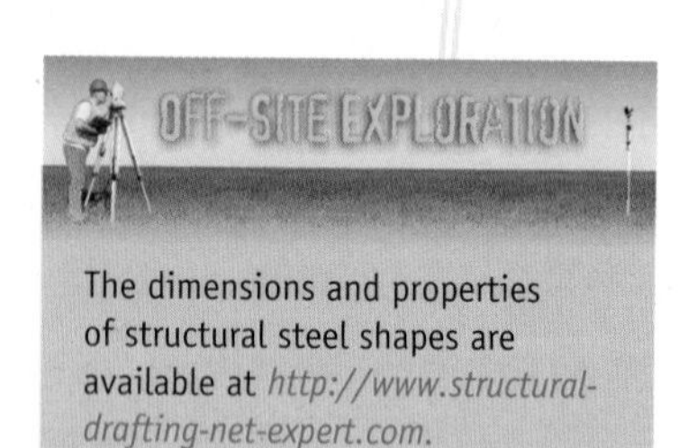

The dimensions and properties of structural steel shapes are available at *http://www.structural-drafting-net-expert.com.*

We will not calculate these properties here, but instead will use published values for the sections we use.

STRUCTURAL DESIGN

As an introduction to structural design we will consider three types of simple structural components: tension members, compression members, and bending members. You have learned how to determine the loads applied to a structural member. In the case of beams, you have learned how to calculate the internal shear and bending moment that result from the applied loads. However, in order to choose a structural member to carry the loads, we must be able to determine loads that a structural member can safely carry. The goal is to choose the most cost-effective member that has a safe load capacity that is greater than the applied load. Of course, there are other considerations that will also affect the structural design such as serviceability, durability, aesthetics, energy efficiency, fire resistance, as well as others. For the purpose of this introduction to structural design we will focus on choosing cost-effective members that provide sufficient strength.

Allowable Stress/Strength

You know from experience that if you apply enough force to an object, say a transverse force to a wooden ruler, you can make it bend. If you gradually apply more force, you can eventually break the ruler. How much force does it take to break the ruler? The answer depends on many factors, but the bottom line is that, on a microscopic level, the material (wood in this case) can only resist so much stress before it ruptures, or fails. The failure point is indicated for steel in the stress-strain curve shown in Figure 14-46.

Based on the discussion in the last section, each structural member should remain within the elastic region of the material in order to safely carry the applied load. But is it safe to assume that every member made from the same material will yield or fail at exactly the same stress? Probably not. There are many factors that can adversely affect the load-carrying capacity, referred to as **strength**, of a structural member, including flawed material, poor fabrication, or incorrect installation. There is also the possibility that unexpected loads will be applied to the structure. Whether it is due to a decrease in material strength or an increase in loads, the possibility exists that a structural member may become overstressed even if it is properly designed.

Structural engineers want to make every effort to ensure that a structural member will never reach a stress level that yields the material—flaws or not. To accomplish this, engineers provide a **factor of safety** in structural designs: a level of comfort that each structural member will not be overstressed. One way to provide a factor of safety is by limiting the stress in a structural member to a value well below the yield stress. Engineers use this reduced stress value, called the **allowable stress**, when designing structural components by ensuring that the maximum stress in the member resulting from the applied load is less than the allowable stress. Another way to express allowable stress is using **allowable strength**, which limits the maximum internal force (such as tension force, shear force, or bending moment) instead of the maximum stress. Allowable stress can easily be converted to allowable strength using the appropriate section property as you will see in later sections.

Factor of Safety:
the ratio of the maximum strength of a member to the probable maximum load to be applied to it.

Allowable stress/strength is dictated by building codes and depends on many factors, the most important of which is the structural material used (such as wood, concrete, or steel). Allowable stress will also vary depending on the type of stress under consideration—that is, the allowable tension stress will typically be different than the allowable shear stress, compression stress, or bending stress. Loading and support conditions can also affect the allowable stress/strength of a member. Rather than specifying allowable stress/strength, building codes often refer to other specifications and documents that are written by outside organizations or trade associations with expertise in a particular material, such as the American Forest & Paper Association (AF&PA), the American Concrete Institute (ACI), and the American Institute of Steel Construction (AISC). These outside organizations publish extensive design guidelines and specifications for a particular material; for instance, the AISC Steel Construction Manual (SCM) presents guidelines and specifications for the design of structural steel that specify allowable stresses.

ALLOWABLE STRESS (STRENGTH) DESIGN VERSUS LOAD RESISTANCE FACTOR DESIGN Within the various structural design guidelines, there are two distinct methods for designing structural components: **Allowable Stress Design** (ASD, also known as *Working Stress Design* or *Allowable Strength Design*) and **Load Resistance Factor Design** (LRFD, also known as *Strength Design*). In some cases, the design guides specify the use of one or the other method. In other cases, as is true in the AISC Steel Construction Manual (SCM), designers are given an option as to which method to use. Although the intent of each method is to provide a reasonable factor of safety to the design of a structure, the two methods use somewhat different approaches.

A great source of information on the comparison of Allowable Stress Design and Load Resistance Factor Design is T. Bart Quimby's online textbook *The Beginner's Guide to the Structural Engineering* at *http://www.bgstructuralengineering.com/Index.html*. Follow the links to Design>Basic Design Concepts>ASD vs LRFD.

Because structural steel is a common commercial structural material and because steel design is less complicated than the design of other structural materials—such as wood or concrete—we will introduce structural design using structural steel shapes. We will apply the ASD design method in our discussion because it is more commonly used than LRFD. However, structural engineers should be well versed in many common structural materials and both ASD and LRFD design methods. With the thirteenth edition of the Steel Construction Manual, AISC changed the method terminology such that ASD now stands for Allowable *Strength* Design, which focuses on limits to the internal forces rather than internal stresses. As a result of the altered terminology, those that are familiar with the ASD method for steel design will find that the formulas and symbols look slightly different from previous editions of the SCM, but the method philosophy remains the same.

STRUCTURAL STEEL

The Steel Construction Manual specifies that the force within a structural member must be less than or equal to the **nominal strength** of the member divided by a factor of safety. The nominal strength of a member is the force that will result in **yielding** of the entire cross section. This requirement is easily expressed and applies to every type of load.

$$F_a \leq \frac{F_n}{\Omega} \quad or \quad F_n \geq \Omega F_a$$

where $\frac{F_n}{\Omega}$ = Allowable Strength F_a = applied load

F_n = nominal strength

Ω = safety factor

The form of this equation will be repeated for each force considered in the design of structural members using the ASD method.

Although structural steel shapes may be available in a variety of steel grades, the preferred grades for common shapes are shown in Figure 14-47.

Figure 14-47: **Preferred grade for structural steel shapes.**

Shape	Preferred Grade of Steel	Minimum Fy (ksi)
W	A992	50
C	A36	36
L	A36	36
Pipe	A53 Grade B	35

Tension Members

Tension members are structural components that carry axial tension forces. Cables in a pulley system are tension components, as are certain truss members, rod hangers, and diagonal braces in a building. Although the connections between the tension member and other structural members are an important consideration in the design of a tension member, connection design is beyond the scope of this textbook. We will consider only the tension strength of the member here.

Allowable Bending Strength

$$P_a \leq \frac{P_n}{\Omega_t} \qquad or \qquad P_n \geq \Omega_t P_a$$

where $\frac{P_n}{\Omega_t}$ = Allowable Tensile Strength
P_a = applied tensile load
P_n =nominal tensile strength
Ω_t = 1.67 = safety factor for tension

If the tensile load is applied in the center of the cross section so that no moment is created, the axial stress is considered to be evenly distributed across the entire cross section. Therefore, the axial stress will theoretically increase until the entire cross section reaches the material yield stress. The load that causes this universal yielding is the nominal tensile strength and is simply the product of the yield stress and the cross-sectional area of the member.

Nominal Tensile Strength

$$P_n = F_y A_g$$

where P_n = nominal tensile strength
F_y = the yield stress $\qquad A_g$ = gross cross-section area of member

Example: Design a Tension Member

Steel bars will be used to laterally brace a structural steel framed building. Each brace will carry a tensile live load of 17 k (17,000 lb). Choose an economical (lightest weight) A36 square steel bar (F_y = 36 ksi) sized to the nearest 1/16 in.

Step 1: *Determine the applied load.*

$$P_a = 17\ k$$

Step 2: *Calculate the required nominal tensile strength.*

$$P_n \geq \Omega_t P_a$$
$$P_n \geq (1.67)(17\ k)$$
$$P_n \geq 28.4\ k$$

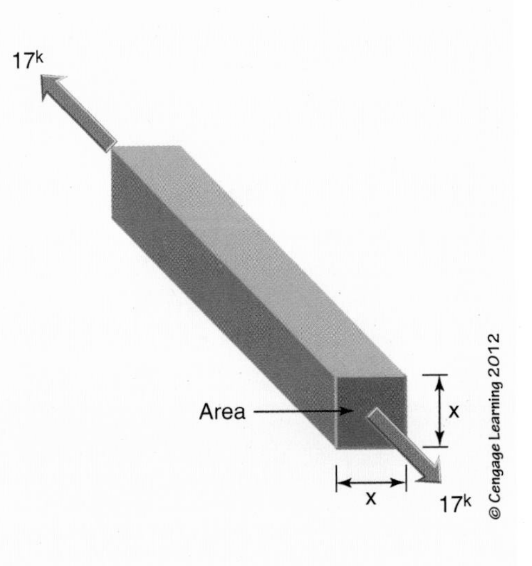

Step 3: *Find the cross-sectional area required to provide $P_n \geq 28.4$ k.*

Since, $P_n = F_y A_g$, substitute

$$F_y A_g \geq 28.4\ k$$
$$A_g \geq \frac{28.4\ k}{F_y}$$
$$A_g \geq \frac{28.4\ k}{36\ ksi} = 0.79\ in^2$$

Step 4: ***Select an economical (light in weight) size.***

For a square section $A = x^2$ where x is the length of each side.

Therefore, $x = \sqrt{A} = \sqrt{0.79\ in^2} = 0.89\ in$

$\frac{14}{16}\ in = 0.875\ in$ – DO NOT USE

$\frac{15}{16}\ in = 0.9375\ in$ – GOOD

Use $\frac{15}{16}\ in$ square A36 steel bar for the lateral braces.

Compression Members

Compression members are structural components that resist axial compression loads. Typical compression members include braces (Figure 14-49), truss elements, and columns (Figure 14-48). Columns are often used in buildings to transfer vertical loads, usually from an upper floor or roof to a lower floor or foundation. Figure 14-49 shows an elevation view of a braced frame. The brace could act in compression or tension depending on the direction of lateral load. In most cases, compression will control over tension in the design of braces.

The design of compression members is similar to the design of tension members in that the stress created by the axial load must be kept below yield stress. In fact, the allowable compressive stress is the same for compression as it is for tension, which results in a similar formula for nominal strength.

Allowable Compressive Strength

$$P_a \leq \frac{P_n}{\Omega_c} \quad or \quad P_n \geq \Omega_c P_a$$

where $\frac{P_n}{\Omega_c}$ = Allowable Compressive Strength

P_a = applied compressive load
P_n = nominal compressive strength
$\Omega_c = 1.67$ = safety factor for compression

Figure 14-48: Columns are vertical compression members that transfer loads from a floor or roof above to a lower floor or foundation.

©iStockphoto.com/Veres Loan.

The major difference between designing for tension and designing for compression is the tendency for compression members to buckle. **Buckling** refers to the sudden sideways bending of a compression member under load and can occur at loads that are much smaller than you might predict based on allowable stress. See Figure 14-50.

You can simulate buckling failure of a column by compressing a metal ruler along its long axis. First, place the metal ruler in tension by pulling on each end. Even if you pull as hard as you can, you will most likely not be able to cause the ruler to fail in tension. By contrast, if you place the ruler in compression by placing it on a desk in a vertical orientation and then compressing the ruler along its long axis, you will probably find that the ruler can support only a small compressive load before it buckles.

Figure 14-49: A braced frame. The brace can act in compression or tension depending on the direction that the lateral load is applied.

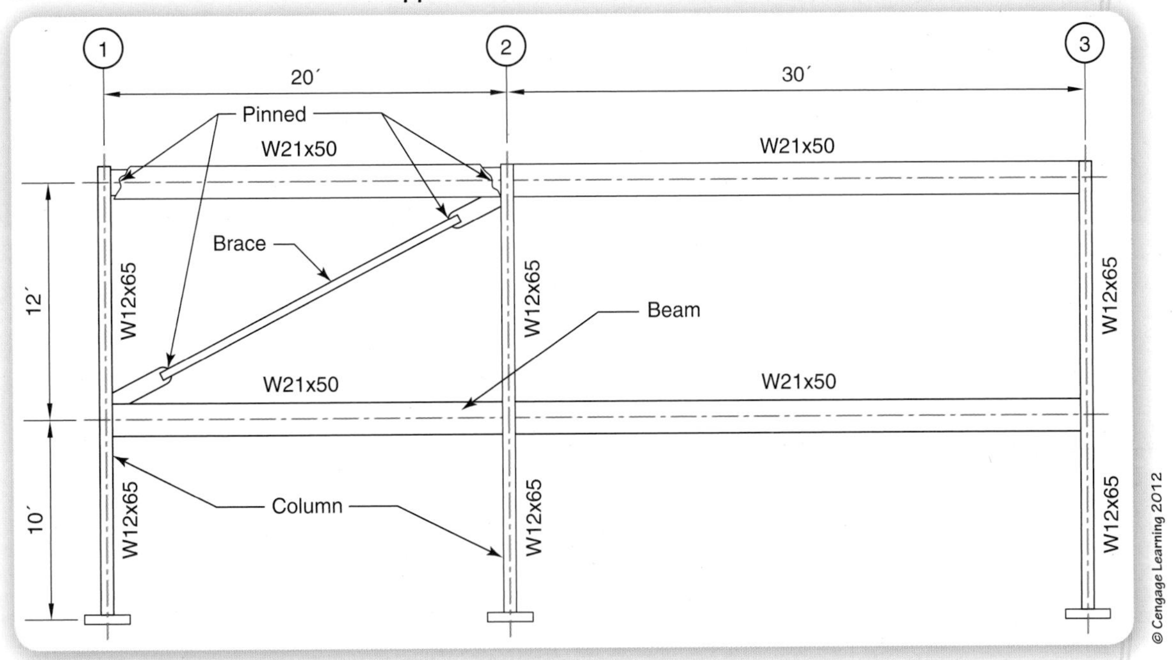

You will also notice that, when the column buckles, the ruler bends in its weak direction—the direction of least thickness. As you can imagine, the thickness, or slenderness, of the column will also affect the buckling load; a thicker ruler will require a larger compressive load to cause buckling.

Engineers often use structural steel columns in the design of buildings. Pipe shapes are commonly used for residential and small commercial facilities. Larger, heavier, steel-framed buildings are often designed with wide flange columns. Because pipes are symmetric about a point, they do not have a strong or weak axis. However, wide flange shapes have two distinct axes, the x-x axis and the y-y axis, as shown in Figure 14-51. The section properties about these two axes are different

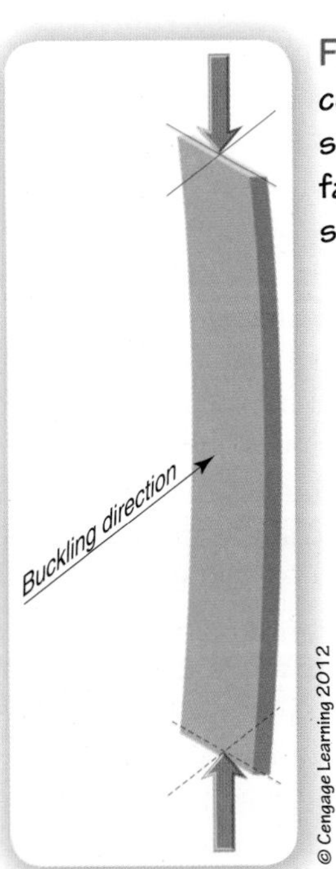

Figure 14-50: Long slender compression members are susceptible to buckling failure before they reach yield stress.

Figure 14-51: Compression members tend to buckle about the weak axis (Y-Y) unless the column is braced in the weak direction.

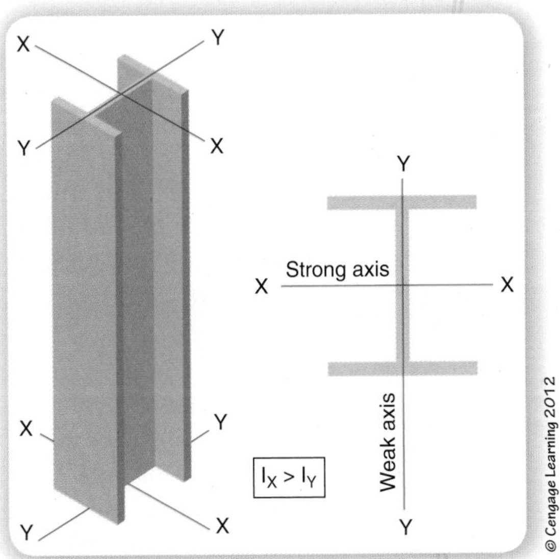

Figure 14-52: Strong axis buckling (left) and weak axis buckling (right).

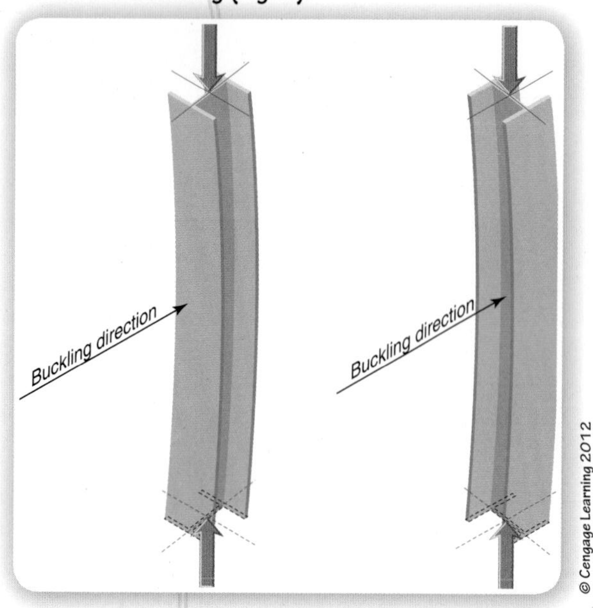

Figure 14-53: Column nominal strength can be increased by bracing a column, thereby decreasing the unbraced length.

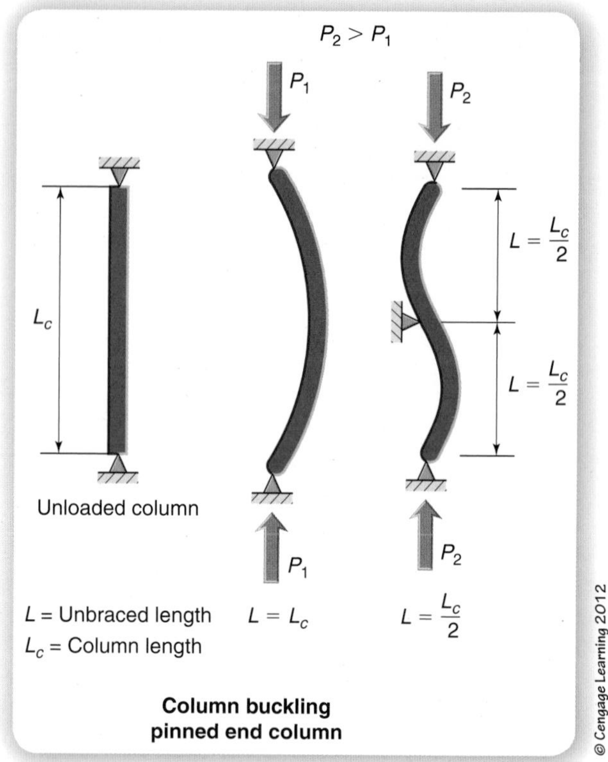

due to the different distribution of cross-sectional area about each axis. The y-y axis is referred to as the weak axis because, all things being equal, a column will buckle about this axis—this is referred to as *weak axis buckling* (see Figure 14-52).

Now, back to the ruler column, you will find that if a classmate supports the ruler at mid-height by holding the midpoint and not allowing the ruler to shift out of its vertical orientation, the ruler will be able to support substantially more load—it may still buckle between the top or bottom and the middle support, but it will take more compressive force to make that happen. What do you think would happen if the column were supported at two or three positions along the ruler? You can probably guess that the column would be able to carry even more load before buckling occurs. The distance between the points of lateral support of a column is referred to the **unbraced length** (see Figure 14-53). The longer the unbraced length, the less load the column can resist before buckling.

It seems apparent from our ruler experiments that the strength of a column cannot be based on yielding of the material alone because columns may buckle before they reach yield stress. Engineers must consider the potential for the column to buckle, and, when buckling is possible, use a nominal strength that addresses the lower buckling failure load. Since the magnitude of the load that causes buckling depends on the unsupported length and the slenderness of the column, the calculation of the nominal compressive load should incorporate these variables. The unsupported length, *L*, and the slenderness of the column are taken into consideration by means of the slenderness ratio. The slenderness ratio indicates the potential for buckling using the unbraced length and the radius of gyration. The radius of gyration represents the distribution of the cross-sectional area around the center axis—the higher the radius of gyration, the more resistance to buckling.

A wide flange column will buckle in its weak direction first unless it is intentionally supported against weak axis buckling. Therefore, in most cases, you will consider the slenderness ratio in the weak direction.

Slenderness Ratio

Assuming pinned end conditions,

$$\text{slenderness ratio} = \frac{KL}{r}$$

where K = effective length factor = 1.0 for pinned support condition

L = laterally unbraced length, in

$r_y = \sqrt{\frac{I}{A}}$ = radius of gyration about weak axis, in

where I = moment of inertia, in^4

A = cross-sectional area, in^2

A lower slenderness ratio provides a higher nominal compressive strength. Therefore, engineers can increase the nominal compressive strength of a column by decreasing the unbraced length (which can be accomplished by providing support along the column) or by choosing a member with a larger radius of gyration. If the slenderness ratio is larger than about 200, the member is susceptible to damage during transport or construction. The SCM restricts the slenderness ratio of steel structural members to $(KL/r)_{max} \leq 200$.

Nominal Compressive Strength

$$P_n = F_{cr} A_g$$

where F_{cr} = flexural buckling stress, psi (ksi)

A_g = gross cross section area of member, in^2

Given that $F_e = \dfrac{\pi^2 E}{\left(\frac{KL}{r}\right)^2}$

$$F_{cr} = \left[0.658^{\frac{F_y}{F_e}}\right] F_y \quad \text{when } F_e \geq 0.44\, F_y, \text{ or}$$

$$F_{cr} = 0.877 F_e \quad \text{when } F_e < 0.44 F_y$$

Example: Determine if a Column Is Adequate

An A582 Grade 50 (F_y = 50,000 psi) W10 × 33 column extends a distance of 28 ft from the first floor to the roof of an atrium and must support a dead load of 12,000 lb and a live load of 36,000 lb. Assuming pinned end supports, is the W10 × 33 adequate to support the applied load? Note: For W10 × 33, r_y = 1.94 in, A_g = 9.71 in^2, and K = 1.0 for pinned connections.

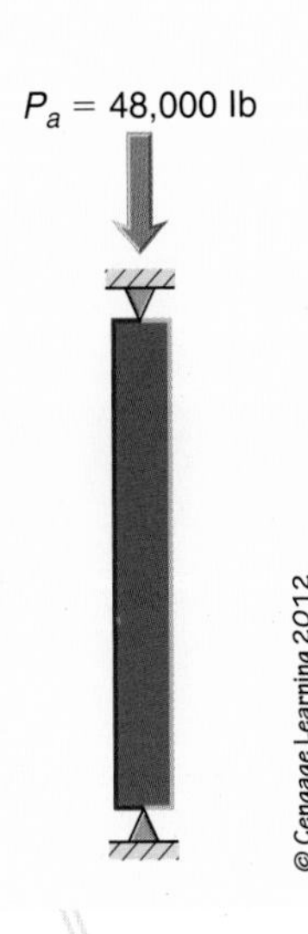

Step 1: *Determine the applied load.*

$$P_a = D + L = 12{,}000\ lb + 36{,}000\ lb = 48{,}000\ lb$$

Step 2: *Calculate the required nominal compressive strength.*

$$F_e = \frac{\pi^2 E}{\left(\frac{KL}{r}\right)^2} = \frac{\pi^2 \left(29 \times 10^6 \frac{lb}{in^2}\right)}{\left(\frac{(1.0)(28\ ft)\left(\frac{12 in}{ft}\right)}{1.94\ in}\right)^2} = 9542\ psi$$

Since $F_e < 0.44\ F_y = 0.44(50{,}000\ psi) = 22{,}000\ psi$

$F_{cr} = 0.877\ F_e = 0.877\ (9542\ psi) = 8368\ psi$

So, $P_n = F_{cr}A_g = (8368\ \frac{lb}{in^2})\ (9.71\ in^2) = 81{,}253\ lb$

Step 3: ***Divide nominal strength by factor of safety. This is the allowable compressive strength.***

$$\frac{P_n}{\Omega_c} = \frac{81{,}253\ lb}{1.67} = 48{,}654\ lb$$

Step 4: ***Compare applied force to allowable strength.***

$$P_a \leq \frac{P_n}{\Omega_c}$$

$$48{,}000\ lb \leq 48{,}654\ lb \text{ GOOD}$$

The W10 × 33 column is adequate.

Students will evaluate a design solution using conceptual, physical, and mathematical models at various intervals of the design process in order to check for proper design and to note areas where improvements are needed.

Beams

A beam is a structural member that carries a transverse load in bending and transfers the load to other supporting elements. Usually beams are oriented so that the length of the beam is horizontal and the load is applied vertically, as a gymnast applies load to a balance beam (see Figure 14-54). However, beams can be oriented in any direction. For instance, a flagpole acts as a cantilever beam when the wind exerts a force on the flag (Figure 14-55).

Figure 14-54: Like a balance beam, beams are typically oriented horizontally such that the load is applied vertically.

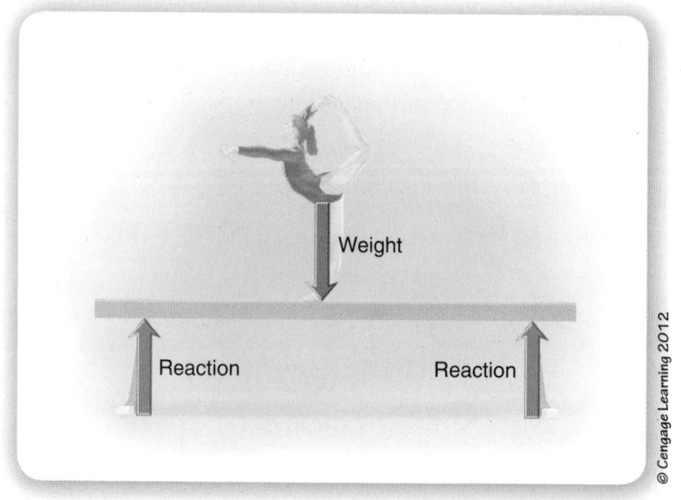

Beam design is based on four important considerations:

- Bending Moment
- Shear
- Deflection
- Cost

The first two considerations take into account the strength of the beam. Is the beam strong enough to carry the applied loads? Deflection, on the other hand, is a serviceability condition. Will the beam perform within acceptable limits of movement? Although excessive deflection can be a symptom of imminent failure, large deflections do not necessarily indicate that the beam is not strong enough to carry the loads, but it can make a structure uncomfortable or unusable for humans or equipment. In order to provide a safe and servicable beam design, the structural engineer must ensure that bending moment, shear, and deflection are within allowable limits per code requirements. However, as previously mentioned, cost is very often a key consideration in structural design. Because the cost of structural steel components varies directly with member weight—the heavier the beam, the more expensive—you should choose the lightest beam that can safely carry the applied loads and perform within acceptable deflection limits.

Again, we will use ASD as specified in the AISC Steel Construction Manual (SCM) to determine allowable bending strength and allowable shear strength.

Figure 14-55: When the wind blows, a flagpole acts as a vertical cantilever beam.

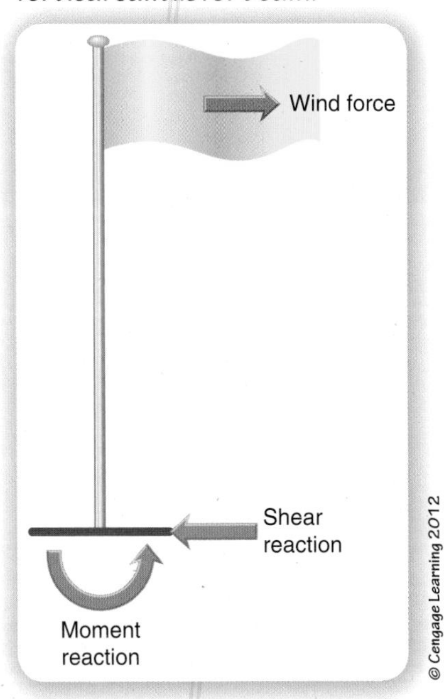

Allowable Bending Strength

$$M_a \leq \frac{M_n}{\Omega_b} \qquad or \qquad M_n \geq \Omega_b M_a$$

where $\frac{M_n}{\Omega_b}$ = Allowable Bending Strength

M_a = bending moment due to applied loads, ft · lb (ft · k)

M_n = nominal bending moment strength, ft · lb (ft · k)

Ω_b = 1.67 = safety factor for bending

The nominal moment is the maximum bending moment that a beam can carry without failure and occurs when the entire cross section is fully yielded. That is, both the tension and compression regions of the cross section are stressed consistently to F_y. Figure 14-56 illustrates the stages through which a beam moves as it is loaded to its maximum moment capacity, M_n.

Nominal moment strength is directly related to the plastic section modulus, *Z*, of the beam. Recall from our discussion earlier in the chapter that the magnitude of the plastic section modulus is an indication of the strength of the member in bending. In general, more cross-sectional area further away from the neutral axis will result in a larger plastic section modulus. If the section is square or circular, the plastic section modulus will be the same about both the x-x and y-y directions. However, most structural steel shapes will provide a

Figure 14-56: Nominal moment graph. The stress distribution within the beam is shown at various stages of loading. As the moment increases, more of the member cross section is subjected to yield stress. The nominal moment is the moment at which the entire cross section experiences yield stress.

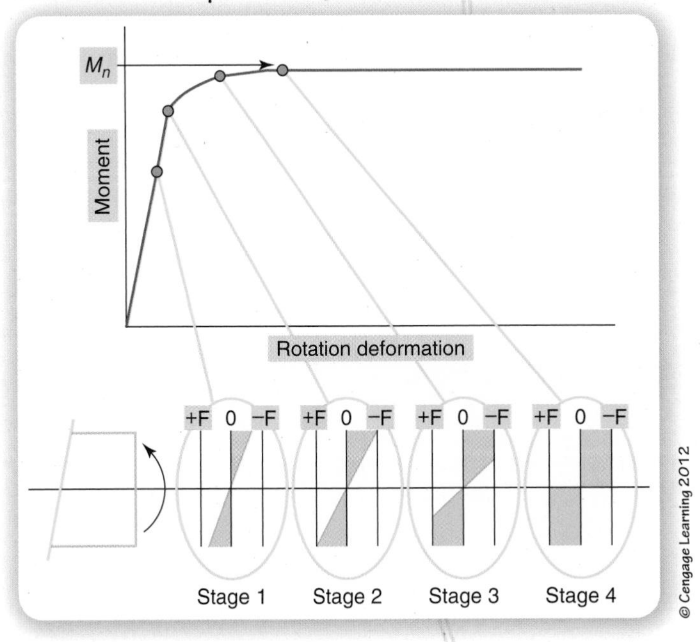

larger plastic section modulus about the x-x axis. Therefore, beams should generally be oriented with the x-x axis parallel to the floor. I-shaped sections (like wide flanges) are good choices for beam design because the flanges provide a larger portion of the cross-sectional area located at a significant distance from the neutral axis resulting in a larger plastic section modulus.

Nominal Moment

$$M_n = F_y Z_x$$

where F_y = the yield stress, psi (ksi)

Z_x = plastic section modulus about the x-axis, in^3

When a beam is oriented so that it bends in its strong direction, it may tend to twist sideways, which is called *lateral torsional buckling.* This occurs because the compression flange of the beam buckles and displaces laterally, but the tension flange does not. The SCM requires a reduction in the nominal moment strength of the beam when this failure mode is likely. Buckling can be avoided by bracing the beam laterally—that is, by constructing the beam so that the part of the beam in compression (usually the top) cannot move laterally. This can be accomplished by nailing floor sheathing to wood joists, for example, or embedding the top of a steel beam in the concrete floor it is supporting. For simplicity, and because many beams are continuously braced, we will assume that the compression region of each beam under consideration is continuously braced so that the beam will not undergo lateral torsional buckling and the nominal moment is not reduced.

Another consideration for beam design is that the web or a flange of a steel beam can fail locally (within a small region of the beam) under a given load even if the member itself is theoretically strong enough to carry the load. This occurs because the thin plates that make up steel sections may be too slender to carry the compression stress resulting from beam bending. Engineers must check this condition when sizing beams; however, most wide flange shapes typically used for beams are **compact sections**, which means that the shape dimensions preclude local buckling. For simplicity, we will not check for local buckling, but, again, structural engineers must ensure that this will not be a failure mode for each beam they design.

Allowable Shear Strength

$$V_a \leq \frac{V_n}{\Omega_v} \quad or \quad V_n \geq \Omega_v V_a$$

where $\frac{V_n}{\Omega_v}$ = Allowable Shear Strength

V_a = actual shear, lb (k)

V_n = nominal shear strength, lb (k)

Ω_v = 1.5 = *safety factor for shear*

Nominal shear strength depends on the area of the cross section that resists shear. In general, the flanges of a wide flange beam provide little shear resistance. Therefore, only the web area is considered for shear resistance.

$$V_n = 0.6F_y A_w$$

where F_y = the yield stress, psi (ksi)

A_w = area of the web

DEFLECTION Although we concede that structures will naturally move when loads are applied—beams bend when loaded and buildings sway when the wind blows—we typically want to limit the movement of the building to minimize the negative effects of motion on building inhabitants and equipment. We describe the displacement of a beam when it bends as **deflection** (see Figure 14-57). Deflection is limited by codes, but unlike strength requirements, deflection limits are not based on safety but on comfort and ease of use. The IBC limits on deflection are shown in Figure 14-58.

Figure 14-57: Beam deflection.

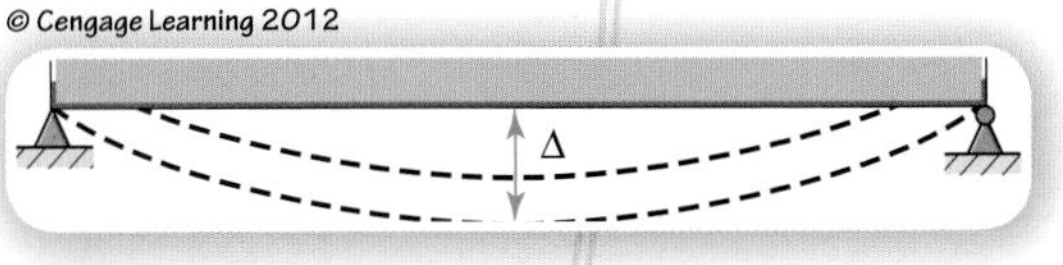

How do you determine the deflection of a beam under a given loading? Beam deflection depends on several factors including the span of the beam, the loading condition, the modulus of elasticity of the beam material, and the moment of inertia of the beam shape. Although beam deflection can be derived using mathematical techniques (beyond the scope of this textbook), engineers typically use a deflection formula to calculate a beam's deflection. Several common deflection formulas are given in Figure 14-59. It is very important that you pay close attention to units when substituting values into a formula to ensure that the resulting deflection magnitude is in appropriate units. It is often easiest to convert all input values to pounds (or k) and inches—that way the resulting deflection is expressed in inches.

$$\Delta_{max} = \frac{Pab(a + 2b)\sqrt{3a(a + 2b)}}{27\ EIL}$$

Figure 14-58: Deflection limits. International Code Council, (2009) International Building Code. Table 1604.3, p 305.

TABLE 1604.3
DEFLECTION LIMITS[a, b, c, h, i]

CONSTRUCTION	L	S or W^f	$D + L^{d,g}$
Roof members:[e]			
Supporting plaster ceiling	$l/360$	$l/360$	$l/240$
Supporting nonplaster ceiling	$l/240$	$l/240$	$l/180$
Not supporting ceiling	$l/180$	$l/180$	$l/120$
Floor members	$l/360$	—	$l/240$
Exterior walls and interior partitions:			
With brittle finishes	—	$l/240$	—
With flexible finishes	—	$l/120$	—
Farm buildings	—	—	$l/180$
Greenhouses	—	—	$l/120$

For SI: 1 foot = 304.8 mm

a. For structural roofing and siding made of formed metal sheets, the total load deflection shall not exceed $l/60$. For secondary roof structural members supporting formed metal roofing, the live load deflection shall not exceed $l/150$. For secondary wall members supporting formed metal siding, the design wind load deflection shall not exceed $l/90$. For roofs, this exception only applies when the metal sheets have no roof covering.

b. Interior partitions not exceeding 6 feet in height and flexible, folding and portable partitions are not governed by the provisions of this section. The deflection criterion for interior partitions is based on the horizontal load defined in Section 1607.13.

c. See Section 2403 for glass supports.

(Table notes continued)

Figure 14-59: **Deflection formula.**

Concentrated load at center P · A · B · $\frac{L}{2}$ · L	**Deflection**	$\Delta_{max} = \frac{PL^3}{48EI}$ (at point of load)
Uniformly distributed load ω · A · B · L	**Deflection**	$\Delta_{max} = \frac{5\omega L^4}{384EI}$ (at center)
Uniformly distributed load and concentrated load at center P · ω · A · B · $\frac{L}{2}$ · L	**Deflection**	$\Delta_{max} = \frac{PL}{4} + \frac{5\omega L^4}{384EI}$ (at point of load)
Two equal concentrated loads (symmetrically placed) P · P · A · B · a · a · L	**Deflection**	$\Delta_{max} = \frac{Pa}{24EI}(3L^2 – 4a^2)$ (at center)
Two equal concentrated loads (symmetrically placed) and a uniformly distributed load P · ω · P · A · B · a · a · L	**Deflection**	$\Delta_{max} = \frac{5\omega L^4}{384EI} + \frac{Pa}{24EI}(3L^2 – 4a^2)$ (at center)
Concentrated load (asymmetrical) P · A · B · a · b · L	**Deflection**	$\Delta_{max} = \frac{Pab(a + 2b)\sqrt{3a(a + 2b)}}{27EI}$ (at $x = \sqrt{\frac{a(a + 2b)}{3,}}$ when $a > b$)

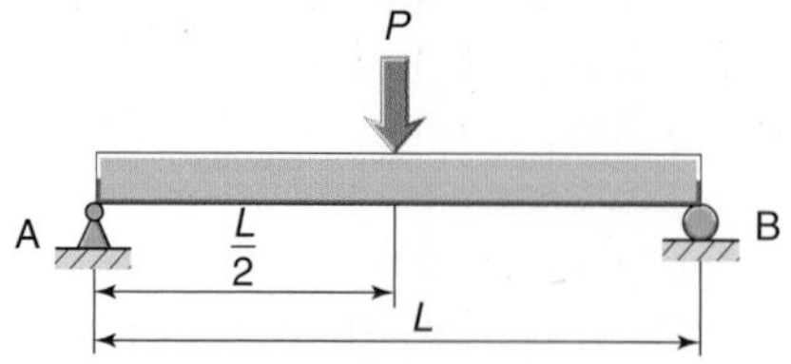

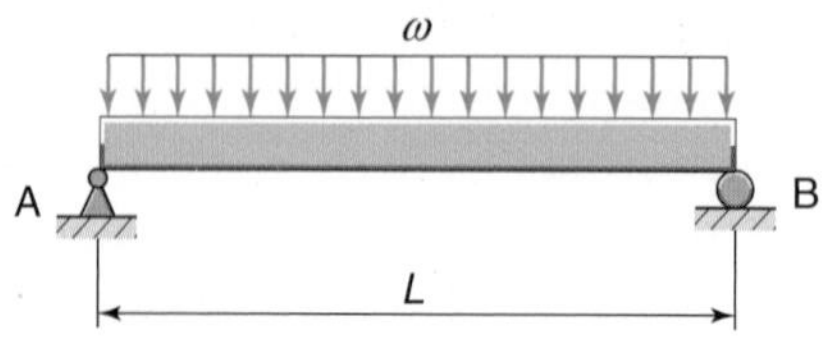

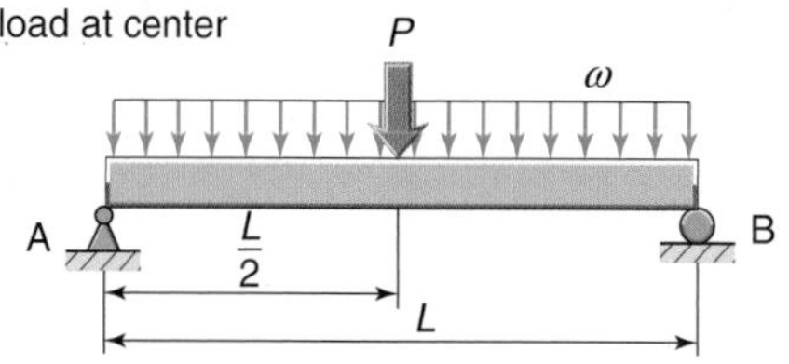

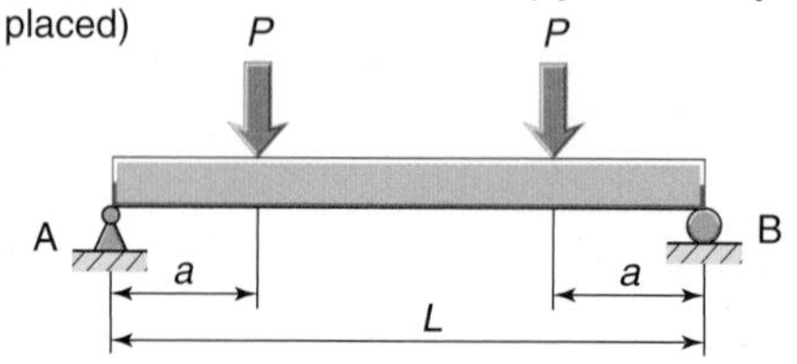

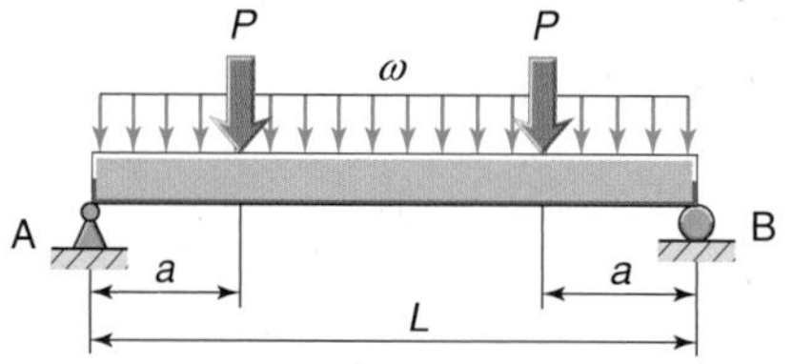

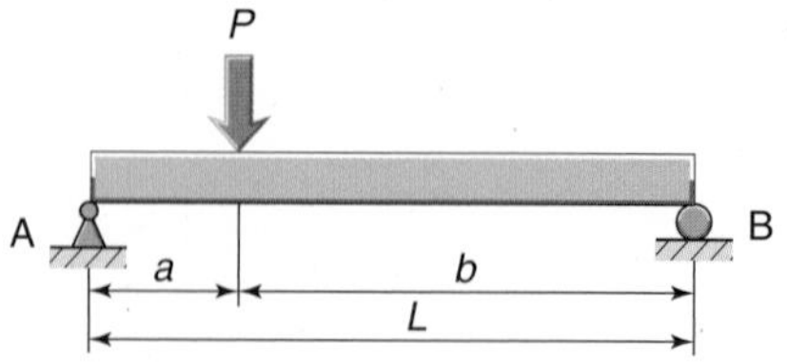

Example: **Select a Structural Steel Beam**

Choose an efficient structural steel section to support the loads shown in the beam diagram. Assume the yield stress of the steel is 50 ksi.

W_{DL} = 800 lb/ft
W_{LL} = 1000 lb/ft
32′

Step 1: ***Find the maximum moment and maximum shear.***

Use the beam formula to find

$$M_{max} = M_a = \frac{wL^2}{8} = \frac{(1800\,\frac{lb}{ft})(30\,ft)^2}{8} = 202{,}500\,ft \cdot lb = 202.5\,ft \cdot lb$$

$$V_{max} = V_a = \frac{wL}{2} = \frac{(1800\,\frac{lb}{ft})(30ft)}{2} = 27{,}000\,lb = 27.0\,ft \cdot lb$$

Step 2: ***Find the required nominal moment.***

$$M_n \geq M_a\Omega_b = (202\,ft \cdot k)\,(1.67) = 338.2\,ft \cdot k$$

Step 3: ***Determine the required plastic section modulus.***

$$M_n = F_yZ_x \qquad \text{therefore} \qquad F_yZ_x \geq 338.2\,ft \cdot k$$

$$\geq Z_x \frac{338.2\,ft \cdot k}{F_y} = \frac{(338.2\,ft \cdot k)\left(\frac{12\,in}{ft}\right)}{50\frac{k}{in^2}} = 81.2\,in^3$$

Step 4: ***Find an economical steel shape with the required plastic section modulus.***

Visit *http://www.structural-drafting-net-expert.com.* Follow the links to Steel Sections and Wide Flange Beams to view the dimensions of the steel wide flange shapes. Choose the Properties button and find the steel shape of least weight that has a plastic section modulus of 81.2 in^3 or larger.

W21 × 44 has the following properties and dimensions:

$Z_x = 95.4$ in^3
$I_x = 843$ in^4
$t_w = 0.350$ in
$d = 20.66$ in

Step 5: ***Check the shear strength.***

$$V_n \geq V_a\Omega_v = (27{,}000\,lb)(1.5) = 40{,}500\,lb = 40.5\,k$$

$$V_n = 0.6\,F_ydt_w = 0.6\left(50{,}000\,\frac{lb}{in^2}\right)(20.66\,in)(0.350\,in) = 216{,}930\,lb = 216.93\,k$$

$$V_n = 216.93\,k \geq 40.5\,k \text{ GOOD}$$

The W21 × 44 is adequate to resist the shear stress.

Step 6: ***Calculate deflection limits.***

$$(\Delta_{DL+LL})_{max} = \frac{L}{240} = \frac{(30\ ft)(\frac{12\ in}{1\ ft})}{240} = 1.5\ in$$

$$(\Delta_{LL})_{max} = \frac{L}{360} = \frac{(30\ ft)(\frac{12\ in}{1\ ft})}{360} = 1.0\ in$$

Step 7: ***Find the deflection for dead and live loads and for live load alone.***

$$\Delta_{LL} = \frac{5w_{LL}L^4}{384\ EI} = \frac{5\left(1000\ \frac{lb}{ft}\right)(30\ ft)^4\left(\frac{12\ in}{1ft}\right)^3}{384\left(29{,}000{,}000\ \frac{lb}{in^2}\right)(843\ in^4)} = 0.75\ in \leq 1.0\ in \text{ GOOD}$$

$$\Delta_{DL+LL} = \frac{5wL^4}{384\ EI} = \frac{5\left(1800\ \frac{lb}{ft}\right)(30\ ft)^4\left(\frac{12\ in}{ft}\right)^3}{384\left(29{,}000{,}000\ \frac{lb}{in^2}\right)(843\ in^4)} = 1.34\ in \leq 1.5\ in \text{ GOOD}$$

Note that when evaluating the deflection expression, the units should cancel such that the final unit for deflection is inches. Without the conversion of feet to inches, the deflection units would be cubic feet over square inches. Therefore, the conversion factor must be cubed in order to obtain a deflection in inches.

Students will use unit analysis to check measurement computations.

Note that if the deflection exceeds the limits, a new steel shape should be chosen. The new shape should possess a plastic section modulus greater or equal to the required section modulus and a larger moment of inertia, *I*, than the original section selected. The final beam selection should meet all of the requirements noted above. In this case, use a W21 × 44.

Structural Analysis Software

As a student in the twenty-first century, you have come to rely on technology to help you solve problems. As you probably have already guessed, structural analysis and design can be aided by the use of software packages that allow engineers to input the details of a structure, analyze the forces and stresses to which each structural element is subjected, and select efficient and economical designs. Some packages allow analysis and design of only relatively simple structural components, such as individual beams or columns. Other analysis packages are quite sophisticated and can accomplish accurate finite element analyses of large structural designs. Figure14-60 shows the output of a beam analysis performed by structural analysis software showing the beam diagram with the corresponding shear and moment diagrams. It is important that you are comfortable with and understand the concepts and equations used in the analysis of structural members so that you can assess the accuracy of the output from structural analysis software packages. When used properly analysis software can help you quickly design efficient and cost-effective structures.

Figure 14-60: This is an example of a beam analysis performed using engineering software.

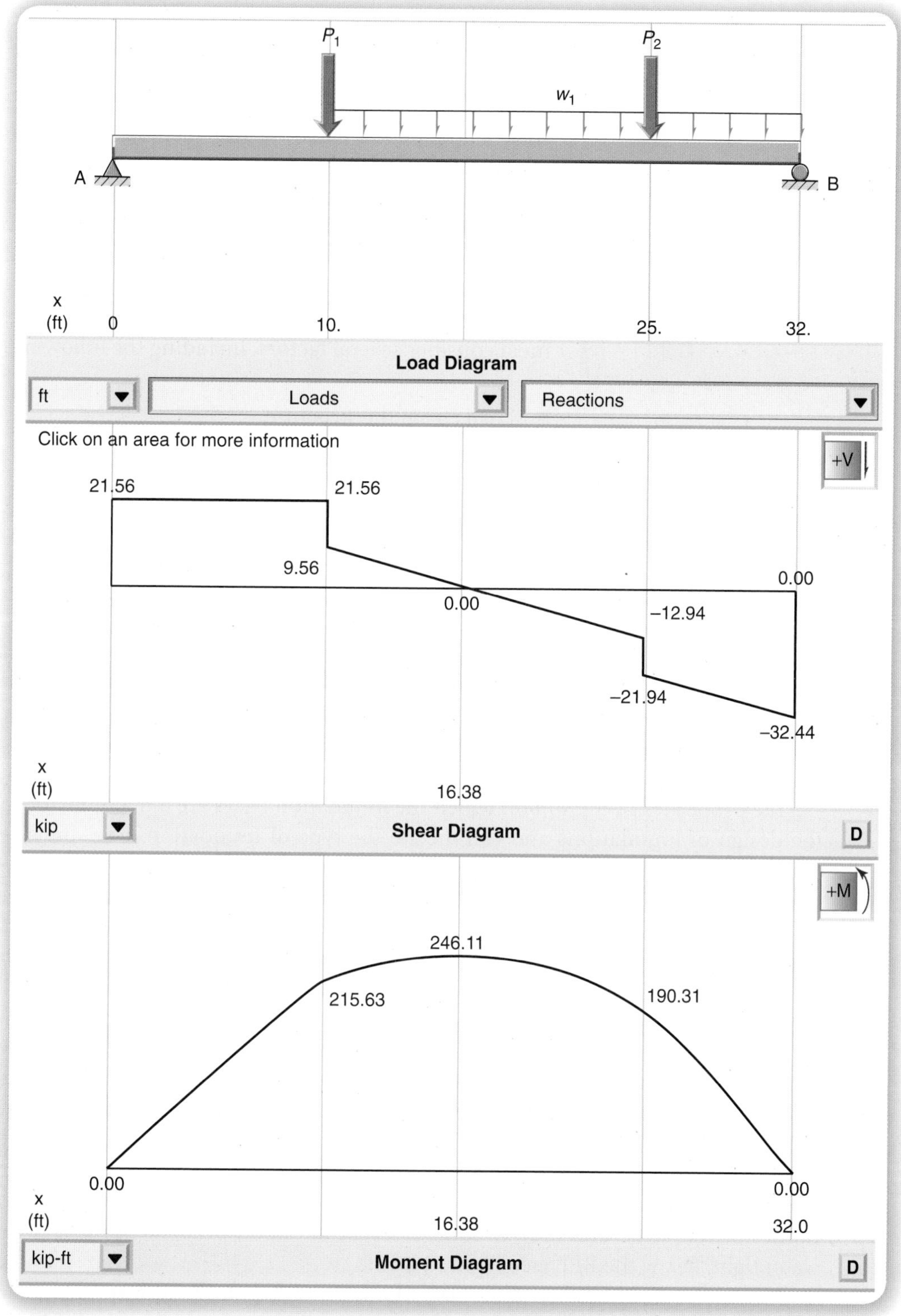

P_1 P_2 w_1
A B
x (ft) 0 10. 25. 32.
Load Diagram
ft Loads Reactions
Click on an area for more information
+V
21.56 21.56 9.56 0.00 0.00 −12.94 −21.94 −32.44
x (ft) 16.38
kip Shear Diagram D
+M
246.11 215.63 190.31
0.00 0.00
x (ft) 16.38 32.0
kip-ft Moment Diagram D

FOUNDATIONS

The foundation acts as the legs of a structure. Just as your legs transfer your weight and the weight of objects you carry or lift to the ground, the foundation transfers the weight of the structure as well as any additional loads that are applied to the structure. And just as your legs can transfer both vertical (for instance the weight

Figure 14-61: Foundation failure can cause cracking in walls.

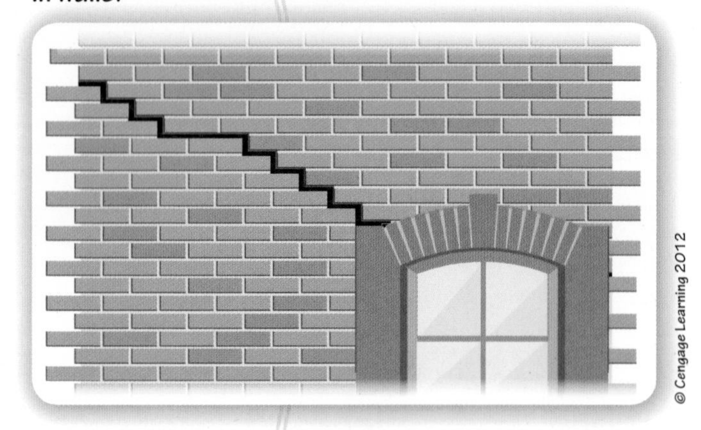

of a box you are carrying) and horizontal (caused, for example, by a collision on a soccer field) forces to the ground, a building foundation must be able to support both vertical and lateral loads and safely transfer the loads to the supporting soil. In order to achieve this goal, a foundation must be designed such that the structure remains stable, with minimal movement or settlement. An inadequate foundation will allow the structure to shift and deform and can result in an array of structural issues from cracked walls and misaligned windows to structural collapse (Figure 14-61).

In order to design an adequate foundation, you must consider several factors, including the following:

- Loads from the structure
- Bearing capacity of the soil
- Flood elevation
- Water table
- Drainage
- Frost depth
- Cost

All loads applied to the structure must eventually be transferred to the soil through the foundation. We have already discussed the types of loads to which a structure may be subjected and the paths that these loads may take. Remember that, by code, the structure must be designed to resist several load combinations. This requirement applies to the foundation design as well.

The properties of the soil on which the foundation bears are a major factor in the design of foundations and can dictate the type of foundation used, the size and depth of a foundations, and the materials from which the foundation is constructed. As discussed in Chapter 6, the type of soils present will also have a significant effect on the drainage characteristics, susceptibility to settlement and frost heave, and the ability of the soil to support loads. Much of the soil information necessary to design a foundation is available in the soils report (see Chapter 6).

Water can also impact foundation design: it can alter loading conditions; it can change the behavior of soil; and it can physically move soil, which can potentially reduce soil support or impose unexpected soil pressures. To any building in its path, moving water can impose lateral loads from water pressure and velocity and uplift forces especially when the level of the water rises above the enclosed building causing the building to float. For that reason, a typical design strategy is to elevate the building above the base flood level or base flood elevation (BFE). Alternatively, nonresidential structures may, in some cases, be dry flood-proofed, or made to be water tight, below the BFE (see Figure 14-62).

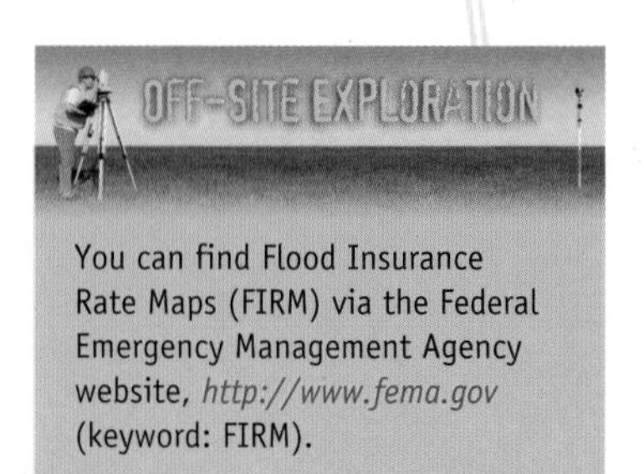

You can find Flood Insurance Rate Maps (FIRM) via the Federal Emergency Management Agency website, *http://www.fema.gov* (keyword: FIRM).

Remember from Chapter 5 that the BFE is provided on Flood Insurance Rate Maps (FIRM) produced by the Federal Emergency Management Agency (FEMA). Be aware that local building regulations may require that design be based on a design flood elevation (rather than the BFE) that is higher than the BFE. Always check all applicable codes and regulations.

Building codes and local regulations address the design of structures located in flood zones. The IBC specifies that the design and construction of buildings and structures located in flood hazard areas must meet the requirements of ASCE 24, *Flood Resistant Design and Construction*. This document complements the National Flood

Figure 14-62: Dry flood-proofing techniques.

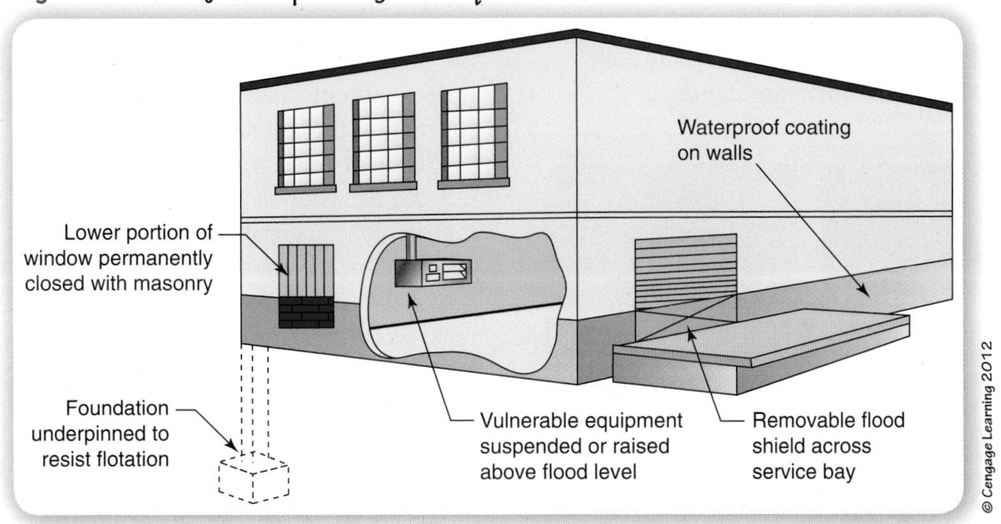

Insurance Program (NFIP) regulations (discussed in Chapter 5) and includes additional requirements and limitations (Figure 14-63). For the purpose of this textbook, we will highlight the following NFIP regulations that will affect foundation design. These regulations will dictate the elevation at which the foundation must support the structure.

For buildings within an A-zone:

- Residential buildings must have the lowest floor (including basement) elevated to or above the BFE.
- Nonresidential buildings must have either the lowest floor (including basement) elevated to or above the BFE or **dry flood-proofed** to the BFE (see Figure 14-64).

For buildings within a V-zone:

- All buildings must be elevated on piles and columns so that the bottom of the lowest horizontal structural member of the lowest floor is at or above the BFE.
- All buildings must be anchored to resist flotation, collapse, and lateral movement.
- The area below the lowest floor must either be free of obstruction, or an enclosure must be constructed with nonsupporting/non–load-bearing breakaway walls and used only for the parking of vehicles, building access, or storage.

Figure 14-63: National Flood Insurance Program (NFIP) regulations.

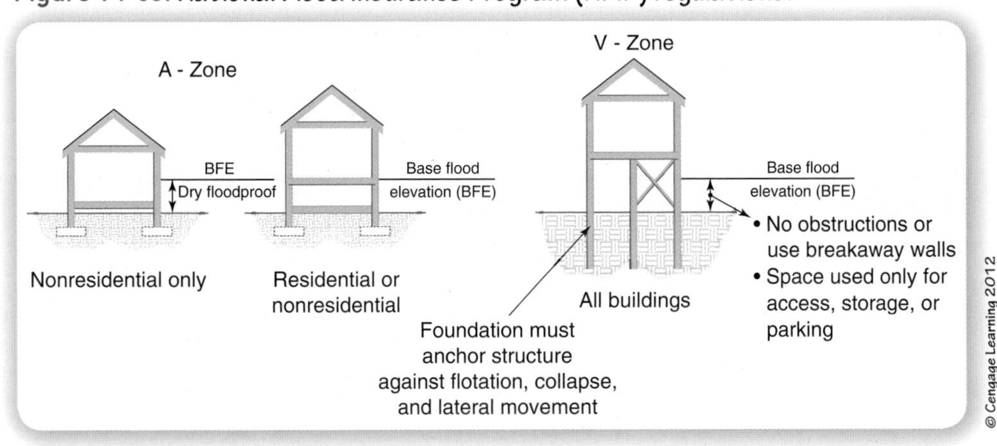

Figure 14-64: The head librarian, Jean Bogun, of the then new elevated library shows the level of flood waters caused by Hurricane Isabel (2003) on the foundation. Because the building was built above the flood elevation, it was saved from major damage.

Photo by Mark Wolfe/FEMA News Photo.

In addition to imposing significant loads to the building, water can reduce the strength of soil. In general, as the water content of a soil increases, the strength of the soil decreases.

Stagnant water may also drastically affect the behavior of soil. A phenomenon that is often associated with earthquakes or vibration is **liquefaction**. Liquefaction can occur in sandy soils that are saturated so that the water fills all of the voids between the grains of the soil. During an earthquake, the water pressure increases to the point that the soil particles can easily slide against each other reducing the bearing capacity of the soil (see Figure 14-65). Therefore, a high water table in sandy soils could pose concern in areas that are susceptible to earthquakes.

Water also causes dry soils (especially expansive clay) to swell and lift up foundations. In addition, as you are probably aware, water expands when it freezes; that is why ice floats in water—ice is less dense. If a foundation rests on wet soil that is susceptible to freezing, the ice could potentially move the foundation as it forms. This phenomenon is referred to as **frost heave**. To avoid water accumulation near the foundation, the site should be graded to direct water away from the building and a drainpipe, or footing drain, should be installed at the foundation level (see Figure 14-66). In addition, the bottom of the foundation should be constructed below the frost depth unless frost protection is provided to the foundation. Figure 14-67 provides a map showing the frost depth for the continental United States, but local regulations may provide more accurate local information and required foundation depths.

Figure 14-65: Many buildings supported on shallow foundations were damaged in the 1999 Adapazari, Turkey, earthquake, which measured 7.4 on the open ended Richter scale. A major cause of damage was the liquefaction of soil resulting in bearing capacity failure and excessive settlement.

AFP/Getty Images.

The soils report will generally provide the soil strength in terms of the **allowable soil bearing pressure** or allowable foundation pressure. The allowable soil bearing pressure is the maximum pressure to which the soil may be subjected for design purposes and includes a factor of safety. In some cases, allowable soil bearing pressures are specified in local codes and ordinances, or local codes may require that the design meet the requirements of the International Building Code (IBC) or International Residential Code (IRC). Figure 14-68 provides recommendations for the maximum allowable foundation pressures. The allowable soil bearing pressure used in design should reflect the soil classification found below the bottom of the footing. Alternatively, allowable soil bearing pressures can be estimated based on an on-site soils investigation.

The IBC permits the use of higher foundation pressures if data is submitted (typically in a soils report) to substantiate the use of a higher value, and the higher value is approved. Be aware that unusual circumstances, such as landfills, expansive soils, or organic deposits, could also cause the load capacity of the soil to be less than specified in the table. Although soil

Figure 14-66: To avoid water accumulation at the foundation and frost heave, slope the ground away from the building and use a footing drain.

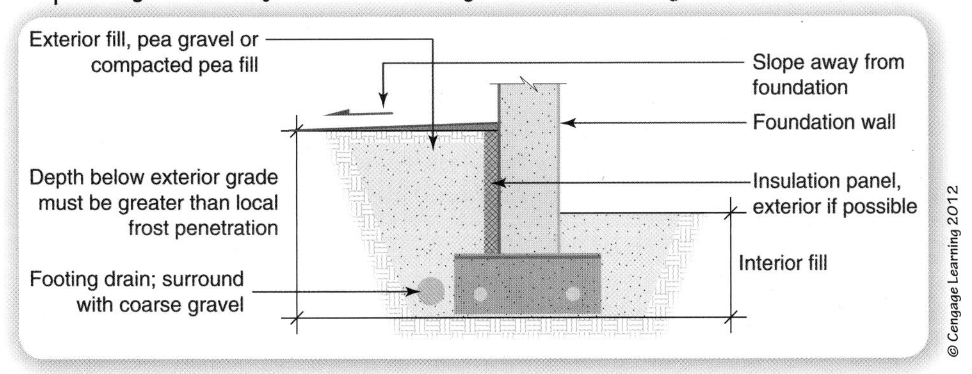

investigations are rarely needed for residential construction due to the relatively light loads, a soils report is often required by local municipalities in order to obtain a commercial building permit. At this point, you may wish to review the discussion on soil classifications in Chapter 6.

Of course, the cost of construction is almost always a factor in design. The cost of a foundation will be affected by all of the factors just addressed: loads, flood elevation, water table, drainage, frost depth, and soil bearing capacity. The most cost-effective foundation for residential and low-rise commercial buildings is typically a shallow foundation. The loads associated with low-rise construction are generally small, and the foundation design is based more on environmental conditions (flood elevation, water table, frost depth) and constructability than on load capacity. You will generally get a good indication of cost-effective foundation types by investigating the foundations in the vicinity of the project.

Figure 14-67: Frost depth penetration for the Continental United States. From JEFFERIS AND MADSEN. Architectural Drafting and Design, 5e.

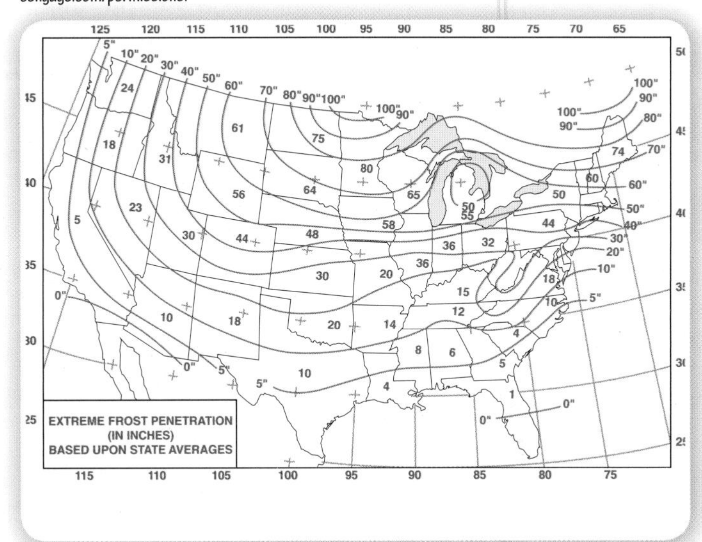

Figure 14-68: Allowable soil bearing pressure.

Type of Material	Allowable Soil Bearing Pressure (psf)
Bedrock	12,000
Sedimentary and foliated rock	4,000
Gravel and sandy gravel (GW and GP)	3,000
Gravel with fines (GM, GC)	2,000
Sand, silty sand, clayey sand (SW, SP, SM, SC)	2,000
Clay, sandy clay, silty clay (CL, CH)	1500
Silt, clayey silt and sandy silt (ML, MH)	1500

Preliminary information on soil conditions throughout the United States can be obtained via the U.S. Department of Agriculture (USDA) Web Soil Survey available online at *http://websoilsurvey.nrcs.usda.gov* (or keyword search: Web Soil Survey).

Foundation design for larger commercial, institutional, and industrial facilities must be based on a careful analysis of the specific project circumstances before a decision on the type of foundation can be made. A shallow foundation is preferable if it will provide adequate support because it is typically less expensive than a deep foundation. Deep foundations may be indicated if the site has weak soils near the surface but satisfactory soils at a greater depth or the structure needs to be elevated above grade. Often geotechnical engineers make recommendations regarding the types of foundation to be used for a particular project but more than one type of foundation should be investigated so that costs can be compared.

SHALLOW FOUNDATIONS **Shallow foundations** are used when the soils near the ground surface are relatively stable and possess adequate strength to support the structure. Although shallow foundations exist in many forms, the basis of support from the soil is the same: the foundation transfers the building loads directly to the soil directly under the foundation by vertical pressure. We typically assume that the load is evenly distributed over the area of the foundation in contact with the soil as illustrated in Figure 14-69. The interface between the shallow foundation and the soil is the critical load area: the soil must resist the highest pressure at this point. As the depth below the bearing surface increases, the load spreads out over a larger and larger area and the soil bearing pressure decreases. As a result, shallow foundations are designed based on the allowable bearing pressure at the bottom of the foundation.

Often, a slab on grade is thickened under walls and columns to provide more strength where the vertical loads from the structure above are concentrated as shown in the section views in Figure 14-70. A typical slab for light construction is 3.5 or 4 in thick. Welded wire steel mesh is typically embedded into the slab to resist cracking.

Figure 14-70: Typical sections of a thickened slab on grade foundation at an exterior and an interior wall.

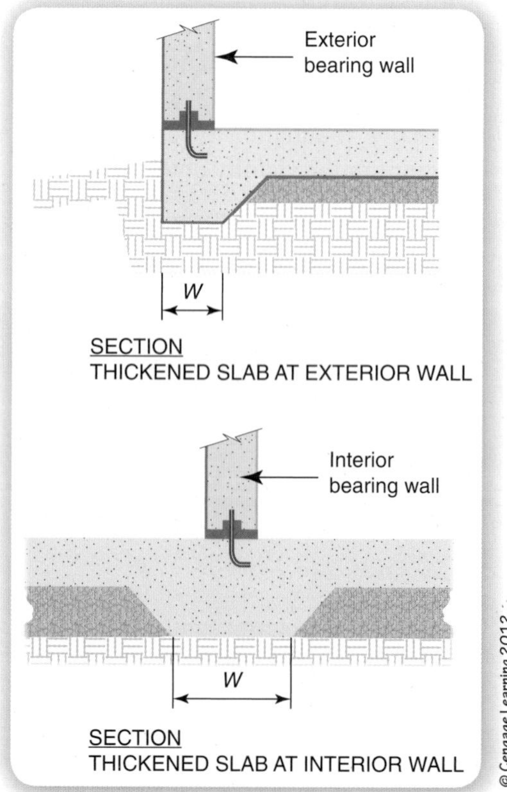

Figure 14-69: Soil bearing pressure caused by a shallow foundation decreases as the depth below the foundation increases.

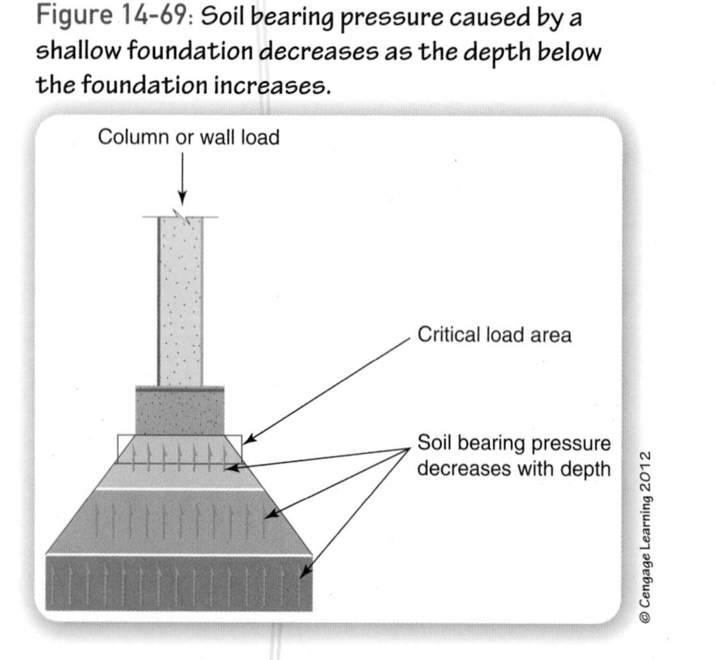

Spread footings are often used to accommodate a large frost depth, a need to support the building below grade because of a crawl space or basement, or inadequate surface soils. If the spread footing is designed to continuously support a foundation wall the footing is referred to as a **strip footing**, or a continuous footing. Continuous footings subjected to relatively light loads that have a reliable bearing capacity of 1500 psf or greater are typically sized to accommodate the width of the foundation wall with a projection on either side of the wall to provide allowance for some construction misalignment.

The IRC stipulates that the thickness (T) of a strip footing for residential construction shall be at least 6 in, and that the footing width, W, shall be based on the load-bearing value of the soil in accordance with the table in Figure 14-71. In addition, the projection, P, must be at least 2 in but not more than the footing thickness.

A spread footing, designed to support an individual column load is called a column footing or **isolated footing** (see Figure 14-72). Isolated footings are generally rectangular, reinforced concrete pads that act to disperse the column loads over an area of soil sufficient to maintain a soil bearing pressure less than the allowable soil pressure. In many cases a building column does not rest directly on the footing, but bears on a short concrete column called a pier. This arrangement allows the column to remain protected from the damp and possibly corrosive environment below grade.

A mat foundation, also called a raft foundation, is a large, thick, heavily reinforced slab that serves as the foundation for many columns or an entire building (see Figure 14-73). When the building loads are relatively high and the soils are relatively weak such that the total area of the isolated column footings will require a large percentage of the building footprint, it may be more economical to create one large mat foundation than many smaller footings.

Once you have determined the building loads and the allowable soil pressure, it is a relatively straightforward calculation to determine the required bearing area for a shallow foundation.

Figure 14-71: The IBC requires that the thickness, T, of a spread footing for residential construction shall be at least 6 in, and that the footing width, W, shall be based on the load-bearing value of the soil. In addition, the projection, P, must be at least 2 in but not more than the footing thickness.

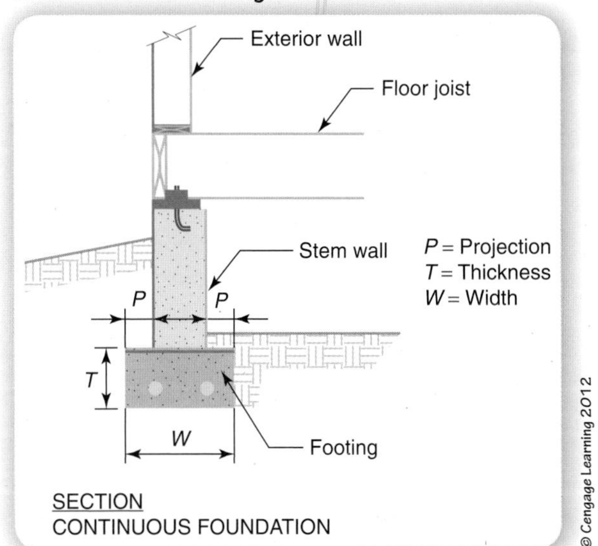

Figure 14-72: Column footing.

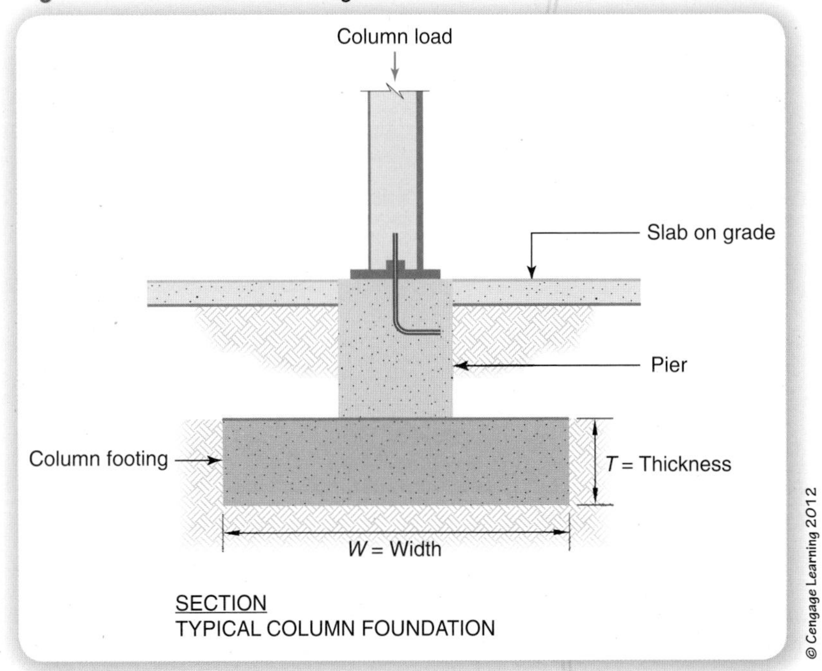

Bearing Area of a Shallow Foundation

$$A_{required} = \frac{F}{q}$$

where $A_{required}$ = required bearing area of the foundation

F = total applied load = P (column load) + W (weight of foundation)

q = allowable soil bearing pressure

Figure 14-73: Mat foundation.

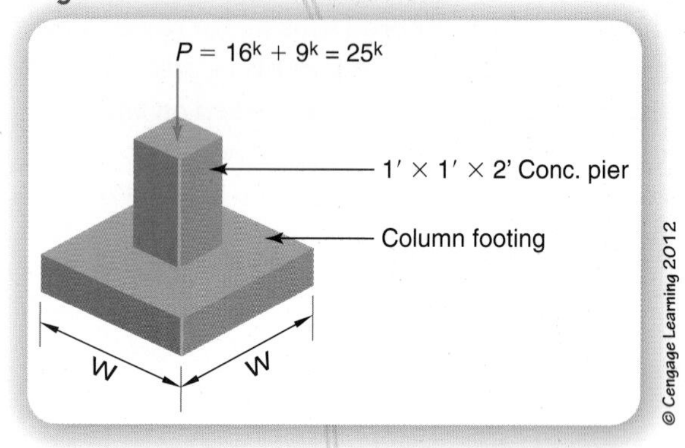

© Cengage Learning 2012

Because it is difficult to calculate the weight of the foundation itself if you have not yet determined an appropriate size, it is often easier to reduce the allowable soil bearing capacity to account for an assumed thickness of foundation. Although the overall size of the foundation is not known until the required area has been determined, if we assume a footing thickness, we can determine an approximate weight of the footing per square foot of area—that is, we can approximate the pressure that the footing will exert on the soil. If we subtract the pressure due to the footing (p) from the allowable soil pressure (q), we get the magnitude of allowable soil bearing pressure remaining to support the other building loads. This is referred to as the *net allowable soil bearing pressure.*

Alternate Equation for the Bearing Area of a Shallow Foundation

$$A_{required} = \frac{P}{q_{net}}$$

where $A_{required}$ = required area of foundation bearing
P = column load
q_{net} = net allowable soil bearing pressure $= q - p$
q = allowable soil bearing pressure
p = pressure due to weight of footing

Example: Size a Spread Footing

Size an economical square column footing to transfer to the supporting soil a column dead load of 16,000 lbs (including the column weight) and a live load of 9000 lbs. The column rests on a 1 ft × 1 ft× 2 ft tall pier. Assume a footing thickness of 15 in. The footing rests on a siltly gravel soil—soil tests indicate that the allowable soil bearing pressure is 2500 psf.

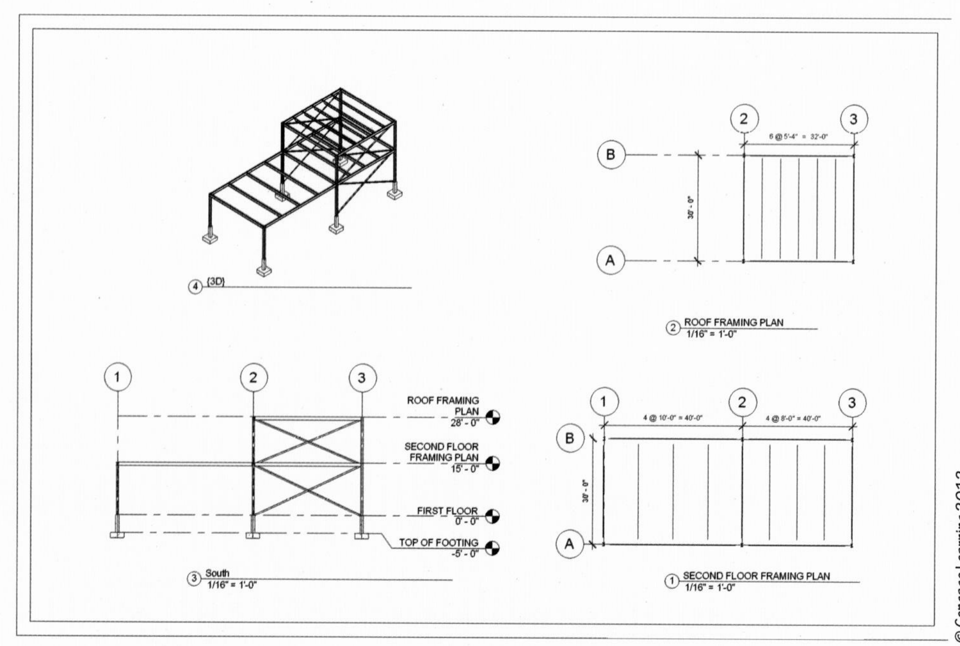

© Cengage Learning 2012

Step 1: ***Calculate the total column load.***

$$Pier\ DL = (1\ ft)(1\ ft)(2\ ft)\left(150\ \tfrac{lb}{ft^2}\right) = 300\ lb$$

$$P = DL + Pier\ DL + LL = 16{,}000\ lb + 9{,}000\ lb = 25{,}000\ lb$$

Step 2: ***Calculate the soil pressure due to the footing weight. Since concrete weighs 150 psf***

$$P_{footing} = \left(150\ \tfrac{lb}{ft^2}\right)(15\ in)\left(\frac{1\ ft}{12\ in}\right) = 187.5\ \tfrac{lb}{ft^2}$$

Step 3: ***Reduce the allowable soil bearing pressure to account for the concrete pressure.***

$$q_{net} = 2500\ \tfrac{lb}{ft^2} - 187.5\ \tfrac{lb}{ft^2} = 2312.5\ \tfrac{lb}{ft^2}$$

Step 4: ***Calculate the required footing area.***

$$A_{required} = \frac{P}{q_{net}} = \frac{25{,}000\ lb}{2312.5\ \frac{lb}{ft^2}} = 10.81\ ft^2$$

Step 5: ***Size the footing.***

Since the footing will be square

$$A_{required} = w^2\ or\ w = \sqrt{A_{required}}$$

$$w = \sqrt{10.81\ ft^2} = 3.29\ ft$$

This is approximately equal to 3 ft – 4 in. In practice, footings are generally dimensioned to the next larger 3 or 6 in because it is less expensive to pour a little more concrete than to spend extra time forming the foundation to very accurate measurements. Therefore, use a 3 ft 6 in × 3 ft 6 in square isolated footing.

Once a shallow foundation has been sized, the foundation components (which may include a wall, pier, and footing) must be designed. These components are typically constructed of concrete or concrete masonry units. The structural engineer will finalize the design of the components and then compare the final design to the loads assumed in the initial steps of sizing the footing. If the final design is heavier than that assumed, the calculations should be revisited to ensure that the additional load does not overstress the concrete or the soil.

Students will test solutions against needs and criteria it was designed to meet.

Deep Foundations

Although shallow foundations are often less expensive, many site, soil, or environmental conditions may necessitate the use of **deep foundations**. As discussed in Chapter 13, deep foundations include structural elements, such as piles, that extend well below the ground surface. In some cases, deep foundations are necessary because the soils near the surface are unstable or weak, but deeper soil layers provide better support characteristics. In other cases, deep foundations provide a more efficient method of elevating structures above flood levels and anchoring building during earthquake, wind, and flood events. Deep foundations also provide protection against foundation undermining that occurs when rushing water erodes soil from under the foundation.

Figure 14-74: End bearing pile group.

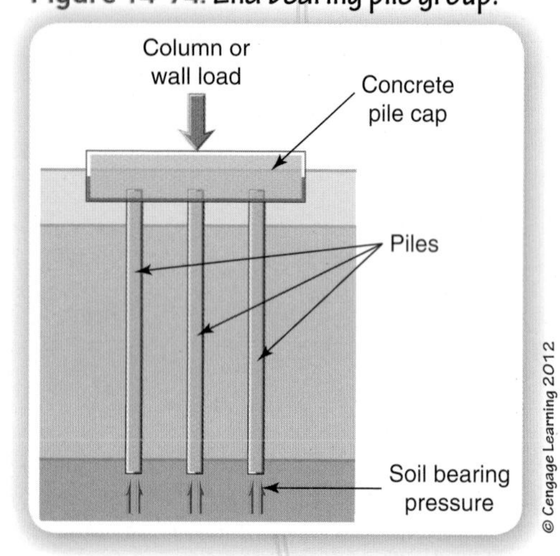

Deep foundations can carry load in two distinctly different ways. If a deep layer of soil is sufficiently strong to support the building, deep foundations are often designed to bear on the deeper layer and transfer load through soil bearing pressure at the bottom of the foundation, as is the case with shallow foundations. Imagine pushing a straw into a jar of mayonnaise. The mayonnaise (or weak soil) cannot adequately resist the load that you apply to the straw. However, when the straw reaches the bottom and bears on the interior surface of the jar (the strong soil), the straw is well supported. Piles that transfer load in this manner are referred to as *bearing piles* (see Figure 14-74).

Now imagine trying to force a straw through a jar of extra thick peanut butter. It takes much more effort to push the straw through this "soil." In fact, a few straws strategically placed in the jar and driven into the peanut butter (without touching the bottom) could potentially hold up a textbook. These straws are transferring load through friction between the surface of the straw and the surrounding peanut butter. This same phenomenon occurs when piles are driven into stiff cohesive soil. Piles that transfer load to surrounding soil through friction are referred to as *friction piles* (see Figure 14-75). When designing friction piles, the coefficient of friction that is exhibited by the soil is used to determine the number of piles needed to support a given load and should be provided in the soils report.

Piles are typically driven or bored into the ground and are most often installed vertically to resist vertical loads. Piles can also be installed at an angle in order to better resist lateral loads from the structure. These non-vertical piles are referred to as *battered piles* (see Figure 14-76).

Figure 14-75: Friction pile group.

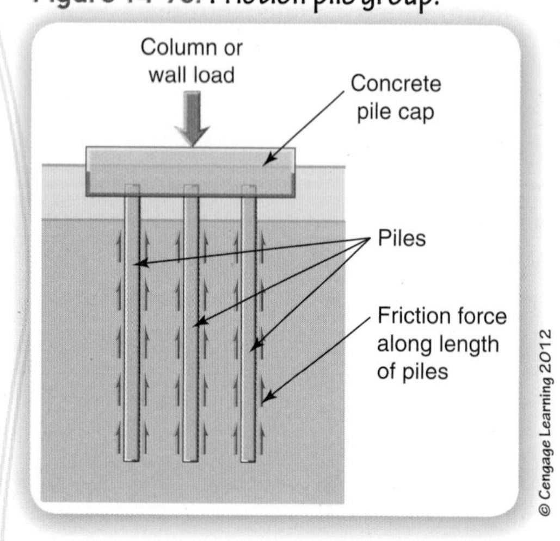

Figure 14-76: Battered piles can provide more resistance to lateral loads.

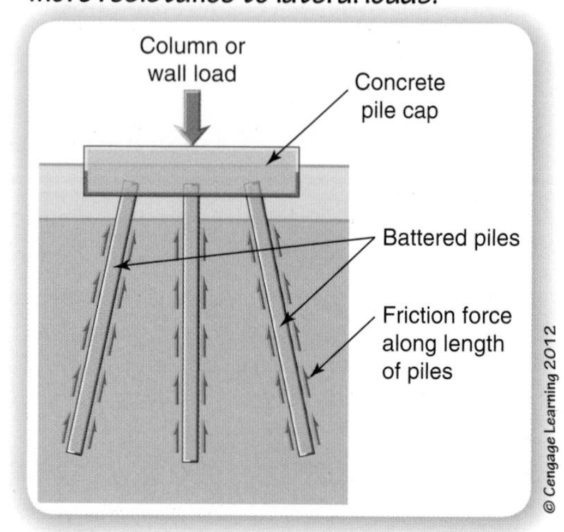

SUMMARY

One of the most important aspects of building design is the design of the structural systems that support the building. Without an adequate structure, a building could not withstand the loads that are imposed by the weight of the building itself; the people, furnishing, and equipment that occupy the building; and the natural environment. Potential loads and the likelihood of different loads affecting the building concurrently must be carefully considered. Building codes dictate the loads and the combinations of loads that must be considered in building design. The type of occupancy will most affect the imposed live loading inside a building, but the location of the building will significantly affect the snow, wind, and earthquake loads that must be included in the design process for the structure.

The building structure typically consists of a variety of structural components that can include flat plates (like floors and walls), beams, columns, braces, and foundations. Each of these structural components must not only carry the loads that are applied directly to it, but must also be securely connected to other structural members within the system. The connections allow loads to be transferred from one structural element to another so that the building loads can be safely transferred along a continuous load path to the ground. Each structural member will experience some combination of tension, compression, bending, shear, and torsion forces that cause reaction forces and internal stresses within each member.

The capability of a structural member to resist internal stress depends on the material from which it is made and the shape of the member as well as how the loads are applied. A factor of safety is always used in the design process to determine the allowable strength of each member. Another important consideration in structural design is the potential movement of the building and its individual components when load is applied. The amount of deflection of structural members is limited by codes to ensure that the building is comfortable for human occupants and provides an appropriate operating environment for equipment. It is the responsibility of structural engineers to determine reaction forces and select structural members to safely resist internal stress while ensuring that the building performs within certain serviceability limits.

BRING IT HOME

1. Assume that you are reviewing the design of a hotel in the upper peninsula of Michigan. The building has two floors. The elevated floor is designed to be built with 6 in lightweight, hollow-core pre-cast concrete panels with a hardwood floor finish, and will support private rooms. The flat roof will be constructed with a 3 in. cast in place concrete slab, 5 in of rigid insulation with a 5-ply membrane. Determine the magnitude of the floor live and dead load, the roof live and dead load, and the ground snow load that should be used in the design of the building.
2. A multistory office building will be supported with a braced steel frame. Each floor will be constructed using a 4.5 in concrete slab and will support approximately 5 psf of lighting and HVAC equipment for the floor below. Floor framing will include 25 ft long steel beams spaced at 7 ft apart. Assume the weight of the beam itself is 30 plf.
   a. Calculate the total uniformly distributed load (in plf) that will be applied to the beam.
   b. Sketch a beam diagram to represent the loading condition for the beam.
   c. Draw shear and moment diagrams and indicate the maximum shear and bending moment. Be sure to show all of your work.
3. The layout of the structural steel framing for the second floor of a retail store is shown. Assume the total second floor dead load is 70 psf, including the weight of the steel framing. The first floor is a slab on grade.

Second-floor framing plan.

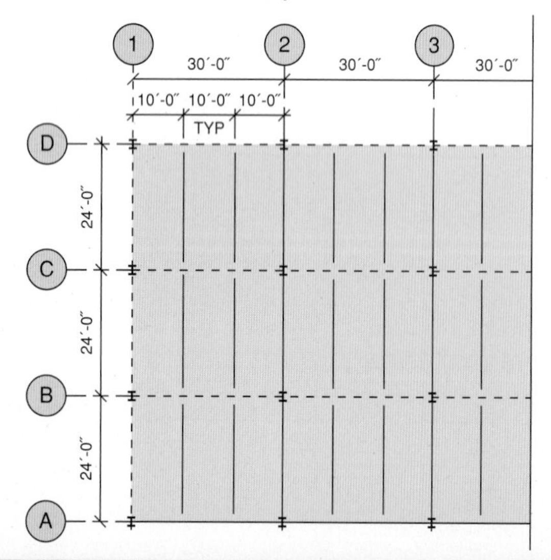

   a. What are the IBC required uniformly distributed live load and concentrated load?
   b. Will the concentrated live load or the uniformly distributed live load control the beam design? Justify your answer.
   c. What is the tributary width for the beams on column line 2?
   d. What is the total uniformly distributed load (dead and live) applied to the beams on column line 2?
   e. What is the maximum shear and bending moment for the beams on column line 2?
   f. Choose the most efficient structural steel shape for the beams on column line 2. Be sure to check deflection.
   g. Choose the most efficient structural steel shape for the girders on column line B. Be sure to check deflection.
4. A W12 × 50 column has been chosen to support both the roof and the elevated floor of a building. Assume that the total roof load (dead and live) supported by the column is 55 k. And the total elevated floor load is 100 k. If the unbraced length of the column is 18 ft, is a W12 × 50 column adequate to support the floor plus the roof loads?
5. Design a square spread footing to support the column from the previous problem. The W12 × 50 column is 30 ft long (ground floor to roof) and the footing thickness is 2 ft. The soil on which the foundation will bear is well-graded gravel.
   a. What is the minimum depth at which the foundation may bear based on the frost depth?
   b. Assume that the base of the column will be supported at grade. Will a pier be needed to transfer the load from the column base to the foundation? If so, what size should it be? How much will it weigh?
   c. What is the total load that will be transferred to the foundation at column B-2? Be sure to include the roof load, the second floor load, the column weight, and the pier weight (if needed).
   d. What is the allowable soil bearing pressure reduced for the footing weight?
   e. What is the minimum footing size that should be used to support column B-2?
   f. Sketch a section view of your foundation design.

EXTRA MILE

Design a structural steel frame for a 100 ft × 75 ft two-story retail store. Size the structural steel floor and roof framing, columns, and spread footings for the building and create a second floor framing plan and a roof framing plan showing the correct sizes for the framing members. Make the following assumptions for the design:

- The yield strength of the steel is 50 ksi.
- First floor dead load is composed of a 3½ in concrete slab on a steel deck. In addition the floor will support a suspended ceiling. Allow 10 lb per square foot for plumbing, electrical, and mechanical equipment.
- The roof will be constructed of a steel deck with 4 in of rigid insulation and a built-up roof. Allow 10 lb per sq ft for the plumbing, electrical, and mechanical equipment.
- The floor-to-floor height and floor-to-roof height is 20 ft.
- The columns are braced at the first floor, second floor, and roof.
- The allowable soil bearing capacity is 2500 lb per sq ft.
- The spread footings are 2 ft thick.

PART VI
Mechanicals: The Building Comes Alive

CHAPTER 15
Planning Electric Codes

GPS DELUXE

Menu

START LOCATION | DISTANCE | END LOCATION

Before You Begin

Think about these questions as you study the concepts in this chapter:

1. What are some of the common terms and properties of electricity?
2. How do we determine a building's electrical needs?
3. What are some of the considerations for determining the electrical load for a building?
4. How are different types of electrical systems communicated to building professionals?
5. What are some of the code requirements for electrical systems?
6. What are some of the considerations in lighting design?

INTRODUCTION

Courtesy to come

It's hard to image what buildings today would be like without electricity. It is the preferred power source for lighting, heating, air conditioning, and appliances and other types of electronic equipment throughout a home or workplace. Since the 1980s, electrical usage in offices has increased from 3 W per sq ft to 5 W per sq ft and to 10 W per sq ft in the 1990s. With the continuing increase in demand for power, new technologies focused on energy conservation have brought current power usages to below the 1990s levels. One such example is the refrigerator. Today's energy-efficient fridges use as little as 250 to 600 kWh per year and rack up $50 or less in annual energy bills. By comparison, a typical 1983 brand, according to the Energy Star online calculator (*http://www.energystar.gov/index.cfm?fuseaction=refrig.calculator&screen=1*), uses 1500 kWh and costs $153 a year.

In this chapter, areas for discussion will include common terms and properties of electricity, determining a building's electrical needs and, loads, and some of the code requirements for electrical systems. The chapter will also discuss how are these electrical systems communicated in plans and diagrams as well as the multiple aspects of lighting design, especially the types of considerations and drawings necessary to develop a complete system.

COMMON TERMS AND PROPERTIES OF ELECTRICITY

Electrical systems are normally designed by electrical engineers. However, every member of the building design profession must have a fundamental knowledge of electricity. Architects, engineers, contractors, and building managers should understand electrical terms commonly associated with building electrical systems. Some of these terms include the following listed in Table 15-1.

Table 15-1

Common Electrical System Terms		
1. Auxiliary systems	9. Feeders	17. Panelboards
2. Branch circuit	10. Fuses	18. Single phase
3. Circuit breakers	11. Ground fault interrupters (GFIs)	19. Switches, 3-way, 4-way
4. Diagrams, connection	12. Ground	20. Three-phase
5. Diagrams, one line	13. Lighting fixtures (luminaries)	21. Transformers
6. Diagrams, riser	14. Motor	22. Voltage
7. Diagrams, schematic	15. Overcurrent	
8. Energy, watt-hour (Wh)	16. Overcurrent protection	

Table 15-2

Common Units of Measure and Quantities	
1. Ampere (A)	7. Volt (V)
2. British thermal unit (Btu)	8. Watt (W)
3. Horsepower (hp)	9. Ohm
4. Kilovolt-ampere (kVA)	10. Megawatt (MW)
5. Kilowatt (kW)	11. Watt-hour (Wh)
6. Volt-ampere (VA)	12. Kilowatt-hour (kWh)

Some of the most common units of measure for electrical systems include the following Table 15-2.

Electric charge is expressed in **watt/hour** (Wh) or **kilowatt/hour** (kWh). The flow of electricity in an electrical circuit is referred to as a **current** (*I*) (see Figure 15-1). The unit for current is given in **ampere** (amp) and is defined as the amount of electric charge passing through a given point per unit of time (see Figure 15-1).

Voltage (*V* or *E*) is the measure of electric force available to cause the movement of electrons. The unit for voltage is given in volts.

The rate at which electricity flows through a material is an internal property known as **resistance** (*R*). A material with low resistance is called a **conductor**. Materials with low resistance are usually pure metals such as copper, aluminum, silver, and gold. A material with high resistance is called an **insulator**. Materials like paper, rubber, neoprene, and other synthetic materials are good insulators.

Capacitance (*C*) is the amount of electrical charge stored. Capacitance is in every form of circuitry with a nonconductive material in the component such as plastic. **Capacitors** are devices designed to store electric charge.

Figure 15-1: Electric circuit.

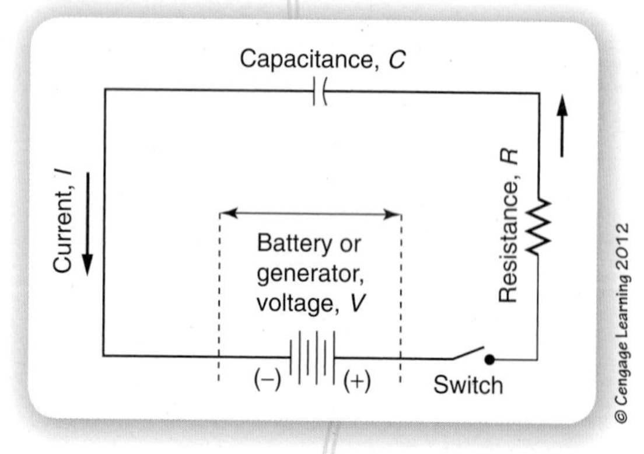

Resources are the things needed to get a job done, such as tools and machines, materials, information, energy, people, capital, and time.

Direct and Alternating Current

There are two basic types of electrical current: direct current and alternating current. **Direct current** is an electric current that flows through a circuit in one direction regardless of the flow rate. Direct current was the first form of electricity used in providing power to mechanical systems and is still in use today. Systems such as batteries, fuel cells, and solar cells produce current that always flows in the same direction, originating in the negative terminal and flowing toward the positive terminal.

Alternating current (AC) is an electrical system in which voltage and current are reversed periodically in the circuit. Nearly all power provided by electric utilities in the United States is through alternating current systems. Some of the advantages of alternating current are lower generating costs, the ability to transform AC to either higher or lower voltages, and easier long-distance voltage transmissions. AC systems have easier voltage transformations because of their ability to have large amounts of electrical power at higher voltages transferred over long distances while maintaining relatively low transmission loads. For example, as Figure 15-2 illustrates, a single utility power company usually generates between 4000 to 25,000 V at the plant per hour. The voltage is then transmitted at 10,000 to 70,000 V for distribution, and for longer distance it may be even higher. These transmission voltages are then stepped down to around 3000 to 4000 V for local distribution through substations. At the end-users' locations, further transformation of the voltage may be stepped down to levels such as 120, 208, 240, 277, and 480 V, depending on the building's needs.

Direct Current:
an electric current that flows through a circuit in one direction regardless of the flow rate. Direct current was the first form of electricity used in providing power to mechanical systems and is still in use today.

Alternating Current:
(AC) is an electrical system in which voltage and current are reversed periodically in the circuit.

POWER AND ENERGY

While there are many forms of energy, and while one form of energy can be converted to another, all energy eventually degrades into heat energy. Of all forms of energy, electricity is perhaps the most convenient to use. In terms of buildings, electrical energy can be easily converted into mechanical energy through a motor, lighting energy through a lamp, or heat energy through a resistant heater. How does this impact the power needs of a particular situation? By determining the various systems and their usage we can determine the power loads of a facility. Recalling **Ohm's Law**, we know that V (voltage) $= I$ (current) $\times R$ (resistance) and E (energy) is the product of P (power) $\times T$ (time) usually given in units of Wh or kWh. While in most cases the power is given by the manufacturer, the energy is simply determined by the power times the usage.

Some of the basic electrical design procedures fall into five steps: determining the building needs, calculating the electrical loads, determining the types of electrical systems, coordinating other design aspects in the building, and preparing electrical plans and specifications.

Ohm's Law:
developed by George Simon Ohm in 1827, describes the relationship between current, voltage, and resistance of a DC circuit: V (voltage) $= I$ (current) $\times$ R (resistance)

Determining the Building Needs

In most cases, the needs of a building are determined by several factors, including building occupancy, cost factors, architectural factors, building environments, lighting requirements, and other mechanical and electrical systems in the building, with allowances for future growth.

Each professional involved in the building design needs to know particular electrical information. Architects need to know the type and size of all needed electrical equipment rooms. Structural engineers need to know the loads of various

Figure 15-2: **Electrical power distributions.**

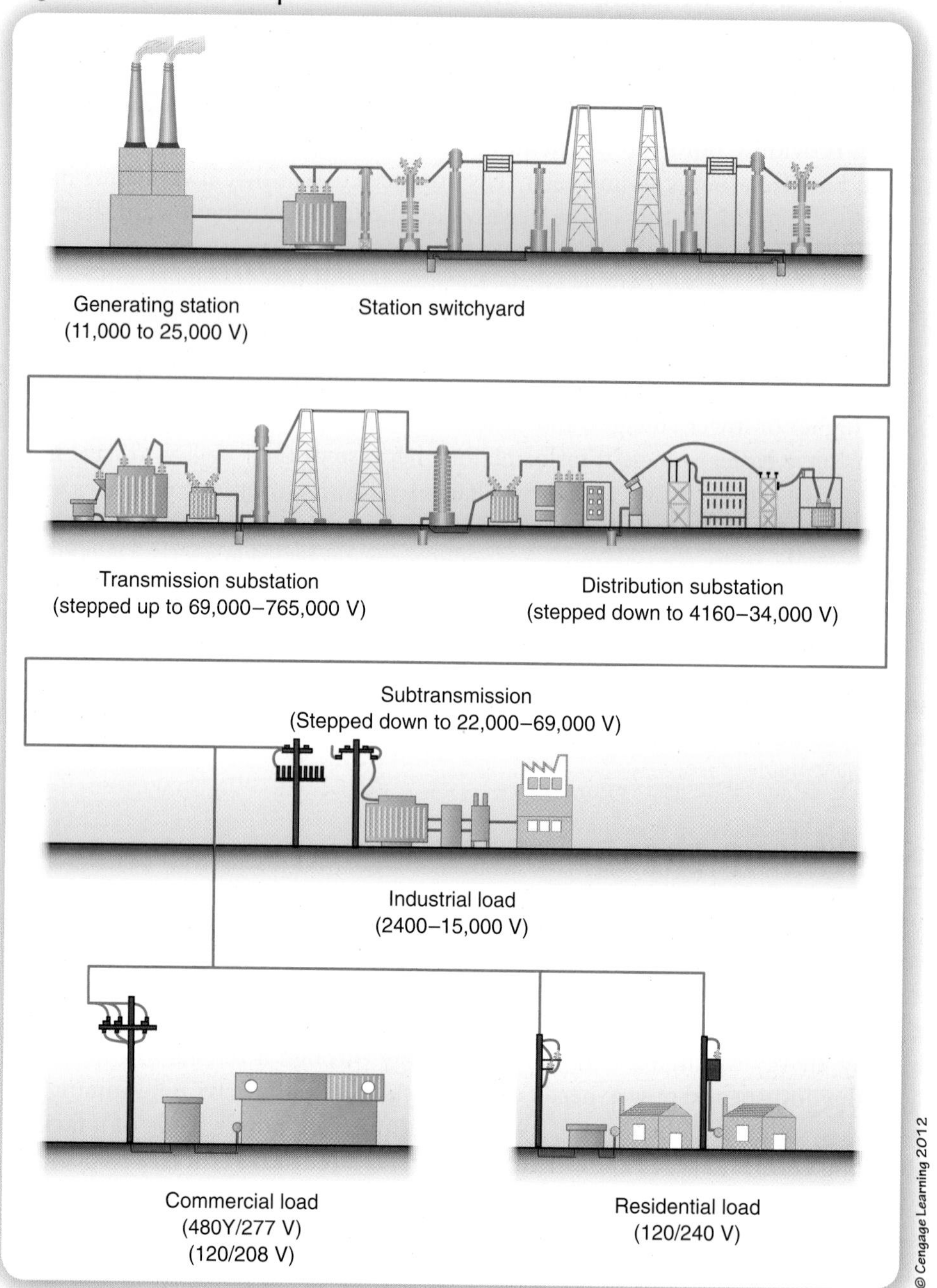

electrical systems to be used. Mechanical engineers need to know system options so they may determine types of HVAC equipment best suited to the available power in a building.

Designing electrical systems may also require input from the utility company to know the types of services available to the building site.

Calculating the Electrical Loads

Calculating the electrical loads of a building is dependent upon several factors. One such factor is determining the building type and its use not only in the present but also in the future planning. Cost is always a key factor and the quality as well as the safety of the occupants in the building. Architecture influences the electrical needs by determining the size of the building, the number of floors, and the building's overall footprint. The building environment considers what types of mechanical systems are needed to condition the space. Additional needs such as interior and exterior lighting, and emergency backup systems will also need to be considered. Mechanical systems include building equipment such as transportation systems, food preparation, recreational equipment, and processing equipment. Building management systems such as clocks, fire alarms, telecommunications, radio, and TV antenna are auxiliary systems. Other types of electrical loads include power provided for plug-in equipment such as household appliances, personal computers, office equipment, laboratory instruments, service equipment, portable lights, and audio and video equipment—commonly referred to as **convenience power.** In addition to the previously mention considerations the need to allow for future expansion of the building is included in the calculations by factoring a minimum of 25 percent spare capacity.

Figure 15-3: *Coffee Shop.*

As an example, let us determine the energy needs for a proposed coffee shop (Figure 15-3) with 600 ft^2 of building area. The hours of operation will be from 7:00 a.m. till 7:00 p.m.

Our first step is to determine the power needs for each of the components listed below. The second step is to determine the amount of energy the coffee shop needs per day. Several assumptions will need to be made when considering the hours of operation.

The following items will need to be considered in determining the coffee shops energy needs:

- Commercial refrigerator – 800 W × 18 h/day (12 h of operation during shop operating hours and 6 additional hours when shop is closed for a combined total 18 h/day)/1000 W/kWh = 14.4 kWh/day.
- Coffee grinder – 100 W × 6 h/day (every hour of store operation the equipment is grinding fresh grounds for a half hour for a combined total 6 h/day)/ 1000 W/kWh = 0.6 kWh/day.
- Commercial microwave – 1500 W × 6 h/day (every hour of store operation the equipment is grinding fresh grounds for a half hour for a combined total 6 h/day)/ 1000 W/kWh = 9 kWh/day.
- Commercial cappuccino maker – 1250 W × 6 h/day (every hour of store operation the equipment is grinding fresh grounds for a half hour for a combined total 6 h/day)/1000 W/kWh = 7.5 kWh/day.
- Commercial coffee maker – 1200 W × 6 h/day (every hour of store operation the equipment is grinding fresh grounds for a half hour for a combined total 6 h/day)/1000 W/kWh = 7.2 kWh/day.
- Commercial toaster oven – 1500 W × 6 h/day (every hour of store operation the equipment is grinding fresh grounds for a half hour for a combined total 6 h/day)/1000 W/kWh = 9 kWh/day.
- Interior lighting – (ASHRAE/IESNA Standard 90.1 (2004) for a restaurant) 1.5 W/ft^2 1.5 W/ft^2 × 600 ft^2 × 16 h/day (2 h of operation before and after hours and 12 h during operating hours for a combined total 16 h/day)/ 1000 W/kWh = 14.4 kWh.

- 40-gallon water heater – 5500 W = 4 h/day (3 h of operation during shop operating hours and 1 additional hours for when shop is closed for a combined total 4 h/day)/1000 W/kWh = 22 kWh/day.
- Central air conditioning system – 1500 W × 12 h/day (assume 9 h of operation during shop operating hours and 3 additional hours for when shop is closed for a combined total 12 h/day)/1000 W/kWh = 18 kWh/day.
- Hand dryer (×2) for restrooms – 1200 W × 2 h/day (assume 30 s of operation per use times 120 customers/restroom times 2 restrooms for a combined total 2 h/day)/1000 W/kWh = 2.4 kWh/day.
- Computer system – 550 W × 16 h/day (2 h of operation before and after hours and 12 h during operating hours for a combined total 16 h/day)/1000 W/kWh = 8.8 kWh/day.

By adding together the kWh/day for each piece of equipment we have determined the total energy needed for a single day of operation to be 112.7 kWh/day. If we assume the voltage of all the appliances in the shop operates on a 120 V system then the amperage for the coffee shop is (112.7 kWh/day × 1000 W/kWh) / 120 V = 939 amps per day. A factor of 120 percent is needed to compensate for future expansion and other considerations. Essentially a service capacity of approximately 1127 amps per day is needed for the coffee shop. We must remember that several assumptions have been made in this example. First, we assumed that all systems and appliances were working at the same time; rarely does this occur. Secondly any ancillary equipment such as blenders, clocks, or any other plug-in devices was not considered in our calculation thus the allowance for additional needs was assessed in the final calculations.

Engineers and architects identify constraints and external factors that may impact the outcome.

Your Turn

Based on a new residence, try to determine the amount of electricity that would be consumed per month of average use. The new residence is a ranch style home that has the following parameters:

1. 3 Bedrooms – ceiling fan with interior lighting in each bedroom
2. 2 Baths – interior lighting
3. Kitchen with stove, refrigerator, dishwasher, microwave, trash compactor
4. Living Room – ceiling fan with interior lighting
5. Dining Room – ceiling fan with interior lighting
6. Family Room – a complete entertainment center including TV, stereo, CD/DVD player
7. Central Air Source Heat Pump
8. Washer and Dryer
9. Garage with Door Opener
10. Upright vacuum cleaner
11. Heat Pump for heating water
12. Home Office computer, all-in-one scanner and printer
13. Ancillary equipment assume 5 percent of subtotal
14. Projected future growth of 25 percent

Calculate the electrical loads per month of this residence based on recommended appliances at U.S. Environmental Protection Agency Energy Star program (*http://www.energystar.gov*) under "Products" and choose an appliance to meet the homes needs. Click on one of the appliance links located below "Qualified Products," such as Clothes Washer. Then select "Find a Product" and "Find Clothes Washer." You can narrow your search by selecting a name brand. Repeat the same steps for all the appliances. If an appliance is not listed on the Energy Star website then perform a search on the appliance type to determine it power consumption. Check your units!

Determining the Types of Electrical Service

When determining the best electrical services for a building, a designer must take into account a variety of aspects including the cost of the system, the cost of maintenance, the reliability of equipment, space allocation, and the overall electrical power needs of the building.

In the case of residential power most household appliances operate as a **single-phase system**. A single-phase system supplies 120 or 240 V three wire connection including a neutral ground from the utilities to the home.

A **three-phase system** is the standard form of electricity distribution from the power station to the powerpole. Three wires carry three currents that are identical except for their cycles being shifted one third of a cycle from each other. In the case of three-phase systems, it can have three different load voltage systems: 120/208 V, 277/480 V, and a 480 V system.

Major areas of commercial loads typically stem from elevators, food service equipment, mechanical equipment, and appliances. Elevators range from 15–200 hp depending on their speed and capacity. They are usually equipped with a three-phase motor to provide the power necessary to run such a system. Three phases provide a smoother current flow than a single phase, and reduced harsh running in large machines. Induction motors that use three phases are smaller, cheaper, efficient, and have a high starting torque than single-phase motors.

Figure 15-4: *Centralized system.*

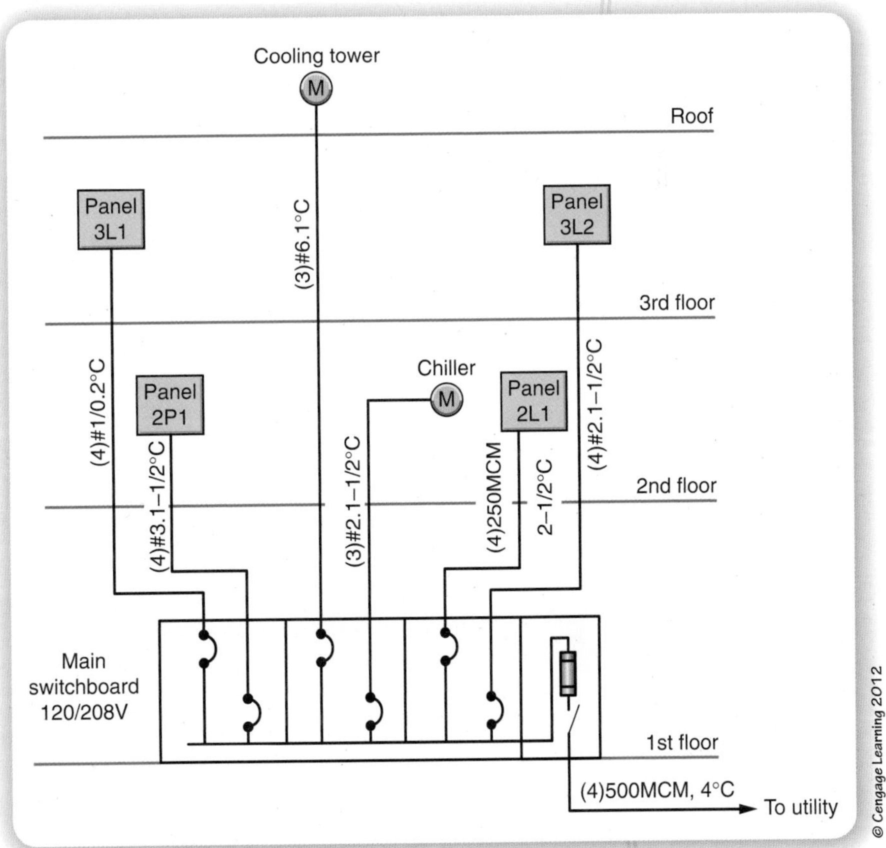

Coordination of Other Design Aspects in the Building

In the case of interfacing of systems in a building, the distinction between a centralized system that supplies the building's power needs versus multiple single power supplies can influence the overall layout and power needs of the building itself. As an example, a multistory office building (Figure 15-4) serviced by one

Figure 15-5: Electrical plan.

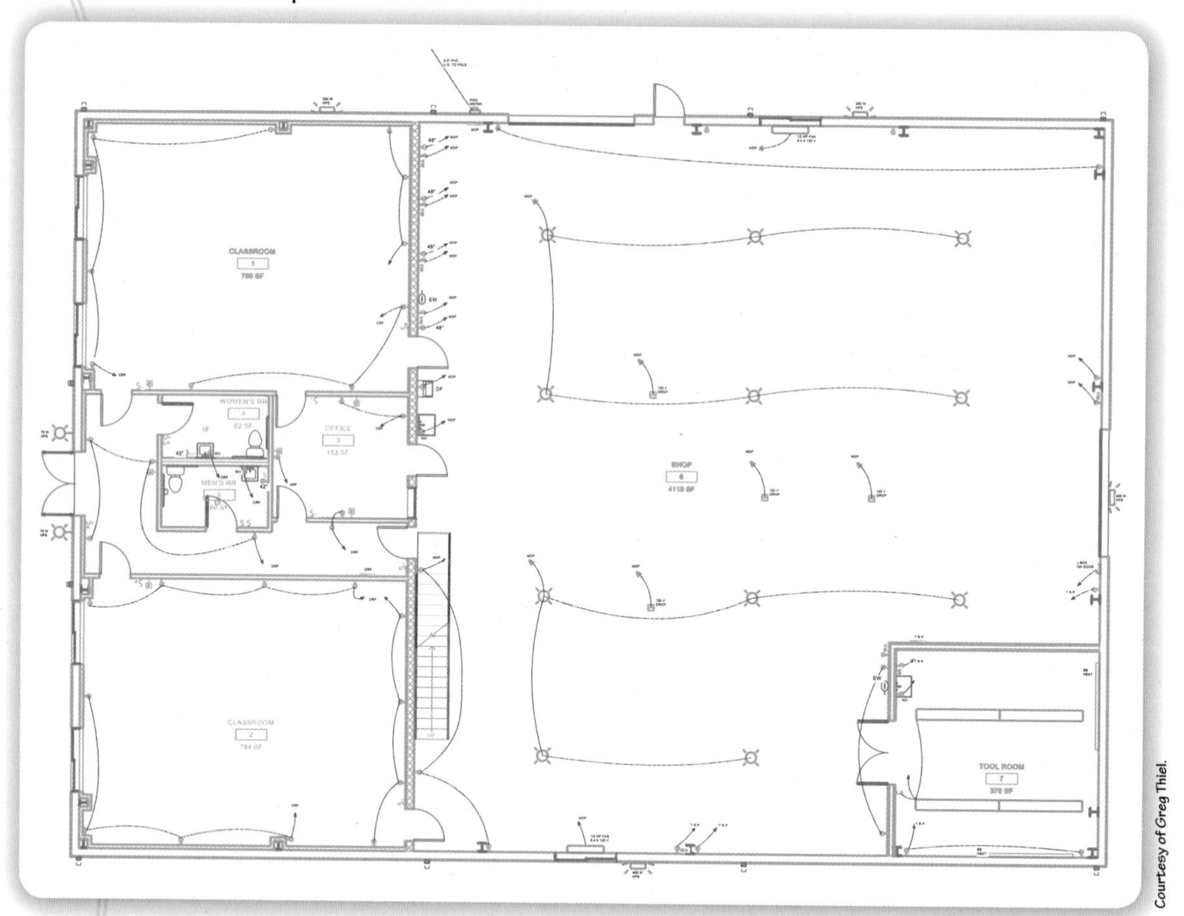

Courtesy of Greg Thiel.

Figure 15-6: Schematic electrical diagram.

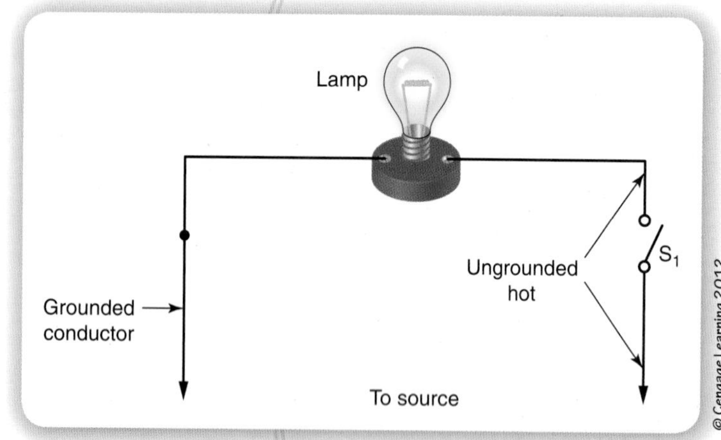

© Cengage Learning 2012

heating/cooling system for the entire building could require a three-phase electrical power system at 480 V, but if that same building contains individual cooling/heating system for each floor it may operate on a single-phase system.

In the case of space planning, decisions made early on pertaining to the size of the mechanical/electrical rooms are often difficult to change later. In terms of designing space for electrical and mechanical equipment, additional consideration is given toward accessibility and safety.

Preparation of Electrical Plans

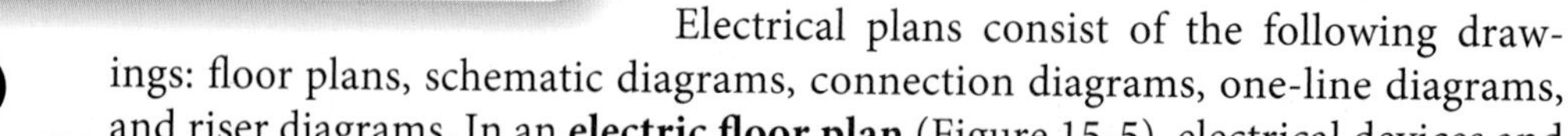

Electrical plans consist of the following drawings: floor plans, schematic diagrams, connection diagrams, one-line diagrams, and riser diagrams. In an **electric floor plan** (Figure 15-5), electrical devices and equipment are superimposed on an architectural background. The plan also indicates wiring between light fixtures and switches and may indicate the types of wall switches and circuit loads. **Schematic diagrams** (Figure 15-6) may be used to illustrate the circuitry of the system in particular, with special illustrations for clarity purposes. **Connection diagrams** (Figure 15-7), also known as wiring diagrams, may provide additional instruction as to how wiring terminals and equivalent are connected along a circuit pattern. **One-line diagrams** (Figure 15-8)

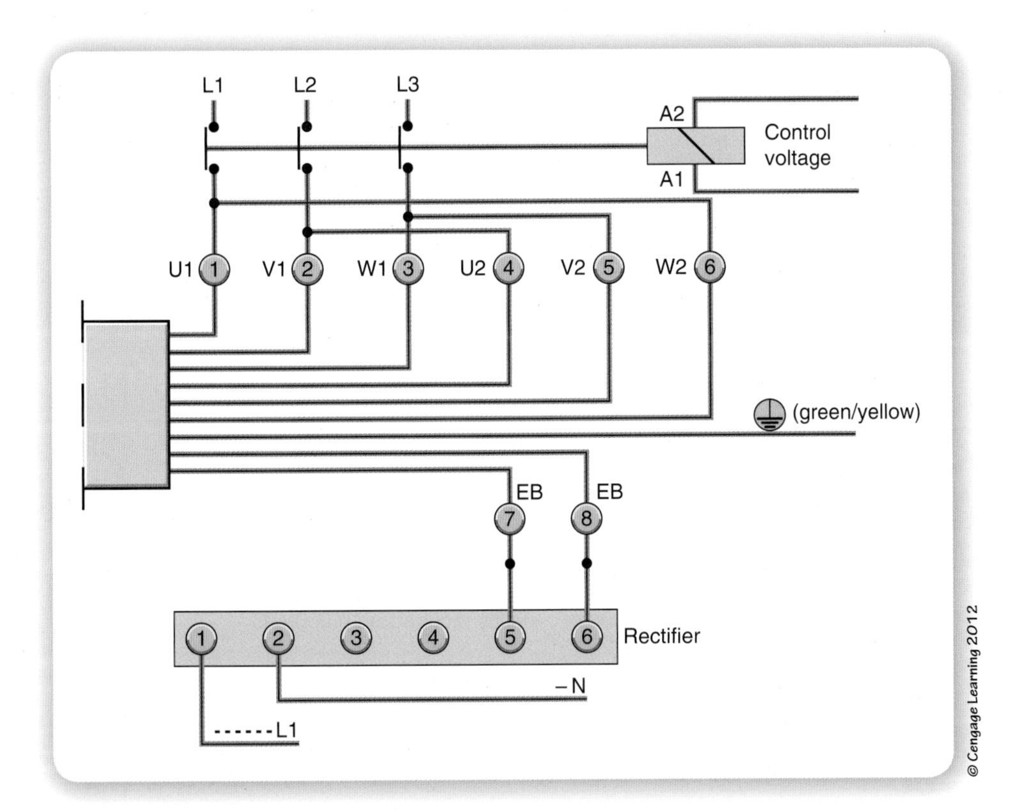

Figure 15-7: *Connection diagrams.*

indicate principal relationships between major equipment, while **riser diagrams** (Figure 15-9) typically express the physical relationship between floors in larger buildings.

Graphic symbols (Figure 15-10) are used to indicate various electrical designs including equipment devices, wiring, and other types of hardware. Standardized symbols are used as a common language to communicate between architects, engineers, and other building professionals. The figure below provides numerous types of electrical symbols commonly used.

Vocabulary, visual symbols, and icons are commonly used in design and construction to be successful in workplace/jobsite communications.

National Electric Code

The National Electric Code (NEC) is prepared by the National Fire Protection Association (NFPA) with the intent to protect the building occupants from any potential hazards from electrical systems. The NEC has been adopted as national and state building codes in the United States and covers methods of connection to

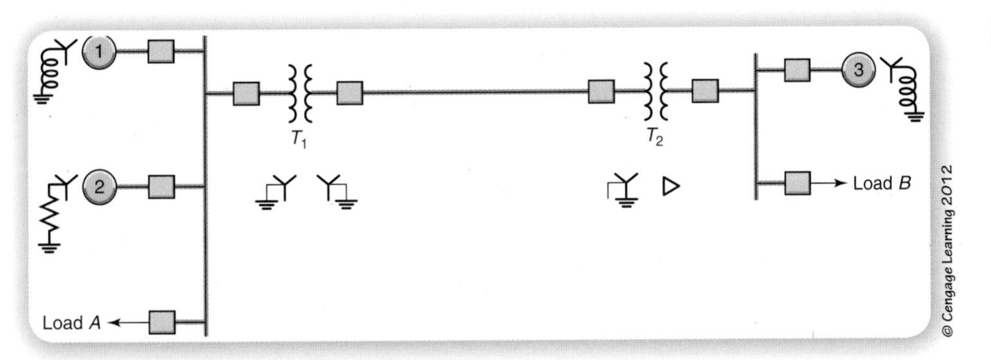

Figure 15-8: *One-line diagrams.*

Figure 15-9: *Electrical riser diagrams.*

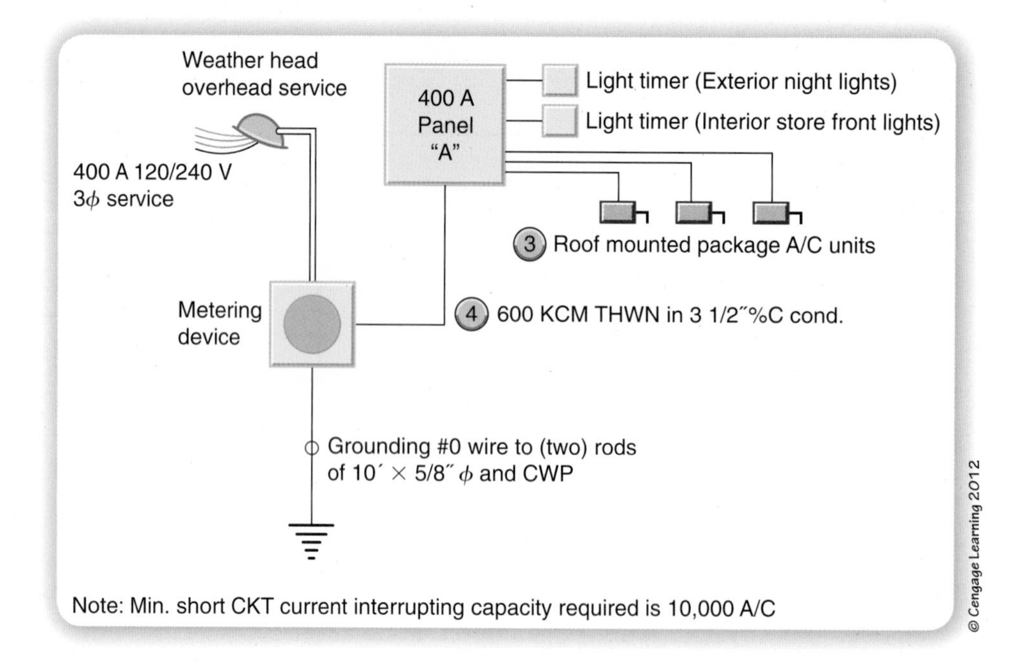

Figure 15-10: *Commonly used electrical symbols. From FLETCHER. Residential Construction Academy House Wiring, 2e.*

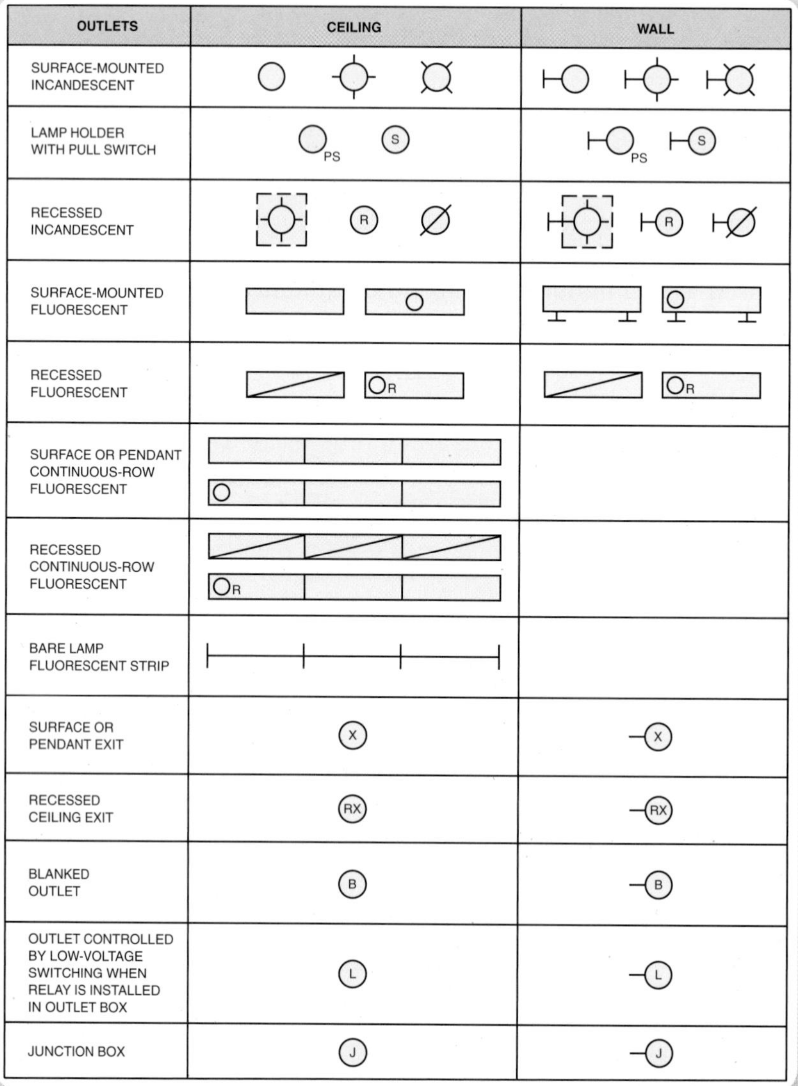

OUTLETS	CEILING	WALL
SURFACE-MOUNTED INCANDESCENT		
LAMP HOLDER WITH PULL SWITCH	PS S	PS S
RECESSED INCANDESCENT	R	R
SURFACE-MOUNTED FLUORESCENT		
RECESSED FLUORESCENT	R	R
SURFACE OR PENDANT CONTINUOUS-ROW FLUORESCENT		
RECESSED CONTINUOUS-ROW FLUORESCENT	R	
BARE LAMP FLUORESCENT STRIP		
SURFACE OR PENDANT EXIT	X	X
RECESSED CEILING EXIT	RX	RX
BLANKED OUTLET	B	B
OUTLET CONTROLLED BY LOW-VOLTAGE SWITCHING WHEN RELAY IS INSTALLED IN OUTLET BOX	L	L
JUNCTION BOX	J	J

Figure 15-10: *(continued)*

RECEPTACLE OUTLETS	
SINGLE-RECEPTACLE OUTLET	D — ELECTRIC CLOTHES DRYER OUTLET
DUPLEX-RECEPTACLE OUTLET	F — FAN OUTLET
TRIPLEX-RECEPTACLE OUTLET	C — CLOCK OUTLET
DUPLEX-RECEPTACLE OUTLET, SPLIT CIRCUIT	FLOOR OUTLET
DOUBLE-DUPLEX RECEPTACLE (QUADPLEX)	X" — MULTIOUTLET ASSEMBLY; ARROW SHOWS LIMIT OF INSTALLATION. APPROPRIATE SYMBOL INDICATES TYPE OF OUTLET, SPACING OF OUTLETS INDICATED BY "X" INCHES.
WP — WEATHERPROOF RECEPTACLE OUTLET	FLOOR SINGLE-RECEPTACLE OUTLET
GFCI — GROUND FAULT CIRCUIT INTERRUPTER RECEPTACLE OUTLET	F — FLOOR DUPLEX-RECEPTACLE OUTLET
R — RANGE OUTLET	FLOOR SPECIAL-PURPOSE OUTLET
DW — SPECIAL-PURPOSE OUTLET (SUBSCRIPT LETTERS INDICATE SPECIAL VARIATIONS: DW = DISHWASHER. A, B, C, D, ETC., ARE LETTERS KEYED TO EXPLANATION ON DRAWINGS OR IN SPECIFICATIONS).	

SWITCH SYMBOLS	
S OR S_1	SINGLE-POLE SWITCH
S_2	DOUBLE-POLE SWITCH
S_3	THREE-WAY SWITCH
S_4	FOUR-WAY SWITCH
S_D	DOOR SWITCH
S_{DS}	DIMMER SWITCH
S_G	GLOW SWITCH TOGGLE—GLOWS IN OFF POSITION
S_K	KEY-OPERATED SWITCH
S_{KP}	KEY SWITCH WITH PILOT LIGHT
S_{LV}	LOW-VOLTAGE SWITCH
S_{LM}	LOW-VOLTAGE MASTER SWITCH
S_{MC}	MOMENTARY-CONTACT SWITCH
M	OCCUPANCY SENSOR—WALL MOUNTED WITH OFF-AUTO OVERRIDE SWITCH
M, P	OCCUPANCY SENSOR—CEILING MOUNTED "P" INDICATES MULTIPLE SWITCHES WIRE-IN PARALLEL
S_P	SWITCH WITH PILOT LIGHT ON WHEN SWITCH IS ON
S_T	TIMER SWITCH
S_R	VARIABLE-SPEED SWITCH
S_{WP}	WEATHERPROOF SWITCH

electrical power as well as the installation of electrical conductors and equipment in public and private facilities. It includes general guidelines, wiring and protection, wiring methods and materials, equipment for general use, special occupancies, special equipment, special conditions, and communication systems. While the NEC has several hundred pages of rules and guidelines, some of the most common have been summarized as follows:

- Electrical power systems shall have a minimum capacity for various types of building occupancies in the case of a single family dwelling, in the United States, should have electrical service not less than 60 amps for a 120/240 V service panel, three wire service and not less than 100 amps for initial loads of 10 kilovolt amps (kVa).
- The feeder capacity and the number of branch circuits shall not be lower than either the actual demand load or the calculated demand load.
- All wires and equipment shall be protected from abnormal situations with the use of over current protection devices.
- Motors and electrical equipment shall be provided with a means of disconnection and shall be in sight of the equipment or shall be capable of being locked in an open position for safety reasons.

Figure 15-11: Lighting and appliance branch-circuit panelboard.

Courtesy of Schneider Electric.

Figure 15-12: Single-phase three-wire branch-circuit panel.

Courtesy: From MULLIN. Electrical Wiring Commercial. © 2005 Delmar Learning, a part of Cengage Learning, Inc. Reproduced with permission. *http://www.Cengage.com/permissions.*

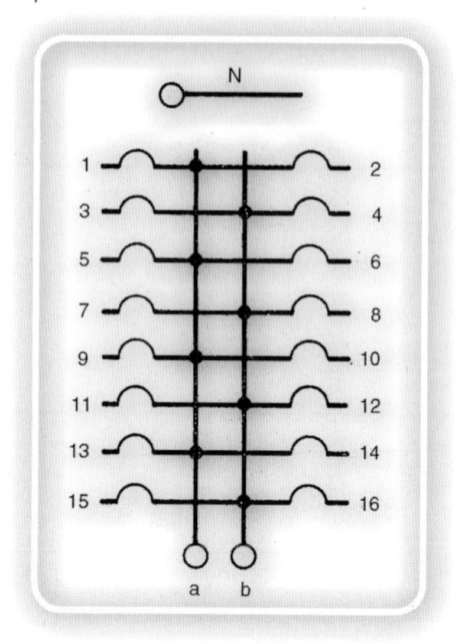

Figure 15-13: Commercial or industrial panel.

Courtesy of Schneider Electric.

The above-mentioned rules are just some of the many rules of the national electric code. The code also considers other types of systems including branch circuits. A **branch circuit** is the portion of the wiring system that extends beyond the overcurrent protection device (Figure 15-12). Some of the design considerations include:

- No wire smaller than 14 gauge shall be used in a dwelling application.
- No wire smaller than 12 gauge shall be used in commercial type structures.
- For two wire 120 V circuits the continuous load shall be limited to 1200 W for 15 amp circuits and 1500 watts for 20 amp circuits.
- Branch circuits shall not exceed 80 percent of its load capacity.
- Inductive lighting such as fluorescent or high-intensity discharge (HID) lamp fixtures shall not exceed 70 percent of branch circuit capacity.
- Fixed wire appliances shall not exceed 50 percent of branch circuit capacity.
- The maximum number of convenience outlets shall be based on the following: outlets supplying specific appliances should be rated accordingly, outlets supplying heavy duty lamp holders shall be considered 5 amps per receptacle, general purpose outlets use 3 amps per duplex receptacle.

ELECTRICAL PANEL LOCATION

All buildings need electrical panels for branch circuits to receptacles and lighting. A panel is powered by a single feeder and circuit breakers connected to the plus feed branch circuits. The electrical panel is often considered the location where power is evenly distributed throughout the building and typically it is located within 100 ft of the utility service provider. In the case of residential buildings, panels can be located in the basement and range in size and shape (Figure 15-13). In the case of commercial buildings, a main service panel may be located in a mechanical room with several sub panels serving multiple floors. Panels (Figures 15-11 thru 15-14) may be located in spaces where appearance is not as an important. Clearance to serviceability is required by the NEC (Figure 15-15).

Figure 15-14: Three phase four-wire branch circuit panel.

From MULLIN. Electrical Wiring Commercial. © 2005 Delmar Learning, a part of Cengage Learning, Inc. Reproduced with permission. http://www.Cengage.com/permissions.

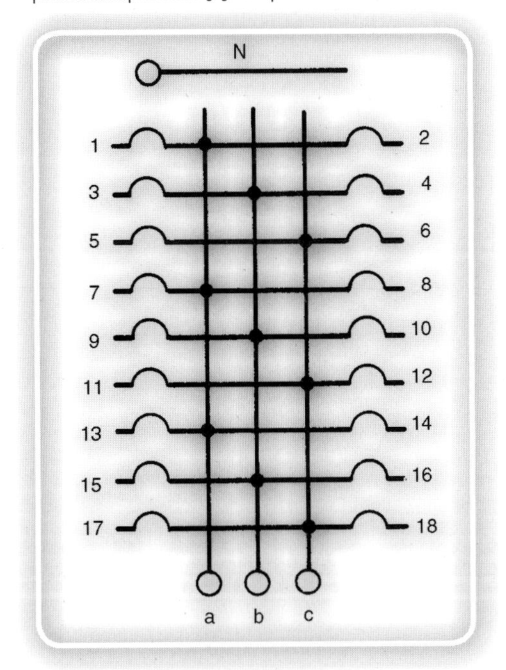

Figure 15-15: Minimum working space clearances in front of service panel.

Courtesy: From FLETCHER. Residential Construction Academy House Wiring, 2e. © 2008 Delmar Learning, a part of Cengage Learning, Inc. Reproduced with permission. http://www.Cengage.com/permissions.

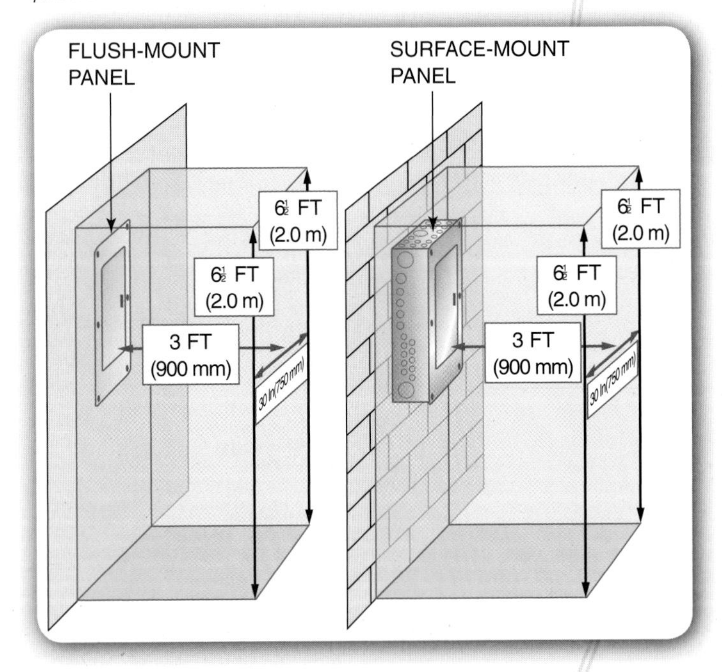

In the case of most commercial buildings today, electrical closets containing electrical panels are preferred for ease of maintenance and flexibility of working on a system without disruption of activities to occupied space. In addition to serviceability, the electrical closets also provide security. Other types of electrical equipment that might be found in mechanical or electrical closets are **transformers.** Transformers step down the incoming power to the building from 480 V or higher to a useable power supply of 120–240 V, or in the case of lighting receptacles to 120–208 V.

As a professional, you will need to create and implement project plans considering available resources and requirements of a project/problem to accomplish realistic planning in design and construction situations.

LIGHTING

It has been said that some of the best engineering designs in a building are behind the walls. For lighting systems, this is generally not the case. For all practical terms and conditions, lighting systems are very visible to the public and can be described as a design area that falls into both fields of art and science. While electric lighting was invented and evolved relatively recently (Figure 15-16), it is hard to imagine what our lives would be like without it.

The importance of lighting is everywhere, and we depend on it to provide functional service, security, and a suitable environment to meet our needs. Lighting creates the glow of the U.S. Capitol (Figure 15-17), the sparkle in a candlelit dinner (Figure 15-18), or the mood of a concert (Figure 15-19).

Figure 15-16: Tree of light–125 years of electric lighting.

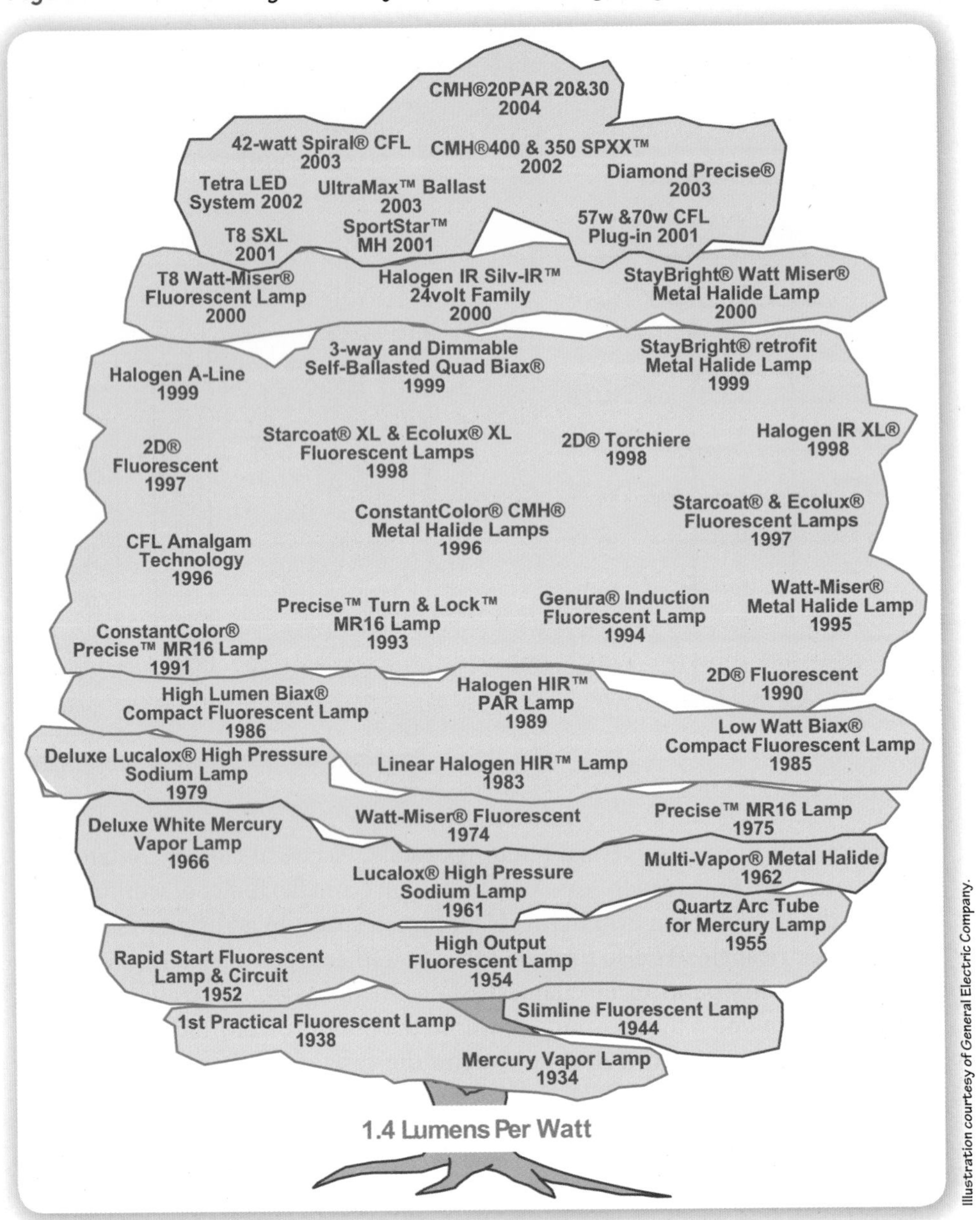

Illustration courtesy of General Electric Company.

Figure 15-17: United States Capitol.

© iStockphoto.com/Qingwa LLC.

Figure 15-18: Festive dinner.

© iStockphoto.com/Paul Piebinga.

Of these scenarios, lighting quality is an important and complex design issue. A quality lighting design is one that meets or exceeds the design criteria appropriate for the tasks that occur within a space; meets the aesthetic objectives in the space (including both light distribution and general appearance); and, finally, meets any energy, cost, and maintenance criteria that apply. The relevant criteria in the design of a lighting system must take into consideration the many lighting quality issues that confronts a particular space. These will be discussed in terms of the design process and include the steps one should follow in the development and execution of a lighting design.

Figure 15-19: Music set.

The Design Process

The design process should provide you with the information necessary to develop appropriate design criteria and allow for the easy implementation of these concepts. For this reason, a clear outline providing a general design process should be used. This allows for easy implementation that lighting designers can follow. When beginning the design, the first step is the **programming**, where a designer researches and develops appropriate design criteria. The second stage, known as the **design development**, is when individual equipment that appears to conform to the original design criteria is selected; this equipment is then tested to determine if it meets the design. Finally, when a final design is reached, the design is documented in the **construction document**, which includes drawings, details, schedules, and written specifications.

Examine how the roles and responsibilities among trades/professions work in relationship to complete a project/job.

Programming

A good place to start in the design of a lighting system is with a process referred to as *programming*. This part of the design process generally involves two steps: collecting important information regarding the spaces to be designed and the tasks performed within these spaces and then establishing goals for one's lighting design.

DATA COLLECTION The relevant information that one should collect in this initial programming step includes the following:

- **Space dimensions**, meaning the width, length, and height of the room. In addition, you should have a good understanding as to what height will certain tasks be performed. This is important in determining the spacing and type of lighting needed to accomplish a set task.
- **Furnishings**, in terms of finishes, which are important to determine reflectances. The impact of a dull finish versus a high gloss finish will impact the lighting requirements if both conditions require a similar lighting level. See Table 15-3 for approximate reflectances below.
- **Surface reflectances,** typically describe the ceiling, wall, and floor finishes. A room with white painted ceilings, dark wood paneling, and a dark carpet would have a reflectance factor of 80-20-15, respectively. These factors are used to

Table 15-3: Approximate Surface Reflectances of Typical Interior Finishes

Building Finishes	Approximate Reflectances (%)
Ceilings	
White pain (plain plaster surfaces)	80
White paint on acoustic tile	70
White paint on smooth concrete	60
White paint on rough concrete	50
Walls	
White paint on plaster	80
Medium blue-gray, yellow-gray	50
Light gray concrete	40
Bricks (Other than rough gray)	30
Unfinished cement, rough tile	25
Wood panel (light)	25
Wood panel (dark)	20
Rough brick	15
Floors	
Light wood	35
Medium wood	25
Dark wood	20
Light tile	30
Dark tile	20
Light carpet (gray, orange, medium blue)	20
Dark carpet (dark gray, brown)	15

determine the amount of lights needed to illuminate a room to the recommended levels. Some approximate reflectances (Table 15-3) include the following:

- **Space activities**, which determine the use of or classify a room or space as, for example, an art room or a classroom.
- **Visual tasks,** describing the actual task to be performed; in the case of an art room, for example, the tasks will include sculpture or oil painting.
- **Occupant type,** (in particular, their ages), specifically the age category of occupants. As we age, our eyes began to require more and more light to perform the same task. As a rule of thumb the visual illuminance of a 66-year-old is one-third that of a 20-year-old due to aging eyes and therefore requires higher illuminance levels.
- Other considerations may include any special geometry or architectural elements that may impact your lighting design solution, the existing lighting conditions that the occupants of this space are currently experiencing, and lastly the client expectations related to image, budget, maintenance, and any other relevant issues.

Once you collect all of this information, you can begin to develop design goals and target design criteria for your lighting system.

DESIGN DEVELOPMENT Generally, there are three types of needs in the design of quality lighting systems. These are human needs, needs related to economics and the environment, and needs related to the architecture of the space. To properly meet these needs, a designer must develop appropriate criteria on which to base the design. Within each of the above categories, there are a variety of aspects that are important (Figure 15-20). A quality lighting design must strike a balance across these areas that are appropriate for the space, and tasks at hand.

In the human needs category, the issue of visibility is the central need in most lighting designs. This is because the primary purpose of a lighting system is to provide for vision. This may be vision to perform a task such as reading, inspecting parts, hitting a ball, locating an elevator, walking down a flight of stairs, or it may be vision that provides for a particular atmosphere or mood (see Figures 15-21 through 15-23).

CONSTRUCTION DOCUMENTS A series of scaled construction documents are prepared indicating the lighting design and its various systems configurations. A reflected ceiling plan (Figure 15-24) usually indicates the types of lighting fixtures

Figure 15-20: The issues that a lighting design must address.

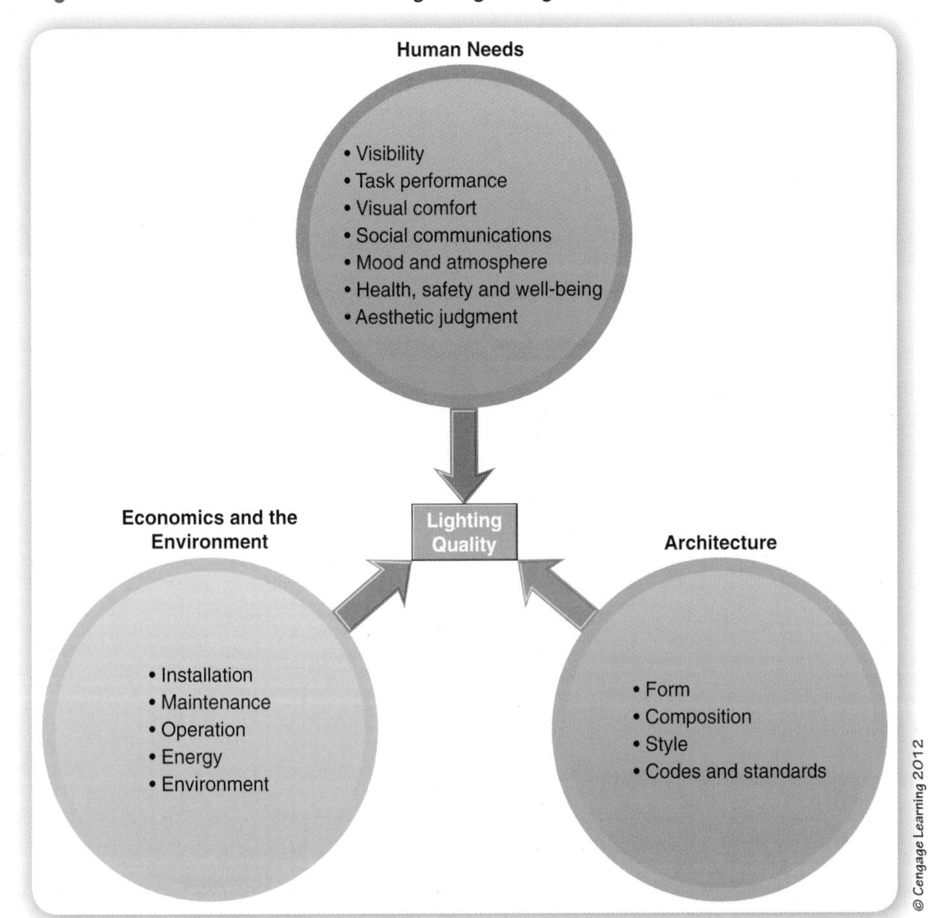

Figure 15-21: Lighting between the stacks.

or luminaries, ceiling heights, ceiling finishes, diffusers, return air grilles, fire protection equipment (smoke detectors, sprinkler heads, and alarm devices), exit signs, and sound and communication systems. A **luminaire** refers to a complete lighting fixture including the housing, bulb, lenses, and ballast in the case of high-intensity discharge lamps. A **ballast** a piece of equipment required to control the starting and operating voltages of electrical gas discharge lights. Electrical circuits,

Luminaire:
a complete lighting fixture that includes the housing, bulb, lens, and ballast (for fluorescent or high-intensity discharge lamps).

Figure 15-22: Lighting for sports.

Figure 15-23: Lighted staircase.

Figure 15-24: Reflected ceiling plan.

Courtesy of Leo Peirick.

Figure 15-25: Electrical plan.

Courtesy of Leo Peirick.

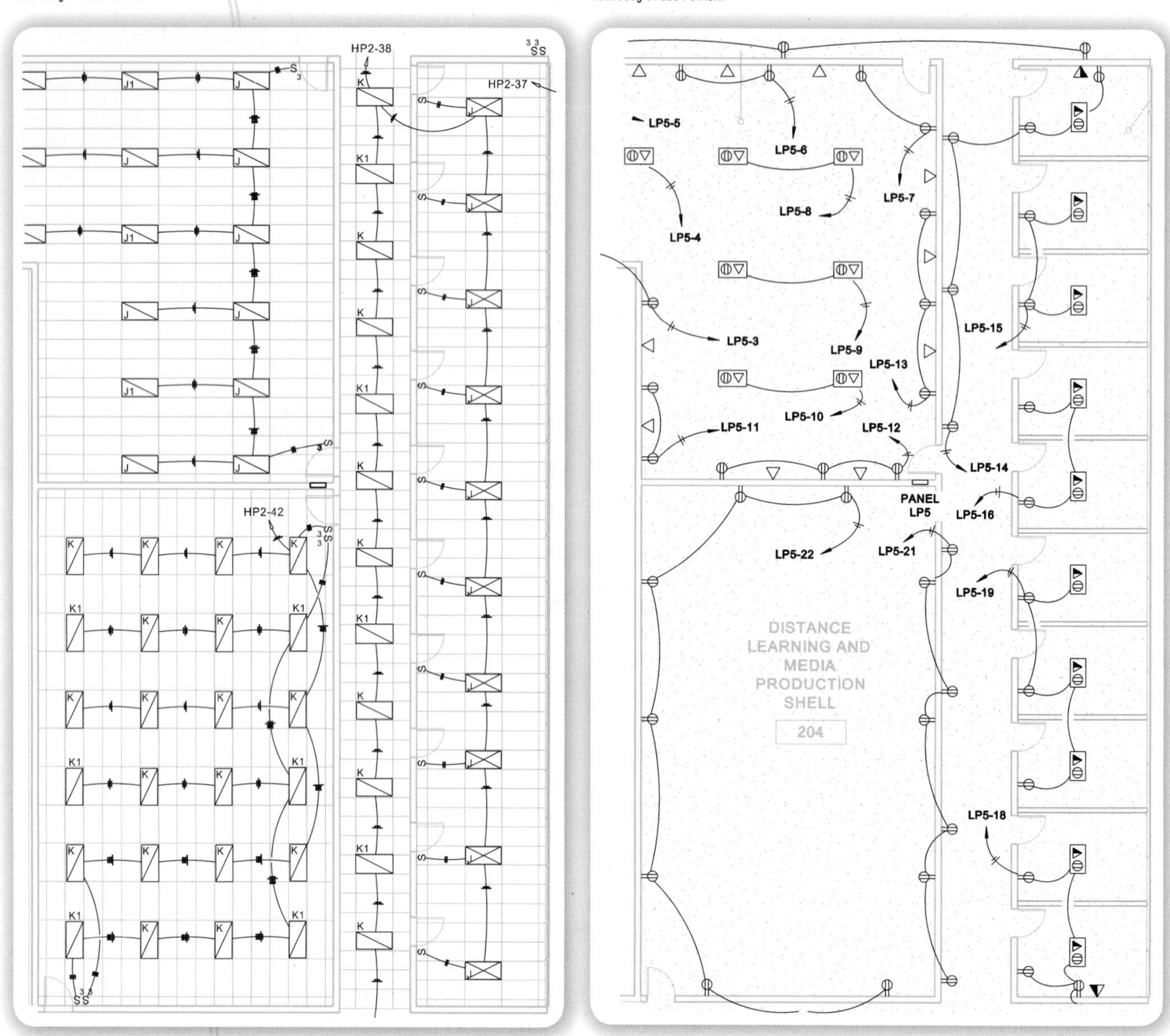

controls, and luminaires are indicated on an electrical plan (Figure 15-25). The electrical drawings will indicate a luminaire type designation that corresponds with a luminaire schedule.

The **reflected ceiling plans** indicate details as to a luminaires layout including key dimensions and notes to assist the contractors with installation. Additional interior luminaire information such as sconces or chandeliers may be indicated in building, room and wall sections, and construction details. Exterior luminaires such as ground mounted accent lights maybe indicated on landscape or site plans with information pertaining to their location and mounting height.

Lighting Types

Lighting sources typically fall into four different groups: incandescent, low-pressure discharge sources, high-pressure discharge sources, and a miscellaneous category. The different light sources and the categories are shown in Table 15-4.

Table 15-4: *Categories of Light Source Types*

Courtesy of OSRAM SYLVANIA Inc., and Paul Kevin Picone/PIC Corp.

Category	Type	Characteristic Features	Typical Application
Incandescent Lamps			
	General service and reflector	Easy to install and use, large variety, low upfront cost, allow concentrated light beam	General and non-decorative lighting, localize, accent and decorative lighting
	Halogen	Compact, high light output, easy to install, long life compared to incandescent	Accent lighting and flood lighting
Fluorescent Lamps			
	Tubular	Wide choice of light colors, high lighting levels possible, economical in use	Commercial, industrial, street and home lighting
	Screw in base	Energy-efficient, easy replacement for incandescent lamps	Most applications where incandescent was used
	Compact fluorescent	Compact; long-life, energy efficient	General lighting, signs, security, orientation lighting
Gas-Discharge Lamps			
	Self-ballasted mercury	Long-life, good color rendering, easy to install, low efficacy but better than incandescent lamps	Small industrial and public light projects; plant irradiation
	High-pressure mercury	High efficacy, long-life, reasonable color quality	Residential area, factory, and landscape lighting
	Metal halide	Very high efficacy, excellent color rendering, long-life	Industrial, street and retail; plant irradiation
	High pressure sodium	Very high efficacy, extremely long-life, high color rendering	Public spaces, street and industrial lighting; plant irradiation
	Low-pressure sodium	Extremely high efficacy, long-life, high visual acuity, poor color rendering, monochromatic	Wherever energy cost effectiveness is important and color rendering is not; tunnel lighting

Figure 15-26: Analogy of lumens to gallons per minute.

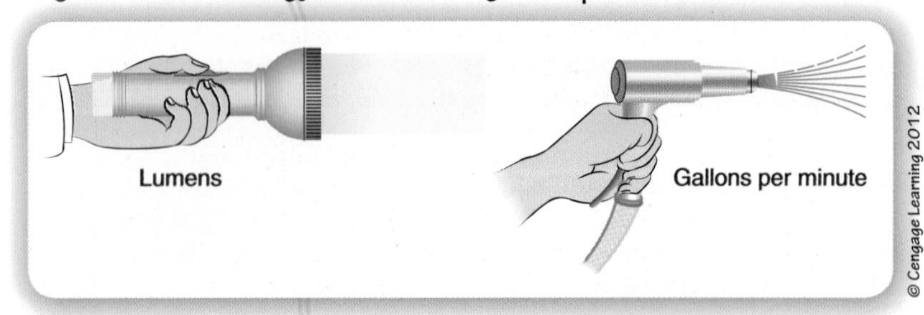

Figure 15-27: Comparison of lumen output to watt.

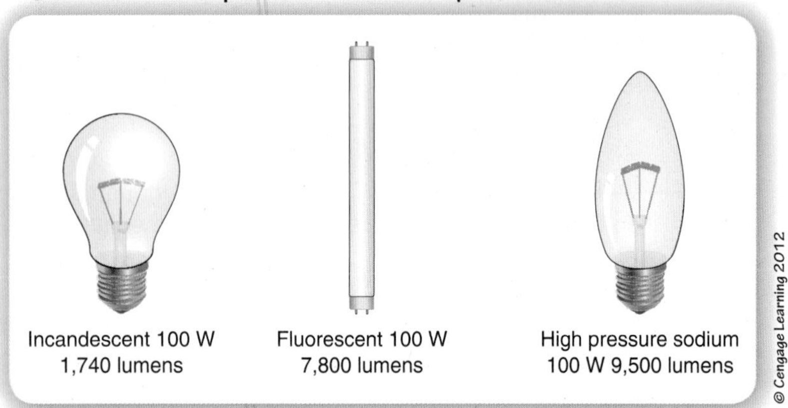

Selecting Light Sources and Equipment

Several factors should be considered in light selection, including light output, efficacy, luminaire efficiency, rated lamp life, correlated color temperature, color-rendering index, and brightness.

Light output is typically expressed in **lumens** (Figure 15-26) and given in two states: initial lumens and mean lumens. Initial lumens are the measurement of light output for a relatively new lamp. This measure is taken after the lamp has been in operation for 100 h. The means lumens is the amount of light output measured at 40 percent of the rated lamps lives.

Efficacy (Figure 15-27) is defined as the light output per unit of electrical power (watts) input or lumens/watt. Referring to the figure below the efficacy for the incandescent bulb is 17.4 lpw versus the fluorescent is 78.0 lpw or the high pressure sodium at 95.0 lpw.

Lumen:

A **lumen** is a rate at which light source emits light energy.

Efficacy:

the efficiency of light energy emitted by a bulb per unit of power output and is given in units of lumens/watt.

Candlepower:

a term used to express levels of light intensity in terms of the light emitted by a candle of specific size and constituents. Today it is expressed as a candela.

Luminaire efficiency: the ratio of total light output from a luminaire. It is often used in comparing varying luminaires of similar **candlepower** distribution curves.

The rated life of a lamp is determined by a point in time when 50 percent of a group of lamps remain illuminated. Rated lamp life is important in determining the longevity of a lighting system.

Lumen depreciation: the loss of light over time. The range varies between 10 percent to 40 percent of a lamp's initial output. As with rated lamp life, lumen depreciation is a factor when determining the service life of a lighting system.

Correlated color temperature (CCT) refers to the color appearance of the light source itself; that is, the color of the light that the source emits (i.e., how it appears when lighting a true white surface) or the color of the source itself (as you look directly at it). CCT is measured in units of degrees Kelvin. A light source with a cooler color appearance (bluer appearance) possesses a CCT of a higher color temperature. Likewise, sources that emit "warm" light, that is, that emit light with a slight red or yellow cast to it, have a lower color temperature (because they correspond to the color of light emitted from a material heated at a lower temperature).

Color-rendering index (CRI) refers to how well a lamp renders color as compared to reference light source of the same correlated color temperature. The change in one's skin tone under fluorescent light as opposed to incandescent light is an example of this. It is not the skin tone that has changed, but rather the type of light illuminating it causing the change in appearance. This is because fluorescent light has a different spectral power distribution than does incandescent light. Since an object reflects a fixed fraction of the radiation incident upon it, skin will reflect a different distribution of light when illuminated by fluorescent light as opposed to incandescent light. Since incandescent contains more emitted radiation at the red end of the visible spectrum, skin will likely possess a "warmer" tone than when it is lighted by fluorescent light.

Brightness is often a condition of lighting that causes irritation to a recipient as a result of direct or indirect reflection of light. A common occurrence of direct brightness is found when one drives down a highway late at night and is confronted by an oncoming vehicle with the high beams on. The irritation that is caused by the lights is quickly resolved by the driver looking at the side of the road and safely passing the oncoming traffic. Indirect brightness from glare or reflection of a glossy surface is usually minimized with the use of shielding and lenses.

Table 15-5: Correlated Color Temperature of Typical Light Sources

1500 K	Candlelight
2680 K	40 W incandescent lamp
3000 K	200 W incandescent lamp
3200 K	Sunrise/sunset
3400 K	Tungsten lamp
3400 K	1 h from dusk/dawn
5000–4500 K	Xenon lamp/light arc
5500 K	Sunny daylight around noon
5500–5600 K	Electronic photo flash
6500–7500 K	Overcast sky
9000–12,000 K	Blue sky

Lighting Controls

Lighting controls prior to the Energy Policy Act (EPACT) of 1973 were simply limited to manual switches and rheostat dimmers. Since EPACT the government required new standards ultimately resulting in greater energy conservation and efficiency. The lighting industry responded by providing new types of lighting technology including lighting control devices. In combination with previous control devices, current lighting systems are also operated by occupancy sensors, photosensors, and automated building systems all in effort to provide buildings with greater energy efficiency (see Table 15-5).

Manual Switching

Manual switching involves **"single-pole, single-throw"** and **"three-way" switching**, as well as standard relay systems. In the development of a manual switching scheme, it is important to address the flexibility of the control system and the potential to achieve energy savings through switching.

SINGLE-POLE, SINGLE-THROW SWITCHES Single-pole, single-throw switches control a group of luminaries from a single switch. When the switch is on, the circuit is closed and power is supplied to the load. When the switch is off, the circuit is open and power is cut off from the load. The switch is connected to one of the phase wires of the power system (a "hot" wire, which is also referred to as a *line wire*), and delivers a voltage to the load when the switch is on (i.e., when the circuit is closed and operating). The circuit diagram for a single-pole, single-throw switch is provided in Figure 15-28.

Figure 15-28: A wiring diagram for a single-pole switch. From FLETCHER. Residential Construction Academy House Wiring, 2e.

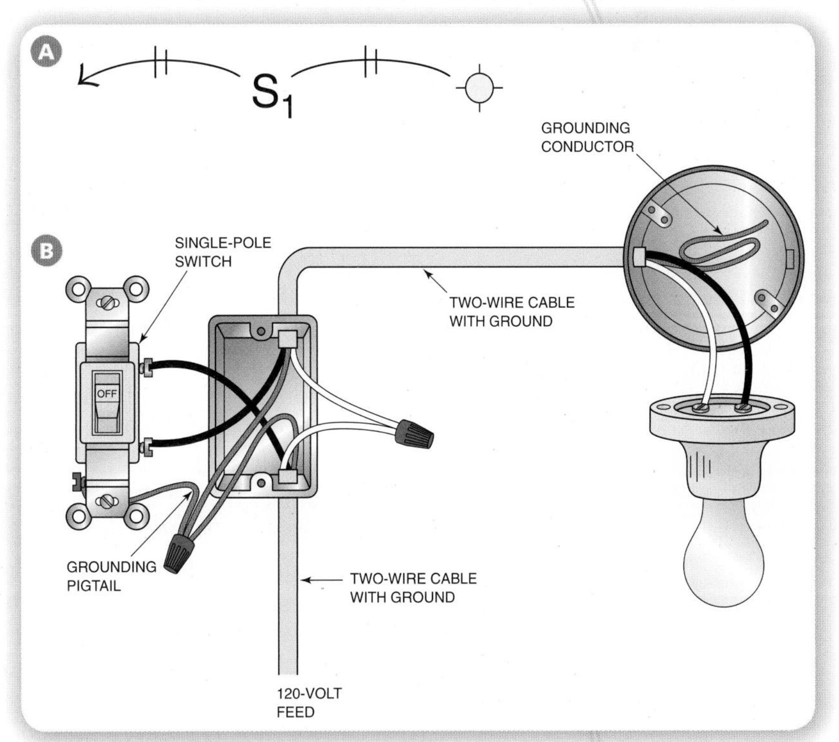

THREE-WAY SWITCHING Three-way switching is applied to control a group of luminaires from two different locations through the use of two three-way switches. This is particularly convenient in the lighting of large spaces, where a switch controlling a group of luminaires may be placed at two different entrances or at opposite ends of

Figure 15-29: Wiring diagram for a three-way switch. From FLETCHER. Residential Construction Academy House Wiring, 2e.

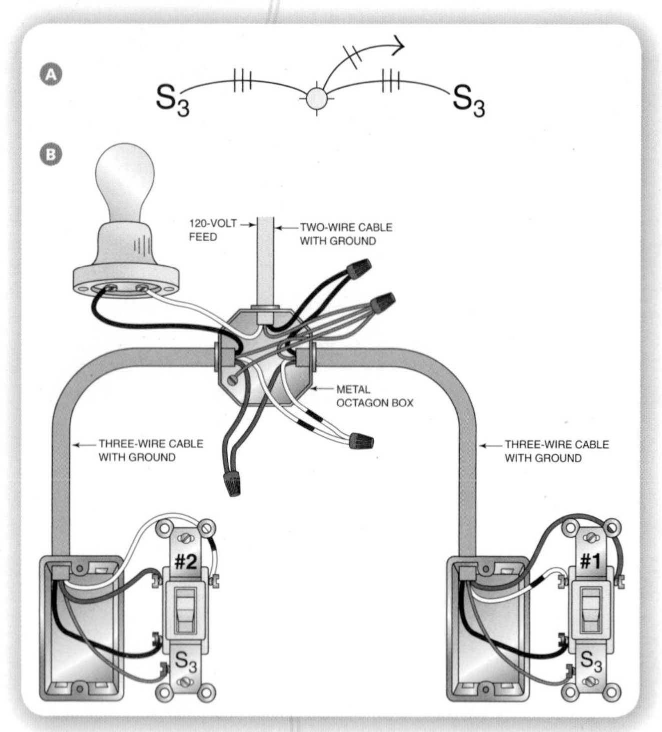

Figure 15-30: Drawing designating multilevel switching in a fluorescent troffer.

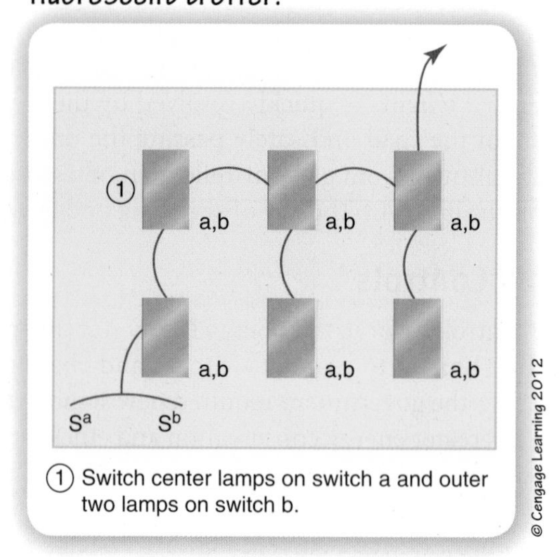

the space. This type of switching operates as follows: If the lights are off, the occupant may turn the lights on by flipping either of the two three-way switches. Likewise, when the lights are on, the occupant can turn them off at either switch. A typical three-way switching (Figure 15-29) are connected by two wires. These two wires are referred to as the **traveling wires**. The power passes through the first switch and then through one of the two traveling wires to the second switch (the wire that it travels in depends on the position of the switch). If the second switch is switched to the energized traveling wire (as opposed to the non-energized one) the power will reach the load and the circuit is therefore in an "on" configuration.

Three-way switches are commonly used in hallways or any rooms that may be entered from two or more locations such as a dining room.

Simulate a lighting design/layout for a preselected business facility by using a web program called "Virtual Lighting Designer." You can go to *http://www.gelighting.com/eu/resources/solutions_advisor/wizard/virtualdesigner/index.html* or simply type the title into your search engine. Note the options to develop lighting in three areas: ambient, task, and accent lighting. Select a choice of light fixtures and select the "Light It" option. Further options may be selected in areas of energy efficient lights, light output, product life, light quality, and environmentally friendly lights. With each change, select "Light It" to see how the space is lit with varying choices of lamps. Learn about the different types of lighting systems available, what lamps were selected including information about the wattage of the bulb, its rated life, correlated color temperature, and color-rendering index.

Multilevel Switching

In some cases, **multilevel switching** (Figure 15-30) is applied by switching every other luminaire on one of two switches. The advantage is the flexibility it allows as compared to a lighting system completely controlled by a single setting. The disadvantage of this arrangement is that uniformity of the resulting work plane illumination is likely to be compromised. The lighting level may be reduced significantly at some positions in the space, while it is reduced very little at other positions.

It is important to note that in multilevel switching applications, "tandem wiring" of ballasts is often applied. Since separately switched lamps require separate ballasts, two ballasts are required to serve one three-lamp luminaire switched as described above. In this case, the center lamp is generally tandem wired, where every other luminaire has two ballasts and one of these ballasts powers the center lamp in two different luminaires. Ballast wires must travel between the two luminaires in this arrangement. Usually, this can only be done if the distance between the two luminaires is less than about 10 ft. It is possible to purchase luminaires with this wiring factory installed and the luminaires are then shipped in pairs of two three-lamp luminaries (Figure 15-31) in which the center lamp in each luminaire is switched separately from the other two. Hence, the center lamps may share one two-lamp ballast. One advantage of this arrangement is that the center lamps can be wired and switched separately from the other lamps. Two ballasts would be located in one luminaire and a single ballast in the other.

Figure 15-31: Schematic diagram showing tandem wiring.

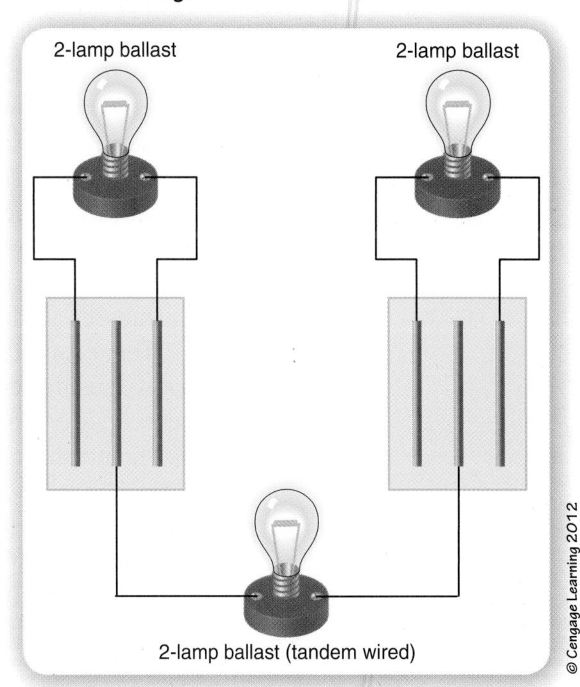

Zoning

Another control option in the layout of a switching system is to "zone" (Figure 15-32) the lighting. When zoning is applied, a room is generally divided into separate zones, and the luminaires belonging to one zone are controlled separately from the luminaires in another zone. This is particularly useful in deep rooms, where the entire space may not be used at once. In these situations, only the luminaries in the zones over or near the occupied portion of the space could be turned on. Zoning is also useful in classrooms, where the area near the front of the room could be zoned separately from the rest of the room. This permits the luminaires that might cast direct light on a projector screen to be turned off, permitting students to better view the projected images. The students are therefore provided with sufficient illumination on their paper tasks, while the projection screen receives very little light from the room lighting system.

Another good application of zoning is in natural lighted spaces that are illuminated by windows. Here, the lighting system should be divided into zones that run parallel to the window wall. During the daytime, the zones nearest the window may be turned off or dimmed, while the zones away from the window are still operated at full power. Before applying this type of zoning, you should analyze the daylight conditions to determine if the daylight zone should consist of one or two rows of luminaires.

Figure 15-32: Layout of two control zones in a space with windows.

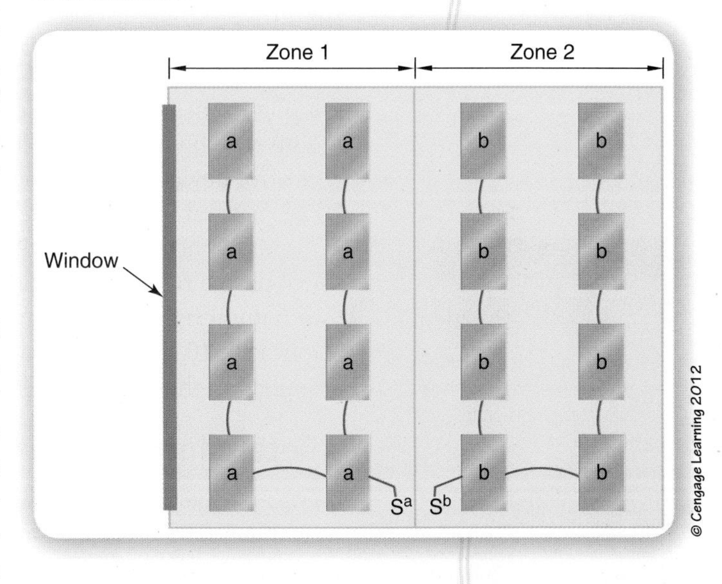

Dimming

Dimming is another form of lighting control that provides additional flexibility. A **dimmer switch** is a device that is used to regulate the amount of power dispersed into a light to control its brightness. Two common forms of dimmer technologies used are rheostat, sometimes referred to as the *turn knob button,* and thyristor, which operates using a semiconductor device. Dimming is easiest to perform with incandescent systems, since it only requires a reduction in the voltage delivered to the luminaires. Until recently, fluorescent fixtures used a ballast to control the amount of light being emitted, and therefore a change in the amount of voltage will ultimately discontinue the power to the bulb. At the turn of the century, a new type of ballast was becoming available to the public providing dimming capabilities.

Today, it is also possible to have dimming capability at more than one controller location, usually a main dimming controller and a remote (secondary) wall control station. Handheld dimmers operating through the use of an infrared signal, similar to that of a TV remote, are also available. With some special systems, the occupants can control their lighting through their computer system.

Learn more about dimming ballasts at Lutron Fluorescent Control Lighting Systems or go to *http://www.lutron.com/product_technical/fluorescent.htm.*

Occupancy Sensors

An **occupancy sensor** is an electron device that detects the presence or absence of people and turn lights on and off accordingly. Occupancy sensors offer an opportunity to save energy that otherwise would be wasted. Occupancy sensors operate in one of two ways. One form of occupancy sensor is the infrared family of sensors. These sensors survey their field of view and look for changes in the location of radiant heat sources within their environment. Our bodies are warm and emit heat, and these sensors operate primarily by sensing body motion. When movement is detected in a space, the lighting in the space remains on. Infrared sensors can only detect movement they can directly see, so they must be mounted where they have a clear view of the space, and the space must be free of obstructions.

The second type of occupancy sensor is the ultrasonic family of sensors. Ultrasonic sensors send out a sound wave whose frequency is above the threshold of hearing. Movement within the space will alter these sound waves (through reflection), and these effects are detected by the sensor. Their operation is similar, in some respects, to that of a police radar gun.

Occupancy sensors can be specified to turn the lights on automatically when you walk into a space or can require that the occupant switch on the lights. Their main purpose, however, is to switch the lights off when a space is not occupied. Adjustments to the sensor generally permit one to adjust both the sensitivity of the sensor and the unoccupied time period required before shutoff occurs.

Photosensors

A photosensor is an electron controlling device that adjusts the lighting levels based on the amount of natural light coming into a room. While some sensors are designed to switch lights on or off, others are used for dimming the lights in a room to acceptable levels. The National Lighting Product Information Program defines a photosensor as a complete unit that houses a photocell and the circuitry to that converts the electrical unit into a signal to be used by a dimming ballast or other control device. The photocell is a silicon chip that converts radiant energy in this case natural light into an electrical current.

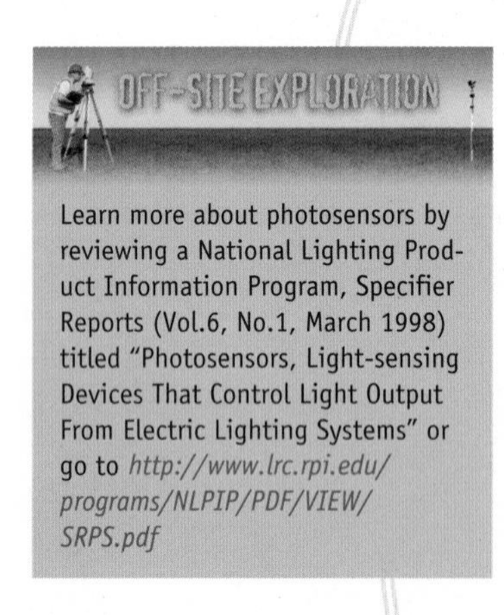

Learn more about photosensors by reviewing a National Lighting Product Information Program, Specifier Reports (Vol.6, No.1, March 1998) titled "Photosensors, Light-sensing Devices That Control Light Output From Electric Lighting Systems" or go to *http://www.lrc.rpi.edu/programs/NLPIP/PDF/VIEW/SRPS.pdf*

Most photosensors are working by modulating current through the input control device ranging between 0 and 10 V. When a photosensor detects low amount of natural light, it limits current into its control device, causing its voltage output to rise to 10 V and thus causing full light output in the room. The opposite can be applied in the case where high amounts of natural light are detected.

Determining the Proper Quantity of Light

For many years, designers reference the Illuminating Engineering Society of North America (IESNA) Handbook for recommended illuminance levels to apply to a space. *Illuminance* refers to the amount of light reaching a given square foot surface area; the unit is in footcandles. As a lighting designer, one can adjust these levels up or down when it is appropriate. Adjustments may be necessary due to the nature of the task, its criticality, contrast, or size; due to the preferences or age of the space occupants; to balance the distribution of light in a space due to the presence of abundant daylight in other parts of the space; and for a variety of other reasons.

Task Lighting

For energy efficiency, a lighting designer should select an appropriate distribution and locate the lighting equipment accordingly to maximize the delivery of light to the task while providing appropriate lighting conditions elsewhere in the space.

Locating the source of illumination closer to the task can generally save energy, since the light can be directed more efficiently to where it is needed (with less light directed elsewhere within the space). This system of focusing light on a specific area to make the visual task easier is what is referred to as **task light.**

Task-ambient lighting is typically used in open office environments where overhead cabinet lighting or adjustable task lighting is used to provide appropriate levels on principal work surfaces. A general overhead system also referred to as an **ambient light** system provides sufficient lighting for circulation and other general space activities. The task-ambient approach to lighting design generally requires fewer total lumens to meet all design criteria and adequately illuminate the space, which provides energy savings in comparison to a uniform overhead system that provides all of the light at the task. Similar task lighting arrangements are often applied in industrial settings where higher levels are provided on work surfaces, such as in assembly areas, through the use of localized task lighting.

Again, to provide these luminance levels from overhead would expend significantly more energy and spread the light well beyond the work surface where it is needed.

One of the problems with under-cabinet task lighting is that it generally places the light source in the location most likely to place reflectances on a horizontal desk task. Special optical configurations are often applied in an attempt to direct the light to the work surface from the side or from the back wall with these types of luminaires.

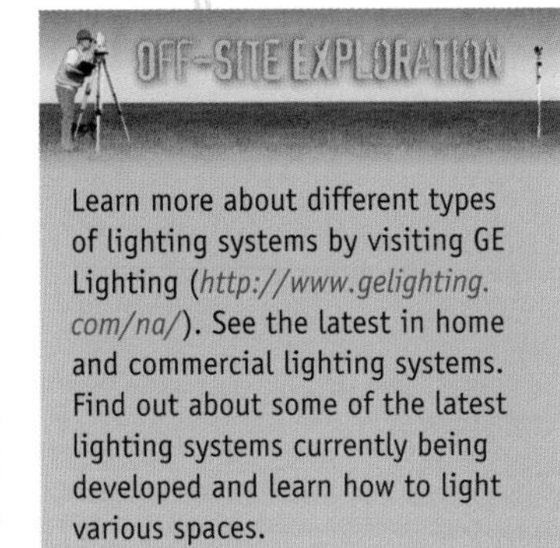

Learn more about different types of lighting systems by visiting GE Lighting (*http://www.gelighting.com/na/*). See the latest in home and commercial lighting systems. Find out about some of the latest lighting systems currently being developed and learn how to light various spaces.

Your Turn

Simulate a lighting design/layout with controls for a classroom by using a web program called "Lighting for Learning: Classroom Design Tool" or go to *http://www.lightingforlearning.com/Products.aspx.* You will be prompted in three easy steps with the first step to select light fixtures, then select controls, and lastly create a design package. The first step, selecting a light fixture, will prompt you for the lighting requirements needed in a classroom you are designing. Let's consider the following conditions for a typical classroom. In accordance with the American Society of Heating Refrigeration and Air Conditioning Engineers and the Illuminating Engineering Society of North America (ASHRAE/IESNA) Standard 90.1, 2007 the maximum power density for a classroom (or reading room) is 1.2. The recommended illuminance ranges between 30 and 50 fc. Select any of the ballast types given. Each refers to specific characteristics based on the manufacturer specifications. Excluding or including illumination of the room's whiteboard or chalkboard is left for the designer's discretion. Choose a light fixture that meets the needs set forth and proceed to next step. The second step is to add controls. The options provided allow you to select the type of daylighting controls, the number of switches needed in the room, and how the controls should interface with the remaining electronics in the room. Note that as options are selected the program updates the controls diagram. Select the options you prefer and proceed to the next step. At this point, you, the designer, have created a complete lighting system with controls package and may now have the opportunity to have the information sent to the your e-mail along with additional information pertaining to the system selected.

Careers in Civil Engineering and Architecture

AN ADVENTURE IN IRAQ

Timothy Ernster is a civil engineer in a dangerous place. He works for the United States Army Corp of Engineers in Iraq, where violent attacks are common. Ernster is helping to rebuild the country's infrastructure following the war that began in 2003. When he visits a site, he dons body armor and is accompanied by a security force. He must sometimes stop work because of rocket and mortar fire.

Still, Ernster is happy to be where he is. "When I joined the Army Corps of Engineers, I wanted to go to Iraq," he says. "I was looking for an adventure."

Timothy Ernster
Project engineer for the Loyalty Resident Office of the United States Army Corps of Engineers, Gulf Region Central District

On the Job

Ernster works with Iraqis on solving the country's infrastructure problems. One of his projects is building an electrical substation that will increase the distribution of electricity to a region near Sadr City. Another is building a large police college in Baghdad to train Iraqi recruits. Ernster oversees the construction of four barracks and four classrooms that can handle 2000 people. He makes sure that construction is up to standard and that the contractors are on schedule. He also signs off on any changes to the contract.

Inspirations

Ernster was always attracted to math and science, taking as many classes as he could. As a high school senior, he ran out of math classes offered at his school, so he began taking classes at a community college. "I liked math and science because they're challenging subjects, and easy to apply to everyday life," he says.

A relative suggested that Ernster pursue engineering in college. He liked the fact that the major was well respected. "I also liked the idea of trying to make things better," he says.

Education

Ernster attended Gonzaga University, where he studied electrical engineering and found the program to be intense and rigorous. He became interested in civil engineering when he joined the Army Corps of Engineers. "Electrical engineering would lead you to an office job, or to a manufacturing plant," he says. "I liked the idea of going into a field where stuff was being built, and where I could see it happen."

Advice for Students

Ernster recommends that high school students interested in engineering take as much math and science as they can. "It can also be helpful to get a job with a construction company doing some kind of labor, to give you a feel for the work," he says.

While the engineering curriculum can be difficult, Ernster says it's worth sticking with even if you feel like quitting.

"There's a shortage of technical professionals," he says. "When you get your degree, there's a whole spectrum of opportunities out there. With a background in engineering, you won't ever have to worry about being employed."

SUMMARY

Today's electrical systems provide the power for state of the art technology and current architectural design. These systems not only supply the power to condition the air, connect people and computers throughout the world, and provide warm meals through the use of microwave ovens and stove tops, but also allow for the illumination of spaces so that we may function with individual freedom 24-h a day, seven days a week. This chapter has focused on various types of power distribution, some of the common terms associated with electrical systems, designing of electrical systems for various types of establishments and how professionals in the building industry communicate the needs and functions of these power systems to their target audiences.

BRING IT HOME

Designing a building's electrical system is an integral part of the overall building design process. Nearly all mechanical equipment including air conditioning and heating systems, pumps, elevators, and appliances are electronically operated. The process of determining the power needs and then selecting the electrical power system is often based on the previously mentioned determinants. This is further complicated with the need to supply power for ancillary equipment such as computers, printers, and other modern conveniences. The procedure for designing an electrical system has been described as analyzing the building needs, determining the electrical loads, then selecting the electrical systems to meet the electrical loads, coordinate with other design requirements, and communicating the decisions to various contractors and builders through the development of drawings and specifications.

In the design of a lighting system, one must address many different conditions. These conditions generally can be divided into those that affect the viewing and performance of the task and those that affect the general appearance of a space. Both sets of issues are important and must be simultaneously addressed in the resulting design. In carrying out the lighting design for a space, you need to also address a number of issues, including lighting quality, energy efficiency, space appearance and its operation, and the needs of the occupants within the space. In most cases, there are not one but multiple acceptable designs that will adequately meet these objectives. The designs that best meet these objectives, however, will often have many of the same qualities in common.

CHAPTER 16
Planning for Plumbing

GPS
DELUXE

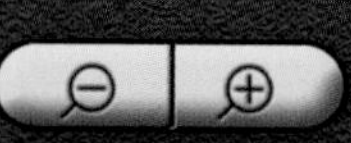

Menu

START LOCATION | DISTANCE | END LOCATION

Before You Begin

Think about these questions as you study the concepts in this chapter:

1. What factors determine the quality and quantity of water supply?
2. How does the location of the water supply impact the building's design and plumbing?
3. Why are plumbing codes used to govern materials and process?
4. What determines the plumbing process and materials of a building project?
5. Why are symbols and abbreviations used to communicate plumbing specifications?
6. How can the plumbing plan support water conservation?
7. What are the two distinctly different systems in a plumbing plan?

INTRODUCTION

Can you imagine what it would be like to live or work in a building without plumbing? We often take for granted the planning and infrastructure behind the plumbing fixtures we use everyday. Although small residential buildings are not deemed large or complex enough to require a plumbing plan, commercial building plans usually include a plumbing drawing set of several sheets. The plumbing plans clearly communicate the components and specifications desired for the plumbing system. The three main categories of a plumbing plan include water supply, waste removal (also known as the *drain waste and vent system* or *DWV*), and planning for efficiency. This chapter will highlight the main features of a plumbing plan and include standardized practices that utilize commonly recognized symbols, abbreviations, and line type standards. A plumbing plan provides a diagram of two separate and distinct systems: water supply and wastewater removal (see Figure 16-2).

To keep this introduction simple and easy to understand, a residential plan will be used to guide your plumbing plan development. In addition to water supply and waste removal, a plumbing plan may specify water and energy efficient fixtures or even detail a plan for the creative use of rainwater.

Figure 16-1: *Planning is essential for a successful plumbing outcome. From JOYCE. Residential Construction Academy: Plumbing.*

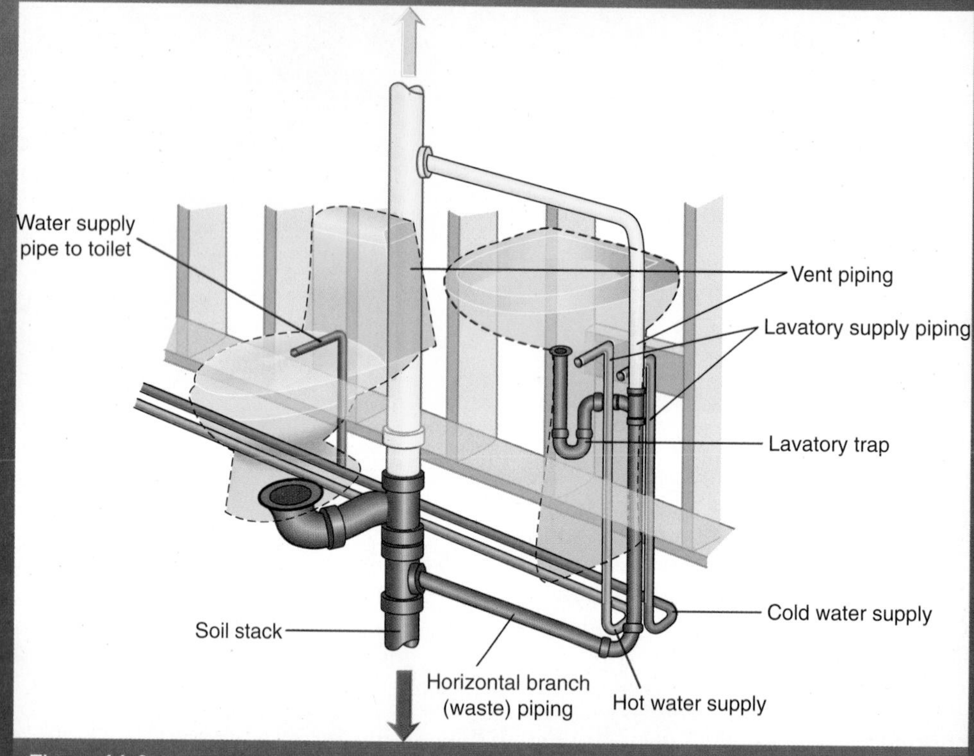

Figure 16-2: *Water supply and wastewater removal are two separate systems.*

QUALITY AND QUANTITY OF WATER SUPPLY

Potable:
suitable for drinking.

Backflow:
water traveling back into the main distribution system, usually by siphoning.

The plumbing plan communicates the desired outcome of two separate systems, water supply and waste removal. Water supply refers to the water, both **potable** and nonpotable, required by the occupants of the building.

Potable water is safe to drink, free of harmful bacteria and other contaminants. Many municipal water districts require installation of a **backflow** prevention device to be installed to block the flow of foreign liquids, gases, or substances into the distribution pipelines of a potable water supply.

The Atmospheric Vacuum Breaker (AVB) is one example of a backflow prevention device. Take note of the right side of Figure 16-3: when the water is not flowing in, the unit holding the seal drops down, preventing backflow. The arrows on the left of the diagram show the path water flows from the water supply to the residence when water is needed. The figure on the right shows how air channels through the top prevent backflow when the residence is not drawing water.

Most residences use only potable water from a well or public watershed. However, buildings designed to be green, sustainable, and resilient may employ systems that increase water efficiency through collection and use of nonpotable water. One such example is a rainwater collection system for flushing toilets, washing

Figure 16-3: An atmospheric vacuum breaker (AVB) is a simple and inexpensive mechanical backflow prevention assembly.

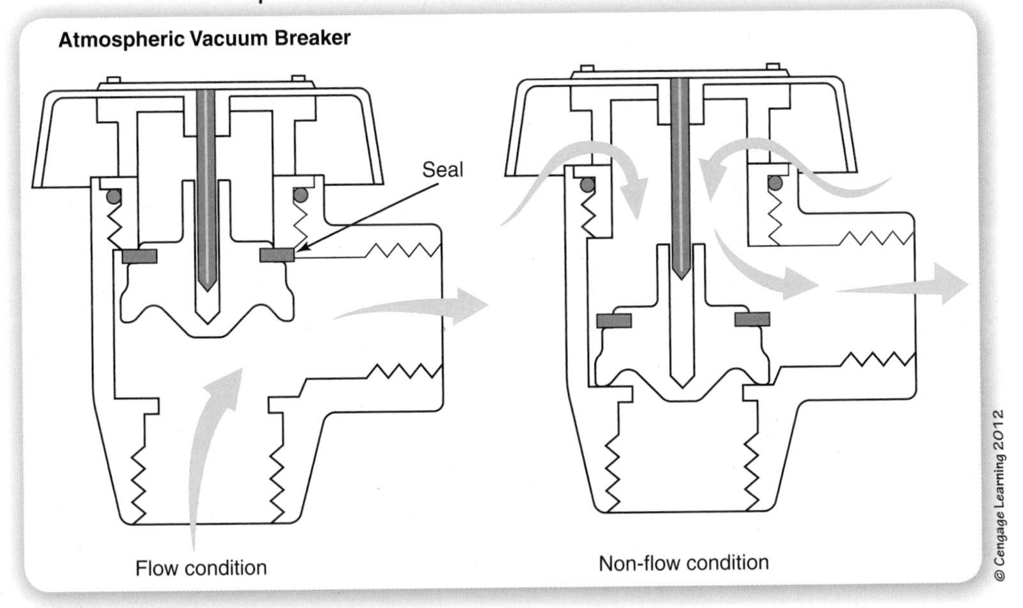

cars, or watering a garden. A nonpotable water supply is more commonly found in commercial applications such as manufacturing facilities where it can be used for cleaning and cooling operations.

Two of the most important factors relating to water supply are pressure and flow rate. Imagine what it would be like to take a shower without adequate pressure or stable flow rate. Water pressure is derived from pressure from the community main or a private well pump. In Chapter 6: Site Planning, you learned to calculate water supply pressure by taking into consideration exterior factors that impact the pressure and flow rate before the water reaches the building. In addition to exterior conditions, water pressure and flow at the faucet is affected by interior conditions. There are three main interior conditions that create resistance to water flow and reduce pressure. First, if the water enters the building at ground level, it must overcome gravity to reach fixtures at higher elevations. This explains in part why the upper floors of apartment buildings often suffer lower water pressure than lower floors. Second, as the water moves from the building's entry to the various fixture locations, resistance occurs due to the water pipes' rough interior surface; this happens as water flows through fittings, regulating devices, and other components as well. Finally, the operation of the actual fixtures will add resistance to water flow. You can determine the interior water pressure using a procedure similar to that used in Chapter 6 to calculate the water pressure of the water supply system. On a commercial project, the civil or piping engineer will calculate the resistance of the internal building conditions to insure that the water pressure entering the building is equal or greater than the total resistance. The water pressure must be sufficient to overcome the static head and minor loss to provide adequate pressure and flow at each fixture.

LOCATION OF THE WATER SUPPLY

Planning and analysis of the site completed early in the land development determines the source of water supply. This location may influence the building design and be re-examined in development of the plumbing plan. In the case of

Figure 16-4: An inside look at a residential plumbing system.

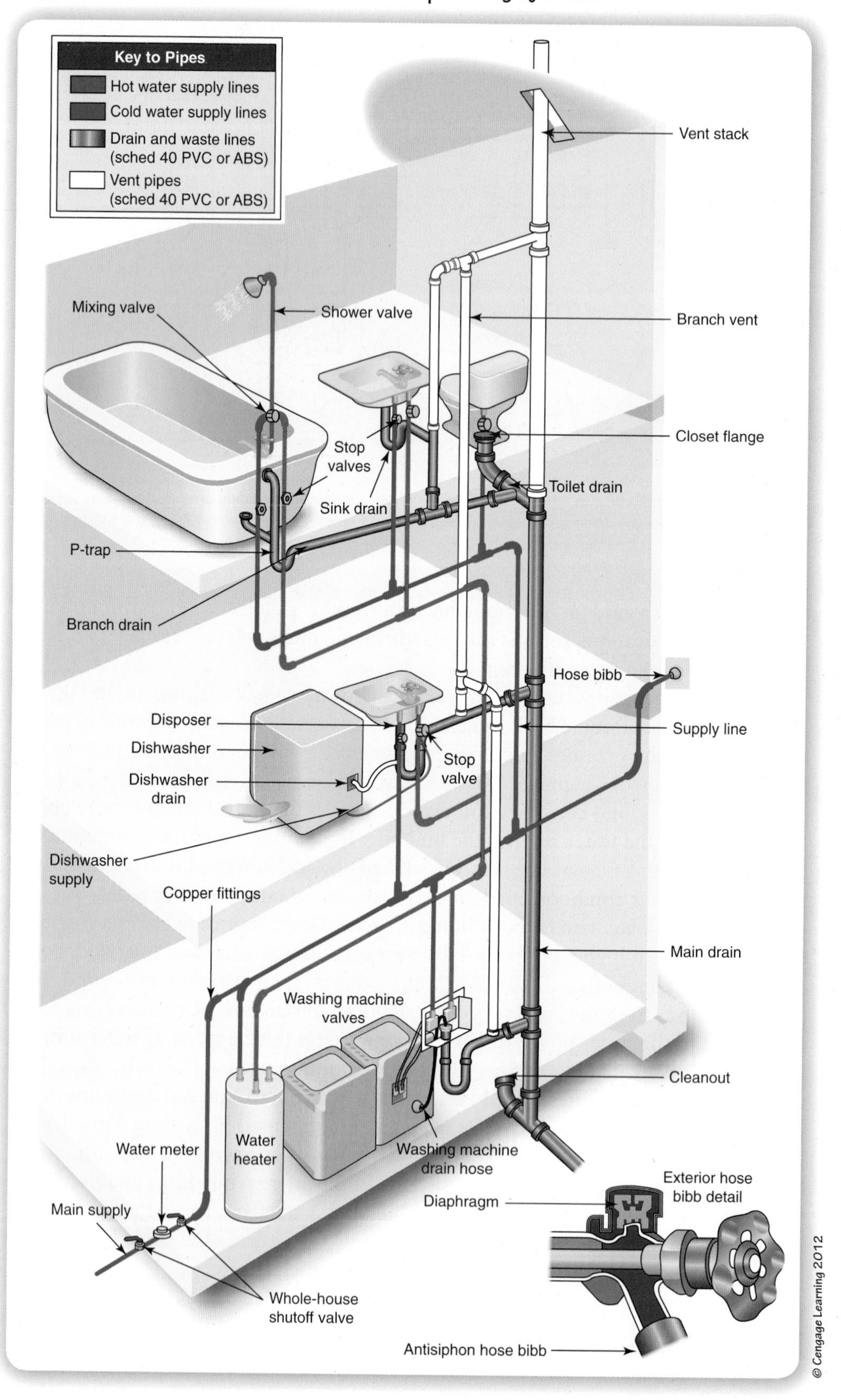

Key to Pipes
Hot water supply lines
Cold water supply lines
Drain and waste lines
(sched 40 PVC or ABS)
Vent pipes
(sched 40 PVC or ABS)
Vent stack
Mixing valve
Shower valve
Branch vent
Closet flange
Stop
valves
Toilet drain
Sink drain
P-trap
Branch drain
Hose bibb
Disposer
Supply line
Dishwasher
Stop
valve
Dishwasher
drain
Dishwasher
supply
Copper fittings
Main drain
Washing machine
valves
Cleanout
Water meter
Water
heater
Washing machine
drain hose
Exterior hose
bibb detail
Diaphragm
Main supply
Whole-house
shutoff valve
Antisiphon hose bibb

public utilities, the water lines run parallel to the street and may enter near the front of the building. Several factors determine water pressure of public water delivery. Water pressure is calculated in pounds per square inch. Public water is generally supplied from a water tower. To begin thinking about water pressure, let's look at using a simple example to calculating water pressure. We will use the formula p = h × d or pressure = height × density. We will use a large water tank 50 ft in height as an example. To calculate the water pressure at a water tank faucet 2 ft above the bottom of the tank, you would first subtract the height of the faucet from the water level in the tank. After subtracting the faucet height from the height of the faucet take the distance of 48 ft (height) and multiply by the water density, which is 62.5 lb per sq ft; the result is 3000 lb per sq ft. Most water companies identify water pressure in pounds per square inch, so you will need to divide your answer by one square foot, which is 144 sq in. Diving 3000 by 144 will identify that the water pressure at the faucet is 20.8 psi. If the faucet were higher on the tank, would the water pressure be greater or less? You are right if you said less. The principle seems simple enough, but there are many factors that impact water pressure, such as the elevation of your building in relationship to the water level of the water tank. The water company will try to keep the water in the tower at a constant level to maintain adequate water pressure for customers. Another factor that impacts water pressure is the route the water must travel to get to the building. Water pressure is reduced by friction loss as it flows through pipe and fittings. The Hazen-Williams formula is commonly used to calculate head (pressure) loss in pipes due to friction.

The **Hazen-Williams** formula:

$$P_d = \frac{4.52\, q^{1.85}}{(c^{1.85}\, d_h^{4.8655})}$$

where

P_d = pressure drop (psi/ft pipe)

c = design coefficient determined for the type of pipe or tube

q = flow rate (gpm)

d_h = inside hydraulic diameter (inch)

It is important to remember that Hazen-Williams formula is only valid for water flowing at ordinary temperatures between *40 to 75 °F; the higher the factor, the smoother the pipe or tube.*

For example, a Hazen-Williams coefficient—*c*—for 40-year-old cast-iron pipes is 64–83, whereas a new unlined cast-iron pipe has a coefficient of 130. The higher the design coefficient factor, the smoother the pipe. You can find charts of Hazen-Williams coefficient by searching the Internet (see Figure 16-5). You may want to record the coefficient copper pipe, which has a 130–140 coefficient.

The amount of public supply water used by a building is generally measured by a meter at the point of entry. In a rural setting where the water supply is derived from a well, the water will most likely enter the building on a side toward the well. During the building's design phase, the architect considers this information to plan a convenient location for the utility room, which contains mechanicals such as water heaters and water treatment systems. Well water is tested to insure drinkability. Even after the water is deemed potable, some well water has undesirable features that impact smell, taste, or performance.

Figure 16-5: Hazen-Williams coefficient and equations.

Roughness Coefficient *C* Values for Hazen-Williams Equation

Values of C Type of Pipe	Range	New Pipe	Design C
PVC	160–145	150	150
Polyethylene	150–130	140	140
Asbestos-Cement	160–140	150	140
Cement-Lined Steel	160–140	150	140
Welded Steel	150–80	140	100
Riveted Steel	140–90	110	100
Concrete	150–85	120	100
Cast Iron	150–80	130	100
Copper, Brass	150–120	140	130
Wood Stave	145–110	120	110
Vitrified Clay		110	100
Corrugated Steel		60	60

Above values of *C* for use with Hazen-Williams Equation, friction head losses in feet per foot of pipe length for fresh water at 50 degrees Fahrenheit.

$$Hf = \frac{4.55}{C^{1.852}} \times \frac{Q^{1.852}}{D^{4.871}} \times L$$

Where Hf = Friction Head Loss in psi

C = Roughness Coefficient

Q = Flow Rate (gpm)

L = Pipe Length (ft)

D = Pipe Inner Diameter (inches)

Water treatment systems can be added to counteract undesirable features. For example, a water softener can be installed to improve the performance of hard water.

The architectural designer or space planner will also consider minimizing the cost of plumbing by grouping plumbing areas and locating them in close proximity to the utility room. This arrangement provides quick response of hot water from the heater to the shower or tub. With the increasing cost of plumbing materials, such as copper, careful planning can also provide major cost savings. Main shut-offs are generally located near the meter. Inside the building, additional shut-offs may be installed at several places along the water lines and at individual fixtures to ease the repair or replacement process. After the meter, union, shut-offs, and backflow protection device, a water line is traditionally divided to feed a hot water heating unit. The hot water line leaving the heating unit then runs parallel with the cold water line to plumb the building.

Codes

Acceptable plumbing materials and practice may vary from one locale to the next based on local governance. Each state or community either adopts or bases their own codes on nationally recognized codes. The two main plumbing codes are the International Plumbing Code® and the Uniform Plumbing Code (UPC). Plumbing codes identify acceptable practice, safety requirements, and the number and types of fixtures required based upon occupancy. Plumbing systems in commercial buildings must also accommodate individuals with disabilities. In Chapter 10: Commercial Space Planning, you read about accessibility guidelines for restroom facilities established by the Americans with Disabilities Act. The International Plumbing Code® for residential and commercial buildings was developed by the International Code Council, a membership association dedicated to building safety and fire prevention. The Uniform Plumbing Code, provided by the International Association of Plumbing and Mechanical Officials, is the product of a collaborative effort and consensus of industry, government, and consumers.

Figure 16-6: Diagram of residential public water supply and building entry details.

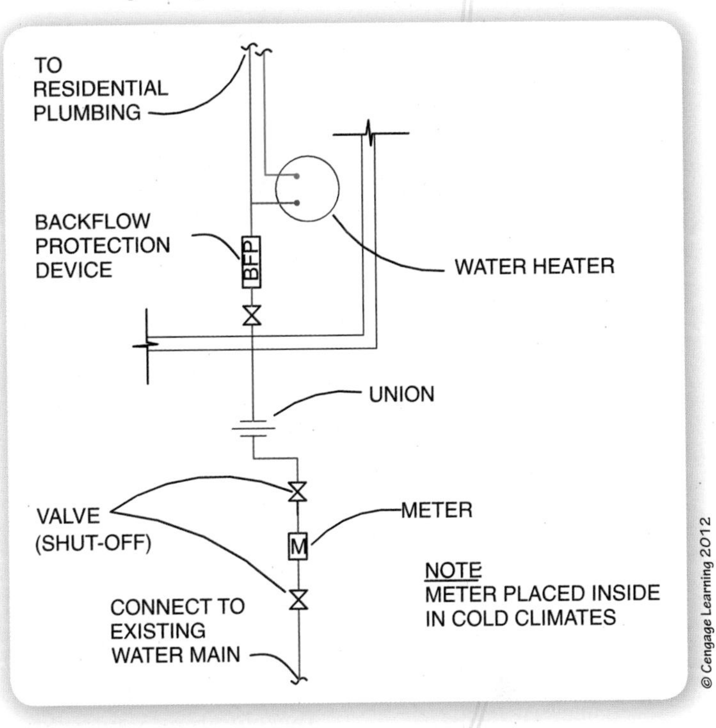

Your Turn

Use the Internet to learn more about the International Plumbing Code® and the Uniform Plumbing Code. Try to determine which plumbing code your state or local government has adopted. Research to learn the difference between the two codes and why there are opponents and proponents of each. List this information in your notebook.

http://www.iccsafe.org International Code Council
http://www.boma.org The Building Owners and Managers Association
http://www.iapmo.org/ The International Association of Plumbing and Mechanical Officials

PLUMBING MATERIALS AND PROCESS

The plumbing materials for a building project are identified on the plumbing plan. Many materials are available, but not all meet code required by local governance (Figure 16-7). For example, although plastic plumbing pipe and fittings

Figure 16-7: Acceptable plumbing materials are specified by local governance.

© iStockphoto.com/Branko Miokovic.

made of chlorinated polyvinyl chloride (CPVC) are suitable for potable water and are readily available, some locations may only accept copper plumbing. Materials often impact the plumbing process. CPVC, copper, and galvanized steel are ridged materials requiring various fittings, such as Ts and elbows to complete the system. Another system uses flexible cross-linked polyethylene (PEX) with a **manifold** to distribute water into several lines that can be easily pulled through the building framework (see Figure 16-8). Advantages of using PEX include reduced installation time, quick delivery of hot and cold water, and more stable pressure to each fixture when fixtures are operating simultaneously.

COMMUNICATION OF THE WATER SUPPLY SYSTEM

Manifold:
a chamber with ports for receiving and distributing a fluid or gas.

How is a building's water supply system communicated on the plumbing plan? The plumbing plan is basically a diagram of specifications of the buildings plumbing needs, including the type and location of each fixture.

You do not have to specify every fitting or show where to run the pipes. Instead, the plumbing contractor is interested in details such as whether a refrigerator needs a cold-water line to supply an icemaker, or where the owner wants the outside **hose bibb** placed in a particular location to wash the car or water plants.

Figure 16-8: A PEX-based plumbing system uses hot and cold manifolds to distribute water.

Photo: Courtesy of Zurn®

Figure 16-9: Communication of hot and cold water needs for back-to-back bathrooms.

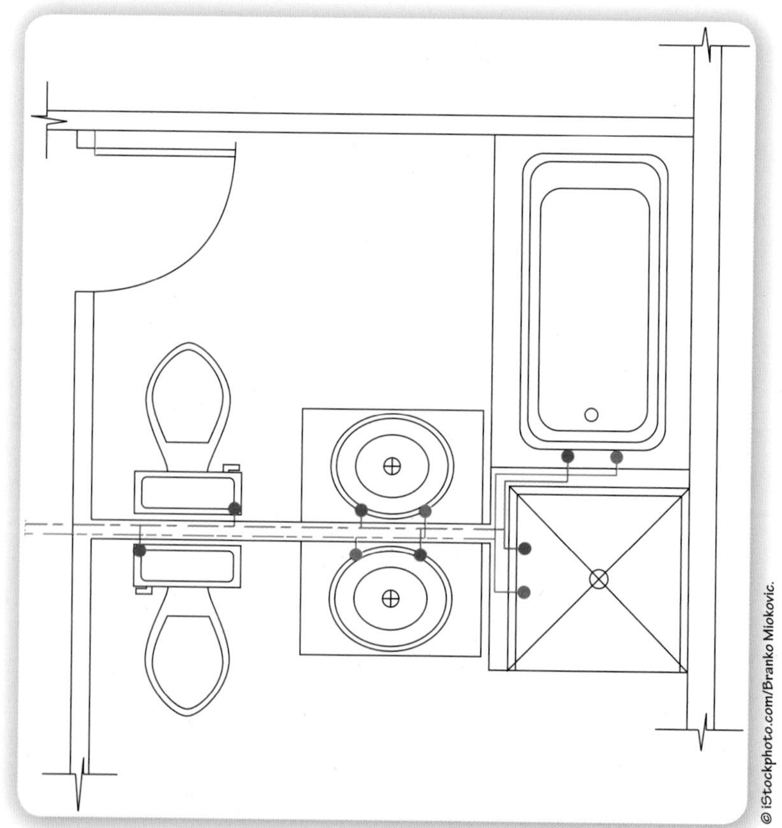

© iStockphoto.com/Branko Miokovic.

These details and others are communicated using symbols, abbreviations, and line types (Figure 16-10).

General notes and schedules provide procedures and specifications that cannot be conveniently shown in the drawing itself. For example, a general note might read: "Maintenance labels shall be affixed to all plumbing equipment and a maintenance manual shall be provided to the owner." It is through general notes that important information is conveyed to the plumbing contractor.

On a small residential project the plumbing contractor may be expected to determine plumbing needs without a plumbing plan. Information about the desired outcome is gathered from the floor plan and conversations with the owner and general contractor. If a set of residential plans includes a plumbing plan it is generally placed on a copy of the floor plan. Plumbing plans for commercial projects often include a plumbing set of several drawing sheets. Multilevel commercial buildings may require riser diagrams of the elevation view showing how the water or waste system connects to various levels. Figure 16-11 shows part of a sanitary riser diagram showing the connection of floor drains on the first and second level to the sanitary sewer. Floor drains allow floors to be washed easily.

Isometric plumbing drawings show a three-dimensional view of the drain, waste, and vent (DWV) system (Figure 16-12). This drawing is sometimes referred to as the *plumbing riser diagram*. The piping layout for the DWV system is shown using conventional line types. Plumbing fixtures, major components, and pipe diameters are labeled.

Figure 16-10: **Partial list of plumbing symbols and line types.**

SYMBOLS	
– – – – –	SANITARY PIPING BELOW GRADE
....	WASTE
– – – – –	VENT
——- ——	COLD WATER PIPING
——- - ——	HOT WATER PIPING
	SHUT-OFF VALVE
B F P	BACKFLOW PREVENTER
M	METER
	HOSE BIBB
	PIPE BREAK
	DIRECTION OF FLOW
	CLEANOUT
	"P" TRAP
	PIPE DROP
	PIPE UP

Hose Bibb: an outdoor faucet, also used to supply washing machines.

Figure 16-11: **A partial plumbing sanitary riser diagram.**

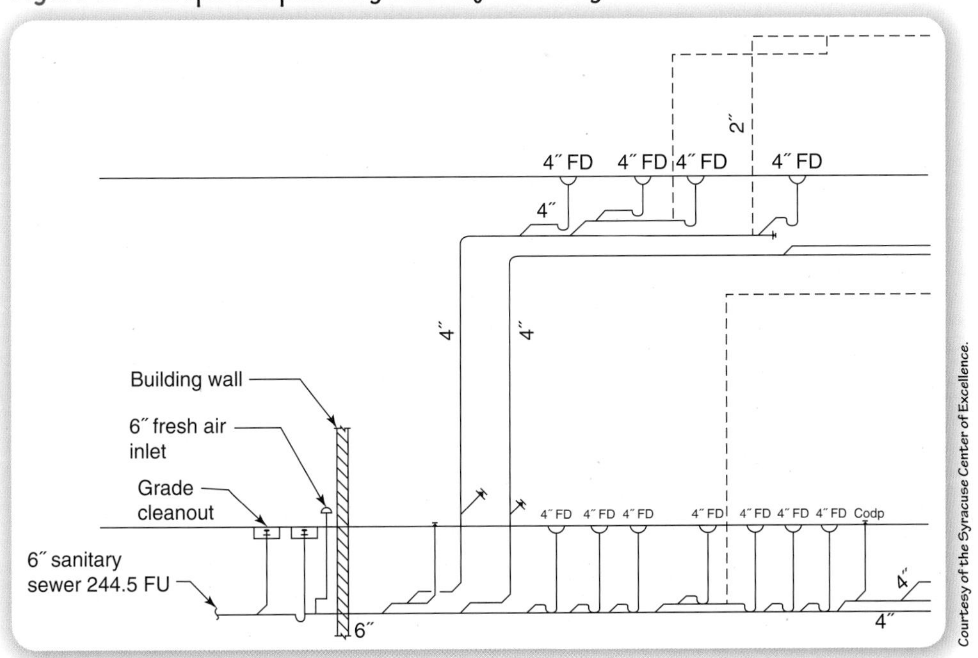

Courtesy of the Syracuse Center of Excellence.

Figure 16-12: An isometric plumbing drawing of a drain/, waste and vent system.

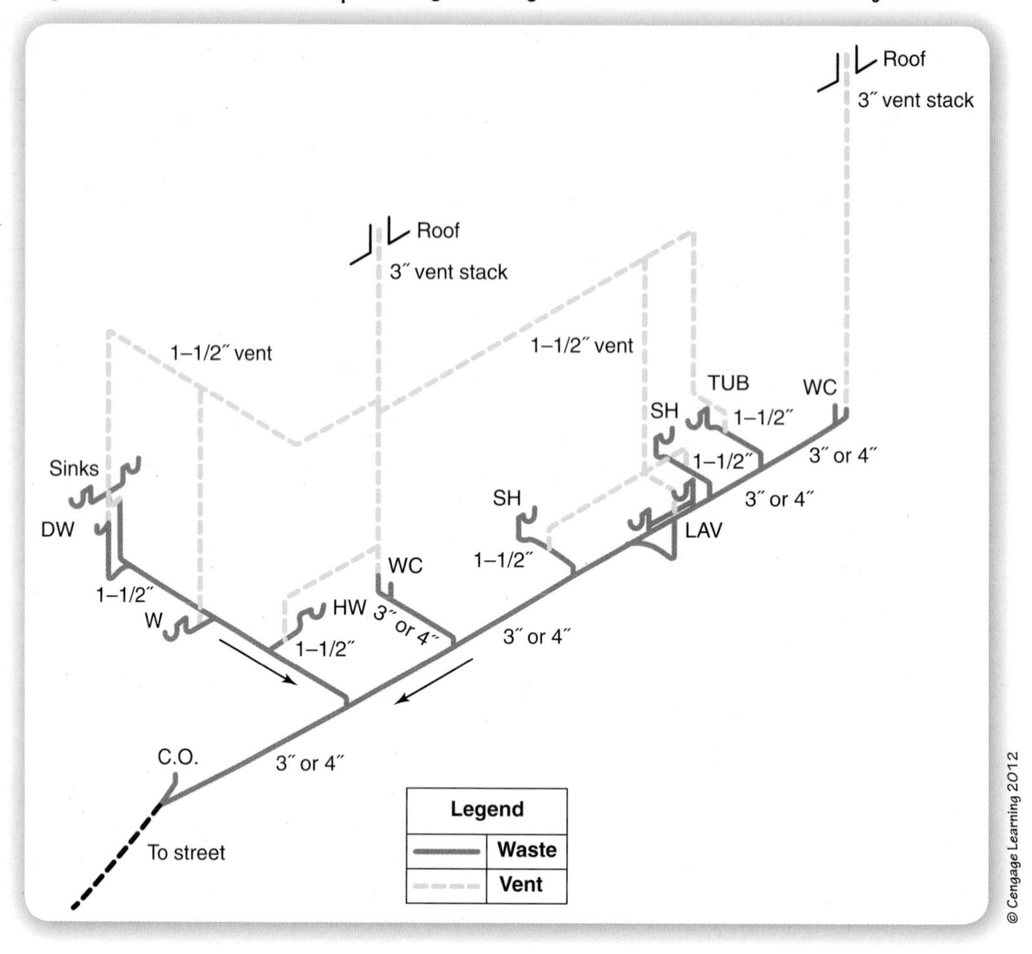

Diagramming the Water System

Let's begin with the steps used to show the water system on a residential plumbing plan. For clarity, our technique will use color to show the line types of our plumbing system. On large commercial plans, the plumbing lines are identified only by line type and line weight. We will use a copy of the floor plan and confirm that all the fixtures and appliances requiring water are shown. To begin the plumbing plan we will place small colored dots at the location of each fixture or appliance needing water. A blue dot, used to identify cold water, is placed at the right side when viewed from the front of each sink, lavatory, tub, shower, and washing machine. In contrast, a blue dot is placed at the *left* side of all water closets, beneath where the flush handle would be located. Don't forget the refrigerator's automatic icemaker and the outside hose faucet. Once we have located all the cold water needs, we will use red dots to indicate where hot water is desired (Figure 16-13).

Next, use the symbols and line type for cold water to indicate where the water is entering the building. Use a tee to divide the line and connect to the water heater. Draw the cold water line through the building and connect to the locations previously marked with blue dots. Drawing the lines parallel and perpendicular to the walls will create an easy to read plan. The actual water lines will be placed in a similar fashion running parallel and perpendicular to the floor joists. Applying the same process, start at the water heater and draw the hot water lines

Figure 16-13: Begin the plumbing plan by locating the hot and cold water needs.

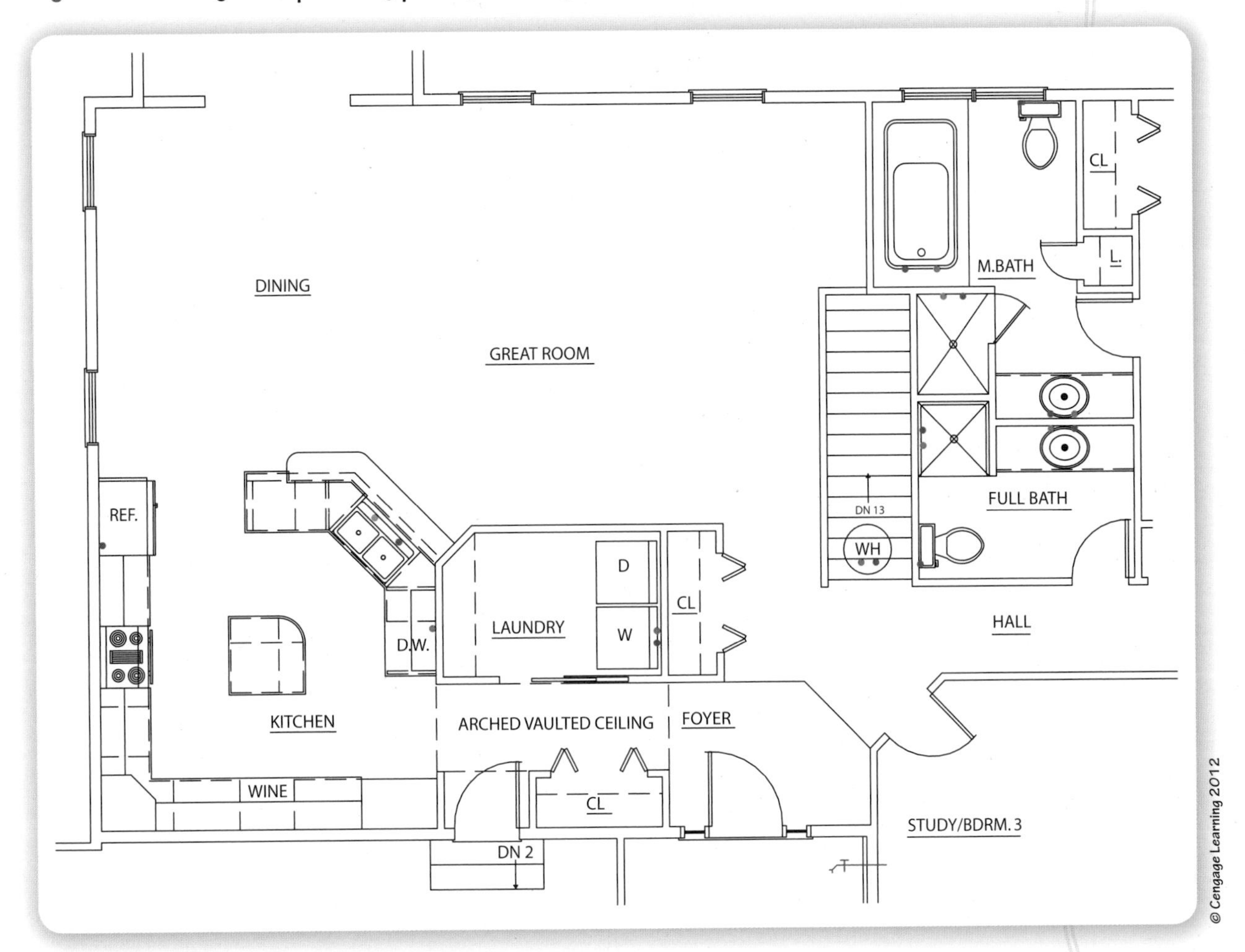

parallel to the cold water line to feed the hot water needs. Remember, these lines may not be the actual location of the pipes. The plumber will determine the best location while working between floor joists and up through walls (Figure 16-14).

The goal of this plan is to clearly communicate the plumbing needs and desired outcome. The plumbing plan shows a water heater located on the lower level with the use of hidden lines. When a building has more than one level, the plan should indicate the wall in which the pipes will be located to provide water to the next level. Add supplemental information with notes and leaders (Figure 16-15).

Strategies for Water Conservation

Water is a limited resource. Water shortages occur as a result from a combination of urban sprawl, population increase, drought, rising temperatures, excessive use, contamination, and waste. A traditional strategy to increase water conservation is through the selection of fixtures that operate with minimal water consumption. A water-saving showerhead or faucet restrictor will significantly reduce water consumption (Figure 16-16).

Figure 16-14: The plumber will determine actual piping location.

Figure 16-15: A completed water supply layout.

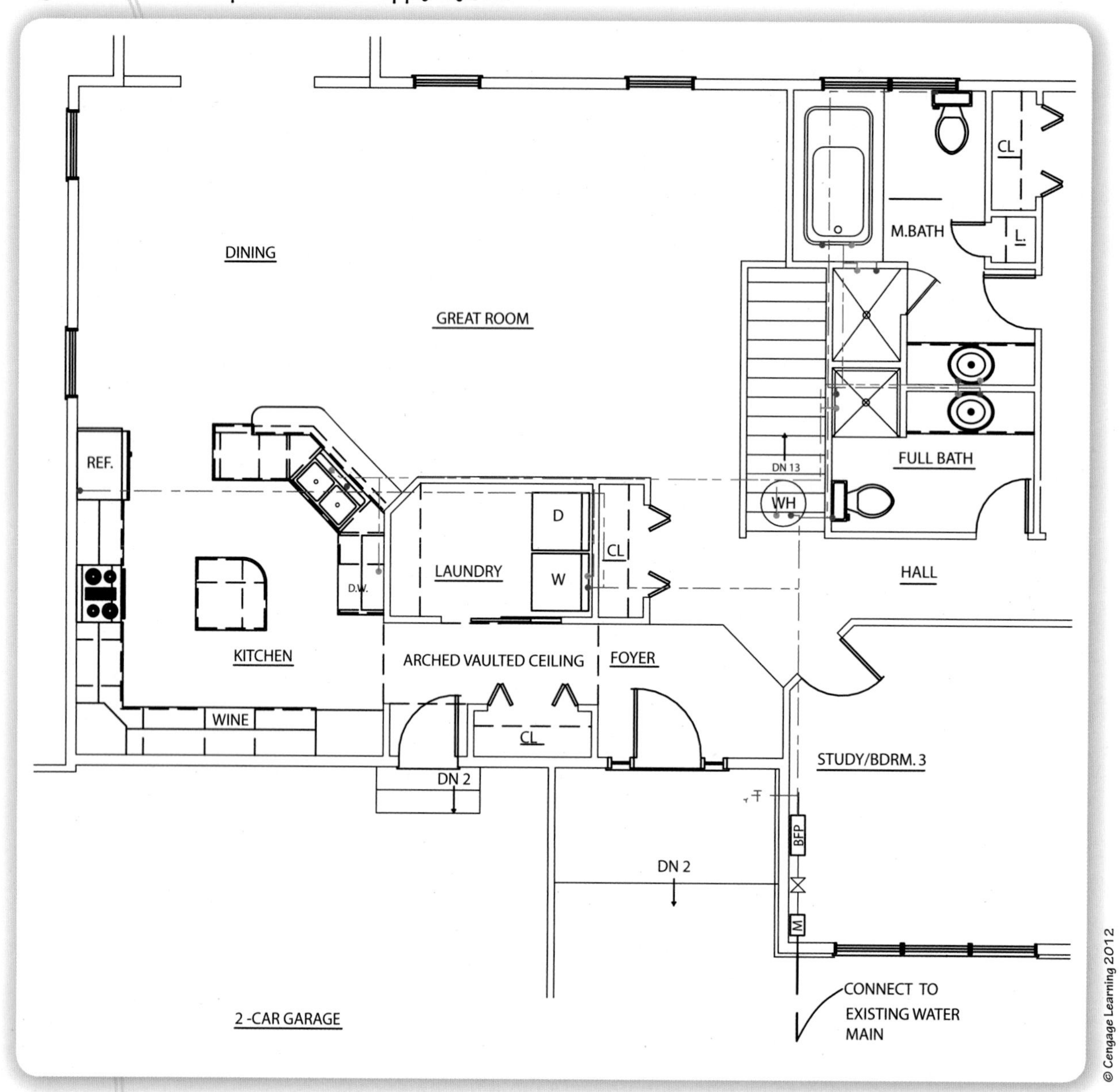

Figure 16-16: Water saving faucets and showerheads greatly reduce water consumption.

ACTIVITY	Water Usage Comparison	
	GALLONS USED (Conventional faucet or showerhead)	GALLONS USED (Water saving faucet or showerhead*)
Shower (water running)	7–10 gallons per minute	2–4 gallons per minute
Brushing teeth	10 gallons tap running	2–3 gallons tap running
Washing hands	2 gallons tap running	1–2 gallons tap running
Dish washing by hand	30 gallons tap running	10–20 gallons tap running

* The rate of consumption for water saving devices varies by product and manufacturer.

Case Study

Low flow water fixtures at CityCenter will save more than 76 million gallons of water per year over conventional fixtures. That is a savings of between 30 and 40 percent in each building (CityCenter, 2007, p. 1).

But most exciting is the fact that green practice is becoming contagious at the CityCenter development site. Recently, an engineer observed a water truck filling up with fresh water for use in dust abatement. Thinking this was necessary, but would be wasteful to use fresh water for this purpose, he found an alternative source of water at the nearby Monte Carlo resort that could be captured and re-used for this purpose (CityCenter, 2007, p. 1).

Household toilets are often identified as large users of water. To address this concern, several toilets are available that conserve water by employing a small tank design or dual-cycle controls. A dual-cycle control allows the user to choose how much water is needed for the flush (see Figure 16-17).

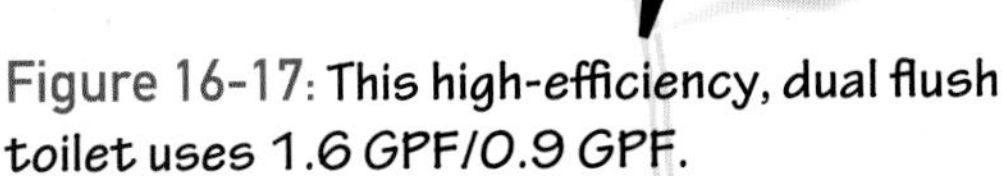

Figure 16-17: This high-efficiency, dual flush toilet uses 1.6 GPF/0.9 GPF.

Courtesy of TOTO USA, INC.

Another option is an Ultra Low Flow (ULF) toilet. The tank is higher and narrower than traditional toilets. Although the tank holds 1.6 gal of water, the flush valve only releases a portion of the water, using the remaining water for increased pressure. The creative shape and operation of this toilet provides both flush power and water conservation. Newer washing machines and dishwashers are more efficient, and use less energy, detergent, and water to achieve the desired results.

Water efficiency may also be achieved through the harvesting and use of rainwater. This strategy will require research of local government codes to determine if this is an option. Some areas may have a specialty code that provides explicit parameters to gather, store, and use rainwater. The code may require the use of specific types of holding tank, piping, and pumps, and may be subject to inspections and maintenance. Large building projects look at harvesting rainwater as a way to increase a building's sustainability. Rainwater is often harvested from green roofs. (See Figure 16-18). The additional weight and infrastructure of a garden or green roof is calculated into the original building plan. A green roof can also reduce stormwater runoff. The amount of stormwater is determined by local climate, depth of soil, plant types, and other variables. Stormwater collected from green roof systems can be used for nonpotable purposes, thus reducing utility needs. Both water consumption and sewer loading can be reduced by using rainwater to flush toilets or water vegetation (see Figure 16-18).

Another non-traditional approach is the use of **gray water**. An example of gray water would be the water from the shower or tub. It generally contains soap.

Gray Water: waste water from sinks, showers, and bathtubs.

If rainwater or gray water collection is allowed by code, it can be used to flush toilets, saving upwards of 20 gal of water per person each day. One fixture that has been used in Japan since the early 1990s is a sink bowl that fits over a toilet tank. Upon flushing, the fresh water is fed through a gooseneck spigot,

Figure 16-18: *One Bryant Park in New York City harvests rainwater to reduce water consumption and sewer system load.*

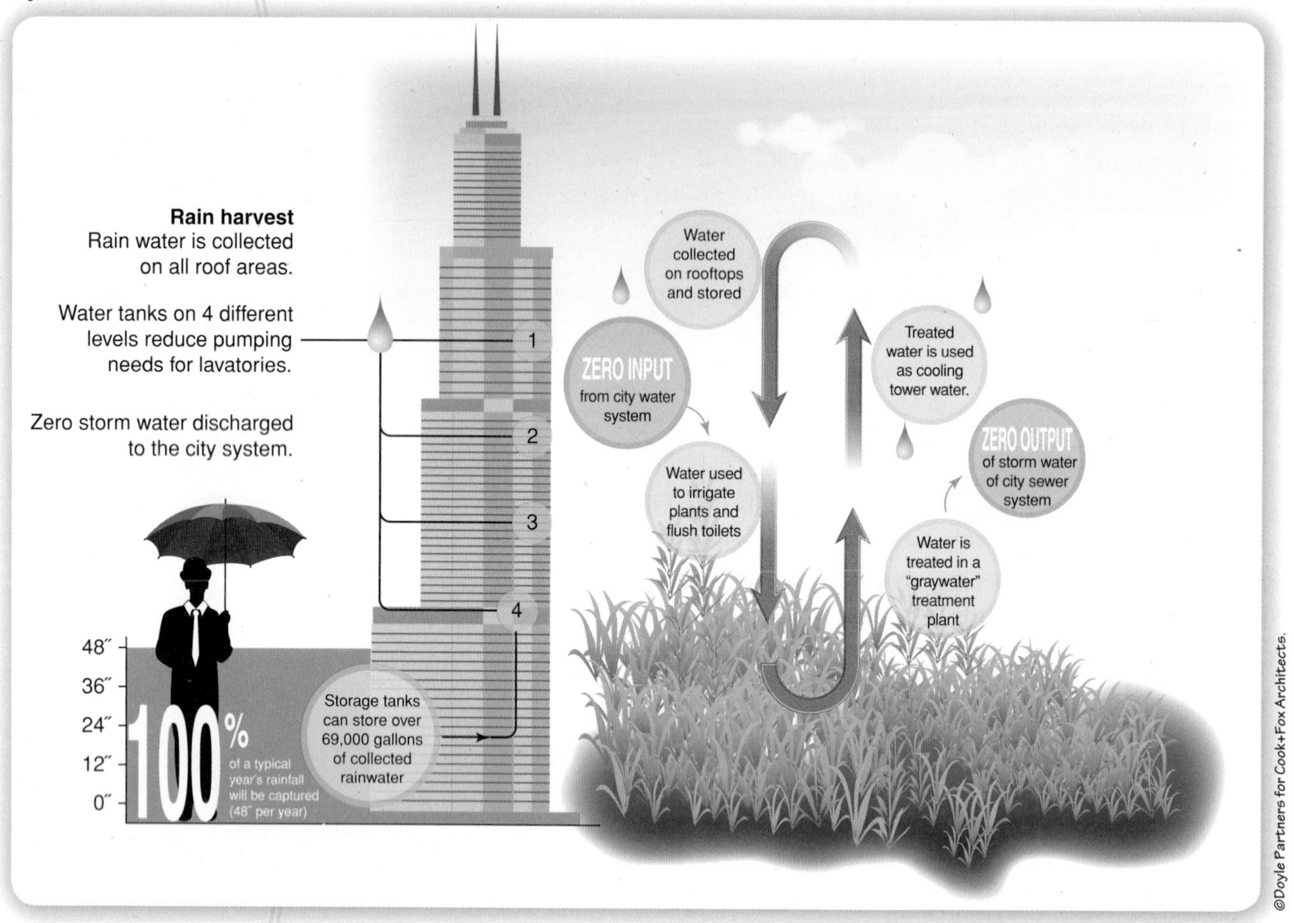

©Doyle Partners for Cook+Fox Architects.

Your Turn

Use the Internet to search for ways to conserve water through the use of water-saving fixtures. As a class, see who can find the most water-efficient toilet, dishwasher, washing machine, and showerhead. Determine the price of water in your area and calculate how much it would cost to use the new appliances over the period of a year for a household of three people. Assume the daily consumption of one load of wash, one load of dishes, six toilet flushes per person, and three showers per person at seven minutes each.

Calculate and compare your energy efficient household to a household of three with the following appliances and fixtures:

- washing machine: 55 gal of water per load
- dishwasher: 15 gal a load
- toilet: 3 gal per flush
- shower head: 5 gal per minute

What is the yearly cost difference between the two households for wash, shower, and flushing operations? Write a brief description in your notebook about how you would convince clients to choose water-efficient devices for their new residence.

allowing handwashing before it fills the toilet tank (see Figure 16-19). With this toilet, the water you wash your hands with is fresh. After handwashing, it becomes gray water and enters the tank ready for the next flush.

Our case study, the Syracuse Center of Excellence, will collect rain and water from snow on the roof. This water will be stored in a 5000 gal tank and will be used to flush toilets, reducing the consumption of potable water and the amount of water discharged to the sewer. If unconventional conservation systems such as rainwater harvesting are desired, the architect and engineer must plan early in the design process to insure the systems are accommodated within the building's design.

Figure 16-19: A toilet lid sink provides fresh water for hand washing before filling tank.

Courtesy of realgoods.com

COMMUNICATION OF THE WASTE REMOVAL SYSTEM

The wastewater removal system is entirely separate system from the water supply. It is often referred to as the DWV (drain, waste, and vent) system. Materials used in this system include polyvinyl chloride/PVC, acrylonitrile-butadiene-styrene/ABS, cast iron, and galvanized steel. These materials are rigid and require fittings such as elbows, tees, and **wyes**. The components of a waste system include **traps**, vents, vertical stacks, horizontal branch pipes, and horizontal drainage pipe (see Figure 16-20).

Trap: curved section of drain line that prevents sewer odors from escaping into the residence.

Each component of the waste systems performs a unique function. The traps hold 2 to 4 in of water, creating a seal that blocks gasses and odors from the drainage system from entering the building. Codes require traps be installed on or near each fixture. Usually, a P-shaped trap is placed below the fixture drain. Sometimes, adjacent lavatories located within a short distance from each other, usually 6 ft or less, may share the same trap. A dishwasher located adjacent to the kitchen sink commonly shares the sink trap. The water closet has a trap formed inside the fixture and does not require an additional trap.

Once the wastewater leaves the trap, it travels down through a vertical waste stack and across horizontal branch piping to reach the main soil stack. From there it moves down to the horizontal drainage piping and exits the building (see Figure 16-19). A watertight cleanout fitting is located at the bottom of the soil stack where it connects to the horizontal drainpipe. The cleanout provides access to the pipe if a blockage occurs. The horizontal drainage pipe should slope downward at least ¼ in per every 1 ft run toward the public sewage system or private septic system.

Venting of the Waste System

The waste system must be vented to allow sewer or septic gas to move upward and out of the building. Vents also enhance the waste flow, allowing air in to prevent siphoning and backpressure.

The main stack, which is located behind the water closet, is generally vented through the roof. Other drains are connected to the vent stack through a series of horizontal and vertical pipes. Three-dimensional illustrations such as Figure 16-20 are not included in plumbing drawings because the plumbing contractor already knows how a plumbing system works. Instead, the communication of the water supply and

Figure 16-20: Vent and waste piping.

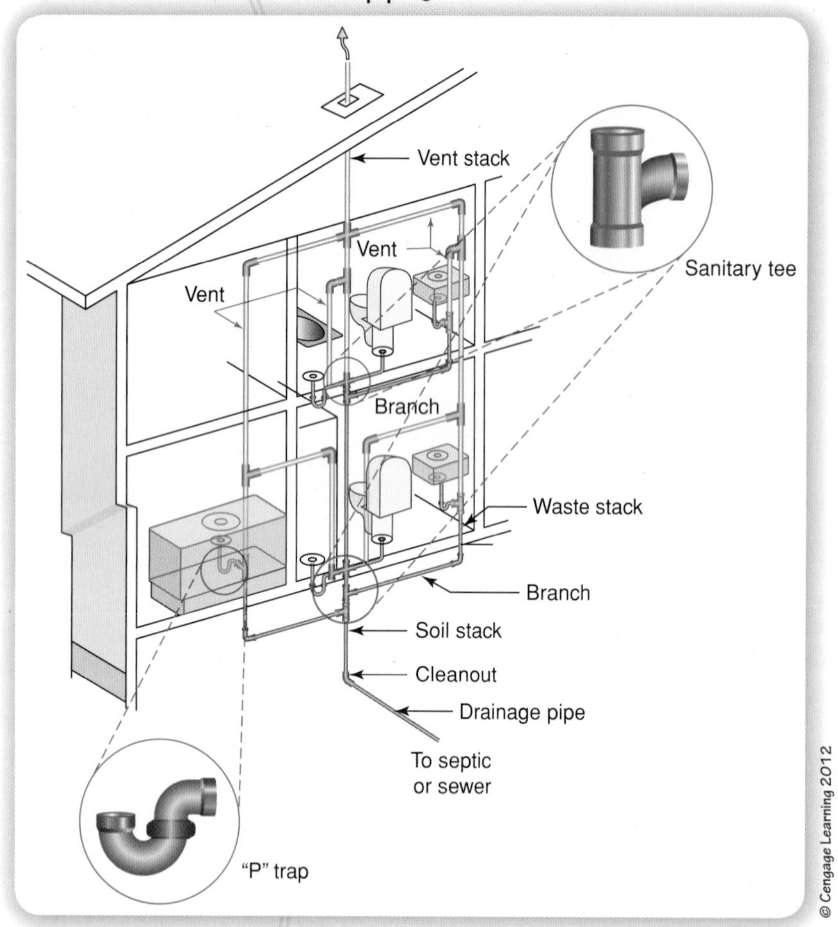

waste removal plan is shown through plumbing plans with symbols, abbreviations, and line types, supported with general notes, schedules, and legends.

Diagramming the Waste Removal System

To explain the waste removal system in a plumbing plan, we will use the same residential plan, Figure 16-23. We will begin by re-examining the location of the public sewage system or private septic system that was determined during the site planning. In an urban setting, the public sewage system is at a set location. In rural areas, codes will impact the location of a septic system, such as placement no less than 100 ft from the well and a specified setback from adjacent properties, lakes, or waterways. The water closets generally connect into the main soil stack with a sanitary tee that directs the waste downward. The soil stack connects to the drain pipe, which should form a fairly direct path to the sewage system or septic system, with minimal bends to reduce the chance of blockage. Other pipes carry mostly liquid and are secondary in nature. The horizontal bridging, from the sink, lavatory tub, and other drains, connects to the main soil stack. Using the same technique as the water system layout, we will draw a brown dot at each water closet and fixture drain. We need to include places that need drains or overflow protection, such as a dehumidifier or water heater. Draw the main drainage pipe line from the soil stack to where it exits the building. Draw connections from the fixture drains to the main drainage pipe line using thick solid lines. When joining into a horizontal branch or drainage piping, show the direction of the waste flow by placing an angled line to represent a sanitary tee or **wye** connection (see Figure 16-22). When drawing the waste lines, think about direction of the flow: down and out.

Wye:
a Y-shaped fitting with three openings allowing one pipe to be joined to another at a 45-degree angle.

In a two-level building, a 6 in plumbing wall on the lower level will contain the soil stack for distribution of waste. The lower-level soil stack is one of the first things you will draw and label on your upper level plan. You will then draw the branch lines connecting the upper level fixtures to the soil stack. These lines are usually drawn parallel and perpendicular matching the wall and floor system.

Figure 16-21: The waste system is vented through the roof.

VENTING THE WASTE SYSTEM Once waste lines are drawn, it is time to locate the main vents. They are generally located above the soil stack and extend past the roof of the building. When two water closets (toilets) share the same soil stack, one vent is used. However, when water closets are placed at different ends of a building, code will require that each soil stack be vented through the roof (Figure 16-21). To show vents on your plan, label with a leader indicating that it is a vent through the roof of a specific diameter. Codes will specify the vent size required; generally a residential vent stack range is 2 to 3 in in diameter, whereas commercial vent stacks are generally 4 in. Often fixtures in close proximity can share the same vent stack (see Figure 16-24).

It is helpful when placing the vent lines to remember that the vent pipes will go up through an adjacent wall to above the ceiling joist and connect to the roof vent. With that in mind, examine each fixture to determine a wall to place a vertical vent pipe referred to as a vent stack, see figure 16-25. Draw a circle to represent the stack.

Using yellow dashed lines, show a connection from the fixture trap to the vent stack. Once above the ceiling joist, it can branch (horizontal) to the main vent stack (vertical). It may help to refer back to Figure 16-20 to visualize the plumbing in 3D. When the waste lines of all fixtures have been vented, the plan will appear similar to Figure 16-26. When plans for water supply and waste removal are combined your plan should resemble figure 16-28.

Unique circumstances, such as an island sink, may require the use of special venting methods or an air admittance valve (Figure 16-27).

Figure 16-22: A sanitary tee (top) and wye direct the wastewater toward the exit.

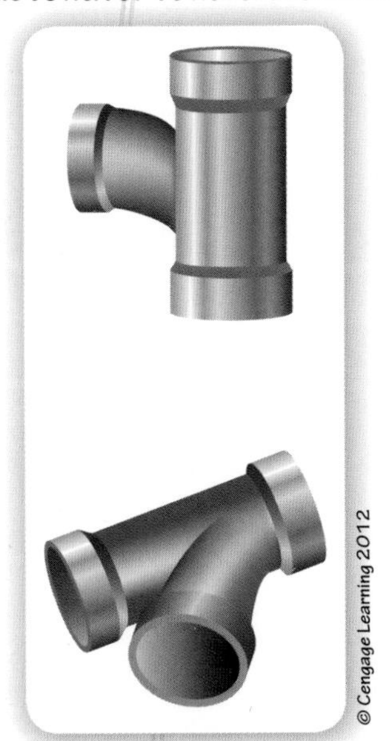

Figure 16-23: The lines representing the wastewater piping illustrate the direction of flow.

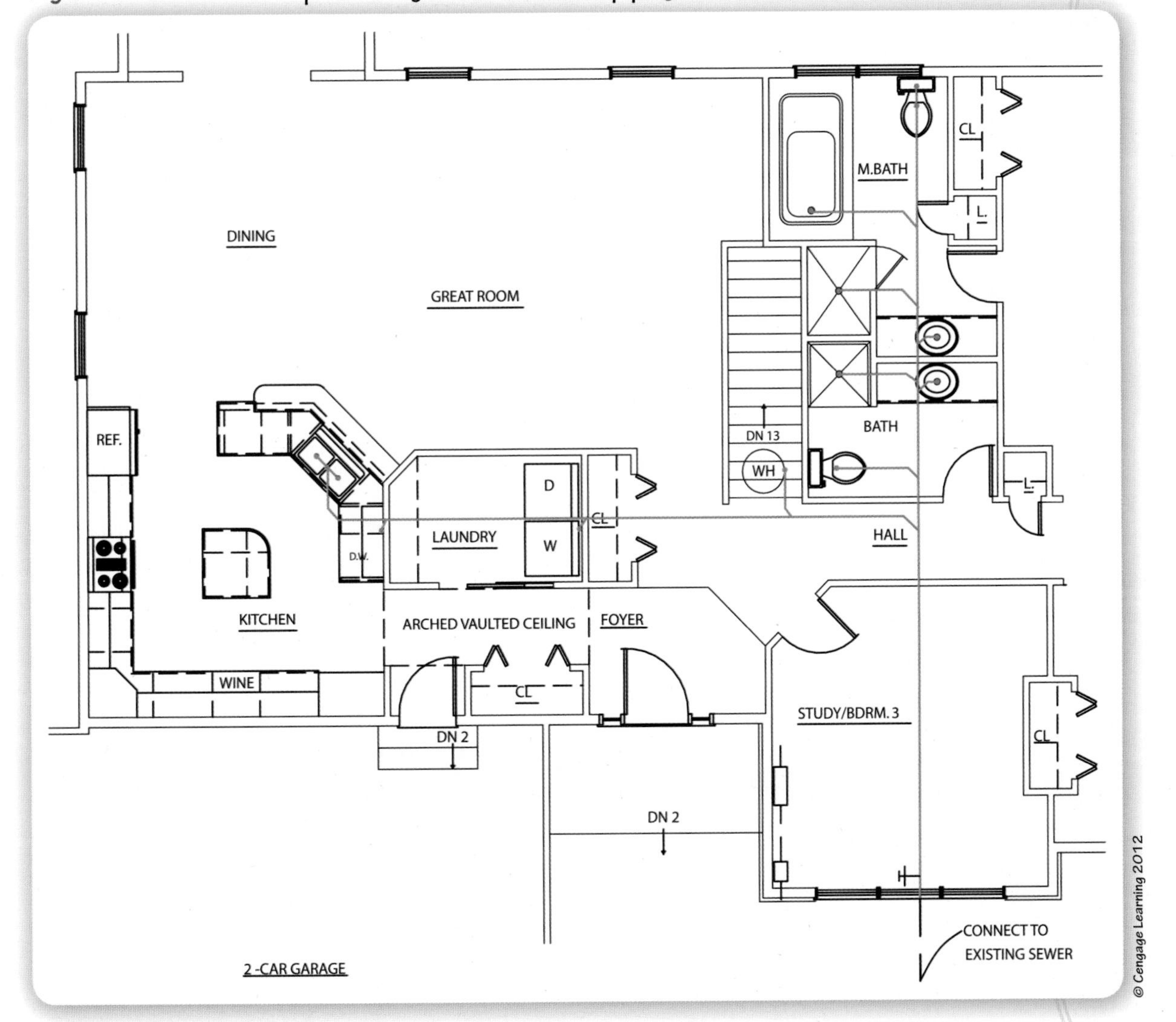

Figure 16-24: Plumbing of sinks placed back to back.

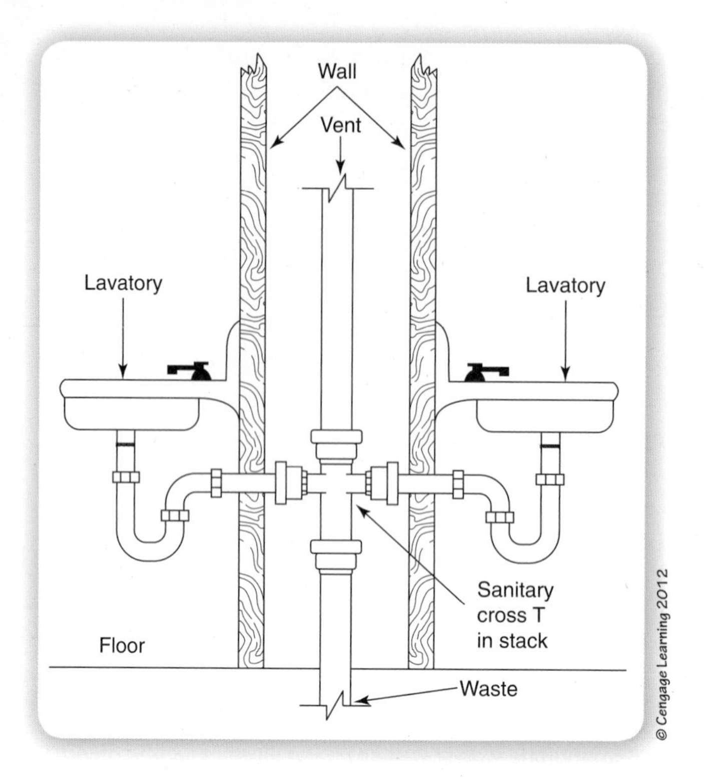

Figure 16-25: Vent piping is placed in a wall adjacent to the plumbing fixture.

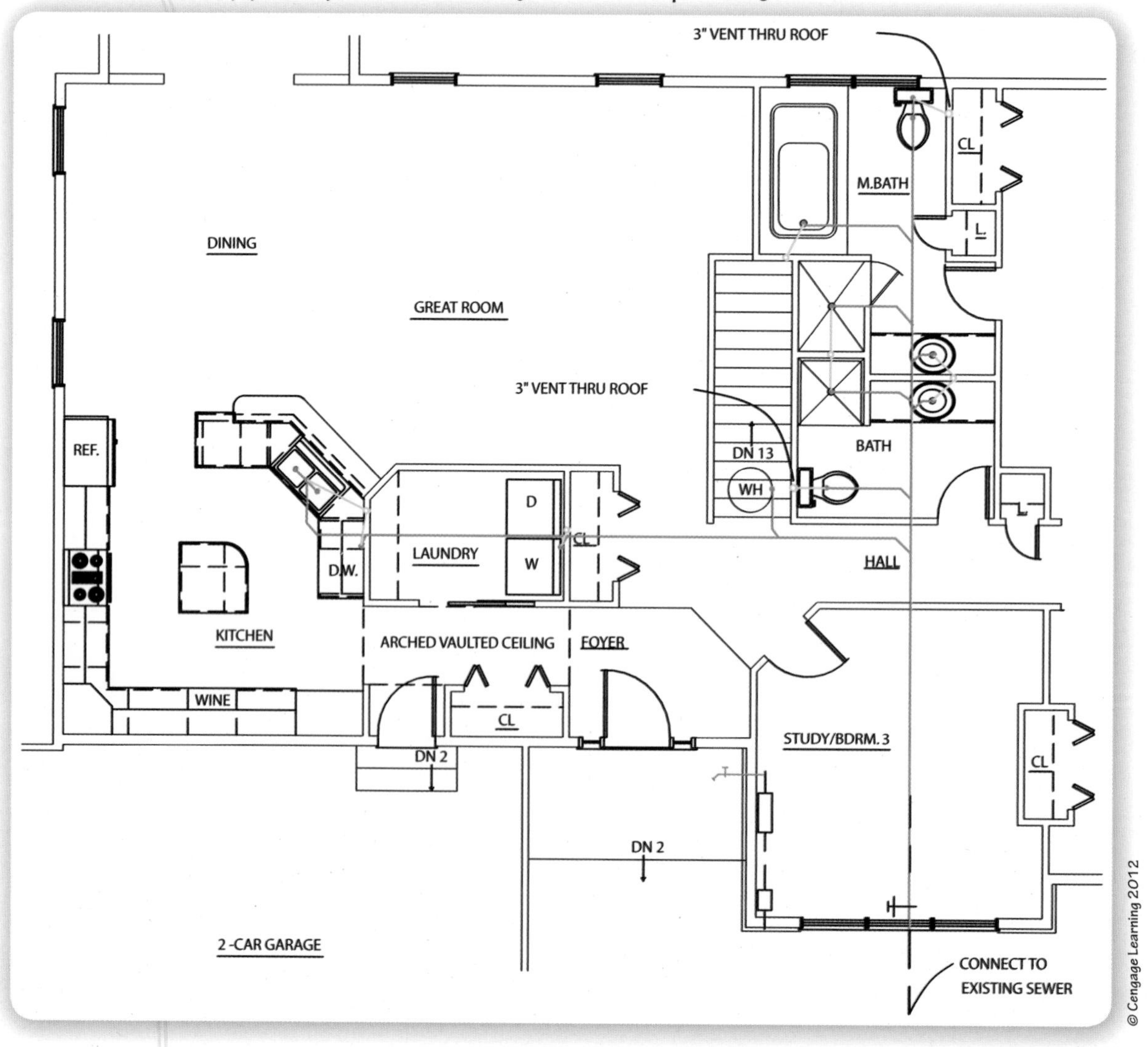

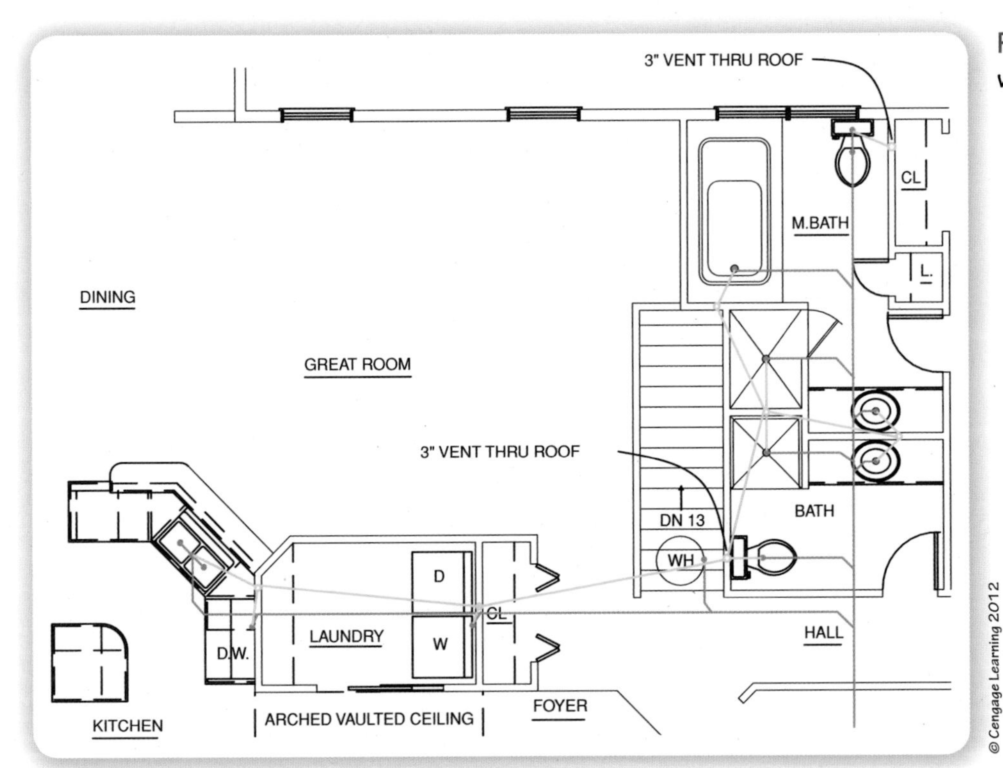

Figure 16-26: Drains are vented to the vent stacks.

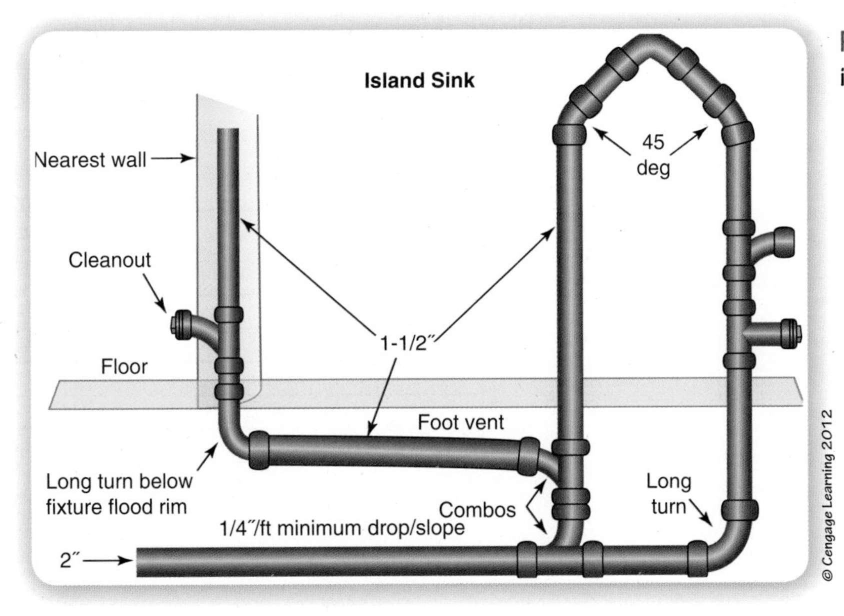

Figure 16-27: Venting of an island sink.

STORMWATER DRAINAGE

Commercial buildings will need to show a plan to disperse stormwater. In Chapter 6: Site Planning you read about stormwater runoff and calculated stormwater runoff for a paved parking lot. Many commercial buildings have fairly flat roofs that must be drained of stormwater. To determine the amount of drainage needed, you will need to calculate the roof area and the maximum rainfall rate for that location. A storm chart will help you identify the amount of rain expected to fall during 60 minutes of a once-in-100-years storm. The building plans must illustrate how the stormwater will be conveyed to the sewer system. In some

Figure 16-28: *Completed plumbing plan.*

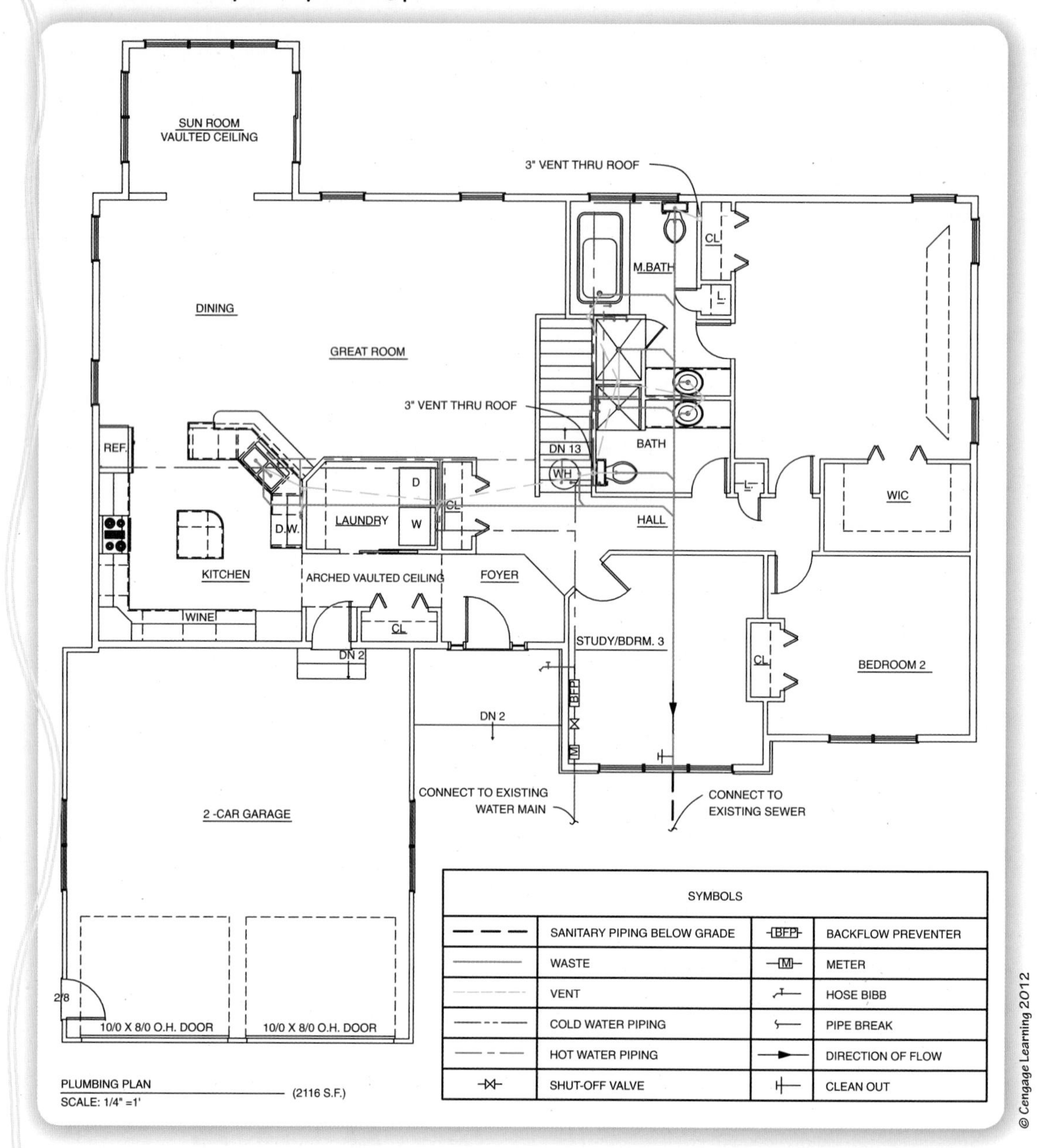

locations, storm sewers are separate from sanitary sewers. Rainwater is often directed through roof drains and channeled down through the building. A pitched roof may use gutters, downspouts, and drainage pipes to direct stormwater away from the building. This method is often seen on residential buildings.

Case Study

There are 14 sheets of plumbing plans for the Syracuse Center of Excellence. The first sheet in the set provides the general notes, legend, and drawing list. The general notes communicate expectations to the plumbing contractor (see Figure 16-29).

A vast amount of information is communicated through the use of symbols and abbreviations. The legend on the front sheet shows the standard line type for waste, vent, and cold and hot water that was used on your residential plan. Plumbing contractors can easily recognize standard plumbing symbols and abbreviations. If you reference the schedule to read the plumbing plan in Figure 16-30, you will be able to identify water and wastewater pipe sizes required to accommodate each fixture. What information is provided directly on the plan?

Plumbing fixture information is clearly communicated through the use of a schedule. The schedule can be as detailed or general based on the scope of the project. Small residential plans may not include a schedule for plumbing fixtures. Instead the fixtures may be chosen by the owner or contractor and placed in the room to be plumbed. In our case study of the Syracuse Center of Excellence, the plumbing schedule describes the code for each fixture and the water supply and waste removal needs. Which of the fixtures listed on the schedule shown in figure 16-30 will require only cold water?

Figure 16-29: *General notes on sheet P001 communicate important information for the plumbing installation.*

GENERAL NOTES
1. All materials and workmanship shall conform to NYS plumbing code and all applicable local codes and regulations.
2. Drawings are diagrammatic and indicate general arrangement of systems and work included. Follow drawings in laying out work and check drawings of other trades relating to work to verify space in which work will be installed. Maintain headroom and space conditions at all times.
3. Coordinate plumbing systems with work of all other trades prior to any fabrication or installation. Provide all fittings, offsets, and transitions as required for a complete workable installation.
4. Platforms, pits, and flashings for plumbing equipment shall be as indicated on the structural and architectural plans, unless noted otherwise. Coordinate exact sizes of required openings and supports for furnished equipment.
5. All equipment shall be installed in strict accordance with the equipment manufacturer's recommendations and applicable codes. Provide all fittings, transitions, valves, and other devices required for a complete workable installation.
6. Maintenance label shall be affixed to all plumbing equipment and a maintenance manual shall be provided to owner.
7. The contractor shall visit the job site for submission of a bid proposal and submission of a cost proposal. Bid will be judged as evidence that site examination has been made. Claims for extra costs for labor, equipment, or materials required, or for difficulties encountered, which could have been foreseen had such examination been made, will not be recognized.
8. Contractor shall refer to all the architectural drawings for plumbing related work.
9. Contractor shall be responsible for obtaining all necessary permits, including but not limited to, entering manholes, use of water from low pressure hydrants, etc., prior to commence of work.

Case Study

Figure 16-30: A schedule on P001 is used to provide information to supplement the plan on P402.

PLUMBING FIXTURE SCHEDULE						
CODE	DESCRIPTION	WASTE	VENT	CW	HW	REMARKS
P–1	WATER CLOSET	4"	2"	1 ¼"*	–	WALL MOUNTED
P–1A	WATER CLOSET	4"	2"	1 ¼"*	–	WALL MOUNTED, ADA COMPLIANT
P–2	URINAL	2"	2"	–	–	WATERLESS
P–2A	URINAL	2"	2"	–	–	WATERLESS, ADA COMPLIANT
P–3	LAVATORY	1 ½"	1 ½"	½"	½"	UNDERCOUNTER
P–3A	LAVATORY	1 ½"	1 ½"	½"	½"	UNDERCOUNTER, ADA COMPLIANT
P–3B	LAVATORY	1 ½"	1 ½"	½"	½"	WALL MOUNTED
P–4	KITCHENETTE SINK	1 ½"	1 ½"	½"	½"	TWO COMPARTMENT
P–4A	KITCHENETTE SINK	1 ½"	1 ½"	½"	½"	SINGLE COMPARTMENT
P–5	SHOWER	2"	1 ½"	½"	½"	
P–6	SERVICE SINK	3"	1 ½"	¾"	¾"	WALL MOUNTED
P–7	ELECTRIC WATER COOLER	1 ½"	1 ½"	½"	–	ADA COMPLIANT, TWIN RECEPTORS
P–7A	ELECTRIC WATER COOLER	1 ½"	1 ½"	½"	–	
P–8	EYE WASH STATION	2"	–	¾"	¾"	TEMPERED WATER
HB	HOSE BIBB	–	–	½" OR ¾"	–	
WH	WALL HYDRANT	–	–	¾"	–	NON–FREEZE TYPE
GH	GROUND HYDRANT	–	–	¾"	–	NON–FREEZE TYPE

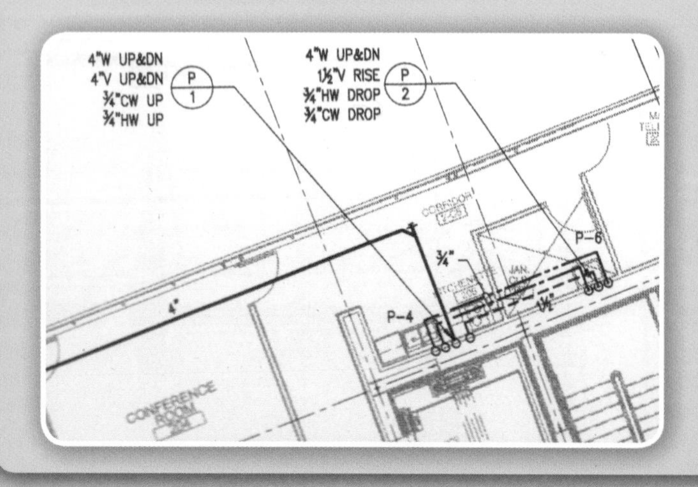

Source: Courtesy of the Syracuse Center of Excellence.

SUMMARY

The occupants of a building need safe and dependable water supply and waste removal systems. These systems require thoughtful planning. In small residential buildings, the plumbing contractor may work directly with the general contractor or owner to achieve the desired outcome. In large commercial projects information may be provided through a set of detailed plumbing plans. In both cases, the materials and the installation process must adhere to local governance and codes. General notes, symbols, abbreviations, and schedules provide clear communication of the desired outcome to the plumbing contractor. The water supply system and waste removal are two distinct and separate systems. The quality of water is determined by its level of purity. The term potable is used to identify water that is safe to drink. Because water is a limited resource, conscious planners often recommend installation of low consumption fixtures. In some areas, water conservation is achieved through harvesting rainwater to use as a nonpotable water source. Nonpotable water can be used to flush toilets or in industrial applications such as cleaning and cooling. A building's water system is generally described by its pressure and flow rate, which are affected by gravity and resistance.

The plumbing contractor uses a plumbing plan to gather information about the desired plumbing outcome. Notes, schedules, and legends provide details not easily communicated in the drawing. A plumbing plan shows the location of all fixtures and major items including the meter, main shut-off, hose bibb, cleanout, roof drain, and vents. It does not specify the location of each and every fitting or show how to run the pipes. A plumbing plan for a multilevel building will identify the plumbing wall where pipes and stacks are located for water and waste distribution between floors. The wastewater removal system of a building is often referred to as the DWV (drain, waste, and vent) system. The main components of a DWV system include traps, vents, vertical stacks, horizontal branch pipes, and horizontal drainage pipe. The waste piping must be vented to permit sewer or septic gas to move up and out of the building. Venting also enhances the waste flow, allowing air in to prevent siphoning and backpressure.

BRING IT HOME

1. Using a copy of the floor plan for a building you have designed, draw the water supply system. Use standard line types, symbols, abbreviations, and notes to show the hot and cold water needs. Note the piping material.
2. Add a waste removal system to the plumbing plan started in #1. Use standard line types, symbols, abbreviations, and notes to show the waste and vent needs. Note the piping material.
3. Create a cover sheet that includes a symbol legend of all line types, symbols, and abbreviations used in your plumbing plan.
4. Using the Internet, research and choose efficient water and energy fixtures for your building. Provide details and specifications in a plumbing fixture schedule.
5. Using the Internet, research and select an energy-efficient hot water heater, washing machine, and dishwasher for your building. Create a short description for each appliance, highlighting the energy or water-saving feature.

EXTRA MILE

Rainwater Resources

Use the Internet to find out more about rainwater use across the country. Start with a search for Portland Oregon Bureau of Environmental Services to examine their rainwater harvesting code guide.

http://www.portlandonline.com/bes/

The website for the Office of Sustainable Development in Portland offers tips on calculating your rainwater harvesting potential. You can also read about case studies and tutorials.

http://www.portlandonline.com/osd/

After gathering information from other sources, make a plan to include a rainwater collection system in one of your own building designs.

CHAPTER 17
Indoor Environmental Quality and Security

GPS DELUXE

Menu | START LOCATION | DISTANCE | END LOCATION

Before You Begin

Think about these questions as you study the concepts in this chapter:

1. How is energy defined?
2. What factors influence building's energy consumption?
3. What main considerations must be taken into account when selecting an indoor environmental quality (IEQ) strategy?
4. How are heating and cooling loads calculated?
5. What are the methods of heat transfer?
6. How do security measures protect building occupants?
7. What are the advantages of an intelligent building system?

INTRODUCTION

Did you know that your home heating and cooling cost is most likely the major portion of your energy bill? According to the U.S. Energy Information Agency, residential and commercial buildings represent 72 percent of U.S. electricity consumption (see Figure 17-1). As you read in Chapter 8: Energy Conservation and Design, most older buildings are not energy efficient because of a poor building envelope or inefficient Heating, Ventilation and Air Conditioning (HVAC) systems. In the United States, HVAC accounts for nearly 50 percent of the energy used in buildings.

Most older buildings have inefficient HVAC systems which operate on nonrenewable fossil fuel. Unfortunately, America's supply of nonrenewable fossil fuel

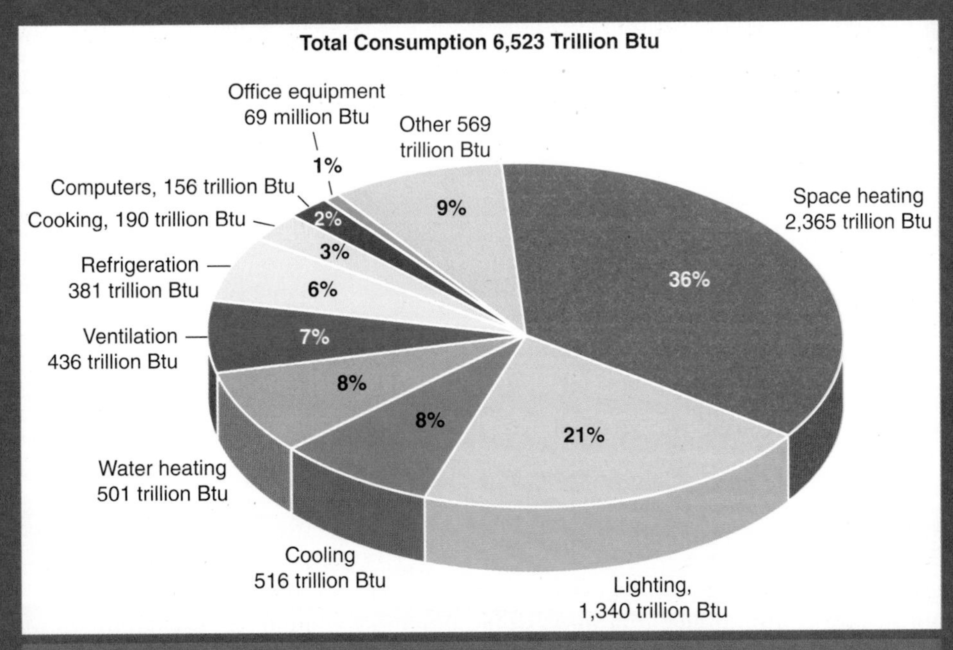

Figure 17-1: *In 2003, more than half of energy consumed by commercial buildings was used for space heating, ventilation, and air conditioning.*

INTRODUCTION

Energy-neutral: a building designed for low energy consumption, often significantly lower than required by building regulations.

Net-zero: a building requires zero energy from a utility provider because it is able to produce as much or more energy then it needs.

is dwindling and energy consumption and greenhouse gases are rising. Buildings are one of the largest consumers of natural resources, and account for 38 percent of all CO_2 emissions. To address these issues, architects are designing high performance buildings that employ renewable energy and fuel efficient HVAC systems. Building designers work together with contractors, Indoor Environmental Quality (IEQ) planners, and government agencies to determine factors that will support **energy-neutral** or **net-zero**-energy building design. These factors most often include building envelope design and solar and geothermal technology. The American Society of Heating Refrigerating and Air Conditioning Engineers (ASHRAE) defines net-zero-energy buildings as using 50 to 70 percent less energy than traditional buildings, and uses no more from the utility grid than is generated and replaced by on-site renewable energy sources.

Indoor environmental quality (IEQ) also affects occupants' health, comfort, and productivity. We have all been in buildings that are hot, cold, damp, or lacking good air quality. IEQ is best addressed early in the building design process by the whole-building integrated design team. The team will look at the building operation as a whole to plan a safe, comfortable, and sustainable indoor environment. The American Society of Heating, Refrigerating, and Air-Conditioning Engineers (ASHRAE) and other organizations provide standards for indoor environmental quality. The energy efficiency of a building is impacted by many factors: exterior environment, climate, building envelope, construction materials, window choices, number of occupants, occupant activities, and interior features and components. A correctly sized HVAC system will reduce energy consumption, provide a more comfortable environment, and may last longer and require fewer repairs. In this chapter, you will learn a how to calculate interior heating and cooling loads to support the sizing of an HVAC system. You will also learn about building features for occupant safety and building security. Building security consists of systems, staff, and procedures that will monitor and respond to prevent damage. Systems must address a range of damage types from fire and flood to unlawful entry and vandalism. Occupant safety is supported by adequate lighting in parking areas and corridors, alarm systems, and security personnel. In commercial buildings, security features and indoor environmental quality are often controlled by an *intelligent building system*. Buildings that employ intelligent building systems to provide a comfortable and safe environment are commonly referred to as smart buildings. An intelligent building system integrates multiple components to control and operate building processes. The computer control of an intelligent building system enhances efficiency and sustainability through feedback, analysis, and custom programming.

Figure 17-2: Renewable energy consumption in the nation's energy supply, 2003.

Source: US Government Energy Information Administration, Office of Coal, Nuclear, Electric and Alternate Fuels.

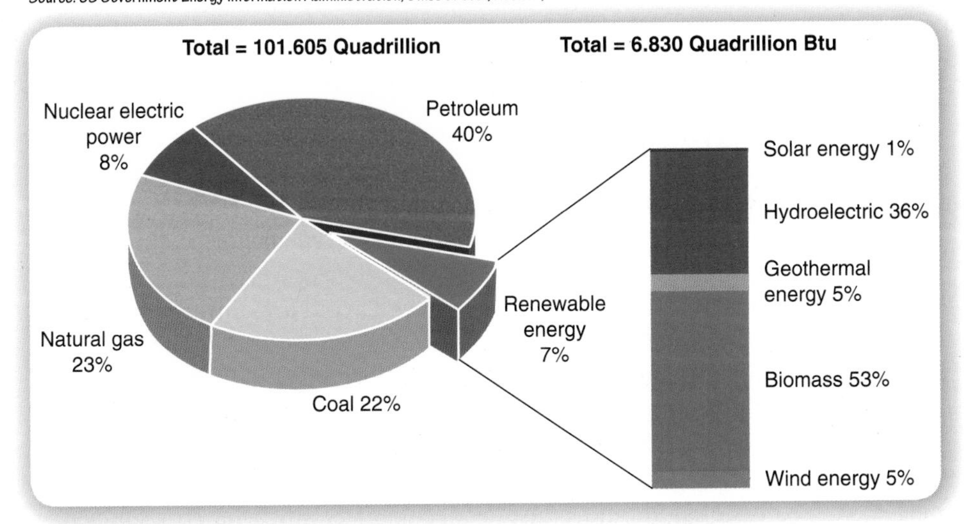

ENERGY

America's supply of fossil fuels is dwindling and the impact of greenhouse gases on world climate is rising. We must find ways to reduce energy load, increase system efficiency, and utilize more renewable energy resources. Fossil fuels (petroleum, coal, and natural gas) are considered nonrenewable resources because they take millions of years to form and are depleted much faster than they are replaced. Non-fossil fuel options include hydroelectric, nuclear, geothermal, solar, and wind (see Figure 17-2).

The earth does not have infinite resources; human consumption negatively impacts the natural processes that renew some resources, and depletes nonrenewable resources.

The Environmental Protection Agency (EPA) recognizes generation of **green power** to be more environmentally friendly than traditional electricity generation. Generating green power emits little or no air pollution and leaves behind no radioactive waste, unlike nuclear power generation. Most importantly, green resources are naturally replenished by the earth and sun (see Figure 17-3).

Green Power: electricity generated from renewable sources (solar, wind, geothermal, biomass, or small hydro).

Figure 17-3: The EPA describes green power as electricity generated from renewable sources.

Green Power	
Solar	Converting energy from the sun into electricity using photovoltaic panels and solar thermal plants
Wind	Harnessing the power of the wind using turbines (wind power is the fastest growing renewable energy technology)
Geothermal	Use of steam that lies below the earth's surface to generate electricity
Biomass	Releasing solar energy stored in plants and organic matter by burning agricultural waste and other organic matter to generate power
Small Hydro	Use of flowing water to power electric turbines (small hydro plants are less than 30 MW in size)

Source: http://www.epa.gov/greenpower.

FACTORS INFLUENCING ENERGY CONSUMPTION

Remember reading about the whole building integrated design approach which involves a team of experts in the design process? The best indoor environmental quality plans are created using this process. An energy analyst, indoor environmental quality (IEQ) planner, and systems professional will discuss with the architect how individual building components, walls, windows, and HVAC systems will act as a whole to provide performance and sustainability. The HVAC system will be sized by both **load** and consideration of building and site factors such as natural light, temperature, climate humidity, shading, and envelope.

Load:
demand, the amount a system can handle.

A correctly sized high-performance system will avoid uncomfortable changes in temperatures, excessive moisture, drafts, and excessive noise. The American Society of Heating, Refrigerating, and Air conditioning Engineers (ASHRAE) is accredited by the American National Standards Institute (ANSI) and follows ANSI's requirements for standards development. ASHRAE recommends that when selecting a strategy and system to provide indoor environmental quality the HVAC professional should consider occupant comfort, energy efficiency, ease of use, service life, initial cost, lifecycle cost, and value-added features. To reduce loads, building designs maximize daylight, include ventilation and moisture control, and avoid materials with high **Volatile Organic Compounds** (VOCs). VOCs are emitted as gases that may have adverse health effects. The United States Environmental Protection Agency's Total Exposure Assessment Methodology (TEAM) study found levels of approximately a dozen common organic pollutants to be two to five times higher inside homes than outside, regardless of whether the homes were located in rural or highly industrial areas" (*http://www.epa.gov/iaq/voc.html*).

Energy Codes

The indoor environmental quality (IEQ) planner must research energy codes, specific to building type and location, to define an environmentally friendly, cost-effective, efficient, and dependable air quality and control system. The IEQ planner will gather the latest recommendations from sources such as ASHRAE, Air Conditioning Contractors of America (ACCA), International Code Council (ICC), International Energy Conservation Code (IECC), Leadership in Energy and Environmental Design (LEED), and the EPA. Model energy codes for residential and commercial energy efficiency are developed by the ICC. The International Residential Code (IRC) specifies energy requirements for residential buildings. A separate energy code, the IECC, was created by the ICC for commercial buildings: The IECC provides specifications for insulation, lighting, mechanical, heating and air conditioning systems, service water heating, and electrical power usage. Many states or local jurisdictions have adopted a version of the IECC (see Figure 17-5).

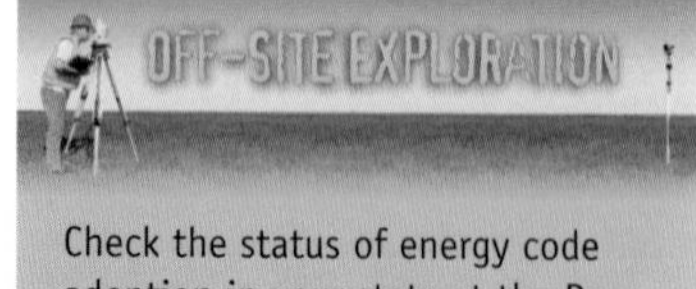

Check the status of energy code adoption in your state at the Department of Energy (DOE) Building Energy Codes Program website: (*http://www.energycodes.gov/states/*).

The IEQ planner performs extensive research to determine the amount of cooling and heating needed for occupant comfort. This is not an easy task due to the many unique building and occupancy factors that impact the calculation.

In addition to calculating the number and activity of the occupants, the IEQ planner will investigate the building composition, building site, and environmental conditions. The site's landscape and adjacent structures can provide windbreak and shading, thus reducing the amount of heat or cooling needed. Environmental studies identify ground temperatures, climate and weather factors, humidity, wind, and sun intensity and direction at various times of the year. Remember that

Your Turn

Take a look at the table shown in Figure 17-4 that shows household energy consumption for the years 1978 to 2005. Identify the end use with the largest consumption gain. What factors do you think contributed to the increase? Make a list of building design options that would reduce the amount of air conditioning needed.

Figure 17-4: **Household Energy Consumption (quadrillion Btu) by *end use and energy*.**

	Space Heating[1]				Air Conditioning[2]	Water Heating				Appliances[3,4]			Total				
Year	Natural Gas	Electricity[5]	Fuel Oil[6]	LPG[7]	Electricity[5]	Natural Gas	Electricity[5]	Fuel Oil[6]	LPG[7]	Natural Gas	Electricity[5]	LPG[7]	Natural Gas[2]	Electricity[5]	Fuel Oil[4–5]	LPG[7]	Wood[8]
	Consumption (quadrillion Btu)																
1978	4.26	0.40	2.05	0.23	R0.31	1.04	0.29	0.14	0.06	0.28	R1.46	0.03	5.58	2.47	2.19	0.33	NA
1979	NA	NA	NA	NA	NA	NA	NA	NA	NA	NA	NA	NA	5.31	2.42	1.71	0.31	NA
1980	3.41	0.27	1.30	0.23	0.36	1.15	0.30	0.22	0.07	0.36	1.54	0.05	4.97	2.48	1.52	0.35	0.85
1981	3.69	0.26	1.06	0.21	0.34	1.13	0.30	0.22	0.06	0.43	1.52	0.05	5.27	2.42	1.28	0.31	0.87
1982	3.14	0.25	1.04	0.19	0.31	1.15	0.28	0.15	0.06	0.43	1.50	0.05	4.74	2.35	1.20	0.29	0.97
1984	3.51	0.25	1.11	0.21	0.32	1.10	0.32	0.15	0.06	0.35	1.59	0.04	4.98	2.48	1.26	0.31	0.98
1987	3.38	0.28	1.05	0.22	0.44	1.10	0.31	0.17	0.06	0.34	1.72	0.04	4.83	2.76	1.22	0.32	0.85
1990	3.37	0.30	0.93	0.19	0.48	1.16	0.34	0.11	0.06	0.33	1.91	0.03	4.86	3.03	1.04	0.28	0.58
1993	3.67	0.41	0.95	0.30	0.46	1.31	0.34	0.12	0.05	0.29	2.08	0.03	5.27	3.28	1.07	0.38	0.55
1997	3.61	0.40	0.91	0.26	0.42	1.29	0.39	0.16	0.08	0.37	2.33	0.02	5.28	3.54	1.07	0.36	0.43
2001	3.32	0.39	0.62	0.28	0.62	1.15	0.36	0.13	0.05	0.37	2.52	0.05	4.84	3.89	0.75	0.38	0.37
2005	2.95	0.28	0.75	0.32	0.88	1.41	0.42	0.14	0.15	0.43	2.77	0.05	4.79	4.35	0.88	0.52	0.43

1 Wood used for space heating is included in "Total Wood."
2 A small amount of natural gas used for air conditioning is included in "Total Natural Gas."
3 Includes refrigerators.
4 A small amount of distillate fuel oil and kerosene used for appliances is included in "Fuel Oil" under "Total."
5 Retail electricity. One kilowatthour = 3412 Btu.
6 Distillate fuel oil and kerosene.
7 Liquefied petroleum gases.

Source:, US Government Energy Information Administration / *Annual Energy Review 2008.*

Windows provide day lighting and ventilation, but if poorly constructed or improperly installed, they can be the source of heat loss and excessive moisture. High performance windows cut energy consumption, condensation, and pollution. In cold climates the sun's energy transmitted through a window can contribute heat to a building. Passive solar heating is based on this principle. Use the Internet to research the National Institute of Building Sciences Whole Building Design at *http://www.wbdg.org/resources/windows.php* and search for "solar heat gain coefficient" (SHGC). Record in your notebook the specifications and characteristics of high-performance windows. For example, record the *U*-factor range of a high-performance multi-paned window with low-emissivity coatings and insulated frames.

Figure 17-5: Floor section illustrates the thoughtful planning of IEQ.

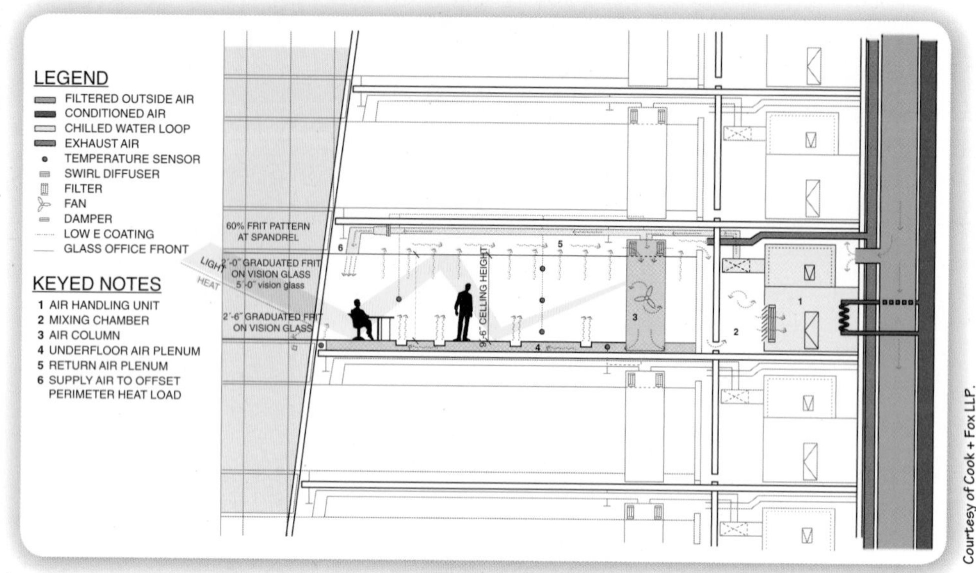

Courtesy of Cook + Fox LLP.

climate is impacted by a location's terrain, altitude, and latitude. Climates vary in temperature, humidity, wind, rainfall, atmospheric pressure, and atmospheric particles. Other factors that affect climate include relationship to the ocean, ocean currents, and persistent snow and ice cover. All these factors must be considered when planning for IEQ.

An HVAC system is matched to the building's energy consumption needs. IEQ planners often work with the architect to address energy impact of a building's size, shape, and orientation to the environmental factors. Even the building construction itself will affect energy consumption. As you read in Chapter 8: Energy Conservation and Design, architectural designers can reduce energy consumption through building envelope, windows, doors, insulation, minimization of exterior surface areas, and inclusion of green IEQ methods, such as passive solar. A building's cooling needs can be reduced with night ventilation, large overhangs on the west side, and shading from **deciduous** landscaping.

Deciduous Plants: trees and bushes that shed their leaves.

Glazing: glass coating or treatment, also used to describe the process to secure glass into the frame.

Low-e: low-emissivity coating that reduces radiant heat transfer by blocking infrared wavelengths.

The choice of window and **glazing** properties will impact the building's energy needs. For example, low-e glazing allows visible light to pass though but shades heat carrying infrared radiation. **Low-e** glazing on windows facing west will reduce heat gain and thus reduce cooling needs (see Figure 17-7).

Window properties are described by Visible Transmittance (VT), Solar Heat Gain Coefficient (SHGC), Shading Coefficient (SC), Visible and Solar Reflectance, UV Transmittance, and *U*-Value (see Figure 17-8).

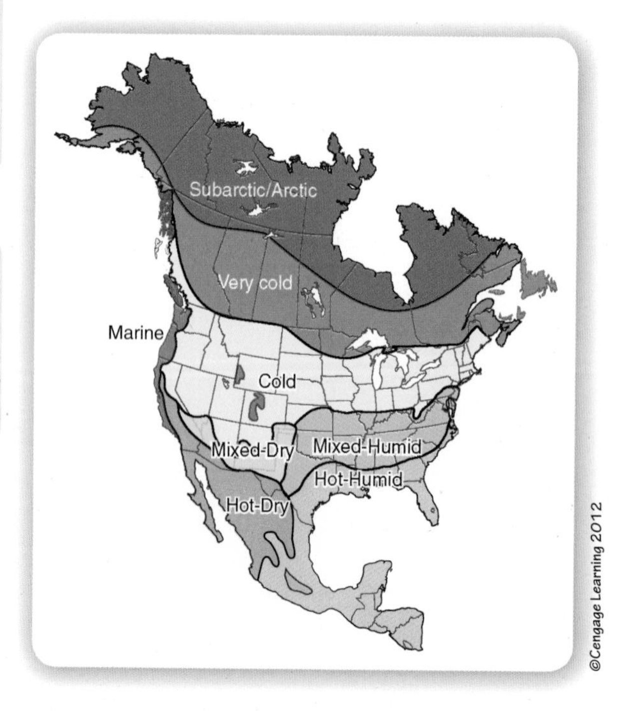

©Cengage Learning 2012

Figure 17-6: Map for indentifying cold, hot, and humid climatic zones of the continental North America.

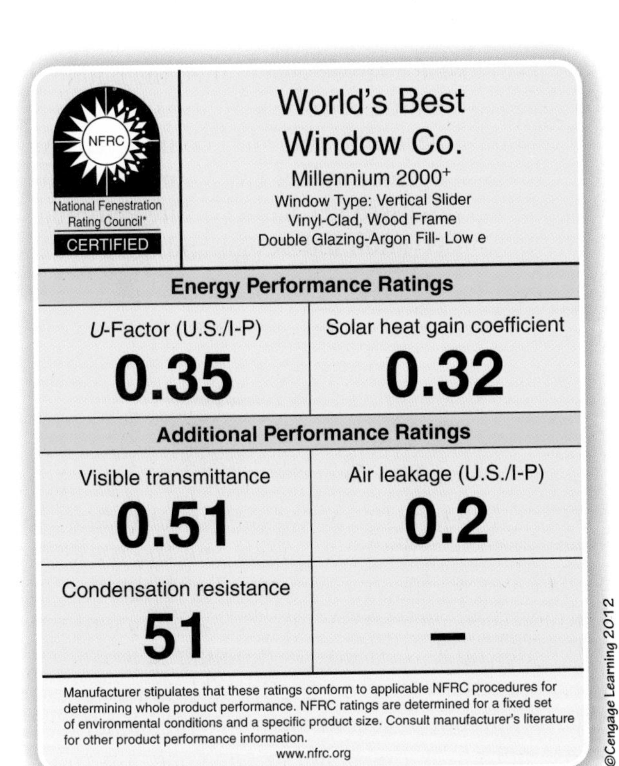

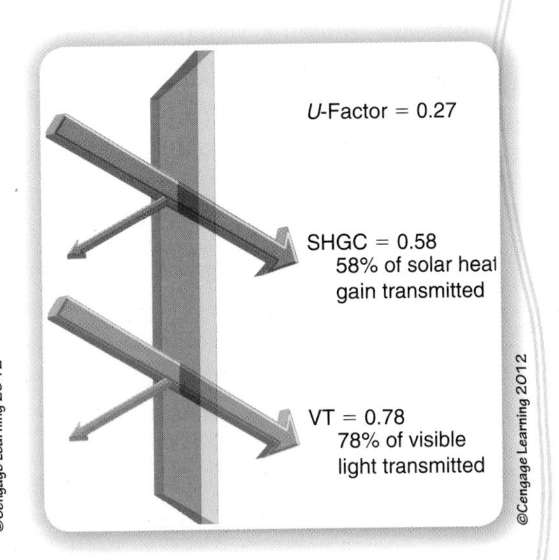

Figure 17-7a: Most windows are labeled for energy performance. b: Characteristics of a typical double-glazed argon/krypton gas filled window with a moderate solar gain low-E glass.

Figure 17-8: Glazing properties will impact a building's energy needs.

Source: Los Alamos National Laboratory Sustainable Design Guide.

Window Properties	
Visible Transmittance (VT)	Percent of the visible spectrum striking the glazing that passes though the glazing
Solar Heat Gain Coefficient (SHGC)	Ratio of total transmitted solar heat to solar energy
Shading Coefficient (SC)	Ratio of solar gain of a specific glazing compared to the solar gain of clear single and double pane glazing
Visible and Solar Reflectance	Percent visible light or solar energy reflected from the glazing
UV Transmittance	Percent transmittance of ultraviolet-wavelength solar energy
U-value	A measure of the rate of conductive heat transfer through the glazing due to temperature change between inside and outside surface

CALCULATING HEATING AND COOLING LOADS

4

After the factors that impact the energy use of a building are determined, the IEQ planner can calculate loads and determine HVAC size.

Mathematics is essential for accurately measuring change. Measurement of change provides evidence of performance and sustainability.

The windows of many new commercial buildings have spectrally selective glazing to reduce energy cost yet still provide daylight and view. Spectrally selective glazing permits only some of the solar spectrum to enter a building, while preventing the transmission of heat-generating ultraviolet and infrared radiation. To learn more about spectrally selective glazing use the Internet to search the U.S. Department of Energy for "energy efficiency" and "renewable energy" at: *http://www.eere.energy.gov/*.

Read the article Spectrally Selective Glazing published in the Federal Technology Alert, DOE/EE-0173 at *http://www1.eere.energy.gov/femp/pdfs/FTA_Glazings.pdf*.

In the past, a common mistake was to install a larger system than the building really needed. Current practice, to improve building performance and sustainability, matches the system size to the building's specific heating and cooling needs.As you read this chapter in preparation of planning your indoor environmental quality, you may need to revisit the information discussed in Chapter 6: Site Planning, Chapter 7: Site Design, Chapter 8: Energy Conservation and Design, and Chapter 12: Building Materials and Components.

Your Turn

An important factor that impacts building energy performance and indoor air quality is the relationship between the building envelope and the HVAC system. The building envelope includes all the components of the exterior wall systems, roofs, and foundations that enclose the interior environment. Use the Internet to research the National Institute of Building Sciences Whole Building Design Guide and read the report on the HVAC Integration of the Building Envelope. In your notebook, record their six recommendations to minimize energy consumption at the building envelope. Go to *http://www.wbdg.org/resources/env_hvac_integration.php?r=envelope*.

Let's try a basic technique to calculate heating and cooling demands of a small building.

We will use the formula $Q' = UA\,\Delta T$ to calculate the total heat transmission load or heat loss in British thermal units per hour (Btu/h) (see Figure 17-9).

Btu/h is commonly used to measure the heating and cooling power of a system.

We will begin by determining the ΔT which means Delta T. Delta T is the temperature difference between the desired interior temperature and the exterior temperature. For this example let's say your outside temperature is 28 degrees and you desire an interior temperature of 68 degrees. The ΔT would be 40 degrees.

Now let's calculate the A, or area, part of the formula. We will practice the formula with a 28 ft by 42 ft single-level structure with a wall height of 9 ft. You could begin by calculating the area of the floor: 28 ft × 42 ft = 1176 sq ft. You can use this same number for the ceiling area. The four walls would calculated next, first by total perimeter, and then multiplied by wall height: 28 ft + 42 ft + 28 ft + 42 ft = 140 (perimeter) × 9 ft = 1260 sq ft. Next, because the windows and doors have a different R-value than the wall, we must subtract out the window and door areas and calculate the wall separately. For this exercise let's say there are 10 windows, each 3 ft × 5 ft, or 15 sq ft. The total area of windows would be 150 sq ft. The two 3 ft × 7 ft doors are 21 sq ft each, for a total of 42 sq ft. Now we can subtract the area of the doors and windows from the wall area: 1260 sq ft − (150 sq ft + 42 sq ft) = 1068 sq ft (wall area minus the windows and doors). When you are done your area list should look like Figure 17-10.

Next will identity the thermal properties of the materials used in the building walls, floors, roof to determine the **R-value**.

R-value: resistance to heat flow.

Figure 17-9: Formula for calculating the heat loss per hour.

$Q' = UA\Delta T$	
Q'	heat loss in BTUs per hour (BTU/h)
U	reciprocal of the R-value ($U = 1/R$)
A	area of the wall, floor, or ceiling
ΔT	difference in temperature between exterior and desired interior

Every material used in the building has an *R*-value rating. You can use the Internet to research the *R*-value for specific materials or use the *R*-value material list in Figure 17-32. You will add up the *R*-values for all the materials used in the wall system. For example the wall may include siding, exterior sheeting, insulation, and gypsum or sheetrock (see Figure 17-11).

After the *R*-values for the wall systems have been determined, they are converted to **U-values** and applied to the formula. This is a fairly easy process; however, when *R*-values are converted to *U*-values, at least the first three decimal places are shown. Do not round the third digit. The *U*-value is the reciprocal of the *R*-value ($U = 1/R$). You will want to write this in your notebook. When you convert *R* 21 of the walls to a *U*-value, take 1, divided by 21. Your answer will be 0.047*U*. Now convert the other *R*-values shown in Figure 17-11 and see if you come up with the same *U*-values (see Figure 17-12).

We now have all the information we need to complete the formula: $Q' = UA\,\Delta T$ and determine the heat loss per hour (Btu/h) or Q'. Starting with the wall calculation we fill in the numbers: 1068 sq ft (area) × 0.0476 (*U*-value) × 40° (ΔT/ temperature differential) = 2033.472 Btu/h (the total heat transmission load per hour for the building under the given conditions). Calculate each system separately then add the Btu/h together to determine the total heat loss. Check you numbers against the chart shown in Figure 17-13.

Remember that heat loss calculations are only part of the IEQ planning. Interior spaces are assessed for heat gain from solar, lighting, equipment, and the number of people that occupy the space. The activity of the occupants will also impact the thermal temperature. For example, ASHRAE lists people in an office setting with moderate activity at 450 Btu/h per person. When the actual number of occupants is unknown, the number can be estimated per ASHRAE Standard 62. Estimates are based on people per a given area and type of activity. For example, an office of 1076 sq ft may have an average of seven workers. ASHRAE Standard 62 gives us the estimate that a person could produce 115 W when seated during light office work. Lighting can be estimated using 1.0–2.0 W/ft^2 or 1.3 W/ft^2 in an office setting (ASHRAE Standard 90.1).

Figure 17-10: **Area calculations of 28 ft by 42 ft single-level practice structure.**

System or component	Areas
Floors	1176 sq ft
Ceiling/Roof	1176 sq ft
Walls minus windows & doors	1068 sq ft
Windows	150 sq ft
Doors	42 sq ft

U-value:

(coefficient of heat conductivity) the measure of the flow of transmittance through a material given a difference in temperature on either side.

Figure 17-11: **Material R-values are added to determine the R-value of the wall system.**

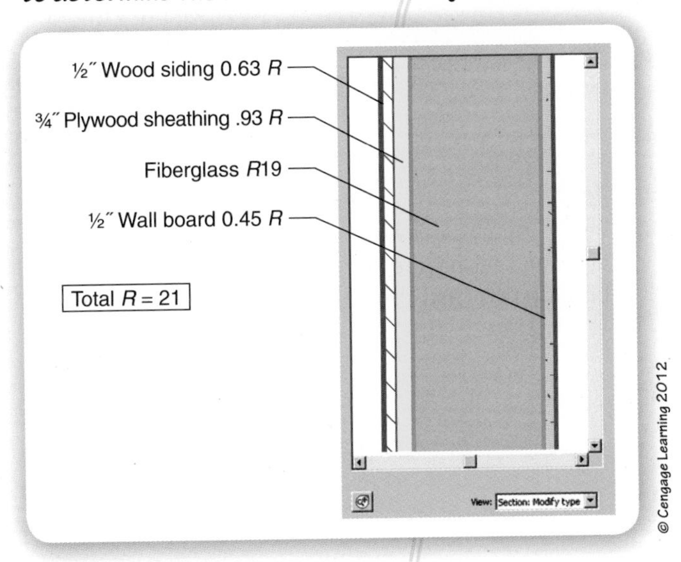

Figure 17-12: **The formula $U = 1/R$ is used to convert the R-value to U-value.**

System or component	Areas	R-value	U-value
Floors	1176 sq ft	21	0.0476
Ceiling/roof	1176 sq ft	42	0.0238
Doors	42 sq ft	3.03	0.3300
Walls minus windows & doors	1068 sq ft	21	0.0476
Windows	150 sq ft	3.23	0.3095

Watt:

(symbol: W) the International System of Units "unit" of power; 1 watt is equal to 3.4 Btu/h.

OFF-SITE EXPLORATION

One Bryant Park features its own state-of-the-art on-site 5.1-megawatt (MW) co-generation plant, providing a clean, efficient power source for a portion of the building's energy requirements. Additional features include filtered under-floor displacement air ventilation, advanced double-wall technology, extremely transparent low iron, low-e insulating glass in floor-to-ceiling windows, and planted roofs (commonly referred to as green roofs) to reduce the urban heat island effect. Taking advantage of energy from the co-generation plant, a thermal storage system at the cellar level produces ice in the evening when electricity rates are lowest to reduce peak daytime demand loads on the city. LED lights reduce electric usage while carbon dioxide monitors automatically introduce more fresh air when necessary. The 20-person project team for One Bryant Park included an energy environmental consultant, a solar design consultant, a wind consultant, and a mechanical engineer.

Zone: one or more rooms within a building that require similar air quality and thermal temperature.

Figure 17-13: Heat loss chart.

System or component	Areas	*U*-value	Δ*T*	*Q*' heat loss BTU/h
Floors	1176 sq ft	0.0476	40°	2239.104
Ceiling/roof	1176 sq ft	0.0238	40°	1119.552
Doors	42 sq ft	0.3300	40	554.4
Walls minus windows & doors	1068 sq ft	0.0476	40°	2033.472
Windows	150 sq ft	0.3095	40°	1857
				Total heat loss: 7803.528 BTU/h

Remember heat gain is not constant as people come and go, lights and equipment are turned on and off. Desired temperatures often vary from one space to another, which will require building HVAC systems to accommodate multiple temperature **zones**.

The HVAC system is sized to meet building demands and will include monitors and controls to insure the system performs properly, reliably, and efficiently. Smaller HVAC systems can be utilized in well-insulated, high-performance buildings. High-performance buildings generally harness the benefits of solar lighting or daylighting and may also include geothermal, heat recovery, or co-generation. Some large commercial buildings, such as One Bryant Park, contain their own co-generation plant. During the building design process, the architect collaborated with an energy analyst and IEQ planner to discuss the possible inclusion of passive solar, photovoltaic panels, geothermal, roof pond, roof radiator, or energy recovery device.

Your Turn

Use the Internet to research urban heat island effect. Record your findings in your notebook.

METHODS OF HEAT TRANSFER

Heating is achieved through either heating of the air or heating of the objects and occupants. Heat transfer mechanisms include conduction, convection, and radiation.

Heat moves in predictable ways, flowing from warmer objects to cooler ones, until both reach the same temperature.

Conduction describes thermal energy transferred from one solid material by direct contact with another. Convection occurs when thermal energy is carried through the movement of gas or liquid, such as the rising of warm air. Radiation is the exchange of thermal energy by electromagnetic waves between bodies or surfaces. The most common example of radiation is from the sun.

In conduction, heat flows from the warmer object to the cooler object, which explains some of the heat loss though the walls, ceilings, and floors a building. An example of convection is the circulation of natural rising of warm air and falling of cooler air, as seen in Figure 17-14. Through the convention transfer method, warm air is added to the room through supply ductwork or registers generally located under the windows. To avoid unequal pressurization of spaces, return ductwork removes the cooler air, which is then mixed with outside air and reheated for supply (see Figure 17-15).

Radiation transfers thermal energy by electromagnetic waves between bodies or surfaces. You experience heat gain when solar radiation enters through building windows. Passive solar uses this heat gain to heat a thermal mass, which then releases the heat through radiation when the temperature in the room drops below the temperature of the thermal mass.

Figure 17-14: **Heat is transferred though conduction, convection, and radiation.**

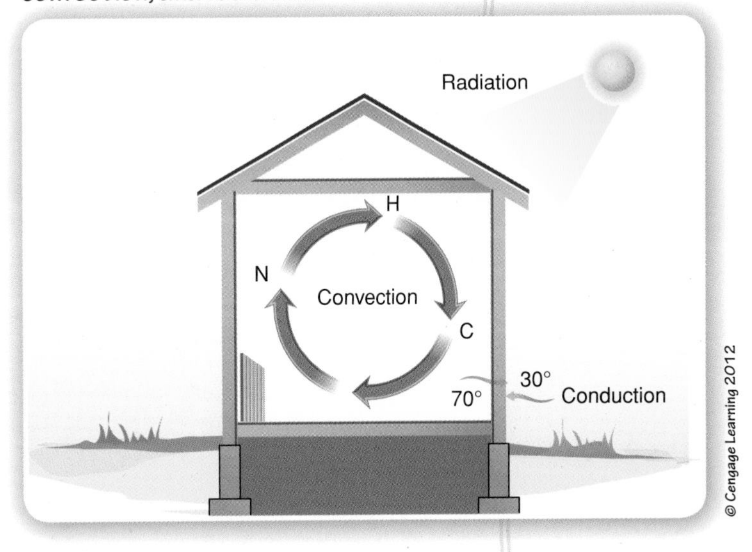

Figure 17-15: **When warm air is supplied to the space it naturally rises, then cools and falls.**

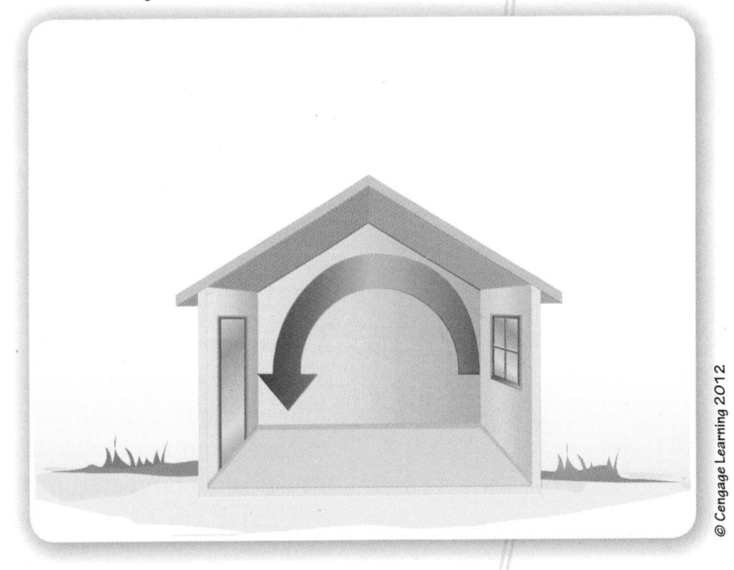

HEATING SYSTEMS

There are many choices of HVAC systems available. The HVAC system is determined during the building design to insure effective performance and space accomodation. A correctly sized, controlled, and maintained HVAC system can reduce energy consumption and provide a comfortable, healthy environment for many years to come. Traditionally, small buildings of less than 5000 sq ft are matched with furnaces, packaged heating, heat pumps, or central air conditioning. Medium-sized buildings of 5000 to 50,000 sq ft may use boilers for heating and chillers for cooling. To reduce energy consumption, many buildings employ earth-friendly systems such as geothermal and solar. If available, buildings may choose to purchase their heat from a central plant or utility. Hospital complexes or universities commonly have a central plant that produces and distributes steam to the separate buildings. Some commercial buildings have **co-generation** plants to provide their own electricity, hot water, and steam.

Co-generation: generation of electricity, hot water, and steam from the same fuel source.

Heat from Solar Radiation

In passive solar systems, heat flows by natural means. Unlike active solar systems, passive systems do not require mechanical equipment. Passive solar systems often include a **trombe wall** that stores energy during the daytime, which is

Trombe Wall: a solid mass, generally of 8 to 16 in thick masonry, for the purpose of absorbing solar heat gain.

Figure 17-16: **Passive solar heat does not require mechanical equipment.**

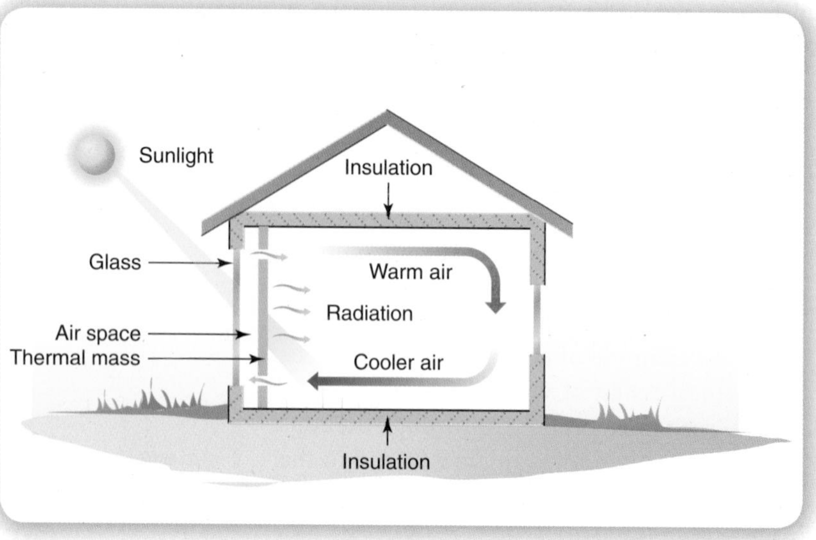

Sunlight
Insulation
Glass
Warm air
Radiation
Air space
Thermal mass
Cooler air
Insulation
© Cengage Learning 2012

then radiated into the room at night, when the room becomes cooler (see Figure 17-16). Passive solar systems reduce the demand for space heating and have no operating costs.

Some active solar heating systems use a collector to absorb and collect solar radiation and fans or pumps to circulate the heated air or fluid. Active systems often include some type of energy storage system. Active solar systems may also use photovoltaic (PV) devices, commonly called *solar cells,* that change sunlight into electricity. Photovoltaic devices reduce the amount of electricity needed from a utility provider to operate HVAC equipment, lighting, and appliances (see Figure 17-17).

Heat Pumps

A heat pump includes an evaporator, compressor, and condenser and works similar to your refrigerator or air conditioner. Electric air-source heat pumps use the difference between outdoor air temperatures and indoor air temperatures to cool and heat your home. Ground source heat pumps (GSHP) use ground temperature, instead of air temperature, to provide heating and air conditioning. The three parts of a GSHP are a closed loop pipe buried in the ground (filled with water and antifreeze), a heat pump, and a heat distribution system. Energy Star recognizes geothermal heat pumps to be among the most efficient and comfortable heating and cooling technologies currently available because they use the earth's natural temperature, Figure 17-18. One unit of electricity to operate a geothermal heat pump can produce three to four units of heat. Heat pumps should be rated at a minimum of 7.7 Heating Season Performance Factor (HSPF) (see Figure 17-18).

Figure 17-17: **Solar cells change sunlight into electricity to reduce the buildings energy consumption.**

© iStockphoto.com/acilo.

Traditional water-based HVAC systems use a chiller plant, or chiller plus boiler, to produce water, which is distributed to coil units in each room. A fan blows air over the coil unit to cool or heat the room. In air-based systems, air is heated or cooled in a central plant and transferred through ductwork to rooms. Refrigerant is used as a final cooling medium. Central air conditioning systems distribute compressed refrigerant in a closed pipe from the compressor to the room-based evaporators, where it expands causing a refrigeration effect. Air is then blown across this coil, causing cooling. Refrigerant gains the rejected room heat and is pumped back to the outdoor unit where it is condensed, giving off the heat to the outside air. The refrigerant is then compressed into liquid and redistributed back to the room based evaporator units.

Figure 17-18: Closed loop piping of a GSHP is buried in the ground either vertically or horizontally.

Boilers

A boiler burns gas or fuel oil to heat water or make steam, which is then fed through a piping system and distributed to radiant floor panels and tubes, radiators, baseboard units, or finned tube units. Once the heat is expended, the cooled fluid is returned to the boiler to be reheated. Boilers do not use a duct system for heat distribution. One example of a boiler-based distribution device uses a series of fluid-filled tubes that run in radiant floor panels or under the floor. The fluid is heated at the boiler and sent to zones radiating heat to objects and occupants of the room. In this radiant floor heat scenario, spaces are zoned, or grouped, by the desired thermal comfort level, and are thermostatically or computer controlled.

Energy Star qualified boilers have minimum annual fuel utilization efficiency (AFUE) ratings of 85 percent. Steam boilers can generally provide from 60,000 to over 100,000,000 Btu/h. Standard size water boilers range from 35,000 to 100,000 Btu/h. HVAC systems for commercial buildings in North America are rated in tons. One ton is equal to 12,000 Btu/h.

Furnaces

Furnaces are the most commonly used residential-heating system in the United States. They use natural gas, propane, fuel oil, and electricity to produce and deliver heat through a duct system to individual rooms in the house. The furnace should be sized to match the purpose. Remember, bigger is not better. The on/off cycling of an oversized system will result in an uncomfortable thermal environment. Some furnaces and applications may require power or fan-forced ventilation. In very tight homes, the furnace may require intake of outside air to mix with the return air. Utilizing existing ductwork some furnaces are fitted with central air conditioning.

Manuals have been developed to help contractors a correctly size a system. Many contractors use manuals commonly referred to as Manual J (calculating load), Manual S (sizing equipment), and Manual D (sizing ductwork). Highly efficient furnaces have a seasonal efficiency of 90 percent or better and create a low-temperature exhaust that must be power vented to the outside. Energy Star labeled oil and gas furnaces have an annual fuel utilization efficiency (AFUE) of 90 percent or better. When drawing a mechanical layout of a furnace system and ductwork, annotations and duct dimensions should be included (see Figure 17-21).

Standard 62.1-2007 **Ventilation for Acceptable Indoor Air Quality**

Standard 62.2-2007 **Ventilation and Acceptable Indoor Air Quality in Low Rise Residential Buildings**

Standard 90.1-2007 **Energy Standard for Buildings Except Low Rise Residential Buildings**

Standard 90.2-2007 **Energy Efficient Design of Low Rise Residential Buildings**

Case Study

Our case study, the Syracuse Center of Excellence, will include radiant panels. Take a look at Figure 17-19a, which shows the key for the west end of the fourth level that will be used for monitoring/control systems, and then compare the key to the diagram shown in Figure 17-19b. Trace the diagram see how the radiant pipe is run for separate zones. Why do you think the center of the space is a cooling only zone?

Figure 17-19a: Radiant Panel Zone Key for the fourth level of the Syracuse Center of Excellence.

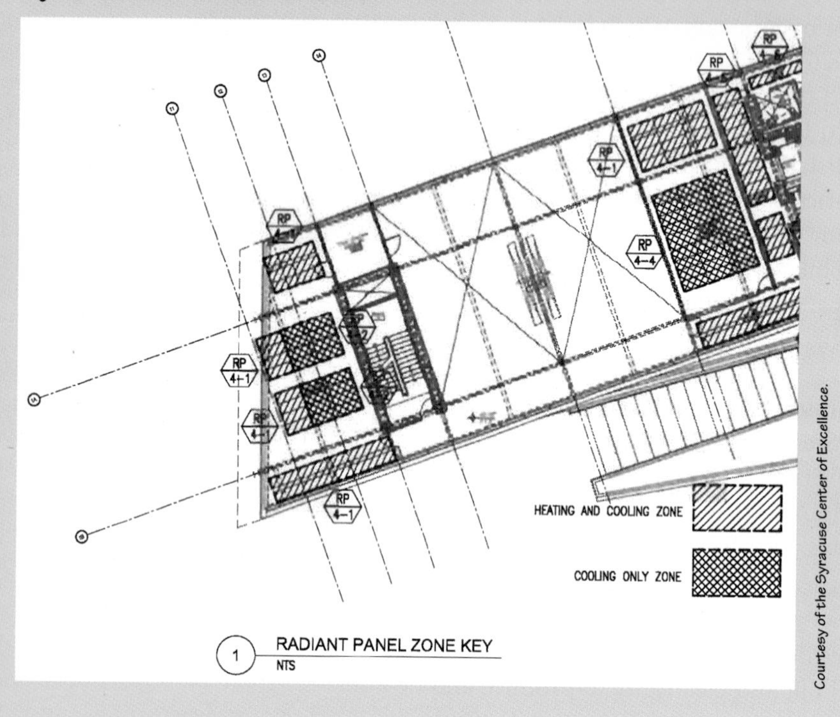

Courtesy of the Syracuse Center of Excellence.

Figure 17-19b: Radiant tubing layout for the radiant panels for the fourth level of the Syracuse Center of Excellence.

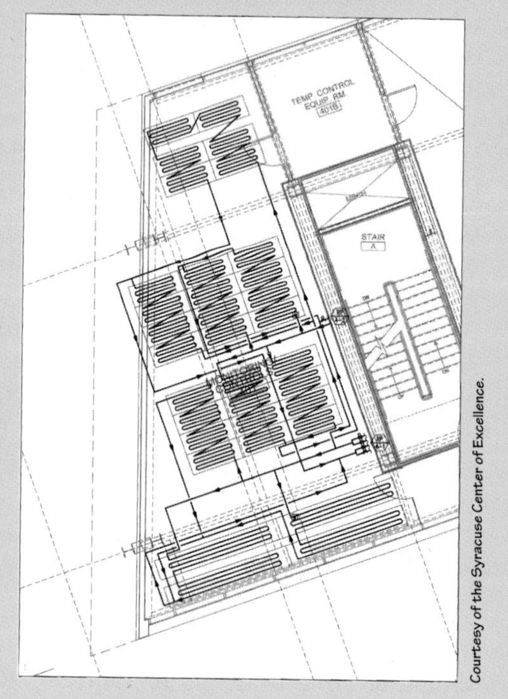

Courtesy of the Syracuse Center of Excellence.

Your Turn

The American Society of Heating, Refrigerating and Air-Conditioning Engineers (ASHRAE) is accredited by the American National Standards Institute (ANSI) and follows ANSI's requirements for due process and standards development. Use the Internet to log on to ASHRAE's website and preview popular ASHRAE standards at *http://www.ashrae.org/*. Minimum efficiencies of HVAC units are specified under ASHRAE 90.1: *http://www.ashrae.org/technology/page/548. At the bottom of the webpage you will have an opportunity to view the popular standards listed below*. HVAC systems for commercial buildings in North America are rated in tons. One ton is equal to 12,000 Btu/h. Popular ASHRAE Standards:

Standard 62.1-2010
Standard 62.2-2010
Standard 62.2-2007
Standard 90.1-2007
Standard 90.2-2007
Standard 189.1-2009

Figure 17-20: A traditional boiler system. Supply pipes are shown in red, and return pipes in blue.

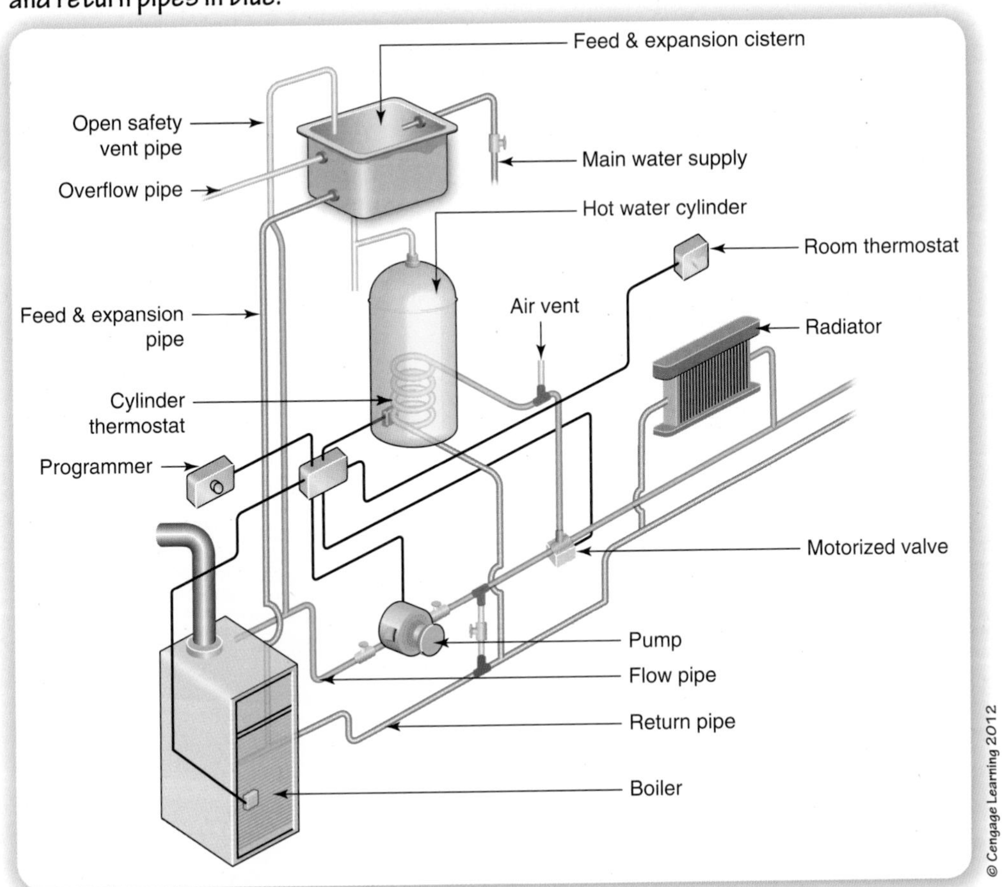

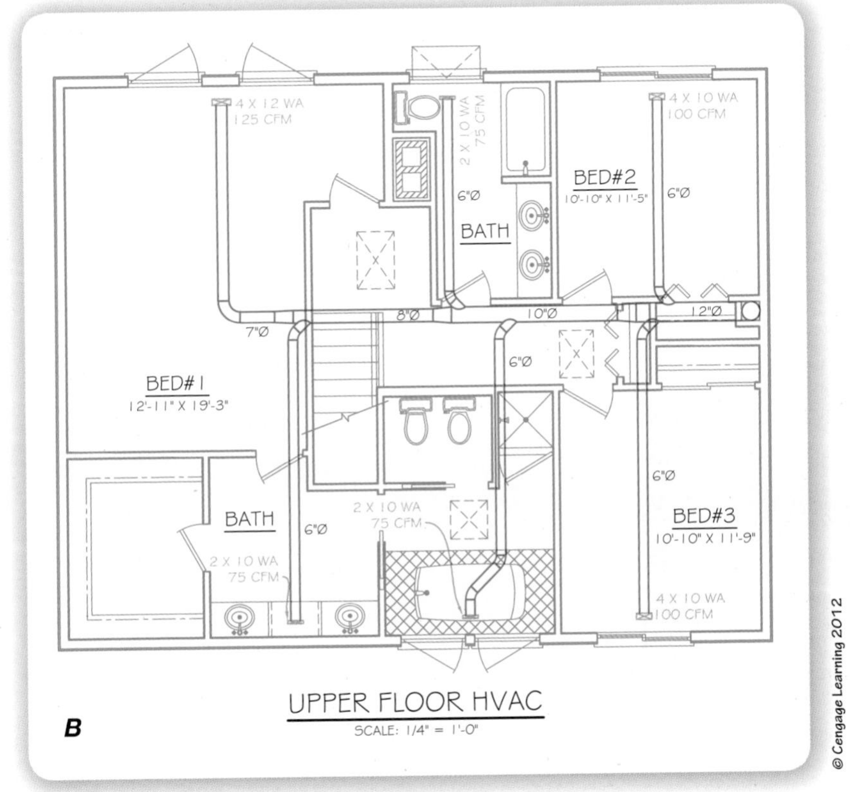

Figure 17-21: Layout of ductwork for forced air furnace.

Case Study

Our case study, the Syracuse Center of Excellence, selected a pressurized floor plenum heat system with individually controlled outlets labeled DD-A (8 in floor displacement diffusers). The outlets at the desk locations provide each worker with custom climate control, Figure 17-22. Notice the dimensions placed on the ductwork that supplies air to the floor plenum. After close examination of the plan, can you locate the supply and return duct that vertically connects to the other levels of the building? Notice how the symbols differ.

Figure 17-22: A partial mechanical plan for the third level of The Syracuse Center of Excellence, sheet M-403W.

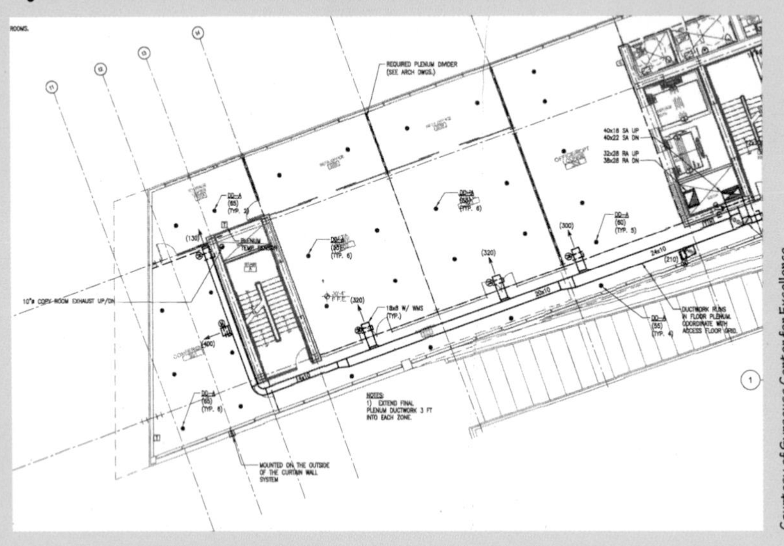

Courtesy of Syracuse Center for Excellence.

Air Handling Units (AHUs) and Roof-Top Units (RTUs)

Air handling units (AHUs) are normally associated with commercial buildings. AHUs collect and mix outdoor air with return air from the building, which is then cooled or heated and discharged into the building through a duct system. ATUs are placed in large mechanical rooms or located on the roof. You have probably seen a commercial roof-top unit (RTU) made specifically for outdoor installation. RTUs usually contain their own internal heating and cooling devices and are often chosen for single-level commercial buildings.

Figure 17-23: Air handling units include fans, heating and cooling coils, air-control dampers, filters, humidifiers, and silencers.

Air Conditioning

Air conditioning can be defined as the sensible and latent cooling of air. Sensible cooling controls the air temperature, while latent cooling controls the humidity. Cooling can be achieved with several systems including induction units such as central air, fan-coil units, and chillers. The size and type of building is a major determinant of the type of cooling equipment used. A chiller produces chilled water to cool air. The chilled water or cold air is distributed throughout a building via pipe or ducts. Chillers are generally available in various sizes from 15 to 1000 tons rated to supply 180,000 to 12,000,000 Btu/h of cooling power. Air conditioner size should be matched to the space. Air Conditioning Contractors of America (ACCA) has published a simple computerized method of determining loads ("Manual J") and sizing ductwork ("Manual D"). Over-sizing an air conditioning system in hot humid climates can increase condensation and cause mold problems. Guidelines set by the U.S. Environmental Protection Agency and the U.S. Department of Energy recommends central air conditioners to be rated at a minimum of 13 Seasonal Energy Efficiency Ratio (SEER). The higher the SEER, the greater the level of efficiency. Choosing air conditioning equipment with SEER rating of up to 20 will increase savings. High-humidity climates can be uncomfortable for building occupants. Excess humidity and moisture from condensation can also create problems with materials and equipment. Mold and corrosion are two common problems that occur. Air conditioning can be used to solve the problem, but a dehumidifier, which removes water vapor from the air, may provide a comfortable alternative in rooms without air conditioning. Dehumidifiers may also reduce the amount of air conditioning needed.

OFF-SITE EXPLORATION

If every room air conditioner sold in the United States were Energy Star qualified, it would prevent 1.3 billion pounds of greenhouse gas emissions—the equivalent emissions from 115,000 cars (*http://www.energystar.gov*). Energy Star qualified room air conditioners use at least 10 percent less energy than conventional models and may include timers for better temperature control. The Energy Star qualification indicates that the product has met strict energy efficiency guidelines set by the U.S. Environmental Protection Agency and the U.S. Department of Energy. A list of Energy Star–rated appliances can be found at *http://www.energystar.gov/index.cfm?c=products.pr_find_es_products*.

Ventilation

Acceptable air quality is defined by ASHRAE as "air in which there are no known containments at harmful concentrations and a substantial majority (80 percent or more) of the people exposed do not express dissatisfaction" (ASHRAE Standard

Figure 7-24: The air inside One Bryant Park in New York City is cleaner than the outside air.

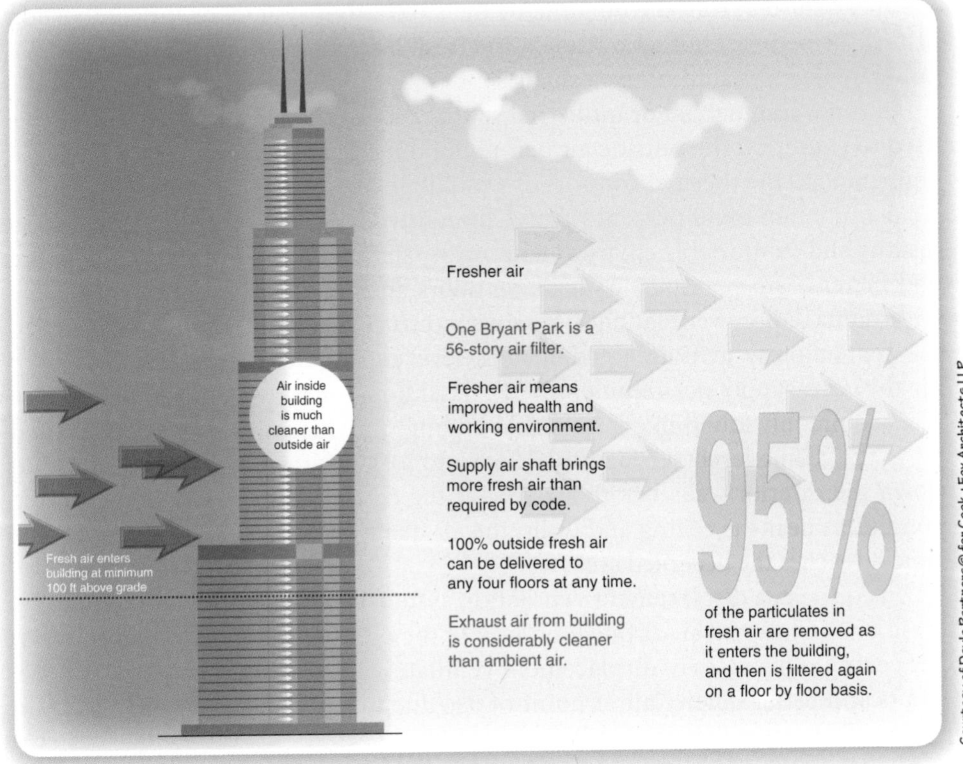

Courtesy of Doyle Partners© for Cook +Fox Architects LLP.

Figure 17-25: Exhausts can be collected together and routed through an "in-line" high-efficiency fan located in the attic.

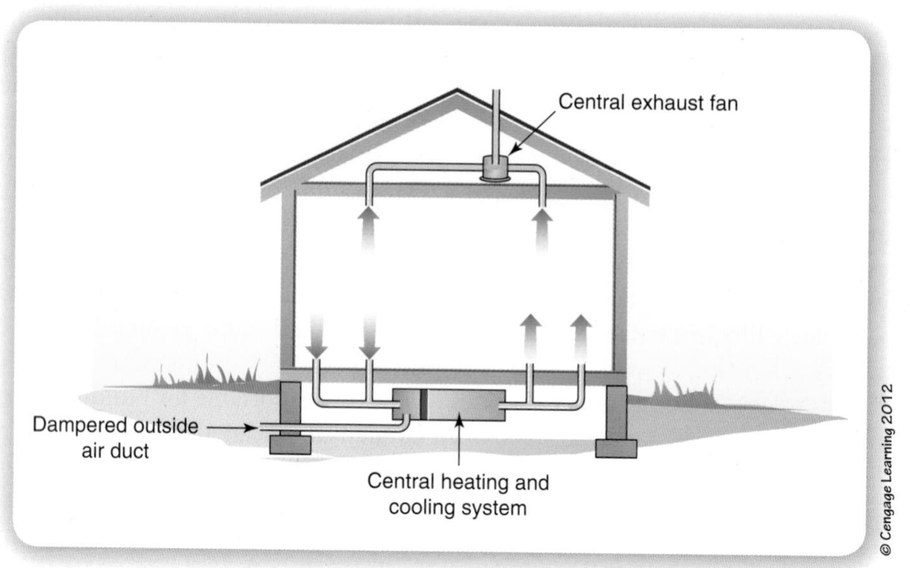

Infiltration: uncontrolled and unintended flow of outdoor air into a building.

Ventilation: purposeful air exchange with the outside environment.

62.1-2007). Changes in air quality are impacted by equipment operation, occupancy activity, material off-gas, or **infiltration** of outside air or contaminants. The intake of fresh air can be achieved through natural means like the opening of windows, or actively supplied with fan powered air distribution systems. Ventilation of fresh air is described as air changes per hour (ACH), the fraction of room air volume exchanged with outside air in a given hour. For example, an ACH of 1.0 would mean each hour the entire air volume of a space is replaced with outside air. In some states, residential building codes require whole house ventilation to provide a controlled amount of fresh filtered outside air. Fresh air is needed to replace the air exhausted from the building by the **ventilation** system, or for combustion of fuel. Remember, not all outside air is fresh air. Some outside air may contain pollutants such as automotive exhaust.

In some commercial or industrial settings, a makeup air unit (MAU) may be used to condition the outside air before building use. Fresh air and air flow also contribute to the thermal comfort of a building.

A minimum of 40 percent relative humidity is generally recommended for a healthy and comfortable environment. Adequate ventilation can reduce moisture and increase air quality by controlling odors and reducing contaminants. Most bathrooms require ventilation to remove moisture and odor (see Figure 17-25).

Ventilation systems are rated by noise level and by the amount of cubic feet of air they can supply per minute (CFM). Traditional mixing-type ventilation systems commonly found in commercial buildings provide and remove conditioned air at the ceiling level (see Figure 17-26 a and b). Air circulates across ceiling and down the exterior walls before arriving at the occupied area. Depending on the conditions near the ceiling and walls, the air quality may be compromised by the time it reaches the occupied area.

In contrast, a displacement ventilation system uses a floor plenum to provide fresh or conditioned air at the floor level of the occupied area and removes air at the ceiling. Advocates of displacement ventilation cite the advantages of reduced drafts and better quality air at point of use due to less circulation of germs and contaminates (see Figure 17-27).

Figure 17-26a and b: Traditional mixing-type air distribution system.

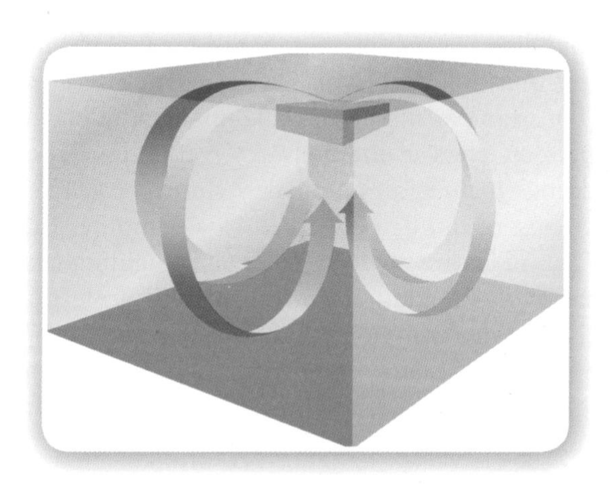

SECURITY AND PROTECTION

If you rented an apartment in a multilevel building in New York City, what types of security measures would you need in order to feel safe? Would you have a different list if you were elderly or had children? What if you were disabled? Security systems provide protection, for both occupants and their assets, against hazards, natural disasters, and criminals.

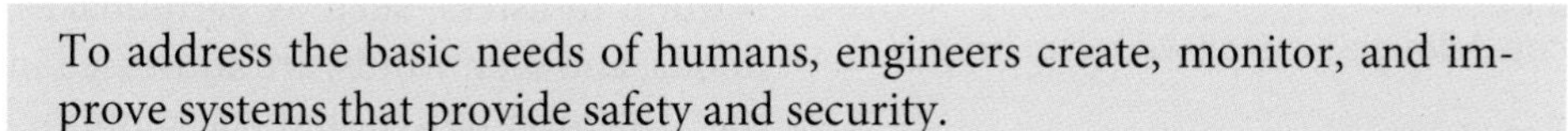

To address the basic needs of humans, engineers create, monitor, and improve systems that provide safety and security.

Security devices for sensing, notification, illumination, monitoring, and control are addressed during whole building design. When you think of security you might first think of a simple locking system for the door to your new apartment, but what about the lobby, the front entry, and elevator? You should feel secure from the moment you drive into the parking garage or parking lot.

Fire and Smoke Protection

Security and protection systems are computer integrated to alert and protect the building and its occupants from smoke and fire. Building plans include the locations of

Figure 17-27: Ventilation by displacement improves air quality at the point of use.

Figure 17-28: Sprinkler systems are specified by code.

© iStockphoto.com/Kenneth Schulze.

control rooms, smoke and fire alarms, lighted exits, fire doors, safe rooms, and fire escapes for a quick exit from the building. Fire suppression devices such as automatic sprinklers, wall-mounted hose bibs, and fire extinguishers are also included. The International Building Code (IBC), International Fire Code (IFC), and National Fire Protection Association (NFPA) Codes specify criteria, equipment, and installation methods for sprinkler systems. For example, for a 2000 sq ft area code might require specific sprinkler heads and a system able to provide a minimum of 0.45 gallons of water per minute, per square foot (see Figure 17-28).

A commercial building is generally monitored for fire and has strict rules occupants must follow to avoid creating conditions that might start a fire. For example, a building rule might prohibit the attachment of papers or flyers in the hallway or common areas. Most modern buildings include an automated control system to protect occupants. Sensors can be networked to a computer that is programmed to sound alarms, activate emergency lighting, and notify the fire department. In addition to security and protection systems, multilevel buildings commonly have a safe room for occupants who cannot navigate stairs. These safe rooms can also be used to protect occupants from natural disasters and hazards such as toxic gas or assault.

Command Center, Monitor and Control

The whole building design team will discuss the desired security of the proposed building. Designers of commercial buildings may recommend a command center with an integrated security system monitored by security staff. Depending on the building's purpose, the security system could also include card scanners, cameras, special lighting, and communications systems. Some schools and government buildings incorporate metal detectors, video surveillance, and motion detectors. Natural disasters, such as earthquakes, floods, and high winds, are addressed in the structural design, but the monitoring, notification, and response plan is normally part of the security plan. The type and amount of security is determined by the building's purpose, location, applicable codes, and occupant or owner requests (see Figure 17-29).

Figure 17-29: U.S. General Services Administration website: http://www.gsa.gov.

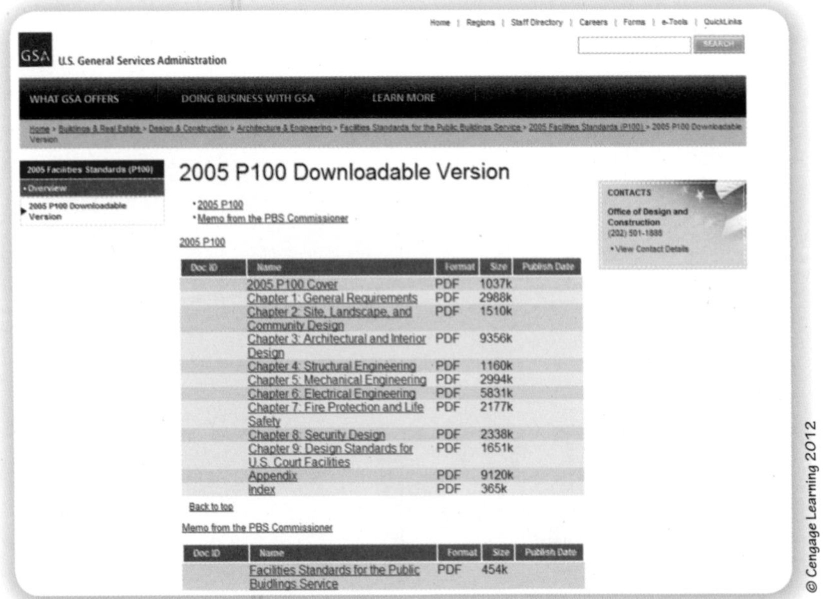

© Cengage Learning 2012

To further protect the occupants of commercial buildings, the architect and security consultant will discuss options for building devices to prevent unauthorized entry that could lead to threat, injury, fire, vandalism, explosions, or the release of chemical, biological, or radiological (CBR) agents. Intruders can be detected, deterred, and denied access by monitored gates, fences, or locked entries and alarms activated by motion, sound, or infrared sensors. Authorized access to buildings can be provided with various personal identification and communication systems. Commercial architects realize that all this equipment requires space, electricity, and networking and must be considered early in the design phase. Obviously, it is impossible to protect a building and its occupants against all harm. This is especially true for buildings with public access. For example, a hotel with a small restaurant will allow public access to the restaurant but must restrict public access to hotel rooms. Restricted access is often achieved with key card systems at elevators and side entrances. Businesses in an office building may require protection for expensive equipment and confidential computer data. Computers or storage devices containing vital or highly sensitive data may need protection from both unauthorized access and physical or electronic damage.

The U.S. General Services Administration (GSA) mission is to deliver superior workplaces, quality acquisition services, and expert business solutions supported by stewardship, best value, and innovation. Use the Internet to access the GSA website and review the chapters on fire protection and security design Fire Protection Engineering can be found at *http://www.gsa.gov/portal/category/21056*.

Security design can be found at *http://www.gsa.gov/portal/category/21057*.

Your Turn

To plan protection for your building, begin listing the various activities to be performed in the building. If possible, survey the security and protection interests of potential occupants. Now use the Internet to research security and protection codes for your area. For example, an Internet search of the keyword NYC crime identified the website for crime statistics: *http://www.nyc.gov/html/nypd/html/crime_prevention/crime_statistics.shtml*.

To deter burglary, further research locates NYC security code for use of security grills to secure commercial property when unoccupied. The information desired was located at: *http://home2.nyc.gov/html/dob/downloads/pdf/ll75of2009.pdf*.

Continue research to determine the area's history and identify events that could be prevented with better security.

Record in your notebook a list of concerns that your security and protection plan will address.

Intelligent Buildings

Intelligent buildings integrate multiple components into one control system to improve energy efficiency, safety, protection, telecommunications, and automation. Mechanical, electrical, and communication devices are controlled and monitored by a computer system. These systems may be referred to as building automation system (BAS), energy management system (EMS), energy management and control system (EMCS), central control and monitoring system (CCMS), or facilities management system (FMS). Integrated systems provide communications between a variety of systems allowing them to work cooperatively and respond to data supplied from different sensors. The result is an efficient, comfortable, safe, and desirable building environment. Let's look at one example. A sensor can detect a fire and instantly communicate with the main computer control. Almost simultaneously, the system can notify the fire department, activate sprinklers, alarms, warning lights, fire doors, and broadcast announcements, perform emergency shut-offs, and activate back-up generators. Depending on the complexity of the restricted access, the system may even be able to report the exact number and location of building occupants.

Figure 17-30: Intelligent building systems allow multiple components to interact and function as a whole.

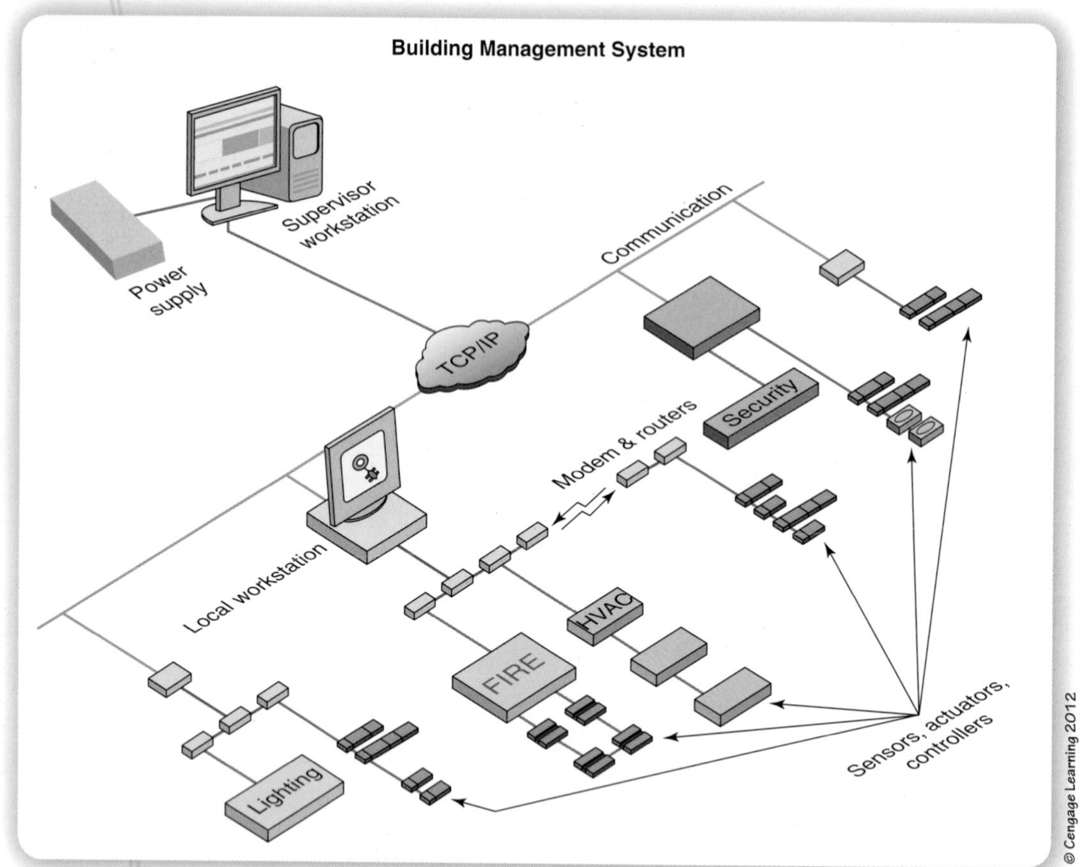

Technology advances the invention of new systems and processes to solve problems.

One of the most common applications of intelligent building systems is to monitor changes in air quality and heat gain and control the HVAC system to provide a constantly comfortable and healthy environment. Intelligent building systems can be programmed to monitor equipment emissions and automatically provide filtration and ventilation when needed. Intelligent buildings can be programmed to anticipate the number of people that enter a building at a specific time each day and adjust the HVAC to prepare an optimal environment for their arrival. During the day, the intelligent building system can monitor the number of visitors to each area of the building, determine the change in environment, and automatically adjust controls to insure a comfortable temperature and adequate supply of fresh air. Intelligent buildings are programmed with schedules that indicate HVAC and lighting changes for different times of the day and week. Cost-effective adjustments can be based on these schedules. Intelligent buildings can also include programs to detect and prevent unauthorized access to building computers or any of the integrated systems. Although there are many advantages to including an intelligent building system in the building plan, one of the most desirable is the reduction of energy (see Figure 17-30).

SUMMARY

Every building differs in climate, exterior environment, construction, occupancy, and activity. An indoor environmental quality (IEQ) analyst will gather this unique building information and consult the latest technologies and recommendations from organizations such as the EPA, ASHRAE, and LEED. ASHRAE suggests the IEQ professional consider occupant comfort, energy efficiency, ease of use, service life, first and lifecycle cost, value-added features, and indoor environmental quality when selecting a strategy and system. The IEQ professional will research factors that influence the building's energy consumption, such as exterior environment, solar orientation, air temperature, barometric pressure, wind speed, wind direction, rain, snow, landscape, ground reflectance, and temperature. These factors play a significant role in determining the thermal loads and energy consumption of a building. The building envelope also impacts how building will respond to the exterior environment. In addition to building and site considerations, the IEQ planner will determine the building's interior heating and cooling loads. To calculate loads one must consider the number of building occupants, occupant activity, lighting, and equipment. The goal is to specify a correctly sized HVAC system that will provide indoor environmental quality for buildings occupants and efficiency and sustainability for the building owner. Security must be provided to protect the building, occupants, equipment, and furnishings. Security and protection measures are determined by the building purpose, location, applicable codes, occupant needs, and owner requests. Security and protection equipment requires space, electricity, and networking and is generally considered early in the design phase. Intelligent buildings monitor and control multiple systems with one computer system programmed to improve energy efficiency, security, protection, communications, and automation. One of the most common applications of an intelligent building system is to monitor changes in occupancy and IEQ and adjust the HVAC systems to provide a constantly comfortable and healthy environment.

BRING IT HOME

1. As a class, work in small teams and select from the list of energy sources. Research and report to the class the advantages and disadvantages of the energy, including efficiency, cost, environmental impact, and sustainable factors. Discuss with the other teams the application of your energy source to HVAC of a commercial building. As a class, make a chart listing the positive and negative features of each energy source.

Electricity
Natural Gas
Propane
Coal
Fuel Oil
Solar
Wind
Steam
Water
Co-generation
Geothermal
Heat Recovery

Figure 17-31a: *R-value per climate for residential building.*

Climate Zone	Ceilings: Attic Space, Wood, Cavity	Ceilings: Attic Space, Steel, Cavity	Ceilings: Without Attic Space (Cathedral or Flat Roof), Wood, Cavity	Ceilings: Without Attic Space (Cathedral or Flat Roof), Steel, Cavity	Walls: Above-Grade Frame, Wood, Cavity	Walls: Above-Grade Frame, Wood, Cont. Ins.	Walls: Above-Grade Frame, Steel, Cavity	Walls: Above-Grade Frame, Steel, Cont. Ins.	Walls: Frame Adjacent to Unconditioned Space, Cavity	Walls: Above-Grade Mass Exterior Insulation, Continuous Insulation	Walls: Above-Grade Mass Interior Insulation, Interior Insulation	Walls: Below-Grade Exterior Insulation[a], Continuous Insulation	Walls: Below-Grade Interior Insulation[a], Interior Insulation	Walls: Unvented Crawlspace, Interior Insulation	Floors: Frame Over Exterior, Wood, Cavity	Floors: Frame Over Exterior, Steel, Cavity	Floors: Frame Over Unconditioned Space and Vented Crawlspace, Wood, Cavity	Floors: Frame Over Unconditioned Space and Vented Crawlspace, Steel, Cavity	Floors: Slab-on-Grade, Perimeter Insulation	Doors: Non-Wood	Fenestration: Vertical Glazed Assemblies	Fenestration: Vertical Glazed Assemblies	Fenestration: Skylights	Fenestration: Skylights
No.	R	R	R	R	R	R	R	R	R	R	R	R	R	R	R	R	R	R	R[b]	U	U	SHGC	U	SHGC[b]
1	30	30	13	19	13	0	15	0	0	0	0	0	0	0	15	22	13	15	NR	0.39	0.67	0.37	1.60	0.4
2	30	30	22	19	15	0	21	0	0	0	0	0	0	0	19	22	13	15	NR	0.39	0.67	0.37	1.05	0.4
3A,B	30	30	22	22	15	0	15	7.5	11	0	4	0	0	13	19	30	19	30	NR	0.39	0.47	0.40	0.90	0.4
3C	30	30	22	22	15	0	15	7.5	11	0	4	0	0	13	19	30	19	30	NR	0.39	0.47	0.40	0.90	NR
4	38	38	22	22	15	5	15	7.5	13	3	4	0	0	21	21	38	19	30	NR	0.39	0.35	NR	0.60	NR
5	43	43	26	30	21	0	21	10	13	3	4	5.4	11	30	25	38	25	38	NR	0.39	0.35	NR	0.60	NR
6	49	49	38	38	15	10	21	10	15	6	15	8.1	11	30	25	38	25	38	NR	0.39	0.35	NR	0.60	NR
7	49	49	38	38	21	10	21	10	15	6	15	10.8	11	30	38	38	30	38	NR	0.39	0.35	NR	0.60	NR
8	52	52	38	38	21	10	21	10	15	6	21	10.8	11	30	38	38	30	38	NR	0.39	0.35	NR	0.60	NR

[a] Either the below-grade exterior insulation or the below-grade interior insulation requirements must be met. Insulation must extend the full height of the wall.
[b] NR = No requirement.

From ANSI/ASHRAE Standard 90.2-07. Courtesy to come.

Figure 17-31b: *Climate map.*

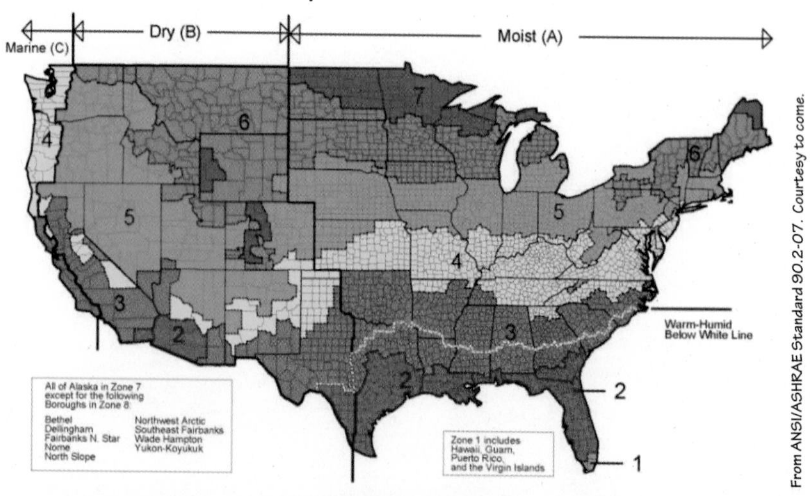

From ANSI/ASHRAE Standard 90.2-07. Courtesy to come.

2. Refer back to Figure 17-6 to determine your climate zone. Use the table in Figure 17-31a and b to determine the ASHRAE recommended *R*-value for ceilings, walls, and floors of a residential building in your climate zone. Record the findings in your notebook for future reference (see Figure 17-32).

3. In your notebook, make a list of store security measures that protect workers and customers. Think about the safety features that are present both during the day when the store is open and after hours. Add to your list safety features for disabled persons. If possible interview a store employee to identify what building features makes them feel safe. Make a plan to provide security features for a small clothing store. Make a list of suggestions and briefly explain to the class how your plan would insure occupant safety and property security.

Figure 17-32: *R*-value chart.

Wall Assembly *R*-Value	
Component	***R*-value**
Wall - Outside Air Film	0.17
Siding - Wood Bevel	0.80
Plywood Sheathing - 1/2"	0.63
3 1/2" Fiberglass Batt	11.00
1/2" Drywall	0.45
Inside Air Film	0.68
Total Wall Assembly *R*-Value	**13.73**

***R*-Value Table**		
Material	***R*/Inch**	***R*/Thickness**
Insulation Materials		
Fiberglass Batt	3.14	
Fiberglass Blown (attic)	2.20	
Fiberglass Blown (wall)	3.20	
Rock Wool Batt	3.14	
Rock Wool Blown (attic)	3.10	
Rock Wool Blown (wall)	3.03	
Cellulose Blown (attic)	3.13	
Cellulose Blown (wall)	3.70	
Vermiculite	2.13	
Autoclaved Aerated Concrete	3.90	
Urea Terpolymer Foam	4.48	
Rigid Fiberglass (> 4lb/ft^3)	4.00	
Expanded Polystyrene (beadboard)	4.00	
Extruded Polystyrene	5.00	
Polyurethane (foamed-in-place)	6.25	
Polyisocyanurate (foil-faced)	7.20	
Construction Materials		
Concrete Block 4"		0.80
Concrete Block 8"		1.11
Concrete Block 12"		1.28
Brick 4" common		0.80
Brick 4" face		0.44
Poured Concrete	0.08	
Soft Wood Lumber	1.25	
2" nominal (1 1/2")		1.88
2 × 4 (3 1/2")		4.38
2 × 6 (5 1/2")		6.88
Cedar Logs and Lumber	1.33	

(*continued*)

R-Value Table		
Material	***R*/Inch**	***R*/Thickness**
Sheathing Materials		
Plywood	1.25	
1/4"		0.31
3/8"		0.47
1/2"		0.63
5/8"		0.77
3/4"		0.94
Fiberboard	2.64	
1/2"		1.32
25/32"		2.06
Fiberglass (3/4")		3.00
(1")		4.00
(1 1/2")		6.00
Extruded Polystyrene (3/4")		3.75
(1")		5.00
(1 1/2")		7.50
Foil-faced Polyisocyanurate (3/4")		5.40
(1")		7.20
(1 1/2")		10.80
Siding Materials		
Hardboard (1/2")		0.34
Plywood (5/8")		0.77
(3/4")		0.93
Wood Bevel Lapped		0.80
Aluminum, Steel, Vinyl (hollow backed)		0.61
(w/ 1/2" Insulating board)		1.80
Brick 4"		0.44
Interior Finish Materials		
Gypsum Board (drywall 1/2")		0.45
(5/8")		0.56
Paneling (3/8")		0.47
Flooring Materials		
Plywood	1.25	
(3/4")		0.93
Particle Board (underlayment)	1.31	
(5/8")		0.82
Hardwood Flooring	0.91	
(3/4")		0.68
Tile, Linoleum		0.05
Carpet (fibrous pad)		2.08
(rubber pad)		1.23
Roofing Materials		
Asphalt Shingles		0.44
Wood Shingles		0.97
Windows		
Single Glass		0.91
w/storm		2.00
Double Insulating Glass (3/16" air space)		1.61
(1/4" air space)		1.69
(1/2" air space)		2.04
(3/4" air space)		2.38
(1/2" w/ Low-E 0.20)		3.13
(w/suspended film)		2.77
(w/2 suspended films)		3.85

(continued)

Material	R/Inch	R/Thickness
(w/ suspended film and low-e)		4.05
Triple Insulating Glass (1/4" air spaces)		2.56
(1/2" air spaces)		3.23
Addition for Tight Fitting Drapes or Shades, or Closed Blinds		0.29
Doors		
Wood Hollow Core Flush (1 3/4")		2.17
Solid Core Flush (1 3/4")		3.03
Solid Core Flush (2 1/4")		3.70
Panel Door w/ 7/16" Panels (1 3/4")		1.85
Storm Door (wood 50% glass)		1.25
(metal)		1.00
Metal Insulating (2" w/ urethane)		15.00
Air Films		
Interior Ceiling		0.61
Interior Wall		0.68
Exterior		0.17
Air Spaces		
1/2" to 4" Approximately		1.00

R-Value Table

EXTRA MILE

Some builders guarantee their buildings to be energy efficient. They inform buyers of the amount of energy the home will use and guarantee a consistent, comfortable room temperature that will not vary more than 3 degrees at the center of any room served by a thermostat. Building America, sponsored by the U.S. Department of Energy, assisted in the development of the energy guarantee programs. See *http://www1.eere.energy.gov/buildings/building_america/about.html.*

Research your local area to identify builders that guarantee their buildings to be energy efficient. Make a list of the energy features that are incorporated into their buildings. If possible, determine the added initial cost, time needed for payback, environmental benefits, lifecycle, and sustainability of the features. Use the internet to research the following programs and make a comparison of the most common energy features.

Environments for Living – *http://www.environmentsforliving.com/*

Ten sustainable buildings – *http://curiosity.discovery.com/topic/green-living/10-sustainable-buildings.htm*

The Energy Use and Comfort Guarantee – *http://www.artistichomessw.com/guarantee.html*

Green Building – *http://www.epa.gov/greenbuilding/*

Energy Star Buildings – *http://www.energystar.gov/index.cfm?c=business.bus_index*

Sustainable Whole Building Design – *http://www.wbdg.org/design/sustainable.php*

PART VII
Curb Appeal

CHAPTER 18
Landscaping

GPS
DELUXE

Menu

START LOCATION | DISTANCE | END LOCATION

Before You Begin

Think about these questions as you study the concepts in this chapter:

1. What effect does landscaping have on a project?
2. What site characteristics are important to consider when designing the landscape of a project?
3. What functions does landscaping provide?
4. How are the principles of visual design incorporated into landscape design?
5. How can you plan an environmentally friendly landscape?
6. How do you choose and design appropriate landscape elements?
7. What should be included in a final landscape plan?

INTRODUCTION

As the saying goes, you never get a second chance to make a first impression; a home, place of business, or property that produces a favorable first impression is often said to have **curb appeal**. Curb appeal is based on the visual attractiveness of the building and its surrounding landscape. Careful landscape design will increase curb appeal resulting in more personal enjoyment for homeowners and a better psychological and emotional reaction by commercial customers, renters, and employees. Curb appeal can also translate into economic gain because effective landscape design can increase the value of the project.

By incorporating principles of design, such as emphasis, repetition, balance, and unity, a designer can create a landscape that is aesthetically pleasing. However, in order to create a truly *effective* landscape, the design must also improve the property in other ways. If properly designed, landscaping can provide privacy and security, improve the energy efficiency of a building, and buffer the property from conflicting land uses. It can keep parking lots cool and guide human and vehicular circulation within and around a site. In many ways, landscaping can improve the quality of life. In some special cases, landscaping can transform a piece of land into a place of exceptional beauty that speaks to our emotions (Figure 18-1).

Figure 18-1: *Millenium Park, in the heart of Chicago, demonstrates a variety of landscape treatments and uses. Lurie Gardens, a three acre landscape built on top of a parking garage (center of the photo) was designed by Gustafson Guthrie Nicol LTD and has won numerous landscape design awards. The band shell, adjacent to Lurie gardens, was designed by Frank O. Gehry & Associates.*

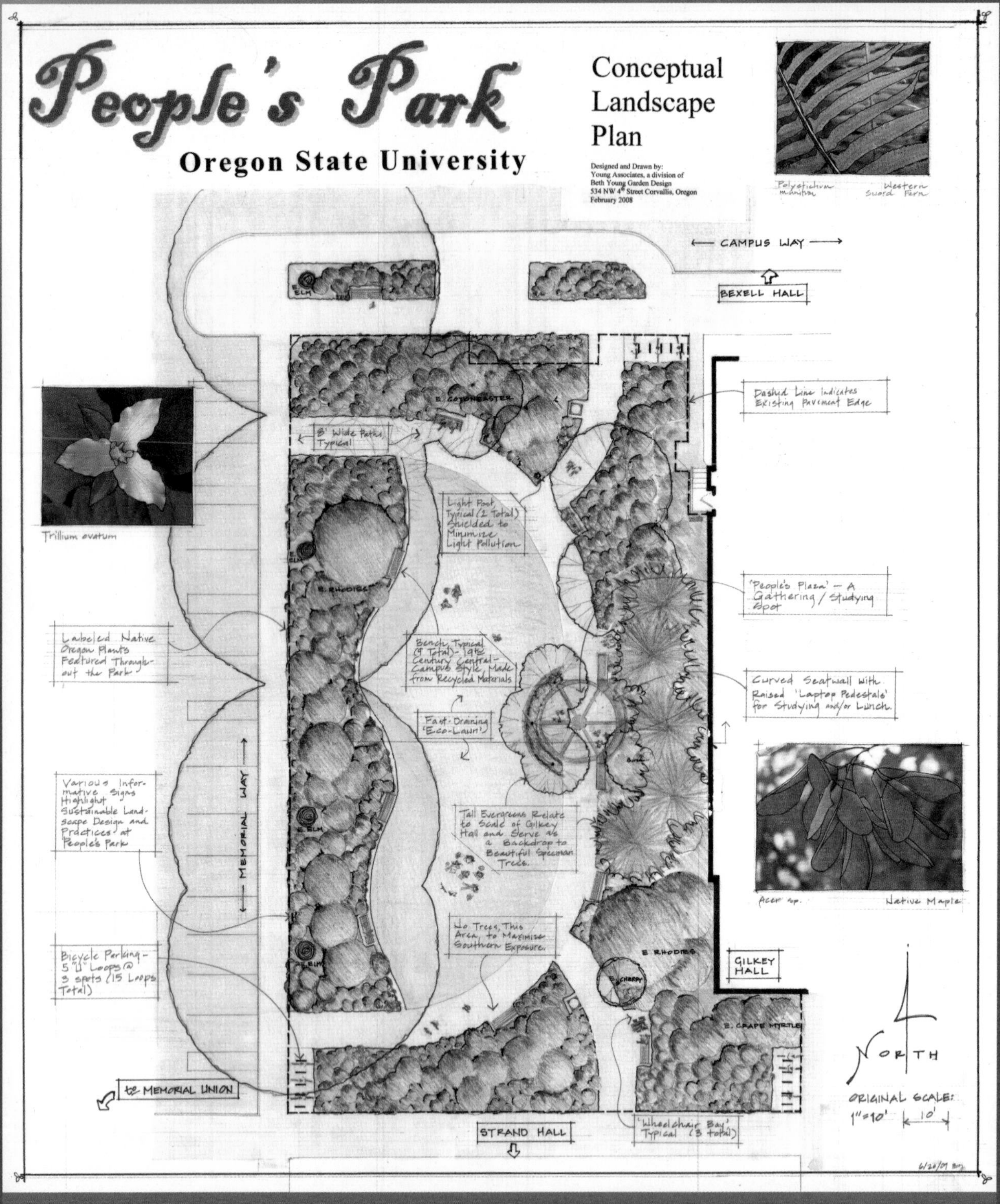

Figure 18-2: *A conceptual landscape plan for the People's Park at Oregon State University. Principles and goals for the park include creating a quiet contemplative space for use by small groups and individuals, encouraging diverse species of plants and animals (including pollinators and beneficial insects), introducing visitors to plants and materials that have a minimal lifecycle impact on the earth, and giving the park an identity and sense of place.*

Courtesy of Oregon State University.

INTRODUCTION

The landscape design process is similar to that used for site or building design. First, data collection regarding the end-users' needs, the site, and the applicable regulations is necessary in order to identify the constraints and criteria for the project (Figure 18-2). Next, a conceptual design is created to address these constraints and criteria. The conceptual plan is reviewed and revised as necessary based on input from the project stakeholders. Once a concept is accepted, the specific landscape materials (including plant and hardscape elements) are selected and a final landscape plan is created.

PROBLEM IDENTIFICATION

Before you can solve a problem, you must first know the **criteria,** or characteristics of a successful solution, and the **constraints,** or restrictions on the design. The criteria and constraints are based on the planned uses of the site, the needs of the end-users, the rules and regulations controlling the project design, and the existing site conditions. A "good" design will meet the needs of the end-users while incorporating the characteristics of the site, meeting regulations, and minimizing the cost of construction, maintenance, and operation.

Needs Analysis

Information on the needs of the end-users should be gathered during the initial needs analysis (Chapters 6 and 8) and should include information on the planned site uses and landscaping requirements. Be aware that requirements for a successful residential landscape are different than those of an effective commercial landscape. An experienced landscape designer will be able to identify many of the potential uses of the site and potential landscape elements that may work for the project, but he or she should have input from everyone who will use the property—the owner, employees, customers, and service personnel—in order to determine the project criteria. Because the landscape treatment of each will be different, consideration should be given to three distinctly different use categories:

1. **Public areas** are areas that are open to the general public. Entryways and parking areas are primarily public spaces.
2. **Private areas** restrict access to the general public and may include areas for outdoor recreation, eating, entertaining, or relaxing.
3. **Service areas** provide space for service equipment, such as HVAC equipment and utility meters, and service activities such as trash disposal, delivery, and storage.

As part of the needs analysis, be sure to get specific information on the activities for each area. For example, how many people should be accommodated in an outdoor eating area? Which recreational activities will take place? Sunbathing? Badminton? Soccer? How large is the service equipment and where is it located? What "feeling" should be projected in the public spaces? Calming? Exciting? Professional?

In addition to identifying the types of uses and activities, find out the likes and dislikes of the client and end-users. What colors and textures are desired? Are any particular plant species favored? What level of maintenance can be tolerated? What special features (such as garden benches or fountains) would appeal to the end-users? If you take these factors into consideration, you will be able to create a more effective design that better addresses the needs of the client.

Requirements involve the identification of the criteria and constraints of a design and the determination of how they affect the final design and development.

Site Analysis

Many of the site characteristics that we discussed in Chapters 5 and 6 with respect to site design are also important to the landscape design of a project. Before a landscape architect or designer can create an effective and efficient landscape plan, he or she will need to know the solar, wind, terrain, view, and noise orientations as well as climate and soil conditions of the site. Existing vegetation, structures, and utilities will also affect the landscape and should be documented. Before you begin the landscape design process, visit the site, get a feel of the property in its natural (or existing) state, and try to identify strengths and weaknesses of the site that may affect the landscape design. In addition, you will need to gather the additional information including,

1. Climate. (See Chapter 5)
2. Prevailing Wind Speed and Direction. (See Chapter 5)
3. Solar Orientation. (See Chapter 5)
4. Terrain. Document the rises and falls of the land and the drainage patterns that exist on the site.
5. Soil Analysis. Several soil characteristics may affect the selection of plant material or indicate the need to amend the soil. (See Figure 18-3)
6. Existing Vegetation. Document the types of plants that naturally thrive (or have been successfully established with previous landscaping attempts). In addition, map the existing trees on the site and indicate location, species, and size.
7. Noise, Smells, and Lights. Unpleasant noise, smells, or harsh lights may need to be screened from the property.
8. Surrounding Property Landscaping. In most cases, one goal of the project design is to integrate the project into its surroundings so that it appears consistent with the design of the area or neighborhood.

You should record the findings of your site analysis on a copy of the project site plan. The site plan should include the property lines, easements, and setbacks;

topography of the site; the location and size of structures (including septic systems) and man-made surfaces; utility line locations; and circulation patterns—all of which must be considered during the landscape design process. Figure 18-4 shows an example of a residential site plan marked up to indicate existing conditions and opportunities for the landscape design.

Figure 18-3: Soil sampling and testing provides a scientific basis for making decisions related to maintaining optimal soil conditions for crops, home gardens, fruit trees, ornamentals, and lawns.

Rules and Regulations

Landscape design, like all aspects of a building project, will be controlled by rules and regulations. Many communities have adopted a community landscape code but other regulations (including state regulations, local, or municipal ordinances and zoning, restrictive covenants, and environmental regulations) can also affect the landscape design. Requirements will differ from location to location, and it is important that you are familiar with, and comply with, these regulations for your proposed site. In most cases, your landscape design must show compliance with the regulations before you can obtain a building permit. Compliance in the construction phase is required before an occupancy permit is issued.

Some common code requirements that may affect the landscape design of your project include the following:

OFF-SITE EXPLORATION

The National Arbor Day foundation provides an online Tree ID guide at *http://www.arborday.org/trees/whattree/* that can help you identify trees on your property.

1. Stormwater Drainage. As discussed in Chapter 6, communities often restrict the stormwater runoff from a site to pre-development levels. Landscaping will have a major impact on the volume of the post-development runoff—the landscape design must take into consideration the change in the runoff characteristics of the site resulting from landscaping. Note that if stormwater ponds are used (see Chapter 6) they can (and are sometimes required by code to) be incorporated into the landscape design to provide an attractive water feature.
2. Tree Ordinances. In order to preserve mature, native trees, many municipalities have adopted **tree ordinances** to control the destruction or cutting of trees. Tree ordinances often specify the size and/or species of trees that are protected and the consequences for destroying

Figure 18-4: Site analysis map showing existing conditions and opportunities for landscape design.

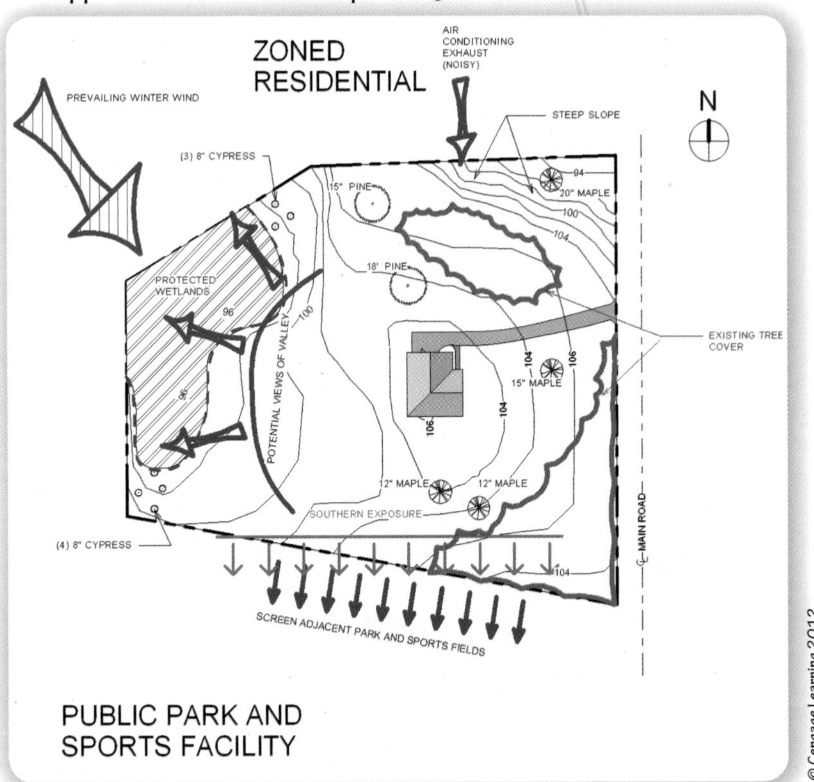

protected trees. Typically an owner must obtain a permit in order to remove a protected tree and may be required to pay fees or plant replacement trees (Figure 18-5).

3. Street Trees. Some communities require owners of properties that are adjacent to a public roadway to provide, plant, and maintain trees within a median in front of their property or within the right-of-way, the area between the property line and the street curb. See Figure 18-6.
4. Street Yard. A *street yard* is an area within the property line between the property line adjacent to a street and the building. If required, the size and location of the street yard as well as the required landscaping elements may vary from

Point of Interest

An Example of a Tree Ordinance

Tree ordinance requirements vary among municipalities across the United States. However, a tree ordinance will typcially define protected trees by the diameter at breast height (DBH) and species. DBH is often measured at 4.5 ft above grade. The following are examples of requirements that may appear in your local codes:

- A tree having a DBH of 8 in or more is a protected tree.
- Trees of all species that meet the size requirements are protected trees, except Hackberry, Cedar, and fruit trees.
- It is unlawful for any person to remove a protected tree without first obtaining a tree removal permit.
- An application for a Tree Removal Permit shall indicate the size of the property on which the tree(s) are located and must be accompanied by documentation showing the following for each tree:
  - Location
  - DBH
  - Dripline
  - Species and common name
  - Reason for requested removal
  - Tree Replacement Plan
- When protected trees are removed, tree replacement shall be required and a tree replacement fee shall be collected.
- Tree replacement shall be completed as per the tree replacement schedule provided below. The total diameter inches of replacement trees in a single category shall be calculated by multiplying the total diameter of trees in that category by the tree replacement ratio for that category.

Category Based on DBH of Existing Tree	Tree Replacement Ratio
I. 8 to 17.99 in	1.0
II. 18 or more in	3.0

Figure 18-5: *Sample tree inventory and tree protection plan for improvements to a residential property.*

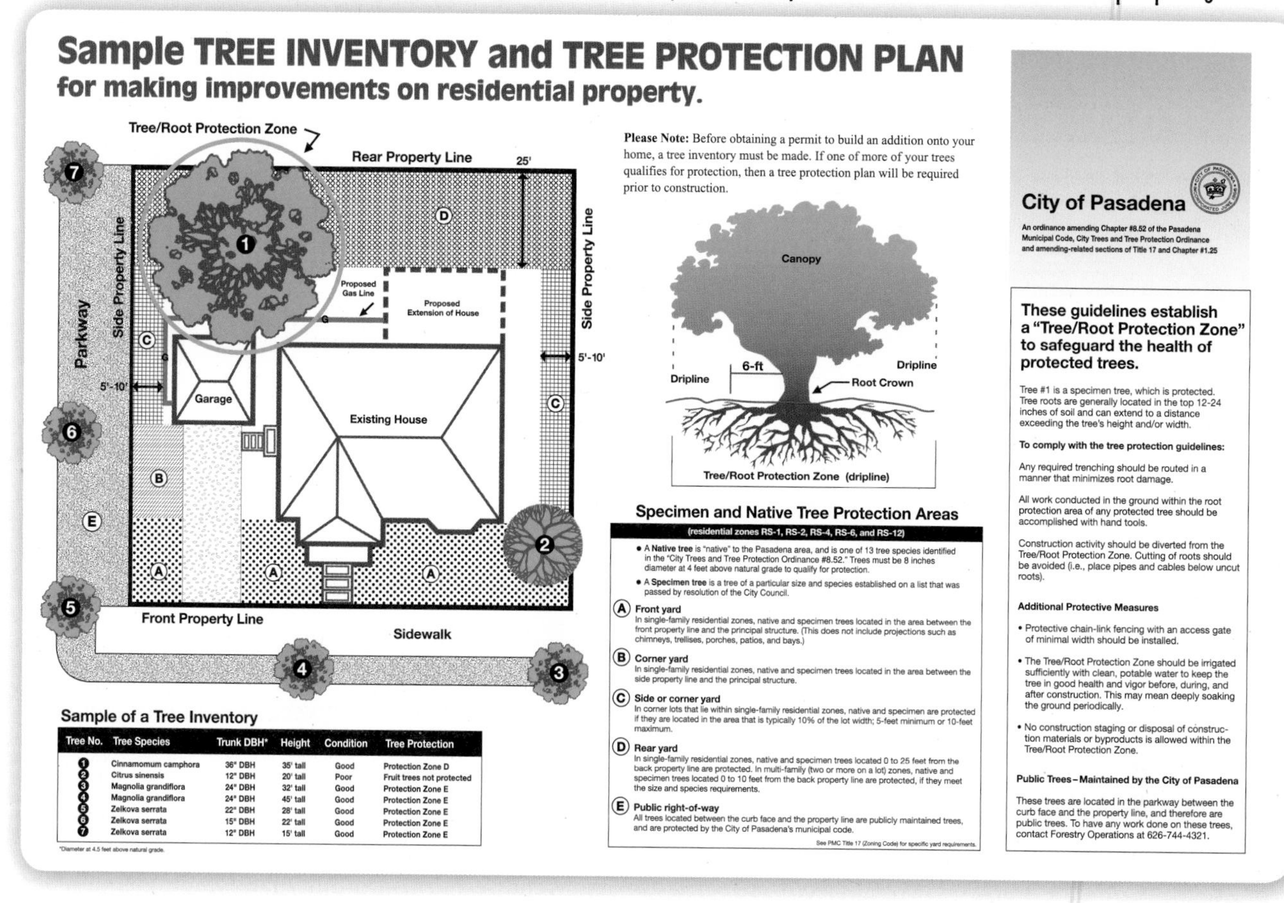

Sample TREE INVENTORY and TREE PROTECTION PLAN

for making improvements on residential property.

Please Note: Before obtaining a permit to build an addition onto your home, a tree inventory must be made. If one of more of your trees qualifies for protection, then a tree protection plan will be required prior to construction.

Sample of a Tree Inventory

Tree No.	Tree Species	Trunk DBH*	Height	Condition	Tree Protection
1	Cinnamomum camphora	36" DBH	35' tall	Good	Protection Zone D
2	Citrus sinensis	12" DBH	20' tall	Poor	Fruit trees not protected
3	Magnolia grandiflora	24" DBH	32' tall	Good	Protection Zone E
4	Magnolia grandiflora	24" DBH	45' tall	Good	Protection Zone E
5	Zelkova serrata	22" DBH	28' tall	Good	Protection Zone E
6	Zelkova serrata	15" DBH	22' tall	Good	Protection Zone E
7	Zelkova serrata	12" DBH	15' tall	Good	Protection Zone E

*Diameter at 4.5 feet above natural grade.

Specimen and Native Tree Protection Areas

(residential zones RS-1, RS-2, RS-4, RS-6, and RS-12)

- A **Native tree** is "native" to the Pasadena area, and is one of 13 tree species identified in the "City Trees and Tree Protection Ordinance #8.52." Trees must be 8 inches diameter at 4 feet above natural grade to qualify for protection.
- A **Specimen tree** is a tree of a particular size and species established on a list that was passed by resolution of the City Council.

(A) **Front yard**
In single-family residential zones, native and specimen trees located in the area between the front property line and the principal structure. (This does not include projections such as chimneys, trellises, porches, patios, and bays.)

(B) **Corner yard**
In single-family residential zones, native and specimen trees located in the area between the side property line and the principal structure.

(C) **Side or corner yard**
In corner lots that lie within single-family residential zones, native and specimen are protected if they are located in the area that is typically 10% of the lot width; 5-feet minimum or 10-feet maximum.

(D) **Rear yard**
In single-family residential zones, native and specimen trees located 0 to 25 feet from the back property line are protected. In multi-family (two or more on a lot) zones, native and specimen trees located 0 to 10 feet from the back property line are protected, if they meet the size and species requirements.

(E) **Public right-of-way**
All trees located between the curb face and the property line are publicly maintained trees, and are protected by the City of Pasadena's municipal code.

See PMC Title 17 (Zoning Code) for specific yard requirements.

City of Pasadena

An ordinance amending Chapter #8.52 of the Pasadena Municipal Code, City Trees and Tree Protection Ordinance and amending-related sections of Title 17 and Chapter #1.25

These guidelines establish a "Tree/Root Protection Zone" to safeguard the health of protected trees.

Tree #1 is a specimen tree, which is protected. Tree roots are generally located in the top 12-24 inches of soil and can extend to a distance exceeding the tree's height and/or width.

To comply with the tree protection guidelines:

Any required trenching should be routed in a manner that minimizes root damage.

All work conducted in the ground within the root protection area of any protected tree should be accomplished with hand tools.

Construction activity should be diverted from the Tree/Root Protection Zone. Cutting of roots should be avoided (i.e., place pipes and cables below uncut roots).

Additional Protective Measures

- Protective chain-link fencing with an access gate of minimal width should be installed.
- The Tree/Root Protection Zone should be irrigated sufficiently with clean, potable water to keep the tree in good health and vigor before, during, and after construction. This may mean deeply soaking the ground periodically.
- No construction staging or disposal of construction materials or byproducts is allowed within the Tree/Root Protection Zone.

Public Trees – Maintained by the City of Pasadena

These trees are located in the parkway between the curb face and the property line, and therefore are public trees. To have any work done on these trees, contact Forestry Operations at 626-744-4321.

location to location. In some communites, all existing trees must be preserved within this defined area. In other communities, specific planting requirements are imposed and often require a combination of large and small trees, shrubs, and ground cover. See Figure 18-7.

5. Buffer Yards. *Buffer yards,* or *buffer zones,* are areas, typically at side property lines, that separate different land uses or adjacent properties. Screening materials, such as fences and plants, may be required in the buffer to provide a visual barrier and sound break. Again, a combination of trees, shrubs, and groundcover may be specified by regulations, and a recommended species list provided. See Figure 18-8.
6. Street Wall. Communities that have developed community landscaping guidelines often prevent continuous paving up to the wall of a building and require a planting strip between the building and parking areas. The width of the strip and the types, sizes, and density of plantings are often specified.
7. Parking Screen and Islands. Many communites recognize the negative aesthetic and environmental effect of vast areas of paving and require

Figure 18-6: *Street tree area diagram.*

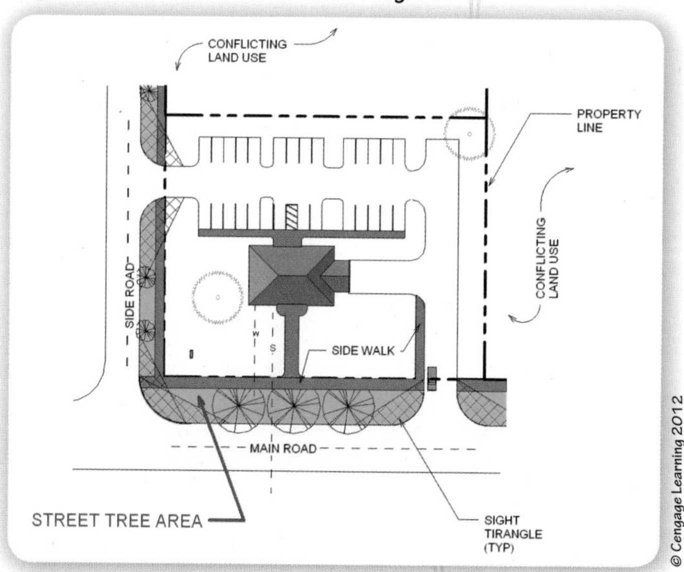

Figure 18-7: Street yard diagram.

© Cengage Learning 2012

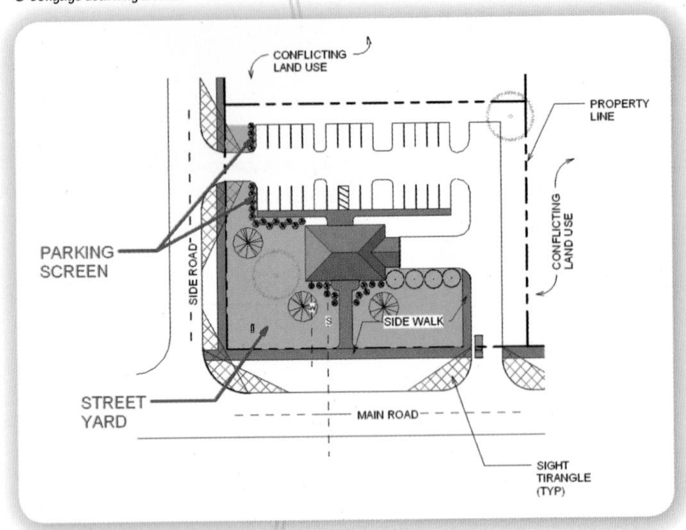

Figure 18-8: Buffer yard diagram.

© Cengage Learning 2012

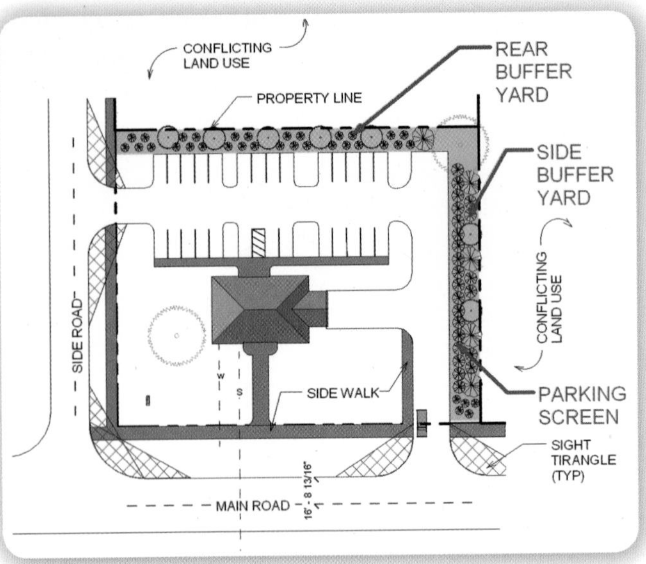

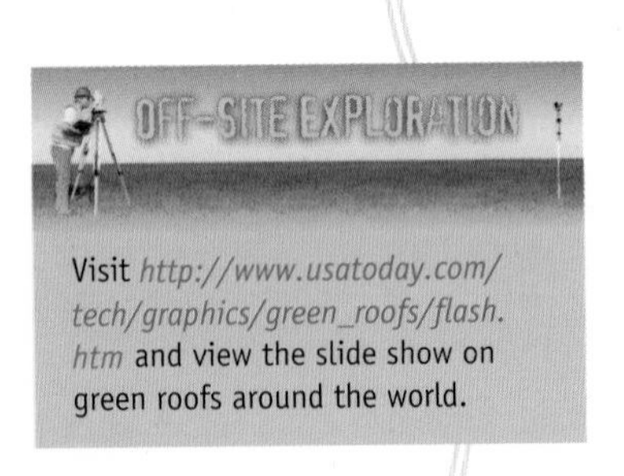

Visit *http://www.usatoday.com/tech/graphics/green_roofs/flash.htm* and view the slide show on green roofs around the world.

parking screens—which provide a visual break between parking lots and adjacent streets or properties—and parking islands, which provide interior landscaped breaks in parking lots. See Figures 18-8 and 18-9.

8. Trash Screening. Screening unsightly elements, such as trash recepticals, loading docks, and mechanical equipment areas from public view or from the view of adjacent properties is often required in community landscaping codes. See Figure 18-10.

Figure 18-9: Parking islands diagram.

© Cengage Learning 2012

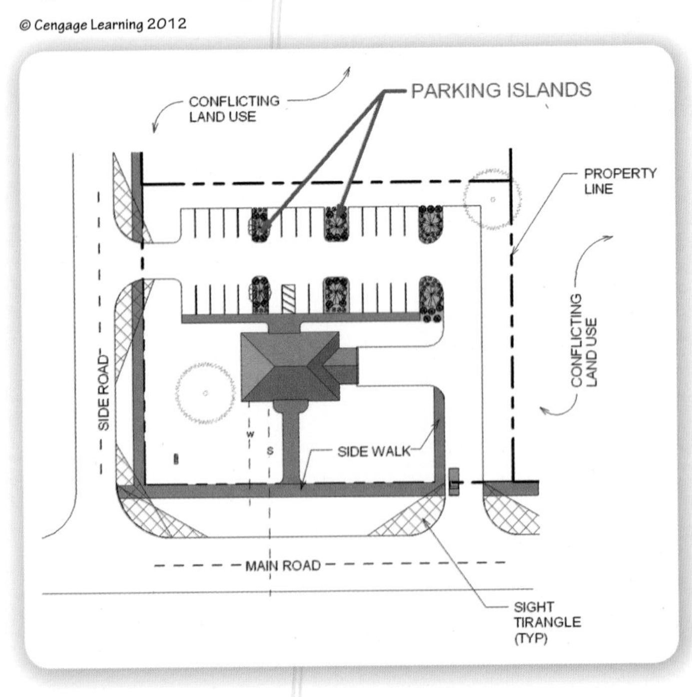

Figure 18-10: Trash screening may include fences or planted screens.

© Cengage Learning 2012

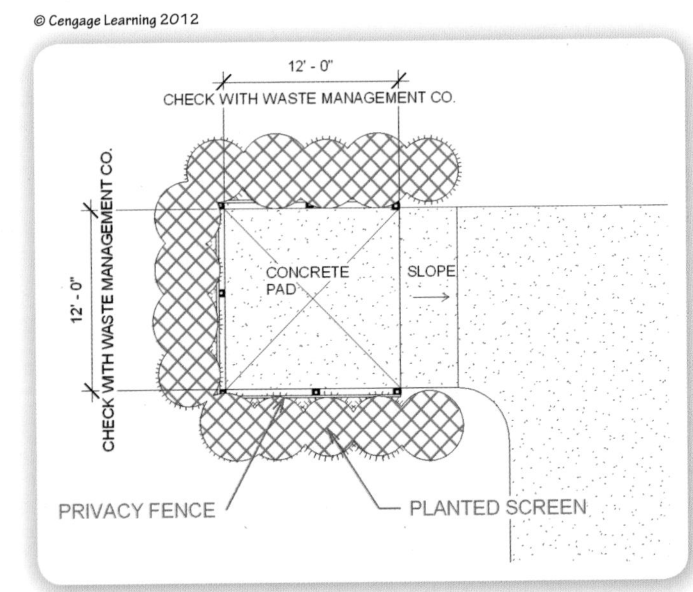

Point of Interest

Ford Motor Company Going Green

The Ford Motor Company Dearborn Truck Assembly Plant in Dearborn, Michigan, was recognized by Guinness World Records in 2004 as having the largest green roof in the world—more than 10 ac. The plant material had to be ordered a year in advance so that enough cuttings could be cultivated during a growing season. The growing medium is only one inch thick and is composed of porous stone, sand, and organic material. When saturated, the roof weighs less than 10 lb per sq ft. It is estimated that the roof will retain 50 percent of the water that falls on the roof.

Figure 18-11: **The green roof of Ford Company Dearborn Truck Assembly Plant in Dearborn, Michigan.**

Photo courtesy Green Roofs for Healthy Cities (http://www.greenroofs.org). William McDonough + Partners, ARCADIS.

Your Turn

Arbor Laws

The Municipal Code Corporation website at *http://www.municode.com/* and the American Legal Publishing Corporation website at *http://www.amlegal.com/library/* each provides access to many municipal codes. Each provides a link to a library of municipal codes. Visit one or both of these sites and search for the municipal code for your city, county, or another local municipality. Then research the ordinances to identify specific landscaping requirements for your project. Be aware that you may need to access several sections of the code to identify landscaping requirements. You may want to begin by researching requirements for the project zoning designation (for example, heavy industrial, general business, etc.). Identify which of the landscaping topics listed above are addressed by your local code and record specific requirements.

In addition to local code regulations, covenants or restrictions for the neighborhood or development in which the project is located may include landscaping requirements. Be sure to obtain a copy of any covenants that will affect your project and comply with the requirements.

Meeting societal expectations is the driving force behind the acceptance and use of products and systems.

The Function of the Landscape

Once you have completed a needs analysis, gathered information on the site conditions, and researched the applicable rules and regulations, you are ready to determine the criteria for your landscape design. The form of the landscape design will depend on the desired functions of the landscape. What specific functions should the landscaping provide? At this point, it is important to understand that landscape architecture includes not only plantings, sometimes called **softscape,** but also man-made elements like fences, decks, walkways, walls, and terraces, which are referred to as **hardscape** (Figure 18-12). Other landscape elements, like ponds, fountains, and lighting, are also used by landscape architects to enhance the design and accomplish specific functions.

Softscape: refers to the live, horticultural elements of a landscape—that is, plants.

Hardscape: refers to the inanimate elements of landscaping such as walls, walks, patios, wooden decks, arbors, and sculptures.

Common functions of landscaping include the following:

1. Aesthetic value and enhanced livability
2. Conservation and environmental protection
3. Solar heat control
4. Wind control
5. Sound control
6. Slope stabilization

Figure 18-12: **Retaining walls, walkways, stone terraces, and other hardscape features combine with a variety of plants (softscape features).**

AESTHETIC VALUE AND ENHANCED LIVABILITY The most obvious function of landscaping is to improve the appearance of a property. On a large scale, landscaping can soften the sometimes hard and massive structural forms of a building and help integrate the building with the landscape. On a smaller scale, the use of plants and landscape elements can hide unattractive areas (such as service areas or a neighbor's storage shed) or emphasize desirable features of a property or structure (like fountains or impressive vistas). When the landscaping provides an attractive outdoor space, people are more inclined to use the site for relaxing, dining, entertaining, or playing. Landscaping can also improve livability of outdoor spaces beyond improving the visual appeal. Landscape elements can provide shade and privacy, reduce noise and pollution, or separate areas of the site that are intended for different uses. In addition, with careful design, landscaping can create a pleasant psychological mood—whether calming and serene or exciting and invigorating.

CONSERVATION AND ENVIRONMENTAL PROTECTION Plants can provide habitat for wildlife and can be used to attract or discourage specific species. Butterflies, hummingbirds, song birds, and small mammals can enhance residential and commercial developments and decrease the "urban" feel of an area. Landscaping techniques that minimize the disturbance of natural habitats and use native plants can reduce the use of water and energy needed to power mechanical equipment for landscape installation and maintenance. Native plants also require less fertilizers and pesticides and generate less waste (such as tree trimmings and grass clippings). Low Impact Development (LID) techniques (discussed in Chapter 6) encourage the use of landscape elements to reduce the volume of storm water runoff and improve the quality of the water leaving the site, which will also improve the health and biodiversity of watersheds and ecosystems (Figure 18-13).

Did you know that you can have your yard certified as a wildlife habitat by the National Wildlife Federation? Check out the requirements on their website (keyword search: create a certified wildlife habitat) at *http://www.nwf.org/backyard/certify.cfm*.

If human destruction of habitats is not addressed, ecosystems will be irreversibly affected.

Figure 18-13: Conceptual landscape plan and completed Ariel Rios Building South Courtyard LID demonstration project at EPA headquarters in Washington, DC. The project incorporates several LID techniques including permeable paving, bioretention cells, a cistern, and sustainable landscaping.

Courtesy of U.S. Environmental Protection Agency, EPA http://www.epa.gov/greeningepa/stormwater/ars_perspective.htm.

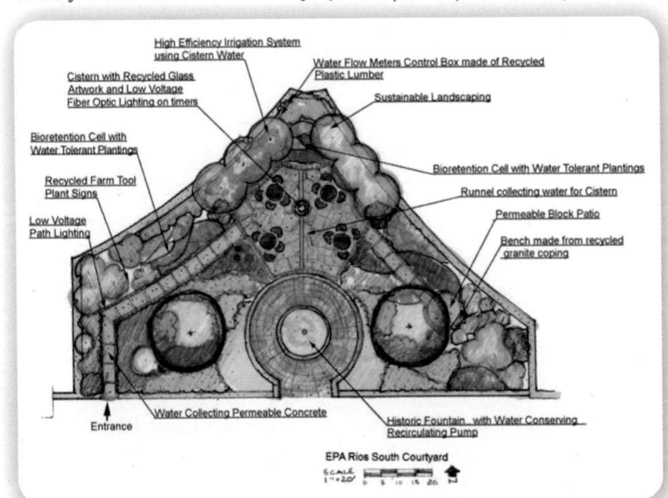

Figure 18-14: The National Wildlife Federation recognizes efforts to incorporate sustainable gardening practices and provide food, water, cover and a place for wildlife to raise their young with the Certified Wildlife Habitat program.

Photo courtesy Angela Fierro, Emily's Bed and Breakfast, Lattimer, PA.

Your Turn

The USDA Natural Resources Conservation Service (NRCS), in cooperation with other organizations and agencies has published "Backyard Conservation–Bringing Conservation from the Countryside to Your Backyard." This booklet details conservation practices that can be used by homeowners to improve the environment, help wildlife, and make an area more attractive and enjoyable. An example of a Certified Wildlife Habitat is shown in Figure 18-14. Access a copy of the publication online (keyword: backyard conservation). In your project notebook list species of plants that are attractive to hummingbirds, butterflies, and other wildlife. Then research each plant species to determine if it can be successfully used in your area. Consider using these plants in your landscape design.

Wildlife Friendly Gardens

For examples of wildlife friendly garden landscape designs, check out the American Beauties Landscape plans at *http://www.nwf.org/backyard/americanbeauties.cfm* (keyword: American Beauties garden plans).

Solar Heat Gain: the increase in temperature in an object when exposed to direct sunlight that results from solar radiation.

SOLAR HEAT CONTROL Objects, including building materials and humans, that are exposed to direct sunlight will absorb heat—we call this **solar heat gain.** When building materials, such as roofs and pavement, absorb heat, the surrounding air temperature will also increase. Of course, solar heat gain is appreciated in the winter when the added warmth makes exterior spaces more pleasant and can significantly reduce the interior heating load of a building. But in the summer, solar heat gain in hardscape surfaces may create an uncomfortably hot outdoor environment, and the heat absorbed by exterior building surfaces, such as the roof and exterior walls, will increase air conditioning loads.

Landscaping can reduce solar heat gain and therefore air temperature in two ways. First, vegetation can reduce the solar heat gain on surfaces by shading the surface. Careful landscape design can reduce solar heat gain up to 20 percent

(EPA) by shading the portions of the building or pavement that are exposed to direct sunlight. Second, plants and the surrounding soil reduce temperatures by *evapotranspiration*, a process that uses heat to evaporate water. More vegetation means more heat is dissipated by evapotranspiration.

Planting **deciduous trees** in a location that will provide shade during the summer but allow solar heat gain during the winter improves the usefulness and livability of the outdoor spaces and can decrease the cost of energy to heat or cool a building. Shading windows and air conditioning units in the summer is also effective (see Figure 18-15).

The best solar orientation is obtained when trees are planted to shade the morning and especially the afternoon sun when the sun's rays are received at a much shallower angle (see Figure 18-16). Using **deciduous** trees, which lose their leaves in the winter, will allow the sun to heat the building surfaces and decrease the energy demand in the winter.

WIND CONTROL If a building is not protected, wind can increase the heating demand for a building. Because buildings typically are not airtight, heat is lost when air leaks out of the building through cracks, poor seals, and other opening. When air blows on a building, it forces cold air into the openings on the windward side and sucks warm air out on the leeward side; we refer to this as **infiltration.** By creating a barrier to the wind, a windbreak (see Figure 18-17) can reduce the wind forces thereby reducing the amount of cold air forced into the building. In open areas well designed windbreaks can cut heating costs by 20 percent or more (Univeristy of Nebraska Extension).

Coniferous Trees: trees that have leaves that look like needles or small scales (resembling fish scales), as well as cones. Most coniferous trees are **evergreen,** meaning they do not lose their leaves during the winter. Examples of coniferous trees include cedars, pines, spruce, and hemlock trees.

Deciduous Trees: often called *hardwoods*, trees that lose their leaves during the cold or dry season. Examples of deciduous trees include oaks, maples, dogwoods, and birch trees.

Broadleaf Trees: trees that have fruits, flowers, and leaves that are thin and flat. Most broadleaf trees are **deciduous**, meaning they lose their leaves in the fall.

Figure 18-15: **In the northern hemisphere, the altitude of the sun at midday in the summer is much higher than the altitude of the sun at midday in the winter. Planting trees at a safe distance to the south often does not provide shade to a building. However, south- facing windows can be effectively shaded by appropriately designed overhangs.**

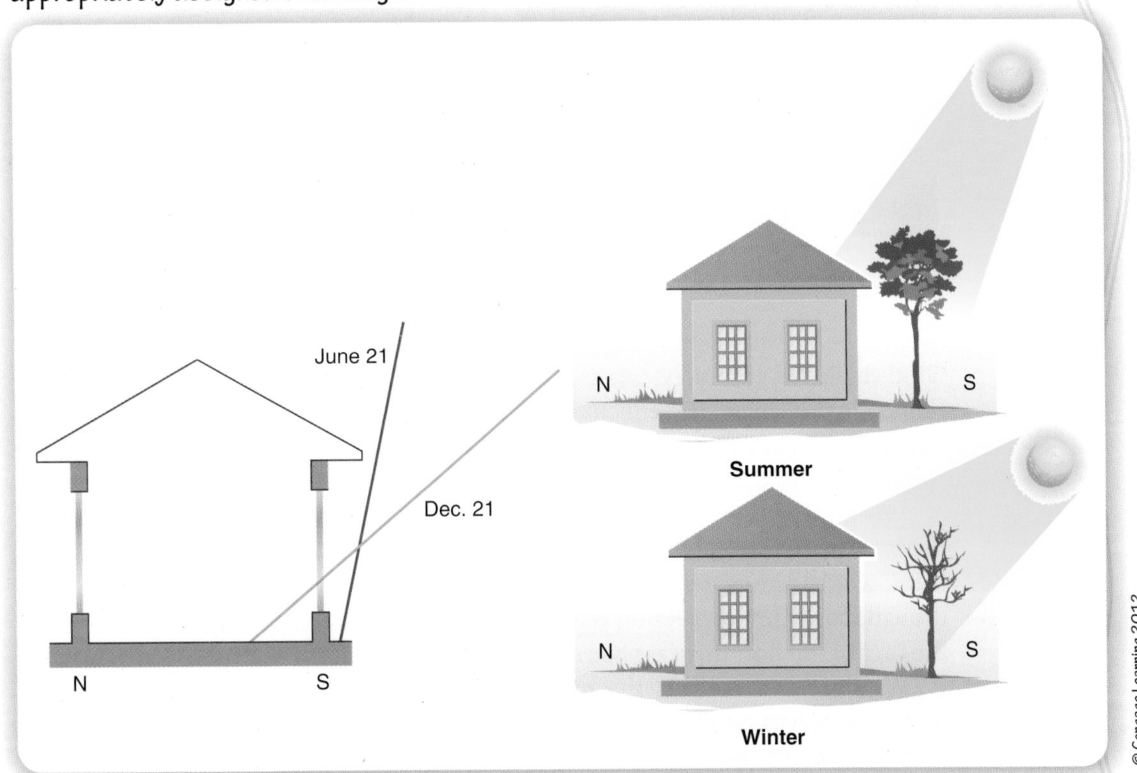

Figure 18-16: Use deciduous trees to shade the morning and especially afternoon sun when the sun's rays are received at a shallow angle.

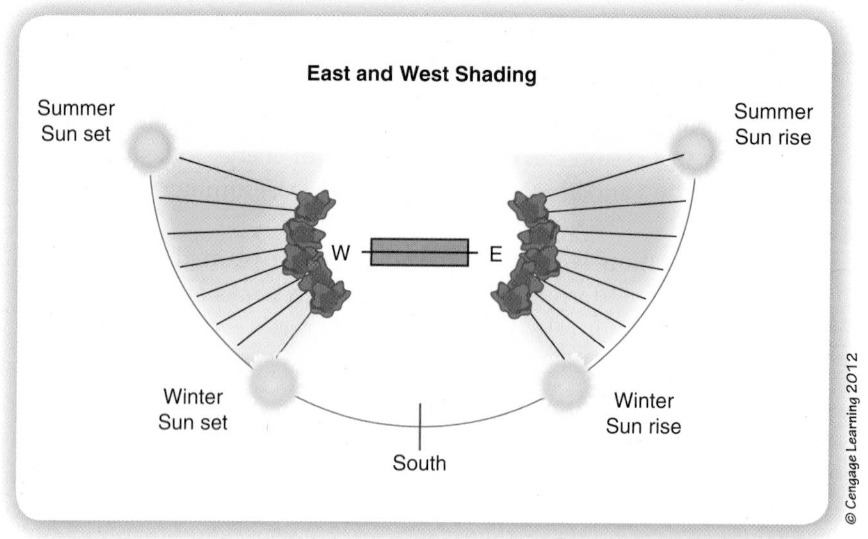

Figure 18-17: A windbreak conserves energy and improves the livability of a site.

Fences or walls can be used as a windbreak, but plants tend to provide a better wind shield because they create a screen through which some air can pass, which reduces turbulence. The effectiveness of the windbreak depends on a number of barrier characteristics: the height, length, number of rows of plants, orientation, species of plants, and the density of the plants.

Wind speed is generally decreased for a distance of up to 10 times the height of the barrier on the leeward, or downwind, side. So, for instance, a 20 ft tall windbreak would reduce wind speed for a distance of up to 200 ft (= 10 × 20 ft), although maximum protection is provided within a distance of 5 times the height of the break, 100 ft in this case. (see Figure 18-18).

Windbreak density is the ratio of the solid portion of the barrier (trunks, branches, leaves, fence pickets, etc.) to the total area of the windbreak. Hedges often use plants with dense foliage— one cannot see through the row of plants.

Figure 18-18: A windbreak provides maximum protection from the wind for a distance of five times the height of the windbreak.

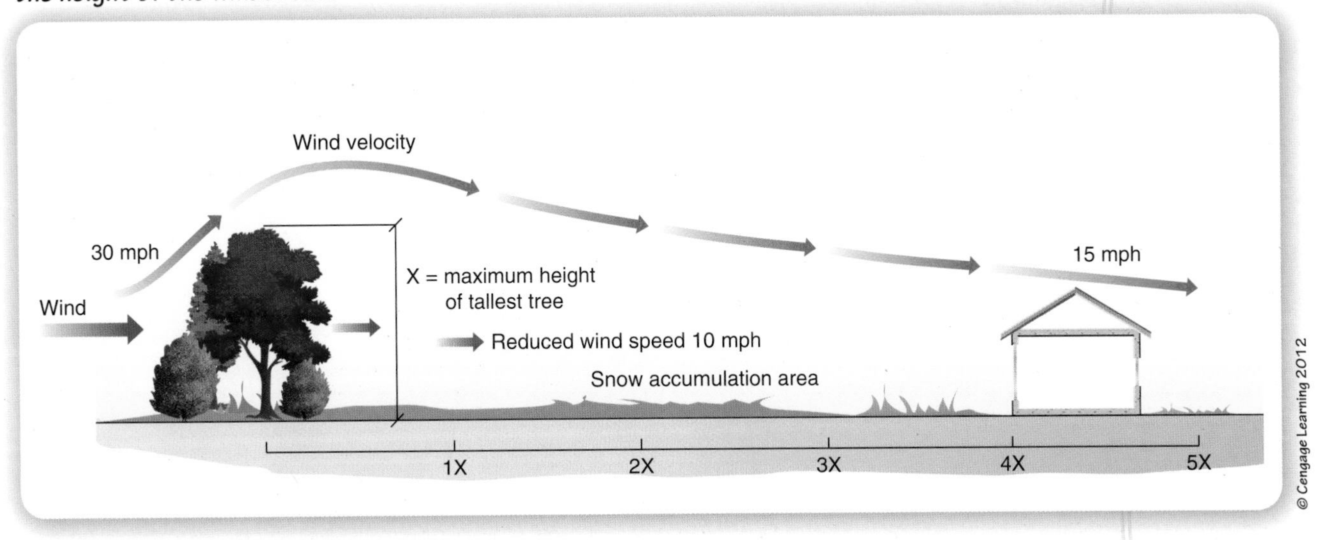

Density can be increased by planting multiple rows of plants (Figure 18-19). Using two or three rows is much more effective than using only one.

Figure 18-19: Windbreak planting plan. For residential and commercial applications where there is not enough space for five rows, fewer rows can be used. For a two-row windbreak, use rows two and three. For a three-row windbreak, use rows one through three with row one on the windward side.

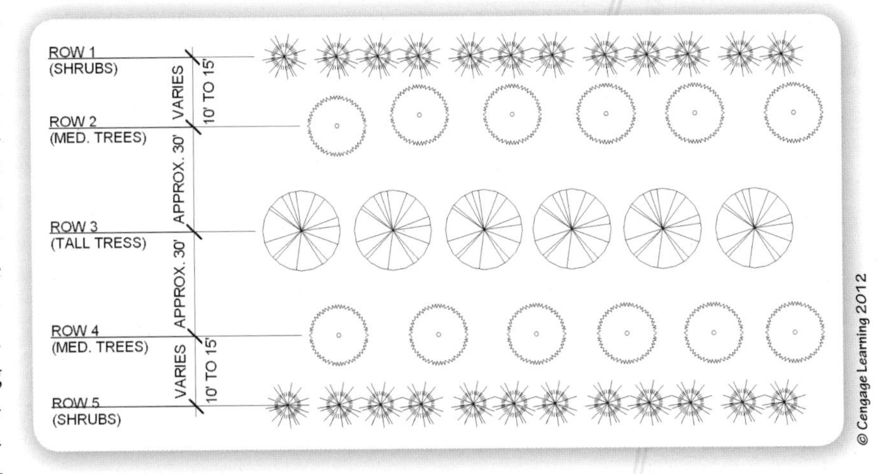

A variety of plant sizes and shapes will create the best barrier, particularly if taller trees are placed toward the center and shorter shrubs toward the outer edges of the break. It is important to break the wind at ground level; therefore, one should choose trees that have branches that reach to the ground at maturity. Coniferous trees will provide an effective windbreak year round, whereas deciduous trees are only effective in the summer.

You can probably guess that placing the windbreak perpendicular to the direction of the wind will provide the largest area of protection if you use a straight barrier. Using an L-shaped break located so that the prevailing wind blows toward the corner of the L will provide protection against winds of variable direction and a much greater area of protection (see Figure 18-20).

The form of an object, such as a windbreak, is frequently related to its use, operation, or function.

SOUND CONTROL Often in urban development, noise pollution will adversely affect a project. A nearby road, a neighbor, or a business next door may create noise that is consistent and annoying. Although landscape plants can provide a small amount of noise reduction, a large mass—such as a large mound of soil, called an earthen **berm**, or a concrete or masonry wall—is much more effective at deflecting or absorbing sound. Ideally, to provide significant sound reduction, the wall or berm

Figure 18-20: Windbreaks should be located at a distance of between two and five times the height (H) of the tallest row within the windbreak. Windbreaks are most effective when oriented perpendicular to the wind direction. L-shaped barriers provide a larger protected area and protection against variable winds.

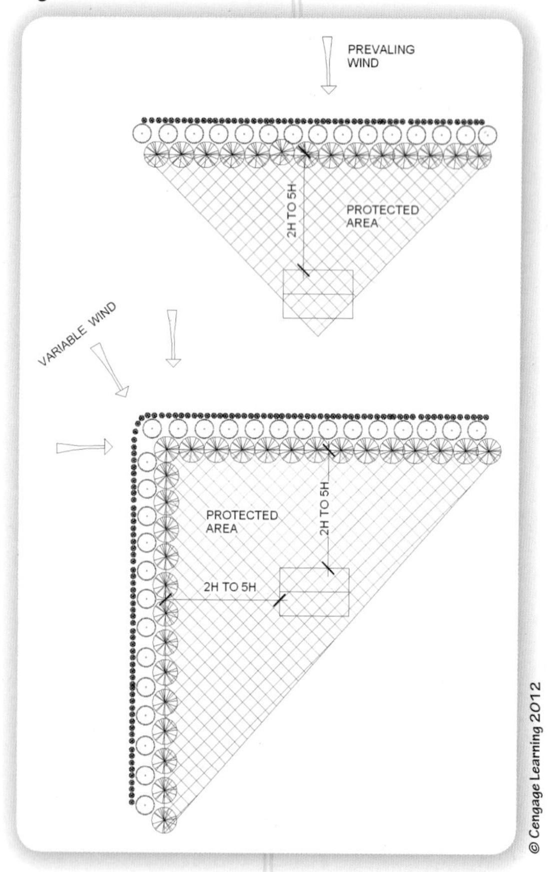

Figure 18-21: An earthen berm can provide noise reduction. When combined with plantings, the noise barrier can also provide privacy, wind reduction, and a visual screen.

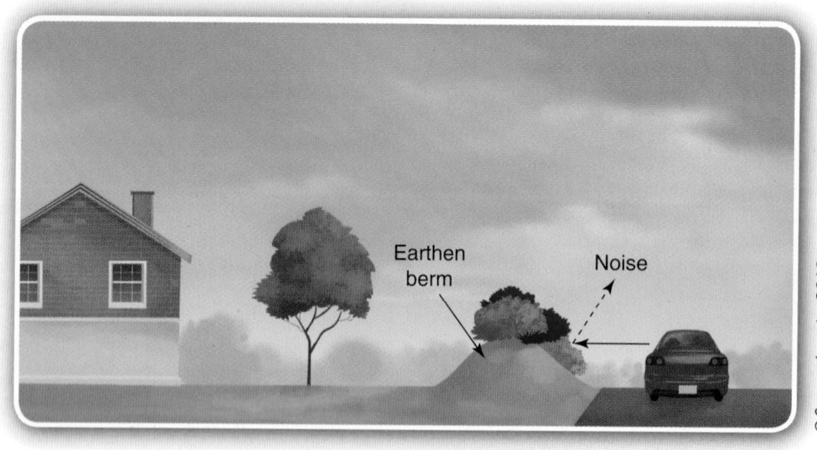

Figure 18-22: Heavy masonry or concrete walls can deflect or absorb sound.

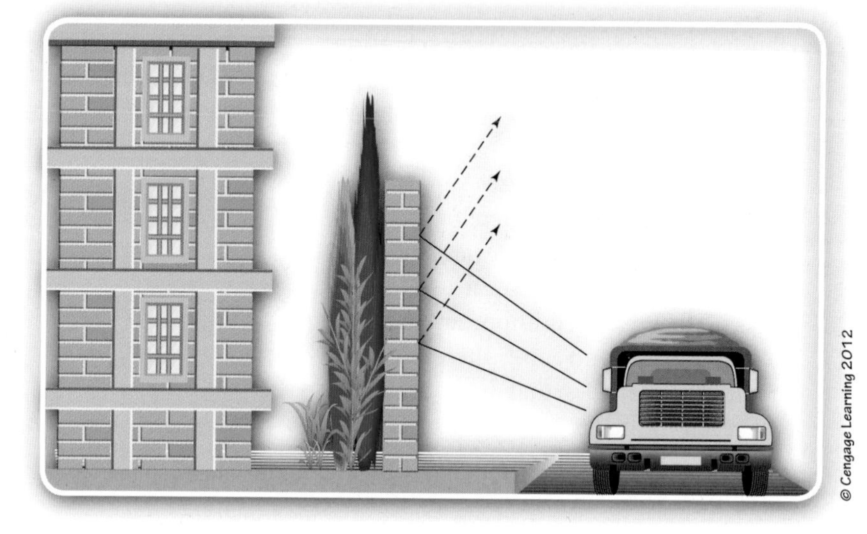

Figure 18-23: A sound barrier wall can add to the landscape. This sound attenuation wall separates the Miami International Airport from two adjacent neighborhoods and is part of the Miami-Dade Art in Public Places program. The mile-long wall was designed by Martha Schwartz Partners and is constructed from six different precast concrete panels with inset colored glass panes.

must be high enough to block the direct path from the source of the sound. Vines or plantings near the wall on the noisy side may help dampen the reflected sound (Figures 18-21,18-22,18-23).

Figure 18-24: Terraces can be used to stabilize steep slopes. These terraces are supported by stone retaining walls.

© iStockphoto.com/laughingmango.

SLOPE STABILITY Water and wind can cause soil erosion on steep slopes. Vegetation can effectively stabilize slopes with a slope gradient of 50 percent (or about 25 degrees). Ground covers and leafy deciduous and evergreen plants are particularly helpful in controlling erosion due to splashes of rain; turf and other plants with fibrous roots can reduce erosion. Mulch can also reduce erosion around the roots of the plants as well as retain moisture and reduce weeds, which will reduce maintenance on the slope. Terraces can also be used to reduce erosion by creating a series of flat areas that "step" down the slope. Hardscape elements such as timber, concrete, masonry, or stone can be used to retain the soil at each terrace face. See Figure 18-24. **Geotextiles,** which are permeable fabrics designed to stabilize soil, can also be used in conjunction with landscaping to help protect slopes from erosion.

Requirements for a design include such factors as the desired elements and features of a design and the limits that are placed on the design.

CONCEPTUAL PLANNING

Now that you are familiar with your site, know the regulations that will affect your landscape design, have determined the desired functions that you want the landscaping to perform, and have reviewed the basic principles of landscape design, you are ready to explore possible landscaping design concepts. Conceptual design of the landscape should begin with a site plan of the property that shows the building, existing trees, and proposed structures including the building, pavement, septic system, wells, and drainage structures and should include important site conditions such as prevailing wind direction, views, sources of noise or smells that need to be screened, steep slopes, and poor soil conditions. An example of a site plan that has been annotated to include site conditions is shown in Figure 18-4.

Models, such as landscape plans, help people understand how things work.

Principles of Design

No matter how much energy is saved or habitat created, a landscape must be appealing in order to be successful. Designs that are simply functional but have no aesthetic value are generally viewed as ineffective because they have not met the most obvious purpose of landscaping—to improve the visual impact of the property. It is not difficult to identify landscape designs that are not visually successful, but you may not be able to identify the specific reasons that one landscape is more attractive than another—you simply have a sense that something is either "right" or "wrong" with the design.

Landscape designers, like architects, engineers, and artists, know that following basic principles of design provide the foundation for an effective visual design that will appeal to human sensibilities. Although you may not recognize the fact, your impression that a landscape design is attractive can usually be attributed to the designer's consideration of these principles including unity, repetition, balance, and emphasis. Effective landscape design also requires consideration of the surrounding environment—a sense of place.

Established design principles are used to evaluate existing designs and to guide the design process.

UNITY **Unity** is perhaps the most important principle to consider when designing a landscape. Unity refers to the impression that the separate parts of the landscape are tied together and are part of a whole. To achieve unity, you must link the different parts of your landscape in some way. Some effective unifying techniques include the following:

- Use a consistent style. Once you have selected a style, stick with it—use it throughout the landscape to create a unified design. The landscape style should relate to the building and the site, but can also be used to express personality or create a psychological impression. Styles can be derived from many different themes—a culture, a natural ecosystem, a geographic region. There are a multitude of styles to choose from: English, Oriental, Mexican, formal, woodland, and so on See Figure 18-25.

Figure 18-25 a and b: Examples of landscaping styles. (a) English cottage style. (b) Oriental style. (c) Geometric formal style.

- Create visual pathways. Visually linking separate areas of the design with a ribbon of walkway or mulched beds can help unify a design if you are consistent with the type of material used.
- Use a consistent color scheme. Choose plants with colors from the same palette—for instance reds, pinks, and whites—scattered throughout the landscape. Yellow and orange enliven the design and create a cheerful feeling, whereas blues and purples can convey a relaxed mood (Figure 18-26). But mixing too many colors or including disjointed areas of distinctly different colors creates confusion. Remember that green is a color too and comes in a great variety of shades. Too many different greens can be distracting. And don't forget the hardscape elements. The colors of the walkways, decks, walls, and mulch should be harmonious with the plantings and be used consistently.
- Repeat basic lines, plants, and hardscape elements. For example, use the same stone walls, boxwood hedge, or curved planting beds consistently and repeatedly. We will talk more about repetition in the following section.

Figure 18-26: A consistent use of color will create unity in the landscape. Blues and purples in the landscape provide a sense of calm.

Your Turn
It's All about Style

Using library resources and the Internet, identify at least five different landscape styles and list characteristics of each style. Then choose two styles that may be appropriate for your project. Collect at least three images that provide examples of each of the two selected styles and insert them into your project notebook; be sure to document the image source. Under each image state the landscape style, list the characteristics that help identify the style, and discuss the use of the principles of design in the landscape design shown.

REPETITION **Repetition** refers to the use of the same elements throughout the landscape design. Like the refrain of a song where words and melody are repeated at intervals throughout the music, repeating elements within a landscape creates continuity and unity. Repetition in landscape is achieved when the same line, shape, color, texture, plant, or material is used over and over. For example, using only curved lines for walkways and planting beds throughout the site or using the same groupings and species of plants scattered across the landscape will create a strong cohesive design. Of course, if you repeat the same chorus of a song without the interceding verses, the music becomes boring; the same is true for landscaping. Some variety is

Figure 18-27: This landscape plan demonstrates repetition of many design elements: the square shape, the hardscape material (wood, river rock), plant species, and colors (the muted red color of several plant species and the grey of the retaining walls) in the landscape. Through repetition, the landscape is unified.

Courtesy of Acres Wild – Landscape and Garden Design, Horsham, West Sussex, UK.

appealing—a little contrast will introduce interest—but including too many different elements is chaotic and disjointed. Figure 18-27 shows a landscape that successfully uses repetition.

Figure 18-28: This view of the White House shows formal balance. The landscape can be divided into two identical halves.

© iStockphoto.com/narvikk.

BALANCE **Balance** refers to a sense of equality in a landscape design and can be either symmetric or asymmetric. **Formal balance,** like twins sitting on opposite ends of a teeter-totter, indicates a symmetric design that can be divided into two identical halves. In landscaping, formal balance is achieved through the use of the same species of plants in the same geometric arrangement on both sides of an imaginary axis. Gardens with formal balance are often referred to as *formal gardens*. Figure 18-28 shows an example of a formal garden. Notice that the two halves of the design contain the same number, size, and species of plants.

Informal, or *asymmetrical,* balance exists when the landscape cannot be divided into two identical halves but the visual weight of the overall design provides a sense of equilibrium (Figure 18-29). Consider the teeter-totter analogy again. If you were to sit on the end of a teeter-totter opposite a small child, the teeter-totter would not balance unless you were closer to the pivot point than the child. Because you are larger and weigh more than the child, you must be closer to the center

Figure 18-29: **Informal balance is achieved when the landscaping elements are not symmetrically placed but achieve a sense of equilibrium. In this example, the small flowering tree on the right is balanced by the cluster of flowering shrubs and small coniferous plant on the left.**

of the teeter-totter than the child in order to create equilibrium. Different species, sizes, and numbers of plants can be arranged in a similar fashion to create the appearance of balance—a large tree can be balanced with several smaller plants, a large flowing shrub can be balanced with a bed of flowers of a similar color.

Geometric and spatial relationships and symmetry can be used to analyze and solve problems.

EMPHASIS Emphasis is the result of focusing attention on one aspect of a design. In landscaping, emphasis is created by a **focal point**, an element that draws visual attention. A focal point can be any important visual element. In a residential landscape, the focal point might be a fountain, a striking plant, or a view. In a commercial design, it may simply be the front entrance of the building. A small garden should have one focal point; larger areas may have several. Too many focal points can be confusing, so try to limit the number of focal elements that are visible from any given spot to three.

The key to creating an effective design that emphasizes the focal point is creating a visual path to the object from a vantage point. A **vantage point** is a place from which the object is viewed. The vantage point in residential design may be a patio, a garden bench, or a window. In commercial design, the vantage point may be the roadway or parking lot. Every focal point needs a direct line of sight from at least one vantage point. The visual path is then created by framing the focal point with plantings and hardscape elements. Be sure to avoid plants that will grow to block the visual path. Figure 18-30 shows a residential landscape plan for a design that includes several focal points.

Figure 18-30: Emphasis is placed on several landscape elements in this design. The water feature is prominent when entering the yard via the walkway around the home, from the terrace area, and from inside the house. The three potted plants and the unidentified focal point in the rear of the yard are also afforded visual paths from the terrace and inside the house.

Courtesy of Acres Wild – Landscape and Garden Design, Horsham, West Sussex, UK.

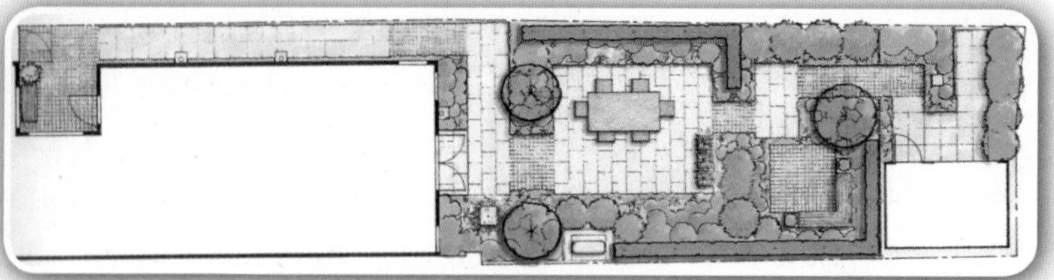

GENIUS LOCI: THE SPIRIT OF PLACE In Roman mythology, **Genius loci** referred to the protective spirit, or god, of a place. Today it is an important principle in landscape design that refers to the subtle and unique quality of a place. Genius loci, also called the "spirit of place," involves not only physical characteristics of a location, but also the internal perceptions that a person experiences while there. Landscape architects try to capture this subtle aspect of a place by being sensitive to and harmonious with the existing natural environment. Of course, this principle, because it is more of

Point of Interest

IMAGINE

Strawberry Fields, designed by landscape architect Bruce Kelly (1948–1993), is a naturalistic 2.5 ac tear-dropped-shaped section of Central Park. It is named after the song "Strawberry Fields Forever" in memory of John Lennon (1940–1980), the musician and member of the internationally famous Beatles. While living near Central Park, Lennon and his wife, artist and performer Yoko Ono, chose this small landscape as their favorite section of the park. The park was named in 1981 after Lennon's death. Ono later donated $1 million to the Central Park Conservancy to re-landscape and maintain Strawberry Fields. It was dedicated on October 9, 1985, what would have been Lennon's forty-fifth birthday.

The park entrance lies across the street from the Dakota Apartments, where Lennon and Ono lived. The focal point in the park is a circular mosaic set in the main pathway, a gift from the city of Naples, Italy. The mosaic displays a single word, "IMAGINE," the title of a famous Lennon song. A nearby bronze plaque lists 121 countries that have endorsed Strawberry Fields as a "Garden of Peace."

(continued)

Point of Interest (Continued)

Figure 18-31: This mosaic of inlaid stones is the focal point of Strawberry Fields, a 2.5 ac section of Central Park named in honor of slain musician John Lennon.

an individual perception, is difficult to quantify and describe, but at the very least, landscaping should respond to the qualities of a site; in other words, one shouldn't try to create a tropical landscape in New England or a woodland design in a desert. A good design will consider the spirit of the place and elevate the spirit of all who visit.

Bubble Diagram

Use a bubble diagram to brainstorm the proposed locations for specific functional areas of the landscape. Use a separate bubble to indicate each area identified in the needs analysis or required by regulations. For residential sites, this may include a patio, play area, water feature, vegetable garden, pool, or screening from neighbors, and so forth (see Figure 18-32). For commercial facilities, your bubble diagram might show a windbreak, a detention pond, screening for a dumpster, and required landscaping buffers, yards, and parking islands (see Figure 18-33). The size of each bubble should represent the relative size of each area.

Since the first design idea is rarely the best, sketch another bubble diagram to illustrate a different potential layout. Then, think outside the grass and make a few more bubble diagrams. Once you have exhausted your creative thinking and created several bubble diagrams, visit the site and try to visualize each potential landscape idea in place. Based on your observations and knowledge of the site, choose the bubble diagram that represents the best potential plan, or combines several of the ideas into a final proposal.

Figure 18-32: Bubble diagram for a residence showing planned use areas.

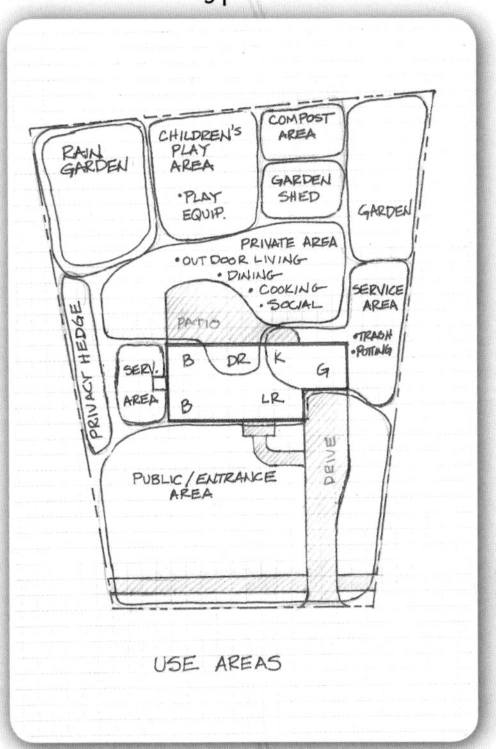

Figure 18-33: *Commercial bubble diagram.*

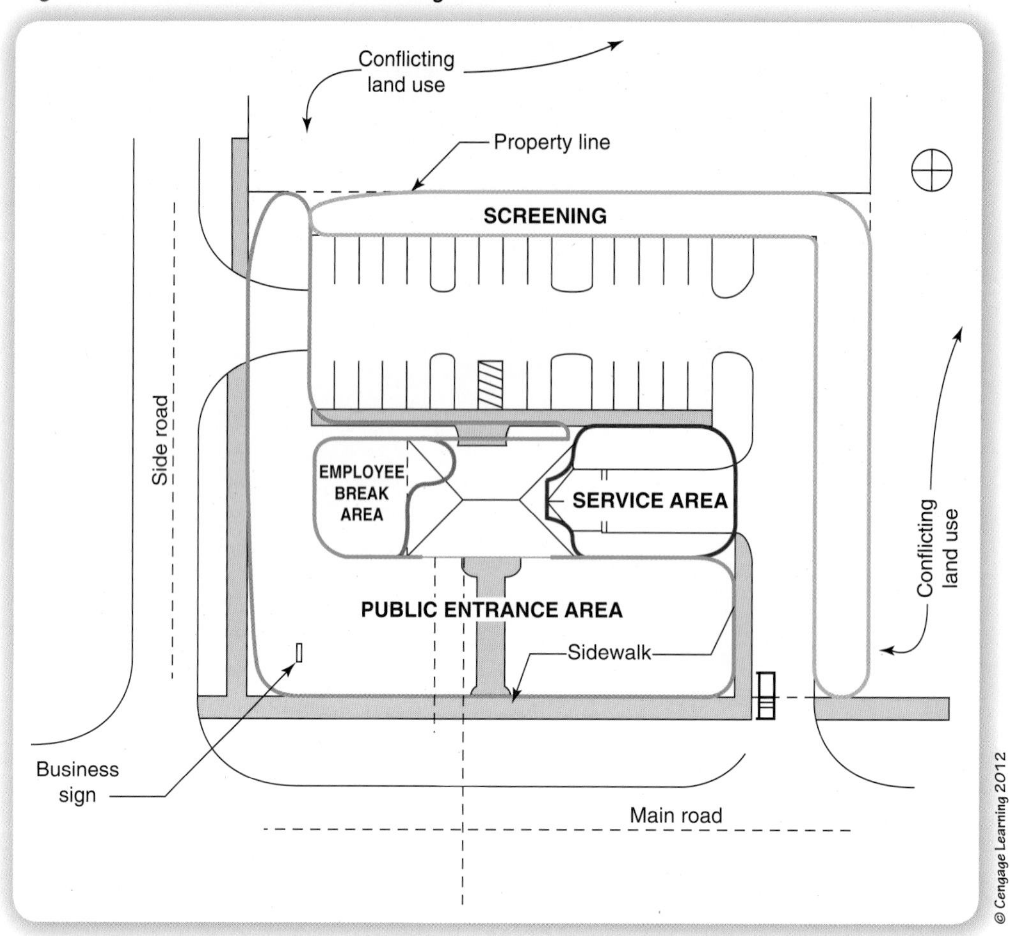

Representations, such as bubble diagrams and landscape plans, can help model physical objects and help solve problems.

Concept Plan

A *concept plan* will provide the details to support the landscape ideas that you have outlined in your bubble diagram. It will show relative sizes and locations of landscape elements that will be included in your design such as walkways, patios, and plants. In this way, you will determine the overall layout of the landscape and insure that it functions in a way that will enhance the project. Typically, the landscape architect will indicate individual elements but will not show dimensions or indicate the specific material to be used. For example, the concept plan may show large trees on the southwest corner of the building to provide shade but not call out a specific species; or the concept plan may show a rectangular patio but not indicate the finished size or surface. The main purpose of the concept plan is to provide an overall sense of the final landscape. An example of a residential concept plan is shown in Figure 18-34a, and a commercial concept plan is shown in Figure 18-35.

Thoughtful planning is necessary to devise a workable solution.

Figure 18-34: Residential landscape concept plan. The photograph shows one of the focal points in the design—the fountain.

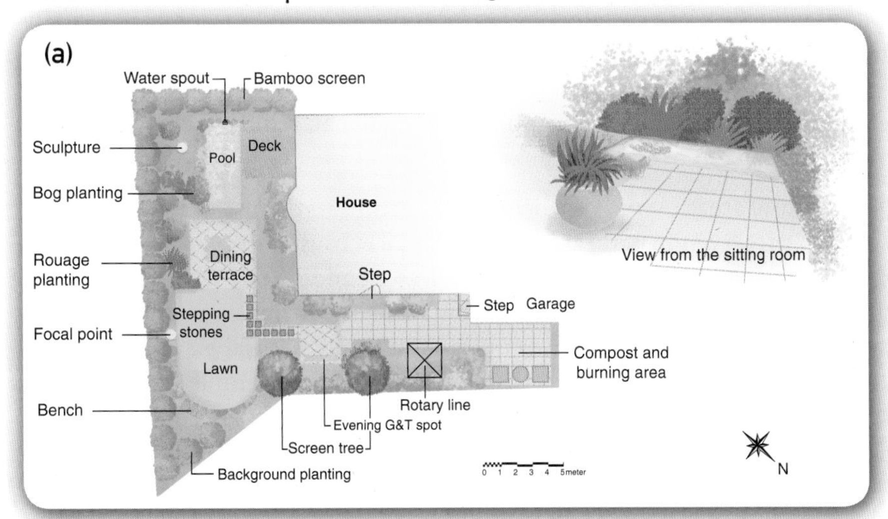

Courtesy of Acres Wild – Landscape and Garden Design, Horsham, West Sussex, UK.

Figure 18-35: Commercial landscape concept plan.

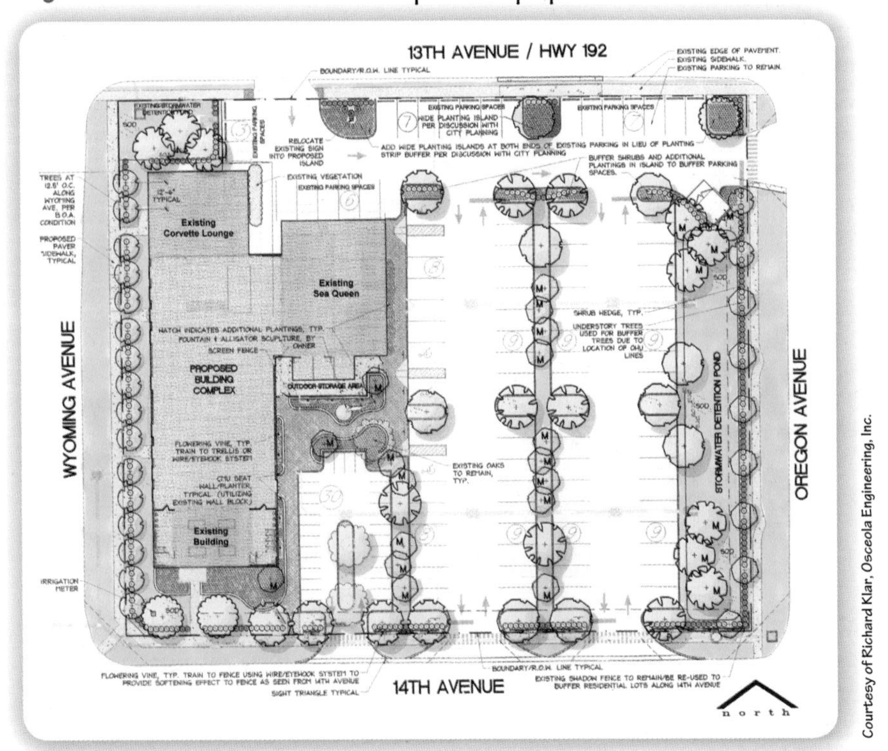

Courtesy of Richard Klar, Osceola Engineering, Inc.

SUSTAINABLE LANDSCAPING

In recent years, concern for the environment has increased, and the concept of creating sustainable landscapes has come to the forefront of landscape design. *Sustainable landscaping* is an approach that seeks to create a beautiful and functional landscape without causing harm to the environment. By reducing the need for

A Native Plant: of a given area is a plant that grew in that area prior to European contact and has co-evolved with local animals and environment.

Invasive Species: are species that are not native to an area whose introduction or spread threatens the existing ecosystems by causing damage to native species and their habitats.

OFF-SITE EXPLORATION
You can check out invasive plants, insects, and other species at Center for Invasive Species and Ecosystem Health website at http://www.invasive.org/.

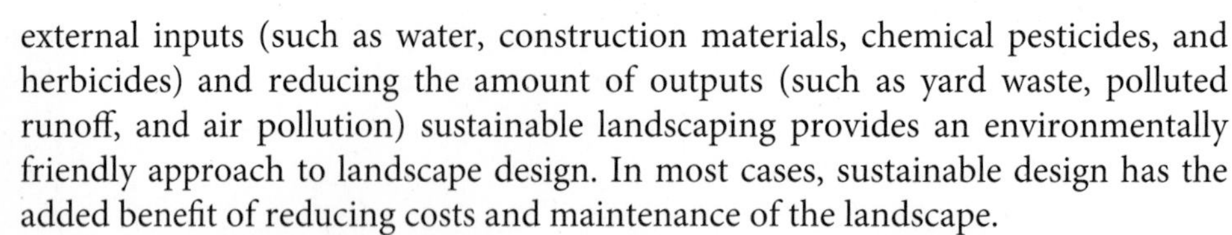

external inputs (such as water, construction materials, chemical pesticides, and herbicides) and reducing the amount of outputs (such as yard waste, polluted runoff, and air pollution) sustainable landscaping provides an environmentally friendly approach to landscape design. In most cases, sustainable design has the added benefit of reducing costs and maintenance of the landscape.

The following sustainable landscaping techniques can reduce the environmental impact of a project:

1. Protect natural areas. Avoid disturbing woodlands, wetlands, and watercourses in order to preserve the natural balance of local ecosystems.
2. Reduce the use of turf. Grass lawns require chemicals and frequent maintenance that produce waste products, contaminate stormwater, and result in toxic emissions from maintenance equipment. Install natural plantings instead of turf areas. If lawns are needed (for sports areas, for example) consult your local cooperative extension for the least harmful species.
3. Mulch planted areas. Mulch helps reduce water loss and, as it breaks down, provides nutrients to the soil thereby reducing the need for fertilizers. Using yard wastes as mulch is even better.
4. Use native plant species. **Native plants** generally requires less fertilizer, less water, and less effort in pest control. In addition, native plants are of great importance to the ecology of a place, and native wildlife may depend on specific species. Some non-native plants, sometimes called **invasive species**, can negatively affect the environment and can even eliminate native species (Figure 18-36).
5. Use landscape elements to reduce energy consumption. As previously discussed, trees and other landscaping elements can help control solar heat gain and reduce wind effects to buildings, which can significantly reduce heating and cooling costs. In addition, landscaping that requires less maintenance will require less fossil fuel due to a reduced need to run maintenance equipment.
6. Use *xeriscape* (pronounced *zera-scāp*) techniques. Xeriscape, often referred to as a water-wise landscape, is a landscaping approach that minimizes the use of water through careful plant selection, landscape design, and irrigation (Figure 18-37).

6
S T
E M

Conservation includes controlling soil erosion, reducing sediment in waterways, conserving water, and improving water quality.

THE DETAILS: CHOOSING LANDSCAPE ELEMENTS

An attractive and effective landscape requires integrating the plantings with the man-made elements on the site. The hardscape and softscape must work together according to the principles of design in order to provide the functional value intended.

5391732

Figure 18-36: **Kudzu, native to Asia, is an invasive species that was widely planted throughout the eastern United States to control erosion. It is a climbing, deciduous vine that can grow to 100 ft long and often grows over, smothers, and kills all other vegetation, including trees.**

Photograph: James H. Miller, USDA Forest Service, http://www.bugwood.org.

Figure 18-37: Xeriscape landscape planning involved identifying water use areas and minimizing high water zones. The plan on the left identifies water use areas for an apartment complex. The plan on the right shows the final xeriscape landscape plan.

Courtesy U.S. Department of Agriculture, Rural Development Colorado State Office.

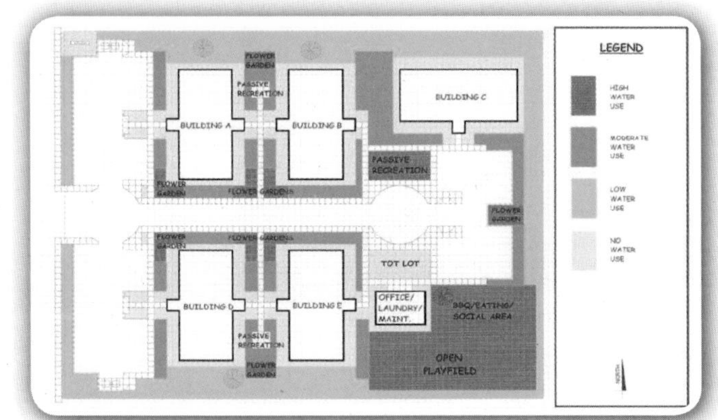

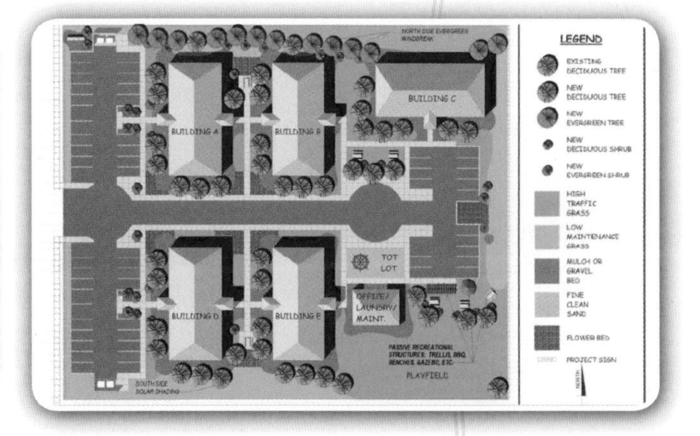

Your Turn

Back to Nature

Research native plants for your project. A good place to start is the National Wildlife Federation website (*http://www.nwf.org/backyard/food.cfm*). Review the list of top ten most popular native plants in your state, and record the common and scientific name for each in your notebook. Then browse the Native Gardening and Invasive Plants Guide for your state (*http://www.enature.com/native_invasive/*). Identify at least two native conifers, two native deciduous trees, and two native shrubs that you might consider using in your project landscape design. Print images of each plant and insert the images into your notebook. Then record important descriptive information about each species such as height, diameter, spread, color, fruit, and so on.

Xeriscape Techniques

Find out more about xeriscaping using the Sustainable Sources website at *http://xeriscape.sustainablesources.com/*.

Hardscape

Hardscape elements can provide the skeleton of a landscape design and will often be required to perform a specific structural function—for instance, provide a driving or walking surface, retain soil, or divide space. Therefore, the choice of materials must depend on both the material properties and aesthetic value of the hardscape material. Although each landscape design will pose different challenges, some questions that may need to be answered concerning hardscape include the following:

- What type of pavement will surface the parking area, gathering places, and walkways?
- Will you need steps or ramps? How will they be constructed?
- If you need retaining walls or fencing, what materials will provide the strength required?
- Do you plan to create a covered area? What structural elements and finishes will be required?

Check out the Virtual Tour of Central Park at *http://www.centralparknyc.org/site/PageNavigator/virtualpark_main*.

Point of Interest

Fredrick Law Olmsted

Frederick Law Olmsted (1822–1903) was an American landscape architect before landscape architecture was a profession–in fact, Olmsted was the first (with partner Calvert Vaux) to use the term *landscape architecture*. Although his work encompassed a wide range of projects including residential, institutional, and government landscape designs, he became famous as an American park maker, and his work has had a significant impact on urban design. Before urban sprawl became a reality, Olmsted envisioned the need for urban parks to provide recreation and to allow the city dweller to become immersed in nature. He felt that the purpose of landscape design was to affect the emotions.

His firm designed many well-known urban parks, including the Emerald Necklace in Boston, the 1893 Chicago World's Fair grounds, the landscape surrounding the U.S. Capitol building, and the grounds of the Vanderbilt Mansion, Biltmore Estate, in Asheville, North Carolina. Perhaps Olmsted's most notable landscape design is that of Central Park in New York City, the first public park built in America. Along with partner, Calvert Vaux, Olmsted created an oasis of nature, originally called Greensward, in the middle of the concrete, glass, and steel that is the city.

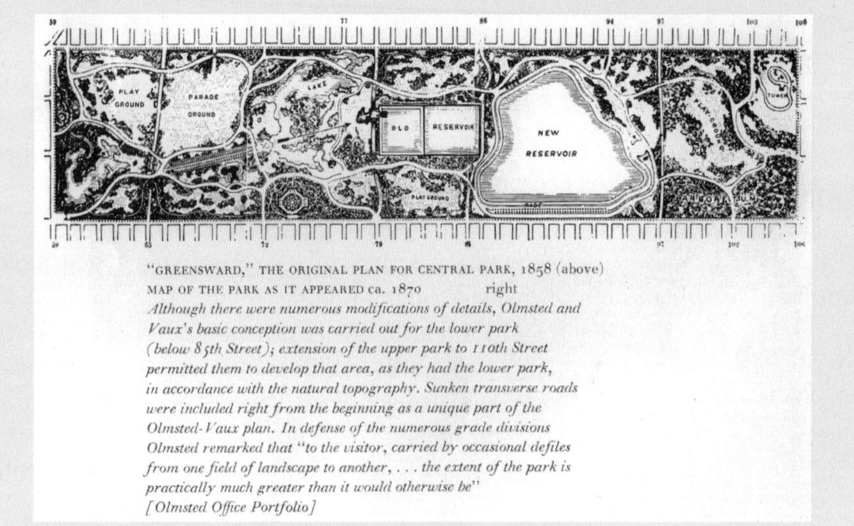
PLAY GROUND
PARADE GROUND
LAKE
OLD
RESERVOIR
NEW RESERVOIR
PLAY GROUND
"GREENSWARD," THE ORIGINAL PLAN FOR CENTRAL PARK, 1858 (above)
MAP OF THE PARK AS IT APPEARED ca. 1870 right
Although there were numerous modifications of details, Olmsted and Vaux's basic conception was carried out for the lower park (below 85th Street); extension of the upper park to 110th Street permitted them to develop that area, as they had the lower park, in accordance with the natural topography. Sunken transverse roads were included right from the beginning as a unique part of the Olmsted-Vaux plan. In defense of the numerous grade divisions Olmsted remarked that "to the visitor, carried by occasional defiles from one field of landscape to another, . . . the extent of the park is practically much greater than it would otherwise be"
[Olmsted Office Portfolio]

Figure 18-38: The Greensward Plan for Central Park.

Courtesy of the National Park Service, Frederick Law Olmsted National Historic Site.

Figure 18-39: Fredrick Law Olmsted, with partner Calvert Vaux, designed Central Park in New York City as an urban oasis of nature.

The variety of hardscape materials is extensive, and you should research all of the options before making selections. Some possible construction materials for parking and paving include poured concrete, asphalt, pervious concrete (which has voids that absorb water), paver blocks, gravel, and urbanite (old, broken concrete recycled into stone-like paving). Decks can be constructed of traditional wood decking or plastic lumber made from recycled plastic and wood. Garden walls or structural *retaining walls,* which hold back soil on slopes (see Figure 18-24), can be built from poured-in-place concrete, concrete masonry units, brick, stone, or heavy timber. Fences may be built from wood, metal, vinyl, or plastic—but don't forget the possibility of creating a living fence with plants.

For each material choice, there may be a multitude of alternatives. For instance, a concrete mixture may include shells that are visible on the surface; it can be colored or it can be finished in patterns that resemble stone or other designs. Stone comes in many shapes, colors, and textures. Hundreds of species of domestic and exotic woods are available. When choosing hardscape material, be sure to keep in mind the principles of landscape design and understand how the material will impact the overall design. Become educated on the sustainability of each alternative material so that you can make an informed decision with the least environmental impact possible. And, of course, consider the cost of the alternatives.

OFF-SITE EXPLORATION

Check out the variety of decorative concrete finishes available at the ConcreteNetwork.com (*http://www.concretenetwork.com/concrete/decorative/*).

Selecting resources, such as landscape elements, involves trade-offs between competing values such as desirability, availability, and cost.

Softscape

Do you have any plants in your yard or neighborhood that are constantly creating maintenance issues? Maybe a shrub requires frequent trimming so that it does not block a stop sign or pathway or does not scrap the siding on your house. Maybe you have to reseed lawn areas under dense trees every year because the turf doesn't get enough sunlight. Or, maybe a group of plants in your backyard begins to look distressed during moderate spells of dry summer weather and needs daily watering. Most landscape maintenance issues and functional failures are a result of a lack of respect for the natural characteristics of the vegetation. Plants are genetically programmed to grow to a certain height and width, in a certain shape, at a certain speed, under certain climate conditions. In order to create a successful landscape design, you must know and understand the natural characteristic of the plants you specify—humans rarely win when they try to fight nature.

Your concept plan should indicate general locations and types of plants (e.g., deciduous trees, low growing shrubs, screening plants). Based on the intended function of each plant and the environmental constraints at each location, you can determine the species characteristics necessary to ensure that the plant will thrive. For instance, trees planted in a low boggy area that are intended to provide a windbreak from cold winter winds should be coniferous and thrive under wet soil conditions. A plant used to provide privacy to a patio from neighboring properties should provide thick foliage near the ground to a height of at least 6 ft.

Irrigation

Mother Nature doesn't always supply the optimal amount of rain for plants. All landscaping, even xeriscapes, will need supplemental water, at least to get them started. Choosing low water-use plants and using mulch to retain moisture will

Figure 18-40: Hardscape elements are an important part of landscape planning.

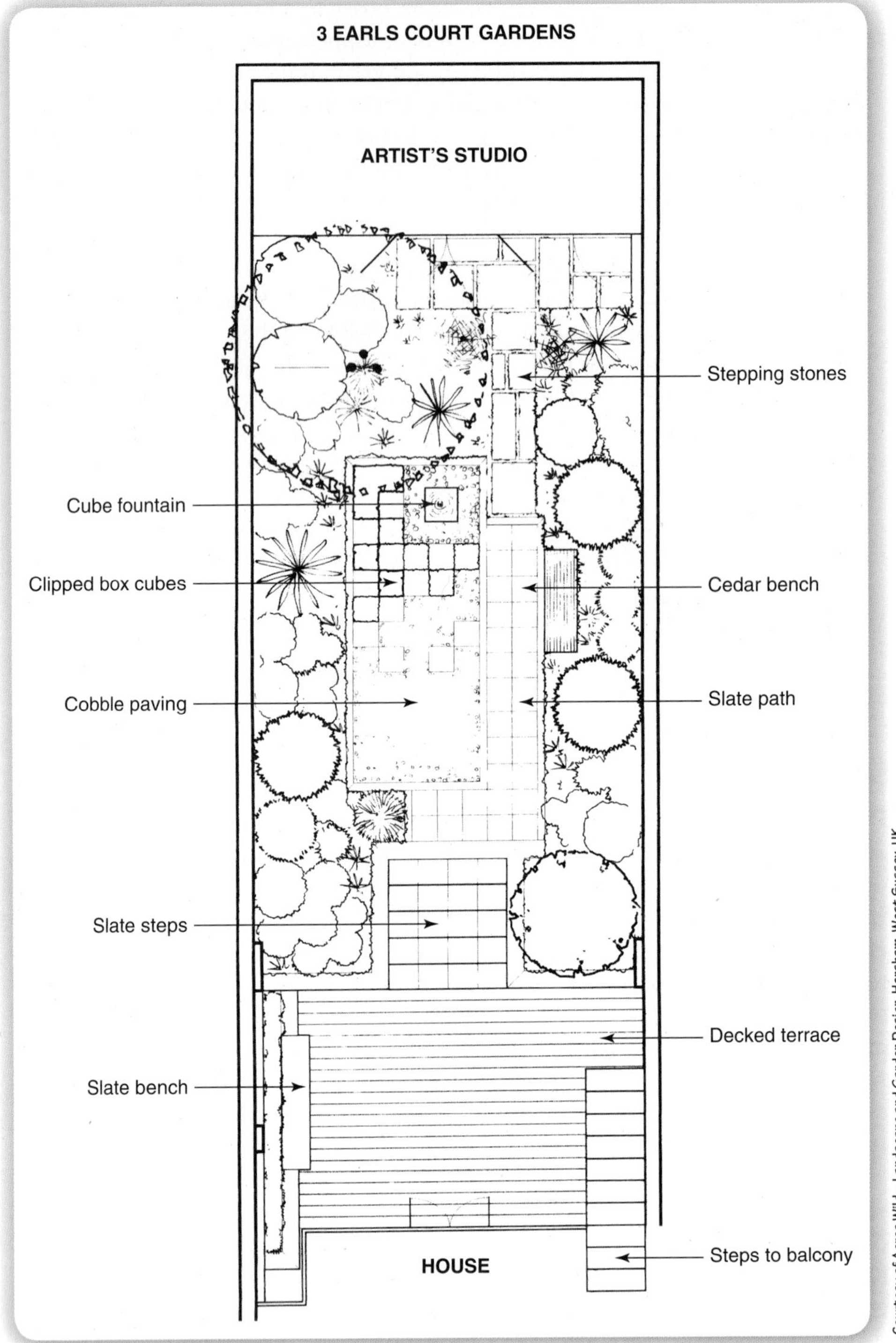

Courtesy of Acres Wild – Landscape and Garden Design, Horsham, West Sussex, UK.

minimize the amount of supplemental water needed, but you must consider the need for irrigation in the landscape design. Several options are available for irrigation of your landscape. The best option for a given property will depend on cost, the water needs of the landscape plants, and the maintenance effort required by each system. Some options for irrigation include the following:

Your Turn

Get Expert Advice

Identify potential trees for your project using the Arbor Day Foundation website at *http://www.arborday.org/*. Search the site for the Hardiness Zone Map. Using your zip code, identify the hardiness zone for your location and find the ten most popular trees for your hardiness zone. Then search for the Tree Guide and use the search feature to find each of these ten trees in the data base. Print a picture of at least two of these trees that you may consider using in your landscape plan and insert them into your notebook. Next to each tree image record the following: name, height, spread, growth rate, soil, sun, shape, and whether the tree is deciduous or coniferous.

Now visit the Lady Bird Johnson Wildflower center Recommended Species page (*http://www.wildflower.org/collections/*). Print out the printer friendly species list of native plants. Browse the list and highlight at least one species each of tree, shrub, herb, and vine that will thrive in each of the water use zones (high, moderate, and low) using a different color highlighter for each water use zone. If you identified potential native trees in the previous *Your Turn*, you may want to include these species in your selections. Print an image of each species that you highlighted and document the plant characteristics and growing conditions.

Greywater: non-industrial wastewater from domestic processes such as dishwashing, laundry, and bathing (but not toilet use) that can be recycled for use in flushing toilets, irrigation, and other non-drinking purposes.

Specifications: the technical requirements of a project that provide a detailed description of materials and quality of work required for something to be built. Specifications are typically published in a written document that accompanies the working drawings. The working drawings and specifications together are referred to as construction documents.

- Harvesting rainwater. Some of the water that falls on your property may not be absorbed by the vegetation—it may seep into the groundwater system or run off onto a neighboring property or into a stormwater system. By collecting the rainwater, in rain barrels (see Figure 18-41), cisterns, or ponds and redirecting it to the landscape plants, you can reduce the need for supplemental water.
- Greywater systems. **Greywater** is generally considered to be untreated household waste water from bathroom sinks, showers, bathtubs, and clothes washing machines—water that can safely be used again for irrigation. This water can be piped to a storage tank, filtered, and later used for outdoor watering. The definition of greywater and restrictions on its use varies among municipalities, so check your local codes before incorporating a greywater system into your irrigation plans.

Humans can devise technologies to conserve water through such techniques as reusing, reducing, and recycling.

FINAL LANDSCAPE PLAN

Now you're ready to finalize your design. The final landscape plan must include all aspects of the landscape design and is typically documented on working drawings and in **specifications.** Because these documents are used by the landscape contractors to estimate the cost as well as install the landscaping, the drawings and specifications must convey the information necessary to completely describe the design requirements. Depending on the complexity of the design and permit requirements, the landscape plan may be as simple as a single drawing indicating

Figure 18-41: Rain barrels can be used to collect and store rainwater that can later be used to water plants.

© iStockphoto.com/Sebastian Santa.

plant species and locations for a residential project, or it may include a series of working drawings and lengthy specifications for a large commercial project. Ideally, the final landscape plan should provide information on the types of plantings, types and location of hardscape elements and beds, irrigation plans, details needed to construct the landscape, and specifications. A partial example of a final commercial landscape plan is given in Figure 18-42.

Figure 18-42: Partial final commercial landscape plan.

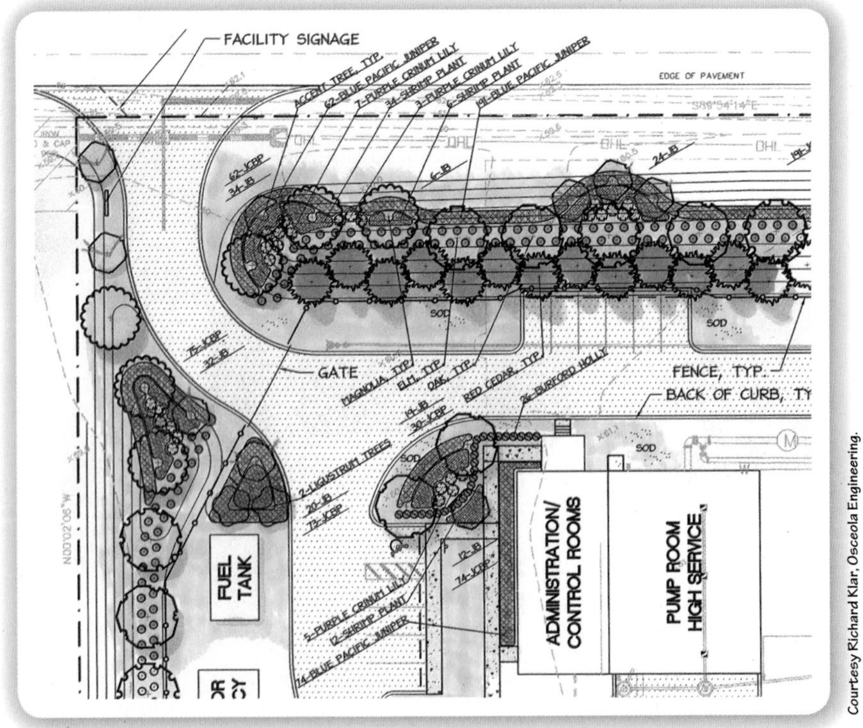

Courtesy Richard Klar, Osceola Engineering.

Careers in Civil Engineering and Architecture

LANDSCAPE ARCHITECT / PLANNER, SEAMONWHITESIDE + ASSOCIATES

The Man with a Plan

As a landscape architect, Ben Liebetrau encounters tremendous variety in his work. He designs commercial, residential, mixed-use, and institutional sites of all sizes, from a single lot up to hundreds of acres. He lays out the streets, the orientation of the buildings, the utilities, and the plantings. While engineers typically handle the technical elements of a site, he takes care of the artistic side.

"We do the layout and design of the site," Liebetrau says. "We determine where the elements will go and how the land will be used."

Ben Liebetrau, RLA, LEED® AP

On the Job

Liebetrau is currently designing a park on South Carolina's Daniel Island. He has to figure out where to put the trails, the buildings, the sports fields, and the parking lot. The city of Charleston will eventually own the park, so Liebetrau has to coordinate with city planners, consulting architects, and the city's Recreation Department and incorporate their ideas. He works closely with the civil engineers in his office to make sure the site drains correctly.

"We concentrate on maintaining the existing vegetation," he says. "We want to minimize disturbance of the site as much as possible on all projects that we design."

Inspirations

As a child, Liebetrau always liked art and design. He took advanced art classes in middle school and did layout and design for the high school newspaper. He was also fascinated by real estate—and that, combined with his interest in art, is a classic combination for a career in landscape architecture.

Education

Liebetrau didn't decide on a landscape architecture major until he got to college at Clemson University. He got the idea during orientation and entered the program his first semester.

During the five-year program, Liebetrau did a lot of drawing and got a lot of hands-on experience in the studio on a wide range of projects. He learned neighborhood design and commercial site design. "It's a very design-oriented major," he says.

For his fifth-year thesis project, Liebetrau spent a whole semester designing a neighborhood for Summerville, South Carolina, taking the project from start to finish.

Advice to Students

Liebetrau suggests that students interested in landscape architecture take a tour of a firm to get a feel for the profession. He emphasizes that landscape architecture is not just about plantings, as some people think. The profession spans many other disciplines.

"An interest in plants can be a big part of it, but it's not essential," he says. "We have a wide range of responsibilities. That's what makes it interesting."

Liebetrau says that anyone with an interest in conservation and the environment might like landscape architecture. "Landscape architects have the great responsibility of being stewards of the natural environment," he says. "It is up to us to ensure that our natural land is developed in a sensitive and sustainable way."

In terms of preparation, he stresses drafting, design, and math, as well as writing.

"If you can write well, that's a bonus," he says. "There are times when you have to explain yourself, not only through drawings but through words."

SUMMARY

You are probably familiar with the popular saying "form follows function," which means that the visual appearance of a designed object or feature should conform to the intended function of that object or feature. This philosophy can be applied to the design of consumer products, buildings, and landscapes alike. While a primary function of the landscape may be to improve the aesthetics of the site, the most effective landscape designs are achieved when the landscape is used to provide desired benefits beyond curb appeal while meeting all regulations and minimizing the negative impacts of development on the environment. In fact, landscaping can improve a property in a number of ways, such as by reducing energy costs, decreasing noise pollution, protecting a property from excessive light and undesirable odors, stabilizing slopes, and conserving the environment.

In order to design a landscape that will provide multiple functions, you must understand the constraints of the site—research the surrounding developments and the local climate and analyze existing environmental conditions. This information will be the basis for your conceptual design. By applying the visual principles of design—such as repetition, balance, emphasis, and unity—to your work, you can create a special place that appeals to the senses, inspires the mind, and touches the heart.

Of course, any type of development, including landscaping, will disturb the natural environment, so an important goal of landscaping should be to reduce the negative impacts of the project. There are many landscaping techniques that can be used to create a Low Impact Development (LID). Many of the environmentally friendly techniques involve making informed decisions on specific landscape elements. For instance, using native species of plants, reducing turf areas, mulching planting bed, using pervious hardscape elements, and recycling rain water for irrigation (to name a few) can all improve the health of the landscape, the environment, and the animals and humans that enjoy the landscape.

All of the details and specifics of your design will be delineated in the landscape construction documents. These documents will describe exactly what will be included in the landscape, where it will be located, and how it should be installed. Details on grading, plantings, hardscape, irrigation, and installation instructions should be specified. In this way, you can be sure that your design intent is clearly conveyed to the landscape contractor and that the design you envisioned is carried out. With continued care and proper maintenance, the resulting landscape can continue to have a positive impact for many years to come.

Creativity, imagination, and a good knowledge base are all required in the work of science and engineering.

BRING IT HOME

1. Assume that your parents, grandparents, or another adult couple are planning to build a small retirement home in an established development in the mountains near Ashville, North Carolina, but have not yet selected a specific site. They have asked you to research the area to gather information that will help them design their landscaping once they have chosen a specific lot. Research and document the climate, wind, sun orientation, existing soil, and native plants for Asheville.
2. Study the site analysis map shown in Figure 18-4. As a landscape designer, describe your approach to providing each of the following functions for this property. Use sketches to explain your solutions.
3. Study the landscape design in Figure 18-32. Describe how the principles of unity, balance, emphasis, and repetition are used in the design.
4. Study the landscape design in Figure 18-26. Discuss at least three functions that the landscape provides for the property and list the specific landscape features that contribute to the performance of each function.
5. Sketch a plan of the existing landscaping at your home or a small nearby commercial facility. Pretend that you have been hired to create a new sustainable landscape for the site. What improvements would you suggest in order to make the landscaping more sustainable and environmentally friendly?
6. Visit the American Society of Landscape Architects Green Roof Education site (*http://www.asla.org/greenroofeducation/*) and explore the benefits of a green roof. Address each of the following items in your notebook:
   a. How is a green roof constructed? Sketch a section of a green roof in your notebook and label the components.
   b. What are the two kinds of green roofs? Describe each.
   c. What functions does a greenroof perform?
7. The Jones family has just purchased a rustic lake cabin on a lake near Williamsburg, Virginia. The property (shown below) was originally cleared and turf planted across the entire property. There are no paved surfaces, and no landscape planting has ever been installed. The Joneses would like to create a landscape design that enhances the property and returns it to a more natural state. Of course, they also want to minimize the required landscape maintenance, since they would prefer not to spend their vacations maintaining the yard.

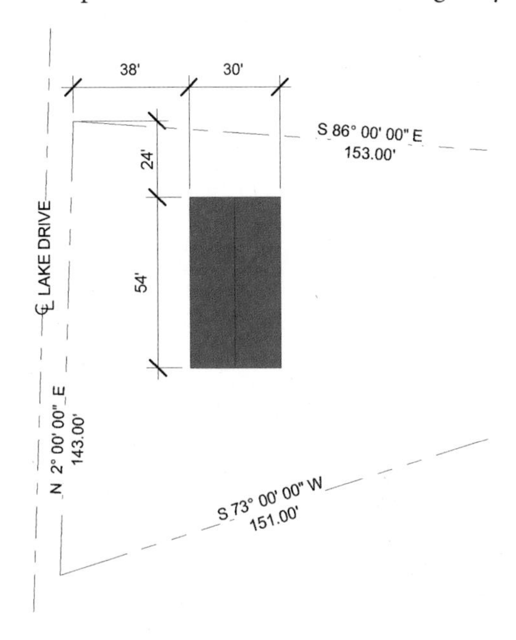

The family of five includes the parents, a teenage girl, and twin preschool-age boys. They own a boat and, while visiting the house during warm weather, will spend a significant amount of time outside, either on the lake or within view of the lake. During the winter, the family likes to visit the lake to cross country ski and ice skate. Your job is to provide a landscape plan that is natural, attractive, and low maintenance.

   a. Using the site plan for the property, create a bubble diagram to identify specific areas of use which should incorporate public areas, private areas, and service areas.
   b. Research the site conditions and sketch a site analysis map for the property.
   c. Create a conceptual landscape plan for the property.

8. Visit the site of a local fast food restaurant. Approximate distances and sketch the site/landscape plan in your notebook indicating the building location, pavement, plantings, turf, etc. Take pictures of the different species of plants used in the landscape.
    a. Identify each planted area as a street yard, buffer yard, street wall, parking island, vegetative screen, an area containing street trees, or another type of planting.
    b. Research the plant species used in the landscape and identify each plant.
    c. Create an existing planting plan for the property.

ABOVE AND BEYOND

In 2007, nine firefighters were killed while battling a fire in a Sofa Super Store building in Charleston, South Carolina. Assume that a 100 ft × 180 ft parcel of land has been donated on which to construct a memorial park (see the preliminary plot plan). Charleston is in an area of South Carolina known as the *low country*—it is close to sea level and flat; therefore, assume insignificant changes in elevation across the site. A secondary road is adjacent to the property. Design an *environmentally friendly* memorial park that will attract wildlife to honor these fallen firefighters. The design should include walkways, benches, a small restroom facility, parking for 10 cars, and a memorial. Write a design narrative that addresses the following questions:

- How does your design memorialize the firefighters?
- How did existing site conditions affect your design choices?
- How did you incorporate the principles of design into your landscape?
- What environmentally friendly landscaping techniques did you use?
- Which elements of your design are intended to attract wildlife?
- Why did you choose the hardscape materials that you used?

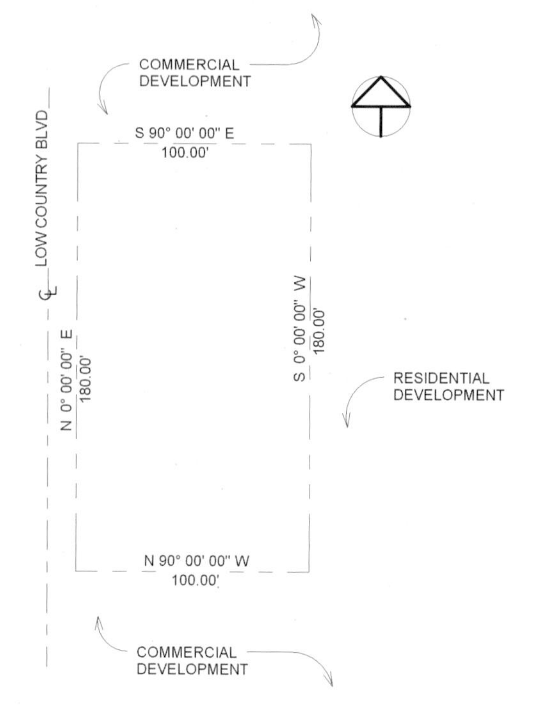

PART VIII
Selling the Plan

CHAPTER 19
Visual Communication of Design Intent

GPS DELUXE

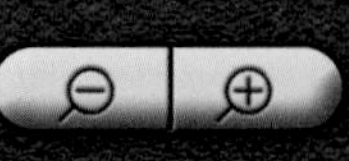

Menu

START LOCATION | DISTANCE | END LOCATION

Before You Begin

Think about these questions as you study the concepts in this chapter:

1. Why are pictorial drawings used for design visualization?
2. How does the audience influence presentation graphics?
3. What is the purpose of presentation graphics?
4. How do presentation drawings differ from working drawings?
5. How does building design software enhance design development and presentation?
6. What techniques can be used to communicate interior design?
7. How do models enhance design visualization?

INTRODUCTION

If a new building was planned for your neighborhood, what information would you, a resident, need and want to know? You might question the purpose of the building, what it will look like, and how will it impact your neighborhood. The architectural firm hired for the project is usually responsible for clear and concise communication of this information. The architectural team assigned to project communication will thoughtfully devise a strategy. What communication method would best convey general understanding of the building plan and projected outcome? Would dozens of complex working drawings inspire design visualization? Probably not. Most large building projects are presented to stakeholders and community members without the background necessary to visualize building design from just looking at working drawings. To help the **layperson** understand the building design and intent, the architectural firm develops promotional materials, slideshows, and displays of presentation drawings and

Layperson: a person without background or expertise in a subject.

Courtesy of CityCenter, Las Vegas.

Figure 19-1: *Display the CityCenter Pavilion of the proposed Veer Towers.*

INTRODUCTION

models, Figure 19-1. Presentation drawings include floor plans with furniture layouts, rendered elevation views, and various interior and exterior pictorials.

Models may be physical or virtual. In addition to communication of building size, shape, and appearance, models show how the structure will fit into and enhance the surroundings. You may have seen a model of a future building project on display at a bank or other community building. These displays are often combined with illustrations or computer presentations with computer generated images.

Virtual models provide the viewer an opportunity to see how the building looks or performs under different conditions. Date, time of day, and seasonal conditions can be set in building design software to simulate how light and shadows will impact the building or surroundings. Adding images or models of people, cars, and accessories provide a sense of realism and building proportion. A walkthrough of a virtual model can greatly enhance design visualization and intent. After reading this chapter, you will understand various methods that may be chosen to communicate the building plan.

PRESENTATION GRAPHICS SUPPORT DESIGN VISUALIZATION

Presentation drawings are much different than working drawings. The purpose of presentation drawings is to communicate general building plan information to project stakeholders, community members, or other interested individuals such as potential investors. Instead of detailed specifications found in working drawings needed for construction, presentation drawings focus on design visualization. The intended audience's interests are important when planning presentation graphics.

Drawings are the common language for clear communication of ideas.

Pictorial Drawings and Design Visualization

Pictorial drawings are commonly used to enhance design visualization. When used to describe architecture, they provide a three-dimensional look at the interior and exterior of a building. Common pictorials include isometric, oblique, and perspective views (see Figure 19-2).

Pictorial Drawing: a drawing that shows a three-dimensional view of the object.

Principles and elements of design are used in pictorial drawings to illustrate a buildings aesthetic quality. Before the computer age, all presentation drawings were completed by hand.

Some architectural firms still produce architectural illustrations using ink and markers, whereas the majority now uses building design software. Most building design software has the ability to view or print a virtual model as a simple shaded

Figure 19-2: Common pictorial drawings used to describe architecture.

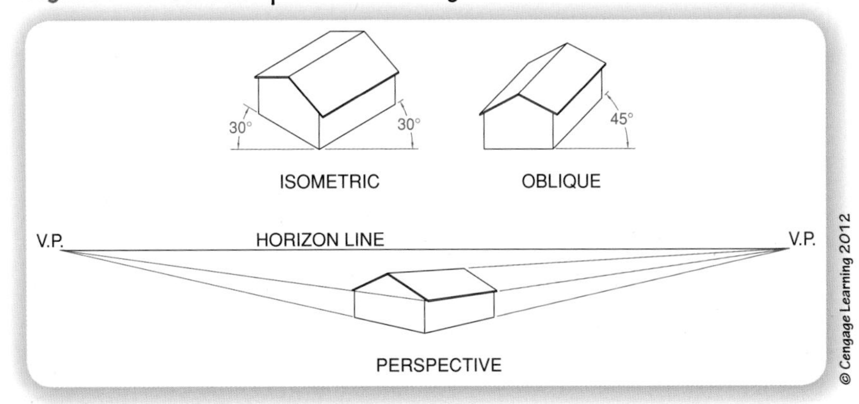

Render:
to portray something artistically (Encarta).

image or a more detailed **rendering.** Applicable design and illustration software can produce multiple renderings of the building at different times of the day or show how the building will appear in different seasons.

A rendering provides a photorealistic, or near photo-like image, of the proposed building. Some advanced software programs even offer an option to represent the model as a hand-drawn illustration (Figure 19-3).

AUDIENCE AND PURPOSE INFLUENCE PRESENTATION GRAPHICS

The choice of visual representation is based on its application. For example, a quick, less expensive shaded image may be appropriate for use in a graph, chart, or small brochure, whereas a detailed rendered image may be a better choice for a large poster display (Figure 19-4). Realistic drawings and illustrations help the layperson visualize the design outcome. A rendered image of a virtual model can depict how the building will look when finished, complete with light and shadows, landscaping, and adjacent properties. To enhance image realism and show

Figure 19-3: A hand-drawn perspective illustration of a proposed mixed-use development for downtown Milwaukee, Wisconsin.

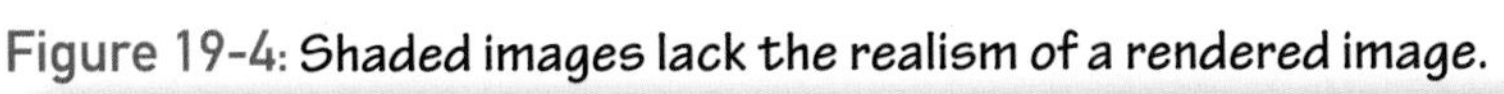

Figure 19-4: Shaded images lack the realism of a rendered image.

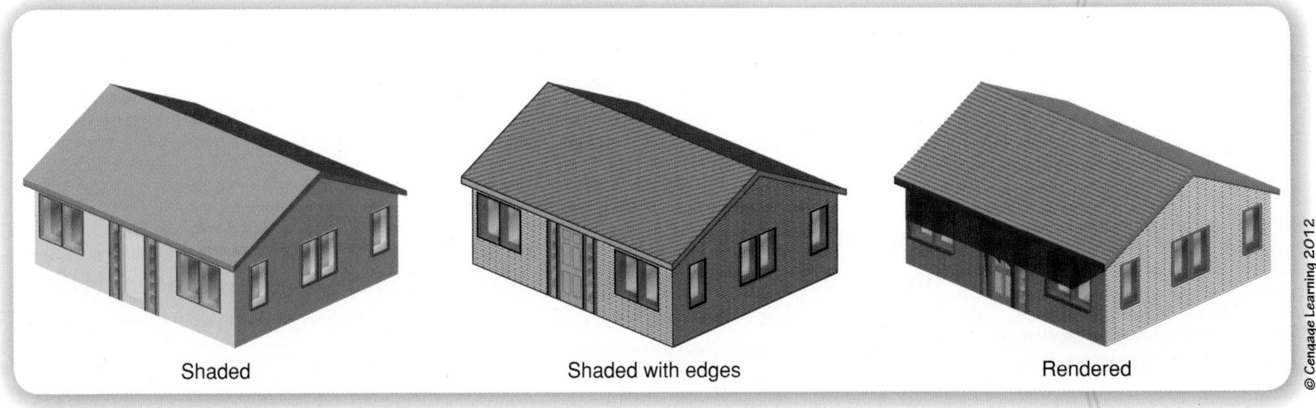

the building's scale, virtual people, cars, pets, and decorative objects are often included. These three-dimensional computer objects are similar to other component families, such as windows, doors, and cabinetry.

Your Turn

Create a pictorial sketch of a building similar to Figure 19-5. Develop the sketch into a detailed hand-drawn illustration. Use colored pencils to enhance the illustration. Create a virtual model of the same building and print out a shaded isometric view. Compare the drawings and record your thoughts about the advantages and disadvantages of hand drawn and computer-generated illustrations.

Figure 19-5: ***Sketches can be used to plan presentation graphics.***

Courtesy of Mike Elliott, Rich Salamone, and Wendy Stearns, PLTW Teachers.

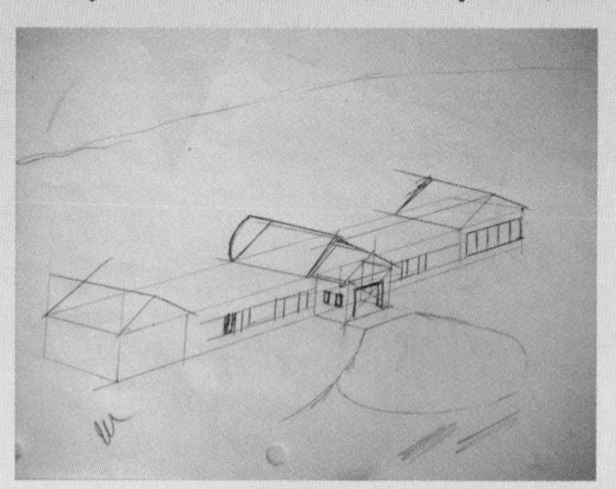

Colorful Elevation Views

Elevation views show the front, back, and sides of a building (Figure 19-10). When elevation views are included in a set of building plans, they are generally line drawings with notes and level dimensions. When used in presentation drawings, elevations are often shaded or rendered to take on a realistic appearance.

Perspective View: a pictorial drawing that shows depth and proportion of the object as it would appear to the human eye.

Enhancing the Realism of Presentation Graphics

The purpose of presentation graphics is to promote design visualization. Highly realistic graphics will maximize visualization of the design intent. Many consider a drawing of a **perspective view** to be an effective way to communicate how the building will actually appear.

Case Study

Let's take a look at some of the presentation graphics for the Mandarin Oriental Hotel at CityCenter (Figure 19-6).

Look at Figure 19-7 to determine the location of the Mandarin Oriental to the other buildings of the urban plan. Notice that the buildings are shown in color and in three dimensions to help the viewer understand the buildings' size, shape, and relationship to one another.

Locate the Mandarin Oriental on the model shown in Figure 19-8. The lighted buildings of the model represent CityCenter. Can you find the adjacent Monte Carlo with the three building sections extending from a center hexagon?

Now examine the perspective illustration of the Mandarin Oriental in Figure 19-9. If you look closely in the lower left quadrant, you can see a small portion of the Monte Carlo Hotel. From this information where would you physically stand to see this view? Notice that the size and proportion of the building is conveyed with images of people, trees, and cars.

An object's location can be described by its relationship to other objects.

Figure 19-6: *A computer rendering of CityCenter.*

Courtesy of CityCenter, Las Vegas.

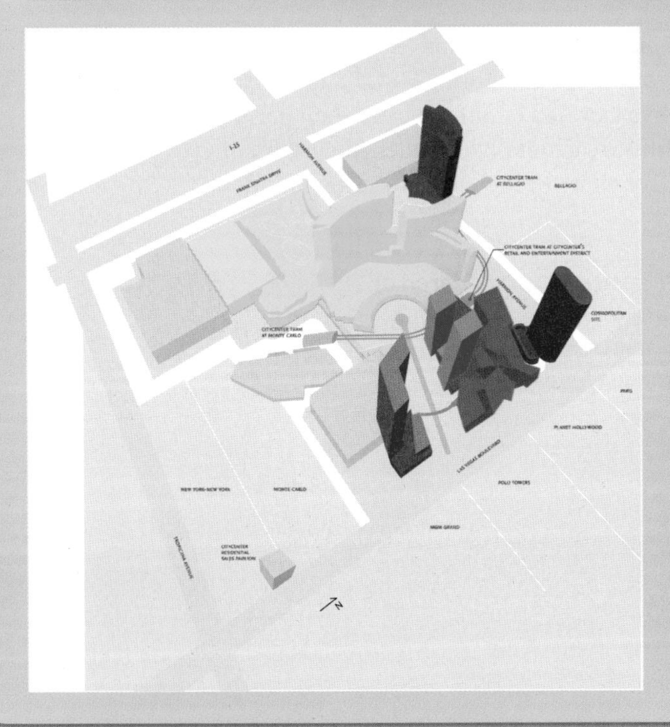

Figure 19-7: *This CityCenter graphic used three-dimensional images and color chart to identify building location.*

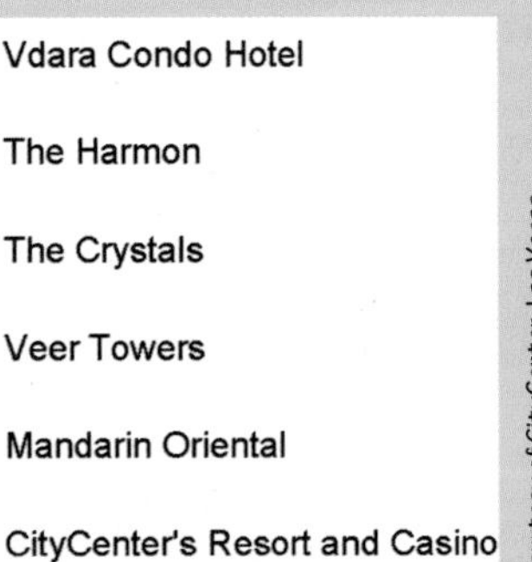

Courtesy of CityCenter, Las Vegas.

Case Study (continued)

Figure 19-8: *The size and proportion of CityCenter compared to adjacent properties are shown in this lighted model.*

Courtesy of CityCenter, Las Vegas.

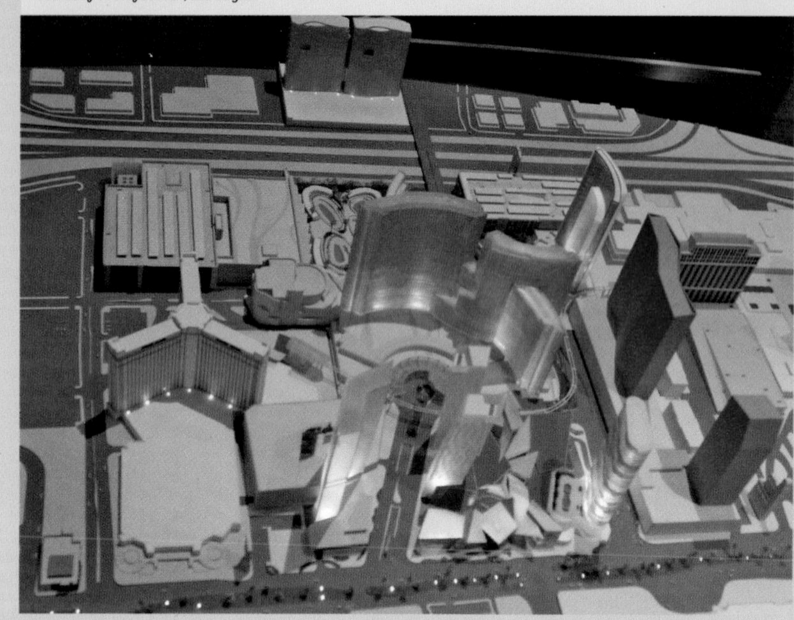

Figure 19-9: *The architectural perspective illustration of the Mandarin Oriental is enhanced with details to support design visualization.*

Courtesy of CityCenter, Las Vegas.

Perspective Drawing: a pictorial drawing that provides the illusion of depth by converging all horizontal lines and represents the object's depth to a single vanishing point on the horizon (for one-point perspectives) or to two vanishing points on the horizon (for a two-point perspective). *PLTW CEA Curriculum*

A perspective view is a pictorial drawing that uses the horizon line and vanishing points to create an image of how the building would appear from a specific viewpoint. Drawing a perspective view is a laborious task when performed by hand, but today's building design software generally includes the option of viewing, rendering, and printing perspective views of the virtual building model.

Figure 19-10: **Elevation views are often rendered or shaded for presentation graphics.**

Courtesy of Mike Elliott, Rich Salamone, and Wendy Stearns, PLTW Teachers.

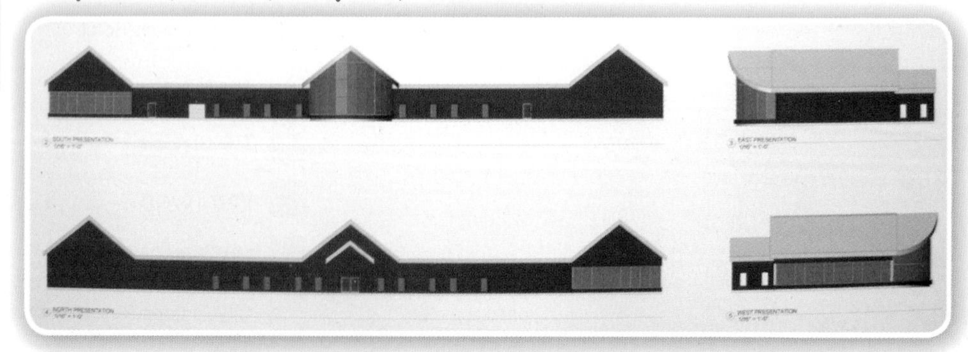

A perspective view is easy to identify by looking at the lines forming the floor, ceiling, or top of windows and doors. If projected to the horizon, the lines would converge to a vanishing point. In the interior perspective shown in Figure 19-11 b, the lines draw you into the room providing a realistic perception of depth.

Actual photographs can add realism to presentation graphics. Cropped photographs of people may be scaled proportionally and placed onto the image, as in Figure 19-11b. This technique can be used to provide a view from a window, or show an existing adjacent building. Some building design programs offer the option of setting the location and time of year to achieve a realistic rendering of foliage or show solar gain on a specific date. (See Figure 19-12.)

Figure 19-11 a: Computer-rendered perspective view of a proposed sustainable housing design.

Courtesy of LiveWorkHome, designed for the From the Ground Up competition, a project sponsored by the Syracuse University School of Architecture, the Center of Excellence in Environmental and Energy Systems and Home HeadQuarters, Inc.

Building Design Software Illustrates Climate Conditions

Building design software offers many advantages over hand drawing. One advantage is the ability to easily show the natural light and shadows one might expect on a specific day and time at a given location, Figure 19-12. This ability not only enhances the presentation graphics, but is also relevant during the design phase to determine optimal building orientation, window features, and placement to minimize energy consumption and improve indoor climate and comfort.

Figure 19-11 b: Computer rendering of the LiveWorkHome interior design.

Courtesy of LiveWorkHome, designed for the From the Ground Up competition, a project sponsored by the Syracuse University School of Architecture, the Center of Excellence in Environmental and Energy Systems and Home HeadQuarters, Inc.

Technology can be used to simulate environmental conditions to inform decision making.

COMMUNICATION OF INTERIOR DETAILS

When communicating the interior layout of a building, artistic and colorful representations of floor plans are used. They show room adjacencies and furniture layouts to help the viewer visualize proportion and scale. A presentation drawing of the floor plan often includes general room size instead of the detailed dimensions found on working drawings (see Figures 19-14 a and b).

Figure 19-12: The advanced model graphic settings in Revit Architecture 2009 ® provide realistic renderings for presentation and analysis.

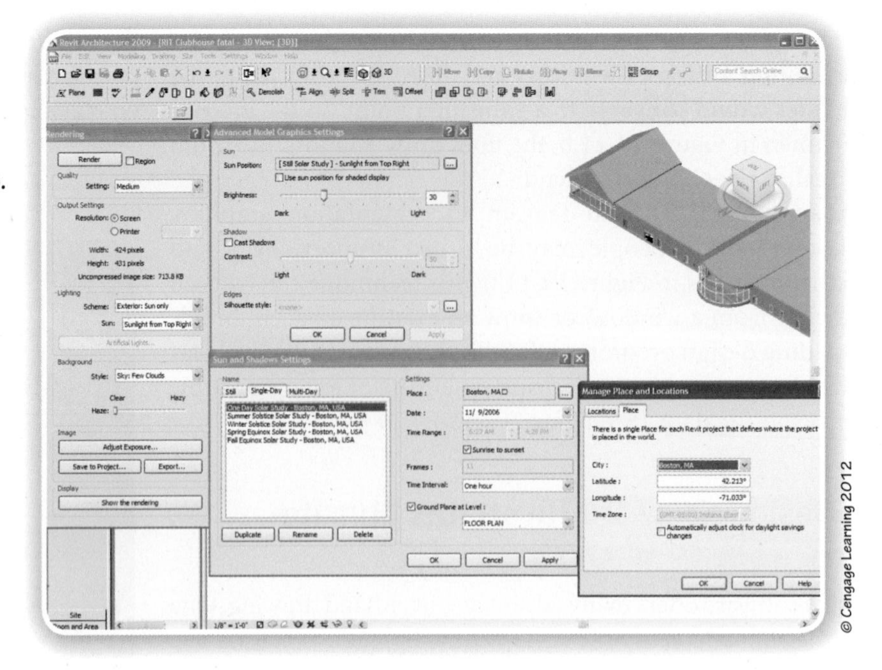

© Cengage Learning 2012

Using building design software, building features and decorative objects can be customized with materials, texture, and color to enhance design visualization. Virtual interior lighting can be added and manipulated for realism.

A virtual walkthrough provides the illusion of walking from room to room. The goal is to provide the viewer an opportunity to experience the look and feel of the proposed building. A walkthrough is created by placing a camera path on the floor plan of a virtual model, as in Figure 19-15. When played, the walkthrough provides an animated virtual tour of the building.

Your Turn

Examine the renderings shown in Figures 19-17 and 19-18 and make a list of the architectural features that might be desirable to a prospective buyer.

Figure 19-13: Computer generated day-to-night time lapsed view of One Bryant Park, New York City.

Courtesy of © dbox for Cook + Fox Architects LLP.

Figure 19-14: A presentation floor plan often includes furniture and color.

(a)

(b)

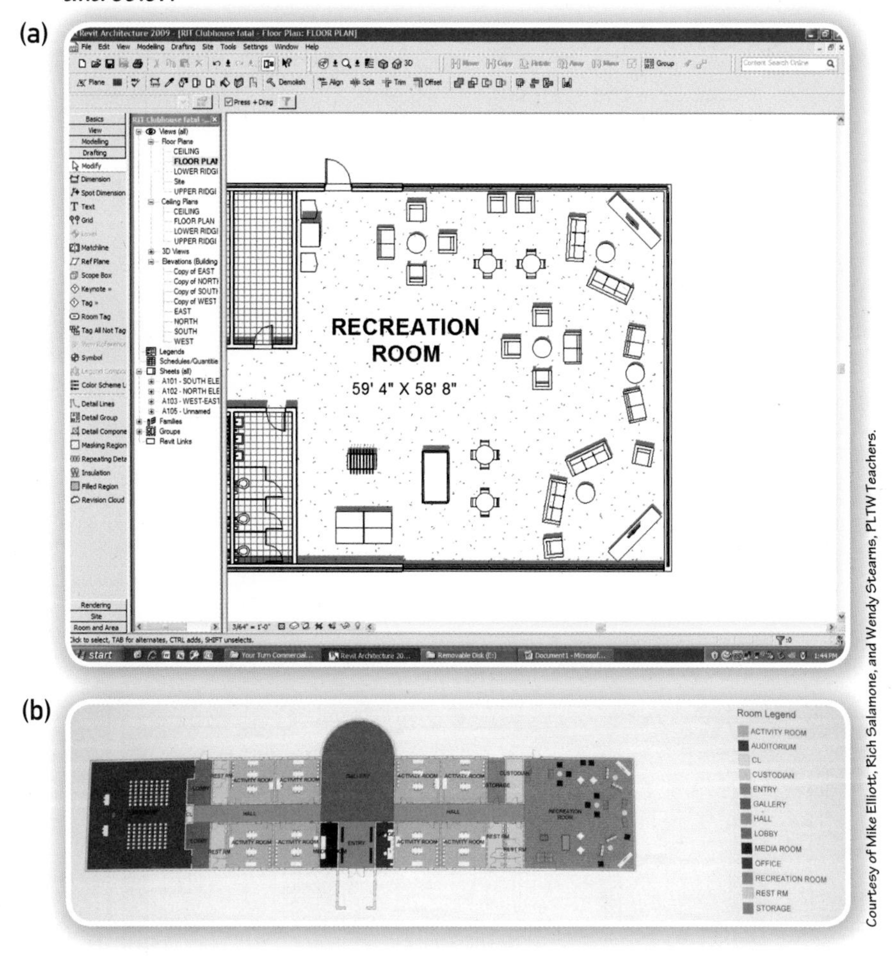

Courtesy of Mike Elliott, Rich Salamone, and Wendy Stearns, PLTW Teachers.

Figure 19-15: A walkthrough in Revit® Architecture is created by placing a camera path on the floor plan of the model.

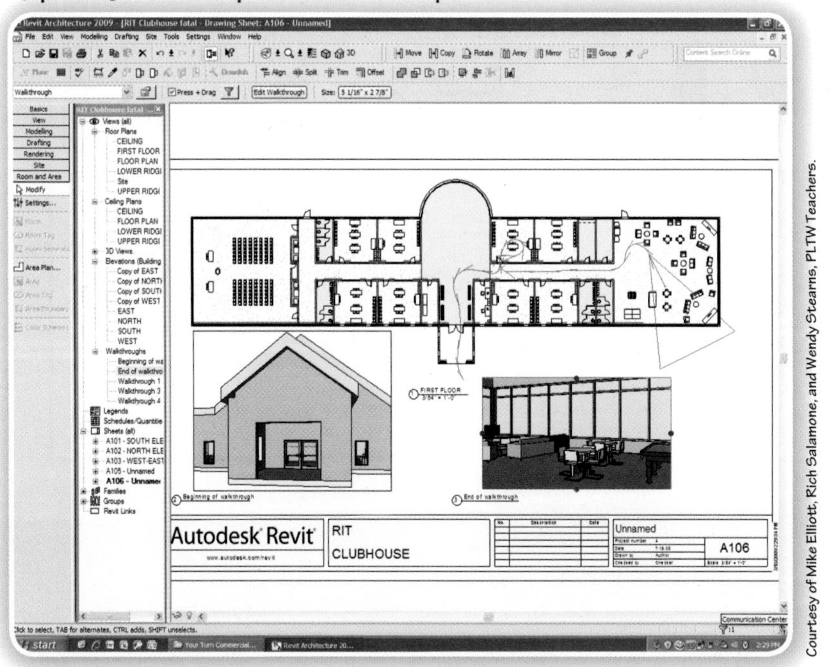

Courtesy of Mike Elliott, Rich Salamone, and Wendy Stearns, PLTW Teachers.

Case Study

Let's take a look at one of the buildings in our case study: CityCenter. The Mandarin Oriental Hotel and Residences has several floor plans available. Figure 19-16 is a photograph of a presentation poster showing the floor plan and possible furniture layout of the two-bedroom penthouse plus den.

Notice that on the bottom right of the poster shown in Figure 19-16 you can see the location of the residence in relationship to the rest of the building. The north direction symbol shows that windows face northeast. Why would prospective owners want to know this information?

Figure 19-16: *A presentation display board showing a floor plan option in the Mandarin Oriental Hotel and Residences.*

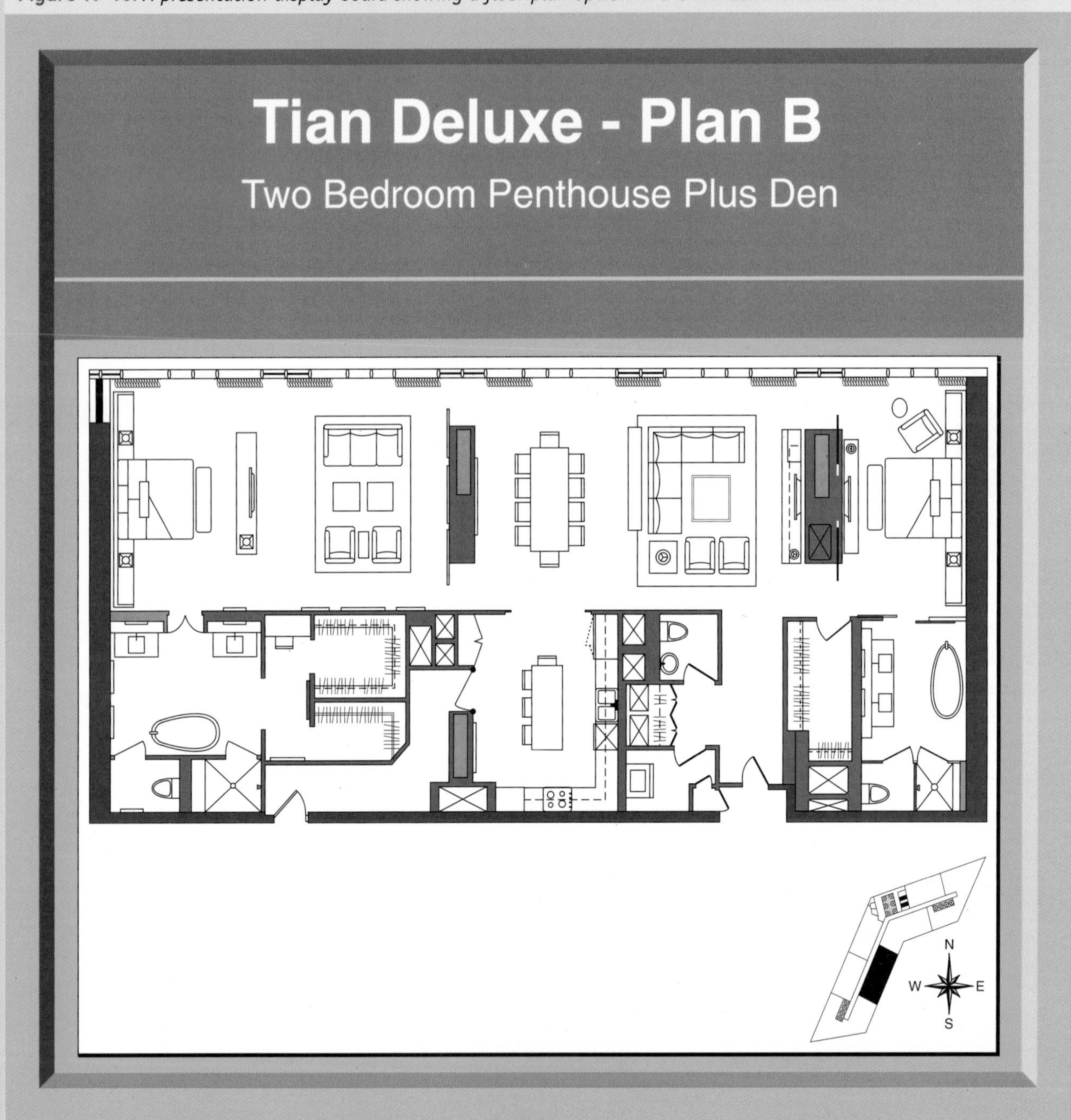

Courtesy of CityCenter, Las Vegas.

Case Study

(*continued*)

Now let's take a look at some interior renderings of the Mandarin Oriental Hotel and Residences in Figure 19-17.

Notice how the table setting, centerpiece, and lighting provide realism to the image. What other information can you gather from viewing Figure 19-17? Now look at the living room shown in Figure 19-18. Where is the living room located in relationship to the dining area and kitchen? Can you picture yourself in this room? Quality graphics inspire that feeling. Lights, shadows, and reflections enhance design visualization. Can you imagine looking out of the window at the scenery below?

Figure 19-17: *A rendering of the proposed kitchen and dining area at the Mandarin Oriental.*

Courtesy of CityCenter, Las Vegas.

Figure 19-18: *A rendering of the proposed living room at the Mandarin Oriental.*

Courtesy of CityCenter, Las Vegas

Figure 19-19: Models convey architectural details and building proportion.

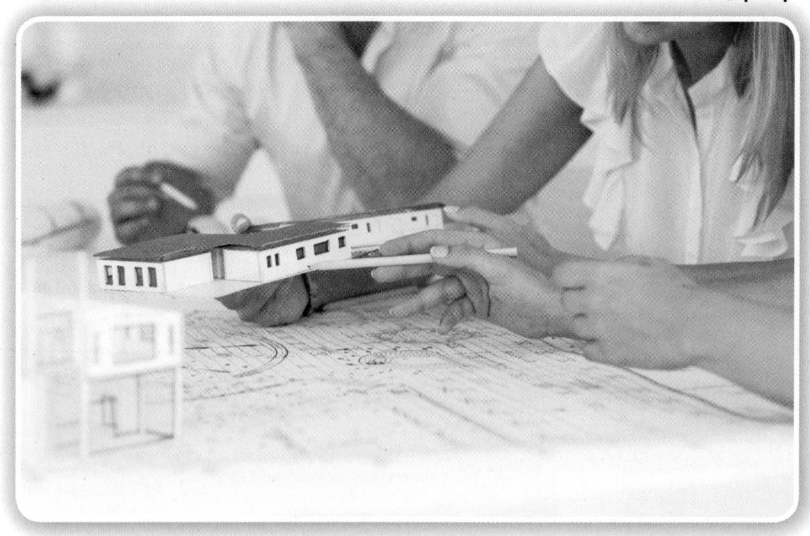

© iStockphoto.com/londoneye.

MODELS PROVIDE PERSPECTIVE AND ANALYSIS

Presentation graphics are often enhanced with physical or virtual architectural models, Figure 19-19.

Virtual models can be used for visual presentations, animation videos, or **simulations.**

Simulation: a computer-generated scenario of programmed conditions and effects.

Virtual models created with building design software can be rendered and printed for presentation graphics and slideshows or animated for video. They are also used for simulations or analysis. One example is to determine solar gain and project shadows from natural light, allowing the designer opportunity to manipulate variables to improve indoor air quality. During the design phase, virtual models can be tested to determine the effect of extreme weather conditions such as wind, rain, or snow.

Figure 19-20: A scaled model is often made of foam core material.

Courtesy of Rich Salamone, PLTW Teacher.

Figure 19-21: Star William Shatner stands next to a model of the Star Trek Enterprise. Time & Life Pictures/Getty Images.

Models have been used for decades in filmmaking. Did you know that the Star Trek Enterprise you see on television is only 5½ ft long? Use the Internet to access The Future Channel® website at *http://www.thefutureschannel.com* and select Design and Architecture > Movies. (See Figure 19-21.) You can watch Greg Jein describe the process of creating the *Star Trek* models at *http://www.thefutureschannel.com/dockets/science_technology/models_for_movies/*.
While at the site, watch the five-minute video titled *Space Architecture,* in which University of Houston architecture student Candy Feuer presents her plan for a Martian greenhouse complete with presentation graphics, physical models, and animated virtual models: *http://www.thefutureschannel.com/dockets/hands-on_math/space_architecture/*.

Physical models are constructed using a variety of materials. Depending on their purpose, some are exquisitely detailed, while others may be quite plain. Models are constructed to precise scale to communicate the building shape and proportion.

Scaled objects are achieved with precise measurement and accurate calculation.

Physical models are created by hand, using model-building tools, or with computer numerical control (CNC) or three-dimensional printing devices. Foam core board is a common material used in creating architectural models, Figure 19-20.

Careers in Civil Engineering and Architecture

THE GREENEST SKYSCRAPER

Rick Cook's architectural firm strives to design buildings that are both beautiful and environmentally responsible. Cook's firm operates as a studio, where architects at all levels of experience work collaboratively on designs. He reviews their work and presents the studio's ideas to clients.

Rick Cook, partner, Cook + Fox Architects

"One of the things I love about being an architect is that it's a way of being an artist while working with many other people and hearing their viewpoints," says Cook. "So every day is a learning experience."

On the Job

Cook's most famous project with Cook + Fox Architects is the Bank of America Tower, the second tallest building in New York City. It was designed to be America's greenest skyscraper.

"We had a blank slate: an empty site for creating a new iconic skyscraper for New York City," says Cook. "We sprang into action by doing research, gathering the best talent, and leading a brainstorming process to gather input. Then we started drawing and building models of the form that made sense for the site and for the client."

The Bank of America Tower generates about two-thirds of its annual energy from a super-efficient on-site power plant. It uses only about half of the water that a conventional building would, thanks to innovations like waterless urinals. And it provides highly filtered air.

"It's designed to be the healthiest workplace in America," says Cook.

Inspirations

At age 12, Cook visited local businesses with his father so he could decide what he wanted to do when he grew up. His revelation came when he met a sign painter with a huge wall-to-wall easel. Cook was fascinated.

"From then on," he says, "I wanted to do something where I faced a blank sheet of paper and created something new every day."

Education

Cook graduated from a five-year architecture program at Syracuse University. He got a job working in the library so he could absorb all the architecture books. During every school vacation he worked for an architecture firm to learn all he could about the profession.

In Cook's fourth year of college, he studied architecture in Florence, Italy. "That's where I learned to love cities and the nuances of culture," he says.

When Cook returned to Syracuse for his fifth year, he entered a competition among all the northeastern schools of architecture. He designed a skyscraper for midtown Manhattan and won. One of the competition's jurors worked at an architectural firm called Fox and Fowle, and after graduating college, Cook was offered a job there.

"I had a chance to work in New York City on skyscrapers and learn how to do it in real life," he says.

Advice for Students

The only way to be an architect, Cook says, is if you love it. "You must find out if you *have* to do it," he says.

Cook suggests that students visit architectural firms to learn what they can. At college, he himself went to every nearby firm and asked for a five-minute tour. "At almost every one of them, somebody stopped what they were doing and showed me around," he says. "All of us architects tend to be very generous with advice. It's a very collegial and supportive environment."

SUMMARY

Presentation graphics and models are created to enhance communication of the building design to stakeholders, community members, and other interested parties. Because these parties are interested in general information, such as the appearance, location, and purpose of the building, they may not have the technical expertise to interpret working drawings. Design visualization is stimulated with presentation drawings that feature realistic pictorial drawings and artistic interpretations of floor plans and elevations. Pictorial drawings provide three-dimensional views of the interior and exterior of proposed buildings. Images of furniture, people, cars, landscape, and adjacent properties are added to enhance realism. Building design software provides the ability to manipulate factors such as color, texture, light, and shadow. Models are used to enhance design visualization by showing a three-dimensional view of architectural details and building proportion. Building models may be physical or virtual. Virtual models provide the opportunity to create animated walkthroughs or computer simulations of weather and solar conditions. Thoughtful selection and preparation of presentation graphics and models enhance communications and maximize design visualization.

BRING IT HOME

1. Draw an interior pictorial view (by hand or using a computer) of a room of your proposed building that you believe will be a key selling point. Include furniture, appliances, and two or three decorative items. Add color and texture to the floors, walls, and ceilings.
2. Draw an exterior pictorial view, by hand or computer, showing how your proposed building will appear when viewed from the street. Include color and texture to the siding, windows, and doors. Add landscaping and small details to provide "curb appeal" to your design.
3. Develop an 8½ × 11 in trifold promotional brochure to communicate your general building plans to community members. Include a presentation quality floor plan to describe the interior layout.
4. Create a poster, suitable for display at a public building, to stimulate design visualization of your building project.

EXTRA MILE

Create a scaled physical model of your building. Build the model on a 1/8 in = 1 ft scaled floor plan print attached to a base of foam core. After constructing walls of foam, create a removable roof that allows you to view the interior layout, refer to Figure 19-20. Use natural or found materials to enhance the model. Work with other students to design a neighborhood or urban setting where all the models can be displayed together. Design roads, shared spaces, and areas for future development. Propose a list of buildings you would like to add to your new urban setting.

CHAPTER 20
Formal Communication and Analysis

GPS DELUXE

Menu

START LOCATION | DISTANCE | END LOCATION

Before You Begin

Think about these questions as you study the concepts in this chapter:

1. What is included in a formal presentation of a building proposal?
2. How is a building proposal described and justified?
3. How is support documentation organized for building proposal presentation?
4. Why are building proposals presented by a team?
5. How do presenters defend a building proposal?
6. Why do presenters request presentation feedback?
7. What happens if the desired outcome of the presentation is not achieved?

The success of a building project often hinges on the effectiveness of the formal proposal presentation (Figure 20-1). A team of professionals, knowledgeable in the details of the building design, are usually assigned to perform the task. A successful building proposal presentation requires research planning, development, practice, and reflection.

The presenters must also possess strong communication skills. They must clearly communicate the building plan and convince stakeholders of the project's value. The purpose of a formal presentation may be to inform, persuade, or gain support. Stakeholders may include clients, financial backers, local government officials, and community members. These are often the people who will determine whether or not the project continues and becomes a reality. Understanding the detrimental consequences of a weak or ineffective presentation, an architectural firm takes the formal presentation very seriously.

Figure 20-1: *Effective presentations are the result of thorough preparation and practice.*

Preparation is a key component to an effective presentation that clearly explains and justifies the building plan. The team must thoroughly research the presentation purpose, audience, and logistics (PAL) to determine the most effective support documents, graphics,

➢ INTRODUCTION

and visuals. This supporting documentation is carefully developed and organized to enhance audience or stakeholder understanding. Verbal communication is just as important. Early in the presentation, the audience is informed of the presentation's purpose, which may be to gain support, funding, or feedback. The presentation will then provide evidence of the building project's value. The presentation team must be prepared to discuss the project in detail and answer any questions that might arise. During the presentation, the team will visually survey the audience for clues or non-verbal signals that may indicate how the presentation is being received. Upon closure, presenters will summarize key points and may request audience action to support the desired outcome. Presenters will gather audience feedback, of strong and weak points, for post-presentation reflection. The presentation team must be prepared to successfully defend their project proposal when faced with opposition or criticism. Sometimes a formal presentation does not lead to the desired outcome. When that happens, supplemental information, modifications, or additional presentations may be necessary.

FORMAL PRESENTATION OF A BUILDING PROPOSAL

Formal communication of the building proposal includes several steps. The process usually begins with a team being assigned to research the presentation purpose, audience, and logistics (Figures 20-2 and 20-3). Next, the presentation team will brainstorm and select optimal content and format. Comprehensive research and planning will enhance the development of an effective presentation (Figure 20-4).

Figure 20-2: The presentation team identifies the presentation purpose, audience, and logistics.

© iStockphoto.com/Jacob Wackerhausen.

The presentation team creates an outline to help guide presentation development and determine necessary support documentation and visuals. In large architectural firms, graphic designers may be assigned to develop the desired support materials. To insure effective communication, the presentation is practiced several times. Evidence is provided to support the desired outcome or action, which may be to inform, persuade, or acquire feedback or support. Following a formal presentation, the audience outcome is compared to the desired outcome. If the desired outcome is not realized, the team may need to modify content, strategy, or develop additional materials to address audience concerns. Post-presentation analysis and reflection will help the presenters improve their presentation development and implementation skills.

PRESENTATION PURPOSE, AUDIENCE, AND LOGISTICS

Let's take a close look at each of the steps for a successful building proposal presentation. The process begins by identifying the desired outcome. What is the purpose or goal of the presentation? Is it to inform, convince, gain support, or promote action? This outcome should be clear and precise, as it will guide the presentation development.

Figure 20-3: Research of purpose, audience, and logistics (PAL) is the first step in preparing for a formal presentation.

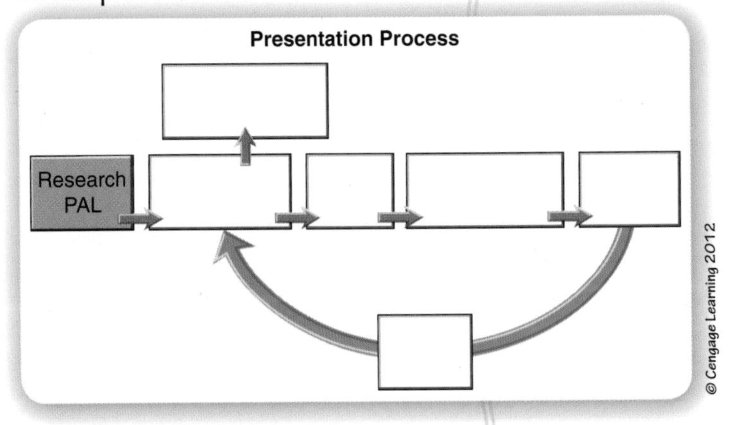

Students select and use research and information technologies to effectively communicate design intent.

Once the goal is determined, the architectural team will identify audience details, such as how many will attend and each individual's expertise. Each audience member may have a unique connection to the project and may be looking for particular information or answers. This information may be difficult to uncover, but is vital to achieving the desired outcome.

The professional background or technical expertise of the audience can determine appropriate terminology for the presentation. For example, if presenting to local community residents, presenters will avoid using unfamiliar technical terms or acronyms without explanation.

Acronym:
a word formed from initials, such as LEED (which stands for Leadership in Energy and Environmental Design).

In addition to using appropriate terminology, presentation content and format is also chosen to match the purpose and intended audience. For example, a presentation to a group of financial investors will be different than a presentation to a local community group. A presentation to local government officials, reviewing the proposal for a building permit, would take on yet another format. The presentation team must also consider the logistics of the presentation including the amount of time allotted and room features, such as layout, furniture, equipment, acoustics, and lighting. Once the presentation purpose, audience, and logistics are clearly defined, the team can begin to discuss strategies to effectively communicate and justify the building project.

Figure 20-4: Research is a vital step in preparing an effective presentation.

Courtesy of W. Stearns, PLTW Teacher.

Your Turn

Figure 20-5: Assigned presentation location.

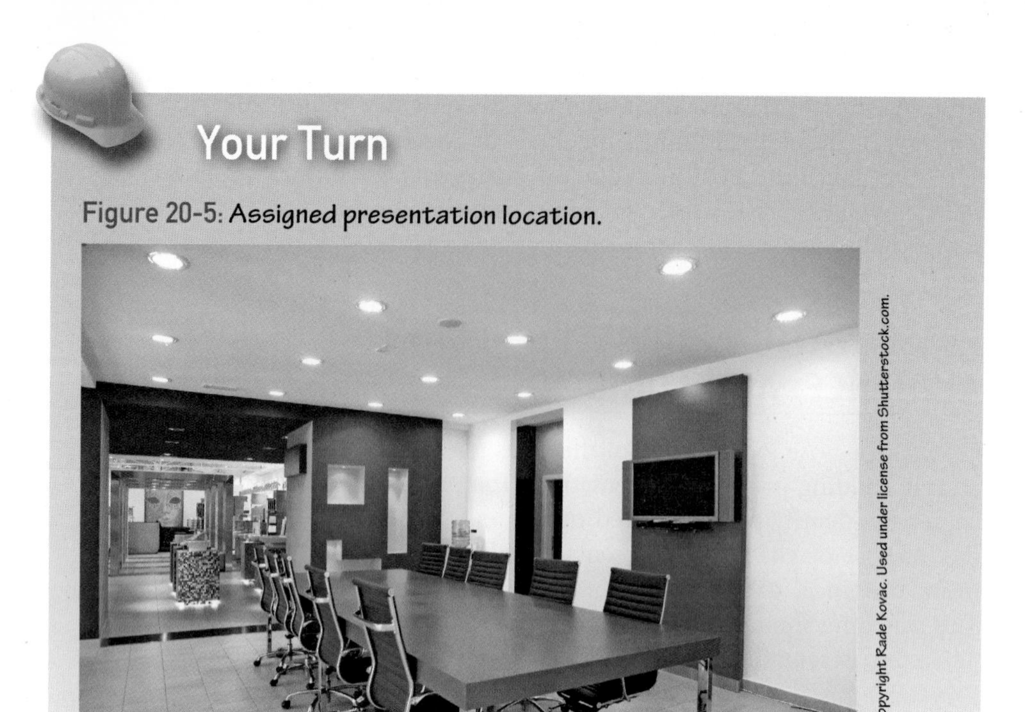

Look closely at the room shown in Figure 20-5. If this room was assigned for your building proposal to seven potential investors, what additional items would be needed? List the desired items, concerns, and questions you must address to effectively use this facility.

BRAINSTORMING CONTENT AND FORMAT

An architectural firm will dedicate many hours to discussion and analysis before selecting the content and format to communicate the project's description and justification (see Figure 20-6). Diligence in the planning phase is vital to achieving the desired outcome. Whether the goal is to inform or persuade, the project description and justification is tailored to fit the situation.

Relevant information can be gathered from the project program. The project program is described in Chapter 3: Research, Documentation, and Communication as a written document highlighting the project goals, design objectives, constraints,

Figure 20-6: Following research, the presentation team brainstorms content and format.

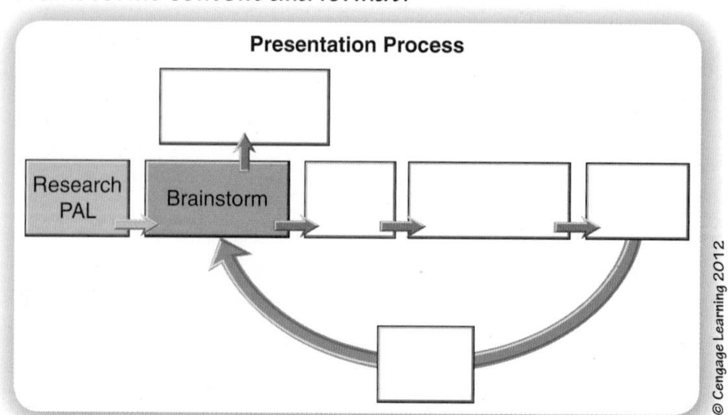

Figure 20-7: Presentation content is chosen to achieve a specific goal.

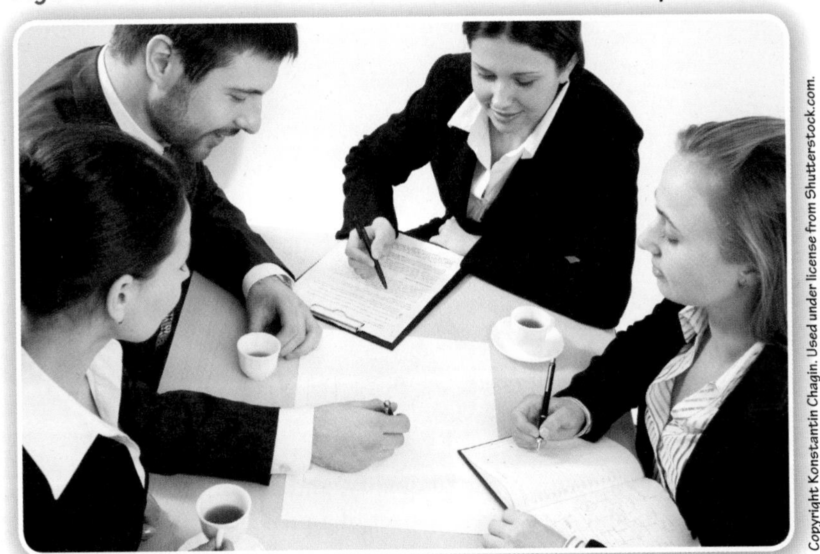

and specifications. The program generally includes detailed drawings and schedules necessary for the accurate and timely completion of the project. Although the program may contain more information than desired, this is a good resource for support documentation as relevant facts can be simplified or refined to fit the formal presentation, Figure 20-7.

Figure 20-8: Following the brainstorming process, the team prepares the presentation and documentation.

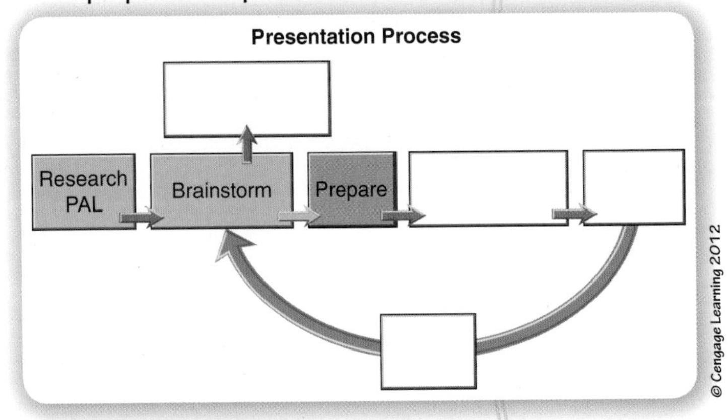

GATHER AND PREPARE PRESENTATION AND SUPPORT MATERIALS

Although a formal presentation is tailored for a specific audience, building proposals may include information about the project concept, goals and objectives, project need, permit requirements, business strategy, building and site design, environmental and community impacts, timeline, schedule, budget, sustainability, resiliency, building lifecycle, and future expansion options. The project proposal may utilize a **value-added** approach to justify the building project. Value-added examples include new jobs, greater tax base, a desired attraction or service, or area rehabilitation.

The team must prepare a realistic agenda to communicate relevant information and key points. If a **call to action** is the goal, then presenters must identify the desired action, and explain the importance of the action. An experienced presenter will make every minute of the presentation count by selecting only relevant information, and communicating an efficient, easy to understand format.

Visual media, models, and support documents stimulate design visualization and provide evidence to justify the project proposal (see Figures 20-9 a and b). Presentation graphics are designed to communicate to a specific audience. For example, a presentation of a private office building project to community members might include graphics of the building's exterior appearance and surrounding landscape as well as a study of the expected change in street traffic. A presentation to future office staff might also include visuals of the office layout, floor plans,

Value-Added: an increase in benefit, worth, or importance as a result of project development.

Call to Action: a formal request or challenge requiring a response.

Figure 20-9: *Graphics, documents, and models provide clarity to the presentation.*

Courtesy of M. Elliott, R. Salamone, and W. Stearns, PLTW Teachers.

security features, indoor environmental quality, and parking. You may think that building proposals should cover every aspect of a project, but due to time limitations and short attention spans, most presentations provide only information that is relevant to the audience and will support the desired goal.

Your Turn

Select one type of audience from the list below and identify several concerns they might have about a building proposal. You may select from the list of concerns below or add your own. Brainstorm how you might best communicate relevant information to address the concerns. Record your ideas in your notebook.

Presentation Audience (choose one)

Members of the city or town planning board
Zoning and codes officials
Community members
Prospective builders
Future building occupants
Building owners
Financial lenders
Environmental rights organization
Historical association
Chamber of commerce
Federal funding commission
Community development agency

Concerns

How will the building enhance the community?
What is the life expectancy of the building?
Does the building style fit the area?
What is the projected construction cost?
When is the projected completion date?
How is the building oriented to the site?
How much parking is available?
Is the building handicapped accessible?
What is the size and shape of the building?
What are the buildings green and sustainable features?
What is the building occupancy?
What utilities and maintenance will the building require?
What are the project risks?
What materials will be used?
Is the land zoned for this project?
Who will be insuring quality control?
What are the buildings intelligent systems and safety features?
What is the estimated operational cost?
What type of climate conditions can the building withstand?
How much natural light enters the building?
How convenient is the floor plan and traffic pattern?

Figure 20-10: Supporting documentation is selected and organized to enhance presentation clarity.

Thomas Barwick/ Photographer's Choice/Getty Images

© Cengage Learning 2012

Organization of Documentation

Have you ever had trouble finding a specific piece of information in a textbook? Without an index, table of contents, or page-numbering system, it could be time consuming and frustrating. To address this issue, experienced presenters put significant thought into organizing the presentation documentation to make it easy for the audience to navigate. Supporting documentation is usually organized in a logical sequence that correlates with the presentation agenda (see Figure 20-10).

Clear and legible documentation is vital to effective communication. Page setup, headings, paragraphs, line spacing, alignment, text font, and font size are carefully tested and selected to allow for optimum readability. Clarity is supported by a consistent, unified format throughout the presentation materials. If the proposal's purpose is to persuade, gain support, or cause action, the documentation will help the audience to make an informed decision. Documents can enhance the presentation by providing details or evidence not easily communicated verbally.

Figure 20-11: Individual document folders provide building project details and evidence.

© iStockphoto.com/Baris Simsek

Handouts are organized, numbered, and placed into folders or binders, and may be supplemented with illustrations, small presentation drawings, maps, charts, graphs, and tables, Figure 20-11. During the presentation, presenters can quickly direct the audience to a specific document by referencing a page number in the binder. Larger drawings are referenced by sheet number

Figure 20-12: **Simplicity of text font and layout enhances legibility of visual aids.**

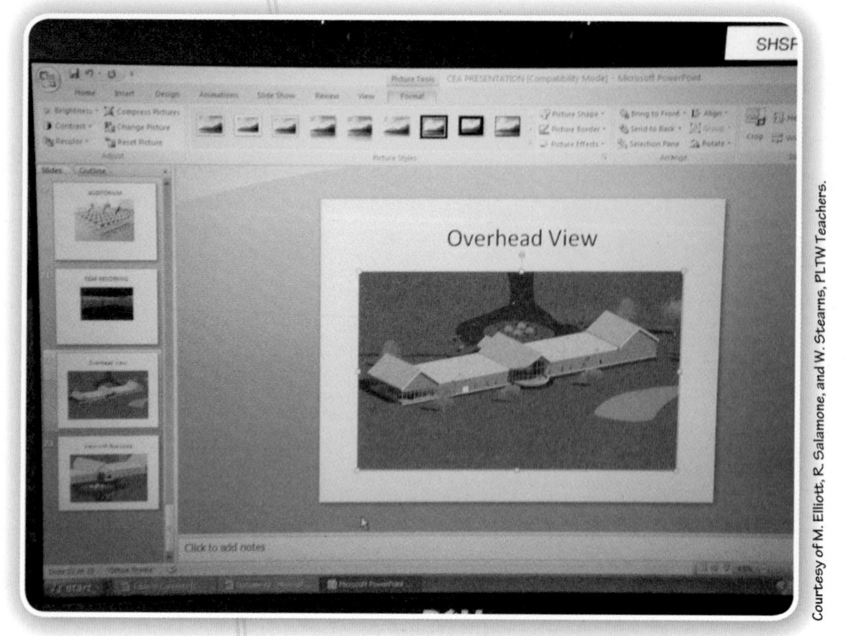

Courtesy of M. Elliott, R. Salamone, and W. Stearns, PLTW Teachers.

and are usually separate from the document binder. A formal presentation of a project proposal is often enhanced with an electronic slideshow or website demonstration. As with documents, effective slideshows and websites are consistent in format for clarity and readability, Figure 20-12. Font, size, alignment, and color are considered along with image selection, slide background, and layout.

Several formats may be tested to determine what will work best with the audience seating and room conditions. Remember, visual aids enhance the presentation and provide greater clarity of key information. Slideshows and websites may include videos, photographs, drawings, illustrations, maps, charts, graphs, tables, and bulleted lists. The advantage of a website is the ability to provide detailed information, links to related websites, and the opportunity for the audience to revisit the project information at a later time (see Figure 20-13).

Figure 20-13: **An Internet site provides the opportunity to revisit the project details.**

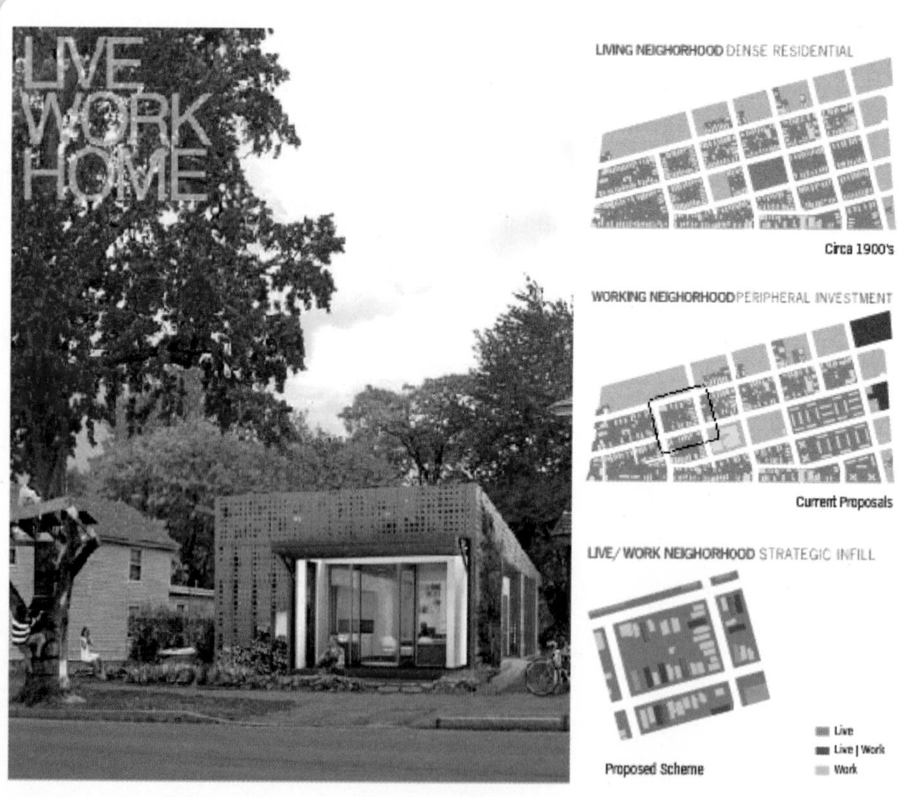

http://www.liveworkhome.com/ Courtesy of Cook+Fox ©.

Visualization of project intent can be supported by computer created drawings, three-dimensional images, and walkthroughs. Charts and graphs can communicate the construction timeline, LEED (Leadership in Energy and Environmental Design) credits, cost estimates, and the results of feasibility studies or other calculations.

Mathematical tools and models provide clarity to evidence and enhance the communication of proposals and justifications.

A slideshow is commonly used to highlight key concepts and support verbal information with pictures, drawings, graphics, models, or videos. Before the presentation, the slideshow is previewed and tested to insure legibility and correlation with the presentation. Videos are also tested to confirm they will run properly and project to the audience as desired. Other visual aids include posters, flip charts, display boards, models, project drawings, and maps (see Figure 20-14).

Visual aids are used to do the following:

- Gain interest
- Increase presentation effectiveness
- Highlight key points
- Clarify a concept or idea
- Support design visualization
- Show data and study results
- Enhance audience understanding and retention

Figure 20-14: Display boards support visualization of project intent.

COMMUNICATION/IMPLEMENTATION OF THE PRESENTATION

The Rehearsal

The presentation team usually rehearses the presentation several times to optimize delivery, content, and continuity. Experienced presenters who take the time to memorize the information and rehearse, portray confidence. Well-practiced presenters are able to focus on the audience instead of the script or notes. Presentation notes are not to be read, but instead used for reference. Notes provide brief memory joggers for key points, names, affiliations, or statistical data (Figure 20-15).

Figure 20-15: Following thorough preparation, the presentation is ready for implementation.

Presentation Process

Research PAL → Brainstorm → Prepare → Implement →

The Presentation

The actual presentation occurs after the team has practiced and refined their presentation to clearly and convincingly communicate the project within the allotted time.

Figure 20-16: Presentation materials and equipment should be set up prior to the audience arrival.

Courtesy of M. Elliott, R. Salamone, and W. Stearns, PLTW Teachers.

The day of the presentation, the presenters will arrive early to set up materials and test their equipment (see Figure 20-16). The time remaining between setup and the presentation may be used to informally welcome the audience as they arrive. This is generally referred to as meet and greet. Once seated, the audience is welcomed and presenters are formally introduced. The introduction includes a brief overview of each presenter's experience, expertise, and project involvement to establish credibility and invoke audience confidence. Occasionally the presenters may ask each member of the audience to briefly introduce themselves. This technique can be time consuming and is generally only used with a small audience. Because the presentation has been rehearsed several times, the team has predetermined who will perform supporting roles such lighting and equipment control. To keep within a given timeframe and avoid being side-tracked, most presenters request that questions be held until the end. This is respectfully requested of the audience before the presentation begins.

Presentations often begin with a short story or joke to put the audience at ease. It is important to select stories or jokes that are not offensive and do not reflect bias. The story should make a clean transition to the project proposal. For example, one presenter began his presentation with this story:

> Yesterday I ran into my friend Bob who asked me about a building I designed a few years ago that had its own co-generation plant. "So how does it work?" he asks. Excited about the project I took the next 20 minutes, without taking a breath, to explain all the details about how the co-generation plant worked. When I was done, Bob stared at me and said: "'Working well' would have answered my question." After a brief pause for laughter the presenter continued: "As we present our project this evening I'll try to be sensitive to why you are here and what you want to know. If you are interested in specific information such as how a co-generation plant operates, I'll be happy to meet with you after the presentation."

Figure 20-17: **Presenters emphasize building project benefit and value.**

This brief story put the audience at ease and communicated respect for audience time and interest in the project. Another option is to tell a story that helps the audience envision a future enhanced by the building project. If done effectively it will both excite the audience and provide a mental vision of the completed project.

Communicating Features and Benefits

Most presenters will emphasize project benefit and value, both at the beginning and end of the presentation (see Figure 20-17). If the presentation goal is to request a specific action, the presenter should communicate this at the beginning, so the audience will be looking for information to support their decision. For example, if you are presenting a building proposal in competition with another architectural design firm, you might communicate to the audience that your proposal offers something unique or distinctly advantageous and worthy of their support. Whether the advantage is flexibility, sustainability, or a cost-effective design, the presenter would give a preview of what the audience should expect to see. One example is the Live-WorkHome building proposal, designed by Cook+Fox architects. LiveWorkHome was a winning entry in the From the Ground Up Competition and featured a modern style, incorporating flexibility and nature on small, single-family urban lot (see Figure 20-18). Unique features of their proposal included a dozen skylight tubes to increase natural lighting and a perforated wrap, patterned to create the illusion of sunlight filtering through the trees.

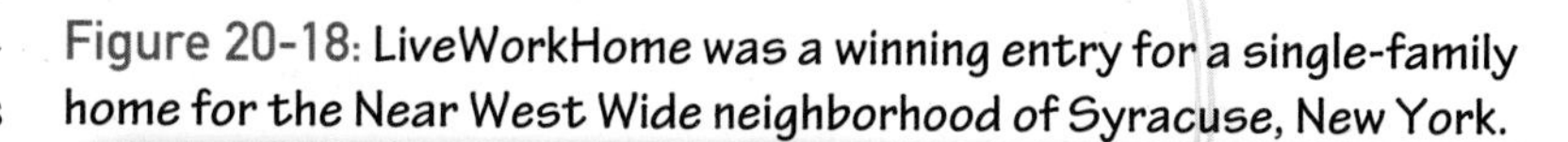

Figure 20-18: **LiveWorkHome was a winning entry for a single-family home for the Near West Wide neighborhood of Syracuse, New York.**

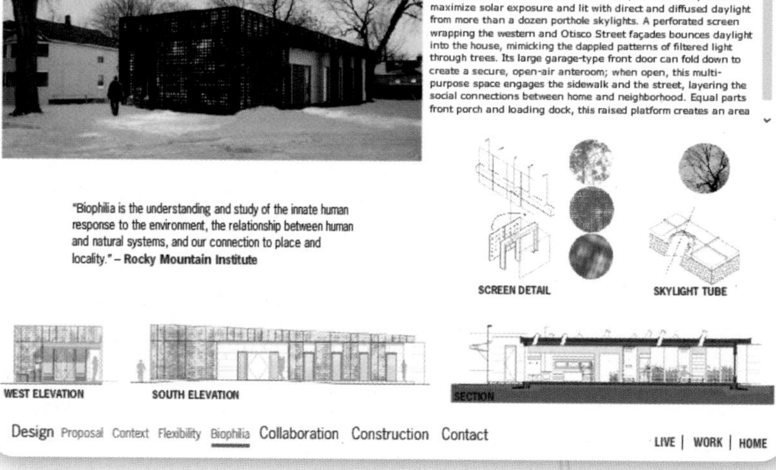

Figure 20-19: Experienced presenters look for clues to determine audience reaction.

© iStockphoto.com/Jacob Wackerhausen.

Presentation Components

The presentation uses an introduction, body, and closure to highlight key factors that substantiate the proposed building's value, advantage, or claim. The **introduction** identifies the claim, presentation goal, and request for action; the **body** provides proof and justification to support the claim, goal, or action; and the **closure** summarizes, insures the claim's benefit and value, and requests audience support or action.

The body of a formal presentation provides evidence for the audience to consider in support of the pre-established objective or desired outcome. It is important to present the information in an organized and sequential manner to avoid confusion. Team members plan and practice their presentation to communicate a cohesive message. During the presentation, team members are attentive and supportive of each other, giving attention to the presenter and not distracting the audience. The presenter uses eye contact to establish a connection with the audience. During the presentation, the team will look at audience body language and other non-verbal clues that may signal comprehension or confusion. Audience behaviors can also signal physical discomfort or restlessness resulting from a room that is too warm or cold or from the audience's inability to see or hear what is being presented (Figure 20-19).

The building proposal presentation is presented in a logical format to make a case for the desired outcome. The desired outcome may be to inform, persuade, seek approval, gather feedback, or promote a specific action. Simple visuals highlight key points to support verbal information. All statements or claims should be supported by evidence, such as examples, comparisons, testimonials, quotations, estimates, and other statistical information. Charts, graphs, and tables, used to present evidence, can be added to slideshows or document folders. During the presentation, avoid disagreements or conflicting information that may confuse or negatively impact the audience.

How does the presentation team defend their building proposal? Sometimes a presentation will face opposition based on audience bias, faulty or limited information, or misconceptions. Some presenters take a proactive approach and address anticipated opposing views and concerns before they are voiced, while other presenters provide prepared answers only if questions arise.

6 PRESENTATION CLOSURE

The presentation closure will revisit key points and objectives, answer any questions, and request audience action (see Figure 20-20). The beginning of the closure should be clear to the audience. Some presenters use cues such as "In summary" or "In closing," while others revisit their claim and provide a synopsis or conclusion. A strong conclusion can support a positive outcome.

Questions should be answered by the presenter who is most experienced with the topic. Some questions may require additional research for a complete and

Figure 20-20: The presenters gather audience feedback during closure.

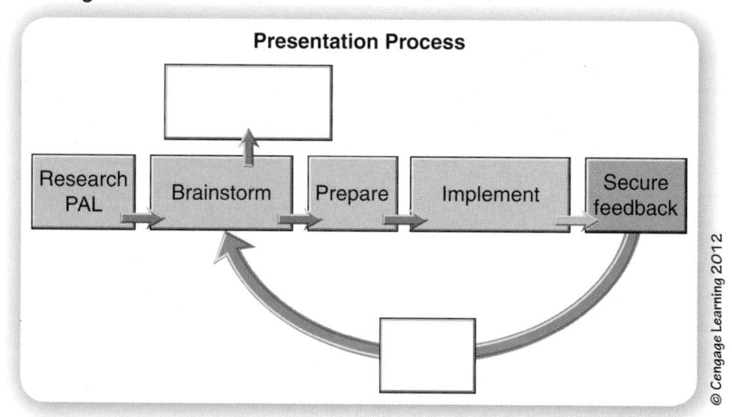

Figure 20-21: Unanticipated questions may require additional research and time.

© iStockphoto.com/Joshua Hodge.

accurate answer. If this is the case, it is best to request additional time and follow up with reliable and substantial information, rather than provide weak, or possibly inaccurate, information (Figure 20-21).

The presenters should portray confidence that the information presented to support the proposal or claim is accurate and based on current or accepted practice. Once the key elements and evidence are summarized, and questions have been adequately answered, the presenters will generally communicate a statement of confidence. An example of a confidence statement to a community group might be something like this: "As you can see from the research presented, the building project will enhance your community by providing much needed affordable housing to seniors." A confidence statement to a group of investors might be: "As presented, our projected cost analysis of the senior housing building project predicts a 20 percent profit the first year at only 75 percent occupancy." A presentation for a senior housing design competition might end with the following confidence statement: "After viewing the other proposals, we believe you will select our plan for its cost-effective design, senior friendly floor plans, amenities, and the many green and sustainable features."

Following review of the main points, a question-and-answer period, and a convincing confidence statement, the presenters will communicate what action they expect from the audience.

If the goal is to provide information and gather feedback, a hard-copy or online questionnaire can be used. If the goal is to persuade project support, then the audience may be encouraged to take a specific action such as approve financing or vote accordingly. The final piece of closure is to thank the audience and acknowledge appreciation for their time and attention (Figure 20-22).

Figure 20-22: Closure should be evident to the audience.

© iStockphoto.com/AVAVA.

FEEDBACK SUPPORTS PRESENTATION ANALYSIS

Even though it may take several days before learning the results of the audience action, presenters generally discuss the presentation as soon as possible. A discussion of presenters' observations and perceptions provides feedback for analysis and may enhance future presentations (see Figure 20-23). The team can view a video or audio recording of the presentation to determine strong and weak points. The presenters will look for audience reactions and note specific questions and who posed them. If the desired outcome is not achieved, the team may refer back to these notes to identify attributing factors to address in future presentations (Figure 20-24).

Figure 20-23: **Following a formal presentation, the architectural team will analyze the outcome to determine presentation effectiveness.**

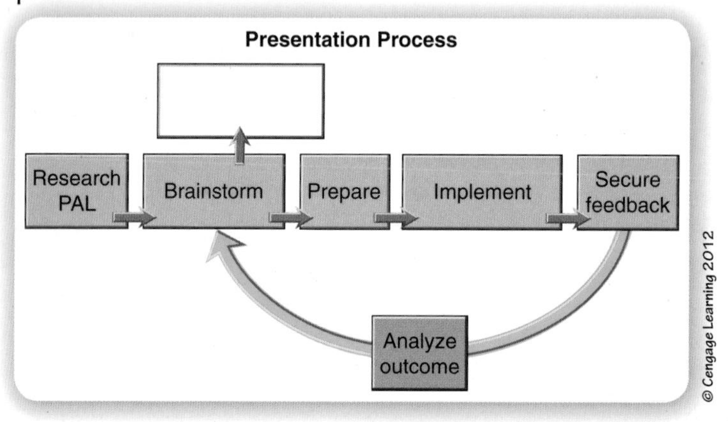

Figure 20-24: **Questionnaires can be a good source of feedback.**

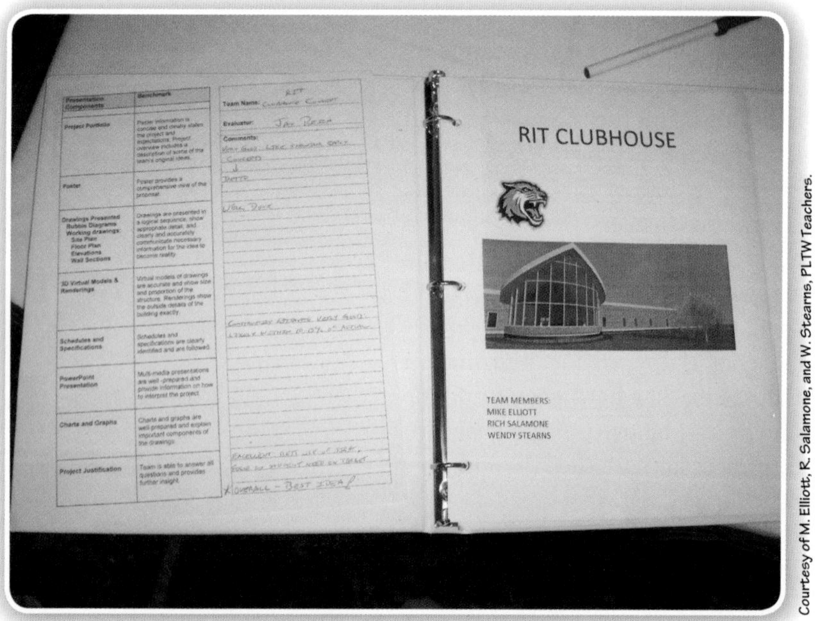

Figure 20-25: Modification or supplementation may be necessary if the desired outcome is not achieved.

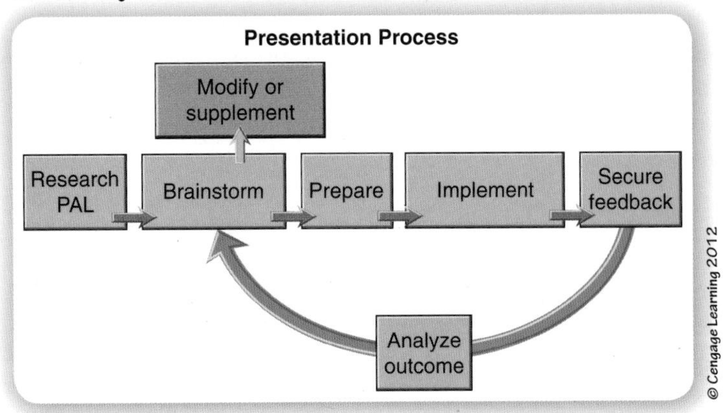

MODIFICATION OR SUPPLEMENTATION TO ACHIEVE DESIRED OUTCOME

7

Even a well-prepared and implemented presentation may require modification or supplementation (see Figure 20-25). Sometimes the audience may pose a question or concern that is not anticipated. Additional questions may surface following the meeting. All questions need to be answered. Let's look at the senior housing example. After the presentation, one member of the audience wondered if the building would be pet friendly, and if so, what pet accommodations would be available. After the architectural team confirmed the plan for a pet-friendly building, they highlighted the convenience of a nearby dog park. This led to other questions about shared spaces in the building, how they would be kept free of pet dander and allergens, and how tenants would dispose of pet waste.

Occasionally a building plan may need to be modified to produce the desired outcome (Figure 20-26). Financial backers might insist on a better cost-profit

Figure 20-26: Supplemental information or modifications may be necessary to achieve the desired outcome.

ratio, requiring a change in design, building material, or other features. Local governance might specify additional green space, parking, or special curbing before a permit is issued. When major modifications to the building project are made, the team may be asked to come back for a follow-up presentation to show how the new plan has addressed specific concerns. It is not unusual for an architectural firm to determine and propose options or trade-offs to achieve a satisfactory compromise or resolution.

All technological solutions have trade-offs, such as safety, cost, efficiency, and appearance.

Careers in Civil Engineering and Architecture

PUTTING IT ALL TOGETHER

Life is never dull for architect Pamela Campbell. She works on multiple projects at once, all of them interesting. She sees them through various stages and meets with a wide range of people.

"Architects are coordinators to a great extent," Campbell says. "We pull different bodies of knowledge together to make a complete physical creation. It's very satisfying to put it all together for a project."

Pamela Campbell
Senior Associate, Cook + Fox Architects

On the Job

For six years, Campbell has worked on Henry Miller's Theatre in New York City's Broadway. She was part of the team that developed the building, and she has followed it through to construction. She works with the construction manager, the owner of the project, the tradespeople, and the theater company that will occupy the site. She also communicates with the press and others interested in the project.

"In this stage, I spend about a third of my week on the construction site for the theater," Campbell says. "I answer questions, do sketches, and try to resolve issues that come up during construction."

Campbell is also working on a house in Syracuse, New York, called the LiveWorkHome. It's a prototype intended to provide low-cost housing for single families. Part of the challenge is incorporating an environmentally sustainable element into a tight budget.

"It's been a fun process," Campbell says. "It's a lot smaller and a lot faster than the Henry Miller's Theatre project."

Inspirations

Campbell always liked to make things, even as a child playing in the yard or building sand castles.

"There's a cliché with architecture students that you start playing with Legos and you suddenly know you're meant to be an architect," she says.

Campbell enjoyed art classes in school, as well as math and science.

"Architecture is one of those professions that bring together lots of fields," she says. "It's a combination of the creative side of things and practical considerations, such as physics and measurement."

Campbell toured universities and found herself attracted to the atmosphere of architectural classes. "The creative aspect is what drew me to it," she says. "I liked all the little models and drawings. You are given a project and have to create something in physical form."

Education

Campbell enrolled in the Mackintosh School of Architecture in Glasgow, Scotland. She liked the fact that the program wasn't classroom-based.

"It was about self-motivation and discovering for yourself," she says.

In one of her first projects at school, Campbell had to build a children's playhouse. She measured her "clients," figuring out how big their hands were so she could create a structure that suited them. "It was a small project," she says, "but a fun starter place. As we went on, the projects became larger and more complex."

Advice for Students

Campbell suggests that high school students find an architect to shadow for a few days.

"I would try to get a sense of what the daily work is like before you start university and commit to it completely," she says.

SUMMARY

Architectural firms realize the importance of a well-planned and implemented presentation. Successful presentations are the result of diligent planning, practice, implementation, and assessment. The process begins with research to identity the presentation purpose or desired outcome. The goal may be to inform, convince, persuade, or gain support. The presentation team continues their research to determine audience expertise and connection to the project. Additional research will uncover presentation logistics, including parameters such as facility design and amount of time allotted. Once the purpose, audience, and logistics are thoroughly analyzed, the presentation team brainstorms to determine content and optimal presentation format. Next, a presentation outline is developed and supporting documentation, drawings, graphics, models, computer simulations, and other visual aids are created. The presentation will have an introduction, body, and closure. Once all the pieces are ready, the team will practice the presentation and prepare answers to address anticipated questions. During the presentation, team members are attentive and supportive of one an other and work together to control lighting, run equipment, and visually survey the audience. The closure revisits key points and objectives, answers questions, and requests audience action. A strong conclusion and statement of confidence will support a positive outcome. Audience feedback will occasionally identify concerns requiring modifications or additional information. If this is the case, a follow-up presentation may be necessary.

BRING IT HOME

1. Identify the desired outcome or purpose of your formal building proposal presentation. Are you looking to inform, persuade, seek approval, or promote a specific action? Are you competing against other proposals? What feedback do you hope to acquire? Discuss these questions with your team and record the answers in your project notebook. Write a short paragraph that describes the desired presentation outcome and how you want the audience to react and respond to your presentation.
2. Gather as much information about the audience who will be viewing your presentation. Determine the size of the audience and identify each member's company, title, technical experience, and connection to the project. This information can be obtained through a pre-presentation questionnaire or contact by phone, e-mail, or person-to-person. Record the team findings in your project notebook.
3. Research the presentation logistics. Find out how much time your team is allowed for setup, presentation, question and answer, feedback, and cleanup. Take a tour of the presentation room and sketch in your notebook the room layout and other details. Sketch and label the number and size of tables and chairs. Make a list of other items such as a podium, microphone, screen, and computers. Add the location of electrical outlets and light switches to your sketch. Before you leave the room, have your team discuss the following: Do you have internet access? Will you need an extension cord? If the room is too light, can you close the blinds? Are there enough tables for your displays and models? Will you need fasteners to attach drawings to a bulletin board? Can the chairs be arranged so everyone will be able to see and hear? Make a list of items not readily available in the room that you will need for the presentation, such as an extension cord, laser pointer, or whiteboard markers.
4. In your notebook write a short paragraph about how the goal, audience, and logistics will guide your presentation preparation.
5. Create a presentation outline to guide the development of your project proposal presentation. Include the main points to be covered in the introduction, body, and closure.
6. Plan an introduction to your formal presentation that includes one of the following:
   a. a story or appropriate joke that transitions to the project proposal
   b. the audience envisioning a future enhanced by your building project

EXTRA MILE

Attend a project proposal for your local community and take notes of how the information was communicated, what supporting documentation was provided, and the overall outcome. It may help to research local news articles for public forums explaining building proposals. Ask if your local school district has plans for an architectural firm to present a building proposal or renovation to the community.

GLOSSARY

A

100-year flood: a flood that has a one percent chance of occurrence in any given year.

Accessibility: degree in which the environment or structure is suitable for people with physical challenges.

Accessible: meets requirements of federal regulations for access including the Americans with Disabilities Act, the Architectural Barriers Act, and the Fair Housing Amendments Act.

Accessible routes: paths of travel that are designed to be easily navigated by persons in a wheelchair.

Acronym: a word formed from initials, such as LEED (which stands for Leadership in Energy and Environmental Design).

Adjacency matrix: a technique used by space planners to determine where rooms should be placed identifying essential space relationships, compatible zones, and shared space options.

Aesthetics: appearance or artistic quality pleasing to the eye or senses.

Allowable soil bearing pressure: the maximum pressure that a foundation may induce on the underlying soil for design purposes. The allowable soil bearing pressure includes a factor of safety. Also known as allowable foundation pressure.

Allowable strength: nominal strength divided by the safety factor.

Allowable stress: allowable strength divided by the appropriate section property, such as section modulus or cross-section area.

Alternating current (AC): an electrical system in which voltage and current are reversed periodically in the circuit.

Altitude: an angle between 0 and 90 degrees which is measured up from the horizon; zero degrees altitude falls exactly on the local horizon and 90 degrees altitude is straight up.

Ambient light: light that is already present in a space before any additional lighting is added; usually refers to natural light but can also refer to artificial lights.

Ampere (amp): the amount of electric charge passing through a given point per unit of time.

Annotations: explanatory or essential comments or notes used to provide more information.

Anthropometrics: man (anthro) measurements (metric), the study of human body measurements.

Applied force: an external force that loads a structure or other object.

Architects: "licensed professionals trained in the art and science of building design." (U.S. Department of Labor Bureau of Labor Statistics Occupational Outlook Handbook)

Architectural style: the classification of a structure based on distinguishing characteristics and design elements.

Architecture: responsible and purposeful design of functional, sustainable, and aesthetically pleasing buildings.

Architrave: the principle beam and the lowest member of the entablature, resting directly on the capitals of a series of columns.

Allowable stress design: a method of designing structural elements such that the allowable strength is greater than or equal to the strength necessary to support the required load combinations.

Asymmetry: a lack of symmetry; a composition in which both halves are not identical.

Atterberg limits: basic measures of water content at important points of transition in the state of fine grained soils used to distinguish amoung types of silts and clays; include the liquid limit and plastic limit.

Axial load (force): a load (force) that acts along the length of a member, either tension or compression.

Azimuth: an angle represented by the compass direction from which sunlight is coming. At solar noon, the sun is always directly south in the northern hemisphere and directly north in the southern hemisphere.

B

Backflow: water traveling back into the main distribution system, usually by siphoning.

Backsight (BS): a rod reading taken when "looking back" to a point of known elevation.

Balance: a principle of design describing the relationship among the various aspects of a structure, landscape or other object as they relate to an imaginary centerline. There are three types of visual balance: symmetric, asymmetric, and radial.

Ballast: a piece of equipment required to control the starting and operating voltages of electrical gas discharge lights.

Baluster: a small, decorative post that supports an upper rail; a row of balusters is called a balustrade.

Baseline: one of the principal east-west lines used for survey control in the rectangular survey system that divides townships between north and south; the baseline meets its corresponding principal meridian at the point of origin.

Beam: structural member that carries a load that is applied transverse to its length.

Beam diagram: graphical representation of a beam indicating the type of beam supports, applied loads, and beam span.

Bearing: a direction indicated by an angle measure, less than 90 degrees, measured from either due north or due south in an easterly or westerly direction, e.g., the bearing North 45 degrees East (N 45° E) indicates a direction 45° from due north toward the east (clockwise, in this case).

Benchmark: a permanent mark that establishes the exact elevation of a point; used by surveyors as a starting point for surveys to establish elevations of other locations.

Bending moment: the algebraic sum of all moments at a cross-section caused by forces applied between the cross-section and one end of the structural member.

Bending stress: an internal tensile and compressive stress developed in a beam to resist the bending moment. Also known as flexural stress.

Berm: a mound of soil created to provide privacy, protect against water, insulate against the elements, provide sound reduction and/or for visual appeal.

Best practice: proven performance or process of highest standard.

Biophilic: in tune with nature.

Bottom plate: the horizontal framing member connecting the bottom of a wood stud frame to the floor of a building's structure; also known as a sole plate.

Braced frame: a structural frame which includes diagonal braces designed to transfer lateral building loads to the foundation.

Branch circuit: the portion of the wiring system that extends beyond the service panel to the devices to which power is delivered.

Bridging: framing members placed in between spanning floor or roof members to prevent twisting and buckling.

Broadleaf trees: plants that have leaves that are thin and flat and that produce fruit and flowers; most broadleaf trees are deciduous.

Brownfield: an abandoned or underused industrial or commercial site that is available for re-use; often have environmental contamination or are perceived to be contaminated.

Buckling: the sudden displacement out of alignment of a structural member subjected to compression stress.

Buffer zone: a vegetated zone adjacent to a structure or group of buildings, used to minimize the effects of the development; often used to separate different land use.

Buildable area: the area within a site on which building is not restricted by codes or regulations.

Building area method: method in lighting design for determining the overall power usage based on square footage, building use, and activity.

Building codes: regulations, ordinances or statutory requirements established or adopted by a governing unit relating to building construction and occupancy.

Building footprint: the shape of the area that a building covers, easily identified by the floor plan.

Building Information Modeling (BIM): the collaborative process of generating and managing building data using electronic three dimensional, real-time, dynamic building models linked to a database of project data. Data includes quantities and properties of building components, geographic information, llight analysis, and test sumulation results predicting how the building will perform under projected conditions.

Building Integrated Photovoltaic (BIPV): photovoltaic units that have been joined directly into building components.

Building life cycle: all stages of a building over time including construction, use, maintenance, preservation, revitalization, deconstruction, and demolition.

Building optimization: a building designed to function at its best or most effective.

Building regulations: set of guidelines designed to uphold standards of public safety, health, and construction; controls the quality of buildings.

Built environment: man-made surroundings created to accommodate human activity, ranging from personal residences to large urban developments.

C

Call to action: a formal request or challenge intended to illicit a response.

Candlepower: a description of the level of light intensity in terms of the light emitted by a candle of specific size and constituents.

Capacitance (C): the amount of electrical charge stored.

Capacitors: devices designed to store electric charge.

Capital: the crowning member of a column or a pilaster.

Carbon dating: common method used by archeologists to estimate the age of organic remains and other substances rich in carbon up to 60,000 years in age.

Cartographic survey: a survey that combines information from both a topographic survey and a hydrographic survey.

Cast in place concrete: concrete members formed and poured at the building site in the locations where they are needed.

Ceramics: a group of materials made from non-metallic minerals fired at high temperatures.

Charette: a work session of about 12 to 30 people involved in the design, construction, and operation of a building project.

Check valve: pipe fitting that allows fluid to flow in one direction only, closing when fluid reverses direction; a check valve may be of the swing, lift, or ball type.

Chord: a structural member that is part of an assembly; i.e., an open web steel truss is made up of top and bottom chords held in place by an open web system.

Civil engineering: The profession of designing and providing specifications for public works such as roads, bridges, harbors, water systems, and buildings.

Cladding: external finish covering the base material on a wall.

Clerestory: a vertical extension beyond the single-story height of a room, generally with windows.

Closed plan: a design that generally features individual offices, separate work and break areas, and full height walls.

Coarse-grained soil: soil in which 50 percent or more, by weight, of the soil is retained on the no. 200 sieve; alternatively 50 percent or more of the sample is composed of sand and/or gravel.

Cogeneration: generation of electricity, hot water, and steam from the same fuel source.

Color: integral part of design and decorating, referring to the dye or tint, that helps distinguish materials and accent shapes.

Color-rendering index (CRI): referes to how well a lamp renders color as compared to a reference light source of the same correlated color temperature.

Column: a vertical member that is responsible for supporting compressive loads and transferring those loads to the foundation of a structure.

Commercial structure: any building intended for carrying out a business or service.

Compact section: a steel shape that is capable of developing yield stress through the entire cross-section before local buckling will occur.

Compliance: a state of being in accordance with established guidelines, specifications, rules, codes, or regulations.

Composite deck: a combination floor framing system composed of a corrugated metal sheet fastening to the beams and then strengthened with steel reinforcing rods and concrete.

Composite materials: engineered materials made from two or more constituent materials with significantly different physical or chemical properties.

Composite runoff coefficient: a value equal to the weighted average of the runoff coefficients of the various surfaces that occur in a given area; coefficients are weighted by the percentage of total area covered by each surface type

Compression: a force tending to press together or shorten an object.

Compression member: structural componants that resist axial compression.

Concentrated load: a load that acts on a relatively small area and is often represented as a point load.

Conductor: material with low resistance to the flow of electricity; copper aluminum, silver and gold are all considered good conductors.

Coniferous trees: trees that have leaves that look like needles or small scales (resembling fish scales), and cones. Most coniferous trees are evergreen, meaning they do not lose their leaves during the winter. Examples of coniferous trees include cedars, pines, spruce, and hemlock trees.

Connection diagram: a graphic representation that provides instruction as to how wiring terminals (and equivalent) are connected along a circuit pattern; also known as wiring diagrams.

Constraints: a consideration that limits a decision, design, or project.

Constructability: the ease with which a material can be cut, shaped, and joined to other materials.

Construction document(s): a set of drawings, details, schedules, and written specifications that represent a design and that can be universally understood within the construction industry.

Construction survey: a survey that locates exact points and elevations for civil engineering and architectural projects; also known as an engineering survey.

Contour interval: the change in elevation between adjacent contour lines on a topographic map.

Contour line: a line on a map indicating points of equal elevation.

Contrast: a noticeably different appearance from something else when placed side by side, often achieved with color, shade, or texture.

Control survey: a survey used to establish precise horizontal and vertical positions of points that serve as a reference framework for other types of surveys.

Corbelling: a construction technique in which rows of stone or brick wall or chimney are gradually offset outward.

Cornice: a decorative projection or crown along the underside of the roof where the roof hangs over the wall.

Correlated color temperature (CCT): the color appearance of a light source; that is, the color of the light that the source emits when lighting a true white surface or the color of the source itself when looking directly at it.

Covenant: a contract in which a homeowner agrees to abide by rules and regulations for property use often created for a planned development.

Criteria: needs, rules, standards, or tests used to evaluate a decision, design, or project used to determine its effectiveness.

Crown elevation: the elevation at the top of a pipe at a given location.

Cupola: a small structure located above the roof for the purpose of adding light and air, or for ornamentation; historically used as a lookout.

Curb appeal: attractive external appearance of a commercial or residential property.

Current (I): the flow of electricity in an electrical circuit; the unit is given in amperes (amps).

Curtain wall (systems): metal framework, typically filled with glass, spanning several floors.

Cut: the removal of naturally occurring earth material.

D

Dead load: the weight of construction materials used in the building, such as the roof, walls, floors, cladding, etc., and the weight of fixed service equipment, such as plumbing, electrical feeders, heating, ventilating, and air-conditioning systems and fire sprinkler systems.

Decentralized wastewater treatment system: independent onsite sanitary wastewater treatment system that collects, treats, and disperses or reclaims wastewater.

Deciduous trees: trees that lose their leaves during the cold or dry season, often called hardwoods; examples include oaks, maples, dogwoods, and birch trees.

Deed: a written document that provides the transfer of ownership of real property from one person(s) to another and must include a description of the property.

Deed restriction: a clause in a deed that limits the use of a property.

Deep foundation: a foundation that transfers structural loads to the earth well below the ground surface and includes long foundation members such as piles or caissons.

Deflection: the movement of a structural member that results from loading.

Deformation: a change in the shape of a structure or structural member caused by a load or force acting on the structure.

Demand charge: an additional charge imposed by an electrical utility for electricity used during periods of peak demand.

Demographic data or demographics: population characteristics such as race, age, household income levels, mobility (in terms of travel time to work or number of vehicles available), educational levels, home ownership, and employment status, etc.

Dentil: a small rectangular block in horizontal series often applied at the cornice.

Design development: a phase that defines and describes all important aspects of the project so that all that remains is the formal documentation step of construction contract documents.

Design load: the load applied to a structure or structural component during design determined by the required load combinations.

Design storm: a selected storm event, described in terms of the probability of occurring once within a given number of years, for which drainage or flood control improvements are designed and built.

Details: magnified drawing views enhanced with additional graphics and annotations to clarify detailed, atypical, or unusual construction.

Detention pond: a pond that collects, temporarily stores, and then slowly releases stormwater into the municipal stormwater system; also known as a "dry pond."

Diagonal brace: an inclined structural member that is part of a braced frame and carries axial forces.

Diaphragm: a roof, floor, or other flat plate that transfers lateral forces applied to a building (such as wind or earthquake loads) to the lateral force resisting system.

Differential leveling: surveying technique that involves establishing the elevation of a horizontal plane of sight in order to determine the differences in elevation between two or more points with respect to a datum.

Differential settlement: a condition in which the rate of settlement differs across a structure.

Dimmer switch: an electrical device used to regulate the amount of power dispersed to a light controlling its brightness.

Direct current (DC): an electric current that flows through a circuit in one direction regardless of the flow rate.

Distribution box: part of a septic system where liquid wastewater is dispersed into perforated pipes buried under the ground.

Dormer: a structure with a window, added to a sloping roof.

Drawing conventions: widely accepted drawing techniques intended to speed the drawing process and enhance communication and understanding.

Dry flood-proofing: creating a watertight seal below flood level to protect habitable space in a structure from inundation of floodwater. Dry flood-proofing includes sealing the exterior envelope, penetrations, and openings.

Ductile metals: metals capable of being drawn into a wire or hammered thin.

Duration: the period of time over which rain is measured. For example in the case of annual rainfall measurements the duration is one year.

E

Easement: the privilege to pass over the property of another for a specific purpose.

Efficacy: the efficiency of light energy emitted by a bulb per unit of power output; given in units of lumens/watt.

Egress: exit or the means of exiting; may also refer to the right to exit.

Elastic limit: the maximum stress a solid can sustain without undergoing permanent deformation.

Elastic modulus: the ratio of axial stress to strain within the elastic range of a material. It is a fundamental material property and is an indication of the stiffness of the material.

Elastic region: on a stress-strain curve, the portion of the graph that is linear and to the left of the elastic limit. The material acts elastically in this region and returns to its original size and shape when the load is removed.

Electrical (floor) plan: a plan that indicates wiring between light fixtures and switches and may indicate the types of wall switches and circuit loads, electrical devices, and equipment overlayed on an architectural floor plan

Elements of design: features including lines, space, form, shape, color, value, and texture that define the appearance of a product or structure.

Elevation: the height of a point above an adopted datum, such as mean sea level (MSL).

Elevation drawings: drawing views that show an orthographic projection of a building and indicate vertical dimensions, materials, architectural design, and construction details not apparent in the floor plan.

Embodied energy: the energy consumed by all processes associated with the production and use of a material, beginning with the acquisition of natural resources and ending with the final demolition of the product.

Emphasis: principle of design describing stress or prominence given to an element of a design creating points of attention that attract the viewer's eye.

Endangered species: a species in danger of extinction.

Energy charge: basic rate at which electricity is charged.

Energy neutral: a building design that incorporates low-energy consumption practices that are often significantly lower than required by building regulations.

Engineered wood: a material made by bonding wood strands, veneers, lumber, or fibers to produce larger composite units that are stronger than solid wood members.

Entablature: an order of horizontal moldings and bands that rest directly on the capitals and columns.

Environmental Protection Agency (EPA): an independent regulatory agency of the United States responsible for establishing and enforcing environmental protection standards .

Environmentally Preferred Products (EPP): materials that have lesser or reduced effect on human health and the environment when compared to competing products that serve the same purpose.

Equilibrium: a physical state in which the forces acting on an object counteract each other and do not cause motion.

Equivalent length: length of straight pipe that creates the same head loss as a particular pipe fitting.

Ergonomics: the science of designing to meet the physical needs of the occupant or user.

F

Façade: the exterior "face" of a building.

Factor of safety: the ratio of the maximum strength of a member to the probable maximum load to be applied to it.

Ferrous metals: iron-based metallic materials.

Fill: the addition of soil, rock, or other materials; can also refer to the material being deposited.

Fine-grained soil: soil in which more than 50 percent, by weight, of the soil passes the no. 200 sieve. In other words, more than 50 percent of the soil is composed of silt and/or clay.

Finish: final treatments that are applied to a design or construction element.

Finish grade: final elevation of the ground surface after excavating or filling.

Fixed support: support that resists translation and rotation.

Floor framing systems: the primary way that horizontal loads are carried to the beams or columns; also provide lateral support for walls.

Focal point: a design element that most strongly draws a viewers attention.

Foresight (FS): rod reading taken when "looking ahead" to a point of unknown elevation.

Form: a principle of design described by geometric shape, outline, or structure.

Formal balance: a sense of equilibrium or stability achieved by symmetry in a design such that elements are arranged equally on either side of a central axis.

Foundations: part of the structural system that transfers the load of a building to an area of the ground.

Framing systems: basic framing systems help support the building structure not only from the exterior but from the interior side of the building by transferring loads both horizontally and vertically down to the foundation of the building.

Free-body diagram: a graphical representation of an object showing all external constraints and forces acting upon it and all geometric measurements necessary to model the body.

Frieze: the middle section of the order; below the cornice of a wall and above the architrave.

Frost heave: movement of a foundation or structure caused when water in the voids of the supporting soil freezes and expands.

Function: an object or structure's purpose or performance.

Furnishings: furniture and other items that are deemed necessary or useful for comfort or convenience.

G

Gable: vertical triangular portion of a side building wall with a pitched roof; fills the space created by the roof pitch.

General notes: notations that pertain to the entire drawing set.

Genius loci: subtle and unique quality of a place; the atmosphere that characterizes a place.

Geodetic survey: a land survey that takes into consideration the curvature of the earth's surface.

Geographic Information System (GIS): a computer based mapping system referenced to a common spatial coordinate system that integrates geographic data with other relevant information such as transportation networks, hydrology, population characteristics, economic activity, political jurisdictions, environmental characteristics, etc.

Geotechnical investigation: below-ground investigation by boring, sampling, and testing the soil strata to establish its compressibility, strength, and other characteristics likely to influence a construction project.

Geotextile: a fabric or textile used as an engineering material in conjunction with soil, foundations, or rock to provide a range of functions including soil reinforcement, moisture barriers, and erosion control.

Glazing: hard shiny coating or type of glass.

Global Positioning System (GPS): a constellation of satellites that orbit the Earth and continuously beam radio waves that can be used to determine the location of the receiver.

Grading: the process of changing the topography of a property for a purpose.

Gravity load: a load, such as a dead load or a live load, that results from the affect of gravity on an object and acts in a downward direction.

Gravity sewer: a sanitary sewer pipe that slopes downward away from a building toward the sewer main such that the flow within the pipe is caused by gravity.

Gray water: used water from sinks, showers, and bathtubs (but not toilets) that can be recycled for use in flushing toilets, irrigation, and other non-drinking purposes.

Green power: electricity generated from renewable sources (solar, wind, geothermal, biomass, or small hydro).

Groundwater: water located below the earth's surface in soil pore spaces and in rock fractures.

H

Habitable space: space used for living, sleeping, cooking and eating, i.e., bedrooms, living rooms, dining rooms, kitchens, and family rooms.

Hardscape: the permanent, man-made features of a landscape.

Harmony: the quality of forming a pleasing and consistent whole.

Header: a structural member that spans the distance above a framed opening, such as a door or a window.

Heat-island effect: built-up areas that are hotter than nearby rural areas; commonly associated with urban areas due to heat storage capabilities of man-made infrastructure thus causing temperature to be higher than rural areas.

Height of instrument (HI): elevation of the line of sight of a level.

Highest and best use: the property use that would produce the most profit.

Hipped roof: a roof style that slopes down to the eaves on all sides.

Horizontal datum: a collection of reference points on the earth that have been identified according to their precise latitude and longitude.

Hose bibb: a faucet angled downward with male hose threads on the spout; used for outdoor water supply.

Hydraulic length: the longest drainage path within the drainage area to the point at which the water exits the site.

Hydrographic survey: a survey that provides data for mapping shorelines and the bottom of bodies of water; often conducted using SONAR (Sound, Navigation, and Ranging), which uses sound waves to determine depth and to find objects in the water.

I

Illuminance: the amount of light energy reaching a given square foot_area, the unit is in footcandles.

Impervious: incapable of being penetrated.

Infiltration: uncontrolled and unintended flow of outdoor air into a building through cracks and other openings.

Informal balance: a sense of equilibrium or stability achieved in an asymmetrical design such that visual elements of varying weight balance one another around a fulcrum point.

Infrastructure: (1) the basic facilities, services, and installations needed for the functioning of a community or society such as transportation and communication systems; water, sewer, and power lines; and public institutions including schools, post offices, and prisons; (2) the underlying foundation or basic framework of a system such as the internal systems of a building including the electrical, plumbing, HVAC, and telecommunications systems.

Ingress: entrance or the means of entering; may also refer to the right to enter.

Insulator: material with high resistance to the flow of electricity; paper, rubber, neoprene, and other synthetic materials are good insulators.

Integrative Design: recently adopted as an ANSI standard, an Integrative Design Process is defined as: "A discovery process optimizing the elements that comprise all building projects and their interrelationships across increasingly larger fields in the service of efficient and effective use of resources." (American National Standards Institute)

Intelligent building: a self-regulating computer-controlled building.

Invasive species: a species of plant or animal that is not native to an area and whose introduction or spread threatens the existing ecosystems by causing damage to native species and their habitat.

Isolated footing: a spread footing that supports an individual load or column. Isolated footings under a single column are often referred to as column footings.

J

Jack arch: a brick or stone element supporting openings in the masonry. Decorative patterns and elements are often incorporated to visually frame architectural features; also known as flat arch or straight arch.

K

Keystone: the central stone generally placed at the top profile of arched elements.

Kilowatt/hour: a unit of energy output per hour use. A kilowatt is equal to 1000 watts.

L

Land Information System (LIS): a special type of Global Information System most often used by municipal agencies that manage information related to land ownership such as parcels, land use, zoning, and infrastructure.

Land surveying: the science of determining the relative positions of points on or near the earth's surface.

Landing: level area at entry doors or between flights of stairs.

Lateral bracing: diagonal bracing, shear walls, or other structural components that transfer building lateral loads down toward the foundation.

Lateral load: a force or pressure that is applied horizontally or perpendicular to a surface.

Lavatory: bathroom sink.

Layperson: a person without background or expertise in a subject. (Encarta)

Leach field: part of a septic system comprised of the area onto which wastewater is distributed; also known as an infiltration field.

LEED: leadership in Energy and Environmental Design; a voluntary national rating system developed by the U.S. Green Building Council for construction of high performance, sustainable buildings that reduce negative environmental impact and improve occupant health and well-being (now includes LEED for Homes with slightly different categories).

Legal description: a written passage or statement that defines property and is descriptive enough so that the property can be differentiated from other properties and located without other evidence.

Lending institution: an organization that provides financing for individuals or companies in exchange for a fee. Typically lending institutions charge interest on the money provided.

Leveling rod: a graduated pole or stick used with surveying equipment to measure differences in elevation.

Life Cycle Analysis (LCA): a detailed procedure for compiling and analyzing inputs and outputs of resources, energy, and associated environmental impacts over the lifecycle of a product or structure.

Light wells: openings or shafts that allow light and natural ventilation to enter buildings.

Line: an element of design; lines can be vertical, horizontal, diagonal, or curved.

Lintel: a beam supporting the weight above.

Liquefaction: the process through which soil, when loaded, suddenly transitions from a solid state to a liquefied state.

Liquid limit: moisture content of a fine-grained soil at the boundary between the liquid and plastic states.

Live load: a load produced by the use and occupancy of the building including, but not limited to, the forces produced by people and movable equipment; but is not an environmental load such as a wind load, snow load, rain load, earthquake load, flood load, or dead load.

Load: demand, the amount a system can handle.

Load-bearing wall: wall that carries a load other than its own weight.

Load path: the chain of structural components through which a load is transferred.

Load Resistance Factor Design (LRFD): an alternate method of designing structural members such that the loads are increased by load factors and the nominal strength is determined by reducing the maximum strength of a member by a resistance factor.

Lot and block: a method of legal description which provides for subdivision of a tract of land into lots of various shapes and sizes within larger blocks of land.

Low Impact Development (LID): stormwater management approach that uses green space, native landscaping, and techniques that mimic a site's pre-development water cycle.

Low-emissivity (Low-e) glass: a treated glass that allows visible light to pass through but reduces radiant heat transfer by reflecting infrared wavelengths

Lumen: the SI unit of luminous flux (luminous power) indicating the perceived power of light.

Lumen depreciation: the loss of light over time typically between 10 to 40 percent of a lamp's initial output.

Luminaire: a complete lighting fixture that includes the bulb, housing, ballast, and lens.

Luminaire efficiency: the ratio of light output from a luminaire to the total light output by the lamp or lamps in the luminaire.

M

Manifold: a chamber with ports for receiving and distributing a fluid or gas.

Masonry: construction made of stone, brick, and/or concrete units, including decorative and customized blocks.

MasterFormat TM: trademarked title of a uniform system for indexing construction specifications published by the Construction Specifications Institute and Construction Specifications Canada.

Material property: a property that depends only on the material from which an object is made, e.g., elastic modulus, yield stress, elastic limit.

Mechanicals: plumbing, electrical, HVAC (Heating, Ventilation, and Air Conditioning), and protection systems.

Metal track: horizontal framing member connecting the bottom or top of a metal stud wall assembly.

Metes and bounds: a method of legal description in which the boundaries of the property are described by a distance and direction between property corners (metes) or by less accurate descriptors (bounds).

Mitigation: the restoration, creation, enhancement, or preservation of a wetland, stream, or critical habitat area that offsets expected adverse impacts to similar nearby ecosystems.

Modillion: an ornamental bracket, placed under eaves, in series, for support or decoration. An S-shape or scroll easily identifies classic modillions. Later modillions are in the form of a plain block.

Modulus of elasticity: see elastic modulus.

Moisture content: the ratio of the weight of water to the weight of the dried soil in a sample expressed as a percent; values of more than 100 are possible.

Moment (of a force) about a point: the tendency of the force to cause rotation of a body about a point. It is equal to the product of the magnitude of the force and the perpendicular distance to the force.

Mortar: a plastic mixture of cementitious materials, water, and a fine aggregate used in masonry construction.

Mortgage: a document that gives a lender rights to a property as security for a loan and is considered a property lien against the lot.

Movement: a principle of design, a feeling of action or flow when viewing a structure's form.

Multilevel switching: a system that uses two or more separate light circuits each of which is controlled by a different switch.

N

Native plant: a plant that existed in a place prior to European contact and has co-evolved with local animals and environment.

Needs analysis: the activities involved with the collection of information about the goals, needs and function of the project, design expectations, and available budget.

Net-zero building design: a building requiring zero energy from a utility provider because it is able to produce as much or more energy then it needs.

Neutral axis: the plane along the length of a member on which there is no tension or compression.

Nominal strength: the force in a member that will result in yeilding of the entire cross-section.

Nonferrous metals: metallic materials in which iron is not a principal element.

Non-habitable space: spaces that are not prime living areas, i.e., bathrooms, laundry rooms, closets, pantries, utility rooms, kitchenettes.

Nonpoint source pollution (NPS): pollution that originates from many diverse sources and is picked up and carried by rainfall or snowmelt over and through the ground to ground or surface water.

Normal stress: a stress that acts perpendicular to the cross-sectional area of a member.

O

Occupancy sensor: electronic device that detect the presence or absence of people and turn lights on and off accordingly.

Occupant type: describe the age category of a group of occupants.

Off-gassing: the process by which solid materials evaporate at room temperature causing chemically unstable substances to slowly release contaminants.

Ohm's Law: developed by George Simon Ohm in 1827; describes the relationship of current, voltage, and resistance of a DC circuit. V (Voltage) = I (Current) × R (Resistance).

One-line diagrams: a drawing that indicates principal relationships between major pieces of electrical equipment.

Open plan: generally features informal workspaces grouped by commonality with low or no partitions.

P

Palladian window: a large, multi-paned window unit with a large center arched section and two short, narrow side windows.

Parapet: a low wall along the edge of a roof, balcony, platform, or terrace.

Partition wall: wall that serves to separate spaces but does not carry structural loads.

Pattern: elements that repeat in a predicable manner.

Pediment: a decorative gable end extension that emphasizes a building's width giving it a sturdy appearance; pediments are often visually supported by columns.

Percolation test: technique used to estimate the rate of water infiltration into the soil.

Permeability: measure of the rate that water or moisture will flow through a material.

Perspective drawing: a pictorial drawing that provides the illusion of depth by converging horizontal lines and represents the object's depth to a single vanishing point on the horizon (for one-point perspectives) or to two vanishing points on the horizon (for a two-point perspective). (PLTW CEA Curriculum)

Perspective view: a pictorial drawing that shows depth and proportion of the object as it would appear to the human eye.

Photosensors: electronic controlling device that adjusts the lighting levels based on the amount of natural light coming into a room.

Photovoltaic: a device for converting radiant energy into electric energy. Also known as a solar cell.

Pictorial drawing: a drawing that shows a three-dimensional view of the object.

Pilasters: a rectangular or half-round column attached to a wall to provide the appearance of columns without the expense of actual columns.

Pile foundations: a deep foundation used to carry and transfer the load of the structure to the earth well below the ground surface and includes long foundation members such as piles or caissons.

Pinned support: a support that will allow rotation about the support but resists translation in all directions.

Pipe fitting: a joint or connector used in a pipe system; i.e., an elbow, union, or tee.

Plane survey: a land survey that ignores the curvature of the earth's surface and assume a flat planar surface.

Plastic: a group of man-made polymers that contain carbon atoms covalently bonded with other elements.

Plastic limit: the moisture content of a fine-grained soil at the boundary between the non-plastic state and the plastic state.

Plastic region: on a stress-strain curve, the portion of the graph that is between the elastic limit and the point of fracture. The material acts plastically in this region and will not return to its original size and shape when the load is removed.

Plasticity index: the mathematical difference between the liquid limit and the plastic limit.

Plat: a plan or map of an area of land intended to be filed for record that shows boundaries, geographical features, some easements, and rights of use over the land. Be aware that not all easements are shown on the plat.

Plenum: a closed chamber with higher pressure than the surrounding atmosphere.

Poorly graded soil: soil that does not contain a good representation of all particle sizes. A poorly graded soil may contain a narrow range of particle sizes (uniformly graded) or not contain one or more ranges of particle sizes (gap graded).

Portico: a porch-type structure with a roof supported by columns or walls that leads to the entrance of a building.

Potable: suitable for drinking.

Potable water: raw or treated water that is considered safe to drink.

Pre-cast concrete: concrete cast in a form and cured before it is lifted into its intended position.

Precipitation intensity: the average amount of rain that falls per hours of a storm; precipitation intensity is calculated by dividing the amount of rainfall that falls during a storm (in inches) by the duration of the storm (in hours).

Presentation drawings: drawings that communicate the general ideas for the development in an colorful easy to understand format.

Pressure: the force applied to a unit area of surface.

Principal meridian: one of the principal north-south lines used for survey control in the rectangular survey system and which divides townships between east and west. The principal meridian meets its corresponding baseline at the point of origin.

Principles of design: the basic rules or standards that describe a design : balance, emphasis, movement, pattern, repetition, proportion, rhythm, variety and unity.

Private area: area of a site to which access is restricted and not open to the general public.

Profile: a vertical section showing the ground surface; may show existing and new grade lines and or soil strata.

Program: a written document highlighting the project goals, design objectives, constraints, and specifications.

Property lien: a legal state applied to a property that gives the mortgage company the right to collect payment of a debt from the sale of the property.

Property survey: a land survey that establishes property lines; also known as a boundary survey or cadastral survey.

Proportion: a principle of design describing comparative relationships between elements in a design with respect to size.

Public accommodation: a facility operated by a private entity, whose operations affect commerce and are used by the public such as places of lodging, restaurants, movie theaters, retail stores, schools, and many more.

Public area: area of a site open to the general public

Publicly Owned Treatment Works (POTW): a state or municipally owned system that is involved in storage, treatment, recycling, and reclamation of municipal sewage or industrial wastes of a liquid nature. It includes sewers, pipes, and other conveyances if they carry wastewater to the POTW treatment plant.

Q

Quoins: bricks or stones placed at the corners of a building for visual emphasis or support.

R

Rainfall depth: the total amount, reported in inches, of rainfall during a storm

Rainfall intensity: the ratio of the total amount of rain (rainfall depth) falling during a given period to the duration of the period. Also known as precipitation intensity.

Range lines: lines parallel to a principal meridian marking off the land into 6-mile strips; known as ranges in the rectangular survey system.

Ratchet clause: a provision under which the demand charge for a period is based on the highest measured demand over the previous year.

Raw material: any natural resource that is used to make finished products.

Reaction force: a resistive force that results from an applied force.

Reflected ceiling plan: a drawing indicating all components that appear on the ceiling of an interior room or space and ceiling heights. It shows the ceiling as if reflected onto a mirror on the floor.

Regrading: the process of changing the topography of the land by adding or removing soil.

Regulation: a rule adopted by a regulatory agency.

Render: to portray something artistically. (Encarta)

Renewable resource: resource or raw material that can be grown and replaced.

Repetition: a principle of design that describes reproduced or reoccurring elements or a pattern.

Resistance (R): the rate at which electricity flows through a material; is an internal material property.

Restrictive covenant: see Covenant.

Retention pond: a permanent on-site pond used to manage stormwater in which pollutants are allowed to settle out or be removed by biological activity; also known as a "wet pond."

Return period: the length of time, on average, over which an event (or an event of greater magnitude) is expected to occur not more than one time. For example, a Category 3 hurricane may have a return period of 100-years which means that a Category 3 hurricane (or stronger) is expected to occur no more than one time, on average, within 100 years.

Rhythm: a principle of design that describes the illusion of flow or movement created by having a regularly repeated pattern of lines, planes, or surface treatments.

Right-of-way (ROW): the strip of land granted through an easement.

Rigid frame: a structural frame which includes rigid connections between members designed to transfer lateral building loads to the foundation.

Rise: vertical distance.

Riser diagrams: a drawing that expresses the physical relationship of the components of an electrical (or plumbing) system.

Roller support: structural support that allows translation in the direction of the roller and allows rotation about the support, but resists translation perpendicular to the direction of the roller. Also know as a rocker support.

Rolling blackouts: the intentional interruption of service by utilities affecting small areas in succession as a means of conserving electricity when supply is low.

Roof framing systems: a structural system used to support the roof of a building.

Rotation: the act or process of turning about a center or an axis.

Route survey: a land survey used to map existing routes or layout new projects with long extents such as highways, pipelines, canals, etc.

Run: horizontal distance.

Runoff coefficient: a number that represents the percentage of rainwater that is not absorbed or does not infiltrate a given surface.

R-value: resistance to heat flow.

S

Sanitary wastewater: liquid waste containing animal or vegetable matter in suspension or solution, including liquids containing chemicals in solution.

Sash window: window where one or both of the glass panels (sashes) overlap and slide up and down inside the frame; often called single or double-hung windows.

Scale: (1) on a drawing, the ratio of a distance on the drawing to the corresponding distance on the actual object; (2) the proportions or size of one part of a design in relationship to another part.

Schedules: a clean, concise way to itemize the features of building components or applications.

Schematic diagram: a drawing that provides a simplified illustration of the parts of a system.

Section: an area of land 1 mile square, containing 640 acres. Normally, 36 sections make up a township in the rectangular survey system.

Section property: a property of a structural member that depends only on the cross-sectional size and shape of the member. Examples: Area moment of inertia, plastic section modulus, elastic section modulus.

Section view: a view, created from a cutting plane passed through an object or building portion, to show internal structure or components.

Septic tank: a large underground concrete box that is designed to hold wastewater for approximately two days.

Service area: area of a site used for service equipment (e.g., HVAC equipment and utility meters) and service activities (e.g., trash disposal and storage) necessary for the operation of the facility.

Setback: (1) the minimum legal distance from a property line or street where improvements to a site can be built or the minimum distance from the property lines to the front, rear, and sides of a structure; (2) the actual distance from a building to a road or property line.

Shallow foundation: (1) a foundation that supports lighter loads and distrubutes them to the ground; (2) a foundation that supports a structure on soil at or near the ground surface. Slabs-on-grade, strip footings, spread footings, and mat foundations are examples of shallow foundations.

Shape: an element of design referring to the contour, profile, or silhouette of an object.

Shear and bending moment diagrams: graphical representations that indicate the internal shear force and bending moment (respectively) at every point along the length of a beam.

Shear modulus: the ratio of shear stress to strain within the elastic range of a material.

Shear stress: an internal stress in which the material on one side of a plane pushes on the material on the other side of the plane with a force parallel to the plane.

Shear wall: a wall that transfers a lateral load down toward the foundation.

Sick Building Syndrome (SBS): a term used to describe building-related health complaints such as headaches; dizziness; nausea respiratory problems; eye, nose, or throat irritations; and skin problems.

Sieve analysis: a procedure commonly used to assess the particle size distribution of soil in which the soil sample passes through a series of sieves of decreasing opening size; also known as a "gradation test."

Sight distance: the distance along a roadway visible to a driver.

Sill plate: a structural member that spans the distance below a framed opening such as a window.

Simple beam: a beam that is supported on one end by a roller support and on the other end by a pinned support.

Simulation: a computer-generated scenario of programmed conditions and effects.

Single-phase system: producing, carrying, or powered by a single alternating voltage; supplies 120 or 240 V three-wire connection including a neutral ground from the utilities to the home.

Single-pole, single-throw switch: a switch that turns on a light, receptacle, or device from a single location by closing the circuit; also known as a "single-throw switch."

Site: an area of land, typically one plot or lot in size, on which a project is to be located.

Site discovery: a process of research and investigation aimed at gathering information about a site in order to make informed decisions about the viability of development on the site.

Site orientation: placement of a structure on a site with consideration to the environmental and physical conditions present.

Site plan: a drawing, or set of drawings, representing a view of the project site when looking down from above, highlighting important site elements.

Slope gradient: the steepness of the slope, or rate of change of the elevation represented by the difference in elevation between two points, given as a percentage of the distance between those points.

Softscape: the part of the landscape consisting of plants.

Solar collector: a device used for extracting the energy from the sun directly into an alternate use.

Solar constant: the average density of solar radiation measured outside earth's atmosphere and at earth's mean distance from the sun, equal to 0.140 watt per square centimeter.

Solar heat gain: the increase in temperature of an object exposed to direct sunlight that results from solar radiation.

Solar orientation: placement of a structure on a site with consideration given to the orientation of the sun throughout the day and throughout the seasons.

SONAR: a system using transmitted and reflected underwater sound waves to detect and locate submerged objects or measure the distance to the floor of a body of water; originally an acronym for Sound, Navigation, and Ranging.

Sound orientation: placement of a structure on a site with consideration to the potential affect of sound and noise.

Space: (1) the dimensions of height, depth, and width within which structures are defined and built; (2) an element of design describing the areas composing a structure; positive or negative visual space.

Space activities: a general determination or classification of a room or space function.

Space adjacencies: proximity to other room.

Space dimensions: refer to the width, length, and height of the room.

Space planning: the design of safe, comfortable, and efficient spaces.

Space-by-space method: method in lighting design for determine the overall power use based on square footage of the building spaces and their uses.

Specific notes: notations that refers to a certain item.

Specifications: a set of technical requirements that provide a detailed description of materials and quality of work required for something to be built. Specifications are typically published in a written document that accompanies the working drawings. The working drawings and specifications together are referred to as construction documents.

Spread footing: the part of a shallow foundation that transfers loads from a building to the underlying soil and is usually constructed of reinforced concrete.

Stakeholder: a person or entity that has some interest in a project; stakeholders can be community residents, businesses, construction and design professionals, funding sources and/or government agencies.

Standard: a level of quality or excellence that determines a level of attainment that is acceptable. It is not a law or an enforceable code unless it is adopted by a governmental entity.

Static head (at a point): the vertical distance between the point and the free water surface; also the level to which a liquid will rise in a piezometer.

Statically determinant beam: a beam for which the reactions can be calculated using only the equations of equilibrium.

Statically indeterminate beam: a beam for which the reactions can not be calculated using only the equations of equilibrium.

Statics: a branch of physical science that deals with forces on bodies at rest.

Stop valve: a pipe fitting used to shut off the fluid flow by an outside force (like turning a knob) and can be one of four general types: gate, globe, butterfly, and ball valves.

Stormwater wetlands: a permanent shallow pool of diverted rainwater that incorporates wetland plants and where pollutants are removed through settling and biological activity; also known as "constructed wetlands."

Strain: the change in size or shape of a material caused by the application of external forces, also called deformation.

Strength: the load carrying capacity of a structural member.

Stress: a measure of the magnitude of the internal forces acting between the particles of the body exerted as a reaction to external forces on the body. Stress is expressed as the average amount of force exerted per unit area of the surface on which internal forces act.

Strip footing: a spread footing that continuously supports a wall; also knowns as a continuous footing.

Subcontractor: a person or business hired by the general contractor to perform a specific task.

Subdivision: a neighborhood created by breaking up property into smaller lots.

Surface reflectances: the ratio of the amount of electromagnetic radiation, usually light, reflected from a surface to the amount originally striking the surface; typically describe the ceiling, wall, and floor finishes.

Surface water: water that is located on the earth's surface in streams, rivers, ponds, lakes, wetlands, and oceans.

Sustainability: building performance, operation, and maintenance that saves both money and natural resources.

Symbol: a written or printed sign used to represent something else; used on working drawings to indicate drawings, rooms, levels, building orientation, components, materials, etc., that are quickly identifiable.

Symmetry: a balanced or harmonious appearance achieved by locating similar shapes at equal distances from the center.

Synergy: a result achieved when a unit or team becomes stronger than the sum of its individual members.

T

Target market: a specific group of consumers for which a company designs or produces its products and services.

Task light: lighting which is focused on a specific area to make the completion of visual tasks easier.

Tension: a force tending to stretch or elongate an object.

Terrain orientation: the placement of a structure on a site with consideration to the characteristics of the land on which the structure is built.

Texture: element of design referring to the surface look or feel.

Theodolite: a precision surveying instrument having a telescopic sight for establishing horizontal and vertical angles.

Threatened species: a species that is likely to become endangered.

Three-way switching: controling a single or group of lumminaires from two different locations through the use of 2 three-way switches

Time of concentration: the travel time of a particle of water from the most hydraulically remote point in the contributing area to the point under study.

Title: a written document providing evidence of the right to legally possess or dispose of property.

Tongue and groove: a type of joint that has one edge with a protruding thinner piece of wood that is inserted into a groove-cut edge of an adjoining piece to form a tight uniform fit.

Topographic survey: a land survey used to gather data to prepare topographic maps that show the location of natural and man-made features and the contours and elevation of the ground; often used by engineers and architects when designing improvements to a site.

Topography: (1) the features on the surface of an area of land; (2) the configuration of a surface including its relief and the position of its natural and cultural features.

Total dynamic head (TDH): the actual pressure of moving water in a pipe measured in feet of water; the difference between static head and head loss due to friction and pipe fittings.

Township line: a line parallel to a baseline, marking off the land into 6-mile strips, know as townships, in the rectangular survey system.

Township: a square unit of land, 6 miles on a side. Each 36 square mile township is divided into 36 one-square mile sections. Townships are numbered north or south of a baseline in the rectangular survey system.

Trade association: an organization of manufacturers and businesses involved in the production and supply of materials and services within a specific business or industry formed to promote common interests.

Traffic pattern: the projected movement of people for a given floor plan.

Transformer: a device used to transfer electric energy from one circuit to another, especially a pair of multiply wound, inductively coupled wire coils that affect such a transfer with a change in voltage, current, phase, or other electric characteristic.

Translation: the motion of a body in a straight line.

Transom: framed glass typically placed above doors to allow light into the entranceway.

Trap: curved section of drain line that prevents sewer odors from escaping into the building.

Traveling wires: connected by two wires to complete a three-way switch circuit.

Tread: the horizontal part of a step; width of a tread is measured from front to back.

Tree ordinance: laws adopted by municipalities to control the destruction of trees in order to preserve mature native trees.

Tributary area (of a structural member): the area of a structure that will contribute load directly to the structural member.

Tributary width (of a structural member): the width of a supported floor or roof, along the length of the structural member that will contribute load to the structural member.

Trombe wall: a solid mass, generally constructed of 8 to 16 in thick masonry, for the purpose of absorbing solar heat gain.

Turret: a small tower attached to a larger building.

U

U Factor: the measure of the rate of heat loss through a material given a difference in temperature on either side; the coefficient of heat conductivity.

Unbraced length: the distance between braced points of a member.

Uniform load (or uniformly distributed load): a load that is applied uniformly, or evenly, over a given area or along a structural member.

Unity: an element of design that provides a measure of how well the elements of a design fit together to create a whole; can be achieved by applying consistent use of design elements.

V

Value: an element of design referring to the relative lightness or darkness of a color.

Value-added: an increase in benefit, worth, or importance resulting from project development.

Vanity: bathroom base cabinet.

Vantage point: a position giving a good view of an object.

Variance: a waiver from the governing body that allows you to deviate from the specifics of an ordinance.

Ventilation: purposeful air exchange with the outside environment.

Vernacular architecture: a classification of architecture used to categorize building traditions that address local needs and construction methods which use locally available resources.

Vertical datum: collection of reference points with known heights above or below mean sea level (MSL).

Viability: practicality or feasibility.

Viability analysis: a site evaluation process that considers the factors that will affect the success of a proposed development on the site.

View orientation: the placement of a structure on a site with consideration to the potential for visually appealing and unappealing views from the structure.

Visual tasks: an activity of which performance requires that details and objects be seen.

Volatile organic compounds (VOCs): gaseous emissions from certain materials that may cause adverse health effects

Voltage (V or E): the measure of electric force available to cause the movement of electrons; the unit for voltage is volts.

W

Wastewater: water that is discharged from a residential, commercial, industrial or agricultural facility and has been adversely affected in quality by contaminants.

Water closet: a toilet.

Water table: (1) the level below which the ground is saturated with water; (2) a horizontal row of specially molded bricks that projects from a wall and deflects water running down the face of the wall.

Watt (symbol: W): the International System of Units "unit" of power; 1 watt is equal to 3.4 Btu/h.

Watt/hour (Wh) or kilowatt/hour (kWh): the rate at which electricity flows.

Well-graded soil: soil that displays a good representation of all particle sizes; a well graded sand will contain a fairly even distribution of coarse, medium, and fine sand.

Wetlands: lands where water saturation is the dominant factor. The resulting bogs, marshes, swans, and fens provide a habitat for many species of plants and animals.

Widow's walk: a railed walkway built on a roof, traditionally for viewing the sea while waiting for fishing boats to return.

Wind orientation: placement of a structure on a site with consideration given to the wind speed and direction experienced on the site throughout the year.

Working drawings: technical drawings drawn to scale for the purpose of communicating specific information necessary for construction.

Wye: a Y-shaped fitting with three openings allowing one pipe to be joined to two others at a 45 degree angle.

Y

Yield stress: the stress at which deformation will begin to change more quickly per unit of stress.

Yielding: the condition when the entire cross-section of a member has reached yield stress.

Young's modulus: see elastic modulus.

Z

Zone: one or more rooms within a building that require similar air quality and thermal temperature.

Zoning ordinances: laws for property development that govern the design and use of buildings, structures, and utilities within designated "zones" within a municipality.

INDEX

Page numbers followed by *f* and *t* indicate figures and tables, respectively.

A

B

C

D

E

F

G

H

I

M

N

O

P

Q

R

S

T

U

V

W

Y

Z